Science and Technology of Fruit Wine Production

Science and Technology of Fruit Wine Production

Edited by

Maria R. Kosseva
University of Nottingham Ningbo Campus (UNNC)
Ningbo, Zhejiang, China

V.K. Joshi
Dr. Y.S. Parmar University of Horticulture and Forestry
Nauni, Solan, India

P.S. Panesar
Sant Longowal Institute of Engineering and Technology
Longowal, Sangrur, India

ELSEVIER

AMSTERDAM • BOSTON • HEIDELBERG • LONDON
NEW YORK • OXFORD • PARIS • SAN DIEGO
SAN FRANCISCO • SINGAPORE • SYDNEY • TOKYO

Academic Press is an imprint of Elsevier

Academic Press is an imprint of Elsevier
125 London Wall, London EC2Y 5AS, United Kingdom
525 B Street, Suite 1800, San Diego, CA 92101-4495, United States
50 Hampshire Street, 5th Floor, Cambridge, MA 02139, United States
The Boulevard, Langford Lane, Kidlington, Oxford OX5 1GB, United Kingdom

Notices

Knowledge and best practice in this field are constantly changing. As new research and experience broaden our understanding, changes in research methods, professional practices, or medical treatment may become necessary.

Practitioners and researchers must always rely on their own experience and knowledge in evaluating and using any information, methods, compounds, or experiments described herein. In using such information or methods they should be mindful of their own safety and the safety of others, including parties for whom they have a professional responsibility.

To the fullest extent of the law, neither the Publisher nor the authors, contributors, or editors, assume any liability for any injury and/or damage to persons or property as a matter of products liability, negligence or otherwise, or from any use or operation of any methods, products, instructions, or ideas contained in the material herein.

Library of Congress Cataloging-in-Publication Data
A catalog record for this book is available from the Library of Congress

British Library Cataloguing-in-Publication Data
A catalogue record for this book is available from the British Library

ISBN: 978-0-12-800850-8

For information on all Academic Press publications
visit our website at https://www.elsevier.com/

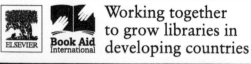

Working together
to grow libraries in
developing countries

www.elsevier.com • www.bookaid.org

Publisher: Nikki Levy
Acquisition Editor: Nancy Maragioglio
Editorial Project Manager: Billie Jean Fernandez
Production Project Manager: Caroline Johnson
Designer: Victoria Pearson

Typeset by TNQ Books and Journals

Contents

List of Contributors...xvii
Preface...xix
Introduction...xxi

CHAPTER 1 Science and Technology of Fruit Wines: An Overview 1

V.K. JOSHI, P.S. PANESAR, V.S. RANA AND S. KAUR

1. Introduction...1
2. Origin and History of Wine 2
    2.1 Origin of Wine, Yeast, Barrels, and Chips 3
    2.2 History of Wine 5
    2.3 Alcoholic Fermentation 6
3. Role of Wine as Food and Its Health Benefits 6
4. Fruit Wines, Their Types and Diversity 8
    4.1 Classification of Wine...8
    4.2 Types of Fruit Wines...9
5. Fruit Cultivation Practices and Their Varieties 10
    5.1 Nongrape Fruits Used for Winemaking...11
    5.2 Cultural Practices Affecting Wine Quality 13
6. Role of Genetic Engineering in Wine...25
    6.1 Genetic Modification of Plants 25
    6.2 Genetic Engineering of Wine Yeast 26
    6.3 Specific Targets for Wine Yeast Genetic Engineering 28
    6.4 Concerns Associated With the Use of Genetically Modified Yeasts...................32
7. Technology of Fruit Wine Production 33
    7.1 General Aspects and Problems in the Production of Fruit Wines.......................33
    7.2 Required Raw Materials 33
    7.3 Fruit Composition and Maturity 34
    7.4 Screening of Suitable Varieties 34
    7.5 Microbiology of Fermentation...34
8. General Methods for Fruit Winemaking...35
    8.1 Preparation of Yeast Starter Culture 35
    8.2 Preparation of Must 35
    8.3 Fermentation..35
    8.4 Siphoning/Racking 36
    8.5 Maturation 37

8.6 Clarification 37
8.7 Pasteurization 37

9. Technology of Wine Production From Various Fruits 37
10. Special Wines 38
10.1 Vermouth 38
10.2 Sparkling Wine 46
11. Nongrape Fruit Wine Industry: Global Status 46
11.1 Factors Influencing Fruit Wine Production 46
11.2 Global Production of Fruit Wines 49
11.3 Country-Wise Status of Fruit Wines 54
12. Summary and Future Strategies 57
References 58

CHAPTER 2   Microbiology of Fruit Wine Production 73

F. MATEI AND M.R. KOSSEVA

1. Introduction 73
2. Microbial Biodiversity Detected During Fruit Wine Production ...74
2.1 Yeasts 74
2.2 Killer Yeast in Wine Fruits 76
2.3 Lactic Acid Bacteria 78
3. Selection of Yeast as Starter Cultures for Production of Fruit Wines ...79
3.1 Inoculation With Mixed Yeast Cultures 80
3.2 Inoculation With Pure Yeast Cultures 81
3.3 Procedures for Preparation of the Inocula for Fruit Wines 82
4. Factors Affecting the Yeast Growth During Fruit Wine Fermentation ...84
4.1 Temperature 84
4.2 Acidity 84
4.3 Sugar Concentration 84
4.4 Aeration 84
4.5 Ethanol Concentration 85
5. Malolactic Fermentation in Fruit Wines 85
6. Use of Immobilized Biocatalysts in Winemaking 86
6.1 Methods of Immobilization 86
6.2 Applications of Immobilized Biocatalyst Technology
in Enology 88
7. Spoilage of Fruit Wines 90
7.1 Must Spoilage 91
7.2 Spoilage During Malolactic Fermentation 91
7.3 Postfermentative Spoilage 91
7.4 Fruit Wine Spoilage Prevention 93

8. Microbiological Analysis in Fruit Winemaking.................................93
    8.1 Monitoring Yeast Population Development During Alcoholic
        Fermentation 93
    8.2 Detecting Spoiling Microorganisms During Fruit Winemaking 94
    8.3 Mycotoxins Detection..96
9. Genetically Modified Microorganisms for Fruit Winemaking 96
    9.1 Methods for Wine-Yeast Development..................................97
    9.2 Genetically Engineered Wine-Yeast Strains 97
    9.3 Legislation and Consumer Behavior on Genetically Modified
        Microorganisms 98
10. Conclusions...98
    References 99

CHAPTER 3 Chemistry of Fruit Wines................................... 105

H.P. VASANTHA RUPASINGHE, V.K. JOSHI, A. SMITH AND I. PARMAR

1. Introduction...105
2. Types of Fermentation 106
3. Chemistry of Winemaking..107
    3.1 Carbon Metabolism 109
    3.2 Nitrogen Metabolism...126
    3.3 Metabolism of Sulfur: Chemistry of the Production of
        Off-Flavor..131
    3.4 Acetic Acid Fermentation...132
4. Role of Enzymes in Winemaking 133
5. Use of Antimicrobials 133
6. Malolactic Fermentation 133
7. Fermentation Bouquet and Yeast Flavor Compounds....................135
8. Chemical Changes Occurring During Fermentation of Sparkling and
    Fortified Wines...137
    8.1 Sparkling Wines...137
    8.2 Biologically Aged Wines..138
9. Toxic Metabolites of Nitrogen Metabolism.................................139
    9.1 Amines...140
    9.2 Ethyl Carbamate 141
10. Chemistry of Wine Spoilage..142
    10.1 Spoilage by Acetic Acid Bacteria......................................142
    10.2 Spoilage by Lactic Acid Bacteria 142
11. Composition and Nutritional Significance of Wine.......................143
    11.1 Compositional Parameters..143
    11.2 Phenolic Characterization of Fruit Wines.............................147

11.3 Aroma and Volatile Compounds..153

12. Summary and Future Outlook 157

References 157

**CHAPTER 4   Composition, Nutritional, and Therapeutic Values
of Fruit and Berry Wines 177**

V. MAKSIMOVIĆ AND J. DRAGIŠIĆ MAKSIMOVIĆ

1. Composition of Fruit and Berry Wines..178
   1.1 Alcohols..178
   1.2 Sugar Content 180
   1.3 Organic Acids 181
   1.4 Aldehydes, Esters, and Other Volatile Constituents 182
   1.5 Vitamins..183
   1.6 Carotenoids 184
   1.7 Minerals 184
   1.8 Dietary Fiber 185
   1.9 Histamine...185

2. Main Classes of Phenolic Compounds From Fruit and Berry Wines
   With Health Benefit Potential 186
   2.1 Anthocyanins and Anthocyanin-Derived Compounds 189
   2.2 Phenolic Acids 191
   2.3 Flavonoid Compounds...192

3. Nutritional Facts 195
   3.1 Caloric Value 195
   3.2 Antioxidants...195
   3.3 Alcohol 196

4. Enzymatic Transformations of Phenolic Compounds During Vinification 196

5. Bioavailability of the Major Health Benefit Components of Fruit Wines 197
   5.1 Bioavailability of Anthocyanins 198
   5.2 Bioavailability of Flavonoids...200

6. Health Benefit Potential of Different Fruit and Berry Wines 203
   6.1 Cardioprotective Potential 203
   6.2 Antioxidative Effects 206
   6.3 Prevention of Various Types of Cancers, or It's Suppression.....................212
   6.4 Cognitive Support..213
   6.5 Prevention of Gastrointestinal Disorders...................................213
   6.6 Other Health Benefits 214

7. Conclusions..214

References 215

CHAPTER 5 **Methods of Evaluation of Fruit Wines**...**227**

D.R. DIAS, W.F. DUARTE AND R.F. SCHWAN

1. Introduction...227
2. Physicochemical Analysis 227
   2.1 Titrimetry 231
   2.2 Potentiometry...235
   2.3 Densimetry..236
3. Chromatographic Analysis...237
   3.1 Liquid Chromatography 237
   3.2 Gas Chromatography...241
4. Microbiological Analysis 242
   4.1 Classical Techniques..243
   4.2 Molecular Techniques..244
5. Sensory Analysis...244
6. Future Prospects..246
   References 246

CHAPTER 6 **Chemical Engineering Aspects of Fruit Wine Production** **253**

M.R. KOSSEVA

1. Introduction...253
2. Emerging Methods for Fruit Juice Extraction 254
   2.1 Microwave Heating for Improved Extraction of Fruit Juice 254
   2.2 Case Studies on Cider Production 257
   2.3 Ultrasound-Assisted Enzymatic Extraction of Fruit Juice 260
   2.4 Pulsed Electrical Field Technology 262
3. Development of Membrane Technologies Applied to Fruit Winemaking 263
   3.1 Must Correction 266
   3.2 Clarification of Fruit Juice and Wine..268
   3.3 Alcohol Removal From Fruit Wines 268
   3.4 Reduction of Malic Acid in Must 269
   3.5 Membrane Bioreactors for Fruit Wine Processing 270
   3.6 Productivity of Membrane Bioreactors 274
4. Racking Process and Transport of Wine 275
   4.1 Pumps 275
5. Preservation Processes Applicable to Wine Production..............................279
   5.1 High Hydrostatic Pressure Treatment of Fruit Wine 279
   5.2 High-Pressure CO_2 Sterilization...281
   5.3 Application of Ultrasound in Must Treatment for Microbial Inactivation 283

5.4 Pasteurization of Fruit Juice Using Microwaves 284
5.5 Pulsed Electric Fields Technology for Wine Preservation..............285
6. Conclusions...287
    References 288

CHAPTER 7   Specific Features of Table Wine Production Technology 295

7.1 Pome Fruit Wines: Production Technology................................ **295**

V.K. JOSHI AND B.L. ATTRI

1. Introduction..295
2. Technology for the Preparation of Apple Wine 296
    2.1 Apple Wine..296
    2.2 Apple Tea Wine 312
    2.3 Apple Wine With Medicinal Value 313
3. Pear Wine/Perry 315
    3.1 Composition of Pear Fruit 315
    3.2 Process for Making Perry 316
4. Cider 318
    4.1 Definition and Characteristics of Cider 318
    4.2 Flavor-Affecting Factors in Cider..................................319
    4.3 Technology of Cider Production.....................................320
    4.4 Cider Quality 329
    4.5 Spoilage of Cider 337
5. Loquat Wine..338
6. Medlar Wine 338
7. *Pyracantha* Wine 339
8. Toyon Wine 339
9. Quince Wine 339
10. Mixed Fruit Wines 340
11. Future Outlook...342
    References 342

7.2 Stone Fruit Wines **348**

V.K. JOSHI, P.S. PANESAR AND G.S. ABROL

1. Introduction..348
2. Production of Stone Fruit Wines: General Aspects 348
    2.1 Production...348
    2.2 Composition of Fruit 349
    2.3 Problems in Wine Production 350

3. General Method of Wine Preparation 350
    3.1 Preparation of Yeast Starter Culture 350
    3.2 Preparation of Must 350
    3.3 Fermentation 351
    3.4 Siphoning/Racking 351
    3.5 Maturation 351
    3.6 Clarification 351
    3.7 Blending...352
    3.8 Pasteurization..352
4. Table Wine 352
    4.1 Plum Wine 352
    4.2 Peach Wine 362
    4.3 Apricot Wine..368
    4.4 Cherry Wines 370
5. Summary and Conclusions 377
    References 378

7.3 Berry and Other Fruit Wines...382

V.K. JOSHI, S. SHARMA AND A.D. THAKUR

1. Introduction..382
2. General Aspects: Production of Berry Wine...383
    2.1 Problems of Berry Wines...383
    2.2 Raw Materials..384
    2.3 Composition and Maturity of Fruits 384
3. Methods of Preparation of Table Wine 384
    3.1 Strawberry Wine 384
    3.2 Red Raspberry Wine 389
    3.3 Sea Buckthorn Wine 390
    3.4 Pumpkin Wine 392
    3.5 Pumpkin-Based Herbal Wine 392
    3.6 Blackberry *Jamun* Wine 395
    3.7 Red Wine Made by Blending of Grape (*Vitis vinifera* L.) and *Jamun*
        (*Syzygium cumini* L.) Juices 398
    3.8 Persimmon Wine...401
    3.9 Fermented Garlic Beverage 403
4. Lychee Wine 403
5. Papaya Wine 404
6. Blended Passion Fruit 405
7. Future Trends 406
    References 406

7.4 Citrus Wines...**410**

S. SELLI, H. KELEBEK AND P.S. PANESAR

1. Introduction..410
2. Orange Wine...410
   2.1 Orange Winemaking...411
   2.2 The Chemical Composition of Orange Wine 414
3. Mandarin Wine..424
   3.1 Chemical Composition of Mandarin Juices and Wines.......425
   3.2 Flavor Composition of Mandarin Wine.............................430
   3.3 Phenolic Composition of Mandarin Juice and Wine 431
   3.4 Antioxidant Activity of Mandarin Juice and Wine.............434
4. Conclusions...435
   References 435

7.5 Production of Wine From Tropical Fruits............................**441**

L.V. REDDY, V.K. JOSHI AND P.S. PANESAR

1. Introduction..441
2. Types of Fruit Wine...442
   2.1 Mango Wine..442
3. Pineapple Wine..451
4. Cashew Apple...451
5. Lychee Wine...452
6. Coconut Wine...454
7. Sapota Wine...454
8. Palm Wine..455
9. Conclusions and Future Perspectives......................................457
   References 458

CHAPTER 8   Technology for the Production of Agricultural Wines **463**

N. GARG

1. Introduction..463
2. Mahua Wines..463
   2.1 Mahua Tree..463
   2.2 Mahua Flower..464
   2.3 Mahua Flower: Composition 464
   2.4 Mahua Liquor..465
   2.5 Mahua Flower Wine 466

2.6 Flower Collection and Processing 466
2.7 Addition of Nutrients...466
2.8 Fermentation Temperature 468
2.9 Addition of Tannins 468
2.10 Flavor Masking 468
2.11 Phenolic Profiling 469
2.12 Mahua Vermouth 470
3. Mead 471
3.1 Method for Mead Production 471
4. Rhododendron Wine 474
4.1 Rhododendron 474
4.2 Tree 474
4.3 Flower 475
4.4 Wine 476
5. Sweet Potato Wine 477
5.1 Sweet Potato 477
5.2 Wine 477
6. Tomato Wine 478
6.1 Tomato 478
6.2 Wine 479
7. Whey Wines 480
7.1 Whey Composition 480
7.2 Other Fermented Beverages 480
7.3 Whey Beer 481
7.4 Whey Wine 481
8. Cocoa Wine 481
8.1 Cocoa 481
8.2 Composition of Cocoa Pulp 482
8.3 Wine 482
9. Regulations for Making Agricultural Wines 482
References 484

**CHAPTER 9 Technology for Production of Fortified and Sparkling
Fruit Wines 487**

P.S. PANESAR, V.K. JOSHI, V. BALI AND R. PANESAR

1. Introduction..487
2. Vermouth...487
2.1 Types of Vermouth 488
2.2 Technology of Vermouth Production 488

2.3 Vermouth Production From Nongrape Fruits 494
2.4 Commercial Production of Vermouth 500
3. Sparkling Wine 500
3.1 Introduction 500
3.2 Technology of Production 502
3.3 Sparkling Plum Wine 504
3.4 Methods of Secondary Fermentation 509
3.5 Malolactic Fermentation in Sparkling Wine Production 512
3.6 Production of Sparkling Plum Wine 513
3.7 Production of Sparkling Apple Wine and Cider 520
3.8 Sparkling Mead 524
3.9 Other Sparkling Fruit Wines 524
4. Conclusions and Future Trends 524
References 525

CHAPTER 10 Fruit Brandies...531

F. LÓPEZ, J.J. RODRÍGUEZ-BENCOMO, I. ORRIOLS AND J.R. PÉREZ-CORREA

1. Introduction 531
1.1 Brief History of Distillation 531
1.2 Alcoholic Beverages 531
1.3 Importance of Distilled Beverages 531
1.4 Classification of Fruit Alcoholic Beverages 532
1.5 Factors Affecting the Quality of the Brandies 534
2. Distillation Systems 534
2.1 Discontinuous Distillation 534
2.2 Continuous Distillation 536
3. Pome Fruit Brandy 537
3.1 Pear Brandy 537
3.2 Apple Brandy 539
4. Stone Fruit Brandy 541
4.1 Cherry Brandy 542
4.2 Plum Brandy 543
4.3 Apricot Brandy 545
4.4 Other Stone Fruit Brandies 546
5. Berry Fruit Brandy 547
6. Other Fruit Brandy 549
6.1 Kiwi Brandy 549
6.2 Melon Brandy 550
6.3 Orange Brandy 551
6.4 Banana Brandy 552
7. Conclusions..552
References 552

CHAPTER 11 Waste From Fruit Wine Production................................**557**

M.R. KOSSEVA

1. Introduction..557
    1.1 Defining Food and Fruit Waste.................................557
2. Unavoidable Solid Food and Fruit Waste......................559
    2.1 Apple Pomace 560
    2.2 Mango Peels..561
    2.3 Citrus Peels...562
    2.4 Berry Peels, Pulp, and Seeds.................................564
    2.5 Coconut Waste...565
3. Valorization of Fruit By-Products and Juices 568
4. Cider Lees 570
    4.1 The Characteristics of Crude Cider Lees 571
    4.2 Valorization of Wine Lees 572
5. Liquid Stream and Wastewater...................................574
    5.1 Characterization of Winery Liquid Effluents 574
    5.2 Methods of Wastewater Treatment 576
6. Ecotoxicity 585
7. Sustainability in the Winemaking Sector 586
    7.1 Life Cycle Assessment in Fruit Production...............587
    7.2 Life Cycle Assessment of Waste Management in
        Cider Production 588
    7.3 Life Cycle Assessment of Wine Packaging..............588
8. Conclusions..589
    References..591

CHAPTER 12 Biorefinery Concept Applied to Fruit Wine Wastes **599**

M.-P. ZACHAROF

1. Introduction..599
    1.1 Generation of Energy and Products From Alternative Sources:
        The Biorefinery Concept 600
    1.2 Waste as a Renewable Source for Energy and Resource
        Recovery...602
    1.3 Biorefinery Feedstock: The Fruit Winery Waste 605
2. Biotechnological Conversion of Fruit Wine Waste to Platform
    Chemicals and Energy 609
    2.1 Case Studies 609
    2.2 The Use of Apple Pomace as Feedstock 611
    2.3 The Use of Mango Waste as Feedstock....................612
3. Conclusions..612
    References..612

CHAPTER 13   Innovations in Winemaking **617**

   R.S. JACKSON

   1.   Introduction 617
   2.   Basic Winemaking 622
   3.   Innovations in the Vineyard/Orchard 625
   4.   Winery Innovations 631
   5.   Sparkling Wines 644
   6.   Fortified Wines 647
   7.   Sensory Evaluation 649
   8.   Authenticity 652
   9.   Future Prospects 653
        References 655

CHAPTER 14   Technical Guide for Fruit Wine Production **663**

   F. MATEI

   1.   Introduction 663
   2.   Fruit Wine Types and Styles 664
   3.   Methods for Fruit Wine Production 665
        3.1 Crushing the Fruit and Must Preparation 666
        3.2 Must Fermentation 671
        3.3 Yeast Inoculation 673
        3.4 Clarifying/Racking 674
        3.5 Aging Fruit Wines 675
        3.6 Bottling Fruit Wines 675
   4.   Traditional Recipes of Fruit Wines 676
        4.1 Berries Fruits Wines 677
        4.2 Stone Fruits Wines 681
        4.3 Tropical and Exotic Fruit Wines 686
        4.4 Citrus Wines 691
        4.5 Pome Fruits Wines 692
   5.   The Fruit Wines in the Market 695
        5.1 Niche Market Segments 698
   6.   Conclusions 699
        References 700

Index 705

List of Contributors

G.S. Abrol   UUHF, Bharsar, Uttarakhand, India
B.L. Attri   ICAR-Directorate of Mushroom Research (DMR), Solan, HP, India
V. Bali   Sant Longowal Institute of Engineering and Technology, Longowal, Punjab, India
D.R. Dias   Federal University of Lavras, Lavras, Minas Gerais, Brazil
J. Dragišić Maksimović   University of Belgrade, Belgrade, Serbia
W.F. Duarte   Federal University of Lavras, Lavras, Minas Gerais, Brazil
N. Garg   ICAR-CISH, Lucknow, India
R.S. Jackson   Brock University, St. Catharines, ON, Canada
V.K. Joshi   Dr. Y.S. Parmar University of Horticulture and Forestry, Nauni, Solan, HP, India
S. Kaur   Sant Longowal Institute of Engineering and Technology, Longowal, Punjab, India
H. Kelebek   Adana Science and Technology University, Adana, Turkey
M.R. Kosseva   University of Nottingham Ningbo Campus, Ningbo, China
F. López   Universitat Rovira i Virgili, Tarragona, Spain
V. Maksimović   University of Belgrade, Belgrade, Serbia
F. Matei   University of Agronomic Sciences and Veterinary Medicine of Bucharest, Bucharest, Romania
I. Orriols   Instituto Galego da Calidade Alimentaria, Leiro, Spain
P.S. Panesar   Sant Longowal Institute of Engineering and Technology, Longowal, Punjab, India
R. Panesar   Sant Longowal Institute of Engineering and Technology, Longowal, Punjab, India
I. Parmar   Dalhousie University, Truro, NS, Canada
J.R. Pérez-Correa   Pontificia Universidad Católica de Chile, Santiago, Chile
V.S. Rana   Dr. Y.S. Parmar University of Horticulture and Forestry, Nauni, Solan, HP, India
L.V. Reddy   Yogi Vemana University, Kadapa, AP, India
J.J. Rodríguez-Bencomo   Universitat Rovira i Virgili, Tarragona, Spain
S. Selli   Cukurova University, Adana, Turkey
R.F. Schwan   Federal University of Lavras, Lavras, Minas Gerais, Brazil
S. Sharma   Shoolini University, Solan, HP, India
A. Smith   Dalhousie University, Truro, NS, Canada
H.P. Vasantha Rupasinghe   Dalhousie University, Truro, NS, Canada
A.D. Thakur   Shoolini University, Solan, HP, India
M.-P. Zacharof   Swansea University Medical School, Swansea, United Kingdom

Preface

Winemaking technologies are among the oldest known to man. Wine production is one of the alternatives for value addition and waste minimization of the fruit-processing industries. Wine has been extolled as a therapeutic agent besides serving as an important adjunct to the human diet, having polyphenols and other bioactive compounds with antioxidant activities. The nongrape fruits, including apple, citrus, kiwi, mango, peach, pear, plum, strawberry, and others, comprise a considerable portion of all the fruits grown worldwide compared to grapes. Being highly perishable commodities, they have to be either consumed fresh or processed into various products. Fruit and fruit-related products also form an important part of human nutrition and are essential both to improve health and as a source of natural energy. Wines derived from the nongrape fruits have functions like those of the grapes. Furthermore, they are amazingly delicious and full of flavors. Consequently, fruit wine production is gradually changing its profile from a limited regional industry to a sector in drink manufacturing in many parts of the world, above all in Asia and South America, where fruit diversity is represented by over 500 different species. At present, interest in functional foods is rising rapidly, and fruit wines, being considered a functional drink, are increasingly receiving the attention of winemakers and consumers. Various fruits can be used for winemaking, but tropical or "exotic" fruits are especially sought. Their unique flavor and color appeal have made these fruits very popular in Europe, North America, and Asia. Tropical and subtropical fruits have high and diverse vitamin and mineral contents that can form an essential part of a nutritionally balanced diet.

Considering all these aspects, a manuscript entitled *Science and Technology of Fruit Wine Production* comprising of 14 chapters was planned. It aims to present consolidated information on the state of art of the science and technology of fruit wine production, composition, chemistry, role and quality of raw materials, medicinal value, quality factors, bioreactor technology and engineering aspects, production, optimization, standardization, preservation, and evaluation of various wines and brandies, as well as on the sustainability of fruit winemaking technologies. To our knowledge, this is the first manuscript covering the science involved and technology employed to produce and evaluate fruit wines. It will stimulate the development of new products and their entry into the market.

The book consists of six major sections, the first of which is an introductory chapter on the origin, history, and role of wine. The second section deals with the science of fruit wines, covering microbiology and chemistry, composition, nutritional, therapeutic values, and methods for the evaluation of wines. The third section focuses on traditional technologies highlighting the specific features of table, agricultural, fortified, and sparkling fruit wine and fruit brandy production technologies. The fourth section is devoted to innovations in winemaking; it also presents engineering aspects such as pulsed electric fields and ultrasound treatment, application of microwave, and membrane separation considered as part of fruit wine processing, which could be successfully scaled up to the industrial level. Membrane bioreactor technology and various modern preservation techniques are discussed in this section. The penultimate section presents the utilization of waste generated from fruit wine production, including the biorefinery concept for valorization of solid and liquid wine by-products. The final section is a technical guide containing traditional recipes for homemade fruit wines. It is hoped that the volume will be highly useful to students, academicians, researchers, industrialists, and amateur winemakers.

Key features of the book include its focus on producing nongrape wines, highlighting their microbiology and chemistry, or composition, flavor, taste, and other quality attributes, including therapeutic properties. The book is well illustrated and have a number of references to quench the thirst of the interested readers. It provides a single-volume resource consolidating the research findings, technology, and equipment employed in making wines from nongrape fruits. It explores options for reducing postharvest losses as well.

This book was written by experts from around the globe, providing the latest information on international research and development of novel technologies in producing wine from fruits other than grapes and strategies for treatment of wine wastes. It includes both theoretical and practical information providing, we hope, inspiration for additional research and applications to create novel niche products. Finally, the volume is expected to contribute to the state of the art of fruit wine manufacturing and the valorization of fruit by-products by providing novel concepts in winemaking.

We would seize to take this opportunity to acknowledge and thank the contributors to this book for their excellent contribution in bringing out a comprehensive range of topics together in a single volume. We would also like to thank Nancy Maragioglio, the senior acquiring editor for the Food Science and Technology Book Program at Academic Press; Billie Jean Fernandez, editorial project manager; and Johnson Caroline and her production team at Elsevier for their helpful assistance throughout this project. Last, but not least, we are grateful to our families for their current and continued support. Considering the scope of production of fruit wines, one of the purposes of this book is to bring together information scattered in various research and review papers available randomly worldwide.

M.R. Kosseva
V.K. Joshi
P.S. Panesar

Introduction

V.K. Joshi, P.S. Panesar, M.R. Kosseva

Winemaking technologies are among the oldest known to humans, as archeological excavations have uncovered many sites with sunken jars indicating the existence of wine for more than 7500 years (McGovern et al., 1996). The spontaneous fermentation of grape juice conducted by natural microflora gave a product with exhilarating properties that humans consumed and then never looked back (Amerine et al., 1980; Joshi et al., 2011). The scientific journey began (1632–1723), some 200 years before Pasteur, when Antonie Van Leeuwenhoeck built a new microscope to observe tiny living creatures that he called "animalcules," now called microbes (Stanier et al., 1970). In 1810, Guy-Lussac summarized the process with the famous equation that $C_6H_{12}O_6$ yielding $C_2H_5OH + 2CO_2$. Proof of the living nature of yeast appeared between 1837 and 1838, when three publications by Cagniard de la tour, Swann, and Kuetzing, appeared, each of whom independently concluded that yeast was a living organism that reproduced by budding (Stanier et al., 1970). Later developments revealed the involvement of yeast (*Saccharomyces cerevisiae* var. *ellipsoideus*) in the production of wine or any other alcoholic beverage (Amerine et al., 1980). The scientific breakthroughs to unravel the mysteries of fermentation started in the 1830s, primarily by French and German chemists. During the 1800s, the making of wine and beer was refined into the techniques known today. Later on, in 1857, Louis Pasteur solved the problem of wine spoilage when Napolean III referred the same to him. He found that heating the wine after it was fermented prevented spoilage, thus opening the way for the aseptic conditions used even today in winemaking: a process called pasteurization. Thus the history of wine is the history of microbiology and biochemistry. The process of winemaking is unique in the sense that it is multidisciplinary in its approach, and nearly all the physical, chemical, and biological sciences contribute to its production.

Wine has been extolled as a therapeutic agent. It is an important adjunct to the human diet, having polyphenols and other bioactive compounds that have antioxidant activities. In addition, the compounds bonded to insoluble plant compounds are released into the aqueous ethanolic solution during the winemaking process, which makes them more biologically available for absorption during consumption (Shahidi, 2009). Wines, because they are not distilled, have more nutrients such as vitamins, minerals, and sugars than distilled beverages like brandy and whisky (Joshi et al., 1999a), especially the polyphenolic compounds that act as antioxidants and antimicrobials. Moderate alcohol and/or wine consumption protects against the incidence of many diseases of modern society, like cardiovascular diseases, dietary cancers, ischemic stroke, peripheral vascular disease, diabetes, hypertension, peptic ulcers, kidney stones, and macular degeneration, in addition to stimulating resistance to infection and retention of bone density (Jindal, 1990; Joshi et al., 1999a; Stockley, 2011). The medicinal or therapeutic value of wine has also been highly acclaimed in the scientific literature (Catherine, 1996; Klatsky and Armstrong, 1993; Gronbaek et al., 1995). With respect to their therapeutic value, the wines from nongrape fruits do not lag behind the grape wine.

Grapes are the main source of raw material for the production of wine by the fermentation process the world over, though the percentage contribution of grapes to total fruit production is only 15.53%. Fig. 1 shows the amount of annual world production of fruits in 2013. The quantities of fruit produced have risen very fast since developed and developing countries achieved equal fruit supply figures in

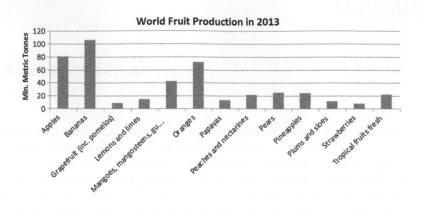

FIGURE 1

World fruit production in 2013. http://www.geohive.com/charts/ag_fruit.aspx.

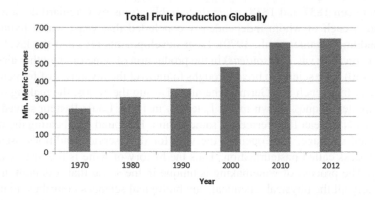

FIGURE 2

Variation in annual world total fruit production from 1970 to 2012. http://www.geohive.com/charts/ag_fruit.aspx.

2009–2011 (FAO, 2011), shown in Fig. 2. The top six fruit producers, in declining order of importance, are China, India, Brazil, United States, Italy, and Mexico. China, India, and Brazil account for almost 30% of the world's fruit supply. In the Southern Hemisphere, Chile, South Africa, and New Zealand have become major suppliers in the international trade of fresh fruit commodities. The amplified production of fruits leads to larger variety and availability of fruit-derived products, which could satisfy the global market and meet various requirements of consumers'.

Therefore, based on fruit production there is ample scope for preparation of wine from fruits other than grapes. However, the utilization of nongrape fruits for wine production generally depends upon the method of preparation, the raw material used, and the vinification practices (Joshi, 2009). The nongrape fruits used in various parts of the world for the production of wine include apples, berries, cherries, wild apricots, kiwifruit, plums, peaches, and strawberries, and one of the most widely produced nongrape fruit wines is cider, or apple wine, made and consumed throughout England, Germany, France, Spain, Ireland, Argentina, and Australia (Joshi et al., 1999b, 2011). Though the production of wine from nongrape fruit is beset with several problems, the challenges have been accepted by the scientists and winemakers, alike who have solved the major hurdles with appropriate solutions.

To develop and optimize bioprocesses involving living systems, viz., fruit and microorganisms, there is a constant need for genetic improvement to achieve or keep high production rates and efficient fermentation with improved quality. Using genetic engineering, considerable progress has been made in the development of genetically improved plants and microorganisms, especially yeast strains, and in exploring the possibilities of introducing new improved quality characteristics into fruits and wine yeast (Kaur et al., 2011). In the case of yeast, modification have been done by selection of variants, mutagenesis, hybridization, and transformation to obtain consistent wine flavor and predictable quality (Pretorius et al., 1999; Soni et al., 2011). Sequence data of the complete *S. cerevisiae* genome (Goffeau et al., 1996) and advances in wine technology have enabled the researchers to comprehensively assess the modifications required for genetic improvement of yeast cells. Winemakers generally require cost-competitive wine production with minimum resource input, which in turn depends upon specific targets, i.e., improvements in fermentation performance, wine wholesomeness, sensory qualities, processing efficiency, and resistance to antimicrobial compounds (Pérez-Torrado et al., 2015). The future use of genetically modified yeasts will be dependent on the ability to assess potential or theoretical risks associated with their introduction into natural ecosystems (Schuller and Casal, 2005), in addition to the improvements in the product currently made. It is likely that the changed parameters will be sensed by the complex regulatory mechanisms that exist within any living cell, and will lead to a specific molecular response. How much of these changes are accepted by the consumers would determine the future course of development.

The techniques used for the production of nongrape fruit wines are basically similar to those for the production of wines made from white and red grapes (Joshi et al., 1999b; Jagtap and Bapat, 2015). However, the major problems associated with nongrape fruit wine are first, that it is quite difficult to extract the sugar and other soluble materials from the pulp of some fruits, and second, that the juices obtained from most fruits are lower in sugar content and higher in acids than grape juice (Vyas and Gandhi, 1972; Amerine et al., 1980; Joshi et al., 1999b; Joshi et al., 1990; Swami et al., 2014), and even lack a nitrogenous source, essentially required for alcoholic fermentation. To overcome these problems, sugar may be added to accelerate the fermentation process and the higher acidity in some fruits can be diluted with water. Malic acid can be biologically converted by malolactic fermentation by applying immobilized cells (Kosseva et al., 1998), or the use of malate-utilizing yeast like *Schizosaccharomyces pombe* in the wine, to an acceptable level (Vyas and Joshi, 1988; Joshi et al., 1991a; Joshi and Attri, 2005), and the addition of a nitrogen source (Joshi et al., 1990) can be used to accelerate the fermentation process, or the addition of enzyme (Joshi and Bhutani, 1991) can be made to hasten the process of clarification.

The raw materials used for the production of fruit-based alcoholic beverages are naturally the fruits, sweetening agent, fruit concentrate, sugar, acid, nitrogen source, clarifying enzyme, filter aid, etc., whereas spices, herbs, or their extracts constitute an essential raw material to make fortified wines like vermouth (Joshi et al., 1991b, 1999; Panesar et al., 2011). Like grape wine, the juice or the pulp of the fruit to be used for winemaking is made into must and is generally prepared depending upon the fruit used and the type of wine to be made. The addition of ammonium sulfate with thiamin and biotin as source of nitrogen and growth regulators gives a greater increase in the rate of fermentation. The must is allowed to ferment at a suitable temperature (20–25°C) after inoculation with yeast culture. Other operations are similar to the preparation of grape wine. A few studies have been conducted on the maturation of wine, the period of which may extend from 6 months to 2–3 years and which makes the wine mellow in taste and fruity in flavor, in addition to clarifying it. Clarification of fruit wine is done in a way similar to that of grape wine, and then, pasteurization is done for bottling of wines with or without addition of preservatives as per requirements. Like grapes, special types of wines, like vermouth and sparkling wines, have also been developed from the nongrape fruits (Joshi, 1997, 2009; Joshi et al., 1999a; Panesar et al., 2011). The dismissal picture of fruit and vegetable utilization by most developing countries has revealed that as of this writing, staggering postharvest losses range from 20% to 30%, which is a colossal loss to the economy of these countries (Joshi, 2001). Growth of the fruit processing industry is, thus, of utmost significance, and production of wine is an integrated component of this industry.

Wine production, as an industry, is one of the alternatives for value addition and waste minimization of nongrape fruit processing also (Joshi et al., 2011). So in brief, unless the processing industry is linked with the horticultural industry, it is unlikely to achieve any worthwhile results, either for the farmers or for the consumers. A great advantage of production of fruit wines is that there are virtually no differences in the manufacturing plants required for the production of nongrape wines and grape wines except for minor modifications, and the manufacturer can make use of the same facility.

As a result, in Europe one can find a significant production of nongrape fermentation products, where the United Kingdom and Germany represent attractive markets. The United Kingdom has a long tradition of fruit and other nongrape wines and is one of Europe's largest markets for fruit wine, with annual production of 40–50 million liters a year, whereas the export of nongrape wine to other countries (mainly the United States) from Canada was recorded to be higher in comparison to grape wines.

The production of cider, an example of a typical fruit wine, is an important economic resource in Europe, and it is a popular drink, with consumption rates of over 14 million hectoliters per year (Association of the Cider and Fruit Wine Industry of the European Union, 2010). The affiliated members of the European Cider and Fruit Wine Association represent over 180 cider and fruit wine manufacturing companies in the EU. Their largest producers are situated in the United Kingdom, France, Spain, Germany, Ireland, and Belgium (http://www.aicv.org).

The United States already has an important tradition of fruit wine production. Michigan is one of the foremost US states in the production of fruit wine, where apple wine and cherry wine are produced in the highest volume. New World regions like Argentina, Chile, South Africa, Australia, and New Zealand have important fruit industries and established wine industries as well.

Compared to the horticulturally advanced countries, wine production in India is almost negligible (Joshi, 2001). Most of the fresh fruits produced in India is marketed for table purposes and only a small fraction is processed into wine, juice, and raisins, which is true for other fruits like apple, plum, peach, pear, etc. But a silver lining in the cloud is the production of grape-based champagne near Nasik, vermouth from grapes in Maharashtra, feni in Goa, and apple cider, wine, and vermouth (at Badhu in Mandi) in Himachal Pradesh (Joshi, 1997; Joshi and Attri, 2005). The famous fruit wines of the Chinese market are wolfberry wine, cherry wine, lychee wine, mulberry wine, pomegranate wine, kiwifruit wine, berry wine, and blueberry wine, whereas *umeshu* or plum wine is one of the most popular fruit wines produced in Japan from a traditional Asian stone fruit.

The ever-increasing demand for alcoholic beverages in the national and international market reflects a considerable scope for alcoholic beverages from nongrape fruits. However, the variable global demand for fruit wines and lack of actual data on the import and export of fruit wines in different countries are some of the hurdles in marketing and export of the wines (Rivard, 2009).

Wine made from fruits other than grapes is diverse, and its cascading flow of possibilities opens up a whole new world to the wine drinkers. The utility and scope of the fruit wines, especially from nongrape fruits, reflects that in depth systematic research on the various facets of enology needs to be strengthened. It could certainly be a potential area of research, especially in those countries and regions where grape cultivation is not practiced. Needless to say, research currently in progress will continue to document the healthful properties of wine and, in addition, the industry will need to play a highly visible role in the promotion of sound and sustainable environmental stewardship as a strong motivator in the purchase of nongrape fruit wines, thus contributing more to the economy of the fruit wine industry.

The nongrape fruits, constitute a considerable portion of all the fruits grown all over the world. Being highly perishable commodities (Kosseva and Webb, 2013), they have to be either consumed fresh or processed into various products. Moreover, fruit and fruit-related products form an important part of human nutrition and are essential both to improve health being a source of natural energy. As an alternative, fruit or nongrape wines are amazingly delicious, full of flavors, and have functions like those of grapes. Consequently, fruit wine production is gradually changing its profile from a limited regional industry to a sector in drink manufacturing in many parts of the world, above all in Asia and South America, where fruit diversity is represented by over 500 different species. At present, interest in functional food is rising rapidly, and fruit wines, being considered a functional drink, are attracting the attention of winemakers and consumers. Various fruits can be used for winemaking, but tropical or "exotic" fruits are especially sought. Their unique flavor, attractive fragrance, and color appeal. These fruits have high and diverse vitamin and mineral contents that can form an essential part of a nutritionally balanced diet. As fruits tend to have a substantial amount of potassium, phosphorus, calcium, and, frequently, iron and magnesium, they are particularly important in providing the building blocks of healthy muscles, bones, teeth, and brain in children, as well as aiding protein digestion, cellular metabolism, and a fully functional nervous system (FAO, 2003, 2011).

World production of tropical fruits will reach 82 million tons in 2014, according to the estimates by the United Nations' Food and Agriculture Organization (FAO, 2013). Seventy-eight percent corresponds was expected to major fruits (mango, pineapple, avocado, and papaya) and 22% to the secondary ones (lychee, rambutan, guava, and so on). Ninety percent of tropical fruits are produced in countries that are developing. The major mango producers are India, Thailand, and Mexico, and for pineapple the Philippines, Thailand, and China. Papaya is produced in India, Brazil, and Mexico, mostly. Regarding the tropical fruits considered secondary, their production is concentrated in the Philippines, Indonesia, and India. Pineapple production is expected to reach more than 20 million tons in 2014, representing 23% of the global harvest of tropical fruits. Asia–Pacific countries account for 46% of the total exotic fruit produced. The booming South Asian region and other tropical area, shown on the world map, is very likely to form a belt for production of the fruit wine (Fig. 3), where production of nongrape wines will be concentrated, because of the great potential for growing fruits. One of the reasons to focus nongrape wine manufacturing in this region is the excessive production of tropical fruits, nearly 50% of the total world production. The important role played by the Asian markets is even more evident from the production trends since 2005; whereas America, Europe, Africa, and Oceania recorded fairly constant fruit production, in Asia it increased by about 55%, making China (Fig. 4) and India the highest producers of fruits in the world, with 20.06% and 13.92% of world production, respectively (Cerutti et al., 2014).

Further development of global fruit wine manufacturing is aligned with the current achievements in agriculture and global transportation making raw materials available year-round, fresh, frozen, or as a concentrate. Other reasons are based on the cost of fruit wine production, which can be lower, and the manufacturing process, which can be faster than grape winemaking, and the space efficiency and higher potential profit that can be achieved because of the consumer demand for fruit wine (Rivard, 2005).

The concept of manuscript of *Science and Technology of Fruit Wine Production* was created to meet the needs of the growing customer demand for healthy drinks, because interest in functional food is rising rapidly. It aspires to compile knowledge and present techniques concerning various nongrape

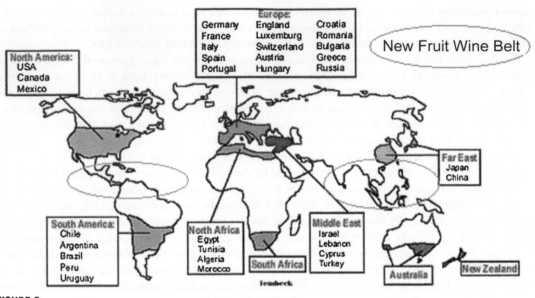

FIGURE 3

The world wine map including the proposed new fruit wine belt.

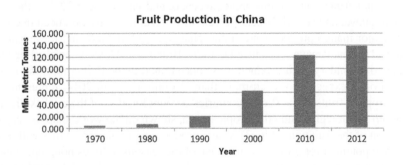

FIGURE 4

Growth of annual fruit production in China since 1970.

wines, to facilitate researchers: scientists and engineers working in the field of wines to develop new methods and technologies. Thus it will provide an opportunity to industrialists and entrepreneurs to set up a wine industry in nongrape-fruit-growing regions intended for novel products or to improve the existing wines. Additionally, the waste by-products from various stages of wine manufacturing can be potentially recycled into the fermentation processes, reducing the postharvest losses, which are quite high in developing countries (Kosseva and Webb, 2013).

In conclusion, such attractive products as fruit wines have great potential for further development, because various fruits as such or the combinations of fruits with grapes can serve as raw starting material for the fermentation process and that will create an innovative niche market product.

REFERENCES

Amerine, M.A., Kunkee, R.E., Ough, C.S., Singleton, V.L., Webb, A.D., 1980. Technology of Wine Making. AVI Publ. Co, Westport, Connecticut.

Association of the Cider and Fruit Wine Industry of the European Union (AICV)., 2010. http://www.aicv.org/pages/industry-data/production-and-sales.html.

Catherine, S., 1996. Wine and health. Medically is wine just another alcoholic beverage? In: Conference Summary of International Wine & Health Conference, 12–13 June.

Cerutti, A.K., Beccaro, G.L., Bruun, S., Bosco, S., Donno, D., Notarnicola, B., Bounous, G., 2014. LCA application in the fruit sector: state of the art and recommendations for environmental declarations of fruit products. Journal of Cleaner Production 73, 125–135.

FAO, 2011. Fruit Products for Profit. Diversification booklet number 16, by Clarke, C., Schreckenberg, K., Haq, N.N., Rome.

FAO Diversification Booklet on No. 21 Traditional Fermented Food and Beverages for Improved Livelihoods. Food and Agriculture Organization of the United Nations. FAOSTAT., 2013. http://faostat3.fao.org/faostat-gateway/go/to/browse/Q/QC/E.

FAO, 2003. Agricultural Services Bulletin 149. Handling and Preservation of Fruits and Vegetables by Combined Methods for Rural Areas. (Technical manual. Barbosa-Cánovas, G.V., Fernández-Molina, J.J., Alzamora, S.M., Tapia, M.S., López-Malo, A., Chanes, J.W.)

Goffeau, A., Barrell, B.G., Bussey, H., Davis, R.W., Dujon, B., Feldmann, H., Galibert, F., Hoheisel, J.D., Jacq, C., Johnston, M., Louis, E.J., Mewes, H.W., Murakami, Y., Philippsen, P., Tettelin, H., Oliver, S.G., 1996. Life with 6000 genes. Science 274, 563–567.

Gronbaek, M., Deis, A., Sorensen, T.I.A., Beckear, U., Schnohor, P., Jensen, G., 1995. Mortality associated with moderate intake of wines, beers and spirits. British Medical Journal 310, 1165–1169.

Jagtap, U.B., Bapat, V.A., 2015. Wines from fruits other than grapes: current status and future prospectus. Food Bioscience 9, 80–96.

Jindal, P.C., 1990. Grape. In: Bose, T.K., Mitra, S.K. (Eds.), Fruits, Tropical and Sub-tropical. Naya Prakashan, Calcutta, p. 85.

Joshi, V.K., 2001. Technologies for the postharvest processing of fruits and vegetables. In: Marwaha, S.S., Arora, J.K. (Eds.), Food Processing: Biotechnological Applications. Asia Tech Publishing Co., New Delhi, pp. 241–263.

Joshi, V.K., Attri, D., 2005. A Panorama of research and development of wines in India. Journal of Scientific and Industrial Research 64 (1), 9–18.

Joshi, V.K., 2009. Production of wines from non-grape fruit. In: Natural Product Radiance Special Issue, July–August. NISCARE, New Delhi.

Joshi, V.K., Sharma, P.C., Attri, B.L., 1991a. A note on the deacidification activity of Schizosaccharomyces pombe in plum musts of variable composition. Journal of Applied Bacteriology 70, 386–390.

Joshi, V.K., Attri, B.L., Mahajan, B.V.C., 1991b. Studies on preparation and evaluation of vermouth from plum. Journal of Food Science and Technology 28, 138.

Joshi, V.K., Bhutani, V.P., 1991. The influence of enzymatic clarification on fermentation behaviour and qualities of apple wine. Sciences Des Aliments 11, 491.

Joshi, V.K., Bhutani, V.P., Sharma, R.C., 1990. Effect of dilution and addition of Nitrogen source on chemical, mineral and sensory qualities of wild apricot wine. American Journal of Enology and Viticulture 41 (3), 229–231.

Joshi, V.K., Thakur, N.S., Anju, B., Chayanika, G., 2011. Wine and brandy: a perspective. In: Joshi, E.V.K. (Ed.), Handbook of Enology, vol. 1. Asia Tech Publishers, Inc., New Delhi, pp. 3–45.

Joshi, V.K., 1997. Fruit Wines, second ed. Directorate of Extension Education. Dr. YS Parmar University of Horticulture and Forestry, Nauni, Solan, HP, p. 255.

Joshi, V.K., Bhutani, V.P., Thakur, N.K., 1999a. Composition and nutrition of fermented products. In: Joshi, V.K., Pandey, A. (Eds.), Biotechnology: Food Fermentation, vol. I. Educational Publishers and Distributors, New Delhi, pp. 259–320.

Joshi, V.K., Sandhu, D.K., Thakur, N.S., 1999b. Fruit based alcoholic beverages. In: Joshi, V.K., Pandey, A. (Eds.), Biotechnology: Food Fermentation. Microbiology, Biochemistry and Technology, vol. II. Educational Publishers and Distributors, New Delhi, pp. 647–744.

Kaur, R., Kumar, K., Sharma, D.R., 2011. Grapes and genetic engineering. In: Joshi, V.K. (Ed.), Handbook of Enology, vol. 1. Asia Tech Publication, New Delhi, pp. 266–286.

Klatsky, A.L., Armstrong, M.A., 1993. Alcoholic beverage choice and risk of coronary heart disease mortality: do red wine drinkers fare best. American Journal of Cardiology 71, 467–469.

Kosseva, M., Kennedy, J.F., Lloyd, L.L., Beschkov, V., 1998. Malolactic fermentation in Chardonnay wine by immobilised *Lactobacillus casei* cells. Process Biochemistry 33, 793–797.

Kosseva, M.R., Webb, C., 2013. Food Industry Wastes: Assessment and Recuperation of Commodities. Academic Press, Elsevier, USA.

McGovern, P.E., Glusker, D.L., Exner, L.J., Voigt, M.M., 1996. Neolithic resinated wine. Nature 381, 480.

Panesar, P.S., Joshi, V.K., Panesar, R., Abrol, G.S., 2011. Vermouth: technology of production and quality characteristics. In: Advances in Food and Nutritional Research. 63. Elsevier, Inc., London, UK, pp. 253–271.

Pérez-Torrado, R., Querol, A., Guillamón, J.M., 2015. Genetic improvement of non-GMO wine yeasts: strategies, advantages and safety. Trends in Food Science & Technology 45, 1–11.

Pretorius, I.S., 1999. Engineering designer genes for wine yeasts. Australian and New Zealand Wine Industry Journal 14, 42–47.

Rivard, D., 2005. Professional Winemaker, in Consultation With AAFRD Alberta Fruit Winery Project.

Rivard, D., 2009. The Ultimate Fruit Winemaker's Guide, second ed. Bacchus Enterprises Ltd.

Schuller, D., Casal, M., 2005. The use of genetically modified *Saccharomyces cerevisiae* strains in the wine industry. Applied Microbiology and Biotechnology 68, 292–304.

Shahidi, F., 2009. Nutraceuticals and functional foods: whole versus processed foods. Trends in Food Science and Technology 20, 376–387.

Soni, S.K., Sharma, S.C., Soni, R., 2011. Yeast genetics and genetic engineering in wine making. In: Joshi, V.K. (Ed.), Handbook of Enology: Principles, Practices and Recent Innovations, vol. III. Asia Tech Publishers, New Delhi, pp. 441–501.

Stanier, R.Y., Doudoroff, M., Adelberg, E.A., 1970. General Microbiology. Prentice Hall, Inc, Engelwood Cliffs, NJ, Macmillan, New Delhi, pp. 2–10.

Stockley, C., 2011. Therapeutic value of wine: a clinical and scientific perspective. In: Joshi, V.K. (Ed.), Handbook of Enology: Principles, Practices and Recent Innovations, vol. 1. Asiatech Publishers, Inc., New Delhi, pp. 146–208.

Swami, S.B., Thakor, N.J., Divate, A.D., 2014. Fruit wine production: a review. Journal of Food Research and Technology 2, 93–94.

Vyas, K.K., Joshi, V.K., 1988. Deacidification activity of *Schizosaccharomyces pombe* in plum musts. Journal of Food Science and Technology 25, 306–307.

Vyas, S.R., Gandhi, R.C., 1972. Enological qualities of various grape varieties grown in India. In: Proc Symp Alcohol Beverage Ind India, Present Status and Future Prospects, Mysore, pp. 9–11.

SCIENCE AND TECHNOLOGY OF FRUIT WINES: AN OVERVIEW

V.K. Joshi[1], P.S. Panesar[2], V.S. Rana[1], S. Kaur[2]

[1]Dr. Y.S. Parmar University of Horticulture and Forestry, Nauni, Solan, HP, India; [2]Sant Longowal Institute of Engineering and Technology, Longowal, Punjab, India

1. INTRODUCTION

Wine has a rich history dating back thousands of years, with its earliest traces so far discovered in 6000 BC in Georgia (Robinson, 1994). It is regarded as a gift from God and has also been described as a divine fluid in Indian mythology from ancient times (Vyas and Chakravorty, 1971; Amerine et al., 1980; Joshi et al., 1999a,b). It has been prepared and consumed by humans since antiquity (Joshi et al., 2011a). Throughout the millennia, wine, more than most of the foods, has captured the imagination of poets and philosophers. Admittedly, except for water and milk, no other beverage has earned such universal acceptance and esteem throughout the ages as wine.

Wine is regarded as a food. The word "food" has many definitions, but it is reasonable to quote that provided by the Codex Alimentarius Commission: food means any substance, whether processed, semiprocessed, or raw, which is intended for human consumption, and this includes drink (Burlingame, 2008). The esthetic postures of wine can also be gauged by quotations like "wine is the most healthful and the most hygienic beverage" (Louis Pasteur) and "wine is a chemical symphony" (Amerine et al., 1980).

Wine is a completely or partially fermented juice of the grape, but fruits other than grapes, like apple, plum, peach, pear, berries, strawberries, cherries, currants, apricots, etc., have also been utilized for the production of wines (Amerine et al., 1967, 1980; Amerine and Joslyn, 1973; MAFF, 1980; Jackson and Schuster, 1981; Jackson, 1994; Joshi, 1997; Brand et al., 2001). Fruit wines are fermented alcoholic beverages made from a variety of base ingredients and can be made from virtually any plant matter that can be fermented.

The fruits used in winemaking are fermented using yeast and aged in wood barrels to improve the taste and flavor quality. A typical wine contains ethyl alcohol, sugar, acids, higher alcohols, tannins, aldehydes, esters, amino acids, minerals, vitamins, anthocyanins, and flavoring compounds (Joshi and Kumar, 2011). Being fruit based, fermented, and undistilled, wines retain most of the nutrients present in the original fruit juice (Joshi and Kumar, 2011; Swami et al., 2014). The yeast *Saccharomyces cerevisiae* var. *ellipsoideus* is the microorganism on whose activity the production of wine or any other alcoholic beverage depends (Rebordinos et al., 2010). The process of winemaking is unique in the sense that nearly all the physical, chemical, and biological sciences, especially microbiology and biochemistry, contribute to its production.

Science and Technology of Fruit Wine Production. http://dx.doi.org/10.1016/B978-0-12-800850-8.00001-6

The focus of this chapter is to present an overview of fruit wines other than grape wine, various fruits used, problems in production of fruit wines, technologies employed in production, global status, and future strategies contributing to the popularization of and marketing of these wines for their uses in human nutrition.

2. ORIGIN AND HISTORY OF WINE

It is established that wine is the oldest fermented product known to humans. Rather, the history of the alcoholic beverage is as old as that of humans. Among the alcoholic beverages, wine was the first to be made and has been used as a food adjunct by humans ever since their settlement in the Tigris–Euphrates Basin. In addition, it has a long history as a therapeutic agent. A peep into the history of humankind would clearly reveal that the preparation of fermented products like wine might have started accidently as a means of storage of perishables, but became an important method of preservation and preparation of products with appealing qualities, even today (Amerine et al., 1980; Joshi, 1997; Joshi et al., 2011a).

Ancient scriptures like the Rigveda and the New Testament have referred to wine, whereas literary writings have described wine profusely. The actual birthplace of wine is unknown, although it had been prepared by the Assyrians by 3500 BC. The qualities of wine were praised in pre-Christian as well as post-Christian times (Vine, 1981). In the past, the addition of spices and herbs to wine was a common practice, both for flavor and for medicinal or aphrodisiacal properties. Some other aspects related to wine history are summarized in Table 1.1.

Table 1.1 Salient Features in the History of Wine

- The Romans organized grape and wine production and during that era the flavor of wine was developed.
- The Greeks stored wine in earthenware amphoras, and the Romans extended the life of their wines with improved oak cooperage.
- Both Greek and Roman civilizations drank almost all of their wines within a year of vintage and disguised spoilage by adding flavoring agents.
- After the collapse of the western Roman Empire in the 5th century AD, the survival of viticulture depended on the symbolic role that wine played in Christianity.
- Muslims destroyed the wine industry of the countries they conquered.
- The need for wine for religious ceremonies led to the development of wine in central Europe.
- In western Europe, vineyards were developed.
- Early colonial fermented beverages were from sugar-rich fruits.
- The more aristocratic colonists preferred imported wines.
- By the 17th century, coopers were building more and better casks and barrels for longer and safe aging of wine.
- Wooden barrels remained the principal aging vessels until the 17th century, when mass production of glass bottles and the invention of the cork stopper allowed wines to be aged for years in bottles.
- The trend of wine consumption shifted toward distilled beverages in the 18th century after the discovery of distillation.
- By the 19th century, the scientific work of Pasteur revolutionized the wine industry by recognizing the roles of yeasts and bacteria.
- Pasteur also identified the bacteria that spoil wine and devised a heating method (called pasteurization) to kill the bacteria.
- In the 1960s, mechanization (grape harvesters and field crushers) in the vineyards contributed to better quality control.
- Advances in plant physiology and plant pathology also led to better vine training and less mildew damage to grapes.
- Stainless steel fermentation and storage tanks that could be easily cleaned and refrigerated to precise temperatures improved the quality of wine.
- Automated, enclosed racking and filtration systems reduced the contact with bacteria in the air, thereby preventing spoilage.

Adapted from Joshi, V.K., Thakur, N.S., Bhat, A., Garg, C., 2011a. Wine and brandy: a perspective. In: Joshi, V.K. (Ed.), Handbook of Enology: Principles Practices, vol. 1. Asia Tech Publisher Inc., New Delhi, pp. 3–45; Vine, R.P., 1981. Wine and the history of western civilization. In: Commercial Wine Making, Processing and Controls. The AVI Publishing Co., Westport, CT.

In the following sections we discuss the origin of wine, yeast, barrels, and chips and the discovery of microorganisms, their physiology, and their biochemistry.

2.1 ORIGIN OF WINE, YEAST, BARRELS, AND CHIPS

The origin of wine as mentioned earlier might have been accidental, when the juice of some fruit might have transformed itself into such a beverage having exhilarating or stimulating properties. When humans became civilized, wine and brandy were at the top of the list of requirements (Joshi et al., 2011a). Their consumption induced euphoria and pleasing relaxation from the strains of life, so it eventually gained social importance and was used for religious feasting and celebration as well as for entertaining guests. Starting about 1000 BC, the Romans made major contributions by classifying grape varieties and observing color and charting ripening characteristics, identifying diseases, and recognizing soil type preferences. Pruning skills and increased yields through irrigation and fertilization were also acquired by the Romans. The Greeks introduced viticulture to France, northern Africa, and Egypt, whereas the Romans exported the vines to Bordeaux, to the valleys of the Rhone, Marne, Seine, etc., and to Hungary, Germany, England, Italy, and Spain.

Evidence of the existence of grape farming in India has also been documented. The ancient Aryans possessed the knowledge of grape culture as well as preparation of beverages from it (Shanmugavelu, 2003). Grapes have been known in India since the 11th century BC. The famous Indian scholars Sushruta and Charaka, in their medical treatises entitled *Sushruta Samhita* and *Charaka Samhita,* respectively, written during 1356–1220 BC, mentioned the medicinal properties of grapes. However, the information about other fruits or fruit wines is totally lacking except that of cider, which was made and consumed widely (Vine, 1981), especially in England and France, well before the 12th century.

The existence of alcoholic beverages like wine in ancient times has been amply proven by paintings, articles, and writings of historic themes in various parts of the world. Most of the civilizations that had their characteristic wine or other alcoholic beverages had myths about the origin of winemaking and attributed its discovery to divine revelation. But the beginning of the art of winemaking was shrouded in prehistoric darkness. There is evidence to suggest that the process of winemaking existed even long before the chronicles found in Egyptian hieroglyphics. In the very first chapter of the Old Testament, it is described how Noah landed his ark on Mount Ararat and promptly planted a vineyard to make wine (Vine, 1981). It is also certain that wine drinking had started by about 4000 BC and possibly as early as 6000 BC. Texts from tombs in ancient Egypt amply prove that wine was in use around 2700–2500 BC when priests and royalty were using it. Archeological excavations have also uncovered many sites with sunken jars (Plate 1.1) indicating the existence of wine for more than 7500 years (McGovern et al., 1996). Evidence of winemaking first appeared in representations of wine presses (Fig. 1.1) that date back to the reign of Udimu in Egypt, some 5000 years ago (Petrie, 1923). The hypothesis of the Near Eastern origin and spread of winemaking is also supported by the remarkable similarity between the words meaning "wine" in most of the Indo-European languages (Renfrew, 1989). In addition, most eastern Mediterranean myths locate the origin of winemaking in northeastern Asia Minor (Stanislawski, 1975). The Phoenicians from Lebanon are well known for the introduction of wine and its secrets to the Romans and Greeks, who subsequently, propagated the art of winemaking.

Wine came to Europe with the spread of the Greek civilization around 1600 BC. Homer's *Odyssey* and *Iliad* both contain excellent and detailed descriptions of wine. It was an important article of Greek commerce, and Greek doctors, including Hippocrates, were among the first to prescribe it. The Romans'

PLATE 1.1

The oldest bottle of wine.

Reproduced from https://www.google.co.in/search?q=oldest+bottle+of+wine&client=firefox-b&tbm =isch&imgil
=BAiK19WA1QyPUM%253A%253BQ.

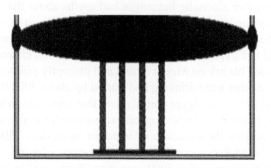

FIGURE 1.1

A diagrammatic view of a wine press (hieroglyph of Shemw, god of the wine press).

Redrawn from Gasteineau, F.C., Darby, J.W., Turner, T.B., 1979. Fermented Food Beverages in Nutrition. Academic Press, New York.

technology of winemaking was highly developed though it was lacking in the preparation of medicinal wines and methods of wine preservation. The export of Italian wine to Gaul in exchange for slaves was also in practice.

A breakthrough in the understanding of wine fermentation however came at the end of the 17th century, when Van Leeuwenhoek described the occurrence of yeasts in grape musts and beer worts. The

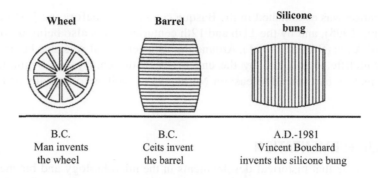

Wheel	Barrel	Silicone bung
B.C.	B.C.	A.D.-1981
Man invents the wheel	Ceits invent the barrel	Vincent Bouchard invents the silicone bung

FIGURE 1.2

Depiction of some of the important events in the history of humans.

first scientific work on fermentation was published by Lavoisier, and in 1836 Cagniard-Latour proved the role of yeasts as living organisms, which cause biochemical transformations, as summarized earlier (Goyal, 1999; Rana and Rana, 2011; Joshi et al., 2000). The wine yeast (*S. cerevisiae*) apparently is not an indigenous member of the grape skin flora; the natural habitat of the ancestral strains of *S. cerevisiae* may be the bark and sap exudates of oak trees (Phaff, 1986). The three major events in the history of humans including the origin of barrels are depicted in Fig. 1.2. The origin of barrels dates back to the prehistoric era of wine (Dennison, 1999).

2.2 HISTORY OF WINE

The earliest biomolecular archaeological evidence for plant additives in fermented beverages dates from the early Neolithic period in China and the Middle East, when the first plants and animals were domesticated and provided the basis for a complex society and permanent settlements (McGovern et al., 2009). In ancient China, fermented beverages were routinely produced from rice, millet, and fruit (McGovern et al., 2004). In earlier years, in Egypt, a range of natural products, specifically, herbs and spices, were added to grape wine to prepare herbal medicinal wines (McGovern et al., 2009). Information about nongrape wines is, however, lacking.

In the crypts of Pyramids, numerous grapes have been found (Vine, 1981). *The Periplus of the Erythraean Sea,* written toward the end of the first century AD and translated in 1912, noted that wine was produced in southern Arabia, particularly in the vicinity of Muza, modern Al Mokha. The oldest bottle of wine, which is from approximately 325 BC, was found in 1867 in an excavation work in a vineyard near the town of Speyer, Germany (Plate 1.1). In addition to bottles, the Romans also developed wooden cooperage for wine storage. Wine amphoras may also have been employed to store on their sides or upside down, thus keeping the cork wet with wine (Addeo et al., 1979; Grace, 1979; Koehler, 1986). Amphoras, cork-sealed and containing wine remnants, have been excavated on several occasions from the Mediterranean (Cousteau, 1954; Frey et al., 1978). Queen Nefertiti is reported to have used wine as a base for her perfume (Vine, 1981). Ancient relics of wine growing are proudly displayed at the wine museum in Beaune, the capital city of Burgundy in France. In the north of Beaune is the famous Clos de Vougeot, established during the mid-1300s by the monks of Citeaux Abbey, where much of the winery remains are intact. In Germany, a few miles west of Wiesbaden near the Rhine, one can see the magnificent Kloster Eberbach, which was used for making wine (Vine, 1981).

Cider preparation was established in the Basque country well before the 12th century (Forbes, 1956; Jarvis et al., 1995), and by the 11th and 12th centuries it was also being produced in Cotentin and in Pays d' Auge (Braudel, 1981). Around the 13th century, it was also being made in southeast England (Sutcliffe, 1934) and by the end of the 15th century and beginning of the 16th century, its production had spread to eastern Normandy as well as to Brittany as reviewed (Joshi et al., 2011a).

2.3 ALCOHOLIC FERMENTATION

It is interesting to note that historical developments in the microbiology and biochemistry of alcoholic fermentation paralleled the developments in wine fermentation. Some of the important events in the history of alcoholic fermentation are summarized here. Jan Baptist van Helmont (1577–1644) explained some of the concepts of fermentation and the chemistry involved in it. Antoine Lavoisier (1743–94) restored the term "alcohol" and quantitatively determined the amount of carbon dioxide and ethanol produced during fermentation of grape juice and gave us the equation of ethanolic fermentation (Areni et al., 1999). The "Father of Microbiology" Leeuwenhoek, in 1680, observed yeast cells with his ground lens, for the first time, in fermenting beer, and Schwann, in 1837, recognized it as a fungus and gave it the name "Zuckerpilz" (sugar fungus). This was perpetuated in the generic term *Saccharomyces* (Lafar, 1910). From 1855 to 1876 the idea that fermentation was a physiological action associated with the life processes of yeast (Comant, 1952; Pasteur, 1866) was considered as a milestone in the development of the biochemistry of fermentation. Later on, the substances responsible for fermentation were named "enzymes," which means "in yeast," coined by Wilhelm Kühne in 1878; however, in 1897 Buchner obtained an enzyme from cell-free juice from yeasts that was not capable of fermentation (Harden, 1924). Neither the filtrate nor the residue from the yeast cell dialysis was capable of fermenting glucose. But on combination of the two, the fermentation took place, indicating that fermentation required the presence of another substance, "coenzyme," which was dialyzable and thermostable. Thus, the study of biochemistry and biochemical reactions has evolved from research on yeast and alcoholic fermentation (Dubos, 1950; Krebs, 1968; Rose et al., 1971). For more information, readers are referred to the literature cited (Leicester, 1974; Amerine et al., 1980; Goyal, 1999) and chapters in this book on the microbiology of wine fermentation (Chapter 2) and the chemistry of winemaking (Chapter 3).

3. ROLE OF WINE AS FOOD AND ITS HEALTH BENEFITS

Wine has been a part of the human diet ever since the settlement of the Tigris–Euphrates Basin, as reviewed earlier (Joshi, 1997; Gasteineu et al., 1979; Vine, 1981). It has been used as a therapeutic agent and served as an important adjunct to the human diet by increasing satisfaction. The medicinal or therapeutic value of wine has been highly acclaimed in the scientific literature (Catherine, 1996; Klatsky and Armstrong, 1993; Gronbaek et al., 1995). Wine contains some minerals, vitamins, sugars, acids, phenols, and small quantities of the B vitamins such as B_1 (thiamine), B_2 (riboflavin), and B_{12} (cyanocobalamine), but is devoid of vitamins A, D, and K (Anon and Barid, 1963; Gasteineau et al., 1979; Ibanezo et al., 2008; Soni et al., 2011a).

Because wines and beers are not distilled, they have more nutrients, such as vitamins, minerals, and sugars, than distilled beverages like brandy and whisky (Soni et al., 2011a). Clearly, wines and beers are more nutritious than distilled liquors, as advocated earlier also (Gasteineau et al., 1979; Joshi et al., 1999a). The mineral composition of wine would also support this contention (Table 1.2). The presence of various phenolic compounds (catechin, epicatechin, quercitin, and ellagic acid) has been identified by thin-layer chromatography in wine of various cultivars of strawberry (Joshi et al., 2009a). The winemaking process releases many of these bioactive components into the aqueous ethanolic solution, thus making them more biologically available for absorption during consumption (Shahidi, 2009).

In ancient times, wine was also used in wound dressing, as a nutritious dietary beverage, as a cooling agent for fevers, as a purgative, and as a diuretic by Hippocrates of Cos (460–370 BC)

Table 1.2 Mineral Contents of Various Fruit Wines

Type of Beverage	Minerals (mg/L)							
	Na	K	Ca	Mg	Cu	Fe	Mn	Zn
Vermouth								
Grape vermouth Quina	111.64	735.64	89.25	62.18	0.53	6.95	0.58	–
Aperitivo vini	45.56	297.62	57.06	53.03	0.46	5.13	0.47	–
Vermut blanco	58.70	225.00	59.50	37.17	0.48	4.31	0.38	–
Vermut rojo	73.65	524.72	54.96	57.57	0.42	7.13	0.34	–
Plum vermouth	41.00	973.00	101.00	17.0	1.07	1.30	1.07	0.82
Sand Pear vermouth	45.0	967.00	43.00	15.0	1.23	7.11	1.23	2.39
Wine								
Grape wine	51.00	803.00	106.00	88.00	3.00	0.13	0.66	0.70
Apricot wine (Newcastle)	11	1481	18	71	2.72	0.96	1.92	0.88
Wild apricot wine (chulli)	43	2602	25	94	5.97	0.50	2.69	0.99
Apple wine (Golden Delicious)	18	1044	11	144	3.68	0.21	0.76	0.84
Cider (HPMC apple juice concentrate)	61	1900	23	137	4.31	0.32	1.54	1.01
Hard cider (Golden Delicious)	19	1069	17	97	3.03	0.19	0.91	0.82
Pear wine (Sand Pear)	87	1906	37	122	8.91	0.16	0.80	1.10
Plum wine (Santa Rosa)	20	1008	18	82	12.73	0.20	1.04	0.95

Compiled from Amerine, M.A., Kunkee, R.E., Ough, C.S., Singleton, V.L., Webb, A.D., 1980. Technology of Wine Making. AVI Publ. Co., Westport, Connecticut; Joshi, V.K., Sandhu, D.K., Thakur, N.S., 1999b. Fruit based alcoholic beverages. In: Joshi, V.K., Pandey, A. (Eds.), Biotechnology: Food Fermentation (Microbiology, Biochemistry and Technology), vol. II. Educational Publishers and Distributors, New Delhi, pp. 647–744; Bhutani, V.P., Joshi, V.K., Chopra, S.K., 1989. Mineral contents of fruit wines produced experimentally. Journal of Food Science Technology 26 (6), 332–333; Bhutani, V.P., Joshi, V.K., 1995. Plum. In: Salunkhe, D.K., Kadam, S.S. (Eds.), Handbook of Fruit Science and Technology, Cultivation, Storage and Processing. Marcel Dekker, New York, USA, pp. 203–241.

(Lucia, 1963; Seward, 1979). The healthful benefits of wine are also associated now with the antimicrobial activities of ethanol and antioxidant properties of phenolic components and flavonoids (Nijveldt et al., 2001; Kinsella et al., 1993). Glucose tolerance factor, a chromium-containing compound that is synthesized by yeast and considered beneficial in the cure of diabetes, is also found in wine (Schwarj and Mertz, 1959; Offenbacker and Sunyer, 1980; Tuman and Doisy, 1977; Sandhu and Joshi, 1995).

Moderate alcohol and/or wine consumption protects against the incidence of many diseases of modern society like cardiovascular diseases, dietary cancers, ischemic stroke, peripheral vascular disease, diabetes, hypertension, peptic ulcers, kidney stones, and macular degeneration, in addition to stimulating resistance to infection and retention of bone density (Leake and Silverman, 1966; Jindal, 1990; Joshi et al., 1999a; Stockley, 2011).

Most of the antimicrobial effects of wine have been attributed to phenolic compounds such as *p*-coumaric acid, which is particularly active against Gram-positive bacteria, whereas other phenols inhibit Gram-negative bacteria, for example *Escherichia coli, Proteus,* and *Vibrio* (Masquelier, 1988), which are known to cause various forms of diarrhea. Apple wine has also been found to have antimicrobial activity against pathogens and microbes of public significance (Joshi and John, 2002). Wine has also been found to be a tranquilizer, to be a diuretic, to reduce muscle spasms and stiffness associated with arthritis, to delay the development of some forms of diabetes and cardiovascular diseases, to have antioxidant effects, and to inhibit platelet aggregation, among other effects (Jackson, 2000). It is a particularly rich dietary source of flavonoid phenolics, including resveratrol. Resveratrol could reduce coronary heart disease (CHD) mortality by its ability to inhibit platelet aggregation, eicosanoid synthesis, and oxidation of human low-density lipoproteins. Resveratrol can reduce serum lipid levels and can prevent or inhibit cellular events associated with tumor initiation, promotion, and progression, and may help in prevention of cardiovascular diseases and cancer (Joshi et al., 1999a, 2011a; Joshi and Preema, 2009). Resveratrol concentration is higher in red wines than in white wines (Jeandet et al., 1993; Okada and Yohotosuha, 1996; Langcake and Carthy, 1979). The relationship of alcohol and mortality has been described as a J-shaped curve, which successfully explains the French Paradox, whereby people consume a large quantity of fat as well as wine, but have reduced incidence of heart diseases (Stockley, 2011). With regard to the fruit wines, there is scarcity of such type of research and consequently a lack of relevant literature.

4. FRUIT WINES, THEIR TYPES AND DIVERSITY

4.1 CLASSIFICATION OF WINE

Unless otherwise specified, the term "wine" is applied here to the product made by the alcoholic fermentation of grape or grape juice by the wine yeast and a subsequent aging process (Amerine et al., 1980; Jackson, 1994, 2000, 2003; Joshi, 1997; Joshi et al., 2004). It is a completely or partially fermented juice of the grape, but fruits other than grapes, viz., apple, plum, peach, pear, berries, strawberry, cherry, currant, apricot, rhubarb, banana, pineapple, cashew nut, pomegranate, lemon, tangerine, orange, date, fig, etc., have also been utilized for the production of wines (Amerine et al., 1967, 1980; Amerine and Joslyn, 1973; Jackson, 1994, 1999; Sandhu and Joshi, 1995; Joshi, 1997; Joshi et al., 1999b, 2000, 2004).

4.2 TYPES OF FRUIT WINES

Generally, there is no accepted system of classification of wines (Amerine et al., 1980; Jackson, 2000, 2003), though they may be classified broadly based on geographic origin, grape variety used, fermentation, or maturation process or may also bear generic names. Based on alcohol content, wines can be classified as table wines and fortified wines. The system already employed for the classification of wine from grapes is made use of in the classification of nongrape wines, as described here. The broad classification of wine is depicted in Fig. 1.3 and is described here briefly, and for more detail the reader may consult the literature cited (Karagiannins, 2011).

Similar to the wines from grapes, there are different types of fruit wines depending upon the method of preparation, with or without distillation; the raw material used; and the vinification practices. Most of the wines are still wines, as they retain no carbon dioxide produced during the fermentation, in contrast to sparkling wines, which contain a considerable amount of carbon dioxide artificially. Champagne in France is the sparkling wine made in the Champagne region (Amerine et al., 1980; Jeandet et al., 2011; Joshi et al., 1995). It is prepared either by bottle or by tank fermentation for carbonation of the base wine by secondary fermentation.

Wines usually contain 11–16% alcohol but can be as low as 7%. Fortified or dessert wines contain added brandy and can be red or white, with alcoholic content ranging from about 16% to 23%. These wines are usually sweet and have as high as 18–20% alcohol. Wine can be classified based on the carbon dioxide content, as those without it are still wines and those with it are sparkling wines.

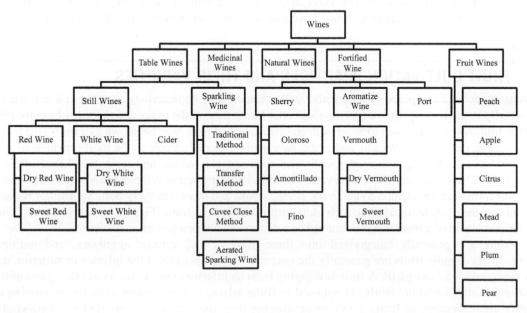

FIGURE 1.3

Various types of wines.

Dry·wines have negligible or no sugar content, whereas those with perceptible levels of up to 8% sugar are termed as sweet wines. Table wines have a comparatively low alcoholic content and little or no sugar, whereas dessert wines are fortified sweet wines. The table wines generally have an alcohol content of 10–11%, whereas that of cider and perry is usually 2–8% (Amerine and Joslyn, 1973; Joshi, 1997).

Fortified wines, to which a distillate of wine called brandy is added, may contain 19–21% alcohol.

In Great Britain and France, the term "cider" (or cidre) means apple wine, hard cider, or fermented apple juice, but in the United States it may mean fermented or unfermented apple juice based on the definitions and available products; broadly it can be classified as soft cider, with 1.5% alcohol content, or hard cider, with 5–8% alcohol content. Apple wine contains alcohol above 8% but may go up to 14%.

Boukha is wine made from figs. Perry is a wine prepared from pear juice. It can be sweet or dry. Because the pear is more astringent, the same characteristic is imparted to the wine also. The mead type of wine used to be prepared by the ancient people of India and was known for excellent digestive qualities. Honey is also used for sweetening the musts for other fermentation.

Medicinal wines are flavored wines containing extracts of several herbs of medicinal importance and may even include bitter compounds, quinine, for example. These wines are usually sweet and have as high as 18–20% (v/v) alcohol.

Natural wines are the products obtained by the use of approved formulas with a natural wine base, herbs, spices, fruit juices, aromatics, essences, and other flavoring materials and are intended to attract the non-wine-drinking public. Many of these wines are under 14% alcohol.

Vermouth is a wine with 15–21% (v/v) alcohol, flavored with a characteristic mixture of herbs and spices, some of which impart an aromatic flavor and odor, and others a bitter flavor. It may be sweet or dry.

5. FRUIT CULTIVATION PRACTICES AND THEIR VARIETIES

Fruits are defined as ripened ovary together with whatever may be intimately attached to it at maturity. Normally in the angiosperms, a fruit develops after fertilization of the ovary and the seeds lie embedded in the fruit. An enlarged pistil together with closely associated parts may be known as a fruit. True fruits, like orange, persimmon, avocado, and grape, have well-developed carpels as the most conspicuous portion. In fruits developing from epigynous flowers, such as apple, banana, and cucumber, the receptacle is a part of the fruit structure. There are a variety of fruits based on their development and organs associated in their formation. In addition to the ovary, petals, sepals, and other accessory parts sometimes become fused with the ovary to form a fruit; such fruits are botanically false fruits (Fig. 1.4A). Horticulturally, fruit is the edible part of a plant that is consumed as a dessert and not as a part of main diet.

Fruits are generally categorized into three groups, viz., simple, aggregate, and multiple (Fig. 1.4B). Simple fruits are generally the outcome of the ripening of the inferior or superior, dry or fleshy ovary of one pistil. A fruit developing from an inferior ovary consists of other parts of the flower in addition to the ovule, as opposed to fruits arising out of a superior ovary, consisting of ovary alone. Aggregate fruits are those developing from the apocarpous pistil (free carpels) of a flower. It refers to the aggregation of simple fruits borne by a single flower (etaerio), e.g., etaerio of achenes (strawberry), etaerio of berries (*Annona* group, e.g., custard apple, bullock heart,

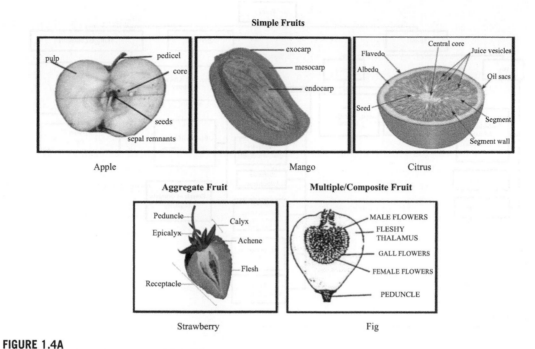

FIGURE 1.4A

Examples of simple, aggregate, and composite fruits.

Adapted from https://www.google.co.in.

atemoya, and cherimoya), and etaerio of drupes (raspberry, blackberry, and loganberry). Multiple or composite fruits are those developed as a result of the fusion of a number of flowers, e.g., sorosis (developing from a spike or spadix, e.g., pineapple, jackfruit, breadfruit, and mulberry) and syconus, i.e., developing from a hypanthodium, e.g., fig (Singh, 2000).

5.1 NONGRAPE FRUITS USED FOR WINEMAKING

A number of fruits are cultivated throughout the world with a large production to convert into alcohol during the fermentation process. The percentage contribution of grapes to the total fruit production is nearly 15.53% (Fig. 1.5). Therefore, there is ample scope for preparation of wine from other fruits also. Several nongrape fruits are used in many parts of the world for the production of wine, such as apples, berries, cherries, wild apricots, kiwifruit, plums, peaches, and strawberries (Joshi, 2009). One of the most widely produced nongrape fruit wines is cider, or apple wine, which is made from fermented apples throughout England, Germany, France, Spain, Ireland, Argentina, and Australia. While ciders can be made from any apple variety, they are typically produced from specific cultivars that are high in sugar content and phenolic content.

The Japanese, Koreans, Chinese, and Taiwanese have their own unique versions of fruit wines made from plums. Whereas ciders can range from dry to sweet, most plum wines available in the United States tend to be sweeter in style. Plum wines can also be used in cocktails, either with soda water in spritzers or

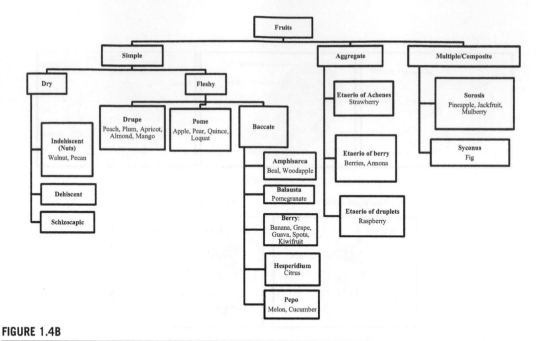

FIGURE 1.4B

Classification of fruits.

Adapted from Sandhu and Chattopadhyay, 2001.

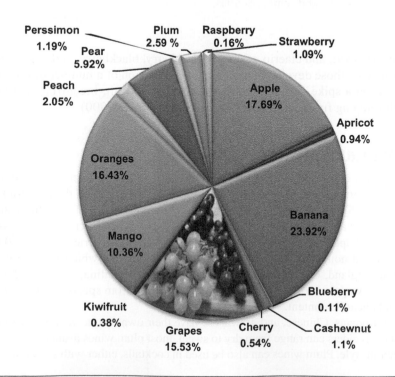

FIGURE 1.5

Percentage production of major fruit crops suitable for winemaking.

Based on FAOSTAT, 2012. http://www.fao.org.

as a complement to *shochu*, which is a spirit made from distilled rice, barley, or sweet potatoes. Furthermore, by blending apples, peaches, pears, and autumn berries in England, well-balanced, complex dry white wines are made. Many of the red wines are prepared from blends of blueberries, blackberries, currants, and other dark fruits growing wild. These are aged in oak up to 2 years. When dealing with fruits other than grapes, sugar may be added to accelerate the fermentation process if the fruit does not contain enough natural sugar to ferment on its own in the presence of yeast. Some fruits, such as cherries, raspberries, strawberries, and pineapples, are also very high in acid, which can translate into a very sour tasting wine. In such cases, sucrose and/or water can be added to counter the fruit's tart acidity. All these aspects are described in various technological Chapters 7–10 of this book and in the literature cited too.

5.2 CULTURAL PRACTICES AFFECTING WINE QUALITY

Quality is perhaps the most important factor in the production of many edible fruit species especially grown for wine production. Wine quality is primarily determined by the fruit quality, mainly its chemical composition (Table 1.3). Despite the recent improvements in winemaking technologies, wine quality is determined primarily by the chemical composition of the fruit. Some of the chemical constituents of fruit do not undergo changes during the fermentation process, whereas others serve as precursors for newly formed compounds. These aspects have been further developed in Chapter 3 of this book. Basic perceptions such as bitterness, sweetness, sourness, and astringency (Joshi, 2006) are elicited by groups of compounds including glycosides, dipeptides, sugars, acids, alkaloids, and phenols characterized by low volatility (Goyal, 1999; Rana and Rana, 2011).

Fruit quality is affected by both environmental and plant factors. The environmental factors known as "terroir" consist of the interaction between soil and climate. In fact, this interaction also includes the crop load as determined by the reproductive-to-vegetative ratio as summarized earlier (Kumar et al., 2011a). The effect of soil on fruit and wine quality is most important in the nonirrigated old orchards where the water and mineral supply is greatly dependent on the soil type. In a dry climate, drip irrigation enables absolute control of the water supply to the plants as well as an efficient mineral supply through the irrigation water. The climate is an important factor. Day and night temperatures, solar radiation, air humidity, and wind velocity at the plantation site are functions of the macroclimate, whereas the microclimate in the immediate vicinity of the clusters is affected by canopy management and the training system. Cultural practices such as irrigation and fertigation can also affect the microclimate through manipulation of vegetative growth. Some of the aspects are discussed here as regards fruit. However, no detailed discussion in depth can be made considering the scope of this volume.

5.2.1 Apple

Table 1.3 Approximate Composition of Major Fruits Grown in the World (per 100 g of Edible Portion)

Fruit	Water (%)	Energy (kcal/100 g)	Protein (%)	Fat (%)	Carbohydrate (%)	Calcium (mg)	Phosphorus (mg)	Iron (mg)	Magnesium (mg)	Ascorbic Acid (mg)
Apple	83.9	59.0	0.19	0.6	15.3	7.0	10.0	0.3	8.0	4.0
Pear	87.7	61.0	0.7	0.4	15.8	13.0	16.0	0.3	7.0	4.0
Peach	89.1	38.0	0.6	0.1	9.7	9.0	19.0	0.5	10.0	7.0
Plum	81.1	66.0	0.5	0.2	17.8	18.0	17.0	0.5	9.0	5.0
Apricot	85.3	51.0	1.0	0.2	12.8	17.0	23.0	0.5	12.0	10.0
Strawberry	89.9	37.0	0.7	0.5	8.4	12.0	21.0	1.0	12.0	59.0
Kiwifruit	81.2	60.0	0.11	0.07	17.5	16.0	22–64	0.51	30.0	105
Orange	86.0	49.0	1.0	0.2	12.2	41.0	20.0	0.4	11.0	50.0
Mango	83.4	59.0	0.5	0.2	15.4	12.0	12.0	0.8	0.063	53.0
Grape	81.6	67.0	1.3	1.0	15.7	16.0	12.0	0.4	13.0	4.0
Banana	75.7	85.0	1.1	0.2	12.6	8.0	26.0	0.7	33.0	10.0
Pineapple	85.4	52.0	0.4	0.2	13.7	18.0	8.0	0.5	8.0	61.0
Pomegranate	83.0	83.0	1.67	1.17	18.7	10.0	36.0	0.3	12.0	10.2
Blueberry	83.4	57.0	0.74	0.33	14.49	6.0	12.0	0.28	6.0	9.7
Blackberry	88.5	43.0	1.39	0.49	9.61	29.0	22.0	0.62	20.0	21.0
Raspberry	87.0	53.0	1.2	0.65	11.94	25.0	29.0	0.69	22.0	26.2
Jamun	84.75	60.0	0.99	0.23	14	11.65	15.6	1.41	35.0	11.85
Lychee	83.0	66.0	0.83	0.44	16.53	5.0	31.0	0.13	10.0	71.5
Loquat	58.0	47.0	0.43	0.2	12.14	16.0	27.0	0.28	13.0	1.0
Persimmon	80.0	70.0	0.58	0.19	18.59	8.0	17	0.15	9.0	7.5

Apple is a mature or ripened ovary/ovaries fused together with many closely associated parts whereby five ovaries of the flower are embedded in tissue, which, along with the thalamus or carpel tissue, becomes fleshy and edible.

Apple is grown in most of the temperate parts of the world and higher hills of subtropical regions. It is a deciduous tree in the rose family and is best known for its sweet, pomaceous fruit. Apple wine is a product made from apple juice by alcoholic fermentation (11–14% alcohol). Apple juice or concentrate is the basic raw material but as the alcohol content of wine is more than that of cider, addition of sugar or a sugar source has to be made. Growing cider apples is very similar to growing dessert apple varieties (Downing, 1989; Joshi et al., 2011c). They have similar cultural requirements of climate, soils, site selection, nutrition, irrigation, and pest and disease control. Cider is made by pressing apples to produce juice and then fermenting that juice to make cider. Most commercial ciders available in Australia are a blend of dessert and cider apple varieties, although they can be made exclusively from cider apple juice. Cider and other apple wines are traditional products in some countries (Amerine et al., 1980). In some European countries such as England, France, and Germany, apple wine is produced in significant quantities, although a considerable proportion is also distilled to make apple brandy (Calvados) (Amerine et al., 1980). Cider is also made in most other European countries, in the United States, and in Canada and is now available in India also.

5.2.2 Apricot

Apricot is a delicious fruit having a characteristic, pleasing flavor and originated in western China but grows in many parts of hilly temperate countries including India.

It requires cool weather to break dormancy and dry sunny weather or spring and warm summer for fruit maturity. Hot weather with temperature above 38°C, however, injures the fruit. The fruit, because of its high flavor, holds promise for conversion into wine. Wild apricot (chulli) in India is used locally by tribes in the hills to make liquor. Wild apricot trees are widely distributed in many areas including Europe, western and central Asia, Baluchistan, the northwest Himalayas, and western Tibet up to 12,000 ft. In India the fruit is found growing in Jammu and Kashmir, Utrrakhand, Himachal Pradesh, and some parts of the northeastern states. Chulli oil has medicinal value and is said to cure arthritis and joint pains.

5.2.3 Banana

Banana is one of the oldest fruits known to humans as the "apple of paradise." Frequent mention has been made of this fruit in the great Indian epics, *Ramayana* and *Mahabharata*. Banana is a tall herb producing a pseudo stem, which is the aerial stem made of a number of leaf sheaths completely encircling the axis of the stem. The fruit is a berry and is edible. Bananas develop by vegetative parthenocarpy. The fruit is variable in size, color, and firmness, but is usually elongated and curved, with soft flesh rich in starch covered with a rind, which may be green, yellow, red, purple, or brown when ripe. Production of banana wine is carried out mostly on a small scale, though there are some commercial producers of banana wine also (Kundu et al., 1976).

5.2.4 Berries

Berries are considered soft fruits and include botanically different types of fruits such as blackberries, blueberries, strawberries, cranberries, gooseberries, currants, and raspberries.

The genus *Vaccinium* contains over 100 species of deciduous and evergreen shrubs and small trees. Many of these have considerable economic importance, particularly in North America, where blueberries and cranberries are produced on arid land, much of which would otherwise be rather barren (Jackson and Looney, 1999).

The genus *Rubus* includes raspberries, subgenus *Idaeobatus*, which are distinguished from blackberries, subgenus *Eubatus*. The European raspberry is *Rubus idaeus*, one of over 400 species of the genus found in most European countries and the North American continent. Two related species *Rubus occidentalis*, with black or yellow fruit, and *Rubus strigosus*, with light red and sometimes white and yellow fruit, have provided much of the genetic material for development of many of the American cultivars of raspberry.

5.2.5 Cashew Nut

The cashew tree (*Anacardium occidentale*) is a tropical evergreen that produces the cashew nut and the cashew apple. It can grow as high as 14 m (46 ft), but the dwarf cashew, growing up to 6 m (20 ft), has proved more profitable, with earlier maturity and higher yields. The cashew apple is a fruit, whose pulp can be processed into a sweet, astringent fruit drink or distilled into liqueur. In Goa, the cashew apple (the accessory fruit) is mashed, and the juice is extracted and kept for fermentation for a few days. The fermented juice then, undergoes a double distillation process. The resulting beverage is called *feni* or fenny and has about 40–42% alcohol. The single-distilled version is called *urrac*, which has about 15% alcohol. In the southern region of Mtwara, Tanzania, the cashew apple (*bibo* in Swahili) is dried and stored. Later on it is reconstituted with water and fermented and then distilled to make a strong liquor often referred to by the generic name of *gongo*. In Mozambique, cashew farmers commonly make strong liquor from the cashew apple, *aguaardente* (burning water). According to *An Account of the Island of Ceylon* by Robert Percival, alcohol had been distilled in the early 20th century from the juice of the fruit and had been manufactured in the West Indies.

5.2.6 Cherry

Cherries are eaten either as dessert fruit or more conveniently in a brined, frozen, or canned state. De Candole (1959) reported that the cherry first grew wild in northern Persia and the Russian Provinces south of the Caucasus, but it spread rapidly because of its attractiveness to birds, hence the name *Prunus avium* L. or bird cherry.

The sour or tart cherry, *Prunus cerasus* L., is characterized both by the taste of its fruit and by the spreading characteristics of the tree (Romani and Jennings, 1971). The sweet cherry (*P. avium*) and the sour (red) cherry (*P. cerasus*) have their commercial origin in Europe. Duke cherries are probably hybrids between sweet and red tart cherries. To prepare wine from cherries, sour-type cultivars are preferred more (Schanderl and Koch, 1957), but a blend of currants and the table variety of cherries may also serve the purpose. To enhance the flavor, about 10% of the pits may be broken down while crushing the cherries. However, the production of hydrogen cyanide from the hydrolysis of amygdalin, present in the pit of cherries, has been reported (Bauman and Gieschner, 1974; Benk et al., 1976; Missehorn and Adam, 1976). Cherry fruit has been found more suitable for preparing a dessert wine than a table wine. The alcohol content of such wines may range from 12% to 17%.

5.2.7 Jamun

Jamun is an important minor fruit of Indian origin, which grows widely in various agroclimatic conditions and is one of the hardiest fruit trees (Singh et al., 1967). It is an evergreen tropical tree in the flowering plant family Myrtaceae. *Syzygium cumini* is native to Bangladesh, India, Nepal, Pakistan, Sri Lanka, Malaysia, the Philippines, and Indonesia.

Jamun is also found in Thailand, the Philippines, the West Indies, and many other tropical and subtropical countries (Morton, 1968). The fruit syrup is very useful for curing diarrhea. It is stomachic, carminative, and diuretic, apart from having cooling and digestive properties (Thaper, 1958).

Jamun seeds are also known for their medicinal properties and are used to treat diabetes, diarrhea, and dysentery. Bhargva et al. (1969) have shown that consumption of jamun markedly lowers blood pressure. The fruit can be used for making dry wine of acceptable quality, and high-quality alcoholic beverage can also be prepared from jamun juice (Joshi et al., 2012a). In Goa and the Philippines, jamun fruits are an important source of wine, somewhat like port and the distilled liquors. Brandy and jambava have also been made from the fermented fruit.

5.2.8 Mandarin Orange

Mandarin orange is the most common citrus species cultivated in India and occupies nearly 50% of the total citrus area in the country. Its cultivation is spread in certain belts, for instance, the Nagpur mandarin is chiefly grown in the Vidarbha region of Maharashtra, *Khasi* mandarin is common in the northeastern region, Darjeeling mandarin in the Darjeeling district of West Bengal, and the kinnow mandarin in the states of Punjab and Haryana. Kinnow is a hybrid of two mandarin cultivars, King (*Citrus nobilis*)×Willowleaf (*Citrus deliciosa*), first developed by Howard B. Frost at the University of California Citrus Experiment Station. After evaluation, the kinnow was released as a new citrus hybrid for commercial cultivation in 1935. The factors that have contributed to the success of this fruit are its beautiful golden-orange color, abundant juice, excellent aroma, and taste. Kinnow has many superior characteristics including heavy bearing and good quality, a source of vitamin C and potassium. But the kinnow juice, like other citrus juices, becomes bitter (Joshi et al., 1997) and this is carried on to the wine.

5.2.9 Kiwifruit

The kiwifruit or Chinese gooseberry is the edible berry of a woody vine in the genus *Actinidia*. The fruit is the size of a large hen's egg, 5–8 cm in length and 4.5–5.5 cm in diameter.

It has a fibrous, dull greenish-brown skin and bright green or golden flesh with rows of tiny, black, edible seeds. The fruit has a soft texture and a sweet but unique flavor, and today it is a commercial crop in several countries, such as Italy, New Zealand, Chile, Greece, and France. In India, its performance in Himachal Pradesh is very good.

Kiwifruit is a rich source of vitamins C and K, and a good source of dietary fiber and vitamin E. The fruit and skin contain flavonoids, actinidain, and adhered pollen, which may produce irritation in the mouth and throat of some allergic individuals. A wine of good quality can be prepared from kiwi fruit pulp using two varieties, i.e., Bruno and Hayward (Lodge, 1981; Gautam and Chundawat, 1998). Kiwi juice has high acidity and low sugar, which necessitates juice amelioration for the production of a balanced wine. The wine produced from the fruit is described as of unusual composition and Riesling Sylvan character (Heatherbell et al., 1980).

5.2.10 Lychee

The lychee is a subtropical fruit known for its delicious flavor and sweet juicy aril. It originated in southern China, where it has been cultivated for over 4000 years. The edible portion of the fruit is a white- to cream-colored translucent pulp surrounding a glossy brown seed. The pulp is grapelike in texture, very succulent and aromatic, and is characterized by a sweet and acid taste (Cavaletto, 1980). Lychee was introduced to the tropical and subtropical world from China at the end of the 17th century and is now found situated within 15–30° latitude in most of those countries (Menzel and Simpson, 1993). Lychee is extensively grown in China, Vietnam, and the rest of tropical southeast Asia, the Indian subcontinent, and more recently in South Africa, Brazil, the Caribbean, Queensland, California, and Florida. It requires a warm subtropical to tropical climate that is cool but also frost-free or with only slight winter frosts not below −4°C, and with high summer heat, rainfall, and humidity. Growth is best on well-drained, slightly acidic soils rich in organic matter. Lychee fruit has plenty of flavor and is a good source of minerals and vitamins. The fruit is used for the preparation of alcoholic beverages in China. A low-alcohol high-flavored beverage using the techniques of partial osmotic dehydration has been made as described earlier (Joshi et al., 1999b; Vyas et al., 1989b).

5.2.11 Loquat

The loquat (*Eriobotrya japonica*) is a species of flowering plant in the family Rosaceae, native to south-central China. It is a large evergreen shrub or a small tree, grown commercially for its yellow fruit, in addition to being an ornamental plant. It can also be used to make light wine by fermentation. Being low in saturated fat and sodium, and high in vitamin A, dietary fiber, potassium, and manganese, it is a nutritious fruit. Both loquat seeds and apricot kernels contain cyanogenic glycosides, but the drinks are prepared from varieties that contain only small quantities (such as Mogi and Tanaka), so there is no risk of cyanide poisoning.

5.2.12 Mango

The mango is native to south and southeast Asia, from where it has been distributed worldwide to become one of the most cultivated fruits in the tropics. The center of diversity of the *Mangifera* genus is in India; thus it is essentially a tropical plant but grows equally well in frost-free subtropical regions and its range of climate varies from subhumid equatorial to subarid frost-free subtropical. It has a high amount of total soluble solids, proteinaceous substances, vitamins, and minerals and is thus, suited to conversion into wine. A preliminary screening of 10 varieties of mango for winemaking has successfully been made as reported earlier (Kulkarni et al., 1980).

5.2.13 Passion Fruit

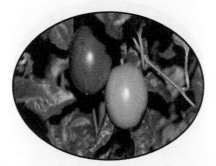

Its common names include passion fruit (United States), passionfruit (UK and Commonwealth countries), and purple granadilla (South Africa). It is cultivated commercially in tropical and subtropical areas for its sweet, seedy fruit and is widely grown in several countries of South America, Central America, the Caribbean, Africa, southern Asia, Israel, Australia, Hawaii, and other United States.

The passion fruit originated in the tropical highlands in South America, but has escaped from cultivation to become endemic in tropical and subtropical wooded upland areas. The purple passion fruit is the form most commonly grown commercially throughout the world but generally on a small scale. A yellow form is also popular and tends to be larger and more globular, whereas the purple passion fruit is ovoid in shape. Hybrids between the two forms are, however, grown commercially in Australia, and are a vine species of passion flower that is native to Brazil, Paraguay, and northern Argentina.

5.2.14 Peach

The peach (*Prunus persica*) is a deciduous tree, native to northwest China in the region between the Tarim Basin and the north slopes of the Kunlun Shan mountains, where it was first domesticated and cultivated. It bears an edible juicy fruit. China is the world's largest producer of peaches. The fruit is also produced in India.

The peach fruit is also utilized for the production of wine. It has less acid than the plum or apricot, but being pulpy has to be diluted with water to make it a palatable wine (Joshi et al., 2005a).

5.2.15 Pear

The pear is shrub species of the genus *Pyrus* in the family Rosaceae. It is also the name of the pomaceous fruit of these trees. Several species of pear are valued for their edible fruit, whereas others are cultivated as ornamental trees.

Wine of good quality can be made from the pears with high tannin contents such as the Bartlett pear. Perry is a drink similar to cider but it is made from pears called snow pears (*Pyrus nivalis*), which are a different species from European pears (*Pyrus communis*) or *nashi* pears (*Pyrus pyrifolia* and *Pyrus bretschneideri*). Perry varieties have a high quantity of tannins in the juice.

5.2.16 Persimmon

Persimmons are the edible fruit of a number of species of trees in the genus *Diospyros. Diospyros* is in the family Ebenaceae. The most widely cultivated species is the Asian persimmon, *Diospyros kaki.*

In color, the ripe fruit of the cultivated strains ranges from light yellow-orange to dark red-orange depending on the species and variety. They vary in size from 1.5 to 9 cm (0.5–4 in.) in diameter, and the varieties may be spherical, acorn, or pumpkin shaped. The calyx generally remains attached to the fruit after harvesting, but becomes easy to remove once the fruit is ripe. The ripe fruit has high glucose content but the protein content is low, though it has a balanced protein profile.

5.2.17 Plum

Plum is a fruit of the subgenus *Prunus* of the genus *Prunus.* The subgenus is distinguished from other subgenera (peaches, cherries, bird cherries, etc.) in the shoots having a terminal bud and solitary side buds (not clustered).

The flowers grow in groups of one to five together on short stems, and the fruit has a groove running down one side and a smooth stone. The taste of the plum fruit ranges from sweet to tart; the skin itself may be particularly tart. It is a juicy fruit and can be eaten fresh or used in jam or other products. Its juice can be fermented into fruit wine (Amerine et al., 1980; Bhutani and Joshi, 1995; Joshi et al., 2011a). In central England, a cider-like alcoholic beverage known as plum jerkum is made from plums. The fruit gives wine with an appealing color and acceptable quality, which is quite popular in many countries, particularly in Germany and Pacific Coastal states.

5.2.18 Pomegranate

The pomegranate is an erect deciduous spreading shrub or tree, with a woody and thorny stem with fruits having a globular shape, crowned by a persistent calyx possessing a hard outer rind.

The flowering season of the wild pomegranate is mid-April to the end of May. The major use of this fruit is in preparation of *anardana*. Its dried rind yields a fast yellow dye. The rind contains up to 30% tannins, which could be used for tanning leather and also in making jet black ink. The bark contains an alkaloid called punicine, which is highly toxic to tapeworms.

Pomegranate (*Punica granatum*) is a fruit-bearing deciduous shrub or small tree growing to between 5 and 8 m (16–26 ft) tall. In the Northern Hemisphere, the fruiting takes place typically in the season from September to February and in the Southern Hemisphere from March to May. As intact arils or juice, pomegranates are used in cooking, baking, meal garnishes, juice blends, smoothies, and alcoholic beverages, such as cocktails and wine. Pomegranate juice can be sweet or sour, but most of the fruits are moderate in taste, with sour notes from the acidic tannins contained in the juice. Pomegranate juice has long been a popular drink in Armenian, Israeli, Persian, and Indian cuisine, and now is widely distributed in the United States and Canada. To prepare wine from pomegranate (*P. granatum*), the fruit is pressed without crushing to avoid excessive astringency in the wine. Sugar is added to the juice to bring its total soluble solids to 22–23°Bx.

5.2.19 Strawberry

Berries have been used for the preparation of wine. Strawberry juice is made from the fruits or its concentrate and is added to cocktails.

Strawberries contain a modest amount of essential unsaturated fatty acids in the achene (seed) oil. The pigment extract of the fruit can be used as a natural acid/base indicator because of the different colors of the conjugate acid and conjugate base of the pigment. Strawberry wine of good quality has the appealing color of premium rosé wine (Joshi et al., 2005b). When frozen berries are used to make wine, they are first thawed and the juice is extracted (Amerine et al., 1980). The juice is ameliorated to 22°Bx by the addition of cane sugar.

6. ROLE OF GENETIC ENGINEERING IN WINE

Considerable progress has been made in the development of genetically improved plants and microorganisms, especially yeast strains, for exploring the possibilities of introducing new improved quality characteristics into fruits and wine yeast. To illustrate both of these aspects, the respective information has been summarized and the reader may consult the literature cited (Pretorious, 1997, 1999, 2001, 2002, Pretorious et al., 2003; Kaur et al., 2011; Srivastva and Raverkar, 1999) and standard texts on the basic aspects of the subject matter.

6.1 GENETIC MODIFICATION OF PLANTS

The present-day fruit industry is based on a few genotypes because the intense selection and fixation of genotypes by clonal propagation led to a narrow germplasm base (Janick, 1992). Thus, it was necessary to change the cultivars, which can be achieved by conventional breeding. However, this is hampered by several biological problems. Chemical controls and various management practices have been used practically in the commercial production of various fruits to overcome disease infection, improve shelf-life, and generate dwarf trees (Kumar et al., 2011a,b). However, these practices are either inefficient or unfriendly to human health and the environment. In contrast, genetic engineering offers a better possibility of improving plant traits (Kaur et al., 2011; Zhu et al., 2004).

6.1.1 Improvement of Traits

Since about 1995, genetic engineering has been successfully used to improve tolerance to biotic and abiotic stresses, increase fruit yield, improve shelf-life of fruit, reduce generation time, and produce fruit with superior nutritional value (Kaur et al., 2011). Selectable marker genes are widely used for the efficient transformation of crop plants. Antibiotic or herbicide-resistance marker genes are generally preferred because they are more efficient. But because of consumer concerns, considerable efforts are being put into developing strategies (homologous recombination, site-specific recombination, and cotransformation) to eliminate the marker gene from the nuclear or chloroplast genome after selection. The absence of a selectable marker gene in the final product and the introduced gene(s) derived from the same plant will increase the consumer's acceptance (Krens et al., 2004). This can be fulfilled by new genetic engineering approaches like cisgenesis or intragenesis. In a study, performance and long-term stability of the barley hordothionin gene in multiple transgenic apple lines has been evaluated (Krens et al., 2011). Still the development of transgenic fruit plants and their commercialization is be hindered by many regulatory and social issues.

6.1.2 Methods of Genetic Transformation

Tissue culture plays a critical role in the genetic transformation of a plant species because the first requirement for most gene transfer systems is efficient regeneration of plantlets from cells carrying a foreign gene. Other than plant tissue culture, various methods of gene transfer are also necessary to obtain transgenic plants (Kaur et al., 2011). In brief, the successful generation of transgenic plants requires the combination of the following:

- a cloned gene with elements, such as promoters, enhancing and targeting sequences for its regulatory expression;
- a reliable method for delivery and stable integration of DNA into cells;
- a suitable tissue culture method to recover and regenerate intact plants from transformed organs or tissues.

As of this writing, a number of strategies have been developed for identification, isolation, and cloning of desirable gene sequences from almost any organism. The next step is to deliver the desired gene sequence into cells of the target organism. This can be achieved by using one of the methods listed in Table 1.4.

Because the methods listed above have proved successful and are applicable to several plant species, they are being used extensively. For more information, the readers are referred to the literature cited (Kaur et al., 2011). It may, however, be pertinent to point out that with respect to winemaking most of the work is on grapes, and nongrape fruits have been neither researched nor documented. However, it could be a hot area of research considering the quantity of nongrape fruits produced.

6.2 GENETIC ENGINEERING OF WINE YEAST

Wine fermentation is mainly carried out by the microorganisms present on the grapes or the winery equipment or as inoculated fermentation. During wine fermentation, a succession of microorganisms (*Kloeckera*, *Hanseniaspora,* and *Candida*) dominates the early stages of alcoholic fermentation, followed by several species of *Metschnikowia* and *Pichia* in the middle stages, when the ethanol rises to 3–4% (Henschke, 1997; Grossmann and Pretorius, 1999). Yeasts, mainly of the species *S. cerevisiae*, invariably take over the latter stages of the process, when alcohol concentration is high, and *S. cerevisiae* is therefore universally known as the "wine yeast" (Amerine et al., 1980; pretorius and Bauer, 2002; Bauer et al., 2003). In modern wineries, reliable and selected starter culture strains of *S. cerevisiae* are used to obtain consistent wine flavor and predictable quality in contrast to spontaneous fermentation. In addition to the primary function (fermentation), they can also have a range of specialized properties for value addition to the final product (Pretorius et al., 1999). This quest for optimized specific tasks set by winemakers has led to genetic engineering of wine yeast strains.

With the emergence of modern molecular genetics, the industrial importance of *S. cerevisiae* has continuously extended beyond traditional fermentation. Moreover, sequence data of the complete *S. cerevisiae* genome (Goffeau et al., 1996) and advances in wine sciences now enable researchers to comprehensively assess the modifications that are engineered into the yeast cells. This interest has also evoked the potential of numerous research laboratories worldwide to engineer yeast strains. Some of the desirable characteristics of wine yeast for genetic engineering are listed in Table 1.5.

Genetically engineered wine yeast strains are being developed in most wine-producing countries, including Canada, France, Germany, Italy, Portugal, South Africa, Spain, Sweden, and the United States (http://www.saasta.ac.za/Media-Portal/download/bio_fs15.pdf). Several studies have already shown the feasibility of improving wine yeast through genetic engineering for traits such as the secretion of specific enzymes that improve wine clarification and juice yield, the production of aroma compounds, and the

---

Table 1.4 Methods of Genetic Engineering Used in Plant Modification

- Agrobacterium-mediated gene transfer
- Particle bombardment or biolistic delivery
- Electroporation
- PEG-induced DNA uptake
- DNA delivery via silicon carbide fibers
- Laser-mediated DNA transfer
- Microinjection
- Liposome-mediated gene transfer, or lipofection
- Diethylaminoethyl-dextran-mediated transfection
- Agroinfiltration
- Vacuum infiltration

Based on Kaur, R., Kumar, K., Sharma, D.R., 2011. Grapes and genetic engineering. In: Joshi, V.K. (Ed.), Handbook of Enology, vol. 1. Asia Tech Publication, New Delhi, pp. 266–286.

Table 1.5 Desirable Characteristics of Wine Yeast

Characteristic	Specific Targets
Fermentation properties	• Rapid initiation of fermentation • High ethanol tolerance • High osmotolerance • Low temperature optima • Moderate biomass production
Flavor characteristics	• Low sulfite–dimethyl sulfide–thiol formation • Low volatile acids production • Low fusel (higher) alcohols production • Liberation of glycosylated flavor precursors • High glycerol production • Hydrolytic activity • Enhanced autolysis • Modified esterase activity
Technological stability	• High genetic stability • High sulfite tolerance • Low sulfite binding activity • Low foam formation • Flocculation properties • Compact sediment • Resistance to desiccation • Zymocidal • Genetic marking proteolytic activity • Low nitrogen demand
Metabolic properties with health implications	• Low sulfite formation • Low biogenic amine formation • Low ethyl carbamate potential

Summarized from Pretorius, I.S., June 15, 2000. Tailoring wine yeast for the new millennium: novel approaches to the ancient art of winemaking. Yeast 16 (8), 675–729; Soni, S.K., Sharma S.C., Soni, R., 2011b. Yeast genetics and genetic engineering in wine making. In: Joshi, V.K. (Ed.), Handbook of Enology, vol. III. Asia Tech Publishers, New Delhi, pp. 441–501.

decrease of ethanol production (Pretorius, 2002; Byaruagaba-Bazirake et al., 2013). However, only two genetically modified yeast strains have been so far released for commercial use as of this writing, i.e., malolactic yeast, ML01, and a yeast strain that produces lower amounts of ethyl carbamate, i.e., a carcinogen. Both of these yeast strains have been authorized for use in the North American wine industry (http://www.capewineacademy.co.za;http://www.saasta.ac.za/MediaPortal/download/bio_fs15.pdf).

Winemakers generally require cost-competitive wine production with minimum resource input, which in turn depends upon improved fermentation performance, efficient nitrogen assimilation, and resistance to antimicrobial compounds (Pérez-Torrado et al., 2015). Methods employed for the genetic engineering of microorganisms have been reviewed extensively in the past few decades, and gene technology has opened up new doors to adding new and improved characteristics to the target cell (Srivastva and Raverkar, 1999; Pretorius, 2000; Schuller and Casal, 2005). The information obtained from the analysis of the entire genomes, transcriptomes, proteomes, and metabolomes of wine yeasts will undoubtedly increase the specificity of the various methods used for modification of wine yeast (Fig. 1.6) in which the starter strains are genetically selected and tailored for the production of particular types and styles of wine. However, the classical strain selection and modification methods, such as variant selection, mutagenesis, and hybridization (mating, spore-cell mating, rare mating, cytoduction, and spheroplast fusion), are based mainly on a "shotgun" approach (Pretorius and Van der Westhuizen, 1991) in which large genomic regions or entire genomes are recombined or rearranged. These methods, therefore, are not specific enough to modify wine yeasts in a well-controlled manner, although they might improve some of the properties of the yeast strain by compromising other desired traits. The only advantages of these methods are that they can be used to improve and combine the various traits under polygenic control (Srivastva and Raverkar, 1999; Soni et al., 2011b) and that they do not give rise to products that are included in the statutory definition of genetically modified organisms, i.e., GMOs (Pretorius, 2002), like in the transformation method. However, genetic engineering remains the only reliable method for modifying an existing property, introducing a new characteristic, and eliminating an unwanted trait without adversely affecting other desirable properties. Several transformation methods and plasmid vectors, in addition to expression and secretion cassettes for the expression of heterologous genes and secretion of their encoded proteins, have been developed for *S. cerevisiae* (Pretorius and Van der Westhuizen, 1991; Pretorius, 2000). This has offered wider applicability and a higher degree of specificity in the development of improved wine yeasts.

6.3 SPECIFIC TARGETS FOR WINE YEAST GENETIC ENGINEERING

Various targets for the direct genetic improvement of wine yeast strains have also been identified (Figs. 1.7 and 1.8), all relating to the improvement of the winemaking process and wine quality by enhancing the characteristics, which are discussed here.

6.3.1 Improvement in Fermentation Process

Many factors affect the fermentation performance of wine yeasts, as reviewed earlier: increased resilience and stress resistance of active dried yeast cells; improved grape sugar and nitrogen uptake and assimilation; enhanced resistance to ethanol and other microbial metabolites and toxins; resistance to sulfite, heavy metals, and agrochemical residues; and reduced foam formation (Srivastva and Raverkar, 1999; Soni et al., 2011b; Snow, 1983; Pretorious, 2000; Pretorius and Bauer, 2002; Pretorius et al., 2003). An imbalance in the high levels of carbon and low levels of freely available nitrogen in grape must is the most common cause of poor fermentative performance. Another thrust

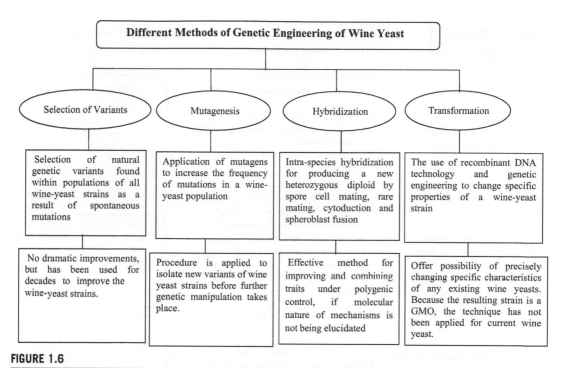

FIGURE 1.6

Various methods used for wine yeast tailoring and their applications. *GMO*, Genetically modified organism.

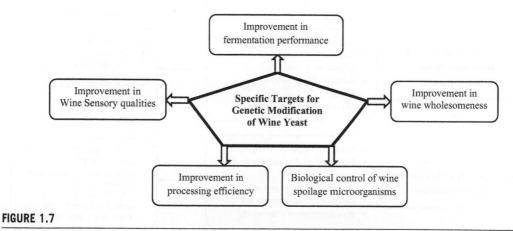

FIGURE 1.7

Various targets for the direct genetic improvement of wine yeast strains.

to improve the fermentation performance of wine yeasts is to increase their resistance to toxic microbial metabolites (e.g., ethanol, acetic acid, and medium-chain fatty acids) and zymocins by modification of the expression of certain genes resulting in increased sterol accumulation and cell-membrane ATPase activity, thereby increasing the resistance to ethanol (Pretorius, 2000, 2001, 2002). Attempts have also been made to tailor wine yeasts to secrete proteolytic and polysaccharolytic enzymes that

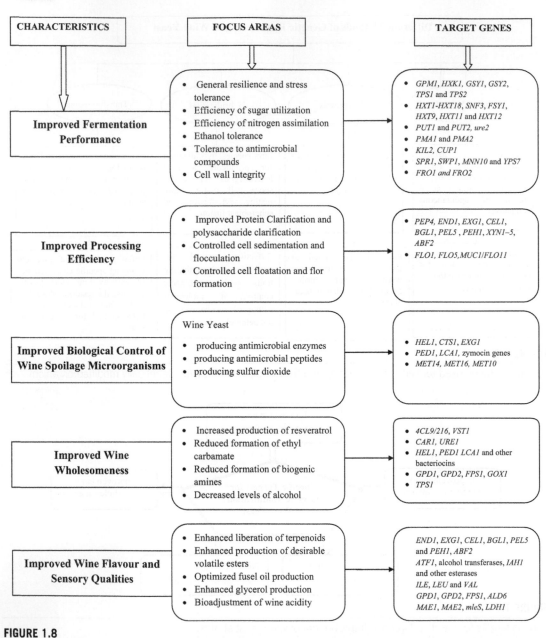

FIGURE 1.8

Focus areas and target genes used for tailoring of wine yeast.

Adapted from Pretorius, I.S., du Toit, A., van Rensburg, P., 2003. Designer yeasts for the fermentation industry of the 21st century. Food Technology and Biotechnology 41, 3–10.

would remove haze-forming proteins and filter-clogging polysaccharides, respectively. Thus, over-expression of several bacterial, fungal, and yeast genes has resulted in proteolytic, pectinolytic, glucanolytic, and xylanolytic wine yeast strains (Laing and Pretorius, 1993; Pretorius, 1997; Van Rensburg and Pretorius, 2000). A hybrid yeast strain (*S. cerevisiae*) that simultaneously overproduces mannoproteins and presents good fermentative characteristics has been developed. This hybrid shares genes (*SPR1*, *SWP1*, *MNN10*, and *YPS7*) related to cell wall integrity with the parental Sc1 and some glycolytic genes (*GPM1* and *HXK1*) with the parental Sc2, as well as the genes involved in hexose transport, such as *HXT9*, *HXT11*, and *HXT12* (Pérez-Traves et al., 2015).

6.3.2 Improvement in Processing of Wine

A second target, which is the regulated expression of the flocculation genes, is important to guarantee a high suspended yeast count for a rapid fermentation rate during the fermentation process, whereas efficient settling is needed to minimize problems with wine clarification at the end of sugar conversion (Henschke, 1997; Barre et al., 1993; Dequin, 2001). The expression patterns of the main genes involved in flocculation (*FLO1*, *FLO5*, and *FLO8*) have been studied both in a synthetic medium and in the presence of ethanol stress in 28 flocculent wine strains, which could be useful for studying the relationship between genetic variation and flocculation phenotype in wine yeasts (Tofalo et al., 2014).

6.3.3 Control of Microbial Spoilage

Various types of chemical preservatives, such as sulfur dioxide, dimethyl dicarbonate, benzoic acid, fumaric acid, and sorbic acid, are generally added to control the growth of unwanted microbial contaminants for maintaining the sensory properties of appearance, aroma, and flavor of wine (Amerine et al., 1980; Du Toit and Pretorius, 2000). However, excessive use of these chemical preservatives reduces the quality of the wine and faces mounting consumer resistance. But, the other approach, i.e., use of purified antimicrobial enzymes and bacteriocins, is expensive and could be circumvented by expressing effective antimicrobial enzymes and peptides in wine yeast starter-culture strains (Du Toit and Pretorius, 2000). To overcome the problems associated with these methods, the genes encoding hen egg white lysozyme (*HEL1*), *Pediococcus acidilactici* pediocin (*PED1*), and *Leuconostoc carnosum* leucocin (*LCA1*) have been used to engineer bactericidal yeasts (Schoeman et al., 1999). The antifungal yeast *CTS1*-encoded chitinase and *EXG1*-encoded exoglucanase have also been expressed in *S. cerevisiae*.

6.3.4 Improvement in Wine Wholesomeness

Another concern is the presence of some unwanted compounds, including suspected carcinogens (e.g., ethyl carbamate), neurotoxins (e.g., biogenic amines), and asthmatic chemical preservatives (e.g., sulfites), in wine. When developing wine yeast strains, a focus on the health aspects and developing yeasts that could enhance the benefits (production of resveratrol and carnitine) and minimize the risks (eliminating ethyl carbamate and biogenic amines and reducing the levels of alcohol) associated with moderate wine consumption is of utmost importance. With regard to the production of resveratrol during fermentation, progress has already been made by constructing a wine yeast that expresses the *4CL9/216* coenzyme A ligase and *VST1* stilbene synthase genes (Becker et al., 2003). In addition, bactericidal yeasts might be developed, producing compounds toxic to unwanted bacteria (bacteriocins), carrying deletions of the *CAR1* arginase gene (blocking the secretion of urea, the precursor for the formation of ethyl carbamate), or containing heterologous urease genes (enabling the degradation of urea). Such yeasts would reduce the levels of added sulfite and reduce yeast-derived ethyl carbamate and bioamines formed by bacterial contaminants (Pretorius, 2000, 2001, 2002; Becker et al., 2003). Similarly, co-transformation has resulted in yeast that constitutively express the urease-encoding gene *Dur1,2*, which led to 90% reduction in the ethyl carbamate content of the wines produced compared with those produced using the parent strain. The commercial name of this strain is EMLo1 (Coulon et al., 2006). Reducing alcohol levels in wines is a key issue to tackle microbiological, technical, sensory, and economic challenges faced during wine fermentation (Tilloy et al., 2014). The specific yeasts capable of performing such function have been modified successfully. Enhancing glycerol production by increasing expression of the glyceraldehyde-3-phosphate dehydrogenase gene, *GPD1*, was the most efficient strategy to lower ethanol concentration (Varela et al., 2012). Similarly, overexpression of the *TPS1* gene, encoding trehalose-6-phosphate synthase, in a commercial wine yeast strain (VIN13) resulted in a moderate increase in trehalose biosynthesis with reduced ethanol yield and fermentation rate accompanied by higher residual sugar levels (Rossouw et al., 2013).

6.3.5 Improvement in Sensory Qualities

The organoleptic quality parameters (appearance, aroma, and flavor) of wine are generally determined by the presence of a well-balanced ratio of flavor compounds and other metabolites (Lambrechts and Pretorius, 2000). Significant progress has been made in the construction of yeasts producing color and aroma liberating enzymes (pectinases, glycosidases, glucanases, and arabinofuranosidases); ester modifying enzymes (alcohol acetyl transferases, esterases, and isoamyl acetate-hydrolyzing enzyme); optimized levels of glycerol through the overexpression of *GPD1*, *GPD2*, and *FPS1F*; and phenolic acids (through the modified expression of the yeast *PAD1* phenyl acrylic acid decarboxylase gene and the expression of bacterial *pdc* *p*-coumaric acid decarboxylase and *padc* phenolic acid decarboxylase genes) (Van Rensburg, and Pretorius, 2000; Laing and Pretorius, 1993; Smit et al., 2003).

6.4 CONCERNS ASSOCIATED WITH THE USE OF GENETICALLY MODIFIED YEASTS

It is apparent that the future use of genetically modified yeasts will be dependent on the ability to assess potential or theoretical risks associated with their introduction into natural ecosystems (Schuller and Casal, 2005). It is likely that the changed parameters will be sensed by the complex regulatory mechanisms that exist within any living cell and will lead to a specific molecular response. This response may result in some unforeseen, indirect consequences regarding the metabolic activity of the cell. In several cases of attempted metabolic engineering of wine yeast strains, such unforeseen consequences have been described, an example being the increased production of acetate in strains with increased levels of glycerol-synthesizing enzymes. Modern biotechnological tools however allow the systematic assessment of any biological process on a "global" level, by analyzing the entire transcriptome (all mRNAs present in a cell), and in the near future the proteome (all proteins) and metabolome (all metabolites). The transcriptome can be analyzed by microarrays, which monitor the transcription of all the protein-encoding genes present in the genome of yeast. Several studies have already been conducted to compare transcriptional regulation in parental and genetically modified yeast strains. By monitoring the transcriptome, possible unexpected side effects of the genetic modification can be revealed and analyzed (Bauer et al., 2003). It is expected that in the future tailor-made yeast would be available to make wine with specific characteristics. For more details on grape and wine biotechnology see Pretorius and Høj (2005), though with respect to other fruits with respect to wine biotechnology there is no information.

7. TECHNOLOGY OF FRUIT WINE PRODUCTION

7.1 GENERAL ASPECTS AND PROBLEMS IN THE PRODUCTION OF FRUIT WINES

For the preparation of wine from grapes, the technology is well standardized and wine production is an established industry in the grape-producing countries of the world. The techniques used for the production of fruit wines are basically similar to those for the production of wines made from grapes (Joshi, 2009). However, differences arise from two facts: (1) it is quite difficult to extract the sugar and other soluble materials from the pulp of some fruits compared to grapes and (2) the juices obtained from most fruits are lower in sugar content and higher in acids (Vyas and Gandhi, 1972; Joshi et al., 1999b; Amerine et al., 1980; Joshi, 2009; Swami et al., 2014). The higher acidity in some fruits makes it more difficult to prepare wine of acceptable quality (Joshi et al., 2011e), whereas in the case of citrus fruits, when the juice is extracted by a rack and cloth press or screw expeller, it contains so much essential oil from the peel that fermentation is reduced considerably. As a solution to these problems, the use of specialized equipment to thoroughly chop or disintegrate the fruit, such as berries, followed by pressing to extract the juice from the pulp, solves the first problem. The fermentation of some fruits is very slow or may even terminate before completion because of a lack of certain nitrogenous compounds or other yeast growth factors in some fruit juices such as pear or other fruit juices. The addition of nitrogen source to such must have solved this problem (Joshi et al., 1990b, 1991a; Amerine et al., 1980). The second problem is solved by the addition of water to dilute the excess acid and the addition of sugar to balance the sugar deficiency. For reduction of acidity, the use of deacidifying yeast like *Schizosaccharomyces pombe* (Vyas and Joshi, 1988; Joshi et al., 1991b) or the malolactic acid bacteria has been successful (Bartowsky, 2011). The basic technique for production of fruit wines is essentially the same, involving routine alcoholic fermentation of the juice or pulp but modifications with respect to the physicochemical characteristics, depending upon the type of wine to be prepared and the fruit used need to be made (Table 1.3). In citrus fruits, bitterness of the juice is another problem, in addition to darkening of the wine. The bitterness of the juice is also carried on in the wine (Amerine et al., 1980; Joshi and Thakur, 1994; Joshi et al., 2012a), so to overcome the same, debittering with the XAD-16 adsorption technique has been successfully done using kinnow juice (Joshi et al., 1997).

Like grape wine industry, the waste from fruit wine production is a useful source of several components which could be a source of value added products (Joshi et al., 2011d).

7.2 REQUIRED RAW MATERIALS

For the production of fruit-based alcoholic beverages, raw materials would naturally be fruits, sweetening agent, fruit concentrate, sugar, acid, nitrogen source, clarifying enzyme, filtering aid, etc. Spices and herbs or their extracts constitute an essential raw material to make fortified wines like vermouth. In some instances, concentrates are also used, e.g., in apple or grape wine and vermouth (Vyas and Gandhi, 1972; Kundu et al., 1980; Joshi et al., 1999b; Panesar et al., 2000; Marwaha et al., 2004).

Sugar is the second most important raw material, which also dictates the final cost of production of all types of wines, vermouth, and brandy. For almost all the fruits, their juice or extract is low in sugar and thus, has to be ameliorated to the sugar level needed for producing a table wine. Even when the sugar level is satisfactory, the high acidity demands dilution, consequently requiring more addition of sugar to the must (Vyas and Gandhi, 1972; Joshi et al., 1990a,b; Suresh and Ethiraj, 1987; Joshi et al., 2001). Generally, the pectin-splitting pectin esterase is used in the clarification of wines such as apple, plum, peach, pear, etc. (Joshi et al., 1990a,b, 2011b; Joshi and Bhutani, 1991).

7.3 FRUIT COMPOSITION AND MATURITY

The composition of fruit and its maturity depend upon the cultivation practices followed. The primary environmental factor influencing the quality of fruit is temperature (maximum, minimum, mean), whereas secondary factors are rainfall, sunshine or cloudiness, humidity, wind, soil, and the combination of all these factors. In addition, different cultivation practices also affect the composition of fruit. The type and amount of fertilization affect maturity, especially sugar content, which ultimately defines the ethanol content in the wine.

7.4 SCREENING OF SUITABLE VARIETIES

Impressive progress has been made in the development of technologies for preparation of wines from grape, mango, apple, peach, pear, plum, cashew apple, pomegranate, banana, ber, strawberry, kinnow (Venkataramu et al., 1979; Kochar and Vyas, 1996; Sandhu and Joshi, 1995; Joshi et al., 2011b), and carrot (Lingappa and Chandershekha,. 1997). Screening of cultivars of grape (Bardia and Kundu, 1980), peach (Joshi et al., 2005b, 2011b), plum (Joshi et al., 2011e), mango, apple (Joshi et al., 2011c), wild and cultivated apricot (Joshi et al., 2011e), lychee, sand pear, and strawberry (Joshi et al., 2005a) has also been done for wine preparation.

The suitability of eight peach cultivars for preparation of table wine was judged and it was revealed that Redhaven, Sunhaven, JH Hale, Flavorcrest, and July Elberta are better than others (Joshi et al., 2005b). A wine with acceptable organoleptic properties could be prepared from custard apple (*Annona squamosa*) (Kotecha et al., 1995; Kumar et al., 2011b). Among the varieties of mango screened (Kulkarni et al., 1980), Fazli, Langra, and Chausa produced good quality wines, and the wine from Dashehari had a characteristic fruity flavor. More information on fruits other than grapes has been elaborated considerably in earlier sections of this chapter.

7.5 MICROBIOLOGY OF FERMENTATION

The entire gamut of winemaking involves microbiology using mainly *S. cerevisiae*, the wine yeast. Several efforts have been made to select yeast capable of producing wine with an optimum rate of fermentation, ethanol content, and sensory qualities (Amerine et al., 1980; Jackson, 2000, 2003; Nigam-Singh, 2011). With respect to grape wines, several reports are available (Kundu et al., 1980). Based on the production of alcohol and aldehyde and the fermentation rate, *S. cerevisiae* var. *ellipsoideus* from France was the most satisfactory yeast, producing a wine with flowery and fruity flavor. These microflora multiply during the early stages of fermentation and produce special flavors but are inhibited later when the alcohol level reaches 4–6% or more. In nongrape fruit, however, there is scant information, with scattered reports on plum and apple (Joshi et al., 2000, 2009a,b).

Spontaneous or natural fermentation of fruits is also carried out, especially for cider (Amerine et al., 1980; Joshi, 2016) or indigenous fermented alcoholic beverages. Evaluation of the quality of naturally fermented alcoholic beverages prepared in tribal areas reveals a higher quantity of methanol and wide variations in ethanol, pH, and volatile acidity among the products (Joshi and Sandhu, 2000a). The source of fermentation normally employed to produce such beverages contains *Aspergillus flavus*, *Aspergillus oryzae*, *Mucor* spp., *Rhizopus* spp., and bacteria like *Pediococcus*, *Leuconostoc*, and *Pseudomonas*. *S. cerevisiae* was, however, found to be predominant in the natural sources of fermentation. More of such information could be helpful in the selection of superior strains of yeast to make alcoholic beverages with desirable qualities.

8. GENERAL METHODS FOR FRUIT WINEMAKING
8.1 PREPARATION OF YEAST STARTER CULTURE

A good strain of the wine yeast *S. cerevisiae* is a prerequisite and needs to be procured for making quality wine. The yeast in the form of slants, tablets, or compressed yeast can be used for this purpose. Before the same is added to the must for fermentation, the yeast culture is activated in the juice/pulp intended for winemaking (Vyas and Gandhi, 1972; Amerine et al., 1980; Joshi, 1997). The container with the juice is plugged and kept in a warm place (25–30°C) and the culture is ready after 24 h. The amount of active culture is added at 2–5% to the must. The yeast, when used in the tablet form, is activated in sterilized water at the optimum temperature for yeast growth. However, compressed yeast can be added directly to the must.

8.2 PREPARATION OF MUST

The juice or the pulp of the fruit to be used for winemaking is made into must. The must is prepared depending on the fruit used and the type of wine to be made. Either the juice is extracted or the fruit is made into pulp. Juice from fruits like apple is extracted first by grating, followed by pressing in a hydraulic press (Plate 1.2A). Usually, pectinol is added to the must to help clarify the wine, as its addition has been found to enhance the quality of wine. SO_2 is added to the must at 100–150 ppm as potassium metabisulfite. The addition of ammonium sulfate with thiamin and biotin gives a greater increase in the rate of fermentation. In apple juice, ammonium sulfate with 4 mg/L thiamine gives a much greater effect on the fermentation rate (Joshi et al., 1991a,b,c; Joshi, 1997). Fining with gelatin has been found to be a good remedy for juices like pear, which have high tannin contents (Joshi, 1997). Further, heating of the juice for 35 min at 122°F can also increase the fermentation rate. In the case of fruits like plum or peach, an appropriate dilution is made and then, amelioration is done.

8.3 FERMENTATION

The must is allowed to ferment at a suitable temperature (20–25°C) after inoculation with the yeast culture. A temperature higher than 25°C should be avoided as it causes the loss of volatile compounds

Basket press   Hydraulic press

PLATE 1.2A

Presses used for juice extraction for wine preparation.

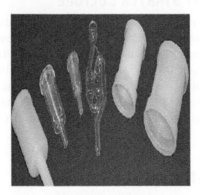

PLATE 1.2B

Air locks used with the container to conduct wine fermentation.

and alcohol. The container in which fermentation is carried out has to be equipped with an air lock (Plate 1.2B). The sugar content or °Brix is measured regularly to monitor the progress of fermentation. Normally, the fermentation is allowed to proceed until all the sugar is consumed completely (usually Brix reading of about 8°B). When the fermentation is completed, the bubbling due to the production of CO_2 is stopped. At the industrial scale, fermentation is carried out in fermenters of various shapes and sizes (Plate 1.2C).

8.4 SIPHONING/RACKING

After the completion of fermentation, the yeast and other materials that settle at the bottom of the container, with the clear liquid separating out, is siphoned/racked or, in the case of pulpy must, it is filtered through a cheese/muslin cloth, followed by siphoning. At the industrial scale, a filter press is used. Two or three rackings are usually done after 15–20days. During the interracking period, no headspace is kept in the bottle or

PLATE 1.2C

Fermenters used for fermentation of wine.

container, which is closed tightly to prevent any acetification. With the help of a vacuum pump, racking can also be practiced in industrial fermentation with an advantage over not doing racking.

8.5 MATURATION

The fresh wine is harsh in taste and has a yeasty flavor. The process of maturation makes the wine mellow in taste and fruity in flavor in addition to the clarification. The period may extend from 6 months to 2–3 years. The process of maturation is complex and the formation of esters takes place thus, improving the flavor of such beverages (Vyas and Gandhi, 1972; Amerine et al., 1980).

8.6 CLARIFICATION

If the wine is not clear after racking and maturation, it is clarified using filtering aids such as bentonite or celite or by tannin/gelatin treatment followed by filtration in a filter press (Plate 1.2D). These treatments usually make the wine crystal clear.

8.7 PASTEURIZATION

Wines, being low-alcohol beverages, are pasteurized at 62°C for 15–20 min, after keeping some headspace in the bottle and crown corking the same. Heating the wines helps the precipitation of tannins or other such materials that are heat sensitive in addition to the preservation. Pasteurized wines however once opened have to be kept at low temperature to prevent their spoilage. Alternatively, table wines can be preserved by the addition of preservatives like SO_2, sodium benzoate, sorbic acid, etc. (Vyas and Gandhi, 1972; Amerine et al., 1980; Joshi, 1997).

9. TECHNOLOGY OF WINE PRODUCTION FROM VARIOUS FRUITS

To make the best wine from available raw materials, it is essential to control the processing technique from the fruit to the glass. This is made possible by modifying the processing techniques. Some of the techniques that have been developed for the purpose of production of wine from various fruits, like

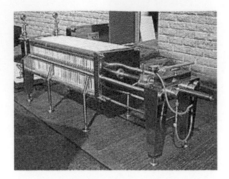

PLATE 1.2D

Filter press used for clarification of wine.

apple, apricot, cacao, cagaita, cherry, cupuaçu, custard apple, elderberry, gabiroba, guava, jabuticaba, jackfruit, jamun, kinnow, kiwifruit, lychee, mango, muskmelon, orange, palm, papaya, peach, pear, pineapple, plum, pomegranate, raspberry, sapota, soursop, strawberry, sweet potato, and umbu (Joshi, 1997; Joshi et al., 1999b; Joshi and Attri, 2005; Joshi and Kumar, 2011; Jagtap et al., 2015), have been summarized in Table 1.6. It is apparent that the focus has been mainly on optimizing the conditions of must making, the effects of the nitrogen source, the nutritive value, the maturation period, screening of varieties, etc. Thus, more in-depth work on various aspects needs to be carried out so as to take fruit winemaking up to the industrial level.

Detailed information about the technology for the production of wines from various fruits is extensively given in Chapter 7, so the readers may consult the same.

10. SPECIAL WINES
10.1 VERMOUTH

Quality dessert and Madeira-style wines have been made from several mango varieties (Onkarayya and Singh, 1984). Madeirization of dessert wines reduces acidity and increases volatile acidity, color, and overall quality. Madeirized wines are more acceptable than the corresponding dry and dessert wines. A rapid madeirization process for mango dessert wine *via* the addition of 0.1% ascorbic acid and madeirization at 50°C for 7 days has been developed (Onkarayya, 1986). Madeira prepared from Totapari is more acceptable than those from Raspuri, Mulgoba, Dashehari, and Langra. Preparation of a new alcoholic beverage, mango vermouth, has been attempted by using a suitable herb mixture for preparing acceptable grades of dry and sweet mango vermouths (Onkarayya, 1985). The suitability of plum for the preparation of vermouth reveals that sweet vermouth with 15% alcohol is of commercial acceptability (Joshi et al., 1991c). Vermouth prepared from plum was found to be a good source of minerals (Bhutani and Joshi, 1996). Tamarind fruits have also been used to prepare, commercially acceptable vermouth (Lingappa et al., 1993). Dry and sweet vermouths with variable alcohol levels are prepared from sand pear base wine (Attri et al., 1994). Sweet vermouth having 15% alcohol was considered the best (Attri et al., 1993).

Table 1.6 Summary of Research Work Done on Nongrape Wines

Name of Fruit	Details of Research Findings	References
Apple	• Production of cider and brandy from Indian apples has been reported. Cider containing 4.5–8.23% alcohol was produced from apple varieties Red Delicious, Rich-a-Red, Kesari, Golden Delicious, and Maharaji. Further, Golden Delicious and Red Delicious apples were suitable for cider production.	Singh Nagi and Manjrekar, (1976) and Patel et al. (1977)
	• The process of sweet cider production was standardized.	Rana et al. (1986)
	• After studying the effects of yeast strains, sugar, and nitrogen on cider production from Red Delicious and Maharaji, the authors found that these varieties were suitable for this purpose.	Karni et al. (1977)
	• Preparation of cider from scabbed apple fruit. The processed apple juice from scabbed fruits was found to be free of patulin, aflatoxin, and microflora. The cider prepared from scabbed fruit juice, too, was comparable in all respects, including fermentation behavior, to normal juice.	Azad et al. (1987)
	• Apple honey wine made by blending apple juice with honey was adjudged to be the best wine among other fruits and the effect of insoluble solids addition on the quality of apple wine was also studied.	Joshi et al. (1990a,b) and Joshi and Sharma (1994)
	• Studies on cider preparation from apple juice concentrate revealed that fortification with DAHP as nitrogen source was essential for rapid fermentation and the must prepared from diluting the concentrate fermented faster than the amelioration done with sugar and the prepared cider showed a high acceptability.	Joshi et al. (1991a)
	• Initial sugar concentration of 20 and 24°Bx was found to be optimum for preparation of cider and apple wine, respectively. Preparation of hard cider was also attempted and successfully done.	Joshi and Sandhu (1997, 1994) and Joshi and Sandhu (2000c)
	• The flavor profile of apple wine fermented with various yeast strains and natural sources of fermentation revealed distinctness among commercial and natural yeast strains. The flavor profiles of wine with and without a nitrogen source were different.	Joshi et al. (2002a) and Joshi and Sandhu (2003)
	• Besides, effect of addition of insoluble solids on physico-chemical and sensory quality especially flavour profile on apple wine was determined.	Joshi et al. (2013a,b)
	• Successful attempt to prepare cider ginger beverage was made.	Panesar et al. (2000)
	• Antimicrobial activity was shown against Escherichia coli, Staphylococcus, Aspergillus, Candida, and Bacillus by adding extract of garlic, hops, honey, etc., to apple wine.	Joshi and John (2002)
	• Nonlinear kinetic model applied to predict the consumption of various sugars as a substrate, during apple wine yeast fermentation with Saccharomyces cerevisiae strain CCTCC M201022 at the flask-scale level. Owing to the high fructose content in apple juice, the evaluation and selection of a fructophilic yeast strain could be significant to the apple wine industry.	Wang et al. (2004)
	• Effects of oak chips having diverse geographical origins (French, American, and Chinese), with different toast levels (light, heavy, and medium), dosages, and aging times on volatile compounds of apple cider were studied. The medium toasting level oak chips released the highest concentrations of volatile components into the ciders. Therefore, the aroma compounds of cider can be increased by using mature oak chips.	Fan et al. (2006)
	• Flavor profiling of apple vermouth with various alcohol, sugar, and spice levels revealed that a product with 4% sugar, 2.5% spice extract, and 15% ethanol gave a desirable flavor profile.	Joshi and Sandhu (2009)
	• Effects of various nitrogen concentrations in apple wine fermentation were studied and revealed that nitrogen content in the apple must was an important factor for growth and fermentation velocity.	Alberti et al. (2011)
	• Attempts to make cider with medicinal value were also made successfully.	Joshi et al. (2014)

Continued

Table 1.6 Summary of Research Work Done on Nongrape Wines—cont'd

Name of Fruit	Details of Research Findings	References
Apricot	• Wild apricot (*Prunus armeniaca*) wine was prepared by dilution of fruit pulp in a 1:1 ratio with water, raising TSS to 24°B, addition of DAHP at 0.1%, 0.5% pectinase, followed by ameliorating to 12°B with sugar.	Joshi et al. (1990a)
	• The influence of wood chip treatments on fruit wines indicated that *Albizia* and *Quercus* produced wines of highest sensory qualities and apricot wine was the best among all the wines prepared.	Joshi et al. (1994)
	• Preparation of apricot wine from Newcastle variety and wild apricot (chulli) was reported. Vermouth production from wild apricot was optimized.	Joshi et al. (1990a,b), Joshi and Sharma (1994) and Joshi et al. (2011b,d,e)
Banana	• Fermentation of banana pulp diluted in 1:1, 1:2, or 1:3 ratio produced acceptable quality wines but the wine flavor was lost after 6 months storage. The wine recovery ranged from 60% to 76% in different varieties.	Kundu et al. (1976)
	• Fermenting banana juice is considered to be an attractive resource for utilizing surplus and overripened bananas. Sensory evaluation results showed that there were no significant differences in flavor, taste, clarity, and overall acceptability between banana wine and a reference wine.	Akubor et al. (2003)
	• The pretreatment of banana must with pectinase followed by treatment with α-amylase enhanced the hydrolysis of complex carbohydrates like pectin and starch, which leads to increased clarity, decreased viscosity, 2.7-fold increase in the amount of extracted juice, and 15% and 39% increase in TSS and reducing sugars in extracted juice, respectively.	Cheirslip and Umsakul (2008)
	• Banana wine can be produced by adopting various methodologies and immediate consumption within 48h without addition of chemical preservatives.	Idise and Odum, 2011
	• Recombinant yeast strains were found able to degrade various polysaccharides instead of commercial enzyme preparations for the process of winemaking to improve wine processing and wine quality.	Byaruagaba-Bazirake et al. (2013)
Ber	• It was found that 50–100ppm SO₂, 2–5% inoculum, and pH 3.5–4.0 of the must were the optimum conditions for preparation of good quality ber (*Ziziphus mauritiana* Lamk) wine.	Patil et al. (1995)
Blueberry	• High-bush blueberry (*Vaccinium corymbosum*) fruits were a rich source of antioxidant phenolic compounds having α-amylase and α-glucosidase inhibitory capacity compared to acarbose, a known antidiabetic drug.	Johnson et al. (2011)
	• Capillary electrophoresis analysis of blueberry wine led to the determination of kaempferol, ferulic acid, vanillic acid, caffeic acid, gallic acid, and protocatechuic acid, thus suggesting that fruit wines made from blueberries may have potential health applications and therefore could contribute to the economy of the wine industry.	Li et al. (2011) and Johnson and Mejia (2012)
Cacao	• The fermentation of cacao fruit pulp juice with yeast UFLA CA 1162 suggested that the use of cacao fruits in the production of wine is a viable alternative that allows the use of harvest surpluses and other underused cacao fruits, thus introducing a new product into the market.	Durate et al. (2010a)
Cagaita	• A comparative study was carried out to evaluate fermentation conducted with free and calcium alginate-immobilized cells of *S. cerevisiae* (UFLA CA11 and CAT-1) during preparation of wine from cagaita fruit pulp.	Oliveira et al. (2011)
	• The cagaita fruit pulp fermentation occurred faster (4 and 8days, respectively) with immobilized UFLA CA11 and CAT-1 yeast strains compared to fermentation (10 and 12days, respectively) with UFLA CA11 and CAT-1 free cells.	
Cherry	• Cherries of the Early Richmond variety were fermented using various *S. cerevisiae* strains to study their effect on the production of volatiles and polyphenols.	Sun et al. (2011)
	• The cherry wines all contained polyphenols (chlorogenic and neochlorogenic acids), anthocyanins (cyanidin 3-glucosylrutinoside and cyanidin 3-rutinoside) in higher amounts.	
Cupuaçu	• The fermentation of cupuaçu pulp juice with *S. cerevisiae* UFLA CA1162 and analysis of minor and major compounds in cupuaçu wine were carried out by using GC/MS and GC/FID.	Durate et al. (2010a)
	• Cupuaçu fruit wine was subjected to a sensory analysis and showed the highest percentage of acceptance for aroma (68%) compared to appearance (61%) and taste (58%).	

		References
Custard apple	• Studies revealed that a wine with organoleptic properties comparable to grape wine could be prepared from custard apple (*Annona squamosa*).	Kotecha et al. (1995)
	• Custard apple (*A. squamosa*) wine was prepared using various dilutions (1:2, 1:3, and 1:4 dilution with and without DAHP). Among all the dilutions 1:4 dilution with DAHP had higher value for physicochemical characteristics, sensory scores, and acceptability in comparison to other dilutions.	Kumar et al. (2011b)
	• The potential of custard apple in the production of a beverage fermented using *S. cerevisiae* (NCIM 3282) yeast was studied, and the antioxidant capacity, total phenolic content, and DNA damage–protecting activity of the wine were assessed.	Jagtap and Bapat (2014)
	• Custard apple fruit wine was also able to protect against γ-radiation (100Gy)-induced DNA damage in pBR 322 plasmid DNA, suggesting that it may have potential health applications and therefore could contribute to the economy of the wine industry.	
Elderberry	• The color, chemical composition, and antioxidant capacity of elderberry must and wine were studied. Elderberry wine produced was an intense red, having pH 3.9 with moderate ethanol concentration (13.2% v/v).	Schmitzer et al. (2010)
	• Elderberry wine showed higher antioxidative potential than elderberry must.	
Gabiroba	• The pH and sugar content of gabiroba fruit pulp make it a good substrate for winemaking.	Durate et al. (2009, 2010a)
	• Compounds with higher aromatic volatiles contribute to the aroma of fruit wines to a greater extent.	
	• Gabiroba fruit wine was subjected to sensory evaluation and had 63%, 60%, and 52% acceptability for appearance, aroma, and taste attributes, respectively.	
Guava	• Wines made from guava juice were highly acceptable because of their low tannin content, optimum color, and flavor. Pectinase treatment of guava pulp prior to fermentation gave about 18% increase in wine yield.	Bardiya et al. (1974)
	• Guava juice was fermented with two different *S. cerevisiae* strains, NCIM 3095 and NCIM 3287, and optimization of guava wine fermentation with respect to osmotolerance, alcohol tolerance, inoculum size, initial pH of the medium, amount of SO_2, amount of diammonium phosphate, and incubation temperature was studied.	Sevda and Rodrigues (2011)
	• GC/FID and GC/MS analysis of guava wine revealed the presence of many volatile constituents like esters, alcohols, ketones, acids, aldehydes, terpenes, phenol derivatives, lactones, sulfur compounds, and other miscellaneous compounds.	Pino and Queris (2011b)
	• Wine was prepared from guava and ber fruit juices using *S. cerevisiae* var. HAU 1. The maximum alcohol yields in guava and ber were 10.653% and 2.246%, respectively.	Younis et al. (2014)
Jabuticaba	• Minor and major compounds in jabuticaba wine prepared by fermenting the fruit juice with yeast UFLA CA 1162 were studied.	Durate et al. (2010a)
	• This fruit wine had a composition similar to other beverages, demonstrating that this fruit has the potential to be used to produce fermented beverages.	
Jackfruit	• Wine from jackfruit pulp was prepared for the evaluation of total phenolic content, flavonoid contents, and antioxidant properties.	Jagtap et al. (2011)
	• The antioxidant and DNA damage–protecting properties of jackfruit wine confirmed health benefits when consumed in moderation and could become a valuable source of antioxidant-rich nutraceuticals.	
	• Jackfruit can be a very important source for commercial wine production. Various process parameters like pH, temperature, and inoculum concentration play important roles during the fermentation of the wine. Moreover, a sweet smell gives an added advantage to its sensory properties.	Sharma et al. (2013)
Jamun	• Jamun varieties, viz., Pharenda, Jamun, and Kathjamun, were used for wine preparation by diluting the whole fruit in the ratio of 1:1 with water and treatment with pectinase enzyme was adjudged to be the best treatment for preparation of wine from these varieties.	Shukla et al. (1991)
	• Anthocyanin-rich jamun fruits were fermented with *S. cerevisiae*, which resulted in a sparkling red wine having an acidic taste and low alcohol (6%) concentration.	Chowdhury and Ray (2007)
	• Jamun wine also possessed medicinal properties, antidiabetic properties, and the ability to cure bleeding piles.	
	• The major antioxidants found were a complicated mixture of hydrolyzable tannins and fruit acids.	Nuengchamnong and Ingkaninan, (2009)
	• The effects of dilution and maturation on jamun wine were documented.	Joshi et al. (2012b)

Continued

Table 1.6 Summary of Research Work Done on Nongrape Wines—cont'd

Name of Fruit	Details of Research Findings	References
Kinnow	• Of cyclodextrin and amberlite X AD-16, amberlite-16 was considered better with respect to the extent of bitterness and changes in composition in kinnow juice for winemaking and it was concluded that debittering the juice either prior to or during fermentation improved the sensory quality.	Joshi et al. (1997)
	• Optimization of various parameters revealed that with 5% inoculum, the fermentation was completed in 10 days with kinnow mandarin juice as substrate and 8 days when a kinnow–cane juice blend was the substrate, indicating a positive effect of addition of cane juice. The by-product (pulp) obtained was processed into a squash-like product for value addition.	Khandelwal et al. (2006)
	• An investigation was carried out to find the optimal conditions for the efficient conversion of kinnow juice into wine using response surface methodology.	Panesar et al. (2009) Singh et al. (1998)
	• The suitability of kinnow fruit for winemaking was also determined.	
Kiwifruit	• The high natural content of ascorbic acid in kiwifruit allowed a large reduction in sulfur dioxide concentration, required to stabilize white table and dessert wines. However, kiwi wine produced from fruit picked in May and fermented immediately was found to be less acceptable.	Withy and Lodge (1982)
	• Chemical composition and sensory properties of kiwifruit wine were studied.	Soufleros et al. (2001)
	• The pretreatment of ripe kiwifruit pulp with pectinase increased the yield of wine production from 63.35% to 66.19%. The kiwifruit wine made from overripened fruits treated with pectinase was superior in sensory value, alcohol, total phenolic content, antioxidant activity, minerals, and production yield.	Towantakavanit et al. (2011)
Lychee	• Method of preparation of high-flavor low-alcohol wine from lychee using osmotic technique was optimized.	Vyas et al. (1989a,b)
	• Among the yeast strains screened for alcoholic fermentation, S. cerevisiae MTCC 178 was found to be the potent strain. Moreover, sensory evaluation revealed the production of lychee wine with attractive aroma and harmonious taste.	Singh and Kaur (2009)
	• The surplus of lychee fruit is thus made into wine, generating additional revenue for the grower and reducing postharvest losses.	Alves et al. (2011)
	• Evaluation of the volatile components and aroma profiles of lychee wines by using HS–SPME–GC/MS showed that the majority of terpenoids derived from lychee juice decreased, or even disappeared, during alcoholic fermentation, whereas terpenol oxides, ethers, and acetates increased.	Wu et al. (2011)
Mango	• Mango varieties were screened and Fazri, Langra, and Chausa produced good-quality wines. Sweet wines from Dashehari had a characteristic fruity flavor.	Kulkarni et al. (1980)
	• Mango juice fermentation using a selected yeast strain (S. cerevisiae 101) was performed. wine produced from the Totapuri cultivar had more volatiles than wine from the Banginapalli cultivar.	Reddy and Reddy (2010)
	• Carotenoid composition of wine made from seven mango cultivars was studied. The xanthophyll percentage in the wines decreased in the range of 69.3–89.7%, and a 480% degradation was noted in Banginapalli, Neelam, Sindhura, and Totapuri mango varietal wines along with15.3–26.5% degradation of β-carotene.	Varakumar et al. (2011)
	• The addition of β-glucosidase could accelerate the release of volatiles and was effective in intensification, diversification, and balancing of the mango wine aroma profile.	Li et al. (2013)
	• The interactions between three S. cerevisiae strains individually and in combination with Oenococcus oeni during the process of malolactic acid fermentation in mango wine were studied.	Varakumar et al. (2013)
	• Mango wine treated with γ-irradiation (3 kGy) resulted not only in an increased total phenolic content and total flavonoid content but also in a decreased microbial load in a dose dependent-manner, leading to improvement in the quality of wine.	Kondapalli et al. (2014)
	• The wine produced from mango was analyzed for its chemical properties and major volatile compounds. The ethanol content of the wines produced was between 8.9 and 9.5% (v/v), the range acceptable for table wine. The sensory evaluation indicated that mango wine exhibited sensory characteristics similar to those of grape wine.	Musyimi et al. (2015) Pino and Queris (2011a)

Muskmelon	• Muskmelon fruits unfit for the table purpose were converted into alcoholic beverages with 6.5% (w/v) alcohol and exhibited a very good sensory quality.	Teotia et al. (1991)
Orange	• Effects of bottle color (clear white, green, and brown), storage temperature, and storage time on the browning of orange wine showed that the use of brown bottles and a short storage time reduced the browning; however, storage at two different temperatures did not significantly affect the browning index.	Selli (2002)
	• A total of 64 volatiles were analyzed in blood orange wine by GC/FID and GC/MS. Similarly, 63 aroma-active compounds were detected by GC–olfactometry.	Selli (2007) and Selli et al. (2008, 2003)
	• Orange wine produced with S. cerevisiae var. ellipsoideus contains the highest ethanol (90.38%), whereas S. cerevisiae from sugarcane molasses produced wine with the least ethanol concentration (81.49%).	Okunowo and Osuntoki (2007)
	• The antioxidant capacity of orange juice was found to be higher than that of orange wine.	Kelebek et al. (2009)
	• Acidity greatly influences the taste, color, and aromatic profile of orange wine as well as the stability and microbiologic control of the wine quality by stopping/retarding the growth of harmful microorganisms that would spoil the wine.	Selli and Kelebek (2015)
Palm	• Palm wine was sensorally evaluated and the key odorants were investigated.	Lasekan et al. (2007)
	• The great yeast diversity and the presence of lactic acid and acetic acid bacteria during the palm wine tapping process indicated that a multistarter fermentation occurs in a natural, semicontinuous fermentation process.	Stringini et al. (2009)
	• Biochemical and microbiological properties of different week-old fresh palm saps were assessed during storage of the fermenting saps to check their quality. The results showed that during storage, the accumulation of alcohol occurred in all palm wine samples with concurrent lactic and acetic acid fermentation taking place as well.	Karamoko et al. (2012)
Papaya	• It has been proposed as a potential renewable energy resource for industrial alcohol production because of its low cost and easy availability.	Sharma and Ogbeide (1982)
	• Fusel oil (by-product of alcohol distillation industry) addition to papaya wine fermented with the yeast Williopsis saturnus var. mrakii NCYC2251 produced a wide range of volatile compounds.	Lee et al. (2010)
	• The impact of amino acid addition on aroma compound formation in papaya wine fermented with yeast W. saturnus var. mrakii NCYC2251 was studied.	Lee et al. (2011)
	• Effects of sequential inoculation of yeasts W. saturnus var. mrakii NCYC2251 and S. cerevisiae var. bayanus R2 on the volatile profiles of papaya wine were investigated.	Lee et al. (2012)
Pear	• Wine of good quality can be made from pears with high tannin content, e.g., Bartlett.	Verma and Joshi (2000)
	• Dry and sweet vermouth with variable alcohol levels are prepared from sand pear base wine.	Joshi and Attri (2005)
	• Blending of plum base wines with different proportions of sand pear juice revealed that it not only reduced the alcohol content but also gave the blend acceptable sensory qualities.	Joshi et al. (2014)
Peach	• Effect of wood treatment on the chemical composition and sensory quality of peach wines revealed significant changes compared to the control. Wine aged with Quercus wood had higher total phenols, aldehyde, and ester contents than the control. The wine treated with Quercus was rated as the best in sensory qualities.	Joshi and Shah (1998)
	• Peach pulp was fermented at 25°C for 2 weeks using S. cerevisiae KCCM 12224, aged at 15°C for 14 weeks.	Chung et al. (2003)
	• Suitability of eight peach cultivars (Sunhaven, Redhaven, Kateroo, JH Hale, Flavorcrest, July Elberta, Stark Early Giant, and Rich-Haven) for preparation of table wine revealed that all cultivars were suitable, but Redhaven, Sunhaven, JH Hale, Flavorcrest, and July Elberta were adjudged as better others.	Joshi et al. (2005b)
	• The wine prepared from Redhaven peaches contained lower alcohol content and higher pH compared with white wine. The main phenolic compounds found in peach wine were chlorogenic acid, caffeic acid, and catechin. It was accepted by consumers as well as sensory panelists.	Davidovic et al. (2013)
	• Fermentation properties were monitored, and fermentation conditions (initial sugar concentration, temperature, and time) were optimized by response surface methodology for production of peach wine using peach juice.	Lee (2015)
Pineapple	• S. cerevisiae species isolated from the fermenting sap of Elaeis guineansis (palm wine) was suitable for making wine from pineapple. This yeast isolate gave a high ethanol yield of 10.2% (v/v) compared to the control wine yeast (7.4%, v/v).	Ayogu (1999)
	• The turbidity of pineapple wine was reduced and a clear product with bright yellow color was obtained after microfiltration. A negative effect of gas sparging, a loss of alcohol content in the wine, was also observed.	Youravong et al. (2010)

Continued

Table 1.6 Summary of Research Work Done on Nongrape Wines—cont'd

Name of Fruit	Details of Research Findings	References
	• Ethyl octanoate, ethyl acetate, and ethyl 2-methylpropanoate compounds strongly contributed to the pineapple wine aroma.	Pino and Queris (2010)
	• Freshly crushed pineapple juice collected from Thailand and Australia shows the presence of *Hanseniaspora uvarum* and *Pichia guilliermondii* yeast strains during the fermentation process. *P. guilliermondii* was dominant and was consistently present during the early stage of the fermentation, whereas *H. uvarum* populations increased from an initial level through the 6-day fermentation period.	Chanprasartsuk et al. (2010)
Plum	• Standardization of plum wine production was done earlier with 1:1 ratio of water and whole-fruit pulp, which reduced the acidity by dilution and produced a wine of acceptable quality.	Vyas and Joshi (1982)
	• Production of acceptable quality wines from plums by reducing the acidity by using *Schizosaccharomyces pombe* was reported.	Vyas and Joshi (1988)
	• Deacidification activity of yeast was rapid at pH 3–4.5 but adversely affected when pH was lowered to 2.5 in plum must. It was also influenced by higher concentration of ethanol SO_2 and addition or not of nitrogen source.	Vyas and Joshi (1988) and Joshi and Bhutani (1991)
	• *S. pombe* was employed to produce plum wine whose acidity can be adjusted as a dry wine. This technique may serve as an alternative to dilution of plum must practiced at present to produce wine of desirable acidity.	Joshi et al., (2002b)
	• Of various yeast strains studied, UCD 595 was the best for wine preparation from plums. • Initial sugar in plums affected the quality of plum wine.	Joshi et al. (2009b)
	• Effect of alcohol concentration and lowering of temperature of incubation on viability of UCD 595 and UCD 522 was determined to acclimatise the yeast for sparkling wine production from plum.	Sharma and Joshi (1996) Bhardwaj and Joshi (2009)
	• Foam formation using different cultivars, yeast culture type, and addition of yeast extract were also determined. • Preparation of sparkling wine from plum was studied to see the effect of deacidifying yeast as such or after immobilization in different plum cultivars.	Joshi and Bhardwaj (2011)
Pomegranate	• Pomegranate wines were rich in anthocyanins, which were responsible for the antioxidant properties and color of the wine.	Meena et al. (2011)
	• The presence of melatonin (N-acetyl-5-methoxytryptamine) was found in the wine made from pomegranate cultivars Wonderful and Mollar de Elche. Melatonin is a neurohormone related to a broad array of physiological functions and has proven therapeutic properties.	Meena et al. (2012)
Prickly pear	• *Saccharomyces* and non-*Saccharomyces* strain combinations can be used to obtain high-quality fermented beverages from prickly pear juice. • GC/MS analysis shows the presence of nine major volatile compounds that are considered to be essential for a fine wine flavor.	Rodriquez-Lerma et al. (2011)
Raspberry	• Korean black raspberry juice and wine were rich sources of phenolic compounds and raspberry wines have been popular as traditional alcoholic beverages in Korea.	Ku and Mun (2008)
	• *Rubus coreanus* Miq. fruits fermented with *S. cerevisiae* had a higher total phenolic content and a better DPPH radical scavenging activity than the unfermented material.	Ju et al. (2009)
	• Black raspberry wine prepared with inclusion of seeds showed higher antioxidant activity.	Jeong et al. (2010)
	• Strain UFLA FW 15 was the yeast that produced a raspberry wine with the best overall chemical and sensory quality.	Durate et al. (2010b)
	• Fruit extract and wine prepared from fruits along with seeds led to increase in functional activities of the juice and wine.	Jeong et al. (2010)
	• Immune-stimulating polysaccharides were isolated and characterized from Korean black raspberry wine by size-exclusion chromatography.	Hwang and Shin (2011)

Fruit	Description	Reference
Sapota	Changes in the physicochemical properties and key compounds of Korean black raspberry wines made from juice, juice–pulp, and juice–pulp–seeds were studied. The results revealed that the color intensity of the wine made from juice supplemented with pulp and seeds was increased compared with the wine made from only juice. The content of proanthocyanidins in raspberry wine was two to three times greater compared to commercial grape wine, which explained the bitter and astringent character of raspberry wine.	Lim et al. (2012)
	Sapota wine can be made from either clarified or nonclarified sapota juice by raising the TSS to 25°Bx, adding 0.7% citric acid, adding 30 ppm SO_2, heating to 80–85°C for 10 min, pressing the pulp for 4 days, adding 0.1% pectinase, siphoning, etc.	Gautam and Chundawat (1998)
Sea buck-thorn	Effects of dilution and deacidification were determined to optimize the process of winemaking from sea buckthorn.	Joshi et al. (2011f)
Soursop	Soursop wine was prepared by using mycoflora associated with various parts of fresh and rotten fruits along with indigenous yeast flora and commercial yeast extract. *Botryodiplodia theobromae* was isolated from the rotten fruits (skin), whereas *Trichoderma viride* was isolated from the fresh fruits. However, *Penicillium sp.* was the most dominant in all parts of fresh soursop fruit, whereas *Rhizopus stolonifer* occurred at a high percentage in rotten fruits.	Okigbo and Obire (2009)
	Annona muricata is a good source for wine production and single-cell protein because of the high nutritional composition of soursop juice, high alcoholic content, and palatability of the wine.	
Strawberry	Strawberry (*Fragaria×ananassa* Duchesne) wine was prepared with various cultivars. Wine from cultivar Camarosa had higher values for physicochemical characteristics, sensory scores, and acceptability in comparison to two other cultivars (Douglas and Chandler).	Joshi et al. (2005a) and Sharma et al. (2009)
	Effects of maturation and the flavor profile of strawberry wine of various cultivars and treatments were determined.	Joshi and Sharma (2003) and Sharma and Joshi (2004)
Sweet potato	An herbal purple sweet potato wine was prepared from purple-fleshed sweet potato rich in anthocyanin pigment and 18 medicinal plant parts, by fermenting with wine yeast, *S. cerevisiae*.	Panda et al. (2012)
	The herbal purple sweet potato wine was a novel product with ethanol content of 8.61% (v/v) and a rich source of antioxidant anthocyanin, which offers remedies for colds, coughs, skin diseases, and dysentery.	
Umbu	Umbu fruit wine showed the presence of geranic acid along with monoterpenes.	Lira Júnior et al. (2005)
	Sensory analysis of wine showed the highest percentage of acceptance (74%) for aroma compared to appearance (63%) and taste (57%). The results indicated that using these tropical fruits in the production of fruit wines was a viable alternative that allows the use of harvest surpluses, resulting in the introduction of new products into the market.	Duarte et al. (2010a)
Multiple fruits	Multifruit wines of high quality from locally available fruits (tomato, almond, orange, lemon, and African star apple extract) have been produced and compared well with standard wines with respect to organoleptic and physicochemical attributes.	Agbor et al. (2011)
	Combinations of sweet potato with grapes and sweet potato with beetroot and banana were effective in the production of white wine.	Karthik et al. (2012)

DAHP, di-ammonium hydrogen phosphate; DPPH, 2,2-diphenyl-1-picrylhydrazyl; FID, *flame ionization detector*; GC, *gas chromatography*; HS–SPME, *head space solid phase microextraction*; MS, *mass spectrometry*; TSS, *total soluble solids*.

Vermouth from apple fruit having 4% sugar, 2.5% spice extract, and 15% alcohol has been made successfully (Joshi and Sandhu, 2000b). Addition of extract of spices and herbs increases the total esters and aldehyde content of vermouth. Increasing the acid content further increases the acceptability of the product with 19% alcohol content. Similar attempts at making wild apricot vermouth have been successful (Joshi et al., 2014). Detailed information can be obtained from Chapter 9 of this book and the literature cited (Panesar et al., 2009, 2011a,b).

10.2 SPARKLING WINE

Like grapes, sparkling wines have been developed from plums (Joshi et al., 1995, 1999b, 2011b; Joshi and Sharma, 1995; Sharma and Joshi, 1996; Joshi et al., 1999c). Six plum cultivars have been evaluated for the production of sparkling wine. Addition of immobilized *S. cerevisiae* followed by *S. pombe* was found as the best wine on account of chemical and sensory analysis (Joshi and Bhardwaj, 2011). The bottled fermented (Methode Champenoise) wine made from base wine with sodium benzoate had a desirable level of CO_2, low aldehydes, higher esters, more crude proteins, better color, and higher sensory qualities than others. In the production of sparkling plum wine, UCD 595 gave better response to ethanol content and lower temperature than UCD 505, whereas a sugar concentration of 1.5% and di-ammonium hydrogen phosphate (DAHP) of 0.2% plum wine at $15\pm2°C$ gave optimum pressure in bottle fermentation (Joshi et al., 1995).

Detailed information about the technology for the production of vermouth and sparkling wines is given in Chapter 9.

11. NONGRAPE FRUIT WINE INDUSTRY: GLOBAL STATUS

Considering the quantity of fruit production, especially of the nongrape fruits, there is considerable scope for wine production. It could lead to a diversification of the products available to consumers. Wine production is an important tool for reducing postharvest losses, which as of this writing range from 20% to 30%. In developing countries, production of wine could soak excessive production of fruits (Joshi et al., 2000, 2011b). It is undisputable that wine is the most important tool for value addition to horticultural produce like the sand pear, which as such commands a small market, but its wine (vermouth) fetches many times higher price. From the dismissive scenario of fruit and vegetable utilization by most developing countries, it is apparent that unless the processing industry is linked with the horticultural industry, it is unlikely to achieve any worthwhile results either for the farmers or for the consumers. So, wine production is an alternative for value addition and waste minimization of the nongrape fruit processing industries (Joshi, 2009; Joshi et al., 2011b).

11.1 FACTORS INFLUENCING FRUIT WINE PRODUCTION

Before advocating the production of fruit wines on a large scale there are several factors that need to be considered, such as availability of raw materials, cost of production and profitability for commercialization, production technology for different types of wines, research and development support for troubleshooting, state policies including tax structure (Nanda, 2015), industrial prioritization, prohibition, advertisement, consumer response, market behavior, and export potential (Joshi and Attri, 2005). The suitability of various fruits for preparation of wines of various types is outlined in Table 1.7.

Another important element is the proper quality of raw materials and for this the proper harvesting date of the fruits is a must. Fruits of proper maturity are important because fine-quality wine can be made from fruits with a proper sugar–acid balance. Even the best varieties give inferior quality of wine if not harvested at proper maturity (Joshi et al., 1999b; Jagtap and Bapat, 2015). The quality of a fermented product like wine is also determined by the characteristics of the fermentation agent, i.e., the yeast strain. It is known that the aroma of wine depends largely on the variety of fruit such as grapes, but yeast strains modify the sensory qualities by forming esters and higher alcohols at various concentrations. Yeasts are also capable of producing hydrogen sulfide, which is undesirable as it imparts a different flavor to the wine (see Chapter 3 of this book for details). The production technology of wine is definitely a major contributory factor to the introduction of wine production commercially. From the overview of research conducted (discussed separately), the establishment of wineries for the production of nongrape wines should not pose any problem as a good base for the technology generation has already been set up. Thus, not only grapes, but other fruits can be converted into wine (Joshi and Attri, 2005; Jagtap and Bapat, 2015). Moreover, there is virtually no difference in the factories required for the production of nongrape wines and grape wines. The manufacturer can make use of the same facility for the production of nongrape wines.

Many times, it is argued that there may be few consumers of nongrape wines, but the variety of apple wine, cider, kiwi wine, and plum wine clearly spells out a big market for nongrape wines, provided these are of proper quality and priced within an acceptable range for the consumer (Joshi and Parmar, 2004; Joshi et 2011a). The explosion in information technology and introduction of satellite communications via the Internet and television have already eliminated the boundaries between countries and regions and have resulted in the globalization of wine. The ever increasing demand for alcoholic beverages in national and international markets reflects a considerable scope for such products. Even efforts are being made to make wine with that medicinal values such as that from Amla or that from apple with additional extract from plants having medicinal value (Soni et al., 2009; Joshi et al., 2014). Thus, an appraisal of these and other factors reveals them to be

Table 1.7 Suitability of Nongrape Fruits for Various Types of Wines

Wine Type	Fruits
Table wine (dry and sweet)	Apple, peach, plum, pear, kiwi, orange, kinnow, strawberry, lychee, mango, apricot, wild apricot
Vermouth	Apple, plum, sand pear, mango, wild apricot, tamarind
Sparkling wine	Plum, orange, apple, honey, orange
Fortified wine	Mango, plum, pear, apricot
Low-alcohol beverages	Apple (cider), sand pear, lychee
Wines with medicinal/antimicrobial properties	Apple, peach

Adapted from Joshi, V.K., Attri, D., 2005. Panorama of research and development of wines in India. Journal of Scientific and Industrial Research 64 (1), 9–18.

quite favorable for the production of nongrape wine in various producer countries, including India (Joshi and Parmar, 2004; Joshi and Kumar, 2011).

The high rate of taxes and excise duties on imported red and white grape wines has evoked interest in producing these wines from tropical fruits (Tables 1.8 and 1.9). Generally, low sugar content and low amounts of extractable pigments limits the use of tropical fruits for red and white wine production, yet efforts have been made toward producing red wine (Okoro and Emeka, 2007). These wines may also contain some additional flavors taken from flowers and herbs (Swami et al., 2014). The most commonly used fruits for the production of red wines are plum and berries, whereas for white wines, peach, pear, apple, etc., are used. Nongrape fruit-based red and white wines are famous in Italy, California, France, Australia, New Zealand, etc. But most of the wines have their own specific color, so the exact terminology of white and red wine is not specifically applicable.

Table 1.8 List of Red Wines Obtained From Nongrape Fruits and Their Region of Origin

Type of Wine	Fruits Used	Region of Origin
Barbera	Blackberry, black cherry, raspberry, plum	Italy, Argentina, California
Cabernet Sauvignon	Dark berries	Italy, France, Australia, California
Malbec	Plum, blackberry	France, Argentina
Merlot	Plum, cherry	Italy, France, Washington, California, Chile
Syrah/Shiraz	Blackberry	Italy, France, Washington, California, South Africa
Pinot Noir	Baked cherry, plum	France, Australia, California, New Zealand
Zinfandel	Blackberry, boysenberry, plum	California

www.foodservicewarehouse.com.

Table 1.9 List of White Wines Obtained From Nongrape Fruits and Their Region of Origin

Type of Wine	Fruits Used	Region of Origin
Chardonnay	Tropical fruit, pineapple, apple	New York, France, Washington, California, South Africa, New Zealand, Australia
Gewurztraminer	Tropical fruit	Germany, New York, France, Washington, California
Muscat/Moscato	Peach, pear	New York, France, Washington, California, South Africa, New Zealand, Australia, Oregon
Pinot Grigio	Grape, pear	Northern Italy, Spain, France, California
Riesling	Peach, citrus, apple	Italy, France, Germany, Oregon, New York, Australia, Idaho
Sauvignon Blanc	Green fruit, gooseberry	France, New York, Australia, California, New Zealand, Oregon, Washington
Semillon	Ripe fruit, orange peel	France, Washington, Australia
Viognier	Apricot	France, Oregon, California

www.foodservicewarehouse.com.

11.2 GLOBAL PRODUCTION OF FRUIT WINES

The market size for fruit wines is difficult to determine because these products are not categorized into separate groups. However, most of the fruit wines are sweet in taste, they are generally grouped into the category of dessert wine, and their market is estimated to be only 2% of the total wine market. The global demand of fruit wines in different countries is largely variable, based on the source of fruit (Table 1.10; Table 1.11) as well as consumer demand (Rivard, 2009).

11.2.1 Apple Wine/Cider

Globally, cider is popular, with different names such as cidre (France), sidre (Italy), sidra (Spain), and apfel wein in Germany and Switzerland (Downing, 1989; Sandhu and Joshi, 1995; Joshi et al., 2011c; Swami et al., 2014). The production rate of cider is highly variable in different countries as shown in Fig. 1.9. Cider is popular in the United Kingdom, especially in the West Midlands, southwest England, and East Anglia, and is available in most corners of the country.

The Fourth Annual Cider Conference held in Chicago, Illinois, on February 8, 2014, highlighted that hard cider sales in the United States jumped over 60% in 2012 (http://expert.msue.msu.edu). The beverage is also popular and native to other European countries like Ireland, northern France, northern Spain, and central Europe (Karagiannis, 2011). Argentina is also a cider-producing country, especially the provinces of Río Negro and Mendoza. Australia also produces cider, particularly on the island of Tasmania, which has a strong apple-growing tradition. Cider is popular in various countries, though with different brand names (Table 1.12).

11.2.2 Plum Wine

Plum wine is quite popular in Germany and Pacific Coastal states, Japan, Korea, and China (Amerine et al., 1980). In China, plum wine is called meijiu, umeshu is a Japanese alcoholic drink, and in Korea, maesil ju is marketed under various brand names, including Mae Hwa Su, Mae Chui Soon, and Seol Joong Mae (http://self.gutenberg.org/articles/non-grape_based_wines). Both the Japanese and the Korean varieties of plum liquor are available with whole *Prunus mume* fruits contained in the bottle. In Taiwan, Japanese-style plum liquor is wumeijiu (smoked plum liquor), which is made by mixing *P. mume* liquor, *Prunus salicina* liquor, and oolong tea liquor. A similar drink known as plum jerkum is made from fermented plums in a manner similar to the use of apples for cider production.

11.2.3 Pineapple Wine

Pineapple wine is made from the juice of pineapples, with a strong pineapple flavor (http://self.gutenberg.org/articles/non-grape_based_wines). It is mainly made in Hawaii by MauiWine and in Nigeria by Jacobs Wines (the first pineapple winery in Africa). It is also made in the Dominican Republic and in Okinawa, Japan.

11.2.4 Cherry Wine

Cherry fruit wine is produced from cherries, usually tart cherries that provide sufficient acid. Michigan winemakers (United States) produce several varieties of cherry wine, including spiced versions and cherry–grape blends. Cherry Kijafa is a fortified fruit wine that is produced in Denmark from cherries with added natural flavors (http://en.wikipedia.org/wiki/Fruit_wine). It usually contains 16% alcohol by volume and is exported to many countries in Europe and North America.

Table 1.10 Description of Fruits Along With Leading Growing Countries

Fruit Crop	Scientific Name	Family	Major Varieties	Leading Countries
Apple	*Malus domestica* Borkh	Rosaceae	Tydeman's Early, Michael, Molies Delicious, Schlomit, Starking Delicious, Vance Delicious, Top Red, Red Chief, Oregon Spur, Red Spur, Red Gold, Golden Delicious, Granny Smith	China, United States, Turkey, Poland, India
Apricot	*Prunus armeniaca*	Rosaceae	Newcastle, Nugget, Charmagaz, Early Shipley, Sufeda, Shakarpara, Kaisa	Turkey, Iran, Uzbekistan, Algeria, Italy
Banana	*Musa paradisiaca*	Musaceae	Poovan, Gros Michel, Lacation, Robusta, Giant Cavendish, Dwarf Cavendish	India, China, Uganda, Philippines, Ecuador
Blackberry	*Rubus* spp.	Rosaceae	Black Knight, Black Treasure, Bountiful, Giant, Forever Amber, Jet, Prestige, Black Hawk, Bristol, Cumberland, Glencoe, Jewel, Munger, Ohio Everbearer, Scepter	North America, Korea, Japan, Brazil, China
Blueberry	*Vaccinium* spp.	Ericaceae	Elliott, Dube, Bluejay, Nelson, Tifblueplus Rose, Denise, Blue Chip	United States, Germany, Turkey, Chile, Australia
Cashew nut	*Anacardium occidentale*	Anacardiaceae	Vengurla 1, Vengurla 4, Vengurla 6, Dhana, Priyanka	Vietnam, Nigeria, India, Benin, Philippines
Sweet cherry	*Prunus avium* L.	Rosaceae	Bing, Stella, Van, Sue, Sam, Lambert, Black Tartarian, Napoleon, Emperor, Black Republican	Turkey, United States, Iran, Italy, Spain
Sour cherry	*Prunus cerasus* L.	Rosaceae	Montmorency	Turkey, Russia, Poland, Ukraine, Iran
Jamun	*Syzygium cumini* (L.) Skeels.	Myrtaceae	Rajamun, Paras, Narendra Jamun	Bangladesh, India, Nepal, Pakistan, Sri Lanka, Malaysia, Philippines, Indonesia
Mandarin orange	*Citrus reticulata* Blanco	Rutaceae	Kinnow, Nagpur, Khasi Coorg	India, Pakistan
Kiwifruit	*Actinidia deliciosa*	Actinidiaceae	Abbott, Allison, Bruno, Monty, Hayward, Hort-16	Italy, New Zealand, China, Greece, Chile, Germany
Lychee	*Litchi chinensis* Son.	Sapindaceae	Early Seedless, Rose-Scented, Dehradun, Gulabi, Calcutta, Shahi, China, Swaran Roopa	
Loquat	*Eriobotrya japonica* Lindl.	Rosaceae	Golden Yellow, Large Round, Pale Yellow, Fire Ball, Mammoth, Matchless, California Advance	
Mango	*Mangifera indica* L.	Anacardiaceae	Alphonso, Kesar, Dashehari, Langra, Chausa, Amrapali, Malika, Ratna, Sindhu, Neelum, Bangalra	

Table 1.10 Description of Fruits Along With Leading Growing Countries—cont'd

Fruit Crop	Scientific Name	Family	Major Varieties	Leading Countries
Passion fruit	*Passiflora edulis* Sims.	Passifloraceae	Purple, Yellow, Noel's Special	
Peach	*Prunus persica* Batsch	Rosaceae	Sharbati, Saharanpu, Prabhat, Florida Sun, Florida Gold, Florida Red, Florida Prince, July Elberta, Sun Red (nectarine), Red Haven	China, Italy, United States, Greece, Spain
Pear	*Pyrus communis* L.	Rosaceae	Bartlett, Max-Red–Bartlett, Laxton's Superb, Comice, Conference, Beurre Hardy, Patharnakh, Baggugosha	China, United States, Argentina, Italy, Turkey
Persimmon	*Diospyros kaki* L.	Ebenaceae	Fuyu, Hachiya, Flat Seedless, Hyakume	China, Korea, Japan, Brazil, Azerbaijan
Plum	*Prunus salicina* Lindl.	Rosaceae	Kala Amritsari, Sutlej Purple, Santa Rosa, Mariposa, Red Beaut, Burbank, Beauty	
Pomegranate	*Punica granatum* L.	Punicaeae	Ganesh, Dholka, Kandhari, Jalore Seedless, Muskat	
Raspberry	*Rubus ellipticus*	Rosaceae	–	Poland, United States, Serbia, Ukraine
Strawberry	*Fragaria × ananassa*	Rosaceae	Pusa Early Dwarf, Tioga, Torrey, Chandler	United States, Turkey, Spain, Egypt, Mexico

Table 1.11 Production of Fruit Wines in Various Countries

Fruit Wine	Countries Involved in Production
Apple	France, Italy, Spain, Germany, India, United States, Spain, Ireland, Australia
Mango	India, Philippines
Plum	United States, Japan, China, Korea
Banana	East Africa, India
Perry	United States, Japan, Sweden, Australia, New Zealand, France, England
Kiwi	New Zealand, India, California, Italy, Chile, France, Australia, Israel
Kinnow	India, United States
Orange	Australia
Strawberry	England, Philippines
Guava	India, United States
Cherry	United States
Blueberry	United States, Canada, England
Pineapple	Thailand, Japan, Mexico, southeast Asian countries
Lychee	China, India, France, Vietnam

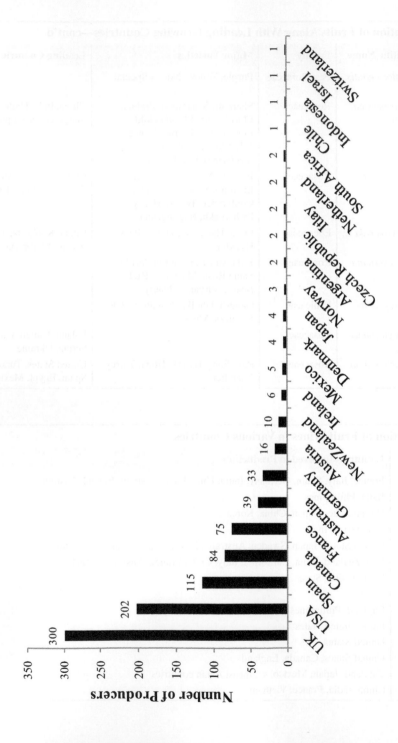

FIGURE 1.9

Country-wise production of cider in August 2013.

Based on the data from http://www.statista.com/statistics/300837/global-number-of-cider-manufacturers-by-country/.

Table 1.12 List of Cider Brands Popular in Various Countries

Country	Cider Brands
Argentina	Real, La Victoria, Del Valle, La Farruca, Rama Caida
Australia	Mercury Cider, Strongbow, Three Oaks Cider, Pipsqueak, Magners, Weston's, Monteith's, Kopparberg, Rekorderlig, Somersby
Canada	Bulwark Cider, Ironwood Hard Cider, Mystique, Okanagan Cider, Rock Creek Dry Cider, Sea Cider, Thornbury Cider, William Cider
China	Apple Vinegar
Finland	Fizz
France	La Chouette Cider
Germany	Elbler Cider, Possmann Pure Cider
India	Tempest, Bragg, Fleischmann
Japan	Asahi Nikka Cidre, Kirin Hard Cidre
Ireland	Finbarra Cider, Stonewell Cider
New Zealand	Speight, Isaac, Old Mout Cider, Zeffer
South Africa	Hunters, Savanna Dry
United Kingdom	3 Hammers, Ashridge Cider, Aspall Cider, Blackthorn Cider, Strongbow Cider, Thatchers Cider
United States	Alpenfire Cider, Angry Orchard, Bantam Cider, Bold Rock Hard Cider, Citizen Cider, Eagle-mount Cider, Finnriver Cider, Fox Barrel Cider, Glider Cider, Gunga Din, Seattle Cider

11.2.5 Jamun Wine

Jan (*S. cumini*) fruit is an important member of the family Myrtaceae. It is considered to be indigenous to India and the West Indies, but is also cultivated in the Philippines, West Indies, and Africa. The pulp of jamun contains good amounts of fermentable sugar, which can be used for alcoholic fermentation and are utilized for the production of red wine, which offers a lot of health benefits by acting as an effective medicine (Patil et al., 2012).

11.2.6 Strawberry Wine

Strawberries, being rich in nutrients, are very delicious but highly perishable; thus they demand immediate utilization as desserts or other valuable products like purees, juices, jams, wine, etc. Chandler, Douglas, and Camarosa are the commonly used cultivars of strawberry used for wine production (Sharma et al., 2009; Joshi et al., 2005b).

11.2.7 Banana Wine

In banana wine production, Cavendish bananas are used, whereas in informal production, a variety of cultivars and apple bananas are used. Commercially produced banana wine is a clear, slightly sparkling alcoholic beverage with a longer shelf-life than banana beer, which spoils easily and therefore is not stored for long periods. Depending on the strain of yeast and amount of sugar added, the sweetness and alcohol level in the final product are variable. Commercial brands of banana wine are Malkia and Meru (http://www.wikiwand.com/en/Banana_wine).

11.2.8 Kiwi Wine

A considerable proportion of kiwifruit grown in New Zealand, California, Italy, Chile, France, Australia, and Israel have been utilized for the production of kiwi wine. One of the interesting things related to kiwi wine is that its flavor bears no relationship to its source of production, i.e., kiwi fruit (Craig, 1988).

11.2.9 Citrus Wine

Citrus fruits are generally known to have poor shelf life and face the problem of postharvest losses (Panesar et al., 2009). Apart from the utility of citrus fruits in juice production, efforts have been made for the development of alcoholic beverages by fermentation. Many investigations have been undertaken to explore the potentials of citrus fruits for the production of wine (Gupta et al., 2009). Wines from various citrus fruits (mandarin, kinnow, galgal, and orange) were compared for their different physicochemical properties and the higher fermentation rates belonged to kinnow wine followed by orange and galgal (Joshi et al., 2012b).

11.3 COUNTRY-WISE STATUS OF FRUIT WINES

The popularity of fruit wines on the global scale is highly variable depending upon the weather conditions and regional fruit and vegetable preferences. Tropical fruit wines are produced in tropical regions globally, using a wide range of tropical fruits by adopting somewhat similar winemaking processes. The actual data for import and export of fruit wines in various countries is still difficult to analyze as they fall within a broader grouping of products in the "fermented beverages" category of most countries' fact sheet statistics (Rivard, 2009).

11.3.1 India

Compared to the horticulturally advanced countries, wine production in India is almost negligible. Most of the fresh grapes in India are marketed as table grapes and only a very small portion is processed into wine, juice, and raisins (Ethiraj et al., 1993), which is true for other fruits like apple, plum, peach, pears, etc. (Joshi, 2001). But a silver lining in the black cloud is the production of grape-based champagne near Nasik, vermouth from grapes in Maharashtra, feni in Goa, and apple cider, wine, and vermouth (at Badhu in Mandi), in Himachal Pradesh (Joshi, 1997). Application of research-generated information can further improve the quality of the product, and hence the market.

India's expanding wine industry is in the midst of a vital transition. As of this writing, there are 93 wineries in India, which are involved in major wine production in the country. In 2014, the country's wine production hit a record of 17 million liters, with export sales rising 40% year on year to reach US$4.4 million in the first 7 months (Grace, 2015). According to the All India Wine Producers Association, the sales of wines also rose by 21% from 9.5 million liters in 2013–14 to 11.5 million liters in 2014–15 (Kasabe, 2015). Since the early 2000s, India has been hyped as an important emerging market for wine. The country has the optimum climate for grape cultivation and its main wine-producing states, Maharashtra and Karnataka, are leading producers of world-class high-quality grapes (Grace, 2015).

Despite huge production, consumption, and the list of wineries in India, the export–import status of wine in India is almost negligible and India stands nowhere in the world map of wine production.

Advancement in the social, educational, and economical status of consumers has led to the shift from hard alcoholic beverages to wine because of the numerous health benefits that the wine offers, especially red wine. Challenges still remain in the industry with regard to consumer culture, importing, and state regulations, which vary according to the state and are very complicated to understand. In addition to these, a ban on advertising alcohol, multiple tax authorities, and a lack of adequate infrastructure

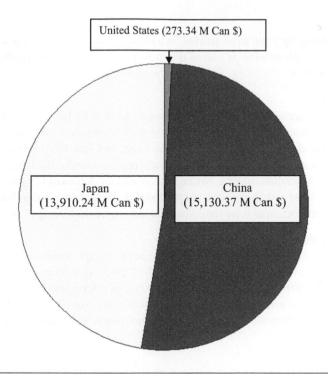

United States (273.34 M Can $)

Japan
(13,910.24 M Can $)

China
(15,130.37 M Can $)

FIGURE 1.10

Sale of nongrape fruit wine in various countries (in Canadian dollars).

Based on Industry, 2013. Market Analysis Report: A Global Export Market Overview for British Columbia's Wine Industry. http://
*www2.gov.bc.ca/assets/gov/farming-natural resources-and-industry/agriculture-andseafood/statistics/exports/global_export_market_*
bcwine_industry.pdf.

are some other challenges that the Indian wine industries face. There are different types of wines that need to be explored by the Indian wine industries, such as alcopops, no alcohol wines, health wines, medicinal/herbal wines, and vermouth. However, with continued estimates of future growth for the wine industry in India and a growing middle class, there are considerable possibilities for opportunity in this market. With a rapidly growing export sector, expanding the domestic consumer market and increasing industry support in major wine-producing states, the Indian wine industry has potential to be a global market competitor (Grace, 2015).

11.3.2 China

Fruit wines (known as *guojiu* in Chinese) are famous in the markets of various cities of China, such as wolfberry wine, cherry wine, lychee wine, mulberry wine, pomegranate wine, kiwifruit wine, berry wine, blueberry wine, and so on. The annual statistics of fruit wine and the medicated liquor industry in China revealed that in 2014 the total profit from these wines decreased despite a decrease in the total tax rate. But in contrast to this, the total sales of fruit wines were recorded to be higher in China compared to other countries (Fig. 1.10). The majority of sales by volume were in the form of nongrape wine (60.6%), followed by still light grape wine (39.4%), and less than 1% in the case of sparkling wine (www.euromonitor.com).

11.3.3 Japan

Japan has a small emerging fruit wine market, which has stabilized after growing consumption in recent years. *Umeshu*, or plum wine, is one of the most popular fruit wines produced in Japan from a traditional Asian stone fruit.

11.3.4 United States

The United States has a substantial market for fruit wines, as well as large markets for cider, perry, and blends of fruit and grape wine and a growing demand for fruity alcoholic drinks. Around 200 small producers and a few large manufacturers produce fruit wine, cider, and fruit-flavored grape wines (Noller and Wilson, 2009). While most fruit wines are from berries and apples, wineries in Hawaii and Florida use tropical fruits, similar to the Australian industry. Michigan is one of the foremost US states in the production of fruit wine, where apple wine and cherry wine are produced in the highest volume, but other fruit juices are also fermented for the production of wine (http://en.wikipedia.org/wiki/Michigan_wine #Fruit_wine).

11.3.5 United Kingdom

The United Kingdom has a long tradition of fruit and other nongrape wines and is one of Europe's largest markets for fruit wine, with an annual production of 40–50 million liters a year, and annual cider and perry consumption of >700 million liters. However, the rate of consumption in all categories has grown in recent years owing to the establishment of small wineries and some large drink manufacturers making a wide range of wines, from stone fruit, berries, apples, pears, and citrus to honey, ginger, and herbs (Noller and Wilson, 2009).

11.3.6 Canada

In Canada, fruit wines and ciders are manufactured throughout the country, mainly in the areas where the weather is milder. Fruit wines are made from a variety of sources including strawberries, raspberries, peaches, pears, apples, and rhubarb. However, most of the fruit wineries are small family businesses and produce locally. The export of nongrape wine to other countries (mainly the United States) was recorded to be higher in comparison to grape wines (Fig. 1.11). However, during past few years, the income from the export of fruit wines from Canada has declined.

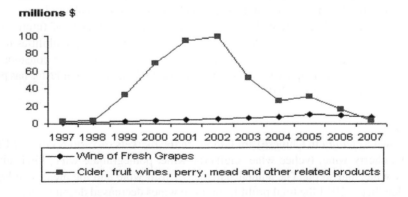

FIGURE 1.11

Exports of grape wine and other related products from Canada to the United States.

The Canadian wine industry (http://www.agr.gc.ca/).

11.3.7 Australia

A small fruit wine industry is emerging in some of the major tropical fruit-growing regions of Australia. The research has been focused on both domestic and international markets, keeping in view the exports and international visitors. During 2002–06, Australian fruit wine products were exported to 26 countries, including Singapore, New Zealand, Japan, Hong Kong, and Vietnam. The two main bodies in Australia representing fruit wine producers are the Association of Tropical North Queensland Wineries, for commercial producers, and Fruit Winemakers (Tas) Inc., for professional and amateur winemakers (Noller and Wilson, 2009).

12. **SUMMARY AND FUTURE STRATEGIES**

Wine constitutes one of the highly acceptable class of fermented alcoholic beverages throughout the world, except where they are forbidden by religion, and make an excellent combination with the diets of the people of the Western world because of their high nutritive value owing to the presence of calories, amino acids, vitamins, minerals, etc. Efforts have been directed to find out ways and means to improve the useful components and eliminate those that are toxic to human health. Similar efforts for nongrape wines are also needed in the future also.

Wine made from fruit other than grapes are just as diverse, and their cascading flow of possibilities opens up a whole new world to wine drinkers. Fruit wine can be sweet and fruity or dry, more complex, and nuanced. Of these, cider and other apple-based wines are produced and consumed in significant amounts throughout the world. Plum wines are quite popular in many areas, particularly in Germany and Pacific Coastal states. Most fruits have numerous varieties with a large range of flavors. The utility and scope of the wines from nongrape fruits reflect that systematic research on different facets of the enology needs to be strengthened.

Several developments in the wine production from grapes have been demonstrated, like the use of enzymes in juice extraction, flavor improvements, and continuous fermentation using bioreactor technology. However, their possible applications in wine production from nongrape fruits need serious consideration, though some efforts made in this direction have been summarized earlier.

In the past few decades, significant advances have been made in the standardization of wine quality and safety by introducing new techniques for the analysis of the various types of phenolics and aromatics that affect the wine's characteristics and in the increased understanding of the factors influenced by vineyard practice and wine aging. There has been a considerable interest in the use of enzymes, notably glucosidases, to release flavor compounds bound in the fruit. There is also an increasing awareness of the disadvantages of overly protecting the juice from the minimal oxidation that occurs during crushing, racking, and other winemaking practices. This is particularly reflected in the reduced use of sulfur dioxide before and after fermentation. In addition, there is growing interest in the use of several yeast strains to induce fermentation, to produce some of the perceived benefits of spontaneous fermentation while retaining the safety of induced fermentation. Inoculation with one or more strains of lactic acid bacteria is also becoming commonplace to induce malolactic fermentation. Considerable interest is also being shown in the use of cell and enzyme immobilization in wine production in batches as well as in continuous fermentation systems. Some of the fruits used in winemaking are highly acidic and the use of biological deacidification (using malolactic bacteria or deacidifying yeasts like *S. pombe*) is being made, but how these practices influence the composition and nutritive value of wines has not been evaluated so far and is an interesting and useful aspect of research in the near future.

From grapes, various types of wines such as vermouth, sparkling wine, and sherry are prepared, but the research on nongrape fruits is scanty. Traditionally, wines are of several types, such as table, sparkling, and fortified wines, in addition to white, red, and rosé and dry or sweet wines, but stress is being put on the production of wines from other fruits. Despite the efforts already made and described here, there are a large number of research gaps for more elaboration in future. It could certainly be a fruitful area of research, especially in those countries and regions where grape cultivation is not practiced. Needless to say, research currently in progress will continue to document the healthful properties of wine and, in addition, the industry will need to play a highly visible role in the promotion of sound and sustainable environmental stewardship as a strong motivator in the purchase of wines.

To monitor and ensure the quality of wine, new techniques have emerged, including the use of molecular biology techniques like polymerase chain reaction. In the future, these techniques are likely to be used more frequently, such as in direct profiling of yeast dynamics in wine fermentation. Use of machinery and equipment as an inseparable component of wine products finds a prominent place. Like in other food industries, the waste from wine and brandy is highly polluting, but rich in several useful constituents, which are recoverable.

With the rise in awareness level of consumers and in health consciousness, the future of the fruit-based wine industry seems to be bright. These wines have been accepted globally by people from various countries. With global players stepping in from Australia, France, India, Italy, etc., the fruit wine market is sure to stay and progress as time passes. Thus, the acceptance and popularity of nongrape fruit wines will increase slowly but steadily with the passage of time.

REFERENCES

Addeo, F., Barlotti, L., Boffa, G., DiLuccia, A., Piccioli, G., 1979. Contituenti acidi di una oleorc sina di conifere rinvenuta in anfore durante gliscavi archeologici di oponti. Annali delta Facolta di Scienze Agararie della degli Studi Napoli, Portici 13, 144.

Agbor, A.A., Ben, U.I., Ubana, E.M., Olayinka, J.C., Okon, A.E., 2011. Production, characterization and safety of wine obtained from a blend of tomato, almond, orange, lemon and African star apple extract. Annals of Biological Research 2, 492–503.

Akubor, P.I., Obio, S.O., Nwadomere, K.A., Obiomah, E., 2003. Production and quality evaluation of banana wine. Plant Foods for Human Nutrition 58, 1–6.

Alberti, A., Vieira, R.G., Drilleau, J.F., Wosiacki, G., Nogueira, A., 2011. Apple wine processing with different nitrogen contents. Brazilian Archives of Biology and Technology 54, 551–558.

Alves, J.A., Lima, L.C.O., Nunes, C.A., Dias, D.R., Schwan, R.F., 2011. Chemical, physical-chemical, and sensory characteristics of lychee (*Litchi chinensis* Sonn) wines. Journal of Food Science 76, S330–S336.

Amerine, M.A., Kunkee, R.E., Ough, C.S., Singleton, V.L., Webb, A.D., 1980. Technology of Wine Making. AVI Publ. Co., Westport, Connecticut.

Amerine, M.A., Berg, H.W., Cruess, W.M., 1967. The Technology of Wine Making, second ed. AVI Publ., Westport, CT, p. 799.

Amerine, M.A., Joslyn, M.A., 1973. Table Wines: The Technology of Their Production, second ed. University of California Press, Berkeley.

Anon, I., Barid, F.D., 1963. The Food Value and Use of Dried Yeast. Bull. Office Inteen. Vin, 39. Cerevisie Yeast Institution, Chicago, Illinois, pp. 1068–1090 (1966).

Areni, C.S., Duhan, D.F., Kiecker, P., 1999. Point of purchase displays, product organization and brand purchase likelihoods. Journal of the Academy of Marketing Science 27 (4), 428.

Attri, B.L., Lal, B.B., Joshi, V.K., 1993. Preparation and evaluation of sand pear vermouth. Journal of Food Science and Technology 30, 435–437.

Attri, B.L., Lal, B.B., Joshi, V.K., 1994. Technology for the preparation of sand pear vermouth. Indian Food Packer 48, 39–44.

Ayogu, T.E., 1999. Evaluation of the performance of a yeast isolate from Nigerian palm wine in wine production from pineapple fruits. Bioresource Technology 69, 189–190.

Azad, K.C., Vyas, K.K., Joshi, V.K., 1987. Some observations on the properties of juice and apples. Indian Food Packer 41, 56–61.

Bardiya, M.C., Kundu, B.S., Daulta, B.S., Tauro, B.S., 1980. Evaluation of exotic grapes grown in Haryana for red wine production. Journal of Research of Haryana Agricultural University 10, 85–88.

Bardiya, M.C., Kundy, B.S., Tauro, P., 1974. Studies on fruit wines-guava wine. Haryana Journal of Horticultural Science 3, 140.

Barre, P., et al., 1993. Genetic improvement of wine yeast. In: Fleet, G.H. (Ed.), Wine Microbiology and Biotechnology. Harwood Academic Publishers, pp. 421–447.

Bartowsky, E., 2011. Malolactic fermentation. In: Joshi, V.K. (Ed.), Handbook of Enology, vol. 3. Asia Tech Publication, New Delhi, pp. 526–563.

Bauer, F., Dequin, S., Pretorius, I., Shoeman, H., Wolfaardt, M.B., Schroeder, M.B., Grossmann, M.K., 2003. The assessment of the environmental impact of genetically modified wine yeast strains. Proceedings of the "Actes de 83ème Assemblée Générale de l'O.I.V".

Baumann, G., Gieschner, K., 1974. Studies on the technology of juice manufacture from sour cherries in relation to the storage of the product. Fluessiges Obst 41, 123.

Becker, J.V., Armstrong, G.O., van der Merwe, M.J., Lambrechts, M.G., Vivier, M.A., Pretorius, I.S., 2003. Metabolic engineering of *Saccharomyces cerevisiae* for the synthesis of the wine-related antioxidant resveratrol. FEMS Yeast Research 4, 79–85.

Benk, E., Borgmann, R., Cutka, I., 1976. Quality control of sour cherry juices and beverages. Fluessiges Obst 43, 17.

Bhardwaj, J.C., Joshi, V.K., 2009. Effect of cultivar, addition of yeast type, extract and form of yeast culture on foaming characteristics, secondary fermentation and quality of sparkling plum wine. Natural Product Radiant 452–460.

Bhargva, U.C., Westfall, A.B., Siehr, D.J., 1969. Preliminary pharmacology of ellagic acid from *Jaglans nigra* (black walnut). Journal of Pharmaceutical Sciences 57, 17–28.

Bhutani, V.P., Joshi, V.K., 1995. Plum. In: Salunkhe, D.K., Kadam, S.S. (Eds.), Handbook of Fruit Science and Technology, Cultivation, Storage and Processing. Marcel Dekker, New York, USA, pp. 203–241.

Bhutani, V.P., Joshi, V.K., 1996. Mineral composition of experimental sand pear and plum vermouth. Alimentaria 272, 99.

Bhutani, V.P., Joshi, V.K., Chopra, S.K., 1989. Mineral contents of fruit wines produced experimentally. Journal of Food Science Technology 26 (6), 332–333.

Brand, G., Millot, J.L., Henquell, D., 2001. Complexity of olfactory lateralization processes revealed by functional imaging: a review. Neuroscience and Biobehaviour Reviews 25, 159.

Braudel, F., 1981. The Structure of Everyday Life: The Limits of the Possible. Civilization and Capitalism 15th–18th Century, vol. I. Collins, London.

Burlingame, B., 2008. Wine: food of poets and scientists. Journal of Food Composition and Analysis 21, 587–588.

Byaruagaba-Bazirake, G.W., van Rensburg, P., Kyamuhangire, W., 2013. Characterisation of banana wine fermented with recombinant wine yeast strains. American Journal of Food and Nutrition 3, 105–116.

Catherine, S., 1996. Wine and health. In: Medically Is Wine Just Another Alcoholic Beverage? Conference Summary of International Wine & Health Conference, 12–13 June.

Cavaletto, C.G., 1980. In: Nagy, S., Shaw, P.E. (Eds.), Lychee, Tropical and Subtropical Fruits. AVI, Westport, CT, p. 469.

Chanprasartsuk, O., Prakitchaiwattana, C., Sanguandeekul, R., Fleet, G.H., 2010. Autochthonous yeasts associated with mature pineapple fruits, freshly crushed juice and their ferments; and the chemical changes during natural fermentation. Bioresource Technology 101, 7500–7509.

Cheirsilp, B., Umsakul, K., 2008. Processing of banana-based wine product using pectinase and α-amylase. Journal of Food Process Engineering 31, 78–90.

Chowdhury, P., Ray, R.C., 2007. Fermentation of jamun (*Syzgium cumini* L.) fruits to form red wine. ASEAN Food Journal 14, 15–23.

Chung, J.H., Mok, C.Y., Park, Y.S., 2003. Changes of physicochemical properties during fermentation of peach wine and quality improvement by ultrafiltration. Journal of the Korean Society of Food Science and Nutrition 32, 506–512.

Comant, J.B., 1952. Pasteur's Study of Fermentation. Harvard Case Histories in Experimental Science, No. 6. Harvard University Press, Cambridge, Massachissetts.

Coulon, J., Husnik, J.I., Inglis, D.L., van der Merwe, G.K., Lovaud, A., Erasmus, D.J., van Vuuren, H.J.J., 2006. Metabolic engineering of *Saccharomyces cerevisiae* to minimize the production of ethyl carbamate in wine. American Journal of Enology and Viticulture 57, 113–124.

Cousteau, T.Y., 1954. Fish men discover a 2200 year old Greek ship. National Geographic 105, 1.

Craig, J.T., 1988. A comparison of the headspace volatiles of kiwifruit wine with those of the wine of *Vitis vinifera* variety Muller-Thurgau. American Journal of Enology and Viticulture 39, 321.

Davidovic, S.M., Veljovic, M.S., Pantelic, M.M., Baosic, R.M., Natic, M.M., Dabic, D.C., 2013. Physicochemical, antioxidant and sensory properties of peach wine made from redhaven cultivar. Journal of Agriculture and Food Chemistry 61, 1357–1363.

De Candolle, A., 1959. Origin of Cultivated Plants. Hafner Publishing Co., New York, NY. 468 pp.

Dennison, C., 1999. Historical note: a version of the invention of barrels and barrel alternatives. International symposium on oak in wine making. American Journal of Enology and Viticulture 50, 539.

Dequin, S., 2001. The potential of genetic engineering for improving brewing, wine-making and baking yeasts. Applied Microbiology and Biotechnology 56 (5–6), 577–588.

Downing, D.L., 1989. In: Downing, D.L. (Ed.), Processed Apple Products. AVI Publishing Co., New York, pp. 168–186.

Du Toit, M., Pretorius, I.S., 2000. Microbial spoilage and preservation of wine: using weapons from nature's own arsenal. South African Journal of Enology and Viticulture 21, 74–96.

Duarte, W.F., Dias, D.R., Pereira, G.V.M., Gervasio, I.M., Schwan, R.F., 2009. Indigenous and inoculated yeast fermentation of gabiroba (*Campomanesia pubescens*) pulp for fruit wine production. Journal of Industrial Microbiology and Biotechnology 36, 557–569.

Duarte, W.F., Dias, D.R., Oliveira, J.M., Teixeira, J.A., Almeida e Silva, J.B., Schwan, R.F., 2010a. Characterization of different fruit wines made from cacao, cupuassu, gabiroba, jaboticaba and umbu. LWT—Food Science and Technology 43, 1564–1572.

Duarte, W.F., Dias, D.R., Oliveira, J.M., Vilanova, M., Teixeira, J.A., Almeida e Silva, J.B., et al., 2010b. Raspberry (*Rubus idaeus* L.) wine: yeast selection, sensory evaluation and instrumental analysis of volatile and other compounds. Food Research International 43, 2303–2314.

Dubos, R.J., 1950. Louis Pasteur: Freelance of Science. Little. Brown, Bosten, Massachusetts, p. 118.

Ethiraj, S., Suresh, E.R., Onkarayya, H., 1993. Controlled de-acidification of Bangalore Blue grape must with *Schizosaccharomyces pombe*. Journal of Food Science and Technology 20, 248–250.

Fan, W., Xu, Y., Yu, A., 2006. Influence of oak chips geographical origin, toast level, dosage and aging time on volatile compounds of apple cider. Journal of the Institute of Brewing 112, 255–263.

FAOSTAT, 2012. http://www.fao.org.

Forbes, R.J., 1956. Food and drink. In: Singer, C., Holmyard, E.J., Hall, A.R., Williams, T.I. (Eds.), A History of Technology, vol. II, the Mediterranean Civilization and the Middle Ages, c.700 B.C. to c. A.D. 1500. Clarendon Press, London.

Frey, D., Hentschel, F.D., Keith, D.H., 1978. Deep water archaeology: the Capsitello wreck excavation, Lipari, Aelian Islands. International Journal of Nautical Archaeology Underwater Explorer 7, 279.

Gasteineau, F.C., Darby, J.W., Turner, T.B., 1979. Fermented Food Beverages in Nutrition. Academic Press, New York.

Gautam, S.K., Chundawat, B.S., 1998. Standardization of technology of sapota wine making. Indian Food Packer 1, 17.

Goffeau, A., Barrell, B.G., Bussey, H., Davis, R.W., Dujon, B., Feldmann, H., Galibert, F., Hoheisel, J.D., Jacq, C., Johnston, M., Louis, E.J., Mewes, H.W., Murakami, Y., Philippsen, P., Tettelin, H., Oliver, S.G., 1996. Life with 6000 genes. Science 274, 563–567.

Goyal, R.K., 1999. Biochemistry of fermentation. In: Joshi, V.K., Pandey, A. (Eds.), Biotechnology: Food Fermentation (Microbiology, Biochemistry and Technology). Educational Publishers and Distributors, New Delhi, pp. 87–172.

Grace, V.R., 1979. Amphoras and the Ancient Wine Trade. Photo 63. American School Classical Studies, Athens, Princeton, NJ.

Grace, T., 2015. Investigating in India's Emerging Wine Industry. http://www.india-briefing.com/news/india-wine-9761.html/.

Gronbaek, M., Deis, A., Sorensen, T.I.A., Beckear, U., Schnohor, P., Jensen, G., 1995. Mortality associated with moderate intake of wines, beers and spirits. British Medical Journal 310, 1165–1169.

Grossmann, M.K., Pretorius, I.S., 1999. Methods for the identification and the improvement of specific properties of *Saccharomyces cerevisiae*: a review. DieWeinwissenschaft 54, 61–72.

Gupta, N., Trivedi, S., Gaudani, H., Gupta, M., Patil, P., Gupta, G., Vamsi, K.K., 2009. Orange: research analysis for wine study. International Journal of Biotechnology Applications 1, 10–15.

Harden, A., 1924. Alcoholic Fermentation. Green Longmans, London.

Henschke, P.A., 1997. Wine yeast. In: Zimmermann, F.K., Entian, K.D. (Eds.), Yeast Sugar Metabolism. Technomic Publishing Co., pp. 527–560.

Heatherbell, D.A., Struvebi, P., Eschenbruch, R., Withy, L.M., 1980. A new fruit wine from kiwifruits. A wine of unusual composition and Reisling Sylvaner character. American Journal of Enology and Viticulture 31, 114.

Hwang, Y.C., Shin, K.S., 2011. Isolation and characterization of immuno-stimulating polysaccharide in Korean black raspberry wine. Journal of the Korean Society for Applied Biological Chemistry 54, 591–599.

Ibanezo, J.G., Alvarej, A.C., Soto, B.M., Casillas, N., 2008. Metals in alcoholic beverages: a review of sources, effect, concentrations, removal, speciation and analysis. Journal of Food Composition and Analysis 21, 672.

Idise, O.E., Odum, E.I., 2011. Studies of wine produced from banana (*Musa sapientum*). International Journal for Biotechnology and Molecular Biology Research 2, 209–214.

Industry, 2013. Market Analysis Report: A Global Export Market Overview for British Columbia's Wine Industry. http://www2.gov.bc.ca/assets/gov/farming-natural resources-and-industry/agriculture-andseafood/statistics/exports/global_export_market_bcwine_industry.pdf.

Jackson, D., Looney, N., 1999. Stone fruits. In: Jackson, D.I., Looney, N.E. (Eds.), Temperate and Subtropical Fruit Production. CABI Publishing, p. 180.

Jackson, D., Schuster, D., 1981. Grape Growing and Wine Making: A Handbook for Cool Climates. Martinborough, Alister Tagcor, New Zealand.

Jackson, R., 1994. Wine Science: Principle and Application. Academic Pres, San Diego.

Jackson, R.S., 1999. Grape based fermentation products. In: Joshi, V.K., Pandey, A. (Eds.), Biotechnology: Food Fermentation (Microbiology, Biochemistry and Technology), vol. 2. Educational Publishers and Distributors, New Delhi, p. 647.

Jackson, R.S., 2000. Wine Science – Principles, Practices, Perception, second ed. Academic Press, San Diego.

Jackson, R.S., 2003. Wines: types of table wine. In: Caballero, B., Trugo, L., Figlas, P.N. (Eds.), Encyclopedia of Food Sciences and Nutrition, second ed. Elsevier Science Ltd., UK.

Jagtap, U.B., Waghmare, S.R., Lokhande, V.H., Suprasanna, P., Bapat, V.A., 2011. Preparation and evaluation of antioxidant capacity of Jackfruit (*Artocarpus heterophyllus* Lam.) wine and its protective role against radiation induced DNA damage. Industrial Crops and Products 34, 1595–1601.

Jagtap, U.B., Bapat, V.A., 2014. Phenolic composition and antioxidant capacity of wine prepared from custard apple (*Annona squamosa* L.) fruits. Journal of Food Processing and Preservation. http://dx.doi.org/10.1111/jfpp.12219.

Jagtap, U.B., Bapat, V.A., 2015. Wines from fruits other than grapes: current status and future prospectus. Food Bioscience 9, 80–96.

Janick, J., 1992. Introduction. In: Hammerschlag, F.A., Litz, R.E. (Eds.), Biotechnology of Perennial Fruit Crops, pp. xix–xxi Wallingford, UK.

Jarvis, B., Foster, M.J., Kinsella, W.P., 1995. Factors affecting the development of cider flavour. Journal of Applied Bacterial Symposium Supplements 79, 55.

Jeandet, P., Bessis, R., Maume, B.F., Sbaghi, M., 1993. Analysis of resveratrol in Burgundy wines. Journal of Wine Research 4, 79–85.

Jeandet, P., Vasserot, Y., Liger-Belair, G., Marchal, R., 2011. Sparkling wine production. In: Joshi, V.K. (Ed.), Handbook of Enology, vol. 1II. Asia Tech Publishers, INC, New Delhi, pp. 1064–1115.

Jeong, J.H., Jung, H., Lee, S.R., Lee, H.J., Hwang, K.T., Kim, T.Y., 2010. Anti-oxidant, anti-proliferative and anti-inflammatory activities of the extracts from black raspberry fruits and wine. Food Chemistry 123, 338–344.

Jindal, P.C., 1990. Grape. In: Bose, T.K., Mitra, S.K. (Eds.), Fruits, Tropical and Sub-tropical. Naya Prakashan, Calcutta, p. 85.

Johnson, M.H., Lucius, A., Meyer, T., de Mejia, E.G., 2011. Cultivar evaluation and effect of fermentation on anti-oxidant capacity and in vitro inhibition of α-amylase and α-glucosidase by Highbush Blueberry (*Vaccinium corombosum*). Journal of Agricultural and Food Chemistry 59, 8923–8930.

Johnson, M.H., Mejia, E.G., 2012. Comparison of chemical composition and antioxidant capacity of commercially available blueberry and blackberry wines in Illinois. Journal of Food Science 71, C141–C148.

Joshi, V.K., John, S., November 2002. Antimicrobial activity of apple wine against some pathogenic and microbes of public health significance. Acta Alimentaria 67–72.

Joshi, V.K., Attri, D., 2005. Panorama of research and development of wines in India. Journal of Scientific and Industrial Research 64 (1), 9–18.

Joshi, V.K., Bhardwaj, J.C., 2011. Effect of different cultivars yeasts (free and immobilized cultures) of *S. cerevisiae* and *Schizosaccharomyces* pombe on physico-chemical and sensory quality of plum based wine for sparkling wine production. International Journal of Food and Fermentation Technology 1 (1), 69–81.

Joshi, V.K., Bhutani, V.P., 1991. The influence of Enzymatic clarification on the fermentation behaviour, composition and sensory qualities of apple wine. Sciences Des Aliments 11 (3), 491–496.

Joshi, V.K., Kumar, V., 2011. Importance, nutritive value, role, present status and future strategies in fruit wines in India. In: Panesar, P.S., Sharma, H.K., Sarkar, B.C. (Eds.), Bio-Processing of Foods. Asia Tech Publishers, New Delhi, pp. 39–62.

Joshi, V.K., Devi, P.M., 2009. Resvertatrol: importance, role, contents in wine and factors influencing its production. Proceedings of the National Academy of Sciences, India, Section B 79 (Pt. III).

Joshi, V.K., Parmar, M., 2004. Present status, scope and future strategies of fruit wines production in India. Indian Food Industry 23, 48–52.

Joshi, V.K., Sandhu, D.K., 1994. Influence of juice contents on quality of apple wine prepared from apple juice concentrate. Research and Industry 39 (4), 250–252.

Joshi, V.K., Sandhu, D.K., 1997. Effect of different concentrations of initial soluble solids on physico-chemical and sensory qualities of apple wine. Indian Journal of Horticulture 54, 116–123.

Joshi, V.K., Sandhu, D.K., 2000a. Quality evaluation of naturally fermented alcoholic beverages, microbiological examination of source of fermentation and ethanolic productivity of the isolates. Acta Alimentaria 29 (4), 323–334.

Joshi, V.K., Sandhu, D.K., 2000b. Influence of ethanol concentration, addition of spices extract and level of sweetness on physico-chemical characteristics and sensory quality of apple vermouth. Brazilian Archives of Biology and Technology 43 (5), 537–545.

Joshi, V.K., Sandhu, D.K., 2000c. Studies on preparation and evaluation of apple cider. Indian Journal of Horticulture 57, 42–46.

Joshi, V.K., Sandhu, D.K., 2003. A note on methanol content in fruit based alcoholic beverages. Processed Food Industry, February 11–17.

Joshi, V.K., Sandhu, D.K., 2009. Flavour profiling of apple vermouth using descriptive analysis technique. Natural Product Radiant 38 (4), 419–425. (special issue).

Joshi, V.K., Shah, P.K., 1998. Effect of wood treatment on chemical and sensory qualities of peach wine during aging. Acta Alimentaria 27 (4), 307–318.

Joshi, V.K., Sharma, S.K., 1994. Effect of methods of must preparation and initial sugar levels on the quality of apricot wine. Research and Industry 39 (4), 255–257.

Joshi, V.K., Sharma, S.K., 1995. Comparative fermentation behaviour, physico-chemical and sensory characteristics of plum wine as affected by the type of preservatives. Chemie, Mikrobiologie, Technologie der Lebensmittel (CMTL) 17 (3/4), 65–73.

Joshi, V.K., Thakur, N.K., 1994. Preparation and evaluation of citrus wines. In: 24th International Congress, Japan Abstract No. 149.

Joshi, V.K., Attri, B.L., Gupta, J.K., Chopra, S.K., 1990a. Comparative fermentation behaviour, physico-chemical characterstics of fruit honey wines. Indian Journal of Horticulture 47, 49–54.

Joshi, V.K., Bhutani, V.P., Sharma, R.C., 1990b. Effect of dilution and addition of nitrogen source on chemical mineral and sensory qualities of wild apricot wine. American Journal of Enology and Viticulture 41 (3), 229–231.

Joshi, V.K., Sandhu, D.K., Attri, B.L., Walia, R.K., 1991a. Cider preparation from apple juice concentrate and its consumer acceptability. Indian Journal of Horticulture 48 (4), 321–332.

Joshi, V.K., Sharma, P.C., Attri, B.L., 1991b. A note on the deacidification activity of *Schizosaccharomyces pombe* in plum musts of variable composition. Journal of Applied Bacteriology 70, 386–390.

Joshi, V.K., Attri, B.L., Mahajan, B.V.C., 1991c. Studies on preparation and evaluation of vermouth from plum. Journal of Food Science and Technology 28, 138.

Joshi, V.K., Mahajan, B.V.C., Sharma, R.K., 1994. Treatment of wines with wood-chip effect on some physicochemical and sensory qualities. Journal of Tree Sciences 13 (1), 27–36.

Joshi, V.K., Sharma, S.K., Thakur, N.S., 1995. Technology and quality of sparkling wine with special reference to plum – an overview. Indian Food Packer 49, 49–63.

Joshi, V.K., Thakur, N.K., Kaushal, B.B.L., 1997. Effect of debittering of kinnow juice on physico-chemical and sensory quality of kinnow wine. Indian Food Packer 50, 5–8.

Joshi, V.K., Bhutani, V.P., Thakur, N.K., 1999a. Composition and nutrition of fermented products. In: Joshi, V.K., Pandey, A. (Eds.), Biotechnology: Food Fermentation, vol. I. Educational Publishers and Distributors, New Delhi, pp. 259–320.

Joshi, V.K., Sandhu, D.K., Thakur, N.S., 1999b. Fruit based alcoholic beverages. In: Joshi, V.K., Pandey, A. (Eds.), Biotechnology: Food Fermentation (Microbiology, Biochemistry and Technology), vol. II. Educational Publishers and Distributors, New Delhi, pp. 647–744.

Joshi, V.K., Sharma, S.K., Goyal, R.K., Thakur, N.S., 1999c. Sparkling plum wine: effect of method of carbonation and the type of base wine on physico-chemical and sensory qualities. Brazilian Archives of Biology and Technology 42 (3), 315–321.

Joshi, V.K., Chauhan, S.K., Bhushan, S., 2000. Technology of fruit based alcoholic beverages. In: Verma, L.R., Joshi, V.K. (Eds.), PostharvestTechnology of Fruits and Vegetables. Indus Publishing, New Delhi, p. 1019.

Joshi, V.K., Sandhu, D.K., Thakur, N.S., Walia, R.K., 2002a. Effect of different sources of fermentation on flavour profile of apple wine by Descriptive Analysis Technique. Acta Alimentaria 31 (3), 211–225.

Joshi, V.K., Attri, D., Bhushan, S., November 2002b. Biological deacidification of Plum (*Prunus saliciana*) must and quality of wine produced with variable acidity. Beverage Food World 33–36.

Joshi, V.K., Sharma, S., Bhushan, S., Attri, D., 2004. Fruit based alcoholic beverages. In: Pandey, A. (Ed.), Concise Encyclopedia of Bioresource Technology. Haworth Inc., New York, pp. 335–345.

Joshi, V.K., Shah, P.K., Kumar, K., 2005a. Evaluation of different peach cultivars for wine preparation. Journal of Food Science and Technology 42 (1), 83–89.

Joshi, V.K., Sharma, S., Bhushan, S., 2005b. Effect of method of preparation and cultivar in the quality of strawberry wine. Acta Alimentaria 34 (4), 339–353.

Joshi, V.K., Sharma, S., John, S., Kaushal, B.B.L., Neerja, R., 2009a. Preparation of antioxidant rich apple and strawberry wine. Proceedings of National Academy of Sciences, Section B 79 (IV), 415–420.

Joshi, V.K., Sharma, S., Devi, P.M., Bhardwaj, J.C., 2009b. Effect of initial sugar concentration on the physic-chemical and sensory qualities of plum wine. Journal of North East Foods 8 (1,2), 1–7.

Joshi, V.K., Thakur, N.S., Bhat, A., Garg, C., 2011a. Wine and brandy: a perspective. In: Joshi, V.K. (Ed.), Handbook of Enology : Principles Practices, vol 1. Asia Tech Publisher Inc., New Delhi, pp. 3–45.

Joshi, V.K., Attri, D., Singh, T.K., Abrol, G., 2011b. Fruit wines: production technology. In: Joshi, V.K. (Ed.), Handbook of Enology. vol. III. Asia Tech Publishers, Inc., New Delhi, pp. 1177–1221.

Joshi, V.K., Sharma, S., Parmar, M., 2011c. Cider and perry. In: Joshi, V.K. (Ed.), Handbook of Enology, vol. 1II. Asia Tech Publishers, Inc., New Delhi, pp. 1116–1151.

Joshi, V.K., Dev, R., Joshi, C., 2011d. Utilization of wastes from food fermentation industry. In: Joshi, V.K., Sharma, S.K. (Eds.), Food Processing Wastes Management. New India Publishing Agency, Pitam Pura, New Delhi, pp. 295–356.

Joshi, V.K., Sharma, R., Abrol, G., 2011e. Stone fruit: wine and brandy. In: Hui, Y.H., Evranuz, E.O. (Eds.), Handbook of Food and Beverage Fermentation Technology. CRC Press, Florida, pp. 273–304.

Joshi, V.K., Sharma, R., Sharma, S., Abrol, G.S., 2011f. Effect of dilution and de-acidification on physico-chemical and sensory quality of seabuckthorn wine. Journal of Hill Agriculture 2 (1), 47–53.

Joshi, V.K., Abrol, G., Thakur, N.S., March–April 2011g. Wild apricot vermouth: effect of sugar, alcohol concentration and spices level on physic-chemical and sensory evaluation. Indian Food Packer 53–62.

Joshi, V.K., Kumar, V., Kumar, A., 2012a. Physico-chemical and sensory evaluation of wines from different citrus fruits of Himachal Pradesh. International Journal of Food Fermentation and Technology 2 (2), 145–148.

Joshi, V.K., Rakesh, S., Aman, G., Ghanshyam, A., 2012b. Effect of dilution and maturation on physico-chemical and sensory quality of jamun wine. Indian Journal of Natural Products and Resources 3 (2), 222–227.

Joshi, V.K., Sandhu, D.K., Kumar, V., 2013a. Influence of Addition of apple insoluble solids, different wine yeast strains and pectinolytic enzymes on the flavor profile of apple wine. International Journal of Food and Fermentation Technology 3 (1), 79–86.

Joshi, V.K., Sandhu, D.K., Kumar, V., 2013b. Influence of addition of insoluble solids, different yeast strains and pectinesterase enzyme on the quality of apple wine. Journal of the Institute of Brewing 119, 191–197.

Joshi, V.K., John, S., Abrol, G.S., 2014. Effect of addition of extracts of different herbs and spices on fermentation behaviour of apple must to prepare wine with medicinal value. National Academy Science Letters. http://dx.doi.org/10.1007/s40009-014-0275-y.

Joshi, V.K., 1997. Fruit Wines, second ed. DTP System, Direcorate of Extension Education. Dr. Y.S. Parmar University of Horticulture and Forestry, Nauni, Solan (HP), p. 226.

Joshi, V.K., 2001. Technologies for the postharvest processing of fruits and vegetables. In: Marwaha, S.S., Arora, J.K. (Eds.), Food Processing: Biotechnological Applications. Asia tech Publishing Co., New Delhi, pp. 241–263.

Joshi, V.K., 2006. Sensory Science: Principles and Application in Food Evaluation. Agro-Tech Academy, Udaipur, pp. 458–460.

Joshi, V.K., 2009. Production of Wines From Non-Grape Fruit. Natural Product Radiance 8 (94), 313–469. Special Issue. NISCARE, New Delhi.

Joshi, V.K.(Ed.), 2016. Indigenous fermented foods of South Asia. Nout, R., Sarkar, P. (Series Eds.), The Fermented Foods and Beverages Series. CRC Press, Roca. FL. pp. 1–849.

Ju, H.K., Cho, E.J., Jang, M.H., Lee, Y.Y., Hong, S.S., Park, J.H., 2009. Characterization of increased phenolic compounds from fermented Bokbunja (*Rubus coreanus* Miq.) and related antioxidant activity. Journal of Pharmaceutical and Biomedical Analysis 49, 820–827.

Karagiannis, S.D., 2011. Classification and characteristics of wines and brandies. In: Joshi, V.K. (Ed.), Handbook of Enology: Principles Practices, vol. 1. Asia Tech Publisher, New Delhi, pp. 46–65.

Karamoko, D., Djeni, N.T., N'guessan, K.F., Bouatenin, K.M.J.P., Dje, K.M., 2012. The biochemical and microbiological quality of palm wine samples produced at different periods during tapping and changes which occurred during their storage. Food Control 26, 504–511.

Karni, P.N., Kala, M., Gupta, S.R.P., 1977. Preparation of cider from commercial apple varieties of Jammu and Kashmir – yeast strain and varietal differences in the physic chemical characteristics. Indian Food Packer 31, 5–13.

Karthik, R.G., Ezhilarasan, V., Yazhini, K.A., Sridhar, S., Chinnathambi, V., 2012. Comparative evaluation of white wine production from different carbohydrate rich substrates using air-lift bio-reactor. International Journal of Pharma and Bio Sciences 3, 392–404.

Kasabe, N., 2015. Maharashtra's Wine Industry Cheers New Liquor Policy. http://www.financialexpress.com/article/markets/commodities/maharashtras-wine-industry-cheers-new-liquor-policy/133870/.

Kaur, R., Kumar, K., Sharma, D.R., 2011. Grapes and genetic engineering. In: Joshi, V.K. (Ed.), Handbook of Enology, vol. 1. Asia Tech Publication, New Delhi, pp. 266–286.

Kelebek, H., Selli, S., Canbas, A., Cabaroglu, T., 2009. HPLC determination of organic acids, sugars, phenolic compositions and antioxidant capacity of orange juice and orange wine made from a Turkish cv. Kozan. Microchemical Journal 91, 187–192.

Khandelwal, P., Kumar, V., Niranjan, D., Tyagi, S.M., 2006. Development of a process for preparation of pure and blended kinnow wine without debittering kinnow mandarin juice. Internet Journal of Food Safety 8, 24–29.

Kinsella, J.E., Frankel, E.N., German, J.B., Kanner, J., 1993. Possible mechanisms for the protective role of antioxidants in wine and plant foods. Food Technology 47, 85.

Klatsky, A.L., Armstrong, M.A., 1993. Alcoholic beverage choice and risk of coronary heart disease mortality: do red wine drinkers fare best. American Journal of Cardiology 71, 467–469.

Kochhar, A.P.S., Vyas, K.K., 1996. Studies on apple brandy from culled apple fruits available in H P. Indian Food Packer 21.

Koehler, C.G., 1986. Handling of green transport amphoras. In: Bulleting de Correspondence.

Kondapalli, N., Sadineni, V., Variyar, P.S., Sharma, A., Obulama, V.S.R., 2014. Impact of y-irradiation on antioxidant capacity of mango (*Mangifera indica* L.) wine from eight Indian cultivars and the protection of mango wine against DNA damage caused by irradiation. Process Biochemistry. http://dx.doi.org/10.1016/j. procbio.2014.07.015.

Kotecha, P.M., Adsule, R.N., Kadam, S.S., 1995. Processing of custard apple: preparation of ready-to-serve beverage and wine. Indian Food Packer 5–7.

Krebs, H., 1968. Introductory remarks. In: Mills, A.K. (Ed.), Aspect of Yeast Metabolism, p. VIII (Davis, Philadelphia, Pennsylvania).

Krens, F.A., Pelgrom, K.T.B., Schaart, J.G., den Nijs, N.P.M., Rouwendal, G.J.A., 2004. Clean vector technology for marker-free transgenic ornamentals. Acta Horticulturae 651, 101–105.

Krens, F.A., Schaart, J.G., Groenwold, R., Walraven, A.E.J., Hesselink, T., Thissen, J.T.N.M., 2011. Performance and long-term stability of the barley hordothionin gene in multiple transgenic apple lines. Transgenic Research 20 (5), 1113–1123.

Ku, C.S., Mun, S.P., 2008. Optimization of the extraction of anthocyanin from Bokbunja (*Rubus coreanus* Miq.) marc produced during traditional wine processing and characterization of the extracts. Bioresource Technology 99, 8325–8330.

Kulkarni, J.H., Singh, H., Chadha, K.L., 1980. Preliminary screening of mango varieties for wine making. Journal of Food Science and Technology 17, 218–220.

Kumar, K., Kaur, R., Sharma, S.D., 2011a. Fruit cultivars for winemaking. In: Joshi, V.K. (Ed.), Handbook of Enology: Principles Practices. vol. 1. Asia Tech Publisher, New Delhi, pp. 237–265.

Kumar, V., Goud, P.V., Babu, J.D., Reddy, R.S., 2011b. Preparation and evaluation of custard apple wine: effect of dilution of pulp on physico-chemical and sensory quality characteristics. International Journal of Food and Fermentation Technology 1, 247–253.

Kundu, B.S., Bardiya, M.C., Tauro, P., 1976. Studies on fruit wines-Banana wine. Haryana Journal of Horticulture Science 5, 160.

Kundu, B.S., Bardiya, M.C., Daulta, B.S., Tauro, P., 1980. Evaluation of exotic grapes grown in Haryana for white table wines. Journal of Food Science and Technology 17, 221–224.

Lafar, F., 1910. Technical Mycology (C.T.C. Salter), vol. 1. Griffin, London, p. 12.

Laing, E., Pretorius, I.S., 1993. Co-expression of an *Erwinia chrysanthemi* pectate lyaseen coding gene (pelE) and an *Erwinia carotovora* polygalacturonase-encoding gene (peh1) in *Saccharomyces cerevisiae*. Applied Microbiology and Biotechnology 39, 181–188.

Lambrechts, M.G., Pretorius, I.S., 2000. Yeast and its importance to wine aroma. South Africa Journal of Enology and Viticulture 21, 97–129.

Langcake, P., Mc Carthy, W.V., 1979. The relationship of resveratrol production to infection of grapevine leaves by *Botrytis cinerea*. Vitis 8, 244–253.

Lasekan, O., Buettner, A., Christlbauer, M., 2007. Investigation of important odorants of palm wine (*Elaeis guineensis*). Food Chemistry 105, 15–23.

Leake, C.D., Silverman, M., 1966. Alcoholic Beverages in Clinical Medicine. Yearbook Publisher, Chicago, Illinois, p. 14.

Lee, P.R., Yu, B., Curran, P., Liu, S.Q., 2010. Effect of Fusel Oil Addition on Volatile Compounds in Papaya Wine Fermented With *Williopsis saturnus* Var. Mrakii NCYC 2251. Food Research International. http://dx.doi.org/10.1016/j.foodres.2010.12.026.

Lee, P.R., Yu, B., Curran, P., Liu, S.Q., 2011. Impact of amino acid addition on aroma compounds in papaya wine fermented with *Williopsis mraeii*. South African Journal for Enology and Viticulture 32, 220–228.

Lee, P.R., Chong, I.S.M., Yu, B., Curran, P., Liu, S.Q., 2012. Effects of sequentially inoculated *Williopsis saturnus* and *Saccharomyces cerevisiae* on volatile profiles of papaya wine. Food Research International 45, 177–183.

Lee, G.D., 2015. Optimization of peach wine fermentation by response surface methodology. Journal of Korean Society of Food Science and Nutrition 44, 586–591.

Leicester, H.M., 1974. Development of Biochemical Concepts From Ancient to Modern Times. Harvard University Press, Cambridge, Massachusetts, p. 182.

Li, Z.C., Chi, L.Z., Zhu, J.K., Zhang, Y.Y., Wang, Q.J., He, P.G., 2011. Simultaneous determination of active ingredients in blueberry wine by CE-AD. Chinese Chemical Letters 22, 1237–1240.

Li, X., Lim, S.L., Yu, B., Curran, P., Liu, S., 2013. Mango wine aroma enhancement by pulp contact and β-glucosidase. International Journal of Food Science and Technology 48, 2258–2266.

Lim, J.W., Jeong, J.T., Shin, C.S., 2012. Component analysis and sensory evaluation of Korean black raspberry (*Rubus coreanus* Mique) wines. International Journal of Food Science and Technology 47, 918–926.

Lingappa, K., Chandershekhar, N., 1997. Wine preparation from carrot (*Daucus carota*). Indian Food Packer 51, 10–13.

Lingappa, K., Padshetty, N.S., Chowdary, N.B., 1993. Tamarind vermouth – a new alcoholic beverage from tamarind (*Tamarindus indica*). Indian Food Packer 47, 23–26.

Lira Junior, J.S., Musser, S., Melo, A., Maciel, S., Lederman, I.E., Santos, V.F., 2005. Caracterizacaofisica e fisicoquimica de frutos de caja-umbu (*Spondias* spp.). Ciencia e Tecnologia de Alimentos 25, 757–761.

Lodge, N., 1981. Kiwi fruit: two novel processed products. Food Technology in New Zealand 16, 35.

Lucia, S.P., 1963. A History of Wine as Therapy. Lippincott, Philadelphia.

MAFF, 1980. Ministry of Agriculture, Fisheries and Food. Grape for Wine. HMSO, London.

Marwaha, S.S., Panesar, P.S., Arora, J., Panesar, R., 2004. Studies on the fermentative production of cider from apple juice concentrate. Indian Food Packer 73–77.

Masquelier, J., 1988. Effects physiologiques du Vin, sa port dans L'alcoolisme. Bulletin de O.I.V 6, 555.

McGovern, P.E., Glusker, D.L., Exner, L.J., Voigt, M.M., 1996. Neolithic resinated wine. Nature 381, 480.

McGovern, P.E., Zhang, J., Tang, J., Zhang, Z., Hall, G.R., Moreau, R.A., 2004. Fermented beverages of pre- and proto-historic China. Proceedings of the National Academy of Sciences of the United States of America 101, 17593–17598.

McGovern, P.E., Mirzoian, A., Hall, G.R., 2009. Ancient Egyptian herbal wines. Proceedings of the National Academy of Sciences of the United States of America 106, 7361–7366.

Meena, P., Gil-Izquierdo, A., Moreno, D.A., Marti, N., Garcia-Viguera, C., 2012. Assessment of the melatonin production in pomegranate wines. LWT—Food Science and Technology 47, 13–18.

Mena, P., Girones-Vilaplana, A., Moreno, D.A., Garcia-Viguera, C., 2011. Pomegranate fruit for health promotion: myths and realities. Functional Plant Science and Biotechnology 5, 33–42.

Menzel, C.M., Simpson, D.R., 1993. In: Macrae, R., Robinson, R.K., Sadler, M.J. (Eds.), Fruit of Tropical Climate – Fruits of Sapindaceae. Encyclopaedia of Food Science. Food Technology and Nutrition. Academic Press, London, p. 2108.

Missehorn, K., Adam, R., 1976. On the cyanide contents in stone-fruit products (Transl). BrauntWein Wirt and Chaft 116 (4), 49.

Morton, J.F., 1968. The jambolan (*Syzgium cuminil* Skeels). Its food, medicinal, ornamental and other uses. Proceedings of the Florida State Horticultural Society 76, 328.

Musyimi, S.M., Sila, D.N., Okoth, E.M., Onyango, C.A., Mathooko, F.M., 2015. Production and characterization of wine from mango fruit (*Mangifera indica*) varieties in Kenya. Journal of Agriculture, Science and Technology 16.

Nanda, K., 2015. Maharashtra's Wine Industry Cheers New Liquor Policy. http://www.financialexpress.com/article/markets/commodities/maharashtras-wine-industry-cheers-new-liquor-policy/133870/.

Nigam-Singh, P., 2011. Microbiology of wine making. In: Joshi, V.K. (Ed.), Handbook of Enology, vol. 2. Asia Tech Publication, New Delhi, pp. 384–406.

Nijveldt, R.J., van Nood, E., van Hoorn, D.E., Boelens, P.G., van Norren, K., van Leeuwen, P.A., 2001. Flavonoids: a review of probable mechanisms of action and potential applications. American Journal of Clinical Nutrition 74, 418–425.

Noller, J., Wilson, B., 2009. Markets for Tropical Fruit Wine Products. Rural Industries Research and Development Corporation, Australian Government.

Nuengchamnong, N., Ingkaninan, K., 2009. On-line characterization of phenolic antioxidants in fruit wines from family myrtaceae by liquid chromatography combined with electrospray ionization tandem mass spectrometry and radical scavenging detection. LWT—Food Science and Technology 42, 297–302.

Offenbacher, E.G., Pi-Sunyer, F.X., 1980. Beneficial effect of chromium-rich yeast on glucose tolerance and blood lipids in elderly subjects. Diabetes 29, 919–925.

Okada, T., Yohotosuha, K., 1996. Trans-resveratrol concentrations in berry skins and wines from grapes grown in Japan. American Journal of Enology and Viticulture 47, 93–99.

Okigbo, R.N., Obire, O., 2009. Mycoflora and production of wine from fruits of soursop (*Annona muricata* L.). International Journal of Wine Research 1, 1–9.

Okoro, C.E., 2007. Production of red wine from roselle (*Hibiscus sabdariffa*) and pawpaw (*Carica papaya*) using palm wine yeast (*Saccharomyces cerevisiae*). Nigerian Food Journal 25, 158–164.

Okunowo, W.O., Osuntoki, A.A., 2007. Quantitation of alcohols in orange wine fermented by four strains of yeast. African Journal of Biochemistry Research 1, 95–100.

Oliveira, M.E.S., Pantoja, L., Duarte, W.F., Collela, C.F., Valarelli, L.T., Schwan, R.F., Dias, D.R., 2011. Fruit wine produced from cagaita (*Eugenia dysenterica* DC) by both free and immobilised yeast cell fermentation. Food Research International. http://dx.doi.org/10.1016/j.foodres.2011.02.028.

Onkarayya, H., Singh, H., 1984. Screening of mango varieties for dessert and madeira-style wines. American Journal of Enology and Viticulture 35, 63–65.

Onkarayya, H., 1985. Mango vermouth – a new alcoholic beverage. Indian Food Packer 39, 85–88.

Onkarayya, H., 1986. A rapid madeirization process to improve mango dessert wines. Journal of Food Science and Technology 23, 175–176.

Panda, S.K., Swain, M.R., Singh, S., Ray, R.C., 2012. Proximate compositions of a herbal purple sweet potato (*Ipomoea batatas* L.) wine. Journal of Food Processing and Preservation. http://dx.doi.org/10.1111/j.1745-4549.2012.00681.x.

Panesar, P.S., Marwaha, S.S., Arora, J., Rai, R., 2000. Fermentative production of cider-ginger beverage. Beverage and Food World 27 (2), 21–22.

Panesar, S.P., Kumar, N., Marwaha, S.S., Joshi, V.K., 2009. Vermouth production technology – an overview. Natural Product Radiant 334–341.

Panesar, P.S., Joshi, V.K., Panesar, R., Abrol, G.S., 2011a. Vermouth: technology of production and quality characteristics. In: Advances in Food and Nutritional Research, 63. Elsevier, Inc., London, UK, pp. 253–271.

Panesar, P.S., Marwaha, S.S., Sharma, S., Kumar, H., 2011b. Preparation of fortified wines. In: Joshi, V.K. (Ed.), Hand Book of Enology: Principles, Practices and Recent Innovations. Asiatech Publisher Inc., New Delhi, pp. 1021–1063.

Pasteur, L., 1866. Etudes su le vin. Impeimerie Imperiala, Paris.

Patel, J.D., Venkataramu, K., Subba Rao, M.S., 1977. Studies on the preparation of cider and brandy from some varieties of Indian apples. Indian Food Packers 31 (6), 5–8.

Patil, S.S., Adsule, R.N., Chavan, U.D., July 1995. Influence of sulphur dioxide, inoculum levels and pH of the must on the quality of Ber wine. Beverage Food World 28–30.

Patil, S.S., Thorat, R.M., Rajasekaran, P., 2012. Utilization of jamun fruit (*Syzygium cumini*) for production of red wine. Journal of Advanced Laboratory Research in Biology 3, 234–238.

Pérez-Torrado, R., Querol, A., Guillamón, J.M., 2015. Genetic improvement of non-GMO wine yeasts: strategies, advantages and safety. Trends in Food Science and Technology 45, 1–11.

Pérez-Través, L., Lopes, C.A., González, R., Barrio, E., Querol, A., 2015. Physiological and genomic characterisation of *Saccharomyces cerevisiae* hybrids with improved fermentation performance and mannoprotein release capacity. International Journal of Food Microbiology 205, 30–40.

Petrie, W.M.E., 1923. Social Life in Ancient Egypt. Methuen, London.

Phaff, H.J., 1986. Ecology of yeasts with actual and potential value in biotechnology. Microbial Ecology 12, 31.

Pino, J.A., Queris, O., 2010. Analysis of volatile compounds of pineapple wine using solid-phase microextraction techniques. Food Chemistry 122, 1241–1246.

Pino, J.A., Queris, O., 2011b. Characterization of odor-active compounds in guava wine. Journal of Agricultural and Food Chemistry 59, 4885–4890.

Pino, J.A., Queris, O., 2011a. Analysis of volatile compounds of mango wine. Food Chemistry 125, 1141–1146.

Pretorius, I.S., Bauer, F.F., 2002. Meeting the consumer challenge through genetically customized wine-yeast strains. Trends in Biotechnology 20, 426–432.

Pretorius, I.S., Van der Westhuizen, T.J., 1991. The impact of yeast genetics and recombinant DNA technology on the wine industry. South African Journal of Enology and Viticulture 12, 3–31.

Pretorius, I.S., du Toit, A., van Rensburg, P., 2003. Designer yeasts for the fermentation industry of the 21st century. Food Technology and Biotechnology 41, 3–10.

Pretorius, I.S., 1997. Utilization of polysaccharides by *Saccharomyces cerevisiae*. In: Zimmermann, F.K., Entian, K.D. (Eds.), Yeast Sugar Metabolism. Technomic Publishing Co., pp. 459–501.

Pretorius, I.S., 1999. Engineering designer genes for wine yeasts. Australian and New Zealand Wine Industry Journal 14, 42–47.

Pretorius, I.S., June 15, 2000. Tailoring wine yeast for the new millennium: novel approaches to the ancient art of winemaking. Yeast 16 (8), 675–729.

Pretorius, I.S., 2001. Gene technology in winemaking: new approaches to an ancient art. Agriculturae Conspectus Scientificus 66, 1–20.

Pretorius, I.S., 2002. The genetic analysis and improvement of wine yeasts. In: Arora, D. (Ed.), Fungal Biotechnology. Marcel Decker.

Pretorius, I.S., Høj, P.B., 2005. Grape and wine biotechnology: challenges, opportunities and potential benefits. Australian Journal of Grape and Wine Research 11, 83–108.

Rana, S.N., Rana, S.V., 2011. Biochemistry of wine preparation. In: Joshi, V.K. (Ed.), Handbook of Enology, vol. 2. Asia Tech Publication, New Delhi, pp. 618–678.

Rana, R.S., Vyas, K.K., Joshi, V.K., 1986. Studies on production and accept-ability of cider from Himachal Pradesh apple. Indian Food Packer 40 (6), 48–56.

Rebordinos, L., Infante, J.J., Rodriguez, M.E., Vallejo, I., Cantroal, J.M., 2010. Wine yeast growth and factor affecting. In: Joshi, V.K. (Ed.), Handbook of Enology, vol. 2. Asia Tech Publication, New Delhi, pp. 406–434.

Reddy, L.V.A., Reddy, O.V.S., 2010. Effect of fermentation conditions on yeast growth and volatile composition of wine produced from mango (*Mangifera indica* L.) fruit juice. Food and Bioproducts Processing. http://dx.doi.org/10.1016/j.fbp.2010.11.007.

Renfrew, C., 1989. The origins of Indo-European languages. Scientific American 261, 106.

Rivard, D., 2009. The Ultimate Fruit Winemaker's Guide, second ed. Bacchus Enterprises Ltd.

Robinson, K., 1994. The Oxford Companion to Wine. Oxford University Press, London.

Rodríguez-Lerma, G.K., Gutiérrez-Moreno, K., Cárdenas-Manríquez, M., Botello-Álvarez, E., Jiménez-Islas, H., Rico-Martínez, R., Navarrete-Bolaños, J.L., 2011. Microbial ecology studies of spontaneous fermentation: starter culture selection for prickly pear wine production. Journal of Food Science 76, 346–352.

Romani, R.J., Jannings, W.G., 1971. In: Hulme, A.C. (Ed.), Stone Fruits. The Biochemistry of Fruits and Their Products. Academic Press, London, p. 411.

Rose, A.H., Harrison, J.S., 1971. The Yeast. Physiology and Biochemistry of Yeasts. Academic Press, New York, p. 1.

Rossouw, D., Heyns, E.H., Setati, M.E., Bosch, S., Bauer, F.F., 2013. Adjustment of trehalose metabolism in wine *Saccharomyces cerevisiae* strains to modify ethanol yields. Applied and Environmental Microbiology 79, 5197–5207.

Sandhu, M.K., Chattopadhyay, P.K., 2001. Classification of fruits and fruit crops. In: Introductory Fruit Crops. Naya Prokash, Calcutta, pp. 20–29.

Sandhu, D.K., Joshi, V.K., 1995. Technology, quality and scope of fruit wines with special reference to apple. Indian Food Industry 14 (1), 24–28.

Schanderi, H., Koch, J., 1957. Die Fruchtwein Be Reitungeugen Ulmer, Stuttgart.

Schmitzer, V., Veberic, R., Slatnar, A., Stampar, F., 2010. Elderberry (*Sambucus nigra* L.) wine: a product rich in health promoting compounds. Journal of Agricultural and Food Chemistry 58, 10143–10146.

Schoeman, H., Vivier, M.A., Du Toit, M., Dicks, L.M., Pretorius, I.S., 1999. The development of bactericidal yeast strains by expressing the *Pediococcus acidilactici* pediocin gene (pedA) in *Saccharomyces cerevisiae*. Yeast 15, 647–656.

Schuller, D., Casal, M., 2005. The use of genetically modified *Saccharomyces cerevisiae* strains in the wine industry. Applied Microbiology and Biotechnology 68, 292–304.

Schwarz, K., Mertz, W., 1959. Chromium (III) and the glucose tolerance factor. Archives of Biochemistry and Biophysics 85, 292–295.

Selli, S., Kelebek, H., 2015. Organic acids. In: Nollet, L.M.L., Toldrá, F. (Eds.), Handbook of Food Analysis. CRC Press (in press).

Selli, S., Canbasi, A., Unal, U., 2002. Effect of bottle colour and storage conditions on browning of orange wine. Nahrung/Food 46, 64–67.

Selli, S., Cabaroglu, T., Canbas, A., 2003. Flavour components of orange wine made from a Turkish cv. Kozan. International Journal of Food Science and Technology 38, 587–593.

Selli, S., Canbas, A., Varlet, V., Kelebek, H., Prost, C., Serot, T., 2008. Characterization of the most odor-active volatiles of orange wine made from a Turkish cv. Kozan (*Citrus sinensis* L. Osbeck). Journal of Agricultural and Food Chemistry 56, 227–234.

Selli, S., 2007. Volatile constituents of orange wine obtained from moro oranges (*Citrus sinensis* [L.] Osbeck). Journal of Food Quality 30, 330–341.

Sevda, S.B., Rodrigues, L., 2011. Fermentative behavior of *Saccharomyces* strains during Guava (*Psidium guajava* L.) must fermentation and optimization of guava wine production. Journal of Food Processing and Technology 2, 1–18.

Seward, D., 1979. Monks and Wine. Mitchell – Breazley, London.

Shahidi, F., 2009. Nutraceuticals and functional foods: whole versus processed foods. Trends in Food Science and Technology 20, 376–387.

Shanmugavelu, G.K., 2003. Grape Cultivation and Processing. Agrobios, Jodhpur (India).

Sharma, S.K., Joshi, V.K., 1996. Optimization of some factors for secondary bottle fermentation for production of sparkling plum (*Prunus salicina*) wine. Indian Journal of Experimental Biology 34 (3), 235–238.

Sharma, S., Joshi, V.K., 2004. Flavour profiling of strawberry wine by quantitative descriptive analysis technique. Journal of Food Science and Technology 41 (1), 22–26.

Sharma, V.C., Ogbeide, O.N., 1982. Pawpaw as a renewable energy resource for the production of alcohol fuels. Energy 7, 871–873.

Sharma, S., Joshi, V.K., Abrol, G., 2009. An overview of strawberry [*Fragaria* x *ananassa* (Weston) Duchesne ex Rozier] wine production technology, composition, maturation and quality evaluation. Natural Product Radiance 8, 356–365.

Sharma, N., Bhutia, S.P., Aradhya, D., 2013. Process optimization for fermentation of wine from jackfruit (*Artocarpus heterophyllus* Lam.). Journal of Food Processing and Technology 4, 204.

Shukla, K.G., Joshi, M.C., Sarswati, Y., Bisht, N.S., 1991. Jambal wine making standardization of methodology and screening of cultivars. Journal of Food Science and Technology 28, 142–144.

Singh, R.S., Kaur, P., 2009. Evaluation of litchi juice concentrate for the production of wine. Natural Product Radiance 8, 386–391.

Singh, S., Krishnamurthi, S., Katyal, S.L., 1967. Fruit Culture in India. ICAR, New Delhi, p. 225.

Singh, M., Panesar, P.S., Marwaha, S.S., 1998. Studies on the suitability of kinnow fruits for the production of wine. Journal of Food Science and Technology 35 (5), 455.

Singh, A., 2000. Fruit Physiology and Production. Kalyani Publishers, p. 177.

Singh Nagi, H.P.P., Manjrekar, S.P., 1976. Studies in the preparation of cider from North Indian apples: I varietal differences in the physico-chemical characteristics. Indian Food Packer 29 (6), 11–14.

Smit, A., Cordero, R.R., Lambrechts, M.G., Pretorius, I.S., van Rensburg, P., 2003. Enhancing volatile phenol concentrations in wine by expressing various phenolic acid decarboxylase genes in *Saccharomyces cerevisiae*. Journal of Agricultural and Food Chemistry 51, 4909–4915.

Snow, R., 1983. Genetic improvement of wine yeast. In: Spencer, J.F.T., Spencer, D.M., Smith, A.R.W. (Eds.), Yeast Genetics: Fundamentals and Applied. Springer, Verlage, New York, pp. 439–459.

Soni, S.K., Bansal, N., Soni, R., 2009. Standardization of condition for fermentation and maturation of wine from amla (*Emblica officinalis* Gaertn.). Natural Product Radiance 8, 436–444.

Soni, S.K., Marwaha, S.S., Marwaha, U., Soni, R., 2011a. Composition and nutritive value of wine. In: Joshi, V.K. (Ed.), Handbook of Enology: Principles Practices, vol. 1. Asia Tech Publisher, New Delhi, pp. 89–145.

Soni, S.K., Sharma, S.C., Soni, R., 2011b. Yeast genetics and genetic engineering in wine making. In: Joshi, V.K. (Ed.), Handbook of Enology, vol. III. Asia Tech Publishers, New Delhi, pp. 441–501.

Soufleros, E.H., Pissa, I., Petridis, D., Lygerakis, M., Mermelas, K., Boukouvalas, G., Tsimitakis, E., 2001. Instrumental analysis of volatile and other compounds of Greek kiwi wine: sensory evaluation and optimisation of its composition. Food Chemistry 75, 487–500.

Srivastava, D.K., Raverkar, K.P., 1999. In: Joshi, V.K., Pandey, A. (Eds.), Genetic Manipulation of Industrially Important Microorganisms, vol. I. Educational Publishers and Distributors, New Delhi, pp. 173–258.

Stanislawski, D., 1975. Dionysus westward: early religion and the economic geography of wine. Geographical Review 65, 427.

Stockley, S, 2011. Therapeutic value of wine. In: Joshi, V.K. (Ed.), Handbook of Enology, vol. 1. Asia Tech Publication, New Delhi, pp. 146–208.

Stringini, M., Comitini, F., Taccari, M., Ciani, M., 2009. Yeast diversity during tapping and fermentation of palm wine from Cameroon. Food Microbiology 26, 415–442.

Sun, A., Kang, Y.H., Gills, J.J., Cuendet, M., 2011. Comparative evaluation of the anticancer properties of European and American elderberry fruits. Journal of Medicinal Food 9, 498–504.

Suresh, E.R., Ethiraj, S., 1987. Effect of grape maturity on the composition and quality of wines made in India. American Journal of Enology and Viticulture 38, 329–331.

Sutcliffe, D., 1934. The vineyards of Northfleet and Teynham in the thirteenth century. Archaeologia Cantiana 46, 140.

Swami, S.B., Thakor, N.J., Divate, A.D., 2014. Fruit wine production: a review. Journal of Food Research and Technology 2 (3), 93–94.

Teotia, M.S., Manan, J.K., Berry, S.K., Sehgal, R.C., 1991. Beverage development from fermented (*S. cerevisiae*) Muskmelon (*C. melo*) juice. Indian Food Packer 49.

Thaper, A.R., 1958. Jamun. ICAR Farm Bulletin, p. 42.

Tilloy, V., Ortiz-Julien, A., Dequin, S., 2014. Reduction of ethanol yield and improvement of glycerol formation by adaptive evolution of the wine yeast *Saccharomyces cerevisiae* under hyperosmotic conditions. Applied and Environmental Microbiology 80, 2623–2632.

Tofalo, R., Perpetuini, G., Di Gianvito, P., Schirone, M., Corsetti, A., Suzzi, G., 2014. Genetic diversity of FLO1 and FLO5 genes in wine flocculent *Saccharomyces cerevisiae* strains. International Journal of Food Microbiology 191, 45–52.

Towantakavanit, K., Park, Y.S., Gorinstein, S., 2011. Bioactivity of wine prepared from ripened and over-ripened kiwifruit. Central European Journal of Biology 6, 305–315.

Tuman, R.W., Doisy, R.J., 1977. Metabolic effects of the glucose tolerance factor (GTF) in normal and genetically diabetic mice. Diabetes 26, 820–826.

Van Rensburg, P., Pretorius, I.S., 2000. Enzymes in winemaking: harnessing natural catalysts for efficient biotransformations. South African Journal of Enology and Viticulture 21, 52–73.

Varakumar, S., Kumar, Y.S., Reddy, O.V.S., 2011. Carotenoid composition of mango (*Mangifera indica* L.) wine and its antioxidant activity. Journal of Food Biochemistry 35, 1538–1547.

Varakumar, S., Naresh, K., Variyar, P.S., Sharma, A., Reddy, O.V.S., 2013. Role of malolactic fermentation on the quality of mango (*Mangifera indica* L.) wine. Food Biotechnology 27, 119–136.

Varela, C., Kutyna, D.R., Solomon, M.R., Black, C.A., Borneman, A., Henschke, P.A., Pretorius, I.S., Chambers, P.J., 2012. Evaluation of gene modification strategies for the development of low-alcohol-wine yeasts. Applied and Environmental Microbiology 78, 6068–6077.

Venkataramu, K., Patel, J.D., Subba Rao, M.S., 1979. Fermentation of grapes with a few strains of wine yeasts. Indian Food Packer 33 (2), 13–14.

Verma, L.R., Joshi, V.K., 2000. An overview of post-harvest technology. In: Verma, L.R., Joshi, V.K. (Eds.), Postharvest Technology of Fruits and Vegetables, vol. I. Indus Publishing Co., New Delhi, pp. 1–65.

Vine, R.P., 1981. Wine and the history of western civilization. In: Commercial Wine Making, Processing and Controls. The AVI Publishing Co., Westport, CT.

Vyas, S.R., Chakravorthy, S.C., 1971. Wines From Indian Grapes. Haryana Agricultural University, Hisar, pp. 1–69.

Vyas, S.R., Gandhi, R.C., 1972. Enological qualities of various grape varieties grown in India. In: Proc Symp Alcohol Beverage Industry in India, Present Status and Future Prospects, Mysore, pp. 9–11.

Vyas, K.K., Joshi, V.K., 1982. Plum wine making: standardization of a methodology. Indian Food Packer 80–86.

Vyas, K.K., Joshi, V.K., 1988. Deacidification activity of *Schizosaccharomyces pombe* in plum musts. Journal of Food Science and Technology 25 (5), 306–307.

Vyas, K.K., Sharma, R.C., Joshi, V.K., 1989a. High flavoured low alcoholic drink from litchi. Beverage Food World 16 (1), 30.

Vyas, K.K., Sharma, R.C., Joshi, V.K., 1989b. Application of osmotic technique in plum wine fermentation. Effect on physico-chemical and sensory qualities. Journal of Food Science and Technology 26 (3), 126–128.

Wang, D., Xu, Y., Hu, J., Zhao, G., 2004. Fermentation kinetics of different sugars by apple wine yeast *Saccharomyces cerevisiae*. Journal of the Institute of Brewing 110, 340–346.

Withy, L.M., Lodge, N., 1982. Kiwifruit wine: production and evaluation. American Journal of Enology and Viticulture 4, 191–193.

Wu, Y., Zhu, B., Tu, C., Duan, C., Pan, Q., 2011. Generation of volatile compounds in Litchi wine during winemaking and short-term bottle storage. Journal of Agricultural and Food Chemistry 59, 4923–4931.

Younis, K., Siddiqui, D.S., Jahan, K., Dar, M.S., 2014. Production of wine from over ripe guava fruit (*Psidium Guajava* L Safada) and ber (*Ziziphus mauritiana* L Umrau) using *Saccharomyces crevices* Var. HAU 1. IOSR Journal of Environmental Science, Toxicology and Food Technology 8, 93–96.

Youravong, W., Li, Z., Laorko, A., 2010. Influence of gas sparging on clarification of pineapple wine by microfiltration. Journal of Food Engineering 96, 427–432.

Zhu, L.H., Li, X.Y., Ahlman, A., Xue, Z.T., Welander, M., 2004. The use of mannose as a selection agent in transformation of apple rootstock M26 via *Agrobacterium tumefaciens*. Acta Horticulture 663, 503–506.

WEBSITES

Saata http://www.saasta.ac.za/Media-Portal/download/bio_fs15.pdf.

http://www.capewineacademy.co.za.

http://www.wikiwand.com/en/Banana_wine.

www.euromonitor.com.

www.foodservicewarehouse.com.

https://www.decanterchina.com/en/?article=868.

statista http://www.statista.com/statistics/300837/global-number-of-cider-manufacturers-by-country/.

http://www.agr.gc.ca/.

MICROBIOLOGY OF FRUIT WINE PRODUCTION

2

F. Matei[1], M.R. Kosseva[2]

[1]*University of Agronomic Sciences and Veterinary Medicine of Bucharest, Bucharest, Romania;*
[2]*University of Nottingham Ningbo Campus, Ningbo, China*

1. INTRODUCTION

Fruit wines are fermented alcoholic beverages made of different fruits, and the basic processes are similar to grape juice production and fermentation. Consequently, it is expected that the microbiology of nongrape wines is more or less the same as the grape wine, while microflora can vary with the type of the fruit. Similarly, the changes in the type of microflora can be different where the fermentation is not conducted naturally.

Both endogenous and exogenous microorganisms are involved in the process of fruit wine fermentation. The endogenous cultures are contributed by fruits and winery surfaces while the exogenous are the selected starter cultures. The quality of the end-product wine depends on the contributions made by all microorganisms present in the whole system including bacteria and yeast. Traditional wine fermentations use the wild microflora present on the surface of fruit-skins and other yeasts available in wineries, which are indigenous.

As terrain products, the fruits get into the processing carrying different types of microorganisms, as bacteria, yeast, and phylamentous fungi (mainly spoilage molds). These microorganisms vary in levels from fruit to must and during the winemaking, but their diversity is very similar to the grape. The final organoleptic quality of the fruit wines depends on the growth and metabolic activities of the microorganisms during fermentation and even in the postfermentation steps.

From a technological point of view, in the case of fruit wines some particularities are rising: the fermentation can be initiated and allowed to proceed for some time prior to pressing; in contrast to grape must, many fruits lack the nutrients necessary to sustain yeast growth. Thus yeast nutrients such as yeast extract or diammonium phosphate may be needed—up to 0.1%. Crushed fruits will have more yeast nutrients than pressed juice, so additions may be unnecessary in the latter.

The aim of this chapter is to describe biodiversity of microorganisms used for fruit wine production and compare it to the grape winemaking microflora. Selection of yeasts for alcoholic fermentation and lactic acid bacteria for malolactic fermentation is considered with emphases on pure and mixed cultures. The factors affecting the yeast growth are also presented. Methods of immobilization of microbial cells are defined; fundamental considerations on selection of the supports for immobilization are provided as well. Applications of immobilized cells in alcoholic fermentation, malolactic fermentation, sparkling wine, and cider manufacturing are given. Currently, spoilage of fruit wines as well as microbial analysis of wine quality has attracted attention of the researchers aiming to satisfy customer needs

Science and Technology of Fruit Wine Production. http://dx.doi.org/10.1016/B978-0-12-800850-8.00002-8

and tastes. Finally, genetically modified microorganisms applicable to fruit winemaking and their legislation for the customer and producer benefits are demonstrated.

2. MICROBIAL BIODIVERSITY DETECTED DURING FRUIT WINE PRODUCTION

In the past two decades researchers have "borrowed" from the grape winemaking field the techniques used for the study of microbial biodiversity in fruit winemaking. Traditionally, the identification and characterization of yeast species have been based on morphological traits and their physiological capabilities (Barnett et al., 1990). The necessity of enrichment, isolation, and characterization of strain, in addition to being time-consuming, is fraught with potential biases. While this approach has proven useful, it often fails to characterize those microorganisms for which culturing is problematic or impossible (Nigam, 2011).

The advances in the field of molecular microbial ecology have brought a variety of new tools to directly assess the microbial diversity present in natural habitats. A common strategy is to sample the DNA (or RNA) of a mixed microbial community and use it as a template for both the assessment of community structure and to reveal individual constituents. Relatively recently, denaturing gradient gel electrophoresis (DGGE) has been employed to differentiate rRNA genes directly purified from complex microbial communities. The polymerase chain reaction (PCR)-DGGE has been employed primarily to examine bacterial diversity (or, less frequently, fungal diversity) in various natural habitats.

Natural grape and fruit fermentation involves a succession of **yeasts**, with *Saccharomyces cerevisiae* as the dominant species (Pretorius, 2000). It is generally accepted that in the first stage of a fruit wine fermentation the non-*Saccharomyces* yeast are predominant. Later on, while the ethanol contents increase, the dominant specie is *Saccharomyces*. This microorganism succession process is a reflection of microbial interactions, competition for intrinsic growth factors such as nutrients, and resistance to inhibitory environmental conditions such as high acidity. As a consequence, microorganisms showing a selective advantage emerge in a given period as the dominant populations during fermentation (Sánchez et al., 2010).

Usually, the genera and species found on the fruit will be recovered in the must. Most of the reports regarding the fruits microbiota are related to the cider production. It has been reported that the apple main genera and species are *Candida malicola*, *Candida krusei*, *Torulopsis famata*, *Debaryomyces*, *Pichia*, and *Kloeckera*; when the fruits are fully ripened, some other species may occur, like *Saccharomyces florentinus*, *Hanseniaspora*, and *Saccharomycodes ludwigii*. If apples come from the soils, different microorganisms have been found: *Hansenula*, *Candida reukaufii*, *Rhodotorula aurantiaca*, *Rhodotorula glutinis*, and *Trichosporon* (Dan, 2011). When the fruits are stored longer before the crushing (about one week), the contents of yeast grow 10 times, especially, *Kloeckera*.

2.1 YEASTS

Different **non-*Saccharomyces* yeasts** were isolated during the fruit wine processing, mainly from the fruit, in the fruit juice, and in the early stages of the fruit winemaking process (Pretorius, 2000; Pando Bedriñana et al., 2010; Chilaka et al., 2010; Matei et al., 2011) and are presented in Table 2.1.

From the indigenous fermentation of gabiroba (*Campomanesia pubescens*), pulp species very similar to those found in the grape fermentation were isolated, respectively *Candida*, *Issatchenkia*, and *Pichia* (Duarte et al., 2009b).

Table 2.1 Non-*Saccharomyces* Yeasts Isolated During Fruit Wine Fermentation

Fruit Wine	Non-*Saccharomyces* Yeast	Source
Gabiroba (*Campomanesia pubescens*)	*Candida, Issatchenkia,* and *Pichia*	Duarte et al. (2009b)
Palm wine	*Kloechera apiculata, Candida krusei* *Candida, Endomycopsis, Hansenula, Kloechera, Pichia, Saccharomycodes, Schizosaccharomyces* *Schizosaccharomyces pombe* *Pichia membrabefaciens* *Saccharomycodes ludwigii; Zygosaccharomyces bailii, Hanseniaspora uvarum, Candida parapsilosis, Candida fermentati,* and *Pichia fermentans*	Amoa-Awua et al. (2007) Enwefa et al. (1992) (Faparusi, 1973; Okafor, 1972) Fahwehinmi (1981) Stringini et al. (2009)
Cider	*Hanseniaspora valbyensis, Hanseniaspora uvarum, Metschnikowia pulcherrima, Pichia guilliermondii,* and *Candida parapsilosis* *Hanseniaspora valbyensis; Hanseniaspora uvarum; Hanseniaspora osmophila; Pichia guilliermondii;* and *Metschnikowia pulcherrima* *Kloeckera apiculata; Candida pulcherima*	Pando Bedriñana et al. (2010) Valles Suarez et al. (2007) Dan (2011)
Pineapple wine	*Hanseniaspora uvarum;* and *Pichia guilliermondii*	Chanprasartsuk et al. (2010)
Papaya	*Schizosaccharomyces pombe; Zygosaccharomyces*	Maragatham and Panneerselvam (2011)
Masau (*Ziziphus mauritiana*)	*Issatchenkia orientali, Pichia fabianii,* and *Saccharomycopsis fibuligera*	Nyanag et al. (2007)

In the case of palm wine, Amoa-Awua et al. (2007) isolated *Kloechera apiculata* and *C. krusei*. Other authors reported the presence of several genera of non-*Saccharomyces* yeast in palm wine, like *Candida, Endomycopsis, Hansenula, Kloechera, Pichia, Saccharomycodes,* and *Schizosaccharomyces* (Enwefa et al., 1992).

In the case of apple wine, the presence of *Hanseniaspora* yeasts was reported in advanced stages of fermentation and can be explained by the minor sugar contents in the apple must compare to the grape juice without addition of SO_2. The species *Candida parapsilosis* were reported for the first time in cider by Pando Bedriñana et al. (2010).

The freshly crushed pineapple juice collected from Thailand and Australia countries showed the presence of *Hansenula uvarum* and *Pichia guilliermondii* yeasts strains during the fermentation process (Chanprasartsuk et al., 2010).

In masau (*Ziziphus mauritiana*) wines, Nyanga et al. (2007) identified species such as *Saccharomyces cerevisiae, Issatchenkia orientalis, Pichia fabianii. S. cerevisiae,* and *I. orientalis, which* were predominant in the fermented fruit pulp but were not detected in the unripe fruits. *S. cerevisiae, I. orientalis, P. fabianii* and *Saccharomycopsis fibuligera* are fermentative yeasts, and they might be used in the future development of starting cultures to produce better quality fermented products from masau fruit.

Some of the non-*Saccharomyces* yeast may lead to undesirable results during fruit wine fermentations. *Hanseniaspora* species may lead to eastery taints due to higher concentration of ethyl acetate and methyl butyl acetate; *Candida, Metschnikowia, Pichia* may develop film formation on wine surfaces and severe flavor taints; *Zygosaccharomyces* can cause wine refermentation during the storage.

Table 2.2 *Saccharomyces* Yeast Species Isolated During Fruit Wine Fermentation

Fruit Wine	*Saccharomyces* Yeast	Source
Palm wine	*Saccharomyces cerevisiae* *S. cerevisiae*	Amoa-Awua et al. (2007) Stringini et al. (2009)
Cider	*S. cerevisiae; Saccharomyces bayanus; Saccharomyces pastorianus; Saccharomyces kudriavzevii; Saccharomyces mikatae* *S. cerevisiae; S. bayanus*	Pando Bedriñana et al. (2010) Valles Suarez et al. (2007)
Citrus fruit (orange, grapefruit)	*S. cerevisiae var ellipsoideus* *Saccharomyces uvarum, S. cerevisiae, Saccharomyces carlsbergensis, S.cer. var ellipsoideus*	Okunowo and Osuntoki (2007) Nidp et al. (2001)
Papaya	*S. cerevisiae; S. bayanus; S. uvarum; Saccharomyces italicus; S. pastorianus*	Maragatham and Panneerselvam (2011)
Masau (*Ziziphus mauritiana*)	*S. cerevisiae*	Nyanga et al. (2007)

In his study, Oafor (1972) isolated 17 yeast strains from the palm wine, four belonging to the species of *Candida*, 12 to the genus of *Saccharomyces* and one to the *Endomycopsis species*.

The presence and permanence of different yeast species throughout fruit fermentation, and consequently, their influence on the final product, is determined by the fermentation conditions, such as inoculum of *Saccharomyces cerevisiae* starter culture, the temperature of the fermentation, and the fruit juice composition. The main species of *Saccharomyces* found during fruit wine fermentation is *S. cerevisiae*; however, some other *Saccharomyces* species have been reported, as *Saccharomyces bayanus*, *Saccharomyces pastorianus*, *Saccharomyces kudriavzevii*, *Saccharomyces mikatae*, *Saccharomyces carlsbergensis* (see Table 2.2). The variability within *Saccharomyces* spp. during spontaneous alcoholic fermentations depends very much on the specific years (Gutiérrez et al., 1999). In the case of cider, Valles Suarez et al. (2007) reported that *S. bayanus*, was the predominant species at the beginning and the middle fermentation steps of the fermentation process, reaching a percentage of isolation between 33% and 41%, whereas *S. cerevisiae* took over the process in the final stages of fermentation.

Nidp et al. (2001) reported that 16 yeast strains isolated from grapefruit (*Citrus paradis*), orange (*Citrus sinensis*) and pineapple (*Ananas comosus*) were characterized using standard microbiological procedures. The species were identified as *Saccharomyces uvarum, S. cerevisiae, S. carlsbergensis*, and *S. ellipsoideus*. Similarly, Matei and collaborators (data not reported) have isolated strains of *Saccharomyces cerevisiae* during pineapple fermentation.

2.2 KILLER YEAST IN WINE FRUITS

Some yeasts strains secrete toxins (proteins) into their environment, killing other sensitive yeast present in that environment. Such strains are known as killer strains (K). The killer factor was discovered in *Saccharomyces cerevisiae* but killer strains have been identified in other yeast genera as well, for example, *Hansenula, Kloeckera, Candida, Hanseniaspora, Pichia, Torulopsis, Kluyveromyces*, and *Debaryomyces*. Generally, the killer toxins described are active at pH values from 3 to 5.5.

FIGURE 2.1

Killer test for yeast on medium with addition of blue methylene (0.03%; the killer positive reaction is recognized by an inhibition areas around the yeast colonies).

The killer yeasts have various degrees of killer factor according to the sensitivity reaction between strains, therefore, killer yeasts have been classified into 11 groups based on the nature and properties of the toxins involved in killer interaction. A medium that contains the toxin exerts a selection pressure on a sensitive enological strain. Stable variants survive this selection pressure and can be obtained in this manner. This is the most simple strategy for obtaining a killer enological strain (Nigam, 2011).

In *Saccharomyces* three different killer toxins (K1, K2, and K28) were clearly identified, which are all genetically encoded by double-stranded M-dsRNA "killer" viruses persistently existing within the cytoplasm of the infected host cell (Wickner, 1996).

Killer tests can be done by the on-plate method recommend by Barre (1980), which is supposed to inoculate the reference strain on surface and the strains to be tested by streaks. The killer positive reaction is recognized by an inhibition area around the yeast colonies after 48–72 h of incubation at 20°C (Fig. 2.1). When incubated at higher temperature the killer toxin may be inactivated. The specific medium contains yeast extract 1%, peptone 1%, glucose 2%, agar 3%, and blue methylene 0.03%, all dissolved in citrate-phosphate buffer pH 4.8.

The majority of the yeast strains in fruit winemaking (around 95%) have a neutral phenotype and only few strains were found to have the killer phenotype. Yeast strains of *Saccharomyces* producing the killer toxin were isolated from Chinese ciders, and three strains were proposed to be used as starter culture on industrial level (Shuang, 2006). Similarly, killer phenotype was reported for the first time in Asturian autochthonous cider yeasts by Pando Bedriñana et al. (2010).

Relatively high inhibitory activity was reported for crude extracts of killer toxins made by *Pichia* sp. to the species as *Candida glabrata*, *Candida stellata*, *Candida vini*, *Hanseniaspora guilliermondii*, *Hanseniaspora uvarum*, *Metschnikowia pulcherrima*, *Debaryomyces hansenii*, and *Zygosaccharomyces bailii* (Satora et al., 2012).

Gulbiniene et al. (2004) isolated killer yeast from fruit and berry wine fermentation. The analyzed strains produced different amounts of active killer toxin, and some of them possessed new industrially significant killer properties. They were recommended for fermentation of particular types of wine—apple, cranberry, chokeberry, or grape. The authors also suggested that strong killers isolated from apple wine lees may be effective in suppressing wild yeast strains during fermentation.

Even if alcoholic fermentation is conducted with natural or starter yeast culture, some **bacteria** can be found in the fermented must and wine.

2.3 LACTIC ACID BACTERIA

Lactic acid bacteria (LAB) are characterized by a unique set of properties, and they have the capability to produce large amounts of lactic acid. Depending on the genus or species, LAB may ferment sugars solely to lactic acid in a homofermentative manner, or the products could comprise of ethanol, carbon dioxide, and lactic acid during the heterofermentation reaction. Homofermentation yields two ATPs per glucose, which is similar to yeast fermentation, whereas only one ATP is produced in heterofermentation. The homofermentative LAB reduces hexose sugars to lactic acid via the Embden–Meyerhof–Pamas (glycolytic) pathway. The formation of D-lactic acid arises from the reduction of pyruvic acid and is performed by homofermentative species of lactobacilli and pediococci. Heterofermentative lactobacilli, *Leuconostoc*, and *Oenococcus* spp. produce D-lactic acid and acetic acid through the 6-phosphogluconate pathway.

The bacteria present in the first steps of winemaking (must and the start of fermentation) belong to different species, generally homofermentative ones. The most abundant belong to the species *Lactobacillus plantarum*, *Lactobacillus hilgardii*, *Leuconostoc mesenteroides*, and *Pediococcus* sp., while to a lesser extent, *Oenococcus oeni* and *Lactobacillus brevis* are also found (Garcia-Ruiz et al., 2012). Bacterial multiplication takes place in the interval between the end of alcoholic fermentation and the start of malolactic fermentation (MLF).

Awe and Nnadoze (2015) reported *Lactobacillus casei* and other lactobacilli in date palm wine when it was inoculated with yeast starter culture. In the case of palm wine, Amoa-Awua et al. (2007) also reported the presence of *L. plantarum* and *L. mesenteroides* as dominated LAB among the yeast biota. These LAB are the only species isolated in the mature samples. While acetic acid bacteria (*Acetobacter* sp.) were isolated only after the third day of fermentation, when levels of alcohol become substantial. *Gluconobacter* has been isolated during palm wine tapping as well.

When using molecular tool in the case of palm wine, the bacterial diversity was found to include also uncultured bacterial clone (Okolie et al., 2013). The 16S rRNA gene fragments were amplified from microbial community and the genomic DNA of the isolates was then sequenced by PCR using universal primers. The analysis revealed that 32 community clones were identified as *Lactobacillus* sp., *L. casei*, *L. plantarum*, *L. mesenteroides* ssp. *dextranicum*, *Leuconcostoc lactis*, *Pediococcus parvulus*, *Acetobacter pomorum*, *Acetobacter pasteurianus*, *Gluconobacter oxydans*, *Acinetobacter calcoaceticus*, *Enterobacterium bacterium*, *Acidovorax* sp., *Comamonas* sp., *Bacillus subtilis*, *Staphylococcus piscifermentans*, and uncultured bacteria clone D1-78. However, the results showed that bacterial diversity in the palm wine sample is dominated by *Lactobacillus* and *Leuconostoc* species as reported by other authors.

In masau wines, Nyanga et al. (2007) identified as predominant strains *Lactobacillus agilis* and *L. plantarum*. Other species identified included *Lactobacillus bifermentans*, *Lactobacillus minor*, *Lactobacillus divergens*, *Lactobacillus confusus*, *L. hilgardii*, *Lactobacillus fructosus*, *Lactobacillus fermentum*, and *Streptococcus* spp.

In terms of **microbial population level**, few studies have counted the yeast and bacteria available on fruits. Most of the reports are related to apples and cider. The apples on the tree have on their surface 10^2 CFU/g of yeast, while on surface of the fruits felt on soil the yeast level rose to 10^4 CFU/g (Dan, 2011). In the case of cider, the initial numbers of LAB may reach 10^5–10^6 CFU/mL, while the yeast population may grow up to 10^6 CFU/mL. The concentrations of acetic acid bacteria population in the initial must usually are about 10^5 UFC/mL, and they decrease to zero about 20 days after the start of the fermentation (del Campo, 2008). During the fermentation, the yeast levels rise to 10^7–10^8 UFC/mL (Pretorius, 2000; Chen et al., 2014). In the case of palm wine with respect to the microbial population, a drastic increase was observed in yeast numbers from average values of 10^5–10^7 CFU/mL; LAB amplified from concentrations of 10^5–10^9 CFU/mL and aerobic mesophiles—from concentrations of 10^7 to about 10^9 CFU/mL during the first 24 h of tapping (Amoa-Awua et al., 2007).

Comparative studies have been done regarding the yeast population during natural and induced fermentation in gabiroba wine. Throughout the process of fermentation, when *S. cerevisiae* UFLA CA 1162 was inoculated, there were stable viable populations around 9 log CFU/mL. During indigenous fermentation, yeast population increased from 3.7 to 8.1 log CFU/mL after 14 days (Duarte et al., 2009b).

3. SELECTION OF YEAST AS STARTER CULTURES FOR PRODUCTION OF FRUIT WINES

Alcoholic fermentation is the biological process in which sugars (glucose, fructose) are converted into cellular energy and into ethanol and carbon dioxide as waste products. Fermentation is considered anaerobic process carried out in the presence of microorganisms (yeasts) and in absence of oxygen.

$$C_6H_{12}O_6 \rightarrow 2C_2H_5OH + 2CO_2 + Energy$$

The fermentation process for elaboration of the beverage depends on the performance of yeast to convert the sugars into alcohol and esters. Different species of yeast that develop during fermentation determine the characteristics of the flavor and aroma of the final product. Due to the differences in fruit composition, yeasts strains used for fermentation have to adapt to different environments (e.g., sugar composition and concentrations, presence of organic acids, etc.). In addition, the applied yeast has to compete for sugar utilization with other microorganisms present in the mashes, e.g., other yeast species or bacteria, depending on the fruit of choice and varying climatic conditions.

Spontaneous fermentation with microbiota that are present on the fruit surface is used in conventional winemaking in different regions of the world. In the case of fruits, must removal of indigenous microbiota occurred by pasteurization or sulfur addition. An alternative for both methods is the inactivation of wild yeasts by killer strains that are introduced into the fermentation medium together with *Saccharomyces* yeasts.

It is generally accepted that adding yeast starters will stabilize the fermentation and will lead to a quality product. It is very important to choose a yeast strain for a specific style of fruit wine. Fruit wines usually have a lower nitrogen content and the yeast can't work properly; in this case one should employ strains that work well in lower nutrient content and accentuate fresh aromatic qualities under less than ideal conditions (Parks, 2006). Toward the end of the fermentation, the yeasts release nitrogenous compounds in the fruit wine, including amino acids and peptides, along with pantothenic acid, riboflavin, and some phosphorous compounds. These nutrients are very important for any subsequent MLF (Waites et al., 2001).

S. cerevisiae eventually dominates most of the wine fermentations. Consequently, it is widely recognized as the principal wine-yeast and these species have been commercialized as starter cultures.

In the case of very low pH, like in berry fruits musts, it is recommended to use strains of yeast resistant to low pH, like *Saccharomyces bayanus* (Heimonen et al., 2013).

For the ciders, the ideal starter yeast should exhibit the following properties: rapid initiation of fermentation; resistance to SO_2 and high ethanol concentrations; low requirements for growth factors; ability to complete attenuation by fermenting all sugars; suitable flocculation characteristics; development of a sound organoleptic profile; and production of polygalacturonase to degrade soluble pectin (Waites et al., 2001).

Very popular for the grape winemaking, the yeast starter selection for nongrape wines has attracted the researchers' interest. In the case of raspberry wine, Brazilian researchers have tested and recommend for industrial use the strain of *S. bayanus* UFLA FW 15, which developed fruit wine characteristics such as "raspberry, cherry, sweet, and strawberry" (Duarte et al., 2010).

In the case of star gooseberry (*Phyllanthus acidus*, L.; Skeels) and carambola (*Averrhoa carambola*, L.), Sibounnavong et al. (2010) reported the shortening of the fermentation process from 30 days to 2 weeks when using a starter yeast of *S. cerevisiae*.

3.1 INOCULATION WITH MIXED YEAST CULTURES

There are reports on usage of **mixed culture of *Saccharomyces* and non-*Saccharomyces*** yeast in the fruit wines. Generally, with an increase in ethanol concentration, the amount of living cells of non-*Saccharomyces* decreases, and *Saccharomyces* strains start to dominate (Lee et al., 2012).

In the case of **cider** fermentation, mixed cultures of *Wickerhamomyces anomalus* and *S. cerevisiae* were used by different teams. Generally, strains of *W. anomalus* are considered to be rather weak fermenters. Sátora et al. (2014) has reported that the application of *W. anomalus* strains together with *S. cerevisiae* yeast as a mixed culture positively influenced the chemical composition and sensory features of produced apple wines. The obtained beverages were characterized by the highest amounts of ethyl alcohol, and generally, more residual sugars were left after fermentation. Similar results were reported by Ye et al. (2014). The mixed cultures produced statistically the same level of ethanol as *S. cerevisiae* monoculture and could produce more variety and higher amounts of acetate esters, ethyl esters, higher alcohols, aldehydes, and ketones. The sensory evaluation demonstrated that ciders obtained from cofermentation with *W. anomalus* gained higher scores than ciders fermented by pure *S. cerevisiae*.

For some fruits, like **lychee**, postharvest problems are often noticed due to rapid browning of its pericarp, which reduces its appeal to consumers and results in significant losses. The production of fruit wine from lychee serves as a way to prolong its shelf life and add economic value. Some non-*Saccharomyces* yeast starter, as *Torulaspora delbrueckii*, *Williopsis saturnus*, or *Kluyveromyces lactis*, were tested for lychee wines production. The fermentation performance of these non-*Saccharomyces* yeasts was significantly different. *T. delbrueckii* PRELUDE had the fastest rate of growth and high sugar consumption and produced the highest level of ethanol (7.6% v/v). *W. saturnus* NCYC22 used the lowest amount of sugars, but consumed the highest amount of nitrogen. However, strain *K. lactis* KL71 and strain *W. saturnus* NCYC22 overproduced ethyl acetate, while *T. delbrueckii* PRELUDE had a better ability to generate high levels of ethanol, isoamyl alcohol, 2-phenylethyl alcohol, ethyl octanoate, and ethyl decanoate and retained high OAVs of lychee aroma-character compounds cis-rose oxide, and linalool, which recommend the strain for further industrial use (Chen et al., 2015).

Lee et al. (2012) have studied the growth kinetics and fermentation performance in **papaya** juice of *W. saturnus* and *S. cerevisiae* added in different ratios (10:1, 1:1, and 1:10) with initial seven-day fermentation by *W. saturnus*, followed by *S. cerevisiae*. The growth kinetics of *W. saturnus* were similar at all ratios, but its maximum cell count decreased as the proportion of *S. cerevisiae* was increased. The findings suggest that ratio of yeasts is a critical factor for sequential fermentation of papaya wine by *W. saturnus* and *S. cerevisiae* as a strategy to modulate papaya wine flavor. For example, the persistence of both yeasts at 1:1 and 1:10 ratios led to formation of high levels of acetic acid.

A mixed culture of *S. cerevisiae* and *W. saturnus* var. *mrakii* have been used by Li et al. (2012) on different **mango** varieties. Both yeasts grew well in all juices and there was no early growth arrest of either yeast. While the changes of major volatiles were similar in all varieties, there were significant varietal differences in the volatile composition of the resultant mango wines; the volatiles, especially most of the terpenes of the juices, decreased drastically, and new volatiles such as β-citronellol were formed.

In the case of pineapple wine, a few studies were reported on the use of mixed starter culture of *S. cerevisiae*, *Sa. ludwigii*, and *Hanseniaspora* sp. The mixed cultures of *Sa. ludwigii* and *Hanseniaspora* sp. produced the highest alcohol content to 14.0% v/v in the final day of fermentation and their fermentation profiles were similar to those of the batch of single *S. cerevisiae* and mixed culture of *S. cerevisiae* and *Hanseniaspora* sp. (Chanprasartsuk et al., 2012).

During grape and fruit wine fermentation, the nitrogen content of juices (musts) is a key factor that regulates yeast metabolism and fermentation rate and ultimately influences the composition of aroma compounds in wine. Nitrogen deficiency of the must may cause sluggish or stuck fermentations in the winery. In general, an increase in the nitrogen content of musts resulted in reduced formation of aliphatic higher alcohols and hydrogen sulfide and in increased formation of esters in wine, although the effects vary with respect to the yeast strain, nitrogen status of musts, and experimental conditions. However, the **addition of large quantities of nitrogen** in the form of ammonium (typically as diammonium phosphate) to the must can result in microbiological instability and increased production of ethyl acetate and acetic acid as well as carcinogenic ethyl carbamate. Alternatively, the **addition of a mixture of amino acids** may enhance the utilization of the respective amino acids and possibly alter the aroma profile of the grape wine produced (Chen et al., 2014). Thus, it may be a better option to supplement must with specific amino acids instead of ammonium or a mixture of amino acids, to positively modulate wine aroma, which has hardly been explored in grape wine or fruit wine fermentation with *S. cerevisiae*.

The combination of selected cryophilic yeast capable of fermenting under conditions of nutrient deficiency and a lower temperature of fermentation (<22°C) could help to reduce anthocyanin degradation like in pomegranate wine, avoiding a dramatic impact on its color and preserving the beneficial effects of these specific bioactive compounds on human health (Berenguer et al., 2016).

3.2 INOCULATION WITH PURE YEAST CULTURES

A good strain of wine-yeast of *S. cerevisiae* is a prerequisite and needs to be procured for making quality wine. Yeast starter culture can be purchased and used slanted, tableted, or compressed. There are many commercial wine yeasts on the market, but for the fruit wines the list has been narrowed by the wine makers and some are listed in the following list. Most of them are recommended by Rivard (2009).

- *Lallemand 71B*: may be used widely, for most off-dry fruit wines. It brings out the fresh fruitiness in most berry fruit wines.
- *Lallemand BA11*: the strain is excellent on tropical fruit wines. It helps develop the aroma and can increase mouthfeel.
- *Lallemand EC1118* (Prise De Mousse): highly used yeast, especially with wines with low pH or starting the fermentation at low temperatures. It is also recommended to ferment the fruit wines to a very dry level, or to make a sparkling wine.
- *Lallemand ICV-K1*: it is recommended for tree fruit wines such as apple or peach; it ferments well, no matter what the pH or temperature of the must is, and it emphasizes the freshness of the product.
- *Lallemand R2*: it is recommended for very sweet fruit wines, cryo-extracted wines, and wines that need to ferment at low temperatures to retain the aromatic qualities.
- *Lallemand VIN13*: it is recommended for wines that need higher alcohol without fortification as it can ferment to almost 17% without any help. It gives good tropical notes and relatively clear flavors and can ferment under cooler temperatures.
- *Bio Springer CKS 102*: it is recommended for aromatic berry wines such as raspberry or delicate strawberry wines. Also works very well on tropical wines such as lychee, pineapple, and passion fruit.
- *Oenoferm Freddo*: it brings out the fresh fruit aroma.
- *Lesaffre UCD 595*: is recommended by Joshi et al. (2011) to obtain high-quality plum wines.
- *Montrachet*: is one of the more neutral yeasts available so that the fruit flavor is what comes through the most; a very good general use wine-yeast for most fruits; when in doubt use this yeast.
- *Pasture Red/RC-212*: works well with raspberry or blackberry.
- *UFLA FW 15*: recommended by Duarte et al. (2010); produces a raspberry wine with the best chemical and sensory quality.

It is important to highlight that trends in food fermentation are focused on the isolation of proper wild-type strains from traditional products, which can be used as starter cultures, with the aim of conducting industrial production processes without losing their unique flavor and product characteristics (Sánchez et al., 2010).

3.3 PROCEDURES FOR PREPARATION OF THE INOCULA FOR FRUIT WINES

If using active dried yeast, some advantages have to be taken into account from a technological point of view:

- avoid contamination during propagation;
- higher efficiency of conversion of sugars to ethanol can be achieved;
- storage of yeast in the dry state is also convenient;
- ease in controlling yeast count:
  - the number of viable yeast cells is approximately constant for a given weight. Thus, the wine maker can easily control the amount of yeast by their weight;
  - large amount of yeast should not be used, since it will accelerate the rate of fermentation, hence forming much heat or energy that will be harmful to the yeast cells;

- ideally, 10^6 cells/mL must be present in the mixture to have a normal fermentation rate. Lower cell count slows down fermentation, while higher cell count will encourage formation of by-products affecting the quality of wine;
- stability of dried yeast starter culture is important; unstable yeast strains have a tendency to lose viability after several transfers. The ability to produce in low concentration of one or more compounds that contribute to the flavor of the wine or spirit is also lost.

When choosing the yeast, it is better to perform smaller-scale bench trials before deciding on a strain to use and making larger quantities of wine with it. The easiest way to experiment with different strains is to split a particular batch of wine into two to four smaller batches and inoculate each with different yeast strains. The obtained wines are better to be evaluated using standard judging system (color, aroma, taste, and finish), used often in wine competitions (Rivard, 2009).

Yeast can be sprinkled directly on top of the must, hydrated separately, or added in a starter solution. The direct add into the must may be easiest, but it takes about two days to be certain the yeast was viable (Keller, 2010). Making a starter solution will help to show within a reasonable time if the yeast is viable before the must is ready for the yeast. Before adding the yeast to the must for fermentation, the yeast culture is activated in the juice/pulp intended for winemaking. The container with the juice is plugged and kept in a warm place (25–30°C) and the culture is ready after 24 h. The amount of active culture is added at a rate of 2–5% (v/v) to the must (Joshi et al., 2011). The yeast from the tablet is activated in sterilized water at optimum temperature of yeast growth; compressed yeast can be used directly in the wine fermentation. However, the recommendations of the yeast producer should be followed.

Two examples of yeast starters used in homemade fruit wines are given here.

- The first one is recommended by Keller (2010) (Box 2.1).
- Another good rehydration procedure was recommended by Monk (1986) (Box 2.2).

BOX 2.1 YEAST STARTERS USAGE FOR HOMEMADE FRUIT WINES BY KELLER (2010)

Dried yeast should be added directly to one cup (240 mL) of 100°F (38°C) tap or spring water in a quart (liter) jar. After stirring gently, they should be covered and allowed to hydrate for at least 30 min, then should be checked if yeast are viable; other 3–4 h should be left for best results. During this time, allow the starter and must to adapt within 10°F (5°C) of one another, and then add to the starter 1/4 cup (60 mL) of strained must or white grape juice (not concentrate). The starter should be re-covered and set in a warm place for about 4 h left and add to it another 1/4 cup (60 mL) of juice or strained must. Cover again and leave for 4 h, then add it to the must or add another 1/2 cup (120 mL) of juice or strained must to increase the yeast population.

BOX 2.2 YEAST STARTERS USAGE FOR HOMEMADE FRUIT WINES BY MONK (1986)

One can use water that is 5 to 10 times the weight of the yeast. For example, for 500 g of dry yeast, 3–5 L of water can be used for rehydration. Rehydrate yeast in warm water, 104° to 113°F (40–45°C). Slowly add yeast to water to obtain even hydration. Allow yeast to remain in warm water for 5–10 min before stirring (a longer duration will reduce yeast activity). The temperature difference between yeast starter and must should not be more than 18°F. To reduce the possibility of cold shock, gradually cool the starter, then add it to the must. The usual rate of inoculation is about 2 lbs/1000 gallons.

4. FACTORS AFFECTING THE YEAST GROWTH DURING FRUIT WINE FERMENTATION

A range of environmental factors influences the production of metabolites and survival of yeasts during fermentations. The main factors are temperature, pH, sugars concentration, acidity of fruit juice (substrate), and content in SO_2. In case of yeasts, temperature and tolerance of ethanol have an important influence on their performance. Initial low contents of nitrogen in grape or apple musts have been associated with the reduction of yeast development, and as a result, with slow and stuck fermentations (Vilanova et al., 2007).

4.1 TEMPERATURE

The **temperature** during fermentation has a direct effect on the yeast activity. The optimum for growth and multiplication of ordinary yeast is 30°C and usually inhibited at temperature higher than 32°C. Temperature increases the yeast growth, speed of enzyme action; also, cell sensitivity to the toxic effect of alcohol increases with temperature, due to increased membrane fluidity. This may partially explain the rapid decline in yeast viability at temperatures above 20°C during wine fermentation (Torija et al., 2002).

Concerning the cider production, it is considered that "normal" fermentation is conducted at 20–25°C and it lasted for 1–4 weeks (Waites et al., 2001).

4.2 ACIDITY

Optimum pH values are also very important for the yeast growth. Many fruits contain a high concentration of acids, so the correction of the acid content during the formulation stage is important (Heimonen et al., 2013). This is normally done by diluting the fruit juice or adding calcium carbonate to have a fermentation mixture with about pH 3.5–4.5. In these types of substrates, acid-tolerant strains must be used as fermenting yeast. Most of the yeasts grow very well between pH 4.5 and 6.5 and nearly all species are not able to grow in more acidic or alkaline media.

At present, it is difficult to make quality wines without the addition of SO_2 as the sulfur is an essential component for yeast growth. SO_2 is used as a preservative agent in wine because it has several functions, for example, preventing enzymatic oxidation of the must. It also inhibits the growth of undesirable microorganisms, which is used for wine preservation.

4.3 SUGAR CONCENTRATION

As a principle, when the sugar concentration is raised, the rate of fermentation and the maximum amount of ethanol produced is decreased. However, special type of yeast strains can be selected and conditioned to grow them at higher sugar concentration (>30% sugar) were adapted.

4.4 AERATION

Aeration is not required during the ethanol production phase, but its presence has many advantages during crushing and yeast proliferation in the production of essential sterols (ergosterol and lanosterol) and unsaturated fatty acids, such as linoleic acid and linolenic acids.

4.5 ETHANOL CONCENTRATION

Ethanol concentration is another important factor. The yield of ethanol varies with the strain of the yeast used, its adaptability to ethanol, and phase of growth when it is added to the fermenting mixture. The maximum ethanol concentration that can be produced by yeasts varies from 0% to about 19% v/v. Those strains that tolerate high concentrations of ethanol store less lipids and carbohydrates compared with less tolerant strains. The researchers have developed a molecular tool using microsatellite typing and RAPD PCR to generate DNA profiles to distinguish the ethanol tolerance in local wine-yeast strains.

Reddy and Reddy (2011) have proposed a systematic approach to study different factors' influence on selected starter yeast used in mango winemaking. It has been proven that temperature had an important effect on yeast growth and on the levels of volatile compounds. It was concluded that the temperature of 25°C, pH of 5, content of 100 ppm of SO_2 in must with initial oxygen were optimum for better quality of wine from mango fruits.

5. MALOLACTIC FERMENTATION IN FRUIT WINES

MLF is known in winemaking as a secondary fermentation. It leads to acid reduction, flavor modification, and also contributes to microbiological stability. MLF can occur spontaneously during or after the alcoholic fermentation (Lonvaud-Funel, 1999; Dharmadhikari, 2002).

The major wine yeasts strains of *Saccharomyces* spp. are unable to degrade malic acid effectively during alcoholic fermentation. Naturally, **LAB** are involved in MLF. Basically, different species of bacteria, like *O. oeni*, *L. plantarum*, or *L. casei* and *Pediococcus* convert the malic acid that is available in a wine into lactic acid and CO_2 (Fig. 2.2). LAB are Gram-positive, catalase-negative, nonmotile, nonspore forming, rod-, and coccus shaped.

There are 17 different species of *Lactobacillus* associated with winemaking either being connected to the beginning of the alcoholic fermentation or the MLF and wine (Du Toit et al., 2011). *Lactobacillus* associated with wine is mainly facultative or obligatory heterofermentative and can adapt to fruit wine conditions such as high ethanol levels, low pH and temperatures, and presence of sulfur dioxide. A special characteristic associated with wine lactobacilli is the production of bacteriocins, especially plantaricins, which would enable them to combat LAB spoilage.

From a genetic point of view, both *Oenococcus* and *Lactobacillus* contains the malolactic enzyme encoding gene, but sequence homology shows that it clusters separately. *Lactobacillus* also possesses

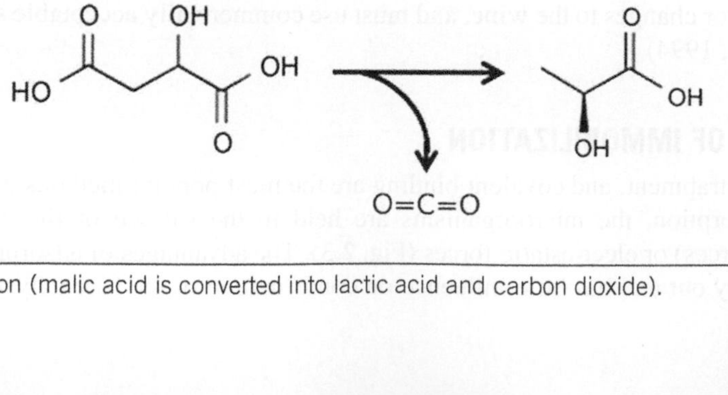

FIGURE 2.2

Malolactic fermentation (malic acid is converted into lactic acid and carbon dioxide).

more enzyme encoding genes compared to *O. oeni*, important for the production of wine aroma compounds such as glycosidase, protease, esterase, phenolic acid decarboxylase and citrate lyase (Du Toit et al., 2011).

MLF is best to be done right after the wine-yeast fermentation has completed. LAB have the ability to consume sugars just like yeast. But instead of turning these sugars into alcohol, they convert these sugars into volatile acids such as acetic acid, which leads to vinegar. That's one of the reasons for proffering the MLF after the end of alcoholic fermentation, when the sugar content is limited. Another reason is the fact that lactic bacteria are more sensitive to the added sulfites than the yeast. Usually, if occurring naturally, it is expected to be done between three and four months after the end of alcoholic fermentation, while if a domesticated culture is added, it may take between 3 and 6 weeks.

Deeper information on the complex microbial ecology of MLF in spontaneous Spanish cider production has been provided by Sánchez et al. (2010). The work it concluded that *Lactobacillus collinoides*, *O. oeni*, and *P. parvulus* were the dominant species involved in the dynamics of the MLF process during spontaneous cider production. However, under the same conditions only *L. collinoides* and *O. oeni* could carry out the MLF. Furthermore, the author's team has isolated strains of *O. oeni* as starters for cider wines (Sánchez et al., 2012).

The interactions between *S. cerevisiae* strains individually and in combination with *O. oeni* (sequentially and simultaneously inoculated) during the process of malolactic acid fermentation in mango wine was studied by Varakumar et al. (2013).

6. USE OF IMMOBILIZED BIOCATALYSTS IN WINEMAKING

The immobilized biocatalyst is defined as "the microbial cells/enzymes physically confined or localized in a certain defined region of space with retention of their catalytic activities, and which can be used repeatedly and continuously" (Chibata, 1978). The use of immobilized cells offers several advantages: improved productivity of fermentations by the high cell density; adaptation to continuous processes that can be better optimized and controlled; simplified systems for removing microbial cells from batch processes; greater tolerance to inhibitory substances; smaller-scale fermentation facilities with reduced capital and running costs; and possibilities of using a variety of microbial strains including genetically modified organisms (GMOs). Some potential disadvantages must also be considered: cell overgrowth, which increases turbidity of the fermented beverage; mechanical stability of the matrix used to immobilize microbial cells; and loss of activity on prolonged operation. To be attractive in wine production, the method must be cheap, easily performed in an industrial situation, not liable to cause oxidation of the wine, robust, not susceptible to contamination, able to impart correct flavor changes to the wine, and must use commercially acceptable supports and organisms (Divies et al., 1994).

6.1 METHODS OF IMMOBILIZATION

Adsorption, gel entrapment, and covalent binding are the most popular methods of biocatalyst immobilization. In adsorption, the microorganisms are held to the surface of the carriers by physical (Van-der-Waals forces) or electrostatic forces (Fig. 2.3). The advantages of adsorption method are that it is simple to carry out and has little influence on conformation of the biocatalyst. However, a major

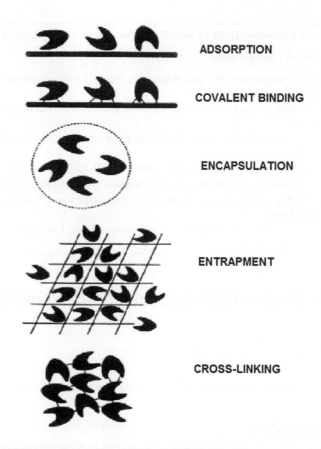

ADSORPTION

COVALENT BINDING

ENCAPSULATION

ENTRAPMENT

CROSS-LINKING

FIGURE 2.3

Illustration of the methods for biocatalyst immobilization.

disadvantage of this technique is the relative weakness of the adsorptive binding forces. The physical entrapment method is extremely popular for the immobilization of whole cells. The major advantages of the entrapment technique is the simplicity by which spherical particles can be obtained by dripping a polymer-cell suspension into a medium containing precipitate-forming counter ions or through thermal polymerization. The major limitation of this technique is the possible slow leakage of cells during continuous long-term operation. However, improvements can be made by using suitable cross-linking procedures. Mass-transfer limitations are a significant drawback for many immobilized cell techniques. Fundamental considerations in selecting a support and method of immobilization are presented in Table 2.3 (Kosseva and Kennedy, 1997).

Among the supports and methods for cell immobilization, the most useful for wine manufacturing is entrapment using natural polymers like alginate, agarose, carrageenan, chitosan, and pectin. Such natural gelling polysaccharides represent an emerging group due to their advantage of being nontoxic, biocompatible, cheap, and offering a versatile technique for biocatalyst preparation (Kosseva et al., 2009, 2011).

Table 2.3 Fundamental Considerations in Selecting a Support and Method of Immobilization

Property	Points of Consideration
Physical	Strength, available surface area, shape/form, degree of porosity, permeability, noncompression of particles, and so on.
Chemical	Hydrophilicity, inertness, available functional groups for modification, and regeneration of support.
Stability	Storage, residual cell activity, cell productivity, maintenance of cell viability, and mechanical stability.
Resistance	Bacterial/fungal attack, disruption by chemicals, pH, temperature, organic solvent, and so on.
Safety	Biocompatibility, toxicity of component reagents, and health and safety for process workers.
Economics	Availability and cost of support, chemicals, special equipment, reagents, technical skills required, and so on.
Reaction	Flow rate, cell loading and catalytic productivity, reaction kinetics, diffusion limitations, mass transfer, and so on.

6.2 APPLICATIONS OF IMMOBILIZED BIOCATALYST TECHNOLOGY IN ENOLOGY

- Alcoholic fermentation;
- MLF;
- "prise de mousse" or Champagne method;
- sparkling drinks;
- enzymatic processing of musts and wines with pectolytic and proteolytic enzymes.

6.2.1 Immobilized Yeast Cells

During the last 25 years, application of immobilized yeast cells in wine production have been explored in a view to reduce labor requirements, to simplify time-consuming procedures, and thereby to reduce costs. In a general manner, the alcoholic fermentation by yeasts immobilized in alginate gel beads is accelerated, which has been related to changes in cell composition and function. Using the same strain of *S. cerevisiae* and 2-mm diameter alginate spheres with a population of 2.109 cells/mL of gel, Divies and Cachon (2005) measured specific ethanol production at 0.6 g/g of cells per hour in the immobilized form and 0.3 g/(g h) for free cells. If smaller spheres are used, even better fermentation activities are possible. In addition, entrapment protects the yeasts against ethanol toxicity as well as against heavy metals, phenols, acidity, and extreme temperatures.

Oliveira et al. (2011) prepared a wine from cagaita (*Eugenia dysenterica*, DC) fruit pulp and carried out a comparative study to evaluate the fermentations conducted with free and Ca-alginate immobilized cells of *S. cerevisiae* (UFLA CA11and CAT-1) in triplicate, at 22°C for 336 h. According to the sensory evaluation, the fruit wine acceptability was greater than 70% for color, flavor, and taste for all cagaita beverages.

Varakumar et al. (2012) proved that yeast-mango-peel immobilized biocatalyst can be a good and effective system for mango wine fermentation at both low and room temperatures, as the wines produced by this procedure had a potentially better aroma and taste than those obtained by free cell fermentation. The biocatalyst is economical, food grade, and does not need special pretreatment before use. Mango peels, which otherwise may pollute the environment, can be beneficially used as an alternative cell immobilization support.

While prospects for the immobilized cells technology are encouraging, further research is needed to optimize reaction variables, to improve the long-term stability of the reactors, and to understand more about secondary metabolite production by yeasts under these conditions.

6.2.2 Control of Malolactic Fermentation by Immobilized Lactic Acid Bacteria

One of the most difficult steps to control in winemaking is MLF, which normally occurs after completion of the alcoholic fermentation. Desired and controlled MLF via the use of selected immobilized LAB can be applied due to the following reasons:

1. Natural MLF takes a long time, and growth limitations of lactic acid microflora affect and depend on the physical–chemical properties and nutritional composition of wine; for example, fatty acids and ethanol may inhibit LAB growth. Therefore, immobilization techniques aim to increase the tolerance of the MLF bacteria.
2. The development of desired flavor by using selected cultures of bacteria.
3. The acceleration of MLF by higher cell densities achieved by immobilization techniques.
4. The feasibility of application and commercialization of the process by lyophilized and immobilized cultures.
5. The reuse of cell for MLF and the application of continuous processes (Kosseva et al., 1998; Kourkoutas et al., 2004).

We conducted a comparative study of MLF in Bulgarian Chardonnay wine by free and encapsulated *L. casei* cells as well as lyophilized cultures of *O. oeni* (Kosseva and Kennedy, 2004). We selected calcium pectate gel as the encapsulation material to accomplish MLF in wine. The application of pectate gel in winemaking holds good prospects due to its good stability at low pH values. Also, pectin is among the polysaccharides permitted by the European legislation as a potential encapsulation "carrier/support" for food additive systems. Encapsulation in pectate gel is a safe, mild, cheap, and versatile technique, which may be used for the preparation of biocatalyst for MLF in wine. Immobilized bacteria of *L. casei* degraded 30% of malic acid in white wine, deacidifying it from pH 3.15 to 3.40, whereas the lyophilized culture of *O. oeni* degraded 48% of malic acid, deacidifying from pH 3.15 to 3.60. The degree of conversion of malic acid in wine was twice as high as that obtained by free cells. The operational stability of calcium pectate gel capsules was 6 months. It has been proven that the encapsulated biocatalyst increases the rate of fermentation and makes the fermentation take place even at high ethanol concentrations (12–13% v/v). The immobilized cells were also not inhibited by SO_2.

O. oeni immobilized onto various natural supports like corn cobs, grape skins, and grape stems was successfully used to conduct consecutive batches of MLF for a total period of around 5 months with the possibility of storage of the biocatalyst for 30 days in wine at 25°C (Genisheva et al., 2013).

By analogy to LAB, the strain of *Schizosaccharomyces pombe* was entrapped in the Ca-alginate gel so that fermentation was carried out under the continuous conditions (Ciani, 2008).

To sum up, the introduction of immobilized bacterial cells for direct inoculation into wine has improved the control of MLF, reducing the possibility of wine spoilage by the other bacteria. The wine makers can now pay more attention to the control of the flavor modification induced by immobilized inocula.

6.2.3 Sparkling Wines

In the case of sparkling wine production by *methode champenoise,* it involves secondary fermentation, producing carbon dioxide and giving a sparkle to the wine, while the prolonged contact of the wine with yeast gives it a characteristic flavor. However, the removal of lees from the bottled wines is the most laborious process. The application of immobilized yeast in the production of sparkling wine has reduced the time and labor involved in the process.

Diviès and Deschamps (1986) used a pressurized batch reactor to produce cider, sparkling wine, or semisparkling grape juice using yeasts immobilized in alginate gel beads. For example, 1,500,000 bottles of sparkling wine could be produced per year by a reactor ($1 m^3$) operating for 220 days. In a continuous system (Fumi et al., 1989), a sparkling wine with composition and sensorial properties comparable with a product obtained conventionally was produced, but with a greatly enhanced productivity.

Sz. pombe has also been employed for simultaneous biological deacidification and secondary fermentation during maturation for production of wine from high malate grape juice and sparkling wines production. It has also been found effective in deacidification of plum must to produce wine. Bhardwaj and Joshi (2009) demonstrated for plum sparkling wine that addition of 0.5% yeast extract to the base wine at the beginning of secondary fermentation improved the foam stability, foam expansion, anthocyanin, total esters, phenols, and soluble proteins.

6.2.4 Immobilized Cells in the Production of Cider

There are limited reports on the usage of immobilized cells in cider-making. The major part of the research concerning the MLF of cider, and therefore immobilization techniques, were considered to increase productivity, accelerate maturation, and improve cider quality and stability. Calcium alginate matrix was used to coimmobilize *S. bayanus* and *Leuconostoc oenos* in one integrated biocatalytic system aiming to perform simultaneously alcoholic and MLF of apple juice and produce cider in a continuous packed-bed bioreactor (Nedovic et al., 2000). Compared with traditional cider-making, they achieved rapid alcoholic and MLF and improved flavor with a reduction of higher alcohols, isoamylacetate, and diacetyl formation. These changes were attributed to altered metabolism of immobilized cells. Soft and dry cider was obtained by adjusting feeding flow rates. However, the profile of volatile compounds in the final product was modified relatively to batch process, especially for higher alcohols, isoamylacetate, and diacetyl. This modification is due to different physiology states of yeast in two processes. Nevertheless, the taste of cider was acceptable for the consumer market (Nedovic et al., 2000).

Likewise, cells of *O. oeni,* isolated from the cellar of a cider industry, were immobilized in alginate beads and used as starter culture for MLF of cider (Herrero et al., 2001). The immobilized system had higher alcohol productivity, and the rates of malic acid consumption were similar to those obtained using free cells while no significant differences of volatile by-products were observed. The immobilized cells produced lower acetic acid and ethyl acetate. The use of immobilized malolactic bacteria for secondary fermentation of cider or coimmobilized bacteria and yeasts for both main and secondary fermentation has proven in all cases to be efficient as far as fermentation rates are concerned. However, further research for application to full-scale production of ciders with improved and controlled flavor profiles is needed.

7. SPOILAGE OF FRUIT WINES

During fruit wine production the spoilage may occur in different steps, from must to alcoholic and MLF, and especially in the postfermentation step.

7.1 MUST SPOILAGE

Different microorganisms may lead to the fruit must spoilage: wild yeasts and lactic acid and acetic acid bacteria. To prevent oxidation and to avoid **must spoilage**, before the fermentation, potassium or sodium metabisulfite should be added. Sánchez (1979) recommended 5 mL of 10% solution for one gallon of must. Dharmadhikari (2004) recommended the addition of sulfur dioxide to must in the range of 50–75 ppm. It is already usual to add SO_2 into the must in the concentration of 100–150 ppm as potassium metabisulfite (Joshi et al., 2011a). There is evidence that the addition of sodium benzoate gave better quality wine than potassium metabisulfite in the plum wine.

7.2 SPOILAGE DURING MALOLACTIC FERMENTATION

The risks involved with spontaneous MLF include the following: undesirable LAB can produce spoilage components (mousiness, bitterness, volatile acidity); it has been proven that spontaneous MLF can lead to significantly higher concentrations of biogenic amines being produced. Biogenic amines such as histamine have health implications and are subjected to more and more strict regulations. The risk of antagonistic interactions between yeast and bacteria increases due to inhibiting products produced by the yeast (alcohol, medium chain fatty acids); uncontrolled timing of the process; and longer MLF duration that results in a longer SO_2-free period with no protection against mold spoilage.

The yeasts of the genus *Schizosaccharomyces* have traditionally been described as wine-spoilage organisms owing to their production of compounds with negative sensorial impacts, such as acetaldehyde, H_2S, and volatile acids.

In the fruit juice where malic acid content can be high, the possible use of non-*Saccharomyces* yeasts, such as *Schizosaccharomyces* spp., to reduce malic acid concentrations by a maloalcoholic fermentation (MAF) is stirring much interest. The industrial use of *Schizosaccharomyces* has been described in the fermentation of cane sugar in rum-making (Pech et al., 1984; Fahrasmaneet al., 1988), the production of palm wine (Christopher and Theivendirarajah, 1988; Sanni and Loenner, 1993) and cocoa fermentation (Ravelomanana et al., 1984). Applications with *Sz. pombe* are fewer than with LAB, mainly because *Sz. pombe* produces off-flavors during MAF. To reduce the production of off-flavors during the long MAF period, many methods, such as immobilized cell fermentation and high cell density fermentation, have been investigated (Ding et al., 2015).

The metabolic properties of the *Sz. pombe* breakdown of malic acid produce the pyruvic acid and also breakdown of ethyl carbamate precursors which is of great interest in modern winemaking. However, its major drawback is its strong acetic acid production at least for the unselected strains commonly used in wine research. The selection of *Schizosaccharomyces* strains with low-strain production of acetic acid could bring a new tool for unbalanced musts. Other fermentation modalities such as mixed and sequential fermentation between *Saccharomyces* and *Schizosaccharomyces* to minimize the levels of acetic acid could be used (Benito et al., 2012).

7.3 POSTFERMENTATIVE SPOILAGE

In the **postfermentative** phase, spoilage of fruit wines may occur due to the presence of yeast and bacteria.

In the case of **yeast**, *Saccharomyces* is regarded as a spoilage organism only if occasionally occurring in wines with residual sugars (semisweet wines) causing refermentation (du Toit and Pretorius, 2000).

Another yeast which may lead to refermentation is *Zygosaccharomyces bailii*; it occurs usually in sparkling wines (with high amounts of CO_2) and will form turbidity and sediment, as well as high levels of acetic acid and esters.

Pellicular yeasts from genera *Candida, Metschnikowia*, and *Pichia*, can form a film layer (pellicle) on top of stored wine, mainly in the presence of oxygen (in partially filled vessels); aside the "cosmetic" problem, these yeast may also be detrimental to the quality of wine, leading to an oxidized flavor due to the production of acetaldehyde (du Toit and Pretorius, 2000).

As no control over LAB is carried out after MLF, a large number of lactic acid bacteria is found and these bacteria are a potential source of different fruit wines alterations. Conventionally, LAB are identified by their morphological and biochemical characteristics. However, results obtained are often ambiguous and the methods involved are time-consuming. Other methods have been applied with success to the identification of wine-associated LAB, including protein fingerprinting, peptidoglycans of the cell walls, and lactate dehydrogenase enzyme patterns. Molecular techniques have been also employed to identify wine LAB by DNA level, and the results obtained are less controversial than for other methods (du Toit and Pretorius, 2000). LAB can increase the acid content of wine by producing lactic acid and acetic acid. The D-lactic acid is associated with spoilage, as the L-lactic acid is produced during MLF.

One of the most widespread is the production of excess of acetic acid by using residual sugars and other compounds such as glycerol or lactic acid. Acetic acid is always present in fruit wines, but an excess leads to the spoilage of the product. Other alterations frequently reported are ropiness, which affect most of the ciders (del Campo et al., 2008).

The nature and the extent of wine spoilage by LAB depends on several factors such as the type of bacteria, composition of the wine, and vinification practices. Depending on the substrate used by the LAB, different types of spoilages have been reported.

Sugars' fermentation. LAB metabolize the residual sugars such as glucose and fructose, and produce lactic acid and acetic acid. The resulting wine acquires a sour vinegar-like aroma due to high volatile acidity levels. It occurs mainly in must with stuck fermentation or wines with higher residual sugars (sweet wines) and the wine cannot be consumed. The positive side of the acetic acid bacteria (AAB) presence is the possibility to produce vinegar of fruits, which are appreciated on the market.

In the case of dry wines, a less serious form of lactic spoilage can occur. In these wines the LAB utilizes pentose sugars, trace amounts of glucose and fructose, and produces lactic and acetic acid as a byproduct. Formation of these acids increases the titratable acidity and lowers the pH. The decrease in pH restricts the growth of those organisms.

Degradation of glycerol. Breakdown of glycerol by LAB results in the formation of lactic acid, acetic acid, and acrolein. The wine smells acetic, butyric, and acquires a bitter taste due to acrolein.

Organic acid fermentation. In this kind of spoilage, the LAB ferments tartaric acid and forms lactic acid, acetic acid, and carbon dioxide. Degradation of tartaric acid occurs especially in wines with low acidity and high pH (pH above 3.5). The titratable acidity is further reduced and the wine acquires an acetic aroma and disagreeable taste. Citric acid degradation has been positively correlated with the formation of diacetyl and acetone as well as acetic acid.

Ropiness. Certain species of *Leuconostoc* and *Pediococcus* have been found to produce dextran slime or mucilaginous substances in wine. The wine appears oily and may not necessarily have high volatile acidity.

7.4 FRUIT WINE SPOILAGE PREVENTION

Removal or inhibition of undesirable bacteria is more challenging in fruits wine than in the grape wines, especially in the case of products with low alcohol content. As an initial barrier, the high ethanol concentrations (up to 16% v/v) and high wine acidity (pH as 2.9) can inhibit development of bacterial populations; however, in wines with lower ethanol concentrations and low acidity (above pH 3.7), it can be challenging to arrest bacterial growth (Bartowsky, 2009). Storage of wine at temperatures below 15°C might assist with minimizing the ability of bacteria to proliferate in wine.

The occurrences of most spoilage scenarios are uncommon and can be avoided with correct hygiene management during the vinification and maturation process. Sulfiting the fruit wine after finishing alcoholic and MLF is a solution which is sometimes performed to eliminate the bacteria that cause undesirable alterations. However, in the elaboration of natural fruit wines, this practice is rarely used (del Campo et al., 2008). It is preferred to keep the addition of chemicals to a minimum to maintain the organoleptic qualities of the final product (du Toit and Pretorius, 2000).

Sulfurous anhydride or sulfur dioxide (SO_2) has numerous properties as a preservative in winemaking; these include its antioxidant and selective antimicrobial effects, especially against LAB. Nevertheless, and due to increasing health concerns, consumer preference, possible organoleptic alterations in the final product, and a tighter legislation regarding preservatives, there is a worldwide trend to reduce SO_2 levels in wine.

While the most employed solution against fruit wine spoilage is the use of sulfite, some authors (Garcia-Ruiz et al., 2012) reported the possibility to inhibit the lactic and acetic bacteria by the use of different phenolic extracts, like eucalyptus extract (2 g/L) which significantly delayed the progress of both inoculated and spontaneous MLF; however, the addition of the phenolic extract as not as effective as $K_2S_2O_5$ (30 mg/L).

8. MICROBIOLOGICAL ANALYSIS IN FRUIT WINEMAKING

Microbiological analysis is aimed at following alcoholic fermentation, detecting microbiological infections, and allowing the detection of any abnormality, not only in the finished product but also during the different phases of manufacturing. All experiments must be carried out under normal microbiological aseptic conditions.

8.1 MONITORING YEAST POPULATION DEVELOPMENT DURING ALCOHOLIC FERMENTATION

The alcoholic fermentation is usually followed by counting the yeast cells/mL or by measuring the turbidity (optical density) by the aid of a spectrophotometer. Finally, the growth curve against the time will be obtained.

Counting cells is a technique to determine the cell concentration in a liquid. The yeast counting can be performed by the aid of a **counting chamber**; there are several counting chambers developed by different manufacturers (Thomas, Neubauer, Mallasez, etc.). A counting chamber is a special glass slide (the size of a normal microscope slide) with engraved squares and a cover slip. Those chambers are usually employed in the clinical analysis for the counting of the blood cells (hemocytometers).

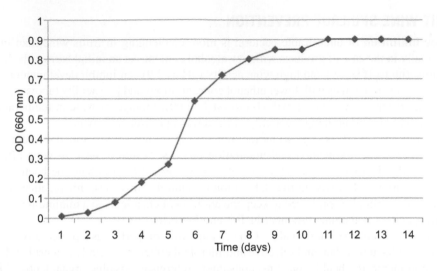

FIGURE 2.4

Typical yeast growth curve during alcoholic fermentation of fruit wines (measured by optical density).

Methylene blue staining allows for a simple way to assess the health of a yeast culture. In theory, dead cells will stain blue while living cells remain colorless. The results will be given in cells of CFU/mL. This method can be time-consuming, and on an industrial level, it is simpler to follow yeast growth by the aid of a spectrophotometer; the turbidity of the suspensions is measured at 600–660 nm and the growth curve will be plotted as optical density against the time (Fig 2.4).

8.2 DETECTING SPOILING MICROORGANISMS DURING FRUIT WINEMAKING

During the fermentation and postfermentation microscopic techniques may be applied for a first glance differentiation between the presence of yeast and bacteria. When differentiating the bacteria, Gram stain may apply (lactic bacteria are Gram positive, while acetic bacteria are Gram negative).

The identification of the spoiling microorganisms is usually performed by the classical method on specific media, by rapid tests, or by molecular tools.

The main tests used to obtain a rough classification of wine-spoilage bacteria include:

- microscopic examination of the cell morphology of cultured colonies—rods or cocci; presence or absence of endospores;
- Gram behavior by means of a potassium hydroxide test;
- oxidase test;
- catalase test;
- aerobic or anaerobic growth;
- formation of acetic acid from ethanol.

Yeasts can be detected under the microscope by their size and shape. In fresh cultures still growing logarithmically, budding is an absolutely reliable indication (unless one is dealing with a representative of the genus *Schizosaccharomyces*).

Specific selective **media** have been designed for LAB. One of the most common medium used to detect acid-tolerant bacteria is orange serum—a medium used to grow primarily the acid-tolerant bacteria (e.g., LAB; Sartorius group). To detect this group, the culture medium should be incubated under anaerobic to microaerophilic conditions because many of the LAB are inhibited by oxygen. AAB can also be grown with orange serum, but should be aerobically incubated. It is also recommended to add 5–8% of ethanol to promote their growth and suppress molds. Yeasts and molds can also grow on media containing orange serum, but they are suppressed if incubated under anaerobic conditions. The *jus de tomate* (tomato juice) culture medium was specifically developed to detect *O. oeni*, it must be incubated anaerobically for the bacterium to grow well.

For liquid enrichment, sterile fruit juice can be used as medium, in which spoilage microorganisms such as yeasts, bacteria, and fungi show growth rates equal to those in the finished product.

Rapid tests, for performing microbial analyses in minutes or a few hours, are available on the market. In this category the use of biosensors, epifluorescent microscopy, and immunochemical tests may be included.

A biosensor transforms a microorganism's metabolic reaction into a signal that usually can be measured physically. Because of the small size, single microbes only produce minute quantities of metabolites. This is why a larger number of microbes is needed to achieve a measurable result. Preincubation for one or more days depending on the type of microorganisms is recommended. The bioluminescence method, which measures the amount of ATP, is restricted to special applications, such as monitoring the fermentation process, or the hygiene inspection of equipment or staff before they start work.

Molecular tools rely mainly on PCR techniques. PCR does not involve propagating the microorganisms, but rather replicating their genetic material, DNA. This process is much faster than growing the microorganisms themselves. A short strand of specific DNA is added, i.e., a targeted search is carried out for a specific microorganism. PCR enables specific target microorganisms, such as product contaminants, to be detected in a sample. Gene probes are complementary to typical sections of the DNA of a specific type of microorganism, its genetic fingerprint, so to speak. Microdroplet-based multiplex PCR on a chip can be used to detect foodborne bacteria producing biogenic amines with great success.

Traditional methods for microbial strain identification have been mostly supplanted in favor of ribosomal RNA-based methods for speciation of cultured yeast and bacterial populations in wine. Moreover, culture-independent molecular methods now allow for more rapid profiling of complex populations, or quantification of targeted species, thereby enhancing the information available to the producer (Mills et al., 2008). A more recent technique that has found wide application in wine fermentations is QPCR. In QPCR the logarithmic amplification of a DNA target sequence is linked to the fluorescence of a reporter molecule; a common reporter used for detection of wine-related organisms is a green fluorescent dye. The fluorescence is read after each round of DNA amplification and may either be compared to an external standard curve known as absolute quantification, or it may be compared to an internal or external control sample in a method known as relative quantification.

Although the high-acid-tolerant strains of *Escherichia coli* can survive for long periods in fruit juice they are extremely sensitive to alcohol and die within a few hours in fermenting. Despite the alcoholic content or low pH in the product, some authors have reported the presence of pathogenic bacteria (coliforms, spore-forming *Bacillus cereus*) in commercially fruit wines (Jeon et al., 2015). This fact should be taken into account when developing manufacturing systems and methods to prolong the shelf life of high-quality **fermented alcoholic beverage (FAB)** products. New strategic quality management plans for various FABs are needed.

8.3 MYCOTOXINS DETECTION

Mycotoxins are secondary metabolites of filamentous fungi and therefore occur naturally in food and beverages. They represent a very large group of different substances produced by different mycotoxigenic species. The mycotoxins are of high concern for human health due to their properties to induce severe acute and chronic toxicity at low-dose levels.

Molds can infect agricultural crops during crop growth, harvest, storage, or processing. The growth of fungi is not necessarily associated with the formation of mycotoxins, and because of the stability of mycotoxins, they may be present in food when fungi are no longer present. Fruit contains natural acids (citric, malic, and tartaric acids) that give the fruits tartness and slow down bacterial spoilage by lowering the pH. The pH of fruit varies from <2.5 to 5.0 and these values are tolerable for many fungal species but less for bacteria (Fernandez-Cruz et al., 2010).

The mycotoxins most commonly found in fruits and their processed products are aflatoxins, ochratoxin A, patulin, and *Alternaria* toxins.

Patulin (PAT) is a toxic metabolite produced by several species of *Penicillium* and *Aspergillus* and it has been detected during apple processing; also it was found in some other fruits like pear, apricot, and peach. The amount of PAT in the juices can be reduced after removal of the rotten or damaged fruit but cannot be eliminated completely as the mycotoxin diffuses into the healthy parts of the fruit.

Aflatoxin (AFL) are a group of closely related metabolites produced by *Aspergillus flavus* and *Aspergillus parasiticus*. Natural aflatoxin contamination has been reported in oranges and apple juices.

Ochratoxin (OTA) was originally isolated from *Aspergillus ochraceus*. Other *Aspergillus* species are also capable of producing OTA. *Penicillium verrucosum* is the best known *Penicillium* species that is able to produce OTA. OTA has been detected in cherry and strawberries and their associated juices. In a large screening, Duarte et al. (2009a) have reported that apple and orange juices were free of OTA, and black currant juices presented levels just above the limit of detection.

The fact that most mycotoxins are toxic at very low concentrations makes it necessary to have sensitive and reliable methods for their detection. A number of different analytical methods have been applied to mycotoxin analysis due to their varied structures. These include widely applicable liquid chromatography (LC) methods with UV or fluorimetric detection (FLD), which are extensively used in research and for legal enforcement of food safety legislation and regulations in international agricultural trade. Other chromatographic methods, such as thin layer chromatography and gas chromatography, are also employed for the determination of mycotoxins, whereas advances in analytical instrumentation have highlighted the potential of LC–mass spectrometric methods, especially for multitoxin determination and for confirmation purposes (Fernandez-Cruz et al., 2010).

Micotoxins can be also detected by the use of RT PCR, which is capable of high throughput analysis, is accurate, and has good sensitivity, it could be used as a centralized control (Heimonen et al., 2013).

9. GENETICALLY MODIFIED MICROORGANISMS FOR FRUIT WINEMAKING

Advances have been made in genetic engineering–enabled gene manipulation and enhanced technologies in applied agricultural and food biotechnology research. These advances in DNA technology have made it possible to clone genes with traits of interest to the wine maker and to manipulate organisms' cells to overproduce the desired protein.

9.1 METHODS FOR WINE-YEAST DEVELOPMENT

Several methods for wine-yeast development have been used (Pretorius and Bauer, 2002):

- Selection of the variants: it is a simple and direct method, based on selection of natural genetic variants found within populations of all wine-yeast strains as a result of spontaneous mutations.
- Mutagenesis: imply the application of mutagens to increase the frequency of mutations in yeast population; the effect of ploidy reduces wine efficiency in diploid or polyploid strains; haploid strains are preferred when inducing mutations.
- Hybridization: intraspecies hybridization entails sporulating diploids, recovering individual haploid ascospores, and mating of haploid cells of opposite mating types to produce a new heterozygous diploid. When sexual reproduction can't be applied special hybridization may apply: spore-cell mating, rare mating, sheroplast fusion, and cytoduction.
- Transformation: represents the use of recombinant DNA technology and genetic engineering to change specific properties of a yeast strain; the resulting strain is a GMO and should apply strict legal regulations.

9.2 GENETICALLY ENGINEERED WINE-YEAST STRAINS

During the past two decades recombinant DNA technology has been used to introduce various desired strains from other organisms into wine strains of *S. cerevisiae*; these would include the genes encoding for Killer K1 toxin to combat wild yeasts; pectinase activity to increase the filterability of wines; glucanase activity to release the bound terpenes and increase the fruity aroma in wine; glycerol phosphatase to increase glycerol levels; lactate dehydrogenase to promote mixed culture fermentation and acidification; malolactic enzyme to promote MLF; and malic enzyme to promote maloethanolic fermentation (Pretorius, 2001; Nigam, 2011).

In the case of grape winemaking in the last two decades several genetically engineered yeast strains have been constructed, and very interesting improvements in the winemaking process or the quality of the wine obtained have been reported, including improved primary and secondary flavors, malic acid decarboxylation by yeast, increased resveratrol, lactic acid, or glycerol contents (Byaruagaba-Bazirake et al., 2013). These improvements can be easily adapted to fruit wine microorganisms.

An efficient technology called **genome shuffling** has been used to improve the industrially important microbial phenotypes. Genome shuffling is the recombination between multiple parents of each generation, and several rounds of genome fusion are carried out. Multiple exchanges and multiple-gene recombinations can occur rapidly and efficiently, resulting in the generation of a large number of mutant strains that can then be tested for the desired phenotypes. Genome shuffling is much easier to carry out using inactivated parental protoplasts than noninactivated ones (Wang et al., 2007), because the parental protoplasts cannot survive without recombination. The advantage of this technique is that genetic breeding can be performed on the tested microorganisms without knowing their genetic background, making it a highly effective method (Ding et al., 2015).

In the case of high level acidity wines, have been proposed the use of shuffled fusants of *Sz. pombe*. The starting mutant population was generated by UV treatment and the mutants with higher deacidification activity were selected and subjected to recursive protoplast fusion (Ding et al., 2015). After three rounds of genome shuffling a strong, the best performing fusant increased (GS3-1) its deacidification activity by 225.2% as compared to that of original strain. This mutant can be recommended in the case of other fruits juices of very high acidity, like cassis or different gooseberries.

S. cerevisiae lacks the ability to produce extracellular depolymerizing enzymes that can efficiently liberate fermentable sugars from abundant, polysaccharide-rich substrates, and this limits *S. cerevisiae* to a narrow range of carbohydrates. Industrial enzyme preparations are widely used to supplement these polysaccharide-degrading activities. Most commercial pectinase and glucanase preparations are derived from *Aspergillus* and *Trichoderma*, respectively.

It has been proven that commercial wine-yeast strains contains the genes responsible for polygalacturonase production, but the enzyme production is inhibited in the presence of glucose (Radoi et al., 2005b). By chemical mutagenesis mutants capable to produce polygalacturonase in high glucose concentrations were obtained, and this led to an improved clarification process in the wines (Radoi et al., 2005a).

Byaruagaba-Bazirake et al. (2013) used recombinant yeast strains able to degrade different polysaccharides (glucans, xylans, pectins, and starch) instead of commercial enzyme preparations for the process of banana winemaking to improve wine processing and wine quality. The banana cultivar Bagoya pulp treated with genetically modified yeast secreting glucanase and xylanase showed a 5.5% v/w increase in wine yield over that of the control yeast (nonrecombinant). The banana cultivar Kayinja pulp which was treated with genetically modified yeast secreting pectinase enzyme showed a 35% increase in wine yield compared to the wine obtained with the control yeast but without altering its physiochemical properties. Therefore, such a recombinant yeast strain could be used in banana wine fermentations as an alternative to commercial enzyme preparations.

9.3 LEGISLATION AND CONSUMER BEHAVIOR ON GENETICALLY MODIFIED MICROORGANISMS

Legislation is an important factor affecting the development of modified fruit wine microorganisms. The need for approval is a major barrier to commercialize these genetically modified yeast strains. These strict requirements of genetically modified food and beverage regulations are rising as a response to perceived consumer concerns about the risk of GMOs (Pretorius, 2001). Former and recent surveys lead to the conclusions that the average consumer is poorly informed about the benefits and risks to use the biotechnology. It is likely that vast potential benefit exists for the wine consumer and producer in the application of gene technology. That benefit will be realized, however, only if the application is judicious, systematic, and achieved with high regard for the unique nature of the product (Pretorius and Bauer, 2002).

10. CONCLUSIONS

Fruit wines are FABs made of different fruits and the basic processes are similar to grape juice production and fermentation. Both endogenous and exogenous microorganisms are involved in the process of fruit wine fermentation. The endogenous cultures are contributed by fruits and winery surfaces while the exogenous are the selected starter culture. Fruits enter into the processing carrying different types of microorganisms, as bacteria, yeast, and phylamentous fungi (mainly spoilage molds). In the first stage of a fruit wine fermentation the non-*Saccharomyces* yeast are predominant (*Candida, Hanseniaspora, Torulopsis, Pichia, Kloeckera*, etc.). Later on, while the ethanol contents increase, the dominant species is *Saccharomyces*. The most abundant LAB isolated during fruit wine fermentation belong to the species *L. plantarum, L. hilgardii, Le. mesenteroides*, and *Pediococcus* sp., while to a lesser extent, *O. oeni and L. brevis* are also found.

Spontaneous fermentation with microbiota that are present on the fruit surface is used in conventional winemaking in different regions of the world. It is generally accepted that adding yeast starters will stabilize the fermentation and will lead to a quality product. The ideal starter yeast should exhibit the following properties: rapid initiation of fermentation; resistance to SO_2 and high ethanol concentrations; low requirements for growth factors; ability to complete attenuation by fermenting all sugars; suitable flocculation characteristics; development of a sound organoleptic profile and production of polygalacturonase to degrade soluble pectin.

MLF occurs also in fruit wines and different species of bacteria, like *O. oeni, L. plantarum* or *L. casei*, and *Pediococcus* degrade the malic acid. A special attention has been given to the potential use of immobilized cells for malic acid degradation as well as application of the immobilized biocatalysts in enology.

Fruit wines are also subject to spoilage, mainly caused by lactic and acid bacteria. Classical food microbiological analysis may be applied, as well as rapid biochemical and molecular tests.

During the past two decades recombinant DNA technology has been used to introduce various desired strains from other organisms into wine strains of *S. cerevisiae*. The improvements can be easily adapted to fruit wine microorganisms. However, the need for approval is a major barrier to commercialize these genetically modified yeast strains. An important issue to be taken into account is that the average consumer is poorly informed about the benefits and risks of the modified microorganism used in winemaking.

REFERENCES

Amoa-Awua, W.K., Sampson, E., Tano-Debrah, K., 2007. Growth of yeasts, lactic and acetic acid bacteria in palm wine during tapping and fermentation from felled oil palm (*Elaeis guineensis*) in Ghana. Journal of Applied Microbiology 102, 599–606.

Awe, S., Nnadoze, S.N., 2015. Production and microbiological assesment of date palm (*Phoenix dactylifera* L.) fruit wine. British Microbiology Research Journal 8 (3), 480–488.

Barnett, J.A., Payne, R.W., Yarrow, D., 1990. Yeasts, Characterisation and Identification, second ed. Cambridge University Press, Cambridge.

Barre, P., 1980. Le mécanisme "killer" dans la concurrence entre souches de levures. Evaluation et prise en compte. Bulletine OIV 56, 345–349.

Bartowsky, E.J., 2009. Bacterial spoilage of wine and approaches to minimize it. Letters in Applied Microbiology 48, 149–156.

Benito, S., Palomero, F., Morata, A., Calderon, F., Suarez-Lepe, J.A., 2012. New applications for *Schizosaccharomyces pombe* in the alcoholic fermentation of red wines. International Journal of Food Science and Technology 47 (10), 2101–2108.

Berenguer, M., Salud, V., Enrique, B., Domingo, S., Manuel, V., Nuria, M., 2016. Physicochemical characterization of pomegranate wines fermented with three different *Saccharomyces cerevisiae* yeast strains. Food Chemistry 190, 848–855.

Bhardwaj, J.C., Joshi, V.K., 2009. Effect of cultivar, addition of yeast type, extract and form of yeast culture on foaming characteristics, secondary fermentation and quality of sparkling plum wine. Natural Product Radiance 8 (4), 452–464.

Byaruagaba-Bazirake, G.W., Rensburg, P.V., Kyamuhangire, W., 2013. Characterisation of banana wine fermented with recombinant wine-yeast strains. American Journal of Food and Nutrition 3, 105–116.

del Campo, G., Berregi, I., Santos, J.I., Duenas, M., Irastorza, A., 2008. Development of alcoholic and malolactic fermentations in highly acidic and phenolic apple musts. Bioresource Technology 99, 2857–2863.

Chanprasartsuk, O., Prakitchaiwattana, C., Sanguandeekul, R., Fleet, G.H., 2010. Autochthonous yeasts associated with mature pineapple fruits, freshly crushed juice and their ferments and the chemical changes during natural fermentation. Bioresource Technology 101, 7500–7509.

Chanprasartsuk, O., Pheanudomkitlert, K., Toonwai, D., 2012. Pineapple wine fermentation with yeasts isolated from fruit as single and mixed starter cultures. Asian Journal of. Food and Agro-Industry 5 (02), 104–111.

Ciani, M., 2008. Continuous deacidification of wine by immobilized *Schizosaccharomyces pombe* cells: evaluation of malic acid degradation rate and analytical profiles. Journal of Applied Microbiology 79 (6), 631–634.

Chen, D., Chia, J.Y., Liu, S.-Q., 2014. Impact of addition of aromatic amino acids on non-volatile and volatile compounds in lychee wine fermented with *Saccharomyces cerevisiae* MERIT. ferm. International Journal of Food Microbiology 170, 12–20.

Chen, D., Yap, Z.Y., Liu, S.-Q., 2015. Evaluation of the performance of *Torulaspora delbrueckii, Williopsis saturnus,* and *Kluyveromyces lactis* in lychee wine fermentation. International Journal of Food Microbiology 206, 45–50.

Chibata, I., 1978. Immobilised Enzymes—Research and Development. John Wiley and Sons, Inc., New York, USA.

Chilaka, C.A., Uchechukwu, N., Obidiegwu, J.E., Akpor, O.B., 2010. Evaluation of the efficiency of yeast isolates from palm wine in diverse fruit wine production. African Journal of Food Science 4 (12), 764–774.

Christopher, R.K., Theivendirarajah, K., 1988. Palmyrah palm wine. 1: microbial and biochemical changes. Journal of the National Science Council of Sri Lanka 16, 131–141.

Dan, V., 2011. In: Galati, A. (Ed.), Microbiological Processes During Cider Preparation in Food Microbiology.

Dharmadhikari, M., 2002. Some issues in malolactic fermentation – acid reduction and flavor modification. Vineyard & Vintage View 17 (4), 4–6.

Dharmadhikari, M., 2004. Wines from cherries and soft fruits. Vineyard & Vintage View, Mountain Grove, MO 19 (2), 9–15.

Ding, S., Zhang, Y., Zhang, J., Zeng, W., Yang, Y., Guan, J., Pan, L., Li, W., 2015. Enhanced deacidification activity in *Schizosaccharomyces pombe* by genome shuffling. Yeast 32, 317–325.

Divies, C., Cachon, R., 2005. Wine production by immobilized cell systems. In: Nedovic, V., Willaert, R. (Eds.), Applications of Cell Immobilization Biotechnology. Springer, Heidelberg, pp. 285–293.

Diviès, C., Deschamps, P., 1986. Procédé et appareillage pour la mise en oeuvre de réactions enzymatiques et application à la préparation de boissons fermentées. French Patent 2601687.

Divies, C., Cachon, R., Cavin, J.-F., Prevost, H., 1994. Immobilized cell technology in wine production. Critical Reviews in Biotechnology 14 (2), 135–153.

du Toit, M., Pretorius, I.S., 2000. Microbial spoilage and preservation of wine: using weapons from nature's own arsenal e a review. South African Journal of Enology and Viticulture 21, 74–92.

du Toit, M., Engelbrecht, L., Lerm, E., Krieger-Weber, S., 2011. *Lactobacillus*: the next generation of malolactic fermentation starter cultures—an overview. Food Bioprocess Technol 4, 876–906.

Duarte, S.C., Pena, A., Lino, C.M., 2009a. Ochratoxin A non-conventional exposure sources—a review. Microchemical Journal 93 (2), 115–120.

Duarte, W.F., Dias, D.R., de Melo Pereira, G.V., Gervásio, I.M., Schwan, R.F., 2009b. Indigenous and inoculated yeast fermentation of gabiroba (*Campomanesia pubescens*) pulp for fruit wine production. Journal of Industrial Microbiology and Biotechnology 36, 557–569.

Duarte Whasley, F., Dias Disney, R., Oliveira José, M., Vilanova, M., Teixeira José, A., Almeida, E.S., Rosan João, B., Schwan, F., 2010. Raspberry (*Rubus idaeus* L.) wine: yeast selection, sensory evaluation and instrumental analysis of volatile and other compounds. Food Research International 43, 2303–2314.

Enwefa, R., Uwajeh, M., Oduh, M., 1992. Some studies on Nigerian palm wine with special reference to yeasts. Acta Biotechnology 12, 117–125.

Fahrasmane, L., Ganou-Parfait, B., Parfait, A., 1988. Yeast flora of Haitian rum distilleries. MIRCEN Journal of Applied Microbiology and Biotechnology 4, 239–241.

Fahwehinmi, M.M., 1981. Studies of Growth of *Saccharomyces Chevalieri* and *Pichia Membranefaciens* Isolated From Palm Wine (B.Sc. dissertation). University of Nigeria.

Faparusi, S.I., 1973. Origin of initial microflora of palm wine from oil palm trees (*Eaeis guineensis*). Journal of Applied Bacteriology 36, 559–565.

Fernandez-Cruz, M.L., Mansilla, M.L., Tadeo, J.L., 2010. Mycotoxins in fruits and their processed products: analysis, occurrence and health implications. Journal of Advanced Research 1 (2), 113–122.

Fumi, M.D., Bufo, M., Trioli, G., Colagrande, O., 1989. Bulk sparkling wine production by external encapsulated yeast bioreactor. Biotechnology Letters 11, 821–824.

García-Ruiz, A., Cueva, C., González-Rompinelli, E.M., Yuste, M., Torres, M., Martín-Álvarez, P.J., Bartolomé, B., Moreno-Arribas, M.V., 2012. Antimicrobial phenolic extracts able to inhibit lactic acid bacteria growth and wine malolactic fermentation. Food Control 28, 212–219.

Genisheva, Z., Mussatto, S.I., Oliveira, J.M., Teixeira, J.A., 2013. Malolactic fermentation of wines with immobilised lactic acid bacteria – influence of concentration, type of support material and storage conditions. Food Chemistry 138, 1510–1514.

Gulbiniene, G., et al., 2004. Killer yeast strains in wine-yeast population. Food Technology and Biotechnology 42 (3), 159–163.

Gutiérrez, A.R., Santamaría, P., Epifanio, S., Garijo, P., López, R., 1999. Ecology of spontaneous fermentation in one winery during 5 consecutive years. Letters in Applied Microbiology 29, 411–415.

Heimonen, F., Lisek, T., Patai, F., Quero, L., Zidar, T., 2013. IP Traditional Food in Combating Food-Borne Pathogens. Berry Wine. University of Applied Sciences, Tampere.

Herrero, M., Laca, A., Garcia, L.A., Diaz, M., 2001. Controlled malolactic fermentation in cider using *Oenococcus oeni* immobilized in alginate beads and comparison with free cell fermentation. Enzyme and Microbial Technology 28, 35–41.

Jeon, S.H., Kim, N.H., Shim, M.B., Jeon, Y.W., Ahn, J.H., Lee, S.H., Hwang, I.G., Rhee, M.S., 2015. Microbiological diversity and prevalence of spoilage and pathogenic bacteria in commercial fermented alcoholic beverages (beer, fruit wine, refined rice wine, and yakju). Journal of Food Protection 78 (4), 812–818.

Joshi, V.K., Rakesh, S., Ghanshyam, A., 2011. Stone fruit: wine and brandy. In: Hue, et al. (Ed.), Handbook of Food and Beverage Fermentation Technology. CRC Press, Florida.

Joshi, V.K., Attri, D., Singh, T.K., Abrol, G.S., 2011a. In: Handbook of Enology: Principles, Practices and Recent Innovations. Asiatech Publishers Inc.

Keller, J., 2010. Stone Fruit Wines. Wine Maker Magazine, p. 937.

Kosseva, M.R., 2011. Immobilization of microbial cells in food fermentation processes. Food and Bioprocess Technology 4, 1089–1118.

Kosseva, M.R., Kennedy, J.F., 1997. Supports and methods for cell immobilization in malolactic fermentation. Comptes Rendus De L'Academie Bulgare Des Sciences 50 (7), 33–36.

Kosseva, M.R., Kennedy, J.F., 2004. Encapsulated lactic acid bacteria for control of malolactic fermentation in wine. Artificial Cells, Blood Substitutes and Biotechnology 32 (1), 55–65.

Kosseva, M., Kennedy, J.F., Lloyd, L.L., Beschkov, V., 1998. Malolactic fermentation in Chardonnay wine by immobilised *Lactobacillus casei* cells. Process Biochemistry 33, 793–797.

Kosseva, M.R., Panesar, P.S., Kaur, G., Kennedy, J.F., 2009. Use of immobilised biocatalysts in the processing of cheese whey. A Review in International Journal of Biological Macromolecules 45, 437–447.

Kourkoutas, A., Bekatorou, A., Banat, I.M., Marchant, R., Koutinas, A.A., 2004. Immobilization technologies and support materials suitable in alcohol beverages production: a review. Food Microbiology 21, 377–397.

Lee, P.R., Chong, I.S.M., Yu, B., Curran, P., Liu, S.Q., 2012. Effects of sequentially inoculated *Williopsis saturnus* and *Saccharomyces cerevisiae* on volatile profiles of papaya wine. Food Research International 45, 177–183.

Li, X., Chan, Li J., Yu, B., Curran, P., Liu, S.-Q., 2012. Fermentation of three varieties of mango juices with a mixture of *Saccharomyces cerevisiae* and *Williopsis saturnus var. mrakii*. International Journal of Food Microbiology 158, 28–35.

Lonvaud-Funel, A., 1999. Lactic acid bacteria in the quality improvement and depreciation of wine. Antonie van Leeuwenhoek 76, 317–331.

Maragatham, C., Panneerselvam, A., 2011. Isolation, identification and characterization of wine-yeast from rotten papaya fruits for wine production. Advances in Applied Science Research 2 (2), 93–98.

Matei, F., Brinduse, E., Nicoale, G., Tudorache, A., Teodorescu, R., 2011. Yeast biodiversity evolution over decades in Dealu Mare-Valea Calugareasca vineyard. Romanian Biotechnological Letters 16 (Suppl. 1), 113–120.

Mills David, A., Phister, T., Neeley, E., Johannsen, E., 2008. Wine fermentation. In: Cocolin, L., Ercolini, D. (Eds.), Molecular Techniques in the Microbial Ecology of Fermented Foods. Springer.

Monk, P.R., 1986. Rehydration and propagation of active dry yeast. Australian Wine Industry Journal 3–5.

Nedovic, V.A., Durieux, A., Van Nedervelde, L., Rosseels, P., Vandegans, J., Plaisant, A.M., et al., 2000. Continuous cider fermentation with co-immobilized yeast and Leuconostoc oenos cells. Enzyme and Microbial Technology 26, 834–839.

Nidp, R.N., Akoachere, J.F., Dopgima, L.L., Ndip, L.M., 2001. Characterization of yeast strains foe wine production: effect of fermentation variables on quality of wine produced. Applied Biochemistry and Biotechnology 95 (3), 209–220.

Nigam, P.S., 2011. In: Joshi, V.K. (Ed.), Microbiology of Wine Making in Handbook of Enology: Principles, Practices and Recent Innovations, vol. II. Asia Tech. Publ., New Delhi, pp. 383–405.

Nyanga, L.K., Nout, M.J.R., Gadaga, T.H., Theelen, B., Boekhout, T., Zwietering, M.H., 2007. Yeasts and lactic acid bacteria microbiota from masau (*Ziziphus mauritiana*) fruits and their fermented fruit pulp in Zimbabwe. International1 Journal of Food Microbiology 120, 159–166.

Oafor, N., 1972. Palm wine yeasts from parts of Nigeria. Journal of the Science of Food and Agriculture 23, 1399–1407.

Okolie, P.I., Opara, C.N., Emerenini, E.C., Uzochukwu, S.V.A., 2013. Evaluation of bacterial diversity in palm wine by 16S rDNA analysis of community DNA. Nigerian Food Journal 31 (1), 83–90.

Okunowo, W.O., Osuntoki, A.A., 2007. Quantitation of alcohols in orange wine fermented by four strains of yeast. African Journal of Biochemistry Research 1, 095–100.

Oliveira, M.E.S., Pantoja, L., Duarte, W.F., Collela, C.F., Valarelli, L.T., Schwan, R.F., Dias, D.R., 2011. Fruit wine produced from cagaita (*Eugenia dysenterica* DC) by both free and immobilised yeast cell fermentation. Food Research International 44, 2391–2400.

Pando Bedriñana, R., Querol Simón, A., Suárez Valles, B., 2010. Genetic and phenotypic diversity of autochthonous cider yeasts in a cellar from Asturias. Food Microbiology 27, 503–508.

Parks, B., 2006. Country Wine-Yeast: Tips From the Pros. Wine Maker Magazine p. 857.

Pech, B., Lavoue, G., Parfait, A., Belin, J.M., 1984. Rum fermentation: suitability of strains of *Schizosaccharomyces pombe*. Science des Aliments 4, 67–72.

Pretorius, I.S., 2000. Tailoring wine-yeast for the new millennium: novel approaches to the ancient art of winemaking. Yeast 16, 675–729.

Pretorius, I.S., 2001. Gene technology in wine-making: new approaches to an ancient art. Agriculturae Conspectus Scientificus 66 (1), 27–47.

Pretorius, I.S., Bauer, F., 2002. Meeting the consumer challenge through genetically customized wine-yeast strains. Trends in Biotechnology 20, 426–432.

Radoi, F., Kishida, M., Kawasaki, H., 2005a. Characteristics of wines made by *Saccharomyces* mutant which produce a polygalacturonase under wine-making conditions. Bioscience, Biotechnology, and Biochemistry 69 (11), 2224–2226.

Radoi, F., Kishida, M., Kawasaki, H., April 2005b. Endo-polygalacturonase in *Saccharomyces* wine yeasts: effects of carbon source on enzyme production. FEMS Yeast Research 5 (6–7), 663–668.

Ravelomanana, R., Guiraud, J.P., Vincent, J.C., Galzy, P., 1984. Study of the yeast flora of the traditional cocoa fermentation in Madagascar. Revue des Fermentations et des Industries Alimentaries 39, 103–106.

Reddy, L.V.A., Reddy, O.V.S., 2011. Effect of fermentation conditions on yeast growth and volatile composition of wine produced from mango (*Mangifera indica* L.) fruit juice. Food and Bioproducts Processing 8 (9), 487–491.

Rivard, D., 2009. Choosing the right fruit wine-yeast. The Daily Fruit Wine On-Line Journal 4.

Sánchez, P., 1979. The prospects of fruit wines in Phillipines. Journal of Crop Science 4 (4), 183–190.

Sánchez, A., Rodríguez, R., Coton, M., Coton, E., Herrero, M., García, L.A., Díaz, M., 2010. Population dynamics of lactic acid bacteria during spontaneous malolactic fermentation in industrial cider. Food Research International 43, 2101–2107.

Sánchez, A., Coton, M., Coton, E., Herrero, M., García, L.A., Díaz, M., 2012. Prevalent lactic acid bacteria in cider cellars and efficiency of *Oenococcus oeni* strains. Food Microbiology 32, 32–37.

Sanni, A.I., Loenner, C., 1993. Identification of yeasts isolated from Nigerian tradicional alcoholic beverages. Food Microbiology 12, 517–523.

Satora, P., Sroka, P., Błaszczyk, U., 2012. The action of *Pichia* killer strains against wine spoilage microorganisms. In: Materials of Laboralim 2012 Conference, Banska Bystrica, Slovakia, pp. 227–229.

Sátora, P., Tarko, T., Sroka, P., Błaszczyk, U., 2014. The influence of *Wickerhamomyces anomalus* killer yeast on the fermentation and chemical composition of apple wines. FEMS Yeast Research 14, 729–740.

ShuangWang, G., 2006. The Isolation, Screening and Application of Killer Yeasts in Cider (Master's thesis). Beijing Forestry University.

Sibounnavong, P., Daungpanya, S., Sidtiphanthong, S., Keoudone, C., Sayavong, M., 2010. Application of *Saccharomyces cerevisiae* for wine production from star gooseberry and carambola. Journal of Agricultural Technology 6 (1), 99–105.

Stringini, M., Comitini, F., Taccari, M., Ciani, 2009. Yeast diversity during tapping and fermentation of palm wine from Cameroon. Food Microbiology 26 (4), 415–420.

Torija, M.J., Roes, N., Pblet, M., Guillamon, M.J., Mas, A., 2002. Effects of fermentation temperature on the strain population of *Saccharomyces cerevisiae*. International Journal of Food Microbiology 80, 47–53.

Valles Suarez, B., Bedriñana, R.P., Tascon, N.F., Simo, A.Q., Madrera, R.R., 2007. Food Microbiology 24, 25–31.

Varakumar, S., Naresh, K., Reddy, O.V.S., 2012. Preparation of mango (*Mangifera indica* L.) wine using a new yeast-mango-peel immobilised biocatalyst system. Czech J. Food Sci 30 (6), 557–566.

Varakumar, S., Naresh, K., Variyar, P.S., Sharma, A., Reddy, O.V.S., 2013. Role of malolactic fermentation on the quality of Mango (*Mangifera indica* L.) wine. Food Biotechnology 27, 119–136.

Vilanova, M., Ugliano, M., Varela, C., Siebert, T., Pretorius, I.S., Henschke, P.A., 2007. Assimilable nitrogen utilisation and production of volatile and non-volatile compounds in chemically defined medium by *Saccharomyces cerevisiae* wine yeasts. Applied Microbiology and Biotechnology 77, 147–157.

Waites, M.J., Morgan, N.L., Rockey, J.S., Higton, G., 2001. Industrial Microbiology. Blackwell Science Ltd.

Wang, Y.H., Li, Y., Pei, X.L., et al., 2007. Genome-shuffling improved acid tolerance and L-lactic acid volumetric productivity in *Lactobacillus rhamnosus*. Journal of Biotechnology 129, 510–515.

Wickner, R.B., 1996. Microbiological Reviews 60, 250–265.

Ye, M., Yue, T., Yuan, Y., September 2014. Effects of sequential mixed cultures of *Wickerhamomyces anomalus* and *Saccharomyces cerevisiae* on apple cider fermentation. FEMS Yeast Research 14 (6), 873–882.

CHEMISTRY OF FRUIT WINES

3

H.P. Vasantha Rupasinghe[1], V.K. Joshi[2], A. Smith[1], I. Parmar[1]

[1]*Dalhousie University, Truro, NS, Canada;* [2]*Dr. Y.S. Parmar University of Horticulture and Forestry, Nauni, Solan, HP, India*

1. INTRODUCTION

Winemaking is among the oldest techniques known to civilization and is one of the most commercially prosperous biotechnological processes, even today (Moreno-Arribas and Polo, 2005; Joshi et al., 2011a), involving alcoholic fermentation (Molinos et al., 2016; Joshi, 2016). Historically speaking, red wine consumption dates back thousands of years to the Egyptians, who prepared medicinal wines made from grapes, using herbs and tree resins (McGovern et al., 2009). Of all the traditional wines, red wine is often touted as being beneficial for human health, rich in many health-promoting compounds such as polyphenols. However, grapes are not the only fruit that can be made into wine. Fruit wines can be made from nearly any fruit; however, some popular fruit sources include blueberries, blackberries, strawberries, cherries, apples, peaches, and plums (Amerine et al., 1980; Joshi et al., 1999a,b, 2011b; Jagtap and Bapat, 2015).

The conversion of a juice into wine is a complex biochemical process involving yeast, predominantly *Saccharomyces cerevisiae*. In brief, winemaking is the result of a number of biochemical transformations brought about by the action of several enzymes from various microorganisms, especially yeast, which carry out the major part of the process of alcoholic fermentation (Amerine et al., 1980; Moreno-Arribas and Polo, 2005). Lactic acid bacteria also contribute to a secondary process, known as malolactic fermentation, responsible for reducing the acidity of the high-acid wines (Bartowsky, 2011). A number of enzymes originating from the fruit, yeast, and lactic acid bacteria or those from contaminating microorganisms are involved in different steps of winemaking. Commercial enzyme preparations such as pectin esterase or amylolytic enzymes are widely used as supplements.

Many changes involved in alcoholic fermentation processes result from various biochemical reactions and the end products formed, which have been studied since the pioneering research of Louis Pasteur about 140 years ago (Amerine, 1985; Amerine et al., 1980; Rana and Rana, 2011). In yeast-based alcoholic fermentation, the yeast utilizes the sugars and other constituents of grape juice, or for that matter, any fruit juice, for its growth, converting them into ethanol, carbon dioxide, and other metabolites (Goyal, 1999; Rebordinos et al., 2011). All the products formed during the process contribute to the chemical composition and sensory quality characteristics of the wine. So, a thorough knowledge of yeast growth and the chemistry of fermentation involved is essential because it is fundamental to winemaking. Wine preparation is a traditional process, yet several advancements have been made on various facets of the process. However, the advances made in the second half of the 20th century have clearly shown that fermentation of fruit must and the production of quality wines is not quite as simple as the process as discovered by Pasteur (Moreno-Arribas and Polo, 2005). Since

2005, great strides have been made in understanding the chemistry and interactions of yeast, lactic acid bacteria, and other microorganisms during the winemaking process that have greatly benefitted winemakers in the production of high-quality wines. It is pertinent to state that the preparation of nongrape wine is basically similar to that from grapes, so naturally the chemistry of alcoholic fermentation, or for that matter, malolactic fermentation, and associated characteristics would remain fundamentally the same. Of course, differences in several components would be there because of the differences in composition of the specific fruits used. Many of these aspects are discussed in this chapter.

While the history of nongrape wines may not be as long or as prestigious as that of grape wines, they are gaining more contemporary notice as more attention is being given to the novel range of commercial opportunities and health benefits fruit wines possess (Rupasinghe and Yu, 2013; Rupasinghe et al., 2012). Studies exploring the health-promoting and disease-preventing potential of foods have shown that, in comparison to grapes, other fruits such as blueberries, blackberries, black currants, cherries, cranberries, crowberries, elderberries, raspberries, lingonberries, and sea buckthorn have similar or even higher flavonoid and phenolic content (Dey et al., 2009; Thilakarathna and Rupasinghe, 2012). Berries contain a wide range of dietary phytochemicals with strong antioxidant capacities. The most common phytochemical constituents are phenolic acids, flavonoids, stilbenes, and carotenoids (Kaur and Kapoor, 2001; Tomas-Barberan et al., 2001; Vinson et al., 2001). In recent years, increased interest in human health, nutrition, and disease prevention has enlarged consumers' demand for functional beverages based on the source of fruit, thereby expanding the market for nongrape fruit wines.

Globally, grape wine is the most popular fruit-fermented alcoholic beverage, followed by apple cider and several other commonly consumed cool climate fruit wines. Such fruit wines have long been served in moderate quantities as dessert wines and recognized as a natural source of essential minerals and many bioactive phytochemicals (Joshi, 1997; Soni et al., 2011; Joshi et al., 1999b). While there are numerous reviews on composition, bioactive phytochemicals, and physiological effects of grape wines, limited studies have been conducted to evaluate health-related major components of fruit wines produced. The published data on phenolic composition, antioxidant capacity, and biological activity in vitro and in vivo for nontraditional cool fruit wines are reviewed here and compared with those found for grape wines.

2. TYPES OF FERMENTATION

Various types of fermentation, viz., alcoholic, lactic acid, acetic, and alkaline fermentation, are involved in the production of fermented foods (Joshi and Pandey, 1999; Molinos et al., 2016) and of these, alcoholic fermentation is one of the most important and oldest processes, involving the production of mainly ethanol and carbon dioxide, resulting in a variety of beverages like wines, beers, and various distilled liquors (Amerine et al., 1980; Joshi et al., 2011b; Rebordinos et al., 2011). In an overview of alcoholic fermentation (AF), the production of wine of any kind follows a series of biochemical processes, which start with the ripening and harvesting of the fruit, carry on through the extraction of juice, and, most importantly, culminate in the microbial interactions facilitated by yeast during the AF process. However, there are some key differences between the winemaking of grapes and that of other fruits, such as in the composition, viz., sugar and acid content, the high acidity of the fruits, and the volume of juice released per mass, as well as the unique chemical fingerprint of each fruit that contributes to its flavors, that need to be acknowledged and compensated for during the winemaking process (Yang, 1953; Amerine et al., 1980; Joshi et al., 2011b).

Table 3.1 Types of Yeast Strains Used for Fruit Wine Fermentation[a]

Fruit Source	Possible Fruit Wine Yeasts	References
Mango	*Saccharomyces cerevisiae* CFTRI 101	Reddy and Reddy (2011)
	Williopsis saturnus var. *mrakii* NCYC500	Li et al. (2012)
Guava	*S. cerevisiae* NCIM 3095	Sevda and Rodrigues (2011)
Kiwifruit	*S. cerevisiae bayanus* LALVIN EC-1118	Towantakavanit et al. (2011)
Apple	*S. cerevisiae* CCTCC M201022	Wang et al. (2004)
Cherry	*S. cerevisiae* BM4x4, RA17, D21, GRE, RC212, and D254	Sun et al. (2011)
Raspberry	*S. cerevisiae* UFLA FW 15	Duarte et al. (2010a)
Cacao	*S. cerevisiae* UFLA CA 1162	Duarte et al. (2010b)
Lychee	*S. cerevisiae* UFLA CA1183	Alves et al. (2011)
Plum	*S. cerevisiae* UCD 595, *Schizosaccharomyces pombe*	Joshi et al. (1991b)

[a]*More strains of yeast are viable for the fermentation of more than one fruit source. Yeast choice may depend on the source of fruit wine being sought.*

By contrast, wine grapes, when fully ripe, have been bred to near perfection for the winemaking process, especially the sugar/acid ratio, low amounts of protein and pectin (leading to a wine that is easily clarified), yield, etc. (Buglass, 2011). Nevertheless, these differences have not limited the broad, and sometimes novel, range of nongrape fruits that have been turned into successful fruit wine beverages (Jagtap and Bapat, 2015).

In the case of fruits with low sugar content, the must can be ameliorated with sugar or a sugar syrup that is used to correct the balance. Water can be used to dilute the acidity to an appropriate level and adjuncts (sugar, pectic enzymes, yeast, and SO_2 or other treatments), though optional, can be used at the discretion of the winemaker to achieve the desired results (Buglass, 2011). Fermentation as a metabolic process remains largely unchanged between the fermentation of grapes and that of other fruit, as described earlier.

Similarly, of the yeast strains appropriate for fruit winemaking, there are a lot to choose from, though all are generally taken from *S. cerevisiae*. The selection of possible yeast strains to ferment fruit wine covers a broad range (Table 3.1) and offers a spectrum of possible flavor profiles through the use of multiple strains yielding unique chemical fingerprints of volatiles and polyphenols, all dependent on the chosen strain (Sun et al., 2011). Thus, fruit wine offers a significant amount of leeway for experimentation in this regard.

3. CHEMISTRY OF WINEMAKING

The most important transformation during the production of wine takes place in the fruit must during vinification, that transformation being the AF of the sugars, especially hexoses (glucose and fructose), producing ethanol and carbon dioxide along with a large number of minor by-products (Amerine et al., 1980; Moreno-Arribas and Polo, 2005). Vinification factors, such as the temperature of fermentation, the pH, and the sugar concentration of the juice, are known to affect the growth and metabolic ability of yeast (Dubois et al., 1996; Charoenchai et al., 1998; Calderoin et al., 2001; Alexandre et al., 1992; Angulo et al., 1993; Santamariia et al., 1995; Rebordinos et al., 2011) and the inhibitory effect of ethanol on the specific growth rates and viability of *S. cerevisiae*. Other known stress factors also occurring during some wine fermentations include high levels of SO_2 and CO_2 (Boulton et al., 1996) and the presence of competing microorganisms (for example, lactic acid bacteria identified as *Lactobacillus kunkeei* spp. nov.; Edwards et al., 1998), as well as some of the killer toxins produced by the killer yeasts (van Vuuren and Jacobs, 1992; Hidalgo and Flores, 1994; Cocolin and Comi, 2011). A pictorial representation of the metabolic events in yeast is shown in Fig. 3.1.

The biochemical events occurring in AF can be broadly classified into the metabolism of carbon, nitrogen, and sulfur compounds, and of these, carbon metabolism is the prime pathway of making wine. But the ultimate concentration of ethanol in a wine depends on the initial concentration of sugars in the must/juice and the conditions prevailing during fermentation (Amerine et al., 1980; Goyal, 1999; Rana and Rana, 2011). Nitrogen is the second important element, required for yeast growth; therefore nitrogen metabolism has an indirect yet considerable impact on the quality of wine. The quantity and type of nitrogen content of the must sometimes also lead to problems other than the flavor of the wine, like sluggish and stuck fermentation (Navarro and Navarro, 2011), the formation of reduced sulfur compounds, and certain other end products like ethyl carbamate (Ough et al., 1988a).

During wine fermentation, many organic acids are either synthesized or degraded by yeast or lactic acid bacteria, which affects the quality of wine considerably (Radler, 1993). The production of various acids, both desirable and undesirable, can be controlled only by understanding the pathways of their

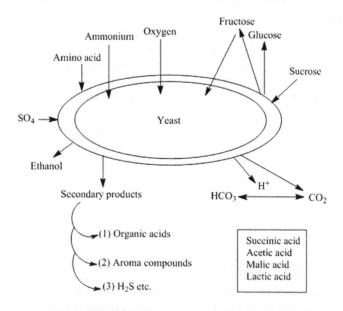

FIGURE 3.1

Diagram showing utilization and formation of various compounds by yeast during alcoholic fermentation such as wine production.

Adapted from Rana, N.S., Rana, V.S., 2011. Biochemistry of wine preparation. In: Joshi, V.K. (Ed.), Handbook of Enology: Principles, Practices, and Recent Innovations, vol. 2. Asiatech Publishers Inc., New Delhi, pp. 618–678.

synthesis and degradation (Radler, 1993). Yeast also utilize sulfur for biosynthesis of several sulfur compounds, including hydrogen sulfide, which imparts an unpleasant aroma during yeast formation.

This chapter deals specifically with the various metabolic events leading to the translocation and metabolism of major ingredients like sugars, nitrogen, organic acids, and sulfur compounds. Changes in fermentation aspects connected with maturation in terms of chemistry, especially with respect to composition and nutrition, are also illustrated, but no attempts are made to discuss the very basics of biochemical aspects here, and for this the reader is referred to the literature cited (Amerine et al., 1980; Goyal, 1999; Rana and Rana, 2011).

3.1 CARBON METABOLISM

When the concentration of sugars is high, such as in the must, *S. cerevisiae* can metabolize sugar only by the fermentation route. A limited number of carbon compounds are utilized by wine yeast (*Saccharomyces* species), such as monosaccharides (glucose and fructose); disaccharides (sucrose, maltose, and melibiose), which can be fermented by most wine yeast strains; and trisaccharides, i.e., raffinose, which can also serve as substrate for a few yeast strains; but pentoses are not utilized by the *Saccharomyces* wine strains (Goyal, 1999). Their utilization involves various pathways but the operation of these pathways depends on the substrate availability and other conditions prevailing during the fermentation, such as oxygen, which plays a very significant role in metabolism. The major pathway employed for glucose and fructose catabolism is glycolysis (Amerine et al., 1980; Rana and Rana, 2011) and is common in yeast and lactic acid bacteria, with the production of pyruvate, which is further metabolized in different microorganisms.

3.1.1 Glycolysis

Broadly, the glycolytic pathway involves the conversion of glucose into pyruvate (Rana and Rana, 2011) by a series of reactions involving a number of enzymes (Fig. 3.2). It is virtually a universal pathway for glucose catabolism and is operational under both fermentative and respiratory modes of metabolism.

Winemaking involves mainly fermentation of the soluble sugars, viz., glucose and fructose, of grape or any other juice into carbon dioxide and ethyl alcohol. The glycolytic pathway includes various steps leading to the conversion of glucose to pyruvate. The process of glucose metabolism for wine production includes three steps, viz., transport of sugars into the cell, the phosphorylation and conversion of glucose into glucose 6-phosphate, and finally its conversion into pyruvate. The first step of glycolysis is the transport of sugars into the cell, for which two types of mechanisms operate, viz., passive transport and active transport. The former does not require an input of metabolic energy, whereas the latter does need energy. For more details see the literature cited (Amerine et al., 1980; Goyal, 1999; Rana and Rana, 2011). Under anaerobic conditions, however, the transport of sugar is a key step for the control of glycolytic flux. Microorganisms growing in sugar-containing media do not usually accumulate intermediates involved in sugar metabolism within the cell, and the sugar transport in *Saccharomyces* is quite complex and is known to involve multiple carriers (Kruckeberg and Bisson, 1990). The glucose is utilized at a faster rate than fructose in most fermentations. Glucose transport has also been examined in an industrial strain of *S. cerevisiae* (UCD 522) during grape juice fermentation (Mc Clellan et al., 1989). At high concentration of glucose, more than one molecule tends to bind simultaneously to the carrier, which inhibits the transporter activity. The next step in the pathway of

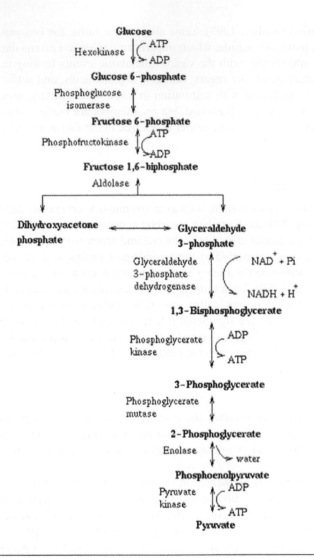

FIGURE 3.2

Flow diagram representation of the glycolytic pathway.

glucose utilization is the conversion of glucose 6-phosphate to pyruvate. As can be seen from Fig. 3.2, in the reaction of glycolysis, one molecule of ATP is used for transfer of a phosphoryl group to the 6-carbon of glucose. Glucose 6-phosphate is then converted into its isomeric form, fructose 6-phosphate, by the enzyme phosphofructokinase (PFK). This is another step of the pathway whereby the second molecule of ATP is consumed and phosphorylation occurs at the 1-carbon position to produce fructose 1,6-bisphosphate. This enzyme also requires Mg^{2+} ions for the reaction to proceed. Fructose 1,6-bisphosphate is cleaved into two triose phosphate sugars, viz., glyceraldehyde 3-phosphate and dihydroxyacetone phosphate (Fig. 3.3), by the enzyme aldolase (Amerine, 1985). Interconversion of triose phosphate sugars is the next very important step from the generation of energy point of view.

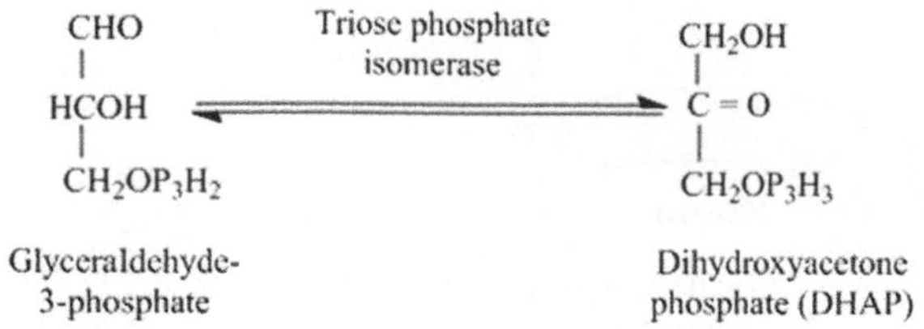

CHO   Triose phosphate CH_2OH
|
HCOH   isomerase $C = O$
|
$CH_2OP_3H_2$ $CH_2OP_3H_3$

Glyceraldehyde-   Dihydroxyacetone
3-phosphate   phosphate (DHAP)

FIGURE 3.3

Interconversion of triose sugars.

Only the glyceraldehyde 3-phosphate formed in the previous reaction, of the two triose sugars, is required for further metabolism, and therefore to utilize the entire glucose molecule for the production of alcohol, the dihydroxyacetone phosphate is isomerized to glyceraldehyde 3-phosphate by the enzyme triose phosphate isomerase in a reversible reaction outlined here. Glyceraldehyde 3-phosphate in the presence of an enzyme, glyceraldehyde-3-phosphate dehydrogenase, is converted into 1,3-bisphosphoglyccrate. The energy released in this oxidation proccss is conscrvcd through phosphorylation of carboxylic groups with inorganic phosphate and with the formation of NADH (Goyal, 1999). The high-energy substrate formed in the previous reaction undergoes transfer of a phosphoryl group to ADP to form ATP, and a low-energy substrate, 3-phosphoglycerate, is generated by the enzyme phosphoglycerate kinase. This phenomenon of generation of ATP from high-energy substrate is called substrate-level phosphorylation (Rana and Rana, 2011).

Conversion of 3-phosphoglycerate to 2-phosphoglycerate is brought about by phosphoglycerate mutase, which catalyzes the transfer of a phosphoryl group from the 3-hydroxyl position to its 2-hydroxyl position to form 2-phosphogycerate. This repositioning of the phosphoryl group brings the molecule to a high-energy state and allows substrate-level phosphorylation to yield another molecule of ATP. The enzyme enolase in this step catalyzes the dehydration of 2-phosphoglycerate and produces phosphoenolpyruvate by the reversible reaction and requires Mg^{2+} ions. This is the last step of the glycolytic pathway, catalyzed by the enzyme pyruvate kinase, and the second reaction that generates ATP by phosphorylating ADP in substrate-level phosphorylation. Mg^{2+} and K^+ ions are essential for the enzyme activity. In addition, the enzyme is also under the control of allosteric effectors (Rhodes et al., 1986). However, the activity of the enzyme is inhibited by ATP, citrate, acetyl-CoA, and long-chain fatty acids and modulated by ADP and P_i. In glycolysis, thus, there is a net production of two molecules of ATP and a NADH molecule.

3.1.2 Metabolism of Pyruvate

The majority of industrially important microorganisms like *S. cerevisiae* metabolize carbohydrates by glycolytic pathway oxidation of NAD^+ and ADP, respectively (Amerine et al., 1980). Both NAD^+ and ADP are present in small quantities in cells and a basic necessity of the cell is to generate NAD^+ and ADP if metabolism is to be continued. Therefore, each living organism is provided with a mechanism to oxidize NADH for the continuity of substrate oxidation. There are three main types of mechanisms that metabolize pyruvate and regenerate NAD^+ during the process. These include fermentation, anaerobic respiration, and aerobic respiration. Fermentation, especially in relation to wine, is described in this section, but only a brief account of other processes is given. A generalized view of pyruvate metabolism in *S. cerevisiae* is depicted in Fig. 3.4.

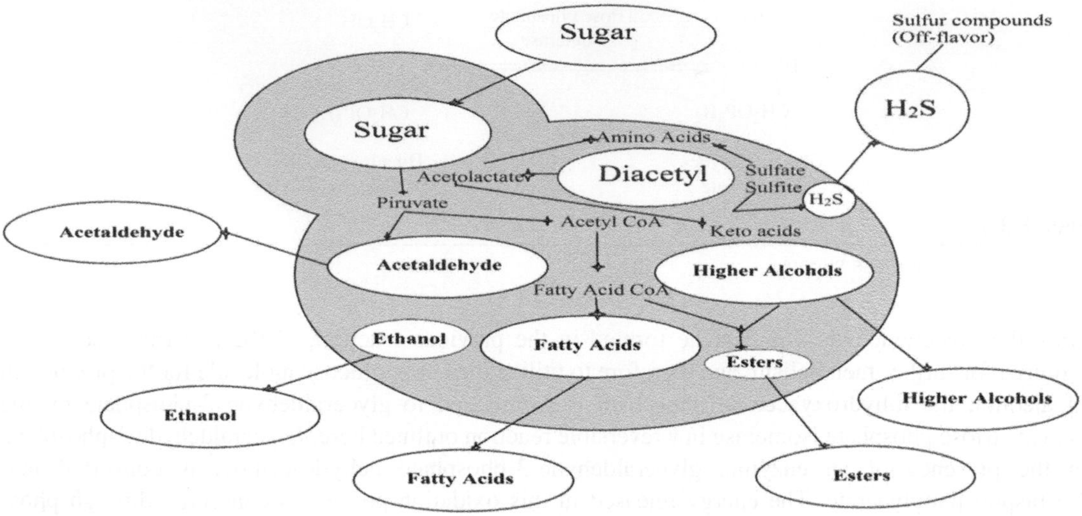

FIGURE 3.4

A generalized overview of metabolism in yeast and the products formed, including the conversion of pyruvate into various products.

Adapted from Moreno-Arribas, M.V., Polo, M.C., 2005. Winemaking biochemistry and microbiology: current knowledge and future trends. Critical Reviews in Food Science and Nutrition 45, 265–286.

3.1.3 Citric Acid Cycle

Glycolysis releases relatively little of the energy present in a glucose molecule; much more energy is released during the subsequent operation of the citric acid cycle [or the tricarboxylic acid (TCA) cycle] and oxidative phosphorylation (Rana and Rana, 2011). Following this route under aerobic conditions, pyruvate is converted into acetyl-CoA by the enzyme pyruvate dehydrogenase and then acetyl-CoA enters into the TCA cycle (Fig. 3.5). When the cellular energy level is high, i.e., ATP is in excess, the rate of the TCA cycle decreases and acetyl-CoA begins to accumulate. Under these conditions, acetyl-CoA can be employed for fatty acid synthesis or for the synthesis of ketone bodies. However, NAD^+ used during glycolysis must be regenerated if glycolysis is to be continued. Under aerobic conditions, NAD^+ is regenerated by the reoxidation of NADH via the electron transport chain. However, when oxygen is limiting, as in muscular contractions, the reoxidation of NADH to NAD by the electron transport chain becomes insufficient to maintain glycolysis. Under these conditions, NAD^+ is regenerated by conversion of pyruvate to lactate by lactate dehydrogenase. The formation of lactic acid from pyruvate by lactic acid bacteria can significantly increase the acidity of the wine. The formation of lactic acid is discussed in detail elsewhere in this chapter, as a minor end product.

After glycolysis, pyruvate is a central metabolite from which a number of products are formed (Fig. 3.4) by various enzymes found in different microorganisms (Goyal, 1999). Various reactions involved in the TCA cycle are depicted in Fig. 3.5. In *S. cerevisiae* pyruvate is converted into ethanol (the main component of alcoholic beverages) and carbon dioxide (Fig. 3.6).

The ethanolic pathway in yeast gives rise to the production of ethanolic beverages, as discussed in other parts of the text. Pyruvate is converted into ethanol via acetaldehyde from pyruvate. In addition

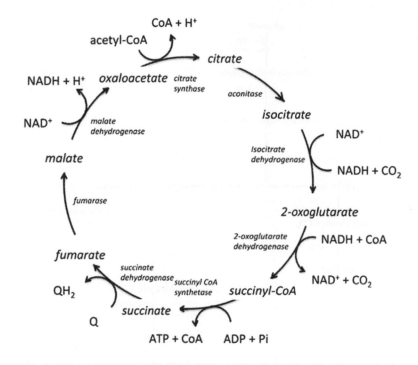

FIGURE 3.5

Various reactions involved in the tricarboxylic acid cycle.

Adapted from Cannon, W.R., 2014. Concepts, challenges and successes in modeling thermodynamics of metabolism. Frontiers in Bioengineering and Biotechnology 2, 1–10. http://dx.doi.org/10.3389/fbioe.2014.00053.

to ethanol, various fermentation products of the various metabolic pathways of pyruvate are formed in different microorganisms (also depicted in Fig. 3.4). In a typical heterolactic acid fermentation, products are formed from pyruvate during lactic acid fermentation and, hence, play an important role in flavor development, such as diacetyl acetoin in fermented milk products. Separate products can be formed from pyruvate by *Acetobacter* or *Gluconobacter*, as is the case with the formation of entirely different products from pyruvate by *Escherichia coli*.

3.1.4 Types of Fermentation

Fermentation is the anaerobic breakdown of pyruvate (Goyal, 1999; Rana and Rana, 2011). Depending upon the end product, fermentation has been classified into various types as discussed here.

3.1.4.1 Neuberg's First Form of Fermentation (Alcoholic Fermentation)

The conversion of pyruvate to ethanol is also called Neuberg's first form of fermentation (Fig. 3.5), named after the German scientist Carl Neuberg. The yeast strains of *Saccharomyces* are the main alcohol producers under anaerobic conditions by converting pyruvate to ethanol and carbon dioxide. NAD^+ molecules reduced during glucose-to-pyruvate conversion are regenerated when pyruvate is metabolized to ethanol and CO_2. The process takes place in two steps.

In the first step, pyruvate decarboxylase catalyzes the irreversible decarboxylation of pyruvate to produce acetaldehyde, for which the enzyme requires thiamine pyrophosphate as a coenzyme and Mg^{2+} ions for the

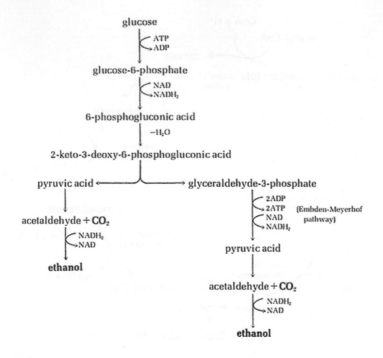

FIGURE 3.6

Production of ethanol from pyruvate in *Saccharomyces cerevisiae*, NAD$^+$ molecule regeneration, and conversion of pyruvate to ethanol during alcoholic fermentation.

Reproduced from Goyal, R.K., 1999. Biochemistry of fermentation. In: Joshi, V.K., Pandey, A. (Eds.), Biotechnology: Food Fermentation, vol. I. Asiatech Publishers Inc., New Delhi, pp. 87–172; Todar, K., 2008. Diversity of metabolism in procaryotes. In: Todar, K. (Ed.), Todar's Online Textbook of Bacteriology. University of Wisconsin. http://textbookofbacteriology.net/metabolism_3.html.

activity. In this step, alcohol dehydrogenase catalyzes the reduction of acetaldehyde to ethanol at the expense of NADH in a reversible reaction. The reaction is favored by pH of 7.0. At higher pH, i.e., 9.5, and in the presence of an excess of NAD$^+$, however, the reaction tends to proceed in the backward direction.

3.1.4.2 Neuberg's Second Form of Fermentation
During the fermentation of glucose by yeast at pH 6 or below, only small amounts of glycerol and other products are formed. The addition of sulfite to the cultures of *S. cerevisiae* enhances glycerol production. Acetaldehyde produced by the action of carboxylase during ethanol production combines with sulfite to form an additional compound. As a result, acetaldehyde is no longer able to act as a hydrogen acceptor for reduced NAD, which accumulates and soon is exhausted unless some other hydrogen acceptor is provided.

3.1.4.3 Neuberg's Third Form of Fermentation
Under alkaline conditions, dihydroxyacetone, an intermediate in the glycolytic pathway, replaces acetaldehyde as a hydrogen acceptor and is reduced to glycerol 3-phosphate, which is dephosphorylated to glycerol with the regeneration of NAD$^+$ (Goyal, 1999) as shown in Fig. 3.7. Many osmophilic yeasts

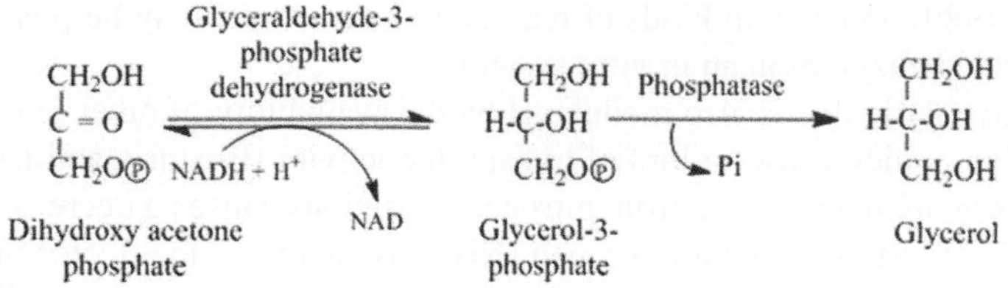

FIGURE 3.7

Neuberg's third form of fermentation.

Reproduced from Rana, N.S., Rana, V.S., 2011. Biochemistry of wine preparation. In: Joshi, V.K. (Ed.), Handbook of Enology: Principles, Practices, and Recent Innovations, vol. 2. Asiatech Publishers Inc., New Delhi, pp. 618–678.

carry out this fermentation corresponding to Neuberg's third form in the absence of alkaline steering agents. The inhibition of alcohol dehydrogenase or pyruvate decarboxylase activity can produce glycerol in proportionally higher yields. The different species of *S. cerevisiae* vary in their ability to produce glycerol, which ranges from 4.2 to 10.4 g/L (Radler and Schutz, 1982).

3.1.5 Yield of Ethanol and By-Products

The yield of ethanol in AF with a surplus amount of fermenting sugars depends on the ethanol tolerance of the yeast, as higher levels of ethanol are inhibitory to microbial enzymes and disintegrate the cell membrane through unfavorable interactions with cell wall components. Ethanol tolerance of yeast is also linked with membrane fluidity. A comprehensive review of this aspect of yeast has been compiled (Casey and Ingledew, 1986). In brief, one molecule of glucose or fructose yields two molecules each of ethanol and carbon dioxide. In a fermentation process, it has been estimated that starting with 22–24% sugars, 95% of the sugars is converted into ethanol and carbon dioxide, 1% is converted to cellular biomass, and the remaining 4% is converted to other end products. The amount of ethanol produced per unit of sugar during wine fermentation is of considerable commercial importance. During AF, one molecule of glucose produces two molecules each of ethanol and CO_2 under anaerobic conditions. In other words, 180 g of glucose (1 mol) should yield 92 g of ethanol (2 mol) and 88 g CO_2 (2 mol). The theoretical yield of ethanol production, therefore, comes out to be 51% under practical conditions and a much higher percentage (i.e., 47%) yield can be achieved. Although the metabolism yields equimolar quantities of CO_2 and ethanol, the actual amount of CO_2 liberated is less than theoretical because of the utilization of CO_2 in anaerobic carboxylation reactions (Ough et al., 1988a,b).

3.1.6 Regulation of Carbon Metabolism and Glycolysis in Yeast

Control of the carbon flux through the glycolytic pathway is a complex process. Glycolysis in *S. cerevisiae* under both aerobic and anaerobic conditions using ^{13}C and ^{31}P NMR spectroscopy combined with the determination of the ratio of various end products using ^{14}C-labeled glucose was investigated (Campbell et al., 1987; Den Hollander et al., 1986; Reibstein et al., 1986). It was also revealed that the major point of control during sugar utilization is at the step of transport or phosphorylation or both. Loss of enzyme activity at any of the three irreversible steps of glycolysis, viz., sugar phosphorylation, PFK, and pyruvate kinase, has an immediate impact on the glucose transporters. There is mounting evidence that the glycolytic enzymes exist as a multienzyme complex in eukaryotic cells (Green et al.,

1965; Walsh et al., 1989) and certain kinds of regulatory interactions may be possible that cannot be duplicated with purified enzymes in an in vitro system.

Control of the glycolytic flux is also modulated by the availability of other nutrients, in particular, by nitrogen limitation, which causes a loss of transporter activity (Busturia and Lagunas, 1986). The inactivation of glucose uptake resulting from nitrogen limitations causes a decrease in the overall gly-colytic rate (Salmon, 1989). Ammonium nitrogen (NH^{4+}) is an allosteric effector of both the PFK and the pyruvate kinase activities, so its effect on transporter activity may be indirect (Rhodes et al., 1986; Sols, 1981). Glucose inactivation refers to the inhibition of activity and subsequent proteolytic destruc-tion of many proteins by an unknown mechanism.

The availability of molecular oxygen also plays a role in the control of sugar metabolism, which is the substrate-level inhibition and the Pasteur effect. The phenomenon of substrate inhibition is observed in winemaking wherein at high sugar concentration, both uptake and consumption of sugar are inhibited, attributable to the inhibition of sugar transport as explained earlier. The Pasteur effect has been described as the inhibition of fermentation in favor of respiration in the presence of oxygen, which operates in *Saccharomyces* only under very special circumstances of nitrogen limitation (Lagunas, 1986).

3.1.7 End Products (Minor) of Sugar Metabolism

AF is associated with the synthesis of a diverse class of compounds, also called secondary metabolites, such as various alcohols, esters, organic acids, fatty acids, aldehydes, ketones, lactones, and terpenes (Fig. 3.4). These compounds determine the characteristic aroma and acceptance of any alcoholic bever-age. The total concentration of aroma substances in wine is about 0.8–1.0 g/L and such metabolites are synthesized mainly during the exponential phase of yeast growth. The yield of ethanol is, however, less during this phase. Some of these compounds are described here.

Higher (fusel) alcohols are collectively referred to as fuel alcohols or fusel oils and constitute a major portion of the secondary products of yeast metabolism (Amerine et al., 1980). Those alcohols that contain more than two carbons are called higher alcohols, such as n-propanol, isobutyl alcohol (2-methyl-1-propanol), 2-methylbutanol (optically active amyl alcohol), isoamyl alcohol, and 2-phenyl-ethanol. Among the various higher alcohols, isoamyl alcohol (3-methylbutanol) accounts for more than 50% of the total (Muller et al., 1993). These higher alcohols have higher boiling points and higher molecular weights than ethanol. The physical properties of higher alcohols and their metabolisms have been reviewed earlier (Webb and Ingraham, 1963). Nitrogenous compounds have a major role in the formation of these alcohols (see the section on nitrogen metabolism). They are mainly produced during amino acid metabolism as well as from the intermediates of carbohydrate metabolism (Fig. 3.8).

The growth of yeasts has been found to be linearly linked with the production of these alcohols (Pierce, 1982) but there are other factors that are responsible for the production of higher alcohols, including fruit, maturation, composition of amino acids, strain of yeast, and fermentation conditions, viz., temperature, pH, aeration (Rankine, 1967; Webb and Ingraham, 1963; Henschke and Jiranek, 1993; Ough and Bell, 1980; Rapp and Versini, 1991; Sinton et al., 1978; Vos, 1981). They can be a major component in wine distillates in which these are more concentrated. The higher alcohols them-selves have little impact on the sensory properties of wine. Generally, a higher alcohol concentration of below 300–400 mg/L reinforces the flavor of alcoholic beverages, whereas a concentration more than this is considered undesirable, being a cause of hangover in the consumer of such beverages (Amerine et al., 1980). These alcohols form an oily layer that imparts a fusel (foul) smell, and hence are called fusel oils (Webb and Ingraham, 1963), and the concentration in white wine has been found to range from 162 to 266 mg/L, and in 130 red wines from 140 to 417 mg/L (Gumyon and Heintz, 1952).

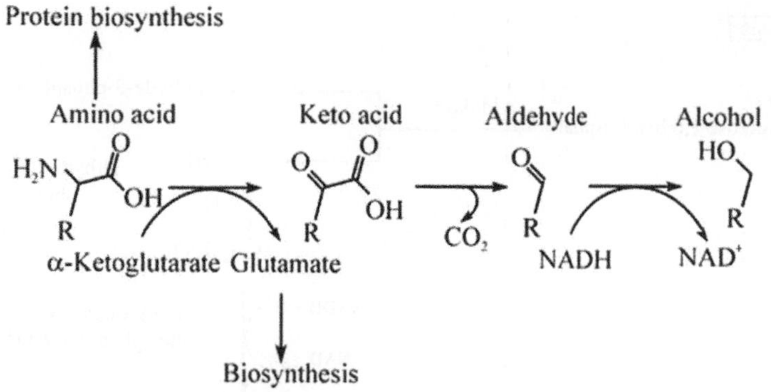

FIGURE 3.8

Pathway of higher alcohol formation from amino acids.

Jansen, M., Veurink, J., Euverink, G.J., Dijkhuizen, 2003. Growth of the salt-tolerant yeast Zygosaccharomyces rouxii *in microtiter plates: effects of NaCl, pH and temperature on growth and fusel alcohol production from branched-chain amino acids. FEMS Yeast Research 3 (3), 313–318.*

However, there is little systematic information on the quantities of higher alcohols in nongrape wines, though there is scattered documented information on this aspect (Joshi et al., 1999b).

In AF by yeast like *S. cerevisiae*, a very important by-product, glycerol, is produced (Fig. 3.9) by the fermentation called Neuberg's fermentation (Goyal, 1999). Neuberg's third form of fermentation is accomplished by the addition of sulfite to the fermentation medium, whereby glyceraldehyde 3-phosphate is converted into glycerol rather than dihydroxyacetone, as is the normal process in the fermentation to produce pyruvate. It is formed from dihydroxyacetone phosphate, which is one of the intermediate products of glycolysis, wherein it is reduced to glycerol phosphate by dihydroxyacetone phosphate reductase enzyme, as described earlier. It is the main by-product of AF. Low temperature, high tartaric acid, and sulfur dioxide favor the production of glycerol in AF, whereas the increase in sugar content decreases glycerol content relative to ethanol. Most of the glycerol is produced during the early stages of fermentation. Glycerol is of considerable sensory importance because of its sweet taste and oiliness, thus improving the taste of alcoholic fermented products, though it is produced only in trace amounts (Goyal, 1999).

It is believed that increased glycerol production will improve wine quality leading to better mouthfeel and enhanced viscosity.

Methanol or methyl alcohol (CH_3OH) is a highly toxic component of alcoholic beverages and is responsible for intoxication, blindness, or death. Eyes have very high sensitivity to formaldehyde, the immediate product of methanol; hence these are affected first. Methanol is formed as a by-product of AF under certain conditions such as the presence of high levels of pectin in a must (Fig. 3.10). The methanol is liberated upon addition of pectin and enzyme action. The pectinases that are employed for clarification and enhanced yield of juice result in the formation of methanol, sometimes beyond the acceptable range.

The stone fruits, like plum and apricot, are rich in pectin and thus have higher amounts of methanol (Woidich and Pfannhauser, 1974). The level of methanol in the distillates can be reduced substantially by discarding the first fraction of the distillate as methanol has a lower boiling point than ethanol (Amerine et al., 1980; Joshi, 1997).

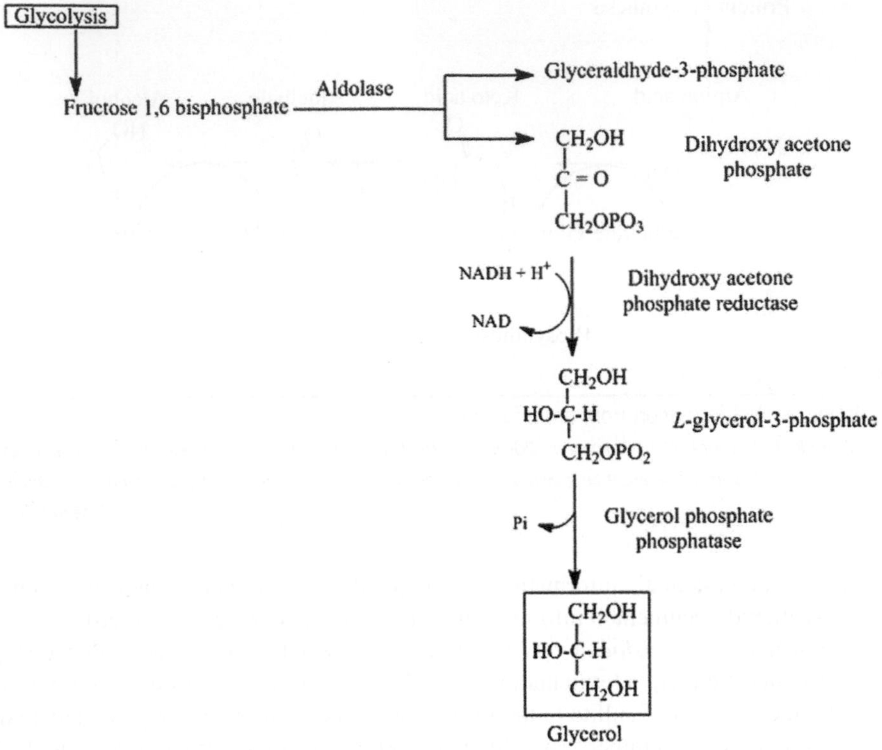

FIGURE 3.9

Pathway showing glycerol production (Goyal, 1999).

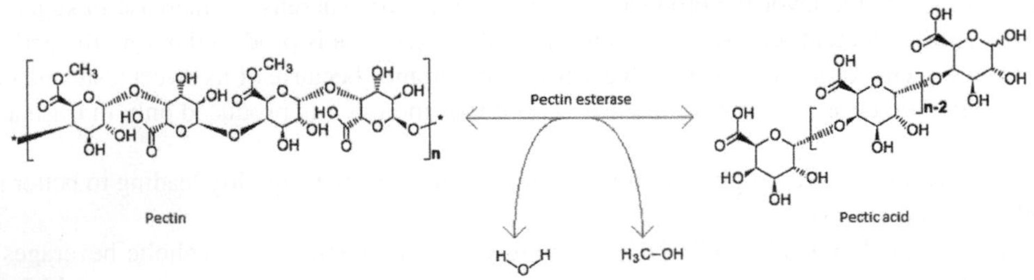

FIGURE 3.10

Formation of methanol in alcoholic fermentation by the action of pectin esterase enzyme on pectin (Goyal, 1999).

3.1.8 Volatile and Nonvolatile Organic Acids

During AF, both volatile and nonvolatile compounds are synthesized, or existing compounds are degraded, by the yeast. The yeast is able to synthesize organic acids by degradation of glutamate (Rana and Rana, 2011).

The concentrations of organic acids can vary greatly and depend on the condition and maturity of the fruit. During AF, the limiting activities of both pyruvate decarboxylase and alcohol dehydrogenase coupled with high sugar concentrations lead to the formation of many organic acids. In respiring cells, pyruvate decarboxylase and alcohol dehydrogenase activities are not highly expressed. Both of these enzymes are inducible by glucose (Denis et al., 1983; Schmitt et al., 1983; Sharma and Tauro, 1986). As a consequence, compounds other than ethanol are produced at the beginning of fermentation. Succinic acid, malic acid, and acetic acid account for the major organic acids formed by yeasts. Succinic acid is the main carboxylic acid produced by yeast during wine fermentation. The ratio of acids, however, varies with different yeast strains.

Succinic acid is the main carboxylic acid produced by yeasts during wine fermentation (Fig. 3.11) and its concentration goes up to 2.0 g/L (Radler, 1993). The addition of sugars to the resting cells of yeasts results in rapid formation of succinic acid. The amino acid glutamate favors the production of succinic acid. When yeasts are grown aerobically on glutamate, an oxidative pathway involving succinyl-CoA becomes operative and succinic acid is synthesized (Goyal, 1999). The important enzyme 2-oxoglutarate dehydrogenase, of the TCA cycle, is present in fermenting yeast, which is responsible for the production of succinate during fermentation by oxidative reactions.

A small amount of succinate can also be synthesized by the reductive pathway from oxaloacetate or malate or aspartate. Malate formed from oxaloacetate may be converted to small amounts of succinate by pyruvate decarboxylase of the reductive pathway (Schwartz and Radler, 1988).

Some yeasts have the capacity to synthesize malate (Fig. 3.12) during fermentation (Goyal, 1999) and thus increase the acidity of the fermented products. Consequently, the pH is decreased, giving additional advantages of preservation of fermented foods like wine. Malic acid formation is dependent upon the yeast strain and cultural conditions and is favored by the concentration of sugars (20–30%), a pH value of nearly 5, a limiting concentration of nitrogen compounds (100–200 mg N/L), and the presence of CO_2 (Radler and Lang, 1982). *S. cerevisiae* forms approximately equal quantities of succinate and malate, whereas the strain *Saccharomyces uvarum* produces about 10 times more malate than succinate.

Acetic acid is produced by many strains of yeast and is also the main product of ethanol oxidation by acetic acid bacteria. It is the main volatile acid in fermented beverages and constitutes more than 90% of the

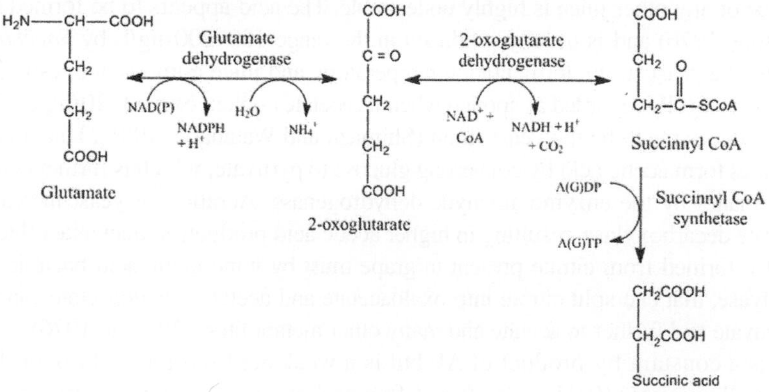

FIGURE 3.11

Oxidative pathway for the synthesis of succinic acid by yeast.

Rana, N.S., Rana, V.S., 2011. Biochemistry of wine preparation. In: Joshi, V.K. (Ed.), Handbook of Enology: Principles, Practices, and Recent Innovations, vol. 2. Asiatech Publishers Inc., New Delhi, pp. 618–678.

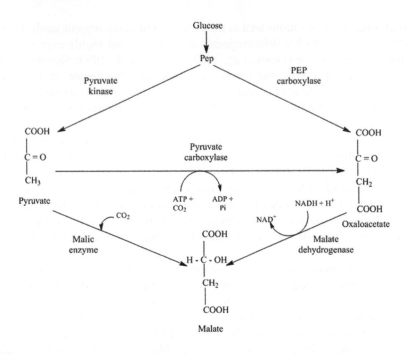

FIGURE 3.12

Synthesis of malate by yeast during fermentation. *PEP*, phosphoenolpyruvate.

volatile acidity in wine (Amerine et al., 1980). Its presence beyond certain limits, however, spoils alcoholic beverages. Some yeast strains, like *Hansenula*, produce a high amount of acetic acid and, hence, are not suitable for brewing (Amerine et al., 1980; Joshi, 1997). Because of the negative sensory attributes of this acid at higher concentrations and its association with *Acetobacter* spoilage, its production during fermentation of grape juice or any other juice is highly undesirable. The acid appears to be formed early in the fermentation (Whiting, 1976) and is usually produced in the range 100–200 mg/L by *Saccharomyces* and is also influenced by the yeast strain, fermentation temperature, and juice composition, especially sugars and nitrogen. The wine is legally regarded as spoiled when its acetate content exceeds 20 mg/L. The major factor in acetate production seems to be the yeast strain (Shimazu and Watanabe, 1981). Under anaerobic conditions, carbohydrates form acetic acid by converting glucose to pyruvate, which is further converted to acetic acid instead of ethanol by the enzyme aldehyde dehydrogenase. Aeration of yeast increases the specific activity of pyruvate decarboxylase, resulting in higher acetic acid production than anaerobic pitching yeast. Acetic acid is also formed from citrate present in grape must by wine lactic acid bacteria, which have an enzyme, citrate lyase, that can split citrate into oxaloacetate and acetate. Oxaloacetate can again be decarboxylated to pyruvate and further to acetate and many other metabolites (Whiting, 1976).

Lactic acid is a constant by-product of AF but is a weak acid with a slight odor. Usually, yeasts produce only small amounts (0.04–0.75 g/L) of free carboxylic acid during fermentation and lactic acid constitutes a small fraction of these acids. The yeasts that produce a large amount of lactic acid belong to the species *Torulaspora pretoriensis* (synonym *Saccharomyces pretoriensis*), which resembles *S. cerevisiae*. Lactic acid is synthesized from sugar by reduction of pyruvate by a lactate dehydrogenase that has been partially purified (Genitsariotis, 1979). The formation of lactic acid is shown in Fig. 3.13.

Yeasts are able to produce several oxoacids, like 2-oxoglutarate, 2-oxobutyrate, 2-oxoisovalerate, 2-oxo-3-methylvalerate, and 2-oxoisocaproate, which are the products of amino acid metabolism. In addition, grape must, or for that matter any fruit must, also contains low concentrations of pyruvate and 2-oxoglutarate. During fermentation, however, their amounts increase and pyruvate is later partially metabolized by the yeast. Both acids may be formed from the corresponding amino acids, alanine and glutamate, but they are also excreted by yeast cells growing in the presence of low concentrations of nitrogen compounds. The oxoacids are important in wine fermentation because they are able to bind to SO_2, thus lowering the content of free SO_2 in wine needed for the safe preservation of wine.

Citric acid is found in grapes only in trace amounts but is an important additive for winemakers. It is active in metal chelation, thus aiding in reducing the probability of iron and copper haziness, and has a stabilizing effect. Consequently, it is usually added to wines that are nearing the end of processing—just before bottling. Although acetic acid is the main volatile acid, comprising about 90%, a variety of other fatty acids are also present, like hexanoic acid, octanoic acid, decanoic acid, propionic acid, and butyric acid present in trace amounts in wine (Sponholz and Dittrich, 1979). The pathway for the conversion of citrate into acetic acid and other metabolites is shown in Fig. 3.14.

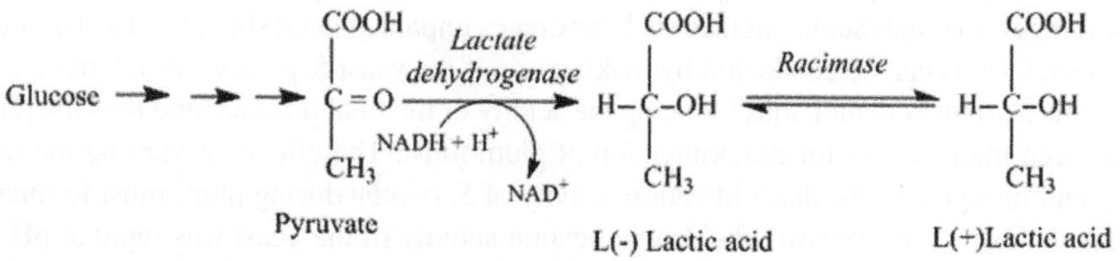

FIGURE 3.13

Mechanism of lactic acid formation (Goyal, 1999).

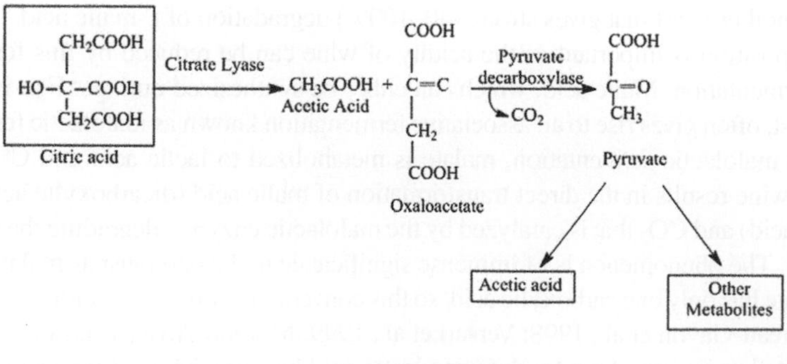

FIGURE 3.14

Pathway showing the conversion of citrate to acetate.

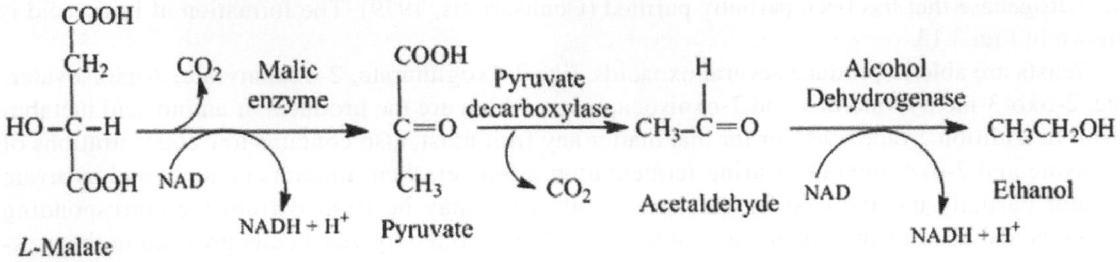

FIGURE 3.15

Pathway showing metabolism of malate to ethanol in *Saccharomyces cerevisiae* and *Schizosaccharomyces pombe*.

3.1.9 Decomposition of Organic Acids

The metabolism of malate by wine yeasts has been studied and documented (Peynaud et al., 1964; Rankine, 1966; Rodriquez and Thornton, 1990; Shimazu and Watanabe, 1981; Wenzel et al., 1982). Wine yeast is capable of metabolizing malate during fermentation, but generally in small amounts, i.e., 3–45%, depending on the strain, as is the case with species of *Kloeckera*, *Candida*, *Pichia*, and *Hansenula*. Thus, these species, as well as *S. cerevisiae*, cause only minor changes to the total acidity of the wine. In contrast, however, strains of *Schizosaccharomyces pombe* and *Schizosaccharomyces malidevorans* can completely degrade malic acid (Rankine, 1966), as shown in Fig. 3.15.

Some fruits, like plums, apples, and grapes, have malic acid as the predominant acid, and in alcoholic fermented beverages, the high acidity makes such beverages unpalatable (Joshi et al., 1991b). Fermentation of juices from such fruits can be conducted by making use of the yeast *S. pombe*, which has the capacity to degrade the malic acid into ethanol, thus reducing the acidity of the final product, and thus the physiology of this yeast has been made use of for deacidification of plum must. The effects of varying the levels of pH, ethanol, SO_2, and nitrogen on the deacidification activity of *S. pombe* during plum must fermentation have been determined (Joshi et al., 1991b). The deacidification activity of the yeast was rapid at pH 3.0–4.5 but was adversely affected at pH 2.5 in the initial stages of fermentation, and 150 ppm SO_2 was effective in enhancing the activity, but the deacidification activity of the yeast was quite susceptible to higher concentrations of ethanol (5–15%). *S. cerevisiae* has low affinity to degrade malate efficiently, but the genetically engineered strain made using *S. pombe* and *Lactococcus lactis* actively metabolized the malate to lactate within 3 days in Cabernet Sauvignon and Shiraz grape must at 20°C (Volschenk et al., 1997). *Zygosaccharomyces bailii* is another yeast that gives strong (40–100%) degradation of L-malic acid.

Malate decomposition is important as the acidity of wine can be reduced by this fermentation, called malolactic acid fermentation. Malic acid, which can either be synthesized during AF or is present in grape, plum, or apple must, often gives rise to an associative fermentation known as malolactic fermentation. As the name suggests, in malolactic fermentation, malate is metabolized to lactic acid and CO_2. The biological deacidification of wine results in the direct transformation of malic acid (dicarboxylic acid) into lactic acid (monocarboxylic acid) and CO_2 that is catalyzed by the malolactic enzyme, degrading the malic acid in wine (Bartowsky, 2011). The phenomenon is of immense significance to the enologist as malate is a dicarboxylic acid, whereas lactate has only one carboxylic acid, so this conversion reduces the acidity of the wine (Henick-Kling, 1993; Ribereau-Gayon et al., 1998; Versari et al., 1999; Moreno-Arribas and Lonvaud-Funel, 2001). Malolactic fermentation is carried out by the wine lactic acid bacteria. Most of the lactic acid bacteria isolated from wine possess the malolactic enzyme (Hinick-King, 1993) responsible for the reaction. Certain lactic acid bacteria, like *Lactobacillus casei* and *Lactococcus faecalis*, also make use of the malic enzyme for the metabolism of malate (Kandler et al., 1973; Renault et al., 1988). The enzyme is linked with NAD(P) coenzyme and requires Mn^{2+} for its activity, and the lactic acid bacteria associated with grape must and wines belong to *Lactobacillus, Leuconostoc, Oenococcus,* and *Pediococcus* (Lonvaud-Funel, 1999; Delaquis et al., 2000). However, in most cases, malolactic fermentation is carried out by *Oenococcus oeni* (formerly known as *Leuconostoc oenos*) (Dicks et al., 1995). In certain wines, malolactic fermentation brings about improvement in flavor by converting malate to pyruvate, which is subsequently metabolized to lactic acid with the help of lactate dehydrogenase (Gambaro et al., 2001; Bertrand et al., 2000; De Revel et al., 1999).

The third pathway of malolactic fermentation is used by *Lactobacillus fermentum,* which elaborates malate dehydrogenase, which catalyzes the reaction that oxidizes malate to oxaloacetate. Among the sensorial changes originated by lactic acid bacteria during malolactic fermentation, diacetyl is considered to produce one of the most important flavors. Carbonyl or acetonic compounds, including diacetyl, acetoin, and 2,3-butanodiol (Fig. 3.16), are formed from citric acid metabolism by lactic acid bacteria via several reactions in which citrate lyase plays a role (Hugenholtz, 1993). At moderate concentrations (of around 5–10 mg/L), diacetyl has a positive effect on the bouquet of the wine because of its buttery taste, whereas at higher concentrations it becomes a defect. The factors influencing their biosynthesis and the buttery diacetyl content of wine have also been reported (Martineau and Henick-Kling, 1995a,b; Nielsen and Richelieu, 1999; Bartowsky and Henschke, 2000). (For more details see a separate section on this aspect.)

The control of lactic acid bacteria during and after vinification is essential to obtain wines of consistently high quality. To meet this objective, lysozyme, an enzyme present in hen egg white with lytic activity against lactic acid bacteria, has been used (Cunningham et al., 1991). An addition of 500 mg lysozyme per liter of grape must results in malolactic fermentation, whereas 250 mg/L to red wine malolactic fermentation promotes microbiological stabilization (Gerbaux, 1997). The fermentation is described in somewhat more detail with respect to microorganisms and quality in a separate section of this chapter regarding malolactic acid fermentation.

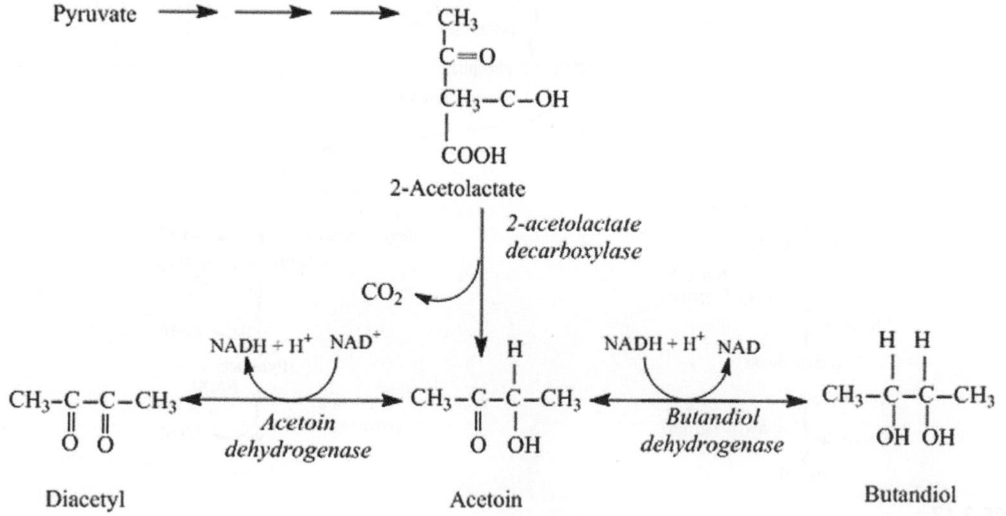

FIGURE 3.16

Metabolism of acetolactate to diacetyl and butandiol by lactic acid bacteria.

3.1.10 Metabolic Pathways of Lactic Acid Bacteria

Simple carbohydrates (hexoses, pentoses) are the main carbon sources used by lactic acid bacteria (LAB), but disaccharides such as lactose, maltose, or sucrose can also be metabolized. A limited number of LAB species (known as amylolytic LAB) can degrade starch. The various LAB groups, however, differ with respect to carbohydrate metabolism, not only concerning the substrates that can be metabolized but also in the final fermentation products. Consequently, each food substrate is usually fermented by a specific group of LAB and, hence, the profile of fermentation products is also distinctive. LAB can obtain ATP only by fermentation, usually of sugars (Axelsson, 2004).

Two main sugar fermentation pathways (homofermentative and heterofermentative) can be distinguished among LAB (Axelsson, 2004; Mayo et al., 2010). The carbohydrate metabolism of various species of LAB and the biochemistry of LAB-produced fermentation products also vary (Figs. 3.17 and 3.18). The reduction of pyruvic acid happens by the action of lactate dehydrogenase, which needs NAD+/NADP+ as a cofactor. In homolactic fermentation the predominant form of lactic acid is dependent not only on the lactate dehydrogenase specificity [*Lactococcus* form L(+)-lactate and *Lactobacillus* D(−)-lactate], but also on the presence of lactate racemase, which, when active in

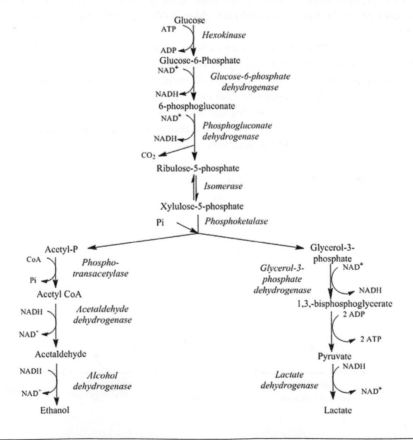

FIGURE 3.17

Formation of ethanol and lactic acid by heterofermentative lactic acid bacteria. *P*, phosphate.

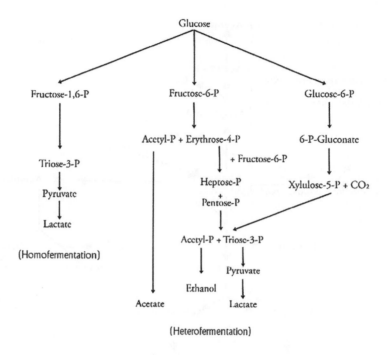

FIGURE 3.18

Formation of different products in the homolactic or heterolactic pathway. *DP*, bisphosphate; *P*, phosphate.

Faria-Oliveira, F., Diniz, R.H.S., Godoy-Santos, R., Piló, F.B., Mezadri, H., Castro, I.M., Brandão, R.L., 2015. The role of yeast and lactic acid bacteria in the production of fermented beverages. In: Amer Eissa, A. (Ed.), South America, Food Production and Industry. InTech. http://dx.doi.org/10.5772/60877.

the microorganism cells, results in a racemic mix (D,L-lactate) by fermentation (Buruleanu et al., 2010). Heterofermentative LAB produce the anaerobic fermentation of glucosides via the pentose phosphate (6-phosphogluconate) pathway. The kind of fermentation products depends on the species involved. Thus, *Lactobacillus brevis* produces, through heterolactic fermentation, lactic acid, acetic acid, and carbon dioxide, whereas *Leuconostoc mesenteroides* produces lactic acid, ethanol, and carbon dioxide. Glycolysis (the EMP pathway) results in almost exclusively lactic acid as the end product under normal conditions, and this type of metabolism is referred to as homolactic fermentation (Neti et al., 2011).

Homofermentative LAB mainly produce lactic acid through glycolysis (homolactic fermentation), whereas heterofermentative LAB produce, in addition to lactic acid, CO_2, acetic acid, and/or ethanol through the 6-phosphogluconate/phosphoketolase pathway (heterolactic fermentation). The utilization of cosubstrates such as oxygen or fructose as electron acceptors by obligate heterofermentative *Lactobacillus* spp. is coupled to an increased production of acetate in dough. In hexose fermentation, facultative and obligately hydrogen phosphate as a nitrogen source is added and the nitrogenous compounds of the must are not utilized, higher alcohols are not produced, and the quality of the alcoholic beverage remains desirable. The formation of various products from pyruvate in LAB is depicted in Fig. 3.19.

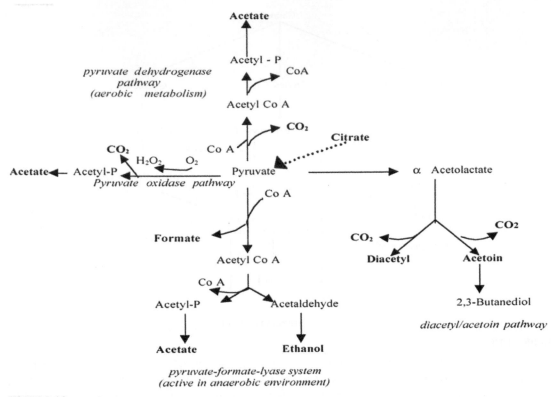

FIGURE 3.19

Formation of various products in lactic acid bacteria. *P*, phosphate.

Caplice, E., Fitzgerald, G.F., 1999. Food fermentations: role of microorganisms in food production and preservation. International Journal of Food Microbiology 50, 131–149.

3.2 NITROGEN METABOLISM

Yeasts require an exogenous source of nitrogen mainly for the synthesis of proteins and nucleic acids and also for the growth and metabolism of the yeast. Nitrogen is the second important nutrient after carbon and is utilized by yeast during the fermentation of any fruit must. Nitrogen has both positive and negative implications in wine production. It is responsible for the production of reduced sulfur compounds (Henschke and Jiranek, 1991; Jiranek and Henschke, 1991) and formation of ethyl carbamate. On the other hand, it is also involved in the biosynthesis of compounds responsible for the fermentation bouquet. So to enhance wine aroma and eliminate sluggish fermentation (Navarro and Navarro, 2011), the nitrogen metabolism is very important. An overview of the complete metabolism of nitrogen is shown in Fig. 3.20.

It is well established that *S. cerevisiae* can grow on a diverse range of nitrogen compounds, including ammonium, urea, amino acids, small peptides, purines, and pyrimidine-based compounds (Cooper, 1982; Large, 1986). Usually, the must contains varying proportions of all these compounds, which are potentially utilized by the cell. However, before their utilization, the nitrogen compound must be transported into the cell and its efficiency depends on the expression, regulation, and efficiency of the transport system and the

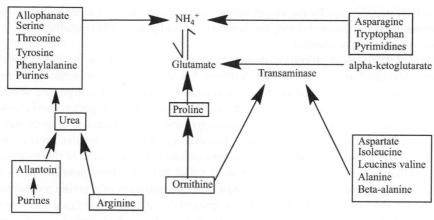

FIGURE 3.20

Major products of nitrogenous compound degradation.

Based on Rana, N.S., Rana, V.S., 2011. Biochemistry of wine preparation. In: Joshi, V.K. (Ed.), Handbook of Enology: Principles, Practices, and Recent Innovations, vol. 2. Asiatech Publishers Inc., New Delhi, pp. 618–678.

regulation and energetics of subsequent catabolic and anabolic processes. Consequently, growth, fermentation rate, and biomass yield depend on the quantity and nature of the nitrogen sources available.

3.2.1 Nitrogen Sources and Nitrogen Supplements

Most musts of fruits contain full complements of nutrients in the form of carbon and nitrogen. The nitrogen is present in a complex range of compounds, i.e., in the form of amino acids, ammonium, peptides, nucleotides, proteins, and vitamins, but the concentrations of these compounds vary according to the type of fruit. However, a great variation has been observed for amino acid concentration and composition in fruits juices. Generally speaking, a preferred yeast nitrogen source is one that is most readily converted biosynthetically into useful nitrogen compounds like ammonia or glutamate diammonium hydrogen source or one that requires the least energy input or cofactors for mobilization of the nitrogen moiety and, more importantly, should not have a toxic effect and be economically feasible (Amerine et al., 1980; Joshi, 1997; Joshi et al., 1990). *S. cerevisiae* can grow on ammonium, urea, and most of the amino acids and does not have any absolute requirement of amino acids. In addition to the natural nitrogen present in fruit juices, must is also supplemented with assimilable nitrogen to avoid the problems associated with nitrogen deficiency during fermentation. For example, diammonium phosphate is added to the must at 200 mg/L prior to inoculation with yeast and a further addition of 100 mg/L is done in response to the production of hydrogen sulfide (Henschke and Ough, 1991). Urea and other commercial yeast foods are other alternate nitrogen supplements (Ingledew and Kunkee, 1985). However, the addition of urea is prohibited in most wine-producing countries because of its involvement in the production of ethyl carbamate (Ingledew et al., 1987; Monteiro et al., 1989; Ough et al., 1988a,b), as discussed in a later section of the chapter.

3.2.2 Amino Acid Utilization Profile

Amino acid composition of the must is of great importance in wine production because amino acids act as a source of nitrogen for yeast during fermentation and have a direct influence on the aromatic composition of wines (Drdak et al., 1993). They have a variety of origins; those indigenous to fruit can be partially or totally metabolized by the yeasts during the growth phase, some are excreted by yeasts at the end of fermentation or released during the autolysis of dead yeasts, whereas others are produced by the enzymatic degradation of proteins.

The pattern of amino acid utilization during fermentation reflects both the initial distribution of nitrogen compounds and strain-dependent preferences. When yeasts were presented with a mixture of amino acids in a model medium, the most important source of nitrogen was arginine, which provided 30–50% of the total nitrogen. However, lysine, serine, threonine, leucine, aspartate, and glutamate were the next most accumulated compounds, and glycine, tyrosine, tryptophan, and alanine were the least utilized. This pattern of nitrogen utilization was, however, quantitatively similar irrespective of yeast strain, the degree of aeration, and the sugar and ammonium concentrations of the medium (Jiranek et al., 1996). Differences in amino acid utilization by yeasts have also been pointed out. In a study on Californian musts, the inability of yeasts to utilize a high concentration of arginine present in nitrogen-rich must was observed (Ough et al., 1991). To understand the relative importance and role of various amino acids as nitrogen sources for yeast metabolism needs to be studied further.

3.2.3 Uptake and Transport of Nitrogen Compounds

In fruit juice fermentations to make wine, nitrogenous compounds with low concentration are taken up very quickly. However, before degradation of compounds for a nitrogen source, the biosynthetic pool of amino acids is filled prior to yeast growth. Once this is achieved and growth has started, various nitrogen compounds are taken up and degraded in a specific order of preference. Ammonium ion, glutamate, and glutamines are the most preferred nitrogen sources and are utilized directly in biosynthesis and these three nitrogen sources are depleted first from the medium before utilization of other nitrogen sources. The next group of nitrogen compounds in terms of preference includes alanine, serine, threonine, aspartate, asparagine, urea and arginine, proline, glycine, lysine, histidine, and pyrimidines. Thiamine, however, is not utilized as a source of nitrogen by most strains of *Saccharomyces*, but it can be readily utilized directly as a biosynthetic precursor.

The metabolism of aromatic amino acids is complex, with some reactions requiring oxygen as a cofactor, which may be limiting during fermentation. This order of preference may change depending upon environmental, physiological, and strain-specific factors. The transport of nitrogen compounds into the cell is the major factor controlling their utilization. Different nitrogen sources are transported by different mechanisms. There are three basic means of transport, viz., simple diffusion, facilitated diffusion, and active transport. Comprehensive reviews of the amino acid transport mechanisms of *S. cerevisiae* have been published (Jiranek et al., 1996; Casey and Ingledew, 1986; Eddy, 1982; Henschke and Rose, 1991; Horak and Kotyk, 1977; Wiame et al., 1985).

The cell's ability to excrete protons by ATPase, a hydrogen ion pump that uses energy from the hydrolysis of one ATP molecule for each hydrogen ion pumped out of the cell, is an important regulatory factor for amino acid uptake (Roon et al., 1975a,b, 1977a,b). Other than this transport, both D and L isomers of neutral and basic amino acids and proline, to a limited extent, are transported by a group-specific transport system called the general amino acid permease (GAP). The GAP appears to operate as a nitrogen scavenger system. During the initial stage of fermentation, permeases are specific for a small number of amino acids, but GAP is the principal transport system for amino acids during the later stages of fermentation (Rose, 1987). Urea is transported via two mechanisms in the yeast. The first is an active transport, which has an apparent K_m of 14 mM and is sensitive to nitrogen repression (Colowick, 1973). The second system is one of passive or facilitated diffusion, which operates at external urea concentrations of greater than 0.5 mM (Cooper and Sumrada, 1975; Cooper, 1982). The utilization of small peptides and proteins depends on the ability of yeast either to transport these compounds or to degrade them extracellularly. The transport of di- and tripeptides is thought to be accomplished by a general peptide transport system, but the utilization of large peptides or proteins depends on the ability of the yeast either to transport these compounds or to degrade them extracellularly. Neither of these mechanisms has, however, been reported in *S. cerevisiae* (Lagace et al., 1990; Rosi et al., 1987; Sturley and Young, 1988). Moreover, the importance of peptides to yeast metabolism remains to be clarified.

3.2.4 Regulation of Nitrogen Transport

The activity of transport systems can be regulated by several mechanisms, which vary in their specificity, response time, and level of influence. Feedback inhibition, *trans*-inhibition, inactivation/reactivation, and repression are the main processes involved to influence transport mechanisms. Feedback inhibition is the process by which an accumulated substrate inhibits further accumulation by the same transport carrier, as reviewed earlier (Rana and Rana, 2011). Feedback inhibition is of least significance when the internal substrate concentrations are kept low by catabolism or compartmentation into subcellular organelles. *Trans*-inhibition refers to the competition for uptake exhibited between substrates that do not share a common transport mechanism (Goyal, 1999). Inactivation/reactivation processes have been described for several ammonium-sensitive permeases of *S. cerevisiae*, including the GAP (Grenson, 1983a,b; Grenson and Acheroy, 1982). During growth on a poor nitrogen source such as proline, the permease is synthesized through the expression of the structural gene, *GAP*. In addition to this mechanism, both ammonium-sensitive permeases and degradative enzymes could be regulated by nitrogen catabolite repression. Repression may act at the level of transcription, translation, or processing of RNA or precursor proteins.

3.2.5 Factors Affecting Nitrogen Accumulation

Many factors influence the accumulation of nitrogen compounds in grape must, such as the cultural conditions, medium composition, and yeast strain factors. The general aspects of nitrogen metabolism in yeasts have also been reviewed earlier (Bisson, 1991; Cartwright et al., 1986, 1989; Cooper, 1982; Davis, 1986; Hinnesbusch, 1988; Jones and Fink, 1982; Large, 1986; Nykanen, 1986; Pierce, 1987; Schwenke, 1991). Depending upon the type of amino acid accumulated, it may be utilized in one of the following ways:

- An amino acid can be incorporated directly into a protein.
- An amino acid can be degraded to liberate nitrogen, which is used for the biosynthesis of other cell nitrogen constituents.
- The carbon component of an amino acid can be released and used for the biosynthesis of other cell carbon constituents.

Thus, amino acid/nitrogen compounds catabolize or anabolize to synthesize various constituents during fermentation through various pathways. Essentially all nitrogen compounds that accumulate are degraded to either of the products, ammonium or glutamate. The nitrogen catabolic pathways of yeasts have been summarized (Large, 1986). The two nitrogen compounds, i.e., NH^{4+} and glutamate, are required by the cell to coordinate biosynthesis of all biologically active nitrogen-containing components. These end products of nitrogen metabolism, however, are interconvertible. In general, two types of glutamate dehydrogenases (GDHs) are expressed, depending upon the available nitrogen source, and rarely they are expressed equally (Cooper, 1982). $NADP^+$-GDH is expressed when ammonium ions are the sole source of nitrogen in the medium; in contrast, NAD-GDH is expressed when glutamate, aspartate, or alanine is the sole source of nitrogen. Different amino acids or nitrogen compounds catabolize to yield glutamate or NH^{4+} used for various biosynthetic processes. Amino acids like lysine, histidine, cysteine, and glycine are considered good sources of nitrogen for many yeasts but none of these compounds is utilized efficiently by *Saccharomyces* as a nitrogen source. Glutamate and NH^{4+} are the two main degradation products of nitrogen metabolism that are directly used for biosynthesis. Glutamine generates both glutamate and NH^{4+} and, therefore, is also a preferred nitrogen source, which is depleted first. The next group of compounds includes alanine, serine, threonine, aspartate, asparagine, urea, and arginine. Proline is the nitrogen source of choice under aerobic conditions. Glycine, lysine, histidine, and pyrimidine are not metabolized by most yeast strains as a source of nitrogen but can be readily and directly utilized as biosynthetic precursors. Metabolism of aromatic amino acids also does not occur during fermentation as it is complex and requires oxygen and other cofactors. The ability of arginine to support high growth rates is due to the fact that it is rich in nitrogen and for transport requires only one proton, making it economical to transport. Some yeast strains excrete urea during arginine metabolism, which is reabsorbed and degraded further when the concentration of assimilable nitrogen becomes low (Henschke and Ough, 1991; Monteiro and Bisson, 1991).

Reduced sulfur is incorporated into carbon compounds to synthesize sulfur-containing amino acids, like cysteine and methionine.

Regarding the formation of higher alcohols, amino acids that are deaminated metabolically to release their nitrogen components leave behind a carbon skeleton, which is regarded as a waste product. Deamination of amino acids can result in the formation of keto acids or higher (fusel) alcohols via a metabolic pathway as shown in Fig. 3.8.

In addition to deamination and decarboxylation, higher alcohols are also produced during the biosynthesis of amino acids from their corresponding keto acids (Nykanen, 1986). Various factors, viz., the yeast strain, growth, temperature, ethanol production, must pH, level of solids, and degree of aeration, affect the production of alcohols. Fruit and maturation also affect their concentration owing to the qualitative and quantitative differences in amino acid composition.

3.2.6 Enological Aspects of Nitrogen Metabolism

Slow or incomplete fermentation and hydrogen sulfide formation are the two serious fermentation problems encountered in winemaking and sporadic research work has been done to tackle these problems (Agenbach, 1977; Eschenbruch et al., 1978). Hydrogen sulfide (H_2S), an unpleasant aroma compound with low sensory threshold (10–100 mg/L), is sometimes formed in excessive amounts by yeast during the fermentation of grape must (Acree et al., 1972; Rankine, 1963). Nitrogen metabolism is one of several mechanisms accounting for its formation (Henschke and Jiranek, 1991; Monk, 1986). There are two broad phases of H_2S production in wineries. The first phase occurs during active fermentation and is responsive to supplementation with assimilable nitrogen. By contrast, the second phase is most frequently observed near to depletion of sugars from the must and proceeds in the presence of assimilable nitrogen. Total production of H_2S in grape juice is inversely related to the levels of assimilable amino acids.

Nitrogen metabolism promotes yeast growth essentially by supplying the precursors of protein and nucleic acid synthesis, so a rich supply of nitrogen will allow high rates of growth and biomass yield and stimulate fermentation activity, but an imbalance in the supply and demand of nitrogen results in the development of fermentation problems as discussed earlier. Nitrogen affects the fermentation rate directly by synthesizing proteins for the glycolytic pathway and NH^{4+} ions serve as an allosteric effector for the activity of PFK, the main regulatory enzyme of the glycolytic pathway. Nitrogen limitation results in the accelerated turnover of glucose permeases and thus, slows down the fermentation rate, resulting in sluggish or incomplete fermentation (Lagunas et al., 1982; Salmon, 1989; Navarro and Navarro, 2011). On the other hand, enhanced wine flavor and urea formation appear to be linked to higher concentrations of assimilable nitrogen and in particular amino acids, but to what extent the nitrogen availability affects the composition and sensory aspects of wine is not clear.

3.3 METABOLISM OF SULFUR: CHEMISTRY OF THE PRODUCTION OF OFF-FLAVOR

In the metabolism of sulfur, the formation of volatile sulfur compounds (VSCs) takes place, which results in the production of off-flavors. It is the main problem in the production of quality wine, and the main components responsible for off-flavors are acetic acids, sulfur-containing volatiles, free amino nitrogen, vitamins, and other factors. The mechanisms of the formation of VSCs during wine fermentation are only partially understood because of the complex nature of the factors involved in winemaking. The production of VSCs varies widely in the composition of model systems with nutrient deficiencies or with residual elemental sulfur. These are spoilage-causing compounds that are very volatile and have unpleasant odors. Although their concentration is very minute (10–100 mg/L), their sensory impact is huge during fermentation and poses a significant problem to winemakers. This group includes compounds that are very volatile and have unpleasant odors, generally described in terms of rotten egg, skunk aroma, garlic or onion, etc.

Among the sulfur-containing volatiles, H_2S creates the biggest problem, with its rotten egg-like flavor. Several factors may contribute to H_2S production (Richard et al., 1997; Paquin and Williamson, 1986). It can be produced by the yeast during fermentation because of several factors like the presence of elemental sulfur in grape skin (Acree et al., 1972; Rankine, 1963; Schutz and Kunkel, 1977; Wenzel and Dittrich, 1978; Thomas et al., 1993), inadequate levels of free α-amino nitrogen (FAN) in the must (Monk, 1986; Vos and Gray, 1979), a deficiency of pantothenic acid (Tanner, 1969) or pyridoxine (Vidal-Carou et al., 1991), or higher than usual levels of cysteine in the juice and yeast strains (Acree et al., 1972; Radler, 1993; Suzzi et al., 1985). The odor threshold value of H_2S in wine varies from 50 to 80 mg/L, 240, and above this concentration, it causes off-flavors. Sulfate uptake and reduction are regulated by a sulfur-containing amino acid, i.e., methionine. The total production of H_2S in eight grape juices with respect to different levels of assimilable amino acids was studied. In general, the highest concentration of H_2S was produced during the rapid phase of fermentation. The deficiency of assimilable amino acids may be a major factor in H_2S production at the later stages of fermentation (Park et al., 2000).

The unavailability or deficiency of FAN in the must also causes H_2S formation by yeast (Henschke and Jiranek, 1991; Monk, 1986). Reduction in H_2S formation in the juices of one cultivar having 100 mg/L at 100 mg/L FAN to essentially zero at 300–400 mg/L FAN has been reported (Vos and Gray, 1979). It has also been suggested that low concentrations of FAN might stimulate proteases, which cause hydrolysis of juice proteins. Deficiency of vitamins, pantothenate or B_6, in the medium can also cause increased H_2S production by yeasts (Wainwright, 1971). A deficiency of pantothenate leads to a deficiency of CoA, which is essential for methionine biosynthesis. Pyridoxine is necessary for several reactions in the pathway of methionine synthesis and a large amount of H_2S is produced if the concentration is below 2 mg/L. Normally, most of the H_2S formed during fermentation is carried away with carbon dioxide. Therefore, in normal faultless wines, H_2S is not detectable despite a detection limit of 1.5 mg/L, as in most cases, H_2S disappears after fermentation slowly. Increased amounts of H_2S can be purged out or oxidized by aeration or oxidized by treatment with sulfite (Tanner, 1969).

Several other sulfur compounds are also detected, which are responsible for off-flavors. These include dimethyl sulfide, at a concentration ranging from 0 to 47 mg/L, synthesized from *S*-methyl-methionine (Loubser and Du Plessis, 1976). Diethyl sulfide, with the flavor of cooked vegetables, onion, and garlic (Fuck and Radler, 1972; Goniak and Noble, 1987; Januik, 1984), and mercaptans, having a rotten egg-like flavor (Fuck and Radler, 1972), formed during fermentation in synthetic media containing cysteine, methionine, or sulfate and thioesters, like *S*-methyl thioacetate. The factors leading to the production of these compounds have not been thoroughly elucidated. Metabolism of sulfur-containing amino acids may be responsible for their synthesis. The production of sulfur compounds that give off-character to the wine results from the amino acid composition of the juice, as discussed earlier. The effect of added threonine and methionine on H_2S production has also been documented (Wainwright, 1971).

Considerable emphasis has been placed on reducing the undesirable H_2S and other VSCs in wine associated with off-flavors (Rauhut et al., 1996; Park et al., 2000; Spiropoulos et al., 2000) and, to a lesser extent, diacetyl and other related carbonyl compounds (Martineau and Henick-Kling, 1995a,b; Romanoand Suzzi, 1996). However, comparatively little effort has concentrated on higher molecular weight sulfur compounds [such as *N*-(3-methylthiopropyl) acetamide and 3-methylthiopropionic acid] that have been shown to be by-products of the metabolism of the amino acids cysteine, methionine, and homomethionine by yeast (Anocibar-Beloqui et al., 1995; Lavigne and Dubourdieu, 1996), which needs attention.

3.4 ACETIC ACID FERMENTATION

The formation of acetic acid during AF, which may be in small amounts, takes place by the pathway shown in Fig. 3.21.

In *Acetobacter* and *Gluconobacter*, however, acetic acid is the major product, and fermentation is called acetic acid fermentation, as discussed here. The first step in the conversion of ethanol to acetic acid is the formation of acetaldehyde according to the following equation:

Step 1

$$CH_3CH_2OH + O_2 \xrightarrow{\text{Alcohol dehydrogenase}} CH_3CHO + H_2O$$

Ethanol   Acetaldehyde

Step 2

$$CH_3CHO + H_2O \longrightarrow H_3C{-}\underset{\underset{H}{|}}{\overset{\overset{H}{|}}{C}}{-}OH + [O] \xrightarrow{\text{Aldehyde dehydrogenase}} CH_3COOH + H_2O$$

Acetaldehyde   Acetic acid

The second step is the formation of acetic acid from acetaldehyde. The latter first reacts with water to yield hydrated acetaldehyde, which in turn is oxidized or dehydrogenated to yield acetic acid.

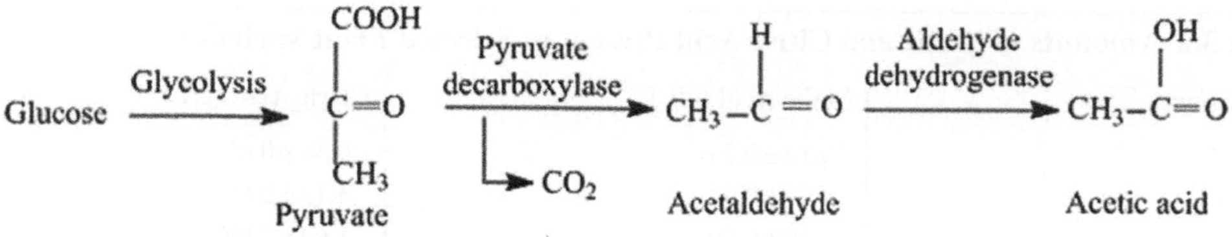

FIGURE 3.21

Mechanism of acetic acid synthesis.

4. ROLE OF ENZYMES IN WINEMAKING

The importance of enzymes involved in winemaking is obvious because grapes and microbes contain and produce divergent kinds of enzymes that relate to the characteristics of wines. The use of commercial enzyme preparations (Bisson and Butzke, 1996) for winemaking arose to address specific purposes such as better juice extraction (Joshi et al., 1991a), clarification of the wine (Lagace and Bisson, 1990; Joshi and Bhutani, 1991), and flavor production (Canal-llauberes, 1993; Sato, 2011). These commercial enzymatic preparations enhance the natural process by reinforcing the fruit's and yeast's own enzymatic activities, giving the winemaker more control over the process. The addition of these commercial enzymes to resolve clarification and filtration problems (pectinases, xylanases, glucanases, proteases), or to release varietal aromas (glycosidases), is a common practice in vinification (Martino et al., 2000). These products are not new, as they were first used in the 1970s.

5. USE OF ANTIMICROBIALS

Sulfur dioxide (SO_2) has been used for centuries in the wine industry as an antioxidant and antimicrobial agent. It is also known that the antimicrobial activity of SO_2 decreases in high-pH wines. However, with increasing health concerns and consumer preferences for more natural foods, there is a worldwide trend to reduce SO_2 (du Toit and Pretorius, 2000) and find some alternative to it. Several studies have described the efficacy of lysozyme in controlling or inhibiting malolactic fermentation (Gerbaux et al., 1997; Pilatte et al., 2000; Bartowsky, 2003). Reacting with oxygen, SO_2 protects vulnerable wine constituents from oxidation (Richard et al., 1997; Clarke and Bakker, 2004). The SO_2 does not simply react with oxygen to protect vulnerable polyphenols from oxidation, its interaction with oxygen is dependent on the concentration of catechol (Danilewiex et al., 2008).

6. MALOLACTIC FERMENTATION

There are two fermentation stages commonly involved in winemaking, AF, in which yeast metabolize sugars into alcohol and carbon dioxide, and malolactic fermentation (MLF), which involves LAB converting L-malate via one of three different enzymatic pathways into L-lactate and carbon dioxide (Radler, 1986). Whereas the former is common to all winemaking, i.e., grape, other fruits having malic acid, or otherwise, MLF can be a spontaneous process instigated by LAB naturally present on the fruit

Table 3.2 Amounts of Malic and Citric Acid Present in Selected Fruit Varieties[a]

Fruit Source	Malic Acid (g/L)	Citric Acid (g/L)
Apple[a]	10.12±0.23	0.36±0.02
Apricot	4.59±0.04	4.13±0.05
Pear[a]	2.49±0.09	1.64±0.07
Kiwi	2.66±0.13	11.00±0.14
Orange	2.13±0.01	11.71±0.17
Strawberry	1.74±0.10	7.13±0.34
Pineapple[a]	1.43±0.09	6.52±0.18

[a]*Results are from one group of fruit when more than one group of the same fruit was available.*
Adapted from del Campo, G., Berregi, I., Caracena, R., Santos, J.I., 2006. Quantitative analysis of malic and citric acids in fruit juices using proton nuclear magnetic resonance spectroscopy. Analytica Chimica Acta 556, 462–468 who used high-resolution NMR spectroscopy.

or can be implemented at the discretion of the winemaker to achieve a specific organoleptic profile, as in the case of Chardonnay and Burgundy white wines and Bordeaux red wine (Bauer and Dicks, 2004). During MLF, additional sugars are fermented and aromatic compounds are produced, which can also play an important role in determining the quality of most red wines, certain white wines, classic sparkling wines, and fruit wines as well. The chemistry involved in MLF has already been discussed in an earlier section of this chapter.

Given the harsh conditions in wine, only strains of *Leuconostoc*, *Lactobacillus*, *Pediococcus*, and *Oenococcus* are able to survive the low pH (<3.5), the high SO_2 (50 ppm), and the ethanol levels of 10% (v/v) (Van Vuuren and Dicks, 1993; Lonvaud-Funel, 1999; Gao and Rupasinghe, 2013). Of these four strains, however, it is typically *O. oeni* that drives the MLF process toward the end of AF (Van Vuuren and Dicks, 1993). The activity of *O. oeni* in wine, among other selected LAB, produces specific compounds such as acetoin and diacetyl (2,3-butanedione), an important compound that imparts a nutty or buttery aroma to wine (Neilsen and Richelieu, 1999; Maicas et al., 1999). The overall softer mouthfeel of wine has also been attributed to the activities of LAB through MLF (Bartowsky and Henschke, 2004).

MLF is typically encouraged in cool climate regions where grapes, and other typical cool climate fruits, possess higher levels of malic acid than in a warm climate or in tropical fruits, as presented in Table 3.2. MLF is also encouraged in wine aging in oak barrels, when long-time bottle aging is part of the process, or when a specific organoleptic profile is required (Bauer and Dicks, 2004). This is in contrast to warm climate grape crops, which, like many varieties of tropical fruit, have lower levels of malic acid, in which case MLF is considered a spoilage of the wine (Bauer and Dicks, 2004). However, it is not only malic acid that is important to MLF; citric acids can be metabolized by LAB alongside residual carbohydrates (Ramos and Santos, 1996), though *O. oeni* cannot grow on citrate alone (Salou et al., 1994; Ramos and Santos, 1996). The important flavor compounds diacetyl and acetoin are actually produced specifically through citrate catabolism (Ramos and Santos 1996; Miranda et al., 1997; Bertrand et al., 2000), highlighting MLF's significance in fruit wines in which these are desired flavors. In fact, the production of both compounds by *O. oeni* is stimulated by increased citrate concentration (Nielsen and Riechelieu, 1999).

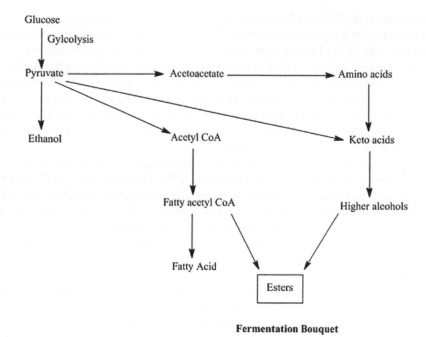

FIGURE 3.22

Pathway of synthesis of the fermentation bouquet.

Current literature regarding the breadth of all fruit wines, cool and warm climate varieties alike, is still significantly sparse compared to that of grape wine research. There are gaps in the literature on the specific effects of MLF on all varieties of fruit wines, and while MLF in grape wine may not be fully understood yet, it is even less understood in fruit wines.

7. FERMENTATION BOUQUET AND YEAST FLAVOR COMPOUNDS

It has been established that, during the fermentation of grape sugars, the yeast produces from pyruvate a low concentration of a range of volatile compounds that make up the so-called "fermentation bouquet" (Fig. 3.22). The main group of compounds, and hence the best studied, are the higher alcohols, fatty acids (Valero et al., 1998; Guitart et al., 1999), aldehydes, and esters (Herraiz and Ough, 1992; Zea et al., 1994; Ferreira et al., 1995; Lema et al., 1996; Vianna and Ebeler, 2001). The yeast strain conducting a clean fermentation without any negative characteristics produces a special fruity aroma in the wine that is called the fermentation bouquet (Singleton et al., 1975). The special fruity odor is primarily due to a mixture of hexyl acetate, ethyl caproate, and isoamyl acetate in a ratio of about 3:2:1.51. Odorwise, the hexyl acetate appears to be the most important and isoamyl acetate the least important to the special fermentation bouquet.

There is no apparent way other than the use of low temperature to stabilize these esters in wine. The other compounds of the fermentation bouquet are the organic acids, higher alcohols, and, to a lesser extent, aldehydes, the formation of which is influenced to various degrees by the nitrogen source (Rapp and Versini, 1991). Lower fermentation temperature, i.e., about 15°C, encourages the production of volatile esters by yeasts (Killian and Ough, 1979). Two major groups of esters are formed during fermentation, the ethyl esters of straight-chain fatty acids and the acetates of higher alcohols. Some of the fatty acids' CoA takes part in ester formation by reacting with alcohol. Its formation is positively correlated with the total nitrogen of the must (Bell et al., 1979). The two important esters of the fermentation bouquet are also positively correlated with the must nitrogen. The concentration of acetate ester, however, is more dependent on the yeast strain and sugar concentration than is the fatty acid ester. The acetate esters of ethanol and the higher alcohols often have a major aroma impact in freshly prepared wines.

The formation of the fermentation bouquet is little influenced by the cultivars involved or the yeast strain conducting fermentation (Houtman and du Plessis, 1981; Houtman et al., 1980). There are, however, some strains that can produce a significantly higher concentration of individual esters (Soles et al., 1982). The fermentation bouquet compounds, such as acetaldehyde, acetic acid, ethyl acetate, higher alcohols, and diacetyl, if present in excess are regarded as undesirable. Esters can also be produced from the alcoholysis of acyl-CoA compounds, which occurs when fatty acid biosynthesis or degradation is interrupted in the cell and generates free coenzymes. Esters can also be formed from the carbon skeleton of amino acids.

The higher alcohols react with acetyl-CoA produced from sugars, amino acids, and sulfates during yeast metabolism. Small acyl-chain esters are typically fruity or floral, whereas the longer acyl chain esters are sweeter or soap-like. If they have more than 12 carbons, then they are not very volatile and hence do not have much effect on the odor. The formation of esters is influenced primarily by temperature, the amino nitrogen content of the juice (Bell et al., 1979; Ough and Lee, 1981), and the yeast strain employed (Soles et al., 1982; Vos et al., 1978). The factors that affect the acyl-CoA production also affect the ester formation.

In fact, a very large number of compounds play roles in determining the flavor of alcoholic beverages. Over 200 compounds that influence flavor have been identified in cider (Jarvis et al., 1995). The number may be higher for other alcoholic drinks. Flavoring compounds as discussed earlier are formed by the products of yeast metabolism during AF. The contributions of several volatile thiols to the aromas of wine made from different grape varieties have been resolved (Tominaga et al., 1998, 2000). Among them, mercapto-4-methyl-2-one, 3-mercaptohexan-1-ol, and 3-mercaptohexyl acetate were found to be considerably higher in certain grape varieties.

Over the past few years research has focused on the role of yeast in the transformation of odorless molecules that are considered to be precursors of the aromatic substances. Several strains of *S. cerevisiae* are marketed as "enhancers of varietal expression," having an ability to hydrolyze conjugated aroma precursors in juice, thus improving the wine aroma (Zoecklein et al., 1997; Mahon et al., 1999). Taking into account that aroma is a wine's most important and distinguishing characteristic, future research into the development of methods and strategies for the accurate quantification of aroma compounds is expected.

Various strains of *S. cerevisiae* winemaking yeasts have been extensively investigated for several years, but the occurrence, biochemical properties, and importance of indigenous yeasts other than *S. cerevisiae* in wine flavor have not been investigated, and the changes that they produce are essentially unknown. In the past, some studies have been carried out on the contributions of "non-*Saccharomyces*" wine yeast to the quality of wine. Non-*Saccharomyces* yeasts have been shown to contribute to the production of esters and other pleasant volatile compounds (Lema et al., 1996; Romano et al., 1997; Ciani and Maccarelli, 1998; Eglinton et al., 2000; Ferraro et al., 2000; Soden et al., 2000). The non-*Saccharomyces* strains of the genera *Kloeckera* and *Hanseniaspora* have been described as producing significant protein protease activity and to affect the protein profile of finished wines (Charoenchai et al., 1997; Dizy and Bisson, 2000). Special interest has been shown in the potential of non-*Saccharomyces* yeast to produce β-glycosidase enzymes involved in the flavor-releasing processes that are of interest in winemaking (McMahon et al., 1999; Yanai and Sato, 1999; Manzanares et al., 2000; Mendes-Ferreira et al., 2001; Cordero-Otero et al., 2003). The acetate esters formed by enzymatic activities of yeast strains belonging to the genera *Hanseniaspora* and *Pichia* have been studied in detail (Rojas et al., 2001, 2003). Clearly, more research on these non-*Saccharomyces* wine yeasts and their chemistry of fermentation can show the potential of these yeasts in improving the flavor profile of the wine produced.

8. CHEMICAL CHANGES OCCURRING DURING FERMENTATION OF SPARKLING AND FORTIFIED WINES

8.1 SPARKLING WINES

The manufacture of sparkling wines is done by the Méthode Champenoise, or traditional method, which is predominantly used in the Champagne region of France and the Cava region in Spain (Amerine et al., 1980). It is characterized by two successive fermentation processes and aging of the wine with the yeasts responsible for the second fermentation in grape wines (Jeandet et al., 2011). However, with respect to sparkling wines from nongrape fruit, there is limited information except for scattered reports on apple wine or cider, plum wine, or the sparkling wine from oranges (Joshi et al., 1995, 1999c, 2011b). Sparkling apple wines are prepared by carbonation in either tank or bottle by secondary fermentation of an apple base wine. However, as the process is similar to that for grape wines, the relevant chemical aspects of the sparkling wines are discussed here taking wine from grapes as an example. The most characteristic aspect of the sparkling wine preparation process is the aging on the lees. The aging period of the wine with yeast varies with the type of wine desired and the legislation of the country, but is not usually shorter than 9 months. During aging, autolysis of the yeast occurs, resulting in enzymatic autodegradation of the cellular constituents that begins immediately after the death of the yeast cells. Consequently, various compounds and enzymes, which bring about important changes in the properties of the wine, are released into the medium and influence the sensorial quality of the resultant wine. For more information on yeast autolysis and its effects on wine quality, structural changes in the yeast cell walls, and the biochemical composition of the wine, see the literature cited (Charpentier and Feuillat, 1993; Martinez-Rodriguez and Polo, 2001).

During autolysis, the polysaccharides of the yeast cell wall are hydrolyzed and the breakdown mainly depends on a strong endo- and exo-β-(1,3)-glucanase enzymatic activity in the yeast cell wall and on the α-mannosidases and end cellular proteases. These enzymes release compounds in several stages, beginning with polysaccharides and short-chain oligosaccharides in the early stages of autolysis, followed by the release of mannoproteins as the cell wall is broken down further (Charpentier and Feuillat, 1993). Changes in the ultrastructure of the yeast also take place, as revealed by electron microscope techniques, with the most of the studies done in the model solution (Kollar et al., 1993; Hernawan and Fleet, 1995). Yeast autolysis results in the release of various products into the wine with special emphasis on proteolysis and the production of nitrogen compounds (amino acids, peptides, and proteins; Moreno-Arribas et al., 1996, 1998a,b; Martinez-Rodriguez and Polo, 2000; Martinez-Rodriguez et al., 2001a,b); the release of polysaccharides, nucleic acids, and lipids (Moreno-Arribas et al., 2000; Pueyo et al., 2000; Aussenac et al., 2001; de la Presa-Owenset et al., 1998); and the formation and degradation of volatile compounds (Pozo-Bayon et al., 2003; Postel and Ziegler, 1991). However, peptides and amino acids are generally considered the majority compounds released into wine during autolysis. These compounds are associated with foam formation in the wine. Of the nitrogen compounds released during the second fermentation and autolysis, amino acids have received the most attention. Other important compounds released during wine autolysis are proteins, and their characteristics in the wines used for making sparkling wines and the changes occurring during secondary fermentation and aging with yeast have been documented (Luguera et al., 1997, 1998).

Lipids play important roles in membrane structure and composition; hence, lipid degradation is an important transformation during autolysis. Research on the contributions of yeast to the release of fatty acids and other lipids and their characterization has been reported (Hernawan and Fleet, 1995; Pueyo et al., 2000; Gallard et al., 1997; Le Fur et al., 1997). Other types of compounds, such as nucleic acids, polysaccharides, fatty acids, and volatile compounds, are also released. A rise in polysaccharide concentration during the aging of sparkling wines in the presence of yeast has been documented (Moreno-Arribas et al., 2000; Luguera et al., 1997; 1998; Gallart et al., 1997; Le Fur et al., 1997; Llauberes et al., 1987; Pueyo et al., 1995; Feuillat et al., 1988; Andreis-Lacueva et al., 1997; Loipez-Barajas et al., 2001; Cavazza et al., 1990).

Similarly, a rise in the major volatile compounds, especially esters, in wine throughout the second fermentation and during aging of the yeast has been recorded (Postel and Ziegler, 1991; Cavazza et al., 1990; Zoecklein et al., 1997; Francioli et al., 2003; Pozo-Bayon et al., 2003). Yeast autolysis is an extremely slow process, partly because of the low temperature of wine storage, which reduces the enzymatic activity required for the same. The addition of yeast autolysate or yeast extract preparations obtained from wines in vitro in order to accelerate the natural process has been attempted (Feuillat, 1987). Another strategy to enhance autolysis during the manufacture of sparkling wines is the use of mixed cultures of killer yeasts and yeasts sensitive to their toxins (Todd et al., 2000; Cocolin and Comi, 2011). Studies on the secondary fermentation of apple wine with low alcohol content (5–6%), as reviewed earlier, showed that hydrolyzed autolysate had a favorable effect on the secondary fermentation of apple wine and was suitable as a nitrogen source.

Phenolic acid is composed mainly of chlorogenic and p-coumaroylquinic acid, together with dihydrochalcones such as phloridzin and phloretin 2-xyloglucosides and flavonols, which are related to bitterness, astringency, and haze in the final product (Picinelli et al., 2000).

Changes in phenolic acid during the sparkling wine process have been documented (Table 3.3). Apple wine produced with a high initial polyphenol level tastes very rough, which could be a consequence of the high content of flavon-3-ols and chlorogenic acid, and a residual amount of malic acid, because an incomplete development of MLF was detected in studies. In the preparation of sparkling wine from plums, the effects of various concentrations of sugar (1%, 1.5%, 2%) and diammonium hydrogen phosphate (0.1–0.3%) and of two strains of *Saccharomyces*, viz., UCD 595 and UCD 505, in two base wines have been reported (Joshi et al., 1999c). More research in these areas is expected in the future.

8.2 BIOLOGICALLY AGED WINES

Biologically aged wines such as sherry are produced using an original method that was developed over several decades (Amerine et al., 1980; Jeandet et al., 2011). These begin as young wines, carefully selected

Table 3.3 Polyphenol Contents of Sparkling Apple Wine on Different Days of Fermentation

Polyphenol	Time (Days)			
	0	4	40	113
(+)-Catechin	12.7	12.8 ± 1.2	11.5 ± 0.9	11.1 ± 1.2
B1	17.1	20.1 ± 0.7	18.7 ± 2.1	16.8 ± 0.7
(−)-Epicatechin	92.2	102.3 ± 3.3	78.5 ± 11.4	66.9 ± 9.5
B2	100.5	113.9 ± 1.2	80.9 ± 13.8	75.6 ± 4.5
C1	15.2	16.9 ± 0.5	11.2 ± 2.4	9.5 ± 1.0
Tetramer phloretin	17.8	22.9 ± 0.6	15.2 ± 1.4	10.6 ± 4.9
2-Xyloglucoside	62.3	72.4 ± 1.0	54.4 ± 7.7	58.0 ± 5.1
Phloridzin	255.9	286.8 ± 7.0	262.4 ± 36.4	260.5 ± 22.1

Picinelli, A., Suarez, B., Garcia L., Mangas J.J, 2000. Changes in phenolic contents during sparkling wine making. American Journal of Enology and Viticulture 51 (2), 144.

soon after completing fermentation, which are fortified by adding vinous alcohol until they reach alcohol contents of 15–15.5% and then transferred to oak barrels before being aged. Wine aging occurs in the so-called "Solera and Criaderas" system under the flor film of yeast that grow on the surface of the wine, with an ethanol content of 15.5% or higher and a low fermentable sugar content. The most significant metabolic change occurring in biological aging is a large acetaldehyde production, considered to be the best marker of the biological aging process, with an important sensory contribution, together with a drastic glycerol and acetic acid consumption and a moderate ethanol metabolism that, in the absence of glucose, is used by the yeast as a carbon source and by certain enzymes (Suarez-Lepe et al., 1990; Suarez-Lepe, 1997; Mauricio et al., 1997). There is also a simultaneous consumption of all the amino acids, especially of proline (Mauricio and Ortega, 1997; Berlanga et al., 2001).

These changes have been the subject of several articles. For example, acetaldehyde has been found to be an intermediate in ethanol metabolism by flor yeast and is the precursor for several other compounds, such as 2,3-butanediol, acetic acid, and acetoin, that are found in sherry wines (Martinez et al., 1995, 1997). Research has also been devoted to understanding the metabolic pathways for the production of these aroma compounds in relation to several factors, such as the period of aging, the flor yeast strain used, and the effect of acceleration conditions applied to shorten the length of this process (Plata et al., 1998; Cortes et al., 1998, 1999; Moyano et al., 2002; Benitez et al., 2003; Peinado et al., 2003). However, there is no information on the preparation of such wines from nongrape fruit and, consequently, no report on the chemical changes can be cited as of this writing.

9. TOXIC METABOLITES OF NITROGEN METABOLISM

One of the most active research areas concerns the study of the metabolic role of LAB in the formation of compounds that are undesirable because of their implications for health, such as ethyl carbamate and biogenic amines (Moreno-Arribas and Polo, 2005; Moreno-Arribas et al., 2003). Ethyl carbamate (also referred to as urethane) is a naturally occurring component of fermented foods and beverages, including wine, that has been shown to be an animal carcinogen (Ough, 1993). Extensive research on ethyl carbamate has been directed at elucidating the origin of this compound in wine (Monteiro and Bisson, 1991; Liu et al., 1996). Ethanol and several compounds that contain a carbamyl group are involved in its formation. In wine, the carbamylic compounds involved are urea, produced by yeast during AF, and citrulline and carbamyl phosphate, produced by LAB during MLF. In both cases, these products are intermediates of arginine metabolism, one of the major amino acids of must and wine (Moreno-Arribas et al., 1998a,b; Pripis-Nicolau et al., 2000). Urea can also react chemically with ethanol during wine storage and produce ethyl carbamate. As of this writing, the addition of commercial preparations of the enzyme acid urease is allowed by the International Organisation of Vine and Wine and EEC legislation to prevent ethyl carbamate production via this pathway.

9.1 AMINES

The other compounds having important repercussions on health are the biogenic amines, which are low-molecular-weight organic bases with a high biological activity and are shown to have undesirable effects when in foods and beverages (Moreno-Arribas and Polo, 2005). Histamine is the best studied biogenic amine and is known to be a cause of headaches, hypotension, and digestive problems, whereas tyramine is associated with migraines and hypertension (Silla-Santos, 1996). During lactic acid fermentation, various amines are synthesized by LAB. Some of these amines are listed in Table 3.4.

In wine, biogenic amines are mainly derived from the activities of LAB that produce enzymes capable of decarboxylating the corresponding precursor amino acids. Although 25 different biogenic amines have been described in wine, the most common ones are histamine, tyramine, putrescine, and cadaverine (Lehtonen, 1996; Gloria et al., 1998; Vazquez-Lasa et al., 1998; Mafra et al., 1999), which are produced by the decarboxylation of histidine, tyrosine, ornithine/arginine, and lysine, respectively. More details about the factors influencing amine formation by LAB in wines can be found elsewhere (Lonvaud-Funel, 2001).

More than 20 amines have been identified in wines (Table 3.4). Histamine plays a special role as an indicator amine to assess the freshness and quality of wine. When there is no histamine, there are no other biogenic amines (Askar and Treptow, 1986). The total concentration of biogenic amines in wine has been reported to range from a few milligrams per liter to about 50 mg/L, depending upon the quality of wine. The main source of ammonia and volatile amines in wine are

Table 3.4 Some of the Amines Found in Wines		
Ammonia	Hexylamine	Phenylamine
Butylamine	Histamine	Phenethylamine
Cadaverine	Indole	Piperidine
1,5-Diaminopentane	Isopentylamine	Propylamine
Diethylamine	Isopropylamine	Putrescine
Dimethylamine	Methylamine	Pyrrolidine
Ethanolamine	2-Methylbutylamine	2-Pyrrolidine
Ethylamine	Morpholine	Serotonin
		Tyramine

the grapes (Mayer and Pause, 1987; Ough et al., 1981; Vidal-Carou et al., 1991). Amines are also formed during MLF by degradation of the parent amino acid. The high amounts of amines may also be due to unsanitary conditions during the winemaking process (Vidal-Carou et al., 1991; Zee et al., 1981). The major physiological effect of biogenic amines are headache, vomiting, and diarrhea (Mayer and Pause, 1987). The recommended upper limit for histamine in wine is 2 mg/L in Germany, 5–6 mg/L in Belgium, 8 mg/L in France, and 10 mg/L in Switzerland (secondary amines like spermine and spermidine from ornithine can react with nitrite to form carcinogenic nitrosamines) (Smith, 1981).

Several factors affect the content of amines in wine, which include type of soil, nitrogen fertilizer, degree of maturation of the grape, elaboration method used for the wine, extension of autolysis, growth of LAB, clarification process, and enological treatment of the wine, as well as the yeast strain that intervenes in the fermentation (Yazquez, 1996). It was found that the strain of yeast directly affects the content of amines, whereas the amino acids used have no relationship with them (Goni and Azpilicueta, 2001).

9.2 ETHYL CARBAMATE

Ethyl carbamate is a carcinogenic compound found in alcoholic beverages (Moreno-Arribas and Polo, 2005). The formation of ethyl carbamate is shown in Fig. 3.23.

The formation takes place from the usage of urea which, by reaction with ethyl alcohol, forms carbamate (Fig. 3.23).

Urea, which is an end product of nitrogen metabolism, may combine with ethanol to form ethyl carbamate, or urethane (Monteiro et al., 1989; Ough et al., 1988a,b). Urea is the principal precursor, and attempts to control the concentration of ethyl carbamate by controlling urea and the concentration of its precursor, arginine, have received significant attention (Ough and Bell, 1980; Stratford and Rose, 1986, 1985). Approximately 0.8 mg ethyl carbamate/L is formed in a model solution of urea and ethanol (12%) under acidic conditions, i.e., at pH 3.2, and temperature of 71°C after 48 h. Freshly prepared wines do not contain significant concentrations of ethyl carbamate, though it is formed at a rate of about 4.8 mg/year at 13°C from an initial concentration of 10 mg urea (Ough et al., 1990). Therefore, it is highly desirable to eliminate or reduce as much as possible the appearance of urea in the fermentation medium. The utilization of adenine by the yeast during grape juice fermentation has been studied as a precursor of urea (Ough et al., 1992). A technique called solid-phase microextraction and gas chromatography has been developed for the determination of ethyl carbamate in wines (Whiton and Zoecklein, 2002).

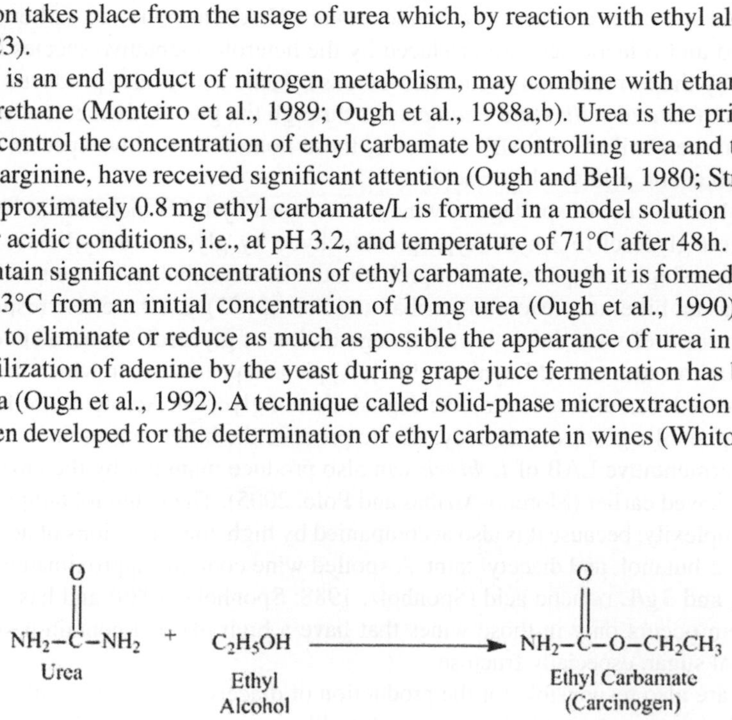

$$NH_2-\overset{\overset{\displaystyle O}{\|}}{C}-NH_2 \ + \ C_2H_5OH \longrightarrow NH_2-\overset{\overset{\displaystyle O}{\|}}{C}-O-CH_2CH_3$$

Urea   Ethyl Alcohol   Ethyl Carbamate (Carcinogen)

FIGURE 3.23

Formation of ethyl carbamate from urea.

10. CHEMISTRY OF WINE SPOILAGE
10.1 SPOILAGE BY ACETIC ACID BACTERIA

The vinegar taint in wines is generally associated with the presence of a high concentration of acetic acid, i.e., above 1.2–1.3 g/L, which is objectionable (Margalith, 1981). Acetic acid bacteria have an anaerobic metabolism and depend on oxygen for their growth (Drysdale and Fleet, 1988). It is generally considered that acetic acid bacteria do not grow during AF but do exhibit exceptionally good survival properties in wines during storage. General exposure of the wine to air or momentary exposure during pumping and transfer operations can quickly stimulate their growth. The main reaction of this growth is the metabolism of ethanol into acetic acid and the chemistry involved has already been described in an earlier section on acetic acid fermentation. The aldehyde dehydrogenase enzyme has low specificity, so that higher aldehydes, with chain lengths up C8, can be oxidized to their corresponding acids. In fully aerated wines, 50–60% of the ethanol could be oxidized by these bacteria for the production of 1.5–3.75 g/L acetic acid (Drdak et al., 1993).

10.2 SPOILAGE BY LACTIC ACID BACTERIA

LAB are known to play a key role in the MLF of wines but they may also cause spoilage in wine, as reviewed earlier (Lonvaud-Funel, 2011). Lactic acid fermentation, i.e., the conversion of carbohydrates to lactic acid and other products, is a characteristic of LAB, as reviewed earlier (Rana and Rana, 2011). Acetic acid and D-lactic acid are produced by the heterofermentative species of *Lactobacillus* and *Leuconostoc* by the fermentation of sugars, whereas D-lactic acid is also produced by the homofermentative species of *Lactobacillus* and *Pediococcus* through the glycolytic metabolism of sugars (see the section on the metabolism of LAB). However, these homofermentative species ferment pentose sugars to give both acetic acid and lactic acid.

Another well-known case of spoilage of wine taste caused by LAB metabolism is due to acrolein. Some LAB convert glycerol to 3-hydroxypropionaldehyde by glycerol dehydratase (Boulton et al., 1996). Acrolein as a single component is not problematic; however, when it reacts with tannins, it can produce an unpleasant bitterness (Moreno-Arribas and Polo, 2005). LAB are also responsible for acrolein taint and bitterness of wine due to bacterial degradation of glycerol. An acrolein concentration of 10 ppm is sufficient to cause taint (Margalith, 1981). Apple and pear wine, in which the growth of LAB is not totally repressed, can undergo complete degradation of glycerol. Acrolein, itself, is not bitter, but it reacts with the phenolic groups of anthocyanins to produce the bitter sensation.

Some heterofermentative LAB of *L. brevis* can also produce mannitol by the enzymatic reduction of fructose, as reviewed earlier (Moreno-Arribas and Polo, 2005). The mannitol taint is a form of spoilage with high complexity, because it is also accompanied by high concentrations of acetic acid, D-lactic acid, *n*-propanol, 2-butanol, and diacetyl taint. A spoiled wine contains approximately 9 g/L mannitol, 3 g/L acetic acid, and 3 g/L D-lactic acid (Sponholz, 1988; Sponholz, 1989) and has a vinegary-estery taste. The problem occurs only in those wines that have a high pH and contain a significantly high amount of residual sugar, especially fructose.

Certain LAB are also responsible for the production of diacetyl. The wines with a high concentration of diacetyl have an undesirable buttery or whey-like aroma and flavor. The wines with 1 mg/L diacetyl content are considered faulty. Normally, wines contain 0.2–0.3 mg/L diacetyl, as a result of yeast activity. The metabolic pathway for the production of diacetyl by LAB is unclear. The intermediate pyruvate, produced during metabolism of hexose and pentose sugars, is converted to acetolactate, which is then converted to diacetyl (El-Gendy et al., 1983; Kandler, 1983).

In addition, it may come from the metabolism of citric acid by citrate lyase, which cleaves this acid into acetic acid and oxaloacetate. The oxaloacetate is then metabolized into diacetyl through pyruvate. Sorbic acid may be added to wines as an antimicrobial agent to control yeast growth. Certain strains of LAB metabolize sorbic acid and impart a geranium off-odor to the wine. The substance responsible for the geranium odor has been identified as 2-ethoxyhexa-3,5-diene. If wines or juices are to be stored in the presence of sorbic acid, they should contain a sufficient amount of free SO_2 to prevent the growth of LAB and development of geranium taint.

Some of the LAB species have also been associated with spoilage of wine (Moreno-Arribas and Polo, 2005). The so-called "taste spoilage," *piqûre lactique*, is one of the most frequent and well-studied forms of spoilage that can occur during any stage of the winemaking process. If the LAB develop before all the must sugar has been transformed into ethanol, they ferment the hexoses, and in addition to ethanol and CO_2, which are also generated by the yeast, they produce acetic acid (Strasser de Saad and Manca de Nadra, 1992). If excess amounts of lactic acid and acetic acid are formed under these conditions, the volatile acidity of the wine increases considerably. The D-isomer of lactic acid is associated with spoilage, whereas L-lactic acid is produced during MLF (Fugelsang, 1997; Sponholz, 1993).

Certain LAB strains can synthesize extracellular polysaccharides from residual sugars, leading to a deterioration of the wine. Wines with an increased viscosity and a slimy appearance are known as "ropy." The genera *Leuconostoc* and *Pediococcus* have been implicated in ropiness (van Vuuren and Dicks, 1993; Manca de Nadra and Strasser de Saad, 1995). Some other strains are able to synthesize an exopolysaccharide characterized as a β-D-glucan (Llauberes et al., 1990). This wine spoilage can occur during vinification and, in most cases, after bottling. Heterofermentative lactobacilli have been implicated in the production of mousy off-flavor in wine, described as an odor similar to "mouse urine" or acetamide (Costello and Henschke, 2002). Moreover, the tendency of some *Dekkera* and *Brettanomyces* yeast to produce a mousy off-flavor has also been described (Grbin and Henschke, 1993). These olfactory defects of wine are attributed to the formation of three heterocyclic volatile bases: 2-acetyltetrahydropyridine, 2-ethyltetrahydropyridine, and 2-acetyl-1-pyrroline.

11. COMPOSITION AND NUTRITIONAL SIGNIFICANCE OF WINE
11.1 COMPOSITIONAL PARAMETERS

Like wine from grapes, the wines from nongrapes in general have water, ethyl alcohol, sugars, acids, minerals, polyphenolic compounds, vitamins, and pigments as major components and aldehydes, esters, volatile acids, higher alcohols, and a number of compounds (Table 3.5) having significance in flavor formation (Amerine et al., 1980; Joshi et al., 1999a, 2011a,b; Soni et al., 2011).

There is very little information pertaining to the total fat and protein of fruit wines; the amounts are nevertheless negligible, as they are in grape wine, contributing little, if anything, to the overall nutritional value of fruit wines. Nevertheless, fruit wines do happen to possess nutritional value by means of their mineral content, alongside other beneficial phytochemical components. Just as their source fruits present natural dietary sources of particular elements, each fruit wine also presents a unique elemental profile (Table 3.6). Among the many areas of interest with wine, mineral content is

Table 3.5 Important Constituents in Alcoholic Beverages

Class	Compounds
Esters	Amyl acetate, butyl acetate, ethyl acetate, ethyl butyrate, ethyl lactate, ethyl benzoate, ethyl hexanoate, ethylguaiacol, ethyl-2-methyl butyrate, ethyl octanoate, ethyl octenoate, ethyl decanoate, ethyl dodecanoate, diethyl succinate, 3-methyl propionate
Alcohols	Ethanol, 2-methylbutan-1-ol (amyl alcohol), methylbutan-1-ol (isoamyl alcohol), heptanol, hexane-1-ol, 2-phenylethanol, 2-methylpropanol, glycerol, 2,3-butanediol
Carbonyls	Decalactone, decan-2-one, acetaldehyde, butyraldehyde, hexanal, nonanal, diacetyl benzaldehyde
Acids	Acetate, butyrate, lactate, malate, succinate, hexanoate, nonanoate, octanoate
Sulfur derivatives	Methional, ethanethiol, methylthioacetate, dimethyl disulfide, ethyl methyl disulfide, diethyl disulfide, 3-methylthiopropyl acetate, 2-methyltetrahydrothiophenone, 2-mercaptoethanol, cis- and trans-2-methylthiophenol, bis(2-hydroxyethyl) disulfide
Phenol compounds	Vinyl phenol, ethyl phenol, ethyl guaiacol, vinyl guaiacol

Based on Rana, N.S., Rana, V.S., 2011. Biochemistry of wine preparation. In: Joshi, V.K. (Ed.), Handbook of Enology: Principles, Practices, and Recent Innovations, vol. 2. Asiatech Publishers Inc., New Delhi, pp. 618–678.

significant for a number of reasons, among them are the contributions to organoleptic properties, toxicological effects (Alvarez et al., 2007a,b), and geographical origin (Fabani et al., 2010; Frías et al., 2002; Moreno et al., 2007; Rodrigues et al., 2011).

With the growing interest in the properties of wines, both grape and otherwise, as functional foods (Dey et al., 2009; Higgins and Llanos, 2015), a greater understanding of the mineral content of fruit wine would not only serve as a means of better understanding the plethora of health benefits inherent in moderate consumption, but also as a means of calculating the toxicological risk factors with regard to heavy metals in fruit wines (Alvarez et al., 2012), thus there is an economic benefit. In a study done by Higgins and Llanos (2015) on traditional wine consumption, they found significant future potential in marketing wines for their health benefits, which could give winemakers a new edge in a functional food-driven market. Likewise, Dey et al. (2009) reviewed the potential for fruit wine as functional food and concluded that fruit wine's antioxidant potential gave it excellent functional food marketing potential; the mineral value could be a facet of the functional aspect of fruit wines (Rupasinghe and Clegg, 2007).

The nongrape wines do not lag behind grape wine in terms of mineral content. Of the five minerals of significant dietary concern—calcium (Ca), potassium (K), phosphorus (P), iron (Fe), and magnesium (Mg)—all but one are present in greater than trace amounts (Table 3.3). Iron, alongside manganese (Mn) and zinc (Zn), was among the least present elements. The elements cobalt (Co), chromium (Cr), copper (Cu), molybdenum (Mo), nickel (Ni), and lead (Pb), all of which could become toxic if accumulated in biological systems, were below the detection limits (Rupasinghe and Clegg, 2007). The most predominant element across the range of fruit wines, as well as the most dominant element present with any single tested wine, was potassium—one of the major necessary dietary minerals in the human diet. Potassium was most abundant in black currant and kiwi wine; however, the levels in other fruit wines were not insignificant, easily comparable to those of grape wine (Table 3.7). Considering that the daily recommended amount of potassium in Canada is around 3500 mg, a single 175-mL glass of black currant wine would provide approximately 6% of daily potassium intake.

Table 3.6 Minerals Present in Select Fruit Wines (mg/L)

Source of Fruit Wine	Ca	K	Mg	Na	P	S	Fe	Mn	Zn	References
Elderberry	89	831	106	62	149	171	1.2	0.9	0.2	Rupasinghe and Clegg (2007)
Blueberry	96	958	82	63	75	113	0.6	14.6	0.5	Rupasinghe and Clegg (2007)
Black currant	108	1202	45	33	102	81	3.2	1.0	0.5	Rupasinghe and Clegg (2007)
Cherry	69	834	50	38	54	136	1.2	0.2	0.1	Rupasinghe and Clegg (2007)
Raspberry	96	742	67	42	64	80	0.5	0.5	0.5	Rupasinghe and Clegg (2007)
Cranberry	643	791	43	48	120	99	4.5	1.5	0.4	Rupasinghe and Clegg (2007)
Apple	45	958	38	31	68	80	0.4	0.2	0.2	Rupasinghe and Clegg (2007)
Kiwi	38.78	1004.43	42.55	10.46	–	–	0.55	–	0.53	Towantakavanit et al. (2001)

Table 3.7 Minerals Found in Select Grape Wines (mg/L)

Wine	Ca	K	Fe	P	Mg	Na	Mn	Zn	References
Cabernet Sauvignon[a]	66.65±8.97	1292.75±169.10	1.63±0.78	220.82±71.13	115.2±10.75	–	–	0.83±0.32	Kondrashov et al. (2009)
Merlot[a]	79.25±14.93	1107.33±51.36	2.07±1.46	120±30.73	102±10.75	–	–	0.55±0.17	Kondrashov et al. (2009)
Chardonnay	62[b]	632	1.3	201	349	52	1.9	0.7	Rupasinghe and Clegg (2007)
Champagne	83±7.0	346±37	1.6±0.5	117±23	84±7.0	11±2	0.8±0.10	0.63±0.08	Jos et al. (2004)
Fino	85.3±27.7	888.3±196.1	3.83±1.80	71.7±13.1	68.0±11.9	31.8±10.5	0.77±0.35	0.63±0.37	Álvarez et al. (2012)

[a]Means and standard deviations presented are calculated from the data presented in Kondrashov et al. (2009) regarding Cabernet and Merlot.
[b]No standard deviation available in data.

Among the other important dietary minerals, calcium or phosphorus tended to be the second most abundant, followed by magnesium. It is interesting to note the unique mineral signature of cranberry wine, whose calcium levels exceed those of all other wines, including grape varieties. With the daily value of calcium being recommended at 1100 mg in Canada, a single 175-mL glass of cranberry wine would serve over 10% of the recommended amount of calcium. On the other hand, a glass of elderberry wine contains phosphorus and magnesium, offering just over 2% and 7%, respectively, of each mineral's daily recommended amount in a single glass.

Like fruit, fruit wines are elementally unique, with each offering its own nutritional advantages. The elemental uniqueness of each wine, aided by distinct compositions in other fields as well, may serve as a means of identifying exclusive categories that specifically endowed wines could serve best (Rupasinghe and Clegg, 2007). Nevertheless, fruit wines, in general, possess mineral contents comparable to those of grape wines (Tables 3.6 and 3.7; Tayler et al., 2003). The vermouths made from plum, pear, and apple have mineral contents comparable to those of their counterparts. While the influence of other factors, such as geography, climate, soil chemistry, and viticulture practices, on mineral content in fruits and wines needs to be better understood, there exists excellent evidence to illustrate that fruit wines can serve a novel dietary role in human nutritional needs.

11.2 PHENOLIC CHARACTERIZATION OF FRUIT WINES

Some of the known bioactive agents present in cool climate berries include vitamins (A, C, E, and folic acid), minerals (calcium and selenium), carotenoids (carotene and lutein), phytosterols (sitosterol and stigmasterol), triterpene esters, and phenolic compounds such as anthocyanins, flavonols, flavan-3-ols, proanthocyanidins, ellagitannins, and phenolic acids. The chemistry of fruit phenolics directly influences their metabolism upon fermentation, bioavailability, and biological effects. The compositional analysis of organic acids has reported malic and citric acids as the predominant organic acids in elderberry must and wine (Schnitzer et al., 2010), whereas malic acid and lactic acid are the most abundant acids found in blackberry wine (Worobo and Splittstoesser, 2005). From a technological standpoint, phenolic compounds are key determinants of several organoleptic attributes of fruit wines, including color, taste, astringency, and bitterness (Rupasinghe, 2008). The concentration of polyphenolic compounds in fruit wines depends on various factors such as the variety, growing season, geographical location, extraction method, transport, and storage, to name a few. In particular, anthocyanins present in fruits and fruit wines are mostly affected by pH, temperature, light, oxygen, enzymes, ascorbic acid, sulfur dioxide, salts, minerals, metal ions, and copigments (Castañeda-Ovando et al., 2009; Francis, 1986). Fermentation is known to increase the phenolic content of berry juice products by increasing extraction from the skins, including anthocyanins (Heinonen et al., 1998; Su and Chien, 2007), and therefore may increase the antioxidant capacity (Martin and Matar, 2005) and potential for health benefits.

The phenolic content obtained from published research for some of the cool climate fruit wines is presented in Table 3.8. By far, the most extensive study that has been conducted for comparing the total phenolic content of berry wines is by Heinonen et al. (1998), who used the Folin–Ciocalteu procedure to compare 27 berry wines. Lehtonen et al. (1999) also quantified several phenolic compounds from eight berry wines using HPLC. Published research reports show a great variation in the total phenolic values of berry wines that range between 592 and 2651 mg GAE/L. The available data vividly show that blackberry, elderberry, black raspberry, and blueberry wines contain the highest phenolic content,

Table 3.8 Phenolic Characterization of Some Common Fruit Wines

Fruit Wine	Total Phenolic Content (mg GAE/L)	Total Anthocyanins (mg C3GE/L)[c]	Total Flavonoids (mg GAE/L)[d]	Total Proanthocyanidins (mg CE/L)[e]	References
Elderberry (Canada)	1753–2004	865	534–1134		Rupasinghe and Clegg (2007) Schmitzer et al. (2010)
Elderberry (Croatia)	5136–8307	2509–5061			
Blackberry	1697–2789 733–2698 1051-3621 (USA)[a] 1607–2836 601–1624 (Andean) 1232 (Turkey) 1548 (Croatia) 2212 (USA) 2755–4044 (Korea)	13–135[b] 1–125[b] 174–465 35–121 (Andean) 16–192 (USA) 39–86 (Korea)	924–1417	3048–3933 (Korea)	Mudnic et al. (2012), Klarić et al. (2011), Johnson et al. (2011), Johnson and de Mejia (2012), Mitic et al. (2013), Arozarena et al. (2012), Yildirim (2006), and Lim et al. (2012)
Sea buckthorn	689 2182				Negi and Dey (2009) and Negi et al. (2013)
Cherry	1080 592–742 1533–2651	18–27[b] 107–281			Heinonen et al. (1998), Sun et al. (2011), and Mitic et al. (2013)
Red raspberry	1050 1052–1490	103–257			Heinonen et al. (1998) and Mitic et al. (2013)
Black raspberry	2755–4044 2400 6.2 mg GAE/g 650–1179	39–86 98 2.4 mg C3G/g		3048–3933	Lim et al. (2012), Cho et al. (2013), Jeung et al. (2010), and Jung et al. (2009)
Black currant	1050 1985 520–1000		10.4–34.5 mg/L		Heinonen et al. (1998), Vuorinen et al. (2000), and Czyzowska and Pogorzelski (2002)
Crowberry and birch sap	776				Heinonen et al. (1998)
Black currant/ red currant and strawberry	775				Heinonen et al. (1998)

Table 3.8 Phenolic Characterization of Some Common Fruit Wines—cont'd

Fruit Wine	Total Phenolic Content (mg GAE/L)	Total Anthocyanins (mg C3GE/L)[c]	Total Flavonoids (mg GAE/L)[d]	Total Proanthocyanidins (mg CE/L)[e]	References
Strawberry mash		130			Klopotek et al. (2005)
Blueberry	1676 966–2510[a] 2197 1090 600–1860 (USA)	11–192 76.8 8 15-162 (USA)			Rupasinghe and Clegg (2007), Johnson et al. (2011, 2013), Ortiz et al. (2013), and Sanchez-Moreno et al. (2003)
Blueberry/ blackberry (50:50)	2932[a]	105.8			Johnson et al. (2013)

[a]Data expressed as mg ellagic acid equivalents (EE)/L.
[b]Data expressed as mg malvidin-3-glucoside equivalents (M3GE)/L.
[c]Data expressed as mg cyanidin 3-glucoside equivalents (C3GE)/L.
[d]Date expressed as mg gallic acid equivalents (GAE)/L.
[e]Data expressed as mg catechin equivalents (CE)/L.

whereas blends of crowberry and birch sap wine, black currant/red currant wine, and strawberry wine contain relatively lower total phenolic amounts. The literature also shows that other fruit wines, from sea buckthorn, raspberry, cherry, black currant, bilberry, and elderberry, have total phenolic content that is comparable to or higher than that of most common red wines (Negi and Dey, 2009; Klopotek et al., 2005; Sanchez-Moreno et al., 2003; Rupasinghe and Clegg, 2007). On the other hand, wines obtained from apple, plum, and peach are found to be lower in phenolic content compared to red wines.

Similarly, values for total anthocyanins vary greatly between 8 and 467 mg cyanidin-3-glucoside Eq/L. It is noteworthy that some of the data reported here are expressed in units other than the ones normally used. From the reviewed studies, it is evident that anthocyanins contribute the major part of the phenolics present in cool climate berry wines. Anthocyanins in most berry fruits are primarily glycosides of cyanidin, delphinidin, malvidin, peonidin, and petunidin (Mazza and Miniati, 1993; Table 3.9). The anthocyanin content of berry wines was the highest for elderberry wine, followed by blackberry and cherry wines.

Another important class is flavonols, which are prominently found in cool climate fruits and thus the wines produced from them. Myricetin and quercetin have been recognized as the main flavonols detected in red berry wines (Vuorinen et al., 2000; Table 3.9). The concentrations of quercetin and myricetin have been shown to be 3.8–22.6 and 2.2–24.3 mg/L, respectively, in red berry wines, which are higher than those found in red grape wines (Vuorinen et al., 2000). For instance, in black currant wines the amount of myricetin ranges from 7.3 to 22.6 mg/L and that of quercetin from 3.1 to 11.9 mg/L. Similar levels have been reported for wines from black currant and red currant, strawberry, or red raspberry, whereas lower values have been registered for crowberry wines (Hakkinen et al., 1999). Also, it has been demonstrated that the polyphenolic content of wines made from black currants (England) is

Table 3.9 HPLC Characterization of Some Common Fruit Wines

Fruit Wine	Phenolic Acids	Anthocyanins	Flavonols	Flavan-3-ols	References
Elderberry	Chlorogenic acid, neochlorogenic acid	Cyanidin-3-sambubioside, cyanidin-3-sambubioside-5-glucoside, cyanidin-3-glucoside, cyanidin-3,5-diglucoside, cyanidin-3-rutinoside	Quercetin-3-rutinoside, quercetin-3-glucoside, kaempferol-3-rutinoside		Schmitzer et al. (2010)
Blackberry	Gallic acid, chlorogenic acid, p-coumaric acid, caffeic acid, ferulic acid	Malvidin-galact, gluc, arabino; delphinidin-gluc, arabin; cyanidin-gluc, arabin; petunidin-gluc, arabin, xyloside, peonidin-gluc	Quercetin, kaempferol		Mitic et al. (2013) and Johnson et al. (2013)
Sea buckthorn			Rutin, myrecetin, quercetin, kaempferol		Negi et al. (2013)
Cherry	Chlorogenic acid, neochlorogenic acid, p-hydroxybenzoic acid, p-coumaric acid, caffeic acid, chlorogenic acid, ferulic acid	Cyanidin-glucoside, rutinoside, glucosylrutinoside, sophoroside, peonidin-rutinoside	Quercetin, kaempferol	Catechin, epicatechin, proanthocyanidin dimer B2, trimer C1	Sun et al. (2011), Czyzowska and Pogorzelski (2002), and Mitic et al. (2013)
Red raspberry	Caffeic acid, p-coumaric acid, ferulic acid	Cyanidin-sophoroside, glucoside, rutinoside	Quercetin		Mitic et al. (2013)
Black raspberry	4-Dihydroxybenzoic acid, 3,4,5-trihydroxybenzoate, gallic acid				Cho et al. (2013)
Black currant	Chlorogenic acid, neochlorogenic acid, caffeic acid, p-coumaric acid, chlorogenic acid, ferulic acid	Cyanidin-rutinoside, glucoside, aglycone; delphinidin-rutinoside, glucoside, aglycone		Gallocatechin, catechin, epigallocatechin, proanthocyanidin dimer B2, epicatechin	Czyzowska and Pogorzelski (2002)
Crowberry and birch sap					
Black currant/red currant and strawberry					
Blueberry		Malvidin-galact, gluc, arabino; Delphinidin-galact, gluc, arabin; cyanidin-gluc, arabin, 6-acetyl-3-gluc, petunidin-gluc, arabin, peonidin-gluc			Johnson et al. (2013)

similar to that of wines produced from Cabernet or Merlot grapes (from Australia, California, or Chile; Vuorinen et al., 2000).

In addition to flavonols, the berry wines contain other classes of phenolics such as flavan-3-ols, hydrocinnamic acids, and other phenolic acids, which are suggested to be major contributors to the observed antioxidant activity and physiological activities in these fruit wines (Lehtonen et al., 1999). The phenolic characterization of elderberry wine has demonstrated the presence of 10 major compounds, including chlorogenic acid, neochlorogenic acid, quercetin-3-O-rutinoside, quercetin-3-O-glucoside, kaempferol-3-O-rutinoside, and five cyanidin-based anthocyanins. Similarly, blackberry wine is also known to have high concentrations of benzoic and cinnamic acids (Siriwoharn and Wrolstad, 2004; Szajdek and Borowska, 2008), especially gallic acid, chlorogenic acid, p-coumaric acid, and caffeic acid (Klarić et al., 2011). Nonetheless, certain berries contain special classes of phenolics other than the most common categories that get transferred to their respective wines. For instance, blackberry wines contain ellagitannins as one of their primary classes of polyphenols in the range between 265 and 1445 mg EAE/L (Arozarena et al., 2012).

Fermentation-induced changes in phenolic composition have also been described in the published literature. For instance, during fermentation of Korean black raspberry juice, the amounts of most phenolic acids, including 3,4-dihydroxybenzoic acid, caffeic acid, gallic acid, and p-coumaric acid, were increased by 27–188%, whereas those of catechin, epicatechin, malvidin-3-glucoside, myricetin, and quercetin were decreased by 5–30% (Lim et al., 2012). This wide variation reflects the phenolic metabolism of the wine-producing yeast strains.

11.2.1 *Antioxidant Capacity of Fruit Wines*

Recently, Yildirim (2006) evaluated the antioxidant capacities of fruit wines and arranged them in the following order: bilberry > blackberry > black mulberry > sour cherry > strawberry > raspberry > apple > quince > apricot > melon. Studies have been performed in relationship to phenolic content and their contribution to the antioxidant activities in various fruit wines. Several studies have shown a strong positive correlation between the total antioxidant capacity of fruit wines and the total phenolics (Rupasinghe and Clegg, 2007; Sanchez-Moreno et al., 2003; Yildirim, 2006). However, Heinonen et al. (1998) demonstrated no such correlation comparing the polyphenolic content of 27 fruit wines to the antioxidant activity. This could be because most berry wines are rich in anthocyanins, which respond poorly in the Folin–Ciocalteu assay, thereby giving a poor correlation (Singleton, 1974).

In a study of three fruit wines and one red wine, Pinhero and Paliyath (2001) found that blackberry, blueberry, and summer cherry wines had 30–40% more superoxide radical scavenging activity than the red wine tested. They also examined hydroxyl radical scavenging activity using dealcoholized phenolic components from the fruit wines and found that summer cherry and blueberry wine extracts showed greater inhibition than blackberry and red wine (with values of 5.98%, 9.07%, 3.57%, and 2.8% inhibition/µg of gallic acid equivalent, respectively). The antioxidant capacity of some fruit wines is presented in Table 3.10.

11.2.2 *Biological Activity of Nongrape Fruit Wines*

While there is an abundance of studies available investigating the phenolic and antioxidative capacities of fruit wines, there is a considerable gap in the literature concerning the possible beneficial bioactivity of fruit wine. That is not to say there is no information regarding the health benefits, but in comparison to the significant body of information regarding the activities of grape wines, fruit wines are severely understudied. One significant difference between fruit wine and grape wine is that fruit wines

Table 3.10 Antioxidant Capacity of Fruit Wines

Fruit Wine	FRAP (mM TE/L)	DPPH (IC_{50}) (mM Trolox/L)	ABTS (mM Trolox/L)	ORAC (mM Trolox/L)	References
Elderberry	146	9.95			Schmitzer et al. (2010)
Blackberry		4–7	4–9 (IC_{50} mg/L)	6–49 (USA)	Klarić et al. (2011), Johnson et al. (2011), and Mitic et al. (2013)
		10–18	23–28 (Korea)		Lim et al. (2012)
Sea buckthorn	3.1	2.6	7.4		Negi et al. (2013)
Cherry		4–17			Mitic et al. (2013)
Red raspberry		4–8			
Black raspberry		187–221	22–28		Lim et al. (2012)
			198–225		Jeung et al. (2010)
Blueberry		5.4		17–32 25.3 16–24 (USA)	Johnson et al. (2011, 2013), Ortiz et al. (2013), and Sánchez-Moreno et al. (2003)
Black currant					
Blueberry/black-berry (50/50)				25.3	Johnson et al. (2013)

encompass a wide array of fruits, whereas traditional grape wine is still made only from grapes, though the range of cultivars of grapes is significant; the study of fruit wine is a decidedly broad topic, including fruits such as blueberry, blackberry, raspberry, strawberry, elderberry, mulberry, cherry, apple, pear, peach, plum, etc. While there are some studies that have focused on one aspect of many fruits, such as quantifying mineral content in a list of fruit wines (Rupasinghe and Clegg, 2007) or otherwise compiling information with regard to a broad range of fruit wines to form an overview (Jagtap and Bapat, 2015), far more common is a single study that focuses on one or two fruit wines in particular.

There is a limited body of research available pursuing the bioactive properties of some increasingly popular fruit wines such as blueberry wine. For instance, blueberry wine possesses some interesting possibilities as a health-promoting wine because there have been a considerable number of studies on health benefits of the whole berry as well as the juice. Scientific evidence exists on the potential effects of whole blueberries as a dietary supplementation for the management of type 2 diabetes (Stull et al., 2010), as well as of neurodegenerative diseases such as Alzheimer disease (Ramassamy, 2006). Johnson et al. (2011) looked at in vitro inhibition of α-amylase and α-glucosidase by blueberry bioactives and their wines, demonstrating that blueberry wines were not only a good source of polyphenols and antioxidants but also a potential source for the development of antidiabetic therapeutics.

Cherries are one of the emerging fruits to enter into the fruit wine market. The fruit is becoming increasingly popular in countries like China, where cherry production is high and it is necessary to circumvent spoilage of the product because of its short shelf life. Therefore, fruit wine itself offers a unique potential in addition to its accepted flavor profile (Sun et al., 2013). Cherries have also attracted interest because of their beneficial health aspects such as in their strong antioxidant activities (Serrano et al., 2005). For example, Yoo et al. (2010) demonstrated that cherry wine, compared to fresh cherries and cherry juice, has enhanced superoxide dismutase and catalase enzyme levels. Cherry wine also exerted significant protective effects on V79-4 cells against oxidative stress induced by hydrogen peroxide, greater than those of juice applied at the same concentration.

While studies thus far have shown promising bioactive potential in the area of fruit wines, the available literature may still be lacking in comparison to the knowledge available for whole fruits or grape wines. Because of the significant gap in the literature with regard to fruit wines and their effects on chronic human diseases such as cancer, cardiovascular disease, and neurodegenerative diseases, the overall benefits of various fruit wines are inconclusive. Further studies with animal models and/or human clinical trials are needed to better investigate and demonstrate the full range of potential health benefits of fruit wines.

11.3 AROMA AND VOLATILE COMPOUNDS

Fruit wines, as with all foods, possess an array of volatile compounds that act separately as well as in concert with one another to produce a wide range of aromas, flavors, and mouthfeels. Fruit wines, like grape wines, are composed of volatiles produced through several means—by the fruit itself, as byproducts of AF and MLF, and through methods of bottling, aging, and storing. In general, the most common volatiles that contribute to the flavor and/or aroma profile of fruit wines can be broadly categorized into five groups: higher alcohols, esters, acetates, organic acids, and other compounds. Nevertheless, there are many minor volatile and nonvolatile components such as aldehydes, ketones, lactones, terpenes, phenols, and glycerol, which contribute to the full bouquet of subtle flavors, aroma, and mouthfeel of wines.

Higher alcohols tend to have a significantly higher aroma threshold compared to the esters, acetates, and organic acids (Table 3.8); their aromas tend toward harsher, "sweet," alcoholic scents, although there are a few particular higher alcohols, such as 2-phenylethanol, that can impart positive sensory attributes to wine with descriptors like "rose-like" and "perfume-like" (Falqué et al., 2001). In the case of higher alcohols, too much can be overpowering; lower concentrations (less than 300 mg/L) can contribute to the overall complexity and fruitiness of the wine, whereas higher concentrations (exceeding 400 mg/L) may be detrimental to wines because of their harsh chemical scents (Lerm et al., 2010). Higher alcohols in wine are formed by decarboxylation and reduction of α-keto acids; amino acid biosynthesis is responsible for most of the higher alcohols formed during fermentation (Lerm et al., 2010).

The production of esters, particularly ethyl esters, is important in determining wine aroma, as they are often associated with sweet, fruity scents (Sweigers et al., 2005; Siebert et al., 2005). Their aroma threshold can range from as high as 14,000 μg/L to as low as 1 μg/L or lower; low concentrations of

Table 3.11 Volatile Compounds, Their Aroma Descriptors, and Their Aroma Thresholds

Volatile	Aroma Descriptor[a]	Aroma Threshold (µg/L)
Ethyl acetate	VA, nail polish	7500[b]
Ethyl propanoate	Fruity	1840[c]
Ethyl 2-methylbutanoate	Sweet fruit	1[b]
Ethyl 3-methylbutanoate	Berry	3[b]
Ethyl octanoate	Sweet, soap	2[b]
Ethyl decanoate	Pleasant, soap	200[d]
2-Methylpropyl acetate	Banana, fruity	1600[e]
2-Methylbutyl acetate	Banana, fruity	160[c]
3-Methylbutyl acetate	Banana	30[b]
2-Phenylethyl acetate	Flowery	250[b]
2-Methylpropanol	Fusel, spiritous	40000[b]
2-Methylbutanol	Nail polish	65000[e]
3-Methylbutanol	Harsh, nail polish	30000[b]
2-Phenylethanol	Roses	10000[b]
Acetic acid	VA, vinegar	280000[b]
Hexanoic acid	Cheese, sweaty	8000[f]
Octanoic acid	Rancid, harsh	8800[g]
Decanoic acid	Fatty	6000[f]

VA, *Volatile acidity.*
[a]*Aroma descriptors listed in Table 3.1 of Siebert et al. (2005).*
[b]*Guth (1997).*
[c]*Etievant (1991). In: Maarse, H. (Ed.), Volatile compounds in foods and beverages.*
[d]*Ferreira et al. (2000).*
[e]*Meilgaard (1975).*
[f]*Amerine and Roessler (1976).*
[g]*Salo (1970).*

esters can still impart significant aromatic qualities to wine (Siebert et al., 2005). While many esters in wine are yeast fermentation derived, MLF does play a part in producing additional esters, affecting the aroma of wines. The two main groups of MLF-derived esters associated with fruitiness in the wine are acetate esters and ethyl fatty acid esters. Acetates are formed by the reaction of acetyl-CoA with higher alcohols in the presence of acetyltransferase (Yoshioka and Hashimoto, 1981; Matthews et al., 2004; Ugliano and Henschke, 2008). The activity of alcohol acetyltransferase is also highly variable based on the specific strain of yeast, and therefore the flavors imbued by the acetate products can vary between wines (Fujii et al., 1996). Ethyl fatty acid esters are formed through the enzymatic esterification of activated fatty acids formed during lipid biosynthesis (Lerm et al., 2010). Acetates of higher alcohols, such as 3-methylbutyl acetate and 2-phenylethyl acetate identified in raspberry wine (Duarte et al., 2010c), impart pleasant aromas in wine, similar to the qualities of esters. 3-Methylbutyl acetate and 2-phenlyethyl acetate can be perceived as "banana" and "fruity" notes in wine (Table 3.11; Seibert et al., 2005). Table 3.11 shows that acetates have a low to moderate aromatic threshold and acetates do not need to be present in large amounts to have an impact on the aroma that can be recognized by consumers. If the former three categories present largely positive attributes in wines, the fourth category, volatile organic acids, does the opposite. The presence of organic acids in large quantities may lower the acceptance of wines because of the negative effects on sensory characteristics (Nikolaou et al., 2006). Organic acids are formed by the hydrolysis of tri-, di-, and monoacylglycerols (Liu, 2002). Compiled in a study done by Seibert et al. (2005), the organic acids happen to be the least aromatically appealing fermentation-derived compound in wine. Hexanoic acid was described as "cheesy" and "sweaty," octanoic acid was described as "rancid" and "harsh," and decanoic acid was described as "fatty." It goes without saying that the presence of large quantities of acetic acid would impart vinegary flavors and would ruin the wine.

For wines that undergo MLF, LAB not only metabolize malic acid into lactic acid but can also affect the final aroma balance by means of alternate metabolic pathways and substrates present in the wine (Nielsen and Richelieu, 1999), modifying fruity aromas as well as producing active aroma compounds (Davis et al., 1985; Henick-Kling, 1995). Diacetyl (2,3-butanedione), a diketone, is considered one of the most important aroma compounds produced (Bartowsky and Henschke, 1995; Lonvaud-Funel, 1999). It is responsible for contributing buttery, nutty, and butterscotch aromas to wine, as well as a yeasty character to sparkling wine (Batowsky and Henschke, 1995, 2004; Martineau et al., 1995; Bartowsky et al., 2002). Diacetyl is an intermediate in the metabolism of citric acid by LAB during MLF (Bartowsky et al., 2002; Bartowsky and Henschke, 2004); however, it is a chemically unstable compound that can be reduced to acetoin, which can then be reduced to 2,3-butanediol (Bartowsky et al., 2002; Costello, 2006). This conversion has a direct effect on wine aroma, because acetoin and 2,3-butanediol have higher odor threshold values compared to diacetyl, approximately 150 (Francis and Newton, 2005) and 600 mg/L (Bartowsky and Henschke, 2004), while contributing the same buttery character to wine. In contrast, the odor threshold of diacetyl is much lower, though it varies between wines: 0.2 mg/L in Chardonnay, 0.9 mg/L in Pinot noir, and 2.8 mg/L in Cabernet Sauvignon (Martineau et al., 1995).

In a study done by Sun et al. (2013) investigating effective combinations among three strains of yeast and two strains of LAB in producing favorable cherry wines, the range of diacetyl produced was as high as 0.11 mg/L to as low as 0.03 mg/L. However, there were many aromas working in concert with one another, produced in different amounts by the differing abilities of the paired strains. The conclusion of the study was that AF is more effective in modulating wine flavors and aroma, though MLF does play a subtle role in manipulating aroma character.

Despite the considerable literature devoted to odor thresholds and aroma descriptors of volatile compounds, particularly as they pertain to red and white wines, there is a limited body of knowledge concerning the volatile composition of fruit wines. With growing interest in the fruit wine industry, more studies are being conducted, but their range is still limited to only a selected few fruit wines, for example, raspberry (Duarte et al., 2010a), cherry (Sun et al., 2013), and pomegranate (Andreu-Sevilla et al., 2013). Table 3.12 illustrates some of the major volatile compounds that contribute to these fruit wines' distinctive characters, including their odor activity value (OAV), indicating the ratio between volatile compound concentrations and their odor threshold. Those with concentrations greater than their odor threshold are considered aroma-contributing compounds (Pino and Queris, 2010). Volatile compounds with the greatest OAVs presumably would contribute the most significant aromas. Esters, particularly ethyl esters, are among the most common significant aroma contributors in a range of fruit wines, especially in pineapple wine, in which esters comprise all of the major components in the wine (Pino and Queris, 2010). Given that there is a limited offering with regard to understanding the full

Table 3.12 Some of the Major Contributing Compounds to Fruit Wine Aroma					
Fruit Wine	**Compound**	**Aroma Descriptor**	**Concentration (mg/L)**	**Odor Threshold (µg/L)**	**OAV**
Raspberry[a]	2-Methyl-1-propanol	Malty[e]	71.0 ± 14.28	550[e]	129.1
	2-Methyl-1-butanol	Malty, solvent-like[e]	34.7 ± 8.09	1200[e]	28.9
	3-Methyl-1-butanol	Malty[e]	167.2 ± 37.56	220[e]	760
	Ethyl hexanoate	Green, apple[f]	447.9 ± 36.30	14[i]	31,992.9
	Ethyl octanoate	Sweet, soap[f]	449.5 ± 15.23	5[j]	89,900
	Ethyl butyrate	Papaya, butter, apple, perfumed[g]	135.9 ± 36.86	20[j]	6795
	3-Methylbutyl acetate	Banana[f]	1927.0 ± 154.39	30[i]	64,233.3
	3-Mercapto-1-hexanol	–	0.0039	–	–
Cherry[b]	β-Damascenone	Sweet, honey-like[b]	6.93 ± 0.47	0.05[i]	138.6[b]
	Hexanoic acid	Green[b]	14.41 ± 0.75	420[k]	34.31[b]
	Ethyl 3-methyl butanoate	Fruity[b]	0.05 ± 0.00	3[j]	17.35[b]
	Ethyl hexanoate	Apple, banana[b]	0.18 ± 0.01	14[j]	12.86[b]
	3-Methylbutyl acetate	Banana, sweet, fruity[b]	0.24 ± 0.01	30[j]	7.86[b]
	Octanoic acid	Candy, perfumey, fruity, caramelized, peachy, strawberry[b]	2.71 ± 0.13	500[k]	5.42[b]
Pomegranate[c]	Ethyl octanoate	Sweet, soap[f]	21.0	5[d]	4.2
	Ethyl acetate	Nail polish[f]	45.1	5[d]	9.0
	Phenethyl acetate	Caramel, grape, rose, wine-like[h]	5.02	–	–
Pineapple[d]	Ethyl octanoate	Sweet, soap[f]	0.0974	5[d]	19.5
	Ethyl acetate	Nail polish[f]	0.0383	5[d]	7.7
	Ethyl decanoate	Pleasant, soap[f]	0.0179	200[d]	<1
	Methyl 2-methylhep-tanoate	–	0.0109	ND	–
	Ethyl 2-methylpro-panoate	Fruity	0.0006 ± 0.03	0.1[d]	6

ND, *not determined;* OAV, *odor activity value.*
[a]*Duarte et al. (2010a). Raspberry wine fermented by UFLA FW 15, which was described as the best strain for raspberry fermentation*
[b]*Sun et al. (2013). Cherry wine SI-E was selected from the six different fermentation possibilities in the study for its pleasant fruity product.*
[c]*Andreu-Sevilla et al. (2013). Selected data from Mollar de Elche pomegranate wine, whose bouquet was described as the finest.*
[d]*Pino and Queris (2010).*
[e]*Czerny et al. (2008).*
[f]*Seibert et al. (2005).*
[g]*Meilgaard (1975).*
[h]*SAFC (2012).*
[i]*Guth (1997).*
[j]*Ferreira et al. (2000).*
[k]*Gonzalez Alvarez et al. (2011).*

range of aromatic characters involved in fruit wines, and that these wines are poised to become increasingly important on the market, there is significant room for further study into volatile aroma compounds contributing to fruit wines (McKellar et al., 2005).

12. SUMMARY AND FUTURE OUTLOOK

The chemistry of wine fermentation and the associated changes have been described in detail, especially the metabolism of major elements, i.e., carbon, nitrogen, and sulfur, during the fermentation process. Most of the catabolic pathways of sugar and nitrogen metabolism in *Saccharomyces* have been well characterized biochemically and should remain the same in nongrape fruit wines as in grape wine. Since 2005, there has been considerable progress in understanding the biochemical changes occurring during winemaking. Classical biochemical, chemical analyses, and molecular techniques have all contributed to the enormous progress made in this field. Basic research, however, is still needed to understand the biochemical behavior of yeast and bacteria in a wine medium. Undoubtedly, some interesting insights could be gained concerning the mechanisms of control of sugar and nitrogen utilization, if this process were thoroughly investigated during the growth and fermentation of various substrates. Focus needs to be put precisely on the metabolism concerning the production of several compounds, such as urea, ethyl carbamate, methanol, hydrogen sulfide, and amines, including nitrosamine, for the safety of the wines for the consumer. The concentrations of such compounds may be increased or decreased by the appropriate concentration of additives and types of metabolites employed during the fermentation. Nearly all the studies to date concern the biochemical activities of *S. cerevisiae* and no attention has been given to the production of different compounds by species other than *Saccharomyces*, which can contribute significantly to the quality of the wine. The global objective of increasing the diversity of wines could then be met. The production of healthier wines could be pursued by increasing the knowledge of the biochemical mechanisms of yeast and LAB that give rise to the different compounds in wine responsible for its sensory and biological characteristics and health-related qualities to enhance the components with positive effects and suppress those with negative ones.

REFERENCES

Acree, T.E., Sonoff, E.P., Splittstoesser, D.F., 1972. Effect of yeast strain and type of sulphur compound on hydrogen sulfide production. American Journal of Enology and Viticulture 23, 6–9.

Agenbach, W.A., 1977. A study of must nitrogen content in relation to incomplete fermentations, yeast production and fermentation. Proceedings of the South African Journal for Enology and Viticulture 66.

Alexandre, H., Berlot, J.P., Charpentier, C., 1992. Effect of ethanol on membrane fluidity of protoplasts from *Saccharomyces cerevisiae* and *Kloeckera apiculata* grown with or without ethanol, measured by fluorescence anisotropy. Biotechnology Techniques 5, 295–300.

Álvarez, M., Moreno, I.M., Jos, A., Cameán, A.M., González, A.G., 2007a. Study of mineral profile of Montilla-Moriles ''fino'' wines using inductively coupled plasma atomic emission spectrometry methods. Journal of Food Composition 20, 391–395.

Álvarez, M., Moreno, I.M., Pichardo, S., Cameán, A.M., González, A.G., 2007b. Metallic profiles of sherry wines using inductively coupled plasma atomic emission spectrometry methods. Science Alimentary 27, 80–89.

Álvarez, M., Moreno, I.M., Pichardo, S., Cameán, A.M., González, A.G., 2012. Mineral profile of "fino" wines using inductively coupled plasma optical emission spectrometry methods. Food Chemistry 135, 309–313.

Alves, J.A., Lima, L.C.O., Nunes, C.A., Dias, D.R., Schwan, R.F., 2011. Chemical, physical–chemical, and sensory characteristics of lychee (*Litchi chinensis* Sonn) wines. Journal of Food Science 76, 330–336.

Amerine, M.A., Roessler, E.B., 1976. Wines—Their Sensory Evaluation. WH Freeman, New York.

Amerine, M.A., Kunkee, R.E., Ough, C.S., Singleton, V.L., Webb, A.D., 1980. Technology of Wine Making. AVI Publ. Co, Westport, Connecticut.

Amerine, M.A., 1985. Winemaking. In: Koprowski, H., Plotin, S.A. (Eds.), Worlds Debt to Pasteur. Alan R. Liss in Incorporated, New York, p. 67.

Andres-Lacueva, C., Lamuela-Raventós, R.M., Buxaderas, S., Torre-Boronat, M.C., 1997. Influence of variety and aging on foaming properties of cava (sparkling wine) – 2. Journal of Agricultural and Food Chemistry 45 (7), 2520–2525.

Andreu-Sevilla, A.J., Mena, P., Marti, N., Viguera, C.G., Carbonell-Barrachina, A.A., 2013. Volatile composition and descriptive sensory analysis of pomegranate juice and wine. Food Research International 54, 246–254.

Angulo, L., Lema, C., Lopez, J.E., 1993. Influence of viticultural and enological practices on the development of yeast population during winemaking. American Journal of Enology and Viticulture 44, 405–408.

Anocibar-Beloqui, A., Guedes de Pinho, P., Bertrand, A., 1995. Importance of N-3-(methylthiopropyl) acetamide and 3-methylthiopropionic acid in wines. Journal International des Sciences de la Vigne et du Vin 29, 17–26.

Arozarena, Í., Ortiz, J., Hermosín-Gutiérrez, I., Urretavizcaya, I., Salvatierra, S., Córdova, I., RemediosMarín-Arroyo, M., José Noriega, M., Navarro, M., 2012. Color, ellagitannins, antho-cyanins, and antioxidant activity of Andean blackberry (*Rubus glaucus* Benth.) wines. Journal of Agricultural and Food Chemistry 60, 7463–7473.

Askar, A., Treptow, H., 1986. Treptow. Biogene Amine in Lebensmitteln. Eugen Ulmer GmbH and Co, Stuttgart, p. 197.

Aussenac, J., Chassagne, D., Claparols, M., Charpentier, C., Duteurtre, B., Feuillat, M., Charpentier, C., 2001. Purification method for the isolation of monophosphate nucleotides from champagne wine and their identification by mass spectrometry. Journal of Chromatography A 907, 155–164.

Axelsson, L., 2004. Lactic acid bacteria: classification and physiology. In: Salminen, A.V., Wright, A.O. (Eds.), Lactic Acid Bacteria, Microbiological and Functional Aspects. Marcel Dekker, New York, pp. 1–66.

Bartowsky, E., 2003. Lysozime and winemaking. Australian Journal of Grape and Wine Research 473a, 101–104.

Bartowsky, E.J., Henschke, P.A., 2000. Management of malolactic fermentation for the 'buttery' diacetyl flavour in wine. The Australian Grapegrower & Winemaker, Annual Technical Issue 58–67.

Bartowsky, E.J., Costello, P., Henschke, P.A., 2002. Management of malolactic fermentation – wine flavour manipulation. Australian & New Zealand Grapegrower & Winemaker 461 (7–8), 10–12.

Bartowsky, E.J., Henschke, P.A., 1995. Malolactic fermentation and wine flavour. Australian Grapegrower & Winemaker 378, 83–94.

Bartowsky, E.J., Henschke, P.A., 2004. The "buttery" attribute of wine – diacetyl – desirability, spoilage and beyond. International Journal of Food Microbiology 96, 235–252.

Bartowsky, E., 2011. Malolactic fermentation. In: Joshi, V.K. (Ed.), Hand Book of Enology, vol. 3. Asia Tech publication, New Delhi, pp. 526–563.

Bauer, R., Dicks, L.M.T., 2004. Control of malolactic fermentation in wine. A review. South African Journal for Enology and Viticulture 25 (2), 74–88.

Bell, A.A., Ough, C.S., Kliewer, W.H., 1979. Effects on must and wine composition, rates of fermentation and wine quality of nitrogen fertilization of *Vitis vinifera* var. Thompson Seedless grape wines. American Journal of Enology and Viticulture 30, 124.

Benitez, P., Castro, R., Garcia-Barroso, C., 2003. Changes in the polyphenolic and volatile contents of 'Fino' sherry wine exposed to ultraviolet and radiation during storage. Journal of Agricultural and Food Chemistry 51, 6482–6487.

Berlanga, T.M., Atanasio, C., Mauricio, J.C., Ortega, J.M., 2001. Influence of aeration on the physiological activity of flor yeast. Journal of Agricultural and Food Chemistry 49, 3378–3384.

Bertrand, A., de Revel, G., Pripis-Nicolau, L., 2000. Sensory evaluation of consequences of malolactic fermentation for white wine in barrels. Bulletin de OIV 831–832, 313–321.

Bisson, L.F., 1991. Influence of nitrogen on yeast and fermentation of grapes. In: Rantz, J. (Ed.), Proceedings of International Symposium on Nitrogen in Grapes and Wine. 136. ASEV, Davis, CA, pp. 172–184. Annual meeting, Seattle WA.

Bisson, L.F., Butzke, C.E., May/June 1996. Technical enzymes for wine production. Agro-Food-Industry Hi-Tech 11–14.

Boulton, R.B., Singleton, V.L., Bisson, L.F., Kunkee, R.E., 1996. Principles and Practices Ofwinemaking. Chapman Hall, New York.

Buglass, A. (Ed.), 2011. The Handbook of Alcoholic Beverages: Technical, Analytical, and Nutritional Aspects, vol. 1. John Wiley & Sons, United Kingdom.

Burulearu, L., Nicolescu, C.L., Bratu, M.G., Manea, J., Avram, D., 2010. Study regarding some metabolic features during lactic and fermentation at vegetable juices. Romanian Biotechnology Letters 15, 5177–5188.

Busturia, A., Lagunas, R., 1986. Catabolite inactivation of the glucose transport system in *Saccharomyces cervisiae*. Journal of General Microbiology 132, 379–385.

Calderoin, F., Varela, F., Navascueis, E., Colomo, B., Gonzailez, M.C., Suarez, J.A., 2001. Influence of pH and temperature in the biosynthesis of malic acid in wines by *Saccharomyces cerevisiae*. Bulletin de OIV 474–486 845846.

Campbell, B., den Hollander, J.A., Alger, J.R., Shulman, R.G., 1987. 31PNMR saturation-transfer and 13C NMR kinetic studies of glycolytic regulation during anaerobic and aerobic glycolysis. Biochemistry 26, 7493–7500.

Canal-Llauberes, R.M., 1993. Enzymes in winemaking. In: Fleet, G.H. (Ed.), Wine Microbiology and Biotechnology. Harwood Academic Publishers, Switzerland, pp. 477–506.

Cannon, W.R., 2014. Concepts, challenges and successes in modeling thermodynamics of metabolism. Frontiers in Bioengineering and Biotechnology 2, 1–10. http://dx.doi.org/10.3389/fbioe.2014.00053.

Caplice, E., Fitzgerald, G.F., 1999. Food fermentations: role of microorganisms in food production and preservation. International Journal of Food Microbiology 50, 131–149.

Cartwright, C.P., Juroszek, J.R., Beavan, M.J., Ruby, M.S., DeMorias, M.F., Rose, A.H., 1986. Ethanol dissipates the proton- motive force across the plasma membrane of *Saccharomyces cerevisiae*. Journal of General Microbiology 132, 369.

Cartwright, C.P., Rose, A.H., Calderbank, J., Keenan, M.H.J., 1989. Solute transport. In: Rose, A.H., Harrison, J.S. (Eds.), The Yeasts, vol. 3. second ed. Academic Press, London, p. 5.

Casey, G.P., Ingledew, W.M., 1986. Ethanol tolerance in yeasts. Critical Reviews in Microbiology 13, 219.

Castañeda-Ovando, A., Pacheco-Hernández, M.L., Páez-Hernández, M.E., Rodríguez, J.A., Galán-Vidal, C.A., 2009. Chemical studies of anthocyanins: a review. Food Chemistry 113, 859–871.

Cavazza, A., Versini, G., Grando, H.W.S., Price, K.R., 1990. Variabilita indotta dai ceppi di lievito nella rifermentazione dei vini spumanti. Industrie delle Bevande 19, 225–228.

Charoenchai, C., Fleet, G., Henscke, P.A., 1998. Effects of temperature, pH and sugar concentration on the growth rates and cell biomass of wine yeasts. American Journal of Enology and Viticulture 49, 283–288.

Charoenchai, C., Fleet, G.H., Henschke, P.A., Todd, B.E.N., 1997. Screening of non-*Saccharomyces* wine yeast for the presence of extracellular hydrolytic enzymes. Australian Journal of Grape and Wine Research 3, 2–8.

Charpentier, C., Feuillat, M., 1993. Yeast autolysis. In: Fleet, G.H. (Ed.), Wine Microbiology and Biotechnology. Harwood Academic Publishers, Switzerland, pp. 225–242.

Cho, J.Y., Jeong, J.H., Kim, J.Y., Kim, S.R., Kim, S.J., Lee, H.J., Moon, J.H., 2013. Change in the content of phenolic compounds and antioxidant activity during manufacturing of black raspberry (*Rubus coreanus* Miq.) wine. Food Science and Biotechnology 22 (5), 1–8.

Ciani, M., Maccarelli, F., 1998. Oenological properties of non-*Saccharomyces* yeast associated with wine-making. World Journal of Microbiology and Biotechnology 14, 199–203.

Clarke, R.J., Bakker, J., 2004. Wine Flavour Chemistry. Blackwell, Oxford.

Cocolin, L., Comi, G., 2011. Killer yeasts in winemaking. In: Joshi, V.K. (Ed.), Handbook of Enology, vol. III. Asia Tech Publishers, New Delhi, pp. 565–590.

Colowick, S.P., 1973. The Hexokinases. In: Boyer, P.D. (Ed.), The Enzymes. 9. New York Academic Press, p. 1.

Cooper, T.C., Sumrada, R., 1975. Urea transport in *Saccharomyces cerevisiae*. Journal of Bacteriology 121, 571–576.

Cooper, T.G., 1982. Nitrogen metabolism in *Saccharomyces cerevisiae*. In: Strathern, J.N., Jones, E.W., Broach, J.R. (Eds.), The Molecular Biology of the Yeast *Saccharmyces Cerevisiae*: Metabolism and Gene Expression. Cold Spring Harbor, New York, pp. 39–99.

Cordero-Otero, R.R., Ubeda-Iranzo, J.F., Briones-Perez, A.I., Potgieter, N., Villena, M.A., Pretorius, I.S., van Rensburg, P., 2003. Characterization of the beta-glucosidase activity produced by enological strains of non-*Saccharomyces* yeasts. Journal of Food Science 68, 2564–2569.

Cortes, M.B., Moreno, J., Zea, L., Moyano, L., Medina, M., 1998. Changes in aroma compounds of sherry wines during their biological aging carried out by *Saccharomyces cerevisiae* races *bayanus* and *capensis*. Journal of Agricultural and Food Chemistry 46, 2389–2394.

Cortes, M.B., Moreno, J.J., Zea, L., Moyano, L., Medina, M., 1999. Response of the aroma fraction in sherry wines subjected to accelerated biological aging. Journal of Agricultural and Food Chemistry 47, 3297–3302.

Costello, P., 2006. The chemistry of malolactic fermentation. In: Morenzoni, R. (Ed.), Malolactic Fermentation in Wine – Understanding the Science and the Practice. Lallemand, Montréal, pp. 4.1–4.11.

Costello, P.J., Henschke, P., 2002. Mousy off-flavor of wine: precursors and biosynthesis of the causative N-heterocycles 2-ethyltetrahydropyridine, 2-acetyltetrahydropyridine, and 2-acetyl-1-pyrroline by *Lactobacillus hildargii* DSM 20176. Journal of Agricultural and Food Chemistry 50, 7079–7087.

Cunningham, F.E., Proctor, V.A., Gootsch, S.J., 1991. Egg-white lysozyme as a food preservative: an overview. World's Poultry Science Journal 47, 141.

Czerny, M., Christlbauer, M., Christlbauer, M., Fischer, A., Granvogl, M., Hammer, M., et al., 2008. Re-investigation on odour thresholds of key food aroma compounds and development of an aroma language based on odour qualities of defined aqueous odorant solutions. European Food Research and Technology 228, 265–273.

Czyzowska, A., Pogorzelski, E., 2002. Changes to polyphenols in the process of production of must and wines from blackcurrants and cherries. Part I. Total polyphenols and phenolic acids. European Food Research and Technology 214, 148–154.

del Campo, G., Berregi, I., Caracena, R., Santos, J.I., 2006. Quantitative analysis of malic and citric acids in fruit juices using proton nuclear magnetic resonance spectroscopy. Analytica Chimica Acta 556, 462–468.

Danilewiex, J.C., Seccombe, J.T., Whelan, J., 2008. Mechanism of interaction of polyphenols, oxygen and sulphur dioxide in model wine and wine. American Journal of Enology and Viticulture 59, 2128–2136.

Davis, C.R., Wibowo, D., Eschenbruch, R., Lee, T.H., Fleet, L.H., 1985. Practical implications of malolactic fermentation: a review. American Journal of Enology and Viticulture 36, 290–301.

Davis, R.H., 1986. Compartmental and regulatory mechanism in the arginine pathways of *Neurospora crassa* and *Saccharomyces cerevisiae*. Microbiological Review 50, 280–313.

Delaquis, P., Cliff, M., King, M., Girard, B., Hall, J., Reynolds, A., 2000. Effect of two commercial malolactic cultures on the chemical and sensory properties of Chancellor wines vinified with different yeasts and fermentation temperatures. American Journal of Enology and Viticulture 51, 42–48.

Denis, C.L., Ciriacy, M., Young, E.T., 1983. MRNA level for the fermentative alcohol dehydrogenase of *Saccharomyces cerevisae* decrease upon growth on a non fermentable carbon source. Journal of Biological Chemistry 258, 1165.

Dey, G., Negi, B., Gandhi, A., 2009. Can fruit wines be considered as functional food?—An overview. Natural Product Radiance 8 (4), 314–322.

Dicks, L.M., Dellaglio, F., Collins, M.G., 1995. Proposal to reclassify *Leuconostoc oenos as Oenococcus oeni*, Gen. Nov., comb. Nov. International Journal of Systematic Bacteriology 45, 395–397.

Dizy, M., Bisson, L.F., 2000. Proteolytic activity of yeast strains during grape juice fermentation. American Journal of Enology and Viticulture 51, 155–167.

Drdak, M., Rajniakova, A., Buchtova, V., Simko, P., 1993. Free amino acid content of various red wines. Die Nahrung 37, 77.

Drysdale, G.S., Fleet, G.H., 1988. Acetic acid bacteria in wine making: a review. American Journal of Enology and Viticulture 39, 143.

Duarte, W.F., Dragone, G., Dias, D.R., Teixeira, J.A., Almeida Silva, J.B., Schwan, R.F., 2010c. Fermentative behavior of *Saccharomyces* strains during microvinification of raspberry juice (*Rubusidaeus* L. International Journal of Food Microbiology 143, 173–182.

Duarte, W.F., Dias, D.R., Oliveira, J.M., Teixeira, J.A., AlmeidaeSilva, J.B., Schwan, R.F., 2010b. Characterization of different fruit wines made from cacao, cupuassu, gabiroba, jaboticaba and umbu. LWT—Food Science and Technology 43, 1564–1572.

Duarte, W.F., Dias, D.R., Oliveira, J.M., Vilanova, M., Teixeira, J.A., Almeidae Silva, J.B., et al., 2010a. Raspberry (*Rubusidaeus L.*) wine: yeast selection, sensory evaluation and instrumental analysis of volatile and other compounds. Food Research International 43, 2303–2314.

Dubois, C., Manginot, C., Roustan, J.-L., Sablayrolles, J.-M., Barre, P., 1996. Effect of variety, year, and grape maturity on the kinetics of alcoholic fermentation. American Journal of Enology and Viticulture 47, 363–368.

Eddy, A.A., 1982. Mechanism of solute transport in selected eukaryotic microorganisms. Advances in Microbial Physiology 24, 1.

Edwards, C.G., Hacy, K.M., Collins, M.D., 1998. Identification and characterization of two lactic acid bacteria associated with sluggish/stuck fermentation. American Journal of Enology and Viticulture 43, 445–448.

Eglinton, J.M., mcWilliam, S.J., Fogarty, M.W., Francis, I.L., Kwiatkowski, M.J., Hoj, P.B., Henschke, P.A., 2000. The effect of *Saccharomyces bayanus*-mediated fermentation on the chemical composition and aroma profile of Chardonnay wine. Australian Journal of Grape and Wine Research 6, 190–196.

El-Gendy, S.M., Abdel-Galil, H., Shahin, Y., Hegezi, F.Z., 1983. Acetoin and diacetyl production by homo and hetero-fermentative lactic acid bacteria. Journal of Food Protection 46, 420.

Eschenbruch, R., Bonish, P., Fisher, B.M., 1978. The production of H2S by pure culture yeasts. Vitis 17, 67.

Etievant, P.X., 1991. In: Maarse, H. (Ed.), Volatile Compounds in Foods and Beverages. Marcel Dekker, New York, pp. 483–546.

Fabani, M.P., Arrúa, R.C., Vázquez, F., Diaz, M.P., Baroni, M.V., Wunderlin, D.A., 2010. Evaluation of elemental profile coupled to chemometrics to assess the geographical origin of Argentinean wines. Food Chemistry 119, 372–379.

Falqué, E., Fernández, E., Dubourdieu, D., 2001. Differentiation of white wines by their aromatic index. Talanta 54, 271–281.

Faria-Oliveira, F., Diniz, R.H.S., Godoy-Santos, R., Piló, F.B., Mezadri, H., Castro, I.M., Brandão, R.L., 2015. The role of yeast and lactic acid bacteria in the production of fermented beverages. In: Amer Eissa, A. (Ed.), South America, Food Production and Industry. InTech. http://dx.doi.org/10.5772/60877.

Ferraro, L., Fatichenti, F., Ciani, M., 2000. Pilot scale vinification process using immobilized *Candida stellata* and *Saccharomyces cerevisiae*. Process Biochemistry 35, 1125–1129.

Ferreira, V., Lopez, R., Cacho, J.F., 2000. Quantitative determination of the odorants of young red wine from different grape varieties. Journal of the Science of Food and Agriculture 80, 1659–1667.

Ferreira, V., Fernandez, P., Pena, C., Escudero, A., Cacho, J.F., 1995. Investigation on the role played by fermentation esters in the aroma of young Spanish wines by multivariate analysis. Journal of the Science of Food and Agriculture 67, 381–392.

Feuillat, M., 1987. Preparation d'autolysats de levures et addition dans les vins effervescents elabores selon la methode champenoise. Revue Francaise Oenologie 28, 36–45.

Feuillat, M., Freyssenet, M., Charpentier, C., 1988. Production de colloides par les levures dans les vins mousseux elabores selon la methode champenoise. Revue Francaise Oenologie 28, 36–45.

Francioli, S., Torrens, J., Riu-Aumatell, M., Loipez-Tamames, E., Buxaderas, S., 2003. Volatile compounds by SPME-GC as age markers of sparkling wines. American Journal of Enology and Viticulture 54, 158–162.

Frías, S., Conde, J.E., Rodríguez, M.A., Dohnal, V., Pérez-Trujillo, J.P., 2002. Metallic content of wines from the Canary Islands (Spain). Application of artificial neural networks to the data analysis. Nahrung Food 46, 370–375.

Francis, F.J., 1986. Analysis of anthocyanins. In: Markakis, P. (Ed.), Anthocyanins as Food Colors. Academic press, New York, USA, pp. 181–207.

Francis, I.L., Newton, J.L., 2005. Determining wine aroma from compositional data. Australian Journal of Grape and Wine Research 11 (2), 114–126.

Fuck, E., Radler, F., 1972. Apfelsaurestoffwechsel bei *Saccharomyces* I. Der abnaerobe Apfelsaureabbau bei *Saccharomyces cerevisiae*. Archiv fur Mibrobiologie 87, 149.

Fugelsang, K.C., 1997. In: Fugelsang, K.C. (Ed.), Wine Microbiology. Chapman & Hall, New York.

Fujii, T., Yoshimoto, H., Tamai, Y., 1996. Acetate ester production by *Saccharomyces cerevisiae* lacking the ATF1 gene encoding the alcohol acetyltransferase. Journal of Fermentation and Bioengineering 81, 538–542.

Le Fur, Y., Maume, G., Feuillat, M., Maume, B.F., 1997. Characterization by gas chromatography/mass spectrometry of sterols in *Saccharomyces cerevisiae* during autolysis. Journal of Agricultural and Food Chemistry 47, 2860–2864.

Gallart, M., Francioli, S., Viu-Marco, A., Loipez Tamames, E., Buxaderas, S., 1997. Determination of free fatty acids and their esters in musts and wines. Journal of Chromatography A 776, 283–291.

Gambaro, A., Boido, E., Zlotejablo, A., Medina, K., Lloret, A., Dellacassa, E., Carrau, F., 2001. Effect of malolactic fermentation on the aroma properties of Tannat wine. Australian Journal of Grape and Wine Research 7, 27–32.

Gao, J., Rupasinghe, H.P.V., 2013. Characterization of malolactic conversion by *Oenococcusoeni* to reduce the acidity of apple juice. Journal of Food Science and Technology 48, 1018–1027.

Genitsariotis, R., 1979. Uber die Bildung von L(+) Lactat bei Sachharomyces pretoriensis und die NAD-abhiangige Lactat-dehydrogenase (Dissertation Mainz).

Gerbaux, V., Villa, A., Monamy, C., Bertrand, A., 1997. Use of lysozyme to inhibit malolactic fermentation and to stabilize wine after malolactic fermentation. American Journal of Enology and Viticulture 48, 49–54.

Gloria, M.B., Watson, B.T., Simon-Sarkadi, L., Daeschel, M.A., 1998. A survey of biogenic amines in Oregon Pinot noir and Cabernet Sauvignon wines. American Journal of Enology and Viticulture 49, 279–282.

Goni, D.T., Azpilicueta, C.A., 2001. Influence of yeast strain on biogenic amines content in wines: relationship with utilization of amino acid during fermentation. American Journal of Enology and Viticulture 52 (3), 185.

Goniak, O.J., Noble, A.C., 1987. Sonsory study of selected volatile sulphur compounds in white wines. American Journal of Enology and Viticulture 38, 223–227.

González Álvarez, M., González-Barreiro, C., Cancho-Grande, B., Simal-Gándara, J., 2011. Relationships between Godello white wine sensory properties and its aromatic fingerprinting obtained by GC–MS. Food Chemistry 129, 890–898.

Goyal, R.K., 1999. Biochemistry of fermentation. In: Joshi, V.K., Pandey, A. (Eds.), Biotechnology: Food Fermentation, vol. I. Asiatech Publishers Inc., New Delhi, pp. 87–172.

Grbin, P.R., Henschke, P.A., 2000. Mousy off-flavor production in grape juice and wine. Australian Journal of Grape and Wine Research 6, 255–262 (233).

Green, D.E., Murer, E., Hultin, H.O., Richardson, S.H., Salmon, B., Brierly, G.P., Baum, H., 1965. Association of integrated metabolic pathways with membranes I. Glycolytic enzymes of the red corpuscle and yeast. Archives of Biochemistry and Biophysics 112, 635.

Grenson, M., Acheroy, B., 1982. Mutations affecting the activity and the regulation of the general amino-acid permease of *Saccharomyces cerevisiae*. Localisation of the cis acting dominant PGR regulatory mutation in the structural gene of this permease. Molecular and General Genetics 188, 261.

Grenson, M., 1983a. Inactivation-Reactivation process and repression of permease formation regulate several ammonia-sensitive permeases in the yeasts *Saccharomyces cerevisiae*. Journal of Biochemistry 133, 135.

Grenson, M., 1983b. Study of the positive control of the general amino-acid permease and other ammonia-sensitive uptake systems by the product of the NPR1 gene in the yeast *Saccharomyces cerevisiae*. Journal of Biotechnology 133, 141.

Guitart, A., HernandezOrte, P., Ferreira, V., Pena, C., Cacho, J., 1999. Some observations about the correlation between the amino acid content of musts and wines of the Chardonnay variety and their fermentation aromas. American Journal of Enology and Viticulture 50, 253–258.

Gumyon, J.F., Heintz, J.E., 1952. The fusel oil content of Californian wine. Food Technology 6, 359.

Guth, H., 1997. Quantitation and sensory studies of character impact odorants of different white wine varieties. Journal of Agricultural and Food Chemistry 45, 3027–3032.

Den Hollander, J.A., Ugurbil, K., Brown, T.R., Bednar, M., Redfield, C., Shulman, R.G., 1986. Studies of anaerobic and aerobic glycolysis in *Saccharomyces cerevisiae*. Biochemistry 25, 203–211.

Häkkinen, S.H., Kärenlampi, S.O., Heinonen, I.M., Mykkänen, H.M., Törrönen, A.R., 1999. Content of the Flavonols quercetin, myricetin, and kaempferol in 25 edible berries. Journal of Agricultural and Food Chemistry 47 (6), 2274–2276.

Heinonen, I.M., Lehtonen, P.J., Hopia, A.I., 1998. Antioxidant activity of berry and fruit wines and liquors. Journal of Agricultural and Food Chemistry 46, 25–31.

Henick-Kling, T., 1993. Malolactic fermentation. In: Flect, G.H. (Ed.), Wine Microbiology and Biotechnology. Harwood Academic Publishers, Switzerland, pp. 289–326.

Henick-Kling, T., 1995. Control of malo-lactic fermentation in wine: energetics, flavour modification and methods of starter culture preparation. Journal of Applied Bacteriology Supplement 79, 29S–37S.

Henschke, P.A., Jiranek, V., 1993. Yeast-metabolism of nitrogen compounds. In: Fleet, G.H. (Ed.), Wine Microbiology and Biotechnology. Harwood Academic Publishers, Australia, pp. 77–164.

Henschke, P.A., Ough, C.S., 1991a. Urea accumulation in fermenting grape juice. American Journal of Enology and Viticulture 42, 317–321.

Henschke, P.A., Rose, A.H., 1991. Plasma memberanes. In: Rose, A.H., Harrison, J.S. (Eds.), The Yeasts, vol. 4., Academic Press, London, pp. 297–345.

Henschke, P.A., Jiranek, V., 1991. Hydrogen sulfide formation during fermentation: effects of nitrogen composition in model grape musts. In: Rantz, J.M. (Ed.), Proceedings of the International Nitrogen Symposium on Grapes and Wine American Society for Enology and Viticulture 172–175 Davis, CA.

Hernawan, T., Fleet, G., 1995. Chemical and cytological changes during the autolysis of yeast. Journal of Industrial Microbiology 14, 440–450.

Herraiz, T., Ough, C.S., 1992. Identification and determination of amino acid ethyl esters in wines by capillary gas chromatography and mass spectrometry. Journal of Agricultural and Food Chemistry 40, 1015–1021.

Hidalgo, P., Flores, M., 1994. Occurrence of the killer character in yeasts associated with Spanish wine production. Food Microbiology 11, 161–167.

Higgins, L., Llanos, E., 2015. A healthy indulgence? Wine consumers and the health benefits of wine. Wine Economics and Policy. http://dx.doi.org/10.1016/j.wep.2015.01.001.

Hinick-King, T., 1993. Malolactic fermentation. In: Fleet, G.H. (Ed.), Wine Microbiology and Biotechnology. Harwood Academic Publishers, Switzerland.

Hinnesbusch, A.G., 1988. Mechanisms of gene regulation in the general control of amino acid biosynthesis in *Saccharomyces cerevisiae*. Microbiological Reviews 52, 248–273.

Horak, J., Kotyk, A., 1977. Temperature effects in amino acid transport by *Saccharomyces cerevisiae*. Experimental Mycology 1, 63–68.

Houtman, A.C., du Plessis, C.S., 1981. The effect of juice clarity and several conditions promoting yeast growth or fermentation rate, the production of aroma components and wine quality. South African Journal for Enology and Viticulture 2, 71.

Houtman, A.C., Marais, J., Du Plessis, C.S., 1980. The possibilities of applying present-day knowledge of wine aroma components: influence of several juice factors on fermentation rate and ester production during fermentation. South African Journal for Enology and Viticulture 1, 27–33.

Hugenholtz, J., 1993. Citrate metabolism in lactic acid bacteria. FEMS Microbiology Reviews 12, 165–178.

Ingledew, W.M., Kunkee, R.E., 1985. Factors influencing sluggish fermentations of grape juice. American Journal of Enology and Viticulture 36, 65–76.

Ingledew, W.M., Magnus, C.A., Patterson, J.R., 1987. Yeast foods and ethyl carbamate formation in wine. American Journal of Enology and Viticulture 38, 323–332.

Jagtap, U., Bapat, V., 2015. Wines from fruits other than grapes: current status and future prospects. Food Bioscience 9, 80–96.

Jansen, M., Veurink, J., Euverink, G.J., Dijkhuizen, 2003. Growth of the salt-tolerant yeast *Zygosaccharomyces rouxii* in microtiter plates: effects of NaCl, pH and temperature on growth and fusel alcohol production from branched-chain amino acids. FEMS Yeast Research 3 (3), 313–318.

Januik, M.T., 1984. The Development of a Technique to Quantify and Identify Wine Sulfur Volatiles by Gas Chromatography (Thesis). University of California, Davis.

Jarvis, B., Foster, M.J., Kinsella, W.P., 1995. Factor affecting the development of cider flavour. In: Board, R.G., Jones, D., Jarvis, B. (Eds.), Microbial Fermentations: Berverages, Foods and Feeds Journal of Applied Bacteriology—Symposium Supplement 79, 55.

Jeandet, P., Vasserot, Y., Liger-Belair, G., Marchal, R., 2011. In: Joshi, V.K. (Ed.), Sparkling Wine Production in: Handbook of Enology, vol. III. Asia Tech Publishers, Inc., New Delhi, pp. 1064–1115.

Jeong, J.H., Jung, H., Lee, S.R., Lee, H.J., Hwang, K.T., Kim, T.Y., 2010. Anti-oxidant, anti-proliferative and anti-inflammatory activities of the extracts from black raspberry fruits and wine. Food Chemistry 123, 338–344.

Jiranek, V., Henschke, P.A., 1991. Assimilable nitrogen: regulator of hydrogen sulphide production during fermentation. Australian Journal of Grape and Wine Research 325, 27.

Jiranek, V., Langridge, P., Henschke, P.A., 1996. Determination of sulphite reductase activity and its response to assimilable nitrogen status in a commercial *Saccharomyces cerevisiae* wine yeast. The Journal of Applied Bacteriology 81, 329–336.

Johnson, M.H., Lucius, A., Meyer, T., de Mejia, E.G., 2011. Cultivar evaluation and effect of fermentation on antioxidant capacity and in vitro inhibition of α-amylase and α-glucosidase by Highbush Blueberry (*Vaccinium corombosum*). Journal of Agricultural and Food Chemistry 59, 8923–8930.

Johnson, M.H., de Mejia, E.G., 2012. Comparison of chemical composition and antioxidant capacity of commercially available blueberry and blackberry wines in illinois. Journal of Food Science 71, 141–148.

Johnson, M.H., Gonzalez de Mejia, E., Fan, J., Lila, M.A., Yousef, G.G., 2013. Anthocyanins and proanthocyanidins from blueberry–blackberry fermented beverages inhibit markers of inflammation in macrophages and carbohydrate-utilizing enzymes in vitro. Molecular Nutrition & Food Research 57, 1182–1197.

Jones, E.W., Fink, G.R., 1982. Regulation of amino acid and nucleotide biosynthesis in yeast. In: Strathern, J.N., Jones, E.W., Broach, J.B. (Eds.), Molecular Biology of the Yeast *Saccharomyces*. Metabolism and Gene Expression. Cold Spring Harbor Laboratory, New York, p. 181.

Jos, A., Morenoa, I., González, A.G., Repetto, G., Cameána, A.M., 2004. Differentiation of sparkling wines (cava and champagne) according to their mineral content. Talanta 63, 377–382.

Joshi, V.K., Bhutani, V.P., Sharma, R.C., 1990. Effect of dilution and addition of nitrogen source on chemical mineral and sensory qualities of wild apricot wine. American Journal of Enology and Viticulture 41, 229–232.

Joshi, V.K., Bhutani, V.P., 1991. The influence of enzymatic clarification on the fermentation behaviour, composition and sensory qualities of apple wine. Sciences Des Aliments 11 (3), 491–496.

Joshi, V.K., Chauhan, S.K., Lal, B.B., 1991a. Extraction of juices from plum, peach and apricot by the pectolytic enzyme treatment. Journal of Food Science and Technology 28 (1), 64–65.

Joshi, V.K., Sharma, P.C., Attri, B.L., 1991b. A note on the deacidification activity of *Schizosaccharomyces pombe* in plum musts of variable composition. Journal of Applied Bacteriology 70, 386–390.

Joshi, V.K., Sharma, S.K., Thakur, N.S., 1995. Technology and quality of sparkling wine with special reference to plum—an overview. Indian Food Packer 49, 49–63.

Joshi, V.K., 1997. Fruit Wines, second ed. Direcorate of Extension Education. Dr. Y.S. Parmar University of Horticulture and Forestry, Nauni, Solan, HP, p. 226. DTP System.

Joshi, V.K., Bhutani, V.P., Thakur, N.K., 1999a. In: Joshi, V.K., Pandey, A. (Eds.), Composition and Nutrition of Fermented Products. In: Biotechnology: Food Fermentation, vol. I. Educational Publishers and Distributors, New Delhi, pp. 259–320.

Joshi, V.K., Sandhu, D.K., Thakur, N.S., 1999b. Fruit based alcoholic beverages. In: Joshi, V.K., Pandey, A. (Eds.), Biotechnology: Food Fermentation, vol. II. Educational Publishers and Distributors, New Delhi, pp. 647–744.

Joshi, V.K., Sharma, S.K., Goyal, R.K., Thakur, N.S., 1999c. Sparkling plum wine: effect of method of carbonation and the type of base wine on physico-chemical and sensory qualities. Brazilian Archives of Biology and Technology 42, 315–321.

Joshi, V.K., Pandey, A., 1999. Biotechnology: food fermentation. In: Joshi, V.K., Pandey, A. (Eds.), Biotechnology: Food Fermentation, vol. I. Asiatech Publishers Inc., New Delhi, pp. 1–45.

Joshi, V.K., Thakur, N.S., Bhatt, A., Chayanika, G., 2011a. Wine and brandy: a perspective. In: Joshi, V.K. (Ed.), Handbook of Enology. vol. 1. Asia Tech Publishers, Inc., New Delhi, pp. 3–45.

Joshi, V.K., Attri, D., Singh, T.K., Abrol, G., 2011b. Fruit wines: production technology. In: Joshi, V.K. (Ed.), Handbook of Enology, vol. III. Asia Tech Publishers, Inc, New Delhi, pp. 1177–1221.

Joshi, V.K., 2016. Indigenous Fermented Foods of South Asia. Rob Nout and Prabir Sarkar, Series Editors. The Fermented Foods and Beverages Series. CRC Press, Roca, FL, pp. 1–849.

Jung, J.W., Son, M.Y., Jung, S.W., Nam, P.W., Sung, J.S., Lee, S.J., Lee, K.G., 2009. Antioxidant properties of Korean black raspberry wines and their apoptotic effects on cancer cell. Journal of Science and Food Agriculture 89, 970–977.

Kandler, O., 1983. Carbohydrate metabolism in lactic acid bacteria. Antonie vanm Leeuwenhoek 49, 209–224.

Kandler, O., Winter, J., Stetter, K.O., 1973. Zur Frage der Beeinflusung der Glucosevergrung durch L-Malat bei Leuconostoc mesenteroids. Archiv fur Mikrobiologie 90, 65.

Kaur, C., Kapoor, H.C., 2001. Antioxidants in fruits and vegetables—the millennium's health. International Journal of Food Science and Technology 36, 703–725.

Killian, E., Ough, C.S., 1979. Fermentation esters—formation and retention as affected by fermentation temperature. American Journal of Enology and Viticulture 30, 301.

Klarić, D.A., Klarić, I., Mornar, A., 2011. Polyphenol content and antioxidant activity of commercial blackberry wines from Croatia: application of multivariate analysis for geographic origin differentiation. Journal of Food and Nutrition Research 50 (4), 199–209.

Klopotek, Y., Otto, K., Bohm, V., 2005. Processing strawberries to different products alters contents of vitamin C, total phenolics, total anthocyanins, and antioxidant capacity. Journal of Agricultural and Food Chemistry 53, 5640–5646.

Kollar, R., Vorisek, J., Sturdik, E., 1993. Biochemical, morphological, and cytochemical studies of enhanced autolysis of *Saccharomyces Cerevisiae*. Folia Microbiology 38, 479–485.

Kondrashov, A., Ševčík, R., Benáková, H., Koštířová, M., Štípek, S., 2009. The key role of grape variety for antioxidant capacity of red wines. e-SPEN, the European e-Journal of Clinical Nutrition and Metabolism 4, 41–46.

Kruckeberg, A.L., Bisson, L.F., 1990. The HXT2 gene of *Saccharomyces cerevisiae* is required for high affinity glucose transport. Molecular and cellular Biology 10, 5903.

Lagace, L.S., Bisson, L.F., 1990. Survey of yeast acid proteases for effectiveness of wine haze reduction. American Journal of Enology and Viticulture 41 (2), 147–153.

Lagunas, R., 1986. Misconceptions about the energy metabolism of *Saccharomyces cerevisiae*. Yeast 2, 221.

Lagunas, R.O.S.A.R.I.O., Dominguez, C., Busturia, A., Saez, M.J., 1982. Mechanisms of appearance of the Pasteur effect in *Saccharomyces cerevisiae*: inactivation of sugar transport systems. Journal of Bacteriology 152 (1), 19–25.

Large, P.J., 1986. Degradation of organic nitrogen compounds by yeasts. Yeast 2, 1–34.

Lavigne, V., Dubourdieu, D., 1996. Demonstration and interpretation of the yeast lees ability to adsorb certain volatile thiols contained in wine. Journal International des Sciences de la Vigne et du Vin 30, 201–206.

Lehtonen, P., 1996. Determination of amines and amino acids in wine—a review. American Journal of Enology and Viticulture 47, 127–133.

Lehtonen, P.J., Rokka, M.M., Hopia, A.I., Heinonen, I.M., 1999. HPLC determination of phenolic compounds in berry and fruit wines and liqueurs. Wein Wissenchaft Viticultural and Enological Sciences 54, 33–38.

Lema, C., Garcia-Jares, C., Orriols, I., Angulo, L., 1996. Contribution of *Saccharomyces* and Non-*Saccharomyces* populations to the production of some components of Albarino wine aroma. American Journal of Enology and Viticulture 47, 206–216.

Lerm, E., Engelbrecht, L., du Toit, M., 2010. Malolactic fermentation: the ABC's of MLF. South African Journal for Enology and Viticulture 31 (2), 186–212.

Li, X., Chan, L.J., Bin, Y., Curran, P., Liu, S., 2012. Fermentation of three varieties of mango juices with a mixture of *Saccharomyces cerevisiae* and *Williopsis saturnus* var. *mrakii*. International Journal of Food Microbiology 158, 28–35.

Lim, J.W., Jeong, J.T., Shin, C.S., 2012. Component analysis and sensory evaluation of Korean black raspberry (*Rubus coreanus* Mique) wines. International Journal of Food Science and Technology 47, 918–926.

Liu, S.-Q., Pritchard, G.G., Hardman, M.J., Pilone, G.J., 1996. Arginine catabolism in wine lactic acid bacteria: is it via the arginine deiminase pathway or the arginase-urease pathway? Journal of Applied Bacteriology 81, 486–492.

Liu, S.-Q., 2002. A review: malolactic fermentation in wine beyond deacidification. Journal of Applied Microbiology 92, 589–601.

Llauberes, R.M., Dubourdieu, D., Villetaz, J.C., 1987. Exocellular polysaccharides from *Saccharomyces* in wine production. Industrie delle Bevande 41, 277–286.

Llauberes, R.M., Richard, B., Lonvaud-Funel, A., Dubourdieu, D., 1990. Structure of an exocellular B-D-glucan from *Pediococcus* sp., a wine lactic acid bacterium. Carbohydrate Research 203, 103–107.

Loipez-Barajas, M., Loipez-Tamames, E., Buxaderas, S., Suberbiola, G., de la Torre-Boronat, M.C., 2001. Influence of wine polysaccharides of different molecular mass on wine foaming. American Journal of Enology and Viticulture 52, 146–150.

Lonvaud-Funel, A., 1999. Lactic acid bacteria in the quality improvement and depreciation of wine. Antonie vanm Leeuwenhoek 76, 317–331.

Lonvaud-Funel, A., 2001. Biogenic amines in wines: role of lactic acid bacteria. FEMS Microbiology Letters 199, 9–13.

Loubser, G.J., Du Plessis, C.S., 1976. The quantatitive determination and some values of dimethyl sulfide in white table wines. Vitis 15, 248–252.

Luguera, C., Moreno-Arribas, V., Pueyo, E., Polo, M.C., 1997. Capillary electrophoretic analysis of wine proteins. Modifications during the manufacture of sparkling wines. Journal of Agricultural and Food Chemistry 45, 3766–3770.

Luguera, C., Moreno-Arribas, V., Pueyo, E., Bartolomei, B., Polo, M.C., 1998. Fractionation and partial characterization of protein fractions present at different stages of the production of sparkling wines. Food Chemistry 63, 465–471.

Mafra, I., Herbert, P., Santos, L., Barros, P., Alves, A., 1999. Evaluation of biogenic amines in some Portuguese quality wines by HPLC fluorescence detection of OPA derivatives. American Journal of Enology and Viticulture 50, 128–132.

Mahon, H.M.M., Zoecklein, B.W., Jasinski, Y.W., 1999. The effects of prefermentation maceration temperature and percent alcohol (v/v) at Press on the concentration of Cabernet Sauvignon grape glycosides and glycoside fractions. American Journal of Enology and Viticulture 50, 385–390.

Maicas, S., Gil, J.V., Pardo, I., Ferrer, S., 1999. Improvement of volatile composition of wines by controlled addition of malolactic bacteria. Food Research International 32, 491–496.

Manca de Nadra, M.C., Strasser de Saad, A.M., 1995. Polysaccharide production by *Pediococcus pentosaceous* from wine. International Journal of Food Microbiology 27, 101–106.

Manzanares, P., Rojas, V., Genoves, S., Valles, S., 2000. A preliminary search for anthocyanin-B-D-glucosidase activity in non-*Saccharomyces* wine yeasts. International Journal of Food Science & Technology 35, 95–103.

Margalith, P.Z., 1981. Flavour Microbiology. Illinois Charles G. Thomas, p. 173.

Martin, J.L., Matar, C., 2005. Increase of antioxidant capacity of the lowbush blueberry (*Vaccinium angustifolium*) during fermentation by a novel bacterium from the fruit microflora. Journal of the Science of Food Agriculture 85, 1477–1484.

Martino, A., Schiraldi, C., Lazzaro, A., Di, Fiume, I., Spagna, G., Pifferi, P.G., De Rosa, M., 2000. Improvement of the flavour of Falanghina white wine using a purified glycosidase preparation from *Aspergillus niger*. Process Biochemistry 36, 93–102.

Martineau, B., Henick-Kling, T., 1995a. Formation and degradation of diacetyl in wine during alcoholic fermentation with *Saccharomyces cerevisiae* strain EC1118 and malolactic fermentation with *Leuconostoc oenos* strain MCW. American Journal of Enology and Viticulture 46, 442–448.

Martineau, B., Henick-Kling, T., 1995b. Performance and diacetyl production of commercial strains of malolactic bacteria in wine. Journal of Applied Bacteriology 78, 526–536.

Martineau, B., Acree, T.E., Henick-Kling, T., 1995. Effect of wine type on the detection threshold for diacetyl. Food Research International 28 (13), 9–143.

Martinez, P., Codon, A.C., Perez, I., Benitez, T., 1995. Physiological and molecular characterization of flor yeasts: polymorphism of flor yeast populations. Yeast 11, 1399–1411.

Martinez, P., Perez Rodriguez, L., Benitez, T., 1997. Evolution of flor yeast population during the biological aging of fino sherry wine. American Journal of Enology and Viticulture 48, 160–168.

Martinez-Rodriguez, A.J., Polo, M.C., 2000. Characterization of the nitrogen compounds released during yeast autolysis in a model wine system. Journal of Agricultural and Food Chemistry 48, 1081–1085.

Martinez-Rodriguez, A.J., Polo, M.C., 2001. Enological aspects of yeast autolysis. In: Pandalai, S.G. (Ed.), Recent Research and Development in Microbiology. Research Signpost, Trivandum, India, pp. 285–301.

Martinez-Rodriguez, A.J., Carrascosa, A.V., Polo, M.C., 2001a. Release of nitrogen compounds to the extracellular medium by three strains of *Saccharomyces cerevisiae* during induced autolysis in a model wine system. International Journal of Food Microbiology 68, 155–160.

Martinez-Rodriguez, A.J., Polo, M.C., Carrascosa, A.V., 2001b. Structural and ultrastructural changes in yeast cells during autolysis in a model wine system and in sparkling wine. International Journal of Food Microbiology 71, 45–51.

Matthews, A., Grimaldi, A., Walker, M., Bartowsky, E., Grbin, P., Jiranelç, V., 2004. Lactic acid bacteria as a potential source of enzymes for use in vinification. Applied Environmental Microbiology 70, 5715–5731.

Mauricio, J.C., Ortega, J.M., 1997. Nitrogen compounds in wine during its biological aging by two flor film yeasts: an approach to accelerated biological aging of dry sherry-type wines. Biotechnology and Bioengineering 53, 159–167.

Mauricio, J.C., Moreno, J.J., Ortega, J.M., 1997. In vitro specific activities of alcohol and aldehyde dehydrogenases from two flor yeasts during controlled wine aging. Journal of Agricultural and Food Chemistry 45, 1967–1971.

Mayer, K., Pause, G., 1987. Amingehalte in Ostschweizer Weinen.Schweiz. Zeitschrift fur Obst—und Weinbau 123, 303.

Mayo, B., Aleksandrzak-Piekarczyk, T., Fernández, M., Kowalczyk, M., Álvarez- Martín, P.P., Bardowski, J., 2010. Updates in the metabolism of lactic acid bacteria. In: Mozzi, F., Raya, R.R., Vignolo, G.M. (Eds.), Biotechnology of Lactic Acid Bacteria: Novel Applications. Wiley-Blackwell, Iowa, USA, pp. 3–33.

Mazza, G., Miniati, E., 1993. Anthocyanins in Fruits, Vegetables and Grains. CRC, Boca Raton, Florida.

Mc Clellan, C.J., Does, A.L., Bisson, L.F., 1989. Characterization of hexose uptake in wine strains of *Saccharomyces cerevisiae* and *Saccharomyces bayanus*. American Journal of Enology and Viticulture 40, 9–15.

McGovern, P.E., Mirzoian, A., Hall, G.R., 2009. Ancient Egyptian herbal wines. Proceedings of the National Academy of Sciences of the United States of America 106, 7361–7366.

McKellar, R.C., Rupasinghe, H.P.V., Lu, X., Knight, K., 2005. The electronic nose as a tool for the classification of fruit and grape wines from different Ontario wineries. Journal of the Science of Food and Agriculture 85, 2391–2396.

McMahon, H., Zoecklein, B.W., Fugelsang, K., Jasinki, Y., 1999. Quantification of glycosidase activities in selected yeasts and lactic acid bacteria. Journal of Industrial Microbiology and Biotechnology 23, 198–203.

Meilgaard, M.C., 1975. Flavour chemistry of beer: Part II: flavor threshold of 239 aroma volatiles. MBAA Technical Quarterly 12, 151–168.

Mendes-Ferreira, A., Climaco, M.C., Mendes-Faia, A., 2001. The role of non-*Saccharomyces* species in releasing glycosidic bound fraction of grape aroma components—a preliminary study. Journal of Applied Microbiology 91, 67–71.

Miranda, M., Ramos, A., Veiga-da-Cunha, M.A.R.I.A., Loureiro-Dias, M.C., Santos, H., 1997. Biochemical basis for glucose-induced inhibition of malolactic fermentation in *Leuconostoc oenos*. Journal of Bacteriology 179 (17), 5347–5354.

Mitic, M.N., Obradovic, M.V., Mitic, S.S., Pavlovic, A.N., Pavlovic, J.L.J., Stojanovic, B.T., 2013. Free radical scavenging activity and phenolic profile of selected Serbian red fruit wines. Revista de Chimie (Bucharest) 64 (1).

Molinos, A.C., Gálvez, A., Chauhan, A., Gupta, A.D., Chye, F.Y., Rapsang, G.F., Abrol, G.S., Bareh, I., Naomichi, I., Ruddle, K., Sim, K.Y., Burgos, M.J.G., Pulido, R.P., ogundele, S.O., Sharma, S., Thokchom, S., Joshi, S.R., Joshi, V.K., Chauhan, V., Jaiswal, V., Chandel, V., Bira, Z.M., 2016. Indigenous fermented foods of south Asia-an overview. In: Joshi, V.K. (Ed.), Indigenous Fermented Foods of South Asia. Taylor and Francis, CRC Press, Boca Raton, Florida, pp. 1–52.

Monk, P.R., 1986. Formation, utilization and excretion of hydrogen sulphide by wine yeast. Australian & New Zealand Wine Industry Journal 1, 10–16.

Monteiro, F.F., Bisson, L.F., 1991. Amino acid utilization and urea formation during vinification. American Journal of Enology and Viticulture 42, 199–208.

Monteiro, F.F., Trousdale, E.K., Bisson, L.F., 1989. Ethyl carbamate formation in wine: use of radioactively labeled precursors to demonstrate the involvement of urea. American Journal of Enology and Viticulture 40, 1.

Moreno, I., González-Weller, D., Gutierrez, V., Marino, M., Cameán, A., González, G., et al., 2007. Differentiation of two Canary DO red wines according to their metal content from inductively coupled plasma optical emission spectrometry and graphite furnace atomic absorption spectrometry by using probabilistic neural networks. Talanta 72, 263–268.

Moreno-Arribas, M.V., Polo, M.C., 2005. Winemaking biochemistry and microbiology: current knowledge and future trends. Critical Reviews in Food Science and Nutrition 45 (4), 265–286.

Moreno-Arribas, M.V., Polo, M.C., Jorganes, F., Muñoz, R., 2003. Screening of biogenic amine production by lacticacid bacteria isolated from grape must and wine. International Journal of Food Microbiology 184, 117–123.

Moreno-Arribas, V., Lonvaud-Funel, A., 2001. Lactic acid bacteria. Involvement in wine quality. In: Pandalai, S.G. (Ed.), Recent Research and Development in Microbiology. Research Signpost, Trivandum, India, pp. 481–504.

Moreno-Arribas, V., Bartolome, B., Pueyo, E., Polo, M.C., 1998a. Isolation and characterization of individual peptides from wine. Journal of Agricultural and Food Chemistry 46, 3422–3425.

Moreno-Arribas, V., Pueyo, E., Polo, M.C., 1996. Peptides in musts and wines. Changes during the manufacture of cavas (Sparkling wines). Journal of Agricultural and Food Chemistry 44, 3783–3788.

Moreno-Arribas, V., Pueyo, E., Nieto, J., Martin-Alvarez, P.J., Polo, M.C., 2000. Influence of the polysaccharides and the nitrogen compounds on foaming properties of sparkling wines. Food Chemistry 70, 309–317.

Moreno-Arribas, V., Pueyo, E., Polo, M.C., Martin-Alvarez, P.J., 1998b. Changes in the amino acid composition of the different nitrogenous fractions during the aging of wine with yeast. Journal of Agricultural and Food Chemistry 46, 4042–4051.

Moyano, L., Zea, L., Moreno, J., Medina, M., 2002. Analytical study of aromatic series in sherry wines subjected to biological aging. Journal of Agricultural and Food Chemistry 50, 7356–7361.

Mudnic, I., Budimir, D., Modun, D., Gunjaca, G., Generalic, I., Skroza, D., Katalinic, V., Ljubenkov, I., Boban, M., 2012. Antioxidant and vasodilatory effects of blackberry and grape wines. Journal of Medicinal Food 15 (3), 315–321.

Muller, C.J., Fugelsang, K.C., Wahlstorm, V.L., 1993. Capture and use of volatile flavour constituents emitted during wine fermentations. In: Gump, B.H. (Ed.), Beer and Wine Production: Analysis, Characterization and Technological Advances American Chemical Society Series 536, 219 Washington, DC.

Navarro, G., Navarro, S., 2011. Stuck and sluggish fermentation. In: Joshi, V.K. (Ed.), Handbook of Enology: Principles, Practices, and Recent Innovations, vol. 2. Asiatech Publishers Inc., New Delhi, pp. 591–617.

Negi, B., Dey, G., 2009. Comparative analysis of total phenolic content in sea buckthorn wine and other selected fruit wines. World Academy of Science, Engineering and Technology 3.

Negi, B., Kaur, R., Dey, G., 2013. Protective effects of a novel sea buckthorn wine on oxidative stress and hyper-cholesterolemia. Food & Function 4 (2), 240–248.

Neti, Y., Erlinda, I.D., Virgilio, V.G., 2011. The effect of spontaneous fermentation on the volatile flavor constituents of durian. International Food Research Journal 18 (2), 635–641.

Nielsen, J.C., Richelieu, M., 1999. Control of flavour development in wine during and after malolactic fermentation by *Oenococcusoeni*. Applied Environmental Microbiology 65, 740–745.

Nikolaou, E., Soufleros, E.H., Bouloumpasi, E., Tzanetakis, N., 2006. Selection of indigenous *Saccharomyces cerevisiae* strains according to their oenological characteristics and vinification results. Food Microbiology 23, 205–211.

Nykanen, L., 1986. Formation and occurrence of flavour compounds in wine and distilled alcoholic beverages. American Journal of Enology and Viticulture 37, 84–86.

Ough, C.S., 1993. Lead in wines—a review of recent reports. American Journal of Enology and Viticulture 44, 464–467.

Ortiz, J., Marın-Arroyo, M.-R., Noriega-Domınguez, M.-J., Navarro, M., Arozarena, I., 2013. Color, phenolics, and antioxidant activity of blackberry (*Rubus glaucus Benth.*), blueberry (*Vaccinium floribundum Kunth.*), and apple wines from Ecuador. Journal of Food Science 78 (7), 985–993.

Ough, C.S., Bell, A.A., 1980. Effects of nitrogen fertilization of grapevines on amino acid metabolism and higher- alcohol formation during grape juice fermentation. American Journal of Enology and Viticulture 31, 122–123.

Ough, C.S., Lee, T.H., 1981. Effect of vineyard nitrogen fertilization level on the formation of some fermentation esters. American Journal of Enology and Viticulture 32, 125–127.

Ough, C.S., Crowell, E.A., Gutlove, B.R., 1988a. Carbamyl compound reactions with ethanol. American Journal of Enology and Viticulture 39, 239–242.

Ough, C.S., Crowell, E.A., Monney, L.A., 1988b. Formation of ethyl carbamate precursors during grape juice (Chardonnary) fermentation. I Addition of amino acid, urea and ammonia effects of fortification on nitrocellular and extracellular precursors. American Journal of Enology and Viticulture 39, 243–249.

Ough, C.S., Huang, Z., Stevens, D., 1991. Amino acid uptake by four commercial yeasts at two different temperatures of growth and fermentation: effect on urea excretion and reabsorption. American Journal of Enology and Viticulture 42, 26–40.

Ough, C.S., Stevens, D., Sendovski, T., Huang, Z., An, D., 1990. Factors contributing to urea formation in commercial fermented wines. American Journal of Enology and Viticulture 41, 68–73.

Ough, C.S., DaudtMonterio, F.F., Bisson, L.F., 1992. Utilization of adenine by yeast during grape juice fermentation and investigation of the possible role of adenine as a precursor of urea. American Journal of Enology and Viticulture 43, 18.

de la Presa-Owens, C., Lamuela-Raventoos, R.M., Buxaderas, S., de la Torre-Boronat, M.C., 1998. Characterization of Macabeo, Xarel.lo, and Parellada white wines from the Penedes region II. American Journal of Enology and Viticulture 46, 539–541.

Paquin, C.E., Williamson, V.M., 1986. Ty insertion at two loci account for most of the spontaneous antimycin A resistance mutations during growth at 15°C of *Saccharmyces cervisiae* strains lacking ADHI. Molecular and Cellular Biology 6, 70.

Park, S.K., Boulton, R.B., Noble, A.C., 2000. Formation of hydrogen sulfide and glutathione during fermentation of white grape musts. American Journal of Enology and Viticulture 51 (2), 91.

Peinado, R.A., Moreno, J.J., Ortega, J.M., Mauricio, J.C., 2003. Effect of gluconic acid consumption during stimulation of biological aging of sherry wines by a flor yeast strain on the final volatile compounds. Journal of Agricultural and Food Chemistry 51, 6198–6203.

Peynaud, E., Domercq, S., Boidron, A., Lafon-Lafourcade, S., Guimberteau, G., 1964. Etude der les levures Schizosachharomyces metabolisant l'acide L-malique. Archiv fur Mikrobiologie 48, 150.

Picinelli, A., Suarez, B., Garcia, L., Mangas, J.J., 2000. Changes in phenolic contents during sparkling wine making. American Journal of Enology and Viticulture 51 (2), 144.

Pierce, J.S., 1982. The Margaret Jones memorial lecture: amino acids in malting and brewing. Journal of the Institute of Brewing 88, 228.

Pierce, J.S., 1987. Horace Brown memorial lecture; the role of nitrogen in brewing. Journal of the Institute of Brewing 93, 378.

Pilatte, E., Nygaard, M., Cai Gao, Y., Krentz, S., Power, J., Lagarde, G., 2000. Etude de l'effect du lisozyme sur differentes souches d' Oenococcus oeni. Applications dans la gestion de la fermentation malolactique. Revue Francaise Oenologie Novembre/decembre N°185.

Pinhero, R.G., Paliyath, G., 2001. Antioxidant and calmodulin inhibitory activities of phenolic components in fruit wines and its biotechnological implications. Food Biotechnology.

Pino, J.A., Queris, O., 2010. Analysis of volatile compounds of pineapple wine using solid-phase microextraction techniques. Food Chemistry 122, 1241–2124.

Plata, M.C., Mauricio, J.C., Millan, C., Ortega, J.M., 1998. In vitro activity of alcohol acetyltransferase and esterase in two flor yeast strains during biological aging of sherry wines. Fermentation and Bioengineering 85, 369–374.

Postel, W., Ziegler, L., 1991. Influence of the duration of yeast contact and of the manufacturing process on the composition and quality of sparkling wines II. Free amino acids and volatile compounds. Wein-Wissenschaft Viticultural and Enological Sciences 46, 26–32.

Pozo-Bayon, M.A., Pueyo, E., Martin-Alvarez, P.J., Martinez-Rodriguez, A.J., Polo, M.C., 2003. Influence of yeast strain, bentonite addition, and aging time on volatile compounds of sparkling wines. Journal of Enology and Viticulture 54, 273–278.

Pripis-Nicolau, L., de Revel, G., Marchand, S., Anocibar-Beloqui, A., Bertrand, A., 2000. Automated HPLC method for the measurement of free amino acids including cysteine in musts and wines: first applications. Journal of the Science of Food Agriculture 81, 731–738.

Pueyo, E., Martinez-Rodriguez, A.J., Polo, M.C., Santa-Maria, G., Bartolome, B., 2000. Release of lipids during yeast autolysis in a model wine system. Journal of Agricultural and Food Chemistry 48, 116–122.

Pueyo, E., Olano, A., Polo, M.C., 1995. Neutral monosacharides composition of the polysaccharides from must, wines, and cava wines. Revista Española de Ciencia y Tecnología de Alimentos 35, 191–201.

de Revel, G., Martin, N., Pripis-Nicolau, L., Lonvaud-Funel, A., Bertrand, A., 1999. Contribution to the knowledge of malolactic fermentation influence on wine aroma. Journal of Agricultural and Food Chemistry 47, 4003–4008.

Radler, F., 1986. Microbial biochemistry. Experientia 42, 884–892.

Radler, F., 1993. Yeasts metabolism of organic acids. In: Fleet, G.H. (Ed.), Wine Microbiology and Biotechnology. Harwood Academic Publishers, Chur, Switzerland, pp. 165–182.

Radler, F., Schutz, H., 1982. Glycerol production form various strains of Saccharomuces. American Journal of Enology and Viticulture 33, 36.

Ramos, A., Santos, H., 1996. Citric and sugar co -fermentation in Leuconostoc oenos, a 13C nuclear magnetic resonance study. Applied Environmental Microbiology 62, 2577–2585.

Rebordinos, L., Infante, J.J., Rodriguez, M.E., Vallejo, I., Cantroal, J.M., 2011. Wine yeast growth and factor affecting. In: Joshi, V.K. (Ed.), Handbook of Enology. vol. 2. Asia Tech publication, New Delhi, pp. 406–434.

Ramassamy, C., 2006. Emerging role of polyphenolic compounds in the treatment of neurodegenerative diseases: a review of their intracellular targets. European Journal of Pharmacology 545, 51–64.

Rana, N.S., Rana, V.S., 2011. Biochemistry of wine preparation. In: Joshi, V.K. (Ed.), Handbook of Enology: Principles, Practices, and Recent Innovations, vol. 2. Asiatech Publishers Inc., New Delhi, pp. 618–678.

Rankine, B.C., 1963. Nature, origin and prevention of hydrogen sulphide aroma in wines. Journal of the Science of Food Agriculture 14, 79.

Rankine, B.C., 1966. Decomposition of L-malic acid by wine yeasts. Journal of the Science of Food Agriculture 17, 312.

Rankine, B.C., 1967. Formation of higher alcohols by wine yeasts and relationship to taste thresholds. Journal of the Science of Food Agriculture 18, 584.

Rapp, A., Versini, G., 1991. Influence of nitrogen compounds in grapes on aroma compounds of wine. In: Rantz, J., Davis, C.A. (Eds.), Proceedings of the International Symposium on Nitrogen in Grapes and Wines, p. 156.

Rauhut, D., Kiirbel, H., Dittrich, H.H., Grossmann, M., 1996. Properties and differences of commercial yeast strains with respect to their formation of sulfur compounds. Viticultural and Enological Science 51, 187–192.

Reddy, L.V.A., Reddy, O.V.S., 2011. Effect of fermentation conditions on yeast growth and volatile composition of wine produced from mango (*Mangifera indica* L.) fruit juice. Food and Bioproducts Process 89, 487–491.

Reibstein, D., den Hollander, J.A., Pilkis, S.J., Shulman, R.G., 1986. Studies on the regulation of yeast phospho-fructo-I-kinase: its role in aerobic and anaerobic glycolysis. Biochemistry 25, 219.

Renault, P.P., Gaillardin, C., Heslot, H., 1988. Role of malolatic fermentation in lactic acid bacteria. Biochimie 70, 375.

Rhodes, N., Morris, C.N., Ainsworth, S., Kinderlerer, J., 1986. The regulatory properties of yeast pyruvate kinase. Biochemical Journal 234, 705.

Ribereau-Gayon, P., Dubourdieu, D., Doneche, B., Lonvaud, A., 1998. Le metabolism des bacteries lactiques. In: Traite d'Oenologie, Microbiologie du vin, Vinifications. Dunod, Paris, pp. 171–193.

Richard, P.V., Ellen, M.H., Theresa, B., Cheri, W., 1997. Enology (winemaking). In: Winemaking, from Grape Growing to Market Place. The chapman and Hall Enology Library, pp. 95–145.

Rodrigues, S.M., Otero, M., Alves, A.A., Coimbra, J., Coimbra, M.A., Pereira, E., et al., 2011. Analysis for categorization of wines and authentication of their certified brand of origin. Journal of Food Composite Analysis. http://dx.doi.org/10.1016/j.jfca.2010.12.003.

Rodriquez, S., Thornton, R., 1990. Factors infuencing the utilization of L-malate by yeasts. FEMS Microbiology Letters 72, 17.

Rojas, V., Gil, J.V., Pinaga, F., Manzanares, P., 2001. Studies on acetate ester production by non-*Saccharomyces* wine yeast. International Journal of Food Microbiology 70, 283–289.

Rojas, V., Gil, J.V., Pinaga, F., Manzanares, P., 2003. Acetate ester formation in wine by mixed cultures in laboratory fermentations. International Journal of Food Microbiology 86, 181–188.

Romano, P., Suzzi, G., 1996. Origin and production of acetoin during wine yeast fermentation. Applied Environmental Microbiology 62, 309–312.

Romano, P., Suzzi, G., Domizio, P., Fatichenti, F., 1997. Secondary products formation as a tool for discriminating non-*Saccharomyces* wine strains. Antonie vanm Leeuwenhoek 71, 239–242.

Roon, R.J., Larimore, F., Levy, J.S., 1975a. Inhibition of amino acid transport by ammonium ion in *Saccharomyces cerevisiae*. Journal of Bacteriology 124, 325.

Roon, R.J., Even, H.L., Dunlop, P., Larimore, F., 1975b. Methylamine and ammonia transport in *Saccharomyces cerevisiae*. Journal of Bacteriology 122, 502–509.

Roon, R.J., Levy, J.S., Larimore, F., 1977a. Negative interactions between amino acid and methyll amine/ammonia transport systems of *Saccharomyces cerevisiae*. Journal of Biological Chemistry 252, 3599.

Roon, R.J., Meyer, G.M., Larimore, F.S., 1977b. Evidence for a common component in kinetically distinct transport system of *Saccharomyces cerevisiae*. Molecular Genetics and Genomics 158, 185–191.

Rose, A.H., 1987. Responses to the chemical environment. In: Rose, A.H., Harrison, J.S. (Eds.), The Yeasts, vol. 2. second ed. Academic Press, London, p. 5.

Rosi, I., Costamagna, L., Bertuccioli, M., 1987. Screening for extracellular acid protease(s) production by wine yeasts. Journal of the Institute of Brewing 93, 322.

Rupasinghe, H.P.V., 2008. The role of polyphenols in quality, postharvest handling, and processing of fruits. In: Paliyath, G., Murr, D.P., Handa, A.K., Lurie, S. (Eds.), Postharvest Biology and Technology of Fruits, Vegetables and Flowers. Blackwell Publishing, pp. 260–281.

Rupasinghe, H.P.V., Yu, L.J., 2013. Value-added fruit processing for human health. In: Muzzalupo, I. (Ed.), Food Industry. InTech—Open Access Publisher, Rijeka, Croatia, pp. 145–162.

Rupasinghe, H.P.V., Thilakarathna, S., Nair, S., 2012. Polyphenols of apples and their potential health benefits. In: Sun, J., Prasad, K.N., Ismail, A., Yang, B., You, X., Li, L. (Eds.), Polyphenols: Chemistry, Dietary Sources and Health Benefits. Nova Science Publishers, Inc., Hauppauge, NY, USA, pp. 333–368.

Rupasinghe, H.P.V., Clegg, S., 2007. Total antioxidant capacity, total phenolic content, mineral elements, and histamine concentrations in wines of different fruit sources. Journal of Food Composition and Analysis 20, 133–137.

SAFC, 2012. Flowers & Fragrances. European Ed. Catalogue 2012. SAFC Specialties, Madrid.

Salmon, J.M., 1989. Effect of sugar transport inactivation in *Saccharomyces cerevisiae* on sluggish and stuck enological fermentations. Applied and Environmental Microbiology 55, 9535.

Salo, P., 1970. Determining the odour thresholds of some compounds in alcoholic beverages. Journal of Food Science 35, 95–99.

Salou, P., Loubiere, P., Pareilleux, A., 1994. Growth and energetics of Leuconostoc oenos during cometabolism of glucose with citrate or fructose. Applied and Environmental Microbiology 60, 1459–1466.

Sanchez-Moreno, C., Cao, G., Ou, B., Prior, R.L., 2003. Anthocyanin and proanthocyanidin content in selected white and red wines. Oxygen radical absorbance capacity comparison with nontraditional wines obtained from Highbush blueberry. Journal of Agricultural and Food Chemistry 51 (17), 4889–4896.

Santamariia, P., Loi pez, I., Gutierrez, R., Garciia-Escudero, E., 1995. Evolution des acides gras totaux pendant la fermentation a differents temperatures. Journal International des Sciences de la Vigne et du Vin 29, 101–104.

Sato, M., 2011. Enzymes in wine production. In: Joshi, V.K. (Ed.), Hand Book of Enology, vol. 3. Asia Tech publication, New Delhi, pp. 702–730.

Schmitt, H.D., Ciriacy, M., Zimmermann, F.K., 1983. The synthesis of yeast pyruvate decarboxylase is regulated by large variations in the messenger RNA level. Molecular and General Genetics 192, 247.

Schmitzer, V., Veberic, R., Slatnar, A., Stampar, F., 2010. Elderberry (*Sambucusnigra L.*) wine: a product rich in health promoting compounds. Journal of Agricultural and Food Chemistry 2010 (58), 10143–10146.

Schutz, M., Kunkee, R.E., 1977. Formation of hydrogen sulfide from elemental sulfur during fermentation by wine yeast. American Journal of Enology and Viticulture 28, 137.

Schwartz, H., Radler, F., 1988. Formation of L (-) malate by *Sachharomyces cerevisiae* during fermentation. Applied Environmental Biotechnology 27, 553.

Schwenke, J., 1991. Vacules, internal membranous systems and vesicles. In: Rose, A.H., Harrison, J.S. (Eds.), The Yeasts, vol. 4. second ed. Academic Press, London, pp. 347–432.

Serrano, M., Guillen, F., Martinez-Romero, D., Castillo, S., Valero, D., 2005. Chemical constituents and antioxidant activity of sweet cherry at different ripening stages. Journal of Agricultural and Food Chemistry 53 (7), 2741–2745.

Sevda, S.B., Rodrigues, L., 2011. Fermentative behavior of *Saccharomyces* strains during guava (*Psidium guajava L.*) must fermentation and optimization of guava wine production. Journal of Food Processing & Technology 2, 118.

Sharma, S., Tauro, P., 1986. Control of ethanol production by yeast: role of pyruvate decarboxylase and alcoholic dehydrogenase. Biotechnology Letters 8, 735.

Shimazu, Y., Watanabe, M., 1981. Effects of yeast strains and environmental conditions on forming of organic acids in must during fermentation (Japanese). Journal of Fermentation Technology 59, 27.

Siebert, T.E., Smyth, H.E., Capone, D., Neuwohner, C., Pardon, K., Skouroumounis, G., Herderich, M., Sefton, M., Pollnitz, A., 2005. Stable isotope dilution analysis of wine fermentation products by HS-SPME-GC-MS. Analytical Bioanalytical Chemistry 381, 937–947.

Siila Santos, M.H., 1996. Biogenic amines: their importance in foods. International Journal of Food Microbiology 29, 213.

Singleton, V.L., 1974. Analytical fractionation of the phenolic substances of grapes and wine and some practical uses of such analyses. In: Webb, A.D. (Ed.), Chemistry of Winemaking. American Chemistry Society, Washington, DC.

Singleton, V.L., Sieberhagen, H.A., De Wet, P., Van Wyk, C.J., 1975. Composition and sensory qualities of wines prepared from white grapes by fermentation with and without grape solids. American Journal of Enology and Viticulture 26, 62.

Sinton, T.H., Ough, C.S., Kisssler, J.J., Kasimatis, A.N., 1978. Grape juice indicators for prediction of potential wine quality. I. Relationship between crop level, juice and wine composition, and wine sensory ratings and scores. American Journal of Enology and Viticulture 29, 267.

Siriwoharn, T.S., Wrolstad, R.E., 2004. Polyphenolic composition of marion and evergreen blackberries. Journal of Food Science 69, 233–240.

Smith, T.A., 1981. Amines in food. Food Chemistry 6, 169.

Soden, A., Francis, I.L., Oakey, H., Henschke, P.A., 2000. Effects of co-fermentation with *Candida stellata* and *Saccharomyces cerevisiae* on the aroma and composition of Chardonnay wine. Australian Journal of Grape and Wine Research 6, 21–30.

Soles, R.M., Ough, C.S., Kunkee, R.E., 1982. Ester concentration differences in wine fermented by various species and strains of yeasts. American Journal of Enology and Viticulture 33, 94.

Sols, A., 1981. Multimodulation of enzyme activity. In: Current Topics in Cellular Regulation19, p. 77.

Soni, S.K., Marwaha, S.S., Marwaha, U., Soni, R., 2011. Composition and nutritive value of wine. In: Joshi, V.K. (Ed.), Handbook of Enology: Principles, Practices and Recent Innovations, vol. 1. Asia Tech Publisher, New Delhi, pp. 89–145.

Spiropoulos, A., Tanaka, J., Flerianos, I., Bisson, L.F., 2000. Characterization of hydrogen sulfide formation in commercial and natural wine isolates of Saccharomyces. American Journal of Enology and Viticulture 51 (3), 233–248.

Sponholz, W.R., 1988. Alcohols derived from sugars and others sources and fullbodiedness of wines. In: Linskens, H.F., Jackson, J.F. (Eds.), Modern Methods of Plant Analysis New Series, vol. 6. Wine Analysis, p. 147.

Sponholz, W.R., 1989. Fehlerhafter und Unerwunschte Erscheinungen in wein. In: Wurdig, G., Woller, R. (Eds.), Chemic des Weines, p. 385.

Sponholz, W.R., 1993. Wine spoilage by microorganisms. In: Fleet, G.H. (Ed.), Wine Microbiology and Biotechnology. Harwood Academic Publishers, Switzerland, pp. 395–420.

Sponholz, W.R., Dittrich, H.H., 1979. Analytiche vergleiche von Mosten und Weinen aus gesunden and essigstichigen traubenbeeren. Wein-Wissenschaft Viticultural and Enological Sciences 34, 279.

Strasser de Saad, A.M., Manca de Nadra, M.C., 1992. Sugar and malic acid utilization and acetic acid formation by *Leuconostoc oenos*. World Journal of Microbiology and Biotechnology 8, 280–283.

Stratford, M., Rose, A.H., 1986. Transport of sulphur dioxide by *Saccharomyces cerevisiae*. Journal of General Microbiology 132, 1.

Stratford, M., Rose, A.H., 1985. Hydrogen sulphide production from sulphide by *Saccharomyces cerevisiae*. Journal of General Microbiology 131, 1427.

Stull, A.J., Cash, K.C., Johnson, W.D., Champagne, C.M., Cefalu, W.T., 2010. Bioactives in blueberries improve insulin sensitivity in obese, insulin-resistant menand women. The Journal of Nutrition and Disease 1764–1768.

Sturley, S.L., Young, T.W., 1988. Extracellular protease activity in a strain of *Saccharmyces cerevisiae*. Journal of the Institute of Brewing California 94, 23.

Su, M.S., Chien, P.J., 2007. Antioxidant activity, anthocyanins, and phenolics of rabbiteye blueberry (*Vacciniumashei*) fluid products as affected by fermentation. Food Chemistry 104, 182–187.

Suarez-Lepe, J.A., 1997. El caracter filmoogeno de las levaduras y otras propiedades de interes para vinificaciones especiales. In: Levaduras vinicas, Funcionalidad y uso en bodega, pp. 171–196 (Mundi-Prensa ed. Mundrid: Mundi-Prensa).

Suarez-Lepe, J.A., Hugo, Leal, B., Salmon, J.M., Sablayrolles, J.M., Rosenfeld, E., 1990. Vinificaciones especiales desde el punto de vista microbioloogico: Los vinos con crianza biolgica. In: Suarez-Lepe, J.A., Hugo, Leal, B. (Eds.), Microbiologia Enologica, Fundamentos de vinific, pp. 501–538 (Mundi-Prensa ed. Madrid: Mundi-Prensa).

Sun, S.Y., Jiang, W.G., Zhao, Y.P., 2011. Evaluation of different *Saccharomyces cerevisiae* strains on the profile of volatile compounds and polyphenols in cherry wines. Food Chemistry 127, 547–555.

Sun, S.Y., Che, C.Y., Sun, T.F., Lv, Z.Z., He, S.X., Gu, H.N., Shen, W.J., Chi, D.C., Gao, Y., 2013. Evaluation of sequential inoculation of *Saccharomyces cerevisiae* and *Oenococcusoeni* strains on the chemical and aromatic profiles of cherry wines. Food Chemistry 138, 2233–2247.

Suzzi, G., Romano, P., Zambonelli, C., 1985. *Saccharomyces* strain selection in minimizing SO_2 requirement during vinification. American Journal of Enology and Viticulture 36, 199.

Swiegers, J.H., Bartowsky, E.J., Henschke, P.A., Pretorius, I.S., 2005. Yeast and bacterial modulation of wine aroma and flavour. Australian Journal of Grape and Wine Research 11, 139–173.

Szajdek, A., Borowska, E.J., 2008. Bioactive compounds and health-promoting properties of berry fruits: a review. Plant Foods for Human Nutrition 63, 147–156.

du Toit, M., Pretorius, I.S., 2000. Microbial spoilage and preservation of wine: using weapons from nature's own arsenal—a review. South African Journal for Enology and Viticulture 21, 74–96.

Tanner, H., 1969. Der Weinbockser, Entstehung und Beseitigung. Zeitschrift fur Obst und Weinbau 105, 78 Jhg. 252.

Taylor, V.F., Longerich, H.P., Greenough, J.D., 2003. Multielement analysis of Canadian wines by inductively coupled plasma mass spectrometry (ICP-MS) and multivariate statistics. Journal of Agricultural and Food Chemistry 51 (4), 856–860.

Thilakarathna, S.H., Rupasinghe, H.P.V., 2012. Anti-atherosclerotic effects of fruit bioactive compounds: a review of current scientific evidence. Canadian Journal of Plant Science 92, 407–419.

Thomas, C.S., Gubler, W.D., Silacci, M.W., Miller, R., 1993. Changes in elemental sulfur residues on Pinot noir and Cabernet sauvignon berries during the growing season. American Journal of Enology and Viticulture 44, 205.

Todd, B.E.N., Fleet, G.H., Henschke, P.A., 2000. Promotion of autolysis through the interaction of killer and sensitive yeasts: potential application in sparkling wine production. American Journal of Enology and Viticulture 51, 65–72.

Tomas-Berberan, F.A., Gil, M.I., Cremin, P., Waterhouse, A.I., Hess-Pierce, B., Kader, A.A., 2001. HPLC-DAD-ESIMS analysis of phenolic compounds in nectarines, peaches, and plums. Journal of Agricultural and Food Chemistry 49, 4748–4760.

Tominaga, T., Gachon, C.P., Dubourdieu, D., 1998. A new type of flavour precursors in *Vitis vinifera* L. cv. Sauvignon Blanc, S-cysteine conjugates. Journal of Agricultural and Food Chemistry 46, 5215.

Tominaga, T., Guyot, R.B., Gachons, C.P., Dubourdieu, D., 2000. Contribution of volatile thiols to the aromas of white wines made from several *Vitis vinifera* grape varieties. American Journal of Enology and Viticulture 51 (2), 178.

Towantakavanit, K., Park, Y.S., Gorinstein, S., 2011. Quality properties of wine from Korean kiwifruit new cultivars. Food Research International 44, 1364–1372.

Ugliano, M., Henschke, P.A., 2008. Yeast and wine flavour. In: Moreno-Arribas, M.V., Polo, C. (Eds.), Wine Chemistry and Biochemistry. Springer, New York, pp. 328–348.

Valero, E., Millan, M.C., Mauricio, J.C., Ortega, J.M., 1998. Effect of grape skin maceration on sterol, phospholipid, and fatty acid contents of *Saccharomyces cerevisiae* during alcoholic fermentation. American Journal of Enology and Viticulture 49, 119–124.

van Vuuren, H.J.J., Dicks, L.M.T., 1993. *Leuconostoc oenos*: a review. American Journal of Enology and Viticulture 44, 99–112.

Van Vuuren, H.J.J., Jacobs, C.J., 1992. Killer yeasts in the wine industry: a review. American Journal of Enology and Viticulture 43, 119–128.

Vazquez-Lasa, M.B., Iniguez-Crespo, M., Gonzalez-Larraina, M., Gonzalez-Guerrero, A., 1998. Biogenic amines in Rioja wines. American Journal of Enology and Viticulture 49, 229.

Versari, A., Parpinello, G.P., Cattaneo, M., 1999. *Leuconostoc oenos* and malolactic fermentation in wine: a review. Journal of Industrial Microbiology and Biotechnology 23, 447–455.

Vianna, E., Ebeler, S.E., 2001. Monitoring ester formation in grape juice fermentation using solid phase micro-extraction coupled with gas chromatography-mass spectrometry. Journal of Agricultural and Food Chemistry 49, 589–595.

Vidal-Carou, M.C., Codony-Salcedo, R., Marine-Font, A., 1991. Change in the concentration of histamine and tyramine during wine spoilage at various temperatures. American Journal of Enology and Viticulture 42, 145.

Vinson, J.A., Xuehui, S., Ligia, Z., Bose, P., 2001. Phenol antioxidant quantity and quality in foods: fruits. Journal of Agricultural and Food Chemistry 49, 5315–5321.

Volschenk, H., Viljoen, M., Grobler, J., Baur, F., Lonvaud-Funel, A., Denayrolles, M., Subden, R.E., Vanvuuren, H.J.J., 1997. Malolactic fermentation in grape must by a genetically engineered strain of *Saccharomyces cerevisiae*. American Journal of Enology and Viticulture 40 (2), 193.

Vos, P.J.A., Gray, R.S., 1979. The origin and control of hydrogen sulfide during fermentation of grape must. American Journal of Enology and Viticulture 30, 187.

Vos, P.J.A., 1981. Assimilabe nitrogen-A factor influencing the quality of wines. In: International Association for Modern Winery Technology and Management. 6th International Oenological Symposium (28–30th April 1981). Mainz, Germany, p. 163.

Vos, P.J.A., Zeeman, W., Heymann, H., 1978. The effect on wine quality of diammonium phosphate additions to musts. In: Proc. S. Afric. Soc. Enol. Vitic.Stellenbosch, Cap Town, South Africa, p. 87.

Vuorinen, H., et al., 2000. Content of the flavonolsmyrecetin, quercetin and kaempferol in Finnish berry wines. Journal of Agricultural and Food Chemistry 48, 2675–2680.

Wainwright, T., 1971. Production of H2S by wine yeast: role of nutrients. Journal of Applied Bacteriology 34, 161.

Walsh, J.L., Keith, T.J., Knull, H.R., 1989. Glycolytic enzyme interactions with tubulin and microtubules. Biochimica et Biophysica Acta 999, 64.

Wang, D., Xu, Y., Hu, J., Zhao, G., 2004. Fermentation kinetics of different sugars by Apple wine yeast *Saccharomyces cervisiae*. Journal of the Institute of Brewing 110, 340–346.

Webb, A.D., Ingraham, J.L., 1963. Fusel oil. Advances in Applied Microbiology 5, 317.

Wenzel, K., Dittrich, H.H., 1978. Zur beeinflussung der schwefelwasserstoff—buildung der hefedurch trub, stickstoffgehalt, molecularen schwefel und kupfer bei der vergarung von traubenmost. Wein-Wissenschaft Viticultural and Enological Sciences 33, 200.

Wenzel, K., Dittrich, H.H., Pletzonka, B., 1982. Untersuchungen zur Beteiligung von Hefen am Apfelsaureabbau bei der Weinbereitung. Wein-Wissenschaft Viticultural and Enological Sciences 37, 133.

Whiting, G.C., 1976. Organic acid metabolism of yeasts during fermentation of alcoholic beverages: a review. Journal of the Institute of Brewing 82, 84.

Whiton, R.S., Zoecklein, B.W., 2002. Determination of ethyl carbamate in wine by solid phase microextraction and gas chromatography/mass spectrometry. American Journal of Enology and Viticulture 53 (1), 58.

Wiame, J.M., Grenson, M., Arst Jr., H.N., 1985. Nitrogen catabolite repression in yeast and filamentous fungi. Advances in Microbial Physiology 26, 2.

Woidich, H., Pfannhauser, W., 1974. Zur Gaschromatographichen Analyse Von Branntweinen: quantitative best mming Von Acetaldenyd WEssigsaureanethylester, Essigsaureanthylester, methanol, Butanol-1, Butanol-2, Propanol-1, 2-methuylpropanol-1, AmylalKoholen und Hexanol-1. Mitt. Hoeschesen Bundesleher-Versuchsanst. Wein- und Obstbau Klosterneuburg 24, 155.

Worobo, R.W., Splittstoesser, D.F., 2005. Microbiology of fruit products. In: Barrett, D.M., Somogyi, L., Ramaswamy, H. (Eds.), Processing Fruits: Science and Technology, 2. CRC Press, Boca Raton, pp. 262–284.

Yanai, T., Sato, M., 1999. Isolation and properties of B-glucosidase produced by *Debaromyces hansenii* and its application in winemaking. American Journal of Enology and Viticulture 50, 231–235.

Yang, H.Y., 1953. Fruit wines: requisites for successful fermentation. Journal of Agricultural and Food Chemistry 1 (4), 331–333.

Yazquez, M.B., 1996. Determinacion de aminas biogenas en vinos de Rioja (Thesis). Universidad del Pais Vasco-San Se Bastain.

Yildirim, H.K., 2006. Evaluation of colour parameters and antioxidant activities of fruit wines. International Journal of Food Sciences and Nutrition 57 (1/2), 47–63.

Yoo, K.M., Al-Farsi, M., Lee, H., Yoon, H., Lee, C.Y., 2010. Antiproliferative effects of cherry juice and wine in Chinese hamster lung fibroblast cells and their phenolic constituents and antioxidant activities. Food Chemistry 123, 734–740.

Yoshioka, K., Hashimoto, N., 1981. Ester formation by alcohol acetyltransferase from brewers' yeast. Agricultural and Biological Chemistry 45, 2183–2190.

Zea, L., Moreno, J., Medina, M., Ortega, M.J., 1994. Evolution of C6, C8 and C10 acids and their ethyl esters in cells and musts during the ageing with three *Saccharomyces cerevisiae*races. Journal of Industrial Microbiology 13, 269–272.

Zee, J.A., Simard, R.E., Roy, A., 1981. A modified automated ion exchange method for the separation and quantitation of biogenic amines. Canadian Institute of Food Science and Technology 14, 71.

Zoecklein, B.W., Hackney, C.H., Duncan, S.F., Marcy, J.E., 1997. Effect of fermentation, aging and thermal storage on total glycosides, phenol-free glycosides and volatile compounds of white Riesling (*Vitis vinifera* L.) wines. Journal of Industrial Microbiology and Biotechnology 22, 100–107.

COMPOSITION, NUTRITIONAL, AND THERAPEUTIC VALUES OF FRUIT AND BERRY WINES

V. Maksimović, J. Dragišić Maksimović

University of Belgrade, Belgrade, Serbia

Although the term *wine* is almost exclusively related to grapes, wines prepared from other sources are starting to be seen as a potential health food. Different cultures from different regions use various "raw materials" for winemaking and each has its own specific "connoisseur troupe." A wide variety of alcoholic beverages are prepared by fermentation of carbohydrates originating from various sources such as fruits, vegetables, cereals, honey, and sap. These different beverages are often known generally as country wines, and those made solely from fruits are often called fruit wines (McKay et al., 2010). Increased interest in maintaining human health through proper and optimized nutrition has opened almost unbounded demand for new functional foods such as various fruit products, including different varieties of fruit and berry wines.

There is a considerable amount of data showing that constituents of fruits, present in their wines, are beneficial to human health and contribute to the prevention of degenerative processes caused by different onsets such as aging, unbalanced diet, oxidative stress, and even some inherited disorders. In this chapter a review of potential beneficiary effects of various fruit wine consumption is given in respect to the type of the fruit and berry used. From the substantial number of their bioactive compounds, only those that are considered as carriers of fruit and berry wines' therapeutic characteristics will be pointed out.

Wines made from other fruits are profitable products and often need good marketing help to persuade consumers to try and enjoy nongrape wines. Many different fruits, such as apples, pears, peaches, plums, and cherries, give sufficient juice after crushing or pressing and have a satisfying balance between acids and sugars, which are the main prerequisites for efficient winemaking. Furthermore, immense production of cider and perry, which are often considered as varieties of apple and pear wine, have become a very profitable product in highly developed markets, such as the United States, with an estimated production of 6 million gallons (McKay et al., 2010).

Berry wine can be produced from almost any type of berry, but the most widespread types are blueberry, blackberry, raspberry, strawberry, and cranberry wines. One excellent study, done for the State of Illinois in the United States, depicts raspberry, blackberry, blueberry, and strawberry as the most common berry wine types (Johnson and Gonzalez de Mejia, 2012). Specific wine consumption is highly connected with source berry production and largely depends upon marketing and health benefit promotions. Having in mind that in 2012 the total world production of blueberries overcame 1 billion pounds (Brazelton, 2013), a consecutive increase in the production of corresponding wines is hardly surprising. Numerous studies show that berries, in general, are rich in various bioactive compounds with proven

Science and Technology of Fruit Wine Production. http://dx.doi.org/10.1016/B978-0-12-800850-8.00004-1

therapeutic capacities (Heinonen et al., 1998; Reed, 2002; Giampieri et al., 2012; Khoo and Falk, 2014). Many of them are present in different berries; some are almost exquisite for specific fruit cultivar. Thus, it is not surprising that all of the just-mentioned (and many more) papers advocate regular consumption of berry fruit products as evidently beneficiary for general health.

1. COMPOSITION OF FRUIT AND BERRY WINES

Wine, in general, is a rather complex and diverse mixture of ingredients, consisting mainly of water, alcohol, and sugar. An excellent overview of wine's chemical complexity is given by Moreno-Arribas et al. (2009). Accordingly, fruit and berry wines consist mainly of water, alcohols, sugars, organic acids, and minor constituents such as polyphenols, higher alcohols, esters, etc. (Johnson and Gonzalez de Mejia, 2012). These minor components such as volatile and nonvolatile acids, esters, higher alcohols, aldehydes, and ketones contribute largely to the final organoleptic character of final product. Furthermore, the finest and the most delicate differences appreciated by wine experts are determined by even larger group of compounds such as volatile terpenoids and complex esters. A very informative summary of the composition of less abundant fruit wines from Brazil given by Duarte et al. (2010a,b) can serve as additional information, and an overview of presence of major constituents in fruit and berry wines is given in Table 4.1.

1.1 ALCOHOLS

The amount of alcohol in wine depends upon technology as well as traditional prevalence and usually ranges between 5% and 15%. Since nongrape fruits do not have enough sugars for satisfactory alcohol levels, additional sugars have to be added for fermentation thus defining the final sweetness of the wine (Table 4.1). Alcohol content in strawberry (Kafkas et al., 2006; Sharma et al., 2009), blackberry, and

Table 4.1 Selected Chemical Characteristics of Fruit and Berry Wines

Wine Type	Ethanol (%)	Methanol (mg/L)	Sugar (°Bx)	Nonvolatile Acids (%)	pH
Apple	5–10	120–250	8–10	0.5	4.3
Cherry	10–13	<100	4–5	0.8	3.3–4
Blackberry	10–12	100–250	10–12	0.5–0.8	3.3–3.6
Blueberry	9–12	150–200	7–12.5	0.5–0.9	2.8–3.7
Strawberry	9–11.5	–	8–9.7	0.6–0.7	3.1–3.3
Cashew	7–9	–	10–12	0.5–0.6	4.6
Plum	6–13.5	>250	11	0.7–0.8	3.3
Kiwi	9.5	–	11	0.5	3.5
Mahua	8.7	–	10	0.7	3.3
Cider	5–20	<180	10–20	0.4–0.6	3.7–4.1
Mango	8	–	6	0.6	3.7
Peach	8.1	–	14.50	0.78	3.9
Orange	12.6	–	11	–	3.6

blueberry wines have around 10–12% ethanol (Johnson and Gonzalez de Mejia, 2012), while around 12% was found in black and red currant wines (Heinonen et al., 1998). Plum (Bhardwaj and Joshi, 2009) and cherry (Niu et al., 2012) wines are shown to be stronger than kiwi (Vaidya et al., 2009), mahua (Yadav et al., 2009), and cashew (Attri, 2009) wines, but specific wine selections from different locations can be much different in alcohol content. Apple musts also require sugar addition so they can reach the acceptable 5% for European wine to a high (15%) alcohol level for Korean wine (Lee et al., 2013b). For peach and orange, wine ethanol content can reach 8.1% and 12.6%, respectively (Kelebek et al., 2009; Davidović et al., 2013). Also, ethanol quantity is crucial for the stability, desirable aging, and final sensory properties of wine. During fermentation, rising alcohol content significantly confines yeast growth. Also, the presence of alcohol is important for the future discussion about the bioavailability of wine ingredients with potential health benefits.

Besides ethanol, methanol is present in measurable content in fruit and berry wines (Table 4.1). Since methanol is toxic to humans, maximal levels of methanol are usually below 250 mg/L. It is reported that plum wine can have very high amount of methanol, over 1 g/L (Miljić and Puškaš, 2014). Nevertheless, commercial wines cannot contain over 200–250 mg/L as regulated by laws in different countries.

Glycerol is one of the major wine constituents that can affect wine's viscosity, sweetness, and clearness. Glycerol is a byproduct formed by *Saccharomyces cerevisiae* during fermentation to maintain yeast cytosolic redox balance under anaerobic conditions (Wang et al., 2001). In raspberry wines, it can reach 10 g/L as one of the most abundant compounds, after water and ethanol (Duarte et al., 2010a,b). In cocoa wine, glycerol concentration is 4.6 g/L (Dias et al., 2007), 6.81 g/L in mango wine (Li et al., 2013), and 5 g/L for plum wine (Miljić and Puškaš, 2014). Glycerol is also present in cider in concentration of 3–6 g/L (Garai-Ibabe et al., 2008). Furthermore, during cider maturation glycerol can be metabolized by lactic acid bacteria, diminishing its final sensorial quality.

Sugar alcohols, such as alditol, arabitol, erythritol, mannitol, myo-inositol, and sorbitol, can be found in smaller amounts in fruit wine (Lee et al., 2013b). Their content varies in wines from different fruits and usually is below 200 mg/L, e.g., 150 mg/L of sorbitol was found in apple wine (Lee et al., 2013b). Apple, pears, cherries, peaches, and plums contain significant amounts of sorbitol, and since it is not fermentable, it can be detected in related wines (Wrolstad, 2012). It is expected that these values could be higher since typical sorbitol content in apple juices range between 3 and 8 g/L. Since sorbitol is not used by microorganisms during fermentation, it keeps up to 4–6 g/L concentration in final cider (Picinelli et al., 2000). Pears are especially rich in sorbitol and its significant concentration in perry is expected to be found. In contrast, berry fruits have much lower content of sorbitol. Therefore, its presence in higher concentrations in berry wines is the sign of deception performed by addition of less expensive apple and peach wine.

Usage of various yeast strains during fermentation contributes considerably to variations in higher alcohol profiles and concentrations in wine (Swiegers et al., 2005). Higher alcohols can have both positive and negative impacts on the aroma and flavor of wine depending on its concentration; they are considered favorable compounds, when their total concentration is lower than 300 mg/L. In raspberry wines five of them were identified, i.e., 1-propanol; 2-methyl-1-propanol; 2-methyl-1-butanol; 3-methyl-1-butanol; and 2-phenylethanol in total concentration of 349 mg/L (Duarte et al., 2010a,b). In mango wines the total amount of higher alcohols ranged from 131 to 343 mg/L, where 2-methyl-1-propanol (103 mg/L), 1-propanol (45 mg/L), and 3-methyl-1-butanol (150 mg/L) were dominant members (Reddy et al., 2009). Similarly, 3-methyl-1-butanol (232.00 mg/L), a major aroma compound, represented 32.16% of the total esters and alcohols in Fuji apple wine (Wang et al., 2004).

1.2 SUGAR CONTENT

The main drawback of nongrape fruits for wine production is lower total sugar content of their juices and musts in comparison to grapes. As a prerequisite for fermentation, sugar content is critical to yeast growth and metabolism. Usually, wine yeasts gain most of their metabolic energy from glucose and fructose catabolism. Unfermented sugars are collectively termed as "residual sugars" consisting primarily of pentose sugars, such as arabinose, rhamnose, and xylose, and small amounts of unfermented glucose and fructose (Jackson, 2008). Other sugars, such as trehalose may be synthesized and released by yeast cells during fermentation (Rossouw et al., 2013). According to the content of total sugars, all wines are classified as dry wines if the residual sugar content is less than 1.5 g/L.

The major sugars present in apple products are fructose, sucrose, and glucose. In apple wines, the consumption rate of sugars is fast in the early phases of fermentation, but becomes slower as the fermentation proceeds. At the end of fermentation, most of the sugar is consumed and the remaining total sugar content is approximately 7.2–7.8 Brix (Lee et al., 2013b). Pears have significantly higher sugar content, and the dominant mono- and disaccharides are the same, but unlike apples, pears have a lot of unfermentable polyol sorbitol rendering almost equal amounts of sugars for fermentation to alcohol. Pears and apples have significant amounts of starch, so prior to crushing, it is obligatory to provide enough time (around 7–14 days) for the hydrolysis action of amylases. Apple and pear wines have low, but satisfactory, intrinsically achieved alcoholic content (5–9% ABV), in contrast to many other fruits that either have too little juice or are deficient in sugar. Such musts are usually adjusted by the addition of sugar to achieve the proper alcoholic content (McKay et al., 2010). As an example for ciders, Asturian cider contains not as many sugars, so only fructose was found in the maximal concentration of 0.9 g/L. Since sweet taste is unwanted in Asturian cider, great concern is taken by cider makers to guarantee the utilization of all sugars (Picinelli et al., 2000). Interestingly, in the peach wine trehalose was the dominant sugar (0.2 g/L), followed by glucose (0.14 g/L) and fructose (0.04 g/L; Davidović et al., 2013). The total amount of sugar in orange wine was 48.78 g/L and the main portion of carbohydrates in citrus fruits are the three simple sugars: sucrose (45 g/L), glucose (1 g/L), and fructose (3 g/L; Kelebek et al., 2009). Cherries have glucose as a major sugar, followed by fructose and sucrose in total concentration from 125 to 265 g/kg FW depending on cultivar tested (Usenik et al., 2008). During winemaking there is a necessity to chaptalize the new cherry wine with sufficient sugar to obtain the desired alcohol content of 12–13% by volume (Archibald, 1997). Also, final amount of sugar largely depends on the yeast strain used, and it is estimated in range from 2.09 to 3.88 g/L of reducing sugar content (Sun et al., 2011).

In berries, sugars are represented mainly by glucose and fructose, with variable values in berries from different families (Milivojević et al., 2011a,b, 2012a,b; Mikulic-Petkovsek et al., 2012). In the case of three blueberry cultivars, Milivojević et al. (2013) estimated total sugar content in a range from 155.2 to 164.7 g/kg. As a comparison, average total sugar content in red grapes (*Vitis vinifera*) is around 200 g/kg or more upon its maturity (Jackson, 2014). Although total sugar content of various berries and grapes are similar, berry musts usually have lower sugar content due to the dilution during the winemaking process, which involves the addition of water and other adjuncts. Additional water is necessary to lower the acidity of berry must, which contributes to unfavorable sensory characteristics of wine.

Hence, the final average concentration of glucose/fructose in blackberry wines is 106.3 and 95.1 g/L for blueberry wines (Johnson and Gonzalez de Mejia, 2012). Since EU Commission Regulation (EC) No 753/2002 defines that all wines with sugar content above 45 g/L have to be classified as sweet wines, the aforesaid berry wines belong to that group. Strawberry and cranberry wines could contain more sugar, since during vinification additional sugar has to be added to obtain proper fermentation and to regulate their intrinsic acidity (Sharma et al., 2009). Regularly, consumers found the wine with high sugar amount and low acidity to be the most acceptable.

1.3 ORGANIC ACIDS

The major organic acid in apples and pears is malic acid, but minor quantities of shikimic, pyruvic, and fumaric acid are present in the fruit, along with sugar acids (e.g., gluconic acid), especially in riper fruit. The major organic acids in apple wines (Table 4.2) are malic, succinic, and citric acid with estimated amounts of 6.2, 1.3, and 0.3 g/L, respectively (Martin et al., 1971). Likewise, malic acid was present as a residual acid in all the analyzed ciders (200 mg/L). This acid is converted into the major acid in ciders, lactic acid (3.3–4.4 g/L), via malolactic fermentation (Picinelli et al., 2000). In the same experiment, Picinelli et al. (2000) found variable amounts of shikimic acid, from 0 to 40 mg/L. In Turkish ciders, a slightly different organic acid profile is estimated: malic acid (1.9–2.7 g/L), succinic acid (0.24–0.43 g/L), and small amounts of fumaric acid (3.88 µg/L; Güçer et al., 2008). Nevertheless, malic acid can be almost completely degraded during fermentation during cider preparation (Herrero et al., 1999).

Table 4.2 Organic Acid Detected in Different Fruit and Berry Wines

Wine Type	Malic Acid	Citric Acid	Tartaric	Succinic Acid	Acetic Acid	References
Blackberry	3.5	4.9	–	0.6	0.1–0.7	Johnson and Gonzalez de Mejia (2012)
Blueberry	2.8	3.8	–	0.8	0.1–0.6	Johnson and Gonzalez de Mejia (2012)
Strawberry	0.28	5.7	–	trace	<1	Martin et al. (1971)
Raspberry	1.5	2.1	–	2.8–7.1	0.2–0.5	Ryan and Dupont (1973)
Red currant	0.45	3.6	–	0.74	0.1–0.3	Ryan and Dupont (1973)
Apple	6.2	0.5	–	1.3	–	Martin et al. (1971)
Cider	0–2.7	–	–	0.25–1.6	4	Güçer et al. (2008)
Cherry	6.82	0.25	1.35	0.75	–	Martin et al. (1971)
Mango	3.8–7.2	1.74	0.26	1.01	–	Li et al. (2013)

Amounts of organic acids are expressed in g/L.

Comparable profiles of organic acids have been observed in sweet cherry cultivars where malic (3.53–8.12 g/kg FW), citric (0.11–0.54 g/kg FW), shikimic (6.56–26.7 mg/kg FW) and fumaric acid (0.97–7.56 mg/kg FW) were detected (Usenik et al., 2008). Evaluation of organic acids present in Danish cherry pure gave similar values with higher amounts of citric acid (2.5 g/kg). In a further analysis, changes in organic acid profile during fermentation to wine are monitored and resulted in high concentration of malic acid (6.82 g/kg) followed by tartaric (1.35 g/kg), succinic (0.75 g/kg), and citric acid (0.25 g/kg; Martin et al., 1971). During mango wine production malic acid content changes into the following pattern: malic acid was the major organic acid in juice (3 g/L), transiently increased to 7.2 g/L after acidification with malic acid and reverting to initial level (3.8–3.3 g/L) after fermentation. Besides malic acid, citric, tartaric, and succinic (1.74, 0.26 and 1.01 g/L, respectively) were prominent organic acids detected in mango wine (Li et al., 2013). A major organic acid detected in orange wine is citric acid (6 g/L), along with minor amounts of malic (0.34 g/L) and ascorbic acid (230 mg/L; Kelebek et al., 2009).

The organic acid composition of berry wines will be largely defined by their content in the source berry. The major organic acids present in different berries (Table 4.2) are citric, isocitric, malic, tartaric, fumaric, succinic, and shikimic and their content varies upon species and developmental stage (Whiting, 1958; Xie et al., 2011; Milivojevic et al., 2012a,b). After fermentation, the organic acid profile of berry wines are represented mainly by malic, citric, tartaric, and less often succinic acid (Martin et al., 1971). Similarly, to fruit wines, the increase of succinic as well as lactic acid and the decrease of malic acid during apple wine fermentation and aging are observed (Lee et al., 2013b). Significant amount of acids necessitate water addition after the crushing to obtain optimal yeast action. Final organic acid concentration is dependent on the yeast strain used for fermentation, especially for succinic, acetic, and lactic acid content, since they are mainly produced by yeast during fermentation (Duarte et al., 2010a,b). As shown in Table 4.2, blackberry and blueberry wines are rich in malic acid (3.5 and 2.8 g/L, respectively; Johnson and Gonzalez de Mejia, 2012), while in contrast, strawberry wines have less malic acid 0.28 g/L but they are rich in citric acid (5.7 g/L; Ryan and Dupont, 1973). In the same report it is estimated that raspberry wines have from 0.5 to 1.5 g/L malic acid and rather high concentration of succinic acid (2.8–7.1 g/L). In all berry wines, volatile acid content (e.g., acetic acid) should be kept at minimum level since their presence over 1.2 g/L (Zoecklein, 1995) can develop an unwanted "vinegar-like" flavor (Swiegers and Pretorius, 2005). In an experiment with different yeast strains used for fermentation, one apparent raspberry wine contained higher amount of acetic acid (2.3 g/L) and consequently expressed an intensive vinegar-like character (Duarte et al., 2010a,b).

For fruit and dessert wines the amount of volatile acids above 1.2 g/L is a sign of wine spoilage, so commercial wines must have less than this quantity (Zoecklein, 1995). Increasing acidity is a sign of deterioration of the stability and quality of wine, so it is important to be monitored.

1.4 ALDEHYDES, ESTERS, AND OTHER VOLATILE CONSTITUENTS

Besides the earlier mentioned compounds, large number of aroma and fragrant compounds that provide organoleptic characteristics of various wine types are also detected (Ferreira, 2010). Lots of them are volatile compounds and they can have a crucial role for final wine quality (Moreno-Arribas et al., 2009; Wang et al., 2012). A descriptive system for the classification of various aroma components is given in research conducted by Noble et al. (1987). For example, the main constituents of cherry flavors include methanol, ethanol, butanol, pentanol, octanol, geraniol, linalool, ethyl acetate, acetic acid, isovaleric acid, octanoic acid, and benzaldehyde (Dharmadhikari, 1996). In blackberry wine, composition of aroma volatiles was ethyl caproate, ethyl acetate, ethyl octanoate, isoamyl alcohol, phenethyl alcohol, and acetic acid (Wang et al., 2012). In pomegranate juices, volatile compounds can be grouped in four main chemical families: terpenes (limonene, myrcene, and γ-terpinene); aldehydes (decanal, nonanal, and octanal); esters (octyl acetate, ethyl acetate, and ethyl decanoate), and alcohols (1-butanol and 2-methyl-1-butanol). After fermentation, in pomegranate wine following fragrance compounds prevail: esters and alcohols make up to 80% of all volatiles, with addition of new counterpart, and organic acids, such as octanoic and decanoic acids (Andreu-Sevilla et al., 2013).

Aldehydes and ketones are produced in smaller amounts during incomplete alcoholic fermentation or oxidation of alcohol, but still very important for the generation of varietal aromas. Nevertheless, in higher concentrations they have undesirable effects, thus being considered as off-odor. In fruit wines, aldehydes have been detected in comparable amounts: 100–150 mg/L for cashew apple wine (Attri, 2009), 100–150 mg/L for plum wine (Joshi et al., 2009), and 2.2–21 mg/L of acetaldehyde in apple wine (Kourkoutas et al., 2001). In pomegranate wines, detailed analysis of volatiles showed presence of decanal, nonanal, and octanal in range from 0.03% to 0.95% (Andreu-Sevilla et al., 2013). Berry wines also contain aldehydes; the most abundant are hexanal, benzaldehyde, and *trans*-2-hexenal, making almost 10% of all aroma compounds (Wang et al., 2012).

Esters are also present in fruit and berry wines as one of the major ingredients defining final wine taste. They are responsible for the fruity bouquet of wines, where mixtures of esters may not possess the same intensity or qualitative attributes as individually (van der Merwe and van Wyk, 1981). In pomegranate wine, ethyl acetate and isoamyl acetate are the dominant forms, followed by minor amounts of other esters such as ethyl propanoate and isobutyl acetate (Kafkas et al., 2006). In cashew, mango and apple wines total amount of esters were 82–102, 15–35, and 242.12 mg/L, respectively (Wang et al., 2004; Attri, 2009; Li et al., 2013). Hexyl acetate, butyl acetate, and 2-methylbutyl acetate are present in ciders but, if thermally treated, ciders lost 30% of their original ester and aldehyde content during storage (Azhu Valappil et al., 2009). Ethyl octanoate, ethyl decanoate, and ethyl hexanoate were the major esters of the strawberry wine, taking almost 90% of all identified aroma compounds measured by head-space gas chromatography (Kafkas et al., 2006). In addition, the total amount of esters estimated in different strawberry wine ranged from 78 to 103 mg/L depending on the wine-making method (Sharma et al., 2009). In blackberry wines, major esters are ethyl caprylate, diethyl succinate, and ethyl caproate, with overall values ranging from 52.37 and 189.63 mg/L depending on the winemaking procedure (Gao et al., 2012).

1.5 VITAMINS

Fruit and berry wines belong to a group of sweeter wines, thus in the case of moderate to high consumption, can significantly contribute to overall caloric intake since 0.2 L of blueberry wine have around 130–150 cal. Wine contains some B vitamins (thiamine, riboflavin, and B12) and, unfortunately, very little amounts of vitamins A, C, D, and K (Fuller et al., 2011). As said before, berries are a significant source of ascorbate but in the final fermentation products there is relatively little ascorbate because it is highly degradable by heat and prone to oxidation, so, in the majority of berry wines there is no measurable vitamin C. However, vitamin C may be added during or after fermentation to some wines to prevent unwanted oxidation of wine components.

Thiamine is present in wine in nutritionally insignificant amounts since it is utilized by yeast during fermentation, as well as the riboflavin. Also, there is no data supporting the presence of nutritionally significant amounts of vitamin B12 in berry wines. Niacin is present in small amounts in grape juices, and from negligible to relatively small concentrations in wines, perhaps only 0.1 mg per serving. Water soluble B vitamin, folic acid, can be found in berries in significant amounts (Beattie et al., 2005) and (according to some ongoing investigation reports) can reach up to 1 mg/100 mL in some berry wines. Folic acid is present in grapes in much smaller amounts, hence, in grape wine, it can be found in no more than 1% of requirements per serving. One important aspect is that in chronic and excessive intake of alcohol, folic acid absorption (as well as other B vitamins) is compromised, leading to substantial B vitamin deficiency causing various and severe disorders (Kaunitz and Lindebaum, 1977; Cook et al., 1998). After taking all of these facts into consideration, it can be concluded that moderate wine consumption presents an insufficient source of vitamins needed for maintaining optimal nutritional status and overall health.

1.6 CAROTENOIDS

Several carotenoids were identified, and their changes were monitored during fermentation of mango puree. Major carotenoids detected in mango puree were β-carotene followed by lutein, violaxanthin, and neoxanthin in a total amount range from 980 to 5810 (μg/100 g). The carotenoid profile was the same in mango wine with total carotenoid content in the range of 578–4330 (μg/100 g), depending of the cultivar used for winemaking (Varakumar et al., 2011). In the same report it was shown that the xanthophylls (oxygenated carotenoids) degraded significantly from mango puree to wine (from 70% to over 80%), while β-carotene (hydrocarbon carotenoid) was more stable, with a maximum of 26.5% degradation. Besides its contribution to antioxidant capacity β-carotene, as provitamin A carotenoid, is a very important constituent of mango wine.

1.7 MINERALS

Mineral elements are present in fruit and berry wines in different quantities. Concentrations of the most mineral elements are significantly different among wines from different fruit sources. Generally, wine contains various minerals in readily available forms, especially potassium and iron, while excessive alcohol consumption can disturb the uptake of calcium, magnesium, selenium, and zinc, and increase the excretion of zinc via the kidneys (Rupasinghe and Clegg, 2007). Potassium is a dominant element present in all categories of wines, while other major elements are calcium, sulfur, phosphorous, and magnesium. Rupasinghe and Clegg (2007) estimated elemental composition of different fruit and berry wines. Potassium was the most abundant element in all wines ranging from 742 to 1201 μg/g, with the highest levels detected in black currant wine. Wine's low-sodium content is an additional benefit for those on a low-sodium diet, such as heart attack survivors. Magnesium and sodium were in the concentration range from 31 to 82 μg/g, while calcium, phosphorus, and sulfur had values from tens to hundreds of μg/g of tested wines. Such low-sodium/high-potassium ratio in berry wines makes them one of the more effective sources of potassium for individuals using diuretics.

Cranberry wines had exceptionally high calcium levels (643 µg/g) in comparison to all wine types tested: five times more than the second-best calcium source, black currant wine (108 µg/g). Red or white wines have a significantly higher magnesium amount than fruit and berry wines, with the exception of elderberry wines, which are also high in magnesium. Iron, manganese, and zinc were among the minor minerals present having the same values both for grape and nongrape wines, varying from 0.2 to 4.5 µg/g. In the case of manganese, a cofactor of very important enzymes for antioxidative defense, blueberry stands out from this range having 14.6 µg/g, approximately 10 times more than the average manganese content in other fruit and berry wines. Microelements cadmium, cobalt, chromium, copper, molybdenum, nickel, and lead were below the detection limits in all of the wine categories (Rupasinghe and Clegg, 2007). It should be emphasized that excessive alcohol intake negatively affects absorption, metabolism, and excretion, hence disturbing homeostasis of a number of microelements essential for well-being.

1.8 DIETARY FIBER

Grape wines contain certain amounts of soluble dietary fiber more present in red (1.37 g/L) than in white wine (0.94 g/L; Díaz-Rubio and Saura-Calixto, 2006). Nevertheless, amount ingested after moderate wine consumption is still insufficient to contribute significantly to the daily recommended intake in the human diet.

1.9 HISTAMINE

One important advantage of nongrape wines is that histamine was detected in few orders of magnitude lower concentration in berry wines (58 and 41 µg/L for black currant and blueberry wine, respectively) as compared to red wines (11,143 µg/L). Even better, the infamous byproduct of malolactic fermentation was not detected in plum, apple, and cherry wine (Rupasinghe and Clegg, 2007). Generally, their content is low in fruit and berry musts and rise during malolactic fermentation. The level of histamine was the highest when Bordeaux yeast strain was used, with decreasing content in wines produced by Burgundy, Malaga, Tokay, and Syrena yeasts (Pogorzelski, 1992). As always, generalization should be avoided since histamine was detected in elderberry wine tested in four successive years, ranging from moderate amount of 3650 µg/L to high (grape red wine equivalent) value of 12,360 µg/L. The investigation of factors affecting the formation of histamine has shown that it is formed during fermentation as a result of the action of intrinsic activity of histidine decarboxylase, an enzyme of the elderberry (Pogorzelski, 1992). In wines prepared from fruits with higher malic acid content, proper yeast strain can keep histamine levels very low. For instance, biogenic amines are found in very low contents in ciders and their possible accumulation during fermentation and storage period can be prevented by microbiological stabilization (Garai et al., 2006; Garai Ibabe et al., 2013).

Although almost all of the previously described components of fruit and berry wines have certain bioactive potential, phenolic compounds will be discussed in more detail since they are nearly a synonym for health benefit food potential of wine (De Beer et al., 2002). The following section will give a detailed overview of the phenolic components from fruit and berry wines that are known for their therapeutic potential. As a starting point, diversity of numerous classes of phenolic compounds will be linked with their distribution in the specific type of berry or fruit.

2. MAIN CLASSES OF PHENOLIC COMPOUNDS FROM FRUIT AND BERRY WINES WITH HEALTH BENEFIT POTENTIAL

Phenolic compounds belong to versatile group of plant secondary metabolites essential for proper growth and development (Table 4.3). Their function is even more divergent and they play many of the "main roles" in plant physiology, from structural components to pathogen defense, or from pigmentation to antioxidative protection (Lattanzio et al., 2009). Many subclasses of phenolic compounds have established antibacterial, antifungal, antiviral, anticarcinogenic, immunomodulating, and antiinflammatory properties. Their beneficial therapeutic values are proven in treatments of various disorders like cardiovascular issues, asthma, allergies, diabetes, and hypertension, presumably because of their antioxidative action.

Table 4.3 Representative Phenolic Compounds Present in Fruit and Berry Wines

Class	Basic Structure	Representative Compounds
Benzoic acid derivatives		Gallic, Vanillic Hydroxybenzoic, Salicylic Protocatechuic
Hydroxycinnamic acids derivatives		Chlorogenic Neo-chlorogenic Caffeic *o,m,p*-coumaric
More complex phenolic acids		Ellagic
Stilbenoids		Resveratrol Piceid Astringin Piceatannol
Flavones		Apigenin Luteolin Chrysin
Flavanones		Naringin Naringenin Pinocembrin
Flavonols		Catechin Epicatechin Epicatechin-gallate Epigallocatechin

Table 4.3 Representative Phenolic Compounds Present in Fruit and Berry Wines—cont'd

Class	Basic Structure	Representative Compounds
Flavonols		Kaempferol Myricetin Quercetin
Anthocyanidins		Cyanidin Delphinidin Malvidin Pelargonidin
Anthocyanins		Cyanidin 3-glucoside Delphinidin 3-glactoside Malvidin 3-galactoside Pelargonidin 3-glucoside
Tannins		Ellagitannins Gallotannins Polymeric proanthocyanidins

Before dispersion of chromatographic methods such as gas chromatography–mass spectrometry and liquid chromatography–mass spectrometry, the most frequent parameter used to describe the potential therapeutic value of selected wine was total phenolic content (TPC). Besides TPC, or as its logical outcome, estimation of total antioxidative activity (TAA) is given as more direct link to bioactive potentials of berry or its product. High intrinsic TPC values of berry fruit will largely determine final TPC values of corresponding wines. In Table 4.4, the overview of TPC values from different berry fruits is given (Kähkönen et al., 1999). TPC values are presented in gallic acid equivalents (GAE) since the different berries has significant qualitative differences in polyphenol content.

In the case of the berry wines, TPC content will also vary according to different fruits of origin. As a reference, in a study where TPC from 10 different types of grape red wine was compared, the highest value was 2149 ± 108.21 mg/L GAE (Feng-mei et al., 2014). Very often, TPC could also be presented in comparable ellagic acid equivalents (EAE) and for six different blackberry wines average value was 2212.5 ± 1090.3 mg EAE/L. Second to them, different blueberry wines had an average TPC of 1623.3 ± 645.5 mg EAE/L (Johnson and Gonzalez de Mejia, 2012). Increase in TPC during fermentation of elderberry must is estimated ranging from 1714.53 ± 71.40 mg GAE/L for must to 2004.13 ± 49.44 mg GAE/L as measured for young wine (Schmitzer et al., 2010). In another report

Table 4.4 Overview of Total Phenolic Content Measured for Different Wines

Wine Type	TPC (mg/L)	References
Blackberry	2212 (EAE)	Johnson and Gonzalez de Mejia (2012)
Grape	2149 (GAE)	Feng-mei et al. (2014)
Elderberry	2004 (GAE)	Schmitzer et al. (2010)
Blueberry	1623 (EAE)	Johnson and Gonzalez de Mejia (2012)
Bilberry	1161 (GAE)	Kalkan Yildirim (2006)
Black mulberry	1081 (GAE)	Kalkan Yildirim (2006)
Cherry	991 (GAE)	Rupasinghe and Clegg (2007)
Plum	555 (GAE)	Rupasinghe and Clegg (2007)
Mango	537 (GAE)	Varakumar et al. (2011)
Apple	451 (GAE)	Rupasinghe and Clegg (2007)
Peach	403 (GAE)	Davidović et al. (2013)
Strawberry	142 (GAE)	Dey et al. (2009)

(Kalkan Yildirim, 2006), estimated TPC values for bilberry, blackberry, and black mulberry wines were 1161, 1232, and 1081 mg/L GAE, respectively. Strawberry wine has lower TPC ranging from 126.8 to 142.3 mg/L GAE, so in terms of TPC, berry fruit wines can be placed in descending order: black-berry > bilberry > black mulberry > strawberry > raspberry (Kalkan Yildirim, 2006; Dey et al., 2009). In comparison to grapevines (Table 4.4) berry fruit wines are not inferior if TPC and its correlation with potential health benefit are discussed.

Other fruit wines have lower TPC content in comparison to grape and berry wines and, according to their TPC content, can be placed in descending order: sour cherry, quince, apple, melon, and apricot (Heinonen and Meyer, 2002; Kalkan Yildirim, 2006). Sweet cherries have been reported to contain various phenolics and anthocyanins which contribute to high TPC and antioxidant activity of cherry wine, resembling ones estimated for berry wines (Rupasinghe and Clegg, 2007). Since cider and perry are prepared mainly from specific sorts of fruits with higher polyphenol content, they are strongly flavored and colored than corresponding apple and pear wines (McKay et al., 2010). Higher polyphenol content of any fruit usually leads to oxidative browning which products strongly contribute to wine's final bouquet. Anyway, TPC of apple wines (451 mg GAE/L) or plum wine (555 mg GAE/L) is significantly lower than in cherry wines (991 mg GAE/L) and almost four times lower than TPC of blueberry wines being 1676 mg GAE/L (Rupasinghe and Clegg, 2007). Similarly, TPC of peach wine (402.53 mg/L GAE) have been found significantly higher in comparison with that of white wines (Davidović et al., 2013). Mango wine has a moderate level of TPC estimated in range from 202 to 537 mg GAE/L depending on mango cultivar used for winemaking (Varakumar et al., 2011). More exotic potent source of polyphenols is sea buckthorn wine and its TPC of 689 mg GAE/L was comparable to grape wine tested in the same batch (647 mg GAE/L). Also, investigation of influence of bottle aging on the TPC of sea buckthorn wine showed a slight decrease (534 mg GAE/L; Negi and Dey, 2009). Keeping in mind that higher TPC has high positive correlation with antioxidant activity, it is clear that the further discussion will stay focused on berry, cherry, and some of the fruit wines.

The real insight in bioactive potential of any plant product and branding it as functional food can be done only upon detailed qualitative analysis for presence of specific compounds. As mentioned before, development of analytical methods obtained very confident tools for "fingerprint" analysis of plant extracts. The final goal of modern food chemistry/biotechnology is giving evidence that the presence of a specific compound is responsible for apparent therapeutic effect. It is a very demanding task, since the bioactivity of the selected sample is (almost) usually a product of synergistic effect of many different compounds present. The full complexity of distribution and variety of phenolic compounds present in food and beverages can be grasped just by reading the title of an immense study (Pérez-Jiménez et al., 2010). Furthermore, there are lots of different ways that phenolic compounds are divided according to their structure. The next subsections will give a detailed outlook of occurrence of specific groups of phenolic compounds in a complex matrix, such as fruit wine, using widely adopted methods of chemical classification (Vermerris and Nicholson, 2006).

2.1 ANTHOCYANINS AND ANTHOCYANIN-DERIVED COMPOUNDS

Anthocyanins are the very disperse group of phenolic constituents present in different fruits and vegetables. They are largely responsible for the distinct color of fruits (especially berries known as potent source of anthocyanins), vegetables, and related products. Although they are all various glycosides of only a few anthocyanidins such as cyanidin, malvidin, delphinidin, petunidin, and pelargonidin, anthocyanins are portrayed by more than 540 different molecules (Davies, 2009). Generally, it is predictable that individual anthocyanins detected in specific berry or fruit can be traced down to its final product: wine. Nevertheless, their final content and qualitative profile will be very much dependent on the winemaking process.

Individual anthocyanins are present in berries in different forms and quantities and their key representatives are displayed in Table 4.5 (Ichiyanagi et al., 2004; Borges et al., 2009; Nour et al., 2011; Giampieri et al., 2012). Anthocyanin stability is largely dependent on temperature, light intensity, pH,

Table 4.5 Major Forms of Anthocyanins in Berry and Fruit Wines

Source Berry or Fruit	Major Anthocyanins
Blackberries	Cyanidin-3-glucoside, Cyanidin-3-rutinoside Pelargonidin-3-glucoside
Blueberries	Delphinidin-3-glactoside, Delphinidin-3-arabinoside, Malvidin-3-galactoside, Malvidin-3-arabinoside, Peonidin-3-glucoside, Petunidin-3-galactoside, Petunidin-3-glucoside, Petunidin-3-arabinoside
Bilberries	Delphinidin-3-galactoside, Delphinidin-3-arabinoside, Malvidin-3-galactoside, Malvidin-3-arabinoside, Petunidin-3-glucoside, Petunidin-3-galactoside, Petunidin-3-arabinoside
Raspberries	Cyanidin-3-glucoside, Cyanidin-3-sophoroside, Cyanidin-3-sambubioside, Pelargonidin-3-glucoside, Pelargonidin-3-rutinoside
Strawberries	Cyanidin-3-glucoside, Cyanidin-3-rutinoside Pelargonidin-3-glucoside, Pelargonidin-3-galactoside, Pelargonidin-3-arabinoside
Black Currant	Cyanidin-3-rutinoside, Cyanidin-3-glucoside, Delphinidin-3-glucoside, Delphinidin-3-rutinoside, Malvidin-3-galactoside, Petunidin-3-rutinoside, Peonidin-3-glucoside, Peonidin-3-galactoside, Peonidin-3-rutinoside
Red Currant	Cyanidin-3-rutinoside, Cyanidin-3-sambubioside
Cranberries	Cyanidin-3-galactoside, Peonidin-3-galactoside, Peonidin-3-arabinoside
Cherry	Cyanidin-3-glucoside, Cyanidin-3-glucosylrutinoside, Cyanidin-3-rutinoside, Delphinidin-3-rutinoside, Pelargonidin-3-glucosylrutinoside

and organic moiety, which can be changeable during the winemaking process (Patras et al., 2010; Bener et al., 2013). Unfortunately, there are not many references that could help in tracking their fate during the fermentation of different berry musts. In blackberry wines, Rommel et al. (1992) observed the presence of several anthocyanins, at the same time pointing out instability of one of the major constituents, cyanidin 3-glucoside. In raspberry wine, cyanidin 3-sophoroside was found as a major stable anthocyanin form, the opposite of degradable cyanidin 3-glucoside (Rommel et al., 1992). Major anthocyanins in a few tested blueberry wines (Sánchez-Moreno et al., 2003) were peonidin 3-glucoside, malvidin 3-galactoside, and petunidin 3-glucoside. Delphinidin 3-glucoside, delphinidin 3-rutinoside, cyanidin 3-glucoside, and cyanidin 3-rutinoside are major monomeric anthocyanidins, followed by their aglycones delphinidin and cyanidin (Czyżowska and Pogorzelski, 2004).

Hence anthocyanins are readily found in berry and some fruit wines, and their content is commonly presented jointly as total anthocyanin content (TAC; Table 4.6). The highest TAC values were found in a number of blackberry and blueberry wines averaging from 75.56 ± 70.44 to 20.82 ± 12.14 mg EAE/L, respectively (Johnson and Gonzalez de Mejia, 2012). Considerably high TAC values were estimated for blackberry wines ranging from 134 ± 3 to 164 ± 3 mg/L, expressed as malvidin 3-glucoside equivalents, comparable to TAC measured in red grape wine (Mudnic et al., 2010). During blueberry fermentation only partial preservation of anthocyanin content of blueberry samples occurred, the net loss of $68.8 \pm 1.3\%$ of TAC was observed after seven days of fermentation. Modified bacterial strain achieved much better preservation of TAC since only $24.1 \pm 2.2\%$ were lost for the same period, probably caused by stronger acidification (Martin and Matar, 2005). Literature data are lacking concerning the TAC of raspberry wines and only provisional values can be considered using those measured in different raspberry extracts, 0.17–57.6 mg of cyanidin 3-glucoside equivalents/100 g (Liu et al., 2002). Strawberry wines had significantly lower values of TAC of 8.4–12.6 mg/100 g FW P3GluE (Table 4.6), having only 20–30% of values detected in fresh fruit (Klopotek et al., 2005). Furthermore, the interactions of anthocyanins, total phenolics, and ascorbic acid within a strawberry species are strongly affected by cultivar (Dragišić Maksimović et al., 2015).

As expected, in fruit wines such as apples and pears, TAC is much lower (Table 4.6), with values below 50 mg/L of EAE/L. Furthermore, fermentation process additionally diminishes their content because of the formation of oligomers with other polyphenols (McKay et al., 2010). As expected, cherry wines contain much more anthocyanins (0.12 ± 0.01 g/L malvidin 3-glucoside equivalents), mainly

Table 4.6 Comparison of Total Anthocyanin Content of Various Wines

Wine Type	TAC (mg/L)	References
Grape	212–287 (M3GluE)	Mudnic et al. (2010)
Grape	80–220 (M3GluE)	Pantelić et al. (2014)
Blackberry	115–191 (EAE)	Johnson and de Mejia (2012)
Blackberry	134–164 (M3GluE)	Mudnic et al. (2010)
Cherry	120 (M3GluE)	Pantelić et al. (2014)
Blueberry	30–50 (EAE)	Johnson and de Mejia (2012)
Strawberry	8.4–12.6[a] (P3GluE)	Klopotek et al. (2005)

[a]mg/100 g.

cyanidin 3-glucosylrutinoside and cyanidin 3-rutinoside (Pantelić et al., 2014). Pantelić et al. (2014) also described seven characteristic molecules in anthocyanin profile of cherry wine: delphinidin 3-rutinoside, cyanidin 3-sophoroside, cyanidin 3-pentosylrutinoside, pelargonidin 3-glucosylrutinoside, cyanidin 3-rutinoside, peonidin 3-rutinoside, and pelargonidin 3-glucoside. In cashew wine, major phenolic compound is leucodelphinidin, an anthocyanidin with many potential health benefits (Attri, 2009).

2.2 PHENOLIC ACIDS

In the current literature, phenolic acids are usually divided on hydroxybenzoic acid derivatives (such as gallic, hydroxybenzoic, salicylic, and protocatechuic) and diverse hydroxycinnamic acids and their derivatives (Table 4.3). Ellagic acid is the more complex member of hydroxybenzoic acid derivatives group as dilactone of hexahydroxydiphenic acid. Ellagic acid can be detected in free form or involved in forming oligomeric molecules such as hydrolysable tannins (ellagitannins) in strawberries, raspberries, and blackberries (Manach et al., 2004). Tannins express antinutritional effect because they can bind to amino groups of peptides obstructing their hydrolysis in the stomach. Ellagic acid derivatives, as starting structure for tannin formation, are found predominantly in seeds and achenes (embryos), but also in lignified tissues. Crushing and pressing these structures during winemaking release tannins, especially in the case of berry fruits like blackberry, raspberry, and strawberry. Since tannins along with water and sugar are the main adjuncts used in the winemaking process, their final content in fruit and berry wine largely depends on amount added and metabolic action of yeast strain used (McKay et al., 2010). Hydroxycinnamic acids (major forms are caffeic, ferulic, *p*-coumaric, chlorogenic, and sinapic acid) seldom exist in their free form; more often they are detected as glycosides formed with one or more different mono and disaccharide units. Also, they often form phenolic acid esters with quinic, shikimic, or tartaric acid. For example, caffeic and quinic acids combine to form chlorogenic acid, which is found in significant quantities in bilberry and blueberry fruit (up to 71.2 μg/g FW; Milivojević et al., 2012a,b).

The most abundant hydroxycinnamic acids in apple wines are chlorogenic and *p*-coumaroylquinic acid. They are prone to very fast enzymatic oxidation, catalyzed by polyphenol oxidase, giving *o*-quinones as the first product, generally responsible for browning of apple musts (Satora et al., 2008). The effect of fermentation on the phenolic profile of apple beverage will be given in the next example. In investigation of phenolic profile of Asturian cider, phenylpropionic acids (hydrocaffeic acid and hydrocoumaric acid), were detected as dominant (80%) forms of all detected phenolic acids (Madrera et al., 2005), instead of chlorogenic acid and *p*-coumaroylquinic acid detected in apple juice (Guyot et al., 2003). During fermentation, these two hydroxycinnamic acids esters are hydrolyzed into caffeic acid/quinic acid and *p*-coumaric acid/quinic acid, respectively. Free-form caffeic acid and *p*-coumaric acid are further reduced into hydrocaffeic and hydrocoumaric acids (Madrera et al., 2005). In orange wine, gallic acid was dominant hydroxybenzoic acid being around 2 mg/L, while hydroxycinnamic acids were represented by caffeic (2.5 mg/L), chlorogenic (4.56 mg/L), *p*-coumaric (1.58 mg/L), ferulic (9.91 mg/L), and sinapic acid (7.78 mg/L; Kelebek et al., 2009). The main phenolic compounds found in peach wine were chlorogenic acid, caffeic acid, and catechin (3.59, 0.87, and 0.60 mg/L, respectively; Davidović et al., 2013).

Cherry wines have a significantly higher content of phenolic acids in comparison to grape wine and almost 10 times for caffeic acid. Protocatechuic acid (23.89 mg/kg), chlorogenic acid (3.57 mg/kg), caffeic acid (13.88 mg/kg), *p*-coumaric acid (23.42 mg/kg), and *p*-hydroxybenzoic acid (6.65 mg/kg) were detected as the main phenolic acids (Pantelić et al., 2014). Gallic acid was detected in much lower amounts (1.10 mg/kg) in contrast to its high content in grape wines (above 20.00 mg/kg for four of five tested red wines). In another popular wine-fruit from India, the custard apple, lower levels of TPC (98 mg GAE/L) was found as compared to red wine (Jagtap and Bapat, 2014). In jackfruit wine, three hydroxybenzoic acids (gallic, protocatechuic acid, and gentisic acid), and two hydroxycinnamic acids, caffeic acid and *p*-coumaric acid were found as the major phenolic constituents (Jagtap and Bapat, 2014).

Distribution of hydroxybenzoic acid derivatives varies significantly among different berries. Ellagic acid is major phenolic acid in raspberries and strawberries representing 88% and 51% of all analyzed phenolic compounds, but in blueberries, bilberries, and red currant the amount of ellagic acid was less than 10% (Häkkinen et al., 1999). Another report confirmed the presence (44 mg/L) of significant amount of ellagic acid in strawberry wine (Lehtonen et al., 1999). Protocatechuic, hydroxybenzoic, and gallic are present in more than 100 μg/g FW (El Gharras, 2009) in raspberry, black currant, and strawberry. Hydroxy-cinnamic acids has attracted attention since its presence is highly correlated with food therapeutic potential (Dávalos and Lasunción, 2009). In six berry types belonging to *Fragaria*, *Rubus*, *Vaccinium*, and *Ribes* genus, chlorogenic acid was detected only in blueberries averaging 44 μg/g FW (Milivojević et al., 2013). In another report, caffeic is marked as the dominant (17.4 ± 1.0 μg/g FW) free-form of hydroxycin-namic acids in blueberries, but for blackberries ferulic acid is detected in the highest amount of 39.0 ± 7.1 μg/g FW (Zadernowski et al., 2005). In the same report, a completely different profile is obtained when total phenolic acid content was analyzed illustrating different isomers of coumaric as the major components. Phenolic acids in free form were the minor fraction constituting only from 1.7% (black cur-rants) to 4.2% (black mulberry) of the total phenolic acids present in these berries. Bound phenolic acids constituted from 67.3% (blackberries) to 79.3% (black mulberries) of all hydroxycinnamic phenolic acids identified (Zadernowski et al., 2005). It is important to point out that the "sea-saw" between free and bound forms of phenolic acids, as consequence of physiological processes such as plant stress defense or fruit maturation, sometimes leads to even incomparable data from numerous phytochemical analysis. Unfortunately, there are only a few papers presenting the presence and distribution of different phenolic acids in berry wines as a consequence of focusing on flavonoid and anthocyanin analysis. Hennig and Burkhardt (1960) detected ellagic, caffeic, *p*-coumaric, and 3,4-dihydroxybenzoic acid in black raspberry wines (Lim et al., 2012), while only ellagic acid was present in loganberry wine.

Although TAC could be higher in fruit juices, bioavailability of these compounds can be much lower than in the wines. Support for wine lovers is shown in this quotation: "Drinking fermented bever-ages is important due to the conversion of the compounds present and increase anthocyanin and poly-meric proanthocyanidins compounds and antioxidant capacity" (Martin and Matar, 2005; Johnson and Gonzalez de Mejia, 2012). On the other hand, Goldberg et al. (2003) suggest that efficient absorption does not require the presence of alcohol, so aqueous media and vegetable suspensions are as effective polyphenol resources as the wine.

2.3 FLAVONOID COMPOUNDS

The flavonoids are a very diverse group of phenolic compounds present in many plants having very distinct physiological functions. They are more stable than anthocyanins, so flavonoids can persist in many food products of plant origin. Higher amounts of flavonoids can be found in various fruits, vegetables, tea, coffee, chocolate, juices, and wine. They can be divided into a few major subclasses of flavonoids such as flavones (e.g., apigenin, chrysin, and luteolin); flavonols (kaempferol, myricetin, and quercetin); flavanones (naringin, naringenin, and pinocembrin); and flavanols (catechin, epicatechin, and gallocatechin; Table 4.3).

2.3.1 Flavones, Flavonols, and Flavanones

Historically, flavonoids were described as health benefit components much earlier than phenolic acids, so they were, up to 1950, classified as vitamin P. This could be the reason for better literature support concerning presence of flavonoids in berry fruit wines. In black currant wine, major flavonols are myricetin (3.8–22.6 mg/L) and quercetin (2.2–24.3 mg/L) with higher content than in concurrently tested Cabernet wine (0–14.6 and 1.2–19.4 mg/L, respectively; Vuorinen et al., 2000). In the white berry wines, Vuorinen et al. had found only quercetin (4.1 mg/L) in white currant wine. They concluded the flavonol content of berry wines were comparable to other reported for, geographically different, quality grape red wines. In a phenolic compound analysis of various Finland wines prepared from black currants, blueberries, cranberries, crowberries, and strawberries, Lehtonen et al. (1999) found similar concentrations of phenolic compounds in fruit but smaller quantities in berry fruit wines. In research on flavonoid occurrence, Mudnic et al. (2012) found 2.1–2.4 mg/L quercetin 4-glycoside in blackberry wines. During alcoholic fermentation of mulberry must, besides anthocyanidins, two quercetin glycosides (rutin and isoquercetin) were identified (Pérez-Gregorio et al., 2011). Myricetin, quercetin, and kaempferol were found in flavonoid profiles of wines made from black currants, blueberries, and crowberries (Ollanketo and Riekkola, 2000). Anttonen and Karjalainen (2006) highlighted that the black currant wine contained considerable amounts of quercetin glycosides rutin and isoquercitrin, which is in correlation with their presence in black currant fruit. Analysis of few Finland black currant commercial juices and wines showed variation in flavonoid content between two matrices. Almost twofold higher content of flavonol isoquercitrin has been found in wine, although myricetin, quercetin, and kaempferol were identified mainly in wine samples (Kivilompolo et al., 2008). Again, the choice of winemaking technology influences significant changes in both qualitative and quantitative polyphenol content during fermentation, as described before for TAC.

The most abundant flavonols of the apple and pear wine are quercetin glycosides, where isorhamnetin 3-glucoside was the major form in apple wine, whereas in pear wine, mainly isorhamnetin diglycosides were detected (Weber et al., 2014). Using extremely selective and reliable novel high performance liquid chromatography mass spectrometry (HPLC/MS) techniques, five flavonoids have been reported for the first time in apple juices: dihydrochalcones (2 phloretin-pentosyl-hexoses); flavonols (isorhamnetin 3-O-rutinoside, isorhamnetin 3-O-arabinopyranoside, and isorhamnetin 3-O-arabinofuranoside; Ramirez-Ambrosi et al., 2013). Last mentioned results are given just to present almost daily mounting of data about more and more new metabolites found in food products. The next and even more demanding step is making strong and highly approved correlation of each metabolite found with its potential health benefits alone or as a whole extract. Main flavonoids from orange wine belong to flavanones and major representatives are hesperidin, narirutin, and apigenin with estimated values of 90.65, 21.67, and 16.12 mg/L, respectively (Kelebek et al., 2009).

In cherry wine, naringenin and apigenin were found in quantities of 0.15 and 0.06 mg/L, respectively. Higher content of caffeoylquinic acids (as previously given in phenolic acids section) lead to formation of quercetin 3-O-hexosides, quercetin 3-O-rutinoside (0.23 mg/kg) and caffeoyl-hexosides, which were found only in cherry wine, and not in tested grape wines (Pantelić et al., 2014).

2.3.2 Flavanols

Flavanols are another diverse group of polyphenolics, and major members with dietary importance for human health are catechin, epicatechin, and epicatechin gallates. Regularly, smaller monomeric flavan 3-ols such as catechin and epicatechin, are readily found as major components in red grape red wines (del Álamo et al., 2004). More often they are aggregated in oligomeric or polymeric pattern forming numerous different types of proanthocyanidins (Blanco-Vega et al., 2014). In a very informative report, Mudnic et al. (2012) compared several grape wines (red and white), with four commercial blackberry wines for their flavanol and procyanidin B_2 content. The largest quantities of catechin (45.2 mg/L) and epicatechin (34.7 mg/L) were observed in two blackberry wines, as well as almost sevenfold higher (77.1 mg/L) procyanidin B_2 content in comparison to the maximal found in red grape wine (12.3 mg/L). On the other hand, catechins are quantified in significantly higher amounts (values between 120 and 390 mg/L) in California red grape wines (Frankel et al., 1995) than Mudnic et al. (2012) reported for Slovenian red wines. At risk of being repetitive, it should be emphasized again that, depending on experimental object, wine variety, or even physiological state of berries, one can find very different values in literature. Much smaller quantities of flavanols and procyanidin B_2, monitored in dependence of winemaking technology, are detected in black currant wines being well below 10 mg/L (for procyanidin B_2 maximal amount was 4.92 mg/L; Czyżowska and Pogorzelski, 2004). In the same report, cherry wines are described as a much more potent source of flavonols with almost 10 times higher values (from 24.2 to 117.81 mg/L for procyanidin B_2). Czyżowska and Pogorzelski (2004) found a decrease in catechin and procyanidin B_2 content during vinification process, with increased catechin trimer concentration in one case.

In apple wine, the most abundant flavanols are the major one epicatechin, followed by significant amounts of catechin and their polymerization product procyanidin B_2 (Satora et al., 2008), whereas in pear wines flavan-3,4-diols (leucoanthocyanins) are predominant forms. Also, apple wines are a potent source of dihydrochalcones phloridzin (phloretin-2-glucoside) and phloretin-β-xyloglucoside. The flavanol profile of cherry wine is similar to grape counterparts, with values up to 3.92 mg/kg for epicatechin, followed by smaller amounts of catechin and epigallocatechin (Pantelić et al., 2014).

2.3.3 Stilbenoids

Resveratrol is found in the seed and skin of grapes and berries and in their juice and wine. Red wine has higher amounts of resveratrol due to the fact that during red winemaking must, grape skin, and seeds are often in contact during the whole fermentation process (Fernández-Mar et al., 2012). Resveratrol is a well-known stilbenoid inevitably found in any red grape wine analysis. In a study of four different blackberry wines, cis- and trans-resveratrol are found in small amounts below 1.5 mg/L. Other important molecules belonging to the mentioned group are astringin, piceid, and piceatannol. Piceid and astringin were present in higher quantities (up to 12 mg/L), but astringin were found in only 50% of samples (Mudnic et al., 2012). In blueberries and bilberries trans-resveratrol content was 140.0 ± 29.9 pmol/g in highbush blueberries from Michigan; 71.0 ± 15.0 pmol/g in bilberries from Poland; and no detectable resveratrol in highbush blueberries from British Columbia (Lyons et al., 2003). Current literature lacks data regarding stilbene content in berry wines but differences in resveratrol content between Vitis and Vaccinium species are reported by Rimando et al. (2004). Resveratrol content was the highest in grapes (6471 µg/g DW), followed by lingonberry 5884 µg/g DW and it is superior to any other berries tested: blueberry, cranberry, and bilberry (amounts ranging from 768 to 1691 µg/g DW). It is important to highlight that resveratrol content was measured from the whole fruit and not just the grape skin (which is the usual extraction protocol in most of the other works), providing much more comparable results with berry composition analysis. Besides resveratrol, different distribution of pterostilbene and piceatannol were detected. Deerberries are rich in pterostilbene (520 μg/g DW) and highbush blueberry "Bluecrop" had piceatannol (422 μg/g DW) instead (Rimando et al., 2004). Although there is not enough evidence for their occurrence in berry wines, an overview of the presence of different stilbenoids in berry fruits is given since they are known for their strong and diverse bioprotective activities. On the other hand, resveratrol was not detected in cherry wines (Pantelić et al., 2014).

3. NUTRITIONAL FACTS

Besides the already-mentioned characteristics of fruit and berry wines, fruit wines possess particular nutritional values. Evidently, different fruit wines have variable nutrient content due to differences in the source material as well as in variations of the production process. Also, specific winemaking processes using specific ingredients lead to production of individual wine types, sometimes much unparalleled. Main constituents of berries belong to diverse classes of organic compounds such as sugars, organic acids, vitamin C, and in general, different polyphenols (Milivojević et al., 2011a,b, 2012a,b; Giampieri et al., 2012; Mikulic-Petkovsek et al., 2012). Ascorbic acid is historically given the largest attention due to its early recognized antioxidant capacities (Ezell et al., 1947). It is present in many berries and its faith during wine production is very dependent on the technology used (Agar et al., 1997). Beside depicted principal components, berries are an excellent source of natural antioxidants, mainly phenolic compounds, with proven valuable therapeutic or preventive characteristics (Heinonen et al., 1998; Bowen-Forbes et al., 2010; Im et al., 2013; Nour et al., 2013).

3.1 CALORIC VALUE

Despite its origin, wine's major nutritional value comes from the high caloric value of its ethanol content. Generally speaking, fermentation processes modify the raw ingredients, strongly affecting the overall chemical composition of the final product. Specific origin and winemaking process, as well as the situation governing its practice, determines the consumption and quantity of alcohol and other wine's constituents that will be taken with food or alone. Yeast converts the fermentable sugars to alcohol at the same time producing certain nutrients, such as various acids, biogenic mines and some members of the B group of vitamins. One serving of alcohol is fully absorbed into the blood stream within 30 min to 2 h after intake. Alcohol needs no digestion and it is rapidly absorbed directly through the intestinal wall and metabolized mainly (95%) in the liver. Its high caloric value made wine (in apparent historical situations) as the major source of metabolic energy for the adult population (Fuller et al., 2011).

3.2 ANTIOXIDANTS

Daily intake of polyphenols can reach as much as 1 g partly taken through the diet reach in fruits and vegetables. Significant share of polyphenols is ingested by consumption of fruit juices, wine, coffee, tea, and similar beverages. Of course, different cuisine and cultural habits define the presence of polyphenols in the common diet for certain populations. For instance, average daily intake of polyphenols for the French population is around 1.2 g (40% of flavonoids, 60% of phenolic acids) slightly higher than 0.82 g/day estimated for Spanish inhabitants. Mutually for both populations, plant beverages (juices and wine), tea, and coffee were dominant sources of polyphenols in the regular diet.

3.3 ALCOHOL

Without any doubt, alcohol consumption is of foremost attention for health, since a vast majority of consequences of alcohol intake are clearly damaging in many ways. Presence of alcohol in any type of wine provokes deterioration of nutrient bioavailability not only of those taken from wine, but also from other food, creating conditions for various deficiencies.

Maybe more as marketing necessitates, questions of whether fruit wines can be considered *functional food* have been raised. After what has been said about nutraceutical compounds present in fruit and berry wines, it would not be unjustified to conclude that the apparent presence of antioxidants, as well as other bioactive compounds, is giving them additional values for branding them as *functional food* (Dey et al., 2009).

4. ENZYMATIC TRANSFORMATIONS OF PHENOLIC COMPOUNDS DURING VINIFICATION

As prevention, plants have developed different defense mechanisms to convert radicals in less harmful products. These mechanisms are based on metabolic compounds (phenolics, ascorbate), as well as enzymatic antioxidant systems that consists of several different proteins including specific and nonspecific peroxidases (PODs), polyphenol oxidases (PPOs), superoxide dismutase (SOD), etc. The content of biologically active compounds is influenced by the enzyme/substrate reactions which define the final quality of fruit wines. Therefore, this section will outline major changes of antioxidative enzymes activity during the fruit ripening or during the process of winemaking that may indicate the maturity stage of the berries.

Accumulation of phenolic compounds in plant tissue during ripening activates the enzymes, primarily PPOs or various PODs. Plant-borne substrates of these enzymes include phenolic acids, flavonoids, and other phenolic compounds (Galletti et al., 1996). A large number of monophenolic and/or diphenolic compounds catalyzed by PPOs may in turn form a variety of products such as quinones and condensation products (Eskin, 1990; Whitaker, 1995; Jiménez and García-Carmona, 1999). Low-molecular metabolites, as well as free radical precursors of lignin and superoxide anion radicals in berries were formed depending on POD activity (Kozlowska et al., 2001). Constitutional PODs could be related to the variety of physiological processes, including cane lignification and periderm suberization, which is quite important for berry fruits, which are characterized by lignifying organs as the main site of phenolic metabolism. Generally, the enzymes were found in a glycosylated form and associated to membranes, as well as soluble isoenzymes (Haard and Tobin, 1971; Thomas et al., 1982). Miesle et al. (1991) observed that peroxidase has various functions in the ripening process, including changes in cell wall plasticity and anthocyanin breakdown (Yokotsuka and Singleton, 1997). Furthermore, PODs belong to a large family of enzymes able to oxidize a wide variety of organic and inorganic substrates in the presence of H_2O_2 (Zhang et al., 2005). In relation to this, activated enzymes in berry fruit disburse phenolic substrates resulting in their decreased concentrations. Inversely proportional enzyme/substrate ratio indicates that berries with higher enzyme activities and lower content of antioxidative components has a great potential for enzymatic browning (Dragišić Maksimović et al., 2013), which makes it less usable as high nutrient quality fruit. A better knowledge of the factors that influence the action of POD and PPO is imperative to control and manipulate its detrimental activity in fruit products, including wines.

PPO and POD activities via phenolic metabolism could modulate the nutrient quality of the berry fruits (Milivojević et al., 2011b). Dragišić Maksimović et al. (2013) demonstrated that the values of berry PPO activity highly exceeded values of POD activity emphasizing their role in phenolic metabolism. Accordingly, Kader et al. (1997) demonstrated that the oxidative browning after crushing of fresh berries is due mainly to PPO. This indicates that berries have a great potential for enzymatic browning, resulting in anthocyanin breakdown during the winemaking process (González et al., 1999).

Protein content and activities of PPO and POD were found to be higher in the beginning of berry development and during the final process of ripening compared to the other periods, most probably due to higher metabolic activity during these development stages (Aydin and Kadioglu, 2001). The incremental PPO and POD activity in the beginning of fruit development may be involved in shikimate and phenylpropanoid pathways, strongly affecting final phenolic profile (Cheng and Breen, 1991). On the other hand, the increase in both enzyme activities during ripening may affect anthocyanin metabolism. Simultaneously with the changes of PPO and POD activities, contents of proteins and other metabolites such as phenolics, sugars, and ascorbic acid are also affected. In particular, the increase in sugar content and PPO and POD activities reduce the astringent taste of berries in the ripening stage. Changes in enzyme activities related to the content of substrates play an important role in nutrient quality definition of fruits (Dragišić Maksimović et al., 2013), which implies that all of these changes jointly contribute to the quality of the final product, i.e., fruit wine.

Anthocyanins as the predominant pigments in wine derive from both fruit skin and flesh to wine in maceration process (Elez Garofulić et al., 2012). In most cases, the winemaking process decreased the PPO and POD activities, since during juice production blanching time of 10 min and bath temperature of 70°C was sufficient to inactivate PPO and POD (Galić et al., 2009). Therefore, in fruit wine production each processing step changes the overall antioxidant system, among others, through its enzymatic component which indirectly can assist in achieving wine quality goals.

Some plant cells, especially those in ripening fruits, produce a certain amount of ethanol where metabolic pathways are perfectly functioning, including POD reactions (Kelly and Saltveit, 1988). A broad range of substrate specificities and high stability in different media (Barceló et al., 2003) justify the presence of POD detected in wines, even at the level of residual activity (Galić et al., 2009). This made POD the best candidate enzyme for testing plant responses to ethanol at the enzymatic activity level. Ethanol alters the activity of POD showing highly significant increases at higher ethanol concentrations (Li et al., 2004). Therefore, genuine enzymatic components should not be treated as negligible material for wine production.

5. BIOAVAILABILITY OF THE MAJOR HEALTH BENEFIT COMPONENTS OF FRUIT WINES

The previous content depicted in detail many of the components which are thought to be the carriers of the health benefits of fruit and berry wines. These effects are very versatile in respect to very different disorders and maladies that they act upon, as well as in the site/organ/tissue where they express the remediation/preventive potential. It is easy to conclude that the first site of their action is within the gastrointestinal system, and the main question rises regarding their stability or transformation during the digestion process. Previously reviewed components of fruit and berry wines have confirmed the health benefit effects on very distant organs and systems such as cardiovascular system or even the brain. Having that in mind, it is clear that collecting as much as possible information about the bioavailability of the ingested flavonoids is essential to understand their real biological effects. Besides numerous investigations on polyphenol bioavailability, gathered results about the absorption, distribution, metabolism and excretion of individual flavonoids in human/animal systems are hard to interpret in a straightforward manner. Thus, before a detailed overview of their sanative values is given, their bioavailability has to be discussed as the main precondition for efficient biomedical performance.

Phenolic compounds in general are not well-absorbed, and it is supposed that the presence of ethanol enhances their bioavailability. This can explain why regular wines have often been found to offer somewhat greater health benefits than phenolic extracts or dealcoholized wine alone. Furthermore, most studies have been designed using phenolic compounds in their parent form rather than its derivatives such as glycosides and esters, which predominate in plants. Hence, although significant quantities of various polyphenols are present in the human diet, they are not proportionally effective, either because they have a lower intrinsic activity or because they are poorly absorbed, highly metabolized, or rapidly eliminated from the body. After absorption, polyphenols and their derivatives are transported in plasma extensively bounded to the human serum albumin molecule. Also, circulating polyphenols are able to disperse into the tissues in which they are metabolized, but their accumulation within specific target spots needs to be further verified (Manach et al., 2004).

Low absorption rates result in their restraint amounts; as a result, the average maximum polyphenol concentration in plasma hardly reaches more than $1\,\mu M$ after the consumption of dietary significant amounts (50 mg) of a single phenolic compound. Also, ingested polyphenols are metabolized variably depending on the polyphenol type with relative urinary excretion ranging from 0.3% to 43% of the ingested dose (Manach et al., 2005). Many different investigations showed that it is almost impossible to show a strict ratio or dependence between the amount of polyphenols present in food and their final bioavailability in human body. The majority of food polyphenols are present in derived form, usually as glycosides, esters, or polymers that have to be hydrolyzed by the gut microflora prior to absorption.

5.1 BIOAVAILABILITY OF ANTHOCYANINS

The available literature is full of evidence that consumption of anthocyanins provides various defense mechanisms (anticancer, cardioprotective, neuroprotective, and antidiabetic) against various disorders, but detailed metabolite tracking pointed out their very low rate of absorption, i.e., very low concentration in distant target tissues (Felgines et al., 2002). Furthermore, they are difficult to quantify in situ, so there is a lot of space for further investigation about their action. After consumption, anthocyanins are rapidly absorbed as glycosides mainly by the absorption through the gastric wall (Talavéra et al., 2003).

5.1.1 Absorption Rate and Main Forms Detected

Generally, it is accepted that glycosylated forms of anthocyanins can reach the bloodstream and target tissues in 10^{-8} to $10^{-7}\,M$ concentration, within minutes after ingestion, and remain detectable for up to several hours after the intake. The problem lies in the fact that a much higher $(10^{-6}–10^{-4}\,M)$ concentration is proven as effective in inhibiting malignant cell survival or intercepting oncogenic signaling events (Cooke et al., 2005). Another unsolved problem is: do anthocyanins retain their chemopreventive efficacy in its present condition or do they have to be previously hydrolyzed to their aglycones? Differences in absorption of purified anthocyanins (cyanidin 3-glucoside, cyanidin 3-galactoside, cyanidin 3-rutinoside, and malvidin 3-galactoside) and anthocyanins present in blackberry (14 and 750 µmol/L) and bilberry (88 µmol/L) extracts were compared after in situ gastric administration for 30 min. In agreement with results presented earlier, differential rates of absorption from the stomach between pure anthocyanin monoglycosides (around 25% for glucosides or galactosides) were observed; whereas absorption of cyanidin 3-rutinoside was lower (Talavéra et al., 2003).

Bilberry anthocyanins were also variably and efficiently absorbed (19–37%), where delphinidin glycosides were absorbed most intensively. Absorption of blackberry anthocyanins was not proportional to their concentrations applied: ingestion of high concentration of blackberry anthocyanins (750 µmol/L) lead to lower absorption of cyanidin 3-glucoside than in the case of 14 µmol/L extract. One possible reason for this discrepancy can be the potential toxicity of such high concentration of flavonoids (Halliwell, 2008), which will be discussed later. After administration of this very high concentration of anthocyanins present in blackberry extracts, parent forms were observed in plasma from gastric vein and aorta, whereas neither aglycones nor metabolites were detected (Talavéra et al., 2003). Furthermore, anthocyanin glycosides were quickly and rapidly excreted into bile as intact and metabolized forms. Only 20 min after ingestion, cyanidin 3-glucoside and its methylated form, peonidin 3-glucoside, was detected in bile together with few unidentified anthocyanin metabolites. As an important conclusion stands the fact that delphinidin glycosides were again more efficiently absorbed than others. Since delphinidin glycosides are the most potent antioxidants among anthocyanins present is bilberries, they could play one of the major roles for health promoting effects of bilberry extracts (Talavéra et al., 2003).

5.1.2 Metabolic Transformations and Excretion

Aforesaid facts are verified with experiments on bioavailability among humans when doses of 150 mg to 2 g total anthocyanins were given to the volunteers, generally in the form of berries, berry extracts, juices, or concentrates. In the meantime, around 1.5 h after intake, anthocyanins reached their maximum concentrations ranging from 10 to 50 nM (Manach et al., 2005). In humans, postprandial plasma levels of anthocyanins are low, around 0.1 µM or less, and most of the anthocyanins and their metabolites are excreted in urine during the first 2–4 h after ingestion. Cyanidin 3-glucoside from black raspberry is absorbed through the human small intestine wall mostly as the intact glucoside, which appears as such in the plasma and urine, along with a small amount of O-methylated metabolite. Other cyanidin 3-glycosides, 3-rutinoside, and 3-xylosylrutinoside, are metabolized and excreted in urine as O-methylated forms (Ichiyanagi et al., 2004). Only small fraction (between 1.5% and 5.1%) of the ingested anthocyanins was detected as parent forms in the urine within 12 h after consumption of 300 mL of wine containing around 218 mg of anthocyanins (Lapidot et al., 1998).

Most research confirmed such a pattern where only 32.5% of the anthocyanins excreted in the urine were the parent compounds, and the rest (67.5%) were detected as conjugated metabolites (Mazza, 2007). For example, after drinking of strawberry reach beverage several anthocyanins were identified in human plasma, of which pelargonidin sulfate, pelargonidin 3-O-glucoside, and pelargonidin 3-O-glucuronide were the most abundant (Edirisinghe et al., 2011). Detected maximal concentrations were rather low being 139±7 and 14±1 nmol/L for pelargonidin sulfate and pelargonidin 3-O-glucoside, respectively, after ingestion of a strawberry beverage containing 10 g of freeze-dried strawberry powder. To simplify metabolite detection, Felgines et al. (2002) kept rats on a blackberry-rich diet since blackberries have cyanidin 3-glucoside as the major anthocyanin form. After 8 days of anthocyanin supplementation they found that the vast majority of ingested cyanidin 3-glucoside is excreted in urine as intact and methylated glycoside form. In another report, more optimistic data are presented, showing that bioavailability of anthocyanins vary largely upon their type: cyanidin 3-sambubioside has 24 times less total bioaccessibility (0.75% of the intake) than cyanidin 3-sambubioside-5-glucoside and cyanidin 3,5-diglucoside bioaccessibility (17.84% of the intake; Lila et al., 2011). Other anthocyanins from maqui berries tested in the cited work had a relatively low total bioaccessibility (1.23–4.28% of the intake).

It was reported that proanthocyanidins are the least well-absorbed polyphenols, followed by galloylated tea catechins, and the anthocyanins, reaching final plasma concentrations usually below micromolar range (Manach et al., 2005). On the other hand, highly sensitive methods (such as HPLC-MS/MS) for the detection of anthocyanins and its known derivatives can be too narrow in means of selectivity, so additional anthocyanin metabolites, although present, could be neglected. Also, some UV–VIS methods can give lower results by hindering certain amounts of anthocyanins, because of aggregating or chemical reactions with other components of the plasma or urine (Manach et al., 2005).

5.2 BIOAVAILABILITY OF FLAVONOIDS

Current literature data point out that anthocyanins possess lower bioavailability in comparison to flavonoids. Moreover, only aglycone and some glucoside forms of flavonoids can be absorbed in the small intestine, whereas flavonoids linked to a rhamnose must be hydrolyzed before absorption by rhamnosidases from microflora present in the colon. Polyphenols linked to arabinose or xylose also has to be metabolized in the same manner, but up to date there is not enough supporting data regarding action of specific enzymes. As absorption occurs less readily in the colon than in the small intestine as a general rule, rhamnosides are absorbed less rapidly and less efficiently than their aglycones and corresponding glucosides (Hollman and Katan, 1999).

5.2.1 Quercetin

Mentioned facts about different rates of absorption can be demonstrated by investigations of quercetin bioavailability. Quercetin can be absorbed by humans from the digestive tract and detected in bile and urine as glucuronide and sulfate conjugates, with much difference in absorption between aglycones and corresponding glycosides (Moon et al., 2000). Gugler et al. (1975) found that less than 1% of quercetin was absorbed into the human body after the oral administration of quercetin aglycone. Glycosides of quercetin were more efficiently absorbed than quercetin itself, whereas rutin (quercetin 3-*O*-rutinoside) was less efficiently and less rapidly absorbed, being only 20% of quercetin glucoside (Manach et al., 2004). Furthermore, Sesink et al. (2001), indicated that quercetin was not absorbed as glycosides, hence quercetin glucuronides are the major metabolites present in human plasma. Another point of quercetin bioavailability is that the elimination of quercetin metabolites is quite slow, with reported half-lives ranging from 11 to 28 h and this could lead to its accumulation in plasma after repeated intakes. After several weeks of supplementation, quercetin plasma concentrations reached 1.5 μmol/L after 28 days of supplementation with high doses of quercetin of 1 g/day (Conquer et al., 1998). Another investigation found 0.63 μmol/L of quercetin in plasma after supplementation of 80 mg quercetin per day during 1 week (Moon et al., 2000). Furthermore, many reports indicate high individual variability in flavonoid absorption due to particular polymorphisms for intestinal enzymes or transporters (Hollman et al., 1999).

5.2.2 Catechins

Red grape and berry wine are some of the main sources of catechins, beside tea, various berries, and some fruits (e.g., apples). Although various forms of catechins have been designated for their health benefits, the mechanisms by which these flavonoids act are not well understood. Flavan 3-ols are more efficiently absorbed than quercetin and, as established for a rat's model system, catechin is further methylated and/or conjugated (Manach et al., 1999). Thus, comprehensive research of the bioavailability of the ingested catechin and its metabolites is essential for proper definition of the palette of its possible biological effects. Predominantly, catechins are rapidly eliminated from humans especially in the form of urine and bile. Furthermore, catechins and leucoanthocyanins interact with proteins, starch, and digestive enzymes to form less digestible complexes, resulting in reduced absorption of these nutrients (Kaume et al., 2011).

After ingestion, catechin and epicatechin are primarily present in plasma as metabolites such as conjugated and/or methylated forms. After consumption of 120 mL of red wine containing exactly 35 mg of catechin, the levels of catechin and its metabolites in plasma of five male and four female volunteers are measured (Donovan et al., 1999). To look up for the influence of alcohol on catechin absorption, regular and dealcoholized red wine were given to the volunteers on separate days. Before wine consumption, plasma levels of catechin and its conjugate 3′-O-methylcatechin (MC) were lower than 2 nmol/L. After 1 h, average levels of catechin and MC increased differently with respect to presence of alcohol in the wine: 91 ± 14 nmol/L (regular red) and 81 ± 11 nmol/L (dealcoholized). Also, within 1 h further modification of catechin occurred, so $21 \pm 1\%$ of the catechin metabolites were methylated and around than 98% of catechin and MC were present as glucuronide and sulfate conjugates. As a conclusion of their investigation, Donovan et al. (1999), suggest that catechin is present mainly in its metabolized forms in concentration insignificantly affected by the presence of ethanol.

Another research indicated different rates of urinary secretion of catechin and epicatechin (Baba et al., 2001). When rats were fed by epicatechin, urinary excretion of the total amount of epicatechin metabolites was significantly higher than the amount of catechin metabolites in the urine from rats fed by catechin. When rats were fed by mixture of both catechins, rates of urinary excretion were always lower than those measured for single catechin form administration. Presented results imply that, in rats, bioavailability of epicatechin is higher than catechin and, when given in mixture, catechins are absorbed in competitive manner (Baba et al., 2001).

When compared, quercetin and catechin have very different fate after ingestion of food containing 0.25% of catechin or quercetin. In plasma, quercetin is present mainly in methylated form in stable amount of 50 μM, whereas the concentration of catechin metabolites dropped from 38.0 to 4.5 μM between 12- and 24-h after an experimental meal (Manach et al., 1999). The concentrations of quercetin and catechin derivatives were lower in the liver than in plasma, and no accumulation of metabolites was observed after 2 weeks of flavonoid-supplemented diet (Manach et al., 1999). It is obvious that the health benefit potential of different flavonoids depends primary on the kinetics of their absorption (and elimination), and secondary on the degree of their metabolic modification.

5.2.3 Resveratrol

Coming around just to scratch complexity of wine matrices, bioavailability of resveratrol, as a very important stilbene compound, will be represented. Goldberg et al. (2003) made an experimental set up for the evaluation of the bioavailability of three different wine phenolics (*trans*-resveratrol, catechin, and quercetin). Flavonoids are given orally to healthy human subjects in three different media: white wine, grape, and vegetable juice in the following quantities: *trans*-resveratrol and catechin (25 mg/70 kg) and quercetin (10 mg/70 kg). Blood was collected immediately and at four intervals over the first 4 h after consumption, while urine was collected at zero time and for the following 24-h. As expected, and in accordance with the data for quercetin and catechin presented in this section, three polyphenols were present in serum and urine predominantly as glucuronide and sulfate conjugates. A smaller amount of resveratrol and catechin remain in the serum in free form, with observed values being only 1.7–1.9% and 1.1–6.5%, respectively, while a much higher percentage (17.2–26.9%) of free quercetin was observed. From all of the tested flavonoids, resveratrol was the most efficiently absorbed (urine 24-h excretion 16–17% of consumed dose) catechin absorption was the lowest (1.2–3.0%), while values for quercetin were somewhere in the middle (2.9–7.0%; Goldberg et al., 2003). Resveratrol gained its highest recorded (8.5 µg/L) concentrations in serum 30 min after consumption in all three matrices, with little difference in mean concentrations regardless of alcohol presence in matrix. In the serum, catechin needed more time (30–60 min) to reach it maximum concentration, with the greatest absorption rate when given in vegetable juice. Wine was the best matrix for absorption of quercetin, and the highest serum concentrations were recorded 30 min after uptake. Next, 2 h values had returned to baseline levels, and in the next 2 h unexplainably fell down to 50% below baseline levels (Goldberg et al., 2003). Also, their beneficiary effects should be under further evaluation since their serum concentrations of 10–40 nmol/L are inadequate to exhibit various antioxidant, anticancer, antiinflammatory, and antithrombotic effects, because verified EC_{50} values for these three polyphenols are much higher (in the range from 5 to 100 µmol/L).

Since all of the mentioned results confirm prevalence of polyphenols in conjugated and derivative forms, vast amounts of data reporting powerful in vitro anticancer and antiinflammatory effects of their *free* counterparts seem to be less significant (Goldberg et al., 2003).

5.2.4 *Action in the Gastrointestinal Tract*

Fruit wine consumption exhibited much stronger sanative effect in direct contact with ill tissue. Strong evidence is given through investigations, working on postprandial distribution of resveratrol in colorectal cancer tissue (Patel et al., 2010). Twenty patients with histologically confirmed colorectal cancer consumed eight daily doses of resveratrol (0.5–1.0 g), before surgical resection. As identified by HPLC/MS technique, parent compound and its metabolites resveratrol 3-*O*-glucuronide, resveratrol 4′-*O*-glucuronide, resveratrol 3-*O*-sulfate, resveratrol 4′-*O*-sulfate, resveratrol sulfate glucuronide, and resveratrol disulfate were identified. Of all detected metabolites, the highest mean values measured were for parent resveratrol and resveratrol 3-*O*-glucuronide in tumor tissue with values of 674 and 18.6 nmol/mL, respectively (Patel et al., 2010). Hopefully, this amount of resveratrol is close to 36 nmol/g, the mean concentration of resveratrol which accomplished reduction of Apc^{Min} adenoma number by 27% compared with mice on the control diet. Since resveratrol was predominantly detected in colorectal tissue the results described here suggest that daily intake of resveratrol between 0.5 and 1.0 g can provide more than sufficient levels in the human gastrointestinal tract to elicit pharmacologic effects (Patel et al., 2010).

Another important fact is that polyphenolics are, to some extent, associated with dietary fiber and as such have significantly lower bioavailability. High impact on the accessibility of the health benefit of polyphenol compounds is shown by the fact that 35–60% of total polyphenols in red grape wine and about 9% in white grape wine are associated with dietary fiber (Saura-Calixto and Díaz-Rubio, 2007).

6. HEALTH BENEFIT POTENTIAL OF DIFFERENT FRUIT AND BERRY WINES

First, although vast amount of data is present regarding many sanative effects of various berries and their products, it is not easy to find a comprehensive review about specific berry wine. Furthermore, comparison of wines from different types of berries has been seldom reviewed. Therefore, the main similarities and differences in bioactive characteristics among different fruit and berry wines will be pointed out. Phytochemicals present in fruit and berry wines can be defined as substances that may exhibit a potential for modulating human metabolism in a manner favorable for the prevention of chronic and degenerative diseases. As suggested by many researchers, moderate wine consumption may reduce the incidence of chronic diseases such as heart disease, hypertension, and metabolic disease, as well as gastrointestinal and cognitive disorders.

6.1 CARDIOPROTECTIVE POTENTIAL

During the last few decades, numerous studies present strong evidence suggesting that moderate wine consumption is associated with a reduced risk of cardiovascular disorders. Numerous studies have suggested that the consumption of polyphenol reach beverages such as cocoa, tea, and red wine can improve endothelial function in patients with manifested symptoms of cardiovascular disorders as well as in healthy volunteers without cardiovascular risk factors. Being closer to our topic, many detailed clinical trials have established a positive correlation between moderate consumption of grape wines and cardiovascular disease. Usually, red wines have been recognized as superior to white wines because of their higher TAC and TPC. This is supported by the fact that supplementation of white wine with grape polyphenolics yielded the same cardioprotective effects as the red wine (Aviram and Fuhrman, 2002). Moreover, one of the most clearly established benefits of moderate wine consumption relates to an odd situation in France where the lower death rate from cardiovascular diseases is observed in spite of higher intake of food with saturated fats. Lower incidence of cardiovascular disorders may be the possible outcome of traditionally high wine consumption, and the phenomenon has been described as a French paradox. Another description of the mentioned phenomena is that regular consumption of *red* wine has been proposed to be the most likely cause for the lower mortality (30–35% reduction) than in other industrialized countries, despite the high incidence of risk factors (Renaud and de Lorgeril, 1992). In the current literature, polyphenol compounds such as anthocyanins, catechins, and proanthocyanidins, stilbenes, and other phenolics are the main carriers of the red wine's beneficial effect (Dell'Agli et al., 2004). Several mechanisms of action have been proposed: improvements in vascular function, reduction of atherosclerosis, and potential reduction in total mortality (Lippi et al., 2010).

Normal vascular functions are mainly unbalanced by atherosclerosis, a form of chronic alteration of the arterial morphology. Development of atherosclerosis is characterized by modification of the functional properties of endothelial cells, oxidation of low density lipoprotein (LDL), migration of vascular smooth muscle cells (VSMC) from arterial media into intima, excessive proliferation of VSMC in the neointima, and increased extracellular matrix deposition (Lee et al., 2013a).

6.1.1 Vasodilatory Effect

When compared to red grape wines, the majority of blackberry wines are less potent vasodilators, but significantly better than white grape and one apparent sample of blackberry wine (Mudnic et al., 2012). Unexpectedly lower vasodilatory potential of that one sample of blackberry wine stands in positive correlation with its 10-fold lower anthocyanin content than in other blackberry wines tested, being more similar to white wine where no anthocyanins were detected. Intriguingly, its 3.5-fold higher TPC than in white grape wine didn't make it a more effective vasodilatator. Since wines characterized by the worst vasodilatory effect had at the same time the lowest TAC values, vasodilatory activity was linked with their anthocyanin, rather than polyphenol, content (Mudnic et al., 2012). The same research group tested various phenolic acids present in wine for vasodilatory activity and they estimated different values for hydroxybenzoic and hydroxycinnamic acids (Mudnic et al., 2010). Maximal vasodilatation was observed for vanillic acid followed by *p*-hydroxybenzoic, protocatechuic syringic, and gallic acid. Maximal vasodilatation induced by derivatives of hydroxycinnamic acid had caffeic acid followed by *p*-coumaric, ferulic, and sinapic acid (Mudnic et al., 2010). Detailed overview of presence of significant amounts of various phenolic acids, given in previous subsections, stands for fruit and berry wines as potential functional food for amelioration of cardiovascular disorders. Literature data for the action of wine phenolics as cardioprotective agents could sound even contradictive: investigations of vasodilatory activity suggest that monomeric catechins and simple phenols are devoid of effect (Dell'Agli et al., 2004). In the same review, more support is given for connotation of anthocyanins and oligomeric proanthocyanidins as vasoactive compounds, which is in accordance with findings of Mudnic et al. (2012).

6.1.2 Inhibition of Platelet Aggregation

Loss of proper endothelial function indicated by proliferation and migration of smooth muscle cells are one of the major factors in the onset of atherosclerosis. Nitric oxide synthesis by vascular endothelium is enhanced after wine consumption, thus promoting vasorelaxation and reducing platelet aggregation (Dell'Agli et al., 2004). Many wine phenolics, such as resveratrol, catechin, epicatechin, and quercetin, have inhibitory effects on platelet aggregation (Pace-Asciak et al., 1995). Furthermore, combined effect of several phenolics was shown to be superior to action of individual compounds (Wallerath et al., 2005). Thus, red wine exerts some direct inhibitory effects on platelet aggregation in vitro, and this effect should be attributed to the phenolic compounds and not to the alcohol itself (De Lange et al., 2003). Platelet aggregation is an important stadium in atherosclerosis development and it is shown that anthocyanin extracts possess acute and long-term beneficial effects on endothelial function as measured by Erlund et al. (2008). They found that berry consumption inhibited platelet function by 11% in comparison to a control group, with more cardioprotective effects such as reduced blood pressure and increased high density lipoprotein (HDL) concentrations, attained via the changes in nitric oxide metabolism. A similar vasoprotective effect of red wine consumption was observed (in vivo and in vitro), by the mechanism of wine polyphenol stimulated release of platelet aggregation inhibitor, prostacyclin (Schramm et al., 1997).

6.1.3 Prevention of Low Density Lipoprotein Oxidation

In addition to prevention of platelet aggregation, phenolic wine constituents can bind directly with LDLs obstructing their oxidation as one of the causes of atherosclerosis. Although associated with several independent factors, it is believed that majority of damage provoking atherosclerosis develops as a result of the oxidation of LDLs. It was shown that consumption of red, but not of white wine, resulted in the decreased oxidation of LDL in plasma. Proposed mechanism of action is oxidative protection of LDL by direct binding of quercetin via glycosidic ether bond (Aviram and Fuhrman, 2002). Thus, red wine consumption acts against the accumulation of oxidized LDL in lesions as a first line of defense and as a second line by removal of atherogenic lesions limiting the migration of smooth muscle cells into the intima of artery walls. Dharmashankar and Widlansky (2010) found that red wine polyphenols can bind to plasma LDL rendering them protected from oxidation by free radicals generated from polyunsaturated fatty acid–generated lipid hydroperoxides enzymatic cleavage (Spiteller, 2005). Other results from different clinical trials showed that red wine consumption, besides improving hypertension symptoms, raised HDL and lowered triglyceride levels, uninfluenced by the high sugar and ethanol content in wines (Opie and Lecour, 2007). The intrinsic cardioprotective effect of grapes has been confirmed by the result showing that grape powder reduced the atherosclerotic lesion area by 41% compared to the control or placebo mice (Fuhrman et al., 2005). Observed antiatherosclerotic effect was the consequence of several different actions: 8% reduction in serum oxidative stress; 22% increase in total serum antioxidant capacity; 33% reduction in macrophage uptake of oxidized LDL; and a 25% decrease in macrophage-mediated oxidation of LDL. Furthermore, some older results denote that red wine more efficiently inhibits the cell mediated oxidation of lipoproteins than white wine, still independent of ethanol content (Rifici et al., 1999). All of the presented data serve as an endorsement for the hypotheses that LDL oxidation is involved in early stages of atherosclerosis and that polyphenols behave as antioxidants preventing or delaying the onset of this infirmity.

6.1.4 Influence on Atherosclerosis Signaling Pathways

On the other hand, one critical point of view argues that even after chronic intake of berry polyphenols their plasma levels are too low to induce significant inhibition of platelet aggregation (Hubbard et al., 2003). Another remark is that the majority of results are not gathered after the wine intake but more from indirect in vitro experiments using rather high polyphenol concentrations. As presented in the bioavailability section, reported plasma levels of anthocyanidins and flavonoids are seldom higher than 0.1 μM, existing in their metabolized forms. Thus, in vitro experiments with parent forms used have arguable legitimacy for describing their real mode of action as excellently reviewed by Habauzit et al. (2014). Cardioprotective effects of such low concentrations of metabolites can be explained only if a completely different mechanism of action of wine polyphenols, like mediation and alternation of cell signaling and transcriptional patterns, are taken into account (Auclair et al., 2009). The main criticism of all of these cardioprotective effect examinations concerns the factual concentrations as well as authentic metabolic forms of polyphenols present in plasma after wine consumption. Using the transcriptomic approach, new insights into the mechanisms underlying the effects of dietary polyphenols in vascular protection are provided. Mauray et al. (2009) reported that a nutritional supplementation with a bilberry anthocyanin-rich extract attenuates atherosclerotic lesion development in apo E$^{-/-}$ mice, which spontaneously develop human-like atherosclerosis. They found that bilberry anthocyanin-rich extract altered the expression of numerous aortic genes encoding proteins that are involved in oxidative stress, inflammation, transendothelial migration, and angiogenesis. It is very important to emphasize that the observed action of anthocyanins on a transcriptional level are effective in much lower plasma concentrations than in the case of direct antioxidative and antiinflammatory mechanisms. In a subsequent report they confirmed that bilberry extracts supplementation significantly improved hypercholesterolemia, whereas the plasmatic antioxidant status remained unchanged (Mauray et al., 2012). Detailed nutrigenomic analysis identified 1261 genes which expression was modulated by bilberry extracts in the aorta. Further bioinformatic analysis revealed that these genes are implicated in different cellular processes such as oxidative stress, inflammation, transendothelial migration, and angiogenesis, processes associated with atherosclerosis development/protection. As an example, polyphenols may prevent cardiovascular disorders by decreasing expression of genes regulating cell adhesion, thus discouraging the incorporation of inflammatory monocytes from blood into the endothelium (Choi et al., 2004). Hence, anthocyanins acting like signal molecules and transcriptional modulators initiate whole cascades of cardioprotective processes, without direct interference with target molecules, like LDL or tissue. Moreover, they confirmed feasibility of such mechanisms in animal models of atherosclerosis, where atheroprotective effects are induced by several fruit phenolics (catechin, anthocyanins, or naringin; Auclair et al., 2009). They reported that these fruit phenolics induced changes in the expression of genes related to inflammation and atherogenesis and a number of other regulatory genes involved in processes that control the initial steps of atherosclerosis development.

These results are the first in vivo evidence of action of a dietary polyphenol administered at a low nutritional level by the modulation of the expression of several genes that may play a key role in the prevention of atherosclerosis. Since all of the investigations have been carried out at the gene expression level only, a functional genomics approach should be used to select most important genes that are involved in the onset of atherosclerosis.

6.1.5 Enhancement in Nitric Oxide Production

From all of its potential health benefit effects, resveratrol is best known as one of the most studied (and praised) cardioprotective nutritional supplements. Leaving its direct antioxidant effect aside, resveratrol action is perceptible by the enhanced synthesis and release of nitric oxide (powerful vasodilator) by endothelial cells. A mechanism of action is studied on the molecular level and it is observed that $0.1 \, \mu M$ of resveratrol significantly increased the expression of genes encoding endothelial NO synthase, responsible for synthesis of the vasodilator molecule NO, and even simultaneously decreased expression of a few vasoconstrictor genes (Nicholson et al., 2008). Of all the other polyphenols tested, only quercetin displayed similar effects to resveratrol and ferulic acid had no effect (Nicholson et al., 2008). These presented facts are supported by the results demonstrating that resveratrol performs cardioprotective effects in a twofold manner: by the inhibition of platelet aggregation and by its antioxidant effects on cholesterol metabolism (Das and Das, 2010). However, it is necessary to emphasize the significant lack of data concerning resveratrol content and cardiovascular protective effects of fruit and berry wines, in general. Occasionally, some data is available about berry products other than wine, such as cranberry juice, which shows favorable effects on cardiovascular risk factors (Basu and Lyons, 2011), and grape juice, but not orange or grapefruit juice, which inhibits platelet activity in dogs and monkeys (Osman et al., 1998). Investigation made by Osman et al. (1998) is particularly interesting since it demonstrates that higher flavanol content in grape products is responsible for more perceived cardioprotective effect than one estimated for these two citrus products.

Anthocyanins are denoted for their beneficial effects on preventing cardiovascular disorders, alleviating symptoms of microcirculation diseases that result from capillary fragility and by preventing cholesterol-induced atherosclerosis. Grape seed's proanthocyanidins enhanced NO generation resulting in endothelium dependent relaxation of blood vessels by interaction with nitric oxide synthase (Kruger et al., 2014). High amounts of anthocyanins present in berry wines should, without any doubt, share the same beneficial properties as just mentioned for red grape wine.

6.2 ANTIOXIDATIVE EFFECTS

Abundant literature data show that reactive oxygen species (ROS) are involved in the early stages of development of diverse pathological processes, cardiovascular diseases, certain types of cancer, inflammation, neurological disorders, diabetes, and more (Halliwell, 2005, 2009). They are generated as a result of normal cell metabolism, and the fine-tuning of free radical generation/quenching balance in tissues is known as antioxidative homeostasis. Redox homeostasis is maintained by antioxidant defense systems, which can be enzymatic (superoxide dismutase, catalase, glutathione peroxidase), and nonenzymatic (glutathione), or dietary (vitamins A, C, and E, and polyphenols). Different external causes (radiation, heavy metal intoxication, other toxins, drugs) can trigger extensive and uncontrolled ROS production. ROS, as highly reactive and mobile (OH· as the most reactive oxygen centered free radical is uncharged and can easily pass through membranes) molecules damage lipids, proteins, DNA, or react with many small metabolites. Besides direct interaction with target molecules, ROS can also trigger a set of redox-sensitive signaling pathways that can have very adverse effects by activating a whole cascade of reactions.

6.2.1 Free Radical Depletion Mechanisms

Interaction of phenolic compounds from fruit and berry wines with free radicals is generally described as radical termination reactions. It is believed that the first step in the reaction of polyphenols with ROS (or other free radicals) is the formation of resonance stabilized phenoxy radicals, usually by H-atom transfer. Subsequently, generation of more stable radical form is making them capable of disrupting free radical reaction propagation, such as lipid peroxidation. In general, wine's phenolic compounds belong to a group of excellent hydrogen or electron donors for generation of the mentioned moderately stable phenoxy radical intermediates, thus exhausting further progress of the harmful free radical chain reaction. Such a mechanism can be responsible for the inhibition of the membrane peroxidation or lipoprotein oxidation observed in various model systems upon treatment with wine extracts or phenolics (Fuller et al., 2011). It was reported earlier that analysis of correlation of the relative LDL antioxidant activity with the concentration of individual phenolic constituents served as an estimation of their sanative potential so that they can be sorted in decreasing order: gallic acid, catechin, myricetin, quercetin, caffeic acid, rutin, epicatechin, cyanidin, and malvidin 3-glucoside. Accordingly, the capacity to protect LDL from oxidation appears to be distributed widely among a large number of phenolic constituents in wine. On the other hand, the resveratrol content had no relation to this antioxidant activity (Frankel et al., 1995).

There are numerous experimental approaches for verification of polyphenol antioxidant properties and revealment of mechanisms concerning the interaction between polyphenols and target tissues. Some excellent review papers illustrate that more than three-fourths of all published data about wine health benefit is given for cardiovascular and oncological disorders (Biasi et al., 2014). The same review showed that 53% of all experimental setups for investigation gastrointestinal disorders are performed as various in vitro tests. As discussed before, the main disadvantage of such an approach is their almost impossible extrapolation to in vivo conditions mainly because the concentrations of phenolic antioxidant compounds used are often a few orders higher than those measured in situ (Frankel and German, 2006). Antioxidants present in wine are rarely present in their parent form in plasma, since they are quickly metabolized and many of these metabolites have been shown to be poor in vitro antioxidants compared with the aglycones or parent compounds. A good example of this is that O-methylation of hydroxyl groups of flavonoids, as one of the common metabolic transformations of wine phenolics, significantly decreases their radical scavenging capacity (Rice-Evans et al., 1996). Furthermore, misleading results from experiments involving direct exposure of cancer cells to polyphenol-rich beverages and extracts can be attributed to the cytotoxic effect of H_2O_2 generated from reactions of polyphenols with cell media (Long et al., 2000).

6.2.2 Methods for the Estimation of Antioxidative Potential

The antioxidant activity of wines has been studied using several free radical scavenging assays such as the trolox equivalent antioxidant capacity (TEAC), 2,2'-azino-bis(3-ethylbenzothiazoline-6-sulphonic acid) (ABTS) radical, 2,2-diphenyl-1-picrylhydrazyl (DPPH) radical, bulk lipid (MeLo), and superoxide anion radical assay. Less abundant are data from lipid peroxidation and LDL oxidation assay used mainly for specific experimental setups. It is easy to find excellent textbooks (Nakajima et al., 2014) and reviews (De Beer et al., 2002) about all of the tests used for investigation of wine, must, or berry/fruit extract, so they will not be discussed in detail.

On the other hand, another very potent biophysical method, electron spin resonance spectroscopy (ESR), for direct assessment of TAA of investigated matrix becomes more extensive in functional food assessment. Since free radicals possess paramagnetic characteristics due to their odd electron, they can be registered in a manner similar to the widespread nuclear magnetic resonance technique. Thus, the main advantage of electron spin resonance spectroscopy is the direct registration of occurrence, scale, and type of specific radical form present or generated de novo. Furthermore, using spin trapping as additional modification, it is possible to distinguish between apparent ROS radicals such as hydroxyl and superoxide anion radical (Bacic and Mojovic, 2005). Free radical quenching by wine antioxidants, i.e., attenuation of their corresponding signals in ESR, will be directly recorded, and by this, possible uncontrolled (unwanted) interference of complex matrix such as wine and assay components (such as DPPH) will be avoided. Results from ESR are more informative since they can reveal possible spots for generation or quenching of free radicals during the interaction of investigated matrix with specific target. Also, the sample doesn't have to be in a liquid state for ESR measurements, so even the small tissue filaments can be tested for changes in antioxidant status. On the other hand, a main drawback of the ESR method is expensive equipment and need for trained personnel, but the same was proclaimed for HPLC-MS/MS methods some years ago.

6.2.3 Differences in Antioxidative Efficiency

In scientific literature, as well as in other popular resources, the antioxidant capacity of grape wines is often placed in descending order: red, rosé, and white wine. In their excellent paper, Heinonen et al. (1998) determined that extracts of dealcoholized fruit and berry wines (made of mixtures of black currants with crowberries or bilberries) or mixtures of black and red currants possess antioxidant activity in a MeLo model. Apart from wine mixtures, raw materials such as apple, arctic bramble, cowberries, cranberries, or rowanberries show antioxidative capacity as well. Following inhibition of formation of MeLo hydroperoxides acquired by addition of different wine extracts were measured for: control red wine (90%), mixture of black currants and crowberries (98%), cranberries (92%), mixture rowanberries and apple (90%), apple (84%), and rather low value (20%) for cranberry wine (Heinonen et al., 1998). These results clearly indicate superiority of black currant, crowberry, and bilberry wines to other fruit and berry wines, being even slightly better than the red grape wines. These results are supported by findings showing that wines from black fruits, such as blueberry, have been shown to have similar TAA values to red wines (Sánchez-Moreno et al., 1999). In accordance with their high TPC and TAC values presented in composition section, dark berries (blackberry, blueberry, blackberry and cornelian, and black cherry), were found to have similar TAA values to those of black wine grapes when tested by the ferric reducing antioxidant power (FRAP) method (Pantelidis et al., 2007). In another report, a similar pattern of distribution of TAA among different fruit and berry wines is observed, hence they can be ranked from the highest to the lowest TAA values as follows: bilberry > blackberry > black mulberry > sour cherry > strawberry > raspberry > apricot > quince > apple > melon (Kalkan Yildirim, 2006).

Almost the same pattern was observed for their corresponding source fruits and berries TPC indicating that the polyphenol abundance is the basis of their successor wine's TAA. Furthermore, verity that the antioxidant activity of any wine is almost exclusively dependent on its phenolic composition was also confirmed by Heinonen et al. (1998), presenting that white wine, due to low TPC of 265 mg/L GAE, had a negligible effect on the oxidation in the MeLo model system. Since deprived of anthocyanins, white wine's hydroxycinnamic acid derivatives and flavonols were specified to be responsible for any observed antioxidant capacity (Makris et al., 2003). Also, most of the fruit and berry wines with a phenolic content <600 mg/L GAE (cloudberries, red raspberries, and strawberries) did not markedly inhibit the oxidation of MeLo. Heinonen et al. (1998) concluded that strong colored and relatively small size berries with a tough skin such as bilberries, blackcurrants, cowberries, cranberries, and crowberries exert significant antioxidant activity.

Another proof that the antioxidant potential of wines is predominantly dependent of phenolic and anthocyanin content is given in comparison of TAA measured in commercially available blueberry and blackberry wines by TEAC assay. Regularly used, TEAC assay designate TAA of wine expressed as TEAC as a measurement based on comparison of unknown matrix with trolox (hydrosoluble analog of vitamin E) antioxidant capacity, expressed in units of trolox equivalents (TE), commonly per liter. TE values are used to standardize antioxidative capacity estimations of complex mixtures (such as fruit and wine), thus making them more comparable. However, the complexity of wine as a matrix often raises a need for the application of two or even more different assays to achieve appropriate results. Average estimated TAA of blackberry wines was not significantly higher than the average one of the blueberry wines (26.39 and 21.21 mmol TE/L, respectively), with strong positive correlation ($r=0.88$) between TPC and TAA, and a positive correlation ($r=0.55$) with TAC (Johnson and Gonzalez de Mejia, 2012). The considerable high antioxidative potential of berry wines is confirmed by another common TAA assay, FRAP. In this assay, TAA is evaluated as matrix reduction potential of Fe^{3+} to Fe^{2+}, which is in turn chelated by 2,4,6-tripyridyl-s-triazine (TPTZ) forming dyed complex with absorption maxima around at 590 nm. TAA of two blackberry wines (15.8 and 13.9 TE/L) were slightly higher than maximal TAA measured for red wine (12.8 TE/L), being at the same instance more than 10 times above TAA of white wines (1.9 TE/L), proving their inferiority regarding antioxidative protection (Mudnic et al., 2012). Another frequently used assay (ABTS) measures the inhibition of formation of $ABTS^+$ cation radical by antioxidant components present in wine, or other matrix. Furthermore, estimated TAA can be normalized to aforementioned trolox or ascorbic acid equivalent units, for easier comparison with other literature data. Investigation of the antioxidant activity of the black raspberry wine extract by ABTS test (IC_{50} 374–407 µg/mL) prove their similarity with antioxidant capacity of red grape wine (Jeong et al., 2010). Taking into account the lower TPC of blackberry wines in comparison with red grape wines, their higher TAA values were described as a consequence of much higher amounts of gallic acid present in blackberry wines in comparison to their grape counterparts. The noteworthy antioxidative benefit potential of gallic acid in vitro is previously confirmed in research by the same group (Mudnic et al., 2010). Not less significant antioxidative potential has been confirmed for elderberry wine also, with increased TAA (9.95 mM TE/L) in comparison to its must (8.18 mM TE/L), although with decreasing trend during 3 years of wine aging (6.13 mM TE/L; Schmitzer et al., 2010). Since the TAA values measured by the same test for red grape wine were from 5.03 to 7.73 mM TE/L it is clear that elderberry wine can provide significant health promotion advantages.

As mentioned a few times before, the final phenolic composition of fruit and berry wines is defined by the type and physiological state of berry as well as the vinification process used. It is essential to evaluate the presence of individual components in different wines, as well as from different vintages, in an attempt to reveal their possible transformation during aging (Frankel et al., 1995).

Fruit wines were also investigated, and significant TAA of pomegranate wine, as well as its must, was estimated by using both ABTS and DPPH in vitro assays. Antioxidant activity of pomegranate wine was similar or even higher than one for red grape wine, with a significant decrease in its value during wine aging, analogous to berry wines (Mena et al., 2012). Furthermore, antioxidant capacity of peach wine was estimated using two methods (DPPH and FRAP) and reported to be 1.55 ± 0.09 and 3.01 ± 0.12 (mMTE/L), respectively (Davidović et al., 2013). In another report, peach wine had two times lower TAA in comparison to plum wine, 395 and 618 mg/L ascorbic acid equivalents, respectively (Rupasinghe and Clegg, 2007).

More exotic fruit wines have certain antioxidant potential, measured by DPPH radical quenching (in percent) as another assay for TAA estimation. As one of the most frequently used colorimetric assays, the DPPH test deserves a very short description. Likewise, previous assays, DPPH method is based on the spectrophotometric measurement of the DPPH concentration change as a result from the reaction with antioxidant: DPPH radical has a deep-violet color in solution (since it has strong absorption at about 520 nm) and becomes colorless upon the reaction with tested antioxidant. So, transmittance of the solution is directly proportional to the antioxidative capacity (i.e., DPPH radical quenching) of the tested compound or matrix. Although very widely used, some concerns regarding the applicability of this assay has been reported (Xie and Schaich, 2014). Custard apple wine attained more efficient DPPH quenching (36.8%) than wines from pineapple (35.61%), lime (20.1%), tamarind (15.7%), garcinia (15.4%), rambutan (15.1%), and star gooseberry (14.8%; Jagtap and Bapat, 2014). In the same report, antioxidant capacity of custard apple wine was demonstrated by the threefold reduction in the number of strand breaks (induced by γ radiation) in custard apple wine–treated plasmid DNA, in comparison to the control plasmid DNA.

6.2.4 Direct Antioxidative Effect on Targeted Tissue

Polyphenol concentrations in the colon can reach several 100 μM, and they play an important local role in the removal or conversion of potential toxins as well as carcinogens, acting as extracellular antioxidants. Boban and Modun (2010) concluded that another aspect of wine-related increase in plasma antioxidant activity is the antioxidative effect of unabsorbed polyphenols that remain in gastrointestinal system. Due to their sizable concentrations, they can scavenge free radicals, prevent lipid peroxidation, or oxidize different antioxidant compounds. In this manner, although acting locally, the polyphenols could enhance plasma concentration of various antioxidants, which in turn influences plasma TAA. The colon is particularly exposed to oxidizing agents, because the majority of other dietary antioxidants (vitamins C and E) are absorbed in the upper parts of the intestinal tract. It has been shown that after 1 h of ingestion of 300 mL of red wine serum antioxidant capacity increased by 18%, which is comparable with a 22% increase after ingestion of 1 g of vitamin C (Whitehead et al., 1995).

6.2.5 Antiinflammatory Effect

The antiinflammatory effect of red wine polyphenols is obtained by direct radical scavenging as well as by suppressing the expression of specific inflammation-related genes involved in cellular redox signaling (Biasi et al., 2014). Thus, the first instance of the protective effects of wine phenolics is prevention of various intestinal diseases caused by inflammation triggered by oxidative stress. In addition, the importance of wine polyphenols has recently been stressed for their ability to act as prebiotics maintaining optimal conditions for perseverance of gut microbiota. Biasi et al. (2014) highlighted results from studies on intestinal cell lines or on experimental animal models showing that wine compounds are able to reduce colonic injury by protecting intestinal redox homeostasis, hence preventing inflammation. Also, polyphenols as good metal chelators can suppress free radical generation catalyzed by transition metal ions (Fe^{2+}, Cu^+, Mn^{2+}), during the Fenton or Haber Weiss reactions (Lopes et al., 1999). Red wine extract (60.7% epicatechin, 18.1% catechins, 17.9% epigallocatechin, and 3.3% epicatechin 3-O-gallate), forming chelate complexes with ferric and ferrous ions, showed potent inhibition of oxidative DNA damage independent of degree of polyphenol polymerization in wine (Lodovici et al., 2001).

6.2.6 Cumulative Antioxidative Effect

For all these results, one can inquire: does the antioxidant activity of tested polyphenols differ markedly when tested separately and in their mixtures, and is it influenced by the type of the assay used? Validity of such a doubt is confirmed by the results showing that the antioxidant activity of each polyphenol is not a constant value as measured individually and when other substances are present (and interact) with this polyphenol in the same antioxidant assay (Kurin et al., 2012). Also, significant discrepancies can be observed by reviewing results of antioxidant activity of individual compounds from different assays since their apparent activities significantly varied upon their structure and by the mechanism of the assay. For example, resveratrol showed the lowest inhibitory activity in each assay used, except the ABTS where it had better results than caffeic acid. Estimated antioxidant activity values for three, maybe mostly studied, wine phenolic compounds, sorted them in following descending order: quercetin > caffeic acid > resveratrol (Kurin et al., 2012). Furthermore, for suitable evaluation of differences in TAA values gathered by different tests, factors such as solubility and partitioning between different phases has to be taken into account. Red wine polyphenols have been reported to show antioxidant activity in water-rich conditions such as in oxidation of human LDL and in an aqueous β carotene-linoleate system (Kanner et al., 1994). Nevertheless, wine extracts are also effective in hydrophobic lipid systems; LDL isolated from plasma treated with red wine procyanidins possesses increased resistance to oxidation in terms of lipid peroxidation (Lourenço et al., 2008).

In addition, possible synergistic or antagonistic interactions of phenolic compounds need to be considered in very complex and divergent mixture such as wine. Therefore, the TAA of the phenolic compounds may not fully correspond to the sum of individual antioxidant capacity estimated for the antioxidants present in the same matrix.

The magnitude and diversity of the antioxidative effects of wine polyphenols indicate existence of additional mechanisms working beyond the direct modulation of oxidative stress. Alongside ROS scavenging, antiinflammatory actions due to modulation of signaling pathways and gene expression as well as epigenetic phenomena are reported. As pointed out in many reviews, the overall protective effect of polyphenols is due to free radical scavenging, metal chelation, enzyme and cell signaling pathways modulation, and gene expression effects (Rodrigo and Libuy, 2014). Similar rationalization of additional mechanisms for CVD prevention can be given since recent findings revealed that polyphenols could interact with cellular signaling cascades regulating the activity of transcription factors that regulate the expression of genes involved in antioxidative protection. A large number of studies have described, in both in vitro and in vivo models, the numerous effects of procyanidin-rich extracts on cell signaling. Despite providing significant new insight in definition of the health benefits of flavanols and procyanidins, they have very limited value when mechanisms of action are discussed, lacking the convincing identification of molecules responsible for the observed effects (Fraga and Oteiza, 2011). Presented evaluation of antioxidative potential of berry wines should persuade the wider population to take advantage of available blueberry and blackberry wines as potential functional food with health benefits.

6.3 PREVENTION OF VARIOUS TYPES OF CANCERS, OR IT'S SUPPRESSION

Literature is almost devoid of data concerning antitumor activities of fruit and berry wines. Lots of reports are made by testing specific components present in wines for their specific bioactivity. So, besides giving few available data of direct testing of nongrape wines on amelioration of cancer disorders, the brief overview of facts supporting health benefit effects of particular compounds has been presented.

Reducing the incidence of colorectal cancer by moderate red wine consumption has been related to the actions of its polyphenols on colon cancer epithelial cells. It was observed that direct action of $50\,\mu M$ of pure wine polyphenols (quercetin, myricetin, laricitrin, and syringetin) as well as 50 of red wine extract is capable of inhibiting the proliferation of colorectal epithelial adenocarcinoma cells. Direct cytotoxic activity of flavonols and their ability to disrupt cell cycle at the end of G2 phase are considered as the main mechanisms that reduce tumor growth (Gómez-Alonso et al., 2012). All of these effects could be possible if the prospective concentration of red wine in the colon reaches $50\,\mu g/mL$. Gómez-Alonso et al. (2012) endorse such outcome by extrapolating phenolic extraction yield (26.5 mg of polyphenols per 125 mL of wine) and a volume of chyme (0.5 L) reaching the large intestine after a meal. Thus, after ingestion of 125 mL of red wine with a meal, $26,500\,\mu g$ of polyphenols solubilized in 500 mL of chyme gives approximately $50\,\mu g/mL$ of wine polyphenols. It was reported that cloudy apple juice was more effective than clear juice or phenolic extracts in the modulation of 1,2-dimethylhydrazine-induced colon cancers in rats (Barth et al., 2007). This implies that the colloidal suspension of proteins, fatty acids, cell wall polysaccharides, pectins, polyphenols, and a host of other compounds, has a greater protective effect than polyphenols alone.

The antiproliferative potential of the black raspberry extracts and wine was shown by the suppression of proliferation of both colon and prostate cancer cells (Jeong et al., 2010). Investigations took account of the presence of crushed seeds in raspberry wine, and measured antiproliferative effects of wines with or without seed were not found to be significantly different. On the other hand, 60% ethanol extracts of black raspberry fruits showed much lower influence on antiproliferation of the prostate cancer cells than the same extracts with crushed seeds. These results suggest that ethanol extraction and/or inclusion of the seeds in the black raspberry juice, but not wine, may increase the antiproliferation effect on the cancer cells (Jeong et al., 2010). Seeram et al. (2006) also demonstrated that the black raspberry fruit extracts inhibit proliferation of human colon and breast tumor cells strongly suggesting black currant extracts as a health benefit food/supplement.

These results indicate that the observed inverse correlation between development of different types of cancer and a moderate red wine intake may be partly mediated by the actions of red wine phenolic constituents. As said before, the wholesome biological effect of wine cannot and should not be attributed to a single phenolic compound. It is much more probable that the mutual actions of different polyphenolics are responsible for various health benefit effects observed in treating different disorders.

6.4 COGNITIVE SUPPORT

Polyphenol compounds found in fruit and berry wines, in particular flavonoids, have been associated with health benefits including improvement in cognition and neuronal function with aging. Although there is no data for the cognitive beneficiary effects of fruit wines, there are only a few reports of sanative action of similar matrices, such as grape juice. Investigation on Concord grape juice supplementation has shown, next to already reviewed wine sanative effects, reduced risk for dementia. In addition, preliminary animal data have indicated improvement in memory and motor function with grape juice supplementation, suggesting its potential for cognitive benefit in aging humans (Krikorian et al., 2010). Twelve older adults with diagnosed memory decline (but not dementia) were included in a randomized, placebo-controlled, double-blind trial with Concord grape juice supplementation for 12 weeks. Significant improvement in a measure of verbal learning and nonsignificant enhancement of verbal and spatial recall was observed in a group supplemented for with 100% Concord grape juice. Further survey of cognitive protecting effects during 16 weeks of Concord grape juice administration demonstrated that participants who consumed grape juice had reduced semantic interference on memory tasks. Also, greater activation in anterior and posterior regions of the right hemisphere was also observed with functional magnetic resonance imaging in the Concord grape juice–treated subjects (Krikorian et al., 2012). Having in mind that grape juice and apparent fruit wines share many bioactive compounds it is very likely that some fruit wines could exhibit a similar effect.

The complexity of model systems and the undefined places of engagement lead to contradictory results. It was found that there was no significant difference in neuroprotective effects between the regular red wine and the red wine supplemented with a tenfold concentration of resveratrol on specific cell lines. The contribution of resveratrol to the antioxidant activity of red wine was less than other tested polyphenols, so it was suggested that resveratrol may be negligible with respect to health benefits of red wine (Xiang et al., 2014). On the other hand, other reports show that cerebral ischemic damage in the rat brain was prevented by pretreatment with resveratrol (Della-Morte et al., 2009).

The importance of grape wine experiments for our topic could be supported by the argument that although (to our best knowledge) no corresponding data is available for fruit and berry wines, there are a number of papers confirming the benefits obtained by berries itself (Seeram, 2011; Miller and Shukitt-Hale, 2012; Tavares et al., 2012). Results showing that the same phenolic compounds, present in grape and nongrape wines, are responsible for neuro/cognitive protection effects provide enough support for the hypothesis that consumption of nongrape wines offer similar sanative benefits.

6.5 PREVENTION OF GASTROINTESTINAL DISORDERS

According to the site of action, two general modes of in vivo action of wine phenolic compounds are present: preabsorption and postabsorption action. For direct gastrointestinal sanative effects, preabsorption action (where polyphenols play their beneficial roles prior to absorption or excretion) is more important. This mode of action is thought to be most likely in the large intestine, where levels of phenolic substances can be comparatively high. Wine components have been proposed as an alternative natural approach to prevent or treat inflammatory bowel diseases. The difficulty remains to distinguish whether these positive properties are due only to polyphenols in wine or also to the alcohol intake, since many studies have reported ethanol to possess various beneficial effects. To date, the real function of wine components in managing human intestinal inflammatory diseases is still quite limited, and further clinical studies may offer more solid evidence of their real beneficial effects (Biasi et al., 2014).

6.6 OTHER HEALTH BENEFITS

Besides antioxidative health benefits of moderate wine consumption, there are other bioprotective effects like the maintenance of cell membrane integrity and function, DNA integrity, antiinflammatory, and antiseptic effects. Polyphenols possess certain antibacterial, antifungal, and antiviral activities. Wine inhibited microbial, especially *Escherichia coli*, growth, and the inhibition increased as the polyphenol concentration increased, and clarified wines were inactive against all bacteria tested (Xia et al., 2014). The antimicrobial activity and the phenolic composition of the tested white and red wine extracts indicate that some phenolic acids have the potential to inhibit growth of certain pathogens such as *Staphylococcus aureus*, *E. coli*, and *Candida albicans* strains (Papadopoulou et al., 2005). All of the mentioned investigations suggest that polyphenolic compounds contained in red wines were responsible for the antimicrobial effects. Also, it was reported that 25% solution of apple wine inhibited growth of *Candida utilis*, *Aspergillus niger*, *S. aureus*, *E. coli*, and *Bacillus subtilis*. Antimicrobial effects were positively correlated with ethanol content, but addition of spices extract, as well as honey, further enhanced the antimicrobial activity of wine (Joshi and John, 2002). Some reports depict modified apple wine (with garlic, anola, mentha, ginger, and hop extract) as a potential functional food with medicinal properties (Joshi and John, 2002; Joshi et al., 2014).

Effect of berry consumption on modulating symptoms of metabolic syndromes as a prologue for serious disorders such as obesity or type 2 diabetes are well documented (Basu and Lyons, 2011). Several clinical trials confirmed that phenolic compounds from berries, especially strawberries, blueberries, and cranberries, may alleviate initial features of the metabolic syndrome, principally via antioxidant, antihypertensive, and antiatherosclerotic activity. Results from population-based study made in Sweden indicated that moderate wine drinkers exhibited a more favorable general health pattern according to both lifestyle factors and metabolic parameters (Rosell et al., 2003). On the other hand, alcohol consumption represents an additional source of calories and contributes to obesity incidence because alcohol consumers usually add alcohol to their customary daily energy intake. Results obtained in this study made on a rather large Mediterranean cohort are very supportive for wine drinkers: from all of the tested alcoholic beverages, only the wine drinking group has not expressed significant association between wine consumption and the incidence of metabolic syndrome onset (Barrio-Lopez et al., 2013).

7. CONCLUSIONS

After all aforementioned, fruit and berry wines have been deservedly placed in the focus of very diverse investigations with the joined goal to link up their composition with their health benefits. Fruit wines undoubtedly possess a broad spectra of sanative potentials that are the result of interactions of their various bioactive ingredients. Variation in fruit sources and winemaking procedures make characterizations of specific wine with apparent bioactive potential rather unreliable. Classifying fruit wines as functional food is not valuable unless very detailed analysis of overall composition has been performed. Minute variations in potent components such as phenolic compounds can have a much larger effect in prevention and treatment of particular disorders. Thus, a very intriguing task is given to various profiles of researchers whose mutual efforts can help the general public to use the utmost of health benefits provided by moderate consumption of fruit and berry wines.

REFERENCES

Agar, I.T., Streif, J., Bangerth, F., 1997. Effect of high CO_2 and controlled atmosphere (CA) on the ascorbic and dehydroascorbic acid content of some berry fruits. Postharvest Biology and Technology 11, 47–55.

Andreu-Sevilla, A.J., Mena, P., Martí, N., García Viguera, C., Carbonell-Barrachina, Á.A., 2013. Volatile composition and descriptive sensory analysis of pomegranate juice and wine. Food Research International 54, 246–254.

Anttonen, M.J., Karjalainen, R.O., 2006. High-performance liquid chromatography analysis of black currant (*Ribes nigrum* L.) fruit phenolics grown either conventionally or organically. Journal of Agricultural and Food Chemistry 54, 7530–7538.

Archibald, D., 1997. Fruit and berry wines. In: Vine, R., Harkness, E.M., Linton, S.J. (Eds.), Winemaking. Springer, New York, pp. 261–269.

Attri, B., 2009. Effect of initial sugar concentration on the physico-chemical characteristics and sensory qualities of cashew apple wine. Natural Product Radiance 8, 374–379.

Auclair, S., Milenkovic, D., Besson, C., Chauvet, S., Gueux, E., Morand, C., Mazur, A., Scalbert, A., 2009. Catechin reduces atherosclerotic lesion development in apo E-deficient mice: a transcriptomic study. Atherosclerosis 204, e21–e27.

Aviram, M., Fuhrman, B., 2002. Wine flavonoids protect against LDL oxidation and atherosclerosis. Annals of the New York Academy of Sciences 957, 146–161.

Aydin, N., Kadioglu, A., 2001. Changes in the chemical composition, polyphenol oxidase and peroxidase activities during development and ripening of medlar fruits (*Mespilus germanica* L.). Bulgarian Journal of Plant Physiology 27, 85–92.

Azhu Valappil, Z., Fan, X., Zhang, H.Q., Rouseff, R.L., 2009. Impact of thermal and nonthermal processing technologies on unfermented apple cider aroma volatiles. Journal of Agricultural and Food Chemistry 57, 924–929.

Baba, S., Osakabe, N., Natsume, M., Muto, Y., Takizawa, T., Terao, J., 2001. In vivo comparison of the bioavailability of (+)-catechin, (−)-epicatechin and their mixture in orally administered rats. The Journal of Nutrition 131, 2885–2891.

Bacic, G., Mojovic, M., 2005. EPR spin trapping of oxygen radicals in plants: a methodological overview. Annals of the New York Academy of Sciences 230–243.

Barceló, A.R., Pomar, F., López-Serrano, M., Pedreño, M.A., 2003. Peroxidase: a multifunctional enzyme in grapevines. Functional Plant Biology 30, 577–591.

Barrio-Lopez, M.T., Bes-Rastrollo, M., Sayon-Orea, C., Garcia-Lopez, M., Fernandez-Montero, A., Gea, A., Martinez-Gonzalez, M.A., 2013. Different types of alcoholic beverages and incidence of metabolic syndrome and its components in a Mediterranean cohort. Clinical Nutrition 32, 797–804.

Barth, S.W., Faehndrich, C., Bub, A., Watzl, B., Will, F., Dietrich, H., Rechkemmer, G., Briviba, K., 2007. Cloudy apple juice is more effective than apple polyphenols and an apple juice derived cloud fraction in a rat model of colon carcinogenesis. Journal of Agricultural and Food Chemistry 55, 1181–1187.

Basu, A., Lyons, T.J., 2011. Strawberries, blueberries, and cranberries in the metabolic syndrome: clinical perspectives. Journal of Agricultural and Food Chemistry 60, 5687–5692.

Beattie, J., Crozier, A., Duthie, G.G., 2005. Potential health benefits of berries. Current Nutrition & Food Science 1, 71–86.

Bener, M., Shen, Y., Apak, R., Finley, J.W., Xu, Z., 2013. Release and degradation of anthocyanins and phenolics from blueberry pomace during thermal acid hydrolysis and dry heating. Journal of Agricultural and Food Chemistry 61, 6643–6649.

Bhardwaj, J., Joshi, V., 2009. Effect of cultivar, addition of yeast type, extract and form of yeast culture on foaming characteristics, secondary fermentation and quality of sparkling plum wine. Natural Product Radiance 8, 452–464.

Biasi, F., Deiana, M., Guina, T., Gamba, P., Leonarduzzi, G., Poli, G., 2014. Wine consumption and intestinal redox homeostasis. Redox Biology 2, 795–802.

Blanco-Vega, D., Gómez-Alonso, S., Hermosín-Gutiérrez, I., 2014. Identification, content and distribution of anthocyanins and low molecular weight anthocyanin-derived pigments in Spanish commercial red wines. Food Chemistry 158, 449–458.

Boban, M., Modun, D., 2010. Uric acid and antioxidant effects of wine. Croatian Medical Journal 51, 16–22.

Borges, G., Degeneve, A., Mullen, W., Crozier, A., 2009. Identification of flavonoid and phenolic antioxidants in black currants, blueberries, raspberries, red currants, and cranberries. Journal of Agricultural and Food Chemistry 58, 3901–3909.

Bowen-Forbes, C.S., Zhang, Y., Nair, M.G., 2010. Anthocyanin content, antioxidant, anti-inflammatory and anticancer properties of blackberry and raspberry fruits. Journal of Food Composition and Analysis 23, 554–560.

Brazelton, C., 2013. World Blueberry Acreage & Production. North American Blueberry Council.

Cheng, G.W., Breen, P.J., 1991. Activity of phenylalanine ammonia-lyase (PAL) and concentrations of anthocyanins and phenolics in developing strawberry fruit. Journal of the American Society for Horticultural Science 116, 865–869.

Choi, J.-S., Choi, Y.-J., Park, S.-H., Kang, J.-S., Kang, Y.-H., 2004. Flavones mitigate tumor necrosis factor-α-induced adhesion molecule upregulation in cultured human endothelial cells: role of nuclear factor-κB. The Journal of Nutrition 134, 1013–1019.

Conquer, J., Maiani, G., Azzini, E., Raguzzini, A., Holub, B., 1998. Supplementation with quercetin markedly increases plasma quercetin concentration without effect on selected risk factors for heart disease in healthy subjects. The Journal of Nutrition 128, 593–597.

Cook, C.C., Hallwood, P.M., Thomson, A.D., 1998. B Vitamin deficiency and neuropsychiatric syndromes in alcohol misuse. Alcohol and Alcoholism 33, 317–336.

Cooke, D., Steward, W.P., Gescher, A.J., Marczylo, T., 2005. Anthocyans from fruits and vegetables – does bright colour signal cancer chemopreventive activity? European Journal of Cancer 41, 1931–1940.

Czyżowska, A., Pogorzelski, E., 2004. Changes to polyphenols in the process of production of must and wines from blackcurrants and cherries. Part II. Anthocyanins and flavanols. European Food Research and Technology 218, 355–359.

Das, M., Das, D.K., 2010. Resveratrol and cardiovascular health. Molecular Aspects of Medicine 31, 503–512.

Dávalos, A., Lasunción, M.A., 2009. Health-promoting effects of wine phenolics. In: Moreno-Arribas, M.V., Polo, M.C. (Eds.), Wine Chemistry and Biochemistry. Springer, New York, pp. 571–591.

Davidović, S.M., Veljović, M.S., Pantelić, M.M., Baošić, R.M., Natić, M.M., Dabić, D.C., Pecić, S.P., Vukosavljević, P.V., 2013. Physicochemical, antioxidant and sensory properties of peach wine made from redhaven cultivar. Journal of Agricultural and Food Chemistry 61, 1357–1363.

Davies, K., 2009. Annual Plant Reviews, Plant Pigments and Their Manipulation. John Wiley & Sons, Boca Raton.

De Beer, D., Joubert, E., Gelderblom, W.C.A., Manley, M., 2002. Phenolic compounds: a review of their possible role as in vivo antioxidants of wine. South African Journal of Enology & Viticulture 23, 48–61.

De Lange, D.W., Van Golden, P.H., Scholman, W.L., Kraaijenhagen, R.J., Akkerman, J.W., Van De Wiel, A., 2003. Red wine and red wine polyphenolic compounds but not alcohol inhibit ADP-induced platelet aggregation. European Journal of Internal Medicine 14, 361–366.

del Álamo, M., Casado, L., Hernández, V., Jiménez, J.J., 2004. Determination of free molecular phenolics and catechins in wine by solid phase extraction on polymeric cartridges and liquid chromatography with diode array detection. Journal of Chromatography A 1049, 97–105.

Dell'Agli, M., Buscialà, A., Bosisio, E., 2004. Vascular effects of wine polyphenols. Cardiovascular Research 63, 593–602.

Della-Morte, D., Dave, K.R., DeFazio, R.A., Bao, Y.C., Raval, A.P., Perez-Pinzon, M.A., 2009. Resveratrol pretreatment protects rat brain from cerebral ischemic damage via a sirtuin 1-uncoupling protein 2 pathway. Neuroscience 159, 993–1002.

Dey, G., Negi, B., Gandhi, A., 2009. Can fruit wines be considered as functional food? – An overview. Natural Product Radiance 8, 314–322.

Dharmadhikari, M., 1996. Wines from cherries and soft fruits. Vineyard & Vintage View 11, 1–9.

Dharmashankar, K., Widlansky, M., 2010. Vascular endothelial function and hypertension: insights and directions. Current Hypertension Reports 12, 448–455.

Dias, D.R., Schwan, R.F., Freire, E.S., dos Santos Serôdio, R., 2007. Elaboration of a fruit wine from cocoa (*Theobroma cacao* L.) pulp. International Journal of Food Science & Technology 42, 319–329.

Díaz-Rubio, M.E., Saura-Calixto, F., 2006. Dietary fiber in wine. American Journal of Enology and Viticulture 57, 69–72.

Donovan, J.L., Bell, J.R., Kasim-Karakas, S., German, J.B., Walzem, R.L., Hansen, R.J., Waterhouse, A.L., 1999. Catechin is present as metabolites in human plasma after consumption of red wine. Journal of Nutrition 129, 1662–1668.

Dragišić Maksimović, J., Poledica, M., Mutavdžić, D., Mojović, M., Radivojević, D., Milivojević, J., 2015. Variation in nutritional quality and chemical composition of fresh strawberry fruit: combined effect of cultivar and storage. Plant Foods for Human Nutrition 70, 77–84.

Dragišić Maksimović, J.J., Milivojević, J.M., Poledica, M.M., Nikolić, M.D., Maksimović, V.M., 2013. Profiling antioxidant activity of two primocane fruiting red raspberry cultivars (*Autumn bliss* and *Polka*). Journal of Food Composition and Analysis 31, 173–179.

Duarte, W.F., Dias, D.R., Oliveira, J.M., Teixeira, J.A., e Silva, J.B.A., Schwan, R.F., 2010a. Characterization of different fruit wines made from cacao, cupuassu, gabiroba, jaboticaba and umbu. LWT – Food Science and Technology 43, 1564–1572.

Duarte, W.F., Dias, D.R., Oliveira, J.M., Vilanova, M., Teixeira, J.A., e Silva, J.B.A., Schwan, R.F., 2010b. Raspberry (*Rubus idaeus* L.) wine: yeast selection, sensory evaluation and instrumental analysis of volatile and other compounds. Food Research International 43, 2303–2314.

Edirisinghe, I., Banaszewski, K., Cappozzo, J., Sandhya, K., Ellis, C.L., Tadapaneni, R., Kappagoda, C.T., Burton-Freeman, B.M., 2011. Strawberry anthocyanin and its association with postprandial inflammation and insulin. British Journal of Nutrition 106, 913–922.

El Gharras, H., 2009. Polyphenols: food sources, properties and applications – a review. International Journal of Food Science & Technology 44, 2512–2518.

Elez Garofulić, I., Kovačević Ganić, K., Galić, I., Dragović-Uzelac, V., Savić, Z., 2012. The influence of processing on phyico-chemical parameters, phenolics, antioxidant capacity and sensory attributes of elderberry (*Sambucus nigra* L.) fruit wine. Croatian Journal of Food Technology, Biotechnology and Nutrition 7, 9–13.

Erlund, I., Koli, R., Alfthan, G., Marniemi, J., Puukka, P., Mustonen, P., Mattila, P., Jula, A., 2008. Favorable effects of berry consumption on platelet function, blood pressure, and HDL cholesterol. The American Journal of Clinical Nutrition 87, 323–331.

Eskin, N.A.M., 1990. Biochemistry of food spoilage: enzymatic browning. In: Eskin, N.A.M. (Ed.), Biochemistry of Foods, second ed. Academic Press, San Diego, pp. 401–432.

Ezell, B.D., Darrow, G.M., Wilcox, M.S., Scott, D.H., 1947. Ascorbic acid content of strawberries. Journal of Food Science 12, 510–526.

Felgines, C., Texier, O., Besson, C., Fraisse, D., Lamaison, J.-L., Rémésy, C., 2002. Blackberry anthocyanins are slightly bioavailable in rats. The Journal of Nutrition 132, 1249–1253.

Feng-mei, Z., Du, B., Peng-bao, S., Feng-ying, L., 2014. Phenolic profile and antioxidant capacity of ten dry red wines from two major wine-producing regions in China. Advance Journal of Food Science and Technology 6, 344–349.

Fernández-Mar, M.I., Mateos, R., García-Parrilla, M.C., Puertas, B., Cantos-Villar, E., 2012. Bioactive compounds in wine: resveratrol, hydroxytyrosol and melatonin: a review. Food Chemistry 130, 797–813.

Ferreira, V., 2010. Volatile aroma compounds and wine sensory attributes. In: Reynolds, A.G. (Ed.), Managing Wine Quality. Woodhead Publishing, Cambridge, pp. 3–28.

Fraga, C.G., Oteiza, P.I., 2011. Dietary flavonoids: role of (−)-epicatechin and related procyanidins in cell signaling. Free Radical Biology and Medicine 51, 813–823.

Frankel, E.N., German, J.B., 2006. Antioxidants in foods and health: problems and fallacies in the field. Journal of the Science of Food and Agriculture 86, 1999–2001.

Frankel, E.N., Waterhouse, A.L., Teissedre, P.L., 1995. Principal phenolic phytochemicals in selected California wines and their antioxidant activity in inhibiting oxidation of human low-density lipoproteins. Journal of Agricultural and Food Chemistry 43, 890–894.

Fuhrman, B., Volkova, N., Coleman, R., Aviram, M., 2005. Grape powder polyphenols attenuate atherosclerosis development in apolipoprotein E deficient (E0) mice and reduce macrophage atherogenicity. The Journal of Nutrition 135, 722–728.

Fuller, N.J., Lee, S.H., Buglass, A.J., 2011. Nutritional and health aspects. In: Buglass, A.J. (Ed.), Handbook of Alcoholic Beverages: Technical, Analytical and Nutritional Aspects. John Wiley & Sons, Ltd., West Sussex, pp. 933–1110.

Galić, A., Dragović-Uzelac, V., Levaj, B., Bursać Kovačević, D., Pliestić, S., Arnautović, S., 2009. The polyphenols stability, enzyme activity and physico-chemical parameters during producing wild elderberry concentrated juice. Agriculturae Conspectus Scientificus 74, 181–186.

Galletti, G.C., Bocchini, P., Smacchia, A.M., Reeves III, J.B., 1996. Monitoring phenolic composition of maturing maize stover by high performance liquid chromatography and pyrolysis/gas chromatography/mass spectrometry. Journal of the Science of Food and Agriculture 71, 1–9.

Gao, J., Xi, Z., Zhang, J., Guo, Z., Chen, T., Fang, Y., Meng, J., Zhang, A., Li, Y., Liu, J., 2012. Influence of fermentation method on phenolics, antioxidant capacity, and volatiles in blackberry wines. Analytical Letters 45, 2603–2622.

Garai, G., Dueñas, M., Irastorza, A., Martin-Alvarez, P., Moreno-Arribas, M., 2006. Biogenic amines in natural ciders. Journal of Food Protection 69, 3006–3012.

Garai-Ibabe, G., Ibarburu, I., Berregi, I., Claisse, O., Lonvaud-Funel, A., Irastorza, A., Dueñas, M., 2008. Glycerol metabolism and bitterness producing lactic acid bacteria in cidermaking. International Journal of Food Microbiology 121, 253–261.

Garai Ibabe, G., Irastorza, A., Dueñas, M.T., Martín Álvarez, P.J., Moreno Arribas, V.M., 2013. Evolution of amino acids and biogenic amines in natural ciders as a function of the year and the manufacture steps. International Journal of Food Science & Technology 48, 375–381.

Giampieri, F., Tulipani, S., Alvarez-Suarez, J.M., Quiles, J.L., Mezzetti, B., Battino, M., 2012. The strawberry: composition, nutritional quality, and impact on human health. Nutrition 28, 9–19.

Goldberg, D.M., Yan, J., Soleas, G.J., 2003. Absorption of three wine-related polyphenols in three different matrices by healthy subjects. Clinical Biochemistry 36, 79–87.

Gómez-Alonso, S., Collins, V.J., Vauzour, D., Rodríguez-Mateos, A., Corona, G., Spencer, J.P.E., 2012. Inhibition of colon adenocarcinoma cell proliferation by flavonols is linked to a G2/M cell cycle block and reduction in cyclin D1 expression. Food Chemistry 130, 493–500.

González, E.M., de Ancos, B., Cano, P.M., 1999. Partial characterization of polyphenol oxidase activity in raspberry fruits. Journal of Agricultural and Food Chemistry 47, 4068–4072.

Güçer, Y., Güven, A., Anlı, R.E., 2008. Determination of the wine quality of different apple cultivars. International Interdisciplinary Journal of Scientific Research 1, 68–76.

Gugler, R., Leschik, M., Dengler, H., 1975. Disposition of quercetin in man after single oral and intravenous doses. European Journal of Clinical Pharmacology 9, 229–234.

Guyot, S., Marnet, N., Sanoner, P., Drilleau, J.-F., 2003. Variability of the polyphenolic composition of cider apple (*Malus domestica*) fruits and juices. Journal of Agricultural and Food Chemistry 51, 6240–6247.

Haard, N.F., Tobin, C.L., 1971. Patterns of soluble peroxidase in ripening banana fruit. Journal of Food Science 36, 854–857.

Habauzit, V., Milenkovic, D., Morand, C., 2014. Vascular protective effects of fruit polyphenols. In: Watson, R.R., Preedy, V.R., Zibadi, S. (Eds.), Polyphenols in Human Health and Disease. Academic Press, San Diego, pp. 875–893.

Häkkinen, S., Heinonen, M., Kärenlampi, S., Mykkänen, H., Ruuskanen, J., Törrönen, R., 1999. Screening of selected flavonoids and phenolic acids in 19 berries. Food Research International 32, 345–353.

Halliwell, B., 2005. Free Radicals and Other Reactive Species in Disease. Wiley Online Library, West Sussex.

Halliwell, B., 2008. Are polyphenols antioxidants or pro-oxidants? What do we learn from cell culture and in vivo studies? Archives of Biochemistry and Biophysics 476, 107–112.

Halliwell, B., 2009. The wanderings of a free radical. Free Radical Biology and Medicine 46, 531–542.

Heinonen, I.M., Lehtonen, P.J., Hopia, A.I., 1998. Antioxidant activity of berry and fruit wines and liquors. Journal of Agricultural and Food Chemistry 46, 25–31.

Heinonen, I.M., Meyer, A.S., 2002. Antioxidants in fruits, berries and vegetables. In: Jongen, W. (Ed.), Fruit and Vegetable Processing. Woodhead Publishing, Cambridge, pp. 23–51.

Hennig, K., Burkhardt, R., 1960. Detection of phenolic compounds and hydroxy acids in grapes, wines, and similar beverages. American Journal of Enology and Viticulture 11, 64–79.

Herrero, M., Garcia, L.A., Diaz, M., 1999. Organic acids in cider with simultaneous inoculation of yeast and malolactic bacteria: effect of fermentation temperature. Journal of the Institute of Brewing 105, 229–232.

Hollman, P.C., Bijsman, M.N., van Gameren, Y., Cnossen, E.P., de Vries, J.H., Katan, M.B., 1999. The sugar moiety is a major determinant of the absorption of dietary flavonoid glycosides in man. Free Radical Research 31, 569–573.

Hollman, P.C.H., Katan, M.B., 1999. Dietary flavonoids: intake, health effects and bioavailability. Food and Chemical Toxicology 37, 937–942.

Hubbard, G.P., Wolffram, S., Lovegrove, J.A., Gibbins, J.M., 2003. The role of polyphenolic compounds in the diet as inhibitors of platelet function. Proceedings of the Nutrition Society 62, 469–478.

Ichiyanagi, T., Kashiwada, Y., Ikeshiro, Y., Hatano, Y., Shida, Y., Horie, M., Matsugo, S., Konishi, T., 2004. Complete assignment of bilberry (*Vaccinium myrtillus* L.) anthocyanins separated by capillary zone electrophoresis. Chemical and Pharmaceutical Bulletin 52, 226–229.

Im, S.-E., Nam, T.-G., Lee, H., Han, M.-W., Heo, H.J., Koo, S.I., Lee, C.Y., Kim, D.-O., 2013. Anthocyanins in the ripe fruits of *Rubus coreanus* Miquel and their protective effect on neuronal PC-12 cells. Food Chemistry 139, 604–610.

Jackson, R.S., 2008. Chemical constituents of grapes and wine. In: Jackson, R.S. (Ed.), Wine Science, third ed. Academic Press, San Diego, pp. 270–331.

Jackson, R.S., 2014. Chemical constituents of grapes and wine. In: Jackson, R.S. (Ed.), Wine Science, fourth ed. Academic Press, San Diego, pp. 347–426.

Jagtap, U.B., Bapat, V.A., 2014. Phenolic composition and antioxidant capacity of wine prepared from custard apple (*Annona squamosa* L.) fruits. Journal of Food Processing and Preservation, 175–182.

Jeong, J.-H., Jung, H., Lee, S.-R., Lee, H.-J., Hwang, K.T., Kim, T.-Y., 2010. Anti-oxidant, anti-proliferative and anti-inflammatory activities of the extracts from black raspberry fruits and wine. Food Chemistry 123, 338–344.

Jiménez, M., García-Carmona, F., 1999. Myricetin, an antioxidant flavonol, is a substrate of polyphenol oxidase. Journal of the Science of Food and Agriculture 79, 1993–2000.

Johnson, M.H., Gonzalez de Mejia, E., 2012. Comparison of chemical composition and antioxidant capacity of commercially available blueberry and blackberry wines in Illinois. Journal of Food Science 77, C141–C148.

Joshi, V., John, S., 2002. Antimicrobial activity of apple wine against some pathogenic and microbes of public health and significance. Alimentaria 67–72.

Joshi, V., John, S., Abrol, G.S., 2014. Effect of addition of extracts of different herbs and spices on fermentation behaviour of apple must to prepare wine with medicinal value. National Academy Science Letters 37, 541–546.

Joshi, V., Sharma, S., Devi, M.P., 2009. Influence of different yeast strains on fermentation behaviour, physicochemical and sensory qualities of plum wine. Natural Product Radiance 8, 445–451.

Kader, F., Rovel, B., Girardin, M., Metche, M., 1997. Mechanism of browning in fresh highbush blueberry fruit (*Vaccinium corymbosum* L.). Role of blueberry polyphenol oxidase, chlorogenic acid and anthocyanins. Journal of the Science of Food and Agriculture 74, 31–34.

Kafkas, E., Cabaroglu, T., Selli, S., Bozdoğan, A., Kürkçüoğlu, M., Paydaş, S., Başer, K., 2006. Identification of volatile aroma compounds of strawberry wine using solid-phase microextraction techniques coupled with gas chromatography–mass spectrometry. Flavour and Fragrance Journal 21, 68–71.

Kähkönen, M.P., Hopia, A.I., Vuorela, H.J., Rauha, J.-P., Pihlaja, K., Kujala, T.S., Heinonen, M., 1999. Antioxidant activity of plant extracts containing phenolic compounds. Journal of Agricultural and Food Chemistry 47, 3954–3962.

Kalkan Yildirim, H., 2006. Evaluation of colour parameters and antioxidant activities of fruit wines. International Journal of Food Sciences and Nutrition 57, 47–63.

Kanner, J., Frankel, E., Granit, R., German, B., Kinsella, J.E., 1994. Natural antioxidants in grapes and wines. Journal of Agricultural and Food Chemistry 42, 64–69.

Kaume, L., Howard, L.R., Devareddy, L., 2011. The blackberry fruit: a review on its composition and chemistry, metabolism and bioavailability, and health benefits. Journal of Agricultural and Food Chemistry 60, 5716–5727.

Kaunitz, J.D., Lindebaum, J., 1977. The bioavailability of folic acid added to wine. Annals of Internal Medicine 87, 542–545.

Kelebek, H., Selli, S., Canbas, A., Cabaroglu, T., 2009. HPLC determination of organic acids, sugars, phenolic compositions and antioxidant capacity of orange juice and orange wine made from a Turkish cv. Kozan. Microchemical Journal 91, 187–192.

Kelly, M.O., Saltveit Jr., M.E., 1988. Effect of endogenously synthesized and exogenously applied ethanol on tomato fruit ripening. Plant Physiology 88, 143–147.

Khoo, C., Falk, M., 2014. Cranberry polyphenols: effects on cardiovascular risk factors. In: Watson, R.R., Preedy, V.R., Zibadi, S. (Eds.), Polyphenols in Human Health and Disease. Academic Press, San Diego, pp. 1049–1065.

Kivilompolo, M., Obůrka, V., Hyötyläinen, T., 2008. Comprehensive two-dimensional liquid chromatography in the analysis of antioxidant phenolic compounds in wines and juices. Analytical and Bioanalytical Chemistry 391, 373–380.

Klopotek, Y., Otto, K., Böhm, V., 2005. Processing strawberries to different products alters contents of vitamin C, total phenolics, total anthocyanins, and antioxidant capacity. Journal of Agricultural and Food Chemistry 53, 5640–5646.

Kourkoutas, Y., Komaitis, M., Koutinas, A., Kanellaki, M., 2001. Wine production using yeast immobilized on apple pieces at low and room temperatures. Journal of Agricultural and Food Chemistry 49, 1417–1425.

Kozlowska, M., Fryder, K., Wolko, B., 2001. Peroxidase involvement in the defense response of red raspberry to *Didymella applanata* (Niessl/Sacc.). Acta Physiologiae Plantarum 23, 303–310.

Krikorian, R., Boespflug, E.L., Fleck, D.E., Stein, A.L., Wightman, J.D., Shidler, M.D., Sadat-Hossieny, S., 2012. Concord grape juice supplementation and neurocognitive function in human aging. Journal of Agricultural and Food Chemistry 60, 5736–5742.

Krikorian, R., Nash, T.A., Shidler, M.D., Shukitt-Hale, B., Joseph, J.A., 2010. Concord grape juice supplementation improves memory function in older adults with mild cognitive impairment. British Journal of Nutrition 103, 730–734.

Kruger, M.J., Davies, N., Myburgh, K.H., Lecour, S., 2014. Proanthocyanidins, anthocyanins and cardiovascular diseases. Food Research International 59, 41–52.

Kurin, E., Mučaji, P., Nagy, M., 2012. In vitro antioxidant activities of three red wine polyphenols and their mixtures: an interaction study. Molecules 17, 14336–14348.

Lapidot, T., Harel, S., Granit, R., Kanner, J., 1998. Bioavailability of red wine anthocyanins as detected in human urine. Journal of Agricultural and Food Chemistry 46, 4297–4302.

Lattanzio, V., Kroon, P.A., Quideau, S., Treutter, D., 2009. Plant phenolics – secondary metabolites with diverse functions. In: Daayf, F., Lattanzio, V. (Eds.), Recent Advances in Polyphenol Research. Wiley-Blackwell, West Sussex, pp. 1–35.

Lee, D.H., Choi, S.S., Kim, B.B., Kim, S.Y., Kang, B.S., Lee, S.J., Park, H.J., 2013a. Effect of alcohol-free red wine concentrates on cholesterol homeostasis: an in vitro and in vivo study. Process Biochemistry 48, 1964–1971.

Lee, J.H., Kang, T.H., Um, B.H., Sohn, E.H., Han, W.C., Ji, S.H., Jang, K.H., 2013b. Evaluation of physicochemical properties and fermenting qualities of apple wines added with medicinal herbs. Food Science and Biotechnology 22, 1039–1046.

Lehtonen, P., Rokka, M., Hopia, A., Heinonen, I., 1999. HPLC determination of phenolic compounds in berry and fruit wines and liqueurs. Die Wein-Wissenschaft 54, 33–38.

Li, X.-Q., Li, S.-H., Chen, D.-F., Meng, F.-R., 2004. Induced activity of superoxide dismutase and peroxidase of in vitro plants by low concentrations of ethanol. Plant Cell, Tissue and Organ Culture 79, 83–86.

Li, X., Lim, S., Yu, B., Curran, P., Liu, S., 2013. Impact of pulp on the chemical profile of mango wine. South African Journal of Enology & Viticulture 34.

Lila, M.A., Ribnicky, D.M., Rojo, L.E., Rojas-Silva, P., Oren, A., Havenaar, R., Janle, E.M., Raskin, I., Yousef, G.G., Grace, M.H., 2011. Complementary approaches to gauge the bioavailability and distribution of ingested berry polyphenolics. Journal of Agricultural and Food Chemistry 60, 5763–5771.

Lim, J.W., Hwang, H.J., Shin, C.S., 2012. Polyphenol compounds and anti-inflammatory activities of Korean black raspberry (Rubus coreanus Miquel) wines produced from juice supplemented with pulp and seed. Journal of Agricultural and Food Chemistry 60, 5121–5127.

Lippi, G., Franchini, M., Guidi, G.C., 2010. Red wine and cardiovascular health: the "French Paradox" revisited. International Journal of Wine Research 2, 1–7.

Liu, M., Li, X.Q., Weber, C., Lee, C.Y., Brown, J., Liu, R.H., 2002. Antioxidant and antiproliferative activities of raspberries. Journal of Agricultural and Food Chemistry 50, 2926–2930.

Lodovici, M., Guglielmi, F., Casalini, C., Meoni, M., Cheynier, V., Dolara, P., 2001. Antioxidant and radical scavenging properties in vitro of polyphenolic extracts from red wine. European Journal of Nutrition 40, 74–77.

Long, L.H., Clement, M.V., Halliwell, B., 2000. Artifacts in cell culture: rapid generation of hydrogen peroxide on addition of (−)-epigallocatechin, (−)-epigallocatechin gallate, (+)-catechin, and quercetin to commonly used cell culture media. Biochemical and Biophysical Research Communications 273, 50–53.

Lopes, G.K., Schulman, H.M., Hermes-Lima, M., 1999. Polyphenol tannic acid inhibits hydroxyl radical formation from Fenton reaction by complexing ferrous ions. Biochimica et Biophysica Acta General Subjects 1472, 142–152.

Lourenço, C.F., Gago, B., Barbosa, R.M., de Freitas, V., Laranjinha, J., 2008. LDL isolated from plasma-loaded red wine procyanidins resist lipid oxidation and tocopherol depletion. Journal of Agricultural and Food Chemistry 56, 3798–3804.

Lyons, M.M., Yu, C., Toma, R.B., Cho, S.Y., Reiboldt, W., Lee, J., van Breemen, R.B., 2003. Resveratrol in raw and baked blueberries and bilberries. Journal of Agricultural and Food Chemistry 51, 5867–5870.

Madrera, R.R., Lobo, A.P., Valles, B.S., 2005. Phenolic profile of Asturian (Spain) natural cider. Journal of Agricultural and Food Chemistry 54, 120–124.

Makris, D.P., Psarra, E., Kallithraka, S., Kefalas, P., 2003. The effect of polyphenolic composition as related to antioxidant capacity in white wines. Food Research International 36, 805–814.

Manach, C., Scalbert, A., Morand, C., Rémésy, C., Jiménez, L., 2004. Polyphenols: food sources and bioavailability. The American Journal of Clinical Nutrition 79, 727–747.

Manach, C., Texier, O., Morand, C., Crespy, V., Régérat, F., Demigné, C., Rémésy, C., 1999. Comparison of the bioavailability of quercetin and catechin in rats. Free Radical Biology and Medicine 27, 1259–1266.

Manach, C., Williamson, G., Morand, C., Scalbert, A., Rémésy, C., 2005. Bioavailability and bioefficacy of polyphenols in humans. I. Review of 97 bioavailability studies. The American Journal of Clinical Nutrition 81, 230S–242S.

Martin, G.E., Sullo, J.G., Schoeneman, R.L., 1971. Determination of fixed acids in commercial wines by gas-liquid chromatography. Journal of Agricultural and Food Chemistry 19, 995–998.

Martin, L.J., Matar, C., 2005. Increase of antioxidant capacity of the lowbush blueberry (*Vaccinium angustifolium*) during fermentation by a novel bacterium from the fruit microflora. Journal of the Science of Food and Agriculture 85, 1477–1484.

Mauray, A., Felgines, C., Morand, C., Mazur, A., Scalbert, A., Milenkovic, D., 2012. Bilberry anthocyanin-rich extract alters expression of genes related to atherosclerosis development in aorta of apo E-deficient mice. Nutrition, Metabolism and Cardiovascular Diseases 22, 72–80.

Mauray, A., Milenkovic, D., Besson, C., Caccia, N., Morand, C., Michel, F., Mazur, A., Scalbert, A., Felgines, C., 2009. Atheroprotective effects of bilberry extracts in apo E-deficient mice. Journal of Agricultural and Food Chemistry 57, 11106–11111.

Mazza, G., 2007. Anthocyanins and heart health. Annali dell'Istituto Superiore di Sanita 43, 369–374.

McKay, M., Buglass, A.J., Chang Gook, L., 2010. Fruit wines and other nongrape wines. In: Buglass, A.J. (Ed.), Handbook of Alcoholic Beverages. John Wiley & Sons, Ltd., West Sussex, pp. 419–435.

Mena, P., Gironés-Vilaplana, A., Martí, N., García-Viguera, C., 2012. Pomegranate varietal wines: phytochemical composition and quality parameters. Food Chemistry 133, 108–115.

Miesle, T.J., Proctor, A., Lagrimini, L.M., 1991. Peroxidase activity, isoenzymes, and tissue localization in developing highbush blueberry fruit. Journal of the American Society for Horticultural Science 116, 827–830.

Mikulic-Petkovsek, M., Schmitzer, V., Slatnar, A., Stampar, F., Veberic, R., 2012. Composition of sugars, organic acids, and total phenolics in 25 wild or cultivated berry species. Journal of Food Science 77, C1064–C1070.

Milivojević, J., Maksimović, V., Dragišić Maksimović, J., Radivojević, D., Poledica, M., Ercili, S., 2012a. A comparison of major taste-and health-related compounds of *Vaccinium* berries. Turkish Journal of Biology 36, 738–745.

Milivojević, J., Maksimović, V., Nikolić, M., Bogdanović, J., Maletić, R., Milatović, D., 2011a. Chemical and antioxidant properties of cultivated and wild *Fragaria* and *Rubus* berries. Journal of Food Quality 34, 1–9.

Milivojević, J., Rakonjac, V., Fotirić Akšić, M., Bogdanović Pristov, J., Maksimović, V., 2013. Classification and fingerprinting of different berries based on biochemical profiling and antioxidant capacity. Pesquisa Agropecuaria Brasileira 48, 1285–1294.

Milivojevic, J., Slatnar, A., Mikulic-Petkovsek, M., Stampar, F., Nikolic, M., Veberic, R., 2012b. The influence of early yield on the accumulation of major taste and health-related compounds in black and red currant cultivars (*Ribes* spp.). Journal of Agricultural and Food Chemistry 60, 2682–2691.

Milivojević, J.M., Nikolić, M.D., Dragišić Maksimović, J.J., Radivojević, D.D., 2011b. Generative and fruit quality characteristics of primocane fruiting red raspberry cultivars. Turkish Journal of Agriculture and Forestry 35, 289–296.

Miljić, U.D., Puškaš, V.S., 2014. Influence of fermentation conditions on production of plum (*Prunus domestica* L.) wine: a response surface methodology approach. Hemijska Industrija 68, 199–206.

Miller, M.G., Shukitt-Hale, B., 2012. Berry fruit enhances beneficial signaling in the brain. Journal of Agricultural and Food Chemistry 60, 5709–5715.

Moon, J.H., Nakata, R., Oshima, S., Inakuma, T., Terao, J., 2000. Accumulation of quercetin conjugates in blood plasma after the short-term ingestion of onion by women. American Journal of Physiology. Regulatory, Integrative and Comparative Physiology 279, R461–R467.

Moreno-Arribas, M.V., Polo, M.C., Carmen, M., 2009. Wine Chemistry and Biochemistry. Springer, New York.

Mudnic, I., Budimir, D., Modun, D., Gunjaca, G., Generalic, I., Skroza, D., Katalinic, V., Ljubenkov, I., Boban, M., 2012. Antioxidant and vasodilatory effects of blackberry and grape wines. Journal of Medicinal Food 15, 315–321.

Mudnic, I., Modun, D., Rastija, V., Vukovic, J., Brizic, I., Katalinic, V., Kozina, B., Medic-Saric, M., Boban, M., 2010. Antioxidative and vasodilatory effects of phenolic acids in wine. Food Chemistry 119, 1205–1210.

Nakajima, V.M., Macedo, G.A., Macedo, J.A., 2014. Citrus bioactive phenolics: role in the obesity treatment. LWT – Food Science and Technology 59, 1205–1212.

Negi, B., Dey, G., 2009. Comparative analysis of total phenolic content in sea buckthorn wine and other selected fruit wines. World Academy of Science, Engineering and Technology 54, 99–102.

Nicholson, S.K., Tucker, G.A., Brameld, J.M., 2008. Effects of dietary polyphenols on gene expression in human vascular endothelial cells. Proceedings of the Nutrition Society 67, 42–47.

Niu, Y., Zhang, X., Xiao, Z., Song, S., Jia, C., Yu, H., Fang, L., Xu, C., 2012. Characterization of taste-active compounds of various cherry wines and their correlation with sensory attributes. Journal of Chromatography B 902, 55–60.

Noble, A.C., Arnold, R.A., Buechsenstein, J., Leach, E.J., Schmidt, J., Stern, P.M., 1987. Modification of a standardized system of wine aroma terminology. American Journal of Enology and Viticulture 38, 143–146.

Nour, V., Stampar, F., Veberic, R., Jakopic, J., 2013. Anthocyanins profile, total phenolics and antioxidant activity of black currant ethanolic extracts as influenced by genotype and ethanol concentration. Food Chemistry 141, 961–966.

Nour, V., Trandafir, I., Ionica, M.E., 2011. Ascorbic acid, anthocyanins, organic acids and mineral content of some black and red currant cultivars. Fruits 66, 353–362.

Ollanketo, M., Riekkola, M., 2000. Column-switching technique for selective determination of flavonoids in Finnish berry wines by high-performance liquid chromatography with diode array detection. Journal of Liquid Chromatography & Related Technologies 23, 1339–1351.

Opie, L.H., Lecour, S., 2007. The red wine hypothesis: from concepts to protective signalling molecules. European Heart Journal 28, 1683–1693.

Osman, H.E., Maalej, N., Shanmuganayagam, D., Folts, J.D., 1998. Grape juice but not orange or grapefruit juice inhibits platelet activity in dogs and monkeys (*Macaca fasciularis*). The Journal of Nutrition 128, 2307–2312.

Pace-Asciak, C.R., Hahn, S., Diamandis, E.P., Soleas, G., Goldberg, D.M., 1995. The red wine phenolics *trans*-resveratrol and quercetin block human platelet aggregation and eicosanoid synthesis: implications for protection against coronary heart disease. Clinica Chimica Acta 235, 207–219.

Pantelić, M., Dabić, D., Matijašević, S., Davidović, S., Dojčinović, B., Milojković-Opsenica, D., Tešić, Ž., Natić, M., 2014. Chemical characterization of fruit wine made from Oblačinska sour cherry. The Scientific World Journal 2014.

Pantelidis, G., Vasilakakis, M., Manganaris, G., Diamantidis, G., 2007. Antioxidant capacity, phenol, anthocyanin and ascorbic acid contents in raspberries, blackberries, red currants, gooseberries and Cornelian cherries. Food Chemistry 102, 777–783.

Papadopoulou, C., Soulti, K., Roussis, I.G., 2005. Potential antimicrobial activity of red and white wine phenolic extracts against strains of *Staphylococcus aureus*, *Escherichia coli* and *Candida albicans*. Food Technology and Biotechnology 43, 41–46.

Patel, K.R., Brown, V.A., Jones, D.J., Britton, R.G., Hemingway, D., Miller, A.S., West, K.P., Booth, T.D., Perloff, M., Crowell, J.A., 2010. Clinical pharmacology of resveratrol and its metabolites in colorectal cancer patients. Cancer Research 70, 7392–7399.

Patras, A., Brunton, N.P., O'Donnell, C., Tiwari, B., 2010. Effect of thermal processing on anthocyanin stability in foods; mechanisms and kinetics of degradation. Trends in Food Science & Technology 21, 3–11.

Pérez-Gregorio, M.R., Regueiro, J., Alonso-González, E., Pastrana-Castro, L.M., Simal-Gándara, J., 2011. Influence of alcoholic fermentation process on antioxidant activity and phenolic levels from mulberries (*Morus nigra* L.). LWT – Food Science and Technology 44, 1793–1801.

Pérez-Jiménez, J., Neveu, V., Vos, F., Scalbert, A., 2010. Systematic analysis of the content of 502 polyphenols in 452 foods and beverages: an application of the phenol-explorer database. Journal of Agricultural and Food Chemistry 58, 4959–4969.

Picinelli, A., Suárez, B., Moreno, J., Rodríguez, R., Caso-García, L.M., Mangas, J.J., 2000. Chemical characterization of Asturian cider. Journal of Agricultural and Food Chemistry 48, 3997–4002.

Pogorzelski, E., 1992. Studies on the formation of histamine in must and wines from elderberry fruit. Journal of the Science of Food and Agriculture 60, 239–244.

Ramirez-Ambrosi, M., Abad-Garcia, B., Viloria-Bernal, M., Garmon-Lobato, S., Berrueta, L.A., Gallo, B., 2013. A new ultrahigh performance liquid chromatography with diode array detection coupled to electrospray ionization and quadrupole time-of-flight mass spectrometry analytical strategy for fast analysis and improved characterization of phenolic compounds in apple products. Journal of Chromatography A 1316, 78–91.

Reddy, L.V., Reddy, O.V.S., Joshi, V., 2009. Production, optimization and characterization of wine from Mango (*Mangifera indica* Linn.). Natural Product Radiance 8, 426–435.

Reed, J., 2002. Cranberry flavonoids, atherosclerosis and cardiovascular health. Critical Reviews in Food Science and Nutrition 42, 301–316.

Renaud, S., de Lorgeril, M., 1992. Wine, alcohol, platelets, and the French paradox for coronary heart disease. The Lancet 339, 1523–1526.

Rice-Evans, C.A., Miller, N.J., Paganga, G., 1996. Structure-antioxidant activity relationships of flavonoids and phenolic acids. Free Radical Biology and Medicine 20, 933–956.

Rifici, V.A., Stephan, E.M., Schneider, S.H., Khachadurian, A.K., 1999. Red wine inhibits the cell-mediated oxidation of LDL and HDL. Journal of the American College of Nutrition 18, 137–143.

Rimando, A.M., Kalt, W., Magee, J.B., Dewey, J., Ballington, J.R., 2004. Resveratrol, pterostilbene, and piceatannol in *Vaccinium* berries. Journal of Agricultural and Food Chemistry 52, 4713–4719.

Rodrigo, R., Libuy, M., 2014. Modulation of plant endogenous antioxidant systems by polyphenols. In: Watson, R.R. (Ed.), Polyphenols in Plants. Academic Press, San Diego, pp. 65–85.

Rommel, A., Heatherbell, D., Wrolstad, R., 1992. Red raspberry juice and wine: effect of processing and storage on anthocyanin pigment composition, color and appearance. Journal of Food Science 55, 1011–1017.

Rosell, M., De Faire, U., Hellenius, M., 2003. Low prevalence of the metabolic syndrome in wine drinkers—is it the alcohol beverage or the lifestyle? European Journal of Clinical Nutrition 57, 227–234.

Rossouw, D., Heyns, E., Setati, M., Bosch, S., Bauer, F., 2013. Adjustment of trehalose metabolism in Wine *Saccharomyces cerevisiae* strains to modify ethanol yields. Applied and Environmental Microbiology 79, 5197–5207.

Rupasinghe, H.P.V., Clegg, S., 2007. Total antioxidant capacity, total phenolic content, mineral elements, and histamine concentrations in wines of different fruit sources. Journal of Food Composition and Analysis 20, 133–137.

Ryan, J.J., Dupont, J.A., 1973. Identification and analysis of the major acids from fruit juices and wines. Journal of Agricultural and Food Chemistry 21, 45–49.

Sánchez-Moreno, C., Larrauri, J.A., Saura-Calixto, F., 1999. Free radical scavenging capacity and inhibition of lipid oxidation of wines, grape juices and related polyphenolic constituents. Food Research International 32, 407–412.

Sánchez-Moreno, C., Cao, G., Ou, B., Prior, R.L., 2003. Anthocyanin and proanthocyanidin content in selected white and red wines. Oxygen radical absorbance capacity comparison with nontraditional wines obtained from highbush blueberry. Journal of Agricultural and Food Chemistry 51, 4889–4896.

Satora, P., Sroka, P., Duda-Chodak, A., Tarko, T., Tuszyński, T., 2008. The profile of volatile compounds and polyphenols in wines produced from dessert varieties of apples. Food Chemistry 111, 513–519.

Saura-Calixto, F., Díaz-Rubio, M.E., 2007. Polyphenols associated with dietary fibre in wine: a wine polyphenols gap? Food Research International 40, 613–619.

Schmitzer, V., Veberic, R., Slatnar, A., Stampar, F., 2010. Elderberry (*Sambucus nigra* L.) wine: a product rich in health promoting compounds. Journal of Agricultural and Food Chemistry 58, 10143–10146.

Schramm, D.D., Pearson, D.A., German, J.B., 1997. Endothelial cell basal PGI2 release is stimulated by wine in vitro: one mechanism that may mediate the vasoprotective effects of wine. The Journal of Nutritional Biochemistry 8, 647–651.

Seeram, N.P., 2011. Emerging research supporting the positive effects of berries on human health and disease prevention. Journal of Agricultural and Food Chemistry 60, 5685–5686.

Seeram, N.P., Adams, L.S., Zhang, Y., Lee, R., Sand, D., Scheuller, H.S., Heber, D., 2006. Blackberry, black raspberry, blueberry, cranberry, red raspberry, and strawberry extracts inhibit growth and stimulate apoptosis of human cancer cells in vitro. Journal of Agricultural and Food Chemistry 54, 9329–9339.

Sesink, A.L.A., O'Leary, K.A., Hollman, P.C.H., 2001. Quercetin glucuronides but not glucosides are present in human plasma after consumption of quercetin-3-glucoside or quercetin-4′-glucoside. The Journal of Nutrition 131, 1938–1941.

Sharma, S., Joshi, V., Abrol, G., 2009. An overview on strawberry [*Fragaria×ananassa* (Weston) *Duchesne* ex *Rozier*] wine production technology, composition, maturation and quality evaluation. Natural Product Radiance 8, 356–365.

Spiteller, G., 2005. The relation of lipid peroxidation processes with atherogenesis: a new theory on atherogenesis. Molecular Nutrition & Food Research 49, 999–1013.

Sun, S.Y., Jiang, W.G., Zhao, Y.P., 2011. Evaluation of different *Saccharomyces cerevisiae* strains on the profile of volatile compounds and polyphenols in cherry wines. Food Chemistry 127, 547–555.

Swiegers, J., Bartowsky, E., Henschke, P., Pretorius, I., 2005. Yeast and bacterial modulation of wine aroma and flavour. Australian Journal of Grape and Wine Research 11, 139.

Swiegers, J.H., Pretorius, I.S., 2005. Yeast modulation of wine flavor. In: Allen, I., Laskin, J.W.B., Geoffrey, M.G. (Eds.), Advances in Applied Microbiology. Academic Press, San Diego, pp. 131–175.

Talavéra, S., Felgines, C., Texier, O., Besson, C., Lamaison, J.-L., Rémésy, C., 2003. Anthocyanins are efficiently absorbed from the stomach in anesthetized rats. The Journal of Nutrition 133, 4178–4182.

Tavares, L., Figueira, I., Macedo, D., McDougall, G.J., Leitão, M.C., Vieira, H.L., Stewart, D., Alves, P.M., Ferreira, R.B., Santos, C.N., 2012. Neuroprotective effect of blackberry (*Rubus* sp.) polyphenols is potentiated after simulated gastrointestinal digestion. Food Chemistry 131, 1443–1452.

Thomas, R.L., Jen, J.J., Morr, C.V., 1982. Changes in soluble and bound peroxidase – IAA oxidase during tomato fruit development. Journal of Food Science 47, 158–161.

Usenik, V., Fabčič, J., Štampar, F., 2008. Sugars, organic acids, phenolic composition and antioxidant activity of sweet cherry (*Prunus avium* L.). Food Chemistry 107, 185–192.

Vaidya, D., Vaidya, M., Sharma, S., Joshi, V., 2009. Enzymatic treatment for juice extraction and preparation and preliminary evaluation of kiwifruits wine. Natural Product Radiance 8, 380–385.

van der Merwe, C., van Wyk, C., 1981. The contribution of some fermentation products to the odor of dry white wines. American Journal of Enology and Viticulture 32, 41–46.

Varakumar, S., Kumar, Y.S., Reddy, O.V.S., 2011. Carotenoid composition of mango (*Mangifera indica* L.) wine and its antioxidant activity. Journal of Food Biochemistry 35, 1538–1547.

Vermerris, W., Nicholson, R., 2006. Families of Phenolic Compounds and Means of Classification. Springer, Netherlands, Berlin.

Vuorinen, H., Määttä, K., Törrönen, R., 2000. Content of the flavonols myricetin, quercetin, and kaempferol in Finnish berry wines. Journal of Agricultural and Food Chemistry 48, 2675–2680.

Wallerath, T., Li, H., Gödtel-Ambrust, U., Schwarz, P.M., Förstermann, U., 2005. A blend of polyphenolic compounds explains the stimulatory effect of red wine on human endothelial NO synthase. Nitric Oxide 12, 97–104.

Wang, L., Xu, Y., Zhao, G., Li, J., 2004. Rapid analysis of flavor volatiles in apple wine using headspace solid-phase microextraction. Journal of the Institute of Brewing 110, 57–65.

Wang, Y., Li, P., Singh, N.K., Shen, T., Hu, H., Li, Z., Zhao, Y., 2012. Changes in aroma composition of blackberry wine during fermentation process. African Journal of Biotechnology 11, 16504–16511.

Wang, Z., Zhuge, J., Fang, H., Prior, B.A., 2001. Glycerol production by microbial fermentation: a review. Biotechnology Advances 19, 201–223.

Weber, F., Schulze-Kaysers, N., Schieber, A., 2014. Characterization and quantification of polyphenols in fruits. In: Watson, R.R. (Ed.), Polyphenols in Plants. Academic Press, San Diego, pp. 293–304.

Whitaker, J.R., 1995. Polyphenol Oxidase. Springer Verlag, New York, USA.

Whitehead, T.P., Robinson, D., Allaway, S., Syms, J., Hale, A., 1995. Effect of red wine ingestion on the antioxidant capacity of serum. Clinical Chemistry 41, 32–35.

Whiting, G.C., 1958. The non-volatile organic acids of some berry fruits. Journal of the Science of Food and Agriculture 9, 244–248.

Wrolstad, R.E., 2012. Food Carbohydrate Chemistry. John Wiley & Sons.

Xia, E., He, X., Li, H., Wu, S., Li, S., Deng, G., 2014. Biological activities of polyphenols from grapes. In: Watson, R.R., Preedy, V.R., Zibadi, S. (Eds.), Polyphenols in Human Health and Disease. Academic Press, San Diego, pp. 47–58.

Xiang, L., Xiao, L., Wang, Y., Li, H., Huang, Z., He, X., 2014. Health benefits of wine: don't expect resveratrol too much. Food Chemistry 156, 258–263.

Xie, J., Schaich, K.M., 2014. Re-evaluation of the 2,2-diphenyl-1-picrylhydrazyl free radical (DPPH) assay for antioxidant activity. Journal of Agricultural and Food Chemistry 62, 4251–4260.

Xie, L., Ye, X., Liu, D., Ying, Y., 2011. Prediction of titratable acidity, malic acid, and citric acid in bayberry fruit by near-infrared spectroscopy. Food Research International 44, 2198–2204.

Yadav, P., Garg, N., Diwedi, D., Joshi, V., 2009. Effect of location of cultivar, fermentation temperature and additives on the physico-chemical and sensory qualities on mahua (*Madhuca indica* JF Gmel.) wine preparation. Natural Product Radiance 8, 406–418.

Yokotsuka, K., Singleton, V.L., 1997. Disappearance of anthocyanins as grape juice is prepared and oxidized with PPO and PPO substrates. American Journal of Enology and Viticulture 48, 13–25.

Zadernowski, R., Naczk, M., Nesterowicz, J., 2005. Phenolic acid profiles in some small berries. Journal of Agricultural and Food Chemistry 53, 2118–2124.

Zhang, Z., Pang, X., Xuewu, D., Ji, Z., Jiang, Y., 2005. Role of peroxidase in anthocyanin degradation in litchi fruit pericarp. Food Chemistry 90, 47–52.

Zoecklein, B.W., 1995. Wine Analysis and Production. Chapman & Hall, New York.

CHAPTER

METHODS OF EVALUATION OF FRUIT WINES

5

D.R. Dias, W.F. Duarte, R.F. Schwan
Federal University of Lavras, Lavras, Minas Gerais, Brazil

1. INTRODUCTION

In addition to grapes, many other fruits have been used for the production of fruit wines (Amerine and Ough, 1980; Joshi et al., 1999; Joshi et al., 2011; Joshi and Attri, 2005; Jackson, 2008). Fruits like apples and cherries are grown and used for the industrial production of fruit wines in several countries of Europe and Asia. Berries, such as blueberries and other berries, are also used in Northern Hemisphere for small-scale or household production of fruit wines (Joshi, 1997; Joshi et al., 2004). Since about 1995, special emphasis has been given to the production of fruit wines from tropical and subtropical fruits such as pineapple, mango, melon, cacao, persimmon, banana, and jabuticaba, among others.

On the way to opening the market for fruit wines there are many challenges such as scientific knowledge, technical aspects of fruit processing, technological development suitable for the transformation of these raw materials into the final product, and analytical methods that should be developed and employed to evaluate the wines before their consumption. Regarding analytical methods, because of the many compounds present in wines and fruit wines inherent to the fruit used to prepare the must, it becomes difficult to establish individual standard parameters for each fruit wine. However, the analytical methods used to evaluate the wine are commonly shared as useful tools to investigate the physicochemical, microbiological, and sensory parameters of fruit wines. There are several components present in the fruit wines that are common to wines from grapes. Some of the most common components are listed in Table 5.1. Table 5.2 shows analytical methods used to evaluate fruit wines around the world.

The quantification of chemical compounds, or groups of compounds; the evaluation of physicochemical parameters; microbial enumeration/detection; and sensorial analysis in wine employ several techniques and methods (Zoecklein et al., 1995, 2011, 1990; Amerine and Ough, 1980; Joshi, 1997). The techniques commonly used for fruit wine analysis are summarized in this chapter. For better content organization, methodologies were grouped into physicochemical analysis (Section 2), chromatographic analysis (Section 3), microbiological analysis (Section 4), and sensory analysis, and are addressed encompassing the classic and current concerns.

2. PHYSICOCHEMICAL ANALYSIS

Fermented beverages are complex solutions of thousands of chemical compounds originating from the fruit itself, from the fermentation process, from the yeast and other microbial metabolism during fermentation, and from postfermentation steps (including secondary fermentations and chemical reactions

Science and Technology of Fruit Wine Production. http://dx.doi.org/10.1016/B978-0-12-800850-8.00005-3

Table 5.1 Analytical Methods Generally Used for Wine and Fruit Wine Analysis

Group of Analysis	Technique	Target/Analysis
Physicochemical	Densimetry	Total dry matter
	Gravimetry	Ash
		Sediments
		Specific gravity
		Tartaric acid
		Total dry matter
	Refractometry	Total soluble solids
	Nephelometry	Turbidity
	Titrimetry	Acidity (total and volatile)
		Alkalinity of ash
		Carbon dioxide
		Ethyl acetate
		Hydroxymethylfurfural
		Potassium
		Sulfates
		Sulfur dioxide
	Potentiometry	pH
		Carbon dioxide
	Flame photometry	Potassium
		Sodium
	Spectrophotometry (includes colorimetry)	Arsenic
		Color
		Iron
		L-Ascorbic acid
		Sorbic acid
		Tartaric acid
	Atomic absorption spectrometry	Arsenic
		Cadmium
		Calcium
		Copper
		Iron
		Lead
		Sodium
		Zinc
	Chromatography	Alcohols
		Aldehydes
		Biogenic amines
		Esters
		Ethyl carbamate
		Furfurals
		Higher alcohols
		Ketones
		Mycotoxins
		Natamycin
		Organic acids
		Pesticides
		Phenolic compounds
		Polyols
		Reducing sugars
		Sugars
		Volatiles
		α-Dicarbonyl compounds

Table 5.1 Analytical Methods Generally Used for Wine and Fruit Wine Analysis—cont'd

Group of Analysis	Technique	Target/Analysis
Microbiological	Microscopy	Direct microscopic examination Gram staining Yeast cell count
	Plating	Detection of preservatives and fermentation inhibitors Tests for pathogenic microorganism Total plate count Yeast and mold counts
	Broth culture (most probable number)	Coliform counts Tests for pathogenic microorganism
	Molecular tests	Denaturing gradient gel electrophoresis Pulsed-field gel electrophoresis RFLP Reverse transcriptase–PCR
	Flow cytometry	Cell counting
Sensory	Composite scoring Consumer rating Fact rating Flavor profiling Hedonic testing Paired comparison Quantitative descriptive analysis Threshold testing	

Table 5.2 Examples of Fruit Wines and the Analytical Methods Used to Evaluate Physicochemical Parameters

Fruit	Analytical Method	Country or Region	References
Apple (*Malus* sp.)	Titrimetry, potentiometry, spectrophotometry, chromatography, molecular tools	England; Singapore	Ball (1946), Williams and Tucknott (1978), Valles et al. (2007), and Aung et al. (2015)
Banana (*Musa* sp.)	Titrimetry, potentiometry, spectrophotometry	Nigeria; Thailand	Akubor et al. (2003) and Cheirsilp and Umsakul (2008)
Black mulberry (*Morus nigra*)	Titrimetry, potentiometry, spectrophotometry, spectrometry	Spain	Darias-Martín et al. (2003) and Pérez-Gregorio et al. (2011)
Black plum (*Vitex doniana* Sweet)	Titrimetry, potentiometry	Nigeria	Okigbo (2003)
Black raspberry (*Rubus occidentalis*)	Spectrophotometry	Republic of Korea	Jeong et al. (2010)
Blueberry (*Vaccinium corymbosum*)	Titrimetry	China; Turkey	Yan et al. (2012) and Celep et al. (2015)
Cacao (*Theobroma cacao* L.)	Titrimetry, spectrophotometry, chromatography	Brazil	Dias et al. (2007), Duarte et al. (2010a), and Dias (2015)
Cagaita (*Eugenia dysenterica* DC.)	Titrimetry, potentiometry, spectrophotometry, chromatography	Brazil	Oliveira et al. (2011)
Carambola (*Averrhoa carambola* L.)	Titrimetry, potentiometry, spectrophotometry, sensory	Trinidad and Tobago; Taiwan	Bridgebassie and Badrie (2004) and Wu et al. (2005)
Cashew (*Anacardium occidentale* L.)	Titrimetry, potentiometry, spectrophotometry, sensory	India	Mohanty et al. (2006)
Cherry (*Prunus cerasus*)	Titrimetry, potentiometry, spectrophotometry, chromatography	China; Turkey	Sun et al. (2014) and Celep et al. (2015)

Table 5.2 Examples of Fruit Wines and the Analytical Methods Used to Evaluate Physicochemical Parameters—cont'd

Fruit	Analytical Method	Country or Region	References
Cupuaçu (*Theobroma grandiflorum* Schum.)	Potentiometry, spectrophotometry, chromatography	Brazil	Duarte et al. (2010a)
Custard apple (*Annona squamosa* L.)	Titrimetry, potentiometry, spectrophotometry, chromatography	India	Jagtap and Bapat (2015)
Elderberry (*Sambucus nigra* L.)	Potentiometry, spectrophotometry, chromatography	Slovenia	Schmitzer et al. (2010)
Gabiroba [*Campomanesia pubescens* (DC.) O. Berg]	Potentiometry, spectrophotometry, chromatography, molecular tools, sensory	Brazil	Duarte et al. (2009, 2010a)
Grapefruit (*Citrus paradis*)	Potentiometry, spectrophotometry, chromatography	Cameroon	Ndip et al. (2001)
Jabuticaba [*Myrciaria jaboticaba* (Vell.) Berg]	Potentiometry, spectrophotometry, chromatography	Brazil	Duarte et al. (2010a)
Jackfruit (*Artocarpus heterophyllus* Lam.)	Spectrophotometry, chromatography		Jagtap et al. (2011)
Kinnow (*Citrus* sp.)	Titrimetry, potentiometry	India	Singh et al. (1998)
Kiwi (*Actinidia deliciosa*)	Titrimetry, potentiometry, spectrophotometry, chromatography	Greece	Soufleros et al. (2001)
Lychee (*Litchi chinensis* Sonn)	Potentiometry, spectrophotometry, chromatography	Brazil; Singapore	Alves et al. (2010, 2011) and Chen et al. (2014)
Mango (*Mangifera indica* L.)	Titrimetry, potentiometry, chromatography	India	Reddy and Reddy (2005, 2011)
Melon (*Cucumis melo* L.)	Potentiometry, spectrophotometry, chromatography	Spain	Gómez et al. (2008)
Mulberry (*Morus alba*)	Titrimetry, potentiometry, spectrophotometry, chromatography	China	Tsai et al. (2004), Butkhup et al. (2011), and Wang et al. (2015)
Orange (*Citrus* sp.)	Titrimetry, potentiometry, spectrophotometry, chromatography	Turkey; Portugal	Selli et al. (2003, 2008) and Coelho et al. (2015)
Papaya (*Carica papaya*)	Titrimetry, potentiometry, spectrophotometry, chromatography, sensory	India; Singapore	Maragatham and Panneerselvam (2011) and Lee et al. (2012)
Passion fruit (*Passiflora edulis* var. flavicarpa)	Chromatography	United States	Muller et al. (1964)
Peach (*Prunus persica*)	Titrimetry, potentiometry, spectrophotometry, chromatography, sensory	India; Serbia	Joshi and Shah (1998) and Davidovic et al. (2013)
Persimmon (*Diospyros kaki*)	Potentiometry, spectrophotometry, chromatography, molecular tools	Spain; China	Hidalgo et al. (2012) and Zhu et al. (2016)
Pineapple (*Ananas comosus* L.)	Potentiometry, spectrophotometry, chromatography, molecular tools	Thailand; Brazil	Chanprasartsuk et al. (2010) and Ribeiro et al. (2015)
Prickly pear (*Opuntia streptacantha*)	Chromatography, molecular tools	Mexico	Rodríguez-Lerma et al. (2011) and Navarrete-Bolaños et al. (2013)
Raspberry (*Rubus idaeus* L.)	Potentiometry, spectrophotometry, chromatography, sensory	Brazil; Portugal	Duarte et al. (2010b)
Strawberry (*Fragaria × ananassa*)	Potentiometry, spectrophotometry, chromatography, sensory	India; Turkey	Joshi et al. (2005) and Kafkas et al. (2006)
Umbu (*Spondias tuberosa* L.)	Potentiometry, spectrophotometry, chromatography	Brazil	Duarte et al. (2010a)

Table 5.3 Some Regulations Regarding Fruit Wine Standards of Identity and Quality

Parameter	Range or Value	Country or Region	References
Alcohol (% v/v)	1.2–14 (up to 22 for fortified wine)	Europe	AICV (2014)
	4–14 (14–18 for fortified wine)	Brazil	Brasil (2009)
	Less than 14 (14–24 for dessert fruit wine)	USA	CFR (2015)
	4–22	UK	WSTA (2015)
Volatile acidity (% w/w, of acetic acid)	0.12–0.14	USA	CFR (2015)

during aging) (Polášková et al., 2008). In most wines, the interactions of these compounds lead to a beverage with pleasant flavor and aroma. In other situations, the production of certain compounds, called off-flavors, in amounts over threshold values can cause changes in flavor and aroma (e.g., sulfur compounds, higher alcohols, ketones, diacetyl, acetic acid, and methanol, among others). These off-flavors are considered defects in the wine.

Depending on the need for physicochemical quality criteria, for some compounds, or groups of chemical compounds, the parameters are established as wine quality standards. These standards of identity and quality are established for grape wine in virtually all producing countries. Concerning fruit wines other than grape wines, regulation is scarce and usually the quality standards for grape wine are used. Specific regulations for fruit wines are available in a small number of countries (Table 5.3), such as the Austrian (Austrian Wine Act, 2009), Brazilian (Brasil, 2009), New Zealand (MAF, 2011), European Cider and Fruit Wine Association (AICV, 2014), British (WSTA, 2015), and American (CFR, 2015) regulations, especially regarding alcohol content.

2.1 TITRIMETRY

Titrimetric analysis comprises a group of traditional analytical methods based on measuring the amount of an exactly known concentration of reagent (titrant) that is required to completely (stoichiometrically) react with the analyte. Based on the type of reaction between the titrant and the analyte, titrimetry can be classified into four methods: complex titration, precipitation titration, redox titration, and acid–base titration.

Titrations are used for wine and fruit wine analysis to determine acidity, ash alkalinity, and total sulfur dioxide, for example (Table 5.1). As mentioned before, specific regulations and methods to be used for fruit wine analysis are scarce and it is a common practice to adopt, or to adapt, the standard methods established for wines as a protocol for fruit wine analysis. For that reason, for the following examples we will use the analytical methods of the International Organization of Vine and Wine (OIV, 2015), except for the use of a specific regulation for fruit wine, as a reference guide because of its international coverage and its availability online in English and French. Some OIV methods are also available in Spanish, German, and Italian.

2.1.1 Acidity

Organic acids are compounds of great importance because of their influence on several sensory properties such as flavor, aroma, and color; rates of synthesis and hydrolysis of esters; anthocyanin polymerization; and protein stability in alcoholic beverages. These compounds are also related to the control of microbiological stability of beverages (Boulton, 1980; Mato et al., 2005). The determination of organic acids in juices, wines, and fruit wines is a parameter related to quality, and it is useful to assess the integrity of the beverage, too, in case of adulteration (Ehling and Cole, 2011).

Organic acids found in beverages are divided into two groups, volatile and nonvolatile acids (Table 5.4). In grape must and wine, nonvolatile acids found in greater amounts are tartaric and malic acids, accounting for approximately 90% of acidity. Citric acid is the major organic acid in citrus, raspberry, strawberry, cocoa, and pomegranate, whereas malic acid is the main acid in peach, apple, and plum (Amerine and Ough, 1980; Dias et al., 2007; Flores et al., 2012).

The volatile acids have shorter carbon chains than nonvolatile acids. The main representative of this group is acetic acid, whose amount may match 90% of the content of volatile acids in wines. Acetic

Table 5.4 Most Common Organic Acids Analyzed in Fruit Wines

Acid	Structure	Weight (g/mol)	Volatility	Source	Method of Analysis
Tartaric		150.09	Nonvolatile	Fruits	HPLC, titrimetry, spectrophotometry
Malic		134.09	Nonvolatile	Fruits	HPLC, spectrophotometry
Citric		192.12	Nonvolatile	Fruits	HPLC, titrimetry
Succinic		118.09	Nonvolatile	Fruits, yeasts	HPLC
Lactic		90.08	Nonvolatile	Lactic acid bacteria	HPLC
Ascorbic		176.12	Nonvolatile	Fruits	HPLC, spectrophotometry
Sorbic		112.13	Nonvolatile	Additive/preservative	HPLC, GC, spectrophotometry
Acetic		60.05	Volatile	Fruits, yeasts, acetic acid bacteria, chemical oxidation	HPLC, GC, titrimetry
Butyric		88.11	Volatile	Bacteria	HPLC, GC

acid can be present in some fruits and also can be synthesized by yeasts during fermentation; nevertheless it can be generated by chemical oxidation of ethanol in the presence of oxygen. Acetic acid bacteria can synthesize it during aging and storage of wines. A high quantity of acetic acid in the wine is related to poor quality. Further, improper or poor storage is the prime reason for acetification (Amerine and Ough, 1980). Thus, the presence of acetic acid is considered a wine defect and its concentration should be monitored and maintained below 1.2 g/L or according to the legal limits for each country. Some other acids, such as lactic acid and succinic acid, synthesized by microorganisms during fermentation, are also found in wines (Ribéreau-Gayon et al., 2006b; Bartowsky and Henschke, 2008).

2.1.1.1 Total Acidity or Titratable Acidity

From a chemical point of view, total acidity and titratable acidity refer to different electronic structures. Total acidity is related to the total number of protons of all organic acids present in the substrate if they were undissociated. Titratable acidity refers to the number of dissociated protons, from organic acids, that are neutralized by a strong base during titration. The value found for titratable acidity is always less than that found for total acidity. Because titratable acidity can be more easily determined than the total acidity, and represents an estimate of the acidity in beverages, it is used as a quality parameter for wines (Boulton, 1980; OIV, 2015).

According to the OIV (2015), the total acidity of a wine is the sum of its titratable acidities when it is titrated to pH 7 against a standard alkaline solution. Carbon dioxide, however, is not included in the total acidity and should be removed from the sample before total acidity determination. Titration, using bromothymol blue as the indicator and comparison with an end-point color standard, or potentiometric titration, using a pH meter calibrated with a pH 7 buffer at 20°C, are the methods used to determine the total acidity in wines (method OIV-MA-AS313-01; OIV, 2015).

The total acidity value found in fruit wines can be expressed in milliequivalents per liter or in grams of the major organic acid present in the fruit, per liter, as showed in Table 5.5.

2.1.1.2 Volatile Acidity

Volatile acidity is the measure of volatile acids of the acetic acid series, which are in wine in the free form or as combined salts. Volatile acidity is measured by titration of the distillate separated from the wine sample by steam distillation (Fig. 5.1), using sodium hydroxide (0.1 mol/L) as the titrant and phenolphthalein as the indicator. To determine volatile acidity in wines the acidity regarding sorbic (added to some wines as a preservative) and salicylic (used in some countries as a wine stabilizer) acids, carbon dioxide, and free and combined sulfur dioxide should be subtracted from the distillate obtained from the wine by steam distillation (method OIV-MA-AS313-02; OIV, 2015).

Table 5.5 Expression Units of Total Acidity in Fruit Wines

	Expression	**Equation**
Total acidity (A)	mEq/L	$A = 10n$
	g tartaric acid/L	$A' = 0.075 \times A$
	g malic acid/L	$A' = 0.067 \times A$
	g citric acid/L	$A' = 0.064 \times A$

n *is the volume, in mL, of sodium hydroxide solution, 0.1 mol/L, added to reach the end point of titration.*

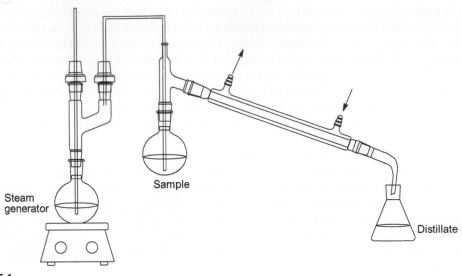

FIGURE 5.1

A schematic of a steam distillation apparatus for volatile acidity determination.

Calculations of volatile acidity are given by:

- grams of acetic acid per liter, $0.300 \, (n - 0.1 \, n' - 0.05 \, n'')$;
- grams of sulfuric acid per liter, $0.245 \, (n - 0.1 \, n' - 0.05 \, n'')$;
- milliequivalents per liter, $5 \, (n - 0.1 \, n' - 0.05 \, n'')$;

where n is the volume (mL) of 0.1 M sodium hydroxide solution used to titrate the distilled wine, n' the volume (mL) of 0.005 M iodine solution used to titrate the free sulfur dioxide, and n'' the volume (mL) of 0.005 M iodine solution used to titrate the combined sulfur dioxide.

2.1.2 Alkalinity of Ash

The residues remaining after the evaporation and ignition of a wine sample are called the ash content. During the ignition, all the organic compounds and inorganic volatile compounds are eliminated from the sample and the residual cations, except for ammonium cations, are converted into carbonates or dehydrated inorganic salts. The ash content varies from fruit to fruit and it is also influenced by the characteristics of the soil, climate, and winemaking technology, e.g., maceration during red wine production leads to an increase in the ash content in red wines compared to white wines, for which maceration is minimal (Huerta et al., 1998; Ribéreau-Gayon et al., 2006a; OIV, 2015).

Alkalinity of the ash represents the amount of cations (other than ammonium cations) combined with organic acids present in the ash that can be determined by titration. The alkalinity of the ash is evaluated by dissolving the ash (from a defined volume of wine, generally 20 mL) in a known excess volume of a hot inorganic acid solution (0.05 M H_2SO_4). The free sulfuric acid that did not react with the ash salts is titrated against 0.1 M sodium hydroxide solution using 0.1% methyl orange as the indicator (method OIV-MA-AS2-01A; OIV, 2015).

Alkalinity of the ash can be calculated and expressed in milliequivalents per liter or, alternatively, in grams per liter of potassium carbonate, as follows:

- milliequivalents per liter, $A = 5(10 - n)$;
- grams per liter of potassium carbonate, $A = 0.345(10 - n)$;

where n, in both cases, is the volume (mL) of 0.1 M sodium hydroxide solution used.

2.1.3 Sulfur Dioxide

Sulfur dioxide (SO_2) is used worldwide in the food and beverage industries as an antioxidant as well as a microbial growth inhibitor. In winemaking, sulfur dioxide inhibits the growth of some yeast species and the majority of bacteria related to wine spoilage (Amerine and Ough, 1980). As an antioxidant in wines, sulfur dioxide reduces oxidative enzyme activities, such as polyphenol oxidase, avoiding enzymatic browning reactions (Ribéreau-Gayon et al., 2006a,b; Isaac et al., 2006; Guerrero and Cantos-Villar, 2015). Sulfur dioxide can also react with carbonyl compounds, forming odorless bisulfite adducts, reducing the off-flavor aroma of free carbonyls (Singleton, 1987; Chinnici et al., 2013). Despite all the beneficial aspects of sulfur dioxide use in the winemaking and food industries, it is known that the sulfites generated from SO_2 can, themselves, cause allergic responses in sensitive consumers. Because of these undesirable effects sulfur dioxide addition to wine and foods is the subject of regulation, and efforts are made to reduce its usage (Santos et al., 2012).

Sulfur dioxide is commonly present in wines in a chemical equilibrium of H_2SO_3 (sulfurous acid) and HSO_3^- (hydrogen sulfite) forms, which is a function of temperature and pH. These chemical forms are dissolved in the wine solution and can be found as free sulfur dioxide (not attracted to or not combined with other wine molecules or constituents) and as combined sulfur dioxide. Wine analyses comprise the evaluation of free and total sulfur dioxide (the sum of free and combined sulfur dioxide forms). There are several methods of analysis of free and total sulfur dioxide. The OIV (2015) describes at least three official methods: Paul's method (OIV-MA-AS323-04A), a modified Ripper's method (OIV-MA-AS323-04B), and the Beech and Tomas method (OIV-MA-AS323-04C). Ripper's method is the simplest and fastest among them, although it is not a precise method. In this method, free sulfur dioxide is determined by direct titration with an iodine solution. The Paul and the Beech and Tomas methods are quite a bit slower, compared to the Ripper method.

2.2 POTENTIOMETRY

Potentiometry is an electroanalytical method based on the measurement of the potential difference of an electrochemical cell in the absence of current. It is a method used to detect the end point of specific titration (potentiometric titration) or for the direct determination of a particular constituent in a sample by measuring the potential of an ion-selective electrode, which is precisely sensitive to the ion under review (analyte) (Zoski, 2006).

Because the equipment is simple and relatively inexpensive, consisting of a reference electrode, an indicator electrode, and a device for reading potential (potentiometer) attached thereto, and the assay dispenses with the use of indicators, which often do not show a detectable color change, potentiometry has become a widespread and reliable method in analytical chemistry. The pH meter is the most common and useful potentiometric-based laboratory apparatus.

2.2.1 pH

Mathematically, pH is the cologarithm of the concentration of H_3O^+ ions (hydroxonium ions) and it ranges from 0 (extremely acid) to 14 (extremely alkaline). This means that lower pH value in wine implies more acidic (true acidity) wine (Boulton, 1980; Ribéreau-Gayon et al., 2006b). In grape wines, the pH values range from 2.8 to 4.0 (Ribéreau-Gayon et al., 2006b). However, in fruit wines these pH normally vary, depending on the fruit must, seeming to be close to that for grape wines. For example, black plum wine has a pH 2.7 (Okigbo, 2003), banana wine has a pH 3.3 (Akubor et al., 2003), pineapple wine has a pH 3.7 (Chanprasartsuk et al., 2012), lychee wine has a pH 4.0 (Alves et al., 2011), and jackfruit wine has a pH of 4.4 (Jagtap et al., 2011).

The pH influences several wine parameters, for instance, high pH (pH higher than 3.6) is associated with the color and a general instability and high potassium (Walker et al., 2004). On the other hand, low pH values in wines are related to better stability. Low pH avoids bacterial growth and spoilage, increases the solubility of iron (ferric casse) and tartrates, and increases the free sulfur dioxide concentration (Ribéreau-Gayon et al., 2006b). pH is measured potentiometrically using a pH meter (method OIV-MA-AS313-15; OIV, 2015).

2.3 DENSIMETRY

Densimetry is a physical method used to determine density. As density is a specific property of matter, it can be used to determine the purity of a material, because it is significantly changed by the presence of contaminants. It also has a relation to temperature. Increasing the temperature of a given mass of matter, will give an increased fixed volume of this matter due to the dilation caused by the separation of atoms and molecules. In contrast, when the temperature is decreased, a decrease in the fixed volume is observed. The amount of mass in a given volume is called its density (Ribéreau-Gayon et al., 2006a; Walker, 2009).

Density can be absolute or apparent. Absolute density (also specific mass or volumetric density) of a body is defined as the quotient of mass and volume of the body. Thus it can be said that the density measures the concentration of mass in a given volume. The symbol for density is ϱ (the Greek letter rho) and the SI unit for density is kilograms per cubic meter (kg/m^3), even though its unit is usually expressed in g/mL or g/L. Apparent density is the ratio of the density of the substance and the density of a reference substance (water is generally taken as the reference). It is a dimensionless magnitude (Hawkes, 2004; Ribéreau-Gayon et al., 2006a).

The density of a liquid or liquid mixture can be determined by mass measurements of the liquid occupying a known volume (pycnometer method) and flotation methods based on the principle of Archimedes. Pycnometry is a process using a pycnometer to determine the relative density of two materials (liquid–liquid or liquid–solid) (Ribéreau-Gayon et al., 2006a; Walker, 2009).

2.3.1 Density and Specific Gravity

In wines, density represents the mass per volume unit of the beverage or must, at 20°C. Its expression is given in g/mL and is denoted by the symbol $\varrho_{20°C}$. Specific gravity (denoted by the symbol $d_{20°C}^{20°C}$) of a wine or must is the ratio of its density to the density of water, both at 20°C. The density and specific gravity at 20°C can be assayed in the sample by pycnometry, by electronic densimetry using an oscillating cell, and by densimetry with a hydrostatic balance (method OIV-MA-AS2-01A; OIV, 2015).

2.3.2 Alcoholic Strength

This parameter is most related to the concentration of ethanol, the major alcohol, in wine. It is defined as the number of liters of ethanol present in 1 hL of wine, at 20°C, and is expressed by the symbol "% vol." Taking into account that one of the determining steps of alcoholic strength is to obtain a distillate, it is reasonable to conclude that ethanol homologs, ethanol esters, and their homologs are included in the alcoholic strength value because of the presence of these compounds in the distillate (OIV, 2015).

According to the OIV (2015), alcoholic strength can be assayed by means of a variety of densimetric methods, including the use of a pycnometer (method OIV-MA-AS312-01A:4A), a frequency oscillator (method OIV-MA-AS312-01A:4B), or a hydrostatic balance (method OIV-MA-AS312-01A:4C).

In addition to these physical methods of measurement, other physical methods are commonly used for determining the alcoholic strength, like refractometry and hydrometry (method OIV-MA-AS312-01B).

3. CHROMATOGRAPHIC ANALYSIS

Since 2005, fruit wine has been highlighted as the new alcoholic fermented beverage. Several fruits like apple, blueberry, elderberry, peach, raspberry, and tropical fruits such as banana, cacao, cagaita, cupuaçu, gabiroba, jabuticaba, guava, lychee, mango, orange, pineapple, and umbu have been used to elaborate wines (Jagtap and Bapat, 2015). The use of these fruits along with the action of microorganisms in the fermentation is necessary to obtain wines with different flavors and aromas. These different flavors are a direct result of the final chemical composition of wines mainly due to the volatile and nonvolatile compounds present. In the context of the great complexity in the composition of fruit wines, the use of robust analytical techniques, such as chromatography, represents an important tool for characterizing these wines (Zoecklein et al., 1990, 1995).

Chromatography has been applied to the analysis of food for many decades, including in fruit wines, to determine the aromatic volatile compounds, organic acids, antioxidants, ethanol, sugars, and other compounds present in the beverage. The complexity of many arrays of fruit wine has required, in some cases, the improvement and/or development of chromatographic methods adapted to these matrices. Generally, the methods employed for grape wine have been a starting point for adjustment or development of chromatographic methods for the analysis of wines of fruits other than grapes. The various techniques that can be used to characterize fruit wines by liquid and gas chromatography are described here.

3.1 LIQUID CHROMATOGRAPHY

In recent years, high-performance liquid chromatography (HPLC) has become one of the more developed analytic techniques and is widely spread and used in various fields, including drink analysis. In the analysis of fruit wines, HPLC has been widely used for the determination of sugars, alcohols, organic acids, and phenolic compounds, among others (Table 5.6).

3.1.1 Analysis of Sugars

Sugar analysis is performed using various separation methods (columns) and may be successful with various detection systems (detectors). The chromatographic columns commonly used in the separation of sugars in

Table 5.6 Columns and Detectors for the Analysis of Sugars and Organic Acids in Fruit Wine by HPLC

Fruit Wine	Compounds	Column	Detection	References
Cherry	Sucrose, glucose, and fructose	Waters Sugar-Pak I cation exchange	RI	Niu et al. (2012)
Gabiroba	Sucrose, glucose, fructose, ethanol, and organic acids	Shim-pack SCR-101H cation exchange	RI for sugars and alcohols; UV–Vis for acids	Duarte et al. (2009)
Cagaita	Sucrose, glucose, fructose, ethanol, and organic acids	Shim-pack SCR-101H cation exchange	RI for sugars and alcohols; UV–Vis for acids	Oliveira et al. (2011)
Gabiroba, cupuaçu, jabuticaba, cacao, and umbu	Sucrose, glucose, fructose, ethanol, and organic acids	Chrompack 67H	RI for sugars and alcohols; UV–Vis for acids	Duarte et al. (2010a)
Raspberry	Sucrose, glucose, fructose, ethanol, and organic acids	Chrompack 67H	RI for sugars and alcohols; UV–Vis for acids	Duarte et al. (2010b)
Papaya	Glucose and fructose	Prevail carbohydrate ES	ELSD-LT	Lee et al. (2010, 2012)
Papaya	Organic acids	Supelcogel C610H	Photodiode array detector	Lee et al. (2010, 2012)
Kiwi	Sucrose, glucose, and fructose	NH_2	UV	Kallithraka et al. (2001)
Pomegranate	Organic acids	Supelcogel C610H	UV	Mena et al. (2012)

ELSD-LT, *low-temperature evaporative light scattering detection;* RI, *refractive index.*

foods and beverages are the NH_2 stationary phase and cation-exchange resins. The mobile phase, water, mixtures of acetonitrile and water (80:20, 75:25), or acidic solutions of sulfuric acid, orthophosphoric acid, and perchloric acid, are used under different conditions of temperature and flow (Fig. 5.2).

After separation on a column, the sugar detection is usually performed using detectors based on refractive index (RI), and evaporative light scattering. Some methodologies from studies conducted using various fruits for which sugars present in wines were determined by HPLC are presented next.

Niu et al. (2012) characterized cherry wine, determining sucrose, glucose, and fructose in five different wines using a Waters Sugar-Pak I cation-exchange column at 85°C with water as mobile phase at a flow rate of 0.4 mL/min; using this methodology, they reported the good separation of those three sugars. In addition to the chromatographic separation with water as the eluent, an acidified mobile phase has been most commonly used for analysis of sugars in fruit wines. Duarte et al. (2009) used a 100 mM perchloric acid solution for the analysis of glucose, fructose, and sucrose in gabiroba wine at a flow of 0.8 mL/min with a Shim-pack SCR-101H cation-exchange column. This column is also reported by Oliveira et al. (2011) for the analysis of sugar in cagaita wine under the same conditions cited previously by Duarte et al. (2009). In both cases, an RI detector was used. In addition to perchloric acid, sulfuric acid has been used as the mobile phase for sugar analysis in wines produced from various fruits. In their work, Duarte et al. (2010a) reported the use of liquid chromatography with detection by RI for sugar analysis in fruit wines produced with tropical fruits such as gabiroba, cupuaçu, jabuticaba,

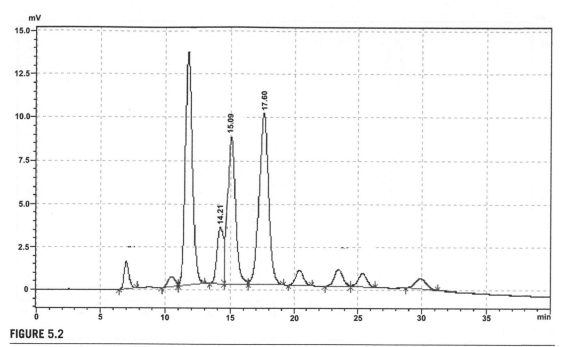

FIGURE 5.2

Chromatogram of organic acids (mix of standards). Peaks: 14.21 min, succinic acid; 15.09 min, lactic acid; 17.60 min, acetic acid. Conditions: UV detector 210 nm; SCR-101H column; 100 mM perchloric acid; 0.6 mL/min flow rate; temperature 50°C.

cacao, and umbu. They used a Chrompack 67H column operated at 37°C with a flow rate of 0.4 mL/min and 5 mmol sulfuric acid. These analytical conditions have been successfully employed in the separation of sugars in raspberry wine (Duarte et al., 2010b).

As mentioned earlier, several studies reported the use of RI detection in sugar analysis. However, some authors (Lee et al., 2012, 2010) have used low-temperature evaporative light scattering detection and acetonitrile:water (80:20) as the mobile phase at a flow rate of 1.4 mL/min at 40°C for the separation of glucose and fructose in papaya wine (Lee et al., 2012, 2010).

Although its use is not as common as the RI, some studies have reported the use of detection with ultraviolet (UV) in the analysis of sugars in fruit wines. According to Kallithraka et al. (2001) glucose, fructose, and sucrose were determined in kiwifruit wine using an NH$_2$ column operated with acetonitrile:water (80:20) with a flow of 0.6 mL/min and UV detection. The sugar separation and detection may be performed using the aforementioned columns and detectors; however, more frequently, owing to cost and efficiency, NH$_2$ columns have been used with acetonitrile:water and RI detection. The choice of detector, column, and mobile phase, however, depends upon various factors such as sample complexity, cost, available equipment, etc.

3.1.2 Analysis of Organic Acids

Analysis of organic acids in fruit wine has been performed, in most cases, employing ion-exchange columns, acidified mobile phases, and detectors based on the use of UV light as shown in Fig. 5.3.

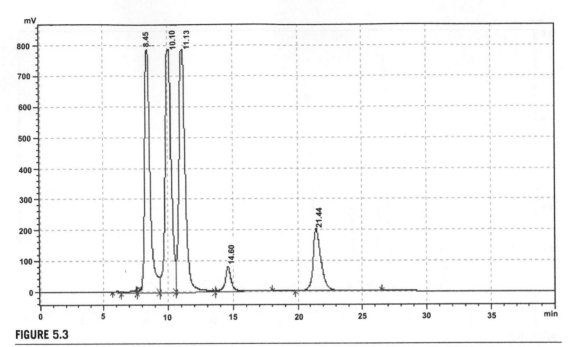

FIGURE 5.3

Chromatogram from sugars and alcohols (mix of standards). Peaks: 8.45 min, sucrose; 10.10 min, glucose; 11.13 min, fructose; 14.60 min, glycerol; 21.44 min, ethanol. Conditions: RI detector; SCR-101H column, 100 mM perchloric acid, 0.6 mL/min flow rate, temperature 30°C.

The organic acid analysis in pomegranate wine was described by Mena et al. (2012), using a Supelcogel C610H column, with mobile phase water:phosphoric acid (99.9:0.1) and a UV detector operated at 210 nm, wherein the authors reported the detection of malic, acetic, tartaric, and citric acids. Similarly, Navarrete-Bolaños et al. (2013) detected lactic acid, acetic acid, tartaric acid, malic acid, and oxalic acid using a Prevail organic acid column, with a mobile phase of potassium phosphate solutions with pH 2.5 adjusted with phosphoric acid and UV detector operated at 210 nm.

In addition to the detection by UV detector operated at a predefined wavelength, fruit wine analysis has been carried out with a diode array detector (DAD), whose main advantage over conventional UV detectors is its operation at variable wavelengths (scan). Lee et al. (2010) analyzed papaya wine using a DAD, Supelcogel C610H column, mobile phase of 1% sulfuric acid at 0.4 mL/min, 40°C, and detected acetic, citric, and malic acids.

The chromatographic analysis of organic acids in fruit wine has been performed, in some cases, together with sugars and other compounds such as glycerol and ethanol. In such cases, the chromatographic system consists of more than one detector connected in series and the separation is performed with a single injection. Duarte et al. (2009, 2010a,b) and Oliveira et al. (2011) analyzed different fruit wines and the chromatographic separation of sugars, ethanol, glycerol, and various organic acids was carried out using a cation-exchange column (Shimadzu SCR-101H and Chrompack 67H column) using an acidified mobile phase (100 mM perchloric acid or sulfuric acid); whereas sugars and alcohols were detected by RI, acids were detected by UV at 210 nm.

In the determination of sugars and acids in fruit wines, most of the workers do not employ specific methodologies for the sample treatment, the samples being diluted in some cases, because some compounds are usually present in relatively high concentrations. The injection of samples into the HPLC is preceded in most cases by centrifuging (for example, twice at 10,000 rpm, 10 min, 4°C) and filtration using 0.22- to 0.45-μm filters.

3.2 GAS CHROMATOGRAPHY

3.2.1 Analysis of Volatile Compounds

The determination of volatile compounds in fruit wine is reported in many studies in which mainly gas chromatography (GC) with flame ionization detection (FID) and mass spectrometry (MS) is used. The chromatographic analyses are preceded in many cases by extraction steps aimed at sample cleanup and concentration of the analytes. The extraction of volatile compounds is mainly performed by liquid–liquid extraction (LLE) and solid-phase microextraction (SPME). As usual for grape wine, in some work with other fruits (Duarte et al., 2010a,b), volatile compounds have been grouped as majority and minority. The major compounds, namely those present in higher concentrations, are analyzed without extraction, whereas those denominated minor compounds are analyzed after an extraction step.

LLE has been used by some authors for the analysis of wines such as orange (Selli et al., 2008). The authors described in their paper the use of 40 mL of dichloromethane added to 100 mL of wine. The mixture was stirred for 30 min at 4°C under nitrogen gas and subsequently centrifuged at 4°C, 9000 g for 4 min. At the end of extraction, the extract was reduced to 0.5 mL with nitrogen gas. LLE with dichloromethane was also used in the analysis of raspberry wine by Duarte et al. (2010b,c). The extraction of volatile compounds in raspberry wine was done using methodology developed by Oliveira et al. (2006) for volatile extraction in grape wine. Four hundred microliters of dichloromethane was added to 8 mL of wine for the extraction, with stirring for 15 min. Then the mixture was cooled for 10 min at 0°C followed by centrifugation at 4°C and RCF (relative centrifugal force)=5118.5 to separate the organic extract. The use of considerable amounts of organic solvents characterizes, in some cases, a disadvantage of the use of LLE.

The extraction of volatile compounds in many studies has been performed by SPME. For this extraction, various kinds of fibers are commercially available (see the following examples). The temperature and extraction time are the parameters determinant of the success or not of the extraction of volatiles in fruit wine samples. Kim and Park (2015) used PDMS/DVB (polydimethylsiloxane/divinylbenzene) fiber for the extraction of volatile compounds from 5-mL samples of blackberry wine in 10-mL vials at 60°C for 30 min. Analyzing strawberry wine, Kafkas et al. (2006) used PDMS fiber with extraction for 30 min at 30°C.

Some authors in the pursuit of optimization of the SPME process have evaluated different fibers for one kind of sample, although most of the manufacturers suggest a specific use for each type of fiber. Bernardi et al. (2014) in their work tested three different fibers, PDMS, PDMS/carboxen (CAR), and PDMS/DVB/CAR, and found that for extraction of minor compounds in jelly palm wine, the PDMS/DVB/CAR fiber showed good results with extraction performed 35°C for 45 min.

The volatile compounds, previously extracted or not, are separated on chromatography columns whose characteristics vary according to the characteristics of the sample and the interest of the researcher. For example, Duarte et al. (2010a,b,c) used a CP-Wax 57 CB (50 m × 0.25 mm i.d., 0.2 μm film thickness; Chrompack), for major compounds, whereas for minor volatile compounds a Factor Four VF-Wax MS Varian, 60 m × 0.25 mm i.d., 0.25 μm film thickness, was used. These columns were employed for separation of volatile compounds in raspberry, cacao, jabuticaba, gabiroba, cupuaçu, and umbu wine. In the three cited studies, the separation was performed using a temperature ramp starting from 50–60 to 220°C with increments of 3°C/min.

Bernardi et al. (2014) and Kim and Park (2015), while analyzing respectively jelly palm and blackberry wine, employed similar capillary column (DB-Wax and ZB-Wax). Bernardi et al. (2014) separated major volatile compounds using a ZB-Wax column (30 m × 0.25 mm × 0.25 μm film thickness), whereas minor compounds were separated by a ZB-Wax 60 m long. In both works, temperature ramps were used to optimize the separations (see original article for details).

The detection of volatile compounds in most studies with fruit wine has been done using FID and MS. Analyses with FID are usually performed at temperatures of 220–250°C for the determination of major volatile compounds as reported by Duarte et al. (2010b) and Bernardi et al. (2014). In analysis with MS, it is commonly operated in electron impact mode (70 eV) with the acquisition of spectra in scan mode with m/z varying from a few tens (e.g., 20 m/z) to a few hundred (450 m/z).

The quantification of volatile compounds is in some cases performed with the use of calibration curves with highly pure reference compounds. However, most commonly an internal standard is used. An internal standard used in work with fruit wine is 4-nonanol as reported by Duarte et al. (2010a,b,c) and Selli et al. (2008).

3.2.2 Methanol

According to the OIV (2015), method OIV-MA-AS312-03A, the content of methanol is determined in distilled wine by GC and detected by flame ionization. To the samples, an internal standard (4-methyl-2-pentanol, 3-pentanol, 4-methyl-1-pentanol, or methyl nonanoate) is added and, after sample preparation, the analysis is performed using a fused silica capillary column coated with a Carbowax 20M-type polar stationary phase (Chrompack CP-Wax 57 CB, 50 m × 0.32 mm × 0.45 μm, DB-Wax 52, 30 m × 25 mm × 0.2 μm). The OIV recommends the use of a carrier gas flow of 7 mL/min, split ratio of 7:50, injection volume of 1 or 2 μL, injector temperature of 200–260°C, detector temperature of 220–300°C, and oven temperature program from 35°C for 2 min to 170°C, at 7.5°C/min.

3.2.3 Ethyl Carbamate

The OIV recommends the following methodology for the analysis of ethyl carbamate with concentration ranging from 10 to 200 μg/L (method OIV-MA-AS315-04). The compound is determined by GC/MS using a capillary fused silica column: 30–50 m × 0.25 mm × 0.25 μm of the Carbowax 20M type. The injector temperature should be 180°C, flow rate of carrier gas 1 mL/min, and oven temperature beginning at 40°C for 0.75 min, programmed to rise 10°C/min to 60°C, then 3°C/min until 150°C, then go up to 220°C, and stay there for 4.25 min. The GC/MS interface should be kept at 220°C.

4. MICROBIOLOGICAL ANALYSIS

In winemaking, during spontaneous fermentation of the must, there is a succession of yeast species. Initially, there is the growth of species susceptible to ethanol, as *Kloeckera* sp., *Hanseniaspora* sp., *Candida* sp., *Metschnikowia* sp., *Pichia* sp., and *Kluyveromyces* sp., followed by some species of *Saccharomyces*. Afterward, with a mean of 3–5 days of fermentation, when the ethanol concentration in the must approaches 10%, the predominant yeast is *Saccharomyces cerevisiae* (Fleet et al., 1984;

Querol et al., 1994; Cocolin et al., 2000). The non-*Saccharomyces* yeasts then lose viability because of their low ethanol tolerance, but they account for some end products characterizing the wine (Ciani and Comitini, 2011). *S. cerevisiae* grows during this period and persists until the end of fermentation because of its high ethanol tolerance.

During the fermentation process, and especially after the alcoholic fermentation, the lactic acid bacteria (responsible for malolactic fermentation) may be present and benefit the final beverage by converting malate to lactate, which reduces the wine acidity, improving the quality and stability in high-acid wines (Davis et al., 1985; Liu, 2002).

Acetic acid bacteria can also be present during fermentation, or grow during wine storage or even in the bottled wine. Acetic acid bacteria are undesirable in winemaking, because these obligate gram-positive bacteria convert ethanol to acetic acid, which is undesirable and spoils the wine (Joyeux et al., 1984; Bartowsky and Henschke, 2008).

Regarding fruit wine spontaneous fermentation, the microbial ecology seems to be similar to that for wine fermentation. Fruit wine fermentation is characterized by a predominance of yeasts, and *Saccharomyces* prevails over non-*Saccharomyces* yeasts, lactic acid bacteria and acetic acid bacteria species being also observed. Spontaneous fermentation of apple must to produce cider showed these characteristic yeast population dynamics. Non-*Saccharomyces* yeast strains of *Hanseniaspora* sp. and *Metschnikowia pulcherrima* in predominated in the early stages, whereas *S. cerevisiae* strains were most abundant at the end of fermentation (Valles et al., 2007). During fermentation of gabiroba (*Campomanesia pubescens*) must, the non-*Saccharomyces* yeasts *Candida quercitrusa* and *Issatchenkia terricola* were found during the early days of fermentation. *S. cerevisiae* prevailed after 7 days of fermentation until the end of the process (Duarte et al., 2009). When studying persimmon wine fermentation, Hidalgo et al. (2012) identified the non-*Saccharomyces* species *Pichia guilliermondii*, *Hanseniaspora uvarum*, *Zygosaccharomyces florentinus*, and *Cryptococcus* sp. throughout the fermentation time course. *S. cerevisiae* strains were higher in population after the middle and late stages, during which it was dominant.

Microbiological analysis is an important tool for the evaluation of the microbial community present in fruit wine fermentation considering the following aspects: studies of microbial ecology and interrelations between microorganisms and their importance in the must ecosystem, to evaluate the performance or stability of a yeast or lactic bacteria during fermentation, and to estimate microbial spoilage of wine (Lonvaud, 2011). In most fermentations in which a pure culture of *S. cerevisiae* is used, the microbiological analysis is performed mostly to determine the purity of the culture and its population. It is similar in cases in which malolactic bacteria carry out fermentation in high-acid musts to reduce acidity through decarboxylation. The spoilage of wine is investigated frequently to identify coliform bacteria, lactic acid or acetic acid bacteria, and commonly non-*Saccharomyces* yeasts in the wine. These kinds of analyses can be carried out though classical techniques or molecular techniques.

4.1 CLASSICAL TECHNIQUES

Official microbiological methods of analysis are focused on the detection of abnormal microbial behavior or populations during fermentation or in the final beverage, commonly using classical microbiological techniques; an exception is made for the molecular identification of *Dekkera/Brettanomyces bruxellensis* (method OIV-MA-AS4-03; OIV, 2015), a spoilage yeast (Oelofse et al., 2009). Classical techniques comprise several routine practices such as microscopic techniques (cell size differentiation,

Gram staining, yeast cell counting, yeast cell viability), plating techniques (counting cell-forming units, colony differentiation, microbial isolation), and culture in liquid medium (most probable number estimation). Detailed information and procedures for these official techniques are described in method OIV-MA-AS4-01 (OIV, 2015).

The phenotypic approach, an important tool in identifying microorganisms, is not mentioned in the official methods. It consists in biochemical and physiological tests for the identification of yeasts and bacteria assessing metabolic characteristics, with respect to fermentative ability of various sources of carbohydrates (mono-, di-, and oligosaccharides), assimilation of carbon sources and nitrogen application vitamins and other growth factors, enzymatic activity, antimicrobial resistance, and other compounds and secretion of metabolites (e.g., acetic acid, starch, enzymes). The phenotypic approach, however, is time-consuming but still useful in identifying microorganisms (Kurtzman et al., 2011; Vos et al., 2011).

4.2 MOLECULAR TECHNIQUES

Phenotypic tests have limitations in the characterization of genera and related species, whereas the use of molecular methods allows verifying a huge differentiation between species. The possibility of identifying species of microorganisms in their native habitat, by molecular techniques, without isolation, has revolutionized studies of microbial ecology and provided new applications in various sectors. Molecular techniques were developed during the 1990s, expanding research in the microbial ecology arena (Kurtzman, 2011).

Among the molecular methods, PCR (polymerase chain reaction) to amplify DNA fragments in substantial quantities is distinguished by simplicity of operation and does not require DNA isolation; it has become accessible for use in routine analysis. Molecular techniques –[pulsed-field gel electrophoresis, random amplification of polymorphic DNA, amplified fragment length polymorphism, restriction fragment length polymorphism (RFLP)] that enable detection and strain differentiation, species identification, and genetic similarity analysis, even in the absence of information of the nucleotide sequence, have been widely used successfully in characterizing yeast in various wine environments (Querol et al., 1992; Querol and Ramon, 1996; Cocolin et al., 2000; Fleet, 2007; Mills et al., 2008; Ivey and Phister, 2011; Cappello et al., 2014; González-Arenzana et al., 2014; Visintin et al., 2016). Regarding fruit wines, there are a few reports on molecular techniques to evaluate microbial ecology and population dynamics of yeasts and bacteria. Among these are the use of RFLP to identify yeasts in spontaneous fermentation of cider (Valles et al., 2007), PCR–denaturing gradient gel electrophoresis to evaluate yeast populations in gabiroba and pineapple musts (Duarte et al., 2009; Chanprasartsuk et al., 2010), and RFLP–PCR to identify yeasts isolated from the fermentation of persimmon must (Hidalgo et al., 2012).

5. SENSORY ANALYSIS

The use of sensory analysis in enology is an ancient practice. It can be used in the screening of fruit for wine production, in the evaluation of yeast and malolactic bacteria for fermentation, and to find out sensory properties related to consumers' acceptance (Joshi, 2006; Narasimhan and Stephen, 2011). Sensory analysis is a very important step in work with beverage characterization. Especially for the

Table 5.7 Methodologies Used in the Sensory Analysis of Fruit Wines

Fruit Wine	Sensory Analysis	References
Gabiroba	Untrained tasters with hedonic scale of 9 points	Duarte et al. (2009)
Gabiroba, cupuaçu, jabuticaba, cacao, and umbu	Untrained tasters with hedonic scale of 9 points	Duarte et al. (2010a)
Cacao	Untrained tasters with hedonic scale of 9 points	Dias et al. (2007)
Jackfruit	Untrained tasters with hedonic scale of 9 points	Asquieri et al. (2008)
Caja	Untrained tasters with hedonic scale of 9 points	Dias et al. (2003)
Raspberry	Quantitative descriptive analysis with 12 trained panelists	Duarte et al. (2010b)
Cherry	Quantitative descriptive analysis with 8 trained tasters	Niu et al. (2011) and Sun et al. (2013)

fruit wines, this analysis provides interesting results because most fruit wines are markedly characterized by aromatic nuances typical from the fruits used. Sensory analysis of fruit wines has been performed with both trained and untrained tasters (Table 5.7).

In the case of analyses with trained panelists, quantitative descriptive analysis (QDA) has been the most commonly cited; whereas for analysis with untrained tasters, the use of the hedonic scale has been reported in several studies on fruit wines.

The QDA was employed by Sun et al. (2013) for sensory evaluation of cherry wine by eight trained tasters with eight sensory terms, floral, fruity, almond, sweet, fusel, green, sweaty, and general aroma, to describe and differentiate the samples. The intensity of each attribute was carried out using a scale from 0 to 5, where 0 indicated the absence of the descriptor and 1–5 respectively corresponding to the intensities very low, low, medium, high, and very high. This 0 to 5 scale was also employed by Kim and Park (2015) when 25 trained tasters described blackberry wine with the following descriptors: herbaceous, sour flavor, fruity aroma, floral aroma, waxy, sweet aroma, and overall acceptability. Niu et al. (2011) described the sensory evaluation of cherry wine using the QDA during five sessions, in which eight panelists used a scale from 0 to 9 (where 0 indicates no descriptor and 9 the highest intensity) to score six descriptors, fruity, sour, woody, fermentation, caramel, and floral, that allowed the characterization of the samples. The QDA was also used by Duarte et al. (2010b) for characterization and differentiation of various raspberry wines, in which 12 trained panelists identified aroma descriptors (sulfide, balsamic resin, yogurt, red fruit, dried fruit, tangerine, pineapple, tropical, flora, blackberry, medicinal, herbaceous), color (rose, orange, strawberry, raspberry, violet, and cherry), and flavor (acid, sweet, salty, and bitter). The authors concluded that the QDA results were directly related to the results obtained by GC/MS and GC/pulsed-flame photometric detection.

Fruit wines have been also evaluated using a hedonic scale of 9 points (where 1 indicates "dislike extremely" and 9, "like extremely") and as key attributes aroma, flavor, appearance, and general aspects. Several authors reported the use of this methodology for various wines of fruits like mango (Sudheer Kumar et al., 2009), gabiroba (Duarte et al., 2009), cagaita (Oliveira et al., 2011), cacao (Dias et al., 2007), jackfruit (Asquieri et al., 2008), and caja (Dias et al., 2003). The number of untrained panelists participating in the sensory evaluation is variable; commonly, samples of 20–25 mL are provided to the tasters. Asquieri et al. (2008) evaluated fermented jackfruit using 73 tasters, whereas the work of

Duarte et al. (2009) and Dias et al. (2007), evaluating gabiroba and cacao wine (respectively), was done by 50 panelists.

In both evaluations, with trained panelists using the QDA or with untrained tasters and hedonic scale, it has been possible to efficiently characterize and distinguish various fruit wines. The results of these analyses are, in most studies, submitted to statistical evaluations using different tools, including mainly nonparametric tests and multivariate techniques.

6. FUTURE PROSPECTS

A critical appraisal of all the methods used to evaluate the quality of fruit wine would clearly reveal that most of the analytical techniques used are the same or small modifications of the methods used for evaluation of wines from grapes. Each fruit has characteristics that give the wine its quirks, making each drink a matrix that requires dedicated analytical methodologies. Thus, as new wines from fruit other than grapes are developed, new analytical techniques for the determination of volatile compounds, nonvolatile compounds, and sensory characterization will be developed and improved. There is also a need to develop protocols that will ensure the safety of the fruit wines. How much of the components that affect human health are provided by the consumption of wine can also be estimated by the appropriate method(s) to determine the active components and the mechanism of action. Similarly, precise methods for determining toxic components in fruit wines need to be developed in the near future.

REFERENCES

AICV, 2014. Code of Practice. http://www.aicv.org/pages/aicv/definitions.htmlhttp://aicv.org/file.handler?f=Cider Trends2014.pdf.

Akubor, P.I., Obio, S.O., Nwadomere, K.A., Obiomah, E., 2003. Production and quality evaluation of banana wine. Plant Foods for Human Nutrition 58 (3), 1–6.

Alves, J.A., de Oliveira Lima, L.C., Nunes, C.A., Dias, D.R., Schwan, R.F., 2011. Chemical, physical–chemical, and sensory characteristics of lychee (*Litchi chinensis* Sonn) wines. Journal of Food Science 76 (5), S330–S336.

Alves, J.A., Lima, L.C.D.O., Dias, D.R., Nunes, C.A., Schwan, R.F., 2010. Effects of spontaneous and inoculated fermentation on the volatile profile of lychee (*Litchi chinensis* Sonn) fermented beverages. International Journal of Food Science and Technology 45 (11), 2358–2365.

Amerine, M.A., Ough, C.S., 1980. Methods for Analysis of Musts and Wines. John Wiley and Sons, New York.

Asquieri, E.R., Rabêlo, A.M.D.S., Silva, A.G.D.M., 2008. Fermented jackfruit: study on its physicochemical and sensorial characteristics. Food Science and Technology (Campinas) 28 (4), 881–887.

Aung, M.T., Lee, P.R., Yu, B., Liu, S.Q., 2015. Cider fermentation with three *Williopsis saturnus* yeast strains and volatile changes. Annals of Microbiology 65 (2), 921–928.

Austrian Wine Act, 2009. Federal Act on the Marketing of Wine and Fruit-Made Wine. Bundesgesetzblattfür die Republik Österreich, Part I, No. 111, November 17, 2009. 35 pp http://faolex.fao.org/docs/pdf/aut97740.pdf (in German).

Ball, E., 1946. The raw material for cider-making. Chemistry and Industry 10 (4), 38–39.

Bartowsky, E.J., Henschke, P.A., 2008. Acetic acid bacteria spoilage of bottled red wine – a review. International Journal of Food Microbiology 125 (1), 60–70.

Bernardi, G., Vendruscolo, R.G., dos Santos Ferrão, T., Barin, J.S., Cichoski, A.J., Wagner, R., 2014. Jelly palm (*Butia odorata*) wine: characterization of volatile compounds responsible for aroma. Food Analytical Methods 7 (10), 1982–1991.

Boulton, R., 1980. The relationships between total acidity, titratable acidity and pH in wine. American Journal of Enology and Viticulture 31 (1), 76–80.

Brasil, 2009. Ministério da Agricultura, Pecuária e Abastecimento. Decreto n° 6.871, de 4 de junho de 2009. Regulamenta a Lei no 8.918, de 14 de julho de 1994, que dispõe sobre a padronização, a classificação, o registro, a inspeção. a produção e a fiscalização de bebidas, Brasília, DF. http://www.planalto.gov.br/ccivil_03/_Ato2007-2010/2009/Decreto/D6871.htm.

Bridgebassie, V., Badrie, N., 2004. Effects of different pectolase concentration and yeast strains on carambola wine quality in Trinidad, West Indies. Fruits 59 (02), 131–140.

Butkhup, L., Jeenphakdee, M., Jorjong, S., Samappito, S., Samappito, W., Chowtivannakul, S., 2011. HS-SPME-GC-MS analysis of volatile aromatic compounds in alcohol related beverages made with mulberry fruits. Food Science and Biotechnology 20 (4), 1021–1032.

Cappello, M.S., De Domenico, S., Logrieco, A., Zapparoli, G., 2014. Bio-molecular characterisation of indigenous *Oenococcus oeni* strains from Negroamaro wine. Food Microbiology 42, 142–148.

Celep, E., Charehsaz, M., Akyüz, S., Acar, E.T., Yesilada, E., 2015. Effect of in vitro gastrointestinal digestion on the bioavailability of phenolic components and the antioxidant potentials of some Turkish fruit wines. Food Research International 78, 209–215.

CFR, 2015. Code of Federal Regulations. Title 27-Alcohol, Tobacco Products and Firearms. Chapter I – Alcohol and Tobacco Tax and Trade Bureau, Department of the Treasury. Subchapter a – Alcohol. Part 4-Labeling and Advertising of Wine. Subpart C – Standards of Identity for Wine. Section 4.21-The Standards of Identity. http://www.ecfr.gov/cgi-bin/text-idx?SID=82a2effe84149c7b68fa58a0b12e4cb8andmc=trueandtpl=/ecfrbrowse/Title27/27cfrv1_02.tpl#0.

Chanprasartsuk, O.O., Prakitchaiwattana, C., Sanguandeekul, R., Fleet, G.H., 2010. Autochthonous yeasts associated with mature pineapple fruits, freshly crushed juice and their ferments; and the chemical changes during natural fermentation. Bioresource Technology 101 (19), 7500–7509.

Chanprasartsuk, O., Pheanudomkitlert, K., Toonwai, D., 2012. Pineapple wine fermentation with yeasts isolated from fruit as single and mixed starter cultures. Asian Journal of Food and Agro-Industry 5 (2), 104–111.

Cheirsilp, B., Umsakul, K., 2008. Processing of banana-based wine product using pectinase and α-amylase. Journal of Food Process Engineering 31 (1), 78–90.

Chen, D., Chia, J.Y., Liu, S.Q., 2014. Impact of addition of aromatic amino acids on non-volatile and volatile compounds in lychee wine fermented with *Saccharomyces cerevisiae* MERIT. ferm. International Journal of Food Microbiology 170, 12–20.

Chinnici, F., Sonni, F., Natali, N., Riponi, C., 2013. Oxidative evolution of (+)-catechin in model white wine solutions containing sulfur dioxide, ascorbic acid or gallotannins. Food Research International 51 (1), 59–65.

Ciani, M., Comitini, F., 2011. Non-*Saccharomyces* wine yeasts have a promising role in biotechnological approaches to winemaking. Annals of Microbiology 61 (1), 25–32.

Cocolin, L., Bisson, L.F., Mills, D.A., 2000. Direct profiling of the yeast dynamics in wine fermentations. FEMS Microbiology Letters 189 (1), 81–87.

Coelho, E., Vilanova, M., Genisheva, Z., Oliveira, J.M., Teixeira, J.A., Domingues, L., 2015. Systematic approach for the development of fruit wines from industrially processed fruit concentrates, including optimization of fermentation parameters, chemical characterization and sensory evaluation. LWT – Food Science and Technology 62 (2), 1043–1052.

Darias-Martín, J., Lobo-Rodrigo, G., Hernández-Cordero, J., Díaz-Díaz, E., Díaz-Romero, C., 2003. Alcoholic beverages obtained from black mulberry. Food Technology and Biotechnology 41 (2), 173–176.

Davidovic, S.M., Veljovic, M.S., Pantelic, M.M., Baosic, R.M., Natic, M.M., Dabic, D.C., Vukosavljevic, P.V., 2013. Physicochemical, antioxidant and sensory properties of peach wine made from red haven cultivar. Journal of Agricultural and Food Chemistry 61 (6), 1357–1363.

Davis, C.R., Wibowo, D., Eschenbruch, R., Lee, T.H., Fleet, G.H., 1985. Practical implications of malolactic fermentation: a review. American Journal of Enology and Viticulture 36 (4), 290–301.

Dias, D.R., 2015. Agro industrial uses of cocoa by-products. In: Schwan, R.F., Fleet, G.H. (Eds.), Cocoa and Coffee Fermentations. CRC Taylor and Francis, Boca Raton, pp. 309–340 (Chapter 8).

Dias, D.R., Schwan, R.F., Lima, L.C.O., 2003. Methodology for elaboration of fermented alcoholic beverage from yellow mombin (*Spondias mombin*). Food Science and Technology 23 (3), 342–350.

Dias, D.R., Schwan, R.F., Freire, E.S., Serôdio, R.D.S., 2007. Elaboration of a fruit wine from cocoa (*Theobroma cacao* L.) pulp. International Journal of Food Science and Technology 42 (3), 319–329.

Duarte, W.F., Dias, D.R., de Melo Pereira, G.V., Gervásio, I.M., Schwan, R.F., 2009. Indigenous and inoculated yeast fermentation of gabiroba (*Campomanesia pubescens*) pulp for fruit wine production. Journal of Industrial Microbiology and Biotechnology 36 (4), 557–569.

Duarte, W.F., Dias, D.R., Oliveira, J.M., Teixeira, J.A., e Silva, J.B.D.A., Schwan, R.F., 2010a. Characterization of different fruit wines made from cacao, cupuassu, gabiroba, jaboticaba and umbu. LWT – Food Science and Technology 43 (10), 1564–1572.

Duarte, W.F., Dias, D.R., Oliveira, J.M., Vilanova, M., Teixeira, J.A., e Silva, J.B.D.A., Schwan, R.F., 2010b. Raspberry (*Rubus idaeus* L.) wine: yeast selection, sensory evaluation and instrumental analysis of volatile and other compounds. Food Research International 43 (9), 2303–2314.

Duarte, W.F., Dragone, G., Dias, D.R., Oliveira, J.M., Teixeira, J.A., e Silva, J.B.D.A., Schwan, R.F., 2010c. Fermentative behavior of *Saccharomyces* strains during microvinification of raspberry juice (*Rubus idaeus* L.). International Journal of Food Microbiology 143 (3), 173–182.

Ehling, S., Cole, S., 2011. Analysis of organic acids in fruit juices by liquid chromatography – mass spectrometry: an enhanced tool for authenticity testing. Journal of Agricultural and Food Chemistry 59 (6), 2229–2234.

Fleet, G.H., 2007. Yeasts in foods and beverages: impact on product quality and safety. Current Opinion in Biotechnology 18 (2), 170–175.

Fleet, G.H., Lafon-Lafourcade, S., Ribéreau-Gayon, P., 1984. Evolution of yeasts and lactic acid bacteria during fermentation and storage of Bordeaux wines. Applied and Environmental Microbiology 48 (5), 1034–1038.

Flores, P., Hellín, P., Fenoll, J., 2012. Determination of organic acids in fruits and vegetables by liquid chromatography with tandem-mass spectrometry. Food Chemistry 132 (2), 1049–1054.

Gómez, L.F.H., Úbeda, J., Briones, A., 2008. Characterisation of wines and distilled spirits from melon (*Cucumis melo* L.). International Journal of Food Science and Technology 43 (4), 644–650.

González-Arenzana, L., López, R., Portu, J., Santamaría, P., Garde-Cerdán, T., López-Alfaro, I., 2014. Molecular analysis of *Oenococcus oeni* and the relationships among and between commercial and autochthonous strains. Journal of Bioscience and Bioengineering 118 (3), 272–276.

Guerrero, R.F., Cantos-Villar, E., 2015. Demonstrating the efficiency of sulphur dioxide replacements in wine: a parameter review. Trends in Food Science and Technology 42 (1), 27–43.

Hawkes, S.J., 2004. The concept of density. Journal of Chemical Education 81 (1), 14–15.

Hidalgo, C., Mateo, E., Mas, A., Torija, M.J., 2012. Identification of yeast and acetic acid bacteria isolated from the fermentation and acetification of persimmon (*Diospyros kaki*). Food Microbiology 30 (1), 98–104.

Huerta, M.D., Salinas, M.R., Masoud, T., Alonso, G.L., 1998. Wine differentiation according to color using conventional parameters and volatile components. Journal of Food Composition and Analysis 11 (4), 363–374.

Isaac, A., Davis, J., Livingstone, C., Wain, A.J., Compton, R.G., 2006. Electroanalytical methods for the determination of sulfite in food and beverages. TrAC Trends in Analytical Chemistry 25 (6), 589–598.

Ivey, M.L., Phister, T.G., 2011. Detection and identification of microorganisms in wine: a review of molecular techniques. Journal of Industrial Microbiology and Biotechnology 38 (10), 1619–1634.

Jackson, R.S., 2008. Wine Science – Principles, Practices, Perception, third ed. Academic Press, San Diego.

Jagtap, U.B., Bapat, V.A., 2015. Phenolic composition and antioxidant capacity of wine prepared from custard apple (*Annona squamosa* L.) fruits. Journal of Food Processing and Preservation 39 (2), 175–182.

Jagtap, U.B., Waghmare, S.R., Lokhande, V.H., Suprasanna, P., Bapat, V.A., 2011. Preparation and evaluation of antioxidant capacity of Jackfruit (*Artocarpus heterophyllus* Lam.) wine and its protective role against radiation induced DNA damage. Industrial Crops and Products 34 (3), 1595–1601.

Jeong, J.H., Jung, H., Lee, S.R., Lee, H.J., Hwang, K.T., Kim, T.Y., 2010. Anti-oxidant, anti-proliferative and anti-inflammatory activities of the extracts from black raspberry fruits and wine. Food Chemistry 123 (2), 338–344.

Joshi, V.K., Shah, P.K., 1998. Effect of wood treatment on chemical and sensory quality of peach wine during ageing. Acta Alimentaria 27 (4), 307–318.

Joshi, V.K., Sharma, S., Bhushan, S., 2005. Effect of method of preparation and cultivar on the quality of strawberry wine. Acta Alimentaria 34 (4), 339–353.

Joshi, V.K., 1997. Fruit Wines, second ed. Directorate of Extension Education, Dr. Y.S. Parmar University of Horticulture and Forestry, Nauni, Solan, India, p. 255.

Joshi, V.K., 2006. Sensory Science: Principles and Application in Food Evaluation. Agro-Tech Academy, Udaipur, p. 527.

Joshi, V.K., Attri, D., 2005. A panorama of wine research in India. Journal of Scientific and Industrial Research 64 (1), 9–15.

Joshi, V.K., Sandhu, D.K., Thakur, N.S., 1999. Fruit based alcoholic beverages. In: Joshi, V.K., Pandey, A. (Eds.), Biotechnology: Food Fermentation (Microbiology, Biochemistry and Technology), vol. II. Educational Publishers and Distributors, New Delhi, p. 647.

Joshi, V.K., Sharma, S., Bhushan, S., Attri, D., 2004. Fruit based alcoholic beverages. In: Pandey, A. (Ed.), Concise Encyclopedia of Bioresources Technology. Haworth Food Product Press, New York, pp. 335–367.

Joshi, V.K., Thakur, N.S., Bhat, A., Garg, C., 2011. Wine and Brandy: a perspective. In: Joshi, V.K. (Ed.), Handbook of Enology, vol. 3. Asia -Tech Publisher, New Delhi, pp. 1–45.

Joyeux, A., Lafon-Lafourcade, S., Ribéreau-Gayon, P., 1984. Evolution of acetic acid bacteria during fermentation and storage of wine. Applied and Environmental Microbiology 48 (1), 153–156.

Kafkas, E., Cabaroglu, T., Selli, S., Bozdoğan, A., Kürkçüoğlu, M., Paydaş, S., Başer, K.H.C., 2006. Identification of volatile aroma compounds of strawberry wine using solid-phase microextraction techniques coupled with gas chromatography–mass spectrometry. Flavour and Fragrance Journal 21 (1), 68–71.

Kallithraka, S., Arvanitoyannis, I.S., Kefalas, P., El-Zajouli, A., Soufleros, E., Psarra, E., 2001. Instrumental and sensory analysis of Greek wines; implementation of principal component analysis (PCA) for classification according to geographical origin. Food Chemistry 73 (4), 501–514.

Kim, B.H., Park, S.K., 2015. Volatile aroma and sensory analysis of black raspberry wines fermented by different yeast strains. Journal of the Institute of Brewing 121 (1), 87–94.

Kurtzman, C., Fell, J.W., Boekhout, T. (Eds.), 2011. The Yeasts: A Taxonomic Study. Elsevier.

Lee, P.-R., Ong, Y.-L., Yu, B., Curran, P., Liu, S.-Q., 2010. Evolution of volatile compounds in papaya wine fermented with three *Williopsis saturnus* yeasts. International Journal of Food Science and Technology 45 (10), 2032–2041.

Lee, P.R., Chong, I.S.M., Yu, B., Curran, P., Liu, S.Q., 2012. Effects of sequentially inoculated *Williopsis saturnus* and *Saccharomyces cerevisiae* on volatile profiles of papaya wine. Food Research International 45 (1), 177–183.

Liu, S.Q., 2002. Malolactic fermentation in wine – beyond deacidification. Journal of Applied Microbiology 92 (4), 589–601.

Lonvaud, A., 2011. Microbial spoilage of wine. In: Joshi, V.K. (Ed.), Handbook of Enology, vol. 3. Asia -Tech Publisher, New Delhi, pp. 1367–1391.

MAF, 2011. Wine Standards Management Plan Code of Practice for Fruit Wine, Cider and Mead. Ministry of Agriculture and Forestry. Manager, Food Standards. New Zealand Standards Group. http://foodsafety.govt.nz/elibrary/industry/wine-standards-management-wsmc-cop-fwcm/cop.pdf.

Maragatham, C., Panneerselvam, A., 2011. Comparative analysis of papaya wine from other fruit wine. Journal of Pure and Applied Microbiology 5 (2), 967–969.

Mato, I., Suárez-Luque, S., Huidobro, J.F., 2005. A review of the analytical methods to determine organic acids in grape juices and wines. Food Research International 38 (10), 1175–1188.

Mena, P., Gironés-Vilaplana, A., Martí, N., García-Viguera, C., 2012. Pomegranate varietal wines: phytochemical composition and quality parameters. Food Chemistry 133 (1), 108–115.

Mills, D.A., Phister, T., Neeley, E., Johannsen, E., 2008. Wine fermentation. In: Cocolin, Ercolini (Eds.), Molecular Techniques in the Microbial Ecology of Fermented Foods. Springer, New York, pp. 162–192.

Mohanty, S., Ray, P., Swain, M.R., Ray, R.C., 2006. Fermentation of cashew (*Anacardium occidentale* L.) "apple" into wine. Journal of Food Processing and Preservation 30 (3), 314–322.

Muller, C.J., Kepner, R.E., Webb, A.D., 1964. Some volatile constituents of passion fruit wine. Journal of Food Science 29 (5), 569–575.

Narasimhan, S., Stephen, N.S., 2011. Sensory evaluation of wine and Brandy wine. In: Joshi, V.K. (Ed.), Handbook of Enology, vol. 3. Asia -Tech Publisher, New Delhi, pp. 1331–1366.

Navarrete-Bolaños, J.L., Fato-Aldeco, E., Gutiérrez-Moreno, K., Botello-Álvarez, J.E., Jiménez-Islas, H., Rico-Martínez, R., 2013. A strategy to design efficient fermentation processes for traditional beverages production: prickly pear wine. Journal of Food Science 78 (10), M1560–M1568.

Ndip, R.N., Akoachere, J.F., Dopgima, L.L., Ndip, L.M., 2001. Characterization of yeast strains for wine production. Applied Biochemistry and Biotechnology 95 (3), 209–220.

Niu, Y., Zhang, X., Xiao, Z., Song, S., Eric, K., Jia, C., Zhu, J., 2011. Characterization of odor-active compounds of various cherry wines by gas chromatography–mass spectrometry, gas chromatography–olfactometry and their correlation with sensory attributes. Journal of Chromatography B 879 (23), 2287–2293.

Niu, Y., Zhang, X., Xiao, Z., Song, S., Jia, C., Yu, H., Xu, C., 2012. Characterization of taste-active compounds of various cherry wines and their correlation with sensory attributes. Journal of Chromatography B 902, 55–60.

Oelofse, A., Lonvaud-Funel, A., Du Toit, M., 2009. Molecular identification of *Brettanomyces bruxellensis* strains isolated from red wines and volatile phenol production. Food Microbiology 26 (4), 377–385.

OIV, 2015. Compendium of international methods of wine and must analysis. Organisation Internationale de la vigne et du vin (OIV). Edition 2015, vols. 1 and 2. OIV, Paris.

Okigbo, R.N., 2003. Fermentation of black plum (*Vitex doniana* Sweet) juice for production of wine. Fruits 58 (06), 363–369.

Oliveira, J.M., Faria, M., Sá, F., Barros, F., Araújo, I.M., 2006. C 6-alcohols as varietal markers for assessment of wine origin. Analytica Chimica Acta 563 (1), 300–309.

Oliveira, M.D., Pantoja, L., Duarte, W.F., Collela, C.F., Valarelli, L.T., Schwan, R.F., Dias, D.R., 2011. Fruit wine produced from cagaita (*Eugenia dysenterica* DC) by both free and immobilised yeast cell fermentation. Food Research International 44 (7), 2391–2400.

Pérez-Gregorio, M.R., Regueiro, J., Alonso-González, E., Pastrana-Castro, L.M., Simal-Gándara, J., 2011. Influence of alcoholic fermentation process on antioxidant activity and phenolic levels from mulberries (*Morus nigra* L.). LWT – Food Science and Technology 44 (8), 1793–1801.

Polášková, P., Herszage, J., Ebeler, S.E., 2008. Wine flavor: chemistry in a glass. Chemical Society Reviews 37 (11), 2478–2489.

Querol, A., Ramon, D., 1996. The application of molecular techniques in wine microbiology. Trends in Food Science and Technology 7 (3), 73–78.

Querol, A., Barrio, E., Ramón, D., 1992. A comparative study of different methods of yeast strain characterization. Systematic and Applied Microbiology 15 (3), 439–446.

Querol, A., Barrio, E., Ramón, D., 1994. Population dynamics of natural *Saccharomyces* strains during wine fermentation. International Journal of Food Microbiology 21 (4), 315–323.

Reddy, L.V.A., Reddy, O.V.S., 2005. Production and characterization of wine from mango fruits (*Mangifera indica* L.). World Journal of Microbiology and Biotechnology 21 (8), 1345–1350.

Reddy, L.V.A., Reddy, O.V.S., 2011. Effect of fermentation conditions on yeast growth and volatile composition of wine produced from mango (*Mangifera indica* L.) fruit juice. Food and Bioproducts Processing 89 (4), 487–491.

Ribeiro, L.S., Duarte, W.F., Dias, D.R., Schwan, R.F., 2015. Fermented sugarcane and pineapple beverage produced using *Saccharomyces cerevisiae* and non-*Saccharomyces* yeast. Journal of the Institute of Brewing 121 (2), 262–272.

Ribéreau-Gayon, P., Dubourdieu, D., Donèche, B., Lonvaud, A., 2006a. Handbook of Enology. The Microbiology of Wine and Vinifications, vol. 1. John Wiley and Sons.

Ribéreau-Gayon, P., Glories, Y., Maujean, A., Dubourdieu, D., 2006b. Handbook of Enology. The Chemistry of Wine Stabilization and Treatment, vol. 2. John Wiley and Sons.

Rodríguez-Lerma, G.K., Gutiérrez-Moreno, K., Cárdenas-Manríquez, M., Botello-Álvarez, E., Jiménez-Islas, H., Rico-Martínez, R., Navarrete-Bolaños, J.L., 2011. Microbial ecology studies of spontaneous fermentation: starter culture selection for prickly pear wine production. Journal of Food Science 76 (6), M346–M352.

Santos, M.C., Nunes, C., Saraiva, J.A., Coimbra, M.A., 2012. Chemical and physical methodologies for the replacement/reduction of sulfur dioxide use during winemaking: review of their potentialities and limitations. European Food Research and Technology 234, 1–12.

Schmitzer, V., Veberic, R., Slatnar, A., Stampar, F., 2010. Elderberry (*Sambucus nigra* L.) wine: a product rich in health promoting compounds. Journal of Agricultural and Food Chemistry 58 (18), 10143–10146.

Selli, S., Cabaroglu, T., Canbas, A., 2003. Flavour components of orange wine made from a Turkish cv. Kozan. International Journal of Food Science and Technology 38 (5), 587–593.

Selli, S., Canbas, A., Varlet, V., Kelebek, H., Prost, C., Serot, T., 2008. Characterization of the most odor-active volatiles of orange wine made from a Turkish cv. Kozan (*Citrus sinensis* L. Osbeck). Journal of Agricultural and Food Chemistry 56 (1), 227–234.

Singh, M., Panesar, P.S., Marwaha, S.S., 1998. Studies on the suitability of kinnow fruits for the production of wine. Journal of Food Science and Technology 35 (5), 455–457.

Singleton, V.L., 1987. Oxygen with phenols and related reactions in musts, wines, and model systems: observations and practical implications. American Journal of Enology and Viticulture 38 (1), 69–77.

Soufleros, E.H., Pissa, I., Petridis, D., Lygerakis, M., Mermelas, K., Boukouvalas, G., Tsimitakis, E., 2001. Instrumental analysis of volatile and other compounds of Greek kiwi wine; sensory evaluation and optimisation of its composition. Food Chemistry 75 (4), 487–500.

Sudheer Kumar, Y., Prakasam, R.S., Reddy, O.V.S., 2009. Optimisation of fermentation conditions for mango (*Mangifera indica* L.) wine production by employing response surface methodology. International Journal of Food Science and Technology 44 (11), 2320–2327.

Sun, S.Y., Che, C.Y., Sun, T.F., Lv, Z.Z., He, S.X., Gu, H.N., Gao, Y., 2013. Evaluation of sequential inoculation of *Saccharomyces cerevisiae* and *Oenococcus oeni* strains on the chemical and aromatic profiles of cherry wines. Food Chemistry 138 (4), 2233–2241.

Sun, S.Y., Gong, H.S., Jiang, X.M., Zhao, Y.P., 2014. Selected non-*Saccharomyces* wine yeasts in controlled multistarter fermentations with *Saccharomyces cerevisiae* on alcoholic fermentation behaviour and wine aroma of cherry wines. Food Microbiology 44, 15–23.

Tsai, P.J., Huang, H.P., Huang, T.C., 2004. Relationship between anthocyanin patterns and antioxidant capacity in mulberry wine during storage. Journal of Food Quality 27 (6), 497–505.

Valles, B.S., Bedriñana, R.P., Tascón, N.F., Simón, A.Q., Madrera, R.R., 2007. Yeast species associated with the spontaneous fermentation of cider. Food Microbiology 24 (1), 25–31.

Visintin, S., Alessandria, V., Valente, A., Dolci, P., Cocolin, L., 2016. Molecular identification and physiological characterization of yeasts, lactic acid bacteria and acetic acid bacteria isolated from heap and box cocoa bean fermentations in West Africa. International Journal of Food Microbiology 216, 69–78.

Vos, P., Garrity, G., Jones, D., Krieg, N.R., Ludwig, W., Rainey, F.A., Whitman, W., 2011. Bergey's Manual of Systematic Bacteriology: Volume 3: The Firmicutes. Springer Science & Business Media.

Walker, J.S., 2009. Physics, fourth ed. United Kingdom: Pearson-Addison-Wesley. 1248 pages.

Walker, T., Morris, J., Threlfall, R., Main, G., 2004. Quality, sensory and cost comparison for pH reduction of Syrah wine using ion exchange or tartaric acid. Journal of Food Quality 27 (6), 483–496.

Wang, L., Sun, X., Li, F., Yu, D., Liu, X., Huang, W., Zhan, J., 2015. Dynamic changes in phenolic compounds, colour and antioxidant activity of mulberry wine during alcoholic fermentation. Journal of Functional Foods 18, 254–265.

Williams, A.A., Tucknott, O.G., 1978. The volatile aroma components of fermented ciders: minor neutral components from the fermentation of sweet cop pineapple juice. Journal of the Science of Food and Agriculture 29 (4), 381–397.

WSTA, 2015. The Code of Practice of the British Wine Producers' Committee of the Wine and Spirit Trade Association. http://www.wsta.co.uk/images/Committees/bwpccode.pdf.

Wu, J.S.B., Wu, M.C., Jiang, C.M., Hwang, Y.P., Shen, S.C., Chang, H.M., 2005. Pectinesterase inhibitor from jelly-fig (*Ficus awkeotsang* Makino) achenes reduces methanol content in carambola wine. Journal of Agricultural and Food Chemistry 53 (24), 9506–9511.

Yan, H.G., Zhang, W.H., Chen, J.H., Ding, Z.E., 2012. Optimization of the alcoholic fermentation of blueberry juice by AS 2.316 *Saccharomyces cerevisiae* wine yeast. African Journal of Biotechnology 11 (15), 3623–3630.

Zhu, W., Zhu, B., Li, Y., Zhang, Y., Zhang, B., Fan, J., 2016. Acidic electrolyzed water efficiently improves the flavour of persimmon (*Diospyros kaki* L. cv. Mopan) wine. Food Chemistry 197, 141–149.

Zoecklein, B.W., Fugelsang, K.C., Gump, B.H., 2011. Analytical techniques in wine and distillates. In: Joshi, V.K. (Ed.), Handbook of Enology, vol. 3. Asia -Tech Publisher, New Delhi, pp. 1287–1330.

Zoecklein, B.W., Fugelsang, K.C., Gump, B.H., Nury, F.S., 1990. Production Wine Analysis. Van Nostrand and Reinhold, New York, p. 475.

Zoecklein, B.W., Fugelsang, K.C., Gump, B.H., Nury, F.S., 1995. Wine Analysis and Production. Chapman and Hall, New York, p. 621.

Zoski, C.G. (Ed.), 2006. Handbook of Electrochemistry. Elsevier.

CHAPTER

CHEMICAL ENGINEERING ASPECTS OF FRUIT WINE PRODUCTION

6

M.R. Kosseva

University of Nottingham Ningbo Campus, Ningbo, China

1. INTRODUCTION

There are two main features, which can distinguish fruit winemaking from the grape wine technologies: difficult extractability of the soluble components from the fruit pulp and different concentrations of sugars and organic acids. These obstacles can be resolved using novel methods and techniques. In this chapter, some of the key chemical engineering aspects of the fruit winemaking processes are reviewed. The application of microwave technology for fruit juice extraction is described as an alternative method to traditional crushing. Emerging technologies such as pulsed electric fields (PEFs), microwaves (MWs), and ultrasound (US) have also been demonstrated to offer a remarkable potential and selectivity for extraction purposes. The employment of these novel processing technologies can reduce food processing wastes and facilitate the production of natural valuable products, which can secure food sustainability and meet consumer demands.

Chemical, physical, enzymatic, or microbiological changes occur within fruit juices as a result of winemaking. The kinetics of chemical changes that take place during processing provides quantitative knowledge, which is a requirement for the design and analysis of winemaking processes. Understanding of the principles that govern some of the unit operations regularly found in the wine production plant is also essential (Singh and Heldman, 2009).

One of the aims of this chapter is to describe novel methods and emerging technologies that can be successfully applied in fruit winemaking processes, such as high hydrostatic pressure (HHP) treatment, PEF, US, and membrane filtration. Application of microfiltration (MF), ultrafiltration (UF), and reverse osmosis (RO) to fruit wine technology is considered with respect to must correction (or malic acid removal), clarification of fruit juice, alcohol removal, and so on. Membrane bioreactors are presented and classified according to the engineering point of view as free biocatalyst membrane bioreactors, or free enzymatic membrane bioreactors, and biocatalytic membrane bioreactors with enzymes or cells immobilized on/in the membrane.

Racking processes for white and red wines are addressed in this work, giving some practical advice to the winemaker. Transportation of juice, must, and wine is described with emphasis on selection of pumps, classification of the pumps, and components of the pumping systems.

Finally, potential applications of the aforementioned enabling technologies and their comparison are projected for the prefermentation processing, fermentation, and postfermentation operations of fruit wines.

Science and Technology of Fruit Wine Production. http://dx.doi.org/10.1016/B978-0-12-800850-8.00006-5

2. EMERGING METHODS FOR FRUIT JUICE EXTRACTION

In principle, the methods used to produce fruit wines mimic those for the production of grape wines (Swami et al., 2014). Two main differences occur:

1. It is more difficult to extract the sugar and other soluble components from the pulp of some fruits (for example, apricots, apples, berries, mango, and plums) than it is from grapes.
2. Sugar contents are lower in the juices obtained from the majority of fruits, but acid contents are higher than in grape juice.

One possible solution to the first problem is to break down the fruit and then use special methods and equipment to extract the juice. The novel equipment includes MW ovens and ultrasonic reactors. By applying PEF technology, one can enhance the extraction of various intracellular compounds, such as sugar, and increase the extraction yield of juice from the fruit (López et al., 2009).

Another solution to the second problem is to use various membrane techniques that can be applied to increase sugar content without the addition of nonfruit components at ambient temperature and to balance the composition of the fruit juice and must.

The aforementioned features, which can distinguish fruit winemaking from the grape wine technologies, necessitate implementation of novel technologies and equipment, which can enhance the extraction yield of the fruit juices and their quality.

2.1 MICROWAVE HEATING FOR IMPROVED EXTRACTION OF FRUIT JUICE

The methods established for juice extraction involve crushing of the fruit prior to pressing. For fruits such as plums or grapes, pectolytic enzymes are used to facilitate pressing and increase yields (Chauhan et al., 2001; Will and Dietrich, 2006). However, endogenous plant enzymes are also released after crushing, along with the ubiquitous polyphenol oxidase (PPO). This enzyme together with oxygen from the ambient air is responsible for browning, which leads to a loss of color quality (Wang et al., 1996).

As an alternate method, Chemat et al. (2010) have patented MW hydrodiffusion to extract antioxidants from plants. In this method, the fruit is directly placed in an MW reactor without added water (Cendres et al., 2011). The internal heating of the in situ water within the fruit until its boiling point expands the cells and leads to their rupture. Heating under MWs thus frees in situ water. This physical phenomenon, known as hydrodiffusion, allows the extract to diffuse outside the fruit and drop by gravity out of the MW reactor through the perforated disk. Cendres et al. (2011) used this new technique to investigate juice extraction from fresh and frozen fruit (grapes, plums, and apricots), particularly because various fruits do not respond well to pressing. Samples (7 plums, 100 grapes, and 5 apricots) were cut into quarters or halves, pooled, weighed, and immersed in liquid nitrogen for freezing (in triplicates for all analyses). MW extraction of fruit juice was performed in a Milestone (Sorisole, Italy) DryDIST MW laboratory oven. This is a multimode 2450-MHz MW oven with a maximum delivered power of 1000 W. The dimensions of the polytetrafluoroethylene-coated cavity were 35 cm × 35 cm × 35 cm. The extraction vessels were made of Pyrex and had a capacity of 1000 mL. During experiments, time, temperature, pressure, and power were controlled.

In a typical procedure performed at atmospheric pressure, 250 or 500 g of fresh or frozen fruit was MW heated using a fixed power without added water. A mixture of hot "crude juice" (in situ water) and steam moved down naturally outside the MW cavity and collected continuously. The extraction was continued until no more fruit juice was obtained or overheating was detected; the kinetics of the extraction was also followed. Cendres et al. (2011) varied the MW power to optimize the yield of extraction. Extraction of juice from minor fruits such as black currant, blueberry, and plums was carried out from frozen fruits because of time constraints during harvesting season. Another factor was the state of the fruit, whether it was fresh or frozen. The impact of MW power, and more exactly of power density, i.e., the ratio of power in watts to fruit weight in grams, was thus established for all three fruits, both fresh and frozen. For all three fruits, plums (*Prunus domestica* L., cv President), grapes (*Vitis vinifera* L., cv Muscat de Hambourg), and apricots (*Prunus armeniaca* L.), the lower the power density tested, the higher the juice yield. At 0.5 W/g, 440–600 mL/kg could be reached for the various fresh fruits, which dropped to 280–340 mL/kg at 1.5 W/g. Cendres et al. (2011) found that maximal flow rates, much higher for frozen fruits, increased with power density. Thus, increasing power density shortened latency and extraction duration and increased the maximal flow rate. The combination of these two factors led to a decrease in the yields with increasing applied power. The process still needs some optimization as the yields did not reach commercial values.

In the classical plum juice process, pectolytic enzyme is added to crushed plums with 3 h incubation at 49°C, before pressing. Chang et al. (1994) reported the yield to vary from 25% (w/w) without enzyme to 79% with pectinase for the red variety of plums. The MW extraction was thus faster. The grape juice yield (after cleaning, crushing, must enzyme treatment to 55°C, dejuicing, and pressing) can reach 80% in 3 h (Hui et al., 2006). The yield of grape juice obtained by MWs was maximum at 620 mL/kg. The sensory evaluation panel was concerned by the very strong acidity of the apricot juice and the plum juice. But the grape juice was rated as sweet by the panel. The color remained a positive point for all juices, as was the "fresh fruit" aroma. The juices were highly viscous (possibly due to the high pectin concentration), which was a positive point for most of the judges. As MW extraction was rapid, the fruits and their vulnerable compounds such as ascorbic acid and anthocyanins were exposed for only a short duration to high temperature.

The reduced cost of extraction is clearly advantageous for the proposed MW extraction method in terms of time and energy. Conventional extraction methods (cloudy juice) required a total processing time of 6 h for pretreatment (heating at 65°C and processing), pressing, centrifugation, and pasteurization (80°C) (Siddiq et al., 1994). The MW processing method required irradiation for 30 min only for extraction and in-line pasteurization for the same material. The energy required to perform the extraction (1 L) was about 3.62 kWh for the conventional method and 1.46 kWh for the MW method. With regard to environmental impact, the calculated quantity of carbon dioxide emitted to the atmosphere is greater for the conventional method (2890 g CO_2/L of fruit juice) than for MWs (1165 g CO_2/L of fruit juice). Thus, MWs are also proposed as an eco-friendly extraction method suitable for aligned extraction and pasteurization of fruit juice. MW extraction of juice from fruit could also be used to produce substantial quantities by using existing large-scale MW extraction reactors. These MW reactors are suitable for the extraction of 100, 200, or 400 kg of fresh plant material per batch (Cendres et al., 2011).

The phenomenon of MW heating and hydrodiffusion for juice extraction from fruit could be empirically modeled (Cendres et al., 2011). The juices were characterized by very bright colors, high acidity for plums and apricots, and a flavor of fresh fruit. Moreover, the highest yields were obtained from frozen fruit and at low power. An advantage of the process is that the juice extracts at 100°C, which is beyond the pasteurization temperature, though for grapes and plums the yields did not reach commercial levels. Fig. 6.1 presents typical results of a juice extraction test run; it was obtained from 500 g of fresh

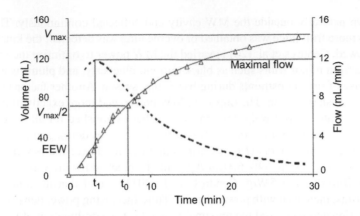

FIGURE 6.1

Empirical model of microwave extraction kinetics, according to Cendres et al. (2011). (*Solid line*) Model curve of microwave extraction (fit of the extraction data); (*dashed line*) derivative curve describing the extraction rate; (*triangles*) experimental points, where t_1, time at maximal flow, and *EEW* (easily exchangeable water) are shown.

plums at a MW power density of 1 W/g of fruit. The extraction data were found to fit well (Fig. 6.1) to an empirical equation Eq. (6.1) where $V(t)$ is the volume collected at time t, t being counted from the appearance of the first drop of juice outside the extraction vessel. The coefficients, which were adjusted to fit the extraction curve, were:

- V_{max}, which represents a theoretical maximum extractable volume;
- t_0, which is the time necessary to obtain half of this maximum extraction volume, $V_{max}/2$;
- b, which represents the general shape of the kinetics of extraction and was found to vary between −1 and −4.

The fitted model gave the following coefficients: $V_{max} = 153$ mL, $b = -1.55$, and $t_0 = 7.6$ min. The inflection point of the extraction kinetics, corresponding to maximal flow rates, could be calculated using the first derivative of Eq. (6.1).

$$V(t) = \frac{V_{max}}{\left(1 + \left(\frac{t}{t_0}\right)^b\right)}, \text{ with its derivative}$$

$$\frac{dV}{dt} = \frac{-V_{max}}{t_0^b} \frac{t^{b-1}}{\left(1 + \left(\frac{t}{t_0}\right)^b\right)^2} \tag{6.1}$$

In this new MW process for extraction of fruit juice, heat was dissipated volumetrically inside the irradiated fruit, and very fast temperature increases could be obtained, depending on the MW power. The pressure created by the boiling water helped to expel the juice. The heat transfer (leading to boiling of internal water) was the driver of the mass transfer, and thus both occurred from the inside to the outside of the fruit.

Cendres et al. (2014) demonstrated that MW hydrodiffusion can generate juices with different compositions during the course of extraction. High concentrations of sugar, acids, and total polyphenols (except in grapes) were markers of the start of extraction with direct hydrodiffusion of these compounds. High concentrations of anthocyanins, some specific polyphenols of plums and sweet cherries, carotenoids, and some volatiles (hexanal, 2-hexenal, linalool) were specifically reached in the middle of extraction. Furfural and benzaldehyde were characteristics of the end of extraction in sweet cherries. Using high-power densities, thus allowing more fruit decomposition, seems useful when compounds are hydrophobic or in the epidermis. Additional examples of MW applications aiming to improve the yields of fruit juice extraction and to enhance the juice properties are given in the following case studies related to cider production.

2.2 CASE STUDIES ON CIDER PRODUCTION

2.2.1 Microwave Heating of Fruit Mash as a Pretreatment Technique

Owing to the health benefits of phenolic compounds and flavonoids, it is desirable to develop methods to increase the extraction of phenolics and flavonoids during fruit juice processing. Heat treatment of fruit mash has proven effective for increasing the concentration of phenolic compounds in fruit juice as well as yield. However, most current heat treatments produce juice with unacceptable analytical and sensory properties. MW energy has the advantage of heating solids rapidly and uniformly, thus inactivating enzymes more quickly to minimize browning. Therefore, Gerard and Roberts (2004) evaluated the effect of MW heat treatment of apple mash on juice yield, quality, and content of total phenolics/flavonoids in the juice. Then, they compared the heat-treated juice to juice produced from unheated apple mash at 21°C. Fuji and McIntosh apple mashes were heated to bulk temperatures of 40, 50, 60, and 70°C in a 2450-MHz MW oven at 1500W. A flow diagram of the juice production and evaluation is shown in Fig. 6.2.

Commercial production of apple cider using rack-and-frame presses results in juice yields around 70g/100g apples when using fresh apples, but juice yield can drop as low as 60g/100g apples when using stored apples that have lost firmness (Rutledge, 1996). A rack-and-frame hydraulic press is shown in Fig. 6.3. The racks on this pilot-scale press measured 0.56×0.56m. Apple mash was made on a large scale using a hammer mill (W.J. Fitzpatrick Co., Model D, Elmhurst, IL, USA) with blunt hammers at 4600rpm. Approximately 70kg of mash was loaded in the pilot-scale rack-and-frame press, and juice yields were determined. Gerard and Roberts (2004) showed that 3kg of apple mash at a depth of 0.016m heated in the MW oven using 1500W for 4.0, 7.1, 10.9, and 16.2min were the optimum parameters to achieve bulk temperatures of 40, 50, 60, and 70°C, respectively. The control mash was not heated. Both the control and the heated apple mashes were loaded into the lab-scale rack and frame. Three 1-kg batches of the mash were placed in cheesecloth and stacked between the racks. The prepared mash was placed in the press (Carver, Inc., Wabash, IN, USA) and 2170kPa of pressure was applied to the mash for 2min. Four replicates at each temperature and for each mash variety were performed.

The results demonstrated that MW heating of mash, as a pretreatment preceding pressing, increased juice yields as high as 7%. MW heat treatment of the mash also increased extraction of phenolics and flavonoids from mash and resulted in juice with increased concentrations of total phenolics and flavonoids. All of the juices produced from heated mash were of a high quality, with comparable pH and titratable acidity. Soluble solids and turbidity increased in the juice with increasing mash temperature. Sensory panelists were unable to detect differences between the cider produced at room temperature and the ciders produced at 40 and 60°C. Therefore, MW heating of apple mash before juice extraction resulted in a high-quality juice with increased phenolic and flavonoid contents as well as increased

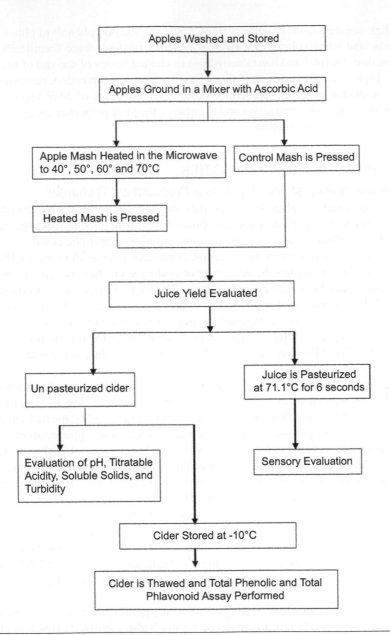

FIGURE 6.2

Example of flow diagram of cider processing, according to Gerard and Roberts (2004).

juice yield. Hence, the optimum temperature for improving juice quality and yield was 60°C (Gerard and Roberts, 2004).

An identical technique can be applied to other fruits such as tropical fruits. Susceptible compounds in fruits will be exposed to heating for a short period of time, because the process of MW

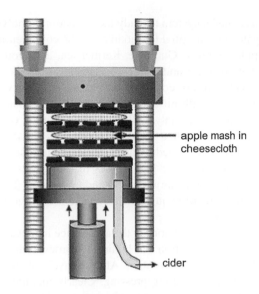

apple mash in
cheesecloth

cider

FIGURE 6.3

Mash racked in a hydraulic press (Gerard and Roberts, 2004).

extraction is rapid. The reduced cost of the extraction process is another advantage of the MW treatment. However, this process still requires some optimization to reach yields with commercial values.

2.2.2 Regulating the Pressing Conditions for Enhanced Properties of Fruit Juice

The first processing step in traditional raw juice production is crushing and pressing of the fruits, during which polyphenols, PPO, oxygen, and cell walls, initially segregated, come into contact and react. This is the step in which the most changes occur in polyphenolic composition, and extraction can be very limited for some polyphenol classes (Guyot et al., 2008). For example, three mechanisms were involved in apple juice production:

1. Noncovalent interactions with cell walls, which can be modulated by modifying the pressing temperature, contribute primarily to retention of procyanidins in the pomace (Le Bourvellec et al., 2007);
2. Oxidation: upon destruction of the apple tissue, PPO oxidizes chlorogenic acid to its o-quinones, which can react in various ways, notably by formation of adducts with flavan-3-ols or transfer of oxidation to these molecules. With oligo or polymeric flavan-3-ols (procyanidins), this subsequently leads mostly to formation of additional intramolecular bonds (Guyot et al., 2008).
3. Flavonols or anthocyanins, located almost exclusively in the apple peel, are physically segregated and diffuse only slowly to the juices.

The same mechanisms may also explain the polyphenolic composition of other juices obtained by pressing, notably for other fruits from the Rosaceae family, which contain the same classes of proanthocyanidins, monomeric flavan-3-ols, phenolic acids, and flavonols.

Cider apples were used as a model system to study the extraction of polyphenols from fruit to juice. Renard et al. (2011) investigated the potential for modulating the polyphenol composition in juice from three distinguished cider apple varieties: Guillevic, Kermerrien, and Douce Moen. The experiments were carried out under inert atmosphere and increased temperature in the range from 5 to 24°C. The crushed apples were also subjected to four conditions of oxidation: preserved from oxidation as above, short contact with air, short contact with air and mixing, and long contact with air and mixing. The impact of these treatments was assessed analytically and, for oxidation of the mash, organoleptically. Apples were pressed at four temperatures in a Speidel-90 pneumatic press (Speidel Tank-und Behälter-bau GmbH, Ofterdingen, Germany), modified by addition of an inox (stainless) steel tube around the press to allow inerting by a heavier-than-air gas. Apples (c. 20 kg/pressing) were left at the chosen temperature for 24 h for temperature equilibration. A standing time of 20 min was observed between crushing and pressing of the apples to allow juice diffusion and procyanidin adsorption. For oxidation, the apples were pressed as above, at 11°C.

As a result, native polyphenol concentrations in the apple juice increased with pressing temperature and decreased with mash oxidation. Retention was highest for procyanidins of a high degree of polymerization and increased with procyanidin oxidation. At high levels of mash oxidation, colored compounds were also retained in the mash during pressing. Clearly maintaining inert conditions during processing could enhance the nutritional properties of premium juices, whereas oxidation of the mash during application of the enzyme could lead to more stable juices for concentrate production. Renard et al. (2011) manipulated the sensory characteristics of the juice (color, bitterness, and astringency) by regulating the pressing conditions. Increased temperatures during pressing led to an increased transfer of bitter and/or astringent procyanidins, whereas the mash could be oxidized to decrease both bitterness and astringency. This can be of great interest to cider producers, for which a moderate level of bitterness and astringency is needed to balance the residual sugars. Bitterness and astringency decreased much less than the concentrations of native polyphenols, and bitterness actually increased for the most oxygenated mash: clearly more efforts are needed to better understand the organoleptic properties of polyphenolic oxidation products. Oxidation decreased the concentrations of native polyphenols in the juices, especially for flavan-3-ols. For the highest oxidation state the color became paler and yellower. Bitterness and astringency decreased upon oxidation, probably because of increased retention of oxidized moieties.

2.3 ULTRASOUND-ASSISTED ENZYMATIC EXTRACTION OF FRUIT JUICE

Acoustic cavitation is the phenomenon of the generation of microbubbles (cavities) in a liquid from the negative pressure produced by concentrating the diffuse energy of sound. If a sufficiently large negative pressure is applied to a liquid, the distance between the molecules increases, producing voids, or cavitation bubbles. The required negative pressure to make cavitation is proportional to the tensile strength of the liquid and thus depends on the type and purity of the liquid, for example, in tap water a few atmospheres of negative pressure is enough for cavitation. An oscillating bubble can accumulate energy from the oscillations in the form of heat. With continuing energy input, the bubble grows until reaching a size (typically tens of millimeters), at which the void structure is no longer stable. The bubble then suddenly collapses, resulting in the rapid release of the stored energy with a heating rate of $>10^{10}$K/s. This transient cavitational implosion is highly localized with associated temperature of roughly 5000K and pressure of about 1000 bars. In summary, the phenomenon of cavitation consists of three distinct steps: (1) bubble nucleation (formation), (2) rapid growth and expansion to a critical size during alternating cycles of compression–rarefaction, and (3) implosion and violent collapse of the bubble in the liquid (Suslick, 1990).

As a result, US enhances mass transfer rates by cavitation forces, whereby bubbles in the liquid/solid extraction can explosively collapse and generate localized pressure causing plant tissue rupture and improving the release of intracellular substances into the solvent (Knorr et al., 2004).

Experiments have shown that US decreases consumption of expensive enzymes, reducing processing times, but does not influence the secondary structure of the enzyme, although the microenvironment and tertiary structure of the enzyme may be disturbed. It was suggested that US enhances physical phenomena such as diminishing the unstirred diffusion layer, enhancing emulsification, decreasing particle size, affording better mixing, and/or generating microstreaming and mechanical stress. In the case of two-phase reactions, it is suggested that US reduces the adsorption of organic compounds on the enzyme surface, thus promising easy processing and recycling of the immobilized enzyme. The use of combined enzyme and US bioprocessing produces cavitation effects that enhance the transport of enzyme to the surface of the substrate, which results in the opening up of the substrate surface to the action of enzymes as a result of mechanical impact of cavitation. To date most of the published research has concentrated on the low-frequency US region, generally 20–40 kHz (Kwiatkowska et al., 2011). US has been used in the extraction of plant materials because of the enhancement of yield and shortening of extraction time (Mason et al., 1996; Toma et al., 2001).

Le Ngoc Lieu and Van Viet Man Le (2010) reported treatment of grape mash in juice processing via application of US and by a combination of US and enzyme. Sonication treatment used to optimize the conditions for grape mash increased extraction yield by 3.4% and reduced treatment time by a factor of 3; combined US and enzyme treatment increased extraction yield slightly, only 2%, but shortened treatment time by a factor of >4. After sonication treatment, enzymatic treatment increased the extraction yield by 7.3% and total treatment time of this method was still less than that of the traditional enzymatic method. In comparison with traditional enzymatic processing, application of US in grape mash treatment enhanced extraction yield and shortened treatment time. The processing in this study was performed as follows. Two liters of grape mash with total solid content of approximately 20% was directly poured into an ultrasonic bath. The height of the mash in the bath was about 4.5 cm. The bath (Elma T 660/H, Singen, Germany) was a rectangular container (300 × 151 × 150 mm) with a maximal volume of 5.75 L, to which 35-kHz transducers were annealed at the bottom. The equipment operated at an US intensity of 2 W/cm^2 and a US power of 360 W. The sonotrode of the bath had a surface area of about 180 cm^2, which was large enough for the US waves to distribute homogeneously in the height of the treated sample. The bath was equipped with a thermostatic system. The treatment temperature ranged from 60 to 80°C and the time from 5 to 15 min. The combined US and enzyme treatment was carried out in the same US bath. A determined amount of Pectinex SP-L (from 0.02% to 0.06% v/v) was added and the mixture was stirred before using. The treatment time ranged from 4 to 12 min. Temperature was maintained at 50°C. At the end of the treatment, enzymes in the sample were inactivated by heating the mash at 90°C for 5 min in a water bath. Summing up, the methods reported by Le Ngoc Lieu and Van Viet Man Le (2010) improved the quality of the grape juice obtained. Sugar content, total acid content, and phenolic content as well as the color of the grape juice were increased.

Thus, US technology can be extensively applied in the fruit and wine processing industries. The advances achieved in the design of US processing equipment can provide industrially robust processing capability (Vilkhu et al., 2008). Enabling design and operational features have included: (1) automated frequency scanning to enable maximum power delivery during fluctuation of processing conditions, (2) nonvibrational flanges on the US horns for construction of high-intensity in-line flow cells, and (3) construction of radial and hybrid sonotrodes to provide greater range in application design and product opportunities. As of this writing, 16 kW is the largest available single US flow cell, which can be configured in series or in parallel modules. Industrial US manufacturers have already implemented industrial processing capability for food extraction applications (Hielscher, 2006).

Several US reactor designs have been described by Chisti (2003), Gogate et al. (2003), and Vinatoru (2001), the last specifically for industrial extraction of plant tissue. These include (1) a stirred sonotrode directly immersed into a stirred bath or reactor, (2) a stirred reactor with US coupled to the vessel walls, and (3) recycling of product from a stirred reactor through an external ultrasonic flow cell. The configurations may provide both intermittent and continuous US exposure, from low intensity in a large-volume reactor ($0.01–0.1 \, W/cm^3$) to high intensity ($1–10 \, W/cm^3$) in an external flow cell. Mixed-frequency reactors have been shown to offer advantages with respect to process efficiency and energy distribution (Moholkar et al., 2000; Tatake and Pandit, 2002; Feng et al., 2002).

The proposed benefits of US-assisted extraction for the fruit wine industry include (1) overall enhancement of extraction yield or rate, (2) improvement of aqueous extraction processes or juice concentrate processing, and (3) enhanced extraction of heat-sensitive components under conditions that would otherwise give insufficient yields (Vilkhu et al., 2008).

2.4 PULSED ELECTRICAL FIELD TECHNOLOGY

Various studies have reported that PEF is an appropriate technology to enhance extraction of various intracellular compounds, such as sugar, and to increase the extraction yield of juices from various fruits (Bazhal and Vorobiev, 2000; López et al., 2009).

Extraction of polyphenols assisted by PEF has been investigated in various vegetal matrices. Corrales et al. (2008) observed that the application of a PEF treatment (3 kV/cm, 30 pulses, 10 kJ/kg) to grape skins increased the total phenolic content and the anthocyanin concentration in the extraction medium by 100% and 17%, respectively, compared with untreated control samples.

The low energy consumption and the short processing times necessary for permeabilization of berry skin cells are key advantages of PEF technology. Thus, wines with a high content of phenolic compounds can be produced, and the duration of maceration during vinification can be reduced (Puertolas et al., 2010).

Corrales et al. (2008) performed a comparative study on the feasibility of various promising methods such as HHP, PEF, and US. They applied those techniques to extract bioactive substances from grape by-products. The effect of heat treatment at 70°C combined with the effects of novel technologies such as US (35 KHz), HHP (600 MPa), and PEF (3 kV/cm) showed a great feasibility and selectivity for extraction purposes. After 1 h extraction, the total phenolic content of samples subjected to novel technologies was 50% higher than in the control samples. The higher yields obtained in extractions carried out by HHP and PEF are of major interest from an industrial point of view, because solvent amounts were reduced and extraction times shortened. Hence, the combination of emerging technologies for extraction purposes and low-cost raw materials is an economical alternative to traditional extraction methods. This can contribute to further sustainable development of the juice and wine industries (Corrales et al., 2008).

3. DEVELOPMENT OF MEMBRANE TECHNOLOGIES APPLIED TO FRUIT WINEMAKING

The membrane technologies used in the wine industry include the pressure-driven membrane processes of MF, UF, nanofiltration (NF), and RO. In this chapter, we consider membrane separation as an essential part of wine processing, which could be applied successfully to fruit wine production on an industrial scale. In a membrane separation system, contact between a fluid, containing two or more components, and a membrane permits some components (for example, water) to permeate more readily than others. The physical and chemical nature of the membrane (e.g., pore size and pore-size distribution) affects the separation of liquid streams. As shown in Fig. 6.4, the membrane in an RO system allows water to permeate, whereas salts and sugars are rejected. UF membranes are useful in fractionating components by rejecting macromolecules. In MF, the membranes separate suspended particulates (Singh and Heldman, 2009).

In cross-flow MF (CFMF, also known as tangential flow filtration) the fluid to be filtered flows parallel to the membrane surface and permeates the membrane by means of a pressure drop. The shear exerted by the feed solution flowing parallel to the membrane surface can remove the deposited particles toward the concentrate so that the cake layer remains relatively thin (El Rayess et al., 2011).

The liquid amount that passes through the membrane is called the permeate and the retained molecules and solvent constitute the retentate, which is concentrated progressively during filtration cycles. Today this standard operation mode is applied in many food industries, namely, wine, vinegar, beer, juice, and so on (Daufin et al., 2001).

Darcy's Law or the general filtration equation governs the permeate flux; it is:

$$J = \frac{\Delta P - \Delta \Pi}{\mu \cdot R_h}$$

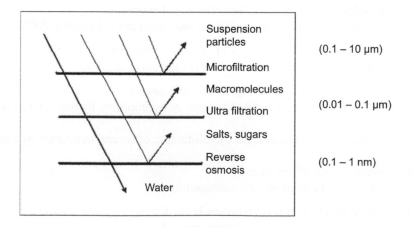

FIGURE 6.4

Use of membrane systems to separate molecules of different sizes.

Adapted from Singh, R.P., Heldman D.R., 2009. Introduction to Food Engineering, fourth ed. Academic Press, Elsevier.

where J (m³/m²/s) is the permeate flux, ΔP (Pa) the applied pressure, $\Delta \Pi$ the osmotic pressure, μ (Pa s) the solvent viscosity, and R_h (m⁻¹) the hydraulic resistance. In wine CFMF, the membranes used are 0.2–0.45 μm average pore size. Thus, they retain only colloids and particles and the solutes and salts pass through the membrane (El Rayess et al., 2011).

Therefore, $\Delta \Pi$ could be considered negligible in this case and Darcy's Law could be given as:

$$J = \frac{\Delta P}{\mu \cdot R_h}$$

Permeation of the selected components is the result of a "driving force." For dialysis, the concentration difference across the membrane is the driving force, whereas in RO, UF, and MF systems, hydrostatic pressure is the key driving force. MF membrane systems require the lowest amount of hydraulic pressure; about 1–2 bars. UF membrane systems operate at higher pressures, on the order of 1–7 bars. These pressure levels are required to overcome the hydraulic resistance caused by a macromolecular layer at the membrane surface. In an RO system, considerably higher hydraulic pressures, in the range of 20–50 bars, are necessary to overcome the osmotic pressures (Singh and Heldman, 2009).

The membranes were initially established as a clarification step after fermentation in winemaking. In traditional winemaking membrane processes can replace several of the separation steps: must correction by RO, clarification by MF/UF, alcohol removal by RO. The use of MF/UF can reduce the number of steps by combining clarification, stabilization, and sterile filtration in one continuous operation. It can eliminate the use of fining substances and filter materials (Lipnizki, 2010).

The application of membrane technology in the beverage industry is has the major advantages of membrane processes compared to traditional separation techniques. They are:

- gentle product treatment throughout processing due to moderate temperature changes;
- high selectivity based on distinctive separation mechanisms, for example, ion-exchange mechanisms;
- compact and modular design for ease of installation and extension;
- low energy consumption compared to condensers and evaporators (Lipnizki, 2010).

In addition to the technological advantages of cross-flow filtration (CFF) in winemaking, there are some operational and economic benefits to consider (El Rayess et al., 2011):

- elimination of labor costs and time savings due to automation;
- elimination of sheets and kieselguhr, which reduces costs, improves hygiene and work safety, and reduces waste;
- reduction of product loss and energy costs by substitution of several treatments by a single operation;
- reduction/elimination of clarifying agents;
- possibility of data recovery (high process automation).

The main disadvantage of membrane filtration is the fouling of the membrane, causing a loss in process productivity in time. This effect can be minimized by regular cleaning cycles. During the plant design, the selection of hydrophilic membranes to reduce fouling by bacteria or membrane modules with an open-channel structure to avoid blockage by particles can be useful. Operating the plant below

Table 6.1 Wine Components and Their Sizes

Component	Dimension
Suspended solids	200–200 µm
Yeast	1–10 µm
Bacteria	0.5–1.0 µm
Polysaccharides	50–200 kDa
Proteins, tannins, polymerized anthocyanins	10–100 kDa
Simple phenols, anthocyanins	500–2000 Da
Ethanol, volatiles	20–60 Da

Adapted from Lipnizki, F., 2010. Cross-flow membrane applications in the food industry. In: Peinemann, K.-V., Pereira Nunes, S., Giorno, L. (Eds.), Membranes for Food Applications, vol. 3, Wiley-VCH, pp. 1–24.

the critical flux or, alternatively, in a turbulent flow regime can act as a preventive measure against fouling. Other limitations to membrane applications can be related to the feed characteristics, like increased viscosity with concentration, or separation mechanisms due to the osmotic pressure increase with concentration (El Rayess et al., 2011).

By analogy to crude grape wine, fruit wines contain three main groups of compounds identified according to their components' sizes:

- solutes (<1 nm)—ions, salts, sugar, glycosides, organic acids, alcohols, and phenolic compounds;
- colloids (1 nm–1 µm)—polysaccharides, proteins, polymerized phenolic compounds, and colloidal aggregates;
- particles (>1 µm)—microorganisms, cell debris, colloidal aggregates, and so on.

These complex wine contents are determined by the fruit juice composition, which depends on the genetic characteristics of cultivars, growing conditions, ripeness of the fruit, and the winemaking practice. The initial must composition also changes during the wine production process. A selection of wine compounds critical for CFMF and their sizes is given in Table 6.1.

The variety of chemical and biological components present in fruit juices/wines and the broad distribution of their dimensions allows application of CFMF processes, which are based on the pore size of the membranes used, as shown in Fig. 6.5.

The clarification of fruit juices by MF has been widely applied in the industry; the analogous process for wine started its rapid growth after 2005 (Ulbricht et al., 2009). Although the first trials of CFMF in enology were conducted at the beginning of the 1980s, since 2005 membrane filtration has found proven use in several production steps. Examples are shown in Fig. 6.5; they include must correction carried out by RO, clarification of wine by MF/UF, wine correction and alcohol removal by RO, and microbial stabilization or sterile filtration via MF.

When operating a membrane system, optimal conditions should be found. Measurement of the process efficiency can be done by the selectivity and the permeate flux. Temperature can affect the flux significantly. Operating at high flux levels means that less membrane area is required and economies can be made in terms of capital, operating, and membrane replacement costs (Pap et al., 2004).

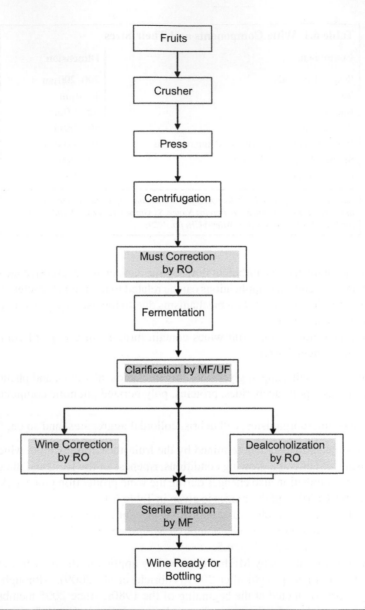

FIGURE 6.5

Membrane processes in the fruit wine production. *MF*, Microfiltration; *RO*, reverse osmosis; *UF*, ultrafiltration.

Adapted from Lipnizki, F., 2010. Cross-flow membrane applications in the food industry. In: Peinemann, K.-V., Pereira Nunes, S., Giorno, L. (Eds.), Membranes for Food Applications, vol. 3, Wiley-VCH, pp. 1–24.

3.1 MUST CORRECTION

In the category of fruit wines, the must is not naturally well balanced for the production of table wines. The sugar content is usually low and the acid content can be either too much or not enough to produce a balanced table wine (Dharmadhikari, 2004). RO can be applied to increase sugar content without the

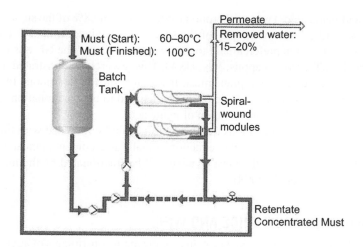

FIGURE 6.6

Batch plant for must correction by reverse osmosis (Lipnizki, 2010).

addition of nonfruit components at ambient temperature and to adjust and balance the composition of the fruit must. The use of RO leads to enrichment in tannin and organoleptic components by water reduction between 5% and 20%. Applying this method to must containing a high concentration of acids (for example, made of red raspberry, blackberry, or strawberry) may not be effective, because apart from sugar, acids are also concentrated (Smith, 2002). In general, the use of this method is limited by legislation in various countries (Lipnizki, 2010).

RO, as an alternative to chaptalization (sugar addition to must) or vacuum evaporation, enables must correction if the sugar concentration needs to be increased (Fig. 6.6). Mietton-Peuchot et al. (2002) used the RO system manufactured by Indagro Ltd., with DESAL RO membranes inside. The RO modules (5.8 m²/module) contained a double spacer. A sieve filter protected the modules against eventual deposit from grape must. The following conclusions were made from the results of RO concentration obtained from various types of must:

- The membranes used enabled an excellent retention of must components.
- The wines so obtained were very close to those produced by bleeding and chaptalization, excluding white wines, because RO concentration of the must slightly raised their acidity.
- The preparation of red wine musts was essential for good operation of the RO.
- The specificity of the product and the strengths relative to the membrane fouling led to definition of the optimum operating conditions: maximum pressure value of about 75×10^5 Pa, temperature between 10 and 15°C, and corresponding maximum reduction volume factor.
- The musts obtained by RO were of good quality, but this technique should not be used in palliation of lack of fruit maturity (Mietton-Peuchot et al., 2002).

Fractionation of sugar and organic acids in grape must and fruit juice by NF was investigated by several research groups (Masson et al., 2008; De Pinho, 2010). Rosa Santos et al. (2008) proposed NF for the simultaneous concentration and rectification of grape must. This was studied via the capability of NF to fractionate sugars from organic acids in grape must using four model solutions of must. In the model solutions, the gap between the rejection coefficients of the sugars (glucose and fructose) and of the acids (tartaric and malic acids) was well pronounced. More than 88% of the sugars was rejected and less than 37% of the acids, so the major part of the sugars was retained in the NF concentrate stream and the organic acids permeated preferentially to the permeate stream. The NF was performed with an NF 270 supplied by FilmTec (Minneapolis, MN, USA). This research demonstrated the capacity of NF for sugar/organic acids fractionation in grape must. It was also shown that among the acids, there was a preferential permeation of malic acid, which could be applied to the fractionation of fruit must, e.g., blackberry must containing a high concentration of malic acid.

In general, numerous other applications of NF and RO could be intended with the aim of bringing some correction to must or wine, such as the reduction of malic acid content in must, pH control, or bad taste elimination. In all these cases, the must or wine will be fractionated by the membranes selected and carefully tasted (Masson et al., 2008).

3.2 CLARIFICATION OF FRUIT JUICE AND WINE

Phenolic compounds, polysaccharides, and proteins are the main constituents of aggregated macromolecules. The formation of colloidal aggregates results in instability, and it is responsible for most physicochemical disorders and deposits in wines.

The traditional fining after fermentation often involves several steps of centrifugation and kieselguhr filtration to remove the aggregated macromolecules and obtain the desired quality. The use of MF/UF can reduce the number of steps by combining clarification, stabilization, and sterile filtration in one continuous operation. This could also eliminate the use of fining substances and filter material. The key to success in the clarification of wine is the membrane selection with regard to fouling behavior and pore size/diameter (Lipnizki, 2010).

De Pinho (2010) reported on the permeate fluxes of MF/UF of a red wine, which were much lower than those of a white wine. At the transmembrane pressure of 1.0 bar, the MF/UF of white wine yielded permeate fluxes of 118 and 129 L/h/m^2, respectively. At the same transmembrane pressure of 1.0 bar, the MF and the UF of red wine yielded final permeate fluxes of 34 and 18 L/h/m^2, respectively. The wine clarification by MF was associated with a small removal of polysaccharides (in the range of 11.4–7.6% at various ΔP from 0.6 to 1.4 bar) and polyphenols (0.9–2.6% at ΔP from 1.0 to 1.4 bar) in the case of white wine. The removal was slightly higher in the case of the red wines, for example, polysaccharides, 24.6–23.1% at ΔP from 0.6 to 1.4 bar, and polyphenols, 9.6–12.6% at ΔP from 0.6 to 1.0 bar). As in the case with MF, the clarification of white wine by UF also led to a low removal of polysaccharides and negligible removal of polyphenols. In contrast, in the case of red wine there was a significant removal of polysaccharides and polyphenols. The membrane regeneration processes in both cases of white and red wines were different as well (De Pinho, 2010).

3.3 ALCOHOL REMOVAL FROM FRUIT WINES

NF and RO are the most popular membrane-based processes used for beverage dealcoholization. According to Commission of European Communities (EC) regulations (Commission Regulation, 2009), the minimum ethanol concentration in wine should not be less than 8.5 vol%, and the reduction of alcohol strength by volume should not exceed 2 vol%. Deeper dealcoholization produces a beverage that cannot be called "wine" (Catarino and Mendes, 2011).

The commonly used alcohol-reduction technique is based on the selective separation of water and alcohol from the wine by RO, which preserves the aromatic compounds (Peuchot, 2010). This process is carried out together with the separation of alcohol from the RO permeate by distillation. The dealcoholization of red wines is carried out after the malolactic fermentation, and in white wines at the end of alcoholic fermentation. This process presents the advantage of being alcohol selective, but treatment capacities are limited because of the low flow rates, which imply a need for high investment and operating costs. In contrast, NF provides substantially higher alcohol flow rates together with greater permeation rates. Working pressures are lower, leading to savings in investment and operating costs. However, the results obtained by NF are very close to those obtained by RO; some examples of permeate and wine analysis results are given by Peuchot (2010).

Diafiltration mode is usually used to perform the dealcoholization. As the permeate (formed mainly by water and ethanol) is withdrawn from the feed, water is added to the retentate at the same flow rate to keep the volume in the system constant. The permeate stream can be subject to a second ethanol removal step, and the dealcoholized stream can be reintroduced into the filtration retentate (Pickering, 2000).

Several authors have reported dealcoholization of alcoholic beverages based on membrane processes. López et al. (2002) applied RO, in batch and diafiltration modes, for producing low-alcohol apple cider, preserving the desirable aroma compounds. Pilipovik and Riverol (2005) reported the advantages and disadvantages of RO for removing ethanol from beverages down to 0.45% ethanol. They found that RO is not economically feasible for producing low-alcohol brews. Catarino and Mendes (2011) investigated a combination of two membrane processes for producing a high-quality low-alcohol wine from a standard alcoholic wine (c. 12 vol% ethanol). Various RO and NF membranes (CA995PE, NF99 HF, NF99, NF97 from Alfa Laval and YMHLSP1905 from Osmonics) were used for removing the ethanol by diafiltration. In the first approach, all membranes were used for removing the ethanol from c. 12 vol% to c. 7–8 vol% (single-step dealcoholization). Next, the most promising membranes (NF99 HF, NF99, and YMHLSP1905) were used for dealcoholizing the original wine down to c. 5 vol%. The resulting dealcoholized wine samples were then blended with the original wine to produce reconstituted wine samples. Finally, 0.3 vol% of an aroma extract obtained by pervaporation of the original wine was added to the wine samples produced by the two methods, single-step dealcoholization and reconstitution, improving the aroma and taste profile of these samples. Concerning all indicators, Catarino and Mendes (2011) showed that the NF membranes YMHLSP1905, NF99, and NF99 HF were the most effective in producing low-alcohol wine. In addition, the incorporation of aroma compounds obtained by pervaporation of the original wine into the dealcoholized samples produced a high-quality low-alcohol wine.

3.4 REDUCTION OF MALIC ACID IN MUST

One typical characteristic of the must produced from cherries and berries is the high total acidity owing to the relatively high content of organic acids in fruit (e.g., English Morello cherry has 1.86% total acidity). In cherries, malic acid is the main organic acid, and small amounts of citric, succinic, and lactic acids have also been reported. Several amino acids also occur, but proline is the main one present in ripe fruit. In soft berry fruits organic acids are the second most abundant soluble solids after sugars. With the exception of blueberry, the soft fruits contain significant levels of organic acids. Citric acid is the key organic acid found in strawberry, raspberry, and blueberry. In blackberry, malic and isocitric acids are the major organic acids; here citric acid is present only in trace amounts (Dharmadhikari, 2004).

Various methods can be applied to reduce malic acid contents in must or fruit wine. Peuchot (2010) described the membrane process performed in two NF stages to reduce the content of malic acid in must. In the first stage, the racked grape must was nanofiltered. The permeate containing mainly water, malic acid, and tartaric acid was neutralized to a pH of about 7 by using potassium hydroxide (Ducruet et al., 2010). In the second stage, the neutralized permeate was nanofiltered through the same membrane. The potassium malate was thus retained by the membrane. The permeate was incorporated into the must. For a continuous process using two membrane units, the permeate flow rates of the two membranes can be kept identical by adjusting the operating pressure. This will permit correct control of the pH during neutralization.

El Rayess et al. (2011) evaluated the main organic and ceramic membranes used in MF of grape wine and their characteristics. Different configurations of ceramic membranes adapted for different types of wines, musts, and lees were also presented. The same authors discussed in detail CFMF fouling mechanisms in enology. For example, adsorption of wine macromolecules to the membrane polymer was studied by Ulbricht et al. (2009). A model solution for wine ("synthetic red wine") has been established by using a commercial red grape marc extract (after fermentation). It was found that polyphenols and polysaccharides are only marginally adsorbed by polypropylene (PP) but strongly adsorbed by polyether sulfonate membranes. Comparisons of data for individual model substances for polyphenols or polysaccharides and their mixtures with data for synthetic red wine support the hypothesis that aggregates of polyphenols and polysaccharides present in red wine have a strong contribution to adsorptive fouling, and that the interaction between polyphenols and the membrane surface is the main driving force. In consequence, the low adsorption tendency of wine ingredients to PP membranes results in higher fluxes and longer service life of the respective filtration modules in wine clarification. As a general rule, membrane fouling depends on the composition of the wine, the operating conditions, and the membrane type. Applications of CFF to the must clarification process require further investigations in the area of membrane fouling. The areas for further study include (El Rayess et al., 2011):

- the physicochemical interactions between wine molecules themselves and between these molecules and the membrane,
- the individual impacts and contributions of wine compounds to fouling,
- mechanisms causing membrane fouling by wine compounds,
- methods to predict the degree of membrane fouling.

3.5 MEMBRANE BIOREACTORS FOR FRUIT WINE PROCESSING

A variety of bioreactor technologies have been applied in the production of wines. This chapter gives emphasis to the high-density microbial cell bioreactors because wine productivity achieved in these systems is enhanced while capital costs can be reduced. Primarily, bioreactors containing high-density microbial cells are classified into two major systems: homogeneous and heterogeneous bioreactors (Divies et al., 1994). Homogeneous bioreactor systems contain a uniform distribution of biomass in the form of free cells in the medium. The repeated use of the microbial biomass can be achieved via centrifugation, flocculation of the yeast cells with a decanter, or retaining the cells in a membrane reactor.

Heterogeneous bioreactor systems have two separate phases: liquid medium and solid particles—usually containing immobilized cells.

Additionally, membrane bioreactors can integrate bioconversion with selective membrane separations leading to continuous, clean, and safe production with low energy consumption, suitable for application in winemaking. The use of a biocatalyst in combination with membrane operations permits drawbacks to be overcome, enabling the bioreaction to be integrated into a continuous production mode, which can be better controlled for increased productivity and economic viability.

Whereas a bioreactor can be defined as an apparatus within which biocatalysts (enzymes or living cells) accomplish biochemical processes, membrane bioreactors can be distinguished as unit operations that combine a bioreactor with a membrane system. The membrane can be used for various tasks in the membrane bioreactor (Calabro, 2013):

- separation,
- selective extraction of reactants,
- retention of the biocatalyst,
- distribution/dosing of reactant,
- biocatalyst support.

Mazzei et al. (2010) described the role of biocatalytic membrane bioreactors (BMRs) in the production of ingredients for functional food and beverages. Based on the role of the membrane, bioreactors are divided into two systems. In the first system, the membrane only controls mass transport, it does not contribute to the reaction. The configuration is referred to as a free biocatalyst membrane reactor (MBR). This is a traditional and commonly used bioreactor containing various types of biocatalysts, e.g., cells and enzymes of different molecular weights. The enzyme membrane reactor is shown in Fig. 6.7 (Calabro, 2013). It consist of a continuously stirred tank reactor with volume V, equipped with a UF membrane and characterized by an exposed area A_m. The membrane is chosen to ensure complete rejection of the enzyme, a specific amount of which is loaded into the reactor. It is

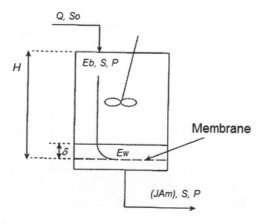

FIGURE 6.7

Diagram of an enzyme membrane reactor with volume V, in which the rejected enzyme profile is shown. E_w is the enzyme concentration at the membrane wall, E_b is the bulk concentration, H is the reactor height, δ is the cake thickness and J is the permeate flux (Calabro, 2013).

then continuously fed with the liquid substrate (or must), which has concentration S_0 and flow rate Q. The reaction takes place and the product with concentration P is removed continuously by permeation through the membrane. A uniform enzyme concentration E_0 can be held before the filtration takes place. Continuous product withdrawal will allow an increase of the reaction yield and minimization of enzyme inhibition. However, the rejected enzyme accumulates on the membrane surface owing to the concentration polarization phenomenon as well as enzyme deactivation. As a consequence, the reaction and the conversion rates will decrease. This MBR system allows simple replacement of the biocatalyst in case of enzyme deactivation. An analogous CFF membrane separation system for cells is shown in Fig. 6.8 (Calabro, 2013).

In the second system, the membrane works as a catalytic and separation unit. In this configuration the reaction also occurs at the membrane level. This second configuration is the BMR (Mazzei et al., 2010). In the BMR the biocatalysts can be gel-entrapped or bound to the membrane as a result of ionic binding, cross-linking, or covalent binding. This configuration is analogous to the heterogeneous high-cell-density bioreactor and it is considered a combination of two unit operations in one step (Calabro, 2013).

Membrane bioreactors contain commercial membranes with various configurations, for example, tubular, hollow fiber, and spiral wound. Characteristics such as the pore size, structure, and construction material are important in the selection of the membrane for a particular application. On the basis of the structural characteristics, membranes are classified as isotropic (with a homogeneous composition) and anisotropic (consisting of a thin layer of membrane supported by a dense layer of porous understructure) (Obradovic et al., 2004). The growth of the cells and formation of a biological film on the membrane, which could create significant resistance to mass transfer and even lead to complete plugging of the membrane, are a main drawback of these systems. The flat module membrane bioreactors have a simpler configuration, which allows easy access to the compartments for cleaning the membrane and, where necessary, its replacement. Hollow fiber counterparts, however, provide a higher surface-to-volume ratio and eliminate the need for a rigid membrane (Nemati and Webb, 2011).

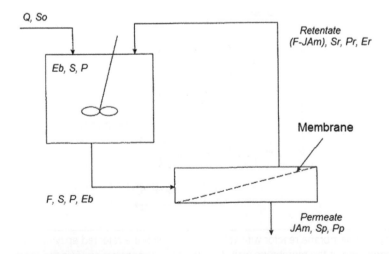

FIGURE 6.8

Diagram of a membrane bioreactor with continuous biocatalyst recirculation in the retentate (Calabro, 2013).

In winemaking, the membrane bioreactors are mainly applied to the production of aromatic compounds, flavors, and stabilizer molecules (Mazzei et al., 2010). Additives are produced through pectinase hydrolysis by glucosidases. The production of stabilizer compounds such as lactic acid via malolactic fermentation is achieved by using lactic acid bacteria (Kosseva et al., 1998). For the purpose of applying these enzymes or microorganisms with enzymatic activity on an industrial scale, the development of methods for immobilization on the membrane or other support is highly important. The methods for enzyme and cell immobilization with applications in the food industry are described elsewhere (Kosseva, 2011, 2013). Regardless of the immobilization technique, the membrane protects the cells from the existing shear forces and bubble bursting. A scheme of a continuous membrane bioreactor with enzymes immobilized in the membrane fibers with recycling of unreacted substrate is illustrated in Fig. 6.9 (Calabro, 2013).

Some examples of coupling enzymes useful in winemaking and membrane reactors are as follows. β-Glucosidase is employed in various winemaking applications, for example, in rosé wine production from red grapes, for the hydrolysis of anthocyanins, and for the hydrolysis of terpene glucosides, which

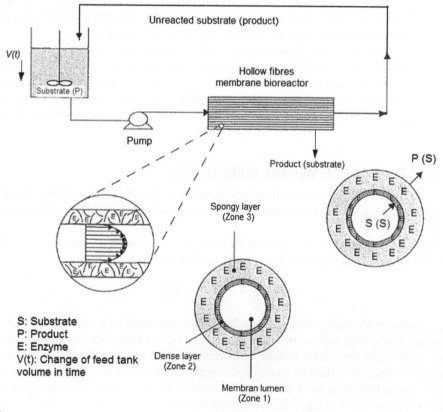

FIGURE 6.9

Diagram of a continuous membrane bioreactor with enzyme immobilized in the membrane fibers (Calabro, 2013).

are well documented by Romo-Sanchez et al. (2014), Martino et al. (1996), and others. Laboratory-scale experiments were carried out in a continuous-flow stirred-tank membrane reactor, which configuration is close to those employed in industrial processes in a model system of wine (Gallifuoco et al., 1998). The experiments proved the advantage of upgrading aroma in wines at the end of the fermentation stage. The components of wine, such as ethanol, fructose, and terpenol, did not affect either the activity or the stability of the enzyme, and immobilization in chitosan was appropriate for the process. This natural polysaccharide is an appropriate support for enzyme and cell immobilization in winemaking (Kosseva and Kennedy, 2004). The mechanical and chemical stability under the conditions investigated was also demonstrated.

Pectinases represent other important enzymatic activities in winemaking, particularly for the clarification process. They are responsible for improving the processability and producing additives. An investigation of the enzyme hydrolysis of pectin, using a free enzyme membrane reactor (EMR) equipped with a Prep/Scale TFF spiral-wound polysulfone membrane (10 kDa), was carried out by Rodriguez-Nogales et al. (2008). Production of crude enzymes (polygalacturonase and pectin lyase) in *Aspergillus niger* cultures grown on an apple pomace as a low-cost inducer was also studied with the aim of wine clarification. According to the retention of induced pectinases by the UF membranes, the best EMR configuration was achieved using the membrane with a 10,000 nominal molecular weight limit, because the biocatalyst was retained with no loss of enzyme activity.

The kinetics and mass transfer limitations occurring in the system largely affect the behavior of membrane bioreactors. Therefore, full consideration of transport phenomena is necessary to optimize their performance. A complete analysis of the effectiveness of the biocatalytic process is required to predict the performance of a membrane bioreactor. Calabro (2013) has discussed mass transfer phenomena occurring in membrane systems and described their behavior using a mathematical modeling approach.

3.6 PRODUCTIVITY OF MEMBRANE BIOREACTORS

Productivity of the BMRs was derived based on the concept of mass balance taking into account the enzyme kinetics, transport phenomena, and effectiveness of the immobilized biocatalysts in the system (Calabro, 2013).

In an enzyme membrane bioreactor, in which the enzyme is homogeneously distributed and no membrane separation happens, productivity can be evaluated as:

$$\Theta_{EMR} = \frac{P \cdot V}{E_0 \cdot V} = \frac{(S_0 - S) \cdot V}{E_0 \cdot V} = \frac{\alpha \cdot S_0}{E_0}$$

where Θ_{EMR} is the productivity of the enzyme bioreactor expressed as a ratio of the total amount of product formed to the total amount of enzyme. The total mass of product can be calculated from its concentration (P) times the reactor volume V; E_0 represents the amount of enzyme fed to the reactor without activity reduction. If the concentration is expressed in terms of weight, the productivity can be calculated, knowing that the mass of the product is equal to the mass of the converted substrate ($S_0 - S$).

For the bioreactor shown in Fig. 6.7, the productivity is expressed as the ratio of the total amount of product formed at time t to the total amount of enzyme, where total permeate flow rate ($J \cdot A_m$) and operating time t are known.

$$\Theta_{EMR} = \frac{P \cdot (J \cdot A_m) \cdot t}{E_0 \cdot V} = \frac{(S_0 - S) \cdot V}{E_0 \cdot V} \cdot \frac{t}{\tau} = \frac{\alpha \cdot S_0}{E_0} \cdot \frac{t}{\tau}$$

$\alpha = \dfrac{(S_0 - S)}{S_0}$ is the degree of conversion and $\dfrac{t}{\tau} = \dfrac{Q \cdot t}{V} = \dfrac{J \cdot A_m \cdot t}{V}$ is the ratio between actual time and characteristics retention time τ, in seconds. This equation can be extended to the continuous reactor shown in Fig. 6.8 with total time of operation t.

For the bioreactor in which a heterogeneous biocatalyst is present (Fig. 6.9), the productivity can be expressed as the ratio of the mass of product formed to the mass of immobilized biocatalyst, or

$$\Theta_{CSMB} = \frac{P \cdot Q \cdot t}{E_{imm} \cdot (1 - \varepsilon) \cdot V} = \frac{(S_0 - S)}{E_{imm} \cdot (1 - \varepsilon)} \cdot \frac{t}{\tau_R} = \frac{\alpha \cdot S_0}{E_{imm}(1 - \varepsilon)} \cdot \frac{t}{\tau_R}$$

In large-scale applications of BMRs, reproducibility of the results obtained requires further research as well as enzyme lifetime and immobilized enzyme stability during the membrane cleaning procedure (Mazzei et al., 2010). There is a need to establish knowledge-based control strategies applicable to these parameters on the industrial scale.

4. RACKING PROCESS AND TRANSPORT OF WINE

Most commercially produced wines undergo a clarification process, to be clear and free of any particles or mist. Kolpan et al. (2010) considered the principal clarification methods used today (from slowest to quickest) as racking, fining, refrigeration or cold stabilization, followed by racking, filtration, centrifugation (or spinning the solids out of the wine).

The young wine is separated or siphoned from the lees (microbial biomass) by transferring the wine to another container, leaving the lees behind. This process is called racking; it is done to accomplish several aims (Dharmadhikari, 1991):

- to clarify the wine,
- to prevent new wines from picking up off-odors,
- to discourage malolactic fermentation and microbial spoilage,
- to help aeration and aging (especially in red wines).

The racking process differs for white and red wines. A key difference in racking red wine as opposed to white is that in red wine limited aeration helps in wine aging. Young red wines are rich in phenolic compounds such as pigments and tannins, which convey a coarse, astringent, and bitter taste to the wine. Aeration (oxidation) facilitates the polymerization of phenolic compounds. Oxygen also participates in numerous complex reactions that cause aging. Because controlled aeration is essential for aging of red wines, it should be done early in the life of a wine, preferably during the first racking. In a modern approach oxygen is introduced on the suction side of the pump and the amount dissolved is monitored on the delivery side. Because each wine requires a specific amount of oxygen, the degree of aeration should be experimentally determined (Dharmadhikari, 1991). A variable-speed pump is used so that the speed can be reduced when the wine level in the bioreactor gets close to the surface of the lees, to prevent the lees from being sucked in during the transfer. The speed of the pump can also be reduced by partially closing the valve on the suction side of the pump.

4.1 PUMPS

During racking and other processes, pumps are used to transport wine, juice, and must. They are devices that expend energy to raise, transport, or circulate fluids. In addition, in the food industry, the transport system must be designed to allow for ease and efficiency of cleaning. The main components of a pumping system are:

- prime movers, such as electric motors or an air system;
- piping, used to carry the fluid;
- valves, used to control the flow in the system;
- other fittings, controls, and instrumentation;
- end-use equipment, which has various pressure and flow requirements (e.g., tanks).

It is essential that all components of the pipeline system contribute to sanitary handling of the product as well. The stainless steel surfaces ensure smoothness needed for cleaning and sanitizing. In addition, proper use of the system provides the desired corrosion prevention (Singh, 2009).

Pumps can be classified according to their basic operating principle as dynamic or positive-displacement pumps. Because positive-displacement pumps are widely used mostly for pumping viscous fluids, in this chapter the accent is on the use of centrifugal pumps.

4.1.1 Centrifugal Pumps

Centrifugal pumps are the most economical and most common pumps used for pumping water in industrial applications. Their components are rotating components, an impeller coupled to a shaft, and stationary components, casing, casing cover, and bearings. The operation of a centrifugal pump is based on a rotating impeller, which converts kinetic energy into pressure or velocity that is needed to pump the fluid. The product enters the pump at the center of the impeller rotation and, owing to centrifugal force, moves to the impeller periphery. At this point, the liquid experiences maximum pressure and moves through the exit to the pipeline. The simple design of the centrifugal pump makes it easily adaptable to cleaning-in-place functions. Most sanitary centrifugal pumps used in the food industry use two vane impellers. Centrifugal pumps are most efficient with low-viscosity liquids such as fruit juices and must, for which flow rates are high and pressure requirements are moderate. The discharge flow from a centrifugal pump is steady. These pumps are suitable for either clean/clear or dirty liquids. Flow rates through a centrifugal pump are controlled by a valve installed in the pipe and connected to the discharge end of the pump. This approach provides an inexpensive means to regulate flow rate, including complete closure of the discharge valve to stop flow. Because this step will not damage the pump, it is used frequently in liquid food processing operations (Singh, 2009).

4.1.2 Positive-Displacement Pumps

By application of direct force to a confined liquid, a positive displacement pump produces the pressure required to move the liquid product. Product movement is related directly to the speed of the moving parts within the pump. Thus, flow rates are accurately controlled by the drive speed to the pump.

In rotary pumps, the displacement is obtained by rotary action of a gear or vanes in a chamber of a diaphragm in a fixed casing. Rotary pumps are further classified as internal gear, external gear, lobe and slide vane, etc. They have the capability to reverse flow direction by reversing the direction of rotor rotation. They deliver a steady discharge flow.

4.1.3 Selection of an Appropriate Pump

Selection of a pump is based on two main considerations: flow rate (capacity) and head required. Matching the pump characteristics with hydraulic requirements is also needed, taking into account process conditions, temperature of the fluid, and presence of solids. General practical considerations are as follows:

- If a high head is required with a low flow rate, use positive-displacement, reciprocating pumps.
- Where high flow rates and moderate heads are desired, use impeller pumps (centrifugal pumps).

The following factors influence the choice of pump for a particular operation:

1. The quantity of liquid to be handled. This primarily affects the size of the pump and determines whether it is desirable to use a number of pumps in parallel.
2. The head against which the liquid is to be pumped. This will be determined by the difference in pressure, the vertical height of the downstream and upstream reservoirs, and the frictional losses, which occur in the delivery line. The suitability of a centrifugal pump and the number of stages required will be determined largely by this factor.
3. A high-speed centrifugal or rotary pump will be preferred as it can be coupled directly to the motor.
4. If the pump is used only intermittently, corrosion problems are more likely than with continuous working.

The cost and mechanical efficiency of the pump must always be considered, and it may be advantageous to select a cheap pump, and pay higher replacement or maintenance costs, rather than installing a very expensive pump of high efficiency.

Pressure is needed to pump the liquid through the system at a certain rate. This pressure has to be high enough to overcome the resistance of the system, which is also called the "head." The total dynamic head is the sum of the static head and the friction head. Static head is the difference in height between the source and the destination of the pumped liquid (Fig. 6.10). The static head at a certain pressure depends on the weight of the liquid and can be calculated with the following equation:

$$\text{Head (in feet)} = [\text{pressure (psi)} \times 2.31]/\text{specific gravity}$$

US DOE (2006).

Friction head is the loss caused by the resistance to flow in the pipe and fittings. It is dependent on size, condition and type of pipe, number and type of pipe fittings, flow rate, and nature of the liquid. The friction head is proportional to the square of the flow rate.

The head and flow rate determine the performance of a pump. As the resistance of a system increases, the head will also increase. This in turn causes the flow rate to decrease and eventually to reach zero. A zero flow rate is acceptable for only a short period before it causes the pump to burn out. The pump operating point (or duty point) is determined by the intersection of the system curve and the pump head curve as shown in Fig. 6.10.

Pumps may be arranged so that the inlet is under a suction head or the pump may be fed from a tank. These two systems alter the duty point curves as shown in Fig. 6.11. In developing such curves, the normal range of liquid velocities is 1.5–3 m/s, but lower values are used for pump suction lines. With the arrangement shown in Fig. 6.11A, there can be problems in priming the pump and it may be necessary to use a self-priming centrifugal pump (Coulson, 1999).

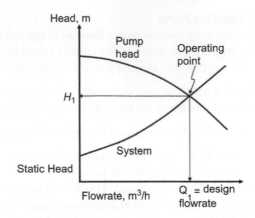

FIGURE 6.10

Pump operating point.

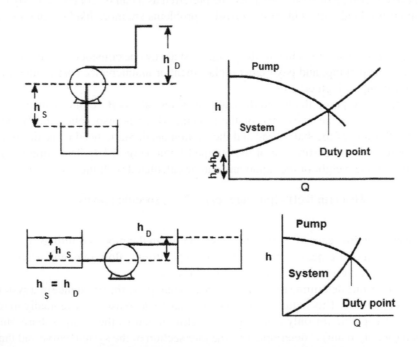

FIGURE 6.11

Pump arrangements that alter the operating or duty points.

Adapted from Coulson, J.M., Richardson, J.F., Backhurst, J.R., Harker, J.H., 1999. Coulson & Richardson's Chemical Engineering.
sixth ed. Fluid Flow, Heat Transfer and Mass Transfer, vol. 1. Butterworth-Heinemann, Oxford (Chapter 8).

    The net positive suction head (NPSH) available indicates how much the pump suction exceeds the liquid vapor pressure, and is a characteristic of the system design. The NPSH required is the pump suction needed to avoid cavitation, and is a characteristic of the pump design. For any pump, the manufacturer specifies the minimum value of the NPSH, which must exist at the suction point of the pump.

This practical information is essential for the selection of an appropriate pump used as the transporting device for fluids such as fruit juice, must, and wine during the winemaking process.

5. PRESERVATION PROCESSES APPLICABLE TO WINE PRODUCTION

It is well known that the most significant factors affecting the growth of microorganisms are temperature and pressure (Isenschmid et al., 1995). Each microorganism has a specific maximum temperature of survival. Above that temperature, proteins denature, cytoplasmic membranes collapse, and cells lyse and are deactivated (Madigan et al., 2002). On the other hand, wine cannot be treated with heat because wine characteristics such as flavor, taste, and color are very sensitive to temperature (Mermelstein, 1998). One common practice in winemaking is the addition of sulfur dioxide (SO_2) to wine to reduce the microbial population of the must and to preserve the final product for a long period of time (Ribéreau-Gayon et al., 2006; Buzrul, 2012). SO_2 can have negative effects on human health (Romano and Suzzi, 1993), so the wine industry is challenged to meet consumer demands of reducing the levels of SO_2 used in wine production (du Toit and Pretorius, 2000). Techniques such as filtration and fining are efficient in controlling microbial growth, but some of these techniques have harmful effects on the sensorial properties of the wine (Buzrul, 2012).

Several novel technologies have been explored to reduce microbial populations in food and drinks without the need of thermal energy. For example, HHP and PEFs are used and are described in review articles published by Puertolas et al. (2010), Oey et al. (2008), and Rendueles et al. (2011).

5.1 HIGH HYDROSTATIC PRESSURE TREATMENT OF FRUIT WINE

HHP treatment is an effective method for increasing food safety and shelf life while preserving the organoleptic properties of food products. HHP treatment could facilitate the wine industry in reducing the levels of SO_2 or could be used in combination with other antimicrobial agents such as nisin (Buzrul, 2012). Although rice wine (nigori sake) was one of the earliest HHP-treated commercial products to appear on the Japanese market (Suzuki, 2002), no HHP-treated alcoholic beverage such as beer or wine has been introduced onto the global market, as of this writing (Buzrul, 2012).

High-pressure-treated foodstuffs could be considered "novel foods." According to the issues covered by Regulation (EC) No. 258/97, 1, novel foods do not have a significant history of consumption within the European Union before May 15, 1997. Regarding the processing aspects, novel foods and food ingredients are those to which a production process not currently used has been applied, particularly, where that process gives rise to significant changes in the composition or structure of the foods or food ingredients, affecting their nutritional value, metabolism, or level of undesirable substances. If a foodstuff is considered novel, a premarket scientific assessment of the product safety and authorization should be carried out in the corresponding member state where the food is marketed (Heinz and Buckow, 2010). Microbiological, toxicological, and nutritional considerations and allergenic aspects should also be part of the risk evaluation process. Several products, such as high-pressure-pasteurized orange juice (France) and fruits (Germany), have been evaluated at national levels. Common agreement was reached that the high-pressure treatment does not cause significant changes in the composition or the structure of the products affecting their nutritional value, their metabolism, or the amounts of undesirable substances (Eisenbrand, 2005). The general aspects of risk assessment applied to HHP-processed foods have been partly covered by some risk evaluation agencies, e.g., the German Senate Commission on Food Safety of the German Research Foundation (Eisenbrand, 2005) and the Spanish Food Safety Agency (Rendueles et al., 2011). The risk is minimal in those industrial processes in which a stabilization of food is required to avoid modification of the sensory characteristics during its storage, distribution, and sale. This process can be useful in some fermented drinks such as beer and wine (Mok et al., 2006).

HHP processes (in which the pressure is in the range of 100–1000 MPa) required to deactivate bacteria have been also reviewed by the Institute of Food Technologists (2000). The development of HHP would be favored if standardized treatment conditions were proposed by international bodies such as the Codex Alimentarius, International Commission on Microbiological Specifications for Foods, or International Life Sciences Institute. The process criteria (combinations of pressure–time–temperature for various foods and pathogens) can be calculated by predictive microbiology procedures and can result from the food safety objectives set for food pathogen combinations. The establishment of these parameters can facilitate the development of procedures for quantitative microbiological risk assessment (Smelt et al., 2002). Food companies can incorporate the stage of HHP processing in Hazard Analysis Critical Control Points plans, as a critical control point, which can be monitored, optimized, and validated, and for which critical limits and corrective actions can be set (Rendueles et al., 2011).

A variety of grape wines, including Sauternes, Moscato, Asti Spumante, white, red, and rosé, have been evaluated after using HHP treatment since 1995. The results were presented and discussed in a review published by Buzrul (2012). The pressure levels used to treat wine were similar to the commercial applications in the fruit juice industry, i.e., 400–600 MPa. As an example, two Moscato wines obtained from Barbera grape must (pH 3.0, sugars 177.2 g/L; Delfini et al., 1995) had the following compositions: (1) 8.22% ethanol, 64 mg/L total SO_2, 146 mg/L acetaldehyde, 50.5 g/L residual sugars; (2) 11.35% ethanol, 102.4 mg/L total SO_2, 168 mg/L acetaldehyde, no residual sugars. A third wine (3) was obtained by adding absolute ethanol to wine (2) up to 15.19%. The SO_2 was produced by yeast during fermentation. A mixture of yeasts, molds, and lactic and acetic acid bacteria was used. HHP treatments (300–600 MPa) were done at 20°C initially. The results showed that HHP (600 MPa for 3 min) had a strong antimicrobial effect: $6 \log_{10}$ (complete inactivation) was observed for all types of wine [(1), (2), and (3)]. At 400 MPa for 2 min all the microorganisms added to the sweet wine (1) died, whereas at 300 MPa for 2 and 4 min the survival rates were 19.67% and 10.57%, respectively.

Mok et al. (2006) studied the pasteurization of low-alcohol red wine (9% ethanol, pH 3.27, acidity 0.068%, total sugar 0.851%) using HHP (100–350 MPa for 0–30 min at 25°C). The original microbial count, 4.15×10^5 CFU/mL, decreased to 2.41×10^3 CFU/mL after 5 min treatment at 250 MPa and to 1.17×10^3, 7.10×10^2, and 2.00×10^2 CFU/mL after 10, 20, and 30 min treatment, respectively. At pressures higher than 300 MPa, the inactivation effects became extraordinary, with the initial microbial count reduced to 2.78×10^2 CFU/mL after 5 min and 6.60×10^1 CFU/mL after 10 min treatment at 300 MPa. Treatment at 300 MPa for 30 min killed all the yeasts in wine, whereas all lactic acid bacteria were inactivated at 300 MPa for 20 min or 350 MPa for 5 min. Sensory evaluation done using 10 trained panels revealed that there were no differences in the aroma, taste, mouthfeel, or overall sensory quality between the HHP-treated (350 MPa for 10 min) and the untreated samples. Puig et al. (2003) also reported that HHP treatment at 5000 atm for 5 min resulted in a 99.99% reduction in bacterial count without resulting in changes in the chemical or organoleptic properties of the wine.

HHP processing is a specific technology, because pressure can result in enhancement or retardation of chemical and biochemical reactions as well as in both desired and undesired modifications of biopolymers [e.g., enzyme (in)activation, gel formation]. Based on current knowledge, elucidation of the HHP effects on the sensory properties of fruit-based food products, such as color, flavor, and texture, is not straightforward because of the presence of various enzymatic and chemical reactions during both processing and storage. Moreover, the effects of HHP treatment on sensory properties cannot be generalized, because (1) study on basic awareness in this subject is still limited and (2) sensory properties are product dependent (Oey et al., 2008).

Moreover, cost analyses of the HHP treatment with respect to the capacity of production and also the product itself have to be performed. The consumer awareness and willingness to pay for HHP treatment of beverages requires further investigation. Future studies can focus on sensory evaluations, because HHP has proved itself able to destroy spoilage organisms in beverages. HHP has a vast potential to eliminate the negative effects of heat on the aroma and flavor of the beverages and also to reduce the SO_2 levels in wine while deactivating microorganisms and sterilizing the product (Buzrul, 2012).

5.2 HIGH-PRESSURE CO$_2$ STERILIZATION

In the late 1980s, Japanese researchers explored the use of high-pressure gas treatment as an alternative for sterilization of heat-labile compounds and for preservation of food products. These studies focused on CO_2 because of several benefits. CO_2 is not flammable and is nontoxic; the main hazard in its use is asphyxiation. Unlike ethylene oxide, CO_2 requires no special handling or ventilation, and leaves no toxic residues. It has a low critical temperature (31.1°C) (Zhang et al., 2006). Moreover, it is one of the by-products of wine fermentation.

When using high-pressure CO_2 for sterilization, two possible mechanisms are described in the literature: mechanical cell rupture and physiological deactivation (Zhang et al., 2006). A diagram of deactivation mechanisms proposed for vegetative bacteria is shown in Fig. 6.12. In addition, Zhang et al. (2006) summarized the status of the research in this area as follows:

1. Vegetative bacteria have been extensively studied and are susceptible to high-pressure CO_2 treatment.
2. The effects of temperature, pressure, and medium on deactivation of vegetative cells can be substantial, but there is not yet a clear understanding of these effects.
3. Several deactivation mechanisms have been proposed.

The highest pressure reported for treatment of orange juice was 33 MPa (Arreola et al., 1991). The pressure requirement for sterilization with CO_2 can be lowered below 20 MPa. The pressure threshold below which no deactivation is observed varies with the bacterial species (Debs-Louka et al., 1999). The D value (the time needed to achieve 1-log reduction) of *Saccharomyces cerevisiae* showed a steep decrease with increase in pressure from 4 to 10 MPa (Shimoda et al., 2001). It is believed that higher temperature enhances deactivation by (1) increasing the fluidity of cell membranes, making them easier to penetrate, and (2) increasing the diffusivity of CO_2 (Hong et al., 1997). Therefore, higher temperatures reduce the duration of the first stage of deactivation, which is thought to be diffusion-controlled. The duration of the earlier stage and the inactivation rate of the second stage have been found to be extremely sensitive to pressure (Zhang et al., 2006).

Enomoto et al. (1997) reported inactivation of food microorganisms by high-pressure carbon dioxide treatment with or without explosive decompression. These authors analyzed proteins released using the Lowry protein assay. Unlike absorbance experiments, which showed release of UV-absorbing

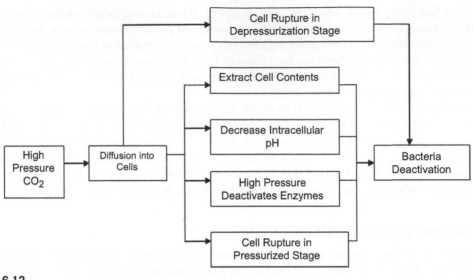

FIGURE 6.12

Diagram of inactivation mechanisms proposed for vegetative bacteria by Zhang et al. (2006).

Table 6.2 Mathematical Models Reported Regarding Microbial Inactivation (Zhang et al., 2006)

Mathematical Model of the Deactivation Kinetics	References
First-order linear	Karaman et al. (2001)
Empirical nonlinear	Peleg (2002)
Sigmoid models	Erkmen (2000)
Empirical equation from the response surface model	Debs-Louka et al. (1999)
Mass transfer model for CO_2 diffusion through the medium	Elvassore et al. (2000)
Thermodynamics pH drop and CO_2 solubility in phospholipid	Spilimbergo et al. (2002)

materials after high-pressure CO_2 treatment, only Enomoto et al. identified protein release from baker's yeast after high-pressure CO_2 treatment (1997). The same scientists observed lower deactivation with pressure cycling. With the same total treatment time of 4 h, the log reduction of baker's yeast decreased from 7-log without pressure cycling to 4-log with a frequency of 1 cycle/h.

The greater part of the publications account for experimental data. Only a few papers propose models of either kinetic or thermodynamic aspects. The kinetics of deactivation are usually characterized by a first-order expression, although various deactivation models have been derived in the articles presented in Table 6.2 (Zhang et al., 2006).

Numerous patents are related to the use of CO_2 in various sterilization applications. Osajima et al. (1996–2003) deposited a series of patents on the use of a supercritical fluid of carbon dioxide to deactivate enzymes and microorganisms in liquids and to deodorize liquid materials, including liquid foodstuffs, with semicontinuous (1996, 1997) or continuous operations (1998, 1999, 2003). At 50°C and

30 MPa, 106 CFU/mL of eight species of *Bacillus* spores, including *Bacillus cereus*, *Bacillus subtilis*, *Bacillus megaterium*, *Bacillus polymyxa*, *Bacillus coagulans*, *Bacillus circulans*, *Bacillus lichenifor-mis*, and *Bacillus macerans* were completely deactivated with 80 min treatment using a semicontinuous microbubble method (Osajima et al., 1997). Sims (2001) designed a new membrane contactor to steril-ize liquids. The liquid or compressed CO_2 was contacted in the pores of the membrane. Wildasin et al. (2002) applied gaseous or liquid CO_2 to reduce microbial and/or enzymatic activity in orange juice. Malchesky (2003) designed a cleaning and antimicrobial decontamination process using subcritical fluids, including carbon dioxide, argon, xenon, nitrous oxide, oxygen, helium, and mixtures of these gases.

The challenges of a high-pressure CO_2 sterilization method were summarized by Zhang et al. (2006) as follows:

1. Accurate sterilization requires killing of spores, but data on the effects of CO_2 on bacterial spores are quite limited. The experimental conditions (high temperature and pressure, long treatment time) reported for killing spores appear to be more extreme in comparison with the alternative sterilization techniques.
2. With regard to the effects of pressure cycling and the use of additives, a small amount of data is available. However, further development of these methods is important for enhancing the deacti-vation process under milder conditions.
3. The weakest links in knowledge are mechanistic studies and mathematical modeling for process design, which are essential in understanding the fundamentals of the high-pressure CO_2 treatment process.

Summing up, CO_2 is inexpensive, nontoxic, nonflammable, and physiologically safe, with a low critical temperature; therefore, a high-pressure CO_2 sterilization technique could be a good option for sterilization of heat-sensitive materials and liquids like fruit wines.

5.3 APPLICATION OF ULTRASOUND IN MUST TREATMENT FOR MICROBIAL INACTIVATION

The manipulation of biological processes by US has became promising since the time Wood and Loomis (1927) studied the influence of US on red blood corpuscles, small fish, and frogs. Frogs were rapidly destroyed and killed, but mice survived for 20 min. Certainly, high-power US, predominantly of a frequency of 20–40 kHz, is destructive to biological molecules both in vivo and in vitro because of the generation of cavitation bubbles, which have a variety of negative physical effects, as well as the production of highly oxidizing hydroxyl radicals (Kwiatkowska et al., 2011). As a result, biologi-cal cell disintegration is a well-established laboratory method to extract intracellular components (Skauen, 1976), because high-intensity cavitation punctures the cell walls and releases the cell con-tent easily (Valero et al., 2007).

5.3.1 Inactivation of Microorganism Via US and Conventional Heating

In response to the desires of consumers for products that are less organoleptically and nutritionally spoiled during processing, and less reliant on additives than previously, alternative approaches to the deactivation of microorganisms in beverages have been developed (Gould, 2001). US in combination with other pres-ervation methods has found many applications (Valero et al., 2007). Destruction of microorganisms by US was studied, for example, for stabilization of wine and as an alternative supplement to traditional sterilization methods (Arena, 1979; Shoh, 1988). The influence of US and conventional heating under different processing conditions on the inactivation and potential subsequent growth of microorganisms in orange juice was investigated by Valero et al. (2007). A limited level of microbial inactivation ($\leq 1.08 \log CFU/mL$) was obtained by selected-batch US treatment at 500 kHz, 240 W, for 15 min. However, microbial growth was observed in the same substrate following 14 days of storage at both refrigeration (5°C) and mildly abusive (12°C) temperatures. The presence of pulp in the juice increased the resistance of microorganisms to US. After continuous US treatments at flow rates of 3000 L/h negligible reductions in microbial counts were found. The quality attributes of the juice such as limonin content, brown pigments, and color were not affected by US. The above treatments were carried out in a semiindustrial unit from Sinaptec (Lezennes, France) developed in collaboration with EDF (Les Renardières, France), working in batch or in continuous-flow mode at changeable temperature. Submerged, stainless steel, high-amplitude horns, with several vibrating equidistant elements and fixed frequency of 23 or 500 kHz, were attached to PP cylindrical treatment chambers. Electronic generators (Nexus N198-N) supplied constant ultrasonic frequencies. For treatments under continuous conditions, a centrifugal pump, two treatment chambers, and a 20-L feeding tank at the appropriate flow rate of 3000 L/h were used. Power levels of 120 and 240 W at high frequency, or 300 and 600 W at low frequency, were used. A 60% US power was applied in all cases. Juice volumes of 1.4 L with various pulp contents (0.1%, 1.0%, and 10.0%) were US-treated under static conditions for 15 min. In addition, 17-L volumes were subjected to treatment with low, high, or a combination of both US frequencies under continuous conditions for 180 min (Valero et al., 2007). For industrial applications, it will be necessary to combine US with other processing techniques with greater antimicrobial potency. This will prevent the development of food-borne pathogens in orange juice. Regarding grape wine, one possible negative effect of US on the organoleptic characteristics can occur from the sedimentation of tannins, potassium bitartrate, and proteins (Arena, 1979). Thus additional research is needed to study the synergetic effects, to improve the processing equipment in conjunction with industrial applications, and for production of more efficient US equipment (Valero et al., 2007). The last will allow the development of combined processes with broad applications in the wine and beverage industries.

5.4 PASTEURIZATION OF FRUIT JUICE USING MICROWAVES

MW pasteurization offers benefits similar to those of the conventional methods, but with an improved product quality and reduced time of exposure to energy (Harlfinger, 1992). Heating by MWs is influenced by the size, shape, and composition of the material, as well as the type of equipment. Canumir et al. (2002) evaluated the effect of pasteurization at various MW power levels (270–900 W) on the microbiological quality of apple juice. Using a home 2450-MHz MW, they obtained data and compared them with those from conventional pasteurization (83°C for 30 s). Apple juice pasteurization at 720–900 W for 60–90 s resulted in a 2–4 log population reduction. At these levels, no significant differences were found between conventional pasteurization and MW treatment. At higher power levels (720–900 W) the D values were comparable to those for thermal inactivation. D values were 0.427 ± 0.03 min at 900 W and 0.4870.10 min at 720 W with temperatures of 70.3 ± 2.1 and 76.2 ± 1.9°C, respectively. It was concluded that the home MW oven was suitable for pasteurization of apple juice in which inactivation of *Escherichia coli* occurred from the heat released (Canumir et al., 2002). Furthermore, the use of MWs has been already developed at the pilot scale and commercial level for citrus juice pasteurization (Schlegel, 1992).

5.5 PULSED ELECTRIC FIELDS TECHNOLOGY FOR WINE PRESERVATION

PEF is considered one of the nonthermal methods for inactivating microorganisms in foods. Electric field strengths in the range of 15–35 kV/cm, generated by the application of short, high-voltage pulses, and specific energies from 50 to 700 kJ/kg induce the formation of pores in the microbial membrane, causing inactivation of microorganisms. Inactivation of the vegetative cells of bacteria and yeast by PEF has been broadly demonstrated (Saldana et al., 2009; Wouters et al., 2001). PEF also enhances mass transfer by electroporation of the cytoplasmic membranes of eukaryote cells. The potential application of PEF for pasteurization has been investigated in a great variety of liquid foods of different physicochemical characteristics, including fruit juices (Molinari et al., 2004; Wu et al., 2005).

Sing and Heldman (2009) described several configurations for the electrodes and the product flow, which have been already developed (Fig. 6.13). They include parallel plate, coaxial, and collinear configurations. Though the parallel plate arrangement supplies the most uniform electric field intensity, the intensity is reduced in boundary regions. At product flow rates for commercial operations, the frequency of pulses increases the product temperature. A primary process variable is the electric field intensity. Depending on the microbial population, the intensities may vary from 2 to 35 kV/cm. For all microorganisms, the rate of reduction in the population will increase as the field intensity is increased. The pulse geometry is an additional factor influencing the process (Sing and Heldman, 2009).

The strict requirements and high costs, which are the main limitations of PEF technology for food pasteurization, would not be necessary for inactivation of wine spoilage microorganisms. In this case, a level of inactivation between 3 and 4 $\log_{10}$ cycles could be enough to facilitate the growth of active dry wine yeast and to avoid microbial spoilage. The specific energy required to obtain this level of inactivation should be around 150–300 kJ/kg. These energy requirements could be translated to an estimated cost of between 4.2 and 8.4 V/ton. Nevertheless, a great effort is required to reduce the cost of the PEF generator before the application of PEF to microbial decontamination in wineries can be considered (Puertolas et al., 2010).

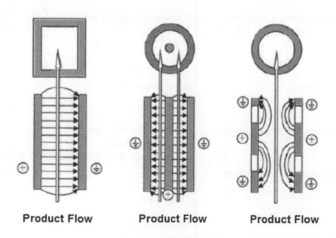

Product Flow   **Product Flow**   **Product Flow**

FIGURE 6.13

Three configurations of continuous-flow pulsed electric field systems (Singh and Heldman, 2009).

Inactivation of wild yeasts in the must by PEF could facilitate the growth of active dry wine yeasts, which are added to grape must to ensure a quick start of the alcoholic fermentation and a reproducible fermentation. Inactivation of the spoilage flora in must before fermentation or in the wine after the fermentation could also be an effective procedure for avoiding contamination of the processing contact surfaces and for controlling the development of alterations throughout the process of aging in barrels and storage in bottles. PEF can also reduce the amount of or replace SO_2 in winemaking (Puertolas et al., 2010).

The potential of PEF to inactivate microorganisms at temperatures that do not cause harmful effects on flavor, color, or nutrient value of must and wine has paved the way for useful applications of this technology in the fruit winemaking process (Puertolas et al., 2010).

Finally, Table 6.3 presents a comparison of the data on numerous applications of alternative methods, such as HHP treatment, PEF, MW, US, and high-pressure CO_2 processing. The experimental

Table 6.3 Comparison of the Alternative Methods Proposed for Fruit Wine Production

Method	Application	Equipment/Parameters	References
HHP	Extraction of bioactive substances	600 MPa	Corrales et al. (2008)
HHP	Pasteurization	100–350 MPa, 30 min 25°C	Mok et al. (2006)
PEF	Extraction of bioactive substances	3 kV/cm	Corrales et al. (2008)
PEF	Inactivation of vegetative cells of bacteria and yeast	Plate, coaxial, and collinear electrode configurations Electrical field intensity 15–35 kV/cm $E = 50$–700 kJ/kg	Sing and Heldman (2009) Saldana et al. (2009) Molinari et al. (2004) Wu et al. (2005)
MW	Juice extraction	MW reactors 100 kg 1.46 kWh	Cendres et al. (2011, 2014)
Pneumatic press	Crushing and pressing	5–24°C	Renard et al. (2011)
US Acoustic cavitation	Enzymatic extraction of juice Extraction of bioactive substances	US bath rectangular container, 5.75 L 35-kHz transducers $I = 2$ W/cm²; $P = 360$ W 50°C 35 kHz + 70°C	Le Ngoc Lieu and Van Viet Man Le (2010) Corrales et al. (2008)
MW	Pasteurization	Home microwave 2450 MHz 720–900 W	Canumir et al. (2002)
US + heating	Destruction of microorganisms	500 kHz, 240 W, 15 min Semiindustrial unit Feeding tank, 20 L, flow rate 3000 L/h	Valero et al. (2007)
High-pressure CO_2	Preservation and sterilization	20–33 MPa Membrane contractor Subcritical fluid system	Zhang et al. (2006) Osajima et al. (2003) Sims (2001) Wildasin et al. (2002) Malchesky (2003)

HHP, *High hydrostatic pressure;* MW, *microwave;* PEF, *pulsed electric field;* US, *ultrasound.*

conditions and equipment used for fruit juice extraction, microbial inactivation of juice and must, and preservation of fruit wines are also highlighted.

6. **CONCLUSIONS**

Whereas the technology for the production of fruit wines has many similarities with grape winemaking processes, there are two main distinctions, which arise from the difficulty of extracting sugars and other soluble compounds from fruits other than grapes. In addition, the contents of the sugar extracted are lower, but the acid contents are higher, than those in the grape juice. These differences necessitate implementation of novel technologies and equipment, which can enhance the extraction yield of fruit juices. MW heating and hydrodiffusion, US-assisted extraction, and PEF technologies can offer various benefits for further industrial development of fruit winemaking processes. The advantages and drawbacks of these applications in relation to fruit wine production are:

1. MW heating and hydrodiffusion allowed rapid extraction of juice from fresh or frozen fruits. The juices are characterized by very bright colors, high acidity for plums and apricots, and a flavor of fresh fruit (Cendres et al., 2011). This process is considered eco-friendly, because the quantity of carbon dioxide emitted to the atmosphere is nearly two times lower than the emission from the conventional method. Energy requirements and cost are also reduced compare to the traditional method, which requires pressing, centrifugation, and pasteurization. One of the serious disadvantages is still the yield of the fruit juice produced by the MW technology, which needs further optimization for commercialization.
2. US-enhanced enzymatic extraction of juice can improve the yields of the fruit juices and their quality. The benefits of US-assisted extraction for the fruit wine industry include improvement of extraction yield or rate and enhancement of fruit mash processing. The quality attributes of the juice (e.g., orange juice) were not affected by the US treatment. In addition, application of US improved grape juice quality because it increased the contents of sugars, total acids, and phenolics as well as the color density of grape juice (Le Ngoc Lieu and Van Viet Man Le, 2010). However, a possible negative effect of US on the organoleptic characteristics of grape wine can occur from the sedimentation of tannins, potassium bitartrate, and proteins (Arena, 1979). For a scale-up of US technique to further industrial applications, vessel geometries, frequency combinations, and frequency modulation need to be explored and optimized (Vilkhu et al., 2008).
3. The main advantages of PEF technology are low energy consumption and short processing times. Thus, fruit wines with a high content of phenolic compounds can be produced, and the duration of maceration time during vinification can be reduced (Puertolas et al., 2010).
4. Various membrane separation techniques can be successfully applied at different stages of industrial fruit wine production. MF found a proven use in the clarification and cold sterilization of wine, UF can be used for clarification of fruit juice and wine, and RO can be applied for must correction, fruit juice concentration, and removal of alcohol. In winemaking, membrane separation processes can satisfy the consumer demand for healthy drinks without artificial additives. State-of-the-art membrane technology methods provide the opportunity to improve safety and decrease energy utilization as well as the environmental impact of wine processing. However, more research is required in the area of membrane fouling mechanisms and its prediction.

5. Membrane bioreactors are already practical for the production of aromatic compounds, flavor, and stabilizers in the wine industry. The immobilized enzyme stability and enzyme lifetime, as well as reproducibility of the experimental results, are important features of this technology to be further developed on a large scale. Knowledge-based control strategies for these applications have to be established in the course of their industrial use.
6. Techniques such as HHP treatment, PEF, MW, US, and high-pressure CO_2 exploitation have been successfully applied for pasteurization and contamination control in the fruit juice and wine industries. These methods offer improved product quality and reduced time of exposure to high temperatures, or eliminate the use of thermal energy altogether, thus decreasing the overall cost.

REFERENCES

Arena, C., 1979. Il processi fisici in enologia. Industrie delle Bevande 6, 175–178.

Arreola, A.G., Balaban, M.O., Wei, C.I., Peplow, A., Marshall, M., Cornell, J., 1991. Effect of supercritical carbon dioxide on microbial populations in single strength orange juice. Journal of Food Quality 14, 275.

Bazhal, M., Vorobiev, E., 2000. Electrical treatment of apple cossettes for intensifying juice pressing. Journal of the Science of Food and Agriculture 80, 1668–1674.

Buzrul, S., 2012. High hydrostatic pressure treatment of beer and wine: a review. High hydrostatic treatment of beer and wine: a review. Innovative Food Science and Emerging Technologies 13, 1–12.

Calablo, V., 2013. Engineering aspects of membrane bioreactors. In: Basile, A. (Ed.), Handbook of Membrane Reactors. Reactor Types and Industrial Applications, vol. 2. Woodhead Publishing, Series in Energy, pp. 3–53.

Canumir, J.A., Celis, J.E., de Bruijn, J., Vidal, L.V., 2002. Pasteurization of apple juice by using microwaves. LWT – Food Science and Technology 35, 389–392.

Catarino, M., Mendes, A., 2011. Dealcoholizing wine by membrane separation processes. Innovative Food Science & Emerging Technologies 12 (3), 330–337.

Cendres, A., Chemat, F., Maingonnat, J.-F., Renard, C.M.G.C., 2011. An innovative process for extraction of fruit juice using microwave heating. LWT – Food Science and Technology 44, 1035–1041.

Cendres, A., Hoerlé, M., Chemat, F., Renard, C.M.G.C., 2014. Different compounds are extracted with different time courses from fruits during microwave hydrodiffusion: examples and possible causes. Food Chemistry 154, 179–186.

Chang, T.S., Siddiq, M., Sinha, N.K., Cash, J.N., 1994. Plum juice quality affected by enzyme treatment and fining. Journal of Food Science 59, 1065–1069.

Chauhan, S.K., Tyagi, S.M., Singh, D., 2001. Pectinolytic liquefaction of apricot, plum, and mango pulps for juice extraction. International Journal of Food Properties 4, 103–109.

Chemat, F., Vian, M., Visinoni, F., 2010. Hydrodiffusion par micro-ondes pour l'isolation de produits naturels. Depositary: Milestone S.r.l. Universite d'Avignon et Des Pays Du Vaucluse. Publication number: 1955749.

Chisti, Y., 2003. Sonobioreactors: using ultrasound to enhance microbial productivity. Trends in Biotechnology 21, 89–93.

Corrales, M., Toepfl, S., Butz, P., Knorr, D., Tauscher, B., 2008. Extraction of anthocyanins from grape by-products assisted by ultrasonics, high hydrostatic pressure or pulsed electric fields: a comparison. Innovative Food Science and Emerging Technologies 9, 85–91.

Coulson, J.M., Richardson, J.F., Backhurst, J.R., Harker, J.H., 1999. Coulson & Richardson's Chemical Engineering, sixth ed. Fluid Flow, Heat Transfer and Mass Transfer, vol. 1. Butterworth-Heinemann, Oxford (Chapter 8).

Commission Regulation (EC) 2009a. No.606/2009 of 10 July 2009 Laying Down Certain Detailed Rules for Implementing Council Regulation (EC) No 479/2008 as Regards the Categories of Grapevine Products, Oenological Practices and the Applicable Restrictions, July 24, 2009. Official Journal of the European Union, p. L193.

Daufin, G., Escudier, J.L., Carrere, H., Berot, S., Fillaudeau, L., Decloux, M., 2001. Recent and emerging applications of membrane processes in the food and dairy industry. Food and Bioproducts Processing 79, 89–102.

Divies, C., Cachon, R., Cavin, J.-F., Prevost, H., 1994. Immobilised cell technology in wine production. Critical Reviews in Biotechnology 14 (2), 135–153.

Debs-Louka, E., Louka, N., Abraham, G., Chabot, V., Allaf, K., 1999. Effect of compressed carbon dioxide on microbial cell viability. Applied and Environmental Microbiology 65, 626.

Delfini, C., Conterno, I., Carpi, G., Rovere, P., Tabusso, A., Cocito, C., et al., 1995. Microbiological stabilisation of grape musts and wines by high hydrostatic pressure. Journal of Wine Research 6, 143–151.

De Pinho, M.N., 2010. Membrane processes in must and wine industries. In: Peinemann, K.-V., Pereira Nunes, S., Giorno, L. (Eds.), Membranes for Food Applications, vol. 3. Wiley-VCH, pp. 105–118.

Dharmadhikari, M., 1991. Racking wines. Publications of the Missouri State Fruit Experiment Station. 6 (12), 2–3. http://guides.library.missouristate.edu/stationpubs.

Dharmadhikari, M., 2004. Wines from cherries and soft fruits. Publications of the Missouri State Fruit Experiment Station 19 (2), 8–15.

Ducruet, J., Fast-Merlier, K.L., Noilet, P., 2010. New application for nanofiltration: reduction of malic acid in grape must. American Journal of Enology and Viticulture 61, 2.

du Toit, M., Pretorius, I.S., 2000. Microbial spoilage and preservation of wine: using weapons from nature's own arsenal—a review. South African Journal of Enology and Viticulture 21, 74–96.

Eisenbrand, G., 2005. Safety assessment of high pressure treated foods. Opinion of the senate Commission on Food Safety (SKLM) of the German research Foundation (DFG). Molecular Nutrition & Food Research 49, 1168–1174.

El Rayess, Y., Albasi, C., Bacchin, P., Taillandier, P., Raynal, J., Mietton-Peuchot, M., Devatine, A., 2011. Cross-flow microfiltration applied to oenology: a review. Journal of Membrane Science 382, 1–19.

Elvassore, N., Sartorello, S., Spilimbergo, S., Bertucco, A., 2000. Microorganisms Inactivation by Supercritical CO_2 in a Semi-continuous Process, 2, p. 773.

Enomoto, A., Nakamura, K., Nagai, K., Hashimoto, T., Hakoda, M., 1997. Inactivation of food microorganisms by high-pressure carbon dioxide treatment with or without explosive decompression. Bioscience, Biotechnology, and Biochemistry 61, 1133.

Erkmen, O., 2000. Predictive modelling of Listeria monocytogenes inactivation under high pressure carbon dioxide. LWT – Food Science and Technology 33, 514.

Feng, R., Zhao, Y., Zhu, C., Mason, J., 2002. Enhancement of ultrasonic cavitation yield by multi-frequency sonication. Ultrasonics Sonochemistry 9, 231–236.

Gallifuoco, A., D'Ercole, L., Alfani, F., Cantarella, M., Spagna, G., Pifferi, P.G., 1998. On the use of chitosan-immobilized β-glucosidase in wine-making: kinetics and enzyme inhibition. Process Biochemistry 33, 163–168.

Gerard, K.A., Roberts, J.S., 2004. Microwave heating of apple mash to improve juice yield and quality. LWT – Food Science and Technology 37, 551–557.

Gogate, P.R., Mujumdar, S., Pandit, A.B., 2003. Large-scale sonochemical reactors for process intensification: design and experimental validation. Journal of Chemical Technology and Biotechnology 78 (6), 685–693.

Gould, G.W., 2001. New processing technologies: an overview. Proceedings of the Nutrition Society 60, 463–474.

Guyot, S., Bernillon, S., Poupard, P., Renard, C.M.G.C., 2008. Multiplicity of phenolic oxidation products in apple juices and ciders, from synthetic medium to commercial products. In: Daayf, F., Lattanzio, V. (Eds.), Recent Advances in Polyphenol Research, vol. 1. Wiley, pp. 278–292.

Harlfinger, L., 1992. Microwave sterilization. Food Technology 46, 57–61.

Hielscher, 2006. Ultrasound in the Food Industry. http://www.hielscher.com/ultrasonics/food_01.htm.

Heinz, V., Buckow, R., 2010. Food preservation by high pressure. Journal für Verbraucherschutz und Lebensmittelsicherheit 5, 73–81.

Hong, S.I., Park, W.S., Pyun, Y.R., 1997. Inactivation of Lactobacillus sp. from kimchi by high pressure carbon dioxide. LWT – Food Science and Technology 30, 681.

Hui, Y.H., Barta, J., Cano, M.P., Gusek, T., Sidhu, J.S., Sinha, N., 2006. Handbook of Fruits and Fruit Processing, first ed. Blackwell Publishing, Ames.

Institute of Food Technologists, 2000. Special supplement: kinetics of microbial inactivation for alternative food processing technologies —IFT's response to task order #1, US Food and Drug Administration: how to quantify the destruction kinetics of alternative processing technologies. Journal of Food Science 4.

Isenschmid, A., Marison, I.W., Vonstockar, U., 1995. The influence of pressure and temperature of compressed CO_2 on the survival of yeast cells. Journal of Biotechnology 39, 229.

Karaman, H., Erkmen, O., 2001. High carbon dioxide pressure inactivation kinetics of *Escherichia coli* in broth. Food Microbiology 18, 11.

Knorr, D., Zenker, M., Heinz, V., Lee, D.-U., 2004. Applications and potential of ultrasonics in food processing. Trends in Food Science & Technology 15, 261–266.

Kolpan, S., Smith, B.H., Weiss, M.A., 2010. Exploring Wine, third ed. John Wiley and Son, Inc., pp. 72–74.

Kosseva, M.R., Kennedy, J.F., Lloyd, L.L., Beschkov, V., 1998. Malolactic fermentation in Chardonnay wine by immobilised *Lactobacillus casei* cells. Process Biochemistry 33, 793–797.

Kosseva, M.R., Kennedy, J.F., 2004. Encapsulated lactic acid bacteria for control of malolactic fermentation in wine. Artificial Cells, Blood Substitutes and Biotechnology 32 (1), 55–65.

Kosseva, M.R., 2011. Immobilization of microbial cells in food fermentation processes. Food and Bioprocess Technology 4, 1089–1118.

Kosseva, M.R., 2013. Use of immobilized biocatalyst for valorisation of whey lactose. In: Kosseva, M.R., Webb, C. (Eds.), Food Industry Wastes: Assessment and Recuperation of Commodities. Academic Press, Elsevier, USA, pp. 137–156.

Kwiatkowska, B., Bennett, B., Akunna, J., Walker, G.M., Bremner, D.H., 2011. Stimulation of bioprocesses by ultrasound. Biotechnology Advances 29, 768–780.

Le Bourvellec, C., Le Quéré, J.M., Renard, C.M.G.C., 2007. Impact of noncovalent interactions between apple condensed tannins and cell walls on their transfer from fruit to juice: studies in model suspensions and application. Journal of Agricultural and Food Chemistry 55, 7896–7904.

Lieu Le, N., Le, V.V., 2010. Application of ultrasound in grape mash treatment in juice processing. Ultrasonics Sonochemistry 17, 273–279.

Lipnizki, F., 2010. Cross-flow membrane applications in the food industry. In: Peinemann, K.-V., Pereira Nunes, S., Giorno, L. (Eds.), Membranes for Food Applications, vol. 3. Wiley-VCH, pp. 1–24.

López, M., Alvarez, S., Riera, F.A., Alvarez, R., 2002. Production of low alcohol content apple cider by reverse osmosis. Industrial and Engineering Chemistry Research 41 (25), 6600–6606.

López, N., Puertolas, E., Condon, S., Raso, J., Alvarez, I., 2009. Enhancement of the solid-liquid extraction of sucrose from sugar beet (*Beta vulgaris*) by pulsed electric fields. LWT–Food Science Technology 42, 1674–1680.

Madigan, M.T., Martinko, J.M., Parker, J., 2002. Brock Biology of Microorganisms. Prentice Hall, USA.

Malchesky, P.S., 2003. Sub-critical Fluid Cleaning and Antimicrobial Decontamination System and Process. United States Patent, US 6558622 B1.

Martino, A., Durante, M., Pifferi, P.G., Spagna, G., Bianchi, G., 1996. Immobilization of β-glucosidase from a commercial preparation. Part 2. Optimization of the immobilization process on chitosan. Process Biochemistry 31, 287–293.

Mason, T.J., Paniwnyk, L., Lorimer, J.P., 1996. The uses of ultrasound in food technology. Ultrasonics Sonochemistry 3, S253–S260.

Massot, A., Mietton-Peuchot, M., Peuchot, C., Milisic, V., 2008. Nanofiltration and reverse osmosis in winemaking. Desalination 231, 283–289.

Mazzei, R., Chakraborty, S., Drioli, E., Giorno, L., 2010. Membrane bioreactors in functional food ingredients production. In: Peinemann, K.-V., Pereira Nunes, S., Giorno, L. (Eds.), Membranes for Food Applications, vol. 3. Wiley-VCH, pp. 201–221.

Mietton-Peuchot, M., Milisic, V., Noilet, P., 2002. Grape must concentration by using reverse osmosis. Comparison with chaptalization. Desalination 148 (1–3), 125–129.

Mermelstein, N.H., 1998. Beer and wine making. Food Technology 52 (84), 86–88.

Moholkar, S., Rekveld, S., Warmoeskerken, G., 2000. Modeling of acoustic pressure fields and the distribution of the cavitation phenomena in a dual frequency sonic processor. Ultrasonics 38, 666–670.

Mok, C., Sonk, K.-T., Park, Y.-S., Lim, S., Ruan, R., Chen, P., 2006. High hydrostatic pressure pasteurization of red wine. Journal of Food Science 71, 265–269.

Molinari, P., Pilosof, A.M.R., Jagus, R.J., 2004. Effect of growth phase and inoculum size on the inactivation of *Saccharomyces cerevisiae* in fruit juices, by pulsed electric fields. Food Research International 37, 793–798.

Nemati, M., Webb, C., 2011. Immobilized cell bioreactors. In: Moo-Young, M. (Ed.), Comprehensive Biotechnology, Engineering Fundamentals of Biotechnology, vol. 2, second ed. pp. 331–346.

Obradovic, B., Nedović, V.A., Bugarski, B., Willaert, R.G., Vunjak-Novakovic, G., 2004. Immobilised cell bioreactors. In: Nedović, V., Willaert, R.G. (Eds.), Fundamentals of Cell Immobilisation Biotechnology. Focus on Biotechnology, vol. 8A. Springer, Netherlands, pp. 411–436. series.

Oey, I., Lille, M., Loey, A.V., Hendrickx, M., 2008. Effect of high pressure processing on colour, texture and flavour of fruit and vegetable-based food products: a review. Trends in Food Science & Technology 19, 320–328.

Osajima, Y., Shimoda, M., Kawano, T., 1996. Method for Modifying the Quality of Liquid Foodstuff. United States Patent, US 5520943.

Osajima, Y., Shimoda, M., Kawano, T., 1997. Method for Inactivating Enzymes, Microorganisms, and Spores in a Liquid Foodstuff. United States Patent, US 5667835.

Osajima, Y., Shimoda, M., Kawano, T., Okubo, K., 1998. System for Processing Liquid Foodstuff or Liquid Medicine With a Supercritical Fluid of Carbon Dioxide. United States Patent, US 5704276.

Osajima, Y., Shimoda, M., Kawano, T., Okubo, K., 1999. System for Processing Liquid Foodstuff or Liquid Medicine With a Supercritical Fluid of Carbon Dioxide. United States Patent, US 5869123.

Osajima, Y., Shimoda, M., Takada, M., Miyake, M., 2003. Method of and System for Continuously Processing Liquid Materials, and the Product Processed Thereby. United States Patent, US 6616849 B1.

Pap, N., Pongrácz, E., Myllykoski, L., Keiski, R., 2004. Waste minimization and utilization in the food industry: processing of arctic berries, and extraction of valuable compounds from juice- processing by-products. In: Pongrácz, E. (Ed.), Proceedings of the Waste Minimization and Resources Use Optimization Conference. June 10, 2004. Oulu University Press, Oulu, Finland, pp. 159–168.

Peleg, M., 2002. Simulation of *E-coli* inactivation by carbon dioxide under pressure. Journal of Food Science 67, 896.

Peuchot, M.M., 2010. New applications for membrane technologies in enology. In: Peinemann, K.-V., Pereira Nunes, S., Giorno, L. (Eds.), Membranes for Food Applications, vol. 3. Wiley-VCH, pp. 119–127.

Pickering, G.J., 2000. Low- and reduced-alcohol wine: a review. Journal of Wine Research 11 (2), 129–144.

Pilipovik, M.V., Riverol, C., 2005. Assessing dealcoholization systems based on reverse osmosis. Journal of Food Engineering 69 (4), 437–441.

Puertolas, E., Lopez, N., Condon, S., Alvarez, I., Raso, J., 2010. Potential applications of PEF to improve red wine quality. Trends in Food Science & Technology 21, 247–255.

Puig, A., Vilavella, M., Daoudi, L., Guamis, B., Minguez, S., 2003. Microbiological and biochemical stabilisation of wines using the high pressure technique. Bulletind de l'OIV 76 (869/870), 569–617.

Renard, C.M.G.C., Le Quéré, J.-M., Bauduin, R., Symoneaux, R., Le Bourvellec, C., Baron, A., 2011. Modulating polyphenolic composition and organoleptic properties of apple juices by manipulating the pressing conditions. Food Chemistry 124, 117–125.

Rendueles, E., Omer, M.K., Alvseike, O., Alonso-Calleja, C., Capita, R., Prieto, M., 2011. Microbiological food safety assessment of high hydrostatic pressure processing: a review. LWT – Food Science and Technology 44, 1251–2160.

Ribéreau-Gayon, P., Dubourdieu, D., Donèche, B., Lonvaud, A., 2006. The Use of Sulphur Dioxide in Must and Wine Treatment. Handbook of Enology. John Wiley & Sons Ltd, Chichester, pp. 193–220.

Rodriguez-Nogales, J.M., Ortega, N., Perez-Mateos, M., Busto, M., 2008. Pectin hydrolysis in a free enzyme membrane reactor: an approach to the wine and juice clarification. Food Chemistry 107, 112–119.

Romano, P., Suzzi, G., 1993. Sulfur dioxide and wine microorganisms. In: Fleet, G.H. (Ed.), Wine Microbiology and Biotechnology. Harwood Academic Publishers, pp. 373–393.

Romo-Sánchez, S., Camacho, C., Ramirez, H.L., Arévalo-Villena, M., 2014. Immobilization of commercial cellulase and xylanase by different methods using two polymeric supports. Advances in Bioscience and Biotechnology 5, 517–526.

Rosa Santos, F., Catarino, I., Geraldes, V., de Pinho, M.N., 2008. Concentration and rectification of grape must by nanofiltration. American Journal of Enology and Viticulture 59 (4), 446–450.

Rutledge, P., 1996. Production of non-fermented fruit products. In: Arthey, D., Ashurst, P.R. (Eds.), Fruit Processing. Chapman and Hall, London, pp. 70–96.

Saldana, G., Puertolas, E., Lopez, N., Garcia, D., Alvarez, I., Raso, J., 2009. Comparing the PEF resistance and occurrence of sublethal injury on different strains of *Escherichia coli*, *Salmonella Typhimurium*, *Listeria monocytogenes* and *Staphylococcus aureus* in media of pH 4 and 7. Innovative Food Science and Emerging Technologies 10, 160–165.

Schlegel, W., 1992. Commercial pasteurisation and sterilization of food products using microwave technology. Food Technology 46, 62–66.

Shimoda, M., Cocunubo-Castellanos, J., Kago, H., Miyake, M., Osajima, Y., Hayakawa, I., 2001. The influence of dissolved CO_2 concentration on the death kinetics of *Saccharomyces cerevisiae*. Journal of Applied Microbiology 91, 306.

Shoh, A., 1988. Industrial applications of ultrasound. In: Suslick, K.S. (Ed.), Ultrasound, Its Chemical, Physical and Biological Effects. VCH Publishers, Inc., New York, pp. 97–122.

Siddiq, M., Arnold, J.F., Sinha, N.K., Cash, J.N., 1994. Effect of polyphenol oxidase and its inhibitors on anthocyanin changes in plum juice. Journal of Food Processing and Preservation 18, 75–84.

Sims, M., 2001. Method and Membrane System for Sterilizing and Preserving Liquids Using Carbon Dioxide. United States Patent, US 6331272 B1.

Singh, R.P., 2009. Fluid flow in food processing. In: Singh, R.P., Heldman, D.R. (Eds.). Introduction to Food Engineering, fourth ed. Academic Press, Elsevier, pp. 65–95.

Singh, R.P., Heldman, D.R., 2009. Introduction to Food Engineering, fourth ed. Academic Press, Elsevier.

Skauen, D., 1976. A comparison of heat production and cavitation intensity in several ultrasonic cell disrupters. Ultrasonics 14, 173–176.

Smelt, J.P., Hellemons, J.C., Wouters, P.C., van Gerwen, S.J., 2002. Physiological and mathematical aspects in setting criteria for decontamination of foods by physical means. International Journal of Food Microbiology 78 (1–2), 57–77.

Smith, C., 2002. Applications of Reverse Osmosis in Winemaking. www.vinnovation.com.

Spilimbergo, S., Elvassore, N., Bertucco, A., 2002. Microbial inactivation by high pressure. Journal of Supercritical Fluids 22, 55.

Suslick, K.S., 1990. Sonochemistry. Science 247, 1439–1445.

Suzuki, A., 2002. High pressure-processed foods in Japan and the world. In: Hayashi, R. (Ed.), Trends in High Pressure Bioscience and Biotechnology. Elsevier Science, pp. 365–374.

Swami, S.B., Thakor, N.J., Divate, A.D., 2014. Fruit wine production: a review. Journal of Food Research and Technology 2 (3), 93–100.

Tatake, A., Pandit, B., 2002. Modelling and investigation into cavitation dynamics and cavitational yield: influence of dual-frequency ultrasound sources. Chemical Engineering Science 57, 4987–4995.

Toma, M., Vinatoru, M., Paniwnyk, L., Mason, T.J., 2001. Ultrasonics Sonochemistry 8, 137–142.

Ulbricht, M., Ansorge, W., Danielzik, I., König, M., Schuster, O., 2009. Fouling in microfiltration of wine: the influence of the membrane polymer on adsorption of polyphenols and polysaccharides. Separation and Purification Technology 68, 335–342.

US Department of Energy (US DOE), Office of Industrial Technologies, 2006. Energy Efficiency Guide for Industry in Asia. www.energyefficiencyasia.org (Pumps and Pumping Systems ©United Nations Environment Programme).

Valero, M., Recrosio, N., Saura, D., Muñoz, N., Martí, N., Lizama, V., 2007. Effects of ultrasonic treatments in orange juice processing. Journal of Food Engineering 80, 509–516.

Vilkhu, K., Mawson, R., Simons, L., Bates, D., 2008. Applications and opportunities for ultrasound assisted extraction in the food industry—A review. Innovative Food Science and Emerging Technologies 9, 161–169.

Vinatoru, M., 2001. An overview of the ultrasonically assisted extraction of bioactive principles from herbs. Ultrasonics Sonochemistry 8, 303.

Wang, H., Cao, G.H., Prior, R.L., 1996. Total antioxidant capacity of fruits. Journal of Agricultural and Food Chemistry 44, 701–705.

Wildasin, R.E., Forbes, J., Robey, R., Paradis, A.J., 2002. Treating Liquid Food and Other Products Using Carbon Dioxide. International Patent, WO 02/03816 A1.

Will, F., Dietrich, H., 2006. Optimised processing technique for colour and cloud stable plum juices and stability of bioactive substances. European Food Research and Technology 223, 419–425.

Wood, R.W., Loomis, A.L., 1927. The physical and biological effects of high frequency ultrasound waves of great intensity. Philosophical Magazine 4, 417–436.

Wouters, P.C., Alvarez, I., Raso, J., 2001. Critical factors determining inactivation kinetics by pulsed electric field food processing. Trends in Food Science and Technology 12, 112–121.

Wu, Y., Mittal, G.S., Griffiths, M.W., 2005. Effect of pulsed electric fields on the inactivation of microorganisms in grape juices with and without antimicrobials. Biosystems Engineering 90, 1–7.

Zhang, J., Davis, T.A., Matthews, M.A., Drews, M.J., LaBerge, M., An, Y.H., 2006. Sterilization using high-pressure carbon dioxide. Journal of Supercritical Fluids 38, 354–372.

SPECIFIC FEATURES OF TABLE WINE PRODUCTION TECHNOLOGY

POME FRUIT WINES: PRODUCTION TECHNOLOGY

V.K. Joshi[1], B.L. Attri[2]

[1]Dr. Y.S. Parmar University of Horticulture and Forestry, Nauni, Solan, HP, India;
[2]ICAR-Directorate of Mushroom Research (DMR), Solan, HP, India

1. INTRODUCTION

A pome is a fleshy fruit consisting of an outer thickened fleshy layer and a central core with usually five seeds enclosed in a capsule (Esau, 1977). The carpels of a pome fruit are fused within the core, where the epicarp and the mesocarp, which form the hypanthial tissue, may be fleshy and difficult to distinguish from one another. The endocarp forms a leathery or stony case around the seeds (see Chapter 1 for more details). The fruits included in the pome group are listed in Table 7.1.1.

Of the pome fruits given in Table 7.1.1, apple and pear are the major ones, whereas the others are minor fruits. Apart from fresh consumption, the fruits are employed to produce various types of wines and their distillates, called brandies (Amerine et al., 1980; Joshi et al., 2004; Vyas and Kochhar, 1993). Compared to the grape wines, most of the other fruit wines are made in countries and regions, where such fruits are produced in abundance. Fruit wines can be made during a less busy time (after grape harvest) of the year, thus permitting efficient use of winery facilities. There are a variety of fruits suited to make a good quality wine. Wines are made from complete or partial alcoholic fermentation of grapes or any other fruit, like apples, plums, peaches, pears, berries, cherries, currants, or apricots (Babsky et al., 1986; Joshi and Bhutani, 1990, 1995; Bhutani and Joshi, 1995; Joshi et al., 1999a,b). However, compared to the quantity of grape wine produced and consumed throughout the world, the quantity of wine produced from nongrape fruits is insignificant (Babsky et al., 1986; Gayon, 1978; Samarajeeva et al., 1985), except for cider and perry, which are produced and consumed in significant amounts throughout the world (Joshi et al., 2011a). Consumption of wine has been associated with the prevention of cardiovascular diseases (Joshi and Thakur, 1995; Delin and Lee, 1991; Joshi, 1997; Stockley, 2011). Apart from production of wines from other pome fruits, the preparation of cider and perry has also been discussed in detail in this chapter.

Table 7.1.1 Name, Botanical Name, and Family of Pome Fruits

S.No.	Name	Botanical Name	Family
1.	Apple	*Malus domestica*	Rosaceae
2.	Pear	*Pyrus communis*	Rosaceae
3.	Quince	*Cydonia oblonga*	Rosaceae
4.	Cotoneaster	*Cotoneaster obovatus*	Rosaceae
5.	Hawthorn	*Crataegus crenulata*	Rosaceae
6.	Loquat	*Eriobotrya japonica*	Rosaceae
7.	Medlar	*Mespilus germanica*	Rosaceae
8.	*Pyracantha*	*Pyracantha crenulata*	Rosaceae
9.	Toyon	*Heteromeles arbutifolia*	Rosaceae
10.	Rowan	*Sorbus aucuparia*	Rosaceae
11.	Whitebeam	*Sorbus aria*	Rosaceae

Esau, K., 1977. Anatomy of Seed Plants. John Wiley and Sons, New York.

2. TECHNOLOGY FOR THE PREPARATION OF APPLE WINE

2.1 APPLE WINE

Apples are used to prepare a table wine that is more nutritious than its distilled liquor counterparts (Bhutani et al., 1989; Goswell and Kunkee, 1977; Joshi and Thakur, 1995). It is a fermented beverage made from fresh or concentrated apple juice. It has a long tradition in Europe and has taken an important place in the global fruit wine industry (Wang et al., 2004). Not only this, it has become the second largest fruit wine industry, with an increasing demand in China. The apple juice fermentation process used to obtain a pleasant alcoholic beverage has been practiced in the eastern Mediterranean for more than 2000 years (Laplace et al., 2001). Fermented apple juice is used as a base for manufacturing apple wine, apple vermouth, and cider (Smock and Neubert, 1950; Sandhu and Joshi, 1995, 2005; Joshi et al., 1995, 2011a,b; Joshi and Sandhu, 2000), a sparkling and refreshing fruit-flavored beverage consumed in many countries throughout the world.

A generalized flow sheet of a method to prepare apple wine is shown in Fig. 7.1.1.

2.1.1 Choice of Fruit

The selection of apple fruit for making wine depends upon several factors such as market demand, availability of raw material, production facilities, and sound economic reasons. Locally grown apples that are surplus after meeting fresh market demands, are used for making juice and wine. It is important that the fruit be sound, i.e., free of decay or rot, and well matured. Unripe or immature fruit should not be used because it is high in starch, acid, and astringency and low in sugar and flavor. On the other hand, overly matured fruits can be low in fresh and fruity flavor, difficult to process, and also difficult to clarify. The amounts of fruit constituents such as sugars, acids, and phenolic compounds, as well as color and flavor, vary considerably among the apple varieties.

The use of apple juice concentrate for wine production has increased considerably because of several advantages offered by it, such as price stabilization, quality maintenance, and storage for longer

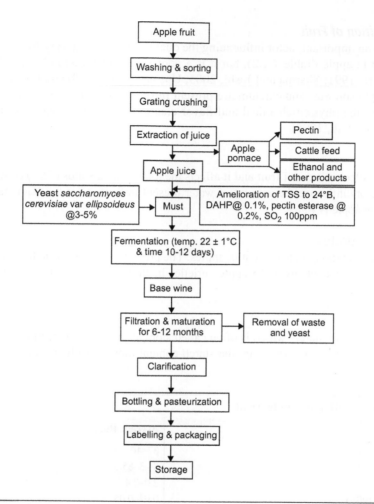

FIGURE 7.1.1

Flow diagram for manufacturing apple wine. *DAHP*, diammonium hydrogen phosphate; *TSS*, total soluble solids.

Adapted from Joshi, V.K., 1995. Fruit Wines, first ed. Directorate of Extension Education. Dr. YS Parmar University of Horticulture and Forestry, Nauni, Solan, HP, p. 155.

times without spoilage (Joshi et al., 1991), though it does lead to a loss of development of specific cultivars for cider making (Beech and Carr, 1977; Labelle, 1980, Downing, 1989; Jarvis et al., 1995). In India, wine production is still in its infancy, and the suitability of Indian varieties for wine production has not been adequately worked out, though Ambri Kashmiri, Red Delicious, Golden Pippin, Golden Delicious, and Rus Pippin apples and Maharaji apples and crab apples have been found suitable for winemaking (Amerine et al., 1980; Rana et al., 1986; Kerni and Shant, 1984; Joshi et al., 1991). A comparative study of scabbed fruits vs normal fruits showed that the fruits with fewer than 15 spots did not affect the fermentation behavior or the physicochemical and sensory qualities of the wine produced (Azad et al., 1987).

2.1.2 Composition of Fruit

This is certainly an important factor influencing the quality of wine. Carbohydrates are the principal food constituent in apple (Table 7.1.2), but it is a poor source of protein (Joslyn, 1950; Gebhardt et al., 1982; Mitra, 1991; Sharma and Joshi, 2005; Upshaw et al., 1978) and fat. Among the minerals, potassium, phosphorus, and calcium are present in significant amounts in apple fruit. There is large variation in the physico-chemical and flavor characteristics of the various cultivars used to make wine (Jarvis et al., 1995).

2.1.2.1 Water

Water is the largest component of fruit and it affects the total soluble solids (TSS) content of the juice. A number of factors influence the amount of water present in fruit at harvest. Generally, 84% of the fresh weight of an apple is water.

2.1.2.2 Carbohydrates

Sugars are the main carbohydrate present in apples. The predominant sugars include fructose, sucrose, and glucose. The sugar contents of 15 apple varieties in Illinois were evaluated, some of which are depicted in Fig. 7.1.2.

2.1.2.3 Starch

The unripe fruit contains starch, which is often called storage sugar. As the fruit matures, the starch is hydrolyzed into sugars. During ripening, the starch content rapidly declines, and at the harvest very

Table 7.1.2 Composition of Apple Fruit	
Constituent	**Average Range**
Calories (kcal/100 g)	37–46
Water (g)	84.3–85.6
Fiber (g)	2.0–2.4
Total nitrogen (g/100 g)	0.04–0.05
Protein (%)	0.19
Lipid (%)	0.36
Sugar (per 100 g flesh)	9.2–11.8
Total sugars (%)	10.65–13.23
Reducing sugars (%)	7.05–10.67
Sucrose (%)	1.95–5.02
Pectin (%)	0.32–0.75
Phenolic compounds (%)	0.15–2.4
Vitamin C (mg/100 g)	3.15–5.7

Compiled from Gebhardt, S.E., Cutrufelli, R., Mathews, R.H., 1982. Composition of Foods. Agric. Handbook. US Dept of Agri., Washington, pp. 8–9; Mitra, S.K., 1991. Apples. In: Mitra, S.K., Bose, T.K., Rathore, D.S. (Eds.), Temperate Fruits. Horticulture and applied Publ., Calcutta, p. 122; Sharma, R.C., Joshi, V.K., 2005. Apple processing technology. In: Chadha, K.L., Awasthi, R.P. (Eds.), The Apple. Malhotra Public. House, New Delhi, p. 445; Upshaw, S.C., Lopez, A., Williams, H.L., 1978. Essential elements in apples and canned apple sauce. Journal of Food Science 43 (2), 449.

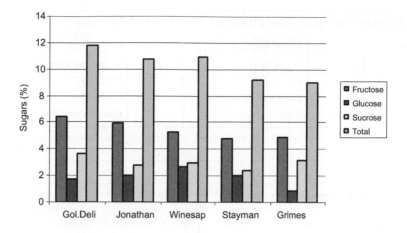

FIGURE 7.1.2

Comparison of the sugar contents of five apple varieties.

Based on data from Loft, R.V., 1943. The levalose, dextrose and sucrose content of 15 Illinois apple varieties. Proceedings of the American Society for Horticultural Science 43, 56.

little, if any, starch may be present in the fruit of the apple. The presence of starch can cause problems in the clarification and filtration of apple wine during processing.

2.1.2.4 Pectic Substances

The pectic substances are complex colloidal carbohydrate derivatives containing large proportions of anhydrogalacturonic acid units. The pectic substances are capable of forming gels and they contribute to the viscosity or body of the juice and often cause cloudiness, thus making the juice difficult to clarify. Pectin-splitting enzymes are used to help the process of juice clarification (Amerine et al., 1980; Joshi and Bhutani, 1991).

2.1.2.5 Organic Acids

A number of organic acids are found in apple fruit, which include malic, citric, quinic, glycolic, succinic, lactic, galacturonic, and citramalic. Malic acid is the principal acid but citric and other acids are present in traces. These acids are present as free or combined forms and the acidity is often expressed in terms of percentage malic acid. The acid composition and pH of the juice are important considerations in apple winemaking. The acid content has an important bearing on taste, pH, fermentation, color, and stability of the wine. The volatile aroma-rich compounds and complex apple aroma are crucial to the quality of apple wine.

2.1.2.6 Astringent Compounds

The astringent compounds in apples include phenolic substances and tannin. These constituents are responsible for the astringency of flavor and darkening of color when sliced fruit or juice is exposed to air. They are present in the range of 0.11–0.34 /100 g of fresh weight. In apples, the phenolic compounds viz; leucoanthocyans, epicatechin, chlorogenic acid, isochlorogenic acid, quinic acid, shikimic acid, coumaryl quinic acid, quercitrin, isoquercitrin, avicularin, rutin, and quercetin xyloside have been reported (Hulme, 1958) (Tables 7.1.3 and 7.1.4).

Table 7.1.3 Bitterness of Apple Wine Produced From Various Cultivars

None or Slight		Moderate		Extreme	
McIntosh	a[a]	Pound Sweet	b[c]	Royal Red Delicious	c[d]
Idared	a	Wayne	c	Winesap	c[d]
Conical Rome	a	Golden Delicious	c	Cortland	d
Roanoke	a[b]	Vance Delicious	c	Cowin Rome	d

All wines fermented to 11–13% ethanol; Cultivars with similar letters are statistically on par with one another.
Based on Buren, J.P.V., Lee, C.Y., Way, R.D., Pool, R.M., 1979. Bitterness in apple and grape wines. HortScience 14 (1), 42.

Table 7.1.4 Phenolic Compounds Present in Wine-Apple Juice

Compound	Example
Phenolic acids	Chlorogenic acid
Phloretin derivatives	Phloridzin
Simple catechins	(−)-Epicatechin
Condensed procyanidins	B^2
Minor constituent	
Flavonol glycosides	Anthocyanins

Adapted from Beech, F.W., Carr, J.G., 1977. Cider and perry. In: Rose, A.H. (Eds.), Economic Microbiology. Alcoholic Beverages, vol. VI. Academic press, London, p. 139.

The amino acid composition of wine apples includes asparagine, aspartic acid, glutamic acid, serine, alanine, and others that are just in traces. Phenolic compounds constitute a significant component of apple and are attributed the role of antioxidants. But the phenolic compounds increase the tendency toward enzymatic browning during juice extraction, and polymers of the catechins and leucoanthocyanins contribute largely to the bitterness and astringency of the wine. Bitterness could be the result of fermentation conditions or the duration of fermentation or could be an effect of variety in apple wine, particularly of yeast nutrients. Some cultivars, like Bittersweet (particularly French and English), have relatively high concentrations of polyphenols, conferring bitterness and astringency to the finished beverage. Because the polyphenols make a major contribution to flavor, color, and pressability and also have some antimicrobial properties, their retention in wine is considered useful. Some changes occur during the storage of apple juice concentrate, as reviewed earlier (Amerine et al., 1980; Beveridge et al., 1986; Wong and Stanton, 1993; Wrolstad et al., 1990).

2.1.2.7 Nitrogenous Compounds

The nitrogenous compounds, such as proteins and amino acids, are important constituents of fruit. The soluble nitrogen content varies from 4 to 33 mg N/100 mL of juice. Asparagine was noted to be the chief amino acid in most apple varieties. Other prominent amino acids were aspartic and glutamic acids. The concentrations of nitrogenous compounds and particularly free amino acids are important to the process of winemaking because they are needed to ensure a sound fermentation. The nitrogen fraction in apple juice comprises the amino acids viz., asparagine, glutamine, aspartic acid, and serine, representing together from 86% to 95% of total amino acids, which are rapidly assimilated by the yeasts. Fruits harvested from orchards with extremely high fertilizer use can contain nitrogen compounds in the juices up to five times higher (Lequere and Drilleau, 1998).

2.1.3 Processing Apples for Juice

2.1.3.1 Grinding

Fresh apple fruit used for extraction of juice should be fully ripened, but are generally stored for a few weeks after harvest so that all the starch is converted into sugar. The fruit should be sorted out to remove decayed fruit and washed to remove dirt and chemical residue. An earlier practice was to empty bulk truckloads or bins of apples on a deleafing screen into a tank of water. Rinsing with clean water is accomplished at the scrubber or after inspection, but routine replacement of holding water is necessary. Various kinds of equipment are available for cleaning apple fruit. A hammer mill is commonly used for grinding apples but other kinds of machines can also be used. In most of the applications, either hammer or grating mills are used and even slicers are required for difficult extraction. A fairly recent development in the production of apple juice for concentrate is diffusion extraction (Buren et al., 1979; Downing, 1989), and the juice can be used for winemaking.

2.1.3.2 Pressing

The crushed apples are pressed to extract the juice. Several types of presses have been developed for pressing and extracting apples, including hydraulic presses, screw presses, basket presses, belt presses, and pneumatic presses (Joshi et al., 2011b). The traditional method, involves the use of a hydraulic press. In this method, a rack is placed in the press rack and then a cloth (usually nylon) is placed on the rack. The pulp is spread in a thin uniform layer on the cloth and the cloth is folded. Another rack is placed on top of the folded cloth and the process is repeated. Several layers of cloth (holding pulp) and racks are stacked and pressure is applied to the stack to squeeze the juice. Initially the pressure is raised to about 500–700 psi, which releases free-running juice from the pulp. Following this, the pressure is slowly raised to 2500–3000 psi to extract the remaining juice. The other presses used for apple juice extraction are the Stoll press, Bucher-Guyer press, Bullmer continuous belt press, and Atlas-Pacific press. Screw presses have been used on hydraulically pressed pomace in France for additional recovery of juice from centrifuged apple pulp containing cellulose fibers (Lowe et al., 1964). Using press aids, the Zenith and Jones presses were among the first continuous presses used for apple juice extraction in the United States, but the amount of apple solids in the juice from screw presses is much greater than that in juice from hydraulic rack and cloth presses (Bump, 1989). A screw press that is well adapted to apple juice production is the Reitz press system. An electro-acoustic dewatering process is one of the newer methods that employs passing of electric current through the pulp prior to pressing, which has been claimed to release a higher yield of juice. Washing and crushing the fruits and adding 50 ppm SO_2 and 10% water in making apple wine are recommended (Vogt, 1977).

2.1.3.3 Juice Treatment

Fresh apple juice is very much susceptible to oxidation and browning, resulting in a decrease in the delicate fruit flavor (Downing, 1989). The oxidative reaction is enzymatically catalyzed but it can also occur without the mediation of enzymes. To obtain the best results with respect to juice clarification, a trial using various enzyme preparations has to be conducted. The action of enzymes depends on pH, temperature, concentration, and reaction time duration. All these factors should be considered in choosing an enzyme treatment suitable for apple juice.

2.1.3.4 Adjusting Sugar and Acid

The correction of raw material to make a product of consistent quality is referred to as amelioration, e.g., adjustment of the sugar and/or acid content of the juice, as regulated by the respective standards (Fig. 7.1.1). Controlling the sugar content of apple juice is required to maintain the proper final alcohol content, which is achieved by the addition of water, juice from the second pressing of the pomace, sugar, or concentrated juice. Fortification of apple juice after dilution from its concentrate with diammonium hydrogen phosphate (DAHP) is essential for rapid fermentation. Must prepared by direct dilution of the concentrate reportedly ferments faster than that ameliorated with sugar (Joshi et al., 1991; Joshi and Sandhu, 1994). Because apple juice does not contain a sufficient amount of sugar to produce a table wine with an alcohol content of 10–12% by volume, the regulations permit the addition of sugar or other sweetening materials, such as syrup or concentrate, to raise the sugar content to the desired level. The addition of sweetening material is also permitted to blend the wine; however, in no case should the volume resulting from the addition of all the sweetening material exceed 35% of the final volume of the wine.

To produce well-balanced wine, the must should contain a sufficient amount of acid. However, two points should also be considered in doing so. First, the addition of sweetening material will dilute the acid level, and second, in apple wine the acidity is expressed in terms of malic acid, which is different from grape wine, in which the acidity is expressed as tartaric acid. Probably the best way to ensure a sufficient level of acidity in the must is to choose medium- to high-acid varieties of apple in the blend to make the wine. As Delicious apples are low in acidity they should be blended with high-acid varieties such as Gravenstein and Jonathan (Downing, 1989).

2.1.3.5 Nitrogen Source

Nitrogen-containing compounds in the must are important to the growth of yeast and, hence, to the fermentation rate and aroma compound production. DAHP at 0.1% has been used as a yeast food in alcoholic fermentation. Addition of a nitrogen source to the musts made from apple juice concentrate for alcoholic fermentation has also been reported (Joshi and Sandhu, 1994). Supplementation with a nitrogen source is also essential because, in its absence, the yeast uses the amino acids of the must, resulting in the formation of higher alcohols (Amerine et al., 1980; Rana and Rana, 2011). Use of a nitrogen source such as ammonium chloride or phosphate is required to produce a wine with 5.6–7.3% alcohol (Amerine et al., 1980). Addition of DAHP improves the fermentation (Fig. 7.1.3) and makes most of the physicochemical characteristics of apple wine desirable (Lagan, 1981; Joshi, 1993). Further, the addition of ammonium salts to the fermenting apple must reduces higher alcohol production by inhibiting the degradation of the amino acids of the must.

2.1.3.6 Must Clarification

The clarification of must is an indispensable factor affecting the overall quality of wine. Pectic enzymes are often added to the juice to hasten and improve the clarification of the cider during and after fermentation. Enzymatically clarified apple wines were rated better in terms of color, appearance, body, and flavor than nonclarified apple wines (Joshi and Bhutani, 1991). So the addition of pectinesterase enzyme is desirable (Joshi and Bhutani, 1991; Downing, 1989) as it increases the rate of fermentation (Fig. 7.1.4), decreases macroelements in the wine, and improves the physicochemical characteristics of the wine produced (Fig. 7.1.5). The addition of increasing levels of insoluble solids to the apple juice leads to the production of undesirable physicochemical characteristics in apple wine. (Joshi et al., 2013a). Thus, to prepare quality apple wine, juice with the insoluble solids removed by presettling and clarification using pectolytic enzyme should be used (Table 7.1.5).

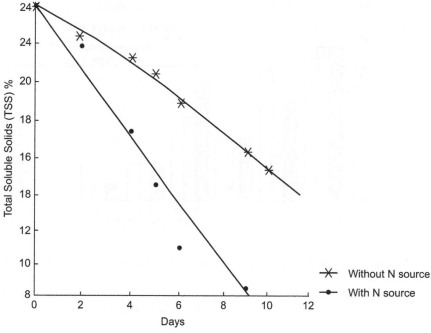

FIGURE 7.1.3

Effect of the addition of a nitrogen source on the rate of fermentation of apple wine. *TSS*, total soluble solids.

Reproduced from Joshi, V.K., 1993. Studies on Alcoholic Fermentation of Apple Juice and Apple Juice Utilization. (Ph.D. thesis submitted to Guru Nanak University Amritsar (Pb), India).

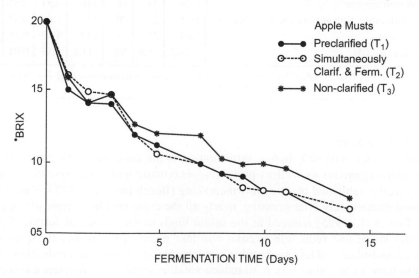

FIGURE 7.1.4

Effect of pectinesterase addition on the rate of fermentation of apple wine.

Reproduced from Joshi, V.K., Bhutani, V.P., 1991. Influence of enzymatic treatment on fermentation behaviour physico chemical and sensory qualities of apple wine. Sciences Des Aliments 11 (3), 491–496.

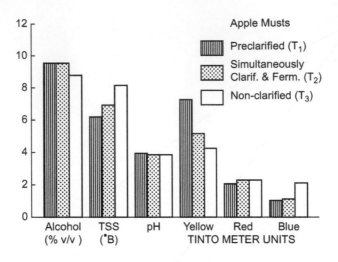

FIGURE 7.1.5

Physicochemical characteristics of apple wine prepared with and without pectinesterase enzyme addition. *TSS*, total soluble solids.

Reproduced from Joshi, V.K., Bhutani, V.P., 1991. Influence of enzymatic treatment on fermentation behaviour physico chemical and sensory qualities of apple wine. Sciences Des Aliments 11 (3), 491–496.

Table 7.1.5 Effect of Addition of Pectinesterase Enzyme on the Mineral Contents of Apple Wine								
Treatment	**K**	**Na**	**Mg**	**Ca**	**Zn**	**Cu**	**Mn**	**Fe**
Preclarification with enzyme	1450	33	18	140	2.51	0.67	0.83	4.26
Simultaneous clarification with enzyme and fermentation	1850	33	20	128	1.64	0.53	0.90	4.26
No enzyme treatment	1381	25	17	114	1.18	0.36	0.80	3.27
Critical difference ($p=.05$)	26.5	5.0	NS	11.8	0.13	0.08	NS	0.08

Reproduced from Joshi, V.K., Bhutani, V.P., 1991. Influence of enzymatic treatment on fermentation behaviour physico chemical and sensory qualities of apple wine. Sciences Des Aliments 11 (3), 491–496.

2.1.3.7 Sulfur Dioxide

The treatment of juice with SO_2 before fermentation is the most common means of controlling undesirable microorganisms as well as preventing enzymatic and nonenzymatic browning reactions and is a well-established practice in winemaking (Beech and Carr, 1977; Rana et al., 1986). If SO_2 is added immediately after pressing, nearly all the color will be chemically bound and will be visually (Lea et al., 1995) reduced as the sulfite binds to the quinoidal forms. If the sulfite is added at a later stage, less reduction in color will take place—presumably the quinones become more tightly cross-linked and less susceptible to nucleophilic addition and reduction. SO_2 has also been shown to have a clarifying action, to reduce volatile acidity, and to exert a solvent effect on anthocyanin pigments (Amerine et al., 1967). Usually, 100–200 ppm SO_2 is added to cider and must (Table 7.1.6).

Table 7.1.6 Effect of SO₂ and Temperature in Wine Preparation

Concentration	Effects	Optimum	Low Temperature	High Temperature
SO₂ (50–200 ppm)	• Controls undesirable microorganisms • Prevents enzymatic browning of the juice • Has clarifying action, i.e., it neutralizes negatively charged colloids • Has solvent effect on anthocyanin pigments • Reduces volatile acidity • Increases glycerol production	15–18°C	• Less activity of bacteria and wild yeast in the must • Less loss of volatile aromatic principles • Greater alcohol yield • More residual carbon dioxide production	• Enhanced growth of thermophilic organisms • More loss of volatile compounds • Slowing down of fermentation • Less carbon dioxide production

Compiled from Amerine, M.A., Berg, H.W., Kunkee, R.E., Qugh, C.S., Singleton, V.L., Webb, A.D., 1980. The Technology of Wine Making. fourth ed. AVI, Westport, CT; Frazier, W.C., Westhoff, D.C., 1995. Food Microbiology. Tata McGraw-Hill Publishing Company Limited, New Delhi; Jarvis, B., 1993a. Chemistry and Microbiology of Cider Making. Encyclopedia. Food Science and Nutrition, first ed. Coleraine campus, Cromore Road, Coleraine, BT52ISA. Belfast Coleraine Jordans town, Magee; Jarvis, B., 1993b. Cider: Cyder; hard cider. In: Encyclopedia Food and Nutrition, first ed. Coleraine campus, Cromore Road, Coleraine, BT52ISA. Belfast Coleraine Jordanstown. Magee.

2.1.3.8 Temperature

The temperature affects the rate of fermentation and the nature of metabolites formed. The effects of temperature on wine fermentation have been summarized in Table 7.1.6.

2.1.3.9 Yeast

Yeast is the most important attribute affecting the quality of the alcoholic beverage. The formation of higher alcohols and esters during alcoholic fermentation is related to the particular yeast strain used (Castelli, 1973). In a traditional fermentation in which no yeast is added and no sulfite is used, the first few days are dominated by the non-*Saccharomyces* species (*Candida pulcherrima*, *Pichia* spp., *Torulopsis* spp., *Hansenula* spp., and *Kloeckera apiculata*), which multiply quickly to produce a rapid evolution of gas and alcohol. They also generate a distinctive range of flavors, characterized by ethyl acetate, butyrate, and related esters. As the alcohol level rises (2–4%), these initial fermenters begin to die out and the microbial succession is taken over by *Saccharomyces uvarum*, which completes the conversion of sugar to alcohol and generation of a more wine-like flavor. The yeast cells become sublethally damaged by the increasing concentration of alcohol, but death does not occur until the concentration exceeds 9% alcohol (v/v). Flow cytometry with the fluorescent dye rhodamine 123 demonstrated that 8 days into the fermentation period the cells could not be distinguished from heat-treated dead cells. However, when oxonol dye was used, the population remained viable for 4 weeks in the fermentation (Dinsdale et al., 1994). The naturally selected yeast ICV-GRE and 71B, malolactic bacteria Lalvin 31, and genetically enhanced yeast ML01 were compared for bio-deacidification of malic acid in the production of Vignoles wine. ICV-GRE yeast consumed 18% of malic acid with no lactic acid production. The number and type of yeast in the juice are also influenced by the method of juice extraction. Consequently, the quality of the wine is also affected.

In a study, various inocula of wine yeast, viz., isolate W, UCD 505, UCD 522, and UCD 595 and a natural source of fermentation (NSF1 and NSF2) were used to make apple wine. The results showed that the source of fermentation significantly affected fermentation behavior and consequently, affected various physicochemical and sensory characteristics of the apple wine. The addition of DAHP as a nitrogen source to the must significantly increased the rate of fermentation and the effect of DAHP addition (Fig. 7.1.6) was visible from the very beginning of fermentation. Except for a slight reduction in the ester contents, all other parameters were improved by the addition of DAHP (increased mineral and ethanol content and decreased titratable acidity, TSS, total phenols, esters, vitamin C, total sugars, total volatiles, amyl alcohol, and methanol contents). It was concluded that most of the characteristics of the wines fermented by standard wine yeasts (W, UCD 522, UCD 595, and UCD 505) were in the desirable range, in contrast to those fermented by the natural source of fermentation. UCD 595-fermented wines had better compositional characteristics than other yeasts (Joshi, 1993). It is interesting to note that a change in vinification practices altered the composition even of wine fermented by a natural source of fermentation or spontaneous fermentation. Application of cluster analysis and principal component analysis (PCA) confirmed the differences in the behavior of the natural source of fermentation, though the differences between the sources of fermentation/inoculum (wine yeast) were stronger than the addition of a nitrogen source. The wine made from UCD 595 was the best in quality (Joshi, 1993). The study confirmed experimentally the findings of abnormal characteristics of naturally fermented alcoholic beverages compared to the wine yeast strains discussed earlier, implicating the source of inoculum or fermentation.

Apple wine produced by various sources of fermentation has also been profiled using quantitative descriptive analysis (QDA) (Joshi et al., 2002). Flavor profiling of apple wines fermented with different *Saccharomyces cerevisiae* strains and a natural source of fermentation with or without nitrogen addition was carried out by descriptive analysis. The natural source of fermentation imparted to the wine flavors like lactic, sharp, acetic, and higher fruity flavor. The W strain gave wines with higher astringency, ethyl acetate, and acetaldehyde-like flavor, and UCD 505- and UCD 522-fermented wines had more apple-like ethanolic, whereas UCD 595 imparted more phenolic, astringent, rose, and ethanolic flavor tones to the wines (Joshi et al., 2002). Application of PCA to the means of flavor scores generated from flavor profiling separated and characterized the wines fermented by different yeasts or sources of fermentation using these descriptors, but did not differentiate clearly the wines fermented with or without a nitrogen source.

In another study, among various yeast strains, insoluble solids, pectinesterase enzyme, rate of fermentation, and physicochemical characteristics of apple wines were examined. The highest rates of fermentation and ethanol production were found in wine fermented by the yeast strain UCD 505, whereas strain UCD 595 gave the smallest amount of methanol. The addition of insoluble solids to the apple juice significantly increased the rate of fermentation and the levels of methanol, vitamin C, amyl alcohol, total volatiles, tannins, color units, and Mn and Zn. The addition of insoluble solids decreased pH, titratable acidity, ethanol, total sugars, total esters, and K, Mg, Ca, Cu, and Fe content of the wine. Addition of pectinesterase significantly increased all the parameters examined except for pH, vitamin C, total esters, and Mn and Mg content (Table 7.1.7). Application of cluster analysis to the results of rate of fermentation, reducing sugars, volatile acidity, and ethanol showed that the influence of the yeast strain was more than the influence of insoluble solids or pectinesterase addition. Further, there was a clear interaction between the yeast strain, the insoluble solids, and the pectinesterase. Addition of insoluble solids to the must led to the production of some undesirable quality characteristics (Table 7.1.8). In contrast, specific yeast strains and enzyme addition improved various physicochemical characteristics of the wine. Presettled or clarified juice was preferred for producing a quality apple wine (Joshi et al., 2013a). The various descriptors as affected by different treatments are shown in a spider web diagram in Fig. 7.1.7.

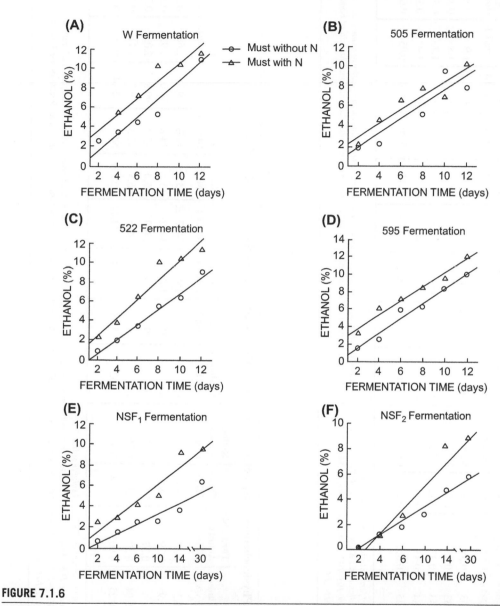

FIGURE 7.1.6

Fermentation behavior of apple wine fermented with different sources of fermentation and with and without a nitrogen.

Reproduced from Joshi, V.K., 1993. Studies on Alcoholic Fermentation of Apple Juice and Apple Juice Utilization. (Ph.D. thesis submitted to Guru Nanak University Amritsar (Pb), India.)

Table 7.1.7 Changes in Physicochemical Characteristics of Apple Wine Affected by Insoluble Solids

Insoluble Solids (%)	RF (°Bx/24h)	Ethanol (% v/v)	Titratable Acidity (%)	TSS (°Bx)	Reducing Sugars (%)	Methanol (µL/L)	Total Esters (mg/L)	Tannins (mg/L)	K (mg/L)	Na (mg/L)	Mg (mg/L)	Ca (mg/L)	Fe (mg/L)	Zn (mg/L)
0.0	0.98	11.63	0.38	6.98	0.38	60	94	143	1513	14	15	98	1.9	0.32
2.5	1.00	11.33	0.38	6.81	0.33	97	88	151	1478	15	18	92	1.8	0.35
5.0	1.03	10.43	0.37	7.42	0.32	84	71	136	1433	18	16	92	1.7	0.39
10.0	1.09	10.84	0.35	7.11	0.37	94	69	153	1400	14	14	85	1.9	0.39
CD (p≥.05)	0.02	0.25	0.017	0.28	0.03	1.26	3.08	3.2	14.4	0.13	5.90	9.94	0.05	0.02

CD, critical difference; RF, rate of fermentation; TSS, total soluble solids.
Reproduced from Joshi, V.K., Sandhu, D.K., Kumar, V., 2013a. Influence of addition of insoluble solids, different yeast strains and pectinesterase enzyme on the quality of apple wine. Journal of the Institute of Brewing, 119, 191–197; Joshi, V.K., John, S., Abrol, G.S., 2013d. Effect of addition of herbal extract and maturation on apple wine. International Journal of Food and Fermentation Technology 3 (2), 103–113.

Table 7.1.8 Changes in Physicochemical Characteristics of Apple Wine as Affected by Different Strains of Yeast

Yeast Type	RF (°Bx/24h)	Ethanol (% v/v)	Titratable Acidity (%)	TSS (°Bx)	Reducing Sugars (%)	Methanol (µL/L)	Total Esters (mg/L)	Tannins (mg/L)	K (mg/L)	Na (mg/L)	Mg (mg/L)	Ca (mg/L)	Fe (mg/L)	Zn (mg/L)
Isolate W	1.03	10.2	0.37	7.08	0.38	95	55	165	1457	13.7	14.6	89.8	1.68	0.33
UCD 505	1.04	11.9	0.39	6.82	0.34	84	91	162	1462	15.2	14.8	88.0	1.92	0.33
UCD 522	1.00	11.6	0.37	7.18	0.32	86	99	154	1369	15.0	17.3	99.8	2.03	0.34
UCD 595	1.04	10.04	0.37	7.17	0.36	77	69	137	1425	16.8	17.3	99.3	1.71	0.40
CD (p≥.05)	0.02	0.25	0.005	0.028	0.03	1.62	3.08	3.20	14.44	0.137	NS	4.44	0.05	NS

CD, critical difference; RF, rate of fermentation; TSS, total soluble solids.
Reproduced from Joshi, V.K., Sandhu, D.K., Kumar, V., 2013a. Influence of addition of insoluble solids, different yeast strains and pectinesterase enzyme on the quality of apple wine. Journal of the Institute of Brewing 119, 191–197; Joshi, V.K., John, S., Abrol, G.S., 2013d. Effect of addition of herbal extract and maturation on apple wine. International Journal of Food and Fermentation Technology 3 (2), 103–113.

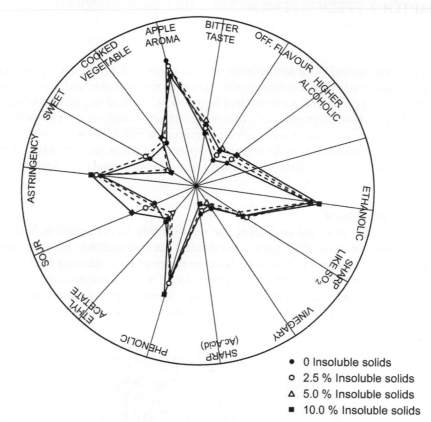

- ● 0 Insoluble solids
- ○ 2.5 % Insoluble solids
- △ 5.0 % Insoluble solids
- ■ 10.0 % Insoluble solids

FIGURE 7.1.7

A spider web diagram showing various descriptors as affected by the addition of insoluble solids in apple wine fermentation.

Reproduced from Joshi, V.K., Sandhu, D.K., Kumar, V., 2013b. Influence of addition of apple insoluble solids, different wine yeast strains and pectinolytic enzymes on the flavor profile of apple wine. International Journal of Food and Fermentation Technology 3 (1), 79–86.

2.1.4 Inoculation With Yeast Culture

The traditional method of apple winemaking is to inoculate the must without any external source of yeast (*S. cerevisiae*), and the indigenous microflora of the apple carry out the fermentation. But the modern practice is to check the indigenous microflora before the fermentation (Lea, 1995). After the juice is sulfited, it is inoculated with the desired yeast culture for fermentation. The growth of yeasts, acetic acid bacteria, and lactic acid bacteria can be excluded by washing and sorting the apples before milling and pressing. High counts of bacteria (including lactic acid bacteria) have been observed during alcoholic fermentation and storage of wine. Fermentation with *Schizosaccharomyces pombe* reduced malic acid in several fruits, including apple (Azad et al., 1986a,b), though with a low rate of alcohol production (Park and John, 1980). *Leuconostoc* has also been employed to reduce the acidity of the fermented product. An ion-exchange sponge with a tailored surface charge for immobilization of *S. cerevisiae* encouraged yeast growth and reduced fermentation (O'Reilly and Scott, 1993). The rate of fermentation may also be controlled by holding the temperature around 16°C, by reducing the yeast population by racking, or by addition of SO_2. However, if the fermentation is too slow, it may be susceptible to sickness, imparting a milky white appearance to the wine and a sweet pungent odor. Industrial wine yeasts *Saccharomyces bayanus* and two interspecies hybrids (*S. cerevisiae* × *S. bayanus*) were checked for their suitability for fermentation of apple musts with different L-malic acid contents (4, 7, and 11 g/L). Based on the fermentation, profiles including main organic acid, acetaldehyde, diacetyl, glycerol, esters, and polyphenols were obtained by the HPLC method (organic acids, acetaldehyde, glycerol, diacetyl), gas chromatography (GC) (esters), colorimetric analysis (polyphenols), and enzymatic analysis (L-malic acid, ethanol). Although the fermentation profiles of wines were characteristic for specific yeast strains, similarities in the organic acid profiles of wines fermented by *S. bayanus* and its hybrid S-779/25 were noted. In all the tested wines L-malic, pyruvic, and citric acids were dominant (Stycztnska and Pogorzelski, 2009).

2.1.5 Fermentation of Must

After SO$_2$ addition, clarification, and sweetening, the must is ready for fermentation. Various kinds of fermenters such as those of wood, plate, plastic, and stainless steel are available; however, stainless steel tanks with temperature control and CO$_2$ venting should be preferred for conducting fermentation. Stainless steel tanks are generally used these days for fermentation of wine, although traditionally oak barrels were used for this purpose. Wooden vats (Plate 7.1.1); vessels of mild steel, with a ceramic or resin lining; bitumen-lined concrete vats; and, more recently, stainless steel and even lined fiberglass–resin vats or

PLATE 7.1.1

Wooden vat traditionally used for the fermentation.

Courtesy: M/S Himachal Fruit Wines and Beverages, Badhu, Mandi, Himachal Pradesh, India.

tanks have been employed commonly for fermentation. Nowadays most companies use vertical stainless steel tanks, whereas other use conicocylindrical vats. One of the fermenters employed in wine fermentation is shown in Fig. 7.1.8. The temperature during fermentation is crucial for preservation of delicate fruit flavors in the resulting apple wine and ranges between 20 and 25°C. To avoid fermentation problems, yeast nutrients such as DAHP and other commercial preparations have to be added to the must at the beginning of fermentation. At cooler fermentation temperatures (13–16°C), the must should reach dryness in 2–3 weeks as tested by analyzing the wine for residual sugar content. After fermentation, the wine should be promptly racked off the lees, sulfited, and stored in full containers. A sweet wine with residual sugar can be produced by stopping the fermentation before reaching dryness or blending the wine later on. To arrest the fermentation, the must needs to be chilled and the yeast should be removed by centrifuging or filtering the cold wine. The chilling temperature will depend on several factors. Favorable results have been obtained by lowering the must temperature to about 29°F. It is important, however, to store the wine

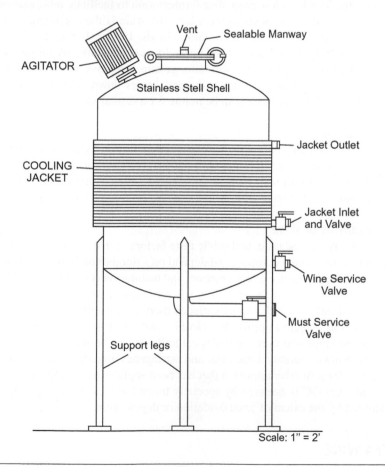

FIGURE 7.1.8

Diagrammatic view of a fermenter used for making wine.

Adapted from Joshi, V.K., 1997. Fruit Wines, second ed. Directorate of Extension Education. Dr. YS Parmar University of Horticulture and Forestry, Nauni, Solan, HP, p. 255.

with residual sugar at cooler, cellar temperatures until bottling. After completion of the fermentation, the wine is stabilized for heat-unstable proteins and clarified. Bentonite is used to achieve protein stability and clarification. Following bentonite treatment, the wine should be fairly clear. Adequate levels of SO_2 should be maintained and the wine should be stored in completely full storage containers, because lactic acid bacteria can metabolize it. In the case of apple wine, malolactic fermentation should be discouraged to preserve acidity and avoid wine spoilage by lactic acid bacteria. The pH of apple wine is generally higher, which makes the wine relatively more susceptible to attack by lactic acid and other bacteria. To avoid microbial spoilage, the wine is processed under scrupulously clean and sanitary conditions using steam, hot water, and cleaning and sterilizing chemicals during processing. Apple wine with a rich and delicate flavor can be prepared for the market after a short aging period of 2–4 months.

2.1.6 Clarification of Apple Wine

Apple wine is left on the lees for a few days after fermentation to facilitate autolyzation of the yeast , thereby adding enzymes and amino acids to the wine. The wine is then separated from the lees and transferred after clarification into storage vats or tanks (Jarvis, 1993a). Initial clarification may be performed by the natural settling of well-flocculated yeast, by centrifugation, by fining, or by a combination of all three. Typical fining agents are bentonite, gelatin, isinglass, or chitosan. Gelatin forms a block with negatively charged tannins in the wine and brings down other suspended materials by entrapment and can also be used together with bentonite for a similar effect.

2.1.7 Aging/Maturation

Apple wine may be stored in bulk or bottled after clarification. Extreme care in the sanitation of storage vessels is necessary to prevent contamination with undesirable microorganisms. Storage temperature can be as low as 4°C but not higher than 10°C. If air is not excluded from the tanks, acetic acid bacteria will produce acetic taints in addition to film, which can also be formed by yeasts.

2.1.8 Sensory Qualities of Wine

The appearance, color, aroma, and taste, and subtle taste factors such as flavor, constitute the quality of wine. The aroma and taste are very complex and depend on a number of factors such as varieties used, agricultural land, vinification practices, fermentation, and maturation (Gayon, 1978). The taste of apple wine is determined more by the fruit composition, whereas odor is governed by technological factors and the yeast employed rather than the apple varieties used to make the wine. The flavor is assessed using both subjective and objective approaches (Joshi, 2006). In the subjective approach, trained or untrained panels can be employed to recognize specific flavors. In human beings, flavor sensation by taste is limited to sweetness, sourness, bitterness, and astringency, together with such tastes as metallic and pungent (Piggot, 1988). Another approach that has been applied to flavor profiling is sniff analysis, whereby the effluent from GC is assessed by specially trained judges (Downing, 1989). The color of the wine is determined by the extent of juice oxidation or degradation.

2.2 APPLE TEA WINE

Tea cider also known as "*Kombucha*" or "tea kvass" or simply "kvass" tasting like sparkling apple cider commonly produced at home by fermentation using a tea fungus (acetic acid bacteria and yeast). A glimpse of literature revealed that Kombucha is a good source of the bioactive compounds which make it as a functional beverage. Based on apples and tea, a tea wine has been developed. With increase in concentrations of tea from 2 to 5 g, a increase in protein content, pH, color, total phenols, epicatechin and caffeine was recorded among the tea leaves extracts with apple juice and apple tea wine fermented by both the fermentations, whereas, a decrease in quercetin and antioxidant activity in the must and wine took place (Kumar et al., 2015). Among the different types of tea, difference for antioxidant activity was non-significant. Method for preparing apple tea wine using different types of tea at different concentrations were optimized and the physico-chemical, antimicrobial and sensory characteristics of the wine analyzed. All characteristics were found to be directly proportional to the concentration of tea fermented naturally or with *S. cerevisiae* var. *ellipsoideus*. In both types of fermentation, CTC (crush, tear and curl) tea-based apple tea wine received significantly higher quality scores ($p \leq 0.05$). Better results in terms of ethanol, higher alcohol concentrations and antimicrobial activity were found with 4 g tea/100 ml apple juice than with other concentrations, particularly 5 g tea/100 ml apple juice. All apple tea wines showed antimicrobial activity (inhibition zone > 7 mm) against *Escherichia coli* (IGMC), *Enterococcus faecalis* (MTCC 2729), *Listeria monocytogenes* (MTCC 839), *Staphylococcus aureus* (MRSA 252) and *Bacillus cereus* (CRI). Our results demonstrated that the best apple tea wine was made with 4 g CTC tea/100 ml apple juice and fermented with *S. cerevisiae* var. *ellipsoideus*, and showed potential as a functional product which also demonstrates the medicinal properties of tea. Among different types of inocula, for most of the parameters there were non-significant differences. However, must inoculated with *S. cerevisiae* var. *ellipsoideus* for the preparation of apple tea wine resulted in the highest fermentation efficiency, ethanol, overall acceptability and lowest volatile acidity. Apple tea wine prepared by ameliorating the apple tea must with apple juice concentrate, DAHP and inoculated with *S. cerevisiae* var. *ellipsoideus* was rated as the best. The product had healthful properties like antioxidant and antimicrobial properties.

2.3 APPLE WINE WITH MEDICINAL VALUE

In an attempt to prepare wine and cider with medicinal value, apple juice was fermented with extracts of *Mentha*, aonla, and garlic with various sources of sugar (Siby and Joshi, 2003). Apple wine having extract of hops, *Mentha*, aonla, ginger, and garlic holds promise as a new product with high consumer acceptability.

Must with varying concentrations of extracts (1, 3, 5, 7, and 9%) were prepared and fermented at $22 \pm 1°C$ using pure culture of *S. cerevisiae* var. *ellipsoideus*. The highest rate of fermentation (RF) was observed in the case of 5% extract of ginger-based must, and the lowest was observed in the case of garlic-based must with 9% extract. The concentration of ethanol in general increased in all the treatments, but was the lowest in the treatments of above 5% garlic-, ginger-, hops-, or *Mentha*-based wines (Fig. 7.1.9). The results showed that a concentration of extract above 5% affected the ethanol production significantly; after this level, the yeast count and the ethanol content registered a decrease (Table 7.1.9). The highest growth inhibition of yeast was observed in garlic-based wine and the lowest was recorded in aonla wine (Joshi et al., 2014).

To impart medicinal value, spices and herb extracts of hops, *Mentha*, aonla, ginger, and garlic each at 5% each were added to the must. The apparent effect of the addition of extract was to delay the fermentation, not to stop it. Physicochemical characteristics of the apple wine before and after 6 months of maturation showed that the addition of extract did not affect the quality of the wine adversely (Joshi et al., 2013a,d). From the sensory quality point of view, extract-treated honey-based, concentrate-based,

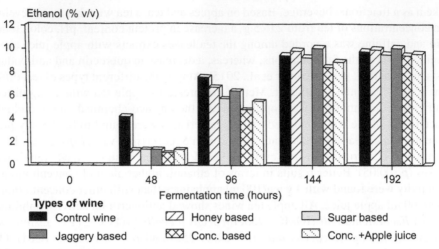

FIGURE 7.1.9

Comparison of ethanol contents of apple wines of various treatments at various intervals of time. *Conc.*, concentrate.

Based on Joshi, V.K., John, S., Abrol, G.S., 2013d. Effect of addition of herbal extract and maturation on apple wine. International Journal of Food and Fermentation Technology 3 (2), 103–113.

Table 7.1.9 Changes in Yeast Population of the Fermenting Must After 48 and 192 h of Fermentation (1x10⁶ CFU/mL)

	Concentration of Extract (%)											
	After 48 h						After 192 h					
Treatment	**0**	**1**	**3**	**5**	**7**	**9**	**0**	**1**	**3**	**5**	**7**	**9**
Apple must + garlic extract	1.8	1.7	1.5	1.4	1.2	1.0	1.4	1.4	1.2	1.1	1.0	0.9
Apple must + ginger extract	1.9	1.8	1.7	1.7	1.4	1.2	1.5	1.4	1.4	1.5	1.2	1.0
Apple must + aonla extract	1.8	2.0	1.8	1.8	1.7	1.9	1.3	1.4	1.5	1.4	1.3	1.2
Apple must + *Mentha* extract	1.7	1.7	1.8	1.8	1.6	1.5	1.5	1.4	1.4	1.5	1.3	1.2
Apple must + hops extract	2.1	1.9	1.8	1.5	1.5	1.4	1.5	1.4	1.5	1.4	1.2	1.1

CFU, *colony-forming units.*
Reproduced from Joshi, V.K., John, S., Abrol, G.S., 2014. Effect of addition of extracts of different herbs and spices on fermentation behaviour of apple must to prepare wine with medicinal value. National Academy Science Letters. http://dx.doi.org/10.1007/s40009-014-0275-y.

or concentrate + apple juice-based wines were superior to the control apple wine in most of the sensory qualities. The highest score was awarded to the honey + herbs and spices extract (5%)-based wine. The addition of extract increased the aldehydes, esters, and total phenols, which are expected to contribute the antimicrobial and antioxidant activities of the wine. The changes during maturation were desirable and, in general, were the same as found in any wine. During maturation, reducing sugars, total esters, titratable acidity, ethanol, and volatile acidity increased significantly, whereas total phenols decreased and there was no effect on TSS and no significant decrease in total sugars or higher alcohols took place.

3. PEAR WINE/PERRY

Pears (*Pyrus communis* L.) are grown in the temperate regions of the world, including India. The fruits are consumed as fresh, and a part of the harvest is used in processing industries also. Pear cultivars fall into two groups: European pears, i.e., having soft-fleshed fruit with inconspicuous grit cells, and oriental pears (*Pyrus pyrifolia*). Bartlett is the most commonly grown cultivar. The cultivars include Giffard, Precoce, Morettini, Seckel, Anjou, Bosc, Conference, Easter, Winter, Nelis, Forelle, Kieffer, Flemish Beauty, etc. (Westwood, 1978). Cultivars such as A-Ri-Rang, Chojuro, Ichiban, Singo, Seuri, Yakumo, and Kosui are the Asian pears. Nutritionally, pears are considered to be a fairly good source of fiber, which malic acid is the major organic acid.

The beverage obtained by fermentation of pears is called perry. The National Association of Cider Makers has drafted a standard for British cider and perry, which reads as "the beverage obtained by the complete or partial fermentation of juice of the pear or a mixture of the juice of pears and apples, with or without the addition of water, sugars or concentrated pears or apple juice, provided that not more than 2.5% of the juice shall be apple juice." Perry is an alcoholic beverage. Its name is derived from the word pirrium, a pear, which in late Latin was pera. In old French, it was rendered as pere or perey. The production of perry can be a promising alternative for the utilization of sand pear fruit, which has a very limited outlet for its direct consumption (Azad et al., 1986a,b; Bhatt and Joshi, 2010). The initial trials on the production of perry from sand pear have been reported (Azad et al., 1986a,b). Both cider and perry are believed to have been produced for over 2000 years. The quantity of cider and perry produced is second only to the wine produced from grapes. Of the two, cider has become an increasingly important commercial product in recent years (Jarvis et al., 1995). Sand pear vermouth is another fortified alcoholic beverage that has been developed from sand pear (see details in another section of this chapter and in the literature cited: Attri et al., 1994; Joshi et al., 1999a,b). The production of perry can be a promising alternative for the utilization of sand pear fruit, which has a very limited outlet for its direct consumption because of its sandy texture (Joshi, 1995).

3.1 COMPOSITION OF PEAR FRUIT

3.1.1 Juice

Perry or pear wine of good quality can be made from pears, such as Bartlett, with high tannin contents.

3.1.2 Sugars

The Asian pears have sweet to sweet tart taste and a fragrant aroma, having 15% natural sugar, although 9–12% is the most typical. Pear contains three major sugars, namely sucrose, glucose, and fructose (Kadam et al., 1995). In pear juice, fructose occurs in the greatest concentration of about 7.0% (w/v), whereas the glucose content is usually low (2–2.5% w/v), with sucrose about 1% (w/v). Sorbitol (sweetener) is also found in perry juice in a concentration ranging from 1 to 5% (w/v). Because this compound is not fermented by yeasts, it remains after fermentation and increases the specific gravity of dry perry. Xylose (0.2% w/v) and other sugars like galactose, arabinose, ribose, and inositol are also present in perry.

3.1.3 Organic Acids
Pears produce juices with pH values within the range of 3.6–3.8. There are a few cultivars below this pH, but there are none with a pH below 3.2. From pH 3.8 upward, there is a similar decline particularly at pH of 4.0–4.2.

3.1.4 Nitrogenous Compounds
The nitrogenous contents in perry do not exceed 10 mg/100 mL, and the amino acid that occurs in greatest quantity in pears is proline. Amino acids like aspartic acid, asparagine, and glutamic acid form a fairly substantial proportion of the amino acids in pears.

3.1.5 Tannins
In perry, only one group of tannins is capable of combining with protein and, more precisely, are called procyanidins. They all contain a phenol structure that is associated with bitterness and astringency.

3.2 PROCESS FOR MAKING PERRY
Perry of good quality can be made from the pears with high tannin contents, such as Bartlett. The fruit used to make perry should be astringent to taste and usually full of stones, but ripe dessert pears are not successful unless the natural esters of the fruits are removed. Most of the steps to make perry are similar to those for cider.

3.2.1 Extraction of Juice and Preparation of Must
The fruits are grated and pressed in a rack and cloth press for preparation of perry. The juice has an original TSS of 14 °Bx and 0.25% acidity. The entire process is shown in Fig. 7.1.10. It is ameliorated to 21 °Bx and 0.5% citric acid along with 100 ppm SO_2. For rapid fermentation of sand pear juice an exogenous addition of nitrogen source has to be made. The procedure used for perry production (Joshi, 1997; Joshi et al., 1999b) includes raising the TSS to 20 °Bx and addition of DAHP at 0.1% (Fig. 7.1.10).

3.2.2 Fermentation
In the very early stages of fermentation, it might be possible to find other weakly fermenting yeasts, such as *Metschnikowia (Candida) pulcherrima, Saccharomycodes ludwigii*, and perhaps even species of *Brettanomyces*. During the latter part of the fermentation, the main fermenting yeast gradually dies out because of depletion of sugars. Addition of SO_2 at the start of fermentation kills off yeasts such as *M. pulcherrima and K. apiculata*. Owing to inversion, sucrose is unlikely to be present in any significant quantity at the beginning of fermentation. The two major sugars are glucose and fructose, which are fermented by yeast. The sugar and sugar-like compounds that the yeast does not metabolize are left in dry perries, viz., xylose can be present in quantities of about 0.05% (w/v) (Whiting, 1961). In dry perries, the sorbitol content can be in excess of 4% (w/v). Probably the most important change that takes place during fermentation or storage is the breakdown of organic acid, including metabolism of citric acids. Sand pear juice does not have sufficient nitrogen for rapid fermentation, for which exogenous addition of a

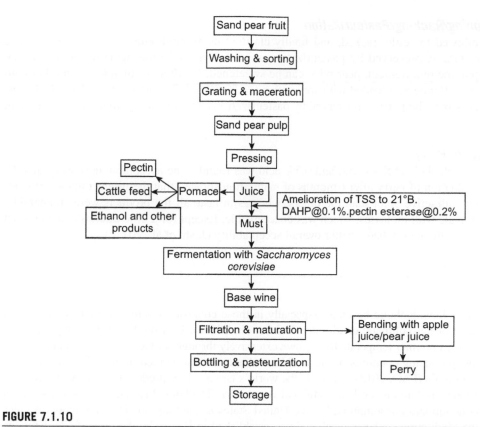

FIGURE 7.1.10

Flow sheet for the preparation of pear wine. *DAHP*, diammonium hydrogen phosphate; *TSS*, total soluble solids.

Based on Joshi, V.K., Sharma, S., Parmar, M., 2011c. Cider and perry. In: Joshi, V.K. (Ed.), Handbook of Enology, Vol. III. Asia Tech Publishers, Inc., New Delhi, pp. 1116–1151.

nitrogen source has to be made. Further, pectic enzyme has been found to hasten the process of clarification in pear wine.

The primary fermentation is followed by a secondary fermentation with *S. cerevisiae* var. *ellipsoideus*. During fermentation nitrogenous compounds tend to disappear and later on reappear in smaller quantities. Alcoholic fermentation at $22 \pm 1°C$ by the addition of yeast culture at 5% is carried out to make perry. The effects of the addition of various sugar sources (sucrose, fructose, glucose, honey, molasses, and jaggery) on the fermentation behavior and sensory quality of perry have been reported (Azad et al., 1986a,b). The highest RF was recorded in must made with jaggery, followed by honey, glucose, fructose, and molasses, and was lowest in the case of sucrose. The temperature affects the fermentation of sugar to ethyl alcohol by the yeast. The maximum ethanol yield with corresponding yeast fermentation efficiency of 87.79% was 8.85% (v/v) in juice fermented at 30°C. Pear juice containing 10% (w/v) pulp (pH 5.0) fermented at 30°C by a pure wine yeast culture, *S. cerevisiae* at 10.0% (v/v), leads to the most efficient alcoholic fermentation.

3.2.3 Siphoning/Racking/Pasteurization

The wine is allowed to settle, racked, and finally clarified using bentonite or pectic enzyme. The prepared pear wine is preserved by pasteurization as in the case of apple wine discussed earlier. Depending upon the requirement, pear wine can be sweetened, fortified, or blended with other fruit wines. Pear wine after being blended with apple juice at a 50:50 ratio becomes a low-alcoholic beverage is called as perry. The perry is preserved by pasteurization and, depending upon the requirement, it can be sweetened.

3.2.4 Quality of Perry

Perry having 5% alcohol, TSS 10 °Bx, and 0.5% acid was found to be acceptable in sensory quality. The sensory evaluation of perry after 6 months of maturation showed that the product made with jaggery had a more attractive color than the others, whereas in taste all the products were comparable, except for that made with molasses, which was unacceptable. Except for a little aftertaste, the product made from must with jaggery had a better overall acceptability (Joshi et al., 2011c).

---

4. CIDER

Cider is a popular low-alcoholic beverage especially in those countries where grape cultivation is not practiced because of agro-climatic conditions (Joshi et al., 2011c). It is produced from apple (*Malus domestica* Borkh.), a premier temperate fruit grown extensively throughout the world, including in India. The major apple-producing countries of the world are China, Russia, United States, Germany, France, Italy, Turkey, and India (FAO, 2012). France is the world's largest cider-producing country, where Normandy and Brittany are famous for their traditional *sweet cidre*. The United Kingdom leads the world in hard cider production and consumption, but the United States is catching up. The production of cider moved from the Mediterranean basin north and was established in France by the time of Charlemagne (9th century) and was probably introduced to England from Normandy well before Duke William's conquest (Revier, 1985; Roach, 1985). At present, cider is produced in several countries.

More recently, low-alcoholic beverages have gained importance in preventing cardiovascular diseases. Wine consumption prevents the formation of low-density lipoproteins (LDL) and increases high-density lipoprotein levels (HDL) (having protective effects against heart disease; Delin and Lee, 1991; Joshi et al., 1999b; Stockley, 2011). The use of hops and spices in cider imparts an antimicrobial activity (Joshi and Siby, 2002) to the product and is considered significant to health from the consumer's point of view.

4.1 DEFINITION AND CHARACTERISTICS OF CIDER

Cider is a low alcoholic beverage made from apple juice by alcoholic fermentation. It has specific characteristics as defined earlier. Cider is an alternative term used for cyder, though both terms have been used at least since 1631, including in Australia. However, in some places, cyder and cider are differentiated: cider is an alcoholic beverage, whereas cyder is usually apple juice or a nonalcoholic beverage (apple juice). The fermented juice is called cider in England but is known as hard cider in the United States. In Europe, fermented apple juice is known as cider (France), sidre (Italy), sidra (Spain), or applewein (Germany and Switzerland) where the name for the corresponding unfermented product is clearly distinguished as apple juice (Nogueira and Wosiacki, 2010). Depending upon the alcohol content, cider could be a soft cider (1–5%) or a hard cider (6–7%) (Downing, 1989; Joshi et al., 2011c; Amerine et al., 1980). It may be made from fresh juice or juice from a single cultivar and accordingly be classified as vintage cider or white cider, which is made from decolorized apple juice or pale-colored juice (Jarvis, 1993a,b). The sweet cider (Pourlx and Nicholas, 1980) has residual sugar from fermentation or is sweetened after fermentation, and still cider has low sugar and is without carbon dioxide. Dry cider is without sugar and with an alcohol content of 6–7%, whereas ciders having alcohol content of not more than 1.2% by volume are made by removing the alcohol from strong cider by thermal evaporation, by reverse osmosis, or generally by adding apple juice to it (Jarvis, 1993a,b). The cider produced by the "Méthode Champenoise" is called champagne cider. Sparkling sweet cider is produced by fermenting apple juice and contains not more than 1% alcohol (v/v), and the natural CO_2 formed during fermentation is retained (see the Chapter 9 on specialty wines). Sparkling cider has lower sugar and a higher alcohol content of 3.5% but with partial retention of the CO_2 formed during fermentation, whereas carbonated cider is charged with commercial CO_2 to produce effervescence. No doubt, a number of apple varieties have been recommended for the preparation of cider, which include bittersweet, bittersharp, sharp, and sweet. But the use of apple juice concentrate for cider making has increased considerably because of various advantages such as price stabilization, quality maintenance, and storage of the concentrate without spoilage (Joshi et al., 1991), though this leads to a loss of development of specific cultivars for the making of cider (Kerni and Shant, 1984; Amerine et al., 1980; Downing, 1989) or wine.

4.2 FLAVOR-AFFECTING FACTORS IN CIDER

Flavor is one of the important sensory attributes that determine the overall quality of any product, including cider. The various factors that influence the flavor of cider (Table 7.1.10) have been discussed in detail by Jarvis et al. (1995).

Table 7.1.10 Factors Influencing the Flavor of Cider	
• Apple juice • Other ingredients in the raw material	• Variety of fruit(s) • Maturity and condition of fruit at pressing • Fresh juice or concentrate • Condition of concentrate • Type of chaptalization sugar(s) • Quantity of SO_2 • Amelioration of pH by addition of acid yeast nutrients
• Yeast	• Natural or inoculated fermentation • Strain of yeast(s) • Condition of yeast(s) when inoculated
• Fermentation • Fermenter design and operation • Secondary fermentation	• Temperature, time of fermentation • Hydrostatic pressure, operational pH/acidity level • Natural or induced malolactic fermentation secondary yeast • Spoilage organisms
• Maturation • Processing factors	• Chemical and enzymatic changes • Decolorization of juice or final cider • Dealcoholization
• Final product makeup	• Carbonation

Summarized from Jarvis, B., Foster, M.J., Kinsella, W.P., 1995. Factors affecting the development of cider flavour. Journal of Applied Bacteriology – Symposium Supplement 79, 55.

4.3 TECHNOLOGY OF CIDER PRODUCTION

4.3.1 Cider Making Methods

The traditional procedure for cider making was more of an art than science and was described as early as 1939. A generalized method of cider making is shown in Fig. 7.1.11. It is a complex process that combines two successive biological fermentations: the first one is the classical alcoholic fermentation of sugar into alcohol conducted by yeast strains like *S. cerevisiae* and the second one is the malolactic

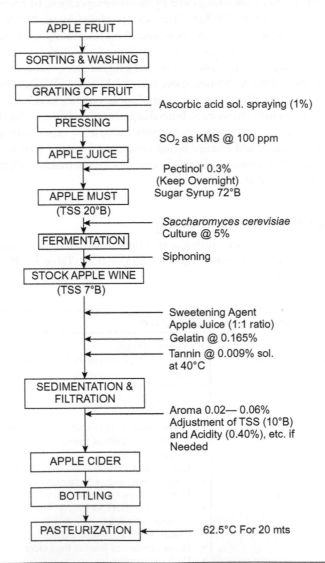

FIGURE 7.1.11

Flow sheet of preparation of sweet cider. *KMS*, potassium metabisulfite; *sol.*, solution; *TSS*, total soluble solids.

Based on Rana, R.S., Vyas, K.K., Joshi, V.K., 1986. Studies on production and acceptability of cider from Himachal Pradesh apples. Indian Food Packer 40, 56-66; Joshi, V.K., 1997. Fruit Wines, Directorate of Extension Education, second ed. Dr. YS Parmar University of Horticulture and Forestry, Nauni, Solan (HP), p. 255.

fermentation that occurs during the maturation process by lactic acid bacteria. The latter is an important manufacturing step to reduce the acidity of the cider and stabilizes it with respect to microbial spoilage through the bacteriostatic effect of the lactic acid produced. Lalvin 31 malolactic bacteria added to wine fermented with ICV-GRE yeast converts the remainder of the malic acid to lactic acid (Gary et al., 2007) and reduces the acidity of wine. The different methods used to make cider have been reviewed in a systematic way (Amerine et al., 1967, 1980; Downing, 1989; Lea, 1978, 1995; Joshi et al., 1995; Joshi et al., 2011c) and are summarized in Table 7.1.11. There are a few reports on the preparation of cider in India also (Joshi and Sandhu, 1997). Cider with 5% alcohol and a TSS/acid ratio of 25 was found to be the most favored at laboratory and consumer survey scales (Rana et al., 1986).

4.3.2 Required Raw Materials

The sweet and low-acid cultivars such as Delicious, Cortland, Ben Davis, and Rome Beauty are recommended for extraction of the base juice. Cultivars like Jonathan, Stayman, Winesap, Northern Spy, Rhode Island Greening, Wayne, and Newtown possess higher acid levels and add tartness to the cider. MacIntosh, Gravenstein, Ribston Pippin, Golden Russet, and Delicious are aromatic and add flavor and bouquet to the cider (Mitra, 1991). The body and flavor can be improved by using astringent apples such as Red Astrachan, Lindel, and crab apples (Downing, 1989; Lea, 1995). A good rule of thumb is to add less than 10% of an astringent cider to an acidic cider, and not more than 20% should be added to any blend (Joshi et al., 1999b; 2011b).

4.3.3 Milling and Pressing

The fresh apples used for juice extraction are stored for a few weeks after harvest so that all the starch is converted into sugar (Downing, 1989; Joshi et al., 2011c). The apples selected for juice processing are then washed and inspected for the presence of any foreign materials or decay, which have an adverse effect on the microbiological status and ultimately the cider quality. Rinsing with clean water is accomplished at the scrubber or after inspection and routine replacement of the holding water is necessary (Joshi et al., 2011c). The fruits are transferred into a mill using a water flume, which provides an additional advantage of washing the fruits (Downing, 1989). The apples are ground to a mash before pressing. Pectinesterase enzyme is also used for better extraction of the juice (Amerine et al., 1980; Joshi and Bhutani, 1991). In most applications either hammer or grating mills are used and even slicers and a grater are required for difficult extraction. A fairly recent development in the production of apple juice for concentrate is diffusion extraction (Bump, 1989), and the juice can be used for making cider. Several types of presses have been developed for pressing and extracting apples, including hydraulic presses (Fig. 7.1.12), screw presses, basket presses, belt presses, and pneumatic presses. The other presses used are the Stoll press, Bucher-Guyer press, Bullmer continuous belt press, and Atlas-Pacific press for apple juice extraction (Downing, 1989). Screw presses have been used on hydraulically pressed pomace in France for additional recovery of juice from centrifuged apple pulp containing cellulose fibers (Lowe et al., 1964). Using press aids, the Zenith and Jones presses were among the first continuous presses used for apple juice extraction in the United States, but the amount of apple solids in the juice from the screw presses is much greater than that in juice from hydraulic rack and cloth presses (Burroughs, 1973). A screw press that is well adapted to apple juice production is the Reitz press system. An electroacoustic dewatering process is also one of the newer methods that employs passing of electric current through the pulp prior to pressing, which has been claimed to release a higher yield of juice. Immediately after pressing, the juice is treated with SO_2, which acts as both an antioxidant, checking browning of the juice, and a preservative by destroying wild yeast and bacteria (Bump, 1989).

Table 7.1.11 Summary of Methods Used in Cider Preparation

Type of Method	Fruit	Juice	Parameters Additive	Fermentation	Maturation	Others
European						
Method 1	(a) Some stored for 3–4 days and others macerated	Extracted as usual, cold stabilized at 0–7.8°C	SO$_2$ 50–100 mg/L	Temperature 4.4–10°C, pure yeast in some, mixed in others	Secondary fermentation in casks for several months	Malolactic fermentation, produces CO$_2$ in bottles
	(b)–	–	Pectic enzymes for clarification			
Method 2	Lower sugar higher acidity	Juice extracted, no maceration, juice centrifuged for bacteria and yeast removal	Lactic acid added to increase the acidity, if needed	Pure yeast such as Steinberg added	–	–
Method 3	Sound fruits separated by flotation	Juice extracted in a hydraulic press	–	Natural fermentation from 1.008 to 1.005 specific gravity	Storage in concrete tanks lined with a coating	Before delivery, cider is sweetened with syrup
Method 4	–	–	–	Fermentation allowed up to specific gravity of 1.025–1.030 (5–7.5 °Bx) filtered or centrifuged	Stored in wooden casks	Carbonated and bottled
American						
Method 1	(a) Sound apples are used for cider making	Juice is extracted in usual press after crushing in a mill	Sulfur dioxide 100–125 ppm added, glucose added to give 13% alcohol	Spontaneous Fermentation may begin during settling	–	Clarified by bentonite treatment, mixed with apple juice blended to give 10 °Bx, filtered
	(b) Instead of juice extraction apple juice concentrate used	Juice made from concentrate	Sweetened	Champagne yeast, 24.4°C temperature was the best	–	–

Summarized from Amerine, M.A., Berg, H.W., Kunkee, R.E., Qugh, C.S., Singleton, V.L., Webb, A.D., 1980. The Technology of Wine Making, fourth ed. AVI, Westport, CT; Joshi, V.K., Sandhu, D.K., Thakur, N.S., 1999b. Fruit based alcoholic beverages. In: Joshi, V.K., Pandey, A., (Eds.), Biotechnology: Food Fermentation, vol. II. Educational Publishers and Distributors, New Delhi, pp. 647–744; Joshi, V.K., Attri, D., Singh, T.K., Ghanshyam, A., 2011b. Fruit wines: production technology. In: Joshi, V.K. (Ed.), Handbook of Enology, vol. III. Asia Tech Publishers, Inc., New Delhi, pp. 1177–1221.

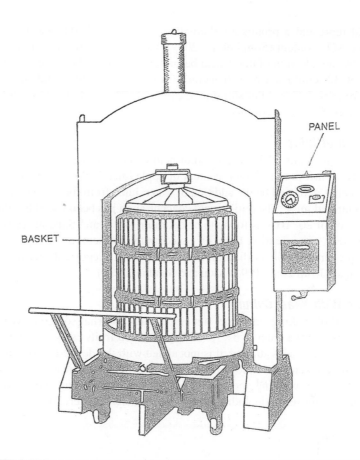

FIGURE 7.1.12

Hydraulic press.

Cider can also be produced from the apple juice concentrate directly after diluting it to a desired brix level (Joshi et al., 1991; Jarvis, 1993a,b; Downing, 1989) (Fig. 7.1.12).

4.3.4 Controlling Microorganisms Before Fermentation

For proper fermentation, the microflora of the juice must be controlled before inoculation with yeast to avoid off-flavor or similar defects in the cider. In northern France, centrifuging or fining of the juice with gelatin and tannin followed by filtration to reduce the RF is practiced. Another approach is to treat the juice with pectin-hydrolyzing enzymes and filter before adding yeast (Jarvis, 1993a,b). SO_2 is used extensively to control bacterial spoilage (Amerine et al., 1980; Rana et al., 1986). The natural fermentation of apple juice depends upon the ability of naturally occurring yeasts in the juice to convert the fruit sugars to ethyl alcohol. Treating juice with SO_2 before fermentation is undoubtedly the most common means of controlling undesirable microorganisms but the amount required depends on the pH of the juice as well as the concentrations of the sulfite-binding compounds that are present in the juice. Addition of lysozyme and enological grade tannins during alcoholic fermentation could represent a promising alternative to the use of SO_2 and for the production of wines with reduced SO_2 content (Sonni et al., 2009). Cider apple juices should always be brought below a pH of 3.8 by the addition of malic acid before SO_2 addition. However, it has been noted that juices with a pH of 3.8 could not be satisfactorily treated within the legal limit of 200 ppm SO_2 (Beech, 1972; Burroughs, 1973). After sulfiting, the juice should be allowed to equilibrate for a minimum of 6 h before free SO_2 is determined.

4.3.5 Amelioration of Juice

Amelioration is the adjustment of the sugar and/or acid content of the juice and is regulated by the respective standards. Controlling the sugar content of apple juice is done to maintain the proper final alcohol content, which is achieved by the addition of water, juice from the second pressing of the pomace, sugar, or concentrated juice. The initial sugar concentration has been found to influence the quality of the cider and a value of 20 °Bx was found to be optimum (Joshi and Sandhu, 1997). Fortification of apple juice with DAHP after dilution from its concentrate is essential for rapid fermentation (Fig. 7.1.13). The must prepared by direct dilution of the concentrate reportedly ferments faster than that ameliorated with sugar (Joshi et al., 1991; Joshi and Sandhu, 1994).

4.3.6 Inoculation With Yeast Culture

The traditional method of making cider does not employ any external source of yeast and the indigenous microflora of apple on the order of 5×10^4 CFU/g of stored fruits carry out spontaneous fermentation (Lea, 1995). After the juice is sulfited, it is inoculated with the desired yeast culture in the case of inoculated fermentation, wherever employed. The growth of yeasts, acetic acid bacteria, and lactic acid bacteria can be excluded by washing and sorting of apples before milling and pressing. As described earlier, high counts of bacteria (including lactic acid bacteria) were observed during alcoholic

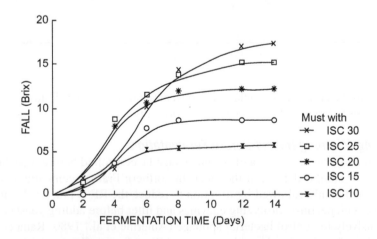

FIGURE 7.1.13

Effect of initial sugar concentration on rate of fermentation in apple must made from apple juice concentrate. *ISC*, initial sugar concentration.

Reproduced from Joshi, V.K., Sandhu, D.K., 1994. Influence of juice contents on quality of apple wine prepared from apple juice concentrate. Research and Industry 39 (4), 250.

fermentation and storage of cider (Table 7.1.12). For more details on microflora development during cider fermentation see Laplace et al. (2001).

For fermentation, the inoculum could be an in-house culture propagated and maintained in flasks or larger containers or it could just be commercially produced dried yeasts, which is a simpler method. Typical strains employed are Uvaferm CM and BC, Lalvin EC 1118, and Siha No 3. The use of a mixed inoculum of *S. uvarum* and *S. bayanus* is a widespread practice, on the grounds that the first yeast provides a speedy start but the second will cope better with the fermentation to dryness to produce a high alcohol level. A small quantity of heat-sterilized juice is inoculated with a dry culture or liquid-nitrogen-frozen culture and after fermentation, the inoculum is added to a larger volume of sulfited juice. The procedure is continued until a final inoculum of 1% or greater by volume is obtained. The yeast species have been isolated during cider making at different periods, i.e., at just the beginning of fermentation (A in Fig. 7.1.14), during active fermentation (B), and in the third period (C) when the fermentation is slow. The categories of yeast present during fermentation were species with strong fermentative metabolism, apiculate yeasts, and species with an oxidative metabolism (Fig. 7.1.14; Deunas et al., 1994). During the initial fermentation *K. apiculata* appeared as the dominant species and later during active fermentation their number decreased. However, rapid growth of *S. cerevisiae* took place.

Ciders were prepared by the traditional and modified methods. A comparison of two methods of cider making has also been made with respect to the evolution of microbial populations and malolactic fermentation (Deunas et al., 1994). The occurrence of malolactic fermentation together with alcoholic fermentation is not considered desirable in French and English ciders (Downing, 1989; Jarvis, 1993a,b; Joshi et al., 2011c) and degradation of malic acid occurs after alcoholic fermentation. However, it does not occur until the population of lactic acid bacteria reaches 10^6 CFU/mL (Deunas et al., 1994) (Fig. 7.1.14). Interestingly, alcoholic fermentation was carried out by *K. apiculata* and *S. cerevisiae* and their

Table 7.1.12 Progress of Lactic Acid Bacteria During Cider Making

Period	During Active Fermentation	During Malolactic Fermentation		During Storage	
Sample	Total LAB Count (CFU/mL)	Species	Total LAB Count (CFU/mL)	Species	Total LAB Count (CFU/mL)	Species
A11	1.2×10^5	*Leuconostoc oenos*	7.4×10^6	*L. oenos*	7.5×10^6	*L. oenos*
A2	1.2×10^5	*L. oenos*	1.2×10^7	*L. oenos*	1.2×10^7	*L. oenos*
B1	5.8×10^5	*L. oenos*	6.8×10^6	*L. oenos* *Pediococcus* sp.	1.0×10^5	*L. oenos*
B2	9.3×10^6	*L. oenos* *Leuconostoc mesenteroides*	1.2×10^7 -1.3×10^6	*L. oenos* *L. oenos*	1.5×10^6 6.3×10^6	*Lactobacillus brevis* *L. brevis*
C1	1.8×10^5	*L. oenos*		*L. brevis*		*L. oenos*
C2	8.6×10^4	*L. oenos*	7.9×10^6	*L. oenos*	7.6×10^5	*L. brevis*

LAB, *lactic acid bacteria.*
Adapted from Salih, A.G., Le Quere, J.M., Drilleau, J.F., Fernandez, J.M., 1990. Lactic acid bacteria and malolactic fermentation in manufacture of Spanish cider making. Journal of the Institute of Brewing 96, 369.

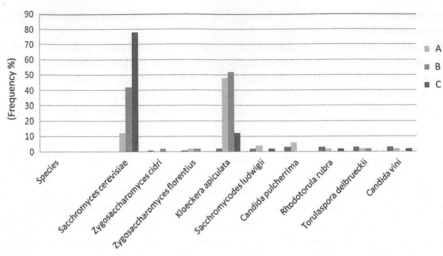

FIGURE 7.1.14

Frequency (%) of yeast species isolated during cider making. (A), after barrel filling, at beginning of alcoholic fermentation; (B), active fermentation (density at 20°C between 1.035 and 1.005); (C), at the end of alcoholic fermentation (density at 20°C below 1.005). 1, yeasts with strong fermentative metabolism; 2, *apiculata* species with low fermentative activity; 3, species with mainly an oxidative metabolism.

Based on Deunas, M. Irastorza, A., Fernandez, A.B., Huerta, A., 1994. Microbial populations and malolactic fermentation of apple cider using traditional and modified methods. Journal of Food Science, 59 (5), 1060.

distribution was similar in both methods. In the traditional method, the malolactic fermentation proceeded at the same time as alcoholic fermentation, but in the modified method no malolactic fermentation occurred, but cider with lower volatile acids was produced. See the cited literature for more details on the microbiology of cider (Lequer and Drilleau, 1998) and malolactic fermentation (Campo et al., 2008).

4.3.7 Temperature

The temperature has been found to affect the RF and the nature of metabolites formed. It takes 3–4 weeks to attenuate cider fermentation at a temperature within the range 20–25°C. Although a temperature of 15–18°C is preferred for flavor development in Germany and France, the optimum temperature for cider fermentation was found to be 15–18°C (Jarvis et al., 1995). The higher temperature increases the RF but enhances the chances of contamination with undesirable thermophilic microorganisms (Flores and Heatherbeth, 1984). Changes in viable cell count, ethanol, glucose, fructose, and sucrose during cider fermentation at 20°C with *S. cerevisiae* have been noted.

4.3.8 Yeast

The yeast is the most important factor affecting the quality of an alcoholic beverage, especially the formation of higher alcohols and esters during alcoholic fermentation, which have been related to particular yeast strains (Castelli, 1973). The desirable characteristics for yeasts used in cider making include producing polygalacturonase to break down soluble pectins; having a rapid onset of

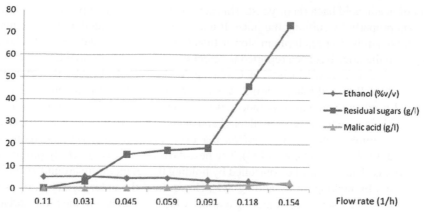

FIGURE 7.1.15

Outlet profile of the cider produced in a thermostable tubular bioreactor loaded with beads containing *Saccharomyces bayanus* and *Leuconostoc oenos*.

Based on the data from Durieux, A., Nicolay, X., Jourdian, J.M., Depaepe, C., Pietercelie, A., Plaisant, A.M., Simon, J.P., 1997. Cider production by immobilized yeast and Leuconostoc oenos: *comparison of different configurations. In: Proceedings of the International Workshop Bioencapsulation VI: From Fundamental to Industrial Applications, Barcelona, Spain, pp. 145–148.*

fermentation; being relatively resistant to SO_2; having a low pH value and high ethanol level; having low requirements for vitamins, fatty acids, and oxygen; able to ferment to "dryness" (i.e., no residual fermentable sugar); not producing excessive foam; utilizing sugars efficiently; producing minimal SO_2; not producing H_2S or acetic acid; and producing the required aroma components, organic acids, and glycerol (Beech, 1993; Beveridge et al., 1986).

Another interesting approach to continuous production of cider was attempted with the use of Ca–alginate material to coimmobilize *S. bayanus* and *Leuconostoc oenos* in one integrated biocatalyst system, which permitted much faster fermentation than traditional cider making, with better flavor fermentation (Duriellx et al., 1997). After completion of fermentation D-lactate was produced, whereas progress of lactic acid bacteria in cider making also took place. Through the immobilization technique the entrapment of yeast cells in alginate beads seems to be simple and effective in the production of good quality cider, yet the packed-bed reactor was found to be the most frequently used type of immobilized cell bioreactor. Its use will definitely improve the industrial process, reducing the time and cost involved (Singh and Sooch, 2009) (Fig. 7.1.15).

4.3.9 Fermentation of Must

Stainless steel tanks are generally used these days for fermentation of cider (Downing, 1989), though traditionally barrels of oak were used for this purpose. Wooden barrels or vats of mild steel, with a ceramic or resin lining; bitumen-lined, concrete vats; and, more recently, stainless steel and even lined fiberglass–resin vats or tanks have been employed commonly for cider fermentation. Nowadays, most companies use vertical stainless steel tanks, whereas others use conicocylindrical vats (Fig. 7.1.8). These tanks may be equipped with temperature-controlled systems, level indicators, and CO_2 venting and blanketing systems. Either juice or cider, if exposed to air during fermentation, will usually develop a surface film of acetic acid bacteria or yeast. This aerobic spoilage can be prevented by excluding the air from the vats properly by sulfiting the juice. It may be noted that heated juice ferments faster than unheated juice, but sulfited juices ferment slower than those not treated with SO_2. The availability of soluble nitrogen in the juice has been reported to affect the RF of cider. Fermentation with *S. pombe* is able to reduce malic acid in several fruits, including apple (Azad et al., 1986b), though with a low rate of alcohol production (Park and John, 1980). *Leuconostoc* has also been employed to reduce the acidity of the fermented product. Simultaneous inoculation of apple juice with *S. cerevisiae* and *S. pombe* produces cider with an acceptable level of alcohol and acidity. An ion-exchange sponge with a tailored surface charge for immobilization of *S. cerevisiae* that encouraged yeast growth and reduced fermentation has been used (O'Reilly and Scott, 1993). The best-flavored cider is generally produced by a slow fermentation process. The RF may be controlled by holding the temperature around 16°C, by reducing the yeast population by racking, or by adding SO_2. However, if the cider fermentation is too slow, it may be susceptible to cider sickness, imparting a milky white appearance to the cider and a sweet pungent odor.

In a study, in the most acidic of four musts, yeast was added to complete the alcoholic fermentation, whereas in rest of the musts, alcoholic and malolactic fermentations took place spontaneously owing to natural microflora, and no chemical was added to control these processes. Malolactic fermentation (MLF) was found to finish before alcoholic fermentation even in the most acidic and phenolic sample (pH 3.18, 1.78 g tannic acid/L). After 4 months, these ciders maintained low levels of lactic acid bacteria ($\leq 10^4$ CFU/mL) and a low content of acetic acid (<0.60 g/L). Both fermentations began simultaneously in the must, but MLF finished 10 days after alcoholic fermentation. Subsequently, the said must maintained a high population of lactic acid bacteria (>10^6 CFU/mL), causing a higher production of acetic acid (>1.00 g/L) than in the other ciders, showing the possible advantages of MLF finishing before alcoholic fermentation (Campo et al., 2008).

4.3.10 Clarification of Cider

The cider is allowed to remain on the lees for a few days after fermentation to facilitate autolyzation of the yeast, thus adding enzymes and amino acids to the cider. The cider is then separated from the lees and transferred after clarification into the storage vats or storage tanks (Jarvis, 1993a,b). Initial clarification may be performed by the natural settling of well-flocculated yeast, by centrifugation, by fining, or by a combination of all the three. Typical fining agents normally used are bentonite, gelatin, isinglass, or chitosan. Gelatin forms a block with negatively charged tannins in the cider and brings down other suspended materials by entrapment and can also be used together with bentonite for a similar effect (Joshi et al., 2011a,b,c).

4.3.11 Aging/Maturation and Secondary Fermentation

The cider may be bulk stored or bottled after clarification but extreme care is taken in the sanitation of storage vessels to prevent contamination with undesirable microorganisms (Downing, 1989). Stored cider should be cultured periodically and removed from storage for special processing if unwanted growth occurs. The storage temperature can be as low as 4°C but not higher than 10°C. If air is not excluded from the tanks, acetic acid bacteria will produce acetic taints. Film yeasts, which may develop, also produce volatile acids in cider. After fermentation, the cider is racked and filtered. The maturation is an important step in cider making, during which most of the suspended material settles down, leaving the rest of the liquid clear, which may be clarified with bentonite, casein, or gelatin followed by filtration. *L. oenos* was revealed as the predominant bacteria in Australian cider (Salih et al., 1990). During maturation, the growth of lactic acid bacteria (LAB) cultures can occur extensively, especially if wooden vats are used (Table 7.1.12), resulting in MLF. Such fermentation would convert malic acid into lactic acid and reduce acidity and impart a subtle flavor that generally improves the flavor of the product. However, in certain circumstances, metabolites of LAB cultures damage the cider flavor by excessive production of diacetyl, with a butterlike taste (Jarvis, 1993a,b). Because malic acid is a predominant acid in the apple, a reduction in acidity due to malolactic acid fermentation might be detrimental to the quality of cider (Salih et al., 1990).

4.3.12 Biochemical Changes During Aging
The production of aldehydes as one of the flavor compounds takes place as a result of autoxidation of polyphenolic compounds and oxidation of ethanol by direct chemical reaction with air (Wildenradt and Singleton, 1974). Alcohols in wine react with organic acids like tartaric, malic, succinic, and lactic acid to form esters, which have been reported to increase with the aging of wine (Amerine et al., 1980). The concentration of total volatile compounds also increases during fermentation as well as in storage. During maturation, a decrease in the tannins due to their complexing with protein and polymerization and subsequent precipitation takes place (Amerine et al., 1980; Joshi et al., 2011c).

4.3.13 Final Treatment and Packaging
The various batches of cider, generally made from mixtures of different juices, are blended to give a specific flavor. To make cider with no haze it is desirable to treat the raw cider with fining agents, organic or inorganic agents, such as bentonite, gelatin, or chitin; silica solution; albumen; casein; isinglass; or tannin, and filter it as reviewed earlier (Sandhu and Joshi, 1995). Cider can be sold as a still or sparkling beverage with varying degrees of sweetness and clarity. The amount of carbonation ranges from saturation for ciders in jars, to 2–2.5 volumes of CO_2 in most bottled ciders, and up to 5 volumes in champagne cider. Carbonation pressure ranges from 2.5 to 3.5 bars, but higher pressure is used in the case of PET bottles (Jarvis, 1993a,b). Sweetening may be from unfermented juice sugars, added juice or concentrate, or sweetening agents, depending upon the appropriate regulations. In terms of clarity, ciders range from turbid farm cider to brilliantly clear ciders (Joshi et al., 2011c). The majority of commercial cider is distributed into kegs, bottles, or cans. Keg cider is carbonated and pasteurized in-line and poured into stainless steel kegs in a plant that rinses, washes, and sterilizes the kegs prior to filling. It can also be distributed in glass bottles that may be carbonated and pasteurized after filling. Common container closures are crown caps and roll-on or plastic stoppers, which have replaced corks. It is then pasteurized at 60°C for 20–30 min or preserved with SO_2 as the best practical approach. Results indicate that UV light is the most effective for reducing pathogens like *Escherichia coli* in cider. The carbonated cider is either sterile-filtered or flash-pasteurized before packaging (Beech and Carr, 1977).

4.4 CIDER QUALITY
4.4.1 Chemical Composition of Cider
The most important compounds formed during fermentation, which are considered key products affecting the organoleptic profile of cider, are ethyl alcohol, higher alcohols, esters, organic acids, carbonyl compounds, sugars, and tannins. Except for extensive hydrolysis by pectolytic and cellulolytic enzymes,

the composition of fermented products, especially the flavor components, remained similar in the products obtained by mechanical or mild enzymatic extraction process (Poll, 1993).

4.4.1.1 Ethyl Alcohol

Various types of ciders are classified according to their ethanol content, which varies from 0.05 to 13.6% (Amerine et al., 1980; Jarvis et al., 1995).

4.4.1.2 Acids

The acids are important in maintaining the pH low enough so as to inhibit the growth of many undesirable bacteria. Like apple must, cider contains a variety of organic acids and their concentration depends on the maturity and fermentation conditions (Beech and Carr, 1977; Labelle, 1980). Sweet cider could have less than 0.45 g acid, but dry ciders made by the traditional method of fermentation, i.e., in which the apples are not washed, have high amounts of volatile acidity (1 g/L) compared to ciders made after washing and blending of apples. In the traditional method of fermentation, malic acid in the must is low (3–3.8 g/L), but in the must made by modern fermentation a high concentration (4.8 g/L) is observed due to acidic apples. The complete degradation of L-malic acid takes place rapidly by LAB in all musts except those made via the modern method of fermentation, where no MLF occurs. It indicates proper development of lactic acid fermentation.

4.4.1.3 Higher Alcohols and Methanol

The formation of higher alcohols is an important criterion to determine the quality of any alcoholic beverage but they vary from strain to strain of yeast, by the cultivars of apple used, and by the fermentation conditions employed (Amerine et al., 1980). The biosynthesis of higher alcohols is generally linked to amino acid metabolism. The higher alcohols are formed as by-products of both anabolic and catabolic metabolism and allow the reequilibrium of the redox balance involving $NAD^+/NADH$ cofactors (Hammond, 1986). Therefore, they may appear via the biosynthesis route using the amino acid biosynthetic yeast or by the deamination and decarboxylation of amino acids present in the substrate. It is also known that higher fusel alcohols are generated from cloudy rather than clear juice fermentation (Beveridge et al., 1986).

Apple eau de vie is a traditional alcoholic beverage produced in France by distillation of fermented apple juice (hard cider). The methanol concentration of hard cider varies from 0.037% to approximately 0.091%, and the methanol content of apple eau de vie ranges from below 200 mg to more than 400 mg/100 mL of 40% ethanol. The US legal limit of methanol for fruit brandy is 0.35% by volume or 280 mg/100 mL of 40% ethanol (Table 7.1.13). Of the four apple cultivars examined, Crispin apples yielded significantly more methanol in hard cider and eau de vie than Empire, Jonagold, or Pacific Rose apples. Pasteurization of Crispin apple juice prior to alcoholic fermentation significantly reduced the methanol content of hard cider and eau de vie (Hang and Woodams, 2010).

4.4.1.4 Tannins

Tannins enhance the sensory qualities of the cider by affecting the astringency level, which varies in cider from 50 to 100 mg/100 mL. The types of polyphenols or tannins found in the bittersweet English pathway of cider are listed in Fig. 7.1.16. No significant change in the phenolic content occurs during fermentation, although the chlorogenic, caffeic, and p-coumaryl acids may be reduced to dehydroshikimic acid and ethyl catechol (Jarvis, 1993a,b). The chlorogenic and caffeic acid in apple juice cultivars and ciders

Table 7.1.13 Influence of Apple Cultivar on the Methanol Content of Hard Cider

Cultivar	Methanol (% v/v)	Methanol (% v/v 40% Ethanol)	Ethanol (% v/v)
Empire			
EP1	0.063	0.361	6.94
EP2	0.060	0.341	7.03
EP3	0.068	0.364	7.44
Mean ± SD	0.064 ± 0.004	0.355 ± 0.012[B]	7.14 ± 0.27
Jonagold			
JG1	0.036	0.222	6.44
JG2	0.039	0.225	6.93
JG3	0.036	0.216	6.66
Mean ± SD	0.037 ± 0.002	0.221 ± 0.004[C]	6.68 ± 0.25
Crispin			
CR1	0.104	0.581	7.08
CR2	0.084	0.446	7.50
CR3	0.085	0.495	6.96
Mean ± SD	0.091 ± 0.011	0.507 ± 0.068[A]	7.18 ± 0.28
Pacific Rose			
PR1	0.076	0.335	9.04
PR2	0.076	0.302	10.12
PR3	0.071	0.277	10.20
Mean ± SD	0.074 ± 0.003	0.305 ± 0.029[B]	9.65 ± 0.76

The values of methanol that have no common superscript are significantly different at the 95% confidence level according to Duncan's multiple range test.
Based on Hang, Y.D., Woodams, E.E., 2010. Influence of apple cultivar and juice pasteurization on hard cider and eau-de-vie methanol content. Bioresource Technology, 101, 1396–1398.

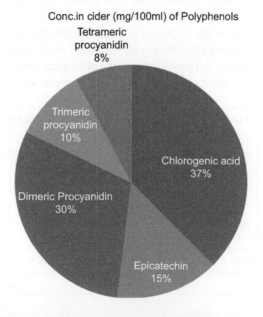

Conc.in cider (mg/100ml) of Polyphenols

Tetrameric procyanidin 8%

Trimeric procyanidin 10%

Chlorogenic acid 37%

Dimeric Procyanidin 30%

Epicatechin 15%

FIGURE 7.1.16

Polyphenols in bittersweet English cider. The concentrations of polyphenols (mg/100 mL) in cider are shown.

Based on the data from Lea, A.G.H., 1984. Colour and tannins in English cider apples. Flussiges Obstetrics 51, 356.

correlated very well with total phenols. The chlorogenic acid constitutes 6.2–10.7% of total phenols and these acids are involved in nonenzymatic autoxidative browning reactions (Cilliers and Singleton, 1989; Cilliers et al., 1990).

The polyphenols have been found to control the quality of ciders, as they predominantly account for astringency, bitterness, color, and aroma. Thirty-two compounds belonging to five groups of phenolic compounds were identified and quantified by reversed-phase liquid chromatography on both fruit extract and juice. The average polymerization degree of flavanols was estimated in fruit by phloroglucinolysis coupled to HPLC. Parental maps were built using SSR and SNP markers and used for quantitative trait locus (QTL) analysis. Sixty-nine and 72 QTLs were detected on 14 and 11 linkage groups of the female and male maps, respectively. The study presented QTLs for the mean polymerization degree of procyanidins, for which the mechanisms involved remain unknown (Table 7.1.14). New markers were designed from sequences of the most interesting candidate genes to confirm their colocalization with underlying QTLs by genetic mapping (Cindy et al., 2014).

Procyanidin B1 and procyanidin B2 were found to be the most powerful antioxidants in Basque cider, and p-coumaric acid, (−)-epicatechin, and hyperin had the greatest capacity to precipitate proteins. Ciders with higher tyrosol concentration will have less reduction potential and higher antioxidant reservoir (Zuriarrain et al., 2014).

The contents of polyphenols and organic acids are in constant flux. After fermentation, the contents of (+)-catechin, (−)-epicatechin, chlorogenic acid, cinnamic acid, p-coumaric acid, gallic acid, caffeic acid, ferulic acid, rutin, and phloridzin decrease by different degrees, whereas protocatechuic acid increases after fermentation (Table 7.1.15). The contents of organic acids are also affected by fermentation. Malic acid, lactic acid, quinic acid, pyruvic acid, and citric acid show various levels of increase, but succinic acid content decreases (Ye et al., 2014), as shown in Table 7.1.16.

4.4.1.5 Carbonyl Compounds

The most important carbonyl compounds formed in cider fermentation are acetaldehydes, diacetyl, and 2,3-pentanedione. Aldehydes, having very low flavor thresholds, tend to be considered off-flavors (green-leaf-like flavor). As intermediates in the formation of ethanol and higher alcohols from amino acids and sugars, the conditions favoring alcohol production also generate small quantities of aldehydes. They are excreted and then reduced to ethanol during the latter stage of fermentation. Diacetyl makes an important contribution to the flavor of cider and its presence is considered essential for correct flavor.

Table 7.1.14 Concentration (mg/L) of Phenolic Compounds Present in the Juice

Parameter	Total Catechins	Total PCA	Total Flavanols	Total HCA	Total Flavonols	Total DHC	Total Polyphenols
Average	249	762	1011	863	47	73	1994
Median	231	716	988	883	43	65	1911
Minimum	61	339	451	137	16	31	740
Maximum	611	1604	2168	1788	177	244	3742

HCA, *hydroxycinnamic acid*; DHC, *dihydrochalcones*; PCA, *protocatechuic acid*.
Reproduced from Cindy, F.V., Sylvain, G., Nicolas, C., Muriel, B., Jean-Marc, C., Sylvain, G., Pauline, L.-Z., Michela, T., David, G., Francois, L., 2014. QTL analysis and candidate gene mapping for the polyphenol content in cider apple. Public Library of Science PLoS One 9 (10), 1–16.

4.4.1.6 Total Esters

Esters are mostly present in smaller concentrations than alcohols (Table 7.1.17), with the notable exception of ethyl acetate and 2- and 3-methyl butyl acetates, which in Yarlington Mill cider apple juice increase 200-fold during fermentation. The esters constitute a major group of desirable compounds, including ethyl acetate (fruity), isoamyl acetate (pear drops), isobutyl acetate (banana-like), ethyl hexanoate (apple-like), and 2-phenyl acetate (honey, fruity, flowery). They are formed by yeast during fermentation in a reaction between alcohols, fatty acids, coenzyme A, and an ester-synthesizing enzyme (Nedovic et al., 2000).

Table 7.1.15 The Effects of Fermentation on the Contents of Polyphenols

Polyphenol	Apple Juice (mg/L)	Apple Cider (mg/L)
Chlorogenic acid	13.29[a]	8.27[b]
Caffeic acid	5.72[a]	4.20[b]
Cinnamic acid	0.65[a]	0.57[b]
Protocatechuic acid	1.06[a]	3.36[b]
Ferulic acid	1.94[a]	1.96[a]
Gallic acid	4.71[a]	3.85[b]
p-Coumaric acid	1.45[a]	1.16[b]
(+)-Catechin	26.37[a]	24.37[b]
(−)-Epicatechin	0.08[a]	0.07[b]
Phloridzin	6.96[a]	5.98[b]
Rutin	2.14[a]	1.84[b]

Results are given as the means of triplicate fermentations, each analyzed in triplicate. Different superscript letters within each column indicate significant differences according to Duncan's test ($p < .05$).
Reproduced from Ye, M., Yue, T., Yuan, Y., 2014. Evolution of polyphenols and organic acids during the fermentation of apple cider. Journal of the Science of Food and Agriculture 94, 2951–2957.

Table 7.1.16 The Effects of Fermentation on the Contents of Organic Acids

Organic Acid	Content (mg/L)	
	Apple Juice	Apple Cider
Malic acid	4938.76[a]	7177.67[b]
Lactic acid	0[a]	130.52[b]
Succinic acid	421.66[a]	233.18[b]
Quinic acid	1202.11[a]	1997.76[b]
Pyruvic acid	31.68[a]	82.11[b]
Citric acid	343.40[a]	536.73[b]

Results are given as the means of triplicate fermentations, each analyzed in triplicate. Different superscript letters within each column indicate a significant difference according to Duncan's test ($p < .05$).
Reproduced from Ye, M., Yue, T., Yuan, Y., 2014. Evolution of polyphenols and organic acids during the fermentation of apple cider. Journal of the Science of Food and Agriculture 94, 2951–2957.

Table 7.1.17 Concentrations of the Major Esters in Apple Juices and Ciders

Major Ester	Yarlington Mill		Content (ppm) Sweet Coppin		Kingston Black		Bramley's Seedling	
	Juice	Cider	Juice	Cider	Juice	Cider	Juice	Cider
Ethyl acetate	2	35	1	20	1	17	2	15
Isobutyl acetate	–	0.2	–	0.003	–	0.1	0.03	0.3
Ethyl butyrate	0.3	–	0.01	0.01	0.2	0.4	0.3	0.1
2- and 3-Methylbutyl acetates	0.15	30	0.02	3	0.2	4	0.1	0.9
Ethyl 2-methyl butyrate	0.006	–	–	–	0.01	–	0.04	–
S-hexyl acetate	0.6	6	0.3	0.1	0.2	1.5	0.3	0.7
Ethyl hexanoate	–	2	–	0.02	–	0.6	–	0.9
2- and 3-Methylbutyl octanoates	–	4	–	0.1	–	0.7	–	0.01

Reproduced from Beech, F.W., Carr, J.G., 1977. Cider and perry. In: Economic Microbiology. In: Rose, A.H. (Eds.), Alcoholic Beverages, vol. VI. Academic press, London, p. 139.

4.4.2 Sensory Quality of Cider

The appearance, color, aroma, taste, and subtle taste factors such as flavor constitute the sensory quality of cider. The aroma and taste are very complex and depend on a number of factors such as varieties used, agricultural land, vinification practices, fermentation, and maturation (Gayon, 1978). The taste of cider is determined more by the apple composition, whereas its odor is governed by technological factors and the yeast employed rather than the apple varieties used to make the cider. The cider flavor is assessed using both subjective and objective approaches (Joshi, 2006; Downing, 1989). In subjective approach, trained or untrained panels can be employed to recognize specific flavors. In human beings, flavor sensation by taste is limited to sweetness, sourness, bitterness, and astringency together with such tastes as metallic and pungent (Piggot, 1988). QDA has also been applied to profile cider flavor analytically (Williams, 1975). Of the various descriptors used, some have greater meaning to characterize cider aroma and perry essence (Salih et al., 1990; Downing, 1989). At a simple level, a number of general descriptors can be used, such as fruitiness, acidity, sweetness, astringency, alcohol, body, bitterness, and sulfur. But at the analytical level, the number of descriptors is kept large to differentiate ciders of different types. A typical cider aroma wheel is shown in Fig. 7.1.17. Another approach that has been applied to flavor profiling is sniff analysis, whereby the effluent from GC is assessed by specially trained judges. It may be mentioned that the color of cider is determined by the extent of juice oxidation or degradation and, in fact, it is possible to make water-white high-tannin ciders if oxidation is completely inhibited (Lea, 1982). During fermentation, however, the initial color diminishes by around 50%, which is presumably because of the strong reductive power of yeast, which readily reduces the keto or carbonyl groups to hydroxyls with consequent loss of the chromophore (Downing, 1989).

The effects of various treatments involving contact with natural lees on the aromatic profile of cider were evaluated. Compared with untreated ciders, contact with the lees brought about a significant increase in the concentrations of most of the volatile compounds analyzed, in particular fatty acids, alcohols, ethyl esters, and 3-ethoxy-1-propanol, in contrast to fusel acetate esters and 4-vinylguaiacol.

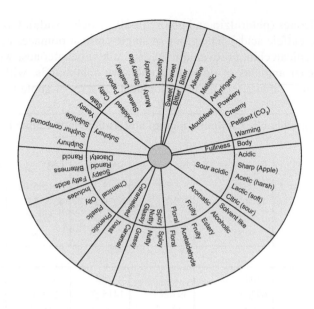

FIGURE 7.1.17

Cider flavor wheel.

Adapted from Joshi, V.K., Sharma, S., Parmar, M., 2011c. Cider and perry. In: Joshi, V.K. (Ed.), Handbook of Enology, vol. 1II. Asia Tech Publishers, Inc. New Delhi, pp. 1116–1151.

The addition of β-glucanase enhanced the content of ethyl octanoate, but produced a decrease in the contents of decanoic acid and all of the major volatiles except acetaldehyde, ethyl acetate, and acetoin, whereas the application of oxygen influenced the rise in the level of 3-ethoxy-1-propanol only. The olfactometric profiles also revealed significant effects of treatment with lees for ethyl propionate, diacetyl, *cis*-3-hexenol, acetic acid, benzyl alcohol, and *m*-cresol, whereas the addition of oxygen significantly influenced the perception of ethyl hexanoate, 1-octen-3-one, 3-methyl-2-butenol, *t*-3-hexenol, and *c*-3-hexenol (Anton-Diaz et al., 2016).

In another study, 90 ciders, including 50 "brut" (near dry), 30 "demi-sec" (half dry), and 10 "doux" (sweet), were analyzed for 180 variables, including processing conditions, sensory descriptors, and physicochemical variables. A multiple factor analysis (MFA) was performed on this data set using the sensory and analytical data as active variables and the processing conditions data were introduced as illustrative variables. Two MFAs were carried out separating the less sweet (brut) from the sweeter (demi-sec and doux) ciders. For both sets, the MFAs showed a strong polarization of the ciders' characteristics: fruity/flowery flavors were associated with cooked flavors and opposed to the descriptors "animal," "sous-bois" (underwood), and "fond de cuve" (vat dregs), themselves associated with bitterness and astringency and also "pomme/cidre" (apple/cider) flavors. The industrial ciders were localized on axis 1 near the fruity and cooked odors, whereas the ciders of small-scale producers were close to the second group of flavors and tastes (Quere et al., 2006).

Eleven different cider apple pomaces (six single-cultivar and five from the cider-making industry) were analyzed for low-molecular-weight phenolic profiles and antioxidant capacity. The Folin index ranged between 2.3 and 15.1 g gallic acid per kilogram of dry matter. Major phenols were flavanols, dihydrochalcones (phloridzin and phloretin-20-xyloglucoside), flavonols, and cinnamic acids (chlorogenic and caffeic acids). The group of single-cultivar pomaces had higher contents of chlorogenic acid, (−)-epicatechin, procyanidin B2, and dihydrochalcones, whereas the industrial samples presented higher amounts of up to four unknown compounds, with absorption maxima between 256 and 284 nm (Table 7.1.18). The antioxidant capacity of apple pomace, as determined

Table 7.1.18 Mean Polyphenolic Contents (mg/kg) and Antioxidant Capacity of Single-Cultivar Cider Apple Pomaces

Polyphenol	P	M	DR	LM	C	DT
Folin (g gallic acid/kg)	6.2[b]	7.2[c]	7.7[c]	5.5[a]	10.9[d]	6.3[b]
TPHPLC (mg/L)	4818.3	5910.0	5230.4	4672.5	5862.6	5198.5
DPPH (g AA/kg)	11.4[a]	11.1[a]	13.5[c]	12.4[b]	15.9[d]	11.1[a]
FRAP (g AA/kg)	9.5[a]	10.8[a]	10.7[a]	9.8[a]	13.8[b]	9.7[a]
Phenolic Acids						
Protocatechuic	20.8[c]	nd	nd	144.7[d]	4.1[b]	1.7[a]
Chlorogenic	693.2[b]	393.2[a]	1415.5[e]	681.5[b]	927.2[d]	836.9[c]
Caffeic	nd	20.7[cd]	16.8[c]	10.5[b]	17.3[c]	25.1[cd]
Other acids	230.6[c]	109.7[a]	110.6[ab]	118.1[b]	278.3[d]	409.5[e]
Sum of phenolic acids	944.7	523.7	1542.9	954.7	1226.9	1273.3
Flavanols						
(−)-Epicatechin	394.9[d]	222.8[c]	314.6[d]	161.1[b]	163.0[b]	136.0[a]
Procyanidin B2	477.0[b]	590.2[c]	437.3[b]	329.1[a]	348.7[a]	553.9[c]
Trimer C1 + tetramer	525.3[b]	372.0[a]	576.0[b]	572.9[b]	585.0[b]	569.6[b]
Other flavanols	664.1[b]	1228.6[e]	944.3[d]	652.1[b]	772.0[c]	619.6[a]
Dihydrochalcones						
Phloretin-20-xyloglucoside	332.8[d]	996.2[f]	136.9[b]	82.9[a]	457.2[e]	228.2[c]
Phloridzin	797.3[c]	1435.4[f]	730.2[b]	587.2[a]	1053.0[e]	875.7[d]
Other dihydrochalcones	38.7[c]	104.0[f]	22.7[b]	18.0[a]	61.7[e]	51.0[d]
Sum of dihydrochalcones	1168.8	2535.7	889.7	688.2	1571.9	1154.9
Flavonols						
Hyperin	223.4[b]	186.7[a]	175.3[a]	377.1[d]	462.9[e]	252.5[c]
Isoquercitrin + rutin	136.0[c]	44.6[a]	63.3[b]	159.6[d]	208.9[e]	218.7[f]
Reynoutrin	61.7[c]	37.5[b]	16.8[a]	144.3[f]	104.4[e]	84.2[d]
Avicularin	129.5[b]	99.2[a]	168.1[c]	364.0[f]	242.9[e]	223.5[d]
Quercitrin	93.0[b]	69.0[a]	96.0[b]	252.0[e]	168.7[d]	103.8[c]
Sum of flavonols	643.6	436.9	525.6	1313.7	1195.0	889.2

Different superscript letters in the same line indicate significant differences at p < .05. DPPH, 2,2-diphenylpicrylhydrazyl; FRAP, ferric-reducing ability of plasma; nd, not detected; TPHPLC, sum of HPLC polyphenols.
Adapted Garcia, Y.D., Valles, B.S., Lobo, A.P., 2009. Phenolic and antioxidant composition of by-products from the cider industry: apple pomace. Food Chemistry 117, 731–738.

by the 2,2-diphenylpicrylhydrazyl (DPPH) and ferric-reducing ability of plasma (FRAP) assays, was between 4.4 and 16.0 g ascorbic acid per kilogram of dry matter, thus confirming that apple pomace is a valuable source of antioxidants. PLSR analysis gave reliable mathematical models that allowed prediction of the antioxidant activity of apple pomace as a function of the phenolic profile (Garcia et al., 2009).

4.5 SPOILAGE OF CIDER

A few ciders with residual sugar or sweet ciders with pH above 3.8 stored at ambient temperature develop a defect called ropiness or oiliness, caused by certain strains of LAB (*Lactobacillus* and *Leuconostoc* spp.) that produce a polymeric glucan (Carr, 1983, 1987) that thickens the consistency. When poured, the cider appears oily in texture with a detectable sheen. At higher concentrations of glucan, the texture thickens so that the cider moves as a slimy "rope" when poured from a bottle. Pasteurization of the affected juice is a simple solution, though it requires blending before use. Another defect is referred to as mousiness (Tucknott and Williams, 1973). Its exact cause is unknown but it occurs in unsulfited cider with a high pH that has necessarily been exposed to air during fermentation. The growth of film-forming yeasts such as *Brettanomyces* spp., *Pichia membranifaciens*, and *Candida mycoderma* also produce "mousy" flavor (1,4,5,6-tetrahydro-2-acetopyridine). *S. ludwigii* is often resistant to SO_2 levels (1000–1500 ppm). These can grow slowly during fermentation and maturation and result in a butyric flavor and the formation of flaky particles, which spoil the appearance of the cider. Contamination of the final product with *S. cerevisiae*, *Saccharomyces bailii*, and *S. uvarum* increases the concentration of CO_2.

Ciders low in acidity, tannins, and nitrogen, but high in mineral matter, occasionally develop an olive green color; the fermentation ceases and starch is deposited. If iron in the cider combines with tannins, a black or greenish black color develops. Bottled cider stored at high temperature sometimes produces a sediment called casse, which is due to the action of peroxidase on tannins. It can be prevented by the addition of SO_2 after fermentation. The classical microbiological disorder of stored bulk cider is known as cider sickness, or "*framboise*" in French (Beech and Carr, 1977), caused by the bacterium *Zymomonas anaerobia*, which ferments sugar in bulk sweet ciders stored at pH values greater than 3.7. The features of the defect are renewed and "almost explosive" fermentation accompanied by a raspberry or banana peel aroma and a dense white turbidity in the beverage due to the production of acetaldehyde at high levels by *Zymomonas*. The acetaldehyde reacts with the tannins to produce an insoluble aldehyde–phenol complex and consequently turbidity. Flavor taints in ciders may also arise from the presence of naphthalene and related hydrocarbons where tarred rope has been stored adjacent to a cider keg. A new taint in ciders is caused by indole and is derived from tryptophan breakdown (Wilkins, 1990) at levels in excess of 200 ppb, whereby its odor becomes increasingly fecal and unpleasant.

The effect of pulsed electric field (PEF) in inactivating naturally occurring microorganisms (yeast and molds) in freshly squeezed apple cider was investigated in a continuous flow system. The microbial count decreased with an increase in applied pulses (17.6–58.7 total) and treatment temperature (45–50 1°C) and a decrease in flow rate (3–10 L/h). At field strength of 27–33 kV/cm (3-mm electrode gap in a concentric chamber), 200 pulses/s, 3 L/h flow rate, and 50 1°C process temperature, there was a 3.10 log reduction in microbial counts. By PEF treatment in the presence of a mixture of nisin and lysozyme (27.5 U/mL for nisin and 690 U/mL for lysozyme), there was an increase in the microbial count from

Table 7.1.19 Effect of Pulsed Electric Field on the Inactivation of polyphenol oxidase in Apple Cider

Reaction Time (min)	Change in Optical Density at 410 nm			PPO Activity Decrease (%)
	Fresh Cider at Room Temperature	50°C Heating for 5 min	50°C and PEF Treatment	
10	0.33a	0.33a	0.22 ± 0.04b	33
20	0.65a	0.65a	0.48 ± 0.01b	27
30	1.00a	0.93a	0.69 ± 0.01b	25

Means with similar letters in a row are not significantly different at the 5% level. PEF, *pulsed electric field;* PPO, *polyphenol oxidase.*
Reproduced from Liang, Z., Cheng, Z., Mittal, G.S., 2006. Inactivation of spoilage microorganisms in apple cider using a continuous flow pulsed electric field system. LWT 39, 350–356.

1.12 to 1.78 log reductions for 10 L/h flow (Table 7.1.19). When cider samples were treated in the presence of clove oil (3 or 5 mL/100 mL), an additive reduction of 1.99 log cycles in microbial counts was observed (Liang et al., 2006).

5. LOQUAT WINE

Loquats are also known as Chinese or Japanese plums. The fruits somewhat resemble apricots in size and color but have large seeds. For preparation of loquat wine, the ingredients include 2 kg loquats, 1 kg sugar, water up to 4 L, 1 Campden tablet (crushed), 5 g pectic enzyme, 10 g acid blend, 1/4 teaspoon grape tannin, 5 g Super Ferment yeast nutrient, and active yeast culture (*S. cerevisiae*). The fruits are washed in cool water and their seeds are removed. The fruits are chopped up and put into the primary fermenter. The pectic enzyme, tannin, acid blend, sugar, and enough water to give a total volume of 4 L are added. The crushed Campden tablet is stirred in, and the container is covered with a plastic sheet and allowed to remain as such for 24 h for activation. The activated yeast is stirred and added to the must. Fermentation in the primary container is allowed to proceed for 7 days, stirring well every day. The pulp is strained by squeezing out as much juice as possible. This liquid is siphoned into another clean container, attached to the air lock, and allowed to ferment for 3–4 weeks. The wine is siphoned into another clean container. The air lock is reattached and the wine is allowed to stand until it is clear. This step is repeated once a month until the product is clear. The clear and stable wine is bottled. If a sweeter wine is preferred, 5 g of potassium sorbate stabilizer per 4 L is used at least 48 h before adding sugar. For better results, the sugar is dissolved in some boiling water and added to the wine to sweeten to taste. After being bottled and corked, the loquat wine is allowed to mature for 6 months (www.google.com).

6. MEDLAR WINE

Medlars (*Mespilus germanica*) are small self-fertile trees originally from southwest Asia and possibly also southeastern Europe. They are pretty trees that seem to be very tough, able to fruit even in the wettest of summers when even many apple varieties struggle. The fruit looks like a cross between a small apple and a rose hip. They do not ripen on the tree and so have to be picked when still hard and allowed to "blet," a softening process like a partial rotting. The fruits are stored in sawdust or bran in a cool, dark place until they go soft and develop an aromatic flavor. The fruits can also be used for making a country wine that tastes rather like sherry. In brief, the bletted fruits are added to sugar and water plus a wine yeast and left to ferment (www.google.com) and the result is a wine.

7. *PYRACANTHA* WINE

Pyracantha, or firethorn, is an evergreen ornamental with small white flowers in the spring. These give way to green berries that persist throughout the summer, turn orange in the autumn, and turn bright red as winter arrives. One of the uses of *Pyracantha* is to prepare a base wine. *Pyracantha* berries are picked when fully ripened. The ingredients are 1 kg *Pyracantha* berries, 1 cup golden raisins, 1 kg sugar, 2 lemons, 1 large orange, 5 g pectic enzyme, 10 g yeast nutrient, 2 L water, and wine yeast. To prepare *Pyracantha* wine, the berries are sorted, destemmed, and washed, discarding any that are not sound or ripe. The berries and raisins are put into boiling water that is adjusted to maintain a simmer for 20 min. The berries and raisins are poured into a blender and chopped. The activated culture of *S. cerevisiae* is added and the must is stirred twice daily for 7 days. The juice is then, separated and the pulp discarded. The remaining sugar is then, added and stirred well and the air lock is attached. The base wine is racked every 30 days for 3 months, refitting the air lock each time. After 6 months the prepared wine is bottled and sweetened to taste (www.google.com). *Pyracantha* wine has an unusual but not unpleasant taste. It ages well and smooths out as it matures, as a dry or a semisweet wine. It is light colored, like an orange-tinted rosé.

8. TOYON WINE

The berries are used for jelly and sometimes in the making of custard or wine. Toyon berries are acidic and contain a small amount of cyanogenic glycosides, which will break down into hydrocyanic acid during digestion, which can be removed by cooking (www.google.com).

9. QUINCE WINE

For preparation of wine from quince, the ingredients required are 25 quinces, 2 lemons, 225 g raisins, 1350 g sugar, water up to 4 L, yeast nutrient, and wine yeast (*S. cerevisiae*). The quince fruit is grated, the core is omitted, and the fruit is boiled for 15 min. After straining, the liquid is poured into a fermentation bucket over the sugar and stirred well to dissolve the sugar. Chopped raisins and lemon juice are added. After the mixture cools, the yeast nutrient and wine yeast are added. The bucket is kept covered for 10 days in a warm place and stirred daily. The mixture is strained through a fine sieve and put into a fermenter and the air lock is attached during the fermentation. When fermentation is finished, the wine is decanted and the clear wine is racked for 6 months (www.google.com).

10. MIXED FRUIT WINES

Various combinations of fruits can be made to produce mixed fruit wines. For this purpose, a combination of grape with other fruits is made so as to have the vinosity of the grape and the flavor of the specific fruit used (Fowles, 1989). Details of the materials required and their proportions for making dry table wines are summarized in Table 7.1.20.

For preparation of the must, the juices are mixed together. The nutrient salts, sugar, and pectolytic enzyme are dissolved in water and then mixed with the juices and the indicated volume is made. The yeast culture is made as discussed earlier. The active yeast culture is added to the must to initiate the fermentation, carried out in a glass container equipped with air locks. The optimum temperature for fermentation is below 20°C. Siphoning and clarification are carried out as per the normal practice. The siphoning is carried out after 3–4 weeks of completion of fermentation. Then, addition of sodium metabisulfite (5 mL of 10% sodium metabisulfite solution) is made. The wine is bottled and sealed with straight wine corks until used. Preparation of mixed wine having a typical flavor of apple, gooseberry, and pineapple has been described (Fowles, 1989). The quantities recommended for dry white table wine (Sancerre type) are shown in Table 7.1.20. The juice is sieved/strained to remove the suspended solids and mixed with juices, sugar, and other ingredients. The yeast is rehydrated and reactivated and added to the must to start the fermentation. The wine improves in quality after storage for 6 months. Five grams of oak chips (commercial), when added to the wine, gives a delicate oak note to the wine, similar to that of white burgundy. For a dry red table wine (Rhone style), red grape concentrate gives flavor, blackberries and elderberries contribute a gusty red color, and black currants provide the elusive flavor and aroma. Pears give the wine body and the desirable acidity. The must is prepared from the juice of blackberries and elderberries with chopped pears with skins. The fermentation is carried out in a plastic container with a lid, on the pulp initially, then the liquid is strained off. The rest of the

Table 7.1.20 Types of Wine and Ingredients for Home-Scale Preparation of Wine Using Grapes in Combination With Other Fruits

Type of Wine and Specifications	Fruit Juice (L/kg)	Sugar (g)	Pectolytic Enzyme	Wine Yeast (Sachet)	Nutrient Salt (g)	Water (Total Volume, L)
1. Simple dry white table wine (ethanol 11–11.5%, acidity 0.55–0.8%)	Apple White grape–1	580	Sufficient	1	1	4.5
2. Dry white table wine (Sancerre style) (ethanol 12.5%, acidity 0.75–0.85%)	Apple–1 White grape–2 Goose berries–0.680 Pineapple–0.200	450	Sufficient	1	1	4.5
3. Dry red table wine (Rhone style) (ethanol 12–12.5%, acidity 0.60–0.65%)	Elderberries–1.600 Blackberries–0.5 Pears–0.5 Black currant–0.100 Red grape concentrate–0.200		Sufficient	1	1	4.5

Adapted from Fowles, G., September 38, 1989. The Complete Home Wine Maker. New Scientist.

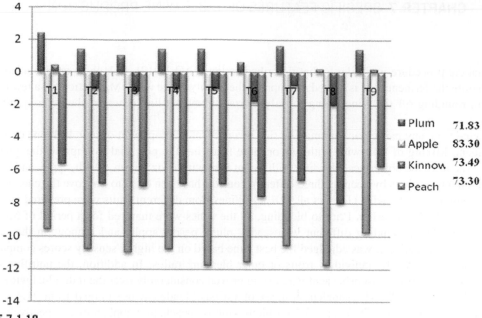

FIGURE 7.1.18

Derivation of suppleness index of various treatments of wine from individual fruits.

Reproduced from Joshi, V.K., Vikas, K., Thakur, N.K., Kumar, P., May–June 2013c. Effect of blending of different fruit wines on the physico-chemical and sensory characteristics of blended wine. Indian Food Packer 40–44.

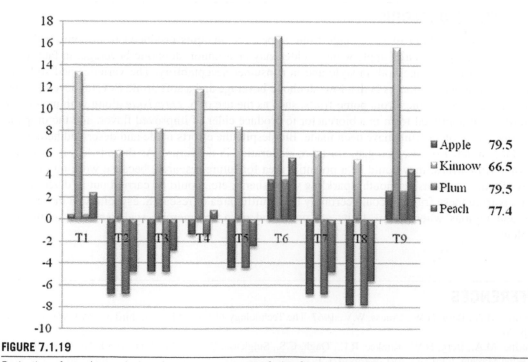

FIGURE 7.1.19

Derivation of suppleness index of various treatments of wine from individual fruits.

Reproduced from Joshi, V.K., Vikas, K., Thakur, N.K., Kumar, P., May–June 2013c. Effect of blending of different fruit wines on the physico-chemical and sensory characteristics of blended wine. Indian Food Packer 40–44.

winemaking procedure is similar to that described earlier (FAO, 2012). Oak chips are included in the must before the fermentation is started, to improve the quality of the wine. Maturation for a few months helps in rounding off the wine owing to polymerization and settling down of the anthocyanins and tannins.

A wine from any particular fruit may be rich in a particular component or may lack one or another constituent. Mixing of fruit wine with one or more fruits has the potential to improve the quality of wine. To achieve the objectives, blending of different fruit wines, viz., plum, peach, apple, and kinnow, in different proportions by keeping three different controls has been done to improve the chemical and sensory qualities of the wine (Joshi et al., 2013c). Wines from different fruits were prepared as per the methods standardized earlier. Prior to blending, all the wines were matured for a period of 6 months. The wine made by blending different wines, viz., plum, peach, apple, and kinnow, in the ratio of 50:15:15:20, respectively, was adjudged the best wine based on the higher sensory scores compared to wines made from all the individual fruits or other blending ratios. In addition, the nutrition of the blended wine in terms of ascorbic acid was also improved considerably over the individual wines from peach, plum, and apple. The suppleness index of various blended wines ranged from 71.4 to 73.7, whereas it was 71.2, 73.5, 79.3, and 83.3 for plum, kinnow, peach, and apple wine, respectively (Figs. 7.1.18 and 7.1.19). It was concluded that to make mixed fruit wine of acceptable quality, plum, peach, apple, and kinnow wines can be blended in the ratio of 50:15:15:20 to improve the nutritional as well as sensory quality characteristics.

11. FUTURE OUTLOOK

Considering the quantity of wine made from grapes, the amounts produced from pome fruits are very low. Although globally apple wine or cider may be a minor alcoholic beverage, in fact it is a well-established beverage both in style and in consumer acceptability. The varieties and styles of cider make it more popular as both a low-alcoholic beverage and a soft drink. Several developments in the production of wines from pome fruits, such as the use of S. cerevisiae along with malolactic bacteria in immobilized form in a bioreactor to produce cider of improved flavor, and the preparation wines from other fruits have been made. But despite the efforts made and described here, there are a large number of research gaps for more elaboration in the future. As there is less information on pome fruit wines, they could be a potential area for future research, because work on different types of wines, such as vermouth, sparkling wine, sherry, etc., could be carried out to diversify the products from these fruits. It could certainly be a fruitful area of research, especially in those countries and regions where pome fruits are cultivated on a commercial scale and marketing is a problem.

REFERENCES

Amerine, M.A., Berg, H.W., Cruess, W.V., 1967. The Technology of Wine Making, third ed. AVI, Westport, CT, p. 76.

Amerine, M.A., Berg, H.W., Kunkee, R.E., Qugh, C.S., Singleton, V.L., Webb, A.D., 1980. The Technology of Wine Making, fourth ed. AVI, Westport, CT.

Anton-Diaz, M.J., Valles, B.S., Mangas-Alonso, J.J., Fernandez-Garcia, O., Picinelli-Lobo, A., 2016. Impact of different techniques involving contact with lees on the volatile composition of cider. Food Chemistry 190, 1116–1122.

Attri, B.L., Lal, B.B., Joshi, V.K., 1994. Technology for the preparation of sand pear vermouth. Indian Food Packer 48 (1), 39.

Azad, K.C., Vyas, K.K., Joshi, V.K., Sharma, R.C., Srivastava, M.P., 1986a. Utilization of Sand pear II. Effect of different sugars on the fermentation behaviour and organoleptic qualities of perry. In: 6th ICFOST, AFST, Mysore, India Abst. 3.30, p. 52.

Azad, K.C., Vyas, K.K., Joshi, V.K., Srivastava, M.P., 1986b. Deacidification of fruit juices for alcoholic fermentation. In: Abst. ICOFOST-86, p. 53.

Azad, K.C., Vyas, K.K., Joshi, V.K., Sharma, R.P., 1987. Observations on juice and cider made from scabbed apple fruit. Indian Food Packer 41, 47.

Babsky, N.E., Toribio, J.L., Lozano, J.E., 1986. Influence of storage on the composition of clarified apple juice concentrate. Journal of Food Science 51, 564.

Beech, F.W., 1972. Quick determination of adequate juice sulfiting. Journal of the Institute of Brewing 78, 477.

Beech, F.W., Carr, J.G., 1977. Cider and perry. In: Rose, A.H. (Ed.), Economic Microbiology. Alcoholic Beverages, vol. VI. Academic press, London, p. 139.

Beech, F.W., 1993. Yeasts in cider making. In: Rose, A.H., Harrison, J.S. (Eds.), The Yeasts, Yeast Technology, vol. 5. second ed. Academic Press, London, p. 169.

Beveridge, T., Franz, K., Harrison, J.E., 1986. Clarified natural apple juice: Product storage stability of juice and concentrate. Journal of Food Science 51, 411.

Bhat, A., Joshi, V.K., 2010. Processing Chapter-29. In: Sharma, R.M., Pandey, S.N., Pandey, V. (Eds.), The Pear: Production, Post-Harvest Management and Protection. IBDC Publishers, Lucknow. xviii, 699, p. ills.

Bhutani, V.P., Joshi, V.K., Chopra, S.K., 1989. Mineral composition of experimental fruit wines. Journal of Food Science and Technology 26, 332.

Bhutani, V.P., Joshi, V.K., 1995. Plum. In: Salunkhe, D.K., Kadam, S.S. (Eds.), Handbook of Fruit Science and Technology, Cultivation, Storage and Processing. Marcel Dekker, New York, USA, pp. 203–241.

Bump, V.L., 1989. Apple processing and juice extraction. In: Downing, D.L. (Ed.), Processed Apple Products. AVI Publishing Co., New York, p. 53.

Buren, J.P.V., Lee, C.Y., Way, R.D., Pool, R.M., 1979. Bitterness in apple and grape wines. HortScience 14 (1), 42.

Burroughs, L.F., 1973. Report, Long Ashton, for 1972. University of Bristol, England, p. 124.

Campo, G., del, Berregi, I., Santos, J.I., Duenas, M., Irastorza, A., 2008. Development of alcoholic and malolactic fermentations in highly acidic and phenolic apple musts. Bioresource Technology 99, 2857–2863.

Carr, J.G., 1983. Microbes I have known. The Journal of Applied Bacteriology 55, 383.

Carr, J.G., 1987. Microbiology of wines and ciders. In: Norris, J.R., Pettipher, G.L. (Eds.), Essays in Agricultural and Food Microbiology. Handbook of Enology: Principles, Practices and Recent Innovations, vol. 1148. John Wiley, London, p. 291.

Castelli, T., 1973. Lecologie Des Levures. Collogue Inter. D's Conolgie Arcensenas. Vignas Vins Mai 19, 25.

Cilliers, J.J.L., Singleton, V.L., 1989. Non-enzymic autoxidative phenolic browning reactions in a caffeic acid model system. Journal of Agricultural and Food Chemistry 37, 890.

Cilliers, J.J.L., Singleton, V.L., Lamuela-Raventos, 1990. Total polyphenols in apples and ciders; Correlation with chlorogenic acid. Journal of Food Science 55 (5), 1458.

Cindy, F.V., Sylvain, G., Nicolas, C., Muriel, B., Jean-Marc, C., Sylvain, G., Pauline, L.-Z., Michela, T., David, G., Francois, L., 2014. QTL analysis and candidate gene mapping for the polyphenol content in cider apple. Public Library of Science PLoS One 9 (10), 1–16.

Delin, C.R., Lee, T.H., 1991. The J-shaped curve revisited: wine and Cardiovascular health update. Australian and New Zealand Wine Industry Journal 6 (1) 15(FSTA 24(5): 5H77, 1992).

Deunas, M., Irastorza, A., Fernandez, A.B., Huerta, A., 1994. Microbial populations and malolactic fermentation of apple cider using traditional and modified methods. Journal of Food Science 59 (5), 1060.

Dinsdale, M.G., Lloyd, D., Jarvis, B., 1994. Membrane potential studies of *Saccharomyces cerevisiae* during cider fermentation. In: Biochemical Society Transactions, 650th Meeting, Cardiff, p. 325.

Downing, D.L., 1989. Apple cider. In: Downing, D.L. (Ed.), Processed Apple Products. AVI Publishing Co., New York, pp. 168–186.

Durieux, A., Nicolay, X., Jourdian, J.M., Depaepe, C., Pietercelie, A., Plaisant, A.M., Simon, J.P., 1997. Cider production by immobilized yeast and *Leuconostoc oenos*: comparison of different configurations. In: Proceedings of the International Workshop Bioencapsulation VI: From Fundamental to Industrial Applications, Barcelona, Spain, pp. 145–148.

Esau, K., 1977. Anatomy of Seed Plants. John Wiley and Sons, New York.

FAO, 2012. Pome Fruits. Food and Agriculture Organizations, Rome. www.fao.org.

Flores, J.H., Heatherbeth, D.A., 1984. Optimizing enzymes and pre-mash treatments for juice and colour extraction from strawberries. Fluessiges Obstetrics 7, 327.

Fowles, G., September 38, 1989. The Complete Home Wine Maker. New Scientist.

Frazier, W.C., Westhoff, D.C., 1995. Food Microbiology. Tata McGraw-Hill Publishing Company Limited, New Delhi.

Gary, L.M., Renee, T.T., Justin, R.M., 2007. Reduction of malic acid in wine using natural and genetically enhanced microorganisms. American Journal of Enology and Viticulture 58, 341.

Garcia, Y.D., Valles, B.S., Lobo, A.P., 2009. Phenolic and antioxidant composition of by-products from the cider industry: apple pomace. Food Chemistry 117, 731–738.

Gayon, P.R., 1978. Wine flavour. In: Charlambous, G., Inglett, G.E. (Eds.), Flavour of Food and Beverage Chemistry and Technology. Academic press. INC, New York, London, p. 335.

Gebhardt, S.E., Cutrufelli, R., Mathews, R.H., 1982. Composition of Foods. Agric. Handbook. US Dept of Agri., Washington, pp. 8–9.

Goswell, R.T., Kunkee, R.E., 1977. Fortified wines. In: Rose, A.H. (Ed.), Alcoholic Beverages. Academic Press, London, p. 477.

Hammond, J., 1986. The contribution of Yeast to beer flavour. Brewers Guardian 115, 27.

Hang, Y.D., Woodams, E.E., 2010. Influence of apple cultivar and juice pasteurization on hard cider and eau-de-vie methanol content. Bioresource Technology 101, 1396–1398.

Hulme, A.C., 1958. Some aspects of the biochemistry of apple and pear fruits. Advances in Food Research 8, 297–413.

Jarvis, B., 1993a. Chemistry and Microbiology of cider making. In: Encyclopedia. Food Science and Nutrition, first ed. Coleraine campus, Cromore Road, Coleraine, BT52ISA. Belfast Coleraine Jordans town, Magee.

Jarvis, B., 1993b. Cider : cyder; hard cider. In: Encyclopedia Food and Nutrition, first ed. Coleraine campus, Cromore Road, Coleraine, BT52ISA. Belfast Coleraine Jordanstown. Magee.

Jarvis, B., Foster, M.J., Kinsella, W.P., 1995. Factors affecting the development of cider flavour. Journal of Applied Bacteriology – Symposium Supplement 79, 55.

Joshi, V.K., Bhutani, V.P., 1990. Evaluation of plum cultivars for wine preparation. In: XXIII Int. Hort. Congress, Held at Italy. Abst. 3336.

Joshi, V.K., Bhutani, V.P., 1991. Influence of enzymatic treatment on fermentation behaviour physico chemical and sensory qualities of apple wine. Sciences Des Aliments 11 (3), 491–496.

Joshi, V.K., Sandhu, D.K., Attri, B.L., Walia, R.K., 1991. Cider preparation from apple juice concentrate and its consumer acceptability. Indian Journal of Horticulture 48 (4), 321.

Joshi, V.K., 1993. Studies on Alcoholic Fermentation of Apple Juice and Apple Juice Utilization. (Ph.D. thesis submitted to Guru Nanak University Ámritsar (Pb), India).

Joshi, V.K., Sandhu, D.K., 1994. Influence of juice contents on quality of apple wine prepared from apple juice concentrate. Research and Industry 39 (4), 250.

Joshi, V.K., Bhutani, V.P., 1995. Peach. In: Salunkhe, D.K., Kadam, S.S. (Eds.), Handbook of Fruit Science and Technology, Cultivation, Storage and Processing. Marcel and Dekker, New York, U.S.A., pp. 243–296.

Joshi, V.K., 1995. Fruit Wines, Directorate of Extension Education, first ed. Dr. YS Parmar University of Horticulture and Forestry, Nauni, Solan (HP), p. 155.

Joshi, V.K., 2006. Sensory Science: Principles and Application in Food Evaluation. Agro-Tech Academy, Udaipur, pp. 458–460.

Joshi, V.K., Thakur, N.K., 1995. Effect of fermentation on nutritive value of foods. Beverage Food World 22 (1), 60.

Joshi, V.K., 1997. Fruit Wines, Directorate of Extension Education, second ed. Dr. YS Parmar University of Horticulture and Forestry, Nauni, Solan (HP), p. 255.

Joshi, V.K., Sandhu, D.K., 1997. Effect of different concentrations of initial soluble solids on physico-chemical and sensory qualities of apple wine. Indian Journal of Horticulture 54 (2), 116–123.

Joshi, V.K., Bhutani, V.P., Thakur, N.K., 1999a. Composition and nutrition of fermented products. In: Joshi, V.K., Pandey, A. (Eds.), Biotechnology: Food Fermentation, vol. I. Educational Publishers and Distributors, New Delhi, pp. 259–320.

Joshi, V.K., Sandhu, D.K., Thakur, N.S., 1999b. Fruit based alcoholic beverages. In: Joshi, V.K., Pandey, A. (Eds.), Biotechnology: Food Fermentation, vol. II. Educational Publishers and Distributors, New Delhi, pp. 647–744.

Joshi, V.K., Sandhu, D.K., 2000. Influence of ethanol concentration, addition of spices extract and level of sweetness on physico-chemical characteristics and sensory quality of apple vermouth. Brazilian Archives of Biology and Technology 43 (5), 537–545.

Joshi, V.K., Sahdhu, D.K., Thakur, N.S., Walia, R.K., 2002. Effect of different sources of fermentation on flavour profile of apple wine by descriptive analysis technique. Acta Alimentaria 31 (3), 211.

Joshi, V.K., Siby, J., November 67, 2002. Antimicrobial Activity of Apple Wine against Some Pathogenic and Microbes of Public Health Significance. Alimentaria.

Joshi, V.K., Sharma, S., Bhushan, S., Devender, A., 2004. Fruit based alcoholic beverages. In: Pandey, A. (Ed.), Concise Encyclopedia of Bioresource Technology. Haworth Inc., New York, p. 335.

Joshi, V.K., Thakur, N.S., Bhatt, A., Chayanika, G., 2011a. Wine and brandy: a perspective. In: Joshi, V.K. (Ed.), Handbook of Enology, vol. 1. Asia Tech Publishers, Inc., New Delhi, pp. 3–45.

Joshi, V.K., Attri, D., Singh, T.K., Ghanshyam, A., 2011b. Fruit wines: production technology. In: Joshi, V.K. (Ed.), Handbook of Enology, vol. III. Asia Tech Publishers, Inc., New Delhi, pp. 1177–1221.

Joshi, V.K., Sharma, S., Parmar, M., 2011c. Cider and perry. In: Joshi, V.K. (Ed.), Handbook of Enology, Vol. 1II. Asia Tech Publishers, Inc., New Delhi, pp. 1116–1151.

Joshi, V.K., Sandhu, D.K., Kumar, V., 2013a. Influence of addition of insoluble solids, different yeast strains and pectinesterase enzyme on the quality of apple wine. Journal of the Institute of Brewing 119, 191–197.

Joshi, V.K., Sandhu, D.K., Kumar, V., 2013b. Influence of Addition of apple insoluble solids, different wine yeast strains and pectinolytic enzymes on the flavor profile of apple wine. International Journal of Food and Fermentation Technology 3 (1), 79–86.

Joshi, V.K., Vikas, K., Thakur, N.K., Kumar, P., 2013c. Effect of blending of different fruit wines on the physico-chemical and sensory characteristics of blended wine. Indian Food Packer May–June 40–44.

Joshi, V.K., John, S., Abrol, G.S., 2013d. Effect of addition of herbal extract and maturation on apple wine. International Journal of Food and Fermentation Technology 3 (2), 103–113.

Joshi, V.K., John, S., Abrol, G.S., 2014. Effect of addition of extracts of different herbs and spices on fermentation behaviour of apple must to prepare wine with medicinal value. National Academy Science Letters. http://dx.doi.org/10.1007/s40009-014-0275-y.

Joslyn, M.A., 1950. Methods in Food Analysis Applied Plant Products. Academic Press, New York.

Kadam, P.Y., Dhumal, S.A., Shinde, N.N., 1995. Pear. In: Salunkhe, D.K., Kadam, S.S. (Eds.), Handbook of Fruit Science and Technology: Production, Composition, Storage and Processing. Marcel and Dekker, INC, New York.

Kerni, P.N., Shant, P.S., 1984. Commerical Kashmir apple for quality cider. Indian Food Packer 38 (1), 78.

Kumar, V., Joshi, V.K., Vyas, G., Tanwar, B., 2015. Effect of different types of fermentation (inoculated and natural fermentation) on the functional properties of apple tea wine. Research Journal of Pharmaceutical, Biological and Chemical Sciences 6 (3), 847–854.

Labelle, R.L., 1980. Apple Cultivars Tested as Naturally Fermented Cider at Geneva. State Agric. Exp. Stn., Memo, N.Y. Geneva, New York.

Lagan, D.F., 1981. Manufacture of Wine by Continuous Fermentation. US patent No NZ 185886.

Laplace, J.M., Jacquet, A., Travers, I., Simon, J.P., Auffray, Y., 2001. Incidence of land and physicochemical composition of apples on the quantitative and qualitative development of microbial flora during cider fermentation. Journal of the Institute of Brewing 107 (4), 227–233.

Lea, A.G.H., 1978. Phenolics of cider – procyanidins. Journal of the Science of Food and Agriculture 29, 471.

Lea, A.G.H., 1982. Analysis of phenolics in oxidizing apple juice by HPLC using a pH shift method. Journal of Chromatography A 238, 253.

Lea, A.G.H., 1984. Colour and tannins in English cider apples. Flussiges Obstetrics 51, 356.

Lea, A.G.H., 1995. Cider making. In: Lee, A.G.H., Piggott, J.R. (Eds.), Fermented Beverage Production. Blackie Academic and Professional, London, U.K, p. 66.

Lequere, J.M., Drilleau, J.F., 1998. Microbiology and technology of cider. Revue des Enologues 88, 17–20.

Liang, Z., Cheng, Z., Mittal, G.S., 2006. Inactivation of spoilage microorganisms in apple cider using a continuous flow pulsed electric field system. LWT 39, 350–356.

Loft, R.V., 1943. The levalose, dextrose and sucrose content of 15 Illinois apple varieties. Proceedings of the American Society for Horticultural Science 43, 56.

Lowe, E., Durkee, E.L., Hamilton, W.E., Moyan, A.I., 1964. Bitter apple juice dejuicing through thick cake extraction. Food Engineering 36 (12), 48.

Mitra, S.K., 1991. Apples. In: Mitra, S.K., Bose, T.K., Rathore, D.S. (Eds.), Temperate Fruits. Horticulture and Applied Publ., Calcutta, p. 122.

Nedovic, V.A., Durieux, A., Van Nedervelde, L., Rossels, P., Vandegans, J., Plainsant, A.M., Simon, J.F., 2000. Continuous cider fermentation by co-immobilized yeast and Leuconostoc oenos cells. Enzyme and Microbial Technology 26, 834.

Nogueira, A., Wosiacki, G., 2010. Sidra. In: Venturini Filho, W.G. (Ed.), Bebidas alcoolicas: Ciencia e Technologia. Blucher, Sao Paulo, pp. 113–139.

O'Reilly, A., Scott, J.A., 1993. Use of an ion-exchange sponge to immobilise yeast in high gravity apple based Cider alcoholic fermentation. Biotechnology Letters 15 (10), 1061.

Park, Y.J., John, C.B., 1980. Decomposition of acid in wine by yeast. Res. Rep. Agricultural Science and Technology 7 (2), 176 Chungnam National University Daejeon, S. Korea.

Piggot, J.R., 1988. Sensory Analysis of Foods, second ed. Elsevier Applied Science, London, New York.

Poll, L., 1993. The effect of pulp holding time and pectolytic enzyme treatment on the acid content of apple juice. Food Chemistry 47 (1), 73.

Pourlx, A., Nicholas, L., 1980. Sweet and Hard Cider. Garden way Publishing Co, Charlotte, V.T.

Quere, J-M. Le, Husson, F., Catherine, M.G.C.R., Primault, J., 2006. French cider characterization by sensory, technological and chemical evaluations. LWT 39 (9), 1033–1044.

Rana, R.S., Vyas, K.K., Joshi, V.K., 1986. Studies on production and acceptability of cider from Himachal Pradesh apples. Indian Food Packer 40, 56–66.

Rana, S., Neerja, Rana, S., Vishal, 2011. Biochemistry of wine preparation. In: Joshi, V.K. (Ed.), Handbook of Enology, vol. 2. Asia Tech Publication, New Delhi, p. 618. 678.

Revier, M. (Ed.), 1985. Le cidre-heir et aujourd'hui. La Nouvelle Libraire, Paris.

Roach, F.A., 1985. Cultivated Fruits in Britain-Their Origin and History. Basil Blackwell, Oxford.

Salih, A.G., Le Quere, J.M., Drilleau, J.F., Fernandez, J.M., 1990. Lactic acid bacteria and malolactic fermentation in manufacture of Spanish cider making. Journal of the Institute of Brewing 96, 369.

Samarajeewa, U., Mathew, D.T., Wijeratne, M.C.P., Warnakula, T., 1985. Effect of sodium metasulphite on ethanol production in coconut inflorescence sap. Food Microbiology 2, 11.

Sandhu, D.K., Joshi, V.K., 1995. Technology, quality and scope of Fruit Wines with special reference to apple. Indian Food Industry 14 (1), 24–34.

Sharma, R.C., Joshi, V.K., 2005. Apple processing technology. In: Chadha, K.L., Awasthi, R.P. (Eds.), The Apple. Malhotra Public. House, New Delhi, p. 445.

Siby, J., Joshi, V.K., 2003. Preparation and Evaluation of Cider with different sugar sources and spices extract. Journal of Food Science and Technology 40 (6), 673–676.

Singh, R.S., Sooch, B.S., 2009. High cell density reactors in production of fruit wine with special reference to cider-An overview. Natural Product Radiance 8 (4), 823–833.

Smock, R.M., Neubert, A.M., 1950. Apple and Apple Products. Interscience Publishers, New York.

Sonni, F., Cejudo, B., Maria, J., Chinnici, F., Natali, N., Claudio, R., 2009. Replacement of sulphur dioxide by lysozyme and oenological tannins during fermentation: influence on volatile composition of white wines. Journal of the Science of Food and Agriculture 688.

Stockley, S.,C., 2011. Therapeutic value of wine. In: Joshi, V.K. (Ed.), Handbook of Enology, Vol. 1. Asia Tech publication, New Delhi, pp. 146–208.

Stycztnska, A.K., Pogorzelski, E., 2009. L-Malic acid effect on organic acid profiles and fermentation by-products in apple wines. Czech Journal of Food Sciences 27, 228–231.

Tucknott, O.G., Williams, A.A., 1973. Report, Long Ashton, for 1972. University of Bristol, England, p. 150.

Upshaw, S.C., Lopez, A., Williams, H.L., 1978. Essential elements in apples and canned apple sauce. Journal of Food Science 43 (2), 449.

Vyas, K.K., Kochhar, A.P.S., 1993. Studies on cider and wine from culled apple fruit available in Himachal Pradesh. Indian Food Packer 47 (4), 15.

Vogt, E., 1977. Der Wein, Seine Beseteurg, seventh ed. Verlag Eugen, Ulmer, Stuttgart.

Wang, W., Xu, Y., Hu, J., Zhao, G., 2004. Fermentation kinetics of different sugars by apple wine yeast *Saccharomyces cerevisiae*. Journal of the Institute of Brewing 110 (4), 340–346.

Westwood, M.N., 1978. Temperate Zone Pomology. W.H. Freeman, San Francisco.

Whiting, G.C., 1961. Annual Report of the Agricultural and Horticultural Research Station, Long Ashton, for 1960. University of Bristol, England, p. 135.

Wildenradt, H.L., Singleton, V.L., 1974. The production of aldehydes as a result of oxidation of polyphenolic compounds and its relation to wine aging. American Journal of Enology and Viticulture 25, 119.

Wilkins, C.K., 1990. Analysis of indole and skatole in porcine gut contents. International Journal of Food Science & Technology 25, 313.

Williams, A.A., 1975. The development of vocabulary and profile and profile assessment method for evaluating the flavour contribution of cider and perry aroma constituents. Journal of the Science of Food and Agriculture 26, 567.

Wong, M., Stanton, D.W., 1993. Effect of removal of amino acid and phenolic compounds on non enzymatic browning of stored kiwi fruit juice concentrate. Lebensmittel-Wissenching und Technologie 26, 138.

Wrolstad, R.E., Spanos, G.A., Durst, R.W., 1990. Changes in Phenolics and Amino Acid Profiles of Apple Juice Concentration during Processing and Storage. Berichte Intenational Fruchtsaft-union, Wissenchaftlich-Technische Kommission, p. 103. www.google.com. www.nhb.org (Hard Honey Cider).

Ye, M., Yue, T., Yuan, Y., 2014. Evolution of polyphenols and organic acids during the fermentation of apple cider. Journal of the Science of Food and Agriculture 94, 2951–2957.

Zuriarrain, A., Zuriarrain, J., Puertas, A.I., Duenas, M.T., Ostra, M., Berregi, I., 2014. Polyphenolic profile in cider and antioxidant power. Journal of the Science of Food and Agriculture 95, 2931–2943.

SUBCHAPTER

STONE FRUIT WINES

7.2

V.K. Joshi[1], P.S. Panesar[2], G.S. Abrol[3]

[1]*Dr. Y.S. Parmar University of Horticulture and Forestry, Nauni, Solan, HP, India;* [2]*Sant Longowal Institute of Engineering and Technology, Longowal, Punjab, India;* [3]*UUHF, Bharsar, Uttarakhand, India*

1. INTRODUCTION

Apricots, plums, peaches, and cherries are the major stone fruits grown the world over (Westwood, 1978; Bhutani and Joshi, 1995; Joshi et al., 2012). These fruits are highly perishable commodities and have to be either consumed immediately or preserved in one or another form. In the developed countries, a considerable quantity is utilized to prepare processed products from these fruits. But in developing countries, lack of proper utilization results in considerable postharvest losses, estimated to be 30–40% (Joshi et al., 2000, 2011a). Conversion of such fruits into wine of acceptable quality, especially in the developing countries, could save these precious resources to a greater extent.

Although production of wine is largely done by the fermentation of grape juice, it has also been practiced widely using fruits such as apples, cherries, currants, peaches, plums, strawberries, etc. (Vyas and Chakravorty,1971; Merine et al., 1980; Joshi et al., 2011a,b). Consumption of wine has assumed a great importance largely due to the presence of phenolic compounds and resveratrol (Joshi and Devi, 2009; Joshi et al., 2011a), which are helpful in preventing cardiovascular disease. Plum wines are also quite popular in many countries, particularly in Germany and Pacific Coastal states (Amerine et al., 1980; Joshi et al., 1999a). The stone fruits, including plum, have many common characteristics, like pulpy nature, appealing color, sugar, high acid, minerals, etc., and can be utilized for the preparation of wine and brandy. Thus, the production of wines in those countries where stone fruits are grown would be advantageous (Joshi and Kumar, 2011). A brief review of the technology of wine production and composition of wine from stone fruits is described in this subunit.

2. PRODUCTION OF STONE FRUIT WINES: GENERAL ASPECTS

2.1 PRODUCTION

Stone fruits are cultivated throughout the world over and occupy an area of 5.07 million ha, with 35.24 million tons produced. In India, about 43,000 ha are planted with apricots, peaches and nectarines, plums, and cherries, with an annual production of about 0.25 million tons (FAO, 2008). The increased production of these fruits can be used profitably, if fruit wines are produced, thus generating employment opportunities and providing better returns to the orchardists (Vyas and Chakravorty, 1971; Sandhu and Joshi, 1995; Joshi et al., 2004).

2.2 COMPOSITION OF FRUIT

Compared to grapes, almost all the stone fruits are low in sugar, the most essential component for alcoholic fermentation, but have to be ameliorated to the sugar level needed for producing a table wine (Table 7.2.1). Even when the sugar level is satisfactory, the high acidity demands dilution, consequently requiring addition of more sugar to the must (Amerine et al., 1980; Bhutani and Joshi, 1995; Joshi and Sharma, 1994). However, the stone fruits are very good sources of anthocyanins; polyphenols; minerals like K, Na, Ca, Mg, and Fe; and vitamins.

Table 7.2.1 Chemical Composition of Some of the Stone Fruits

Parameter	Peach	Plum	Apricot	Cherry
Total sugar (% of fresh weight)	8–9	6–8	6–7	9–13
Major Sugars (% of Total)				
Fructose	10	20	10	55
Glucose	10	55	20	40
Sucrose	80	25	70	5
Acid (% tartaric acid)	0.5–0.8	1.4–1.7	1.1–1.3	0.4–0.6
Major Acids (% of Total)				
Citric	25	-	25	10
Malic	75	95	5	90
pH	3.27	1.4–1.7	1.1–1.3	3.75
Minerals (mg/100 g)				
K	453	120–190	296	-
Na	2.0	0.0–3.0	1	-
Ca	15	6.0–8.0	-	-
Mg	21	4.0–7.0	-	-
Fe	2.4	0.1–0.4	0.54	0.4
Zn	-	0.1	-	-
Vitamins (mg/100 g)				
Vitamin C	1–27	4.0–11.0	10	13–56
Thiamin	0.02	0.02–0.05	-	-
Riboflavin	0.04	0.04–0.05	-	-
Niacin	0.5	0.2–0.9	-	-
Carotene (%)	0.75–0.79	0.26–0.78	-	-
Anthocyanin (mg/100 g)	20.3–178	926	-	-
Phenolic compounds (mg/100 g)	3–14	46.3–57.0	-	-

Compiled from Fowles, G., September 38, 1989. The Complete Home Wine Maker. New Scientist; Bhutani, V.P., Joshi, V.K., 1995. Plum In: Salunkhe, D.K., Kadam, S.S. (Eds.), Handbook of Fruit Science and Technology, Cultivation, Storage and Processing. Marcel Dekker, New York, USA, pp 203–241; Wills, R.B.H., Scriven, F.M., Greenfield, H., 1983. Nutrient composition of stone fruit (Prunus spp) cultivars: apricot, cherry, nectarine, peach and plum. Journal of the Science of Food and Agriculture 34, 1383–1388.

2.3 PROBLEMS IN WINE PRODUCTION

The technique for the production of wine from stone fruits is essentially the same, involving the basic alcoholic fermentation of juice or pulp, but modifications are made with respect to the physicochemical characteristics depending on the type of wine needed and the fruit used. The major differences in the techniques of production of these wines, however, arise from the difficulty in extracting the sugar from the pulp of some of the fruits (Amerine et al., 1980; Joshi, 1997). Not only this juices extracted from most of the stone fruits are lacking in the requisite sugar contents or have poor fermentability. The higher acidity in some of the fruits makes it all the more difficult to prepare a palatable wine. The production of quality wine is affected by several factors, viz., fruit variety, stage of harvest and maturity of the fruit, and total sugars, acid, total phenols, pigments, nitrogen compounds, etc., in the fruit or the must. Additives like nitrogenous compounds, pectinesterase, and sulfur dioxide or other preservatives; fruits yielding pulp or juice; yeast strains and other vinification practices; and postfermentation operations, especially the length of maturation and method of preservation employed, influence the quality of the wine. Fruit, a sweetening agent like fruit concentrate, sugar, acid, nitrogen source, clarifying enzyme, filter aids, etc., are the raw material required to make wine. The pectin-splitting pectinesterase is used mostly in the clarification of wines. Pectinesterase has also been used for extraction of juices from stone fruits like plum, peach, and apricot (Joshi et al., 1991a), which might later on be used for wine production. The nitrogen sources used are several, but diammonium hydrogen phosphate (DAHP) has many advantages (less shifting in pH of the must and low cost); thus it is the most commonly used nitrogen source.

3. GENERAL METHOD OF WINE PREPARATION

In the preparation of wines from stone fruits, many steps are common. Different unit operations involved in the preparation of wine from stone fruits, in general are outlined here.

3.1 PREPARATION OF YEAST STARTER CULTURE

A good strain of the wine yeast *Saccharomyces cerevisiae* is a prerequisite and needs to be procured for making quality wine. The yeast in the form of slants, tablets, or compressed yeast can be employed for this purpose. Before being added to the must for fermentation, the yeast culture is activated in the juice/pulp intended for winemaking. The container with the juice is plugged with cotton and kept in a warm place (25–28°C) and the culture is ready after 24–48 h. The amount of active culture is added at the rate of 2–5% (v/v) to the must. The yeast from a tablet is activated in sterilized water at the optimum temperature for yeast growth. However, compressed yeast can be used directly in the wine fermentation.

3.2 PREPARATION OF MUST

The must is prepared depending upon the fruit used and the type of wine to be made. Either the juice is extracted or the fruit is made into pulp. The juice from fruits like plum and apricot that are highly acidic can also have reduced fermentability, which can be corrected by dilution or by adding a calculated amount of potassium or calcium carbonate. The carbonate is added to the heated mixture at 150–160°F to hasten the reaction to make the calcium citrate less soluble. Ion-exchange treatment can also be used for this purpose. Usually, pectinol is added to the must for clarification of the wine. Its addition has been found to enhance the quality of wine. SO_2 is usually added to the must at a concentration of 100–150 ppm as potassium metabisulfite (KMS). In plum wine fermentation, the addition of sodium benzoate gives better quality wine than KMS (Joshi et al., 1999a). The addition of ammonium sulfate along with thiamine and biotin increases the rate of fermentation.

3.3 FERMENTATION

The must is allowed to ferment at a temperature of 20–25°C after inoculation with yeast culture. Temperatures higher than 25°C are avoided because they cause a loss of volatile components and alcohol. The container in which fermentation is carried out should be equipped with an air lock. The sugar content or degrees Brix is measured periodically to monitor the progress of fermentation. Normally, the fermentation is allowed to proceed until all the sugar is consumed (usually Brix reading of about 8 °Bx). When the fermentation is completed, the bubbling due to production of CO_2 stops.

3.4 SIPHONING/RACKING

After the completion of fermentation, the yeast and other materials settle at the bottom of the container, with a clear liquid separating out, which is siphoned/racked or, in the case of pulpy must, is filtered through a cheese/muslin cloth followed by siphoning. Two or three rackings are usually done every 15–20 days. During the interracking period, no headspace is kept in the bottle or container, i.e., it is tightly closed to prevent acetification of the wine.

3.5 MATURATION

The newly made wine is harsh in taste and has a yeasty flavor. The process of maturation makes the wine mellow in taste and fruity in flavor, in addition to the clarification. The period may extend from 6 months to 2–3 years. The process of maturation is complex and the formation of compounds like esters takes place, thus improving the flavor of such beverages (Amerine et al., 1980). The use of *Quercus* wooden barrels for maturation of table wines is an age-old practice (Quinn and Singeleton, 1985); however, maturation with various wood chips produces characteristics similar to those of barrel-aged table wines (Joshi et al., 1994; Joshi and Shah, 1998).

3.6 CLARIFICATION

If the wine after racking and maturation is not clear, it is clarified using filter aids such as bentonite, celite, or tannin/gelatin treatment in a filter press. These treatments usually make the wine crystal clear.

3.7 BLENDING

The wines from fruits like apricot or plum are acidic and may need some amount of sweetening prior to the final bottling. For apricot and plum wine, sweetening to a total soluble solids (TSS) of 12 °Bx was found to be optimum (Joshi et al., 1999a). The wine is refiltered, if needed.

3.8 PASTEURIZATION

Wines, being low-alcoholic beverages, are pasteurized at 62°C for 15–20 min, keeping some headspace in the bottle and crown corking the same. Heating the wine precipitates tannins or other such materials, which are heat sensitive, in addition to preserving the wine. But pasteurized wines, once opened, have to be kept at low temperature to prevent their spoilage. Alternatively, table wines can be preserved by the addition of preservatives like sulfur dioxide, sodium benzoate, sorbic acid, etc. (Amerine et al., 1980; Joshi,1997). The wines with higher alcohol contents, like fortified wines, need no preservation as the alcohol itself acts as a preservative when above 15%. Carbon dioxide in conjunction with ethyl alcohol and a low level of SO_2 prevents the spoilage of carbonated wines.

4. TABLE WINE

4.1 PLUM WINE

4.1.1 Plum Fruit Cultivars

Among the stone fruits, plum (*Prunus salicina* Linn.) is an important fruit grown throughout the temperate regions of the world (Bhutani and Joshi, 1995). With attractive color and high fermentability, plum has potential for preparation of alcoholic beverages including sparkling and vermouth wine (Vyas and Joshi, 1982; Joshi et al., 1991b,c, 1995). Among the fruit wines, plum is a good source of minerals (Bhutani et al., 1989) and among the cultivars evaluated, viz., Santa Rosa, Methley, and Green Gage, Santa Rosa plum resulted in production of the best quality wine (Joshi and Bhutani, 1990).

4.1.2 Methods of Preparation

The plum is acidic in nature and thus makes an unpalatable wine. To prepare wine from it, basically all the methods involve dilution of the pulp, use of microorganisms to degrade the acids, or osmotic treatment of the fruits to leach out the acids (Amerine et al., 1980; Vyas and Joshi, 1982; Vyas et al., 1989). According to one method to prepare plum wine, for every pound of plum, 1 L of water is added followed by addition of starter culture (Amerine et al., 1980; Joshi et al., 2011b). The mixture is allowed to ferment for 8–10 days before pressing. Because the fruits cannot practically be pressed before fermentation, addition of pectolytic enzyme before fermentation facilitates the processing by increasing the yield of juice and speeding up wine clarification. The sugar may be also added to the partially fermented juice, depending on the type of wine required, i.e., table or dessert. Another method for preparation of wine states that the plums should be fully ripe, diluted with water in a 1:1 ratio (whole-fruit basis), and supplemented with 0.3% pectinol, 150 ppm SO_2, and enough sugar to raise the TSS of the must to 24 °Bx (Vyas and Joshi, 1982). A generalized flow diagram of plum winemaking is shown in Fig. 7.2.1 and the composition of plum wine produced is given in Table 7.2.2.

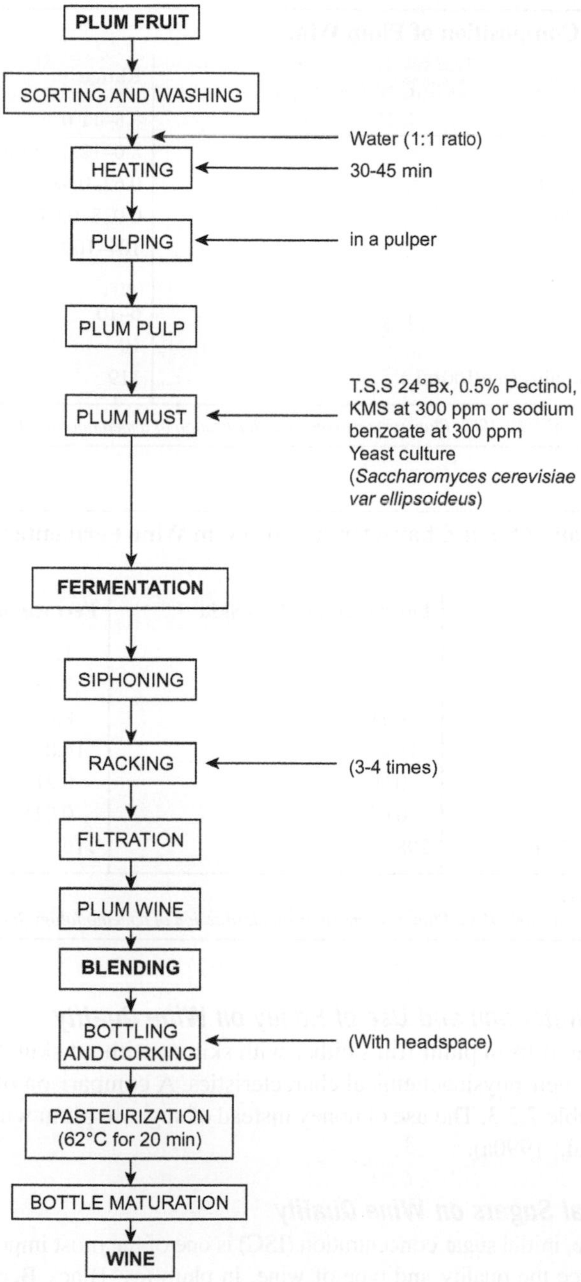

FIGURE 7.2.1

Schematic flow chart for the preparation of plum wine. *KMS*, potassium metabisulfite; *T.S.S.*, total soluble solids.

Reproduced from Joshi, V.K., Sandhu, D.K., Thakur, N.S., 1999b. Fruit based alcoholic beverages. In: Joshi, V.K., Ashok, P. (Eds.), Biotechnology: Food Fermentation, vol. II. Educational Publishers and Distributors, New Delhi, pp. 647–744.

Table 7.2.2 Chemical Composition of Plum Wine

Characteristic	Range
Ethanol (% v/v)	8.8–11.0
Total soluble solids (°Bx)	8.0–12.0 (sweet)
Titratable acidity (% malic acid)	0.62–0.88
Volatile acidity (% acetic acid)	0.028–0.040
Esters (mg/L)	104–109
Color (tintometer color units)	
Red	6–10
Yellow	10
Total coloring matter and tannins (mg/100 mL)	119

Reproduced from Vyas, K.K., Joshi, V.K., 1982. Plum wine making: standardization of a methodology. Indian Food Packer 36, 80–86.

Table 7.2.3 Comparison of Some Characteristics of Plum Wine Fermentation With and Without Skin

Characteristic	Fermentation With Skin	Fermentation Without Skin
Acidity (% MA)	1.43	1.12
pH	3.68	3.75
Alcohol (% w/v)	8.71	8.65
Sugar utilization	16.2	16.0
Color value (OD 540 nm)	0.80	0.21
Volatile acidity (% AA)	0.043	0.041
Total coloring matter (mg/100 ml)	208	116

AA, *acetic acid;* MA, *malic acid.*
Reproduced from Vyas, K.K., Joshi, V.K., 1982. Plum wine making: standardization of a methodology. Indian Food Packer 36, 80–86.

4.1.3 Effect of Skin Retention and Use of Honey on Wine Quality

Plum wine can be prepared from plum fruits either with skin or without skin, but the wines have considerable differences in their physicochemical characteristics. A comparison of such characteristics of plum wine is made in Table 7.2.3. The use of honey instead of sugar in plum wine fermentation has also been reported (Joshi et al., 1990a).

4.1.4 Effects of Initial Sugars on Wine Quality

In the preparation of wine, initial sugar concentration (ISC) is one of the most important quality parameters, as it is known to influence the quality and type of wine. In plum (cv. Black Beauty), an increase in ISC decreased the rate of fermentation, increased the ethanol, and decreased the sugar concentration (Joshi et al., 2009a). Differences in other parameters were also obtained (Table 7.2.4). The sensory quality of the wine was the best from ISC-30 as a sweet table wine. But, ISC-24 was considered suitable as a table wine. However, the required ISC can be chosen depending on the type of wine and the ethanol contents needed.

Table 7.2.4 Effects of Initial Sugar Concentration on Physicochemical Characteristics of Wines From Black Beauty Cultivar of Plum

Characteristic	ISC-20	ISC-24	ISC-30
Total soluble solids	7.9	8.8	12.2
Titratable acidity (% MA)	0.94	0.95	0.88
Fermentation rate (°Bx/24h)	2.13	1.45	1.48
Ethanol (% v/v)	10.8	12.8	14.4
Tannins (mg/L)	565	432	550
Total aldehyde (mg/L)	92	75	72
pH (before blending)	2.71	2.89	2.89
Total sugar (%)	0.248	0.22	2.80
Methanol (μL/L)	277	310	480
Volatile acidity (% AA)	0.050	0.040	0.060
Color Values (Tintometer Color Units)			
Yellow	7.0	7.4	20.8
Red	4.8	13.8	15.8
Blue	0.1	0.1	0.1

AA, *acetic acid*; ISC, *initial sugar concentration*; MA, *malic acid*.
Joshi, V.K., Sharma, S., Devi, M.P., Bhardwaj, J.C., 2009a. *Effect of initial sugar concentration on the physico-chemical and sensory qualities of plum wine. Journal of North East Foods 8, 1–7.*

4.1.5 Effect of Osmotic Treatment of Fruit on Wine Quality

To produce a quality wine, in addition to acids, a proper balance of tannins and good fruity flavor are prerequisites. Application of osmotic techniques after water blanching the fruit increases TSS and decreases acids, though a small loss of anthocyanins and mineral content also occurs. The sensory quality of wine prepared from water-blanched and osmotically treated fruits was the best (Vyas et al., 1989).

4.1.6 Use of Preservatives

The types of preservatives used are reported to affect the quality of wine produced (Joshi and Sharma, 1995). The use of sodium benzoate instead of KMS produces wine with better color and sensory qualities without affecting the physicochemical characteristics. KMS is commonly used in most wine fermentations, but it reduces the color of the wine. Higher ethanol, propanol, and amyl alcohol content is reported in KMS-treated wine compared to sodium benzoate wine. Thus, the use of sodium benzoate does not interfere with normal alcoholic fermentation of plums into wine (Fig. 7.2.2). The effects of various concentrations of sodium benzoate (0–400 mg/L) on fermentability, physicochemical characteristics, and sensory qualities of plum must were also determined. The addition of sodium benzoate decreased fermentation, but the results of a concentration of 100–200 mg/L and the control were statistically on par with each other. Ethanol content decreased proportionately with increase in sodium benzoate concentration in plum must. The titratable acidity and pH remained unaffected. Thus, the addition of 100–200 ppm sodium benzoate to plum must gives the product with the best quality. With respect to sensory qualities, sodium benzoate-treated wine has been found to be superior to KMS-treated wine (Joshi and Sharma, 1995; Joshi and Joshi, 2014).

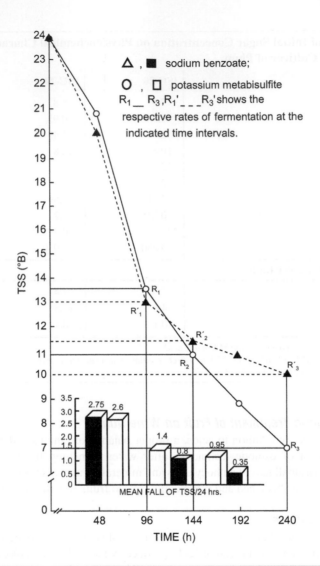

FIGURE 7.2.2

Comparison of the fermentation rates of plum must treated with potassium metabisulfite and sodium benzoate. *TSS*, total soluble solids.

Reproduced from Joshi, V.K., Sharma, S.K., 1995. Comparative fermentation behaviour, physico-chemical and sensory characteristics of plum wine as effected by the type of preservatives. Chemie, Mikrobiologie, Technologie der Lebensmittel 17, 65–73.

The mineral composition of wines has been reported (Joshi and Bhutani, 1990) to be affected by the addition of various preservatives (Fig. 7.2.3).

4.1.7 Microbiology of Fermentation

The qualitative and quantitative composition of the yeast microbiota in fermenting must depends mainly on the type of fruit, its origin, the production procedure, the type of beverage produced,

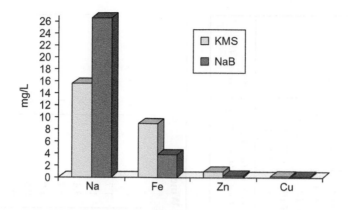

FIGURE 7.2.3

Effects of preservatives on the mineral contents of plum wine.

Based on the data from Joshi, V.K., Bhutani, V.P., 1990. Evaluation of plum cultivars for wine preparation. In: XXIII International Horticulture Congress Held at Italy Abstract 3336.

and the temperature, pH, SO$_2$, and ethanol concentration (Satora and Tuszynski, 2005); but spontaneous fermentation of plum has also been carried out to study the fermentation behavior and quality of the product obtained. Commonly, the wine yeast *S. cerevisiae* var. *ellipsoideus* is employed to produce plum wine. In such fermentation, the early stages of alcoholic fermentation are dominated by non-*Saccharomyces* species, especially *Kloeckera apiculata/Hanseniaspora uvarum* and *Candida pulcherrima*. These yeast strains were identified using killer sensitivity analysis. The activity of these non-*Saccharomyces* species decreased after 2 days, when *Saccharomyces* species took over the fermentation process. This pattern is similar to that of spontaneous fermentation of grapes (Fleet and Heard, 1993). A comparison of the physicochemical characteristics of fresh and fermented plum must using different sources of fermentation/spontaneous fermentation is made in Table 7.2.5.

4.1.8 Effect of Wine Yeast Strain

In commercial production, the wine yeast *S. cerevisiae* var. *ellipsoideus* is used in established practice. However, the effects of different wine yeast strains on the physicochemical characteristics of plum wine have also been studied (Joshi et al., 2009b). The rate of fermentation of all the yeasts showed that the "Tablet" and "W" yeast strains gave the highest reduction in TSS, whereas UCD 505/522 gave the lowest. All the yeast strains recorded rates of fermentation higher than 1.6 and strains UCD 595, W, and Tablet had the same rate of fermentation (Table 7.2.4). The yeasts did not influence the acid production in the wines which is desirable (Joshi et al., 2009b). The plum wines prepared using strains UCD 595, W, and Tablet have better sensory qualities compared to those produced from UCD 522 and UCD 505 (Table 7.2.6).

4.1.9 Effect of Dilution of Pulp/Biological Deacidification

An available alternative to dilution of the pulp with water is the use of *Schizosaccharomyces pombe* as a deacidifying yeast in plum must. The deacidification behavior of *S. pombe* was studied in plum must (Vyas and Joshi, 1988) of varying sugar contents (Table 7.2.7). Clearly, the deacidification activity of the yeast decreased with the increase in TSS of the musts.

Table 7.2.5 Physicochemical Composition of Fresh and Fermented Plum Musts

Plum Must	Extract	Total Sugars	Reducing Sugars	Titratable Acidity*	pH	Ethanol (% Vol)	Fermentation Rate (%)
		(g/L)					
Fresh	175.0 (±1.0)	132.6 (±2.3)	78.7 (±2.5)	7.84 (±0.18)	3.87 (±0.07)	-	-
Fermented							
Distillery Saccharomyces cerevisiae	45.0 (±0.0)	4.8^a (±0.1)	1.2^a (±0.4)	5.40^a (±0.15)	3.77^a (±0.05)	6.5^a (±0.1)	76.9
Wine S. cerevisiae	43.0 (±4.2)	3.8^b (±0.7)	3.4^b (±0.5)	6.73^b (±0.10)	3.78^a (±0.04)	6.5^a (±0.1)	76.2
Indigenous S. cerevisiae	40.0 (±14.1)	4.6^a (±0.1)	3.9^b (±0.2)	6.89^b (±0.18)	3.78^a (±0.03)	6.5^a (±0.1)	76.8
Kloeckera apiculata	45.0 (±2.5)	2.4^c (±0.3)	1.2^a (±0.2)	5.61^a (±0.19)	3.85^b (±0.04)	6.3^b (±0.1)	72.8
Aureobasidium sp.	48.0 (±2.0)	4.8^a (±0.3)	2.4^c (±0.2)	6.11^c (±0.24)	3.75^a (±0.03)	6.3^b (±0.1)	74.1
Spontaneous fermentation	43.0 (±3.0)	1.2^d (±0.2)	0.0^d	6.03^b (±0.23)	3.90^b (±0.01)	8.4^c (±0.0)	96.9

Values in the same row not showing the same superscript letter are different according to the Duncan test. *Titratable acidity expressed as g/L of malic acid. Adapted from Satora, P., Tuszynski, T., 2005. Bio diversity of yeasts during plum Wegierka Zwykla spontaneous fermentation. Food Technology and Biotechnology 43, 277–282.

Table 7.2.6 Effect of Wine Yeast Strain on the Sensory Characteristics of Fermented Plum Wine

Attribute[a]	Yeast Strains				
	UCD 522	UCD 595	UCD 505	W	Tablet
Appearance (2)	1.6	1.9	1.7	1.7	1.8
Color (4)	1.8	1.6	1.8	1.5	1.6
Aroma (2)	2.5	3.4	3.0	3.4	3.4
Volatile acidity (2)	1.0	1.7	1.7	1.7	1.7
Total acidity (2)	1.0	1.6	1.3	1.6	1.6
Sweetness (1)	0.4	0.6	0.5	0.6	0.6
Body (1)	0.5	0.7	0.6	0.7	0.7
Flavor (2)	1.0	1.7	0.8	1.5	1.6
Bitterness (1)	0.4	0.7	0.5	0.7	0.5
Astringency (1)	0.3	0.8	0.4	0.8	0.8
Overall impression (2)	1.3	1.7	1.3	1.7	1.7
Total	**11.8**	**16.4**	**13.6**	**15.9**	**16.0**

The maximum score used for each attribute is shown in parentheses.
[a]Indicates the sensory attributes and the maximum marks for evaluation.
Reproduced from Joshi, V.K., Sharma, S., Devi, M.P., 2009b. Influence of different yeast strains on fermentation behaviour, physico-chemical and sensory qualities of plum wine. Natural Product Radiance 445–451.

Table 7.2.7 Characteristics of Plum Musts Fermented With *Schizosaccharomyces Pombe*

Must	Initial				Final			
	TSS (°Bx)	Titratable Acidity (% MA)	pH	Alcohol (% v/v)	TSS (°Bx)	Titratable Acidity (% MA)	Volatility Acidity (% AA)	pH
P₁	6.8	0.92	3.27	2.7	2.8	0.11	0.04	4.42
P₂	12.0	0.92	3.28	5.5	4.0	0.22	0.09	4.02
P₃	15.0	0.91	3.28	8.2	5.0	0.30	0.09	3.93
P₄	20.0	0.91	3.28	11.6	6.0	0.32	0.13	4.15
P₅	25.0	0.92	3.27	12.6	8.6	0.34	0.20	4.16

AA, *acetic acid;* MA, *malic acid;* P₁ to P₅, *plum must;* TSS, *total soluble solids.*
Reproduced from Vyas, K.K., Joshi, V.K., 1988. Deacidification activity of Schizosaccharomyces pombe in plum musts. Journal of Food Science and Technology 25, 306–307.

The effects of varying levels of pH, ethanol, SO_2, and N_2 source on the deacidification activity of *S. pombe* during plum must fermentation were examined (Joshi et al., 1991c). The deacidification activity of *S. pombe* with different nitrogen sources indicated that the highest amount of acid was degraded in the must containing ammonium sulfate (Table 7.2.8). Higher amounts of acids were metabolized in musts of pH values between 3.0 and 4.5 compared to the control (pH 2.8). The yeast no doubt holds promise as a deacidification method, yet more in-depth studies are needed to utilize the yeast in plum wine fermentation.

4.1.10 Effect of Temperature, Fermentation Time and pH

In another study, the plum variety Cacanska lepotica, the result of a cross between Pozegaca and Wangenheims Fruhzwetsche in 1961, was used to optimize plum wine production. It is one of the most widely grown plum varieties in Serbia and has a high content of natural phenolic phytochemicals, such as flavonoids and phenolic acids (Murcia et al., 2001).

The influence of temperature, pH, and duration of fermentation on plum wine composition and quality and the optimization of these factors by response surface methodology (RSM) were investigated (Miljic and Puskas, 2014). The experimental design and statistical analysis were performed using Stat-Ease software (Design-Expert variables, fermentation temperature; 7.0.0 Trial, Minneapolis, MN, USA). The three dependent factors selected were temperature (X_1), fermentation time (X_2), and pH (X_3) and the analysis was carried out using a full factorial composite design. A 2^3 factorial experiment with six axial points ($a = 1.682$) and six replicates at the center points ($n_0 = 6$) led to a total of 20 experiments. Response parameters were ethanol, methanol, and glycerol contents (Miljic and Puskas, 2014). Values of factors in the central composite design for temperature were 15 and 25°C, for fermentation time were 3 and 7 days, and for pH were 3.0 and 3.6. Second-order polynomial equations, which represent fitted models for investigated responses (Miljic and Puskas, 2014), were shown as adequate ($R^2 > 0.90$ and $p < .05$).

Wine fermentation is usually conducted at relatively low temperature (15–25°C) and pH values (3.3–3.6), despite the risk of slower ethanol production, to decrease the risk of wine spoilage (Jacobson, 2006). Accordingly, optimal temperature and pH values for *S. cerevisiae* activity were in the range of 25–30°C and 4.5–6.5, respectively. Fermentation at 15°C was not completed in the observed time (7 days), according to the ethanol content (3.64–4.35 vol %) and the amount of available sugar in plum pomace (125 g/L), as shown in Fig. 7.2.4A. The maximum ethanol content (6.23%) was obtained during fermentation at 25°C and pH

Table 7.2.8 Final Characteristics of Plum Musts Fermented With *Schizosaccharomyces pombe*

Plum Must	Titratable Acidity (% MA)	Deacidification Activity (%)	TSS (°Bx)	TSS Utilized (%)	pH
Variable SO₂					
50	0.12	90.00	5.5	45	3.92
100	0.12	90.00	4.8	52	4.02
150	0.07	94.76	4.8	52	4.20
200	0.08	92.50	4.9	51	4.44
250	0.08	92.50	5.0	50	4.44
Control	0.11	90.18	4.2	58	4.04
Variable Nitrogen Source					
Yeast extract	0.20	90.19	3.5	65	4.16
Ammonium nitrate	0.44	78.43	3.0	70	3.69
Ammonium sulfate	0.12	94.11	4.0	60	4.15
Tryptone	0.19	90.68	3.0	70	4.18
Peptone	0.16	92.15	3.2	68	4.28
DAHP	0.28	89.21	4.2	58	4.35
Control	0.23	88.23	4.0	60	4.11
Variable pH					
2.5	0.23	90.00	4.2	58	3.72
3.0	0.06	97.20	4.3	57	5.87
3.5	0.07	95.20	4.4	56	6.18
4.0	0.07	95.23	5.0	50	6.38
4.5	0.07	95.23	5.2	48	7.10
Control (2.8)	0.09	92.00	4.1	59	4.63

DAHP, *diammonium hydrogen phosphate; MA, malic acid; TSS, total soluble solids.*
Reproduced from Joshi, V.K., Sharma, P.C., Attri, B.L., 1991c. A note on deacidification activity of Schizosaccharomyces pombe in plum musts of variable composition. Journal of Applied Microbiology 70, 385–390.

3.6. Increase in the temperature of fermentation resulted in an intense increase in ethanol production (Miljic and Puskas, 2014). However, the effect of pH was not so pronounced. Ethanol content was positively affected by the fermentation temperature, time, and pH and negatively by interactions between temperature and time, as well as by the quadratic terms of temperature and time of fermentation (Fig. 7.2.4A).

Most of the methanol was formed in the first 4 days of fermentation, when the activity of pectin methylesterase was the highest (Fig. 7.2.4B), and was increased exponentially with an increase in fermentation temperature. But increase in the pH values of plum pomace caused only a slight increase in its content. The maximum methanol content (1265 mg/L) was documented when the process parameters were 25°C and pH 3.6, after 7 days of fermentation (Miljic and Puskas, 2014). The increase was attributed to high pectin content (2.0–3.5 mass %) and the high degree of esterification (Rop et al., 2009), and the addition of pectinase to the pomace (Craig et al., 1998; Cabaroglu, 2005). Methanol is known to be toxic to humans through ingestion and inhalation and its oxidation leads to production of formic aldehyde and formic acid,

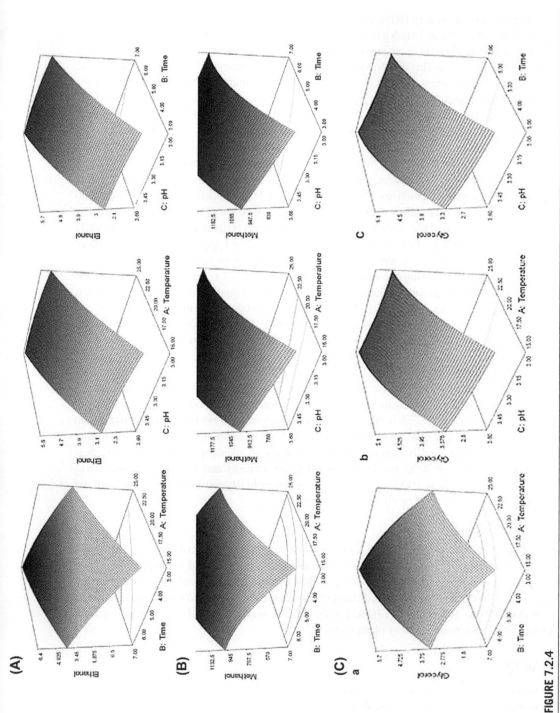

FIGURE 7.2.4

(A) Response surface plots of the interaction of (left) temperature–time (pH 3.3), (middle) temperature–pH (time 5days), and (right) time–pH (temperature 20°C), and their influence on ethanol content. (B) Response surface plots of the interaction of (left) temperature–time (pH 3.3), (middle) temperature–pH (time 5days), and (right) time–pH (temperature 20°C), and their influence on methanol content. (C) Response surface plots of the interaction of (left) temperature–time (pH 3.3), (middle) temperature–pH (time 5days), and (right) time–pH (temperature 20°C), and their influence on glycerol content.

Reproduced from Miljic, U.D., Puskas, S.V., 2014. Influence of fermentation conditions on production of plum (Prunus domestica L.) wine: a response surface methodology approach. Hemijska Industrija 68, 199–206.

both toxic to the central nervous system (Ribereau-Gayon et al., 1999). The lethal oral dose of methanol for humans ranges from 340 ng/kg to 1 mg/kg of body weight (Wang et al., 2004, Anonymous, 2001).

In alcoholic fermentation, glycerol is a nonvolatile compound without aromatic properties, though it significantly contributes to wine quality by providing sweetness and fullness (Ribereau-Gayonet et al., 1999; Remize et al., 2000). Its content increases with increase in fermentation temperature and pH of the pomace (Fig. 7.2.4C). The highest glycerol content (5.72 g/L) was recorded during fermentation at 25°C and pH 3.6 (Miljic and Puskas, 2014). Production of glycerol was more intensive during the first 4 days of fermentation, especially at the higher temperatures (25°C), whereas an increase in pH from 2.8 to 3.7 slightly affected the glycerol yield (Rankine and Bridson, 1971). It was found that higher concentrations of glycerol were obtained in a medium with a higher content of glucose (Radler and Schiitz, 1982), so it is apparent that the glycerol content is positively affected by fermentation temperature and time.

Considering general winemaking practices, it is expected that the fermentation should be optimized to have maximum ethanol and glycerol and minimum methanol yield. Accordingly, the final optimized fermentation conditions obtained with RSM were 18.3°C, pH 3.0, and 7 days fermentation time, which should ensure the production of 4.72 vol % ethanol, 1122 mg/L methanol, and 4.23 g/L glycerol (Miljic and Puskas, 2014). The predicted optimum conditions were verified and the models were proven as adequate after a repeated experiment (triplicate set), using the optimal fermentation conditions. Sensory evaluation of the plum wine produced with optimized fermentation conditions showed good quality and overall acceptability, as per the scores awarded to the color (1.9), clarity (2.0), aroma (3.4), and taste (9.1), in total 16.4 (Miljic and Puskas, 2014). Future studies need to be carried out, however, to find ways to reduce the methanol content in plum wines and determine the suitability of other plum varieties for wine production.

4.1.11 Low-Alcoholic Plum Beverages

Attempts to prepare plum wine with no or low alcohol content were made with variable success. Plum wine was prepared using four different treatments, viz., the conventional method (using *S. cerevisiae*), using *S. cerevisiae* but ameliorating with honey, and the conventional method followed by distillation for removal of alcohol or by deacidification with *S. pombe*, followed by fermentation with *S. cerevisiae*. The products were preserved using different preservation methods (Amandeep et al., 2009), viz., pasteurization (at 80°C for 2 min or 70°C for 10 min or 60°C for 15 min) and chemical preservatives (sodium benzoate, sulfur dioxide at 100 ppm). There was no microbial/yeast count (CFU/mL) in the treatments that were pasteurized at 70 or 80°C (Fig. 7.2.5A and B).

4.1.12 Effect of Maturation

For maturation, it is an age-old practice to use wooden casks for storage of alcoholic beverages produced from grapes. The addition of wood chips of *Albizia* and *Quercus* produced wines, including that of plum, of the highest sensory qualities (Joshi et al., 1994). In general, treatment of wine with wood chips increased the methanol content, and wood chips of *Populus*, *Celtis*, and *Salix* increased the methanol level more than *Albizia* and *Quercus* (Table 7.2.9). In brief, wood chips of *Quercus* and *Albizia* were found to be suitable for maturation of plum wine.

4.2 PEACH WINE

Peach (*Prunus persica*) belongs to the family Rosaceae and is a tasty, sweet, and juicy drupe fruit that originated in China dating back to 1100 BC (Huang et al., 2008). The fruit is processed into products including juice, canned peach halves, wine, jam, and jelly (Joshi and Bhutani, 1995). Peach wines are commercially

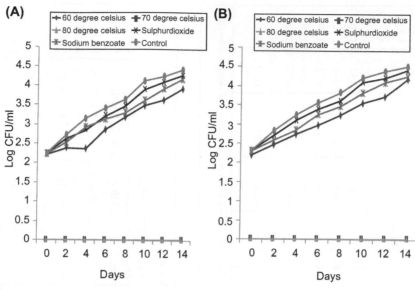

FIGURE 7.2.5

(A) Effects of various preservation methods on total plate count in 20% blended plum wine (T1). (B) Effects of various preservation methods on total plate count in 30% blended plum wine (T1).

Reproduced from Amandeep, G., Joshi, V.K., Rana, N., 2009. Evaluation of preservation methods of low alcoholic plum wine. Natural Product Radiance 8, 392–405.

produced in many countries, but the largest producer is probably the United States (Davidovic et al., 2013). These wines are often characterized by intensively and immediately recognizable peachy aroma, with a pleasant mouthfeel and smooth finish.

4.2.1 Fruit and Cultivars

The peach fruit has less acid than plum or apricot, but, being pulpy, the must has to be diluted with water to prepare for fermentation (Joshi and Bhutani, 1995). But production of wine from dried peaches has not been successful (Amerine et al., 1980). Among the cultivars evaluated for wine production, Redhaven, Sun Haven, Flavorcrest, Rich-a-haven, J.H. Hale and July Elberta were rated suitable for conversion into wine of acceptable sensory quality (Joshi et al., 2005).

4.2.2 Method of Wine Production

The method used for making peach wine consists of dilution of the pulp in the ratio of 1:1, raising the initial TSS to 24 °Bx, and adding pectinol and DAHP at 0.5% and 0.1%, respectively. To the must, 100ppm of SO_2 is added to control the activity of undesirable microflora. Differences in fermentation behavior among peach cultivars have been observed. Some of the physicochemical characteristics of wines produced from various peach cultivars are given in Table 7.2.10. In another study, the peach pulp was fermented at 25°C for 2 weeks using *S. cerevisiae* KCCM 12224 and aged at 15°C for 14 weeks, and it was seen that the yeast count increased to 2.8×10^6 CFU/mL after 2 weeks, but decreased to 7.0×10^3 CFU/mL after 14 weeks (Chung et al., 2003). Benomyl fungicide was demonstrated to be fungistatic to three strains of *S. cerevisiae* at concentrations ranging from 10 to 200 μg/mL in an agar medium prepared from

Table 7.2.9 Effects of Wood Chips on Physicochemical and Sensory Characteristics of Plum Wine

Treatment	Tannins (mg/100 g)	Methanol (µL/L)	Appearance (Max. Score 2)	Color (2)	Aroma (4)	Volatile Acidity (2)	Total Acidity (2)	Sweetness (1)	Body (1)	Flavor (2)
Control	139.00	240.00	1.53	1.51	3.07	1.42	1.32	0.68	0.72	1.20
Alnus nitida	96.00	236.00	1.53	1.49	2.99	1.49	1.41	0.67	0.64	1.35
Populus ciliata	107.00	338.00	1.56	1.49	3.08	1.50	1.52	0.58	0.64	1.45
Celtis australis	122.00	327.00	1.55	1.54	3.08	1.54	1.54	0.61	0.58	1.57
Toona ciliata	148.00	181.00	1.57	1.60	2.99	1.42	1.30	0.61	0.58	1.10
Salix tetrasperma	100.00	229.00	1.50	1.49	2.83	1.25	1.32	0.68	0.61	1.32
Bombax ceiba	122.00	180.00	1.59	1.48	2.99	1.36	1.36	0.68	0.72	1.39
Albizia chinensis	111.00	249.00	1.64	1.61	3.09	1.62	1.64	0.74	0.64	1.58
Quercus leucotrichophora	139.00	198.00	1.74	1.68	3.41	1.61	1.70	0.73	0.78	1.68
Critical difference ($p = .05$)	20.15	11.29	0.10	0.10	0.10	0.10	0.10	0.10	0.10	0.10

Reproduced from Joshi, V.K., Mahajan, B.V.C., Sharma, K.R., 1994. Treatment of fruit wines with wood chips—effect on physico-chemical and sensory qualities. Journal Tree Science 13, 27–36.

Table 7.2.10 Physicochemical Characteristics of Wines From Various Peach Cultivars

Cultivar	Ethanol % (v/v)	Higher Alcohols (mg/L)	Volatile Acidity (% AA)	Total Esters (mg/L)	Total Aldehydes (mg/L)	Total Phenols (mg/L)	Titratable Acidity (% MA)	pH	Total Sugars (mg/L)	Color Intensity
Rich-a-haven	10.73 (3.27)	113.6	0.023 (0.723)	98.23	48.23	241.8	0.66 (0.81)	3.60	1.33 (1.15)	0.53
J.H. Hale	10.67 (3.26)	126.9	0.029 (0.727)	90.92	43.30	259.0	0.71 (0.84)	3.80	1.24 (1.12)	0.33
Redhaven	11.10 (3.33)	121.4	0.023 (0.723)	98.40	45.53	301.7	0.72 (0.85)	3.71	1.26 (1.12)	3.86
Flavorcrest	11.33 (3.36)	134.4	0.029 (0.722)	94.33	49.32	306.1	0.80 (0.89)	3.61	1.19 (1.89)	0.33
Sun Haven	11.67 (3.41)	131.6	0.022 (0.725)	101.50	53.97	278.6	0.63 (0.71)	3.71	1.22 (1.05)	0.36
Stark Early White Giant	10.77 (3.28)	144.2	0.026 (0.725)	97.80	65.73	206.0	0.74 (0.86)	3.81	1.32 (1.15)	0.19
Kateroo	11.83 (3.24)	154.3	0.029 (0.727)	92.37	56.83	230.4	0.61 (0.73)	3.84	1.22 (1.10)	0.23
July Elberta	10.87 (3.14)	151.5	0.020 (0.721)	95.50	63.10	203.3	0.70 (0.83)	0.74	1.36 (1.17)	0.48

AA, acetic acid; MA, malic acid; Values within parentheses are the transformed values; Blue color units = 0 in all the wines.
Reproduced from Joshi, V.K., Shah, P.K., Kumar, K., 2005. Evaluation of peach cultivars for wine preparation. Journal of Food Science and Technology 42, 83–89.

peach pulp. However, heating the peach juice containing 30 μg/mL fungicide for 10 min at 80°C significantly reduced the growth-retarding effects on the three tested strains (Beuchat, 1973).

In a study, the physicochemical, sensory, and health-related characteristics of peach wine produced from the Redhaven variety and those of selected white wines produced from various grape varieties were determined and compared (Davidović et al., 2013). In Table 7.2.11, a comparison of the physicochemical characteristics of peach wine and grape wine is made.

The TSS content of peach must was significantly lower compared with that of white grape must, and therefore the concentration of alcohol in peach wine was also lower. The alcohol content, titratable acidity, and total extract of peach wine were significantly lower compared with those of white wine, whereas its pH value was higher (Davidović et al., 2013). Because of the low alcohol content and relatively high pH value of peach wine, its microbial stability could be expected to be low, so a higher concentration of sulfur dioxide or other type of preservative (e.g., pasteurization) needs to be applied. Volatile acidity is used as an indicator of wine spoilage, and if its value exceeds the limit of 1 g/L, the wine is not marketable and, from this point of view, grape wine is comparable. On the basis of CIELab chromatic parameters of wine samples and the parameters a*, b*, and hue angle, it was concluded that the color of all the samples was yellow with a certain proportion of greenness.

According to the total sugars content (Table 7.2.12), all the analyzed wines could be classified as dry wines (wines in which the residual sugar content is less than 1.5 g/L).

Table 7.2.11 Physicochemical Properties of Peach and Grape Wine Samples

Parameter	Peach Wine	Riesling Italian
TSS in must (°Bx)	14.50±0.10[a]	21.43±0.08[b]
Total extract (%, w/w)	2.87±0.03[a]	3.97±0.01[b]
Alcohol (%, v/v)	8.12±0.03[a]	12.01±0.01[b]
Degree of fermentation (%)	82.22±0.06[a]	83.19±0.08[b]
Calories (kJ/100 mL)	228.94±0.04[a]	334.28±0.06[b]
pH	3.90±0.01[a]	3.03±0.01[b]
Titratable acidity (g/L)	4.47±0.09[a]	6.75±0.08[b]
Volatile acidity (g/L)	0.78±0.02[a]	0.84±0.03[b]

Values represent means of triplicate determinations ± standard deviation. TSS, total soluble solids.
Adapted from Davidovic, S., Veljović, M., Pantelić, M., et al., 2013. Physicochemical, antioxidant and sensory properties of peach wine made from redhaven cultivar. Journal of Agricultural and Chemistry 61, 1357–1363.

Table 7.2.12 Content of Individual and Total Sugars (g/L) in Wines From Peach and Grapes

	Total Sugars	Trehalose	Glucose	Fructose	Sucrose	Maltose
Peach wine	0.20	0.14	0.04	–	–	0.38
Riesling Italian	0.31	0.21	0.55	0.06	0.04	1.17
Riesling Rhine	0.25	0.13	0.17	0.06	0.03	0.64
Chardonnay	0.26	0.16	0.09	0.06	0.05	0.63

Reproduced from Davidovic, S., Veljović, M., Pantelić, M., et al., 2013. Physicochemical, antioxidant and sensory properties of peach wine made from redhaven cultivar. Journal of Agricultural and Chemistry 61, 1357–1363.

Glucose and fructose are the predominant sugars in grapes, and they are the most important source of metabolic energy for wine yeasts. The peach wine had a significantly lower content of total sugars, trehalose, and fructose compared with the white wines. The total phenolic and flavonoid contents and antioxidant activity of the wine samples are presented in Table 7.2.13.

The amounts of phenolic compounds vary markedly in different types of wines, depending on the grape/fruit cultivar, environmental conditions, and winemaking procedure. Peach wine had a significantly higher content of total phenolic compounds (TPC) as well as flavonoids compared with selected white wines. The correlation between TPC and content of flavonoids was very high but was not statistically significant (Table 7.2.14).

The main phenolic compounds found in peach wine were chlorogenic acid, caffeic acid, and catechin (3.59, 0.87, and 0.60 mg/L, respectively). Antioxidant capacities were strongly correlated with total phenolics, with correlation coefficients over 0.99. The highest antioxidant capacity was ascribed to peach wine. A total of 11 phenolics together with *cis*, *trans*-abscisic acid were quantified in wine samples. Results of the UHPLC–MS/MS analysis showed significant differences in the content of individual phenolic compounds found in wine samples (Table 7.2.15). Large amounts of caffeic acid and its derivate, chlorogenic acid (3-*O*-caffeoylquinic acid), were found in peach wine. However, chlorogenic acid was not detected in the white wine samples.

Table 7.2.13 Total Phenolic and Total Flavonoid Contents and Antioxidant Activity of Wine Samples

Samples	TPC (mg GAE/L)	TFC (mg CAF/L)	DPPH (Mm TE)	FRAP (Mm TE)
Peach wine	402.53 ± 3.06^a	332.67 ± 9.75^a	1.55 ± 0.09^a	3.01 ± 0.12^a
Riesling Italian	243.67 ± 1.53^b	134.17 ± 1.44^b	1.20 ± 0.00^b	1.72 ± 0.01^b
Riesling Rhine	319.00 ± 14.73^c	175.17 ± 3.06^c	1.33 ± 0.01^b	2.17 ± 0.01^c
Chardonnay	288.00 ± 8.66^c	129.67 ± 0.76^b	1.30 ± 0.05^b	2.00 ± 0.02^d

Values represent means of triplicate determinations ± standard deviation. Different letters in same column denote a significant according to Tukey's test, p < .05. DPPH, 2,2-diphenylpicrylhydrazyl; FRAP, ferric-reducing ability of plasma; TFC, total flavonoid content; TPC, total phenolic content.
Adapted from Davidovic, S., Veljović, M., Pantelić, M., et al., 2013. Physicochemical, antioxidant and sensory properties of peach wine made from redhaven cultivar. Journal of Agricultural and Chemistry 61, 1357–1363.

Table 7.2.14 Correlation Between TPC, TFC, and Antioxidant Characteristics

	TPC		TFC		DPPH	
Method	r^a	r^b	r	p	r	p
TFC	0.941	0.059				
DPPH	**0.944**[c]	**0.006**	0.955	0.045		
FRAP	**0.981**	**0.010**	0.973	0.027	0.998	0.002

DPPH, 2,2-diphenylpicrylhydrazyl; FRAP, ferric-reducing ability of plasma; TFC, total flavonoid content; TPC, total phenolic content.
[a]*Correlation coefficient.*
[b]*Level of significance.*
[c]*Bolded numbers indicate statistically significant correlation (p < .05).*
Reproduced from Davidovic, S., Veljović, M., Pantelić, M., et al., 2013. Physicochemical, antioxidant and sensory properties of peach wine made from redhaven cultivar. Journal of Agricultural and Chemistry 61, 1357–1363.

A high content of chlorogenic acid in peach wine (3.59 mg/L) is expected, as it is already reported as one of the major phenolic compounds found in peach fruit (Cheng and Crisosto, 1995; Lavelli et al., 2009). Catechin is known as a compound that is present in significant amounts in peach fruit (Cheng and Crisosto, 1995). The concentrations of other identified phenolic compounds in peach wine were several times lower in comparison to the aforementioned phenolic acids. The results of a consumer acceptance test of given wine samples (Fig. 7.2.6) showed that the peach wine was very well accepted by regular consumers of wine and can be a very interesting product on the market.

Peach wine had a remarkably higher antioxidant capacity compared with white wines, and thus, peach wine could be a better source of natural antioxidants than white wines. The results of a sensory analysis indicate that peach wine was very well accepted by regular consumers of wine and can be a very attractive product on the Serbian market.

Table 7.2.15 Content of Some Polyphenols in Various Wine Samples (mg/L Wine)

Compound	Peach	Riesling Italian	Riesling Rhine	Chardonnay
Gallic acid	0.04	0.01	0.00	0.08
Catechin	0.60	0.03	0.02	0.08
Chlorogenic acid	3.59	0.00	0.00	0.00
Caffeic acid	0.87	0.20	0.25	0.16
p-Coumaric acid	0.04	0.05	0.01	0.11
Ferulic acid	0.06	0.16	0.15	0.25
Quercetin	0.00	0.01	0.00	0.02
cis,trans-Abscisic acid	0.09	0.01	0.01	0.01
Naringenin	0.03	0.00	0.00	0.00
Chrysin	0.02	0.01	0.00	0.02
Pinocembrin	0.01	0.00	0.00	0.00
Galangin	0.00	0.02	0.00	0.02

Reproduced from Davidovic, S., Veljović, M., Pantelić, M., et al., 2013. Physicochemical, antioxidant and sensory properties of peach wine made from redhaven cultivar. Journal of Agricultural and Chemistry 61, 1357–1363.

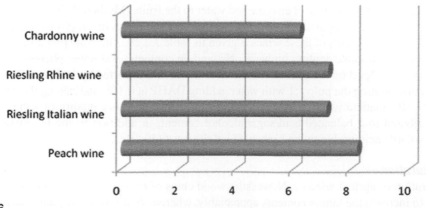

FIGURE 7.2.6

A comparison of the sensory scores for peach wine and grape wine on the hedonic scale (maximum 9.0).

Based on the data from Davidovic, S., Veljović, M., Pantelić, M., et al., 2013. Physicochemical, antioxidant and sensory properties of peach wine made from redhaven cultivar. Journal of Agricultural and Chemistry 61, 1357–1363.

4.2.3 Maturation

Wooden barrels or wood chips have been employed to mature peach wine. A study of the effects of three different types of wood chips (*Quercus*, *Bombax*, *Albizia*) on the chemical composition and sensory qualities of peach wine (eight cultivars) revealed significant changes compared to the control (Joshi and Shah, 1998). An increase in total phenols, aldehydes, and ester contents compared to the control was observed in the wine aged with *Quercus* wood (Tables 7.2.16 and 7.2.17). Further, wines treated with wood chips of *Quercus* gave wine with the best sensory qualities.

4.3 APRICOT WINE

4.3.1 Fruit and Cultivar

Apricot is a delicious fruit grown in many parts of hilly temperate countries, including India (Westwood, 1978). Because of its high flavor the fruit holds promise for conversion into wine. In India, wild apricot fruit grown at higher altitudes naturally is being used locally to make liquor, which completely lacks nutrients. Further, preparation of apricot wine from the Newcastle variety and the wild apricot (*chulli*) has been reported (Joshi et al., 1990b; Joshi and Sharma, 1994). The physicochemical characteristics of wine produced from apricot cv. Newcastle are given in Table 7.2.18.

4.3.2 Method

A method for the preparation of wine from wild apricot has been developed, which consists of diluting the pulp in the ratio 1:2, adding DAHP at 0.1% and pectinol at 0.5%, a TSS of 24 °Bx, and fermentation with *S. cerevisiae* (Fig. 7.2.7). Further, with the increase in the dilution level, the rate of fermentation, alcohol content, and pH of the wines increased, whereas a decrease in titratable acidity and volatile acidity, phenols, TSS, color values, and mineral contents took place. Addition of DAHP at 0.1% enhanced the rate of fermentation. The wine from 1:2 diluted pulp was rated as the best (Joshi et al., 1990b). In another method, honey was used to ameliorate the wild apricot must instead of sugar. Comparison of the physicochemical characteristics of various wines is given in Table 7.2.19. Higher alcohols and total phenols in honey based apricot wine were higher than those in sugar base apricot wine (Ghan Shyam, 2009). Depending upon the ISC, ethanol was produced. However, of three different ISCs, viz., 22, 24, and 26 °Bx, wild apricot wine of 22 °Bx treated with *Quercus* wood chips was rated the best.

For preparation of apricot wine from the Newcastle variety, the extraction of pulp by either the hot method (Method I) or the addition of enzyme and water to the fruits (Method II) has been studied (Joshi and Sharma, 1994). The comparison of rate of fermentation, alcohol content, physicochemical characteristics, and mineral contents of these wines is given in Table 7.2.20. The wine prepared from the latter method had higher titratable acidity; higher K, Na, and Fe contents; and lower phenolic contents. The effect of initial sugar level on the quality of apricot wine was also reported by Joshi and Sharma (1994). It was found that diluting the pulp 1:1 with water, adding DAHP at 0.1%, and raising the TSS to 30 °Bx instead of 24 °Bx made a wine of superior quality. The higher sensory quality of wine with 30 °Bx could be attributed to a balanced acid/sugar/alcohol content, in addition to the production of lower amounts of volatile acidity due to the higher ISC of the must (30 °Bx).

4.3.3 Maturation

In the maturation of apricot wine cv. Newcastle, wood chips of *Quercus*, *Albizia*, *Bombax*, and *Toona* were found to increase the tannin contents appreciably, whereas *Populus*, *Celtis*, and *Salix* wood chips

Table 7.2.16 Effects of Various Wood Chips on Total Phenols, Total Esters, Aldehydes, and Higher Alcohols of Peach Wine of Various Cultivars

Cultivar	Total Phenols (mg/L)				Total Esters (mg/L)				Aldehydes (mg/L)				Higher Alcohols (mg/L)			
	W_0	W_1	W_2	W_3	W_0	W_1	W_2	W_3	W_0	W_1	W_2	W_3	W_0	W_1	W_2	W_3
Rich-a-haven	253.8	264.6	256.0	251.3	115.1	122.2	121.7	120.6	53.26	52.26	51.54	49.11	125.4	123.1	129.9	127.5
J.H. Hale	270.5	279.4	270.2	271.0	109.3	118.1	117.0	114.9	48.81	47.96	51.43	51.41	148.6	147.2	149.0	140.8
Redhaven	315.9	320.4	315.9	313.5	115.2	122.1	119.5	119.3	55.10	56.56	48.91	46.99	144.6	139.1	150.8	164.1
Flavorcrest	317.8	337.5	322.0	321.2	114.9	120.8	119.2	118.2	60.17	56.70	58.87	51.41	126.9	134.3	133.3	129.0
Sun Haven	290.5	307.0	295.0	291.2	120.3	127.3	124.2	125.1	56.91	60.81	55.38	57.91	138.1	130.1	142.5	156.8
Stark Early White Giant	223.2	241.2	227.2	226.2	116.7	121.9	157.4	122.1	68.83	66.80	68.58	68.40	154.6	158.2	161.1	167.4
Kateroo	288.2	309.3	298.9	293.4	109.0	114.3	116.6	113.8	54.04	53.66	50.30	50.56	159.8	197.9	185.6	185.0
July Elberta	214.4	230.6	221.0	218.8	108.0	113.3	111.7	111.8	73.31	73.54	74.31	75.64	215.6	250.8	245.7	256.5

W_0 control; W_1, Quercus chips; W_2, Albizia chips; W_3, Bombax chips.
Joshi, V.K., Shah, P.K., 1998. Effect of wood treatment on chemical and sensory quality of peach wine during aging. Acta Alimentaria Budapest 27, 307–318.

Table 7.2.17 Changes in Total Phenols, Total Esters, Aldehydes, and Higher Alcohols of Peach Wine During Maturation

Cultivar	Total Phenols (mg/L)				Total Esters (mg/L)				Higher Alcohols (mg/L)				Aldehydes (mg/L)			
	M_0	M_1	M_2	M_3	M_0	M_1	M_2	M_3	M_0	M_1	M_2	M_3	M_0	M_1	M_2	M_3
Rich-a-haven	241.8	244.2	256.0	269.1	98.23	100.0	125.6	134.1	113.6	133.2	124.4	121.8	48.23	47.69	51.89	55.43
J.H. Hale	259.0	259.0	273.3	286.1	90.27	92.2	120.1	132.1	126.9	137.6	155.3	146.2	43.30	42.33	50.32	57.02
Redhaven	301.7	302.7	319.2	333.4	98.40	99.9	121.8	135.3	121.4	157.4	140.1	143.4	45.53	45.62	52.75	57.30
Flavorcrest	306.1	306.6	324.9	342.4	94.30	95.6	123.6	135.6	154.4	127.9	138.4	126.3	49.32	48.44	57.82	64.09
Sun Haven	278.6	278.63	299.9	309.9	101.50	100.9	131.8	140.0	131.6	145.3	146.1	141.0	53.97	51.51	56.94	64.83
Stark Early White Giant	206.0	209.0	230.6	247.9	97.80	98.6	130.5	139.4	144.2	164.5	161.8	154.6	65.73	64.77	67.80	71.88
Kateroo	230.4	275.8	297.4	319.1	97.37	93.2	116.2	130.3	154.3	199.1	186.1	176.1	56.83	50.47	50.78	55.16
July Elberta	203.5	203.0	223.5	237.1	92.50	96.3	112.9	124.3	215.6	233.1	252.2	233.6	63.10	72.82	74.51	75.27

M_0 without maturation; M_1, 3 months; M_2, 6 months; M_3, 12 months.
Joshi, V.K., Shah, P.K., 1998. Effect of wood treatment on chemical and sensory quality of peach wine during aging. Acta Alimentaria Budapest 27, 307–318.

Table 7.2.18 Physico-chemical Composition of Apricot wine

Characteristic	Mean ± SD* Apricot (Newcastle)
TSS (°Bx)	8.20 ± 0.07
Titratable acidity (% MA)	0.76 ± 0.02
pH	3.15 ± 0.02
Ethanol (% v/v)	10.64 ± 0.09
Reducing sugars (%)	0.34 ± 0.01
Total sugars (%)	1.11 ± 0.02
Volatile acidity (% AA)	0.025 ± 0.002
Total phenols (mg/L)	253.60 ± 0.8
Total esters	120.6 ± 0.6
Color (units)	
Red	0.70 ± 0.05
Yellow	4.30 ± 0.08
Blue	0.60 ± 0.05

AA, *acetic acid;* MA, *malic acid;* TSS, *total soluble solids.*
Joshi, V.K., Sharma, S.K., 1994. *Effect of method of must preparation and initial sugar levels on the quality of apricot wine. Research and Industry* 39, 255–257.

increased the methanol levels in the treated wine (Joshi et al., 1994). There was significant improvement in the sensory qualities of all the wines except those treated with wood of *Toona, Populus,* and *Alnus.* Of the wood chips tried, *Albizia* and *Quercus* produced wine of the highest sensory qualities (Table 7.2.21). The effects of various wood chips, viz., *Quercus, Bombax,* and *Acacia,* on ethanol, total phenols, and sensory quality characteristics of wild apricot wine have also been reported (Ghan Shyam, 2009). It was revealed that wood chip-treated wines showed more decrease in ethanol content than the control wine after maturation for 6 months, and the lowest ethanol was recorded in wine treated with *Quercus* wood chips (Fig. 7.2.8).

4.4 CHERRY WINES

4.4.1 Fruit and Cultivar

Like other stone fruits, cherries can also be converted into a wine; the sour cultivars are especially preferred (Schanderl and Koch, 1957; Benk et al., 1976). But the cherry has been found more suitable for preparation of dessert wine than table wine. The alcohol content of such wines may range from 12% to 17%. A blend of currants and table varieties of cherries can also be used for winemaking.

4.4.2 Method

To prepare a dessert wine of 16% alcohol, each liter of juice is ameliorated with 430 g of sugar. Addition of KMS before fermentation is advisable. The clarity of the wine is improved considerably if pectolytic enzyme is used. But, the addition of urea to the cherry must did not improve the

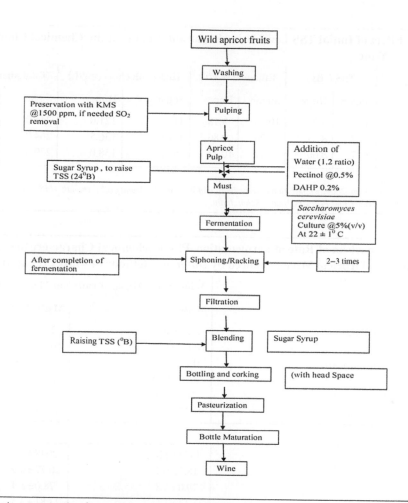

FIGURE 7.2.7

Flow sheet of wine preparation from wild apricot fruits. *DAHP*, diammonium hydrogen phosphate; *KMS*, potassium metabisulfite; *TSS*, total soluble solids.

Joshi, V.K., Sandhu, D.K., Thakur, N.S., 1999b. Fruit based alcoholic beverages. In: Joshi, V.K., Ashok, P. (Eds.), Biotechnology: Food Fermentation, vol. II. Educational Publishers and Distributors, New Delhi, pp. 647–744.

fermentability (Yang and Weigard, 1940). To enhance the flavor, about 10% of the pits may be broken down while crushing the cherries. But production of hydrogen cyanide from the hydrolysis of amygdalin, present in cherry pits, has been detected (Baumann and Gierschner, 1974; Misselhorn and Adam, 1976; Stadel Mann, 1976). Cherry wine does not require a long aging period. To make the wine sweet, sugar can be added prior to bottling. The bottled wine may be preserved either by germ-proof filtration or pasteurization. Yoo et al. (2010) investigated the protective activities of juice and wine produced from tart and sweet cherries. It was found that the processes for making both the juice and the wine reduced the total phenolics to about 30–40% and 42–60%, respectively, and the total

Table 7.2.19 Effect of Initial TSS Level Using Sugar and Honey on the Chemical Characteristics of Wild Apricot Wine

Treatment	TSS (°Bx)		Ethanol (% v/v)		Higher alcohols (mg/L)		Total phenols (mg/L)	
	Sugar	Honey	Sugar	Honey	Sugar	Honey	Sugar	Honey
Initial TSS 22 °Bx	8.33	8.27	10.23	9.18	113.0	121.0	245.0	238.0
Initial TSS 24 °Bx	8.57	8.37	11.70	9.86	131.5	136.5	264.0	255.0
Initial TSS 26 °Bx	8.77	8.60	12.17	10.38	153.0	158.0	279.0	270.0

TSS, *total soluble solids.*
Reproduced from Ghan Shyam, 2009. Preparation and Evaluation of Wild Apricot Mead and Vermouth. (MSc Thesis). Dr YS Parmar, University of Horticulture and Forestry, Nauni, Solan, India.

Table 7.2.20 Comparison of Rate of Fermentation, Physicochemical Characteristics, and Mineral Contents of Apricot Wine Produced by Different Methods of Must Preparation

Parameter	Wines From Musts of Different Methods (Mean ± SD)	
	Method (I)	Method (II)
Rate of fermentation (°Bx/24 h)	2.71 ± 0.08	2.75 ± 0.04
Ethanol (% v/v)	10.17 ± 0.3	10.22 ± 0.3
Titratable acidity (% malic acid)	0.72 ± 0.03	0.82 ± 0.02
pH	3.70 ± 0.05	3.65 ± 0.03
Total phenols (mg/L)	182 ± 3.0	112 ± 8.0
Volatile acidity (% acetic acid)	0.037 ± 0.001	0.039 ± 0.001
Minerals (mg/L)		
Potassium	1463 ± 124	2048 ± 120
Magnesium	18.7 ± 0.2	20.77 ± 0.9
Calcium	70.0 ± 5.8	78.0 ± 1.4
Sodium	11.13 ± 1.6	23.81 ± 4.5
Ferrous	1.69 ± 0.06	7.83 ± 0.04
Zinc	1.94 ± 0.02	1.39 ± 0.01
Copper	1.07 ± 0.01	1.08 ± 0.01
Manganese	0.81 ± 0.01	0.99 ± 0.01

Reproduced from Joshi, V.K., Sharma, S.K., 1994. Effect of method of must preparation and initial sugar levels on the quality of apricot wine. Research and Industry 39, 255–257.

anthocyanin capacity to about 60–77% in juice and 85–95% in wine. However, the radical-scavenging activity and the superoxide dismutase and catalase enzyme levels were higher in cherry wine than in cherry juice (Table 7.2.22).

The Oblačinska sour cherry is the most planted cultivar in Serbian commercial orchards (Fotirić Akšić et al., 2013). The flesh is red, medium-firm, juicy, quite sour, aromatic, and of high quality (Rakonjac et al., 2010). Cherries are considered an important source of polyphenols such as

Table 7.2.21 Effects of Wood Chips on the Physicochemical and Sensory Characteristics of Apricot Wine

Treatment	Tannins (mg/100 g)	Methanol (µL/L)	Max. Score	Appearance 2	Color 2	Aroma 4	Volatile Acidity 2	Total Acidity 2	Sweetness 1	Body 1	Flavor 2
Control	57.00	182.00		1.57	1.45	3.08	1.57	1.34	0.69	0.57	1.00
Alnus nitida	63.00	167.00		1.55	1.62	3.14	1.59	1.32	0.77	0.61	1.06
Populus ciliata	74.00	238.00		1.56	1.56	3.03	1.53	1.28	0.71	0.62	1.38
Celtis australis	104.00	232.00		1.57	1.61	3.019	1.55	1.47	0.60	0.68	1.44
Toona ciliata	118.00	188.00		1.53	1.51	3.03	1.55	1.18	0.52	0.63	1.22
Salix tetrasperma	139.00	220.00		1.54	1.55	3.04	1.54	1.10	0.59	0.53	1.18
Bombax ceiba	126.00	192.00		1.60	1.56	3.33	1.60	1.42	0.66	0.66	1.52
Albizia chinensis	122.00	198.00		1.64	1.48	3.14	1.61	1.33	0.65	0.56	1.50
Quercus leucotri-chophora	83.00	209.00		1.67	1.67	3.57	1.68	1.60	0.67	0.71	1.69
Critical difference (p = .05)	16.42	14.07		0.15	0.15	0.15	0.15	0.15	0.15	0.15	0.15

Reproduced from Joshi, V.K., Mahajan, B.V.C., Sharma, K.R., 1994. Treatment of fruit wines with wood chips—effect on physico-chemical and sensory qualities. Journal Tree Science 13, 27–36.

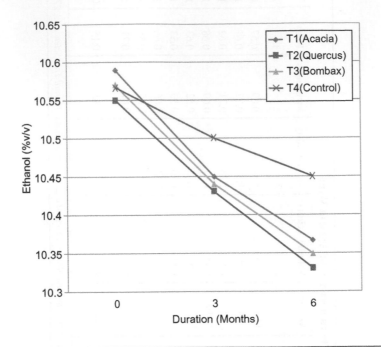

FIGURE 7.2.8

Effects of various wood chips on ethanol content during maturation of wild apricot wine.

Reproduced from Ghan Shyam, 2009. Preparation and Evaluation of Wild Apricot Mead and Vermouth (MSc Thesis). Dr YS Parmar, University of Horticulture and Forestry, Nauni, Solan, India.

anthocyanins, flavan-3-ols, and flavonols (Sun et al., 2012) and have a high phenolic acid content, especially hydroxycinnamic and hydroxybenzoic acid derivates (Gao and Mazza.,1995, Macheix et al., 1990). The phenolic compounds including anthocyanins are associated with antioxidant capacity and nutritional and therapeutic value of cherries (Blando et al., 2004; Bordonaba et al., 2010). This class of compounds has been shown to prevent cardiovascular disease and have strong antiinflammatory and anticarcinogenic activity (He and Giusti, 2010; Kim et al., 2005). In earlier studies, cyanidin 3-glucoside, cyanidin 3-rutinoside, cyanidin 3-glucosylrutinoside, cyanidin 3-sophoroside, pelargonidin 3-glucoside, peonidin 3-rutinoside, and cyanidin 3-arabinosylrutinoside have been found as the main anthocyanins in sour and sweet cherry fruits (Kim et al., 2005; Gao and Mazza, 1995; Giusti and Wrolstad, 2001; Chaovanalikit and Wrolstad, 2004).

Because of their sufficient acid level sour cherries are preferred to prepare wine, using a procedure similar to that used for grape wine (Sun et al., 2012). Two methods based on hyphenated techniques, which combine chromatographic and spectral methods, were used for target search and establishing polyphenolic profiles. UHPLC coupled with a hybrid mass spectrometer, which combines the linear trap quadrupole and the OrbiTrap mass analyzer, was used to identify phenolics. Quantification of phenolics was done with UHPLC coupled with a diode array detector and connected to a triple-quadrupole mass spectrometer (Panetlic et al., 2014). The overall antioxidant activities, TPC, total anthocyanin content (TAC), and mineral content of red wine (three different grape cultivars) and cherry wine, viz., one sample of cherry wine from east Serbia (SCW) and five samples

Table 7.2.22 Contents of Total Phenolics and Total Anthocyanins, Total Antioxidant Capacity, and DPPH Radical-Scavenging Activity of Cherry Fruits and Their Products

Cultivar		Total Phenolics[a]	Total Anthocyanins[b]	Total Antioxidants[c]	DPPH Radical-Scavenging Activity (%)[d]
Danube	Fruit	188 ± 8.1^b	70.1 ± 5.4^b	338 ± 14.0^{bc}	$60.711.0^c$
	Juice	56.7 ± 3.1^e	7.9 ± 2.8^e	202 ± 5.3^e	24.9 ± 0.7^e
	Wine	79.4 ± 4.4^d	29.6 ± 5.9^d	289 ± 2.9^d	53.0 ± 2.9^d
Balaton	Fruit	221 ± 2.0^a	93.5 ± 2.2^a	404 ± 8.1^a	75.2 ± 4.1^a
	Juice	86.8 ± 7.4^d	50.1 ± 4.5^c	311 ± 6.5^c	54.2 ± 5.6^d
	Wine	149 ± 1.5^c	63.4 ± 4.3^c	387 ± 4.7^b	69.0 ± 0.3^b

All mean values are from triplicate determinations. [a,b,c,d,e] Values in the same column with different superscript letters differ significantly ($p < .01$) by Duncan's multiple-range test. DPPH, 2,2-diphenylpicrylhydrazyl.
[a]Total phenolics are expressed in milligrams of gallic acid equivalents (GAE) per 100 g of fresh cherries or mg GAE per liter of liquor.
[b]Total anthocyanins are expressed in milligrams of cyanidin 3-glucoside equivalents (CGE) per 100 g of fresh cherries or mg CGE per liter of liquor.
[c]Total antioxidant activity is expressed as milligrams of vitamin C equivalent antioxidant capacity (VCEAC) per 100 g of fresh cherries or mg VCEAC per liter of liquor.
[d]DPPH radical-scavenging activities of each extract are 100 μg/mL or 100 μL of liquor.
Adapted from Yoo, K.M., Al-Farsi, M., Lee, H., Yoon, H., Lee, C.Y., 2010. Antiproliferative effects of cherry juice and wine in Chinese hamster lung fibroblast cells and their phenolic constituents and antioxidant activities. Food Chemistry 123, 734–740.

of grape wines [Vranac, east Serbia (W1); Cabernet Sauvignon, central Serbia (W2); Cabernet Sauvignon, east Serbia (W3); Cabernet Sauvignon, north Serbia (W4); and Frankovka, north Serbia (W5)], were compared. Cherry wine was made using the procedure used for red wine production, which includes mashing (disintegration), mashed sour cherry maceration, alcoholic fermentation, and removal of skins, seeds, and stalks from the wine (Panetlic et al., 2014). Temperature-controlled fermenters were used for fermentation (18–20°C) with appropriate enzymes facilitating separation of colored and aromatic matters during the maceration process. The TPC, TAC, and radical-scavenging activity of the wines are given in Table 7.2.23.

Table 7.2.23 Total Phenolic Contents, Total Anthocyanin Contents, and Radical-Scavenging Activity in Six Wine Samples

Sample	TPC (g GAE/L)	TAC (g mal-3-glu/L)	RSA (%)
SCW	1.94 ± 0.04	0.12 ± 0.01	34.56 ± 0.18
W1, Vranac, east Serbia	1.76 ± 0.02	0.08 ± 0.01	32.06 ± 0.09
W2, Sauvignon, central Serbia	2.28 ± 0.11	0.17 ± 0.02	46.41 ± 0.20
W3, Cabernet Sauvignon, east Serbia	2.50 ± 0.11	0.10 ± 0.01	48.60 ± 0.13
W4, Cabernet Sauvignon, north Serbia	1.19 ± 0.01	0.17 ± 0.02	21.00 ± 0.05
W5, Frankovka, north Serbia	1.69 ± 0.02	0.22 ± 0.02	31.21 ± 0.24

RSA, radical-scavenging activity; SCW, sample east Serbia wine (cherry); TAC, total anthocyanin content; TPC, total phenolic content.
Reproduced from Pantelić, M., Dabić, D, Matijašević, S., Davidović, S., Dojčinović, B., Milojković-Opsenica, D., Tešić, Ž., Natić, M., 2014. Chemical characterization of fruit wine made from oblačinska sour cherry. The Scientific World Journal, Article ID 454797, 9.

A total of 24 phenolic compounds were quantified using the available standards (Table 7.2.24), in addition to 22 phenolic compounds that were identified based on accurate mass search, and their mean expected retention times (tR), calculated mass, found mass, mean mass accuracy (ppm), and MS/MS fragments for each of the identified compounds and their distribution in wines are summarized in Table 7.2.24.

The most common element in all the samples was potassium (Table 7.2.25), and cherry wine had up to five times higher amounts of this mineral than grape wines. Higher contents of P, Ca, and Mg were found in cherry wine than in grape wine (Pantelić et al., 2014). Toxic elements (As, Cd, and Pb) were found only in small amounts in the tested wines (the allowable levels are 0.2 mg/kg for Pb and As and 0.01 mg/kg for Cd). In general, cherry wine polyphenolics in terms of nonanthocyanins and anthocyanins were shown to be distinctive compared to grape wines. Naringenin and apigenin were characteristic only for cherry wine, and seven anthocyanins were distinctive for cherry wine.

Table 7.2.24 Contents of Phenolics and *cis, Trans*-abscisic Acid (mg/kg) in Wine Samples

Compound	tR, min	SCW	W1	W2	W3	W4	W5
Gallic acid (A)	1.79	1.10	28.57	20.73	30.39	12.56	20.86
Protocatechuic acid (B)	3.73	23.89	7.93	6.11	4.52	5.38	5.19
(−)-Gallocatechin (B)	3.78	ND	ND	4.62	6.02	5.47	2.59
Aesculin (C)	4.78	0.35	0.54	0.37	0.40	0.36	0.45
(−)-Epigallocatechin (D)	4.89	1.01	0.87	0.87	1.07	0.91	0.82
p-Hydroxybenzoic acid (E)	5.10	6.65	4.30	7.45	5.34	0.37	10.45
Gentisic acid (E)	5.12	0.27	0.16	0.42	0.24	1.00	0.14
Chlorogenic acid (F)	5.22	3.57	0.60	0.58	0.58	ND	0.58
(+)-Catechin (G)	5.30	1.31	3.54	5.86	6.13	5.21	14.09
Caffeic acid (H)	5.52	13.88	1.56	2.35	1.25	1.42	1.60
(−)-Epicatechin (I)	5.65	3.92	ND	3.26	3.20	1.86	6.86
(−)-Gallocatechin gallate (J)	5.79	ND	ND	ND	ND	ND	2.79
Rutin (K)	6.08	0.23	0.23	0.23	0.24	0.23	ND
p-Coumaric acid (L)	6.20	23.42	3.62	7.77	3.19	5.86	2.98
Ellagic acid (L)	6.24	ND	2.16	2.73	1.99	2.04	1.79
Naringin (M)	6.46	ND	0.31	0.90	0.90	0.79	0.97
(−)-Epigallocatechin gallate (N)	6.79	ND	2.58	0.90	2.36	1.03	ND
Myricetin (O)	6.93	ND	0.21	0.22	0.28	0.25	0.29
cis,trans-Abscisic acid (P)	7.43	1.06	0.17	0.30	0.21	0.11	0.10
Quercetin (S)	7.60	ND	ND	ND	0.03	ND	ND
Resveratrol (S)	7.65	ND	ND	ND	8.83	ND	ND
Naringenin (T)	8.06	0.15	ND	ND	ND	ND	ND
Apigenin (U)	8.20	0.06	ND	ND	ND	ND	ND
Hesperetin (W)	9.51	ND	0.27	ND	0.39	0.42	ND

ND, *not detected; SCW, sample east Serbia wine (cherry); tR, mean expected retention time. Results are expressed as mg/L.*

Table 7.2.25 The Amounts of Minerals in Cherry Wine and Grape Wine Samples

Mineral	SCW	W1	W2	W3	W4	W5
Al (mg/kg)	0.200	0.070	0.071	0.421	0.055	0.090
As (μg/kg)	0.093	0.060	0.063	0.108	0.042	0.073
B (mg/kg)	2.760	1.364	2.101	1.919	1.690	2.082
Ca (g/kg)	0.084	0.035	0.023	0.041	0.020	0.022
Cd (μg/kg)	0.093	0.002	0.063	0.051	0.042	0.073
Co (μg/kg)	0.577	0.755	0.313	0.501	0.143	0.512
Cr (mg/kg)	0.016	0.006	0.004	0.005	0.002	0.004
Cu (mg/kg)	0.030	0.015	0.041	0.143	0.016	0.041
Fe (mg/kg)	2.192	0.575	0.221	2.415	0.642	1.114
K (g/kg)	1.373	0.246	0.521	0.499	0.283	0.475
Li (μg/kg)	0.678	1.916	2.726	2.794	2.162	1.215
Mg (g/kg)	0.072	0.052	0.052	0.065	0.041	0.054
Mn (mg/kg)	0.632	0.377	0.532	0.489	0.783	0.470
Mo (μg/kg)	0.093	0.060	0.063	0.108	0.042	0.073
Na (mg/kg)	1.650	2.228	2.814	4.227	0.893	4.507
Ni (mg/kg)	0.054	0.027	0.018	0.019	0.007	0.012
P (g/kg)	0.179	0.113	0.137	0.109	0.080	0.129
Pb (μg/kg)	4.404	3.757	0.063	18.490	0.380	28.868
S (g/kg)	0.089	0.093	0.144	0.169	0.066	0.118
Sb (μg/kg)	0.093	0.060	0.063	1.670	0.042	0.073
Se (mg/kg)	0.011	0.009	0.016	0.009	0.012	0.008
V (μg/kg)	6.863	2.515	0.323	0.947	0.498	2.237
Zn (mg/kg)	0.311	0.131	0.142	0.327	0.059	0.156

SCW, *sample east Serbia wine (cherry)*.
Adapted from Pantelić, M., Dabić, D, Matijašević, S., Davidović, S., Dojčinović, B., Milojković-Opsenica, D., Tešić, Ž., Natić, M., 2014. Chemical characterization of fruit wine made from oblačinska sour cherry. The Scientific World Journal Article ID 454797, 9.

5. SUMMARY AND CONCLUSIONS

A crisp overview of the research work carried out on stone fruit wines would reveal clearly that quite good work has been done on wine production from the stone fruits. Despite the efforts made and described here, there are a large number of research gaps, which need more elaboration in the future. Application of yeasts other than *Saccharomyces* in the production of wines as such or in combination could be a potential area of future research. The use of enzymes in juice extraction, flavor improvements, continuous fermentation using bioreactor technology, and their possible applications in wine production from stone fruits need serious consideration. The future could see the usage of biological deacidification using malolactic bacteria or deacidifying yeast like *S. pombe*. From grapes different types of wines, such as vermouth, sparkling wine, and sherry, are prepared, but can such types of wines be prepared from stone fruits also? The role of grape wine and its constituents in preventing coronary heart disease is well established. But there is a lack of information on the stone fruit wines, though there are reports on antioxidant activity with respect to health-related benefits of fruits and wines from stone fruits. This could certainly be a fruitful area of research, especially in those countries and regions where stone fruit cultivation is practiced commercially. In conclusion, the production evaluation of wine especially for healthful properties from stone fruits is certainly an exciting field for future research and development.

REFERENCES

Amandeep, G., Joshi, V.K., Rana, N., 2009. Evaluation of preservation methods of low alcoholic plum wine. Natural Product Radiance 8, 392–405.

Amerine, M.A., Berg, H.W., Kunkee, R.E., Qugh, C.S., Singleton, V.L., Webb, A.D., 1980. The Technology of Wine Making, fourth ed. AVI, Westport, CT.

Anonymous, 2001. Commission regulation (EC) no. 466/2001. Official Journal of the European Communities L77–L79.

Baumann, G., Gierschner, K., 1974. Studies on the technology of juice manufacture from sour cherries in relation to the storage of the product. Flussiges Obstetrics 41, 123.

Benk, E., Borgmann, R., Cutka, I., 1976. Quality control of sour cherry juices and beverages. Flussiges Obstetrics 43, 17–23.

Beuchat, L.R., 1973. Inhibitory Effects of benomyl on *Saccharomyces cerevisiae* during peach fermentation. American Journal of Enology and Viticulture 24, 110–115.

Bhutani, V.P., Joshi, V.K., Chopra, S.K., 1989. Mineral composition of experimental fruit wines. Journal of Food Science and Technology 26, 332.

Bhutani, V.P., Joshi, V.K., 1995. Plum. In: Salunkhe, D.K., Kadam, S.S. (Eds.), Handbook of Fruit Science and Technology, Cultivation, Storage and Processing. Marcel Dekker, New York, USA, pp. 203–241.

Blando, F., Gerardi, C., Nicoletti, I., 2004. Sour cherry (*Prunus cerasus* L) anthocyanins as ingredients for functional foods. Journal of Biomedicine and Biotechnology 5, 253–258.

Bordonaba, J.G., Chope, G.A., Terry, L.A., 2010. Maximising blackcurrant anthocyanins: temporal changes during ripening and storage in different genotypes. Journal of Berry Research 1, 73–80.

Cabaroglu, T., 2005. Methanol contents of Turkish varietal wines and effect of processing. Food Control 16, 177–181.

Chaovanalikit, A., Wrolstad, R.E., 2004. Total anthocyanins and total phenolics of fresh and processed cherries and their antioxidant properties. Journal of Food Science 69, FCT67–FCT72.

Cheng, G., Crisosto, C., 1995. Browning potential, phenolic composition, and polyphenoloxidase activity of buffer extracts of peach and nectarine skin tissue. Journal of American Society and Horticulture Science 120, 835–838.

Chung, J.H., Mok, C.Y., Park, Y.S., 2003. Changes of physicochemical properties during fermentation of peach wine and quality improvement by ultrafiltration. Journal of the Korean Society of Food Science and Nutrition 32, 506–512.

Craig, A., 1998. Comparison of the headspace volatiles of kiwi-fruit wine with those wines of *Vitis vinifera* variety Muller-Thurgau. American Journal of Enology and Viticulture 39, 321–324.

Davidovic, S., Veljović, M., Pantelić, M., et al., 2013. Physicochemical, antioxidant and sensory properties of peach wine made from redhaven cultivar. Journal of Agricultural and Chemistry 61, 1357–1363.

FAO, 2008. FAO Production Statistics. http://faostat.org/default/567.

Fleet, G.H., Heard, G.M., 1993. Yeast-growth during fermentation. In: Wine Microbiology and Biotechnology. Hartwood Academic Publishers, Chur, pp. 27–54.

Fotirić Akšić, M., Rakonjac, V., Nikolić, D., Zec, G., 2013. Reproductive biology traits affecting productivity of sour cherry. Pesquisa Agropecuária 48, 33–41.

Fowles, G., September 1989. The Complete Home Wine Maker. New Scientist, p. 38.

Gao, L., Mazza, G., 1995. Characterization, quantitation, and distribution of anthocyanins and colorless phenolics in sweet cherries. Journal of Agricultural and Food Chemistry 43, 343–346.

Ghan Shyam, 2009. Preparation and Evaluation of Wild Apricot Mead and Vermouth (MSc Thesis). Dr YS Parmar, University of Horticulture and Forestry, Nauni, Solan, India.

Giusti, M.M., Wrolstad, R.E., 2001. Anthocyanins. Characterization and measurement with UV-Visible spectroscopy. In: Wrolstad, R.E. (Ed.), Current Protocols in Food Analytical Chemistry. John Wiley & Sons, New York, USA, pp. 19–31.

He, J., Giusti, M., 2010. Anthocyanins: natural colorants with health-promoting properties. Annual Review of Food Science and Technology 1, 163–187.

Huang, H., Cheng, Z., Zhang, Z., Wang, Y., 2008. History of cultivation and trends in China. In: Layne, D.R., Bassi, D. (Eds.), The Peach: Botany, Production and Uses, first ed. CAB International, London, UK, pp. 4626–4631.

Jackson, R.S., 2004. Wine Science: Principles, Practice, Perception. Academic Press, London, UK.

Jacobson, J.L., 2006. Introduction to Wine Laboratory Practices and Procedures. Springer, New York, pp. 137–179.

Joshi, V.K., Sharma, Rakesh, Ghanshyam, A., 2011b. Stone fruit: wine and brandy. In: Hui, Y.H., Ozgul, E.E. (Eds.), Handbook of Food and Beverage Fermentation Technology. CRC Press, Florida, pp. 273–304.

Joshi, V.K., 1997. Fruit Wines. Dr. Y.S. Parmar University of Horticulture and Foresty, Nauni, Solan (HP), p. 155.

Joshi, V.K., Attri, B.L., Gupta, J.K., Chopra, S.K., 1990a. Comparative fermentation behaviour, physicochemical characteristics of fruit honey-wines. Indian Journal of Horticulture 47, 49–54.

Joshi, V.K., Bhutani, V.P., 1990. Evaluation of plum cultivars for wine preparation. In: XXIII International Horticulture Congress Held at Italy Abstract 3336.

Joshi, V.K., Bhutani, V.P., 1995. Peach. In: Salunkhe, D.K., Kadam, S.S. (Eds.), Hand Book of Fruit Science and Technology. Marcel & Dekker, USA, p. 230.

Joshi, V.K., Bhutani, V.P., Sharma, R.C., 1990b. Effect of dilution and addition of nitrogen source on chemical, mineral and sensory qualities of wild apricot wine. American Journal of Enology and Viticulture 41, 229–231.

Joshi, V.K., Chauhan, S.K., Bhushan, S., 2000. Technology of fruit based alcoholic beverages. In: Verma, L.R., Joshi, V.K. (Eds.), Postharvest Technology of Fruits and Vegetables. Indus Publishing, New Delhi, p. 1019.

Joshi, V.K., Devi, P.M., 2009. Resveratrol: importance, role, content in wine and factors influencing its production. Proceedings of National and Academic Sciences (India) 79, 76–79.

Joshi, V.K., Mahajan, B.V.C., Sharma, K.R., 1994. Treatment of fruit wines with wood chips—effect on physicochemical and sensory qualities. Journal Tree Science 13, 27–36.

Joshi, V.K., Bhutani, V.P., Thakur, N.K., 1999a. Composition and nutrition of fermented products. In: Joshi, V.K., Pandey, A. (Eds.), Biotechnology: Food Fermentation, vol. I. Educational Publishers and Distributors, New Delhi, pp. 259–320.

Joshi, V.K., Sandhu, D.K., Thakur, N.S., 1999b. Fruit based alcoholic beverages. In: Joshi, V.K., Ashok, P. (Eds.), Biotechnology: Food Fermentation, vol. II. Educational Publishers and Distributors, New Delhi, pp. 647–744.

Joshi, V.K., Shah, P.K., 1998. Effect of wood treatment on chemical and sensory quality of peach wine during aging. Acta Alimentaria Budapest 27, 307–318.

Joshi, V.K., Shah, P.K., Kumar, K., 2005. Evaluation of peach cultivars for wine preparation. Journal of Food Science and Technology 42, 83–89.

Joshi, V.K., Chauhan, S.K., Lal, B.B., 1991a. Extraction of juices from plum, peach and apricot by the pectolytic enzyme treatment. Journal of Food Science and Technology 28 (1), 64–65.

Joshi, V.K., Attri, B.L., Mahajan, B.V.C., 1991b. Studies on the preparation and evaluation of vermouth from plum. Journal of Food Science and Technology 28 (3), 138–141.

Joshi, V.K., Sharma, P.C., Attri, B.L., 1991c. A note on deacidification activity of *Schizosaccharomyces pombe* in plum musts of variable composition. Journal of Applied Microbiology 70, 385–390.

Joshi, V.K., Sharma, S., Devi, M.P., Bhardwaj, J.C., 2009a. Effect of initial sugar concentration on the physicochemical and sensory qualities of plum wine. Journal of North East Foods 8, 1–7.

Joshi, V.K., Sharma, S.K., 1994. Effect of method of must preparation and initial sugar levels on the quality of apricot wine. Research and Industry 39, 255–257.

Joshi, V.K., Sharma, S.K., 1995. Comparative fermentation behaviour, physico-chemical and sensory characteristics of plum wine as effected by the type of preservatives. Chemie, Mikrobiologie, Technologie der Lebensmittel 17, 65–73.

Joshi, V.K., Sharma, S.K., Thakur, N.S., 1995. Technology and quality of sparkling wine with special reference to plum – an overview. Indian Food Packer 49, 49–66.

Joshi, V.K., Sharma, Somesh, Bhushan, S., Devender, A., 2004. Fruit based alcoholic beverages. In: Pandey, A. (Ed.), Concise Encyclopedia of Bioresource Technology. Haworth Inc., New York, p. 335.

Joshi, V.K., Sharma, S., Devi, M.P., 2009b. Influence of different yeast strains on fermentation behaviour, physico-chemical and sensory qualities of plum wine. Natural Product Radiance 445–451.

Joshi, V.K., Thakur, N.S., Bhat, A., Garg, C., 2011a. Wine and brandy: a perspective. In: Hand Book of Enology-Principles, Practices and Recent Innovations. Asiatech Publishers, New Delhi, pp. 1–45.

Joshi, V.K., Kumar, V., 2011. Importance, nutritive value, role, present status and future strategies in fruit wines in India. In: Panesar, P.S., et al. (Ed.), Bio-Processing of Foods. Asia Tech Publisher, New Delhi, pp. 39–62.

Joshi, V.K., Sharma, S., Rana, V.S., 2012. Wine and brandy. In: Joshi, V.K., Singh, R.S. (Eds.), Food Biotechnology: Principles and Practices. IK International Publishing House, New Delhi, pp. 471–494.

Joshi, V.K., Joshi, D., 2014. Effect of addition of Sodium Benzoate on the fermentation behaviour, physico-chemical and sensory qualities of plum wine. International Journal of Food and Fermentation Technology 4, 137–142.

Kim, D., Heo, J.H., Kim, Y.J., Yang, H.S., Lee, C.Y., 2005. Sweet and sour cherry phenolics and their protective effects on neuronal cells. Journal of Agricultural and Food Chemistry 53, 9921–9927.

Lavelli, V., Pompei, C., Casadei, M., 2009. Quality of nectarine and peach nectars as affected by lye-peeling and storage. Food Chemistry 115, 1291–1298.

Macheix, J.J., Fleuriet, A., Billot, J., 1990. Phenolic acids and coumarins. In: Macheix, J.J., Fleuriet, A. (Eds.), Fruit Phenolics. CRC Press, Boca Raton, FL, USA, pp. 17–39.

Miljic, U.D., Puskas, S.V., 2014. Influence of fermentation conditions on production of plum (*Prunus domestica* L.) wine: a response surface methodology approach. Hemijska Industrija 68, 199–206.

Misselhorn, K., Adam, R., 1976. On the cyanides contents in stone-fruit products (Transl). Braunt Wein Wirt and Chaft 116, 45–50.

Murcia, J.M.A., Martfnez-Tome, A.M., 2001. Evaluation of the antioxidant properties of Mediterranean and tropical fruits compared with common food additives. Journal of Food Protection 64, 2037–2046.

Pantelić, M., Dabić, D., Matijašević, S., Davidović, S., Dojčinović, B., Milojković-Opsenica, D., Tešić, Ž., Natić, M., 2014. Chemical characterization of fruit wine made from oblačinska sour cherry. The Scientific World Journal 9 Article ID 454797.

Quinn, K.M., Singleton, V.L., 1985. Isolation and identification of ellagitannin from white oak wood and an estimate of their roles in wines. American Journal of Enology and Viticulture 36, 148–155.

Radler, F., Schiitz, H., 1982. Glycerol production of various strains of Saccharomyces. American Journal of Enology and Viticulture 33, 36–40.

Rakonjac, V., Fotirić Akšić, M., Nikolić, D., Milatović, D., Čolić, S., 2010. Morphological characterization of "Oblačinska" sour cherry by multivariate analysis. Scientia Horticulturae 125, 679–684.

Rankine, B.C., Bridson, D.A., 1971. Glycerol in Australian wines and factors influencing its formation. American Journal of Enology and Viticulture 22, 6–12.

Remize, F., Sablayrolles, J.M., Dequin, S., 2000. Re-assessment of the influence of yeast strain and environmental factors on glycerol production in wine. Journal of Applied Microbiology 88, 371–378.

Ribereau-Gayon, P., Glories, Y., Maujean, A., Dubour-dieu, D., 1999. Handbook of Enology. The Chemistry of Wine Stabilization and Treatments, vol. 2. John Willey & Sons, New York.

Rop, O., Jurikova, T., Mlcek, J., Kramarova, D., Sengee, Z., 2009. Antioxidant activity and selected nutritional values of plums (*Prunus domestica* L.) typical of the White Car-patian Mountains. Science and Horticulture 122, 545–549.

Sandhu, D.K., Joshi, V.K., 1995. Technology, quality and scope of fruit wines with special reference to apple. Indian Food Industry 14, 24–34.

Satora, P., Tuszynski, T., 2005. Bio diversity of yeasts during plum Wegierka Zwykla spontaneous fermentation. Food Technology and Biotechnology 43, 277–282.

Schander, H., Koch, J., 1957. Die Fruchtwein Be Reitung. Eugen Ulmer, Stuttgart.

Stadel Mann, W., 1976. Content of hydrocyanic acid in stone fruit juices. Flussiges Obstetrics 43, 45–47.

Sun, S.Y., Jiang, W.G., Zhao, Y.P., 2012. Comparison of aromatic and phenolic compounds in cherry wines with different cherry cultivars by HS-SPME-GC-MS and HPLC. International Journal of Food Science and Technology 47, 100–106.

Vyas, K.K., Joshi, V.K., 1982. Plum wine making: standardization of a methodology. Indian Food Packer 36, 80–86.

Vyas, K.K., Joshi, V.K., 1988. Deacidification activity of *Schizosaccharomyces pombe* in plum musts. Journal of Food Science and Technology 25, 306–307.

Vyas, K.K., Sharma, R.C., Joshi, V.K., 1989. Application of osmotic technique in plum wine fermentation—Effect on physico-chemical and sensory qualities. Journal of Food Science and Technology 26, 126–128.

Vyas, S.R., Chakravorty, S.R., 1971. Wine Making at Home. Haryana Agricultural University Bulletin, Hissar.

Wang, M.L., Wang, J.T., Choong, Y.M., 2004. A rapid and accurate method for determination of methanol in alcoholic beverage by direct injection capillary gas chromatography. Journal of Food Composition and Analysis 17, 187–196.

Westwood, M.N., 1978. Temperate Zone Pomology. W.H. Freeman, San Francisco, pp. 233–235.

Wills, R.B.H., Scriven, F.M., Greenfield, H., 1983. Nutrient composition of stone fruit (*Prunus* spp) cultivars: apricot, cherry, nectarine, peach and plum. Journal of the Science of Food and Agriculture 34, 1383–1388.

Yang, H.Y., Weigard, E.H., 1940. Production of fruit wines in the Pacific North West. Fruit Products Journal 29, 8–12.

Yoo, K.M., Al-Farsi, M., Lee, H., Yoon, H., Lee, C.Y., 2010. Antiproliferative effects of cherry juice and wine in Chinese hamster lung fibroblast cells and their phenolic constituents and antioxidant activities. Food Chemistry 123, 734–740.

BERRY AND OTHER FRUIT WINES

7.3

V.K. Joshi[1], S. Sharma[2], A.D. Thakur[2]

[1]*Dr. Y.S. Parmar University of Horticulture and Forestry, Nauni, Solan, HP, India;* [2]*Shoolini University, Solan, HP, India*

1. INTRODUCTION

Fruits are employed to produce various types of wines and their distillates, called brandies. Wines have been a part of the diet of humans ever since they settled in the Tigris–Euphrates river basin (Amerine et al., 1980; Joshi, 1997). Wines are known to have been prepared by the Assyrians by 3500 BC and have been used as a therapeutic agent (Joshi et al., 2011a). The Rigveda also mentions the medicinal power of wine (Vays and Chakravarty, 1977). Moderate consumption of low levels of alcohol has been associated with lowered mortality from coronary heart disease (Delin and Lee, 1991; Stockley, 2011). Wines also serve as an important adjunct to the human diet by increasing satisfaction, providing the relaxation necessary for proper digestion and absorption of food, besides promoting glucose tolerance factor, which stimulates production of insulin to prevent diabetes (Gasteineau et al., 1979; Joshi et al., 1999a). Production and consumption of fermented beverages like wines is an ancient practice (Amerine et al., 1980), but that of fruit-based distilled alcoholic beverages is a later development. Wines are made from complete or partial alcoholic fermentation using inoculated and natural microflora of grapes or any other fruit, like apple, plum, peach, pear, berries, cherry, currant, apricot, sea buckthorn, blackberry, and pumpkin (Amerine et al., 1980; Joshi et al., 1999b; Joshi and Bhutani, 1990a,b; Sandhu and Joshi, 1995; Joshi and Sandhu, 2000).

Among the colorful fruits, berries such as blackberry (*Rubus* sp.), black raspberry (*Rubus occidentalis*), blueberry (*Vaccinium corymbosum*), cranberry (*Vaccinium macrocarpon*), red raspberry (*Rubus idaeus*), strawberry (*Fragaria × ananassa*), etc., are popularly consumed by humans in fresh (Plate 7.3.1) and processed forms such as beverages, yogurts, jellies, and jams (Navindra et al., 2006). Further, various fruit species are employed for the production of wine and among them berries occupy an important place.

Botanically, a berry is a fleshy fruit produced from a single flower containing one ovary. In everyday English, "berry" is a term for any small edible fruit. These "berries" are usually juicy, round, brightly colored, and sweet or sour, and do not have a stone or pit, although many seeds may be present. Grapes and avocados are two common examples. The berry is the most common type of fleshy fruit, in which the entire ovary wall ripens into an edible pericarp. They may have one or more carpels. The seeds are usually embedded in the fleshy interior of the ovary, but there are some nonfleshy exceptions, such as peppers, that have air rather than pulp around their seeds. In addition, berry extracts are widely consumed in botanical dietary supplement forms for their potential human health benefits, as berries have medicinal properties. Studies have revealed that blackberry, black raspberry, blueberry, cranberry, red raspberry, and strawberry extracts inhibit the growth and stimulate the

PLATE 7.3.1

Various types of berry fruits and developmental stages of the strawberry.

http://grocerytraining.net; Copyright: Thilo Fischer/LMU and Wilfried Schwab/TUM); Modified from www.wzw.tum.de.

apoptosis of human cancer cells in vitro (Navindra et al., 2006). Black raspberry and strawberry extracts showed the most significant proapoptotic effects against the COX-2-expressing colon cancer cell line HT-29 (Seeram et al., 2006).

Compared to the quantity of grape wine produced and consumed in the world, the amount of wine produced from nongrape fruits is insignificant as stated earlier (Amerine et al., 1980; Goswell and Kunkee, 1977), except for cider and perry, which are produced and consumed in significant amounts throughout the world (Joshi et al., 2011b). General principles involved in winemaking have been discussed sufficiently in various chapters of this text, so no attempt will be made to describe them here as is the case with cider and perry, fruit brandy, other wines including brandy and specialty wines which discussed in separate chapters and readers may consult the same. Wines from berries other than grapes are briefly discussed here in this contribution along with their production technologies.

2. GENERAL ASPECTS: PRODUCTION OF BERRY WINE

2.1 PROBLEMS OF BERRY WINES

Winemaking has developed from a haphazard, ill understood, and risky process into a well-defined scientific discipline (Amerine et al., 1980). Basic techniques for the production of fruit wines are essentially the same, involving the routine alcoholic fermentation of the juice or pulp, but modifications with respect to the physicochemical characteristics, depending upon the type of wine to be prepared and the fruit used, are made. The major differences in the techniques of production of these wines, however, are attributable to the difficulty in extracting the sugar from the pulp of some of the fruits (Amerine et al., 1980). The juice/pulp obtained from most of the nongrape fruits is lacking in the requisite sugar content or has poor fermentability. The higher acidity in some of the fruits makes it all the more difficult to prepare wine of acceptable quality. Many tropical and subtropical fruits have low acidity and it is difficult to convert them into attractive and stable wines (Faparusi, 1973; Maldonadd et al., 1975; Nagadowithana and Steinkraus, 1976). The fermentation of such fruits is very slow or may even stop before completion because of the lack of certain nitrogenous compounds or other yeast growth factors

(Lodge, 1981). In the case of blackberries, harvesting maturity is one of the important factors, because the fruits have higher pigment concentration toward the end of the season, leading to pigment deposits in the bottle. To avoid this, blackberries should be picked before full maturity. Among the various products that can be prepared from sea buckthorn, wine is one that is difficult to prepare because of the high acidity of the sea buckthorn pulp (Joshi et al., 2011a).

2.2 RAW MATERIALS

For the production of berry-based alcoholic beverages, the raw material would be the fruit, a sweetening agent, fruit concentrate, sugar, acid, nitrogen source, clarifying enzyme, filter aid, spices and herbs or their extracts (Amerine et al., 1980; Joshi, 1997; Shah and Joshi, 1999). Almost all the non-grape fruits, their juices, or their extracts are low in sugar, the most essential component for alcoholic fermentation, and have to be ameliorated to the sugar level needed for producing a table wine. Even when the sugar level is satisfactory, the high acidity demands dilution, consequently requiring addition of more sugar to the must (Amerine et al., 1980; Bhutani and Joshi, 1995; Joshi and Sharma, 1994). The pectin-splitting enzyme pectinesterase is mostly used in the clarification of wines. The nitrogen sources used are several, but diammonium hydrogen phosphate (DAHP) has many advantages (i.e., less shifting in pH of the must and low cost), so it is used most frequently (Joshi et al., 2011b).

2.3 COMPOSITION AND MATURITY OF FRUITS

The composition of berries (Table 7.3.1) and their maturity, like other fruit, are dependent upon the cultivation practices. The primary environmental factor influencing the quality of the fruit is temperature (maximum, minimum, mean). Secondary factors are rainfall, sunshine versus cloudiness, humidity, wind, soil, and the combination of these. In addition, various cultivation practices affect the composition of fruits. The type and amount of fertilization affects maturity, especially sugar content. For more details, see Chapter 1 and a number of publications on the cultivation of various fruits (Bhutani and Joshi, 1995; Joshi and Bhutani, 1995; Rose, 1992; Kumar et al., 2011).

Unlike grapes, which are a good cultural medium for the growth of wine yeast, needing virtually no supplementation, the other fruit juices/pulps do need supplementation/fortification for successful alcoholic fermentation (Jackson, 2004). Alcoholic fermentation is influenced by a number of factors that affect yeast growth. In addition to the fermentation conditions, the composition and metabolites produced in fermentation exert a profound influence on the yeast and, consequently, the alcoholic fermentation (Amerine et al., 1980; Joshi et al., 2011b).

3. METHODS OF PREPARATION OF TABLE WINE
3.1 STRAWBERRY WINE
3.1.1 Method of Wine Preparation

Strawberry wine of good quality has the appealing color of premium rosé wine. But the attractive color is often short-lived. When frozen berries are employed to make the wine, they are first thawed and the juice is extracted. The juice is ameliorated to 22 °Bx by the addition of cane sugar added in stages, e.g., 25% of the total sugar is added to prepare the must and the rest during fermentation.

Table 7.3.1 Comparative Physicochemical and Nutritional Characteristics of Berries.

Fruits	Moisture (%)	Carbohydrates (%)	Lipids (%)	Protein (%)	Ascorbic Acid (mg/100 g)	Thiamine	Riboflavin	Niacin	Ca (mg/100 g)	Mg	Na	K	P	Fe
Strawberry	89.9	8.4	0.5	0.7	59	0.03	0.07	0.0006	21	12	1.5	161	21	1
Cherry	83.7	–	–	0.7	13–56	–	–	–	–	–	–	–	–	–
Raspberry	–	4.1	–	1.0	38	0.01	0.07	0.0007	35	25	0.7	220	37	1.1
Sea buckthorn	–	6.3	–	0.7	165	0.18	0.07	0.0004	42	30	3.5	133	8.6	0.4
Bilberry	–	6,4	–	0.5	15	0.04	0.07	0.0006	19	9	0.3	110	20	0.6
Cranberry	–	3.5	–	0.4	20	0.05	0.07	0.0002	13	8	0.9	25	10	0.7

Compiled from Benk, E., Borgmann, R., Cutka, I., 1976. Quality control of sour cherry juices and beverages. Flussiges Obstetrics 43, 17–23; Rose, R., 1992. Orange Based Sparkling Alcoholic Beverage. French Patent Application, FR 2657, 878 AL (FSTA 24(2):2H, 88, 1992); Joshi, V.K., Sharma, S., Bhushan, S., 2005. Effect of method of preparation and cultivar on the quality of strawberry wine. Acta Alimentaria 34 (4), 339–355; Baumann, G., Gierschner, K., 1974. Studies on the technology of juice manufacture from sour cherries in relation to the storage of the product. Flussiges Obstetrics 41, 123.

However, addition of sugar after fermentation dilutes the alcohol level. The must is mixed with 1% ammonium phosphate and the fermentation is initiated by the addition of 1% of yeast culture at a temperature of 16°C. The juice is transferred to a jar and the fermentation is continued until 0.1–0.2% reducing sugar contents are obtained. After the completion of fermentation, the wine is racked, bottled, and stored in the dark. The composition and maturity of the fruit and contamination with mold affect the quality of the wine. But overripe fruits with higher anthocyanins and total phenolics give wines with better color than fully ripe fruits. Fruit of the variety Totem have higher anthocyanins, low polyphenol oxidase activity, and lower total phenolics than the Benton variety. Mold contamination increases juice viscosity, reduces the fermentation rate, and accelerates color degradation (Picinelli et al., 2000). Ascorbic acid accelerates the destruction of anthocyanin pigments and also contributes to browning (Pilando et al., 1985). Treatment with enzymes (mainly pectinases) inhibits polymerization and increases color extraction and color density in strawberry wine (Flores and Heatherbeth, 1984; Luthi, 1953). Juice yield, which is affected by the addition of pectic enzyme, is an important cost consideration for wine production, as the price of berry juice is high. The addition of sodium metabisulfite increases ethanol production along with controlling natural microflora (Samarajeewa et al., 1985).

Strawberry fruits of three cultivars, viz., Camarosa, Chandler, and Douglas, were prepared for wine by four different methods, viz., control, thermovinification, fermented on the skin, and carbonic maceration. Among the methods employed (thermovinification, fermented on the skin, and carbonic maceration), the must from the fruit fermented on the skin gave the highest rate of fermentation and ethanol content. Fig. 7.3.1 shows the complete process for making strawberry wine using the thermovinification method. Thermovinified wines have many desirable characteristics, like more total phenols, esters, and color with comparable amounts of higher alcohols, volatile acidity, ethyl alcohol, sugars, and anthocyanins. Thermovinification imparts many desirable qualities to the wine compared to the control or other treatments tried for preparation of wine. Thermovinified wines irrespective of cultivar were scored highest with respect to attractive color, body, flavor, and various other sensory quality attributes. Wine from the Camarosa cultivar registered many desirable characteristics, like more esters, optimum acidity, and more red color units, with comparable other characteristics like alcohol and phenols, whereas that of the Chandler cultivar had higher amounts of ethyl alcohol, phenols, and anthocyanins than other cultivars (Table 7.3.2). The Camarosa wine was rated the best (Joshi et al., 2005). Thermovinification yielded higher content of phenols, followed by wines prepared by fermentation on skin, carbonic maceration, and control. Total phenols (mg/L) in various strawberry wines are shown in Table 7.3.3 and the types of phenolic compounds are listed in Table 7.3.4.

3.1.2 Maturation of Wine

Maturation of strawberry wine for 9 months results in an increase in reducing sugars, total esters, and volatile acidity and a decrease in total soluble solids, total sugars, titratable acidity, color, anthocyanins, and total phenols. Most of these changes are similar to those observed in red grape wine and increase the palatability of strawberry wine and, thus, are desirable. In general, the maturation of wine also improves the sensory quality of strawberry wine (Sharma and Joshi, 2003). The flavor profiling of wines from three cultivars (Camarosa, Chandler, and Douglas) prepared by different methods (control, thermovinification, carbonic maceration, and fermented on the skin) by descriptive analysis was also carried out. Of the 14 descriptors attempted, six had very high intensity (strawberry-like, alcoholic, phenolic, higher alcoholic, astringency, and bitterness) and the remaining (vegetative, yeasty, and earthy) were found to have lower intensity (Fig. 7.3.2). Among the cultivars, Camarosa and Chandler

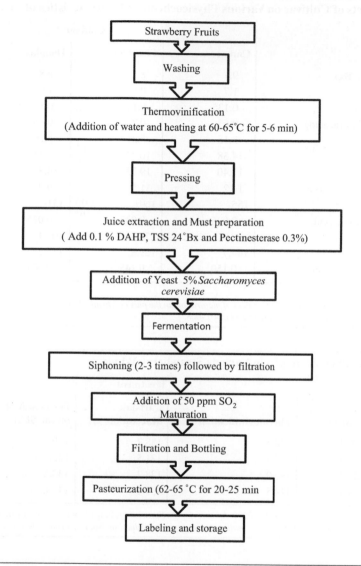

FIGURE 7.3.1

Flow sheet for preparation of strawberry wine by the thermovinification method. *DAHP*, diammonium hydrogen phosphate; *TSS*, total soluble solids.

Adapted from Sharma, S. Joshi, V.K. and Abrol, G., 2009. An overview on Strawberry [Fragaria × ananassa (Weston) Duchesne ex Rozier] wine production technology, composition, maturation and quality evaluation. Natural Product Radiance, 8 (4), 356–365.

had significantly higher flavor intensities than Douglas for alcoholic, vinegary, astringency, sour, phenolic, and strawberry-like but were lower than Douglas for fruity and higher alcoholic attribute. The principal component analysis of the flavor profiling data of these wines revealed a weaker influence of method of preparation compared to that exerted by the cultivar. The flavor profiling characterized the strawberry wines of the different treatments successfully (Sharma and Joshi, 2004). The strawberry wine was also found to be rich in antioxidants (Joshi et al., 2009).

Table 7.3.2 Effects of Cultivar on Various Physicochemical Characteristics of Strawberry Wine

Characteristic	Cultivar			
	Camarosa	Chandler	Douglas	CD > 0.05
Total soluble solids (°Bx)	9.7	8.1	8.8	0.053
Total sugars (%)	1.7	0.6	1.0	0.039
Reducing sugars (%)	0.135	0.124	0.128	0.003
Titratable acidity (% citric acid)	0.65	0.73	0.65	0.030
pH	3.18	3.21	3.26	0.016
Color (red)	14.38	10.72	9.45	0.96
Color (yellow)	17.40	19.33	14.85	0.769
Alcohol (% v/v)	11.2	11.5	9.2	0.366
Higher alcohol (mg/L)	155	169	151	0.672
Volatile acidity (% acetic acid)	0.026	0.032	0.025	0.0006
Esters (mg/L)	90.9	78.3	102.4	0.675
Phenols (mg/L)	144.7	129.8	135.2	3.41
Anthocyanins (OD/mL of wine)	0.150	0.145	0.104	0.022

CD, *critical difference.*
Reproduced from Sharma, S., Joshi, V.K., 2003. Effect of maturation an the phisico-chemical and sensory quality of strawberry wine. Journal of Scientific & Industrial Research 62 (4), 601–608.

Table 7.3.3 Total Phenols (mg/L) in Various Strawberry Wines

Cultivar	Treatment				
	Control	Thermovinification	Carbonic Maceration	Fermentation on the Skin	Mean
Camarosa	139.7	150.2	138.7	150.3	144.7
Chandler	117.0	137.5	131.5	133.2	129.8
Douglas	123.7	146.0	128.2	143.2	135.2
Mean	126.8	144.5	132.8	142.2	

Adapted from Sharma, S., Joshi, V.K., Abrol, G., 2009. An overview on Strawberry [Fragaria × ananassa (Weston) Duchesne ex Rozier] wine production technology, composition, maturation and quality evaluation. Natural Product Radiance, 8 (4), 356–365.

Table 7.3.4 Identification of Phenolic Compounds in Strawberry Wine

Type of Phenol	RF	Control	Thermovinified Wines	Fermented on the Skin	Carbonic Maceration
Catechin	0.48	+1–3	+1–3	+1–3	+1–3
Epicatechin	0.54	+1–3	+1–3	+1–3	+1–3
Quercetin	0.70	+1–3	–	+1–2	+1–3
Ellagic acid	0.68	+1–3	–	+1–3	+1–3

+, present; –, absent. RF, rate of ffermentation.
Reproduced from Sharma, S. Joshi, V.K. and Abrol, G., 2009. An overview on Strawberry [Fragaria × ananassa (Weston) Duchesne ex Rozier] wine production technology, composition, maturation and quality evaluation. Natural Product Radiance, 8 (4), 356–365.

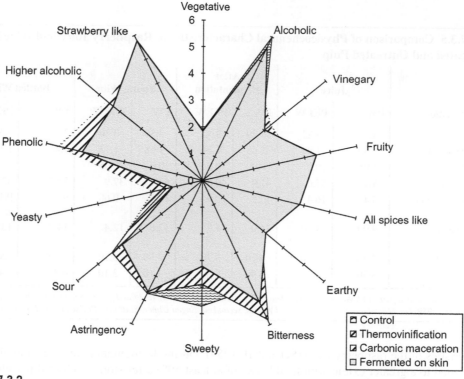

FIGURE 7.3.2

Effects of treatment on the flavor profiling of strawberry wine.

Reproduced from Sharma, S., Joshi, V.K., 2004. Flavour profiling of strawberry wines by quantitative descriptive analysis technique.
Journal of Food Science and Technology, 41 (1), 22–26.

3.2 RED RASPBERRY WINE

Raspberry (*R. idaeus*) is prone to spoilage if not cooled promptly at 0°C but can be preserved for 2–3 days. A significant quantity of juice is used for preparation of the wine. But color loss and deterioration as well as haze and sediment formation during storage of red raspberry wine are the problems encountered by commercial producers. Anthocyanins are mostly responsible for these effects because of their degradation and polymerization.

Red raspberries (Meeker variety) grown and picked near Salemn and commercially block frozen were used for the preparation of wine. The berries were partially thawed at 27°C and were ground through a hammer mill. Addition of 25 ppm SO_2 was made. Lot 1 was fermented at 25°C by adding Champagne yeast (1 g/gal) until about 20 °Bx. Lot 2 was depectinized by addition of 100 ppm liquid pectic enzyme, and in the third set the pulp was pasteurized by high temperature for a short time (85–90°C for 1 min) followed by cooling to 27°C and depectinization similar to the earlier treatment. All the lots had total soluble solids (TSS) 22 °Bx and were fermented at 25°C until 0–0.2 °Bx was reached. The wines after filtration at low temperature were cold stabilized with the addition of 25 ppm SO_2 and 180 ppm potassium sorbate as a preservative, and their sugar contents were raised by 3%. The wines, bottled with corks, were stored in the dark for 6 months (Table 7.3.13). Fermentation of pulp, depectinized juice, and pasteurized juice affected the composition and other physical

Table 7.3.5 Comparison of Physicochemical Characteristics of Raspberry Juice and Wine From the Treated and Untreated Pulp

Characteristic	Juice		After Fermentation		Young Wine		Bottled Wine	
	PW	PEFW	PW	PEFW	PW	PEFW	PW	PEFW
pH	3.29	3.22	3.29	3.32	3.30	3.30	3.34	3.34
Titratable acidity	1.69	1.84	1.68	1.74	1.68	1.73	1.46	1.36
TSS (°Bx)	1.95	10.0	0.00	0.0	0.0	0.05	-	-
Total monomeric	58.1	57.6	41.9	40.0	41.1	37.9	37.5	32.4
Anthocyanin color density	15.4	16.4	13.5	12.0	11.8	9.60	9.5	9.10
Percentage of polymeric color	9.90	4.50	11.0	11.9	12.8	12.4	14.0	13.0
Browning index	4.40	4.70	4.10	3.70	3.60	3.10	3.10	2.90
% Haze	8.50	3.60	4.90	3.50	3.10	3.10	4.80	3.40

PW, *wine from untreated pulp;* PEFW, *wine from pasteurized, pectinesterase-treated, and fined pulp;* TSS, *total soluble solids. Schanderl, H., 1959. Die Mikrobiologie des mestes and Weines, second ed. Eugen Ulmer, Stuttgart (Revised by H.H. Dittrich, 1977).*

characteristics of the raspberry wine (Schanderl, 1959). During fermentation, anthocyanin pigments are degraded to a greater extent with total losses of at least 50% after storage. Cyanidin 3-glucoside was the most unstable anthocyanin, disappearing completely during fermentation, whereas cyanidin 3-sophoroside (the major anthocyanin) was the most stable pigment. It was concluded that pasteurized depectinized wine that had undergone fining had the most stable color and the best appearance after storage (Table 7.3.5).

3.3 SEA BUCKTHORN WINE

Sea buckthorn (*Hippophae rhamnoides* L.) is a hardy, deciduous shrub belonging to the family Elaeagnaceae. It bears yellow to orange berries, which have been used for centuries in Europe and Asia. The natural habitat of sea buckthorn extends widely throughout China, Mongolia, Russia, Finland, Sweden, and Norway. It has attracted considerable attention in North America mainly for its nutritional and medicinal value. The fruit is rich in carbohydrates, protein, organic acids, amino acids, and vitamins and is among the most nutritious and vitamin C-rich fruits known. The fruit, including the seeds, contains large amounts of essential oils. Efforts were made to prepare wine from sea buckthorn (Fig. 7.3.3) by diluting the pulp with water in ratios of 1:5, 1:6, 1:7, and 1:8 in one set and by deacidification of the pulp with sodium bicarbonate at various concentrations (0.6%, 0.8%, 1.0%, and 1.2%) in the second set to reduce the acidity. The pulp was ameliorated with sugar (24 °Bx), 100 ppm SO_2, and 0.5% pectinase enzyme and with or without DAHP (0.1%) and fermented with pure wine yeast culture of *Saccharomyces cerevisiae* var. *ellipsoideus* (5% v/v) at 22 ± 1°C (Joshi et al., 2011c). It has been observed that sea buckthorn must prepared by dilution had better fermentation behavior than that prepared with $NaHCO_3$. Addition of DAHP in general enhanced both the rate of fermentation and the ethanol content. The highest rate of fermentation (RF = 0.80) was recorded in the 1:5 dilution with 0.1% DAHP. After

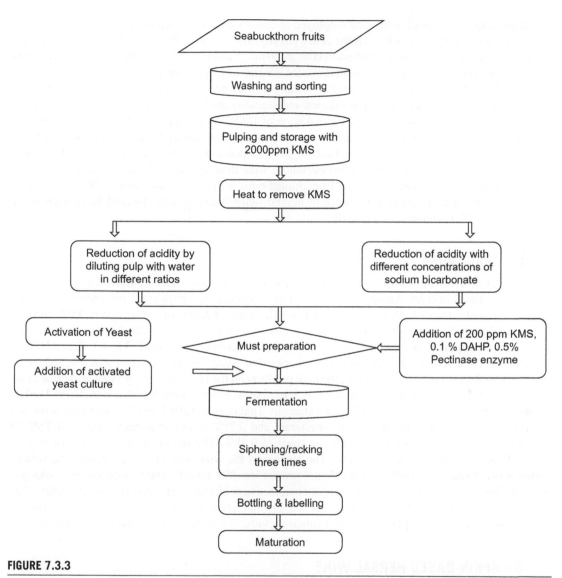

FIGURE 7.3.3

Flow sheet for the preparation of sea buckthorn wine. *DAHP*, diammonium hydrogen phosphate; *KMS*, potassium metabisulfite.

Reproduced from Joshi, V.K., Sharma, R., Sharma, S., Abrol, G.S., 2011c. Effect of dilution and de-acidification on physico-chemical and sensory quality of seabuckthorn wine. Journal of Hill Agriculture 2 (1), 47–53.

fermentation, wines prepared by diluting the pulp had ethanol contents of 9.3–13.18% (v/v), whereas that using NaHCO$_3$ ranged between 8.06% and 10.2% (v/v). The wines prepared with DAHP at 0.1% had higher ethanol content than those without DAHP. Among the physicochemical characteristics, TSS ranged between 6.8 and 10.2 °Bx, whereas titratable acidity (as % citric acid) ranged between 0.96 and

2.48 depending on the level of dilution employed or deacidification carried out by the use of $NaHCO_3$. The total sugars in different wines (Joshi et al., 2011b) ranged from 1.5% to 3.35% and ascorbic acid content was recorded at between 400 and 800 mg/100 mL. Sensory quality of the wine prepared from sea buckthorn pulp by diluting (1:5) with 0.1% DAHP and with an alcohol content of 11.6% (v/v) was adjudged the best on the basis of characteristics like color, aroma, body, and overall acceptability. Joshi et al. (2011c) recommended that out of dilution and deacidification by sodium carbonate treatment, the wine prepared from dilution was superior. Addition of DAHP improved both the RF and the ethanol content. Dilution of must remained superior over neutralizing the acidity by addition of sodium bicarbonate for ethanol content. Even on the basis of physicochemical and sensory attributes pulp dilution remained better for the preparation of sea buckthorn wine than sodium carbonate treatment. Sensory quality characteristics of wines prepared from diluted must were better than those of $NaHCO_3$-treated wines apparently due to salty taste imparted by neutralisation with $NaHCO_3$. Overall the best treatment was sample T1 (1:5 dilution) with DAHP.

3.4 PUMPKIN WINE

Pumpkin (*Cucurbita moschata*) is an important tropical vegetable grown all around the world. Pumpkin is yellow to orange in color. Among the cucurbitaceous vegetables, pumpkin (*C. moschata*) has always been very popular for its high yield, good storage life, longer periods of consumption, high nutritive value, and fitness in transport. *C. moschata* has numerous traditional medicinal uses. The chemical characteristics of pumpkin show that the TSS and titratable acidity range from 3.3 to 4.1 °Bx and 0.11% to 0.13% (as citric acid) in the raw pumpkin. Pumpkin has a very low acid content so it needs addition of some acidulant to improve the taste of the final product and the fermentation by yeast (Amerine et al., 1980). In pumpkin, the pH ranges from 3.9 to 4.3, total sugar content from 0.55% to 0.62%, and reducing sugar content from 0.41% to 0.47%. In a study by Thakur et al. (2014) the best pumpkin wine was that prepared with honey, having 1% pectinesterase and 0.25% acidulant concentration and TSS 7.8 °Bx, with total sugar of 1.81% and reducing sugar of 0.18%. The pH of the wine was 3.24 and TA [titratable acidity (as citric acid)] 0.36%. The alcohol of the wine was 11.37%, whereas the volatile acidity was calculated as 0.03% and the phenol content was 899.1 mg/L. The process of winemaking is shown in Fig. 7.3.4 and the physicochemical characteristics of the wine are presented in Table 7.3.6. The wine had many desirable characteristics, like alcohol, phenols, sensory scores, and other quality characteristics like acidity, pH, sugars, and volatile acidity, that were better than other treatments.

3.5 PUMPKIN-BASED HERBAL WINE

Pumpkin pulp was diluted with water in the ratio of 1:2 by weight. The initial TSS were raised to 24 °Bx with honey and acidity was adjusted with pomegranate extract to 0.25%. Further, pectinol and DAHP were added at the rate of 1.0% and 0.1%, respectively. Various herbs (ginger, *tulsi, mulhati*) and their combinations were added in different concentration (1%, 2%, and 4%) during the must preparation. To each treatment, 50 ppm SO_2 in the form of KMS (potassium metabisulfite) and 5% yeast inoculum were added. Each fermentation was carried out at 27°C and monitored for reduction in sugar and alcohol production. On completion of fermentation the wines were siphoned twice and filtered. Then, 50 ppm KMS was added to the wines and they were distributed into clean sterilized glass bottles. Finally, the filled bottles were pasteurized at 62–65°C for 20 min.

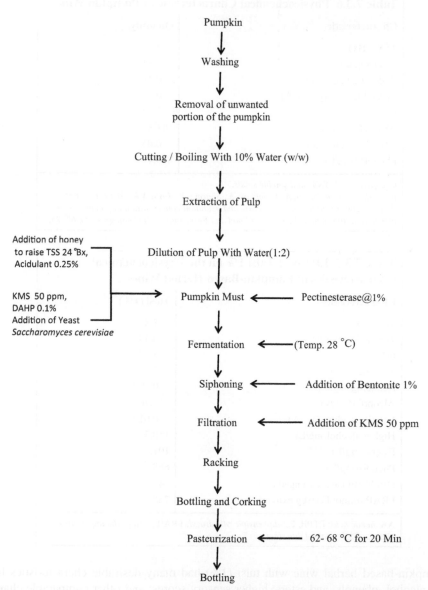

FIGURE 7.3.4

Flow sheet of preparation of pumpkin wine. *DAHP*, diammonium hydrogen phosphate; *KMS*, potassium metabisulfite; *TSS*, total soluble solids.

Adapted from Thakur, A.D., Saklani, A., Sharma, S., Joshi, V.K., 2014. Effect of different sugar sources, pectin esterase and acidulant concentration of pumpkin wine production. International Journal of Food and Fermentation Technology, 4 (1), 67–78.

Table 7.3.6 Physicochemical Characteristic of Pumpkin Wine

Characteristic	Quantity
TSS (°Bx)	7.8
Total sugar (%)	1.81
Reducing sugar (%)	0.18
Titratable acidity (% CA)	0.36
pH	3.24
Alcohol (% v/v)	11.37
Volatile acidity (%)	0.03
Phenols (mg/L)	899.1

CA, *citric acid;* TSS, *total soluble solids.*
Reproduced from Thakur, A.D., Saklani, A., Sharma, S., Joshi, V.K., 2014. Effect of different sugar sources, pectin esterase and acidulant concentration of pumpkin wine production. International Journal of Food and Fermentation Technology, 4 (1), 67–78.

Table 7.3.7 Effects of Tulsi 1% on the Physicochemical Characteristics of Pumpkin-Based Herbal Wines

Characteristic	Tulsi (1%)
TSS (°Bx)	7.8
Titratable acidity (%)	0.48
pH	3.97
Total sugar (%)	2.237
Reducing sugar (%)	0.998
Alcohol (% v/v)	11.70
Volatile acidity (% AA)	0.012
Higher alcohol(mg/L)	158.3
Esters (mg/L)	101.7
Phenols(mg/L)	668
DPPH radicals scavenged (%)	42
FRAP [mmol Fe(II)/g extract]	768

AA, *acetic acid;* DPPH, *2,2-diphenylpicrylhydrazyl;* FRAP, *ferric-reducing ability of plasma;* TSS, *total soluble solids.*

The pumpkin-based herbal wine with tulsi (1%) had many desirable characteristics like higher amounts of alcohol, phenols, and esters; higher sensory scores; and other comparable characteristics like acidity, sugars, higher alcohols, and volatile acidity compared to other treatments as presented in Table 7.3.7. The qualitative estimation of phytochemicals showed that flavonoids, saponins, and reducing sugars were present and alkaloids, tannins, phlobatannins, and cardiac glycosides were absent in various wines. The results of 2,2-diphenylpicrylhydrazyl (DPPH) radical-scavenging assay showed that the scavenging activity was moderate to high in *tulsi* 1% wine (sample T1).

So, the wine prepared with tulsi 1%, having TSS 7.8 °Bx, with total sugar 2.237% and reducing sugar 0.998%, was selected as the best. The pH of the wine was 3.97 and TA 0.48%. The alcohol content was

Table 7.3.8 Qualitative Estimation of Phytochemicals in Tulsi (1%) Pumpkin-Based Herbal Wine

Test	Tulsi (1%)
Fehling	Positive
Alkaloids	Negative
Flavonoids	Positive
Saponins	Positive
Tannins	Positive
Phlobatannins	Negative
Cardiac glycoside	Negative

11.70%, whereas the volatile acidity was calculated as 0.012% and the phenol content was 668 mg/L. The content of esters was 101.7 mg/L and higher alcohols 158 mg/L. The DPPH radicals scavenged were 42%, which was moderate, and ferric-reducing antioxidant power was 768 [mmol Fe(II)/g extract].

The qualitative estimations of phytochemicals in tulsi (1%) pumpkin-based herbal wine is given in Table 7.3.8. The presence of flavonoids, saponins, tannins, and reducing sugars and absence of alkaloids, phlobatannins, and cardiac glycosides are observed. Further, it can be concluded that pumpkin wine with herbal extract has healthful components.

3.6 BLACKBERRY *JAMUN* WINE

Syzygium cumini (Linn. Skeels), commonly known as black plum and *jamun* in Hindi, belongs to the family Myrtaceae. The fruit is gaining popularity among consumers because of its balanced sugar, acid, and tannin contents (Das, 2009). *Jamun* is an important but underutilized fruit crop of India (Prasad and Kumar, 2010). The plant grows naturally in clayey loamy soil in tropical as well as subtropical zones of the Indo-Gangetic plain (Janick and Paull, 2008; Rai et al., 2011). *Jamun* has medicinal value, including antidiabetic, astringent, stomatic, carminative, antiscorbutic, and diuretic actions (Patel et al., 2005). The attractive color, due to anthocyanin pigments (antioxidant), is a major quality attribute in *Jamun* beverages. It is generally consumed fresh and is known to have nutraceutical and therapeutic values (Khurdiya and Roy, 1985). The fruit is an effective food remedy for curing diabetes, heart problems, hemorrhoids, and liver troubles because of its effect on the pancreas. The fruit is a very rich source of anthocyanins and possesses antioxidant properties, too (Chowdhury and Ray, 2007). The berries have a mean weight of about 6.4 g, comprising about 35% seed and 65% pulp, with a seed-to-pulp ratio of 1:1.9 (Table 7.3.9). Further, the berries have a quite high amount of TSS (16 °Bx), which facilitates fermentation, but also have high acid (1.19%) and tannin (345 mg/100 g) contents. Because of the high contents of acids and tannins, the pulp was diluted with water in various proportions for must preparation. Hence, the must was prepared by diluting the pulp in 1:05, 1:1, and 1:2 ratios with water. The concentrations of TSS, DAHP, pectinol, and SO_2 as KMS were kept constant in all the treatments, as reported by Joshi (1997), for wine preparation. The initial TSS was raised to 24 °Bx with sugar syrup of 70 °Bx, and pectinol and DAHP were added at 0.5% and 0.1%, respectively. To each treatment, 100 ppm SO_2 in the form of KMS was also added. The flow sheet for the preparation of *jamun* wine is shown in Fig. 7.3.5.

Joshi et al. (2012) reported that *jamun* must prepared by dilution (1:2) showed better fermentation behavior than the other two treatments. With the increase in dilution level, ethanol content and

Table 7.3.9 Physicochemical Characteristics of *Jamun*

Characteristic	Mean ± SD
Fruit weight (g)	6.4 ± 2.2
Seed weight (g)	2.20 ± 0.50
Pulp weight (g)	4.20 ± 0.60
Seed-to-pulp ratio	1:1.90 ± 0.30
Total soluble solids (°Bx)	16 ± 0.8
Acidity (% citric acid)	1.19 ± 0.2
Reducing sugar (%)	8.60 ± 0.3
Total sugar (%)	12.44 ± 0.2
Ascorbic acid (mg/100 g)	26.80 ± 0.2
Anthocyanins (mL/100 g)	119 ± 0.02
Phenols (as tannic acid, mg/L)	318 ± 0.01
Each value is an average of five replicates	

Reproduced from Joshi, V.K., Attri, D., Singh, T.K., Abrol, G.S., 2011b. Fruit wines: production technology. In: Joshi, V.K. (Ed.), Handbook of Enology: Principles Practices and Recent Innovations, vol. 3. Asia Tech Publisher, New Delhi, pp. 1177–1221.

total esters increased, whereas the total titratable acidity, sugars, and anthocyanin content decreased (Table 7.3.10).

All the wines were matured for a year and changes in the physicochemical characteristics were recorded. It was observed that there was a reduction in TSS, titratable acidity, ethanol content, total phenols, and anthocyanins (Table 7.3.11); however, an increase was observed in reducing sugars and total esters. The maturation of wine for a year improved most of its characteristics. In other study, wine was prepared from *jamun* fruit using two different yeast strains, i.e., *S. cerevisiae* var. *ellipsoideus* and *S. cerevisiae* var. *bayanus*, fermented on three different must types, i.e., juice, pulp + skin, and pulp + skin + seeds. TSS and pH of the must were adjusted to 24 °Bx and 3.2, respectively. Maximum ethyl alcohol of 7.92% and wine recovery of 86.15% were recorded for the treatment in which pulp and skin were used along with *S. cerevisiae* var. *ellipsoideus*. Tannin content was least when juice was fermented with *S. cerevisiae* var. *ellipsoideus* in a study carried out by Lokesh et al. (2014).

3.6.1 Sensory Characteristics

The wines fermented from musts with lower dilutions received lower sensory scores compared to those with higher dilutions. Improper sugar/acid ratio and hazy brown appearance were among the major reasons for the lower scores given to these wines. The wine prepared with the 1:1 dilution was found to be superior (Table 7.3.12) mainly because of better appearance, color, total acidity, sweetness, body, and overall impression.

On the basis of the physicochemical and sensory quality characteristics, *jamun* wine prepared with a 1:1 dilution of the pulp was the best. It had optimum TSS, acidity, ethanol content, appearance, color, sweetness, body, and overall impression. It had balanced sugars, acids, tannins, and ethanol content and thus was found better for the preparation of *jamun* wine of acceptable quality.

Sensory evaluation of *jamun* wine indicated that the wine is acceptable, with a fruity flavor and good color and body. Further, the wine prepared with pulp and skin inoculated with *S. cerevisiae* var. *ellipsoideus* secured the highest score of 15.86 out of 20.0 after 6 months of aging (Lokesh et al., 2014).

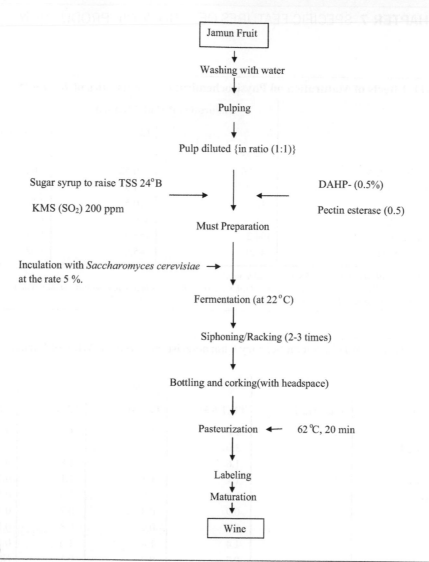

FIGURE 7.3.5

Flow sheet for preparation of *jamun* wine by the method optimized.

Reproduced from Joshi, V.K., Sharma, R., Girdher, A., Abrol, G., 2012. Effect of dilution and maturation on physico-chemical and sensory quality of jamun wine. Indian Journal of Natural Products Radiance 3 (2), 222–227.

Table 7.3.10 Effect of Dilution on Physicochemical Characteristics of *Jamun* Wine

Treatment	TSS (°Bx)	Titratable Acidity (% CA)	Reducing Sugar (%)	Ethanol Content (% v/v)	Total Phenols (mg/L)	Total Esters (mg/L)	Anthocyanins (mg/100 g)
T1 (1:0.5 dilution)	10.8	0.79	5.0	9.9	418	148.1	61.0
T2 (1:1 dilution)	10.2	0.53	4.1	10.4	360	155.3	42.5
T3 (1:2 dilution)	8.8	0.36	2.5	11.8	320	159.2	37.5
CD ($p > .05$)	0.19	0.02	0.22	0.20	2.17	1.26	1.25

CA, *citric acid*; CD, *critical difference*; TSS, *total soluble solids*.
Reproduced from Joshi, V.K., Sharma, R., Girdher, A., Abrol, G., 2012. Effect of dilution and maturation on physico-chemical and sensory quality of jamun wine. Indian Journal of Natural Products Radiance 3 (2), 222–227.

Table 7.3.11 Effects of Maturation on Physicochemical Characteristics of *Jamun* Wine

Treatment	Maturation Period (Months)		CD ($p > .05$)
	0	12	
TSS (°Bx)	9.93	9.20	0.16
Titratable acidity (% CA)	0.56	0.52	0.02
Reducing sugar (%)	3.9	4.3	0.18
Ethanol content (% v/v)	10.7	10.5	0.16
Total phenols (mg/L)	366	360.33	1.77
Total esters (mg/L)	154.2	157.6	1.03
Anthocyanins (mg/100 g)	47.0	45.0	1.02

CA, *citric acid*; CD, *critical difference*; TSS, *total soluble solids.*
Reproduced from Joshi, V.K., Sharma, R., Girdher, A., Abrol, G., 2012. Effect of dilution and maturation on physico-chemical and sensory quality of jamun wine. Indian Journal of Natural Products Radiance 3 (2), 222–227.

Table 7.3.12 Effects of Dilution on Sensory Characteristics of *Jamun* Wine of Various Treatments

Characteristic	Total Marks	Score			CD ($p > .05$)
		T1 (1:0.5)	T2 (1:1)	T3 (1:2)	
Color	2	1.8	1.8	1.8	0.26
Aroma and bouquet	4	3.0	3.4	3.2	0.26
Appearance	2	1.6	1.9	1.8	0.13
Volatile acidity	2	1.1	1.1	1.1	0.13
Total acidity	2	1.8	1.8	1.5	0.23
Sweetness	1	0.9	0.8	0.7	0.13
Body	1	0.6	0.9	0.8	0.13
Flavor	2	1.4	1.8	1.6	0.13
Bitterness	1	0.7	0.7	0.6	NS
Astringency	1	0.8	0.8	0.8	NS
Overall impression	2	1.3	1.8	1.6	0.26
Total sensory score	20	15.0	16.8	15.5	

CD, *critical difference.*
Reproduced from Joshi, V.K., Sharma, R., Girdher, A., Abrol, G., 2012. Effect of dilution and maturation on physico-chemical and sensory quality of jamun wine. Indian Journal of Natural Products Radiance 3 (2), 222–227.

3.7 RED WINE MADE BY BLENDING OF GRAPE (*VITIS VINIFERA* L.) AND *JAMUN* (*SYZYGIUM CUMINI* L.) JUICES

The grape (*Vitis vinifera* L.) is a nonclimacteric fruit belonging to the family Vitaceae and is found growing under a variety of soil and climatic conditions. A significant amount of work on winemaking from minor fruits such as *jamun* is under way because of the importance of these fruits from a therapeutic point of view. *Jamun* beverages are acidic, astringent, and, therefore, not generally preferred for table consumption. The bright and brilliant purple-colored *jamun* juice can, however, be successfully used for blending the beverages (Gehlot et al., 2008). Owing to the presence of high amounts of phenolic compounds (Table 7.3.13) in *jamun* juice, the astringency of *jamun* wine is higher compared to grape wine. A red wine was prepared by blending grape and *jamun* juice in different proportions before fermentation to reduce the astringency and the acceptability of the wines was compared. The juices were extracted as given in Figs. 7.3.6 and 7.3.7.

The fruit juices were ameliorated to 20 °Bx by addition of sugar. The juices were then, supplemented with 0.2% ammonium sulfate to provide additional nitrogen for the growth of yeast. Then,

Parameter	Mean[a]± SD	
	Grape Juice	***Jamun* Juice**
Juice recovery (%)	76.3±2.33	70.3±2.60
Total soluble solids (°Bx)	18.6±0.70	12.3±0.50
pH	4.7±0.04	4.3±0.03
Acidity	0.9±0.00	1.07±0.01
Total sugars (g/L)	151.3±3.50	113.6±2.30
Reducing sugars (g/L)	140±4.80	83±0.92
Phenols (mg/100 mL)	21.8±2.5	100±11.5
Anthocyanins (mg/100 mL)	130±0.45	140±0.64

Table 7.3.13 Chemical Analysis of Grape and *Jamun* Juice

[a]*The values are means of three replicates.*
Reproduced from Chaudhary, C., Yadav, B.S., Grewal, R.B., 2014. Preparation of red wine by blending of grape (vitis vinifera L.) and jamun (Syzygium cuminii L. Skeels) juices before fermentation. International Journal of Agriculture and Food Science Technology 5 (4), 239–348.

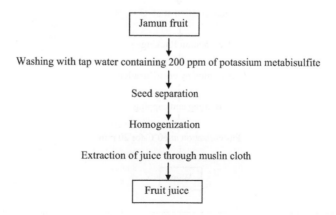

FIGURE 7.3.6

Flow sheet for extraction of *jamun* juice.

Reproduced from Chaudhary, C., Yadav, B.S., Grewal, R.B., 2014. Preparation of red wine by blending of grape (vitis vinifera L.) and jamun (Syzygium cuminii L. Skeels) juices before fermentation. International Journal of Agriculture and Food Science Technology 5 (4), 239–348.

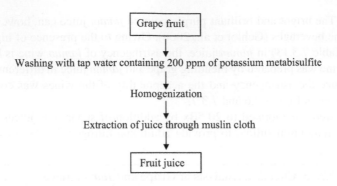

FIGURE 7.3.7

Flow sheet for extraction of grape juice.

Reproduced from Chaudhary, C., Yadav, B.S., Grewal, R.B., 2014. Preparation of red wine by blending of grape (vitis vinifera L.) and jamun (Syzygium cuminii L. Skeels) juices before fermentation. International Journal of Agriculture and Food Science Technology 5 (4), 239–348.

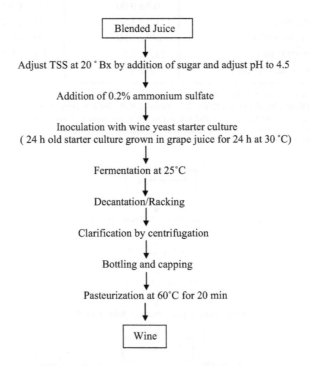

FIGURE 7.3.8

Flow sheet for preparation of *jamun*–grape wine. *TSS*, total soluble solids.

Reproduced from Chaudhary, C., Yadav, B.S., Grewal, R.B., 2014. Preparation of red wine by blending of grape (vitis vinifera L.) and jamun (Syzygium cuminii L. Skeels) juices before fermentation. International Journal of Agriculture and Food Science Technology 5 (4), 239–348.

grape and *jamun* juices were mixed in different proportions (100:0, 75:25, 50:50, 25:75, 0:100). The blended juices were inoculated with 7.5% *S. cerevisiae* strain 4787 and allowed to ferment at 25°C in flasks. The wine was then, clarified by centrifugation at 5700 rpm for 10 min and was bottled and pasteurized at 60°C for 20 min. The entire process of *jamun*–grape wine preparation is given in Fig. 7.3.8.

3.7.1 Extraction of Juices

It is concluded that the blending of grape juice with *jamun* juice before fermentation helps in reducing the astringency, with improvements in taste, body, color, and aroma of the final wine. It was observed that acceptable red wine can be prepared by mixing grape and *jamun* juices, but the acceptability was higher when the juices were blended at a 75:25 (grape/jamun) ratio before fermentation, because the presence of a small amount of *jamun* juice blended with the grape juice enhanced the taste and color of the wine. Considering the neutroclinical value and perishable nature of grapes and *jamuns*, these fruits can be better utilized for preparation of red wine by blending in the appropriate ratio.

3.8 PERSIMMON WINE

Persimmon (*Diospyros kaki*) is an important underutilized fruit crop. Its color varies in different cultivars from yellow and orange to deep red. There are two types of varieties, astringent and non-astringent. The fruits of persimmon are rich in various nutrients and phytochemicals, such as carbohydrates, organic acids, vitamins, tannins, polyphenols, dietary fiber, triterpenoids, and carotenoids, which contribute significantly to their taste, color, and nutritive and medicinal value (Altuntas et al., 2011). Persimmons also have a high antioxidant potential that may have beneficial effects against oxidative damage in humans. The antioxidant activity of the fruit is mainly contributed by condensed tannins, which may reduce the risk of cardiovascular disease, hypertension, diabetes, and a wide range of cancers (Zhou et al., 2010; Thuong et al., 2008; Hwang et al., 2011). Persimmon has an unusual property in that it appears to alter and reduce the rate of alcohol absorption and metabolism and thus, ameliorates the symptoms of a hangover (Srivastava and Das, 2005; Orwa et al., 2009; Chen et al., 2008; George and Redpath, 2008; Hwang et al., 2011; Gorinstein et al., 1998). Apart from the fruits, leaves also possess medicinal properties. The isomeric pentacyclic compounds oleanolic acid and ursolic acid are two common triterpenoids found in *D. kaki* leaves (Zhou et al., 2010; Thuong et al., 2008; Hwang et al., 2011). Both compounds have bioactivities such as antiinflammatory, hepatoprotective, gastroprotective, cardiovascular, antitumoral, anti-HIV, and immunoregulatory effects. Sharma and Mahant (2015) standardized a procedure using mixed strains of yeast for the development of persimmon wine of improved quality with enhanced antioxidant activity. The generalized flow sheet for the same is depicted in Fig. 7.3.9.

The fruit of the Fuyu cultivar was utilized to optimize the additives in persimmon pulp for wine preparation. The pulp was diluted with water in a 1:1 ratio with additives at various concentrations and fermented at a temperature of 27°C. Two concentrations of DAHP as nitrogen source (0.1% and 0.2%) and citric acid (0.3% and 0.4%) were used during the study. The must prepared with addition of 0.1% DAHP and 0.3% citric acid had higher RF and ethanol content (Sharma and Mahant, 2015). Further, the wine had many desirable characteristics such as more alcohol, esters, total phenols, and antioxidant activity (Fig. 7.3.10), with comparable amounts of higher alcohols, total sugars, reducing sugars, and volatile acidity (Table 7.3.14). The wine also had higher scores for various sensory attributes like color, taste and appearance, aroma, volatile acidity, total acidity, sweetness, body, flavor, bitterness, astringency, and overall impression. Hence, from this study it can be concluded that for development of a quality persimmon wine from the Fuyu cultivar, dilution with water in a 1:1 ratio with addition of 0.1% DAHP and 0.3% citric acid is required.

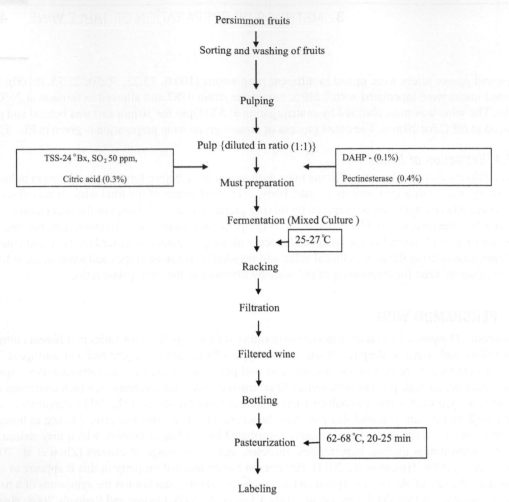

Persimmon fruits
↓
Sorting and washing of fruits
↓
Pulping
↓
Pulp {diluted in ratio (1:1)}

TSS-24 °Bx, SO₂ 50 ppm,
Citric acid (0.3%) → Must preparation ← DAHP - (0.1%)
Pectinesterase (0.4%)
↓
Fermentation (Mixed Culture)
↓ ← 25-27 °C
Racking
↓
Filtration
↓
Filtered wine
↓
Bottling
↓
Pasteurization ← 62-68 °C, 20-25 min
↓
Labeling

FIGURE 7.3.9

Generalized flow sheet for persimmon wine production. *DAHP*, diammonium hydrogen phosphate; *TSS*, total soluble solids.

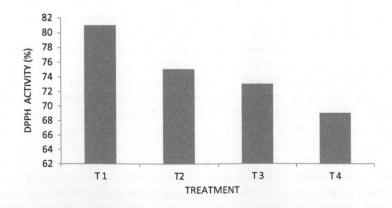

FIGURE 7.3.10

Comparison of 2,2-diphenylpicrylhydrazyl (DPPH) antioxidant activity of various treatments of persimmon wine.

Table 7.3.14 Effects of Additive Concentration on Chemical Characteristics of Persimmon Wine

Characteristic	Treatment (Mean)			
	T1	T2	T3	T4
TSS (°Bx)	7.2[a]	7.4[b]	7.6[c]	8.0[d]
Titratable acidity (% CA)	0.63[a]	0.69[b]	0.62[a]	0.67[b]
pH	3.72[b]	3.52[a]	3.80[d]	3.59[c]
Total sugars (%)	3.25[a]	3.35[b]	3.42[c]	3.49[d]
Reducing sugars (%)	0.222[a]	0.242[b]	0.249[b]	0.260[c]
Ethanol (%)	12.29[d]	12.04[c]	11.89[b]	11.67[a]
Volatile acidity (% AA)	0.024[a]	0.030[b]	0.032[c]	0.036[d]
Total phenol (mg/L)	268[d]	254[c]	229[b]	212[a]
Higher alcohol (mg/L)	170.50[c]	175.26[d]	164.60[a]	167.40[b]
Total esters (mg/L)	59.71[d]	44.34[c]	38.56[b]	31.43[a]

T1, *0.1% DAHP and 0.3% CA;* T2, *0.1% DAHP and 0.4% CA;* T3, *0.2% DAHP and 0.3% CA;* T4, *0.2% DAHP and 0.4% CA.*
AA, *acetic acid;* CA, *citric acid;* DAHP, *diammonium hydrogen phosphate;* TSS, *total soluble solids; superscript letters, numbers across each row having the same superscript letter show no statistical significant difference.*
Mahant, K., Sharma, S., Thakur, A.D., 2016. *Effect of nitrogen source and citric acid addition on wine preparation from Japanese persimmon. Journal of Institute of Brewing (in press).*

3.9 FERMENTED GARLIC BEVERAGE

Garlic is well known for its medicinal value and is being employed as a condiment. It acts as an expectorant, diuretic, antidiabetic, antiseptic and antibacterial, antifungal, antiinflammatory, and antiasthmatic (Grunwald et al., 1993). Allicin obtained by crushing garlic bulbs is soluble in water and ethanol. A combination of three molecules of allicin prevents the aggregation of blood platelets responsible for clotting of blood and arterial blocking (Augusti, 1996). A fermented beverage from garlic has been developed (Tewari et al., 2001). Garlic concentrations of 7.0% and 1.0% (w/v) in the mash after sterilization (15 psi), addition of KMS 100 ppm, TSS 7.5 °Bx, and inoculum at 2.5% were the optimized factors for fermentation conducted by *S. cerevisiae* var. *ellipsoideus*. Attempts were made to prepare a low alcoholic beverage from bitter gourd and apple by standardizing the concentration of apple juice, DAHP and inoculum size by applying central composite design of RSM. On the basis of physico-chemical and sensory characteristics, a run having 40% apple juice, 0.15% DAHP and 2.5% inoculum level were rated as the best. It has the highest TSS, rate of fermentation, ethanol, reducing and, total sugars and total phenols. With the addition of 0.15% DAHP as a nitrogen sources, bitter gourd based wine had the highest TSS, titratable acidity, ethanol, total and reducing sugar content and lowest volatile acidity and higher alcohols. Different concentrations of yeast inoculum used did not significantly influenced most of the physico-chemical characteristics. Must inoculated with 2.5% *S. cerevisiae* var. *ellipsoideus* used for the preparation of bitter gourd based wine had the highest fermentability and ethanol content. However, the product being low alcoholic, need pasteurization at 62°C for 20 min in a glass bottle. Bitter gourd beverage so prepared holds promise as a medicinal drink as bitter gourd is known to be antidiabetic (Joshi and Kumar, 2015).

4. LYCHEE WINE

Lychee fruit is grown in some areas of India. It has plenty of flavor and is a good source of minerals and vitamins, and is used for the preparation of alcoholic beverages in China. This fruit has been used to prepare a low-alcohol high-flavor beverage using the techniques of partial osmotic dehydration. It was found that a product containing 5–6% alcohol, 3–4% sugar, and 0.35% acid can be prepared as an appetizing soft drink instead of an intoxicating liquor. With osmotic dehydration as a pretreatment, the nutrients and flavor are concentrated and this is reflected in the product prepared from untreated fruits. For the same reason, the osmotically treated juice undergoes fast fermentation and the time required for the preparation of wine is considerably less. The detailed procedure involved in preparation of this drink was described earlier (Vyas et al., 1989). Lychee fruits of optimum maturity are washed and peeled, followed by dipping in a sugar solution of 70 °Bx for 4h at 50°C. The treated fruits are taken out and drained, followed by pulping in a pulper. The sugar contents are adjusted to 22 °Bx by dilution with water. A 24-h-old active culture of yeast, *S. cerevisiae*, prepared in the lychee juice, is added at 5% to the sterilized lychee juice to carry out the alcoholic fermentation. The fermentation is carried as a routine practice and allowed to continue until the TSS comes to 7 °Bx (refractometer reading). The wine is matured, followed by blending with an equal quantity of fresh lychee juice, and filtered and distributed into glass bottles, which are closed with crown corks. The bottles are processed in water at 62.5°C for 20min (Vyas et al., 1989). The suitability of lychee juice concentrate was investigated for the production of lychee wine. The large amount of fermentable sugars (85.20%) and high acid content (4.25%) present in the concentrate were found to be suitable for its use in winemaking. Of the four yeast strains screened for alcoholic fermentation of reconstituted lychee juice, *S. cerevisiae* MTCC 178 was the most potent (Singh and Kaur, 2009). The optimal alcoholic fermentation of reconstituted lychee juice by *S. cerevisiae* MTCC 178 was recorded at 25°C, with an initial must pH of 5.0 and TSS of 24 °Bx and an inoculum level of 10% (v/v). The lychee wine produced from reconstituted lychee juice concentrate under the optimized conditions contained 11.60% (v/v) ethanol, 92mg/L total esters, 124mg/L total aldehydes, and 0.78% (v/v) titratable acidity. The sensory evaluation revealed a clean, light amber color, an attractive aroma of natural lychee fruit, and a harmonious wine taste (Singh and Kaur, 2009).

5. PAPAYA WINE

Papaya is a fruit with soluble saccharides in the form of glucose, fructose, and sucrose, and it is widely cultivated in several countries. In tropical climates such as in Nigeria, papaya trees continue bearing fruits throughout the year, and the fruit in turn follows the same pattern of maturity. It displays rapid growth and high yield of 100kg per plant per year or 154,000kg per hectare per year, even from the fourth year of growth. The average yield per hectare is about 22,000 fruits weighing 34 tons. Sugars represent that part of the fruit that is used by microorganisms for wine production (Maragatham and Panneerselvam, 2011a). In a study, the various yeasts such as *S. cerevisiae*, *Saccharomyces bayanus*, *Saccharomyces uvarum*, *Saccharomyces italicus*, *Saccharomyces pastorianus*, *Schizosaccharomyces pombe*, and *Zygosaccharomyces* were isolated from rotten papaya fruit. Their suitability for wine production was tested by using sugar and ethanol tolerance tests. The best biochemically active strain was *S. cerevisiae* for producing wine from papaya variety CO 2. After fermentation for 1month with *S. cerevisiae*, the highest (11.59%) alcohol concentration with corresponding residual sugar concentration of 1.87% was produced from CO 2 papaya. So, *S. cerevisiae* was the best yeast strain, producing wine with the highest acceptable score of 4.8 from CO 2 papaya fruits. The study revealed the possibility of producing wine from locally available fruit using simple, cheap, and adaptable technology with biochemically characterized yeast strains (Maragatham and Panneerselvam, 2011b).

The impact of amino acid addition on aroma compound formation in papaya wine fermented with the yeast *Williopsis saturnus* var. *mrakii* NCYC2251 was studied. Time courses of papaya juice fermentation were determined out using *W. saturnus* var. *mrakii* NCYC2251, with and without the addition of selected amino acids (L-leucine, L-isoleucine, L-valine, and L-phenylalanine). Yeast growth and changes in sugars, °Bx, organic acids, and pH were similar, regardless of amino acid addition. L-Leucine addition increased the production of isoamyl alcohol and some esters such as isoamyl acetate, isoamyl butyrate, and isoamyl propionate, whereas L-isoleucine addition increased the production of active amyl alcohol and active amyl acetate. L-Valine addition slightly increased the production of isobutyl alcohol and isobutyl acetate. L-Phenylalanine addition increased the formation of 2-phenylethanol, 2-phenylethyl acetate, and 2-phenylethyl butyrate, while decreasing the production of most other esters. This study suggests that papaya juice fermentation with *W. saturnus* var. *mrakii* NCYC2251 in conjunction with the addition of selected amino acids can be an effective way to modulate the aroma of papaya wine (Lee et al., 2011).

6. BLENDED PASSION FRUIT

Passion fruit (*Passiflora edulis*), mango (*Mangifera indica*), and pineapple (*Ananas comosus*) are widely cultivated nutritious fruits in the Northern Province of Rwanda, especially in the Gakenke and Musanze districts, where the soil is fertilized with organic manure from a mixture of cow dung and compost. The agriculture is practiced in a green environment and the ripened fruits are harvested, sold fresh, and consumed without postharvest technology or value addition for export or income generation. The fruits were used to make wine. The chemical composition of wine is influenced by soil fertility and climatic conditions. Hence, an attempt to produce a yellow/golden wine was made in the laboratory using a mixture of fruits (33.3% each) as a golden must extracted from passion fruit, mango, and pineapple. The fermentation of the mixture of juices was done at room temperature, i.e., 22°C, using a wild yeast strain of *S. cerevisiae* called "musanzeensis" isolated from local traditional banana wine. Fig. 7.3.11 shows the gradual increase in yeast biomass during must fermentation. The fermentation process greatly reduced the TSS content from 20 to 2°Bx. During substantial must fermentation, the pH decreased from 5.5 of fresh juice to 3.2 of wine, titratable acidity increased from 0.68% to 1.4%, sugar content decreased from 85 to 32 g/L, specific gravity decreased from 1.040 to 1.002, yeast growth increased from 3 to 18 log CFU/mL, and alcohol content increased from 0.0% to 12% by volume.

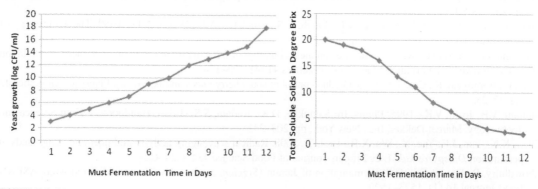

FIGURE 7.3.11

Saccharomyces cerevisiae multiplication and total soluble solids reduction during fermentation of must.

Adapted from Nzabuheraheza, F.D., Nyiramugwera, A.N., 2014. Golden wine produced from mixed juices of passion fruit (Passiflora edulis), mango (Mangifera indica) and pineapple (Ananas comosus). African Journal of Food Agriculture, Nutrition and Development 14 (4), 9104–9116.

After 12 days of fermentation, the color of the wine remained yellow, flavor was enhanced, sweetness was diminished, and acidity (sourness) was increased slightly. These chemical changes could be due to the *S. cerevisiae* activity, which was characterized by a remarkable foam and intensive production of carbon dioxide in the fermenting wine. The mixture of the three juices from *P. edulis*, *M. indica*, and *A. comosus* produced an alcoholic beverage with a wonderful flavor that was generally delicious and acceptable to 40 trained and blinded panelists during sensory evaluation, using the 9-point hedonic scale. Each panelist sipped a 100-mL sample taken from the wine. Thus, the yellow wine obtained should be promoted for value addition to local fruit, imported wine reduction, job creation, income generation, and rural development (Nzabuheraheza and Nyiramugwera, 2014).

7. FUTURE TRENDS

Preparation of wine from nongrape fruits such as berries, persimmon, *Jamun*, litchi, passion fruits, etc. needs serious consideration, though some efforts made in this direction have been summarized in this chapter. But whatever work has been done at the best can be described as preliminary or exploratory research, and more systematic research can only make the production of wines of such fruits as a commercial venture. The role of wine from grapes and their constituents in preventing coronary heart diseases is well established. But there is or no less information on the nongrape fruit wines, which could be a potential area for future research. A critical appraisal of the work done so far, there are a large number of research gaps for more elaboration in the future that could be areas of research, especially in those countries and regions where fruits other than grape are cultivated.

REFERENCES

Augusti, K.T., 1996. Therapeutic values of onion (*Allium cepa* L.) and garlic (*Allium sativum* L.). Indian Journal of Experimental Biology 34, 634–638.

Altuntas, E., Cangi, R., Kaya, C., 2011. Physical and chemical properties of persimmon fruit. International Agrophysics 25, 89–92.

Amerine, M.A., Berg, H.W., Kunkee, R.E., Qugh, C.S., Singleton, V.L., Webb, A.D., 1980. The Technology of Wine Making, fourth ed. AVI, Westport, CT.

Baumann, G., Gierschner, K., 1974. Studies on the technology of juice manufacture from sour cherries in relation to the storage of the product. Fluessiges Obstetrics 41 (4), 123–128.

Benk, E., Borgmann, R., Cutka, I., 1976. Quality control of sour cherry juices and beverages. Fluessiges Obstetrics 43 (1), 17–23.

Bhutani, V.P., Joshi, V.K., 1995. Plums. In: Salunkhe, D.K., Kadam, S.S. (Eds.), Handbook of Fruit Science and Technology. Marcel Dekker, Inc., New York, pp. 203–241.

Chen, X.N., Fan, J.F., Yue, X., Wu, X.R., Li, L.T., 2008. Radical scavenging activity and phenolic compounds in persimmon *Diospyros kaki* L cv Mopan. Journal of Food Science 731, C24–C28.

Chowdhury, P., Ray, R.C., 2007. Fermentation of Jamun (*Syzgium cuminii* L.) fruits to form red wine. ASEAN Food Journal 14 (1), 1523–1529.

Chaudhary, C., Yadav, B.S., Grewal, R.B., 2014. Preparation of red wine by blending of grape (*vitis vinifera* L.) and jamun (*Syzygium cuminii* L. Skeels) juices before fermentation. International Journal of Agriculture and Food Science Technology 5 (4), 239–348.

Das, J.N., 2009. Studies on storage stability on jamun beverages. Indian Journal of Horticulture 66 (4), 508–510.

Delin, C.R., Lee, T.H., 1991. The J-shaped curve revisited: wine and Cardiovascular health update. Australian and New Zealand Wine Industry Journal 6 (1) 15(FSTA 24(5): 5H77, 1992.

Faparusi, S.I., 1973. Original of initial microflora of palm wine from Oil Palm tree (*Elaeis guineensis*). Journal of Applied Bacteriology 36, 559–565.

Flores, J.H., Heatherbeth, D.A., 1984. Optimizing enzymes and pre-mash treatments for juice and colour extraction from strawberries. Fluessiges Obstetrics 7, 327–329.

Gasteineau, F.C., Darby, J.W., Turner, 1979. Fermented Food Beverages in Nutrition. Academic Press, New York, USA, pp. 70–84.

Gehlot, R., Dhawan, S.S., Singh, K., Singh, M., 2008. Nutritious and therapeutic beverages from Jamun. Beverage & Food World 38 (5), 46–47.

George, A.P., Redpath, S., 2008. Health and medicinal benefits of persimmon fruit: a review. Advances in Horticultural Science 22, 244–249.

Gorinstein, S., Bartnikowska, E., Kulasek, G., Zemser, M., Trakhtenberg, S., 1998. Dietary persimmon improves lipid metabolism in rats fed diets containing cholesterol. Journal of Nutrition 128, 2023–2027.

Goswell, R.T., Kunkee, R.E., 1977. Fortified wines. In: Rose, A.H. (Ed.), Alcoholic Beverages. Academic Press, London, p. 477.

Grunwald, J., Schilcher, H., Phillipson, J.D., Loew, D., 1993. Garlic—the best documented medicinal plants in prevention of cardiovascular diseases. Acta Horticulturae 332, 115–119.

Hwang, I.W., Jeong, M.C., Chung, S.K., 2011. The physicochemical properties and the antioxidant activities of persimmon peel powders with different particle sizes. Journal of the Korean Society for Applied Biological Chemistry 54 (3), 442–446.

Janick, J., Paull, R.E., 2008. *Syzygium cuminii* L. Skeels. The Encyclopedia of Fruits and Nuts. Cambridge University Press, Cambridge, p. 551.

Joshi, V.K., 1997. Fruit Wines, Directorate of Extension Education, second ed. Dr. YS Parmar University of Horticulture and Forestry, Nauni, Solan (HP).

Joshi, V.K., Bhutani, V.P., 1990a. Effect of preservatives on the fermentation behaviour and quality of plum wine. In: XXIII Int. Hort. Congress, Held at Italy, 1990. Abst. 2585.

Joshi, V.K., Bhutani, V.P., 1990b. Evaluation of plum cultivars for wine preparation. In: XXIII Int. Hort. Congress, Held at Italy. Abst. 3336.

Joshi, V.K., Bhutani, V.P., 1995. Peach. In: Salunkhe, D.K., Kadam, S.S. (Eds.), Hand Book of Fruit Science and Technology. Marcel & Dekker, USA, pp. 243–296.

Joshi, V.K., Sandhu, D.K., 2000. Influence of ethanol concentration, addition of spices extract and level of sweetness on physico-chemical characteristics and sensory quality of apple vermouth. Brazilian Archives of Biology and Technology 43 (5), 537–545.

Joshi, V.K., Sharma, S.K., 1994. Effect of method of must preparation and initial sugar levels on the quality of apricot wine. Research on Industry 39 (4), 255–257.

Joshi, V.K., Bhutani, V.P., Thakur, N.K., 1999a. Composition and nutrition of fermented products. In: Joshi, V.K., Pandey, A. (Eds.), Biotechnology: Food Fermentation, vol. I. Educational Publishers and Distributors, New Delhi, pp. 259–320.

Joshi, V.K., Sandhu, D.K., Thakur, N.S., 1999b. Fruit based alcoholic beverages. In: Joshi, V.K., Pandey, A. (Eds.), Biotechnology: Food Fermentation (Microbiology, Biochemistry and Technology), vol. II. Educational Publisher & Distributors, Ernakulum, New Delhi, pp. 647–744.

Joshi, V.K., Sharma, S., Bhushan, S., 2005. Effect of method of preparation and cultivar on the quality of strawberry wine. Acta Alimentaria 34 (4), 339–355.

Joshi, V.K., Sharma, S., Bhushan, S., Attri, D., 2004. Fruit based alcoholic beveregaes. In: Ashok, P. (Ed.), Concise Encyclopedia of Bioresource Technology. Haworth Inc., New York, pp. 335–345.

Joshi, V.K., Sharma, S., John, S., Kaushal, B.B.L., Rana, N., 2009. Prepration of antioxidant rich apple and strawberry wines. Proceedings of the National Academy of Sciences of United States of America 79 (IV), 415–420.

Joshi, V.K., Thakur, N.S., Bhat, A., Garg, C., 2011a. Wine and brandy: a perspective. In: Joshi, V.K. (Ed.), Handbook of Enology: Principles Practices and Recent Innovations, vol. 1. Asia Tech Publisher, New Delhi, pp. 146–208.

Joshi, V.K., Attri, D., Singh, T.K., Abrol, G.S., 2011b. Fruit wines: production technology. In: Joshi, V.K. (Ed.), Handbook of Enology : Principles Practices and Recent Innovations, vol. 3. Asia Tech Publisher, New Delhi, pp. 1177–1221.

Joshi, V.K., Sharma, R., Sharma, S., Abrol, G.S., 2011c. Effect of dilution and de-acidification on physico-chemical and sensory quality of seabuckthorn wine. Journal of Hill Agriculture 2 (1), 47–53.

Joshi, V.K., Sharma, R., Girdher, A., Abrol, G., 2012. Effect of dilution and maturation on physico-chemical and sensory quality of jamun wine. Indian Journal of Natural Products Radiance 3 (2), 222–227.

Joshi, V.K., Kumar, N., 2015. Optimization of low alcoholic bitter gourd apple beverage by applying response surface methodology (RSM). International Journal of Food and Fermentation Technology 5 (2), 191–199.

Khurdiya, D.S., Ray, K., 1985. Processing of Jamun (*Syzygium cuminii* L.) fruit into a ready-to-serve beverage. Journal of Food Science and Technology 22, 27–28.

Kumar, K., Kaur, R., Sharma, S.D., 2011. Fruit cultivars for winemaking. In: Joshi, V.K. (Ed.), Handbook of Enology: Principles Practices. vol. 1. Asia Tech Publisher, New Delhi, pp. 237–265.

Lee, P.R., Yu, B., Curran, P., Liu, S.Q., 2011. Impact of amino acid addition on aroma compounds in papaya wine fermented with *Williopsis mrakii*. South African Journal of Enology and Viticulture VWc 32 (2), 220–228.

Lodge, N., 1981. Kiwi fruit: two novel processed products. Food Technology NZ 16 (7), 35–43.

Lokesh, K., Suresh, G.J., Jagadeesh, S.L., 2014. Effect of yeast strains and must types on quality of jamun wine. The Asian Journal of Horticulture 9 (1), 24–27.

Luthi, H., 1953. Gnafu nung and Behandlung der obstwein in kleinbrieh, second ed. Verlag Huber and Co., A.G., Franunfeld.

Mahant, K., Sharma, S., Thakur, A.D., 2016. Effect of nitrogen source and citric acid addition on wine preparation from Japanese persimmon. Journal of Institute of Brewing (in press).

Maldonadd, O., Rolz, C., Schneides, C., 1975. Wine and vinegar production from tropical production from tropical fruits. Journal of Food Science 40, 262–265.

Maragatham, C., Panneerselvam, A., 2011a. Standardization technology of papaya wine making and quality changes in papaya wine as influenced by different sources of inoculums and pectolytic enzyme. Advances in Applied Science Research 2 (3), 37–46.

Maragatham, C., Panneerselvam, A., 2011b. Isolation, identification and characterization of wine yeast from rotten papaya fruits for wine production. Advances in Applied Science Research 2 (2), 93–98.

Nzabuheraheza, F.D., Nyiramugwera, A.N., 2014. Golden wine produced from mixed juices of passion fruit (*Passiflora edulis*), mango (*Mangifera indica*) and pineapple (*Ananas comosus*). African Journal of Food Agriculture, Nutrition and Development 14 (4), 9104–9116.

Nagadowithana, T.W., Stainkraus, K.H., 1976. Influence of the rate of ethanol production and accumulation on the viability of *Saccharomyces cerevisiae* in "rapid fermentation". Applied and Environmental Microbiology 31, 158–162.

Navindra, P.S., Lynn, S.A., Yanjun, Z., Rupo, L., Daniel, S., Henry, S.S., David, H., 2006. Blackberry, black raspberry, blueberry, cranberry, red raspberry, and strawberry extracts inhibit growth and stimulate apoptosis of human cancer cells *in vitro*. Journal of Agricultural and Food Chemistry 54, 9329–9339.

Orwa, C., Mutua, A., Kindt, R., Jamnadass, R., Anthony, S., 2009. Agroforestree Database: A Tree Reference and Selection Guide Version 4.0. http://www.Worldagroforestry.org/sites/treedbs/treedatabases.as.

Patel, V.B., Pandey, S.N., Singh, S.K., Das, B., 2005. Variability in Jamun (*Syzygium cuminii*) accessions from Uttarpradesh and Jharkhand. Indian Journal of Horticulture 62 (3), 244–247.

Picinelli, A., Suarez, B., Garcia, L., Mangas, J.J., 2000. Changes in phenolic contents during sparkling wine making. American Journal of Enology and Viticulture 51 (2), 144–149.

Pilando, L., Wrolstad, R.E., Heatherbell, D.A., 1985. Influence of fruit composition, maturity and mold contamination on the colour and stability of strawberry wine. Journal of Food Science 50, 1121–1125.

Prasad, S., Kumar, U., 2010. A Handbook of Fruit Production. Published by Agrobios, (India), Jodhpur.

Rai, D.R., Chadha, S., Kaur, M.P., Jaiswal, P., Patil, R.T., 2011. Biochemical, microbiological and physiological changes in Jamun (*Syzygium cuminii* L.) kept for long term storage under modified atmosphere packaging. Journal of Food Science and Technology 48 (3), 357–365.

Rose, R., 1992. Orange based Sparkling Alcoholic Beverage. French Patent Application, FR 2657, 878 AL (FSTA 24(2): 2H,88, 1992).

Samarajeewa, U., Mathew, D.T., Wijeratne, M.C.P., Warnakula, T., 1985. Effect of sodium metasulphite on ethanol production in coconut inflorescence sap. Food Microbiology 2, 11–17.

Sandhu, D.K., Joshi, V.K., 1995. Technology, quality and scope of Fruit Wines with special reference to apple. Indian Food Industry 14 (1), 24–34.

Schanderl, H., 1959. Die Mikrobiologie des mestes and Weines, second ed. Eugen Ulmer, Stuttgart (Revised by H.H. Dittrich, 1977).

Shah, P.K., Joshi, V.K., 1999. Influence of different sugar sources and addition of wood chips on physico-chemical and sensory quality of peach brandy. Journal of Scientific & Industrial Research 58 (6), 995–1004.

Sharma, S., Joshi, V.K., 2003. Effect of maturation on the phisico-chemical and sensory quality of strawberry wine. Journal of Scientific & Industrial Research 62 (4), 601–608.

Sharma, S., Joshi, V.K., 2004. Flavour profiling of strawberry wines by quantitative descriptive analysis technique. Journal of Food Science and Technology 41 (1), 22–26.

Sharma, S., Joshi, V.K., Abrol, G., 2009. An overview on Strawberry [*Fragaria* × *ananassa* (Weston) Duchesne ex Rozier] wine production technology, composition, maturation and quality evaluation. Natural Product Radiance 8 (4), 356–365.

Sharma, S., Mahant, K., 2015. Improved Persimmon Wine with Enhanced Antioxidant Activity and Standardized Method of Production. Indian National Patent filling No. 3884/DEL/2015.

Singh, M., Panesar, P.S., Marwaha, S.S., 1998. Studies on the suitability of kinnow fruits for the production of wine. Journal of Food Science and Technology 35 (5), 455–457.

Singh, R.S., Kaur, P., 2009. Evaluation of litchi juice concentrate for the production of wine. Natural Product Radiance 8 (4), 386–391.

Srivastava, K.K., Das, B., 2005. Flowering and fruiting behaviour of persimmon in Kashmir valley. Agricultural Science Digest 25 (4), 287–289.

Seeram, N.P., Adams, L.S., Zhang, Y., Lee, R.S., Scheuller, H.S., Heber, D., December 13, 2006. Blackberry, black raspberry, blueberry, cranberry, red raspberry, and strawberry extracts inhibit growth and stimulate apoptosis of human cancer cells in vitro. Journal of Agricultural and Food Chemistry 54 (25), 9329–9339.

Stockley, C.S., 2011. Therapeutic value of wine: a clinical and scientific perspective. In: Joshi, V.K. (Ed.), Handbook of Enology: Principles Practices and Recent Innovations, vol. 1. Asia Tech Publisher, New Delhi, pp. 146–208.

Thakur, A.D., Saklani, A., Sharma, S., Joshi, V.K., 2014. Effect of different sugar sources, pectin esterase and acidulant concentration of pumpkin wine production. International Journal of Food and Fermentation Technology 4 (1), 67–78.

Tewari, H.K., Chawla, A., Sahota, H.K., Kathuria, S., Julka, T.S., 2001. Fermented beverage from garlic. In: Mann, A.P.S., Munshi, S.K., Gupta, A.K. (Eds.), Biochemistry Environment and Agriculture. Kalyani Publishers, New Delhi, pp. 276–280.

Thuong, P.T., Lee, C.H., Dao, T.T., Nguyen, P.H., Kim, W.G., et al., 2008. Triterpenoids from the leaves of *Diospyros kaki* (Persimmon) and their inhibitory effects on protein tyrosine phosphatase 1B. Journal of Natural Products 71, 1775–1778.

Vyas, S.R., Chakravorty, S.R., 1977. Wine Making at Home. Haryana Agricultural University Bulletin, Hissar.

Vyas, K.K., Sharma, R.C., Joshi, V.K., 1989. High flavoured low-alcoholic drink from litchi. Food Beverage World 16 (1), 30.

Zhou, C., Sheng, Y., Zhao, D., Wang, Z., Tao, J., 2010. Variation of Oleanolic and Ursolic acid in the flesh of Persimmon fruit among different cultivars. Molecules 15, 6580–6587.

CITRUS WINES

7.4

S. Selli[1], H. Kelebek[2], P.S. Panesar[3]

[1]Cukurova University, Adana, Turkey; [2]Adana Science and Technology University, Adana, Turkey; [3]Sant Longowal Institute of Engineering and Technology, Longowal, Punjab, India

1. INTRODUCTION

Citrus is the most economically important tree fruit crop in the world. These fruits may be divided into three botanical species: *Citrus sinensis*, the common orange; *Citrus nobilis*, the mandarin group; and *Citrus documana*, the grapefruit (von Loesecke et al., 1936). Of these, orange is the most commonly grown citrus fruit in the world. In 2012, 68.2 million tons of oranges were grown worldwide, primarily in Brazil, the United States, China, and India (FAO, 2014). The majority of citrus arrives at market in the form of processed products, such as single-strength orange juice and frozen juice concentrate. One possible use of citrus fruits is in the production of fruit wines. Several classes of citrus fruits are available for the preparation of wine and other alcoholic beverages.

Wine is defined as an alcoholic beverage, which is produced by the fermentation of fresh grapes or must, and winemaking is one of the most ancient technologies and is now one of the most commercially prosperous biotechnological processes. Grapes and apples are the crops most widely grown for the production of juice for winemaking. Although grapes and apples are by far the most often used fruits, various other fruits such as oranges, kiwi, peaches, and plums may also be used to make wine. Increasing interest in human health, nutrition, and disease prevention has enlarged the consumer demand for functional food, including fruits and their products such as wine (Rupasinghea and Clegg, 2007). Additionally, the global food industry uses a variety of preservation and processing methods to extend the shelf life of fruits and vegetables so that they can be consumed year round and transported safely to consumers all over the world, not only those living near the growing region (Barret and Lloyd, 2012). Therefore, the utilization of ripe fruit or their juices for wine production is considered an attractive means of utilizing surplus and overripe fruit (Jagtap and Bapat, 2015).

2. ORANGE WINE

Oranges vary in color, flavor, acidity, and sugar profile depending upon the soil character and climate in the regions where they are grown and the methods of horticulture applied in orange orchards. It is recommended that fully ripened, even a bit overripened, oranges be considered to achieve low acidity, good color, and good flavor. Orange juice is well known for its high acid content; producing an orange wine with good balance is a major challenge of utmost importance to the winemaker. Without this balance, orange wines are excessively acidic in character. In addition to high acidity, orange does not have enough natural sugar to reach a sufficient alcohol level after fermentation. Sugar additions can be calculated based on the Brix level of the orange juice. The Brix level of orange juice typically measures between 10 and 12.5 °Bx. To reach 12% alcohol by volume, the orange juice would require approximately 22.5 °Bx. For example, a natural °Bx of 10 would need to be adjusted with sufficient sugar to be increased by 12.5 °Bx.

The scientific literature contains only a limited number of papers on citrus wines. Most of them are related to orange wine (Joslyn and Marsh, 1936; Amerine et al., 1980; Canbas and Unal, 1994; Selli, 2007; Selli et al., 2002, 2003, 2008; Fan et al., 2009a,b). The earliest work was that by Cruess (1914), who prepared orange wine using waste oranges. He fermented clear orange juice with pure yeast, and obtained an orange wine that filtered brilliantly clear and contained between 4.25% and 4.45% alcohol by volume and 1.5% total acid. Then, Joslyn and Marsh (1936) mentioned some suggestions for making orange wine. They recommended sweetening the orange juice by adding sugar before fermenting so as to yield a sweet alcoholic beverage containing about 13% alcohol, and to use pure wine yeast in the sweetened juice. Since 2000, new studies have focused on the phenolic, aroma, and aroma-active compositions of orange wines (Selli et al., 2002, 2003, 2008; Fan et al., 2009a,b; Kelebek et al., 2009). Additionally, Escudero-López et al. (2013) reported that the potential beverage of fermented orange juice would suppose a new manner of orange commercialization and the possibility of providing the consumer with a new healthful alcoholic beverage that widens the supply of other similar drinks with higher alcohol content, such as wine, beer, and cider.

2.1 ORANGE WINEMAKING

2.1.1 Orange Cultivars and Properties

Sound and mature oranges are desired for the production of orange wine. Orange maturity is regularly assessed to determine the harvest date according to variety. Different cultivars of oranges (*C. sinensis* L. Osbeck) have been used in different studies of winemaking, such as Kozan Yerli (Selli et al., 2002), Moro (Selli, 2007), Navel (Escudero-López et al., 2013), Valencia (Canbas and Unal, 1994), Pera Rio (Santos et al., 2013), Jincheng (Li et al., 2010), Qixuecheng (Fan et al., 2009a), and Washington Sanguine (Fan et al., 2009b). The important general properties of orange juices such as color, sugars, flavor, yield of juice, and total soluble solids content of the juice increase with maturity of the juice. Midseason and late-season oranges are better quality for orange winemaking than the early season oranges. Generally, the oranges for winemaking should be mature, but not overmature, because if the oranges are overmature, the juice and produced wine can acquire an off-flavor. Total sugar and acid contents in orange juice are not at the desirable amounts and ratios for winemaking. To guarantee proper fermentation with expected product quality, it is necessary to adjust the sugar and acid content for fermentation.

2.1.2 Extraction of the Juice

Before the extraction of orange juice, the oranges should be sorted to remove rotten fruit, dust, soil, microorganisms, and unfit fruit. They should then be washed thoroughly. The automatic juice extractor has been a main factor in the development of the orange and other citrus juice industry. Orange juice can be extracted using the same machinery for orange winemaking. Some of the new types of orange juice extractors produce juice of low peel oil content due to their better adjustment for fruit size.

For efficient juice extraction, the fruit should be graded into three sizes, as the machines are usually built to work best on fruit of fairly uniform size (Amerine et al., 1980). Mechanically extracted juice contains seeds and segment membranes, which must be removed quickly to avoid adding bitterness to the juice. Therefore, special precautions have to be taken in the extraction of juice. After extraction, the juice is finished in a finisher as quickly as possible. The finisher separates the pulp, peel, and seeds from the stream of liquid. The liquid that leaves the finisher is fresh orange juice. This juice is moved through pipes to stainless steel fermentation tanks, where the juice is fermented to produce orange wine.

2.1.3 Alcoholic Fermentation of the Juice

A general flow diagram of the orange winemaking process is given in Fig. 7.4.1. Orange juice is obtained using an extractor. Then it is passed through the finisher to remove the seeds and pulp, and 50–75 mg/L sulfur dioxide (SO_2) is added. Orange juice does not have enough natural sugar to reach a sufficient alcohol level after fermentation. Brix in orange juice typically measures between 10 and 12.5 °Bx. Each degree of Brix will convert to about 0.535% alcohol during fermentation. Consequently, to

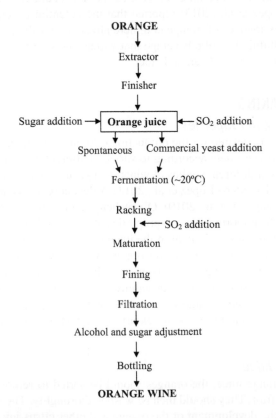

FIGURE 7.4.1

The flow diagram of orange winemaking.

Adapted from Canbas, A., Kelebek, H., 2014. Fermented alcoholic beverages. In: Aran, N. (Ed.), Food Biotechnology: Nobel Akademik. Yayincilik Danismanlik, Ankara, Turkey.

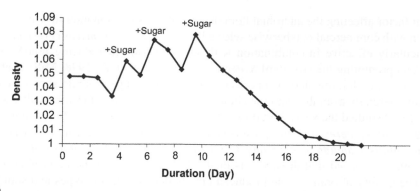

FIGURE 7.4.2

The course of ethyl alcohol fermentation.

Adapted from Selli, S., Canbaş, A., Ünal, Ü., 2002. Effect of bottle colour and storage conditions on browning of orange wine.
Nahrung 46, 64–67.

achieve 12% alcohol by volume, the orange juice would require approximately 22.5 °Bx. For example, a natural Brix of 10 °Bx would need to be adjusted with sufficient sugar to reach a 12.5 °Bx increase. When all the sugar is added at the beginning of fermentation; the concentration levels can be too high for yeast activities, so that the sugar can actually inhibit the fermentation.

Therefore, sugar must be added to the original must to achieve the desired alcohol content in two or three steps (Fig. 7.4.2). The addition of SO_2 is traditionally considered an efficient method to protect and preserve the wine at different stages of its elaboration. It has been used since Roman times for disinfection and cleaning of wine cellars. In wine, this compound has multiple and diverse properties. During winemaking, SO_2 is used on three main occasions. First, in the juice during the prefermentation step, with the prime objective of preventing its oxidation; later, to inhibit microbial growth that can alter the wine, once the fermentation processes have finished and before the aging or storage steps, and, finally, just before bottling, to stabilize the wine and prevent any alteration or accident in the bottles (Pozo-Bayón et al., 2012).

To the fresh orange juice should be added 50–100 mg/L sulfur dioxide. The addition step of this agent is shown in Fig. 7.4.1. The sugar content, fermentation temperature, SO_2 concentration, and yeast strain are important factors in performing a successful fermentative process of fruit wine (Duarte et al., 2009, 2010). Fermentation should be maintained at 18–20°C. Orange juice fermentation is rapid, as it is an excellent culture medium for yeast. Because the fermentation rate of orange juice is very fast, much heat will evolve in the tanks and it may be necessary to cool the fermenting juice so that the temperature does not rise to an undesirable point (Rivard, 2009). Similarly, Von Loesecke et al. (1936) reported that citrus juices could not be fermented at high temperatures because acetic acid microorganisms produced a substance with a distinct aroma of ethyl acetate. Orange wine fermentation is performed either traditionally without inoculation or by the addition of a selected wine yeast into the juice. Ethyl alcohol yield and concentration, tolerance to ethyl alcohol and variations in fermentation temperature, resistance to high sugar contents, and the ability to flocculate and to produce or not certain volatile components are constant sources of interest in the choice of the yeast strain to be employed for the production of various alcoholic drinks (Hammond, 1995).

An important factor affecting the microbial flora in ethyl alcohol fermentation in wines is the practice of inoculation with commercial or otherwise selected strains of *Saccharomyces cerevisiae*. Inoculation can be particularly effective in combination with sulfur dioxide in reducing non-*Saccharomyces* populations and promoting the growth of *S. cerevisiae* (Constanti et al., 1998; Mills et al., 2008). To ensure clean and smooth fermentation, a pure culture of a selected yeast strain should be used. Many strains are commercially available, and the choice of strain depends on the winemaker's preference. Erten et al. (2003) studied the yeast flora during the alcoholic fermentation of orange wine produced from cv. Kozan. Non-*S. cerevisiae* spp. yeast isolated at the beginning of alcoholic fermentation were *Rhodotorula glutinis*, *Candida pulcherrima*, *Candida colliculosa*, and *Cryptococcus laurentii* in this study. The authors reported that these yeasts died off after day 2 and the main wine yeast was *S. cerevisiae*. The quality of orange wine produced greatly depends on the types and sources of yeast strains employed in the fermentation process. *S. cerevisiae* var. *ellipsoideus*, which produced a moderate level of alcohol and an appreciable amount of vitamin C, appears to be the best organism for orange wine production (Okunowo et al., 2005). In another study, the different yeast strains *S. cerevisiae* (isolated from yam), *S. cerevisiae* (from sugarcane molasses), *Saccharomyces carlsbergensis* (from sugarcane molasses), and *S. cerevisiae* var. *ellipsoideus* (from orange juice) were used for fermentation of orange juice and affected the total alcohol concentration in orange wine samples. Based on the results, orange wine obtained from *S. cerevisiae* var. *ellipsoideus* contains the highest ethanol level, whereas *S. cerevisiae* from sugarcane molasses produced wine with the lowest ethanol concentration. The methanol concentration varied between 9.51% with *S. cerevisiae* var. *ellipsoideus* and 14.93% with *S. carlsbergensis* (Okunowo and Osuntoki, 2007).

When ethyl alcohol fermentation is complete, the young wine is transferred to stainless steel tanks, which are completely filled to prevent undesirable bacteria and growth of film yeasts. Film yeasts can grow easily on orange wine exposed to the air and high storage temperature. When the wine is settled in 2 or 3 weeks, it is racked and should be given a dose of 100 mg/L SO_2 to prevent oxidation and other potential problems. Then, the resulting wine should be clarified, filtered, bottled, and sold well within a 1-year production–marketing cycle. Orange wines are not stored for long periods of time in the cellar like normal wine. The contents of ascorbic acid and sugars together with storage temperature, sunlight, and color of the bottles affect browning in orange juice and orange wines. Selli et al. (2002) investigated the effects of bottle color (clear white, green, brown), storage temperature (13–14 and 23–26°C), and storage time (0, 75, and 150 days) on the browning of orange wine. The results showed that orange wine stored in darker brown bottles at low temperature (13–14°C) showed low browning index values compared with wine stored in green and clear white bottles; however, storage at two different temperatures did not significantly affect the browning index.

2.2 THE CHEMICAL COMPOSITION OF ORANGE WINE

2.2.1 General Composition of Orange Wine

The general composition of orange wine is given in Table 7.4.1. The total acid and pH contents of orange wine range from 4.2 to 7.8 g/L and 3.5–4.0, respectively. The principal acid found in oranges, and therefore the wine, is citric acid. Acidity greatly influences the taste, color, and aromatic profile of the wine as well as the stability and microbiologic control of the wine quality by stopping or at least retarding the growth of many potentially harmful microorganisms that would spoil the wine (Selli and Kelebek, 2015).

Table 7.4.1 General Composition of Orange Wine

Analysis	Values	References
Total acidity[a] (g/L)	4.2–7.8	Selli et al. (2003, 2002, 2008), Fan et al. (2009a,b), and Kelebek et al. (2009)
pH	3.5–4.0	Selli et al. (2003, 2002, 2008), Fan et al. (2009a,b), and Kelebek et al., 2009
Ethanol (%, v/v)	10.4–12.6	Selli et al. (2003, 2002, 2008), Fan et al. (2009a,b), and Kelebek et al., 2009
Extract (g/L)	50–76	Selli et al. (2003, 2002, 2008) and Kelebek et al. (2009)
Density (20°C/20°C)	0.9804–1.022	Selli et al. (2003, 2002, 2008), Fan et al. (2009a), and Kelebek et al. (2009)
Volatile acidity[b] (g/L)	0.18–0.32	Selli et al. (2003, 2002, 2008), and Kelebek et al. (2009)
Total sugars (g/L)	3.8–70	Selli et al. (2003, 2002, 2008) and Fan et al. (2009a)
Residual sugars (g/L)	1.5–1.6	Fan et al. (2009a,b)
Ash (g/L)	2.8–3.1	Selli et al. (2003, 2002, 2008) and Kelebek et al. (2009)
Ash alkalinity[c] (g/L)	2.34	Selli et al. (2003, 2002)
Free SO_2 (mg/L)	6–15	Selli et al. (2003, 2002, 2008) and Kelebek et al. (2009)
Bound SO_2 (mg/L)	71–75	Selli et al. (2003, 2002, 2008) and Kelebek et al. (2009)
L*	76.45	Kelebek et al. (2009)
a*	−7.23	Kelebek et al. (2009)
b*	32.40	Kelebek et al. (2009)

[a]As citric acid.
[b]As acetic acid.
[c]As K_2CO_3.

Ethanol is the principal organic by-product of wine fermentation. The level of ethanol in orange wine ranges from 10.4% to 12.5% (Table 7.4.1). Under standard fermentation conditions, ethanol can accumulate up to about 5–8% in orange juice fermentation. Higher levels of alcohol can be obtained by the sequential addition of sugar during fermentation. In addition to its significant physiological and psychological effects on humans, ethanol is crucial to the stability, aging, and sensory characteristics of wine. During fermentation, the increasing alcohol content progressively limits the growth of microorganisms (Jackson, 2000).

SO_2 can exist in wine in free and bound forms. Free SO_2 is the active, effective form of SO_2 in the winemaking process. It has many valuable attributes in wines. At present, it is difficult to make good quality wines without the addition of SO_2, as the sulfur is a key component for yeast growth during fermentation. The free and bound SO_2 levels in orange wine range from 6 to 15 mg/L and 71–75 mg/L, respectively (Table 7.4.1). SO_2 is added to both orange juice and orange wine for protection from oxidation and microbial spoilage.

As shown in Table 7.4.1, the volatile acidity of orange wine ranges from 0.18 to 0.32 g/L in different studies. Volatile acidity corresponds essentially to acetic acid and can play a significant role in wine aroma, and an excessive concentration of this alcoholic fermentation by-product is highly detrimental to wine quality. Its production is carefully monitored and controlled throughout the wine production process. The amount of volatile acidity produced is usually low (0.25–0.50 g/L) but may be higher under certain fermentation conditions (Bely et al., 2003).

2.2.2 Flavor Composition of Orange Wine

The flavor characteristics of foodstuffs are major factors in determining consumer acceptance and preference. Orange is probably the most recognized and accepted flavor in the food and beverage industry worldwide; it is widely used to flavor or aromatize foods and beverages because of its distinctive flavor and aroma. Its fresh and unique flavor is due to complex combinations of several odor components that have interdependent quantitative relationships. Important contributors to orange juice flavor include esters, aldehydes, ketones, terpenes, and alcohols (Nisperos-Carriedo and Shaw, 1990; Shaw, 1991).

The volatile content of orange juice can be changed by ethyl alcohol fermentation, because of the production of yeast volatiles and the metabolism of original fruit volatiles (Cole and Noble, 1995). The fermentation process for the elaboration of the beverage depends on the performance of yeast to convert sugars into alcohol, esters, higher alcohols, and other volatile and nonvolatile compounds. The aroma profile of Turkish Kozan orange wines was determined using liquid–liquid extraction and GC/MS by Selli et al. (2003). In this study, 75 volatile components including terpenes, alcohols, esters, volatile phenols, acids, ketones, aldehydes, lactones, and C13-norisoprenoids were identified in Kozan wine. In another study, Selli (2007) investigated the volatile compounds in blood orange (cv. Moro, *C. sinensis* L. Osbeck) wine.

During alcoholic fermentation, yeasts play a significant role in the formation and modulation of wine aroma compounds by way of releasing varietal volatile components from fruit precursors and, in the meantime, synthesizing de novo yeast-derived volatile components (Duarte et al., 2010; Li et al., 2010; Molina et al., 2009; Sun et al., 2013). Fan et al. (2009a) studied the effects of spontaneous and inoculated (active dry yeast) fermentation on the volatile compounds of orange wine obtained from cv. Qixuecheng using solid-phase microextraction (SPME)–GC/MS. Surprisingly, both sensory and GC/MS analysis showed no differences in volatile compounds of orange wine produced by spontaneous and inoculated fermentation. The same authors also investigated the free and glycosidically bound volatile compounds in orange wine made from *C. sinensis* L. Osbeck cv. Washington Sanguine (Fan et al., 2009b). A total of 19 free and 3 bound volatiles (ethyl 3-hydroxybutanoate, ethyl 3-hydroxyhexanoate, and 3-oxo-α-ionol) were identified in orange wine.

Over 1300 volatile compounds have been identified in alcoholic beverages (Nykanen, 1986). GC/olfactometry (GC/O) has become the most widely used technique for identification of aroma-active compounds in aromatic extracts (Grosch, 1993; Mistry et al., 1997). This technique uses the human nose as a detector to distinguish the single-volatile compounds. Selli et al. (2008) first studied the most aroma-active compounds in orange wine by the two GC/O techniques, detection frequency and time intensity (Table 7.4.2). A total of 63 compounds were identified and quantified in orange wine. The results of the GC/O analysis showed that 35 odorous compounds were detected by the panelists. Of these, 28 aroma-active compounds were identified. Alcohols followed by terpenes and esters were the most abundant aroma-active compounds of the orange wine.

2.2.3 Phenolic Composition of Orange Wine

The quality of orange wine depends on numerous parameters, and phenolic compounds are important among these. The sensory contributions of phenolic compounds mainly concern color and flavor. As in grape wine, the phenolic composition of orange juice and wine depends first on the initial composition of the fruit and then on the extraction and solubilization of phenols during pressing. Also, during the winemaking period, the phenolic content in wine undergoes considerable changes. Little information is available on the chemistry of these changes. Much research on fruit wines has shown that phenolic

Table 7.4.2 Aroma-Active Compounds of Orange Wine

LRI[a]	Compound	Odor Description[b]	Intensity[c]	Number of Judges[d]
1033	1-Propanol	Plastic	5	7
1041	Ethyl butanoate	Fruity, sweet	5	8
1118	Unknown	Plastic	4	8
1166	Unknown	Green, floral	4	6
1217	Isoamyl alcohol	Chemical, harsh	5	7
1254	(E)-β-ocimene	Fruity, minty	5	5
1272	Hexyl acetate	Floral	4	5
1324	3-Methyl-1-pentanol	Roasty	6	8
1354	(E)-3-hexen-1-ol	Green, floral	4	6
1374	(Z)-3-hexen-1-ol	Green	5	6
1398	Unknown	Floral	4	5
1438	Unknown	Bread, roasty	4	7
1452	Furfural	Pungent	3	6
1469	(Z)-furan linalool oxide	Green, sweet	5	5
1524	2,3-Butanediol	Creamy	5	5
1548	Linalool	Floral, citrusy	7	8
1568	1-Octanol	Floral	6	6
1606	Terpinen-4-ol	Rancid	4	6
1647	γ-Butyrolactone	Cheesy, burnt sugar	7	8
1690	α-Terpineol	Green, violet	5	7
1714	3-(Methylthio) propanol	Boiled potato, rubber	5	8
1720	Ethyl acetyl acetate	Fruity	4	4
1748	Citronellol	Floral	6	6
1781	Ethyl-4-hydroxybutanoate	Fruity, floral	5	5
1831	Geraniol	Floral, citrusy	7	8
1864	Benzyl alcohol	Citrusy, sweet	4	7
1908	2-Phenylethanol	Floral, rose	7	8
1974	Unknown	Plastic, roasty	5	7
2029	Diethyl malate	Caramel	7	6
2061	Octanoic acid	Sweaty	3	4
2120	Eugenol	Clove	6	5
2164	Unknown	Cotton candy	5	7
2193	4-Vinyl-2-methoxyphenol	Nutty, spicy	4	6
2285	Ethyl-2-hydroxy-3-phenyl propanoate	Caramel, roasty	5	7
2318	Unknown	Rubber	6	5

[a]*Linear retention index calculated on a DB–Wax capillary column.*
[b]*Odor description as perceived by panelists during olfactometry.*
[c]*Average intensity.*
[d]*Number of judges who detected an odor.*

compounds increase during alcoholic fermentation (Reddy and Reddy, 2009; Sharma et al., 2013). As expected, after alcoholic fermentation and the maturation period, the amounts of these compounds decrease significantly (Fig. 7.4.3). The reasons for the decrease in phenolic compounds in orange juice after alcoholic fermentation probably are precipitation or oxidation during the process; the combination or adsorption of phenolic compounds with solids, proteins, or even yeasts; and polymerization, which induces an important loss of these compounds. The longer the juice remains in contact with the air and pomace, the greater the loss of phenolic compounds.

A total of 13 phenolic compounds, including hydroxybenzoic acids (2), hydroxycinnamic acids (5), and flavanones (6), are reported in blonde orange wines (Table 7.4.3). In addition of these groups, anthocyanins (6) are also identified in blood orange wines. Chemical structures of these compounds are given in Fig. 7.4.4. The total phenolic content in wines of blood orange varieties ranges from 129 to 276 mg/L, whereas in blonde orange wines it is 163–177 mg/L (Kelebek et al., 2008).

Anthocyanins are the major compounds in blood oranges (Arena et al., 2001), as they account for the largest proportion of the total phenolic content, comprising 36–40% of the phenolics in blood orange juice and wine. The concentration of anthocyanins decreases markedly after alcoholic fermentation. It is observed, as in conventional yeast fermentation, that there is a fall in anthocyanin content and an increase in the polymerized forms, which are the products of condensation between phenolics and anthocyanins.

Hydroxybenzoic acids: Orange wines contain 2.38–4.25 mg/L hydroxybenzoic acids, which comprise gallic and protocatechuic acids (Table 7.4.3). The major hydroxybenzoic acid is gallic acid (3,4,5-trihydroxybenzoic acid). Gallic acid is a naturally abundant plant phenolic compound. It is present in foods of plant origin, and since it was found to exhibit antioxidative properties it has attracted considerable interest (Kelebek et al., 2009).

Hydroxycinnamic acids: Caffeic, p-coumaric, ferulic, and sinapic acids have been detected in citrus fruit products in the form of esters, amides, and glycosides (Bocco et al., 1998; Fallico et al., 1996; Manthey and Grohmann, 2001; Mouly et al., 1997; Peleg et al., 1991; Rapisarda et al., 1998;

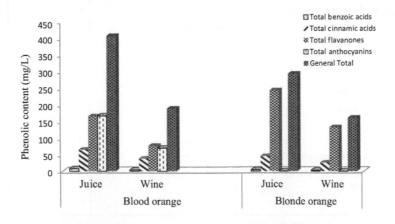

FIGURE 7.4.3

Phenolic distribution of orange juices and wines.

Adapted from Kelebek, H., Selli, S., 2014. Identification of phenolic compositions and the antioxidant capacity of mandarin juices and wines. Journal of Food Science and Technology 51 (6), 1094–1101.

Table 7.4.3 Phenolic Content (mg/L) of Blonde and Blood Orange Wine

Compound	Peak No.	Blonde Orange Wine	Blood Orange Wine	Ref.*
Hydroxybenzoic Acids				
Gallic acid	1	1.93–2.12	2.43–3.33	Kelebek et al. (2009, 2008) and Canbas et al. (2008)
Protocatechuic acid	2	0.45–0.65	0.59–0.93	Kelebek et al. (2009, 2008) and Canbas et al. (2008)
Total		**2.38–2.77**	**3.04–4.25**	Kelebek et al. (2009, 2008) and Canbas et al. (2008)
Hydroxycinnamic Acids				
Caffeic acid	3	2.57–2.83	2.35–3.48	Kelebek et al. (2009, 2008) and Canbas et al. (2008)
Chlorogenic acid	4	4.66–5.21	5.98–7.35	Kelebek et al. (2009, 2008) and Canbas et al. (2008)
p-Coumaric acid	5	1.58–2.16	3.39–4.80	Kelebek et al. (2009, 2008) and Canbas et al. (2008)
Ferulic acid	6	9.91–12.38	12.90–16.94	Kelebek et al. (2009, 2008) and Canbas et al. (2008)
Sinapic acid	7	7.78–8.22	9.40–10.38	Kelebek et al. (2009, 2008) and Canbas et al. (2008)
Total		**26.50–30.80**	**34.01–42.95**	Kelebek et al. (2009, 2008) and Canbas et al. (2008)
Flavanones				
Narirutin	8	21.67–23.56	19.96–18.85	Kelebek et al. (2009, 2008) and Canbas et al. (2008)
Naringin	9	1.29–2.27	1.25–2.53	Kelebek et al. (2009, 2008) and Canbas et al. (2008)
Hesperidin	10	90.65–94.76	38.86–66.22	Kelebek et al. (2009, 2008) and Canbas et al. (2008)
Neohesperidin	11	0.55–0.85	0.14–0.27	Kelebek et al. (2009, 2008) and Canbas et al. (2008)
Didymin	12	3.52–4.01	2.72–3.42	Kelebek et al. (2009, 2008) and Canbas et al. (2008)
Apigenin	13	16.12–17.45	11.23–15.37	Kelebek et al. (2009, 2008) and Canbas et al. (2008)
Total		**133.8–142.90**	**74.15–106.65**	Kelebek et al. (2009, 2008) and Canbas et al. (2008)
Anthocyanins				
Delphinidin 3-glucoside	14	–	1.04–5.13	Kelebek et al. (2008) and Canbas et al. (2008)
Cyanidin 3-glucoside	15	–	6.19–47.64	Kelebek et al. (2008) and Canbas et al. (2008)
Delphinidin 3-(6″-malonyl glucoside)	16	–	0.59–3.19	Kelebek et al. (2008) and Canbas et al. (2008)
Cyanidin 3-(6″-malonyl glucoside)	17	–	6.82–54.05	Kelebek et al. (2008) and Canbas et al. (2008)
Cyanidin 3-*O*-(6″-dioxalyl glucoside)	18	–	3.23–7.89	Kelebek et al. (2008) and Canbas et al. (2008)
Peonidin 3-(6″-malonyl glucoside)	19	–	0–3.86	Kelebek et al. (2008) and Canbas et al. (2008)
Total			**17.86–121.76**	Kelebek et al. (2008) and Canbas et al. (2008)
General total		**162.7–176.5**	**129.01–275.6**	Kelebek et al. (2008) and Canbas et al. (2008)

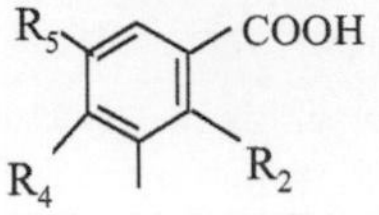

Hydroxybenzoic acids	R₂	R₃	R₄	R₅
Gallic acid	H	OH	OH	OH
Protocatechuic acid	H	OH	OH	H

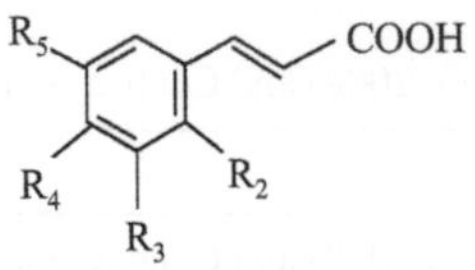

Hydroxycinnamic acids	R₂	R₃	R₄	R₅
Caffeic acid	H	OH	OH	H
p-Coumaric acid	H	H	OH	H
Ferulic acid	H	OCH₃	OH	H
Sinapic acid	H	OCH₃	OH	OCH₃

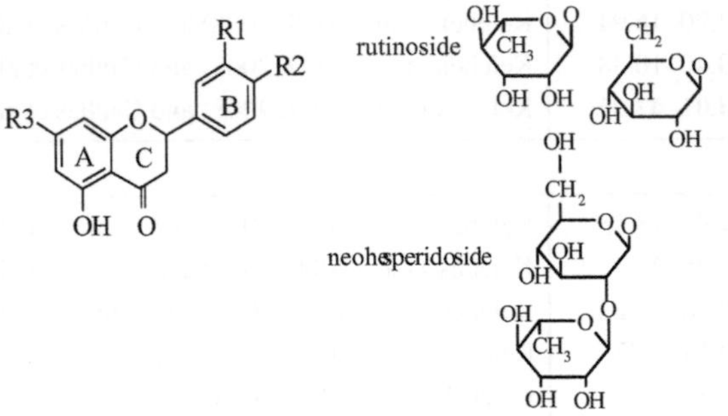

Flavanones	R₁	R₂	R₃
Narirutin	H	OH	ORut[a]
Naringin	H	OH	ONeo[b]
Hesperidin	OH	OMe	ORut[a]
Neohesperidin	OH	OMe	ONeo[b]
Didymin	H	OMe	ORut[a]

[a] rutinoside; [b] neohesperidoside

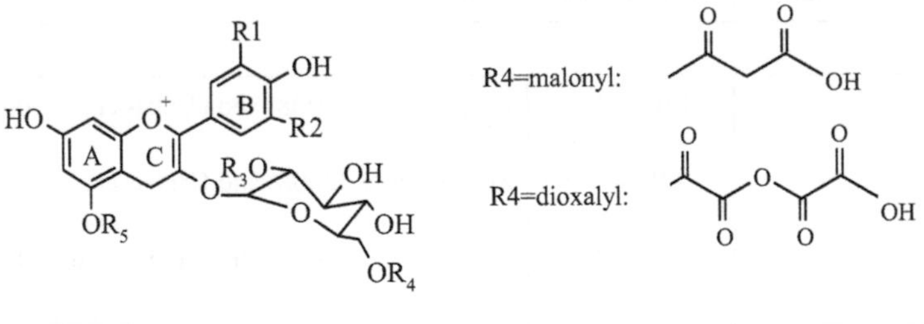

R4=malonyl:

R4=dioxalyl:

Anthocyanins	R1	R2	R3	R4	R5
Delphinidin 3-glucoside	OH	OH	H	H	H
Cyanidin 3- glucoside	OH	H	H	H	H
Delphinidin 3-(6"-malonylglucoside)	OH	OH	H	malonyl	H
Cyanidin 3-(6"-malonyl glucoside)	OH	H	H	malonyl	H
Cyanidin 3-O-(6"-dioxalyl glucoside)	OH	H	H	dioxalyl	H
Peonidin 3-(6"-malonyl glucoside)	OCH3	H	H	malonyl	H

FIGURE 7.4.4

Structures of the hydroxybenzoic acids, hydroxycinnamic acids, flavanones, and anthocyanins.

Risch and Herrmann, 1988). Only very small amounts of free phenolic acids are present (Manthey and Grohmann, 2001; Peleg et al., 1991). Ferulic acid has been reported to be the major phenolic acid in the Shamouti orange fruit (Peleg et al., 1991) and peel and their coproducts (Bocco et al., 1998; Manthey and Grohmann, 2001). Canbas et al. (2008) identified five hydroxycinnamic acids in wine, including caffeic acid, chlorogenic acid, p-coumaric acid, ferulic acid, and sinapic acid. It has been reported that ferulic acid is the most dominant hydroxycinnamic acid in blonde orange wine (9.91–12.38 mg/L) and blood orange wine (12.90–16.94 mg/L), as it accounts for the largest proportion of the total hydroxycinnamic acid contents (Table 7.4.3). Sinapic acid is the second most abundant hydroxycinnamic acid and is followed by chlorogenic, coumaric, and caffeic acids in orange juice and wine. The study of Rapisarda et al. (1998) found that ferulic acid (37.7 mg/L) was the main phenolic acid in Valencia juice.

Hydroxycinnamic acids have been associated with the formation of off-flavors in citrus products. Ferulic and p-coumaric acid esters may be enzymatically hydrolyzed during thermal processing and storage and then decarboxylated to unpleasant compounds such as p-vinylguaiacol and p-vinylphenol (Fallico et al., 1996; Klaren-De Wit et al., 1971; Lee and Nagy, 1990; Naim et al., 1988, 1993; Peleg et al., 1992).

Flavanones: Six flavanones, narirutin, naringin, hesperidin, neohesperidin, didymin, and apigenin, are reported in wines (Kelebek et al., 2009). Table 7.4.3 shows that, of the five flavanones, hesperidin is the most abundant in orange wine. Neohesperidin was the least abundant flavanone in wine. Hesperidin is tasteless and therefore does not contribute to the taste of orange juice and wine (Tomás-Barberán and Clifford, 2000). The level of hesperidin reported for orange wine (38.86–94.76 mg/L) is lower than previously reported by Tomás-Barberán and Clifford (2000), 104–637 mg/L, and Gorinstein et al. (2004), 122–254 mg/L, for orange juice. The reason for the decrease in hesperidin in orange juice after alcoholic fermentation is probably precipitation or oxidation during the process, the combination or adsorption of this compound. Narirutin is the second most abundant flavanone in orange wine. In the literature the narirutin-to-hesperidin ratio has been proposed for quality control of orange juices (Rouseff, 1988). Escudero-López et al. (2013) investigated quantitative changes in the flavanones of orange juice during fermentation. The content of flavanones at day 0 (orange juice), day 11 (peak of total flavonoids), and day 15 (end of fermentation) was evaluated. They reported that naringenin-7-O-rutinoside (363.7 mg/L), hesperetin-7-O-rutinoside (274.9 mg/L), isosakuranetin-7-O-rutinoside (47.9 mg/L), and hesperetin-7-O-glucoside (11.5 mg/L) are the main flavanones in orange juice (Fig. 7.4.5).

Alcoholic fermentation could extract more flavanones, although less soluble ones. Other technological treatments such as extraction pressure or industrial squeezing have produced this effect (Gil-Izquierdo et al., 2002; Escudero-López et al., 2013). Several previous papers have described an increase in the hesperetin absorption rate from orange juice when hesperetin-7-O-glucoside was present in a larger extension thanks to a previous enzymatic treatment of the orange juice (Habauzit et al., 2009). The previous studies showed that flavanone rutinosides from the juice soluble fraction are readily available to the body, whereas precipitates of flavanone rutinosides occurring in the juice cloud are not available for absorption (Gil-Izquierdo et al., 2003; Vallejo et al., 2010).

Anthocyanins: There is no detailed research on the effects of alcoholic fermentation on the anthocyanin profiles of blood orange wine. In our unpublished research, they were identified for the first time ever in the investigated blood orange wines. In that study, a total of six anthocyanin compounds were identified and quantified in Moro and Sanguinello wines. It was observed that the anthocyanin profiles of both Moro and Sanguinello wines were similar (Table 7.4.3). Lee (2002) reported that cyanidin

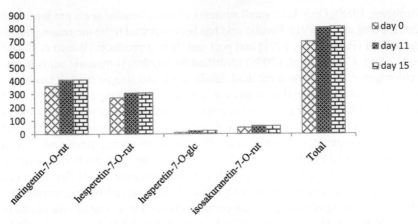

FIGURE 7.4.5

Flavanone content of orange juice during alcoholic fermentation.

Adapted from Escudero-López, B., Cerrillo, I., Herrero-Martín, G., Hornero-Méndez, D., Gil-Izquierdo, A., Medina, S., Ferreres, F., Berná, G., Martín, F., Fernández-Pachón, M.S., 2013. Fermented orange juice: source of higher carotenoid and flavanone contents.

Journal of Agricultural and Food Chemistry 61, 8773–8782.

Table 7.4.4 Sugar and Organic Acid Composition of Orange Juice and Wine

Compound	Juice	Wine
Sugars (g/L)		
Sucrose	59.34	44.68
Glucose	32.30	1.06
Fructose	28.55	3.04
Total	120.19	48.78
Organic Acids (g/L)		
Citric acid	12. 66	6.03
Ascorbic acid	0.49	0.23
Malic acid	1.06	0.34
Total	14.21	6.60

3-(6″-malonyl glucoside) was the most dominant anthocyanin in Moro juice, comprising more than 44% of the total anthocyanin content, followed by cyanidin 3-glucoside. compared with orange juice, the anthocyanin concentration is decreased dramatically.

2.2.4 Sugar Composition of Orange Wine

Orange contains a high concentration of sugar (10–12% w/v) and acids with organoleptic properties. Sucrose, glucose, and fructose are reported as the sugar components in orange juice and wine. Total amounts of sugar are 120.19 and 48.87 g/L in juice and wine, respectively (Table 7.4.4). Together, they represent about 80% of the total soluble solids of orange juice, and the ratio of sucrose/glucose/fructose is generally about 2:1:1 (Lee and Coates, 2000). Sucrose is present in the largest amounts in orange juice (59.34 g/L) and wine (44.68 g/L), accounting for about 49.12% of the total sugar content of orange juice and 91.6% of that of wine (Table 7.4.4).

2.2.5 Organic Acid Composition of Orange Wine

Three organic acids were reported in orange juice and wine: citric, ascorbic, and malic acids. The major organic acid is citric acid in orange juice (12.66 g/L) and wine (6.03 g/L) (Table 7.4.4.). Reported results for citric acid levels in fresh, hand-squeezed Navelina juice from various regions were 8.4–12.6 g/L (Saavedra et al., 2001). Malic acid was the second most abundant organic acid in orange juice (1.06 g/L) and wine (0.34 g/L). The lower concentration of organic acids in the wine compared to the juice can be explained by losses during fermentation. Citric and malic acids of sour orange juice from Antalya, Turkey, were reported by Karadeniz as 48.8 and 2.2 g/L, respectively (Karadeniz, 2004).

Orange wines are rich sources of ascorbic acid, which is an important antioxidant. The concentration of ascorbic acid is a significant indicator of orange wine quality. Kelebek et al. (2009) described vitamin C content in orange juice as decreasing significantly during spontaneous alcoholic fermentation (Table 7.4.4). The influence of fermentation on ascorbic acid content has been evaluated in other foods, obtaining similar results. In this way, fermentation leads to a slight reduction in ascorbic acid content in smoothies. The ascorbic acid content of wine is lower than that of the juice. This decrease in the ascorbic acid level of wine can be explained by the oxidation of these parameters (Pareek et al., 2011). Similarly, Selli et al. (2002) studied the effects of some parameters (bottle color, storage temperature, and storage period) on ascorbic acid degradation in orange wine. The authors reported that the rate of ascorbic acid degradation in orange wine rose with increasing storage time; however, storage at 14 and 24°C did not significantly affect the loss of the ascorbic acid. The influence of fermentation on ascorbic acid content has been evaluated in other foods, obtaining similar results. In this way, fermentation leads to a slight reduction in ascorbic acid content in smoothies (Di Cagno et al., 2011), but this decrease was significant in strawberry juice (Klopotek et al., 2005).

Escudero-López et al. (2013) evaluated the antioxidant capacity of orange juice during fermentation using oxygen radical absorbance capacity (ORAC), ferric-reducing ability of plasma (FRAP), 2,2′-azinobis-(3-ethylbenzothiazoline-6-sulfonic acid) (ABTS), and 2,2-diphenylpicrylhydrazyl (DPPH) assays. They observed a significant increase in ORAC values from day 5 (6517 μM) to day 9 (9355 μM) of the fermentation. Subsequently, this value significantly declined until day 11, remaining constant until the end of the process. FRAP values did not experience significant changes between consecutive measurements during fermentation. In addition, there were no significant differences between the beginning and the end of the process. The trolox equivalent antioxidant capacity (TEAC) value of orange juice was constant throughout the fermentation. There were no significant differences between any pair of measurements. Klopotek et al. (2005) indicated a significant decrease in the TEAC value after fermentation of strawberry mash and press juice. On the other hand, a higher antioxidant capacity was shown for yellow onion after fermentation with *Aspergillus kawachii* (Yang et al., 2012). With regard to the evolution of the DPPH value during fermentation, they obtained changes similar to those for the ORAC values. Kelebek et al. (2009) determined the antioxidant content of orange juice after alcoholic fermentation, but they used spontaneous yeasts, obtaining orange wine with 12.6% ethanol (v/v). Controlled alcoholic fermentation in orange juice to obtain a product with low alcoholic degree was used for the first time in this study; also, the influence of this process on the antioxidant content of orange juice was evaluated. Thus, the references to other works taking into account the impact of fermentation on antioxidant content vary depending on the type of fermentation and the fermentation substrate.

3. MANDARIN WINE

Mandarin (or tangerine) (*Citrus reticulata* Blanco) is classified in the family Rutaceae and largely dedicated to the fresh market, in contrast to oranges, which can be either eaten as a fresh fruit or processed into juice (Plotto et al., 2011). Citrus fruits such as mandarin are also used for wine production (Selli et al., 2004; Kelebek and Selli, 2014). The mandarin winemaking procedure is quite similar to that for orange wine except for extracting the juice. Mandarins, because of the comparatively brittle and thin peel, may be difficult to juice on a reamer and may need to be peeled before the juice is pressed out. There are few studies that have investigated mandarin winemaking in the literature. An et al. (2010) studied the optimal processing parameters for making mandarin wine using the Nanfeng cultivar. The authors reported that the Nanfeng cultivar was suitable for making mandarin wine; the best temperature for fermentation was 25 ± 1°C; and the sugar content of the fruit juice should be adjusted to 22% in two steps during fermentation.

Owing to the poor infrastructure and postharvest losses, kinnow mandarins have a short shelf life. Moreover, the problem of bitterness and delayed bitterness has also been associated with kinnow juice because of the presence of naringin and limonin, respectively (Puri et al., 1996). Efforts have been made with variable success for debittering the kinnow mandarin juice to utilize its immense potentiality in the processed kinnow juice industry (Premi et al., 1995; Amerine and Singleton, 1968; Singh et al., 1998). Apart from using kinnow juice as a processed nonalcoholic beverage, another alternative is its fermentation into wine, which can be a potential value-addition step to this fruit juice (Fig. 7.4.6).

A study on the effects of SO_2 and sugar [to adjust the total soluble solids (TSS) of the juice] added in the preparation of the wine indicated that with an increase in initial TSS, the rate of fermentation decreased, whereas no such effect was observed with respect to the SO_2 level. The alcohol content of the wine during the initial stages of maturation ranged from 8.3% to 10.2% (v/v), and thereafter a slight decrease was observed, which may be due to the conversion of alcohol to esters (Table 7.4.5). Furthermore, during maturation the yellow color units decreased, whereas the red color units increased (Table 7.4.6). The optimum conditions for the production of wine were found to be TSS of 28 °Bx and SO_2 level at 150 ppm (Joshi et al., 2014).

Response surface methodology, a statistical technique, was applied to determine the optimum conditions for maximum alcohol production from kinnow mandarins by varying the TSS, pH, temperature, and inoculum size during fermentation of the wine (Panesar et al., 2009). An ethanol concentration of 11% (v/v) was obtained from the optimum processing conditions of TSS 26 °Bx, pH 5.4, temperature 29°C, and inoculum size 7.5% (v/v) after 5 days of fermentation.

Blending kinnow juice with other juices, especially with cane juice, to improve the quality of the product has been explored by Khandelwal et al. (2006) and Bhardwaj and Mukherjee (2011). The production of blended kinnow wine is depicted in Fig. 7.4.7.

It was observed that during the production of blended kinnow–cane wine, there was no change in the ascorbic acid content nor in the limonin content (Table 7.4.7). The sensory analysis of various blends of kinnow wine was carried on the basis of aroma, taste, appearance, and overall acceptability. The blend of kinnow and cane juice at the ratio of 80:20 showed the highest acceptability followed by 100:0, 50:50, 60:40, 70:30, and 90:10.

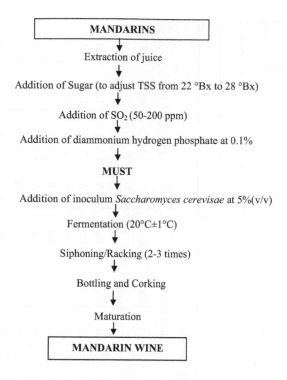

FIGURE 7.4.6

Production of mandarin wine. *TSS*, total soluble solids.

Adapted from Joshi, V.K., Sandhu, N., Abrol, G.S., 2014. Effect of initial sugar concentration and SO₂ content on the physio-chemical characteristics and sensory qualities of mandarin orange wine. International Journal of Food Fermentation and Technology 4 (1), 37–46.

The physicochemical properties (aldehyde, color, pH, TSS, ethanol, and titratable acidity) of wine produced from various citrus fruits, mandarin, kinnow, galgal, and orange, were compared, which indicated higher fermentation rates for kinnow wine followed by orange and galgal (Joshi et al., 2012). Moreover, higher ethanol concentrations were found in kinnow wine compared with the orange and galgal wines (Table 7.4.8).

The bitterness of kinnow wine is mainly due to the presence of limonin and naringin components. The limonin and naringin can be degraded by cyclodextrin and Amberlite XAD-16, of which Amberlite XAD-16 was observed to be the better agent for debittering the wine (Joshi et al., 1997). The limonin content decreased from 206 to 43 mg/L, and a decrease in the naringin content from 350 to 191 mg/L was also observed (Table 7.4.9).

3.1 CHEMICAL COMPOSITION OF MANDARIN JUICES AND WINES

Mandarin is among the most popular citrus fruits. It is a good source of aroma, organic acid, and phenolic compounds. Its nature and concentration largely affect the taste characteristics and organoleptic quality (Peterson et al., 2006; Gattuso et al., 2007). The chemical composition of mandarin juice and

Table 7.4.5 Changes in the Alcohol Concentration (% v/v) of Mandarin Wine During Maturation

Treatment	0 Days				2 Months				4 Months			
	Initial TSS (°Bx)				Initial TSS (°Bx)				Initial TSS (°Bx)			
	22	24	26	28	22	24	26	28	22	24	26	28
T1 (SO$_2$ at 50 ppm)	8.5	8.9	9.3	10.2	8.4	8.7	9.2	10.0	8.3	8.6	9.0	10.0
T2 (SO$_2$ at 100 ppm)	8.4	8.6	9.2	10.1	8.3	8.6	9.0	9.8	8.1	8.4	8.9	9.7
T3 (SO$_2$ at 150 ppm)	8.5	8.6	9.0	9.8	8.3	8.5	8.9	9.7	8.2	8.3	8.7	9.5
T4 (SO$_2$ at 200 ppm)	8.3	8.5	8.9	9.8	8.2	8.3	8.9	9.7	8.0	8.1	8.6	9.7

TSS, total soluble solids.
Adapted from Joshi, V.K., Sandhu, N., Abrol, G.S., 2014. Effect of initial sugar concentration and SO$_2$ content on the physio-chemical characteristics and sensory qualities of mandarin orange wine. International Journal of Food Fermentation and Technology 4 (1), 37–46.

Table 7.4.6 Changes in Color Units of Mandarin Wine During Maturation

Treatment	0 Days				2 Months				4 Months			
	Initial TSS (°Bx)				Initial TSS (°Bx)				Initial TSS (°Bx)			
	22	24	26	28	22	24	26	28	22	24	26	28
T1 (SO$_2$ at 50 ppm)	1.3R	1.0R	1.3R	1.3R	1.0R	1.0R	1.0R	1.0R	1.3R	1.3R	1.5R	1.3R
	1.4Y	2.0Y	1.4Y	1.4Y	2.0Y	2.0Y	2.0Y	2.0Y	1.5Y	1.6Y	2.3Y	1.6Y
	0B	0.2B	0B	0B	0.2B	0.2B	0.2B	0.2B	0.3B	0.4B	0.1B	0.4B
T2 (SO$_2$ at 100 ppm)	1.3R	1.0R	1.0R	1.0R	1.0R	1.0R	1.3R	1.5R	1.3R	1.3R	1.5R	1.5R
	1.4Y	2.0Y	2.0Y	2.0Y	2.0Y	2.0Y	1.4Y	2.3Y	1.5Y	1.6Y	2.3Y	2.3Y
	0B	0.2B	0.2B	0.2B	0.2B	0.2B	0B	0.1B	0.3B	0.4B	0.1B	0.1B
T3 (SO$_2$ at 150 ppm)	1.0R	1.0R	1.3R	1.3R	1.0R	1.0R	1.3R	1.0R	1.3R	1.3R	1.3R	1.3R
	2.0Y	2.0Y	1.4Y	1.4Y	2.0Y	2.0Y	1.4Y	2.0Y	1.5Y	1.5Y	1.6Y	1.6Y
	0.2B	0.2B	0B	0B	0.2B	0.2B	0B	0.2B	0.3B	0.3B	0.4B	0.4B
T4 (SO$_2$ at 200 ppm)	1.0R	1.0R	1.0R	1.3R	1.0R	1.0R	1.0R	1.0R	1.3R	1.3R	1.3R	1.5R
	2.0Y	2.0Y	2.0Y	1.4Y	2.0Y	2.0Y	2.0Y	2.0Y	1.5Y	1.6Y	1.6Y	2.3Y
	0.2B	0.2B	0.2B	0B	0.2B	0.2B	0.2B	0.2B	0.3B	0.4B	0.4B	0.1B

B, blue; R, red, TSS, total soluble solids; Y, yellow.
Adapted from Joshi, V.K., Sandhu, N., Abrol, G.S., 2014. Effect of initial sugar concentration and SO$_2$ content on the physio-chemical characteristics and sensory qualities of mandarin orange wine. International Journal of Food Fermentation and Technology 4 (1), 37–46.

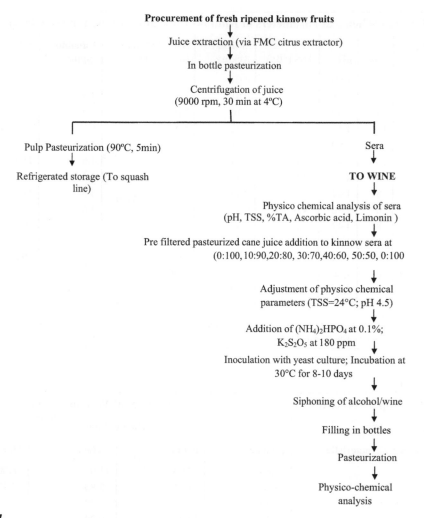

FIGURE 7.4.7

Preparation of pure and blended kinnow–cane wine. *TSS*, total soluble solids.

Adapted from Khandelwal, P., Kumar, V., Das, N., Tyagi, S.M., 2006. Development of a process for preparation of pure & blended kinnow winewithout debittering kinnow mandarin juice. Internet Journal of Food Safety 8, 24–29.

wine reported by Selli et al. (2004) and Kelebek and Selli (2014) is given in Table 7.4.10. In that study, total acidity reported is 1.04%, 0.89%, 1.02%, and 1.03% in Satsuma, Robinson, Fremont, and Clementine, respectively. Citrus juices, especially mandarin juice, are rich sources of ascorbic acid, which is an important antioxidant and also a significant indicator of quality (Selli et al., 2004). The concentration of ascorbic acid in the cultivars Satsuma, Robinson, Fremont, and Clementine was found to be 372.62, 496.20, 398.81, and 441.00 mg/L, respectively (Kelebek and Selli, 2014; Selli et al., 2004). The ascorbic acid content of the wines was lower than that of the juices.

Table 7.4.7 Physicochemical Parameters of Kinnow–Cane Juice Blend for Wine Preparation

Kinnow/Cane Juice Blend	Day	pH	TSS (°Bx)	% Total Acidity	Ascorbic Acid (mg/100 mL)	Limonin (ppm)	Ethanol Content (% v/v)
100:0 (control)	0	4.50	24.0	0.64	22.8	4.6	-
	10	4.41	8.0	0.61	22.7	4.8	12.2
0:100	0	4.50	24.0	0.10	NA	NA	-
	10	4.42	9.2	0.42	NA	NA	10.3
90:10	0	4.50	24.0	0.51	22.3	4.8	-
	10	4.45	9.4	0.54	21.8	4.6	11.1
80:20	0	4.50	24.0	0.45	20.6	4.2	-
	8	4.44	8.8	0.49	20.5	5.6	11.0
70:30	0	4.50	24.0	0.40	20.4	4.6	-
	8	4.42	8.8	0.42	20.1	3.9	10.8
60:40	0	4.50	24.0	0.38	18.6	5.2	-
	8	4.44	9.2	0.39	18.2	4.9	10.0
50:50	0	4.50	24.0	0.34	17.6	4.8	-
	8	4.54	9.2	0.40	17.2	4.9	10.2

TSS, *total soluble solids.*
Adapted from Khandelwal, P., Kumar, V., Das, N., Tyagi, S.M., 2006. Development of a process for preparation of pure & blended kinnow wine without debittering kinnow mandarin juice. Internet Journal of Food Safety 8, 24–29.

Table 7.4.8 Comparison of the Physicochemical Characteristics of Various Citrus Fruit Wines

Physicochemical Characteristic	Citrus Wine			
	Orange	Kinnow	Galgal	Mandarin
TSS (°Bx)	7.80	7.99	11.0	8.00
pH	3.78	3.74	2.80	3.72
Titratable acidity (% CA)	0.70	0.86	1.47	0.86
Ethanol (% v/v)	10.20	12.20	10.20	11.70
Free aldehyde (ppm)	48.0	48.4	48.4	45.0

CA, *citric acid;* TSS, *total soluble solids.*
Adapted from Joshi, V.K., Sandhu, N., Abrol, G.S., 2014. Effect of initial sugar concentration and SO$_2$ content on the physiochemical characteristics and sensory qualities of mandarin orange wine. International Journal of Food Fermentation and Technology 4 (1), 37–46.

The decrease in ascorbic acid level of mandarin wine can be explained by the oxidation of this parameter. Oxidation of ascorbic acid may lead to the formation of dehydroascorbic acid, which cannot be detected by the method used (Puttongsiri and Haruenkit, 2010). The total acidity of mandarin wine is higher than that of normal wine. Sugar is provided partly to balance the taste sensation of high acidity in mandarin wine.

Table 7.4.9 Physicochemical and Sensory Characteristics of Kinnow Wine From Different Debittering Treatments

Treatment	TSS (°Bx)	Ascorbic Acid (mg/100 mL)	Alcohol Content (% v/v)	Limonin content (mg/L)	% Reduction	Naringin Content (mg/L)	% Reduction	Tintometer Color Units Red	Yellow	Sensory Score	Bitterness Rating
T1 (bitter serum + 0.2% cyclodextrin)	8.6	23.7	8.57	115.3	44.1	195	44.2	3.5	60	16	4.0
T2 (bitter serum control)	9.0	24.1	8.53	195.7	5.2	316	9.7	3.5	50	13	3.0
T3 (XAD-debittered serum)	8.0	23.1	9.12	43.8	5.6	191	4.9	4.5	60	17	6.0

Bitterness rating: 9, extremely nonbitter; 8, very nonbitter; 7, moderately nonbitter; 6, slightly nonbitter; 5, neither nonbitter nor bitter; 4, slightly bitter; 3, moderately bitter; 2, very bitter; 1, extremely bitter. TSS, total soluble solids.
Adapted from Joshi, V.K., Thakur, N.K., Kaushal, B.B.L., 1997. Effect of debittering of kinnow juice on physio-chemical and sensory quality of kinnow wine. Indian Food Packer 5–8.

Table 7.4.10 General Composition of Mandarin Juice and Wine

Physicochemical Property	Cultivar			
	Satsuma	Robinson	Fremont	Clementine
Juice Composition				
Density (20°C/20°C)	1.052	1.055	1.057	1.053
Total acidity (TA, %)	1.04	0.89	1.02	1.03
pH	3.50	3.60	3.40	3.50
Ascorbic acid (mg/L)	372.62	496.20	398.81	441.00
Extract (g/L)	111.90	129.30	126.20	98.50
Wine Composition				
Density (20°C/20°C)	1.008	1.015	1.021	1.008
Ethanol (v/v %)	12.70	12.60	12.60	12.50
Total acidity (%)	0.89	0.76	0.85	0.75
pH	3.40	3.60	3.50	3.7
Volatile acidity[a] (g/L)	0.23	0.20	0.18	0.15
Total sugar (g/L)	72.16	71.60	71.33	46.00
Ascorbic acid (mg/L)	203.2	225.6	202.6	330.0
Extract (g/L)	78.91	78.60	77.92	49.00
Free SO_2 (mg/L)	15.20	16.50	15.70	6.00
Bound SO_2 (mg/L)	88.70	89.20	90.30	77.00

[a]As acetic acid.
Adapted from Selli, S., Kürkçüoğlu, M., Kafkas, E., Cabaroglu, T., Başer, K.H.C., CanbaşA., 2004. Volatile flavour components of mandarin wine obtained from Clementine (Citrusreticula Blanco) extracted by head space-solid phase micro extraction. Flavour and Fragrance Journal 19, 413–416; Kelebek, H., Selli, S., 2014. Identification of phenolic compositions and the antioxidant capacity of mandarin juices and wines. Journal of Food Science and Technology 51 (6), 1094–1101.

3.2 FLAVOR COMPOSITION OF MANDARIN WINE

Flavor compounds play an important role in the organoleptic characteristics of alcoholic beverages, as they make a major contribution to consumer acceptance. No scientific studies in the literature discuss the aroma composition of mandarin wine except for Selli et al. (2004). These authors investigated the aroma composition of mandarin wine obtained from the Clementine cultivar using headspace (HS) and immersion (Im)-SPME–GC/MS. Table 7.4.11 shows the volatile constituents of Clementine mandarin wine obtained by HS-SPME and Im-SPME. A total of 15 volatile constituents were identified by HS-SPME, including eight esters, three higher alcohols, three terpenes, and 5-hydroxymethyl furfural. Esters and alcohols were the largest groups and made up more than 99% of the volatiles (Table 7.4.11). These compounds are considered the most important for wines and are formed during alcoholic fermentation by yeasts (Nykanen, 1986; Schreier, 1979). Most volatile compounds have previously been reported for orange wine (Selli et al., 2003, 2008) and orange spirits (Da Porto et al., 2003).

With regard to Im-SPME, 14 volatile constituents were identified, including six esters, four higher alcohols, and four furfural compounds. Furfural compounds and alcohols were the largest groups and made up more than 92% of volatiles determined by Im-SPME. As can be seen from Table 7.4.11, the

Table 7.4.11 Volatile Flavor of Clementine Mandarin Wine

LRI	Compound	HS-SPME (%)	Im-SPME (%)
893	Ethyl acetate	1.8	3.9
1046	Ethyl butyrate	Tr	–
1096	2-Methyl propan-1-ol	Tr	0.7
1136	Isoamyl acetate	3.7	0.9
1188	α-Terpinene	Tr	–
1203	Limonene	0.7	–
1212	Isoamyl alcohol	12.2	24.6
1250	Ethyl hexanoate	4.6	0.5
1290	Terpinolene	Tr	–
1444	Ethyl octanoate	53.3	0.62
1479	Furfural	–	10.7
1586	5-Methyl furfural	–	2.3
1648	Ethyl decanoate	20.4	1.3
1668	Furfuryl alcohol	–	0.2
1674	Ethyl succinate	0.6	0.7
1853	Ethyl dodecanoate	0.4	–
1937	Phenylethyl alcohol	2.1	5.9
1996	4,5-Dimethyl furfural	–	0.4
2515	5-Hydroxy methylfurfural	Tr	47.2

HS, *headspace;* Im, *immersion;* LRI, *linear retention index calculated on polar column (DB–Wax);* SPME, *solid-phase microextraction;* Tr, *trace.*
Adapted from Selli, S., Kürkçüoğlu, M., Kafkas, E., Cabaroglu, T., Başer, K.H.C., Canbaş A., 2004. Volatile flavour components of mandarin wine obtained from Clementine (Citrusreticula *Blanco) extracted by head space-solid phase micro extraction. Flavour and Fragrance Journal 19, 413–416.*

Im-SPME technique showed a lower sensitivity (except for furfural compounds) than the HS-SPME technique for the detection of ethyl esters and terpenes, which are important for wine flavor.

The changes in the composition of the esters in mandarin wine during a maturation period of 4 months were studied, which indicated an increase in the ester composition from 44 to 104 mg/L, having TSS of about 26 °Bx and SO_2 at 150 ppm as depicted in Table 7.4.12 (Joshi et al., 2014).

The flavor profiling of various citrus wines carried out by quantitative descriptive analysis showed that mandarin wine had higher alcoholic flavor (7.0) compared to kinnow wine (Joshi et al., 2012), whereas aromatic flavor was highest in the kinnow wine compared to galgal wine, as shown in Fig. 7.4.8.

3.3 PHENOLIC COMPOSITION OF MANDARIN JUICE AND WINE

A total of 11 phenolic compounds have been identified and quantified in mandarin juice and wine, including hydroxybenzoic acids (3), hydroxycinnamic acids (5), and flavanones (3) (Kelebek and Selli, 2014). In that study, the total amount of phenolic compounds was 36.59–132.63 mg/L for the mandarin juice and 14.08–54.55 mg/L for the wine (Table 7.4.13). The total content of phenolics in several juices

Table 7.4.12 Changes in the Total Esters (mg/L) of Mandarin Wine During Maturation

	Initial TSS (°Bx)				Initial TSS (°Bx)				Initial TSS (°Bx)			
	22	24	26	28	22	24	26	28	22	24	26	28
Treatment	0 Days				2 Months				4 Months			
T1 (SO$_2$ at 50 ppm)	37	39	42	42	62	66	69	72	85	88	93	99
T2 (SO$_2$ at 100 ppm)	39	40	42	44	65	66	71	78	86	86	96	102
T3 (SO$_2$ at 150 ppm)	43	44	46	44	66	71	74	77	82	90	96	104
T4 (SO$_2$ at 200 ppm)	43	46	49	47	69	74	79	82	89	89	100	108

TSS, *total soluble solids.*
Adapted from Joshi, V.K., Sandhu, N., Abrol, G.S., 2014. Effect of initial sugar concentration and SO$_2$ content on the physio-chemical characteristics and sensory qualities of mandarin orange wine. International Journal of Food Fermentation and Technology 4 (1), 37–46.

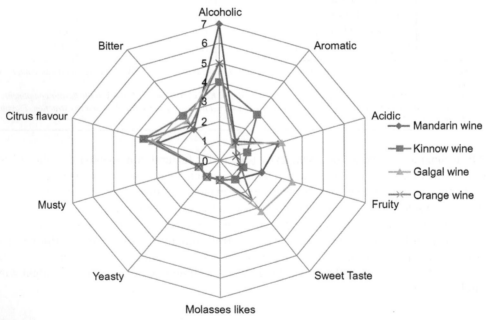

FIGURE 7.4.8

Comparison of the flavor profiles of various citrus wines.

Adapted from Joshi, V.K., Sandhu, N., Abrol, G.S., 2014. Effect of initial sugar concentration and SO2 content on the physio-chemical characteristics and sensory qualities of mandarin orange wine. International Journal of Food Fermentation and Technology 4 (1), 37–46.

Table 7.4.13 Phenolic Composition (mg/L) of Mandarin Juice and Wine

	Satsuma		Robinson		Fremont	
	Juice	Wine	Juice	Wine	Juice	Wine
Hydroxybenzoic Acids						
Protocatechuic acid	0.66	0.26	0.35	0.14	0.44	0.17
p-Hydroxybenzoic acid	0.92	0.35	0.55	0.20	0.68	0.18
Vanillic acid	1.34	0.56	0.74	0.29	0.86	0.37
Total	**2.91**	**1.17**	**1.63**	**0.63**	**1.98**	**0.73**
Hydroxycinnamic Acids						
Caffeic acid	2.72	1.14	0.71	0.22	1.62	0.71
Chlorogenic acid	5.00	2.10	0.88	0.34	2.36	1.03
p-Coumaric acid	2.27	0.90	0.17	0.07	0.37	0.16
Ferulic acid	20.82	8.09	5.41	2.11	6.60	2.89
Sinapic acid	5.73	2.40	2.04	0.66	2.76	1.21
Total	**36.55**	**14.63**	**9.20**	**3.40**	**13.71**	**6.00**
Flavanones						
Narirutin	18.29	7.31	6.56	2.56	17.97	7.86
Hesperidin	74.13	31.13	18.66	7.28	40.50	17.72
Didymin	0.75	0.31	0.54	0.21	0.44	0.19
Total	**93.17**	**38.75**	**25.76**	**10.05**	**58.91**	**25.77**

Adapted from Kelebek, H., Selli, S., 2014. Identification of phenolic compositions and the antioxidant capacity of mandarin juices and wines. Journal of Food Science and Technology 51 (6), 1094–1101.

obtained from different *Citrus* varieties ranged from 37.3 (Liucheng variety) to 52.3 (Kumquat variety) milligrams of gallic acid equivalents per gram (Wang et al., 2007). As can be seen in Table 7.4.13, the total phenolic content of the juices was about twofold higher than that of the wines (Table 7.4.13). These changes were possibly due to the transformation of phenolic compounds into condensed forms that possess slightly different chemical properties.

Hydroxybenzoic acids: Three different hydroxybenzoic acids, protocatechuic, p-hydroxybenzoic, and vanillic acid, were detected and quantified in juice and wine (Table 7.4.13). Satsuma juice presented the highest total hydroxybenzoic acid content (2.91 mg/L). The main hydroxybenzoic acid was vanillic acid, as it accounted for the largest proportion of the total hydroxybenzoic acid content. The highest level of vanillic acid was reported in Satsuma (1.34 mg/L), followed by Fremont (0.86 mg/L) and Robinson (0.74 mg/L). Vanillic acid content varied with the variety of citrus fruit; Satsuma (mandarin) juice had the highest (3.40 mg/L), and Miyou (pummelo) juice had the lowest (0.63 mg/L) (Xu et al., 2008).

Hydroxycinnamic acids: The five hydroxycinnamic acids identified by Kelebek and Selli (2014) are caffeic, chlorogenic, p-coumaric, ferulic, and sinapic acids. The total amount of phenolic compounds ranged from 9.20 to 36.55 mg/L for mandarin juice, and from 3.40 to 14.63 mg/L for the

Table 7.4.14 Changes in Total Phenols of Mandarin Wine During Maturation

Treatment	Initial TSS (°Bx)				Initial TSS (°Bx)				Initial TSS (°Bx)			
	22	24	26	28	22	24	26	28	22	24	26	28
	0 Days				2 Months				4 Months			
T1 (SO$_2$ at 50 ppm)	80	78	80	85	75	72	78	80	74	70	73	78
T2 (SO$_2$ at 100 ppm)	78	78	85	90	74	74	82	88	72	73	79	82
T3 (SO$_2$ at 150 ppm)	87	88	87	96	82	78	84	93	82	76	82	80
T4 (SO$_2$ at 200 ppm)	91	97	94	102	92	96	76	98	90	83	76	91

TSS, total soluble solids.
Adapted from Joshi, V.K., Sandhu, N., Abrol, G.S., 2014. Effect of initial sugar concentration and SO$_2$ content on the physio-chemical characteristics and sensory qualities of mandarin orange wine. International Journal of Food Fermentation and Technology 4 (1), 37–46.

wines. Ferulic acid is the most dominant hydroxycinnamic acid in mandarin juice (5.41–20.82 mg/L) and wine (2.11–8.09 mg/L), as it accounted for the largest proportion of the total hydroxycinnamic acid content (Table 7.4.13). Sinapic acid is also an abundant hydroxycinnamic acid, followed by chlorogenic, caffeic, and *p*-coumaric acids in wine. Juice contains chlorogenic, caffeic, and *p*-coumaric acids in quantities up to 5.00, 2.72, and 2.27 mg/L (Table 7.4.13), respectively, whereas mandarin wine contains about half the quantities of these acids, which amount to 2.10, 1.14, and 0.9 mg/L (Table 7.4.13), respectively (Kelebek and Selli, 2014). Rapisarda et al. (1998) found that ferulic acid (3.77 mg/100 mL) was the main phenolic acid in Liucheng (Valencia late) juice.

Flavanones: Flavanones are the major flavonoids in mandarin juice and wine. Three flavanones, narirutin, hesperidin, and didymin, have been identified in mandarin juice and wine (Kelebek and Selli, 2014). Hesperidin is the most abundant. The levels of hesperidin reported for mandarin juice (18.6–74.13 mg/L) are in good agreement with the report of Gattuso et al. (2007), 8–458 mg/L, whereas levels lower than that were found by Xu et al. (2008). Narirutin is the second most abundant flavanone and its concentration ranges from 6.56 to 18.29 mg/L for mandarin juice, and from 2.56 to 7.86 mg/L for wine. Narirutin levels in mandarin juice have been reported at 1–90 (Gattuso et al., 2007) and 24–288 mg/L in different varieties (Xu et al., 2008).

During the maturation of mandarin wine for 4 months, a decrease in the total phenolic content of the wine was observed. The highest phenol content during the initial maturation period was 102 mg/L at 28 °Bx and SO$_2$ level of 200 ppm. The lowest phenol content, 78 mg/L, was found using an SO$_2$ concentration of 100 ppm at 22 °Bx; and at SO$_2$ of 50 and 100 ppm at 24 °Bx, the levels decreased to 72, 70, and 73 mg/L, respectively, as shown in Table 7.4.14 (Joshi et al., 2014). A decrease in the total phenol content during maturation is essential to increase the palatability of the wine.

3.4 ANTIOXIDANT ACTIVITY OF MANDARIN JUICE AND WINE

The antioxidant potential of mandarin juice and wine measured in terms of its radical-scavenging potential is given in Fig. 7.4.9 (Kelebek and Selli, 2014). DPPH is a stable free radical and the assay can accommodate a large number of samples in a short period of time. The assay is sensitive enough to detect active principles at low concentrations.

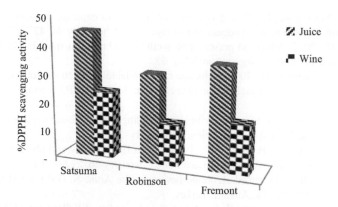

FIGURE 7.4.9

Antioxidant activities of mandarin juice and wine. *DPPH*, 2,2-diphenylpicrylhydrazyl.

Adapted from Kelebek, H., Selli, S., 2014. Identification of phenolic compositions and the antioxidant capacity of mandarin juices and wines. Journal of Food Science and Technology 51 (6), 1094–1101.

The antioxidant activity ranged from 31% to 44% for mandarin juice, and from 14% to 23% for wine (Kelebek and Selli, 2014). As shown in Fig. 7.4.9, Satsuma shows higher antioxidant activities compared to Robinson and Fremont. Antioxidant activities of the juice are about twofold higher than those of the wine. The decreases in the antioxidant activities of the mandarin wine are possibly due to the transformation and oxidation of phenolic compounds and ascorbic acids.

4. CONCLUSIONS

Wines can be made from fruits other than grape, and this subunit is a compilation of studies on wine preparation from citrus fruits especially orange and mandarin fruits. These fruits hold promise to make wine, being a good source of flavor, color, vitamin C, carotenoids, and flavonoids, and the wines made from such fruit could become a good source of such components, being associated with healthful properties. The brief overview of the research work carried out shows that only scattered work on orange and kinnow wine has been carried out so far. So there is a great need to carry out in-depth and systematic studies on various aspects of the winemaking process. The winemaking procedure described here is simple and more systematic research is involving effect of different yeast strains, debittering treatment, maturation with wood or otherwise and antioxidant activity and the associated compounds needed.

REFERENCES

Amerine, M.A., Berg, H.W., Kunkee, R.E., Ough, C.S., Singleton, V.L., Webb, A.C., 1980. The Technology of Wine Making, fourth ed. Avi Publishing Co. Inc, Westport, USA.

Amerine, M.A., Singleton, V.L., 1968. Wine: An Introduction for Americans. University of California Press, Berkeley, USA.

An, D., Sun, A., Meng, C., 2010. Study on the processing technology of citrus fermented wine. China Academic Journal 4, 183–185.

Arena, E., Fallico, B., Maccarone, E., 2001. Evaluation of antioxidant capacity of blood orange juices as influenced by constituents, concentration, process and storage. Food Chemistry 74, 423–427.

Barrett, D.M., Lloyd, B., 2012. Advanced preservation methods and nutrient retention in fruits and vegetables. Journal of the Science of Food and Agriculture 92, 7–22.

Bely, M., Rinaldi, A., Dubourdieu, D., 2003. Influence of assimilable nitrogen on volatile acidity production by *Saccharomyces cerevisiae* during high sugar fermentation. Journal of Bioscience and Bioengineering 96, 507–512.

Bhardwaj, R.L., Mukherjee, S., 2011. Effects of fruit juice blending ratios on kinnow juice preservation at ambient storage condition. African Journal of Food Science 5 (5), 281–286.

Bocco, A., Cuvelier, M.E., Richard, H., Berset, C., 1998. Antioxidant activity and phenolic composition of citrus peel and seed extracts. Journal of Agricultural and Food Chemistry 46, 2123–2129.

Canbas, A., Kelebek, H., 2014. Fermented alcoholic beverages. In: Aran, N. (Ed.), Food Biotechnology: Nobel Akademik. Yayincilik Danismanlik, Ankara, Turkey.

Canbas, A., Kelebek, H., Selli, S., 2008. Colored and Colorless Phenolic Contents of Moro and Sanguinello Orange Juices and Wines. (Unpublished data) Obtained from ZF-2006-BAP-8 project supported by Cukurova University.

Canbas, A., Unal, U., 1994. A study on the evaluation of some orange varieties grown in Adana for wine production. Turkish Journal of Agricultural and Forestry 18, 1–7.

Cole, C.V., Noble, A.C., 1995. Flavor chemistry and assessment. In: Lea, A.G.H., Piggott, J.R. (Eds.), Fermented Beverage Production. Blackie Academic and Professional, London, U.K, pp. 361–385.

Constanti, M., Reguant, C., Poblet, M., Zamora, F., Mas, A., Guillamon, J.M., 1998. Molecular analysis of yeast population dynamics: effect of sulphur dioxide and inoculum on must fermentation. International Journal of Food Microbiology 41, 169–175.

Cruess, W.V., 1914. Orange wine making. California Agricultural Experiment Station Bulletin 244, 157–170.

Da Porto, C., Pizzale, L., Bravin, M., Conte, L.S., 2003. Analyses of orange spirit flavour by direct-injection gas chromatography-mass spectrometry and headspace solid phase microextraction/GC-MC. Flavour and Fragrance Journal 18, 66–72.

Di Cagno, R., Minervini, G., Rizzello, C.G., De Angelis, M., Gobbetti, M., 2011. Effect of lactic acid fermentation on antioxidant, texture, color and sensory properties of red and green smoothies. Food Microbiology 28, 1062–1071.

Duarte, W.F., Dias, D.R., Pereira, G.V.M., Gervásio, I.M., Schwan, R.F., 2009. Indigenous and inoculated yeast fermentation of gabiroba (*Campomanesiapubescens*) pulp for fruit wine production. Journal of Industrial Microbiology and Biotechnology 36 (4), 557–569.

Duarte, W.F., Dragone, G., Dias, R.D., Oliveira, J.M., Teixeira, J.A., Silva, J.B.A.E., Schwan, R.F., 2010. Fermentative behaviour of Saccharomyces strains during micro vinification of raspberry juice (*Rubus idaeus* L.). International Journal of Food Microbiology 143, 173–182.

Erten, H., Cabaroglu, T., Unal, U., Selli, S., Canbas, A., 2003. Yeast flora during the alcoholic fermentation of orange wine produced from cv. Kozan. In: 3th Food Engineering Congres, 2–4 October 2003, Ankara, Turkey, pp. 539–547.

Escudero-López, B., Cerrillo, I., Herrero-Martín, G., Hornero-Méndez, D., Gil-Izquierdo, A., Medina, S., Ferreres, F., Berná, G., Martín, F., Fernández-Pachón, M.S., 2013. Fermented orange juice: source of higher carotenoid and flavanone contents. Journal of Agricultural and Food Chemistry 61, 8773–8782.

Fallico, B., Lanza, M.C., Maccarone, E., Nicolosi Asmundo, C., Rapisarda, P., 1996. Role of hydroxy cinnamic acids and vinyl-phenols in the flavor alteration of blood orange juices. Journal of Agricultural and Food Chemistry 44, 2654–2657.

FAO, 2014. Orange Production Data. The Food and Agriculture Organization.

Fan, G., Lu, W., Yao, X., Zhang, Y., Wang, K., Pan, S., 2009b. Effect of fermentation on free and bound volatile compounds of orange juice. Flavour and Fragrance Journal 24, 219–225.

Fan, G., Xu, X.Y., Qiao, Y., Xu, Y., Zhang, Y., Li, L., Pan, S., 2009a. Volatiles of orange juice and orange wines using spontaneous and inoculated fermentations. European Food Research and Technology 228, 849–856.

Gattuso, G., Barreca, D., Garguilli, C., Leuzzi, U., Coristi, C., 2007. Flavonoid composition of *citrus* juice. Molecules 12, 1641–1673.

Gil-Izquierdo, A., Gil, M.I., Ferreres, F., 2002. Effect of processing techniques at industrial scale on orange juice antioxidant and beneficial health compounds. Journal of Agricultural and Food Chemistry 50, 5107–5114.

Gil-Izquierdo, A., Gil, M.I., Tomás-Barberán, F.A., Ferreres, F., 2003. Influence of industrial processing on orange juice flavanone solubility and transformation to chalcones under gastrointestinal conditions. Journal of Agricultural and Food Chemistry 51, 3024–3028.

Gorinstein, S., Haruenkit, R., Park, Y.S., Jung, S.T., Zachwieja, Z., Jastrzebski, Z., 2004. Bioactive compounds and antioxidant potential in fresh and dried Jaffa sweeties, a new kind of citrus fruit. Journal of the Science of Food and Agriculture 84, 1459–1463.

Grosch, W., 1993. Detection of potent odorants in foods by aroma extract dilution analysis. Trends in Food Science and Technology 4, 68–73.

Habauzit, V., Nielsen, I.L., Gil-Izquierdo, A., Morand, C., Williamson, G., Barron, D., Davicco, M.J., Coxam, V., Chee, W., Offord, E., Horcajada, M.N., 2009. Increased bioavailability of hesperetin-7-glucoside compared to hesperidin results in more efficient prevention of bone loss in adult ovariectomized rats. British Journal of Nutrition 102, 976–984.

Hammond, J.R.M., 1995. Genetically-modified brewing yeasts for the 21st century. Progress to date. Yeast 11, 1613–1627.

Jackson, R.S., 2000. Wine Science: Principles, Practice, Perception. Academic Press, London, UK.

Jagtap, U.B., Bapat, V.A., 2015. Wines from fruits other than grapes: current status and future prospectus. Food Bioscience 9, 80–96.

Joshi, V.K., Kumar, V., Kumar, A., 2012. Physico-chemical and sensory evaluation of wines from different citrus fruits of Himachal Pradesh. International Journal of Food Fermentation and Technology 2 (2), 145–148.

Joshi, V.K., Sandhu, N., Abrol, G.S., 2014. Effect of initial sugar concentration and SO_2 content on the physicochemical characteristics and sensory qualities of mandarin orange wine. International Journal of Food Fermentation and Technology 4 (1), 37–46.

Joshi, V.K., Thakur, N.K., Kaushal, B.B.L., 1997. Effect of debittering of kinnow juice on physio-chemical and sensory quality of kinnow wine. Indian Food Packer 5–8.

Joslyn, M.A., Marsh, G.L., 1936. Suggesting for making orange wine. The Fruit Products Journal and American Vinegar Industry 13, 307–315.

Karadeniz, F., 2004. Main organic acid distribution of authentic citrus juices in Turkey. Turkish Journal of Agriculture and Forestry 28, 267–271.

Kelebek, H., Canbas, A., Selli, S., 2008. Determination of phenolic composition and antioxidant capacity of blood orange juices obtained from cvs. Moro and Sanguinello (*Citrus sinensis* (L.) Osbeck) grown in Turkey. Food Chemistry 107, 1710–1716.

Kelebek, H., Selli, S., 2014. Identification of phenolic compositions and the antioxidant capacity of mandarin juices and wines. Journal of Food Science and Technology 51 (6), 1094–1101.

Kelebek, H., Selli, S., Canbas, A., Cabaroglu, T., 2009. HPLC determination of organic acids, sugars, phenolic compositions and antioxidant capacity of orange juice and orange wine made from a Turkish cv. Kozan. Microchemical Journal 91, 187–192.

Khandelwal, P., Kumar, V., Das, N., Tyagi, S.M., 2006. Development of a process for preparation of pure & blended kinnow wine without debittering kinnow mandarin juice. Internet Journal of Food Safety 8, 24–29.

Klaren-De Wit, M., Frost, D.J., Ward, J.P., 1971. Formation of p-vinyl guaiacol oligomers in the thermal decarboxylation of ferulic acid. Recueil des Travaux Chimiques des Pays-Bas Journal of the Royal Netherlands Chemical Society 90, 906–911.

Klopotek, Y., Otto, K., Bohm, V., 2005. Processing strawberries to different products alters contents of vitamin C, total phenolics, total anthocyanins, and antioxidant capacity. Journal of Agricultural and Food Chemistry 53, 5640–5646.

Lee, H.S., 2002. Characterization of major anthocyanins and the color of red-fleshed Budd blood orange (*Citrus sinensis*). Journal of Agricultural and Food Chemistry 50, 1243–1246.

Lee, H.S., Coates, G.A., 2000. Quantitative study of free sugars and myo-inositol in citrus juices by HPLC and literature compilation. Journal of Liquid Chromatography & Related Technologies 14, 2123–2141.

Lee, H.S., Nagy, S., 1990. Formation of 4-vinylguaiacol in adversely stored orange juice measured by improved HPLC method. Journal of Food Science 55, 162–163.

Li, R., Feng, K., Wu, J., 2010. Effects of *Saccharomyces cerevisiae* strains from different sources on the aromatic composition of orange wine. Food Science 31, 206–213.

Manthey, J.A., Grohmann, K., 2001. Phenols in citrus peel byproducts. Concentrations of hydroxy cinnamates and poly methoxylated flavones in citrus peel molasses. Journal of Agricultural and Food Chemistry 49, 3268–3273.

Mills, D.A., Phister, T., Neeley, E., Johannsen, E., 2008. Wine fermentation. In: Cocolin, L., Ercolini, D. (Eds.), Molecular Techniques in the Microbial Ecology of Fermented Foods. Springer.

Mistry, B.S., Reineccius, T., Olson, L.K., 1997. Gas chromatography olfactometry for the determination of key odorants in food. In: Marsili, R. (Ed.), Techniques for Analyzing Food Aroma. Dekker, New York, pp. 265–292.

Molina, A.M., Guadalupe, V., Varela, C., Swiegers, J.H., Pretorius, I.S., Agosin, E., 2009. Differential synthesis of fermentative aroma compounds of two related commercial wine yeast strains. Food Chemistry 117, 189–199.

Mouly, P.P., Gaydou, E.M., Faure, R., Estienne, J.M., 1997. Blood orange juice authentication using cinnamic acid derivatives. Variety differentiations with flavanone glycoside content. Journal of Agricultural and Food Chemistry 45, 373–377.

Naim, M., Striem, B.J., Kanner, J., Peleg, H., 1988. Potential of ferulic acid as precursor to off flavors in stored orange juice. Journal of Food Science 53, 500–503.

Naim, M., Zuker, I., Zehavi, U., Rouseff, R.L., 1993. Inhibition by thiol compounds of off flavor formation in stored orange juice. II. Effect of L-cysteine and N-acetylcysteine on *p*-vinyl guaiacol formation. Journal of Agricultural and Food Chemistry 41, 1359–1361.

Nisperos-Carriedo, M.O., Shaw, P.E., 1990. Comparison of volatile components in fresh and processed orange juices. Journal of Agricultural and Food Chemistry 38, 1048–1052.

Nykänen, L., 1986. Formation and occurrence of flavor compounds in wine and distilled alcoholic beverages. American Journal of Enology and Viticulture 37, 84–96.

Okunowo, W.O., Okotore, R.O., Osuntoki, A.A., 2005. The alcoholic fermentative efficiency of indigenous yeast strains of different origin on orange juice. African Journal of Biotechnology 4, 1290–1296.

Okunowo, W.O., Osuntoki, A.A., 2007. Quantitation of alcohols in orange wine fermented by four strains of yeast. African Journal of Biotechnology 1 (6), 95–100.

Panesar, P.S., Panesar, R., Singh, B., 2009. Application of response surface methodology in the optimization of process parameters for the production of kinnow wine. Natural Product Radiance 8 (4), 363–373.

Pareek, S., Paliwal, R., Mukherjee, S., 2011. Effect of juice extraction methods and processing temperature-time on juice quality of Nagpur mandarin (*Citrus reticulata Blanco*) during storage. Journal of Food Science and Technology 48 (2), 197–203.

Peleg, H., Naim, M., Rouseff, R.L., Zehavi, U., 1991. Distribution of bound and free phenolic acids in oranges (*Citrus sinensis*) and grapefruits (*Citrus paradisi*). Journal of Agricultural and Food Chemistry 57, 417–426.

Peleg, H., Naim, M., Zehavi, U., Rouseff, R.L., Nagy, S., 1992. Pathways of 4-vinyl guaiacol formation from ferulic acid in model solution of orange juice. Journal of Agricultural and Food Chemistry 40, 764–767.

Peterson, J.J., Beecher, G.R., Bhagwat, S.A., Dwyer, J.T., Gebhardt, S.E., Haytowitz, D.B., Holden, J.M., 2006. Flavanones in grapefruit, lemons, and limes: a compilation and review of the data from the analytical literature. Journal of Food Composition and Analysis 19, 74–80.

Plotto, A., Baldwin, E., McCollum, G.T., Gmitter, F.G., 2011. Sensory evaluation tangerine hybrids at multiple harvests. Proceedings of the Florida State Horticultural Society 124, 260–263.

Pozo-Bayón, M.Á., Monagas, M., Bartolomé, B., Moreno-Arribas, M.V., 2012. Wine features related to safety and consumer health: an integrated perspective. Critical Reviews in Food Science and Nutrition 52, 31–54.

Premi, B.R., Lal, B.B., Joshi, V.K., 1995. Debittering of kinnow juice with amberlite XAD-16 resin. Indian Food Packer 9–17.

Puri, M., Marwaha, S.S., Kothari, R.M., Kennedy, J.F., 1996. Biochemical basis of bitterness in citrus fruit juices and biotech approaches for debittering. Critical Reviews in Biotechnology 16, 145–155.

Puttongsiri, T., Haruenkit, R., 2010. Changes in ascorbic acid, total polyphenol, phenolic acids and antioxidant activity in juice extracted from coated Kiew Wan tangerine during storage at 4, 12 and 20°C. Natural Science 44, 280–289.

Rapisarda, P., Carollo, G., Fallico, B., Tomaselli, F., Maccarone, E., 1998. Hydroxycinnamic acids as markers of Italian blood orange juices. Journal of Agricultural and Food Chemistry 46, 464–470.

Reddy, L.V., Reddy, O.V.S., 2009. Production, optimization and characterization of wine from Mango (*Mangifera indica* Linn.). Natural Product Radiance 8 (4), 426–435.

Risch, B., Herrmann, K., 1988. Hydroxy cinnamic acid derivatives in citrus fruits. ZLebensm Unters Forsch 187, 530–534.

Rivard, D., 2009. The Ultimate Fruit Winemaker's Guide. Bacchus Enterprises Ltd.

Rouseff, R.L., 1988. Liquid chromatographic determination of naringin and neo hesperidin as a detector of grapefruit juice in orange juice. Journal of the Association of Official Analytical Chemists 71, 798–802.

Rupasinghe, H.P.V., Clegg, S., 2007. Total antioxidant capacity, total phenolic content, mineral elements, and histamine concentrations in wines of different fruit sources. Journal of Food Composition and Analysis 20, 133–137.

Saavedra, L., Rupérez, F.J., Barbas, C., 2001. Capillary electrophoresis for evaluating orange juice authenticity: a study on Spanish oranges. Journal of Agricultural and Food Chemistry 49, 9–13.

Santos, C.C.A., Duarte, W.F., Carreiro, S.C., Schwan, R.F., 2013. Inoculated fermentation of orange juice (*Citrus sinensis* L.) for production of a citric fruit spirit. Journal of Institute of Brewing 119, 208–287.

Schreier, P., 1979. Flavor composition of wines: a review. Critical Reviews in Food Science and Nutrition 12, 59–111.

Selli, S., 2007. Volatile constituents of orange wine obtained from Moro oranges (*Citrus sinensis* (L.) Osbeck). Journal of Food Quality 30, 330–341.

Selli, S., Canbaş, A., Cabaroglu, T., 2003. Flavour components of orange wine obtained from a Turkish orange cv. Kozan. International Journal Food Science and Technology 38, 587–593.

Selli, S., Canbaş, A., Ünal, Ü., 2002. Effect of bottle colour and storage conditions on browning of orange wine. Nahrung 46, 64–67.

Selli, S., Canbas, A., Varlet, V., Kelebek, H., Prost, C., Serot, T., 2008. Characterization of the most odor-active and volatiles of orange wine made from a Turkish cv. Kozan (*Citrus sinensis* L. Osbeck). Journal of Agricultural and Food Chemistry 56, 227–234.

Selli, S., Kelebek, H., 2015. Organic acids. In: Nollet, L.M.L., Toldrá, F. (Eds.), Handbook of Food Analysis. CRC press.

Selli, S., Kürkçüoğlu, M., Kafkas, E., Cabaroglu, T., Başer, K.H.C., Canbaş, A., 2004. Volatile flavour components of mandarin wine obtained from Clementine (*Citrusreticula* Blanco) extracted by head space-solid phase micro extraction. Flavour and Fragrance Journal 19, 413–416.

Sharma, N., Bhutia, S.P., Aradhya, D., 2013. Process optimization for fermentation of wine from jackfruit (*Artocarpus heterophyllus* Lam.). Food Process Technology 4 (2), 1–5.

Shaw, P.E., 1991. Fruit II. In: Maarse, H. (Ed.), Volatile Compounds in Food and Beverages. Marcel Dekker, New York, p. 305.

Singh, M., Panesar, P.S., Marwaha, S.S., 1998. Studies on the suitability of kinnow fruits for the production of wine. Journal of Food Science and Technology 35, 455–457.

Sun, S.Y., Che, C.Y., Sun, T.F., Lv, Z.Z., He, S.X., Gu, H.N., Shen, W.J., Chi, C., Gao, Y., 2013. Evaluation of sequential inoculation of *Saccharomyces cerevisiae* and *Oenococcusoeni* strains on the chemical and aromatic profiles of cherry wines. Food Chemistry 138, 2233–2241.

Tomás-Barberán, F.A., Clifford, M.N., 2000. Flavanones, chalcones and dihydrochalcones nature, occurrence and dietary burden. Journal of the Science of Food and Agriculture 80, 1073–1080.

Vallejo, F., Larrosa, M., Escudero, E., Zafrilla, M.P., Cerdá, B., Boza, J., García-Conesa, M.T., Espín, J.C., Tomás-Barberán, F.A., 2010. Concentration and solubility of flavanones in orange beverages affect their bioavailability in humans. Journal of Agricultural and Food Chemistry 58, 6516–6524.

Von Loesecke, H.W., Mottern, H.H., Pulley, G.N., 1936. Wines, brandies, and cordials from citrus fruits. Industrial and Engineering Chemistry 28, 1224–1229.

Wang, Y.C., Chuang, Y.C., Ku, Y.H., 2007. Quantitation of bioactive compounds in citrus fruits cultivated in Taiwan. Food Chemistry 102, 1163–1171.

Xu, G., Liu, D., Chen, J., Ye, X., Maa, Y., Shi, J., 2008. Juice components and antioxidant capacity of citrus varieties cultivated in China. Food Chemistry 106, 545–551.

Yang, E.J., Kim, S.I., Park, S.Y., Bang, H.Y., Jeong, J.H., So, J.H., Rhee, I.K., Song, K.S., 2012. Fermentation enhances the in vitro antioxidative effect of onion (*Allium cepa*) via an increase in quercetin content. Food and Chemical Toxicology 50, 2042–2048.

SUBCHAPTER

7.5

PRODUCTION OF WINE FROM TROPICAL FRUITS

L.V. Reddy[1], V.K. Joshi[2], P.S. Panesar[3]

[1]*Yogi Vemana University, Kadapa, AP, India;* [2]*Dr. Y.S. Parmar University of Horticulture and Forestry, Nauni, Solan, HP, India;* [3]*Sant Longowal Institute of Engineering and Technology, Longowal, Punjab, India*

1. INTRODUCTION

The tropics are generally defined as the regions of the globe that lie between the Tropic of Cancer and the Tropic of Capricorn, and the environmental conditions there are unique, creating a habitat for incredibly diverse animals and plants. Tropical zones on earth are areas where the sun is directly overhead once a year and have only two seasons, namely, wet and dry. The tropics are warm year-round, and they are also very humid, with some areas receiving lots of rain every year. Tropical plants and their fruits have adapted to this climate. Many tropical fruits are large, brightly colored, and very flavorful so that they appeal to the animals they rely on to distribute their seeds. Tropical fruits have been used by humans for centuries, and certain fruits are in high demand all over the world. They are cultivated mostly in countries with warm climates and the only character that they share in common is frost intolerance (Morton, 1987; Reddy et al., 2012).

Some tropical fruits—mango, banana, pineapple, papaya, pomegranate, guava, custard apple, lychee, ber, melon, star fruit (or carambola), kiwi, date, and passion fruit—are well known all over the world. In fact, the banana is one of the highest selling fruits around the world. They can be grown, harvested, and transported easily. Many of these fruits are available in big markets year-round. Other tropical fruit cultivars are more obscure. Although they may be popular in specific regions of the world, they are not familiar to people outside of those areas, and some of them definitely possess an acquired uniqueness in taste. Some more obscure examples of tropical fruit include soursop, cherimoya, sugar apple, jackfruit, durian, acerola, mamey, ackee, breadfruit, lychee, rambutan, and mangosteen. Some of these fruits, like jackfruit and durian, are infamous for their strong odor and flavor, whereas others like mangosteens, lychees, and cherimoyas are quite simply delicious, but difficult to cultivate (Morton, 1987).

Mango, pineapple, avocado, and papaya are known as major tropical fruits. The major mango producers are India, Thailand, and Mexico, and for pineapple the Philippines, Thailand, and China. The avocado is produced in Mexico, Indonesia, and the United States. Papaya is produced in India, Brazil, and Mexico, mostly. In the present chapter, the authors have given ample attention to major tropical fruits. World production of tropical fruit will reach 82 million tons in 2015 (FAO, 2014). Ninety percent of tropical fruits are produced in developing countries. Tropical fruit production helps in creating jobs, increasing farmers' income, food security, and reducing poverty levels. Various aspects of fruit-based alcoholic beverages other than those from grapes have been investigated (Barnett, 1980). The Rigveda amply testifies that wine is perhaps the oldest fermented product known to humans. However, the actual birthplace of wine is still unknown, though it was prepared somewhere in 3500 BC (Joshi and Pandey, 1999; Joshi and Attri, 2005). European explorers in the 16th century introduced wine into the New World (Amerine et al., 1980).

2. TYPES OF FRUIT WINE

2.1 MANGO WINE

Mango, the pride fruit of India, is an important tropical fruit crop occupying about 40% of the total area under cultivation in India. Twenty-five mango cultivars are available in India, and are widely cultivated all over the world. It has a rich, luscious, aromatic flavor and delicious taste in which sweetness and acidity are delightfully blended. It contains a good amount of sugar (16–18% w/v) and many organic acids, and also a good antioxidant, carotene (as vitamin A, 4800 IU). Sucrose, glucose, and fructose are the principal sugars in fully ripened mango with small amounts of cellulose, hemicellulose, and pectin. The unripe fruit contains citric acid, malic acid, oxalic acid, succinic acid, and other organic acids. In contrast, in ripe fruits, the main organic acid is malic acid. Mangoes with higher initial concentration of β-carotene are also reported to be helpful as cancer-preventing agents (Anon, 1962, 1963).

The world's total annual mango fruit production was estimated at 35 million tonnes. Global production of mangoes is concentrated mainly in Asia and more precisely in India, which produces 12 million tonnes per annum. Mangoes are cultivated in 85 countries. Asia and the Oriental countries produce around 80% of the world's total production. Major mango-producing countries are India, China, Thailand, and Pakistan (IHDB, 2013). The crop is significantly important in the fruit economy of India and is the third largest industry in the country. In India mango is grown on 2 million acres and it accounts for 40% of the total fruit production (Fig. 7.5.1). In Andhra Pradesh mango occupies an area of 370,000 hectares, with an annual production of 4,407,000 tonnes, and has placed the state in first position, with a share of 24% of India's production coupled with the highest productivity (APEDA, 2014). Mangoes are a highly perishable tropical fruit, with a shelf life of 2–4 weeks at 10–15°C (Yahia, 1998), limiting their availability in fresh markets. An alternative and profitable method of using mangoes, making wine, could become widely acceptable. Many investigators have carried out much research on mango composition and its cultivation aspects.

2.1.1 Screening of Mango Varieties

The suitability of mango cultivars for wine production is generally based on juice quality, quantity, and other properties. Selected commercially available mango cultivars have been screened for winemaking and the results indicated that the composition of wine depends on the mango variety (Czyhrinciwk, 1966; Kulkarni et al., 1980; Onkarayya and Singh, 1984; Obisanya et al., 1987). The screening of 13 cultivars available in northern India by Kulkarni et al. (1980) indicated that wines made from the varieties Fazri, Langra, and Chausa were good. Another study conducted by Reddy and Reddy (2005) screened 10 mango cultivars, and of them, six varieties showed promising juice yields (450–550 mL/kg mango). Mango contains three types of sugars, glucose, fructose, and sucrose. The total soluble solids (TSS) of mango juice are between 14.2% and 20.5%. The Banginapalli (20.5%) variety has high TSS and titratable acidity as tartaric acid. The pH of the juice is between 3.8 and 4.5 (Table 7.5.1). The authors suggested that the concentrations of ethanol, organic acids, tannins, and aromatic volatile compounds produced differ with mango variety (Table 7.5.2). A study on the quality of mango juice from three varieties suggested that the Nam Doc Mai variety was the best (Li et al., 2012b). The steps involved in mango wine production are depicted in Fig. 7.5.2.

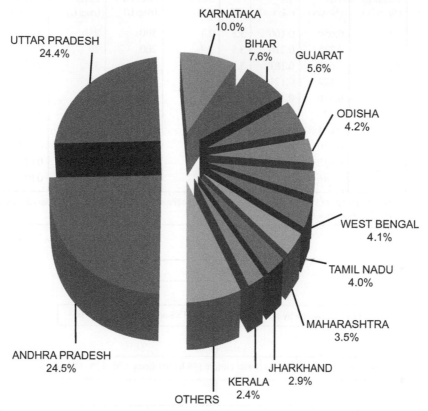

FIGURE 7.5.1

Mango production in the Indian states.

Based on the data presented in Indian Horticulture Database-2013 (IHDB2013).

Table 7.5.1 Physicochemical Characteristics of Mango Pulp

Mango Variety	Juice Yield (mL/kg)	Reducing Sugars (% w/v)	Titratable Acidity (%)	pH	TSS* (%)
Alphonso	570 ± 10	16.3±1.32	0.33	4.1±0.53	16.0±1.2
Raspuri	600 ± 13	15.5±2.21	0.43	3.9±0.86	14.2±1.8
Banginapalli	550 ± 17	18.5±1.24	0.32	4.0±0.6	20.5±0.79
Totapuri	500 ± 22	16.0±1.0	0.31	4.2±1.0	16.5±1.2
A. Banesha	500 ± 15	18.0±0.8	0.32	4.5±0.45	20.1±1.42
Neelam	480 ± 20	15.5±1.7	0.42	4.3±0.8	15.5±1.5
Mulgoa	468 ± 8	14.3±1.4	0.42	4.3±0.5	15.0±1.24
Suvarnarekha	470 ± 12	15.0±0.55	0.40	3.9±1.3	14.4±0.58
Rumani	475 ± 14	14.5±1.0	0.39	4.2±0.72	14.6±1.43
Jahangir	460 ± 10	15.6±1.62	0.46	4.6±0.56	14.2±1.3

TSS, *total soluble solids.*
Reproduced from Reddy, L.V.A., Reddy, O.V.S., 2009a. Effect of enzymatic maceration on synthesis of higher alcohols during mango wine fermentation. Journal of Food Quality 32, 34–47.

Table 7.5.2 Physicochemical Characteristics of Mango Wine Produced at 22 ± 2°C and pH 5

Mango Variety	Ethanol (% w/v)	TA* (% v/v)	VA* (% v/v)	pH	Residual Sugars (g/L)	Higher Alcohols (mg/L)	Total Esters (mg/L)	Tannins (% w/v)	Color OD at 590 nm
Alphonso	7.5	0.650	0.100	3.8	2.1	300	25	0.011	0.22
Raspuri	7	0.735	0.210	3.8	2.4	200	29	0.072	0.18
Banginapalli	8.5	0.600	0.181	3.7	2.0	343	35	0.012	0.23
Totapuri	7	0.622	0.121	4.0	2.0	230	20	0.012	0.17
A. Banesha	8	0.610	0.110	4.0	2.0	320	30	0.013	0.25
Neelam	6.5	0.826	0.234	3.6	2.5	131	15	0.014	0.21
Mulgoa	6.3	0.621	0.109	3.9	3.0	152	18	0.065	0.28
Suvarnarekha	6.8	0.630	0.153	4.1	2.3	175	22	0.025	0.19
Rumani	6.9	0.618	0.125	4.0	2.1	212	15	0.027	0.24
Jahangir	7.1	0.646	0.138	3.8	2.0	256	21	0.042	0.16

Reproduced from Reddy, L.V., Reddy, O.V.S., 2009. Production, optimization and characterization of wine from Mango (Mangifera indica Linn.). Natural Product Radiance 8 (4), 426–435.

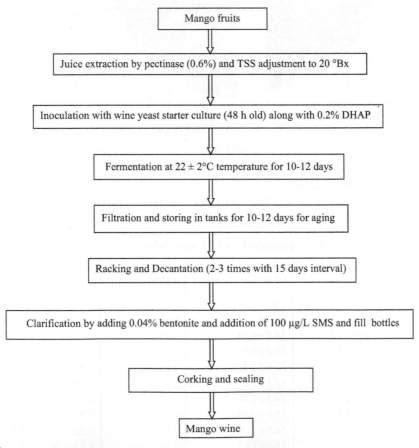

FIGURE 7.5.2

Block diagram of the mango wine production process. *DAHP*, diammonium hydrogen phosphate; *SMS*, sodium metabisulfite; *TSS*, total soluble solids.

2.1.2 Screening of Yeast Strain

The inoculation of must with selected yeasts minimizes the influence of wild yeast on wine quality. However, little is known about the contribution of wild yeasts to the synthesis of volatile compounds during inoculated fermentation. In mango wine fermentation, different investigators used different yeast strains. Fifteen morphologically different groups of yeasts were isolated from fresh, fermenting, and fermented juice of two varieties of mango (Suresh et al., 1982). Onkarayya and Singh (1984) used the *Saccharomyces cerevisiae* strain and Reddy (2005) screened three types of yeast strains isolated from three different sources (palm wine, grapes, and dosa batter). Mixed-culture technology using two different yeasts has been used to improve the wine quality. Li et al. (2012a) conducted a study to compare the chemical and volatile compositions of mango wine fermented with *S. cerevisiae* var. *bayanus* EC1118, *S. cerevisiae* var. *chevalieri* CICC1028, and *S. cerevisiae* var. *cerevisiae* MERIT.ferm. They found that most terpenoids derived from mango juice were retained, unlike mango wine fermented with *S. cerevisiae*, suggesting that mango wine could retain aromatic hints of fresh mango.

2.1.3 Optimization of Mango Wine Production Conditions

Reddy and Reddy (2005) have studied the preliminary processes for the production of wine from mango. Kumar et al. (2009) optimized the fermentation conditions, temperature, pH, and inoculum size, using response surface methodology (temperature 22.5°C, pH 3.8, and inoculum size 11.9%). The results showed a satisfactory production of ethanol and glycerol from mango juice of up to 10% (v/v) and 6.94 g/L, respectively, and it was also observed that the volatile acidity was minimized to 0.29 g/L. Reddy and Reddy (2011) investigated the effects of fermentation conditions like temperature, pH, SO_2, and aeration on mango wine composition and yeast growth.

2.1.3.1 Enzyme Treatment

Mango pulp contains high pectin and needs pectinase treatment for clarification of the juice. A decrease of 50% or more of the relative viscosity of mango pulp resulted from various times of maceration with pectolyase and three of the four commercial enzyme preparations studied (Sakho et al., 1998). Preliminary studies were carried out by Reddy and Reddy (2009a) to optimize the conditions for maximum extraction of juice, using various levels of pectinase enzyme and incubation periods at 28 ± 1°C. Based on their studies, 0.6% pectinase and 8 h of incubation were selected for obtaining juice from the pulp, and treatment increased the juice yield and fermentability of the mango pulp. More recently studies were conducted to evaluate the influence of pulp maceration and β-glucosidase on mango wine physicochemical properties and volatile profiles. β-Glucosidase enhanced terpenols by up to a factor of 10 and acetate esters up to threefold. Furthermore, enzyme treatment mitigated, by up to fivefold, the formation of medium-chain fatty acids and ethyl esters to moderate levels. Sensory evaluation showed pulp contact, and β-glucosidase not only improved the intensity and complexity of the wine aroma but also balanced the odor attributes (Li et al., 2013). Finally, it was suggested that pectinase and other carbohydrate hydrolytic enzyme treatments are good and essential to produce better quality wines from fruits like mango.

2.1.3.2 Temperature

Reddy and Reddy (2009b) reported that the start of fermentation, the stage of maximal cell population, and the decline of the yeast population at 15 and 20°C are slow compared to fermentations at 25, 30, and 35°C. Cell viability was observed to fall as temperature increased. The sum of all the secondary metabolism products increased as fermentation temperature increased from 15 to 35°C. Temperature affects not only the fermentation kinetics (rate and length of fermentation) but also the yeast metabolism, which determines the chemical composition and quality of the wine. Several authors observed increased glycerol levels compared to the control at high temperature (Torija et al., 2002; Kourkoutas et al., 2003; Reddy and Reddy, 2009b, 2011). Increase in temperature increased the volatile composition and also the yeast biomass. Fermentation at low temperatures such as <15°C leads to more aromatic and paler wines. Based on published work it can be suggested that the fermentation temperature has much influence on wine quality and yeast growth compared to all other tested variables.

2.1.3.3 pH

Optimum pH value is necessary for yeast growth and ethanol production. Most yeasts grow very well between pH 4.5 and 6.5 and nearly all species are not able to grow in more acidic or alkaline media. Low or high pH values are known to cause chemical stress on yeast cells. It is also confirmed that the optimum pH is very important to produce good quality wines. Kumar et al. (2009) also used response surface methodology for the optimization of fermentation conditions and suggested that pH 3.8 is optimum for mango wine fermentation.

2.1.3.4 Sulfur Dioxide

Addition of sulfur dioxide (SO_2) does not cause any delay in the onset of alcoholic fermentation except at higher concentrations (200 mg/L). At normal doses (100 mg/L, which is generally used in commercial wine fermentation) a slight stimulatory effect was observed in the initial stages of fermentation (first 2–3 days). Alcoholic fermentation in both cases completed in 7 days. Apart from a slight enhancement in ethanol production, SO_2 also causes some effects on volatile compound synthesis during fermentation (Herrero et al., 2003; Reddy and Reddy, 2011). It has been suggested that SO_2 can stimulate fermentation by *Saccharomyces* in wine by inhibiting the polyphenol oxidase. On the basis of the results obtained, it could be confirmed that the addition of SO_2 induces acetaldehyde formation by yeast in mango wine fermentation.

2.1.3.5 Oxygen

Reddy and Reddy (2011) reported that aeration can improve cell growth and initial fermentation rate. The fermentation is complete sooner (5 days) with aeration and it takes a longer time in the absence of oxygen. The viability of yeast cells is increased (8×10^6 cells/mL) with aeration compared to absence of oxygen (5×10^6 cells/mL). Oxygen appears to be involved in the synthesis of oleic acid and ergosterol, which stimulates yeast growth under anaerobic conditions. The presence of dissolved O_2 increases both the ethanol and the yeast biomass.

Finally, it can be conclude that optimization of fermentation conditions is very crucial in obtaining the best quality wine from mango juice.

2.1.4 Characterization of Mango Wine

Mango wine has been characterized in terms of ethanol and glycerol concentrations and higher alcohols, carotenoids, polyphenols, and volatiles present (Reddy et al., 2009; Kumar et al., 2009; Varakumar et al., 2010; Lee et al., 2010).

2.1.4.1 Analysis of Mango Wine Volatiles by GC—FID and GC—MS

2.1.4.1.1 Ethanol and Glycerol From the fermentation of mango juice, ethanol was produced in higher concentrations than other metabolites. In general, the concentration of ethanol contributes to the whole characteristic quality and flavor of the produced wine. The percentage of ethanol produced in mango wine is between 7% and 8.5% (w/v), when no sugar is added, and comparable with moderate grape wines (Table 7.5.2). Glycerol concentration in mango wines is between 5.7 and 6.9 g/L. Another parameter, which influences the quality of wine highly, is acidity. The main organic acid present in mango must and produced wine is malic acid; and the other acids are less than 1 g/L (Reddy, 2005; Reddy and Reddy, 2011; Reddy et al., 2012, 2014).

2.1.4.1.2 Higher Alcohols and Other Flavor Compounds Mango fruit is known to have good aroma and flavor. The flavor of mango fruit is attributable to the volatile components that are present (Pino and Queris, 2010). The flavor of mango wine depends on many factors, such as fermentation conditions and varietal or fermentative compounds, which are present in highly variable amounts and are mainly alcohols, esters, terpenes, sulfur compounds, acids, and lactones (Reddy et al., 2009; Lee et al., 2010). Volatile aroma compounds are present in fruit juices and many are also synthesized by wine yeast during wine fermentation. The three major compounds (alcohols, esters, and organic acids) are present in different concentrations (Table 7.5.3). Pino and Queris (2010) reported the presence of 102 volatile compounds using advanced headspace solid-phase microextraction–gas chromatography/mass spectroscopy. According to their investigation, mango wine accounted for about 9 mg/L volatile compounds, which included 40 esters, 15 alcohols, 12 terpenes, 8 acids, 6 aldehydes and ketones, 4 lactones, 2 phenols, 2 furans, and 13 miscellaneous compounds. Isopentanol and 2-phenylethanol were the major constituents. Li et al. (2012b) investigated the aroma compounds present in fresh Chok Anon cultivar mango juice and wine fermented with three different yeast strains.

2.1.4.1.3 Polyphenols, Carotenoid Composition, and Antioxidant Activity Phenolic compounds are considered basic components of wine, and over 200 compounds have been identified in grape wine. The concentration of total polyphenols varies with the variety of mango employed in wine fermentation. The Totapuri variety contains the highest concentration, 1050 mg/L, followed by Alphonso, Banginapalli, and Sindhura (725, 610, and 490 mg/L, respectively). Eleven different phenolic compounds were identified in a detailed analysis of Alphonso phenolic compounds using LC/MS.

The total carotenoids in mango wine are in the range of 578–4330 μg/100 g, and the highest amount of total carotenoids was in wine from Alphonso, 4330 μg/100 g, followed by Sindhura, 4101 μg/100 g; Banginapalli, 2943 μg/100 g; Rumani, 2857 μg/100 g; Totapuri, 690 μg/100 g; Raspuri, 634 μg/100 g; and Neelam, 578 μg/100 g (Varakumar et al., 2010). In this mango wine, xanthophylls (oxygenated carotenoids) were degraded more than β-carotene (hydrocarbon carotenoid) (Varakumar et al., 2010). The antiradical activity from 27.57% to 36.70% ascorbic acid equivalents, the antioxidant activity from 73.90% to 85.95% gallic acid equivalents, and the quenching of 2,2-diphenylpicrylhydrazyl were dose dependent (Reddy, 2005; Varakumar et al., 2010; Kumar et al., 2010).

2.1.5 Effect of Storage, Bottle Color, and Temperature on Wine Color

Reddy (2005) examined the change in color of mango wine at 8, 16, and 25°C stored in white, green, and brown bottles based on the browning index. Wine stored in darker brown bottles at low temperature

Table 7.5.3 Composition of Volatiles in Three Mango Varieties (Banginapalli, Alphonso, and Totapuri) Fermented at 25 ± 2°C and pH 4.5 for 10 days.

No.	Retention Time	Name of Compound	Banginapalli (mg/L)	Alphonso (mg/L)	Totapuri (mg/L)
Alcohols					
1	1.271	Ethanol	8.5	7.5	7
2	1.350	Ethyl ether	Solvent	Solvent	Solvent
3	1.492	1-Propanol	54.11	42.32	47.13
4	1.729	Isobutyl alcohol	102.40	115.14	98.87
5	2.581	Isoamyl alcohol	125.2	108.40	140.44
6	2.850	Pentane-2-one	1.43	1.15	1.51
7	4.823	2-Furan methanol	0.123	nd	0.216
8	6.535	Hexane-1-ol	1.42	1.02	nd
9	12.900	Phenylethyl alcohol	22.15	24.15	20.48
10	19.414	Cyclohexane methanol	1.13	nd	1.34
11	42.58	*n*-Pentanedecanol	0.610	nd	Tr
Esters					
12	1.665	Ethyl acetate	35.15	30.42	27.48
13	6.876	Ethyl hexanoate	0.942	0.671	0.552
14	15.92	Ethyl octanoate	1.150	1.06	1.451
15	20.124	Ethyl decanoate	2.34	1.86	1.43
16	33.62	β-Phenylethyl butanoate	nd	0.62	0.92
17	19.67	Dimethyl styrene	1.11	1.34	1.09
Acids					
18	1.950	Acetic acid	0.201	0.163	0.155
19	3.292	Propanoic acid	0.145	0.217	0.184
20	3.829	Butanoic acid	0.932	0.745	0.874
21	12.655	2-Furoic acid	0.910	0.548	0.745
22	15.482	Benzoic acid	1.08	1.21	1.43
23	15.750	Phenyl formic acid	0.643	0.912	0.434
24	16.723	Octanoic acid	0.735	0.427	nd
25	37.99	Decanoic acid	1.18	0.963	Tr
Ketones					
26	2.850	Pentane-2-one	1.43	1.15	1.51
27	6.245	Furanone	1.12	1.51	1.22
28	11.489	Hydroxydimethylfuranone	0.238	0.452	0.331
29	25.967	Phenol 2,6-bis-4-methoxy-one	0.451	0.432	0.312
Unknown					
30	15.165	Unknown	0.183	0.412	0.243
31					
32	23.377	Benzene methane-4-hydroxy	0.531	0.256	0.231
33	35.68	Unknown	0.441	0.131	nd
34	38.86	Unknown	0.12	Tr	Tr
35	46.34	Unknown	Tr	Tr	nd

Tr, *trace*.

Based on the data presented in Reddy, L.V.A., 2005. *Production and Characterization of Wine Like Product from Mango Fruits* (Mangifera indica L.). Thesis Submitted to Sri Venkateswara University, Tirupati, India; Reddy, L.V.A., Sudheer Kumar, Y., Reddy, O.V.S., 2009. Evaluation of volatile flavour compounds from wine produced from mango fruits (Mangifera indica L.) by gas chromatography and mass spectrometry (GC-MS). *Indian Journal of Microbiology* 50, 183–191.

showed low browning indices compared to the wine in a white bottle. The contents of ascorbic acid and sugars together with storage temperature, sunlight, and bottle color affect browning in orange juice and orange wines. Sunlight can cause undesired changes in wines. Therefore, in winemaking the use of bottles that can prevent transmission of short-wave light such as ultraviolet, violet, and blue is recommended, because short-wave light has high energy and can initiate some chemical reactions (Boulton et al., 1996).

2.1.6 Sensory Evaluation of Mango Wine

Mango wine quality was tested based on the main characteristics visual, aroma, taste, and harmony. From Table 7.5.4 it can be concluded that all the different mangoes produced different types of wines, including three yeast strains that produced three wines with different characters. Banginapalli wines with yeast strain CFTRI 101 got the highest score, followed by Alphonso and Totapuri (Table 7.5.4) (Reddy, 2005; Reddy et al., 2009).

2.1.7 Immobilization Studies

Cell immobilization technology was also adopted in mango wine production and a novel yeast biocatalyst was prepared by using watermelon pieces as the immobilizing support for the yeast, *S. cerevisiae* strain 101, for use in wine production. Immobilization was confirmed by electron microscopy (Fig. 7.5.3) (Reddy, 2005; Reddy et al., 2008). Mango peel was used as the support material and produced good quality mango wine (Varakumar et al., 2012). The fermentation rate and other parameters were compared with those of free yeast cells at various temperatures. In all cases, fermentation time was short (22 h at 30°C and 80 h at 15°C) and produced high ethanol productivity (4 g/L/h). The volatile compounds, methanol, ethyl acetate, propanol-1, isobutanol, and amyl alcohol, that formed during fermentation were analyzed with the help of the GC/flame ionization detection. Cell metabolism of the immobilized yeast was not much affected by immobilization. It was reported that the immobilization of yeast on watermelon pieces increased the fermentation rate, vitality, and viability

Table 7.5.4 Sensorial Evaluation of Eight Mango Wines Produced From Eight Mango Varieties

No	Wine Variety	Flavor	Taste	Texture	Appearance	MouthFeel	Overall Acceptability
1	Alphonso	8.90	8.26	8.59[b]	8.23[ac]	8.85[cd]	8.62[cd]
2	Raspuri	7.22[bc]	7.68[ac]	8.14[ac]	7.97[bc]	8[ac]	7.94[ac]
3	Banginapalli	8.56[ac]	8.35[bd]	8.45[bd]	8.76[cd]	8.5[cd]	8.3[bd]
4	Totapuri	7.84[ac]	7.45[ac]	7.98[ac]	8.1[ac]	7.6[ac]	7.4[ac]
5	Sindhura	6.9[ab]	6.4[b]	7.2[ab]	8.45[bd]	7.3[bc]	7.2[ad]
6	Neelam	6.5[b]	6.65[ab]	6.9[b]	6.3[a]	6.7[b]	6.6[b]
7	Rumani	7.1[bc]	6.8[bc]	7.3[bc]	7.5[ab]	6.6[a]	7.0[ab]
8	Imam pasand	6.1[c]	6[a]	6.3[a]	6.7[b]	6.9[ab]	6.2[a]

Values are given as mean ± SD. Values not sharing a common superscript letter differ significantly at p < .05 (DMRT).
Based on the data presented in Reddy, L.V.A., 2005. Production and Characterization of Wine Like Product from Mango Fruits (Mangifera indica L.). Thesis Submitted to Sri Venkateswara University, Tirupati, India; Reddy, L.V.A., Sudheer Kumar, Y., Reddy, O.V.S., 2009. Evaluation of volatile flavour compounds from wine produced from mango fruits (Mangifera indica L.) by gas chromatography and mass spectrometry (GC-MS). Indian Journal of Microbiology 50, 183–191.

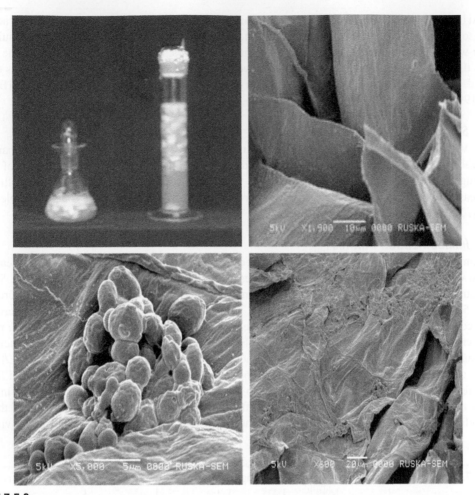

FIGURE 7.5.3

Immobilization of yeast cells on watermelon pieces in mango wine fermentation.

Reproduced from Reddy, L.V.A., 2005. Production and Characterization of Wine Like Product from Mango Fruits (Mangifera indica *L.*).
Thesis Submitted to Sri Venkateswara University, Tirupati, India.

of yeast cells. Preliminary sensory tests suggest a fruity aroma, fine taste, and overall improved quality of the wines produced.

2.1.8 Cost Economics

Approximately 500 mL of juice can be obtained from 1 kg of mangoes. To produce 1 L of wine, it requires about 1250 mL of juice, as there would be fermentation and evaporation losses. This means that 2.5 kg of mangoes is required, costing around Rs. 70.00 as the raw material cost, as of this writing. And about 40% of the raw material cost would be the processing cost. Hence, 1 L of wine would cost approximately Rs. 100.00. However, scale-up studies need to be carried out for an actual assessment of the cost of production of mango wine (Reddy et al., 2014).

3. PINEAPPLE WINE

Pineapple (*Ananas comosus*), a leading member of the family Bromeliaceae, comprises about 2000 species, mostly epiphytic and many strikingly ornamental, and varies from nearly white to yellow in color (Morton, 1987). It is an herbaceous perennial plant that grows to 1.0–1.5m tall with 30 or more trough-shaped and pointed leaves, 30cm long, surrounding a thick stem. It is a multiple fruit, forming what appears to be a single fleshy fruit. Pineapples contain good sugar proportions, which make it suitable for making wine (Adaikan and Ganesan, 2004). Pineapple juice generally has TSS in the range 12–15 °Bx; the sugar content is raised by the addition of sugar up to 22–25 °Bx to produce a wine having 12–13% alcohol. However, the flavor of pineapple is not stable and oxidation can occur easily (Amerine et al., 1980). Wine from two cultivars, Kew and Queen, was prepared and it was observed that nitrogen and phosphate are very important in the production of good quality pineapple wine. The wine produced from the Kew variety inoculated with *Saccharomyces ellipsoideus* 101 recorded the highest percentage of alcohol (8.40%) followed by wine from the Queen variety inoculated with the same strain (8.35%). An organoleptic evaluation was carried out by a selected panel of judges based on the 20-point scale. Wine produced from the Queen variety fermented with *S. ellipsoideus* 101, when supplemented with both N and P sources, recorded the highest score (16.75 of 20.00), followed by the SPQ4 yeast variety (14.75 of 20.00), whereas the lowest score was recorded by wine prepared from the Kew variety inoculated with SPQ3 (13.00 of 20.00) (Patil and Patil, 2006). Chanprasartsuk et al. (2012) produced a pineapple wine using single and mixed non-*Saccharomyces* yeasts and suggested that Queen pineapple juice could be a good substrate for yeast fermentation (Table 7.5.5). The yeast isolates *Saccharomycodes ludwigii* and *Hanseniaspora* used as mixed starter cultures could perform appropriate alcoholic fermentation for Queen pineapple wine production.

4. CASHEW APPLE

Cashew is one of the most economically important plantation crops in India, Brazil, Nigeria, and Vietnam (Mohanthy et al., 2006). Unlike the cashew nut kernel, which has an indisputably exclusive fine taste and a commercial attractiveness of its own, cashew "apple," despite its high nutritive value (high

No.	Characteristic	Juice		Wine	
		Kew	Queen	Kew	Queen
1	pH	3.1	3.2	3.2	3.2
2	Titratable acidity (%)	1.2	1.0	0.9	0.9
3	Tannins (mg/100mL)	22	28		
4	Protein (%)	0.58	0.64	0.4	0.5
5	TSS (°Bx)	12–13	13–15	6.6	7.0
6	Reducing sugars (%)	2.54	2.62	2.1	2.2
7	Alcohol (% v/v)	-	-	8.2	8.1
8	Aldehydes	-	-	57	65
9	Total score	-	-	13.0	14.7

Table 7.5.5 Chemical Composition of Pineapple Juice and Pineapple Wine

Reproduced from Patil, S.K., Patil, A.B., 2006. Wine production from pineapple must supplemented with sources of nitrogen and phosphorus. Karnataka Journal of Agricultural Science 19 (3), 562–567.

Table 7.5.6 Physicochemical Characteristics of Cashew Apple Juice and Wine

Composition	Must	Wine
TSS (°Bx)	17.0	2.0
Reducing sugar (g/100 mL)	6.44	0.9
Titratable acidity (g/100 mL)	0.24	1.21
pH	4.63	2.92
Phenol (g/100 mL)	0.13	0.12
Tannins (mg/100 mL)	2.2	1.9
Lactic acid (mg/100 mL)	nd	2.5
Ethanol	nd	5.0

TSS, *total soluble solids;* nd, *not detected.*

content of vitamin C and minerals, i.e., Ca, P, Fe) and economic potential, is virtually an unknown product on the consumer market. The edible cashew apple is the thick receptacle or "false fruit" to which the cashew nut or true fruit is attached. A popular brandy locally called "fenny" is prepared from cashew apple in Goa and India, having 40% (v/v) alcohol content (Maini and Anand, 1993; Araujo et al., 2011). An attempt was made by Osho and Odunfa (1999) and Mohanthy et al. (2006) to prepare wine from cashew apples after fermenting the juice with the wine yeast strain NCYC 125 and *S. cerevisiae* var. *bayanus* and compare the sensory attributes (taste, aroma, flavor, color/appearance, and aftertaste) with those of commercial grape wine. The composition of fresh cashew apple juice and wine is presented in Table 7.5.6. The basic steps involved in the production of cashew apple wine are depicted in Fig. 7.5.4. The complete utilization of cashew apple for various products through biotechnological processes was investigated by Akinwale (1999). Ethanol, vinegar, and lactic acid production from cashew apple were studied. The maximum production of ethanol reported was 17.6% with a 55% yield (Silva et al., 2007). The effects of initial sugar concentration in the juice on the composition and sensory properties of the wine were studied by Attri et al. (2009) and it was confirmed that fermentation conditions and starting sugar concentration are crucial to the outcome of the ethanol concentration, with different yeast strains behaving differently under certain conditions, and each *S. cerevisiae* strain is unique. In a preliminary sensory evaluation analysis the panelists rated cashew wine as somewhat inferior (except in color/appearance) to commercial grape wine ($p \le .01$), but the attributes taste, aftertaste, and color/appearance were scored at about 3.0 ("like much"). However, the panelists rated aroma and flavor between 2.5 and 3.0 ("like moderately"–"like much"), probably because of the high tannin content in cashew wine, which imparts a somewhat astringent flavor. Nevertheless, the cashew wine was acceptable to all the panelists (Mohanthy et al., 2006).

5. LYCHEE WINE

Lychee wine has a plenty of flavor and is a good source of minerals and vitamins and is used for the preparation of alcoholic beverages in China. This fruit has been used to prepare a low-alcoholic high-flavored beverage using the technique of partial osmotic dehydration. It was found that a product

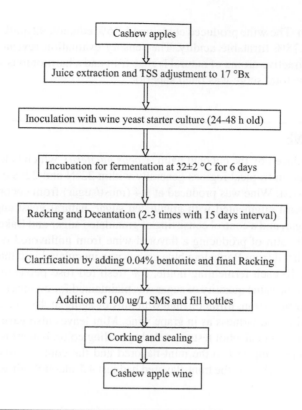

Cashew apples

↓

Juice extraction and TSS adjustment to 17 °Bx

↓

Inoculation with wine yeast starter culture (24-48 h old)

↓

Incubation for fermentation at 32±2 °C for 6 days

↓

Racking and Decantation (2-3 times with 15 days interval)

↓

Clarification by adding 0.04% bentonite and final Racking

↓

Addition of 100 ug/L SMS and fill bottles

↓

Corking and sealing

↓

Cashew apple wine

FIGURE 7.5.4

Flow sheet for the preparation of cashew apple. *TSS*, total soluble solids.

Figure modified and redrawn from Mohanty, S., Ray, P., Swain, M.R., Ray, R.C., 2006. Fermentation of cashew (Anacardium occidental L.) apple into wine. Journal of Food Processing and Preservation 30, 314–322.

containing 5–6% alcohol, 3–4% sugars, and 0.35% acids can be prepared as an appetizing soft drink instead of an intoxicating liquor. With osmotic dehydration as a pretreatment a concentration of nutrients and flavor takes place and is reflected in the product compared to that prepared from untreated fruits. For the same reason, the osmotically treated juice undergoes fast fermentation and the time required for the preparation of wine is considerably less. The detailed procedure involved in preparation of this drink has been described earlier (Vyas et al., 1989; Zeng et al., 2008). Lychee fruit of optimum maturity is washed and peeled, followed by dipping in a sugar solution of 70 °Bx for 4 h at 50°C. The treated fruit is taken out and drained, followed by pulping in a pulper. The sugar content is adjusted to 22 °Bx by dilution with water. A 24-h-old active culture of yeast, *S. cerevisiae*, prepared in the lychee juice is added (5%) to the sterilized lychee juice to carry out the alcoholic fermentation. Singh and Kaur (2009) evaluated the suitability of lychee fruit juice concentrate for wine production. The large amounts of fermentable sugars (85.20%) and acids (4.25%) in lychee juice concentrate were found to be suitable for wine production. Among the four yeast strains screened, MTCC 178 was suitable for wine production from reconstituted lychee juice concentrate. The better quality wine was produced under optimal conditions of 25°C temperature, pH 4.5, TSS 24

°Bx, and 10% inoculum. The wine produced contained 11.6% ethanol, 92 mg/L total esters, 124 mg/L total aldehydes, and 0.78% titratable acidity. The sensory evaluation revealed a clear light amber lychee wine, with an attractive aroma of natural lychee fruit and a harmonious wine taste. The quality of wine as shown by the total score was rated superior.

6. COCONUT WINE

Coconut is the fruit produced by the coconut palm (*Cocos nucifera*), which belongs to the family Arecaceae. Coconut is found in tropical regions generally within 22°N and 22°S of the equator and most commonly near a sea coast. Wine was produced at 1:4 (must/sugar) from coconut (*C. nucifera*) using natural yeast, natural yeast augmented with granulated sugar, natural yeast augmented with Baker's yeast and granulated sugar, and a control consisting of granulated sugar and Baker's yeast. Experiments were conducted with the aim of producing a flavored wine from unflavored tender coconut water. A purified yeast strain isolated from Bangalore blue grapes was inoculated into the sterile tender coconut water with 10% sugar. To each fermenting bottle, 5% fresh red rose petals or 5% mint leaves were added. One control with the same quantity of sugar was maintained for comparison. The tender coconut water after complete fermentation, filtration, and 6 months of aging at 4°C gave a pleasant rose flavor with red color to the wine and tartness as in grape wine. Mint leaves also gave flavor but not color to the wine. The organoleptic and alcohol tests revealed appealing color, better taste, and alcohol content in the rose-flavored wine compared to the mint-flavored and the control. Rose-flavored and control wine had up to 8% alcohol, whereas the mint-flavored had just 4% alcohol (Idise, 2011; Savita D'Souza et al., 2013).

7. SAPOTA WINE

Sapota (*Achras sapota* Linn.) is an important tropical fruit belonging to the family Sapotaceae. At present it is cultivated in all the tropical countries of the world. Sapota can grow to more than 30 m tall with an average trunk diameter of 1.5 m. Fruits are ovoid to globular berries with 1–12 shiny brown or black seeds (frequently 5), surrounded by a brownish skin very much resembling a smooth-skinned potato. The area under sapota cultivation in India is estimated to be 150,000 hectares, with an annual production of 1,117,000 tonnes, and production was 1,238,000 Mt during the season 2007–2008 (www.apeda.com). The fruit has a short shelf life, as the fruit ripens after 8–10 days of harvesting and hence the postharvest loss of this fruit is one of the most pressing problems in tropical countries like India. It has been estimated that the total loss of fruit in India is around 20–30% of the total fruit production. The pulp is yellowish, tender, granular, sweet, juicy, and scented and has a high latex content. Matured sapota pulp is a good source of carbohydrates (21.4 g/100 g), dietary fiber (10.9 g/100 g), and tannins (3.16–3.45%). In view of higher production, nutritional importance, and shorter availability period, researchers have studied the production, composition, and sensory attributes of sapota wine (Reddy, 1959; Gautam and Chundawat, 1998; Parwar et al., 2011; Panda et al., 2014). The steps involved in the production of sapota wine are given in a flow diagram (Fig. 7.5.5). Physicochemical characteristics of the must and wine are summarized in Table 7.5.7. Sensory

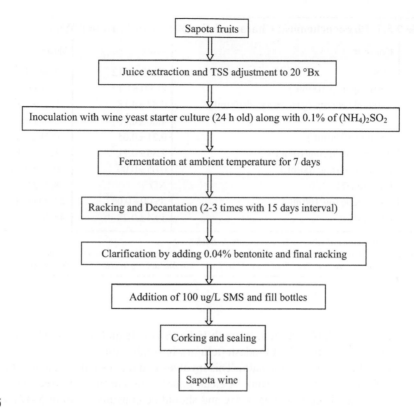

FIGURE 7.5.5

Steps involved in sapota wine production. *TSS*, total soluble solids.

Figure modified and redrawn from Panda, S.K., Sahu, U.C., Behera, S.K., Ray, R.C., 2014. Fermentation of sapota (Achras sapota Linn.) fruits to functional wine. Nutrafoods. http://dx.doi.org/10.1007/s13749-014-0034-1.

evaluation was carried out by 16 panelists on various attributes like taste, aroma, flavor, color/appearance, and aftertaste. The results showed that the flavor, taste, and aroma of sapota wine was strongly liked by the panelists and there was no significant difference ($p < .05$) between the two replicates for most sensory parameters.

8. PALM WINE

Palm wine is the fermented sap of certain varieties of palm trees including raphia palm (*Raphia hookeri* or *Raphia vinifera*), coconut palm (*C. nucifera*), and *Caryota* palm (jeeluga). It is collected by tapping the top of the trunk. It is a cloudy, whitish beverage with a sweet alcoholic taste and a very short shelf life of only 1 day. The wine is consumed in a variety of flavors, from sweet unfermented to sour, fermented, and vinegary. There are many variations of the same product and no individual method or

Table 7.5.7 Physicochemical Characteristics of Sapota Must and Wine

No.	Content	Must	Wine
1	TSS	20.00±0.12	2.38±0.00
2	Total sugar (g/100 mL)	28.00±1.12	3.28±0.77
3	Titratable acidity (g tartaric acid/100 mL)	0.82±0.18	1.29±0.34
4	pH	4.80±0.06	3.02±0.03
5	Phenol (g/100 mL)	0.21±0.05	0.21±0.03
6	Ascorbic acid (mg/100 mL)	2.86±0.21	1.78±0.14
7	Lactic acid (mg/100 mL)	0.05±0.00	0.64±0.08
8	Ethanol (%)	ND	8.23±0.21
9	β-Carotene (μg/100 mL)	35.00±0.05	22.00±0.05
10	DPPH (%)	59.4±0.28	46.00±0.21

DPPH, *2,2-diphenylpicrylhydrazyl*; TSS, *total soluble solids.*
Reproduced from Panda, S.K., Sahu, U.C., Behera, S.K., Ray, R.C., 2014. Fermentation of sapota
(Achras sapota Linn.) fruits to functional wine. Nutrafoods. http://dx.doi.org/10.1007/s13749-014-0034-1.

recipe. Palm wine is particularly common in south India (Shamala and Sreekantiah, 1988). The local name for the product is kallu in Andhra Pradesh (telugu), in south India.

The sap is collected from tapping the palm. This involves making a small incision in the bark of the palm, about 15 cm from the top of the trunk. A clean gourd is tied around the tree to collect the sap, which runs into it. The sap is collected each day and should be consumed within 5–12 h of collection (Fig. 7.5.6). Fresh palm juice is a sweet, clear, colorless juice containing 12–15% sucrose (by weight) and trace amounts of reducing sugars including glucose, fructose, maltose, and raffinose. The sap contains approximately 0.23% protein, 0.02% fat. Half of the total sugars are fermented during the first 24 h and the ethanol content of the fermented palm sap reaches a maximum of 5.0–5.28% (v/v) after 48 h (Sekar and Mariappan, 2005). The sap is not heated and the wine is an excellent substrate for microbial growth. The palm sap fermentation involves alcoholic–lactic–acetic acid fermentation, due to the presence mainly of yeasts and lactic acid bacteria. It has been suggested that *Saccharomyces* spp. are present in the naturally fermented palm sap and are important for the formation of the characteristic aroma of the palm wine (Aidoo et al., 2006). *S. cerevisiae* and *Schizosaccharomyces pombe* have been reported to be the dominant yeast species (Odunfa and Oyewole, 1998). Other yeast species such as other *Candida* spp. and *Pichia* spp. are also present (Atacador-Ramos, 1996). Lactic acid bacteria and other bacteria such as *Lactobacillus plantarum*, *Leuconostoc mesenteroides*, *Acetobacter* spp., and *Zymomonas mobilis* are also present. The microorganisms are reported to originate from the palm tree, the gourd used for sap collection and fermentation, or the tapping equipment. The presence of 17 species of yeasts and seven genera of bacteria in the natural fermented coconut palm sap has been reported (Atputharajah et al., 1986). It is therefore essential that proper hygienic collection procedures are followed to prevent contaminating bacteria from competing with the yeast and producing acid instead of alcohol. Fermentation starts soon after the sap is collected and within an hour or two the sap becomes reasonably high in alcohol (up to 4%). If allowed to continue to ferment for more than a day, it starts

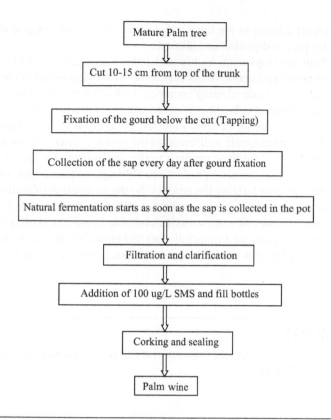

FIGURE 7.5.6

The basic steps involved in the production of palm wine.

Figure modified and redrawn from Atputharajah, J.D., Widanapathirana, S., Samarajeewa, U., 1986. Microbiology and biochemistry of natural fermentation of coconut palm sap. Food Microbiology 3 (4), 273–280.

turning into vinegar. The main control points are extraction of a high yield of palm sap without excessive contamination by spoilage microorganisms and proper storage to allow the natural fermentation to take place. The quality of the final wine is determined by the conditions used for the collection of the sap. Often the collecting gourd is not washed between collections and residual yeasts in the gourd quickly begin the fermentation. This is beneficial as it prevents the growth of bacteria that can spoil the sap. The product should be kept in a cool place away from direct sunlight.

9. CONCLUSIONS AND FUTURE PERSPECTIVES

Wines are made from complete or partial alcoholic fermentation of grapes or any other fruit and contain ethyl alcohol as the intoxicating agent, essential elements, vitamins, sugars, acids, and phenolics. Wines from fruits are preferable to distilled liquors for stimulatory and healthful properties. These beverages also serve as an important adjunct to the human diet by increasing satisfaction and contribute to the relaxation necessary for proper digestion and absorption of food.

Fruits produced from the tropics are highly perishable commodities and have to be either consumed immediately or preserved in one or another form. In the developed countries, a considerable quantity of fruit is utilized, but in developing countries, lack of proper utilization results in considerable postharvest losses, estimated to be 30–40%. The increased production can be used profitably, if fruit wines are produced. Setting up of fruit wineries, in addition to industrialization of the fruit-growing belts, could result in economic upliftment of the people, generating employment opportunities and providing better returns for their produce to the orchardists. The availability of technology is the most important factor determining production, although cost and type of product are also a significant considerations in popularizing the product. So the production of fruit wines in those countries where fruits other than grape are grown would certainly be advantageous. In the future it will be necessary to conduct more studies on more varieties available in India regarding the physicochemical composition, volatile composition, and especially health beneficial effects of mango wine. It is also essential to stimulate and carry out research and development on the production and evaluation of nongrape wines.

ACKNOWLEDGMENTS

The author (Dr. L. V. Reddy) acknowledges the Department of Science and Technology, Government of India, and CSIR, New Delhi, for financial support.

REFERENCES

Adaikan, P., Ganesan, A.A., 2004. Mechanism of the Oxytoxic activity of Comosus proteinases. Journal of Pharmaceutical Biology 42 (8), 646–655.

Aidoo, K.E., Nout, M.J.R., Sarkar, P.K., 2006. Occurrence and function of yeasts in Asian indigenous fermented foods. FEMS Yeast Research 6, 30–39.

Akinwale, T.O., 1999. Fermentation and post fermentation chances in cashew wine. The Journal of Food Technology in Africa 4 (3), 100–102.

Amerine, M.A., Berg, H.W., Kunkee, R.E., Qugh, C.S., Singleton, V.L., Webb, A.D., 1980. The Technology of Wine Making, fourth ed. AVI, Westport; CT.

Anon, 1962. Wealth of India – Raw Materials 6, (L-M.). Publication and Information Directorate, CSIR, New Delhi, India.

Anon, 1963. Mango wine. International Bottler and Packer 37, 157–159.

APPEDA, 2014. http://agriexchange.apeda.gov.in/India%20Production/India_Productions.aspx?hscode=08045020.

Araujo, S.M., Silva, C.F., Moreira, J.J.S., Narain, N., Souza, R.R., 2011. Biotechnological process for obtaining new fermented products from cashew apple fruit by *Saccharomyces cerevisiae* strains. Journal of Industrial Microbiology Biotechnology 38 (9), 1161–1169.

Atacador-Ramos, M., 1996. Indigenous fermented foods in which ethanol is a major product. In: Steinkraus, K.H. (Ed.), Handbook of Indigenous Fermented Foods. Marcel Dekker, New York, pp. 363–508.

Atputharajah, J.D., Widanapathirana, S., Samarajeewa, U., 1986. Microbiology and biochemistry of natural fermentation of coconut palm sap. Food Microbiology 3 (4), 273–280.

Attri, B.L., 2009. Effect of initial sugar concentration on the physico-chemical characteristics and sensory qualities of cashew apple wine. Natural Product Radiance 8 (4), 374–379.

Barnett, J.A., 1980. A history of research on yeasts: work by chemists and biologists 1739–1850. Yeast 14, 1439–1451.

Boulton, R.B., Singleton, V.L., Bisson, L.F., Kunkee, R.E., 1996. Principles and Practices of Winemaking. Aspen, New York, NY, USA.

Chanprasartsuk, O., Pheanudomkitlert, K., Toonwai, D., 2012. Pineapple wine fermentation with yeasts isolated from fruit as single and mixed starter cultures. Asian Journal of Food and Agro Industry 5 (20), 104–111.

Czyhrinciwk, N., 1966. The technology of passion fruit and mango wines. American Journal of Enology and Viticulture 17, 27–30.

FAO Statistics, 2014. Production of Tropical Fruits. Retrieved 26 November 2013.

Gautam, S.K., Chundawat, B.S., 1998. Standardization of technology of sapota wine making. Indian Food Packer 52 (1), 17–21.

Herrero, M., Garcia, L.A., Diaz, M., 2003. The effect of SO_2 on the production of ethanol, acetaldehyde, organic acids and flavour volatiles during industrial cider fermentation. Journal of Agriculture Food Chemistry 52, 3455–3459.

Idise, O.E., 2011. Studies on wine production from coconut (*Cocos nucifera*). Journal of Brewing and Distilling 2 (5), 69–74.

Inndian Horticulture Database, 2013. http://www.nhb.gov.in/area-pro/Indian%20Horticulture%202013.pdf.

Joshi, V.K., Attri, D., 2005. Panorama of research and development of wines in India. Journal of Science and Industrial Research 64, 9–18.

Joshi, V.K., Pandey, A., 1999. Biotechnology. Food Fermentation: Microbiology, Biochemistry and Technology, vol. II. Educational Publishers and Distributors, New Delhi.

Kourkoutas, Y., Douma, M., Koutinas, A.A., Kanellaki, M., Banat, I.M., Marchant, R., 2003. Room and low temperature continuous wine making using yeast immobilized on quince pieces. Process Biochemistry 39, 143–148.

Kulkarni, J.H., Singh, H., Chada, K.L., 1980. Preliminary screening of mango varieties for wine making. Journal of Food Science and Technology 17, 218–221.

Kumar, Y.S., Prakasam, R.S., Reddy, O.V.S., 2009. Optimisation of fermentation conditions for mango (*Mangifera indica* L.) wine production by employing response surface methodology. International Journal of Food Science and Technology 44, 2320–2327.

Kumar, Y.S., Varakumar, S., Reddy, O.V.S., 2010. Evaluation of antioxidant and sensory properties of mango (*Mangifera indica* L.) wine. CYTA Journal of Food 10, 12–20.

Li, X., Lim, S.L., Bin, Yu, Curran, P., Liu, S.Q., 2013. Mango wine aroma enhancement by pulp contact and β-glucosidase. International Journal of Food Science and Technology 48, 2258–2266.

Lee, P.R., Ong, Y.L., Yu, B., Curran, P., Liu, S.Q., 2010. Evolution of volatile compounds in papaya wine fermented with three *Williopsis saturnus* yeasts. International Journal of Food Science and Technology 45, 2032–2041.

Li, X., Chan, L.J., Yu, B., Curran, P., Liu, S.Q., 2012a. Fermentation of three varieties of mango juices with a mixture of *Saccharomyces cerevisiae* and *Williopsis saturnus* var. *mrakii*. International Journal of Food Microbiology 158, 28–35.

Li, X., Bin, Yu, Curran, P., Liu, S.Q., 2012b. Impact of two *Williopsis* yeast strains on the volatile composition of mango wine. International Journal of Food Science and Technology 47, 808–815.

Maini, S.B., Anand, J.C., 1993. Utilization of fruit wastes. In: Chadha, K.L., Pareek, O.P. (Eds.), Advances in Horticulture. Fruit Crops, vol. 4. Malhotra Publishing House, New Delhi, pp. 1967–1992.

Mohanty, S., Ray, P., Swain, M.R., Ray, R.C., 2006. Fermentation of cashew (*Anacardium occidental* L.) apple into wine. Journal of Food Processing and Preservation 30, 314–322.

Morton, J.F., 1987. Fruits of Warm Climates. Miami Printing Press, Miami, pp. 18–28.

Obisanya, M.O., Aina, J.O., Oguntime, G.B., 1987. Production of wine from mango (*Magnifera indica* L.) using *Saccharomyces* and *Schizosaccharomyces* species isolated from palm wine. Journal of Applied Bacteriology 63, 191–196.

Odunfa, S.A., Oyewole, O.B., 1998. African fermented foods. In: Wood, B.J.B. (Ed.), Microbiology of Fermented Foods, vol. 2. Blackie Academic and Professional, London, pp. 713–752.

Onkarayya, H., Singh, H., 1984. Screening of mango varieties for dessert and madeira-style wine. American Journal of Enology and Viticulture 35, 63–65.

Osho, A., Odunfa, S.A., 1999. Fermentation on cashew juice using the wine yeast strain NCYC 125 and three other isolated yeast strains. Advances of Food Science 21 (1/2), 23–29.

Panda, S.K., Sahu, U.C., Behera, S.K., Ray, R.C., 2014. Fermentation of sapota (*Achras sapota* Linn.) fruits to functional wine. Nutrafoods. http://dx.doi.org/10.1007/s13749-014-0034-1.

Pawar, C.D., Patil, A.A., Joshi, G.D., 2011. Quality of sapota wine from fruits of differential maturity. Karnataka Journal of Agricultural Sciences 24 (4), 501–505.

Patil, S.K., Patil, A.B., 2006. Wine production from pineapple must supplemented with sources of nitrogen and phosphorus. Karnataka Journal of Agricultural Science 19 (3), 562–567.

Pino, J.A., Queris, O., 2010. Analysis of volatile compounds of mango wine. Food Chemistry 125, 1141–1145.

Reddy, M.G., 1959. Physico Chemical Investigations on Sapota and its Products (M.Sc. (Food Tech.) Thesis). Central Food Technol. Res. Inst., Mysore (India).

Reddy, L.V.A., 2005. Production and Characterization of Wine Like Product from Mango Fruits (*Mangifera indica* L.). Thesis Submitted to Sri Venkateswara University, Tirupati, India.

Reddy, L.V.A., Reddy, O.V.S., 2005. Production and characterization of wine from mango fruits (*Mangifera indica* L.). World Journal of Microbiology and Biotechnology 21, 1345–1350.

Reddy, L.V.A., Reddy, O.V.S., 2009a. Effect of enzymatic maceration on synthesis of higher alcohols during mango wine fermentation. Journal of Food Quality 32, 34–47.

Reddy, L.V.A., Reddy, O.V.S., 2009b. Production, optimization and characterization of wine from mango (*Mangifera indica* L.). Nature Product Radiance 8, 426–435.

Reddy, L.V.A., Reddy, O.V.S., 2011. Effect of fermentation conditions on yeast growth and volatile composition of wine produced from mango (*Mangifera indica* L.) fruit juice. Food and Bioproducts Process 89, 487–491.

Reddy, L.V.A., Reddy, Y.H.K., Reddy, L.P.A., Reddy, O.V.S., 2008. Wine production by novel yeast biocatalyst prepared by immobilization on watermelon (*Citrullus vulgaris*) rind pieces and characterization of volatile compounds. Process Biochemistry 43, 748–752.

Reddy, L.V.A., Sudheer Kumar, Y., Reddy, O.V.S., 2009. Evaluation of volatile flavour compounds from wine produced from mango fruits (*Mangifera indica* L.) by gas chromatography and mass spectrometry (GC-MS). Indian Journal of Microbiology 50, 183–191.

Reddy, L.V.A., Joshi, V.K., Reddy, O.V.S., 2012. Utilization of tropical fruits for wine production special emphasis on Mango wine Production. In: Satyanarayana, T.S., Johri, B.N. (Eds.), Microorganisms for Sustainable Agriculture and Environmental Development. Springer Publications, USA.

Reddy, L.V.A., Reddy, O.V.S., Joshi, V.K., 2014. Production of wine from mango fruit: a review. International Journal of Food and Fermentation Technology 4, 33–45.

Sakho, M., Chassagne, D., Jaus, A., Chiarazzo, E., Crouzet, J., 1998. Enzymatic maceration: effects on volatile components of mango pulp. Journal of Food Science 63, 975–978.

Savita D'Souza, Kanchanashri, B., Suvarna, V.C., 2013. Development of bioflavoured tender coconut water wine using native grape yeast. International Journal of Biotechnology and Bioengineering Research 4 (6), 553–554.

Sekar, S., Mariappan, S., 2005. Usage of traditional fermented products by Indian rural folks and IPR. Indian Journal of Traditional Knowledge 6 (1), 111–120.

Shamala, T.R., Sreekantiah, K.R., 1988. Microbiological and biochemical studies on traditional Indian palm wine fermentation. Food Microbiology 5 (3), 157–162.

Silva, M.E., Torres Neto, A.B., Silva, W.B., Silva, F.L.H., Swarnakar, R., 2007. Cashew wine vinegar production: alcoholic and acetic fermentation. Brazilian Journal of chemical Engineering 24 (2), 163–167.

Singh, R.S., Kaur, P., 2009. Evaluation of litchi juice concentrate for the production of wine. Natural Product Radiance 8 (4), 386–391.

Suresh, E.R., Onkarayya, H., Ethiraj, S., 1982. A note on the yeast flora associated with fermentation of mango. Journal of Applied Bacteriology 52 (1), 1–4.

Torija, M.J., Roes, N., Pblet, M., Guillamon, M.J., Mas, A., 2002. Effects of fermentation temperature on the strain population of *Saccharomyces cerevisiae*. International Journal of Food microbiology 80, 47–53.

Varakumar, S., Kumar, Y.S., Reddy, O.V.S., 2010. Carotenoid composition of mango (*Mangifera indica* L.) wine and its antioxidant activity. Journal of Food Biochemistry 35, 1538–1547.

Varakumar, S., Naresh, K., Reddy, O.V.S., 2012. Preparation of mango (*Mangifera indica* L.) wine using a new yeast-mango-peel immobilised biocatalyst system. Czech Journal of Food Science 30, 557–566.

Vyas, K.K., Sharma, R.C., Joshi, V.K., 1989. Application of osmotic technique in plum wine fermentation. Effect on physico-chemical and sensory qualities. Journal of Food Science and Technology 26, 126–128.

Yahia, E.M., 1998. Modified and controlled atmosphere for tropical fruits. Horticulture Reviews 22, 123–183.

Zeng, X.A., Chen, X.A., Qin, X.G.F., Zang, l., 2008. Composition analysis litchi juice and litchi wine. International journal of Food Engineering 4 (4), 1–16.

TECHNOLOGY FOR THE PRODUCTION OF AGRICULTURAL WINES

N. Garg

ICAR-CISH, Lucknow, India

1. INTRODUCTION

Wine by definition is a beverage made of the fermented juice of grapes or any other fruit and usually contains 7–15% alcohol by volume (Amerine, 1980). It is also made from vegetables, berries, fruits, herbs, flowers, or honey. Agricultural wines are produced from substrates other than the fruit juices (https://www.law.cornell.edu/cfr/text/27/24.200). Water, sugar, or both, may be added within a limited amount for the production of agricultural wine. Agricultural wine may or may not be flavored or colored; however, hops may be used in the production of such wines. Flower wine has been enjoyed since ancient times. Some examples of agricultural wines are honey wine, flower wines, etc. There are reports of making wine from a number of flowers but that is only at the home scale and just for the sake of hobby (Fessler, 1971; Keller, 2007). In this chapter, some important agricultural wines from mahua, honey, rhododendron, sweet potato, tomato, whey, and cocoa will be discussed.

2. MAHUA WINES

2.1 MAHUA TREE

Madhuca indica (J.F. Gmel. Syn. *Madhuca Latifolia Macb.*), commonly known as mahua, is a large tree found in the dry deciduous forests of India. Mahua, a tree of Indian origin, has been known from prehistoric times (Plate 8.1). Mahua is a multipurpose tree mostly grown on wastelands in the north and central parts of the country where, generally, tropical and subtropical climates prevail (Chateerjee and Parakashi, 2003). It grows well on a wide variety of soils including alluvial soil in Indo-Gangetic plains, rocky, gravelly red, saline soils. A large number of mahua trees are found in the states of Uttar Pradesh, Madhya Pradesh, Orissa, Jharkhand, Chhattisgarh, Gujarat, Andhra Pradesh, Bihar, West Bengal, and Karnataka. The two main species found in India are *Bassia latifolia* Roxb., a deciduous species common to northern parts of the country, and *Bassia longifolia* Koenig, an evergreen species common in southern parts of the country. This most attractive tree with its rounded crown is found everywhere along farm bunds and habitation as well as in the natural forest in India's tribal heartland, for people have maintained it with great love since time immemorial. Due to the economic importance of the mahua tree, British forest administrators historically spared it from felling, a policy still observed today. As a consequence, mahua trees are some of the largest, oldest, and most common in the fields and forests (Anon, 1976).

PLATE 8.1

Mahua tree.

2.2 MAHUA FLOWER

Mahua has a great mythological significance for Hindus. Utilization of mahua in day-to-day life has great nutritional, environmental, and commercial importance. It has special status livelihood systems in different ways. Apart from meeting the food and other requirements, it is also an important source of seasonal income. Its flowers (Plate 8.2) are used to brew liquor which is very popular in the tribal areas. Mahua provides livelihood security to poor households, which collect it both for self-consumption and sale. The rind of the fruit is consumed like a vegetable, and the seed provides edible oil. Most notable of all are its flowers with fleshy petals loaded with sugar.

2.3 MAHUA FLOWER: COMPOSITION

The corollas are a rich source of sugars and contain appreciable amounts of vitamins and calcium. The sugars identified are sucrose, maltose, glucose, fructose, arabinose, and rhamnose (Anon, 1976). The total sugar content of the corollas is maximum, when the flowers are mature and ready to fall. In the growing stage, fructose is present in a greater amount than glucose, and in the ripe stage, the qualities are almost equal; sucrose increases in amount up to the shedding of the corolla; it is later converted in invert sugar. The vitamins present in the corollas are carotene (as vitamin A), 39 IU; ascorbic acid, 7 mg; thiamine, 32 μg; and niacin, 5.2 mg/100 g; folic acid, pantothenic acid, biotin, and inositol are

PLATE 8.2

Mahua flower.

Table 8.1 Composition of Mahua Flower Juice	
Parameters	**Mahua Juice**
Total soluble sugar (°B)	13.0
Titratable acidity (%)	0.11
Ascorbic acid (mg/100 ml)	3.15
Tannins (%)	0.11
Reducing sugar (g %)	1.04
Adapted from Yadav, P., Garg, N., Diwedi, D.H., 2009a. Standardization of pre-treatment conditions for mahua wine preparation. Journal of Ecofriendly Agriculture 4 (1), 88–92.	

present (Anon, 1976). Biochemical analysis of mahua flower juice (Table 8.1) indicated that it is rich in sugar and may be utilized for the production of wine.

2.4 MAHUA LIQUOR

Mahua liquor production is a traditional process being explored since past number of centuries by the ethnic people (Anon, 1976). The estimated production of mahua flowers is more than one million tons in the country. In many states, government organizations like tribal development cooperative corporations purchase the flowers at a minimum support price to set aside the collectors from the exploitation of the middleman. However, due to lack of postharvest processing technologies, most of the flowers get decomposed in the government go downs. At present the only industrial utilization of these flowers in India is in the production of liquor (Yadav et al., 2012). The freshly prepared liquor has a strong, smoky odor, which disappears on aging. It is reported to excite gastric irritation and produce other unpleasant effects. Redistilled and carefully prepared liquors are of good quality and closely resemble Irish whiskey. There has been a fall in the consumption of mahua flowers for the production of alcohol in recent years due largely to the increased use of molasses as raw material. However, the use of mahua flowers for alcohol production is likely to continue in localities where they are available at low cost and where alcohol is intended for use as potable spirits. An average yield of 90 gallons of 95% alcohol is reported from 1 ton of dried flowers. A medium containing mahua flowers in 20% concentration (macerated in a food blender and extracted at 180°F for 30 min) is used for fermentation. The addition of 0.2–0.6% of ammonium sulfate or phosphate, or 5% spent flowers to the medium, gives better yields.

2.5 MAHUA FLOWER WINE

Wine from mahua is a novel product, especially with the view that it will be a flower wine. It is a substantial revenue-generator for the state and for tribals alike. A tremendous scope exists to improve the level of commercialization through standardization and quality control and applying modern fermentation systems to the traditional method. Some consumers have a strong liking for flower wines, and thus mahua wine has a vast export potential. Wine made from the mahua flower may find a good price if prepared as per international standards for wide acceptance and commercialization (Mande et al., 1949; Datta, 2011).

2.6 FLOWER COLLECTION AND PROCESSING

Mahua flowers, which drop naturally on clean, dark-colored polyethylene sheets below the tree in the night making the white flowers clearly visible, are collected/hand-picked. Postharvest treatment of mahua flowers involves cleaning, grading, drying, and storing. Flowers are graded based on quality parameters. The flowers are crushed in a fruit mill and spread on cheese cloth. The juice is extracted by applying pressure of 1500–2000 kg/square inch with a hydraulic press and collected in clean stainless steel utensils. Studies have reflected that KMS treatment (500 ppm) along with heating has been found effective in preparing shelf-stable mahua juice. Yadav et al. (2009a) and Anila et al. (2015) have reported that wine made from mahua flower juice had the least browning and highest sensory acceptance compared to that prepared from dry flowers, which might be due to irreversible changes during the drying of flowers. Yadav et al. (2012) had described a process for the preparation of wine from mahua flowers (Fig. 8.1).

2.7 ADDITION OF NUTRIENTS

The addition of yeast nutrients during winemaking is very important since yeast, a unicellular fungus vital to fermentation, needs carbon, nitrogen, vitamins, and minerals for growth and reproduction (http://www.pennsylvaniawine.com/sites/default/files/Yeast%20Nutrition.pdf). Nutrients added during alcoholic fermentation should contain nitrogen, phosphate, and vitamins, including thiamine and chelated mineral micronutrients. Complex yeast nutrients containing autolyzed yeast cells (dead yeast cells that have been treated so that their cell walls have ruptured) also improve the rate of fermentation. At high ethanol levels, yeast loses the ability to ingest ammonium ions (www.valleyvintner.com/Merchant2/DataSheets/valleyvintnerYeastNutrients_2007.pdf). This is

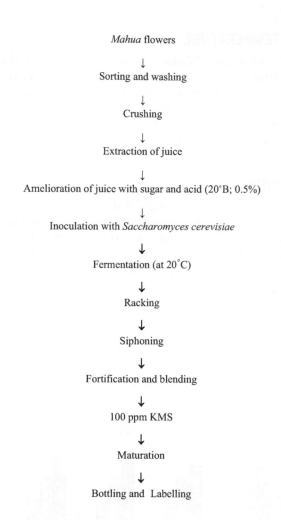

FIGURE 8.1

Flow sheet for the preparation of mahua flower wine.

Adapted from Yadav, P., Garg, N., Dwivedi, D.H., 2009b. Effect of location of cultivar, fermentation temperature and additives on the physico-chemical and sensory qualities on Mahua (Madhuca indica J. F. Gmel.) wine preparation. Natural Product Radiance 8, 406–418.

why it's important to make some nitrogen additions before the fermentation is past the halfway point. Yadav et al. (2009b) and Anila et al. (2015) reported the addition of nutrients in the form of diammonium phosphate, potassium hydrogen phosphate, and yeast extract to mahua must. It was found that nutrient additions effected the fermentation as reflected by higher alcohol and acidity level. Among the three treatments, yeast extract addition was found best for higher alcohol, ascorbic acid, and tannin level. This might be because yeast extracts provide a complex growth medium that best supports the growth of yeast (Jackson, 2008). However, higher ethanol yields were observed in other treatments also (Yadav et al., 2009b).

2.8 FERMENTATION TEMPERATURE

Temperature is perhaps the most critical factor influencing fermentation kinetics. For preparation of indigenous mahua liquor, fermentation is carried out under ambient temperature conditions. In northern India, the temperature during summer rises as high as 40°C. If the wine is made under such conditions the quality of wine will be far inferior. Yadav et al. (2009b) reported that wine fermented at 16°C had better sensory scores compared to that fermented at 25°C. Temperature as low as 13°C decreased both the fermentation and the growth rates (Fig. 8.2).

2.9 ADDITION OF TANNINS

Tannins are an excellent antioxidant that helps in giving the structure, texture, and flavor to wine. Wines that are fermented with the skins (such as red wines) usually contain enough natural tannin. For other wines (such as flower or fruit wines), tannin addition may be desirable. However, in the case of mahua wine, tannin addition was found undesirable since it reduced alcohol levels and increased browning (Yadav et al., 2009b). Sensory evaluation studies further confirmed that tannin addition is not required in mahua wine.

2.10 FLAVOR MASKING

Mahua has musk-scented flowers. Mahua flower wine, though palatable, has a smoky flavor and a burned rice aftertaste. Blending of wines from different varieties for improvement in fragrance and flavor is a long-standing procedure in many wine regions (Jackson, 2008). Fessler (1971) has

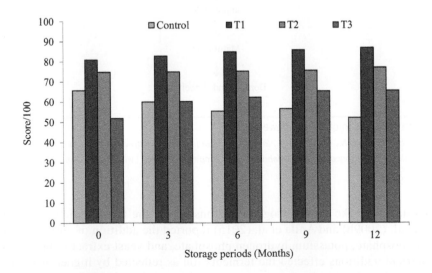

FIGURE 8. 2

Effect of fermentation temperature on sensory quality of mahua wine (T1: 16°C; T2: 20°C; T3: 25°C).

Adapted from Yadav, P., Garg, N., Dwivedi, D.H., 2009b. Effect of location of cultivar, fermentation temperature and additives on the physico-chemical and sensory qualities on Mahua (Madhuca indica J. F. Gmel.) wine preparation. Natural Product Radiance 8, 406–418.

recommended the use of herbs, roots, and spices for development of flavored wines having distinctive character. Yadav et al. (2009b) reported improvement in aroma and acceptability of herbs (lemon, cinnamon, raw mango, and mint) treated in mahua flower wines over the control. However, lemon-treated wine was found superior to other treatments. Lemon is used for flavor improvement of burned rice and vegetables. Citronellal, neral, geranial, 1,8-cineole are the major flavor-producing compounds of lemon (Sawada and Yamada, 1998).

2.11 PHENOLIC PROFILING

High performance liquid chromatography (HPLC) analysis reflected the presence of phenolics *viz.* gallic acid, chlorogenic acid, catechin, epicatechin, caffeic acid, 4-hydroxybenzaldehyde, ascorbic acid, and tannic acid in mahua wine (Fig. 8.3). Lemon treatment of wine resulted in variation in phenolics pattern and quantity (Fig. 8.4) which, perhaps, made it more acceptable. Goldberg et al. (1998) have reported that the concentrations of (+)-catechin and (−)-epicatechin were highest in red Burgundy and Canadian wines. Arts et al. (2000) determined the levels of (+)-catechin, (−)-epicatechin, (+)-gallocatechin, (−)-epigallo-catechin, (−)-epicatechin gallate, and (−)-epigallocatechin gallate in 8 types of black tea, 18 types of red and white wines, apple juice, grape juice, iced tea, beer, chocolate milk, and coffee using HPLC.

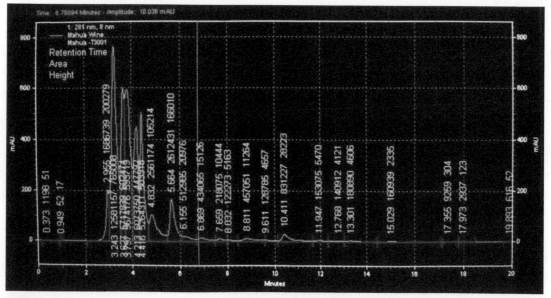

FIGURE 8.3

High performance liquid chromatography analysis of control mahua wine.

Adapted from Kumari, A., Pandey, A., Gupta, A., Raj, A., Sharma, A., Das, A.J., Kumar, A., Chauhan, A., Das, A.J., Ann, A., Neopany, B., Attri, B.L., Panmei, C., Angchok, D., Chye, F.Y., Rapsang, G.F., Vyas, G., Devi, G.A.S., Bareh, I., Kabir, J., Chakrabarty, J., Targais, K., Sim, K.Y., Angmo, K., Palni, L.M.S., Reddy, L.V.A., Swain, M.R., Monika, Devi, M.P., kumar, N., Garg, N., Ningthoujam, S.S., Sharma, N., Yadav, P., Ray, R.C., Deka, S.C., Gautam, S., Thokchom, S., Kumar, S., Khomdram, S., Joshi, S.R., Thorat, S.S., Savitri, Bhalla, T.C., Stobdan, T., Joshi, V.K., Jaiswal, V., Chauhan, V., 2015. In: Joshi, V.K. (Ed.), Indigenous Alcoholic Beverages of South Asia. CRC Press, p. 523 (Chapter 9).

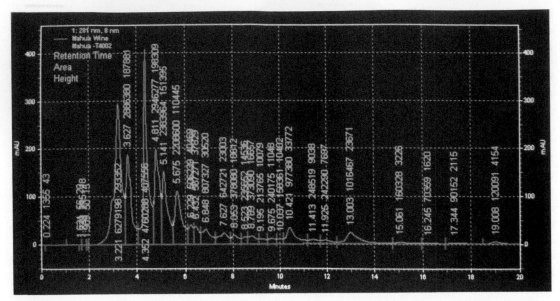

FIGURE 8.4

High performance liquid chromatography analysis in lemon-treated mahua flower wine.

Adapted from Kumari, A., Pandey, A., Gupta, A., Raj, A., Sharma, A., Das, A.J., Kumar, A., Chauhan, A., Das, A.J., Ann, A., Neopany, B., Attri, B.L., Panmei, C., Angchok, D., Chye, F.Y., Rapsang, G.F., Vyas, G., Devi, G.A.S., Bareh, I., Kabir, J., Chakrabarty, J., Targais, K., Sim, K.Y., Angmo, K., Palni, L.M.S., Reddy, L.V.A., Swain, M.R., Monika, Devi, M.P., kumar, N., Garg, N., Ningthoujam, S.S., Sharma, N., Yadav, P., Ray, R.C., Deka, S.C., Gautam, S., Thokchom, S., Kumar, S., Khomdram, S., Joshi, S.R., Thorat, S.S., Savitri, Bhalla, T.C., Stobdan, T., Joshi, V.K., Jaiswal, V., Chauhan, V., 2015. In: Joshi, V.K. (Ed.), Indigenous Alcoholic Beverages of South Asia. CRC Press, p. 523 (Chapter 9).

2.12 MAHUA VERMOUTH

Vermouth is an aromatized wine having added sugar, roots, herbs, spices, and flowers (Fessler, 1971). It contains ethyl alcohol, sugars, acids, tannins, aldehyde, esters, amino acids, vitamins, anthocyanins, fatty acids, and minor constituents like flavoring compounds (Joshi, 1997; Joshi et al., 1999). The additives don't boost the alcohol content, but they do sculpt the flavor of the wine. Fessler (1971) has also recommended the use of herbs, roots, and spices for development of flavored wines having distinctive character. Vermouth can have between 15% and 19% alcohol content by volume. Since the mahua flower has burned, starchy flavors, herbs were used to mask the same and replace with sweet spicy flavor (Yadav et al., 2012). No microbial growth could be observed in mahua vermouth after one year of aging. The sensory results clearly reflected improvement in its aroma and acceptability during storage (Fig. 8.1). The vermouth prepared by mahua had 18.4% alcohol and 1.26 mg/100 g tannin content compared to base wine where it was 10.50% and 0.32%, respectively. Nonsignificant changes were observed in biochemical parameters. It had total soluble sugar (TSS) 6.4°B, acidity 0.57%, and ascorbic acid 1.19 mg/100 mL at the time of bottling, which decreased marginally during storage while the reducing and total sugars increased from a level of 0.27% to 0.33%, respectively. There are reports of loss of ascorbic acid and antioxidants during storage of fruit wines (Patras et al., 2009).

HPLC analysis reflected phenolics compounds viz. gallic acid (122.59 mg/100 g), caffeic acid (10.76 mg/100 g), p–coumaric acid (4.18 mg/100 g) and kaempferol (282.16 mg/100 g) as well as the flavonols (+)-catechins (35.31 mg/100 g) and (−)-epi-catechin (39.51 mg/100 g) in mahua vermouth. Goldberg et al. (1998) have reported high concentrations of (+)-catechin and (−)-epicatechin in red Burgundy and Canadian wines. Arts et al. (2000) reported the presence (+)-catechin, (−)-epicatechin, (+)-gallocatechin, (−)-epigallocatechin, (−)-epicatechin gallate, and (−)-epigallocatechin gallate in 8 types of black tea, 18 types of red and white wines, apple juice, grape juice, iced tea, beer, chocolate milk, and coffee using HPLC. Sirohi et al. (2005) developed a nutritionally rich and therapeutically value-added whey-based mango herbal beverage.

3. MEAD

Mead, or honey wine, is regarded as the ancestor of all fermented drinks, being the oldest alcoholic drink known to man. The invention of making mead has been considered as antedating the cultivation of the soil and as a marker of the passage from nature to culture. The earliest evidence of alcohol in China are jars from Jiahu which date to about 7000 BC containing rice mead produced by fermenting rice, honey, and fruit. The first known description of mead is available in the *Rigveda* dating back to 1700–1100 BC (http://en.wikipedia.org/wiki/Mead). Traditional mead is made by fermenting a mixture of honey and water. However, a number of variations in the protocol are being followed throughout the world. Depending on local traditions and specific recipes, it may be brewed with spices, fruits, or grain mash. It may be still, carbonated, or sparkling; it may be dry, semisweet, or sweet. The amount of alcohol in mead varies from 10% to 18% depending upon the alcohol tolerance capacity of yeast culture being used for fermentation. The types of mead commonly prepared are given in Table 8.2.

3.1 METHOD FOR MEAD PRODUCTION

Honey is diluted with water (Approximately 1 kg with 2.5 or 3.0 L of water). To create the fruit-containing mead, 10–20% fruit juice or purees are added to the honey–water mixture (http://www.honey.com/images/downloads/makingmead.pdf; Joshi et al., 1990). Citric acid is added up to 0.4% acidity. This step balances the flavor and acid sugar blend of product. The mixture may be heated to kill the natural microflora associated with honey. The must is cooled and yeast, yeast nutrition including nitrogen [yeast assimilable nitrogen (YAN)], vitamins (thiamine), and mineral salts (Mg, Zn) are added. Managing nutrient requirements regulates fermentations and enhances sensory quality. Thiamine is a vitamin used as a coenzyme in the alcoholic fermentation pathway. It stimulates yeast growth, speeds up fermentation, and reduces production of SO_2-binding compounds. Minerals are components of the yeast cell membrane and help maintain fermentation metabolism activities. Nitrogen is metabolized by yeast to synthetize proteins. It stimulates yeast multiplication, keeps yeast metabolism active, prevents hydrogen sulphide and mercaptan formation, and stimulates aroma production. YAN is composed of ammonium ions and amino acids (except proline).

The mixture should incubate at a temperature between 16 and 25°C. Lower than this range will inhibit the fermentation, but higher will affect the quality of the product. The mixture should ferment for about 2 weeks and then siphoned into a clean glass container. Almost any spice or herb can be added to mead, either as an extract or directly, at almost any time during the mead-making process. Blends of two or more spices and herbs are commonly used. Adding hops to mead adds a distinctive flavor, clarifies the mead,

Table 8.2 Types of Mead Prepared Throughout the World

Common Name of Mead	Constituents / Procedure	Country Where Produced
Acan	Honey	Mexico
Acerglyn	Honey and maple syrup	–
Bochet	Honey is caramelized or burned separately before adding the water. Gives toffee, chocolate, marshmallow flavors	–
Braggot (also called bracket or brackett)	Originally brewed with honey and hops, later with honey and malt—with or without hops added	Wales
Black mead	Honey and black currants	–
Capsicumel	A mead flavored with chili peppers	–
Chouchenn	Honey	France
Cyser Cider	Honey and apple juice	
Czwórniak	Honey	Poland
Dandaghare	Honey with Himalayan herbs and spices	Nepal
Dwójniak	Water and honey	Poland
Great mead	Mead aged for several years	–
Short mead	Mead aged for a short period; it is effervescent, and often has a cidery taste	–
Gverc or Medovina	Honey and spices	Croatia, Germany
Hydromel	Water-honey; very light or low-alcohol mead	French name for mead
Medica	Honey	Republic of Slovenia, Croatia
Medovina	Honey	Serbia, Czech Republic, Slovakia, and presumably other Central and Eastern European countries
Medovukha	Honey	Eastern Slavic
Melomel	Honey and any fruit	–
Morat	Honey and mulberries	
Pyment	Honey and red or white grapes	
Metheglin	Honey with spices such as such as meadowsweet, hops, vanilla, orange peel, lavender, or chamomile; herbs and spices such as clove, cinnamon, nutmeg, and ginger	–
Mulsum	Unfermented honey blended with a high-alcohol wine	–
Omphacomel	A variety of pyment	–
Oxymel	Blending honey with wine vinegar	–
Pitarrilla	Fermented mixture of wild honey, balche tree bark, and fresh water	Mesoamerica and northern Central America
Póbtorak	Two units of honey for each unit of water	Poland
Rhodomel	Honey, rode hips, petals or rose attar, and water	–
Sack mead	Sweet mead due to the use of comparatively high amounts of honey than usual	–
Sima	Quickly fermented low-alcoholic mead seasoned with lemon	Finland
Tej	Honey fermented with powdered leaves and bark of *gesho*	Ethiopia
Trójniak	Two units of water for each unit of honey	Poland
White mead	Honey, herbs, fruit, or egg white	–

–, Not a location-specific.
Adapted from http://en.wikipedia.org/wiki/Mead.

Table 8.3 Physicochemical Characteristics of Mead

Compound	Minimum	Maximum
Alcohol (%)	12.2	20.8
pH	2.9	3.75
Total acidity (g L-1)	2.20 (36.636 mEq L-1)	7.08(117.902 mEq L-1)
Volatile acidity (g L-1)	0.14(2.331 mEq L-1)	0.779 (12.973 mEq L-1)
Residual sugar (%)	2.5	27.8
Acetaldehyde (mg L-1)	18.2	125.5
Ashes (%)	0.046	0.520
Calcium (%)	0.41	5.11
Magnesium (%)	0.43	2.03
Potassium (%)	8.62	74.19
Sodium (%)	1.24	14.02

Adapted from Steinkraus, K.H., Morse, R.A., 1973. Chemical analysis of honey wines. Journal of Apicultural Research Cardiff 12 (3), 191–195.

and preserves its freshness. Addition of tannins increases astringency and helps in brewing and clarification (http://www.honey.com/images/downloads/makingmead.pdf; Joshi et al., 1990).

When making still mead, the addition of potassium sorbate or wine stabilizer prevents the onset of second fermentation by killing remaining yeast cells (http://www.honey.com/images/downloads/makingmead.pdf). The mead is stored at 16°C for a further 2 months, racked again after 2 months. By this time, yeast and the other heavy impurities will settle down and the mead can be bottled or stored at a low temperature for future bottling. Table 8.3 describes the physicochemical characteristics of mead.

Factors affecting the quality of mead:

- Quality of raw materials affect the final product. Honey, the main raw material for mead production, made by *Apis mellifera*, is the result of two enzymatic reactions inside bee's honey sacks after nectar is collected from flowers. In the first reaction the invertase enzyme converts saccharose to glucose and fructose and in the second, amylase and glucose oxidase converts starch to maltose and glucose to gluconic acid and hydrogen peroxide, respectively (Morales et al., 2013). The main carbohydrates in honey are fructose (38.4%), glucose (30.3%), saccharose (1.3%), and disaccharides as maltose and isomaltose, trisaccharides, and polysaccharides up to 12% (Morales et al., 2013). Water ranges from 0.5 to 0.6 water activity (Aw) and decides the viscosity of honey. The organic acids viz. citric, malic, succinic, formic, acetic, lactic, and piroglutamic acids constitute about 0.57% of honey and make it acidic with pH (3.5–4.8). Mineral content varies between 0.04% and 0.2% with potassium dominating (Anklan, 1998). Proteins constitute about 0.2%. Honey is a natural source of antioxidants. The flavor of honey comes from its acids. The color is also directly linked to the occurrence of these acids, and therefore, the darker the honey, the bigger the presence of mineral salts as well as the more intense the flavor. Physicochemical characteristics of honey affect the quality of mead (Gupta et al., 1992). Different treatments, storage temperature, and period affect the quality of honey (Kaushik et al., 1996). The flavor and color of mead depends upon the age of the honey if the honey is old and dark the product will be dark in color with a strong flavor. If the honey is fresh and light-colored the product will possess this quality.

- The flavor of product will depend upon the floral source of honey.
- The yeast strain used for fermentation is important since different strains exhibit different resistance to acidity and alcohol concentration.
- The initial sugar concentration in the must determines the final alcoholic concentration and the type of mead (dry, medium dry, or sweet). Other ingredients such as water, acid (malic or tartaric), spices and herbs, type of fruit, and sulfites also affect the quality of mead. To avoid stuck fermentation, yeast nutrient should be added.
- Fermentation temperatures between 16 and 25°C is optimal, while temperatures as high as 32.0–38.0°C reduce the quality of the product (Morse and Steinkraus, 1975).
- Pasteurization of must at 80–85°C preserves the delicate flavor of honey; therefore boiling should be avoided to preserve flavor (Steinkraus, 1983; Steinkraus and Morse, 1966).
- To avoid heating at all, sulfiting (60–70 ppm SO_2) is the best option.
- Repeated racking improves the clarity of the mead. Maintaining sanitation is most important for making quality mead (Steinkraus and Morse, 1973; Steinkraus et al., 1971).
- The equipment, containers, and bottles must be thoroughly rinsed with chlorinated water to ensure the long shelf life of the product. Most off flavors are indicative of poor sanitation practices (Steinkraus and Morse, 1973).

4. RHODODENDRON WINE

4.1 RHODODENDRON

The *Rhododendron* is the national flower of Nepal. The hills and mountain-side ranges of Nepal are decorated with different colors and shapes. The Greek translation of *Rhododendron* means "rose tree." The hills of Nepal are colored red, white, or pink during the blooming season of *Rhododendron*. The genus and species, which is dark red in color is *Rhododendron arboreum* which is called as "*Lali Gurans*" in Nepali (Kumar and Srivastava, 2002; Srivastava, 2012). There are more than 30 species of *Rhododendron* in Nepal, with dozens of varieties in all sizes and colors.

4.2 TREE

R. arboreum is an evergreen shrub or small tree with a showy display of bright red flowers. It is called "Burans," "Bras," "Buras," or "Barahke-phool" in the local dialect. It is widely popular for the processed juice of its flowers which have gained market popularity as rhodo juice/sharbat (Srivastava, 2012). The plant is found in the Himalayas from Kashmir eastwards to Nagaland. Various parts of the plant have exhibited medicinal properties and are used for the treatment of various ailments.

R. arboreum is an important plant of the hilly region with extensive medicinal and commercial uses. The plant has a special place in the cultural and economic life of the people. It is offered in temples and religious places for ornamenting and decoration purposes. The wood is used to make tool handles, boxes, and posts and is suitable for plywood. The aesthetic beauty of the fully blossomed flowers burdened on trees in the flowering season attracts the attention of visitors. The plant is of abundant medicinal and economic value. The plant exhibited antiinflammatory, hepatoprotective, antidiarrheal, antidiabetic, antioxidant properties due to presence of flavonoids, saponins, tannins, and other phyto-chemicals (Bhandary and Kuwabata, 2008; Dhan et al., 2007).

4.3 FLOWER

The flowers of *R. arboreum* range in color from a deep scarlet, to red with white markings, pink to white. Bearing up to 20 blossoms in a single truss this rhododendron is a spectacular sight when in full bloom (Plate 8.3, https://en.wikipedia.org/wiki/Rhododendron). It is reported that the bright red forms of this *Rhododendron* are generally found at the lower elevations (Orwa et al., 2009). Flowers are showy and red in dense globose cymes (Chauhan, 1999).

In hilly areas, the flowers of *R. arboreum* with a sweet and sour taste are used in the preparation of squash, jams, jellies, and local brews (Bhandary and Kuwabata, 2008). It is a very common and pleas-ant drink, drunk once daily as a refreshing appetizer and also to prevent high-altitude sickness (Chauhan, 1999). The composition of the *Rhododendron* flower juice is given in Table 8.4.

PLATE 8.3

Rhododendron flower.

Adapted from https://en.wikipedia.org/wiki/Rhododendron.

Table 8.4 Composition of *Rhododendron* Flower Juice	
Carbohydrate	68.5%
Fat	7.6%
Protein	16.3%
Minerals (mg/100 gm)	–
Na	4.8
K	171.5
Ca	2.5

Adapted from http://www.himalayanaturals.in/rhododendron-juice.htm.

The phytochemical screening of the flowers of *R. arboreum* var. nilagiricum showed the presence of secondary metabolites including phenols, saponins, proteins, steroids, tannins, xanthoproteins, coumarins, and carbohydrates which has great medicinal properties. In addition, there are several reports to show *Rhododendron* species for having potent antimicrobial chemicals (Kiruba et al., 2011). Moreover, several species of *Rhododendron* have been widely used as a main ingredient in traditional medicine. Phytochemical analysis has revealed the presence of phenolic compounds, saponins, proteins, steroids, tannins, xanthoproteins, coumarins, and carbohydrates (Kiruba et al., 2011).

4.4 WINE

R. arboreum's nectar is brewed to make wine and is effective in helping with diarrhea and dysentery. The flowers of *R. arboreum* are used for brewing local wine to prevent high-altitude sickness in the Darjeeling hills of eastern Himalayas. Though there is no standard recipe for rhododendron wine, it may be prepared in the way standard flower wines are prepared (http://www.thewinepages.org.uk/flower.htm; http://winemaking.jackkeller.net/wineblog6.asp). The Tinjure Milke Jaljale (TMJ) area of Eastern Nepal is known as the "Capital of *Rhododendron*" where 28 species of *Rhododendron* are available at a single destination.

The preparation and preservation of *Rhododendron* juice among TMJ people is mentioned in the following:

- Plucking: At the time of flowering, when the flowers are matured fully, flowers are plucked from branches carefully. The collection is done in baskets in the forest.
- Grading: The anthers and stigma are removed and only petals are sorted out. These are then cleaned with water.
- Grinding: The common grinder/blenders used for juice extraction are used for the grinding of petals. The extract obtain is filtered to obtain the concentrate form of juice.
- Boiling: About 1300 mL water and 800 gm sugar is boiled on fire. When the solution is about to boil about 1 L of juice is measured and poured in mixture. To serve, mix the juice in water in the ratio of 1:3.
- For making wine, this juice is taken as the raw material. The TSS is adjusted to 22°Brix and inoculated with yeast and yeast nutrients, as discussed in the earlier sections in this chapter. The must is set aside until vigorous fermentation subsides (7–10 days) with daily stirring. Liquid is strained into a secondary fermentation vessel fitted with fermentation trap. Wine is racked after 30 days, then again after an additional 30 days. It is bottled when clear, and then stored in a dark, cool place. It is fit to drink after 6 months but improves enormously after a year. The wine tastes light and flowery upfront, with a hint of a kicky aftertaste (http://ramblingspoon.com/blog/?p=839).

Rhododendron juice, besides being a refreshing drink, also has great medicinal and herbal value. The blood red colored flower, quite appropriately, is beneficial in improving the blood circulation and relieving hypertension. Research also shows that the juice is useful in relieving pain as it relaxes the body (http://winemaking.jackkeller.net/wineblog6.asp; http://ramblingspoon.com/blog/?p=839). Its juice is beneficial in blood pressure, asthma, and heart diseases. All these properties are expected to come in wine. There is a need for exploring these beneficial or harmful (if any) aspects of *Rhododendron* wine.

5. SWEET POTATO WINE

5.1 SWEET POTATO

Sweet potato (*Ipomoea batatas* L.) is a major food crops in the subtropical regions of the world. Its large, starchy, and sweet tuberous roots are rich sources of starch, sugars, vitamin C, pro vitamin A, iron, and minerals. Nutritional facts of sweet potato are given in Table 8.5. Some varieties contain colored pigments such as β-carotene and anthocyanin (Yamakawa, 1998; Hou et al., 2001). These pigments have antioxidant properties that help in fighting against cancer, protect against night blindness, delay aging, and prevent liver injury. Sweet potato is bioprocessed for developing a number of functional foods and beverages such as sour starch, lacto-pickle, lacto-juice, soy sauce, acidophilus milk, sweet potato curd and yogurt, alcoholic drinks, and bioethanol (Coggins et al., 2003; Islam and Jalaluddin, 2004; Wireko-Manu et al., 2010).

5.2 WINE

High starch content coupled with delicious taste and flavor are desirable characters for the production of alcoholic beverage from sweet potato. Good quality, fresh sweet potatoes are collected and washed with water, then peeled, cut, and blended, and the juice is obtained by squeezing the blended material through a muslin cloth. The extracted juice is treated with crude amylase enzyme before the preparation of wine. The TSS of the juice is about 3°Brix. The enzyme-treated sweet potato juice is reconstituted to attain samples of the final TSS and pH. One hundred ppm of potassium metabisulfite is added prior to fermentation. The samples are then inoculated with yeast and allowed to ferment until maximum fermentation is achieved. The percent ethanol produced after the completion of the fermentation of the reconstituted sweet potato juice depends on the combinations of TSS, pH of the reconstituted sweet potato juice, inoculums size, and fermentation temperature. For the inoculum size of 10% v/v, fermentation temperature of 25°C and pH of 4.5 is favorable for the fermentation of sweet potato juice. The increase in temperature above 25°C and pH above 4.5 decreases the percent ethanol produced. The percent ethanol is maximum when the TSS of the juice is in the range of 18–22°Brix and further increase in TSS decreased the ethanol formation (Paul et al., 2014). The flow chart for sweet potato wine preparation is given in Fig. 8.5.

Table 8.5 Nutritional Value of Sweet Potato per 100 g	
Carbohydrates	20.1 g
Starch	12.7 g
Sugars	4.2 g
Dietary fiber	3.0 g
Fat	0.1 g
Protein	1.6 g
Adapted from USDA Nutrient Database.	

Sweet potato

↓

Washing and deskinning

↓

Liquifaction of pulp and sterilization

↓

Enzyme mediated sacharification

↓

Inoculation with 2% actively growing *S. cerevisiae* and nutrient (NH_4SO_4 0.1%)

↓

Fermentation at 25 ± 2 °C for 5 days

↓

Racking at 4 °C

↓

Bottling

FIGURE 8.5

Flow chart for sweet potato wine preparation.

Adapted from Paul, S.K., Dutta, H., Mahanta, C.L., Prasanna Kumar, G.V., 2014. Process standardization, characterization and storage study of a sweet potato (Ipomoea batatas L.) wine. International Food Research Journal 21 (3), 1149–1156.

Purple sweet potato is a special type of sweet potato having high anthocyan in pigment in the root (Panda et al., 2013). Ray et al. (2012) have developed a red wine by fermenting the starchy purple sweet potato roots with *Saccharomyces cerevisiae* and the wine has identical characteristics as that of commercial red (grape) wine. The physicochemical properties of enzyme-treated sweet potato wine are shown in Table 8.6. The red wine produced contains essential antioxidants and acceptable sensory qualities (Ray et al., 2012). The wine has the following proximate compositions: TSS, 2.25°Brix; starch, 0.15 g per 100 mL; total sugar, 1.35 g per 100 mL; TA, 1.34 g tartaric acid per 100 mL; phenol, 0.36 g (caffeic acid equivalent) per 100 mL; anthocyanin, 55.09 mg per 100 mL; tannin, 0.64 mg per 100 mL; lactic acid, 1.14 mg per 100 mL; ethanol, 9.33% (v/v); and pH, 3.61. 2,2-Diphenyl-1-picrylhydrazy (DPPH) scavenging activity of the wine was 58.95% at a dose of 250 µg mL-1.

6. TOMATO WINE

6.1 TOMATO

Tomato, the world's largest vegetable crop, is a rich source of vitamin A and C, minerals like iron and phosphorus, and the pigments lycopene and beta-carotene. It is commonly used in preserved products

Table 8.6 Physicochemical Properties of Enzyme-Treated Sweet Potato Wine

Parameter	Sweet Potato Wine
pH	3.73 ± 0.07
TA (oxalic acid)	$0.83 \pm 0.11\,g/100\,g$
Total sugar	$1.84 \pm 0.23\,g/100\,g$
Reducing sugar	$1.65 \pm 0.16\,g/100\,g$
TSS	$3.15 \pm 0.06°Brix$
Ethanol production	$9.6 \pm 0.28\,mL/100\,mL$
OD	0.22 ± 0.01
L	58.65 ± 0.15
a	0.64 ± 0.09
b	11.6 ± 0.16
H	86.84 ± 0.48
C	11.61 ± 0.15

Values are mean ± standard deviation (n = 3).
Adapted from Paul, S.K., Dutta, H., Mahanta, C.L., Prasanna Kumar, G.V., 2014. Process standardization, characterization and storage study of a sweet potato (Ipomoea batatas L.) wine. International Food Research Journal 21 (3), 1149–1156.

like ketchup, sauce, chutney, soup, paste, puree, etc. (Motamedzadegan and Tabarestani, 2011). There is the need to broaden the processing avenues for tomato to reduce its postharvest losses. Tomato is a potential substrate for wine preparation.

6.2 WINE

The protocol described by Owusu et al. (2012a) includes preparation of yeast inoculums, tomato juice extraction, adjusting TSS, acidity, nutrient addition, and sterilization and inoculation. This is followed by fermentation, filtration, and aging of wine. The schematic representation of protocol is given in Fig. 8.6.

Physicochemical characteristics of fresh wine for pH, titratable acidity, total soluble solids, and ethanol content were recorded as: total soluble solids 7°Brix; pH 3.3; specific gravity 1.035; titratable acidity 0.45 g/100 mL; and alcohol 7.88% has been reported in fresh tomato wine (Mahapati et al., 2010). An increase in acidity during a six-month aging period had been reported by Owusu et al. (2012b). Phytochemical analysis of six-month aged tomato wine indicated the presence of total phenols 0.388%, flavonoids 0.105%, tannins 0.346%, and lycopene 256 mcg/100 mL (Many et al., 2014). The total number of volatile compounds identified in tomato wine before and after storage are 75 (Owusu et al., 2014). These comprise 38 esters, 7 carbonyls, 1 furan, 4 sulfur compounds, 18 higher alcohols, 6 fatty acids, and 1 terpene. Owusu et al. (2015) reported 4.1 pH and 20°C temperature as optimum for higher total antioxidant activity (TAA) levels of tomato wine. TAA was significantly correlated with total phenolics total flavonoid, 1, DPPH, ascorbic acid, and anthocyanin content of wine.

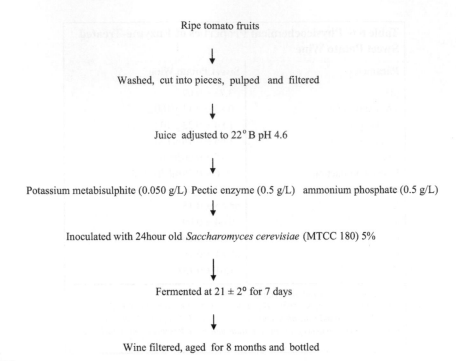

FIGURE 8.6

Flow chart for preparation of tomato wine.

Adapted from Owusu, J., Ma, H., Wang, Z., Afoakwah, N.A., Zhou, C., Amissah, A., 2015.Effect of pH and temperature on antioxidant levels of tomato wine. Journal of Food Biochemistry 39 (1), 91–100.

7. WHEY WINES
7.1 WHEY COMPOSITION

Whey is the liquid byproduct obtained after the precipitation and separation of milk casein during cheese manufacturing. It represents about 90–95% of the milk volume and is approximately 93% water. It contains about 50% of milk nutrients including lactose (4.5–5.0% w/v), soluble proteins (0.6–0.8% w/v), lipids (0.4–0.5% w/v), and mineral salts (8.0–10.0 w/v of dried extract; Kosikowski and Wzorek, 1977; Gandhi, 1996; http://ecoursesonline.iasri.res.in/mod/page/view.php?id=5516). Since lactose is the main constituent (70%) of whey dry matter, whey is a very good material for production of alcoholic beverages.

Alcoholic whey beverages are divided to beverages with low-alcohol content (≤1.5%), whey beer, and whey wine.

7.2 OTHER FERMENTED BEVERAGES

Production of whey beverages with low-alcohol content includes deproteinizing whey, whey concentration, fermentation of lactose (usually by yeast strains *Kluyveromyces fragilis* and *Saccharomyces lactis*) or addition of sucrose until reaching the desired alcohol content (0.5–1%), flavoring, sweetening, and bottling. Thereby, a certain amount of lactose is being transformed to lactic acid which gives a refreshing sour taste to the end product, while the rest ferments to alcohol. Some of the noted beverages belonging to this category are "Milone," obtained by fermentation with kefir culture, and whey sparkling wine "Serwovit," produced in Poland (Jeličić et al., 2008).

7.3 WHEY BEER

Whey beer can be produced with or without the addition of malt; it can be fortified with minerals or can contain starch hydrolyzates and vitamins. Some of the problems that can occur here are the presence of milk fat, which can cause loss of beer foam, undesirable odor, and taste due to the low solubility of whey proteins and inability of beer yeasts to ferment lactose.

7.4 WHEY WINE

Whey is used for production of various types of whey-based wines that contains relatively low-alcohol amount (10–11%) and mostly flavored with fruit aromas. Production of whey wine includes clearing, deproteinization, lactose hydrolysis by β-galactosidase, decanting and cooling, addition of yeasts and fermentation, decanting, aging, filtering, and bottling (Popović-Vranješ et al., 1997).

These have been made from demineralized ultrafiltration permeated or reconstituted acid whey powder. In reconstituted acid whey powder, protein is removed by ultrafiltration and the permeate had a total solids concentration of 26–28% before demineralizing to a mineral level of 1% or less. Any residual whey taints after fermentation are removed with bentonite and charcoal (Popović-Vranješ et al., 1997). In one process for the preparation of sweet whey wine, whey is deproteinized, heated at 82°C for 5 min and approximately 22% dextrose added, depending upon the amount of alcohol desired in the wine (Popović-Vranješ et al., 1997). Fermentation is completed in 7 days at room temperature using the yeast *S. cerevisiae* spp. *ellipsoids*. Wines are clarified by addition of 0.2–0.5% bentonite on a dry weight basis followed by a polishing filtration (https://ir.library.oregonstate.edu/xmlui/handle/1957/26979). As the whey itself contains sufficient nutrients for yeast growth, no additional nutrients are added. Similarly different types of whey-based wines like pop wine, whey champagne, sweet wine etc. can be prepared by using different strains of yeast and processing conditions.

8. COCOA WINE

8.1 COCOA

Cocoa (*Theobroma cacao* L.) is a major cash crop of the tropical world. It is known worldwide for its beans used in the manufacture of chocolate. Use of cacao dates back 2000 years to Central and South America, but is native to the Amazon and is still used today as part of their daily staple diet. Cocoa is the seed of the cacao tree. The seed contains a lot of fat, and is used to produce cocoa butter, which is then used to produce chocolate. Most people will know cocoa as a ground powder they use to produce chocolate. Cocoa beans are the main source of commercial cocoa. The four intermediate cocoa products are cocoa liquor, cocoa butter, cocoa powder and chocolate.

8.2 COMPOSITION OF COCOA PULP

The cacao pulp, white in color, has a much sweeter taste than the bitter seeds within and has similar nutritional benefits similar to cacao bean (Freire et al., 1999; Schwan and Wheals, 2004). The cocoa pulp contains 82–87% water; 10–15% sugar (60% is sucrose and 39% a mixture of glucose and fructose); 2–3% pentoses; 1–3% citric acid; and 1–1.5% pectin. Proteins, amino acids, vitamins (mainly vitamin C), and minerals are also present, being a rich medium for microbial growth (Schwan and Wheals, 2004; Dias et al., 2003; Schwan, 1988).

8.3 WINE

For juice extraction from cocoa pulp, selection of cocoa pods is important for the production of quality juice. The most suitable pods will be those three-quarter ripe and yellowish-green in color. Ripe pods are generally dry and brown in color. This will affect the quality of juice produced. One ton of beans yields 154 L juice (http://www.koko.gov.my/lkm/getfile.asp?id=1204).

Cocoa beans extracted from pods and water is added in the ratio of 1:20 beans/water w/v. Cocoa pulp is extracted from fresh cocoa beans. To obtain a high juice yield (8–10% of the harvest-fresh cocoa beans) and to ensure fast, hygienic processing, it is advisable to open the cocoa fruits in the immediate vicinity of the production facility, i.e., where the cocoa is fermented. The harvest practice common throughout the country, namely breaking open and emptying the cocoa fruits on the plantation, inevitably leads to considerable losses of pulp juice during transport to the production facility, as well as a drop in quality due to contamination of the pulp juice by undesirable microorganisms. The flow chart for cocoa wine preparation is given in Fig. 8.7.

Pulp, diluted with water, ameliorated with sugar to 22°B, treated with pectinolytic enzymes and with SO_2 (100 mg/L) is fermented (Dias et al., 2007). As soon as the cocoa beans have been transferred to the fermentation boxes the white fruit puree adhering to them liquefies, due to the action of pectolytic enzymes contained in the fruit, turning into a milky pulp juice with a very fruity flavor. For this reason, it is advisable to fit a collecting trough to the fermentation boxes so that the juice can flow off hygienically and without loss into a tank.

Cocoa pulp can be readily fermented at 22°C by yeasts such as *S. cerevisiae*, producing an alcoholic beverage (Duarte et al., 2010). At the end of the fermentation, the vats are transferred to a 10°C incubator to aid the sedimentation of solid material from cocoa pulp. After 10 days at this temperature, the wine transfer is carried out with some aeration and the beverage is incubated at 10°C for another 30 days. After this period, another transfer without aeration is carried out and the fruit wine is left for another 10 days at 10°C before filtration. The cocoa wine is then filtered through diatomaceous earth and cellulose filters. The beverage is storage at 8°C in glass bottles fully filled to avoid oxygen entrance.

9. REGULATIONS FOR MAKING AGRICULTURAL WINES

https://www.law.cornell.edu/uscode/text/26/5387.

1. Wines made from agricultural products other than the juice of fruit shall be made in accordance with good commercial practice.
2. No wine spirits may be added to wines produced under this section, nor shall any coloring material or herbs or other flavoring material (except hops in the case of honey wine) be used in their production.

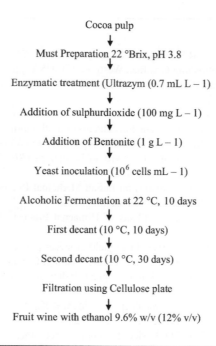

Cocoa pulp

↓

Must Preparation 22 °Brix, pH 3.8

↓

Enzymatic treatment (Ultrazym (0.7 mL L – 1)

↓

Addition of sulphurdioxide (100 mg L – 1)

↓

Addition of Bentonite (1 g L – 1)

↓

Yeast inoculation (10^6 cells mL – 1)

↓

Alcoholic Fermentation at 22 °C, 10 days

↓

First decant (10 °C, 10 days)

↓

Second decant (10 °C, 30 days)

↓

Filtration using Cellulose plate

↓

Fruit wine with ethanol 9.6% w/v (12% v/v)

FIGURE 8.7

Flow diagram of the process of alcoholic beverage production from cocoa pulp.

Adapted from Dias, D.R., Schwan, R.F., Freire, E.S., Serôdio, R.D.S., 2007. Elaboration of a fruit wine from cocoa (Theobroma cacao L.) *pulp. International Journal of Food Science and Technology 42 (3), 319–329.*
http://dx.doi.org/10.1111/j.1365-2621.2006.01226.x.

3. Wines from different agricultural commodities shall not be blended together.
4. In the production of mead:
   a. Water may be added to facilitate fermentation, provided the density of the honey and water mixture is not reduced below 13°Brix.
   b. Hops in quantities not to exceed one pound for each 1000 pounds of honey.
   c. Pure, dry sugar, or honey for sweetening. Sugar may be added only after fermentation is completed.
   d. After complete fermentation or complete fermentation and sweetening, the wine may not have an alcohol content of more than 14% by volume or a total solids content that exceeds 35°Brix (https://www.law.cornell.edu/cfr/text/27/24.203).
5. In the production of wine from agricultural products other than dried fruit and honey, water and sugar may be added to the extent necessary to facilitate fermentation, provided that the total weight of pure dry sugar used for fermentation is less than the weight of the primary winemaking material and the density of the mixture prior to fermentation is not less than 22°Brix, if water, or liquid sugar, or invert sugar syrup is used. Additional pure dry sugar may be used for sweetening, provided the alcohol content of the finished wine after complete fermentation or after complete fermentation and sweetening, is not more than 14% by volume and the total solids content is not more than 35°Brix (https://www.law.cornell.edu/cfr/text/27/24.204).

REFERENCES

Amerine, M.A., Berg, H.W., Kunkee, R.E., Ough, C.S., Singleton, V.L., Webb, A.C., 1980. The Technology of Wine Making, fourth ed. Avi Publishing Co., Inc., Westport, CT, USA, pp. 359–380.

Anklan, E., 1998. A review of the analytical methods to determine the geographical and botanical origin of honey. Food Chemistry 63 (4), 549–562.

Anon, 1976. Wealth India 6, 207–216.

Arts, I.C.W., Putte, B., Hollman, P.C.H., 2000. Catechin contents of foods commonly consumed in the Netherlands. 2. Tea, wine, fruit juices, and chocolate milk. Journal of Agricultural and Food Chemistry 48 (5), 1752–1757.

Bhandary, M.R., Kuwabata, J., 2008. Antidiabetic activity of Laligurans (*Rhododendron arboreum* Sm.) flower. Journal of Food Science and Technology, Nepal 4, 61–63.

Chateerjee, A., Parakashi, S.C., 2003. The Treaties on Indian Medicinal Plants, 4. National Institute of Science Communication and Information Resources, New Delhi, p. 56.

Chauhan, N.S., 1999. Medicinal and Aromatic Plants of Himachal Pradesh. Indus Publishing Company, New Delhi, p. 353.

Coggins, P.C., Kelly, R.A., Wilbourn, J.A., 2003. Juice yield of sweet potato culls. Session 104C, fruit and vegetable products: vegetables (processed). 2003 IFT Annual Meeting. Chicago, USA.

Datta, S., 2011. Sustainable livelihoods through *mahua* plant. Indian Journal of Fundamental and Applied Life Sciences 1 (4).

Dhan, P., Garima, U., Singh, B.N., Ruchi, D., Sandeep, K., Singh, K.K., 2007. Free radical scavenging activities of Himalayan *Rhododendron*s. Current Science 92, 526–532.

Dias, D.R., Schwan, R.F., Lima, L.C.O., 2003. Metodologia para elaboração de fermentado de cajá (*Spondias mombin* L.). Ciência e Tecnologia de Alimentos 23, 342–350.

Dias, D.R., Schwan, R.F., Freire, E.S., Serôdio, R.D.S., 2007. Elaboration of a fruit wine from cocoa (*Theobroma cacao* L.) pulp. International Journal of Food Science and Technology. 42 (3), 319–329. http://dx.doi.org/10.1111/j.1365-2621.2006.01226.x.

Duarte, W.F., Dias, D.R., Oliveira, J.M., Teixeira, J.A., de Almeida e Silva, J.B., Schwan, R.F., 2010. Characterization of different fruit wines made from cacao, cupuassu, gabiroba, jaboticaba and umbu. LWT—Food Science and Technology. 43 (10), 1564–1572. http://dx.doi.org/10.1016/j.lwt.2010.03.010.

Fessler, J.H., 1971. Guideline to Practical Winemaking. P.O. Box 5276, Elmwood Station, Berkelery, CA. , p. 115.

Freire, E.S., Schwan, R.F., Mororo, R.C., 1999. The cocoa pulp agroindustry and the use of its residues in Bahia: progress achieved in the last ten years. In: Proceedings of the 12th International Cocoa Research Conference. Cocoa Producers Alliance, Kuala Lumpur, pp. 1013–1020.

Gandhi, D.N., 1996. Fermented Whey Beverages. NDRI Bulletin, No. 278. , pp. 1–7.

Goldberg, D.M., Karumanchiri, A., Sang, T., Soleas, G.J., 1998. Catechin and epicatechin concentrations of red wines: regional and cultivar-related differences. American Journal of Enology and Viticulture 49 (1), 23–34.

Gupta, J.K., Kaushik, R., Joshi, V.K., 1992. Influence of different treatments, storage temperature and period on some physico-chemical characteristics and sensory qualities of Indian honey. Journal of Food Science and Technology, Mysore 29 (2), 84–87.

Hou, W.C., Chen, Y.C., Chen, H.J., 2001. Antioxidant activities of trypsin inhibitor, a 33 KDa root storage protein of sweet potato (*Ipomoea batatas* (L.) Lam cv. Tainong 57). Journal of Agricultural and Food Chemistry 49 (6), 2978–2981.

Islam, M.S., Jalaluddin, M., 2004. Sweet potato – a potential nutritionally rich multifunctional food crop for Arkansas. Journal of Arkansas Agriculture and Rural Development 4, 3–7.

Jackson, R.S., 2008. Wine Science: Principles and Applications. Published by Academic Press, p. 776.

Jeličić, I., Božanić, R., Tratnik, L., 2008. Whey-based beverages-a new generation of dairy products. Mljekarstvo 58 (3), 257–274.

Joshi, V.K., Attri, B.L., Gupta, J.K., Chopra, S.K., 1990. Comparative fermentation behaviour, physico-chemical characteristics and qualities of various fruit honey wines. Indian Journal of Horticulture 47 (1), 49–54.

Joshi, V.K., Sandhu, D.K., Thakur, N.S., 1999. Fruits based alcoholic beverages. In: Joshi, V.K., Pandey, A. (Eds.), Biotechnology: Food Fermentation, 1. Educational Publisher and Distributors, New Delhi, pp. 647–747.

Joshi, V.K., 1997. Fruit Wines, second ed. Directorate of Extension Education. Dr. Y.S. Parmar University of Horticulture and Forestry, Nauni, Solan, p. 255.

Kaushik, R., Gupta, J.K., Joshi, V.K., 1996. Effect of different treatments, storage temperature and period of storage on reducing sugars, amino acids and diastase activity of Indian honey from *Apis mellifera*. Apiacta 31 (4), 103–106.

Keller, J., 2007. Flower Wines. http://winemaking.jackkeller.net/flowers. asp.

Kiruba, S., Mahesh, M., Nisha, S.R., Miller Paul, Z., Jeeva, S., 2011. Phytochemical analysis of the flower extracts of *Rhododendron arboreum* Sm. ssp. *nilagiricum* (Zenker) Tagg. Asian Pacific Journal of Tropical Biomedicine S284–S286.

Kosikowski, F.V., Wzorek, W., 1977. Whey wine from concentrates of reconstituted acid whey powder. Journal of Dairy Science 60, 1982–1986.

Kumar, S., Srivastava, N., 2002. Herbal research in Garhwal Himalaya: retrospect and prospectus. Annals of Botany 10 (1), 99–118.

Kumari, A., Pandey, A., Gupta, A., Raj, A., Sharma, A., Das, A.J., Kumar, A., Chauhan, A., Das, A.J., Ann, A., Neopany, B., Attri, B.L., Panmei, C., Angchok, D., Chye, F.Y., Rapsang, G.F., Vyas, G., Devi, G.A.S., Bareh, I., Kabir, J., Chakrabarty, J., Targais, K., Sim, K.Y., Angmo, K., Palni, L.M.S., Reddy, L.V.A., Swain, M.R., Monika, Devi, M.P., kumar, N., Garg, N., Ningthoujam, S.S., Sharma, N., Yadav, P., Ray, R.C., Deka, S.C., Gautam, S., Thokchom, S., Kumar, S., Khomdram, S., Joshi, S.R., Thorat, S.S., Savitri, Bhalla, T.C., Stobdan, T., Joshi, V.K., Jaiswal, V., Chauhan, V., 2015. In: Joshi, V.K. (Ed.), Indigenous Alcoholic Beverages of South Asia. CRC Press, p. 523 (Chapter 9).

Mande, B.A., Andreasen, A.A., Sreenivasaya, M., Kolachov, P., 1949. Fermentation of bassia flowers. Industrial and Engineering Chemistry 41, 1451–1453.

Many, J.N., Radhika, B., Ganesan, T., 2014. Nutrient analysis of tomato wine. International Journal of Innovative Science, Engineering and Technology 1 (8), 278–283.

Mathapati, P.R., Ghasghase, N.V., Kulkarni, M.K., 2010. Study of *Saccharomyces cerevisiae* 3282 for the production of tomato wine. International Journal of Chemical Sciences and Applications 1 (1), 5–15.

Morales, E.M., Alcarde, V.E., Angelis, D.F., 2013. Mead features fermented by *Saccharomyces cerevisiae* (lalvin k1-1116). African Journal of Biotechnology 12 (2), 199–204.

Morse, R.A., Steinkraus, K.H., Paterson, P.D., 1975. Wines from the fermentation of honey. In: Crane, E. (Ed.), Honey. Crane, Russak & Company, Inc., New York.

Motamedzadegan, A., Tabarestani, H.S., 2011. Tomato processing, quality and nutrition. In: Sinha, N.K. (Ed.), Handbook of Vegetables and Vegetables Processing. Blackwell Publishing Ltd, Iowa, USA, pp. 739–757.

Orwa, C., Mutua, A., Kindt, R., Jamnadass, R., Simons, A., 2009. Agroforestree Database: A Tree Reference and Selection Guide Version 4.0. Available at: http://www.worldagroforestry.org/af/treedb/.

Owusu, J., Ma, H., Wang, Z., Ronghai, H., 2012a. The influence of pH on quality of tomato (*Lycopersicon esculentum* Mill) wine. International Journal of Advanced Biotechnology and Research 3 (3), 625–634.

Owusu, J., Ma, H., Abano, E.E., Engmann, F.N., 2012b. Influence of two inocula levels of *Saccharomyces bayanus*, BV 818 on fermentation and physico-chemical properties of fermented tomato (*Lycopersicon esculentum* Mill.) juice. African Journal of Biotechnology 11 (33), 8241–8249.

Owusu, J., Ma, H., Wang, Z., Afoakwah, N.A., Zhou, C., Amissah, A., 2014. Volatile profiles of tomato wine before and after ageing. Maejo International Journal of Science and Technology 8 (2), 129–142.

Owusu, J., Ma, H., Wang, Z., Afoakwah, N.A., Zhou, C., Amissah, A., 2015. Effect of pH and temperature on antioxidant levels of tomato wine. Journal of Food Biochemistry 39 (1), 91–100.

Panda, S.K., Swain, M.R., Singh, S., Ray, R.C., 2013. Proximate compositions of a herbal purple sweet potato (*Ipomoea batatas* L.) wine. Journal of Food Processing and Preservation 37, 596–604.

Patras, K., Brunton, N.P., Pieve, S.D., Butler, F., 2009. Impact of high pressure processing on total antioxidant activity, phenolic, ascorbic acid, anthocyanin content and colour of strawberry and blackberry purees. Innovative Food Science and Emerging Technologies 10, 308–313.

Paul, S.K., Dutta, H., Mahanta, C.L., Prasanna Kumar, G.V., 2014. Process standardization, characterization and storage study of a sweet potato (*Ipomoea batatas* L.) wine. International Food Research Journal 21 (3), 1149–1156.

Popović-Vranješ, I., Vujičić, A., Vujičić, I., 1997. Tehnologija Surutke. Poljoprivredni fakultet Novi Sad, Novi Sad.

Ray, R.C., Panda, S.K., Swain, M.R., Sivakumar, P.S., 2012. Proximate composition and sensory evaluation of anthocyanin-rich purple sweet potato (*Ipomoea batatas* L.) wine. Journal of Food Science and Technology 47, 452–458.

Sawada, M., Yamada, T., 1998. The analysis of oxygenated compounds fraction of lemon oils prepared by different extraction methods and application to a beverage. Nippon Shokuhin Kagaku Kogaku Kaishi 45 (2), 134–143.

Schwan, R.F., Wheals, A.E., 2004. The microbiology of cocoa fermentation and its role in chocolate quality. Critical Reviews in Food Science and Nutrition 44, 205–222.

Schwan, R.F., 1988. Cocoa fermentations conducted with a defined microbial cocktail inoculum. Applied and Environmental Microbiology 64, 1477–1483.

Sirohi, D., Patel, S., Choudhary, P.L., Sahu, C., 2005. Studies on preparation and storage of whey-based mango herbal pudina (*Mentha arvensis*) beverage. Journal of Food Science and Technology 42 (2), 157–161.

Srivastava, P., 2012. *Rhododendron arboreum*: an overview. Journal of Applied Pharmaceutical Science 02 (01), 158–162.

Steinkraus, K.H., Morse, R.A., 1966. Factors influencing the fermentation of honey and mead production. Journal of Apicultural Research Cardiff 5 (1), 17–26.

Steinkraus, K.H., Morse, R.A., 1973. Chemical analysis of honey wines. Journal of Apicultural Research Cardiff 12 (3), 191–195.

Steinkraus, K.H., Morse, R.A., Minh, H.V., Mendoza, B.V., Laigo, F.M., 1971. Chemical analysis of honey wines: a brief summary 1900–1971. Bee World, Cardiff 52 (3), 122–127.

Steinkraus, K.H., 1983. Handbook of Indigenous Fermented Foods. Marcel Dekker Inc., USA.

Wireko-Manu, F.D., Ellis, W.O., Oduro, I., 2010. Production of a non-alcoholic beverage from sweet potato (*Ipomoea batatas* L.). African Journal of Food Science 4 (4), 180–183.

Yadav, P., Garg, N., Dwivedi, D.H., 2009b. Effect of location of cultivar, fermentation temperature and additives on the physico-chemical and sensory qualities on Mahua (*Madhuca indica* J. F. Gmel.) wine preparation. Natural Product Radiance 8, 406–418.

Yadav, P., Garg, N., Diwedi, D.H., 2009a. Standardization of pre-treatment conditions for *mahua* wine preparation. Journal of Ecofriendly Agriculture 4 (1), 88–92.

Yadav, P., Garg, N., Dwivedi, D.H., 2012. Preparation and evaluation of mahua (*Bassia latifolia*) vermouth. International Journal of Food and Fermentation Technology 2 (1), 57–61.

Yamakawa, O., 1998. Development of new cultivation and utilization system for sweet potato towards the 21st century. In: Proceedings of International Workshop on Sweet Potato Production System Towards the 21st Century, pp. 1–8.

TECHNOLOGY FOR PRODUCTION OF FORTIFIED AND SPARKLING FRUIT WINES

P.S. Panesar[1], V.K. Joshi[2], V. Bali[1], R. Panesar[1]

[1]Sant Longowal Institute of Engineering and Technology, Longowal, Punjab, India; [2]Dr. Y.S. Parmar University of Horticulture and Forestry, Nauni, Solan, HP, India

1. INTRODUCTION

Wine is a product obtained from partial or complete alcoholic fermentation of grapes or nongrape fruits (Amerine et al., 1980; Joshi, 1997). It is the oldest known fermented product, according to the *Rigveda*, and is one of the most commercially flourishing biotechnological processes throughout the world. The wine can be classified into various types depending upon the characteristics under consideration. On the basis of the amount of alcohol it could be a table or fortified wine; based on the amount of carbon dioxide (CO_2), it could be still or sparkling wine (Joshi et al., 1999a,b). Two types of wines, fortified wine and sparkling wine (with a second fermentation process), are quite distinct and are very popular. Among the fortified wines, vermouth is a wine fortified with alcohol and blended with spices and herbs, which can be sweet or dry (Joshi et al., 1999a). The maceration of herbs and spices constituting different plant parts (seeds, woods, leaves, barks, and roots in dried form termed as botanicals such as cardamom, cinnamon, cloves, ginger, myrrh, sandalwood, etc.) in wine at various stages of fermentation act as natural flavoring and are known for improving the health-related benefits of the wine (Liddle and Boero, 2003). Vermouth is a blend of wine, aromatic plants, sugar, and alcohol. Sparkling wine has excess CO_2 produced either from bottle fermentation or tank fermentation. This wine is very popular all over the world. Champagne is a typical sparkling wine made from grapes (Amerine et al., 1980). Both of these wines are prepared from grapes, so reference to these wines would be made, wherever needed, throughout the chapter, but the focus will be on nongrape wines.

2. VERMOUTH

Vermouth is officially classified as an "aromatized fortified wine," a tongue-twisting term meaning a base white wine fortified and infused with a proprietary recipe of different plants, barks, seeds, and fruit peels, collectively known as botanicals. These types of wines are quite popular in European countries and in the United States in spite of their commercial production in Russia and Poland (Griebel et al., 1955; Amerine et al., 1980). The word vermouth is derived from the German word "wermut" (wer means man; mut means courage, spirit or manhood, or in English, "worm wood," which in Latin stands for *Artemisia absinthium*; Mattick and Robinson, 1960).

Science and Technology of Fruit Wine Production. http://dx.doi.org/10.1016/B978-0-12-800850-8.00009-0

Vermouth is fortified wine with alcohol concentration varying from 15 to 21% with either aromatic or bitter flavor. Vermouth possesses unique flavors and aromas with medicinal qualities. Based on the presence or absence of aromatic plants and differences in sugar content, vermouth is differentiated into two main families, i.e., sweet and dry vermouth. Its origin dates back to ancient Mediterranean history in which the forbearer of vermouth has been ascribed to Hippocrates for obtaining a satisfying and digestive beverage by macerating flowers of plants (aromatic plants and crete) in the Middle Ages, thus called "Hippocratic wine." Romans, following the Greeks, macerated herbs such as celery, myrtle, rosemary, and thyme. Further, in Turin, Florence, and Venice, Italy, herbs and spices were used, such as cardamom, cinnamon, clove, and ginger (from countries of East Africa, India, and Indonesia), during production of the wine (Panesar et al., 2011). Further, in France it was termed "vermouth," and today it is well-liked in the United States and Europe.

2.1 TYPES OF VERMOUTH

2.1.1 Sweet Vermouth

Sweet vermouth have higher sugar levels (around 150 g/L; 12–15% reducing sugar) and their alcohol content varies from 15 to 17% (v/v) with total acidity about 0.45% and tannic acid about 0.04% (Joslyn and Amerine, 1964; Panesar et al., 2011). These fragrant, light Muscat, slightly sweet and nutty flavored drinks, with a mildly sharp aftertaste, have their origin from Italy and are known as Italian vermouth and mostly used in cocktails such as the Manhattan. Caramel/caramelized sugar is mainly added to give it a dark or dusky garnet color. Sweet vermouth has been mainly produced in Italy, Spain, Argentina, and the United States. American vermouths, differentiating from Italian vermouth, have generally higher alcohol but lower sugar content.

2.1.2 Dry Vermouth

Dry vermouth is a flavored white wine with almost 40 or more aromatic herbs, having lower sugar levels (less than 50 g/L; 4–8% reducing sugars) and higher alcohol content (18%, v/v) with total acidity 0.65% and volatile acidity (as acetic acid) 0.053% (Joslyn and Amerine, 1964). This type of vermouth contains comparatively less amounts of botanicals than sweet vermouth but a larger amount of wormwood, bitter orange peel, and aloe (bitter herb) as additional ingredients (Amerine et al., 1967). They are lighter in color but more bitter in flavor than Italian vermouths. They can either be consumed alone or as a base for cocktails (Liddle and Boero, 2003). They were originated from the Marseilles region of France and these are known as French vermouths and mostly used in martinis (The New Encyclopedia Britannica, 1995). Overall, duration of four to five years is needed for the qualitative production of dry vermouth. The differences in composition of sweet and dry vermouth samples from France, Italy, and the United States have been summarized in Table 9.1.

2.2 TECHNOLOGY OF VERMOUTH PRODUCTION

The process of vermouth production includes the preparation of a base wine followed by the distillation of part of it, then flavoring the wine with botanicals extract, fortifying the blend, and finally, maturating the wine. The basic steps involved in the production of vermouth have been depicted in Fig. 9.1.

Table 9.1 Composition of Dry and Sweet Vermouths

Source	Number of Samples	Alcohol (%)			Extract (gm/100 ml)			Total Acid (gm/100 ml)			Tannin (gm/100 ml)		
		Min.	Max.	Avg.	Min.	Max.	Avg.	Min.	Max.	Avg.	Min.	Max.	Avg.
Dry													
France	6	17.4	19.3	18.3	3.7	6.1	4.8	0.55	0.66	0.61	0.05	0.08	0.07
United States	77	15.0	22.0	17.7	1.4	7.9	3.8	0.31	0.66	0.50	0.03	0.07	0.04
Sweet													
Italy	20	15.5	17.1	16.1	14.9	20.7	18.6	0.36	0.52	0.28	0.05	0.11	–
Italy	10	13.7	16.9	15.7	14.0	17.2	15.6	0.36	0.52	0.45	0.05	0.11	0.08
United States	100	14.0	21.0	17.1	10.0	19.0	13.8	0.26	0.63	0.45	0.03	0.10	0.06

Reproduced from Rizzo, F., 1957. La fabricazione del Vermouth. Edizione Agricole, Bologna; Perez, L., Vakarcel, M.J., Ganzalez, P., Domecq, B., 1991. Influence of Botrytis infection of grapes on the biological aging process of fino sherry. American Journal of Enology and Viticulture 42, 58–62.

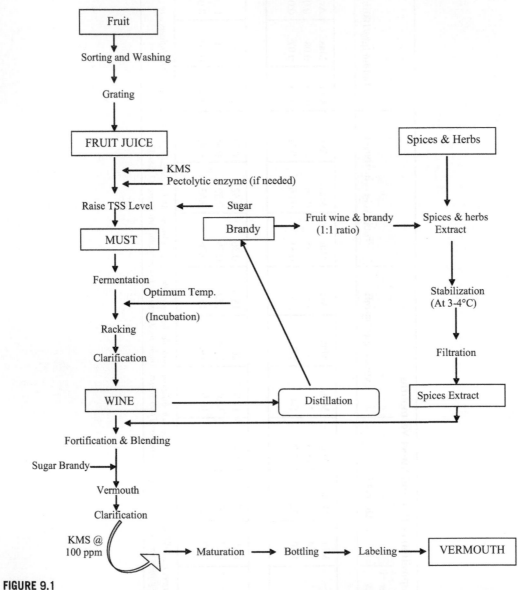

FIGURE 9.1

Basic processes involved in the preparation of vermouth.

Based on Panesar, P.S., Kumar, N., Marwaha, S.S., Joshi, V.K., 2009. Vermouth production technology—an overview. Natural Product Radiance 8 (4), 334–344.

2.2.1 Preparation of Base Wine

Generally, grapes (juice or its concentrate) are used in the preparation of table wine (Amerine et al., 1980; Jackson, 2008). Other fruits, such as apple, mango, pomegranate, and so forth, have also been explored successfully for the preparation of the base wine (Onkarayya, 1985; Joshi and Sandhu, 2000). The most essential feature of base wine for conversion into vermouth is that it should be natural, i.e., no specific flavor notes of variety of grapes should come in the wine. It should be cheap, neutral in flavor and microbiologically sound (Amerine et al., 1980; Joshi et al., 2011a). An appropriate sugar level is required for the preparation of the base wine by fermentation process. Sucrose or grape concentrate is used to maintain the sugar level. Caramel is also an important ingredient, which contributes to the color of the final product; although studies have suggested the use of fruit concentrate, especially grape concentrate (these become darkened on heating), for vermouth coloring (Goswell and Kunkee, 1977). The vermouth prepared from fruits such as pomegranate and plum develops an intensified color naturally. In Italy, producers use refined beet sugar (for sweetening), whereas in France, producers use fortified grape must (*miste-las*). Citric acid is mainly used to regulate the total acidity. The alcoholic content of the base wine is kept higher to compensate the dilution with botanical extracts. Moreover, the base wine has to be analyzed for copper and iron content (Hardisson et al., 1992).

To prepare a base wine, the fruit juice or concentrate is inoculated with active dry yeast culture (minimize undesirable odors) for fermentation of fruit sugars into alcohol. *Saccharomyces bayanus* is generally used for the base wine preparation (Jackson, 2000), although other strains such as *Saccharomyces cerevisiae* have also been preferred for fermentation purposes (Venkataramu et al., 1979). For more details, readers are suggested to see Amerine et al. (1980). The basic step in preparation of the base wine, however, remains the same as any table wine.

Depending upon the vermouth, i.e., dry or sweet, further blending is carried out as will be explained in the subsequent sections of this chapter.

The lower the level of remaining sugar after fermentation, the more dry the wine will be. Overall, the base wine should meet the essential requirements of being sound, neutral flavored, and low in cost (Amerine et al., 1980; Joshi et al., 1999a, 2011a).

2.2.2 Brandy Distillation

The separation of alcohol from wine is the next step after fermentation is complete. Part of the base wine undergoes the distillation process in which three fractions known as "head," "heart," and "tail" are separated. The head and tail fractions of the distillate are discarded, whereas heart fraction is retained and undergo double distillation for increasing the ethyl alcohol content of the brandy (Amerine et al., 1980; Joshi, 1997; Jackson, 2008). The detailed studies on brandy distillation including preparation of the base wine followed by distillation and oak wood maturation have been described in the literature (Amerine et al., 1980; Jaarsveld et al., 2009).

2.2.3 Preparation of Botanical Extract

The extracts of herb and spices, used for flavoring the vermouth, are prepared in brandy or alcohol (Rizzo, 1957) or mixture of wine and brandy (Amerine et al., 1980). The extract of different plant parts (Table 9.2) possessing some medicinal, antimicrobial, or antioxidant properties, are used for flavoring purposes which contributes to provide health benefits to the consumer. Freshly dried botanicals are generally used as their long storage results in low quality due to loss of volatile compounds (Amerine

Table 9.2 List of Herbs and Their Plant Part Used in the Production of Vermouth

Common/Commercial Name	Scientific Name	Portion of Plant Commonly Used
Allspice	*Pimenta dioica* or *P. officinalis*	Berry
Aloe (socotrine)	*Aloe perryi*	Plant
Angelica	*Angelica archangelica*	Root (occasionally seed)
Angostura	*Cuspar febrifuga* or *galipea*	Bark
Anise	*Pimpinella anisum*	Seed
Benzoin, gum benzoin tree	*Styrax benzoin*	Gum
Bitter almond	*Prunus amygdalus*	Seed
Bitter orange	*Citrus aurantium* var. *amara*	Peel of fruit
Blessed thistle	*Cnicus benedictus*	Aerial portion + seeds
Calamus, sweet flag	*Acorus calamus*	Root
Calumba	*Jateorhiza columbo*	Root
Cascarilla	*Croton eleuteria*	Bark
Cinchona	*Cinchana calisaya*	Bark
Cinnamon	*Cinnamomum zeylanicum*	Bark
Clammy sage, common clary	*Salvia sclarea*	Flowers and leaves
Clove	*Syzygium aromaticum*	Flower
Coca	*Erythroxylon coca*	Leaves
Common horehound	*Marrubium vulgare*	Aerial portion
Common hyssop	*Hyssopus officinalis*	Flowering plant
Coriander	*Coriandrum sativum*	Seed
Dittany of crete	*Amaracus dictamnus*	Aerial portion + flowers
Elder	*Sambucus nigra*	Flower (also leaves)
Elecampane, common inula	*Inula helenium*	Root
European centaury	*Erythraea centaurium*	Plant
European meadowsweet	*Filipendula ulmaria*	Root
Fennel	*Foeniculum vulgare*	Seed
Fenugreek	*Trigonella foenum-graecum*	Seed
Fraxinella, gasplant	*Dictamnus albus*	Root
Galangal, galingale	*Alpinia officinarum*	Root
Gentian	*Gentiana lutea*	Root
Germander	*Teucrium chamaedrys*	Plant
Ginger	*Zingiber officinale*	Root
Hart's tongue	*Phyllitis scolopendrium*	Plant
Hop	*Humulus lupulus*	Aerial Portion + flower
Lemon balm, common balm	*Melissa officinalis*	Flowering plant
Lesser cardamom	*Elettaria cardamomum*	Dried fruit
Lungwort, sage of Bethlehem	*Pulmonaria officinalis* or *Pulmonaria saccharata*	Aerial Portion + flower
Lungwort lichen, lung moss	*Styeta polmonacea*	Plant (a lichen)

Table 9.2 List of Herbs and Their Plant Part Used in the Production of Vermouth—cont'd

Common/Commercial Name	Scientific Name	Portion of Plant Commonly Used
Marjoram	*Origanum vulgare*	Aerial Portion + flower
Masterwort, hog's fennel	*Peucedanum ostruthium*	Root
Nutmeg and mace	*Myristica fragrans*	Seed
Orris, Florentine iris	*Iris germanica* var. *florentina*	Root
Pomegranate	*Punica granatum*	Bark of root
Quassia	*Quassia amara*	Wood
Quinine fungus	*Fomes officinalis*	Plant
Rhubarb	*Rheum rhapanticum*	Root
Roman chamomile	*Anthemis nobilis*	Flowers
Roman wormwood	*Artemisia pontica*	Plant
Rosemary, old man	*Rosmarinus officinalis*	Flowering plant
Saffron, crocus	*Crocus sativus*	Portion of flower
Sage	*Salvia officinalis*	Aerial portion + flowers
Savory (summer)	*Satureja hortensis*	Aerial portion of plant
Speedwell	*Veronica officinalis*	Plant
Star anise	*Illicium verum*	Seed
Sweet marjoram	*Marjorana hortensis*	Aerial portion + flower
Thyme, garden thyme	*Thymus vulgaris*	Leaf
Valerian	*Valeriana officinalis*	Root
Vanilla	*Vanilla fragrans*	Bean
Wormwood	*Artemisia absinthium*	Plant
Yarrow	*Achillea millefolium*	Plant
Zedoary, setwell, curcum	*Curcuma zedoaria*	Root

Source: Joslyn, M.A. and Amerine, M.A. (1964). Dessert, Appetizer and Related Flavored wines. University of California. Division of Agricultural Sciences, Berkeley.

et al., 1980). Climatic conditions and harvesting conditions are also known to affect the quality of the botanicals.

Different methods such as direct extraction, concentrate preparation, or maceration with other flavored alcoholic beverages are employed for extraction purposes, depending upon the type of herb or spices used (Pilone, 1954). In direct extraction method, finely ground herbs and spices are directly added to the base wine. This may result in the release of undesirable flavoring agents; therefore, a partial extraction method in which botanicals are placed in cloth bags during extraction is preferred. For concentrate preparation, the base wine is circulated through a vessel containing a concentrated extract of botanicals until the desired flavor and aroma are achieved. In the maceration technique, commercially available brandy or alcohol extract are used to flavor the base wine (Rizzo, 1957). The extraction is generally carried out in one or two in-sealed vessels to minimize the loss of volatiles (Plione, 1954; Jeffs, 1970). The compositions of the extract and extraction method employed, however, differ among various vermouth producers.

2.2.4 Fortification and Blending

The base wine, extract of herbs and spices (flavoring), brandy (higher alcohol content) and sugar (caramel) are blended according to a proprietary formula for each type of vermouth. Alcoholic content (ethanol) and dry extract concentration affects the viscosity of the wine (Nurgel and Pickering, 2005). The young vermouth thus obtained is further matured by aging it for a specific time duration in barrels.

2.2.5 Aging and Finishing

Fortification of vermouth is followed by refrigeration (for stabilization), filtration (clarification), and its maturation. The young vermouth is generally aged for 3–5 years (Valaer, 1950), though a lower maturation period is preferred. It has been reported that further aging can lower the quality of the vermouth (Amerine et al., 1967). Mostly oak barrels but other wooden barrels are used for maturation of the vermouth. Wooden barrels absorb some phenolics and tannins while providing vanillin and other organic molecules during aging. The low pH and addition of sulfur dioxide (SO_2; 50–75 ppm) prevent the microbial spoilage of vermouth during maturation (Plione, 1954).

2.2.6 Bottling

Vermouth undergoes ultrafiltration before bottling and sealing to clarify the wine. SO_2 (above 75 ppm) level is maintained to prevent the spoilage of the vermouth from *Lactobacillus trichodes* (Amerine et al., 1967).

2.2.7 Storage

The storage of vermouth after opening the bottle is not easy; the oxidation of vermouth starts early and it tends to lose refined qualities overtime. Refrigeration makes the oxidation process slower but the flavor starts changing gradually. The shelf life of vermouth after breaking the seal of the bottle is less than a month under refrigerated conditions.

2.3 VERMOUTH PRODUCTION FROM NONGRAPE FRUITS

Vermouth has also been produced from nongrape fruits such as apple, mango, plum, and sand pear with acceptable physicochemical and sensory qualities (Wright, 1960; Onkarayya, 1985; Joshi et al., 1991a, 1999a). There has also been an increasing demand for the utilization of low cost and readily available indigenous fruits for winemaking in the countries where grapes are not abundantly available (Reddy and Reddy, 2005).

2.3.1 Apple Vermouth

Apple is produced and savored throughout the world. To overcome the postharvest losses of the fruit, especially in the developing countries, attempts have been made for its utilization in food processing industry. Preparation of apple vermouth is one such attempt in this direction. A few modifications in the grape vermouth production method are required for the production of apple vermouth due to differences in the type of fruit (Jarczyk and Wrorek, 1997). Joshi and Sandhu (2000) prepared apple-based vermouth by varying alcohol (12–18%), sugar (4–8%) and spices extract (2.5–5%) concentrations and evaluated their effect on physicochemical and sensory attributes (Table 9.3). It has been observed that all the three parameters influenced the sensory quality of the vermouth. Apple vermouth containing 15% alcohol, 4% sugar level, 2.5% spices extract with 0.65% acidity was reported to be a promotable

Table 9.3 Physicochemical Characteristics of Apple Vermouth of Different Alcohol Levels

Characteristics	Alcohol Level (%)		
	12	15	18
Total sugar (%)	9.2	7.8	7.3
Total soluble solids (°B)	16.0	16.2	16.3
Titratable acidity (% MA)	0.43	0.39	0.37
pH	3.36	3.29	3.26
Ethanol (% v/v)	11.9	15.2	19.2
Color (Units)			
Red	3.75	2.92	2.40
Yellow	20.75	20.00	10.60
Total esters (mg/L)	175.7	181.0	246.7
Volatile acidity (%)	0.046	0.040	0.040
Total tannins (mg/L)	633	524	521

Adapted from Joshi, V.K., Sandhu, D.K., 2000. Influence of ethanol concentration, addition of spices extract, and level of sweetness on physico-chemical characteristics and sensory quality of apple vermouth. Brazilian Archives of Biology and Technology 43 (5), 537–545.

commercial product. An increase in the total aldehyde and total ester content of vermouth was observed by adding spices and herb extracts. Further, the flavor profiling of apple vermouth has also been made by using descriptive analysis techniques employing multivariate analysis such as principal component analysis (Joshi and Sandhu, 2009).

2.3.2 Mango Vermouth

The aromatized wine prepared by fermentation of mango juice is known as mango vermouth. Mango (*Mangifera indica* L), the king of the fruits, has quite high carbohydrate content (16–18%, w/v) with affordable market prices, which make it suitable for conversion into commercially profitable fermentation product such as wine (Kumar et al., 2009). It has high amino acids, contains vitamin A, vitamin B_6, and less sodium, cholesterol, and saturated fat (Spreer et al., 2009). Mango wine has been found to contain bioactive molecules with antioxidant activity (Varakumar et al., 2011). Attempts have been made to prepare mango wine from different mango varieties (Kumar et al., 2012). Different botanicals (Table 9.4A) in varying amounts have been reportedly used in the preparation of mango vermouth using Montrachet strain 522 of *S. cerevisiae* for fermentation of cv. Banganpalli (Onkarayya, 1985; Ossa et al., 1987). The mango vermouth obtained was found to be comparable to grape vermouth in terms of pH, total acidity, alcohol, aldehydes, and total phenolic content. According to the sensory evaluation tests (Table 9.4B), mango-based vermouth product had a high acceptability level and good quality, similar to grape-based vermouth (Onkarayya, 1985).

2.3.3 Plum Vermouth

Prunus salicina Linn. (plum) has a short-shelf life (3–4 days) under room temperature as well as cold storage (1–2 weeks) which indeed makes its transportation to far-off places difficult. Therefore, jams, jellies, wine, and other beverages have been prepared for commercial purposes (Gill et al., 2009).

Table 9.4A Spices Used for the Preparation of Sweet and Dry Type Mango Vermouth

| | Amount Used (g/L) of Base Wine | | | |
| | Dry Wine | | Sweet Wine | |
Herbs	**A**	**B**	**C**	**D**
Black pepper	0.75	1.25	2.5	5.0
Coriander	0.70	1.25	2.5	5.0
Cumin	1.25	2.50	3.0	4.0
Bishop's weed	0.50	1.00	1.50	2.0
Clove	0.25	0.50	0.75	1.0
Large cardamom	0.50	1.00	1.50	1.0
Saffron	0.10	0.10	0.10	0.10
Fenugreek	0.50	1.50	2.0	2.50
Nutmeg	0.25	0.50	0.50	0.75
Cinnamon	0.50	1.00	1.50	2.00
Poppy seeds	1.00	1.50	2.0	2.50
Ginger	1.00	1.50	2.0	2.50
Flame of forest	0.25	0.50	0.75	0.75
Lichen	0.25	0.50	1.00	2.00

Reproduced from Onkarayya, H., 1985. Mango vermouth – a new alcoholic beverages. Indian Food Packer 39 (1), 40–45.

Plum offers attractive colors and higher fermentability, and is therefore suitable for wine preparation. The method for the preparation of vermouth is same basically to that of grape vermouth. At industrial scale, the plum brandy is prepared by fractional distillation. Different botanicals have been reported in varying quantities to prepare aromatized plum wine (Table 9.5). The addition of spices and herb extract and an increase in alcohol concentration in the plum vermouth increased the total phenols, aldehyde, and ester content. In addition, an increase in alcohol concentration led to a decrease in acidity and vitamin C. Plum vermouth was evaluated based on its commercial acceptability (Joshi et al., 1991a). The comparison of composition between sweet and dry vermouth has been made in Table 9.6.

2.3.4 Pomegranate Vermouth

Pomegranate (*Punica granatum* L) is a rich source of polyphenols (especially ellagic acid and punicalagins) which act as antioxidants. It contains high sugar levels, and thus, pomegranate juice can act as a good fermentation medium for making wine with good color (Sevda and Rodrigues, 2011). Three types of sweet vermouth with pomegranate (Ganesha cultivar) base were reported to be prepared with different spices (clove, cardamom, and ginger) with three months of maturation. Pomegranate vermouth with cardamom had maximum acceptability in terms of total acidity, sweetness, body, and general quality followed by ginger vermouth and clove vermouth (Patil et al., 2004).

Table 9.4B Physicochemical Characteristics and Sensory Quality of Mango Vermouths

Herbs Mixture Formula and Type of Vermouth	Color (at 420 nm)	pH	Total Acidity Tartaric Acid/100 ml	Volatile Acidity (g. AA/100 ml)	Alcohol (%, v/v)	Total Aldehyde (ppm)	Total Phenols (%)	Organoleptic Scores (Out of 20)
Dry Vermouth								
Formula A	0.420	3.40	0.59	0.088	17.0	15.8	0.055	13.00
Formula B	0.658	3.50	0.60	0.087	17.5	20.9	0.064	11.50
Sweet Vermouth								
Formula C	0.678	3.42	0.59	0.071	17.2	26.4	0.070	15.50
Formula D	0.690	3.50	0.61	0.091	18.0	56.3	0.075	13.60

Reproduced from Onkarayya, H., 1985. Mango vermouth – a new alcoholic beverages. Indian Food Packer 39 (1), 40–45.

Table 9.5 Spices and Herbs Used in the Preparation of Plum Vermouth

Common Name	Botanical Name	Parts Used	Qty/L (g)
Black pepper	*Piper nigrum* L.	Fruit	0.75
Coriander	*Coriander sativum* L.	Seeds	0.70
Cumin	*Cuminum cyaninum* L.	Seeds	0.50
Clove	*Syzygium aromaticum* L.	Fruit	0.25
Large cardamom	*Amomum subulatum roxb.*	Seeds	0.50
Saffron	*Crocus sativus* L.	Flower	0.01
Nutmeg	*Myristica fragrans*	Seed	0.25
Cinnamon	*Cinnamomum zeylanicum Beryn*	Bark	0.25
Poppy seed	*Papaver somniferum* L.	Seed	1.00
Ginger	*Zingiber officinale Rosc*	Dried root	1.00
Woodfordia	*Woodfordia floribunda*	Flower	0.25
Asparagus	*Asparagus* sp.	Leaves	0.10
Withania	*Withania somnifera*	Roots	0.20
Adhatoda	*Adhatoda* sp.	Leaves	0.25
Rosemary	*Rosmarinus officinalis*	Flowering plant	0.10

Reproduced from Joshi, V.K., Attri, B.L., Mahajan, B.V.C., 1991a. Production and evaluation of vermouth from plum fruits. Journal of Food Science and Technology 28, 138–141.

Table 9.6 Physicochemical Characteristics of Dry and Sweet Plum Vermouth

Physicochemical Characteristics	Type of Vermouth	
	Dry	Sweet
Total sugar (%)	ND	4.8
Titratable acidity (% malic acid)	0.81	0.79
Ethanol (% v/v)	15.0	14.5
Volatile acidity (% acetic acid)	0.03	0.04
pH	3.38	3.34
Vitamin C (mg/100 m)	3.5	3.2
Total phenols (mg/L)	417	390
Aldehydes (mg/L)	411	112
Esters (mg/L)	204	219

ND, Not detected.
Based on Joshi, V.K., Attri, B.L., Mahajan, B.V.C., 1991a. Production and evaluation of vermouth from plum fruits. Journal of Food Science and Technology 28, 138–141.

2.3.5 Sand Pear Vermouth

Sand pear (*Pyrus pyrifolia*) vermouth has been successfully prepared by using sand pear for the base wine (Joshi et al., 1999a; Attri et al., 1993). Sweet vermouth with 15% alcohol was considered as the best beverage (Attri et al., 1994). The herbs and spices used for sand pear vermouth were the same as that of plum and apple vermouth (Joshi et al., 1991a; Joshi and Sandhu, 2000). It has been reported that the tannins, aldehyde, and esters are contributed by the botanicals in case of grape, plum, and sand pear vermouths (Amerine et al., 1980; Joshi et al., 1991a; Attri et al., 1994). Therefore, conversion of sand pear wine to vermouth increased aldehydes, phenols, and esters in sand pear vermouth also. Product with 15% alcoholic content was reported to be the best in case of sand pear wine. The physicochemical characteristics of sand pear base wine and vermouth have been given in Table 9.7.

2.3.6 Tamarind Vermouth

Tamarind (*Tamarindus indica*) possesses medicinal properties and has higher levels of tartaric acid, sugar, vitamin B, and calcium. Although higher acidity level is a major hindrance, wine and vermouth from tamarind have been reported in various studies with acceptable quality (Lingappa, 1993; Mbaeyi-Nwaoha and Ajumobi, 2013). Tamarind fruit (50 g/L) with 0.9% acidity and total soluble solids (TSS) 23 °Brix was used for base wine preparation using *S. cerevisiae* var. *ellipsoideus* at 27 ± 1°C. Sweet and dry tamarind vermouths with 17% alcohol have been reported to be of acceptable quality.

2.3.7 Apricot Vermouth

Wild apricot (*Prunus armeniaca* L), a highly acidic, fibrous, and low in TSS content fruit, has been found in the hilly areas of northern India especially in high hilly areas of Himachal Pradesh (India). It is used in a fresh, dry, canned form or preserved as jam, marmalade, or pulp but not on a commercial level. Attempts had been made to produce apricot-based wine and vermouth from cultivated and nondomesticated apricot varieties (Joshi et al., 1990; Joshi and Sharma, 1993; Genovese et al., 2004; Abrol, 2009; Joshi et al., 2011a,b). The apricot distillate aroma has been reported to be constituted by high alcohol concentration and specific terpenes such as linalool, ocimenol, α-terpineol, nerol, and geraniol

Table 9.7 Physicochemical Characteristics of Sand Pear Base Wine and Sweet Vermouth

Characteristics	Wine	Vermouth
Total soluble solids (°B)	6.1	13.0
Titratable acidity (% MA)	0.37	0.43
pH	3.99	3.95
Reducing sugar (%)	–	4.17
Total sugar (%)	–	4.35
Alcohol (% v/v)	10.80	14.95
Volatile acidity (% AA)	0.04	0.04
Ascorbic acid (mg/100 m)	6.6	5.5
Aldehydes (mg/L)	103.21	133.15
Total phenols (mg/L)	226.26	264.46
Esters (mg/L)	197.4	268.04
Optical density	0.64	0.58

Reproduced from Attri, B.L., Lal, B.B., Joshi, V.K., 1993. Preparation and evaluation of sand pear vermouth. Journal of Food Science and Technology 30, 435–437.

Table 9.8 Effect of Alcohol Levels on Chemical Composition of Wild Apricot Vermouth

Characteristics	Base Wine (Mean ± SD)	Alcohol Level (%)		
		15	17	19
TSS (°Brix)	8.20±0.07	17.09	17.23	17.45
Reducing sugars (%)	0.34±0.01	5.55	5.46	5.38
Total sugars (%)	1.11±0.02	10.41	10.16	9.96
Titratable acidity (% MA)	0.76±0.02	0.85	0.78	0.77
Ethanol (% v/v)	10.64±0.09	15.09	17.10	19.04
Total esters (mg/L)	135.40±0.55	260.8	262.7	267.4
Total phenols (mg/L)	253.60±0.8	454.4	451.6	446.6

Adapted from Abrol, G.S., 2009. Preparation and Evaluation of Wild Apricot Mead and Vermouth. (M.Sc. thesis). Dr. Y.S. Parmar University of Horticulture and Forestry, Nauni, Solan, India.

(Genovese et al., 2004). Vermouth with varying sugar (8–12 °Brix), alcohol (15–19%), and spices (2.5–5%) was prepared and evaluated (Abrol, 2009). The effect of varying concentrations of alcohol on apricot vermouth has been given in Table 9.8.

2.4 COMMERCIAL PRODUCTION OF VERMOUTH

The modern vermouth commercial industry (known as Martini and Rossi) was launched in Turin (world's vermouth capital) by King Carlo Alberto of Sardinia–Piedmont in 1840 whereas the first commercial product of vermouth dates back to 1757 (Kauffman, 2001). These wines are mostly popular in United States and European countries, with Russia and Poland being their commercial producers (Amerine et al., 1980; Griebel, 1955). Martini and Rossi (part of Bacardi empire) of Italy and Noilly Prat of France are universally renowned leaders in their respective countries among European producers of uniquely different vermouth followed by Cinzano and Stock of Italy and Dolin *Vermouth de Chambery* of France (Boyd, 2007; Panesar et al., 2009). Martini and Rossi merged traditional methods with modern ones to produce a range of aromatized wine such as Rosso (United States), Extra Dry, and Bianco. Similarly, vermouths are of different types based on color (Table 9.9).

The quality and nature of base wine along with the type, quality, and amount of botanicals used play a significant role on the quality and type of vermouth. Some of the common vermouth with their producer brands have been listed in Table 9.10.

3. SPARKLING WINE
3.1 INTRODUCTION

A wine is called sparkling if it is surcharged with CO_2 (not less than 5 g/L at 20°C) and form bubbles and foam when poured into the flute (Lee and Baldwin, 1988; Buxaderas and López-Tamames, 2010). Carbonic gas is required to have an endogenous origin, obtained via a second fermentation, in the following European categories: sparkling wines and quality sparkling wines (Buxaderas and

Table 9.9 Vermouth Nomenclature on the Basis of Color

	English	Italian	French	Spanish
Red/brown	Red/Sweet/Italian	Rosso	Rouge	Rojo
Clear/colorless	White	Bianco	Blanc	Blanco
Straw	Dry/French	Dry	Dry	Dry
Pink	Rose	Rosato	Rose	Rosado

Table 9.10 Few Examples of Commercially Available Vermouths

Vermouth Region	Type	Example
Chambery	Blanc	Dolin blanc
	Dry	Dolin dry
	Rouge	Dolin rouge
Marseilles	Dry	Noilly prat dry
	Rouge	Noilly prat rouge
Modern	Orange	Sutton cellars
	Rose	Cinzano vermouth rosato
	Ambre	Noilly prat ambre
Torino	Rosso	Martelletti rosso
	Alla vaniglia	Carpano antica formula
	Can bitter	Punt e mes
	Bianco	Martini and Rossi Bianco
	Dry	Cinzano extra dry

López-Tamames, 2012). Elaboration of sparkling wine consists of two phases. In the first phase, the base wine is obtained after applying normal white wine vinification. The second phase consists of refermenting the wine, either in the bottle (champenoise or traditional method) or in isobaric tanks (Charmat method). The second fermentation requires the addition of *liqueur de tirage* to the base wine. The sparkling wines have a special biological aging or aging *sur lies*, i.e., aging on the lies.

Sparkling wine such as champagne from grape is produced by secondary fermentation in closed containers such as bottle or tanks to retain the CO_2 produced (Goldman, 1963). There are four types of sparkling wines that could be made (Amerine et al., 1980; Joshi et al., 2011c; Jeandet et al., 2011):

1. Those with excess carbon produced by fermentation of residual sugar from the primary fermentation (Australian, German, Loire, and Italian, and the Muscato, Ambila of California);
2. Those with excess CO_2 from a malolactic fermentation (Vinho verela wines of northern Portugal);
3. Those that contain excess CO_2 from fermentation of sugar added after the process of fermentation. Most of the sparkling wines of the world are of this type;
4. Those where excess CO_2 is added, including carbonated wines.

Germany consumes the highest sparkling wine per head, followed by France, Italy, and Australia. The major importing country of sparkling wine is Germany (237,000 hl) followed by the United

Kingdom (6.14% of the total volume); other countries include France, Netherlands, Sweden, Norway, the United States, and Japan (Montemiglio, 1992). The export market estimate for Italian sparkling wines has been 663×10^6 hl in 1990.

Out of the nongrape wines, most of the research has been carried out on production of sparkling wines from plum and apple. Since this volume is focused on the production of nongrape fruit wines, this section of the chapter is focused on sparkling plum and apple wine and with description of basic process steps common with grapes.

3.2 TECHNOLOGY OF PRODUCTION

The general steps involved in the production of sparkling wine from the grapes/fruit using the bottle fermentation method, include preparation of a base wine, sugaring, yeasting, bottling, proper blending, secondary fermentation in bottles or tank maturation, finishing, and disgorging (Amerine et al., 1980; Jeandet et al., 2011). The broad step-by-step processes are shown in Fig. 9.2. The maturation is a very important step in the production of sparkling wine and may extend up to three years. Bottle aging results in acquiring the physicochemical and sensory characteristics usually associated with such wines. As sparkling wines remain in contact with the lees, they develop sensory notes such as toasty, lactic, sweet, and yeasty, which can be attributed to proteolytic processes, components that would serve as the substrate for chemical and enzymatic reactions and to causes related with release–absorption between cell walls and the wine (Buxaderas and López-Tamames, 2012). In the production of sparkling wine, the preparation of the base wine is the first and foremost step as the quality of sparkling wine depends largely on the characteristics of the base wine, which in turn are dictated by a number of factors such as cultivar, yeast culture, preservatives, temperatures, nitrogen source, etc., as discussed earlier (Amerine et al., 1980; Joshi et al., 1990).

It has also been reported that glutathione (GSH) in must and wine exerts a protective effect on many desirable wine aromas, limits the formation of off-flavors (Roussis and Sergianitis, 2008; Kritzinger et al., 2013; Webber et al., 2014) and the formation of browning pigments to a certain extent (Hosry et al., 2009; Kritzinger et al., 2013; Sonni et al., 2011). Recently, the effect of the addition of GSH on secondary aromas and on the phenolic compounds of sparkling wine elaborated by traditional method has been reported (Webber et al., 2014). Sparkling wines with addition of GSH to the must showed lower levels of total phenolic compounds and hydroxycinnamic acids. However, the GSH addition to the base wine seems to maintain higher levels of SO_2 in the free form. The variety of the base wine used and the aging time give sparkling wines their characteristic color and aroma (Buxaderas and López-Tamames, 2010). Besides this, yeast autolysis also plays an important role in the development of the yeasty, sweet, and toasty aroma notes.

Several methods are available to transform the base wine into a sparkling wine, including the *methode champenoise*, the *Cremants* from France and Luxembourg, the *methode traditionnelle* (formerly the *methode champenoise*) e.g., used for Cavas, the transfer method, the *methode ancestrale* (Limoux, Gaillac) also including the *Dioise* method and the bulk method (*cuvee close*).

3.2.1 Methode Champenoise

Elaboration of sparkling wines from France and Luxembourg uses the *champenoise methode* for both the base wine and the *prise de mousse* (Lee and Lester, 1985). But differences occur with champagne winemaking, especially when considering the separation of juices after pressing the whole grapes. The second phase includes secondary fermentation and aging in the bottle.

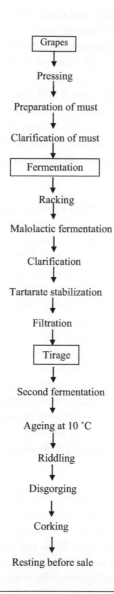

FIGURE 9.2

Steps in the sparkling wine production.

Based on Moulin, J.P., 1986. Champagne: the method of production and the origin of the quality of this French wine. In: Proceedings of 6th Australian Wine Industry Technical Conference, Adelaide, 14–17 July. Australian International Publishers, Adelaide, pp. 218–224.

3.2.2 The Transfer Method

In the transfer method, the wine (obtained by traditional white winemaking) is fermented and aged on lees in the bottle (as for the *methode traditionnelle*) but there are no constraints of riddling (a crucial, expensive, and lengthy operation) and disgorging. After bottle fermentation and proper aging, the bottles are automatically emptied into a steel tank without degassing since the wine is maintained at an isobarometric pressure. At this stage, the dosage can be directly added in the tank, but winemakers generally prefer adding the dosage in another steel tank after having filtered the wine. After standing for several days, the wine is filtered and bottled or just bottled. All operations are carried out under a CO_2 atmosphere using isobarometric bottling. Advantages of the transfer method are overcoming the need of the lengthy operation of *remuage* and making the dosage more uniform. However, this method is expensive, energy-consuming, and a risk of oxidation of the wine does really exist.

3.2.3 Methode Ancestrale

The *methode ancestrale* is an elaborating method though very difficult to control. In this method, the base wine is made from whole grapes (mainly of the Mauzac variety) or through a traditional white winemaking using a semifermented wine. In fact, sugars are used for both the primary alcoholic fermentation and the secondary fermentation. At different steps of the elaboration process, it is essential to stop fermentation each time it starts to accelerate. In this way, refrigeration (until 0°C), sulfiting, depletion of yeast nutrients (using settling, fining, filtration, or centrifugation) are repeated as many times as necessary to regulate or stop the activity of yeast. The wine is then filtered and kept at 0°C until spring time. At this step, the secondary fermentation takes place (2–3 months) in the bottle (at rigorously controlled temperatures) with yeast and remaining sugars of the semifermented wine. Riddling and disgorging (without dosage) then occur, but the wine can also be sold with a slight yeast deposit at the bottom of the bottle. These wines receive the *Blanquette Methode Ancestrale Appellation of Controlled Origin*.

To increase the extraction of aroma compounds, pectinolytic enzymes are added to Muscat grape berries in the crusher. After the second fermentation is stopped by refrigerating the cellar, bottles are emptied into a steel tank maintained at an isobarometric pressure by CO_2 to avoid degassing (as for the transfer method). After filtration, the wine is bottled using isobarometric bottling. The final alcoholic content of the *Clairette de Die* is of approximately 7.5° with 40–50 g/L residual sugars.

3.2.4 Bulk Method

The bulk method (*cuvee close*) is a simpler and more cost-effective technique that has been developed to obtain ordinary wines of low prices. It is also called the tank method or Charmat process. The second fermentation does not take place in the bottle but the base wine is sent to a reinforced steel fermentation tank able to contain several hundred hectoliters of wine. Yeast and sugars are added and the wine is maintained at a temperature of 20–25°C. The *prise de mousse* duration does not exceed 10 days. The secondary fermentation is stopped by a light sulfiting and by refrigerating the wine at −2°C. After having been cold-stabilized at −5°C for several days, the wine is filtered at a low temperature and then bottled using isobarometric bottling. One disadvantage of this method is that there is no aging of the wine as the wine contact with lees is insufficient that affects the quality of wine. The steps involved are depicted in Fig. 9.3.

3.3 SPARKLING PLUM WINE

Production of plum wine involves various steps similar to sparkling wine from grapes. Different steps involved are shown in Fig. 9.4 and discussed here.

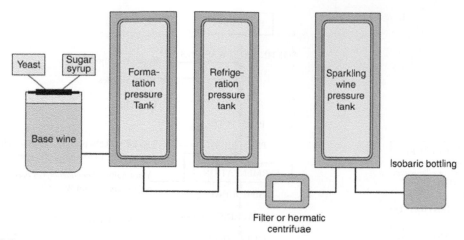

FIGURE 9.3

A schematic presentation of sparkling wine production by Charmat method.

Reproduced from Joshi, V.K., Attri, D., Singh, T.K., Abrol, G.S., 2011c. Fruit wines: production technology. In: Joshi, V.K. (Ed.),
Handbook of Enology: Principles, Practices and Recent Innovations, vol. 3. Asia Tech Publisher, New Delhi, pp. 1177–1221.

3.3.1 Production of the Base Wine

The production of the base wine is the first important step. The quality of the sparkling wine depends largely on the characteristics of the base wine.

3.3.1.1 Method of Base Wine Production

Plum fruit makes wine with appealing colors and acceptable quality. The pulp is fermented to produce the wine. Compared to grapes, the sugar content present originally in plums is not attractive but color and flavor make it acceptable (Vyas and Joshi, 1982).

To prepare plum wine, for every pound of plum, 1 L of water is added followed by the addition of starter culture, allowing to ferment for 8–10 days before pressing as it is practically impossible to press the fruit before fermentation (Amerine et al., 1980). The addition of pectolytic enzyme before fermentation greatly facilitates pressing, increases the yield of juice and hastens the wine clarification (Amerine et al., 1980; Joshi et al., 1990). Aging, filtration, bottling, and further pressing is similar to the grape wine.

Another method of plum winemaking includes the preparation of must by addition of water to the pulp in 1:1 ratio and with initial TSS of 24°B (Vyas and Joshi, 1982). A base wine to support secondary fermentation should have a pH value of less than 3.3, an alcohol content of 8–12% (v/v) and total SO_2 content ranging from 50 to 100 mg/l (Markides, 1986). For the preparation of base wine the plum fruit should be harvested at ripeness. Moulin (1986) reported that to produce a sparkling wine of good quality, the base wine must have a high degree of acidity, a low pH, and low potassium content. From these considerations the plum wine is considered suitable for making sparkling wine.

Cultivar: Among the factors influencing the quality of a wine, choice of proper cultivar is important. Out of the three cultivars of plums tested, the Santa Rosa plum was the best for winemaking (Joshi and Bhutani, 1990b).

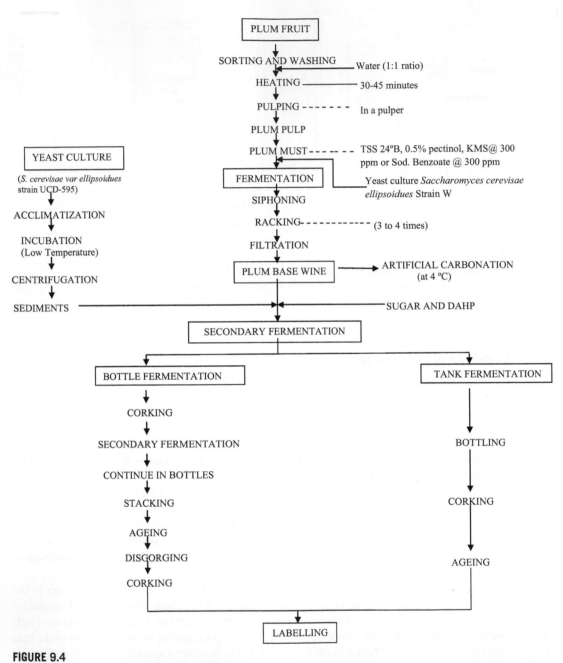

FIGURE 9.4

Flow sheet for the preparation of sparkling plum.

Reproduced from Joshi, V.K., Sharma, S.K., Thakur, N.S., 1995. Technology and quality of sparkling wine with special reference to plum – an overview. Indian Food Packer 49, 49–66.

Yeast culture: For proper fermentation, the addition of yeast and the regulation of temperature are prerequisites (Moulin, 1986). Herraiz et al. (1990) found that the rate of alcoholic fermentation with *S. cerevisiae* varied markedly with the yeast strains used. A good yeast strain should have good fermentability, good sedimentation, low H_2S production, high ethanol tolerance, high sugar consumption, high ethanol productivity, and fermentability to dryness.

Preservatives: Among the preservatives, SO_2 provides a clean fermentation and 100 ppm of SO_2 prevents oxidation and controls undesirable organisms, whereas in the absence of SO_2, the growth of yeasts other than *S. cerevisiae* is possible resulting in variation in the composition of the fermented medium (Herraiz et al., 1990). But the addition of SO_2 to the juice should be low (4–6 g/hl) for sparkling wine production (Randall, 1986; Moulin, 1986). However, the addition of sodium benzoate could be made in place of potassium metabisulfite (KMS) for plum wine preparation (Joshi and Bhutani, 1990a), though the rate of fermentation was found to be higher in KMS-treated wine than that of sodium benzoate in Fig. 9.5.

The type of preservative also affected the physicochemical characteristics, namely, TSS, pH, total sugars, and red color units. Sodium benzoate–treated wine had higher amounts of total phenols, total anthocyanins, Na and K, and lower concentration of aldehyde, Fe, and Zn than the KMS-treated wine (Joshi and Sharma, 1995). Higher loss of color in wine treated with additives such as L-ascorbic acid or sorbic acid than untreated wine took place (Datzberger et al., 1992) but the appropriate use of wine additives did not appear to deteriorate the ruby color of young red wine.

Temperature: Temperature has been found to affect the rate of fermentation and the best temperature for fermentation of wine was 15°C (Atkinson et al., 1959; Moulin, 1986). According to Randall (1986), the juice should be inoculated, after clarification, with the desired yeast strain at 18–20°C and the fermentation should be held for longer, close to 20°C, for optimum anthocyanase action. Casp and Romero (1990) observed that the fermentation at temperatures of less than 20°C produced more esters than at room temperature.

Nitrogen source: The addition of nitrogenous yeast food (urea or ammonium phosphate) is necessary for rapid and complete fermentation of plum juice (Yang and Wiegand, 1949). The assimilable nitrogen supplement stimulated yeast growth and supported higher yeast populations (Vos et al., 1979). According to Vyas and Joshi (1982), diammonium hydrogen phosphate (DAHP) at the rate of 0.1%, can be used as an exogenous nitrogen source in alcoholic fermentation but nitrogen was not a limiting factor in this fruit.

3.3.2 Composition

Titratable acidity and pH: Wine made from undiluted whole plum must was more acidic than that without skin but the pH values of the two wines were almost the same (Vyas and Joshi, 1982). However, titratable acidity of the wines treated with KMS and sodium benzoate were almost comparable (Joshi and Bhutani, 1990b). For the dilution of must with water to produce a wine of lower acidity, deacidifying yeast *Schizosaccharomyces pombe* in plum must has been used (Vyas and Joshi, 1988; Joshi et al., 1991b).

Alcohol: The wine prepared with KMS and sodium benzoate had almost similar alcohol content (Joshi and Bhutani, 1990b). Out of the various strains of *S. cerevisiae* var. *ellipsoideus* tried, UCD 595 produced the highest ethanol content, followed by UCD 522 and W strains (Joshi et al., 1999b).

Color: The wine treated with sodium benzoate was better in color values as compared to that treated with KMS (Joshi and Bhutani, 1990a). According to Stefano and Cravero (1991), the colors in young grape wines were due to anthocyanins, and the colors in aged wines retain a certain sensitivity to SO_2 and pH. In wine aging, anthocyanins are known to pass from low to high molecular weight forms and their oxidation state increases with time.

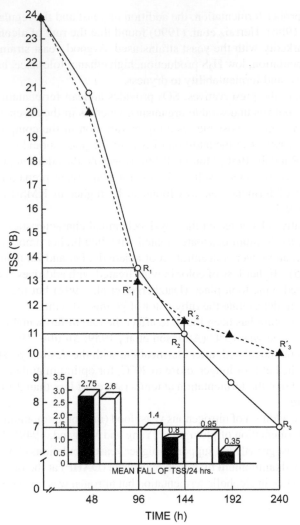

Must treated with :

▲ , ■ sodium benzoate; ○ , □ potassium metabi-sulphite

R_1___R_3, R_1'_ _ _ _R_3' shows the respective rates of fermentation at the indicated time interval.

FIGURE 9.5

Comparison of rate of fermentation of plum musts treated with potassium metabisulfite and sodium benzoate.

Based on Joshi, V.K., Sharma, S.K., 1995. Comparative fermentation behavior and physico-chemical sensory characteristics of wine as effected by the type of preservatives. Chemie Microbiologie Der Labensmittl 17, 45–53.

Minerals: Plum wine was rich in K, Ca, Cu, and Fe, and the values of all the elements were in the acceptable range compared to those reported for grape wines (Bhutani et al., 1989). Higher amounts of K, Na, and Cu were observed in sodium benzoate–treated wine, while increase in Fe and Zn were recorded in wines treated with KMS (Joshi and Bhutani, 1990a).

Biochemical characteristics: The tannin content was found to be higher in plum wine fermented with skin than that without skin, whereas the volatile acidity was found to be almost the same in both the wines (Vyas and Joshi, 1982). The tannin content was found to be higher in wine treated with sodium benzoate than that of KMS, whereas their aldehyde contents were similar (Joshi and Bhutani, 1990a). However, no appreciable differences in total esters and volatile acidity between the two wines were noted. The aldehyde content of a wine has been related to the specific yeast strain used in fermentation. The yeast strain, UCD 595, was found to produce the lowest aldehyde content among those tried (Joshi et al., 1999b).

3.3.3 Sensory Qualities

Temperature has been observed to affect the final flavor of the product (Atkinson et al., 1959). A fermentation temperature of lower than 15°C, produced a grassy aroma while at more than this temperature, the wines lacked in fineness (Moulin, 1986). The wine prepared from pulp with skin had an attractive color but had more astringency than that without skin (Vyas and Joshi, 1982). Further, 1:1 diluted pulp produced a wine of acceptable quality, while that from 1:2 diluted pulp was of a poor quality because of poor body, unbalanced sugar acid blend, and astringency. The wine prepared by the addition of sodium benzoate was judged to be better in terms of sensory qualities than the KMS-treated wine (Joshi and Bhutani, 1990b). The yeast strain used in wine fermentation also influences the sensory qualities. The matured wines prepared by yeast strains of *S. cerevisiae*, UCD 595 and W (a local isolate), had better sensory qualities than that of UCD 522 and UCD 505 (Joshi et al., 1999b). Out of the two types of wines, sodium benzoate–treated wine was rated better mainly due to better aroma, bouquet, astringency, and overall quality (Joshi and Sharma, 1995).

3.4 METHODS OF SECONDARY FERMENTATION

Various methods are available to transform the base wine into a sparkling wine (Jeandet et al., 2011). The base wine produced after proper blending/amelioration undergoes secondary fermentation. Usually, a source of nitrogen and a calculated amount of sugar are added to the base wine in addition to proper blending. The base wine can be fermented either in bottles or a pressurized tank so as to retain CO_2 in the wine (Joshi et al., 1995).

3.4.1 Bottle Fermentation

Bottle fermentation is the most common method of sparkling wine fermentation as described earlier. The name is derived from the fact that secondary fermentation takes place in bottles. It is also known as *methode champenoise* (Amerine et al., 1980). Some factors for producing bottled fermented sparkling wine include the position of the bottle during storage, temperature, excess free SO_2, the combined effects of low temperature, malolactic fermentation in bottles, and so forth (Hardy, 1992). Bottle fermentation occurs in a still wine relatively unfavorable to yeast growth due to its alcohol content (10–11%, v/v), high total acidity (5–7 g/l, H_2SO_4), low pH (3.1–3.2), total SO_2 concentration (60–70 mg/l), and the low cellar temperature (10–12°C) (Juroszek et al., 1987). The secondary fermentation takes about 5–6 weeks at 15°C producing 500–600 KPa pressure, and the yeast must be able to initiate fermentation in the presence of 10–11% alcohol content, pH 2.9–3.1, and SO_2; the fermentation continues under rising pressure (Lee and Baldwin, 1988).

3.4.1.1 Factors Affecting Bottle Fermentation

Other factors on which the quality of sparkling wine depends include yeast culture, nutrient status, temperature, ethanol concentration, preservatives, and carbon dioxide content (Joshi et al., 1995).

The fermentation under carbon dioxide pressure is beneficial if yeast is precultured in the base wine both for the bottle and tank fermentation (Kunkee and Ough, 1966). Although CO_2 pressure delayed onset of fermentation, it did not affect the maximum fermentation rate greatly (Cahill et al., 1980). Further, the ethanol tolerance of the yeast used in the sparkling wine preparation is increased by growing the yeast in wine rather than in a non-ethanolic medium, along with reduction of temperature of incubation (Juroszek et al., 1987; Sa-Correia and Van-Uden, 1983). Cultivation conditions also influence the technological qualities of champagne yeasts used in red sparkling wine production (Tzetanov et al., 1992).

The addition of suboptimal levels of vitamins to the base wine from grapes could produce morphological and biochemical changes, affecting the fermentation rate, the type, and the concentration of fermentation end products (Suomalainen and Oura, 1971). The base wine needs to be supplemented with nutrients such as diammonium phosphate (50–300 mg/l) and vitamin supplements eliminating their deficiencies during the course of secondary fermentation (Vos and Gray, 1979; Markides, 1986).

Differences in survival of various yeast strains of *Saccharomyces* indicated that *S. cerevisiae* var. *ellipsoideus* strain UCD 595 was found to have a higher survival rate than that of UCD 505 (Fig. 9.6) and is considered better for secondary fermentation (Sharma and Joshi, 1996). Temperature has no effect on cell viability but yeast growth rate was slower when temperature was decreased from 20 to 16°C (Nagodawithana and Steinkraus, 1976). A temperature range of 10–15°C was preferred for the steady rate and secondary fermentation (Markides, 1986) to avoid the excessive

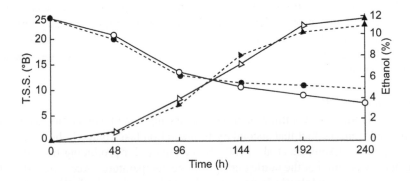

FIGURE 9.6

Changes in total soluble sugar and ethanol concentration during plum must fermentation.

Reproduced from Joshi, V.K., Sharma, S.K., 1995. Comparative fermentation behavior and physico-chemical sensory characteristics of wine as effected by the type of preservatives. Chemie Microbiologie Der Labensmittl 17, 45–53.

production of hydrogen sulfide. For successful second fermentation, the yeast should be acclimatized to the low temperature, prior to inoculation into the base wine for secondary fermentation. It has been successfully used in sparkling plum wine preparation.

Secondary fermentation, increasing ethanol content up to 6% in plum base wine and decreasing the temperature of incubation from 24° to 8°C, reduced the viable count of both yeast strains, though it remained above 80%. Strain UCD 595 was found to give better acclimatization to ethanol and low temperature than did UCD 505 (Fig. 9.7).

The ethanol in a growth medium or a base wine for *methode champenoise* is much less lethal to the yeast cell than the same produced intracellularly as the effect of ethanol on yeast viability was secondary to its inhibitory effect on sugar metabolism. Further, at higher alcohol content in a culture, the growth rates are lower, but growth viability rates remained above 80% remaining unaffected by increasing concentrations of ethanol (Ghose and Tyagi, 1979).

The inhibition due to SO_2 is very severe at low pH and high alcohol content (Amerine et al., 1980). An increase in free SO_2 content from 11 to 31 mg/l in tiraged base wine increased the lag-phase 10-fold and decreased the secondary fermentation rate when free SO_2 content exceeded 18 mg/l (Amerine et al., 1980) with a significant decrease in cell viability (Markides, 1986). Further, higher CO_2 pressure of about 3000 KPa is required for the process to be completed, and during that period, due to yeast autolysis, subtle aromas and flavors are produced (Jordan and Napper, 1986). However, Lee and Baldwin (1988) observed that a contact period of at least four months of bottle fermented wines with lees had a significant influence on the final quality and style of wine. The short-term storage conditions did not significantly affect the results, provided the yeast had no detrimental effect (Moretti and Garafolo, 1992).

3.4.2 Yeast Autolysis

Autolysis is the process whereby the cell constituents are broken down by its own enzymes (Markides, 1986), wherein proteins, peptides, and amino acids are released by the action of proteases. It is considered to be the most important aspect of yeast autolysis. The yeast autolysis is affected by temperatures

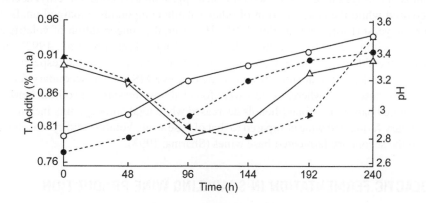

FIGURE 9.7

Changes in titratable acidity and pH during plum must fermentation.

Reproduced from Joshi, V.K., Sharma, S.K., 1995. Comparative fermentation behavior and physico-chemical sensory characteristics of wine as effected by the type of preservatives. Chemie Microbiologie Der Labensmittl 17, 45–53.

at 40°C, where the rate of autolysis has been found to be high but the released components do not have the opportunity to undergo favorable secondary chemical reactions to produce quality champagne (Molnar et al., 1980). The rate of autolysis has been linearly related to temperature in the range of 4–40°C. However, a 10°C increase in temperature corresponds to a 6–7% increase in the rate of autolysis but a temperature of 45°C and above denatured the protease enzymes, preventing further autolysis (Feuillat and Charpentir, 1982). For the aging of wine on yeast maturation, temperature not exceeding 10°C is preferred (Markides, 1986). During the autolysis, a number of changes take place, some of which are discussed here.

3.4.2.1 Biochemical Changes During Yeast Autolysis

The nitrogenous products released into the wine by the action of yeasts might not be the best indicators of autolysis (Moulin, 1986). As autolysis proceeds, many compounds are released into the wine, including amino acids, polypeptides, proteins, protein-carbohydrate complexes (Charpentier and Feuillat, 1989), and lipids (Chen, 1980). At the end of bottle fermentation, the increase in concentration of the total amino acids is due to greater exsorption than the assimilation. During secondary fermentation, total free amino acids decreased followed by an increase during maturation on the yeast (Tracey and Britz, 1989). But the increase was slightly lower for wines made by tank fermentation than those made by the bottle fermentation (Postel and Ziegler, 1991). However, during secondary fermentation, total N-content also decreased but remained constant during subsequent maturation (Postel and Ziegler, 1991). The crude proteins increased in the secondary fermented plum wines and its highest contents were found in bottle fermented, sodium benzoate–treated wines (Joshi and Sharma, 1995). Changes during the secondary fermentation of sparkling plum wine of different treatments (Fig 9.8).

Polyphenol content decreased during secondary fermentation but remained approximately constant during maturation and the composition of the wine was also little affected by maturation. Similarly, a decline in the extractable anthocyanins was observed during alcoholic fermentation, which continued during subsequent aging also (Tercelj, 1992).

The concentration of total volatile compounds increased steadily from secondary fermentation up to 15–18 months of storage on yeast and then remained approximately constant; ethyl acetate and acetaldehyde increased while the concentration of other volatile compounds remained unaffected during maturation on the yeast (Postel and Ziegler, 1991). However, the concentration of volatile compounds in wine remained within the normal limits (aldehydes <84 mg/l; Tzetanov et al., 1992). In secondary fermented wines, the aldehyde content decreased in both wines compared to the respective base wines. The esters, total phenols, and total anthocyanins increased in KMS-treated secondary fermented wine compared to that of base wine, whereas esters were found almost the same in sodium benzoate–treated secondary fermented base wine. Total phenols decreased while total anthocyanins increased in sodium benzoate secondary fermented wine compared to that of base wine. Volatile acidity was, however, comparable in both the secondary fermented base wines (Sharma, 1993).

3.5 MALOLACTIC FERMENTATION IN SPARKLING WINE PRODUCTION

Malolactic fermentation is a very significant stage in sparkling wine production (Beelman and Gallander, 1979; Edwards and Beelman, 1989; Jeandet et al., 2011). The conversion of the dicarboxylic acid, malate, into the monocarboxylic acid, lactate, and carbonic gas by lactic acid bacteria, decreases the acidity of the wine and increases its pH (Henick-Kling et al., 1989; Kunkee, 1967).

This fermentation is particularly important in wines produced from grapes grown in cool climates, which generally have a high organic acid content and low pH. But with respect to plum, such fermentation has not been documented.

3.6 PRODUCTION OF SPARKLING PLUM WINE

The basic methodology for the sparkling plum wine has been depicted in Fig. 9.4. Plum fruits have been evaluated for the preparation of sparkling wine. The wine from plum must was prepared with two preservatives, e.g., potassium metabisulfite and sodium benzoate. It was concluded that both types of wine were suitable for conversion into sparkling wine, though sodium benzoate–treated wine was considered better (Joshi et al., 1995, 2011c). In preparation of sparkling wine from plum, the effect of different concentrations of sugar (1, 1.5, 2%) and DAHP (0.1–0.3%), two strains of *Saccharomyces* viz. UCD 595 and UCD 505 in two base wines have been reported. Secondary fermentation in the inclined bottles is shown in Plate 9.1.

A sugar concentration of 1.5% and DAHP of 0.2% in 200 ml juice bottles at 15+2°C were found to be optimum for bottle fermentation (Sharma and Joshi, 1996). Changes during different secondary fermentation of plum wine with different sugar concentrations are shown in Fig. 9.9. Further, a comparison of physicochemical characteristics of two base wines after secondary fermentation, irrespective of the sugar and DAHP level, is given in Table 9.11.

In the secondary fermented wines, most of the physicochemical characteristics were altered compared to that of artificially carbonated wines except volatile acidity, methanol, propanol, and ethanol. Furthermore, these wines contained lower proteins, minerals, and amyl alcohol than the base wine. In general, the sparkling wines produced by either of the secondary fermentation method had lower sugar, more alcohol, and higher macro elements but lower Fe and Cu contents than the artificially carbonated wines. An overview of the changes occurring in the sparkling wine in comparison to artificially carbonated wine revealed that most of the changes took place due to secondary fermentation. The bottle fermented wine recorded the highest pressure (Plate 9.2), along with low TSS and low sugars.

The secondary bottle-fermented wine was the best in most of the sensory qualities but needed a proper acid-sugar blend of the base wine before conducting secondary fermentation. Sparkling wine made from base wine with sodium benzoate was preferred to that prepared with potassium metabisulfite. The studies showed the potential of plum fruits for production of sparkling wine (Joshi et al., 1999b). A comparison of physicochemical characteristics of sparkling plum wine prepared by bottle and tank fermentation has also been made in Table 9.12. Bottle fermented sparkling wine was rated better both in physicochemical and sensory quality (Joshi et al., 1995). The effect of different cultivars, free and immobilized cultures of *S. cerevisiae* and *Sz. pombe* on physicochemical and on the sensory quality of plum base wine for sparkling wine production has also been evaluated (Joshi and Bhardwaj, 2011).

The preparation and evaluation of base wine for conversion into sparkling plum wine, using two different yeasts and six plum cultivars was studied. The fermentation behavior and chemical composition of base wine was affected differently by various plum cultivars as well as by different combinations of yeasts. The highest rate of fermentation and rate of deacidification (RDA) were given by the must inoculated initially with immobilized *Sz. pombe* followed by *S. cerevisiae* and the combination of both the yeasts in immobilized form (Fig. 9.9).

Deacidification activity of *Sz. pombe* continued even in the medium in which *S. cerevisiae* was inoculated initially. The studies indicated that *Sz. pombe* is useful to reduce the acidity of wine like that

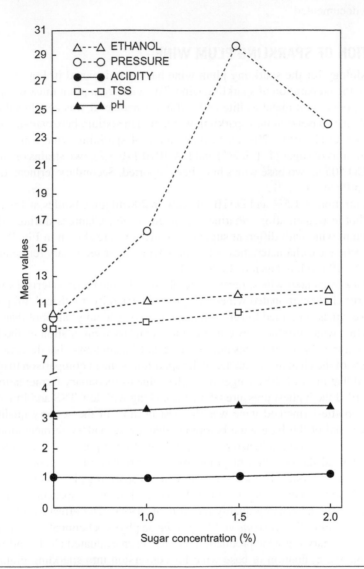

FIGURE 9.8

Effect of different concentration of sugar on physicochemical characteristics in secondary fermentation of plum wine.

Reproduced from Sharma, S.K., Joshi, V.K., 1996. Optimization of some parameters of secondary fermentation for production of sparkling wine. Indian Journal of Experimental Biology 34 (3), 235–238.

PLATE 9.1

Secondary fermentation for production of sparkling wine.

Reproduced from Bhardwaj, J.C., Joshi, V.K., 2009. Effect of cultivar, addition of yeast type, extract and form of yeast culture on foaming characteristics, secondary fermentation and quality of sparkling plum wine. Natural Product Radiance 8 (4), 452–464.

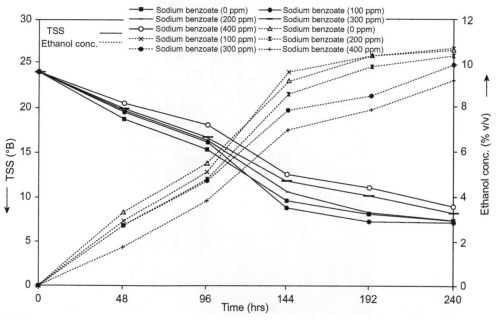

FIGURE 9.9

Effect of treatment on fermentation behavior (until reduction in TSS) and ethanol concentration of plum wines produced by *Schizosaccharomyces pombe* (means of pooled NaB conc.).

Reproduced from Bhardwaj, J.C., 2000. Evaluation of Some Plum Cultivars for Production of Sparkling Wine by Methodechampa-noise Using Biological Deacidification and Foam Stabilisation. A PhD Submitted to Dr Y.S. Parmar University of Horticulture and Forestry, Nauni, Solan, HP, India.

Table 9.11 Comparison of Physicochemical Characteristics of Two Base Wines After Secondary Fermentation

Parameters	Base Wines		CD (<0.05)
	Pot. Metabisulfite	Sod. Benzoate	
Total soluble solids (°B)	9.31	10.94	0.04
Titratable acidity (% MA)	1.03	1.03	NS
CO_2 pressure (lbs/sq. inch.)	20.92	19.04	0.66
pH	3.26	3.38	0.03
Ethanol (% v/v)	11.70	11.33	0.04

Means irrespective of sugar and DAHP levels.
Adapted from Sharma, S.K., Joshi, V.K., 1996. Optimization of some parameters of secondary fermentation for production of sparkling wine. Indian Journal of Experimental Biology 34 (3), 235–238.

PLATE 9.2

Measurement of carbon dioxide pressure in the sparkling wine bottles.

Courtesy: Bhardwaj, J.C., 2000. Evaluation of Some Plum Cultivars for Production of Sparkling Wine by Methodechampanoise Using Biological Deacidification and Foam Stabilisation. A PhD Submitted to Dr Y.S. Parmar University of Horticulture and Forestry, Nauni, Solan, HP, India.

Table 9.12 Comparison of Physicochemical Characteristics of Sparkling Plum Wine		
Characteristics	**Bottle Fermented Wine**	**Tank Fermented Wine**
Aldehydes (mg/L)	27.50	25.45
Esters (mg/L)	121.67	113.33
Total phenols (mg/L)	303.67	293.67
Total anthocyanins (mg/100 ml)	111.50	110.33
Volatile acidity as acetic acid (g/100 ml)	0.020	0.023
Titratable acidity (% MA)	0.94	0.98
Crude proteins (%)	0.47	0.45
Total sugar (%)	2.16	2.52
Ethanol (% v/v)	11.97	11.57
Pressure (lbs/sq inch)	52.50	8.75
Methanol (u/L)	312.13	298.49
Amyl alcohol (u/L)	301.75	327.10
Minerals iron (ppm)	6.70	6.57
Zinc (ppm)	2.16	2.13
Potassium (ppm)	1417.33	1404.33
Sodium (ppm)	29.28	28.52
Manganese (ppm)	0.51	0.48
Copper	0.12	0.14

Reproduced from Joshi, V.K., Sharma, S.K., Goyal, R.K., Thakur, N.S., 1999b. Sparkling plum wine: effect of method of carbonation and the type of base wine on physico-chemical and sensory qualities. Brazilian Archives of Biology and Technology 42 (3), 315–321.

of plum. Out of the various treatments tried, the treatment T_4, where *S. cerevisiae* (immobilized) and *Sz. pombe* in free culture were used, proved to be the best physicochemical and sensory qualities of wine (Figs. 9.10A and B).

Thus, the base wine from acidic fruits like plum can be made by conducting simultaneously deacidification and alcoholic fermentation using *S. cerevisiae* (immobilized) and *Sz. pombe* (free culture). Out of different plum cultivars used, the Santa Rosa plum gave the best wine for sparkling wine preparation, in terms of sensory qualities.

For maintaining foam stability in beer and wine proteins, particularly polysaccharide glycoproteins (hydrophobic) having MW > 5000 act as a surface active agent, contributing to bubble size, low molecular weight proteins reduce these characteristics. The combined effect of cultivars, yeast extract addition, type, and form of yeast culture on the foam characteristics, and physicochemical and sensory quality characteristics of sparkling plum wine was determined (Bhardwaj and Joshi, 2009). It was found that the addition of 0.5% yeast extract to the base wine obtained (*S. cerevisiae* and *Sz. pombe*) to the base wine at the beginning of secondary fermentation improved both the foam stability and foam expansion (Joshi and Bhardwaj, 2011a). The values of foam expansion of the sparkling wine prepared from six plum cultivars (Table 9.13) showed that all three factors affected the foam expansion of the wines significantly. Addition of yeast extract significantly increased the mean expansion values of the sparkling wines, while immobilized yeast culture significantly decreased this parameter. Foam life time

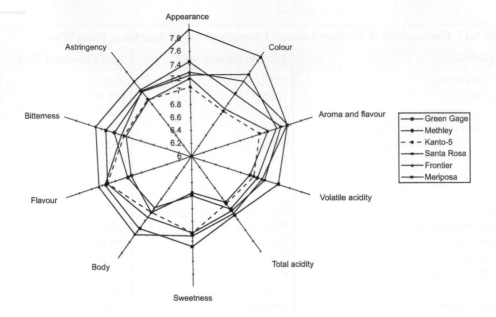

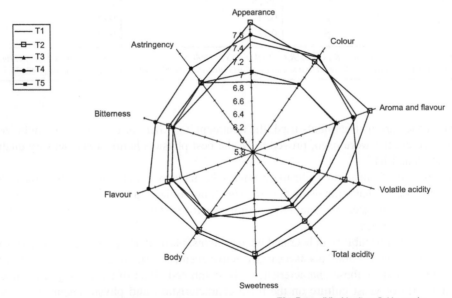

T1 = Free cell liquid culture *S. cerevisae*
T3 = Immobilized *Schiz. pombe* + LC *S. cerevisae*
T5 = Immobilized culture of both yeasts (50+50) (LC = Liquid Culture)

T2 = Free cell liquid culture *Schiz. pombe*
T4 = Immobilized *S. cerevisae*+LC *Schiz. pombe*

FIGURE 9.10

(A) Spider web diagram showing the effect of cultivars on the sensory characteristics of plum base wine.
(B) Spider web diagram showing the effect of treatments on the sensory characteristics of plum base wine.

Courtesy: Bhardwaj, J.C., 2000. Evaluation of Some Plum Cultivars for Production of Sparkling Wine by Methode champanoise Using Biological Deacidification and Foam Stabilisation. A PhD Submitted to Dr Y.S. Parmar University of Horticulture and Forestry, Nauni, Solan, HP, India.

Table 9.13 Effect of Cultivar, Addition of Yeast Extract and Treatment (Yeast Type and Culture Form) on the Foam Expansion (Expanded Volume/Original Volume) of Sparkling Plum Wine

Cultivars	Control					+0.5% yeast					Grand Mean
	Free cell Liquid Culture S. Cerevisiae	Immobilized Cell S. cerevisiae	Free Cell Culture Sz. pombe	Immobilized Cell Sz. pombe	Mean	Free cell Liquid Culture S. cerevisiae	Immobilized Cell S. cerevisiae	Free Cell Culture Sz. pombe	Immobilized Cell Sz. pombe	Mean	
Green gage	1.50	1.42	1.33	1.10	1.34	3.33	1.20	1.65	1.50	1.92	1.63
Methley	1.00	1.10	3.00	1.05	1.54	1.10	1.05	1.58	1.20	1.23	1.38
Kanto-5	1.10	1.54	1.42	1.60	1.41	2.10	1.00	2.58	1.84	1.88	1.65
Santa Rosa	1.20	1.20	1.60	1.20	1.30	2.40	1.05	2.53	1.05	1.76	1.53
Frontier	1.10	1.10	1.60	1.42	1.30	1.54	1.05	1.40	1.05	1.26	1.28
Meriposa	2.50	1.84	1.33	1.20	1.72	1.54	1.42	2.38	1.60	1.73	1.73

Adapted from Bhardwaj, J.C., Joshi, V.K., 2009. Effect of cultivar, addition of yeast type, extract and form of yeast culture on foaming characteristics, secondary fermentation and quality of sparkling plum wine. Natural Product Radiance 8 (4), 452–464.

(L_F) of the sparkling wines by the factor studied (Table 9.14) showed that the addition of yeast extract in the base wine significantly enhanced L_F in the sparkling wine. However, the type of yeast did not influence the L_F values of sparkling wine significantly. The significant effect of cultivar on the L_F values can be attributed to the different surface active agents present in different wines. The studies showed the potential of plum fruits for sparkling wine production by the *methode champenoise*.

3.6.1 Sensory Evaluation

The changes of lipid contents resulting from multiple chemical reactions occur in wines which partly modify the flavor of champagne (Troton et al., 1989). Sensory differences between wines made by the bottle and tank fermentation were found to be significant (Postel and Ziegler, 1991). Similarly, lees significantly affected the sensory quality. Moretti and Garafolo (1992) reported that the type of yeast strain used had a demonstrable effect on the organoleptic characteristics of the sparkling wines. Sensory evaluation indicated that a wine from a highly aerated culture had a lower score, possibly due to products of oxidative metabolism, than a wine obtained by the addition of ammonium nitrogen, increased stirring, and raised temperature (Tzetanov et al., 1992). The production of sparkling wine with immobilized yeast cells resulted in a product with better sensory qualities than the conventional process (Godia et al., 1991). The bottle fermented wine was superior in most of the characteristics except sweetness and body. Out of the two methods tested, bottle fermented wine scored the highest and was found to be the best in all the sensory qualities except sweetness and body. The bottle fermented wine was rated superior to other methods because of better aroma, extent of carbonation, and astringency. The sparkling wine prepared from sodium benzoate base wine was judged to be the best (Joshi and Sharma et al., 1999).

3.7 PRODUCTION OF SPARKLING APPLE WINE AND CIDER

The production of cider and sparkling cider is important in European regions where apple cultivation is viable. A sparkling cider made according to the traditional method characterized by a secondary fermentation in the bottle has been protected by the *European Designation of Origin "Sidra de Asturias"* (EEC Commission Regulation No. 2154/2005). These sparkling ciders must have an alcoholic strength higher than 5% (v/v), a volatile acidity lower than 2 g/L of acetic acid, a total SO_2 content below 200 mg/L and an overpressure in bottle higher than 3 atm (Madrera et al., 2008). Moreover, these ciders will be designated as dry when their content in sugars is lower than 20 g/L, semidry between 30 and 50 g/L, and sweet between 50 and 80 g/L. For information on cider see Chapter 7.1.

The base wine is mixed with a combination of yeast cultures, sucrose, and bentonite to perform the secondary fermentation. It may be noted that fermentation and maturation on lees occur in the same bottles that are subsequently purchased by the consumers. The autolysis of yeasts takes place during aging and several compounds are released that may significantly modify the organoleptic and foaming properties of the cider (Suárez et al., 2005; Picinelli et al., 2005; Madrera et al., 2008).

Sparkling apple wines have been produced by an additional fermentation of apple wines (ethanol content 5–6% vol.) under normal pressure resulting in alcohol content up to 9–10% (v/v). The fermentation proceeds in closed containers, ensuring a sparkling beverage (Yurchenko, 1983; Tchorbanov et al., 1993). The carbon source for secondary fermentation has been introduced as ca. 70% sugar drawing liqueur made on dry apple wine and matured for a suitable time. Ammonium dihydrogen phosphate as well as ammonia solutions have been added as nitrogen sources during the production of sparkling wines.

The production and composition of apple juice and dry apple wine (5–6% vol. ethanol) suitable for secondary fermentation to low-alcohol sparkling beverage has been described (Mitchev et al., 1991a,b).

Table 9.14 Effect of Cultivar, Addition of Yeast Extract, and Treatment (Yeast Type and Culture Form) on Foam Stability (as Seconds) of Sparkling Plum Wine

Cultivars	Control					+0.5% Yeast					Grand Mean
	Free Cell Liquid Culture S. cerevisiae	Immobilized Cell S. cerevisiae	Free Cell Culture Sz. pombe	Immobilized Cell Sz. pombe	Mean	Free Cell Liquid Culture S. cerevisiae	Immobilized Cell S. cerevisiae	Free Cell Culture Sz. pombe	Immobilized Cell Sz. pombe	Mean	
Green gage	20.0	15.0	15.5	16.0	16.6	25.0	30.0	30.0	25.0	28.1	22.4
Methley	16.0	15.0	12.5	15.0	14.6	20.0	15.5	28.0	20.0	20.9	17.7
Kanto-5	12.5	12.5	20.0	15.0	15.0	20.0	19.0	25.0	20.0	18.5	16.7
Santa rosa	12.5	12.5	20.0	13.5	14.6	20.0	15.0	20.0	15.0	20.7	17.7
Frontier	16.0	12.5	20.0	15.5	16.0	20.0	15.0	20.0	25.0	20.0	18.0
Meriposa	17.0	12.5	12.5	15.5	14.4	20.0	25.0	20.0	15.0	20.0	17.2

Based on Bhardwaj, J.C., Joshi, V.K., 2009. Effect of cultivar, addition of yeast type, extract and form of yeast culture on foaming characteristics, secondary fermentation and quality of sparkling plum wine. Natural Product Radiance 8 (4), 452–464.

The secondary fermentation of *Saccharomyces oviformis* (strains Epernet and Varna-1) was studied using dry apple wine with low alcohol content (Tchorbanov et al., 1993). Besides different nitrogen sources, yeast sediment from primary fermentation subjected to autolysis was also used as a nitrogen source. It has been revealed that both the organic and nonorganic amino nitrogen sources are suitable for the sparkling process although the yeast autolyzate demonstrated better results. The maximum CO_2 pressure observed during the sparkling process with hydrolyzed yeast autolyzate as a nitrogen source. The analyses of the dry apple wine and sparkling wine has also been compared (Table 9.15).

To maintain the quality and reproducibility of fermented beverages, the use of isolated strains of *S. cerevisiae* is an interesting strategy (Valles et al., 2008). The use of selected local yeasts is presumed to be more competitive and believed to be much more effective due to their acclimatization to environmental conditions. Moreover, the selection of suitable local yeasts assures the maintenance of the typical sensory properties of the fermented products (Degré, 1993; Melero, 1992; Querol et al., 1992). In addition, the chosen yeast strains to carry out secondary fermentation must satisfy additional characteristics such as the ability to ferment in reducing conditions (bottle), in the presence of ethanol, at a low temperature and under high CO_2 pressure, possess a high flocculation capacity and do not adhere to glass to facilitate their removal from cider when the cider-making process is completed (Bidan et al., 1986; Martínez-Rodríguez et al., 2001).

A methodology for the selection of yeast strains for secondary fermentation of sparkling ciders has been proposed by Valles et al. (2005, 2008). The methodology was divided into consecutive stages, analysis of enological characteristics by means of rapid and nonexpansive tests (plate agar) followed by determination of the technological properties. Various technological properties such as flocculation capacity, ethanol and sulfite tolerance, and production of major volatiles were assessed. Among the 350

Table 9.15 Comparative Analysis of Dry Apple Wine and Sparkling Apple Wine

Component	Dry Apple Wine	Sparkling Apple Wine
Total sugars (g/L)	2.5	3.5
Ethanol (% vol)	5.0	5.6
Methanol (mg/L)	30.0	32.0
Titratable acidity (g/L)	5.5	5.0
Volatile acidity (g/L)	0.26	0.36
pH	3.35	3.25
Ethyl acetate (mg/L)	90	98
Total SO_2	82	80
Free SO_2	6.7	6.5
Total phenolic compounds (mg/L)	700	660
Flavonoid phenolic compounds (mg/L)	660	545
Nonflavonoid phenolic compounds (mg/L)	42	40
Tanninic phenolic compounds (mg/L)	225	190
Catechins (mg/L)	274	209
Pressure (MPa)	–	0.41

Based on Tchorbanov, B., Mitchev, G., Lazarova, G., Popov, D., 1993. Studies on the secondary fermentation of low-alcohol sparkling apple wine. American Journal of Enology and Viticulture 44, 93–98.

isolated yeast colonies, 10 *S. cerevisiae* strains were found as true flocculants and were able to grow in an ethanolic medium and in the presence of 200 mg/l of sulfite.

Four sparkling ciders were produced using the *methode champenoise* with two flocculent strains chosen by applying the selection methodology together with two control strains. The analysis of variance ($p < .01$) among ciders indicated that glycerol, acetaldehyde, ethyl acetate, methanol, propanol, *i*-butanol and 2-phenylethanol were considerably influenced by the secondary yeast strain. The analyses of the base cider and the corresponding sparkling ciders have been shown in Table 9.16. No significant differences among sparkling ciders were observed for alcohol proof, total, and volatile acidities. However, the results of the sensory analysis indicated that all the sparkling ciders were of good quality.

The quality of sparkling wines has been affected by several factors, such as raw material, yeast strains, aging time on lees, and fining treatments. The effect of yeast strain and aging time on the volatile composition of sparkling cider was studied by producing two sparkling ciders from the same base cider using a selected cider yeast strain (*S. bayanus*) and a commercial wine yeast strain (*S. cerevisiae*; Madrera et al., 2008). It has been revealed that the yeast strain and aging time affected the concentration of volatile compounds in cider. It was observed that acetaldehyde and acetoin decreased with time, while higher alcohols, ethyl acetate, ethyl lactate, and ethyl octanoate significantly increased during aging in contact with lees. Moreover, the concentrations of methanol, 2-phenyletanhol, ethyl lactate, and ethyl octanoate were higher in the cider made with the selected yeast strain (*S. bayanus*).

Foam is one of the most important characteristics of cider and the typical visual attributes assessed in sparkling ciders are initial foam, foam area persistence, number of nucleation sites, bubble size, and foam collar (Picinelli et al., 2005). Proteins, polysaccharides, and fatty acids are important molecules in the constitution and stabilization of foam. In sparkling wines, protein concentration, and the presence of

Table 9.16 Analytical Characteristics of the Base Cider and Sparkling Ciders

	Base Cider	Yeast Strains			
		C_6	Levuline	3′	50′
Alcoholic proof (%, v/v)	63 ± 0.01	7.5 ± 0.00	7.4 ± 0.00	7.4 ± 0.01	7.4 ± 0.00
Acetic acid (g/l)	0.8 ± 0.02	0.7 ± 0.02	0.8 ± 0.01	0.8 ± 0.03	0.7 ± 0.01
Glycerol (g/l)	4.7 ± 0.01	5.1 ± 0.01	5.2 ± 0.00	5.2 ± 0.00	5.1 ± 0.01
Acetaldehyde (mg/l)	8 ± 0.35	51 ± 1.13	36 ± 0.39	25 ± 1.19	33 ± 0.83
Ethyl acetate (mg/l)	53 ± 0.63	61 ± 0.62	66 ± 0.50	57 ± 1.59	62 ± 1.61
Methanol (mg/l)	65 ± 2.21	72 ± 0.92	76 ± 2.19	69 ± 1.81	70 ± 3.15
1-Propanol (mg/l)	13 ± 0.30	18 ± 0.06	23 ± 0.46	19 ± 0.10	21 ± 0.56
i-Butanol (mg/l)	39 ± 1.24	46 ± 0.71	51 ± 0.77	49 ± 0.18	48 ± 0.85
1-Butanol (mg/l)	5 ± 0.17	6 ± 0.05	7 ± 0.06	6 ± 0.03	6 ± 0.16
Amyl alcohols (mg/l)	200 ± 5.04	246 ± 3.47	263 ± 2.40	239 ± 2.92	245 ± 5.99
Ethyl lactate + hexanol (mg/l)	213 ± 3.51	244 ± 8.78	253 ± 2.97	235 ± 2.87	239 ± 4.78
2-Phenylethanol (mg/l)	26 ± 0.84	46 ± 1.60	38 ± 0.48	36 ± 0.48	35 ± 0.76

Each value in the table represents the mean value ± SD from three analyses.
Based on Valles, B.S., Bedriñana, R.P., Queipo, A.L., Alonso, J.J.M., 2008. Screening of cider yeasts for sparkling cider production (Champenoise method). Food Microbiology 25, 690–697.

polysaccharide–polypeptide complexes improve the foam stability (Pueyo, 1995; Blanco-Gomis et al., 2009). The chemical and physical characteristics of proteins influence the formation of foam and its stabilization. Thus, for instance, proteins with good foamability tend to be flexible and relatively small. On the contrary, proteins with good foam stability are able to cross-link and aggregate and are resistant to mechanical deformation (Bamforth, 2004). It has also been observed that hydrophobic proteins contribute more to foam constitution than hydrophilic ones (Brissonnet and Maujean, 1993).

The foaming properties such as foam height (FH), foam stability height (FS) are significant parameters in the study of foam of a sparkling beverage (Blanco-Gomis et al., 2007, 2009). Foamability is determined by the FH parameter, FS time characterizes the average lifetime of the bubbles, which is related to collar quality, and ST represents the average lifetime of the foam after gas injection has ceased. The characterization of sparkling ciders from Asturias, northern Spain, by means of the analysis of their protein content and their foam characteristics has been carried out (Blanco-Gomis et al., 2007, 2009). The relationship between polypeptides and foam was demonstrated with a prediction equation of FS time, which was computed using the partial least square regression technique. This mathematical equation confirmed that the polypeptides of high molecular mass are especially related to this foaming parameter.

3.8 SPARKLING MEAD

For preparation of sparkling mead, the honey is dissolved in the warm water and dry mead is made as described earlier (Filipello and Marsh, 1934; Foster, 1967). After storing for six months in champagne bottles, the wine is sweetened with honey and to each bottle is added a small quantity of champagne yeast. The bottles are allowed to stand in a warm place for one week only, then stood on their sides in a cool place for a year. The sparkling wines are served cool (Foster, 1967).

3.9 OTHER SPARKLING FRUIT WINES

Sparkling gooseberry wine is called English champagne and is known to be a very palatable wine. To make such a wine, for every three pounds of green round gooseberries taken in a pan, one gallon of water is poured over them with the addition of four pounds of sugar. It is allowed to be fermented for three weeks by putting into bottles, tying down the corks, and laying the bottles on their sides to carry out secondary fermentation. Sparkling wine from orange has also been prepared (Rommel et al., 1990).

4. CONCLUSIONS AND FUTURE TRENDS

Being rich in medicinal properties, nongrape fruit-based wines and vermouths have a wide scope in the field of enology. Systematic research at both academic and industrial level is required to be strengthened for the commercial productivity of vermouths produced from different fruits. The relationship between the flavor attributes and chemical constituents changes during vermouth production and maturation need to be explored.

Production of sparkling wines has significant potential from some nongrape fruits, especially plum and apple. Different factors, especially the nature of yeast strains and aging time, have to be evaluated and optimized, since these affect the concentration of volatile compounds in wine. Similarly, the foaming properties are significant parameters in the quality of sparkling wine.

It is apparent that the winemaking process can considerably affect the foaming properties of sparkling wines and has to be carefully considered. It is also clear that particles and/or colloids are limiting factors to foam behavior. Lysozyme has a protective effect on wine's foaming properties when added before the bentonite treatment. The combined effect of fruit cultivars, supplement type, and form of yeast culture on the foam characteristics and physicochemical and sensory quality characteristics of sparkling plum wine need to be standardized to make a better quality sparkling wine.

Considering the commercial potential of vermouth and sparkling wines, in depth and systematic research on different facets of enology of these wines can be strengthened.

REFERENCES

Abrol, G.S., 2009. Preparation and Evaluation of Wild Apricot Mead and Vermouth (M.Sc. thesis). Dr. Y.S. Parmar University of Horticulture and Forestry, Nauni, Solan, India.

Amerine, M.A., Berg, H.W., Cruess, W.V., 1967. The Technology of Wine Making, third ed. AVI Publishing Co. Inc., Westport, CT.

Amerine, M.A., Berg, H.W., Kunkee, R.E., Ough, C.S., Singleton, V.L., Webb, A.D., 1980. The Technology of Wine Making. AVI Publishing Company, Inc, Westport, Connecticut, p. 794.

Atkinson, F.E., Bowen, J.F., MacGregor, D.R., 1959. A rapid method for production of a sparkling apple wine. In: Nineteenth Annual Meeting of the IFT. Philadelphia, Pennsylvania, May, 18, pp. 673–675.

Attri, B.L., Lal, B.B., Joshi, V.K., 1993. Preparation and evaluation of sand pear vermouth. Journal of Food Science and Technology 30, 435–437.

Attri, B.L., Lal, B.B., Joshi, V.K., 1994. Technology for the preparation of sand pear vermouth. Indian Food Packer 48 (1), 39–44.

Bamforth, C.W., 2004. The relative significance of physics and chemistry for beer foam excellence: theory and practice. Journal of the Institute of Brewing 110 (4), 259–266.

Beelman, R.P., Gallander, J.F., 1979. Wine deacidification. Advances in Food Research 25, 1.

Bhardwaj, J.C., Joshi, V.K., 2009. Effect of cultivar, addition of yeast type, and form of yeast culture on foaming characteristics, secondary fermentation and quality of sparkling plum wine. Natural Product Radiance 8 (4), 452–464.

Bhardwaj, J.C., 2000. Evaluation of Some Plum Cultivars for Production of Sparkling Wine by Methodechampanoise Using Biological Deacidification and Foam Stabilisation. A PhD Submitted to Dr Y.S. Parmar University of Horticulture and Forestry, Nauni, Solan, HP, India.

Bhutani, V.P., Joshi, V.K., Chopra, S.K., 1989. Mineral contents of fruit wines produced experimentally. Journal of Food Science and Technology 26 (6), 332–333.

Bidan, P., Feuillat, M., Moulin, J., 1986. Les vinsmousseux. Rappot de la France 65e´me. Assemble´e Ge´nerale de l'OIV. Bulletin OIV 59, 563–626.

Blanco-Gomis, D., Mangas-Alonso, J.J., Junco-Corujedo, S., Gutiérrez-Álvarez, M.D., 2007. Cider proteins and foam characteristics: a contribution to their characterization. Journal of Agricultural and Food Chemistry 55 (7), 2526–2531.

Blanco-Gomis, D., Mangas-Alonso, J.J., Junco-Corujedo, S., Gutiérrez-Álvarez, M.D., 2009. Characterisation of sparkling cider by the yeast type used in taking foam on the basis of polypeptide content and foam characteristics. Food Chemistry 115, 375–379.

Boyd, G.D., 2007. Vermouth: The Aromatized Wine. Hotel F & B March/April Issue.

Bravo, F., 1986. Crianza Biologica Del Vino: Procedimientotradicional De Vinos Finos de D. O. Jerez y D.O Montilla. Moriles Enol Enoltec Marzo, pp. 15–19.

Brissonnet, F., Maujean, A., 1993. Characterization of foaming proteins in a Champagne base wine. American Journal of Enology and Viticulture 44 (3), 297–301.

Buxaderas, S., López-Tamames, E., 2010. Managing the quality of sparkling wines. In: Oenology and Wine Quality: A Volume in Woodhead Publishing Series in Food Science, Technology and Nutrition, pp. 553–588.

Buxaderas, S., López-Tamames, E., 2012. Sparkling wines: features and trends from tradition. In: Henry, J. (Ed.), Advances in Food and Nutrition Research, vol. 66. Elsevier Science, Amsterdam, pp. 1–45.

Cahill, I.T., Carroad, P.A., Kunkee, R.E., 1980. Cultivation of yeast under carbon dioxide pressure for use in continuous sparkling wine production. American Journal of Enology and Viticulture 31 (1), 46–52.

Casp, A., Romero, M.P., 1990. Formation of certain volatile compounds of wine. Vignevini 15 (7–8), 59–62.

Charpentier, C., Feuillat, M., 1989. The mechanism of yeast autolysis in wine. Yeasts 5, S181–S186.

Chen, E.C.-H., 1980. Utilisation of fatty acids by yeast during fermentation. Journal of the American Society of Brewing Chemists 38, 148–153.

Datzberger, K., Steiner, I., Washuttle, J., Kroyer, G., 1992. The influence of wine additives on colour and colour quality of young and red wine. Zeitschrift fur Lebensmittel-Untersuchung and Forschung 194 (6), 524–526.

Degre, R., 1993. Selection and commercial cultivation of wine yeast and bacteria. In: Fleet, G.H. (Ed.), Wine Microbiology and Biotechnology. Harwood Academic Publishers, Chur, pp. 421–447.

Edwards, C.G., Beelman, R.B., 1989. Inducing malolactic fermentation in wine. Biotechnology Advances 7, 333–360.

Feuillat, M., Charpentir, C., 1982. Autolysis of yeasts in champagne. American Journal of Enology and Viticulture 33, 5–13.

Filipello, F., Marsh, G.L., 1934. Honey wine. Fruit Wine Journal 14 (40), 42.

Foster, C., 1967. Home Wine Making. Ward Lock, London.

Genovese, A., Ugliano, M., Pessina, R., Gambuti, A., Piombino, P., Moio, L., 2004. Comparison of the aroma compounds in apricot (*Prunusarmeniaca* L. cv Pellecchiella) and apple (*Malus pumila* L. cv Annurca) raw distillates. Italian Journal of Food Science 16, 185–196.

Ghose, T.K., Tyagi, R.D., 1979. Rapid ethanol fermentation of cellulose hydrolysate. I. Batch versus continuous systems. Biotechnology and Bioengineering 21, 1410–1420.

Gill, A., Joshi, V.K., Rana, N., 2009. Evaluation of preservation methods of low alcoholic plum wine. Natural Product Radiance 8 (4), 392–405.

Godia, F., Casas, C., Sola, C., 1991. Application of immobilized yeast cells to sparkling wine fermentation. Biotechnology Progress 7, 468–470.

Goldman, M., 1963. Rate of carbon dioxide formation at low temperatures in bottle fermented champagne. In: Annual Meeting of the American Society of Enologists. Sacramento Inn, California, June, 27–29.

Goswell, R.W., Kunkee, R.E., 1977. Fortified wines. In: Rose, A.H. (Ed.), Alcoholic Beverages. Academic Press, London, pp. 477–534.

Griebel, C., 1955. GemahleneWermutkrauter Z Lebensm. Untersuch u-Forsch 100, 270–274.

Hosry, El-L., Auezova, L., Sakr, A., Hajj-Moussa, E., 2009. Browing susceptibility of white wine and antioxidant effect of glutathione. International Journal of Food Science and Technology 44, 2459–2463.

Hardisson, A., Corrales, J., Gomez-Calcerrada, N., Navarrete, A., 1992. Concentration of iron and copper in wines frequently consumed in the Canary Islands. Anales de Bromatologia 43 (2–3), 231–238.

Hardy, G., 1992. Faults and anomalies encounterd during the development of sparkling wines I. Development of the basic wine. Revue des O Enologues et des Techniques Vitivinicoles et O Enologiques 62, 32–34.

Henick-Kling, T., Sandrine, W.E., Heatherbell, D.A., 1989. Evaluation of malolactic bacteria isolated from Oregon wines. Applied and Environmental Microbiology 55, 2010.

Herraiz, T., Reglero, G., Herraiz, M., Martin-Alvarez, P.J., Cabezudo, M.D., 1990. The influence of the yeast and type of culture on the volatile composition of wines fermented without sulphur dioxide. American Journal of Enology and Viticulture 41 (4), 313–318.

Jaarsveld, F.P., Blom, M., Hattingh, S., Minnaar, P., 2009. Rapid induction of ageing character in brandy products – Part II, influence of type of oak. South African Journal of Enology and Viticulture 30 (1), 16–23.

Jackson, R.S., 2000. Wine Science: Principles, Practice, Perception, second ed. Academic Press, New York.

Jackson, R.S., 2008. Wine Science: Principles and Applications, third revised ed. Academic Press, New York.

Jarczyk, A., Wzorek, W., 1997. Fruit and honey wines. In: Rose, A.H. (Ed.), Alcoholic Beverages. Academic Press, London, p. 387.

Jeandet, P., Vasserot, Y., Liger-Belair, G., Marchal, R., 2011. Sparkling wine production preparation of fortified wines. In: Joshi, V.K. (Ed.), Hand Book of Enology: Principles, Practices and Recent Innovations. Asiatech Publisher Inc., New Delhi, pp. 1064–1115.

Jeffs, J., 1970. Sherry, second ed. Faber and Faber, London, p. 268.

Jordan, A.D., Napper, D.H., 1986. Some aspects of the physical chemistry of bubble and foam phenomenon in sparkling wine. In: Proceedings of Sixth Australian Wine Industry Technical Conference. Adelaide. Australian Industrial Publishers, Adelaide, pp. 237–248.

Joshi, V.K., Bhardwaj, J.C., 2011. Effect of different cultivars yeasts (free and immobilized cultures) of *S. cerevisiae* and *Schizosaccharomyces pombe* on physic-chemical and sensory quality of plum based wine for sparkling wine production. International Journal of Food and Fermentation Technology 1 (1), 69–81.

Joshi, V.K., Bhutani, V.P., 1990a. Evaluation of plum cultivars for wine preparation. In: International Horticulture Congress, Italy, Abstract No. 3336.

Joshi, V.K., Bhutani, V.P., 1990b. Effect of preservatives on the fermentation behavior and quality of plum wine. In: International Horticulture Congress Italy, Abstract No. 2585, p. 727.

Joshi, V.K., Sandhu, D.K., 2000. Influence of ethanol concentration, addition of spices extract, and level of sweetness on physico-chemical characteristics and sensory quality of apple vermouth. Brazilian Archives of Biology and Technology 43 (5), 537–545.

Joshi, V.K., Sandhu, D.K., 2009. Flavour profiling of apple vermouth using descriptive analysis technique. Natural Product Radiance 8 (4), 419–425.

Joshi, V.K., Sharma, S.K., 1993. Effect of method of must preparation and initial sugar levels on the quality of apricot wine. Research on Industry 39 (4), 255–257.

Joshi, V.K., Sharma, S.K., 1995. Comparative fermentation behavior and physico-chemical sensory characteristics of wine as effected by the type of preservatives. Chemie Microbiologie Der Labensmittl 17, 45–53.

Joshi, V.K., Bhutani, V.P., Sharma, R.C., 1990. The effect of dilution and addition of nitrogen source on chemical, mineral and sensory qualities of wild apricot wine. American Journal of Enology and Viticulture 41, 229–231.

Joshi, V.K., Attri, B.L., Mahajan, B.V.C., 1991a. Production and evaluation of vermouth from plum fruits. Journal of Food Science and Technology 28, 138–141.

Joshi, V.K., Sharma, P.C., Attri, B.L., 1991b. A note on deacidification activity of *Schizosaccharomyces pombe*. Journal of Applied Bacteriology 70, 385–390.

Joshi, V.K., Sharma, S.K., Thakur, N.S., 1995. Technology and quality of sparkling wine with special reference to plum – an overview. Indian Food Packer 49, 49–66.

Joshi, V.K., Sandhu, D.K., Thakur, N.S., 1999a. Technology of fruit based alcoholic beverages. In: Joshi, V.K., Pandey, A. (Eds.), Biotechnology : Food Fermentation, vol. 2. Educational Publishers and Distributors, New Delhi, pp. 647–744.

Joshi, V.K., Sharma, S.K., Goyal, R.K., Thakur, N.S., 1999b. Sparkling plum wine: effect of method of carbonation and the type of base wine on physico-chemical and sensory qualities. Brazilian Archives of Biology and Technology 42 (3), 315–321.

Joshi, V.K., Sharma, R., Abrol, G., 2011b. Stone fruit: wine and brandy. In: Hui, Y.H., Evranuz, E.O. (Eds.), Handbook of Food and Beverage Fermentation Technology. Handbook of Plant-Based Fermented Food and Beverage Technology Series. CRC Press, Florida, pp. 273–304.

Joshi, V.K., 1997. Fruit Wine, second ed. Dr YS Parmar University of Horticulture and Forestry, Nauni, Solan, India.

Joshi, V.K., Attri, D., Singh, T.K., Abrol, G.S., 2011c. Fruit wines: production technology. In: Joshi, V.K. (Ed.), Handbook of Enology: Principles, Practices and Recent Innovations, vol. 3. Asia Tech Publisher, New Delhi, pp. 1177–1221.

Joshi, V.K., Ghanshyam, A., Thakur, N.S., March–April 2011a. Wild apricot vermouth: effect of sugar, alcohol concentration and spices level on physic-chemical and sensory evaluation. Indian Food Packer 53–62.

Joslyn, M.A., Amerine, M.A., 1964. Dessert, Appetizer and Related Flavoured Wines. University of California Press, Berkeley.

Juroszek, J.R., Feuillat, M.E., Charpentier, C., 1987. Effect of the champagne method of starter preparation on ethanol tolerance of yeast. American Journal of Enology and Viticulture 38 (3), 194–198.

Kauffman, G.B., 2001. The dry martini: chemistry, history, and assorted lore. The Chemical Educator 6, 295–305.

Kritzinger, E.C., Bauer, F.F., du Toit, W.J., 2013. Role of glutathione in winemaking: a review. Journal of Agricultural and Food Chemistry 61 (2), 269–277.

Kumar, Y.S., Prakasam, R.S., Reddy, O.V.S., 2009. Optimization of fermentation conditions for mango (*Magniferaindica* L.) wine production by employing response surface methodology. International Journal of Food Science and Technology 44, 2320–2327.

Kumar, Y.S., Varakumar, S., Reddy, O.V.S., 2012. Evaluation of antioxidant and sensory properties of mango (*Mangiferaindica* L.) wine. CyTA Journal of Food 10 (1), 12–20.

Kunkee, R.E., Ough, C.S., 1966. Multiplication and fermentation of *Saccharomyces cerevisiae* under carbon dioxide pressure in wine. Journal of Applied Microbiology 14, 643–648.

Kunkee, R.E., 1967. Malolactic fermentation. Advances in Applied Microbiology 9, 235–279.

Lee, T.H., Baldwin, G.E., 1988. Developments in the production and consumption of sparkling wines in Australia. Food Technology Australia 40 (4), 138–140.

Lee, T.H., Lester, D.C., 1986. Production of sparkling wine by the method Champenoise. In: Proceedings of a Seminar, 14, November, 1985, Canberra, ACT, Adelaide, SA, Aus. Sco. Vitic.Oneol, p. 168.

Liddle, P., Boero, L., 2003. Vermouth. In: Encyclopedia of Food Sciences and Nutrition, second ed. , pp. 5980–5984.

Lingappa, K., Padshetty, N.S., Chowdary, N.B., 1993. Tamarind vermouth—a new alcoholic beverage from tamarind (*Tamarindusindica* L.). Indian Food Packer 47 (1), 23.

Madrera, R.R., Hevia, A.G., García, N.P., Valles, B.S., 2008. Evolution of aroma compounds in sparkling cider. LWT Food Science and Technology 41, 2064–2069.

Markides, A.J., 1986. The microbiology of method champenoise. In: Proceedings of the 6th Australian Wine Industry Technical Conference, Adelalde. Australian Industrial Publishers, Adelaide, pp. 232–236.

Martınez-Rodrıguez, A., Carrascosa, A.V., Barcenilla, J.M., Pozo-Bayon, A., Polo, C., 2001. Autolytic capacity and foam analysis as additional criteria for the selection of yeast strains for sparkling wine production. Food Microbiology 18, 183–191.

Mattick, L.R., Robinson, W.B., 1960. Changes in volatile acids during the baking of sherry wine by the Tressler baking process. American Journal of Enology and Viticulture 11, 113–116.

Mbaeyi-Nwaoha, I.E., Ajumobi, C.N., 2013. Production and microbial evaluation of table wine from tamarind (*Tamarindus indica*) and soursop (*Annona muricata*). Journal of Food Science and Technology (Published Online).

Melero, R., 1992. Fermentación controlada y selección de levadurasvı́nicas. Revista Española de Ciencia y Tecnología de Alimentos 32, 371–379.

Mitchev, G.D., Popov, D., Tchorbanov, B., 1991a. Production of Apple Juice for Fermentation to Dry Apple Wine (Engl. Trans., Article in Bulgarian) Lozarstvo I Vinarstvo (Sofia) XXXX (1–2), pp. 15–17.

Mitchev, G.D., Popov, D., Tchorbanov, B., 1991b. Dry Apple Wine Suitable for Production of Low Alcoholic Sparkling Wine (Engl. Trans., Article in Bulgarian) Lozarstvo I Vinarstvo (Sofia) XXXX (3–4), pp. 16–17.

Molnar, I., Oura, E., Suomalainen, H., 1980. Changes in the activites of certain enzymes of champagne yeast during storage of sparkling wine. ActaAlimentaria 9, 313–324.

Montemiglio, L., 1992. Italian sparkling wines: breakdown of the major industrial markets. Enotecnico 28 (1–2), 57–70.

Moretti, S., Garafolo, A., 1992. Production of sparkling wines using various yeasts: effect of wine storage conditions. Enotecnico 27 (12), 69–78.

Moulin, J.P., 1986. Champagne: the method of production and the origin of the quality of this French wine. In: Proceedings of 6th Australian Wine Industry Technical Conference, Adelaide, 14–17 July. Australian International Publishers, Adelaide, pp. 218–224.

Nagodawithana, T.W., Steinkraus, K.H., 1976. Influence of the rate of ethanol production and accumulation on the viability of *Saccharomyces cerevisiae* in "rapid fermentation". Applied and Environment Microbiology 31, 158–162.

Nurgel, C., Pickering, G., 2005. Contribution of glycerol, ethanol and sugar to the perception of viscosity and density elicited by model white wines. Journal of Texture Studies 36, 303–323.

Onkarayya, H., 1985. Mango vermouth – a new alcoholic beverages. Indian Food Packer 39 (1), 40–45.

Ossa, de M., Caro, I., Bonat, M., Perez, L., Domecq, B., 1987. Dry extract in sherry and its evolution in the aging process. American Journal of Enology and Viticulture 38, 293–297.

Panesar, P.S., Kumar, N., Marwaha, S.S., Joshi, V.K., 2009. Vermouth production technology—an overview. Natural Product Radiance 8 (4), 334–344.

Panesar, P.S., Joshi, V.K., Panesar, R., Abrol, G.S., 2011. Vermouth: technology of production and quality characteristics. In: Advances in Food and Nutritional Research, vol. 63. Elsevier, Inc., London, UK, pp. 253–271.

Patil, A.B., Matapathi, S.S., Nirmalnath, P.J., 2004. Pomegranate vermouth – new fermented beverage. Karnataka Journal of Agricultural Sciences 17 (4), 860.

Perez, L., Vakarcel, M.J., Ganzalez, P., Domecq, B., 1991. Influence of Botrytis infection of grapes on the biological aging process of fino sherry. American Journal of Enology and Viticulture 42, 58–62.

Picinelli, A., Ferna´ndez, N., Rodrıguez, R., Suarez, B., 2005. Sensory and foaming properties of sparkling cider. Journal of Agriculture and Food Chemistry 53, 10051–10056.

Pilone, F.J., 1954. Production of vermouth. American Journal of Enology and Viticulture 5 (1), 30–45.

Postel, W., Ziegel, Z., 1991. Effects of duration of yeast contact and the manufacturing process on the composition and quality of sparkling wines. II. Free amino acids and volatile compounds. Wein-Wiss 46, 26–32.

Pueyo, E., Martín-Álvarez, P.J., Polo, M.C., 1995. Relationship between foam characteristics and chemical composition in wines and cavas (sparkling wines). American Journal of Enology and Viticulture 46 (4), 518–524.

Querol, A., Barrio, E., Huerta, T., Ramon, D., 1992. Molecular monitoring of wine fermentations conducted by dry yeast strains. Applied and Environmental Microbiology 58, 2948–2952.

Randall, W.D., 1986. Options for base wine production. In: Proceedings of 6th Australian Wine Industry Technical Conference, Adelaide, 14–17 July. Australian Industrial Publishers, Adelaide, pp. 224–231.

Reddy, L.V.A., Reddy, O.V.S., 2005. Production and characterization of wine from mango fruit (*Magniferaindica* L.). World Journal of Microbiology and Biotechnology 21, 1345–1350.

Rizzo, F., 1957. La fabricazione del Vermouth. Edizione Agricole, Bologna.

Rommel, A., Heatherbell, D.A., Wroslad, R.E., 1990. Red raspberry juice and wine: effect of processingand storage on anthocyanin pigment composition, colour and appearance. Journal of Food Science 55, 1011–1017.

Roussis, I.G., Sergianitis, S., 2008. Protection of some aroma volatiles in a model wine medium by sulphur dioxide and mixtures of glutathione with caffeic acid or gallic acid. Flavour and Fragrance Journal 23 (1), 35–39.

Sa-Correia, I., Van-Uden, N., 1983. Temperature profiles of ethanol tolerance: effects of ethanol on minimum and maximum temperature for growth of the yeast *Saccharomyces cerevisiae* and *Kluyveromyces fragilis*. Biotechnology and Bioengineering 25, 1665–1667.

Sevda, S.B., Rodrigues, L., 2011. The making of pomegranate wine using yeast immobilized on sodium alginate. African Journal of Food Science 5 (5), 299–304.

Sharma, S.K., Joshi, V.K., 1996. Optimization of some parameters of secondary fermentation for production of sparkling wine. Indian Journal of Experimental Biology 34 (3), 235–238.

Sharma, S., 1993. Studies on the Preparation and Evaluation of Sparkling Plum Wine (M.Sc. thesis). Submitted to Dr Y.S.Parmar University of Horticulture and Forestry, Nauni, Solan, HP, India.

Sonni, F., Clark, A.C., Prenzler, P.D., Riponi, C., Scollary, G.R., 2011. Antioxidant action of glutathione and the ascorbic acid/glutathione pair in a model white wine. Journal of Agricultural and Food Chemistry 59, 3940–3949.

Spreer, W., Ongprasert, S., Hegele, M., Wunsche, J.N., Muller, J., 2009. Yield and fruit development in mango (*Mangiferaindica* L. cv. Chok Anan) under different irrigation regimes. Agricultural Water Management 96, 574–584.

Stefano, R., di, Cravero, M.C., 1991. Fractionation of red wine polyphenols. Enotecnico 26 (3), 99–106.

Suarez, B., Palacios, N., Rodrı´guez, R., Picinelli, A., 2005. Influence of yeast strain and aging time on free amino acid changes in sparkling ciders. Journal of Agriculture and Food Chemistry 53, 6408–6413.

Suomalainen, H., Oura, E., 1971. Yeast nutrition and solute uptake. In: Rose, A.H., Harrison, J.S. (Eds.), The Yeasts, vol. 2. Academic Press, London, pp. 3–74.

Tchorbanov, B., Mitchev, G., Lazarova, G., Popov, D., 1993. Studies on the secondary fermentation of low-alcohol sparkling apple wine. American Journal of Enology and Viticulture 44, 93–98.

Terceij, D., 1992. Anthocyanin in Kraski teran wine. Vinid'Italica 33, 33–42.

The New Encyclopædia Britannica, 1995. fifteenth ed. Encyclopædia Britannica, Chicago, IL, vol. 12, p. 323.

Tracey, R.P., Britz, T.J., 1989. The effect of amino acids on malolactic fermentation by Leuconostocœnos. Journal of Applied Bacteriology 67, 589–595.

Troton, D., Charpentier, M., Robillard, B., Calvayrac, R., Duteurtre, B., 1989. Evolution of the lipid content of champagne wine during the second fermentation of *Saccharomyces cerevisiae*. American Journal of Enology and Viticulture 40, 175.

Tzetanov, O., Bambalov, G., Lucvev, S., Tsvetanov, O., 1992. Influence of cultivation conditions upon technological qualities of champagne yeasts for red sparkling wines. In: Proceedings of the Seventh Australian Wine Industry Technical Conference, Adelaide, SA, 13–17 August, vol. 24 (9), p. 9.

Valaer, P., 1950. The Wines of the World. Abelard Press, New York.

Valles, B.S., Garca, N.P., Madrera, R.R., Lobo, A.P., 2005. Influence of yeast strain and aging time on free amino acid changes in sparkling ciders. Journal of Agricultural and Food Chemistry 53 (16), 6408–6413.

Valles, B.S., Bedriñana, R.P., Queipo, A.L., Alonso, J.J.M., 2008. Screening of cider yeasts for sparkling cider production (Champenoise method). Food Microbiology 25, 690–697.

Varakumar, S., Kumar, Y.S., Reddy, O.V.S., 2011. Carotenoid composition of mango (*Magniferaindica* L.) wine and its antioxidant activity. Journal of Food Biochemistry 35, 1538–1547.

Venkataramu, K., Patel, J.D., Rao, M.S.S., 1979. Fermentation of grapes with a few strains of wine yeasts. Indian Food Packer 33 (2), 13–14.

Vos, P.J.A., Gray, R.S., 1979. The origin and control of hydrogen sulfide during fermentation of grape must. American Journal of Enology and Viticulture 30, 187–197.

Vos, P.J.A., Zeeman, E., Heymann, H., 1979. The effect of wine quality on diammonium phosphate additions to musts. South African Journal of Enology and Viticulture 1, 87–104.

Vyas, K.K., Joshi, V.K., 1982. Plum wine making: standardization of a methodology. Indian Food Packer 36 (6), 80–86.

Vyas, K.K., Joshi, V.K., 1988. Deacidification activity of *Schizosaccharomyces pombe* in plum musts. Journal of Food Science and Technology 25 (5), 306–307.

Webber, V., Dutra, S.V., Spinelli, F.R., Marcon, A.R., Carnieli, G.J., Vanderlinde, R., 2014. Effect of glutathione addition in sparkling wine. Food Chem 159, 391–398.

Wright, D., 1960. Factors affecting the colour of dry vermouth. American Journal of Enology and Viticulture 11, 30–34.

Yang, H.Y., Weigand, E.H., 1949. Production of fruit wines in the Pacific North West. Fruits Product Journal 29 (8–12), 27–29.

Yurchenko, L.A., 1983. Biochemistry of the apple wines production (Engl. Trans., book in Russian). In: Vecher, A.S. (Ed.), Nauka I Technika. USSR, Minsk.

FRUIT BRANDIES

10

F. López[1], J.J. Rodríguez-Bencomo[1], I. Orriols[2], J.R. Pérez-Correa[3]

[1]Universitat Rovira i Virgili, Tarragona, Spain; [2]Instituto Galego da Calidade Alimentaria, Leiro, Spain; [3]Pontificia Universidad Católica de Chile, Santiago, Chile

1. INTRODUCTION

1.1 BRIEF HISTORY OF DISTILLATION

Distillation was a technique already used by ancient cultures in China (3000 BC), India (2500 BC), Egypt (2000 BC), Greece (1000 years BC), and Rome (200 BC). Initially, all these cultures produced a distilled liquid, later called alcohol by the Arabs, for the preparation of medicines and perfumes. In the 7th century, Arabs began the invasion of Europe, introducing the technique of distillation there. Later, in Christian Europe, the doctor and theologian Arnau de Vilanova (Valencia, 1238?–1311) published the book *Liber Aqua Vitae*, a treatise on wines and spirits, which was a manual of the period for the production of distillates. In about the year 1520, Theophrastus Paracelsus, professor of chemistry at Basle, developed many liqueurs, which he called the *grand arcanurn* and, among others, the famous *elixir proprietatis*. Initially, these spirits and liqueurs were used as medicines; however, later they were consumed to stimulate the appetite and aid digestion (Paracelsus, Theophrastus Philippus Aureolus Bombastus von Hohenheim, 2008).

1.2 ALCOHOLIC BEVERAGES

Alcoholic beverages are products that have been traditionally obtained, depending on the raw materials and the production area. For example, barley and peat are used in Scotland to produce whiskey; wine is used to produce brandy in the Mediterranean zone; and barley, corn, and rye are used to produce gin in Holland and bourbon in America. In addition, Scandinavians use potato, Mexicans use agave, and the Caribbean countries use sugarcane (Small and Couturier, 2011). The name brandy, used alone, generally refers to the grape product; brandies made from wines or fermented mashes of other fruits are commonly identified by the specific fruit name. With the exception of certain fruit types, known as white types, brandies are usually aged. Fruit brandy usually contains 40–45% ABV (80–90 US proof). It is often colorless (Brandy, 2015).

1.3 IMPORTANCE OF DISTILLED BEVERAGES

From an economic point of view, distillates have always had an important contribution in agro-food zones of production, either by the direct use of agricultural products as raw materials to obtain

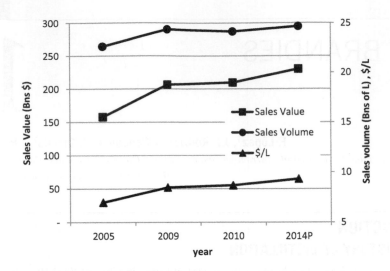

FIGURE 10.1

Global spirit market. Sales volume and sales value during period 2005–2014.

Adapted from Winchester Capital Research, 2015. The Spirits Industry 2015.
http://winchestercapital.com/wp-content/uploads/2014/09/WC-Spirits-Industry-Report_2015-Final.pdf.

distillates, or by profiting from surplus production or residues from the processing of these raw materials. The spirits industry is likely to remain on the steady course of growth shown since 2011 and is poised to capture further market share from wine and beer. Fig. 10.1 shows the increase of sales during the period 2005–2014. The increase of sales volume was 9% while the sales value increased 45%, indicating a higher mean quality of the distillates (Winchester Capital Research, 2015). Leading spirits companies are seeking innovative products and emerging markets for growth. This has the potential to fuel further mergers and acquisitions in 2012. The market will continue to grow for two reasons. First, the spirits-consuming population is growing worldwide as middle classes are expanding in developing nations. Second, it is based upon the expectation that spirits will continue to penetrate the wine and beer markets in developing economies (Winchester Capital Research, 2012). Despite the large variety of spirits, only a few are highly consumed; the greatest economic impact is presented by whiskey, rum, gin, vodka, sake, and brandy (grape origin). Global distilled spirits marketing is also highly concentrated. The 10 largest marketers (by volume) have, since 1991, been responsible consistently for more than half the volume of globalized distilled spirits sold, and the market share of the two largest companies has increased by 65% (Jernigan, 2009).

1.4 CLASSIFICATION OF FRUIT ALCOHOLIC BEVERAGES

The large variety of available fruits and elaboration techniques to produce alcoholic beverages makes their classification extremely complex. In addition, the names of these products are sometimes confusing because a given name refers to different spirits in different countries. For example, the English term cherry brandy can mean either a cherry spirit or more likely a liqueur made by infusing cherries in sweetened brandy. Another confusing name is schnapps, derived from the German Schnapps.

Table 10.1 Legal Restrictions of Main Types of Fruit Spirit Drinks in Europe According to the Definitions Laid Down in Annex II of Regulation (EC) No 110/2008

Spirit Drink Parameter	Fruit Spirits	Macerated Fruit Spirit	Liqueurs	Crèmes
Main characteristic	Can be produced only by the alcoholic fermentation and distillation of fleshy fruit or must of such fruit	Produced by maceration of fruit, whether partially fermented or unfermented	Produced by flavoring, sweetening, and addition of fruits in ethyl alcohol of agricultural origin	Produced as liqueurs with a minimum sugar content of 250 g/L
Alcoholic strength	Distilled at less than 86 vol%	Distilled at less than 86 vol%	–	–
Volatile substances	>200 g/hL pa			
Hydrocyanic acid content (stone fruits)	<7 g/hL pa			
The maximum methanol content	1000 g/hL pa some exceptions			
The minimum alcoholic strength	37.5%	37.5%	15%	15%
Addition alcohol	No	Addition of a maximum of 20 L of ethyl alcohol per 100 kg of fermented fruit	–	–
Flavoring	No	No	Only natural flavoring substances (some exceptions)	Only natural flavoring substances and (some exceptions)
Sugar addition	No	No	>100 g/L exceptions	>250 g/L

In Germany, schnapps is an unsweetened fruit spirit, whereas in the United States, schnapps refers to sweetened spirits flavored with fruit or fruit extracts. These drinks (sometimes known as cordials) are more akin to fruit liqueurs (Buglass et al., 2011b). Nevertheless, these products are normally perfectly defined from a legal point of view through legislation. The spirit drinks in Europe are classified into categories according to the definitions laid down in Annex II of Regulation (EC) No 110/2008. In this Annex, the spirit drinks obtained from fruits are defined from a point of view of the final product; we can distinguish four major groups, and a summary of main legal restrictions are detailed in Table 10.1.

Fruit distillates, normally called brandies, are produced exclusively by the alcoholic fermentation and distillation of fleshy fruit or the must of such fruit, berry, or vegetable, with or without stones and distilled at less than 86% v/v. Macerated fruit spirits are obtained by maceration of authorized fruit or berries, whether partially fermented or unfermented, followed by distillation at less than 86% v/v. Liqueurs, including fruits types, are spirit drinks having minimum sugar content, expressed as invert sugar, of 100 g/L; with a few exceptions. Crèmes are liqueurs, excluding milk products, with a minimum sugar content of 250 g/L. The rules on flavoring substances and preparations are the same of liqueurs. Within the categories of liqueurs and crèmes, the EC regulations also define other fruit sprits drinks, such as Crème de cassis, Guignolet (cherries), Sloe gin, and Maraschino, etc.

1.5 FACTORS AFFECTING THE QUALITY OF THE BRANDIES

Brandies differ widely in quality and composition, depending on the raw material used and the processing procedures. They are especially rich in volatiles in comparison with other types of spirits due to high amount of alcohols and esters.

The quality of distillates is mainly defined by the aroma compounds, which can be classified into four groups: primary aromatic compounds, whose entire aroma appears exactly as in the fruit during ripening; secondary aromatic components, formed during alcoholic fermentation; tertiary aromatic compounds, formed during the distillation process; and quaternary aromatic compounds, formed during the maturation process (Tesevic et al., 2005).

A first condition to obtain quality distillates is that the fruits present adequate characteristics; they must have the correct sugar contents, possess the typical aromas, and be healthy. These characteristics do not have to match the fruits of direct consumption (Tanner and Brunner, 1982; Pischl, 2011). The second is the production process, which consists of four basic stages (Tanner and Brunner, 1982; Claus and Berglund, 2005): fruit preparation (mashing), fermentation, distillation, and storage. The fruit is crushed by mechanical means to extract the juice, but in the case of stone fruits, the cracked stone fruit pits increase the amount of undesirable compounds, such as benzaldehyde (Tanner and Brunner, 1982). The resulting mash is usually fermented in a stirred and cooled fermenter at a controlled temperature. Next the fermented mash is distilled in batch distillation equipment. The clear spirits are stored at high alcoholic strength a minimum of three months prior to alcoholic adjustment with water and bottling.

According to the raw material, three main types of brandies can be distinguished: distillates obtained with pome fruits, of which apples and pears are the most common; those obtained with stone fruits, mainly sweet cherries, sour cherries, plums, apricots, and peaches; and finally the distillates obtained from berries.

While the distillates with pome fruits are those traditionally obtained from pears and apples, new brandies with other fruits are also being developed to valorize surpluses.

2. DISTILLATION SYSTEMS

Distillation is a process of physically separating a mixture into two or more products that have different boiling points, by preferentially boiling the more volatile components out of the mixture. When a liquid mixture of two volatile components is heated, the vapor that comes off will have a higher concentration of the more volatile (i.e., lower boiling point) material than the liquid from which it was evolved. Conversely, if a vapor is cooled, the less volatile (i.e., higher boiling point) material has a tendency to condense in a greater proportion than the more volatile material (Kister, 1992).

2.1 DISCONTINUOUS DISTILLATION

Fig. 10.2 shows the two types of distillation equipment that are commonly used for the production of fruit spirits in batch operations: copper Charentais alembic (French style) and batch distillation columns (German style). Both distillation methods are based on the same theoretical principles, i.e., mass and energy balances, heat and mass transfer, and vapor–liquid equilibrium (Garcia-Llobodanin et al., 2011).

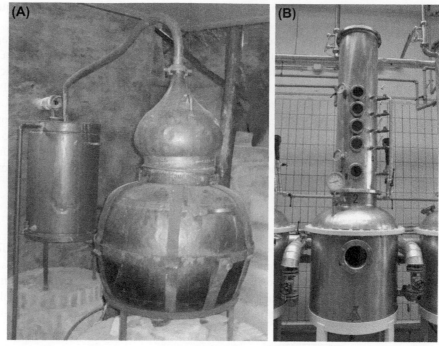

FIGURE 10.2

Typical distillation equipment used for the production of fruit spirits: (A) copper Charentais alembic (French style); (B) batch distillation column (German style).

Both methods differ in their operational principles. Copper pot still distillation presents an almost constant reflux rate, basically determined by the condensation in the swan neck, which in turn is defined by the ambient temperature. In the case of batch column distillation, the reflux rate varies over a wide range. If the column has total and partial condensers, the reflux rate is varied by changing the cooling rate in the partial condenser. If the column has a total condenser only, the flux rate is varied by changing the flow rate of the condensate stream that is returned to the column. In addition, the many equilibrium stages in the rectification column favor the separation, reducing the need for high reflux rates. However, in principle, if a large enough reflux rate can be established in a pot still, the level of separation can be the same as in a distillation column.

The raw materials used to obtain distillates are fermented from agricultural products which acquire a complexity of compounds in the fermentation process. Together, raw material and fermentation compounds will define the aromatic profile of the distillates. In addition, during the distillation, the heat applied in the kettle can produce chemical reactions between the existing compounds, forming other compounds that can increase the complexity of the final distillate. These minority compounds, also known as congeners, may have a different impact on the final product, since there are compounds that are not pleasant, some may even be toxic, and of course, there are compounds that give a positive character. Therefore, by identifying key congeners it will be possible to minimize and remove negative aromas and enhance positive ones. The art of distillation is to obtain distillates with an adequate alcoholic strength and at the same time a pleasant aroma.

In batch distillations, a mixture is heated to its boiling temperature in the boiler, then the vapors pass through the rectification system whether it is a hat in the case of the traditional alembic or a fractionation column. These vapors are then condensed giving rise to the distillate with an alcoholic strength depending on the rectification system. The early fraction of the distillate contains high levels of toxic and unpleasant substances that must be removed. That is why the first fraction collected (head) is rejected. The second fraction collected (heart) contains high levels of the desired aromatic compounds and possesses a high alcoholic strength. When the alcohol content of the distillate begins to decrease (tail fraction), its aromatic quality reduces significantly; the less volatile congeners that have unpleasant tastes are now beginning to distill. The commercial product is elaborated with the heart fraction after adjusting its alcoholic strength.

2.2 CONTINUOUS DISTILLATION

Fig. 10.3 shows the schema of a continuous distillation system. The ferment is fed to the distillation column C from tank F, being previously preheated in the heat exchanger E1 with the vapors obtained (distillate) in the column C, which can be total or partially condensed. Part of the condensate is returned to the column to generate reflux in the zone above the feeding plate and to control the alcoholic degree of the distillate. Next, the distillate passes through the heat exchanger E2, where it is finally cooled with

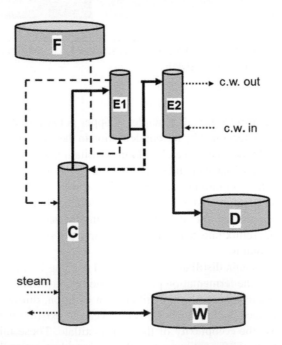

FIGURE 10.3

Schema of a continuous distillation column. *F*, ferment tank; *C*, column; *E1*, preheater ferment heat exchanger; *E2*, distillate condenser; *W*, waste tank; *D*, distillate tank; *C* and *W*, cooling water.

cooling water and sent to distillate tank D. The ferment is fractionated in the column by plates or special packaging, which contacts the falling liquid with the ascending vapor. The vapor is generated in the bottom of the column that is normally heated with steam. The waste (low alcohol content) is extracted from the bottom of the column and sent to tank W. The stability of the system is maintained by controlling the steam inlet, the ferment feed, and the waste removal.

Continuous distillation is faster and achieves higher concentrations of alcohol in the distillate compared with discontinuous distillation (Buglass et al., 2011a). Consequently, the distillate can be produced at lower costs. The main disadvantage of continuous distillation in spirits production is that the distillate is not entirely free of potentially harmful highly volatile components, such as methanol and acetaldehyde, although this situation can be improved by taking the ethanol from just below the top of the rectifier. In this case, most of the acetaldehyde and methanol is taken from the top of the column. Like batch column stills, continuous column stills yield a relatively pure spirit of high ethanolic strength (up to 95.6% ABV, i.e., the strength of the azeotropic mixture) and low in flavor compounds (congeners: acids, higher alcohols, and esters, for example). This is particularly the case with tall-column continuous stills, which are able to supply the highest degree of rectification. However, in the production of many distilled beverages; regulations require the spirit to be of considerably lower ethanolic strength than that of the azeotropic mixture. For example, the EU stipulates that the distillation of fruit spirits must occur at less than 86% ABV, so that the spirit retains some character of the fruit from which it is derived (Regulation (EC) No 110/2008).

3. POME FRUIT BRANDY

The distillates with pome fruits are those traditionally obtained from pears and apples; however, recent studies considered other fruits to valorize surpluses and to develop new brandies.

3.1 PEAR BRANDY

Even though there are many table and juice pear varieties suitable for the production of distillates, two varieties have outstanding distilling qualities: Seckel Sugar pear and Williams or Bartlett pear (Pischl, 2011). Bartlett pear distillates are considered to be the best, because of their pleasant aromas, mainly due to esters (Nikićević, 2005; Buglass et al., 2011b). Willner et al. (2013) have characterized 26 aroma-active compounds in the volatile fraction of Bartlett pear brandy. Sensorial analysis unveiled that ethyl 2-trans, 4-cis decadienoate, and ethyl trans-2-trans-4-decadienoate are key congeners in the overall aroma of Bartlett pear brandies. However, these odorants alone are not able to mimic the overall aroma of a Bartlett pear brandy and, thus, cannot serve as single quality markers. If Bartlett pear spirit is stored in colorless bottles, the 2-trans-4-cis isomers partially isomerize to the 2-cis-4-trans and 2-trans-4-trans isomers, all of which have much less pronounced pear-like odors, so the flavor quality of the spirit decreases. No such isomerization was noted for pear spirit stored in green bottles (Cigic and Zupancic-Kralj, 1999).

The possibility to obtain pear brandies with varieties different from Bartlett has been studied. García-Llobodanin et al. (2007) fermented Blanquilla pear juice concentrate previously diluted to 18° Brix. The pear wine was distilled with and without its lees using three different types of equipment: a glass alembic (a glass pot still coupled to a glass column), a copper alembic, and a glass alembic with the addition of copper shavings to the pot still. The results indicated that methanol, ethyl acetate, and furfural either decreased or showed no change in their concentrations when distilled in the presence of lees and in the copper alembic (see Fig. 10.4). Other compounds (ethyl decanoate and ethyl-2-trans-4-cisdecadienoate) showed increased concentrations in the presence of lees in all equipment tested (see Fig. 10.5).

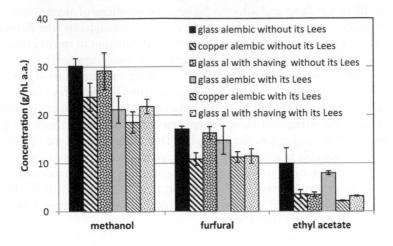

FIGURE 10.4

Effect of the presence of lees and copper in methanol, furfural, and ethyl acetate content for pear distillates.

Adapted from García-Llobodanin, L., Achaerandio, I., Ferrando, M., Güell, C., López, F., 2007. Pear distillates from pear juice concentrate: effect of lees in the aromatic composition. Journal of Agricultural and Food Chemistry 55, 3462–3468.

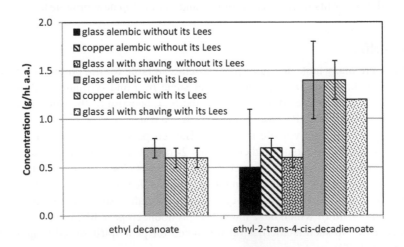

FIGURE 10.5

Effect of the presence of lees and copper in ethyl decanoate and ethyl-2-trans-4-cis-decadienoate content for pear distillates.

Adapted from García-Llobodanin, L., Achaerandio, I., Ferrando, M., Güell, C., López, F., 2007. Pear distillates from pear juice concentrate: effect of lees in the aromatic composition. Journal of Agricultural and Food Chemistry 55, 3462–3468.

It was assumed that the distillation of pear wine in the presence of the lees led to better product quality. García-Llobodanin et al. (2010) studied the pH effect on fermenting Blanquilla diluted pear concentrate. They made two sets of experiments using different fermentation yeasts and different distillation equipment. The results showed, in both experimental sets, that reducing the fermentation pH significantly increased the concentration of most of the higher alcohols and decreased the concentration of ethyl acetate in the spirits. Moreover, pear distillates obtained with the rectification column showed significantly higher concentrations of most of the long-chain ethyl esters (C6–C12) compared to those obtained in the alembic. García-Llobodanin et al. (2011) used Conference pear juice to obtain a pear wine, which was distilled by alembic in double distillation and with a packed column. Pear wine distillations with a packed column produced higher alcoholic degree spirits in just one distillation, compared with two consecutive alembic distillations. In addition, column distillation produced hearts with a lower concentration of toxic compounds such as acetaldehyde and methanol. Furthermore, spirits from column distillations contained significantly more esters and higher alcohols.

Versini et al. (2012) have studied the aroma fraction of Italian distillates of wild (*Pyrus amygdaliformis*, Vill., namely "Pirastru") and cultivated (*Pyrus communis*, L. cvs. "Coscia," "Precoce di Fiorano," and "Butirru de Austu") pear varieties grown in the northern part of the island of Sardinia. They found a wide range of volatile compounds in the distillates, with each varietal product having a specific aromatic profile. Taking into account what has been reported in the literature, the aromatic profiles of these products proved different from the most well-known Bartlett pear distillates: only Coscia distillates are rich in methyl and ethyl unsaturated decanoates, the typical Bartlett pear aroma compounds. Other compounds, such as fatty acid ethyl esters, from hexanoate to decanoate, are also usually at medium-low levels if compared with raw distillates of Bartlett pears. These ethyl esters present differences related to the year of production.

Arrieta-Garay et al. (2013) make a chemical and sensorial comparative examination of pear distillates from the three main varieties grown in Spain (Bartlett, Blanquilla, and Conference) using two distillation systems (copper Charentais alembic and packed column). The Bartlett distillates from both distillation systems possessed higher ethyl ester and acetate and lower cis-3-hexen-1-ol and 1-hexanol concentrations. Despite these differences, a sensory analysis panel could distinguish only the Bartlett alembic distillate from the alembic distillates of the other varieties. In contrast, the panel rated the packed-column distillates equally. Therefore, less aromatic pear varieties can be used to produce distillates with aromatic characteristics similar to those of the Bartlett variety if a suitable distillation process is used.

3.2 APPLE BRANDY

Apple brandy is a spirit obtained by the distillation of cider or apple wine. The apples can be classified as table, commercial, or cider apples, but all are suitable for alcohol production (Pischl, 2011). The sugar content of the average quality fruit is between 8% and 12%. The best known apple brandy, Calvados, is an exception because is an aged product, and it is made by a combination of sweet, tart, and bitter apple varieties that are fermented to produce cider. The cider is doubled-distilled in either pot still, producing a complex brandy that ages well, or a column still, resulting in a fresh brandy with less complex flavors (Small and Couturier, 2011). Calvados is the most important cider spirit bearing the Appelation Controle (AOC) seal, featuring the following varieties: a general AOC Calvados, AOC Calvados du Pays d'Auge, and AOC Calvados Domfrontais; the latter is made from apples and pears, using single column distillation (Burglass et al., 2011). Apple brandies are produced in several other European locations, including northern Spain (especially Asturias), northern Italy (especially Alto Adige and Trentino), Germany, and England, from a wide range of apple varieties—not solely from classic cider varieties that are grown in the major cider manufacturing areas (Vidrih and Hribar, 1999).

The volatile components of freshly distilled Calvados and Cognac have been extensively investigated (Ledauphin et al., 2004), and 331 compounds, of which 162 can be considered as trace compounds, were characterized. Of these, 39 are common to both spirits; 30 are specific to Cognac with numerous hexenyl esters and norisoprenoidic derivatives, whereas 93 are specific to Calvados with compounds such as unsaturated alcohols, phenolic derivatives, and unsaturated aldehydes.

The aroma quality of apple brandy is influenced by the cider maturation (Rodríguez-Madrera et al., 2010). It was found that the most mature cider gave a distillate of superior aroma (with more sweet and spicy character), with higher levels of ethyl acetate, ethyl lactate, and ethyl succinate, and volatiles derived from bacterial metabolism (which is more prevalent in extensively matured cider), such as 2-butanol, 4-ethylguaiacol, eugenol, and 2-propen-1-ol. Moreover, using different yeast species allows the production of spirits with important differences in their aromatic composition, which is certainly interesting from a commercial point of view (Rodríguez-Madrera et al., 2013).

Versini et al. (2009) were able to show that apple distillates made from native Sardinian varieties can be distinguished from those produced from locally grown apple varieties in the traditional cider brandy provinces of Trentino, using similar fermentation and distillation methods. Compounds such as ethyl octanoate, hexyl 2-methylbutanoate, 1-hexanol, benzaldehyde and furfural correlated well with apple varieties (see Fig. 10.6). Ethyl octanoate may be indicative of a more or less favorable yeast activity, while

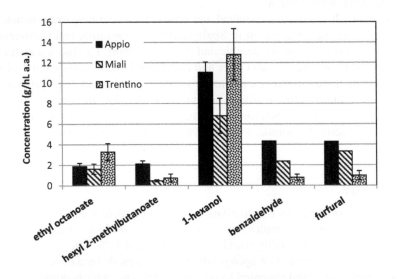

FIGURE 10.6

Comparison of composition of ethyl octanoate, hexyl 2-methylbutanoate, 1-hexanol, benzaldehyde, and furfural for native Sardinian varieties (Appio and Miali) versus Trentino varieties.

Adapted from Versini, G., Franco, M.A., Moser, S., Barchetti, P., Manca G., 2009. Characterisation of apple distillates from native varieties of Sardinia island and comparison with other Italian products. Food Chemistry 113, 1176–1183.

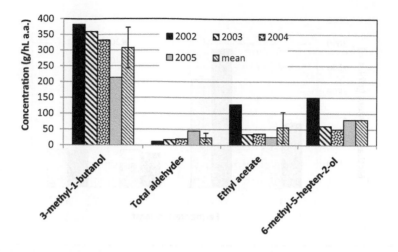

FIGURE 10.7

Effect of year for apple native Sardinian varieties in the volatile composition.

Adapted from Versini, G., Franco, M.A., Moser, S., Barchetti, P., Manca G., 2009. Characterisation of apple distillates from native varieties of Sardinia island and comparison with other Italian products. Food Chemistry 113, 1176–1183.

furfural may be linked to distillation and sugar residues. 3-methyl-1-butanol, total aldehydes, ethyl acetate, and 6-methyl-5-hepten-2-ol are dependent of the year of production (see Fig. 10.7).

Apple pomace (the solid residue generated after juice extraction) is composed of peel, pulp, and seeds, and represents about 25% of the processed apple. Fruit pomaces are usually employed as raw material in the production of spirits following traditional methods, which yields quality products of recognized prestige. Fruit marc spirits, as defined by the European Regulation, are drinks with an alcoholic strength higher than 37.5% (v/v), and a minimum quantity of volatile substances of 200 g/hL pa (EC 110/2008). Due to the seasonality of the raw material, large distillation facilities are required. The use of dry pomace to make apple pomace spirit is a possibility studied by Rodríguez-Madrera et al. (2013). The results of this study show that the treatment with enzymes with pectin methylesterase activity led to excessive levels of distilled methanol, and hence its use is not advisable. In contrast, the indigenous yeasts produced lower concentrations of methanol (see Fig. 10.8). Similar results were obtained by Zhang et al. (2011) in a study with apple mash, juice, and pomace of Crispin apples, which were treated with pectinase. Methanol, ethanol, n-propanol, isobutanol, and isoamyl alcohol were identified as the major alcohols in all the apple spirits.

4. STONE FRUIT BRANDY

Stone fruits such as cherry (Kirschwasser, Cherry, Kirsch), plum (Zwetschgenwasser, Slivovitz), yellow plum, and apricots are used to produce distilled spirits, not only in many regions of Europe, but also in many other parts of the world. Rakia, or rakija, is a brandy, obtained by distillation of fermented fruit, traditionally in Serbia, Slovenia, Croatia, Bosnia, Romania, Czech Republic, Slovakia, Poland, Hungary, and Bulgaria. In the Balkans is called rakia, while in Hungary pálinka; in Czech Republic and

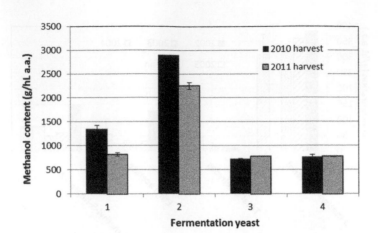

FIGURE 10.8

Methanol content for different fermentation yeasts and enzyme treatment effect. 1, Levuline CHP commercial yeast; 2, Levuline+ β-glucosidase; 3, S.c. 3′ indigenous yeast; 4, H.u. 283 indigenous yeast.

Adapted from Rodríguez-Madrera, R., Pando-Bedriñana, R., García-Hevia, A., Bueno-Arce, M., Suárez-Valles, B., 2013. Production of spirits from dry apple pomace and selected yeasts. Food and Bioproducts Processing 9, 623–631.

Slovakia pálenka. Slivovitz is the most common rakia made from plums. Other stone fruits used to produce rakia are peaches and apricots. The flavor of stone fruit spirits is mostly affected by the aroma compound benzaldehyde, which originates from the enzymatic degradation of amygdalin in the stones of the fruits, passing into the mash during fermentation and later into the distillate at rather high levels. Fermented fruit and beverages frequently contain ethyl carbamate (EC), a potentially carcinogenic compound that can be formed by the reaction of urea with ethanol. It has been regulated in several countries (Lachenmeier, 2005); special care should be taken in the case of stone fruits, since they may contain higher amounts of EC. Some pome fruit brandies, such as quince, may have high EC concentrations as well (Déak et al., 2010). Stone fruit distillates are produced by the yeast *Saccharomyces cerevisiae*, with ethanol as the major product of hexose fermentation and urea as a byproduct in arginine catabolism. In spirit production, EC can also be derived from cyanide found in the stone of the fruit (Schehl et al., 2007). Schehl et al. (2005) studied the influence of the stones on the quality of spirits, cherry, and plum mashes. Although mashes retaining the stones could be clearly distinguished from those where the stones had been removed, no significant preference could be attributed to either spirit, indicating that the aromatic characteristics added by the presence of stones during fermentation are largely a matter of personal taste.

4.1 CHERRY BRANDY

Cherries are classified into sweet cherries (*Prunus avium*) and sour cherries (*Prunus cerasus*). Well-ripened dark-sweet varieties are especially suitable to produce a good brandy. Sour cherries are also a good distilling fruit but not as ideal as sweet cherries (Pischl, 2011). Cherry brandies (kirsch) are produced across the world. Nevertheless, brandies under the name of Kirschwasser are mainly produced in southern Germany, France, and Switzerland by crushing different kinds of sweet cherries, and

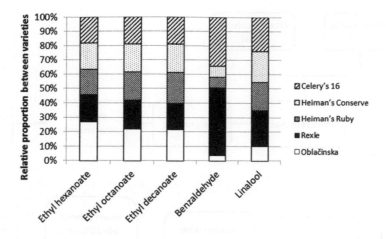

FIGURE 10.9

Relative proportion of volatile compounds in brandies obtained from different cherry cultivars.

Adapted from Nikićević, N., Veličković, M., Jadranin, M., Vučković, I., Novakovi , M., Vujisić, L., Stanković, M., Urošević, I., Tešević, V., 2011. The effects of the cherry variety on the chemical and sensorial characteristics of cherry brandy. Journal of the Serbian Chemical Society 76, 1219–1228.

leaving the mashed mass to ferment for several weeks. The fermented mash is then distilled in copper stills on open fire or vapor, where the head and tails are removed. The resulting distillate has an alcohol content of 60 vol% or more and is sold as a clear and colorless fruit spirit with an alcohol content of 40–50 vol%. Kirschwasser is also used as an additive for different liqueurs (e.g., Curacao, Cherry Brandy, Maraschino, etc.). In Serbia the production of sour cherry brandy has a long tradition. Favorable microclimatic conditions and pedological properties of Serbian soil resulted in Serbia holding fourth place in Europe for the production of this fruit (Nikićević et al., 2011).

Nikićević et al. (2011) studied different varieties of cherries for brandy production, evaluating both their chemical and sensory characteristics. They identified 32 components, including esters, benzaldehyde, terpenes, and acids. Ethyl esters were the most abundant in all samples. All the investigated cultivars yielded brandies of very good to excellent quality. The brandies of two cultivars, Celery's 16 and Rexle, were preferred. These were characterized by high contents of benzaldehyde and linalool (see Fig. 10.9) and significant amounts of aromatic organic acids and esters. In addition, these aromas were present in harmonious proportions.

4.2 PLUM BRANDY

Plum brandies are produced in different areas of Central and Eastern Europe, using different varieties, which are given specific names. In Eastern and Central Europe, plum brandies (slivovitz) matured under appropriate conditions are the most popular fruit brandies prepared from fresh Wegierka plums. This beverage originated in the Balkan Peninsula and the most popular brandy is from Bosnia and Herzegovina (Spaho et al., 2013); it is also quite well-known in Central Europe (Hungary, Poland, Czech Republic, Slovakia, and Romania) and, to a lesser extent, in France (eau de-vie de prunes), Germany (Zwetschgenwasser), and Switzerland (Pflumliwasser; Satora and Tuszynki, 2008). Plum brandy

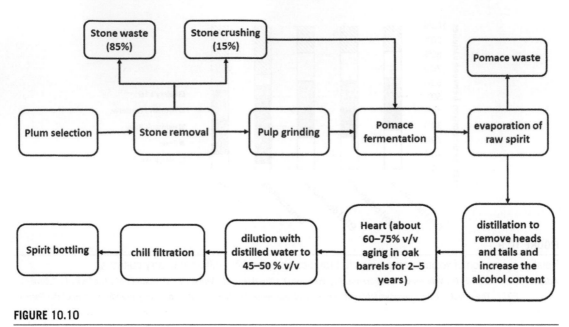

FIGURE 10.10

Traditional plum spirit (Slivovitz) process flow diagram.

is characterized by an intense fruit aroma. They often present a distinctive pungent flavor owing to its artisanal production (Spaho et al., 2013).

Slivovitz production (see Fig. 10.10) includes mill grinding of previously sorted plums (crushing about 15% of the stones), pomace fermentation, evaporation of the raw spirit and, most often, fractional distillation (fractionation) to remove heads and tails and increase the alcohol content. The main fraction of about 60–75% v/v is then subjected to maturation in oak barrels for 2–5 years and, if necessary, diluted with water to reach its consumption strength of 45–50% v/v. The described procedure results in products with various aroma profiles. Apart from its numerous valued components, plum brandy can also contain undesirable ingredients such as hydrocyanic acid (HCN), methanol, and EC (Tesevic et al., 2005). Nevertheless, the fermentation of mashes with selected yeast can reduce the formation of HCN. Balcerek and Szopa (2012), have observed that spontaneous fermentation of fruit pulp resulted in much higher amounts of HCN in the spirits obtained (10.5 mg/L of plum spirit 40%, v/v), in relation to the contents observed in the distillates from the mashes fermented with the addition of *Saccharomyces bayanus* wine yeast (2.80 mg/L of plum spirit 40%, v/v). Satora and Tuszynski (2010) compared spontaneously fermenting plum musts (*Kloeckera apiculata* and *S. cerevisiae*) with musts fermented with different yeasts isolated from fresh blue plum fruits (*Aureobasidium* sp.) and with musts fermented with commercial wine and distillery strains. After spontaneous fermentation, distillates were distinguished by a high content of acetoin, ethyl acetate, and total esters, accompanied by a low level of methanol and fusel alcohols (see Table 10.2). Non-*Saccharomyces* yeasts were responsible for higher concentrations of esters and methanol, while *S. cerevisiae* strains resulted in increased levels of higher alcohols. It was also found that isolated indigenous strains of *S. cerevisiae* synthesized relatively low amounts of higher alcohols compared to commercial cultures. Samples obtained using the distillery

Table 10.2 Chemical Composition of Analyzed Plum Spirit

Sample	Ethanol (% v/v)	SD	Methanol g/hL aa	SD	Total Fusels g/hL aa	SD	Ethylacetate g/hL aa	SD	Total Esters g/hL aa	SD	Acetoin g/hL aa	SD
Distillery, Saccharomyces cerevisiae	69.6	0.6	963.4	29.5	458.5	24.5	46.2	1.1	62.6	0.6	0.0	0.0
Wine, S. cerevisiae	69.5	4.2	760.0	13.5	509.2	38.6	27.7	1.0	39.4	0.8	0.8	0.2
Indigenous, S. cerevisiae	71.0	0.6	893.6	36.7	353.8	5.0	118.3	2.5	138.0	3.0	0.9	0.1
Aureobasidium sp.	69.6	0.0	963.2	67.5	309.7	34.8	85.8	0.9	92.2	1.0	1.0	0.3
Kloeckera apiculata	66.3	0.6	974.4	8.0	250.3	6.1	165.5	3.3	182.9	1.7	0.7	0.2
Spontaneous fermentation	74.3	1.4	755.0	37.8	183.0	13.0	192.5	10.7	247.0	25.6	2.0	0.1
Sig.	**		***		***		***		***		*	

*, **, *** display the significance at 5%, 1%, and 0.1% respectively, by least significant difference.

Adapted from Satora, P., Tuszynski, T., 2010. Influence of indigenous yeasts on the fermentation and volatile profile of plum brandies. Food Microbiology 27, 418–424.

strain of *S. cerevisiae* received the highest score (18.2) during sensory analysis and were characterized by a well-harmonized taste and aroma.

Balcerek et al. (2013) studied the use of intermediate products of plum processing (pulp, concentrate and syrup left after candied fruits processing) as potential raw materials for distillates production. They also studied the supplementation of mashes with sucrose. The resulting distillates presented volatile compounds exceeding 2000 mg/L aa and low content of methanol and HCN. The good taste and aroma make the intermediate products of plum processing very attractive as raw materials for the plum distillates production.

4.3 APRICOT BRANDY

Like many stone fruits, apricots [*Prunus armeniaca* (L.)] are appreciated by consumers all over the world and, consequently, have gained great economic importance. The fruits are presently cultivated in all Mediterranean countries and in South Africa as well as in South and North America, mainly in California. The crop is either marketed fresh, used as canned and dried fruit, or is manufactured into juice (Greger and Schieberle, 2007). Apricot quality consists of a balance of sugar and acidity as well as a strong and characteristic aroma. A significant amount of apricot fruits is processed into brandy, a distillate of apricot fermented must (Puskas et al., 2013).

Genovese et al. (2004) compared aroma compounds in distillates obtained from Pellecchiella apricot (*Prunus armeniaca*, L. cv. Pellecchiella) and Annurca apple (*Malus pumila* L. cv. Annurca). Fifty and 45 volatiles were identified in the apricot and apple distillates, respectively. The aroma volatiles of the apricot distillate were characterized by a high concentration of alcohols and by a specific terpene profile that included linalool, ocimenol, alpha-terpineol, nerol, geraniol, cis-, and translinalool oxide. Gamma-decalactone, gamma-dodecalactone, and ethyl cinnamate were also characteristic of the apricot distillate. Olfactometric analysis showed that volatile compounds, such as beta-damascenone, ethyl 2-methylbutanoate, linalool, methyl anthranilate, ethyl cinnamate, gamma-decalactone, and gamma-dodecalactone, which probably resulted from the original fruit, had a significant odor activity, while 2-phenylethanol was the main odor impact compound.

Urosevic et al. (2014) studied the influence of five commercial yeast strains of *S. cerevisiae* and *S. bayanus* (SB, Top Floral, Top 15, Aroma White, and Red Fruit) and two nutrients, diammonium phosphate and Nutriferm Arom, were examined for their influence on young apricot brandies. Analyses of the major and minor volatiles and sensory analysis of the apricot brandies showed important differences between the samples. The sensory qualities of the assessed apricot brandies indicated that the quality depended on the combination of yeast and nutrients. Nutriferm Arom, as a complex nutrient, gave in all combinations better results than diammonium phosphate, a simple nutrient. The exception was the yeast strain SB that with simple nutrients gave the lowest amounts of some compounds, such as esters and higher alcohols, in the distillate, with better sensory results than the other sample.

The best results were obtained with yeast strain SB with both nutrients and yeast strain Top 15 only with complex nutrients, which gave a high content of linalool. The control sample with no nutrients and selected yeast gave a distillate that was evaluated as having the worst quality with higher concentration of ethyl acetate, 1-butanol, 1-hexanol, and amyl alcohols.

Brandy can be also prepared from wild apricot fruits even with those having high acidity (Joshie et al., 1990; Joshi and Sharma, 2004). The pulp is diluted in a 1:1 ratio with water fortified with a 0.1% 3-Deoxy-D-arabinoheptulosonate 7-phosphate (DAHP) synthase and fermented to completion with total soluble solids of 25° Brix. The wine is then distilled to make brandy. The brandy made from 1:1 diluted pulp was superior to that from 1:2 dilution. The brandy so produced has an intense apricot flavor and high sensory acceptability. The product was treated with slightly roasted oak wood chips during maturation to impart color and flavor. Comparison of the methanol content of brandy made experimentally with that of a locally prepared brandy showed a considerable reduction in the quantity of this alcohol (Joshi and Sadhu, 2000).

4.4 OTHER STONE FRUIT BRANDIES

The interest in using tropical fruits to develop new fermented beverages has grown significantly in the past few years. Reddy and Reddy (2005) studied the viability of fermenting six varieties of mango (*Mangifera indica* L.) produced in India; they also compared the resulting products with fermented grape products. Ethanol concentrations ranged from 7% to 8.5%; the levels of methanol were slightly higher and the concentrations of the other volatile compounds were similar to those of fermented grapes. Alvarenga (2006) studied the effect of enzymatic treatment of mango pulp to increase the extraction efficiency of the juice and to reduce its viscosity in the production of mango spirits. Mango spirits produced under optimum hydrolysis and fermentation conditions showed good quality but presented levels of copper and higher alcohols exceeding those allowed by existing legislation.

Coelho et al. (2011) studied the elaboration of umbu liqueur by maceration of this fruit without stems during 30 days in four types of alcohols: cereals, tubers, grass, and deodorized grass; they tested preferences in relation to flavor, aroma, and purchase intention. After the addition of syrup, a 90-day maturation period was started. The liqueurs were bottled and labeled to perform the sensory analysis through quantitative and acceptance methods. In the analysis, the four kinds of alcohol were found to be suitable for the preparation of umbu liqueur, but the liqueur made with grass alcohol was preferred by the judges in all aspects analyzed, which is an advantage to the small producer because of its low-cost. The liqueur produced with tuber alcohol was less accepted.

Marula is one of the most consumed wild fruits in southern Africa, and since it is drought resistant, it has exceptional yield per tree and both the fruit and the nut are edible. Fundira et al. (2002) studied different enzyme treatments to increase the juice yield and terpenes' recovery in the final distillate. This fruit is also used to produce liqueurs and crèmes.

5. BERRY FRUIT BRANDY

Berries are numerous, both cultivated and wild; the most common are raspberries, blackberries, and blackcurrant. Cranberries and elderberries are rich in tannins and partially poor in nitrogen; hence, their fermentation needs the addition of nutrients (Tanner and Brunner, 1982). Berries are commonly used in the preparation of macerated fruit spirits, liqueurs, and crèmes and are regulated in Europe (Regulation (EC) No 110/2008). These fruits have a significant amount of phenolic compounds and high antioxidant activity (Gorjanović et al., 2010). However, berries are not as widely used for distillation as stone fruits or pome fruits. Cultivated berries possess a relatively low sugar content (4–8%) compared to wild berries (Tanner and Brunner, 1982), resulting in low yields and expensive distillates (Vulic et al., 2012). The diversity of berries allows a wide range of fruit brandies and fruit liqueurs. The production of so-called delicatessen fruit brandies is increasing worldwide. Usual fruits used in these brandies are quince, raspberry, blackberry, cornel berry, currant, and blueberry. If proper technology is applied, they are highly appreciated by the consumer (Vulic et al., 2012).

The suitability of wild fruits to develop new spirits has been explored in several studies. Production of alcoholic beverages from fruits of the forest presents limitations. Sometimes these fruits are not easily fermentable; hence, they should be macerated, leading to liquors with different qualities. In addition, artisanal fermentation using fruit juices, either spontaneous or inoculated, are not reproducible. Solid-state fermentation (SSF) is a good alternative to submerged liquid fermentation (SLF). Even though SLF requires less complicated control systems, SSF yields highly aromatic products, composed of original fruit aromas and those produced during the fermentation process (Alonso-Gonzalez et al., 2010; Santo et al., 2012).

Alonso-Gonzalez et al. (2010) studied the potential of black mulberry and black currant to be used as fermentation substrates for producing alcoholic beverages obtained by distillation of the previously fermented fruits. In the two distillates obtained, the volatile compounds that can pose health hazards are within the limits of acceptability fixed by the European Council (Regulation 110/2008) for fruit spirits. However, the amount of volatile substances in the black currant distillate (121.1 g/hL aa) was lower than the minimum limit (200 g/hL aa) fixed by the aforementioned regulation (see Table 10.3). The mean volatile composition of both distillates was different from other alcoholic beverages, showing the feasibility for obtaining distillates from fermented black mulberry and black currant, which have their own distinctive characteristics. The same authors (Alonso-Gonzalez et al., 2011) obtained two distilled alcoholic beverages from red raspberry and arbutus berry by SSF and subsequent distillation of the fermented fruits. The mean concentrations of ethanol and volatile substances in the distillates from red raspberry and arbutus berry were higher than the corresponding minimum limits fixed by the European Council (Regulation

Table 10.3 Chemical Composition of Analyzed Black Mulberry, Black Currant, Red Raspberry, and Arbutus Berry Spirits

No.	Compound	Black Mulberry Mean	SD	Blackcurrant Mean	SD	Red Raspberry Mean	SD	Arbutus Berry Mean	SD
1	Ethanol (% v/v)	48.3	1.7	38.5	1.2	41.3	1.4	44.3	1.3
2	Methanol	349.6	2.3	167.4	1.3	113.9	1.4	320.5	3.5
3	2-Butanol	nd		0.1	0.0	0.2	0.0	nd	
4	1-Propanol	44.7	0.4	38.2	0.5	36.2	0.2	41.0	0.9
5	1-Butanol	0.9	0.0	0.2	0.0	0.5	0.1	0.8	0.1
6	2-Methyl-1-propanol	39.2	1.3	17.6	0.5	33.4	0.7	36.0	1.9
7	2-Methyl-1-butanol	15.0	2.3	5.6	0.3	11.8	1.2	13.8	0.7
8	3-Methyl-1-butanol	85.9	2.1	33.5	1.2	67.9	1.4	78.7	1.8
9	Allyl alcohol	nd		0.9	0.1	0.0	0.0	nd	
10	1-Hexanol	1.3	0.1	0.1	0.0	1.1	0.0	1.2	0.2
11	Benzyl alcohol	1.5	0.1	3.7	0.2	2.5	0.1	1.4	0.1
12	2-Phenylethanol	nd		0.3	0.0	nd		nd	
13	Ethylacetate	144.7	2.3	7.7	0.4	37.8	0.3	40.7	3.2
14	Ethyl lactate	0.3	0.1	nd		0.3	0.1	0.3	0.1
15	Acetaldehyde	13.9	1.2	12.8	1.5	4.4	0.2	32.7	1.9
16	Acetal	nd		0.3	0.0	4.0	0.3	20.5	1.3
	Total alcohol 3(-8)	185.8	2.7	95.3	1.6	150.0	3.5	170.3	5.3
	Total volatile substances (3-16)	347.6	3.5	121.1	2.2	200.1	4.5	267.1	12.1

nd, *not detected.*

Adapted from Alonso-Gonzalez, E., Torrado-Agrasar, A., Pastrana-Castro, L.M., Orriols-Fernandez, I., Pérez-Guerra, N., 2010. Production and characterization of distilled alcoholic beverages obtained by solid-state fermentation of black mulberry (Morus nigra L.) and black currant (Ribes nigrum L.). Journal of Agricultural and Food Chemistry 58, 2529–2535; Alonso-Gonzalez, E., Torrado-Agrasar, A., Pastrana-Castro, L.M., Orriols-Fernandez, I., Pérez-Guerra, N., 2011. Solid-state fermentation of red raspberry (Rubus ideaus L.) and arbutus berry (Arbutus unedo, L.) and characterization of their distillates. Food Research International 44, 1419–1426.

110/2008) for fruit distillates. In addition, the mean concentrations of methanol in the two alcoholic beverages were much lower than the maximum levels of acceptability that the aforementioned regulation fixed for red raspberry and arbutus berry distillates (see Table 10.3). These results showed that both fruits could be used as fermentation substrates for producing high quality alcoholic beverages.

Aronia berries and chokeberries are excellent fruits to produce strong alcoholic beverages. Balcereck (2010) studied the effect of the quality of raw materials, processing methods, pH of mashes, and yeast strains on the concentration of carbonyl compounds in aronia distillates. Distillates derived from frozen aronia berries contained almost two times more acetaldehydes than those obtained from fresh fruits. An adjustment of pH of aronia mashes from 3.4 to 4.5 resulted in an increase in the concentration of carbonyl compounds (in particular of acetaldehyde). Thermal processing of fruit pulp decreased the concentration of butyraldehyde and increased the concentrations of valeric and isovaleric aldehydes. Spirits produced from mashes fermented by yeast *S. bayanus* contained approximately 43% more acetaldehyde than that obtained from mashes fermented by mixed strains: Burgundy, Bordeaux, and Steinberg.

Galego et al. (2011) studied the preparation of distinct aromatic liquors using deodorized and concentrated fruit distillates. The process was exemplified using fig fruit spirits to prepare myrtle berry liquors. The partial rectification and distillation of used low-quality raw fig spirits lead to an improved final product that kept most of its good aroma properties. This procedure showed high potential to improve other fruit spirits. These improved spirits can be used to prepare new distinctive and highly aromatic liquor drinks. The myrtle berry liquors prepared also showed high levels of polyphenols and anthocyanins.

In the scientific literature, there are some studies about distillates with unusual fruits such as jabuticaba (*Myrciaria jabuticaba*, Berg; Asquieri et al., 2009; Duarte et al., 2011). The quality of the spirits from fermented jabuticaba juice was compared to the parameters established by Brazilian legislation. Asquieri et al. (2009) found that jabuticaba spirits presented values compatible with the parameters established by legislation, with the exception of high ester concentrations (357 mg/100 mL aa). Duarte et al. (2011) concluded that the use of jabuticaba for the production of spirits is a viable alternative usage of this fruit, and showed the potential of jabuticaba spirit as a new product that may be appropriate for a particular niche market.

6. OTHER FRUIT BRANDY

6.1 KIWI BRANDY

Studies on obtaining alcoholic beverages using kiwi fruit are scarce in the literature. Soufleros et al. (2001) evaluated the composition of volatile compounds, organic acids, sugars, and glycerol in a wine of kiwi fruit. They found that kiwi wines had higher concentrations of methanol, lower concentrations of esters, and similar concentrations of higher alcohols than grape wines. Sensidoni et al. (1997) produced kiwi spirits by distilling fermented kiwi juice enriched with rectified grape must and added pectolytic enzymes. In their research, the distillation was performed at two different operating pressures: reduced and atmospheric. The distillates obtained at reduced pressure were aromatically better; however, the characteristic aroma of the kiwi fruit was not detected in either of the distillates. The kiwi aroma is a combination of different volatile compounds such as ethyl butanoate, unsaturated aldehydes, and alcohols of six carbon atoms. The aromatic profile varies with fruit maturity, which increases the fraction of esters. In general, the Hayward variety is aromatically characterized by C6 aldehydes and alcohols, with some esters produced upon ripening (Lopez-Vazquez et al., 2012).

Lopez-Vazquez et al. (2012) study the aromatic composition of kiwi spirits obtained from fermented kiwis of the Hayward variety grown in the southwest of Galicia (Spain); two different strains of *S. cerevisiae* were tested. The spirits obtained were compared with other fruit spirits, in terms of higher alcohols, minor alcohols, monoterpenols, and other minor compounds relevant to quality and taste (see Table 10.4). The yeast strain affected the aromatic profile of the spirits, but the ratio of trans-3-hexen-1-ol and cis-3-hexen-1-ol was unaffected; this ratio in kiwi spirits is different from that found in other fruit distillates. Arrieta-Garay et al. (2014) showed that kiwi spirits obtained with a packed column had the highest concentrations of esters C6–C10 and monoterpenols, while alembic spirits had the highest concentrations of ethyl acetate, methyl acetate, and higher alcohols. Kiwi spirits distilled with a packed column were preferred by consumers. The predominant sensory descriptors in the packed-column kiwi spirits were floral, fruity, and spicy, while burned, smoky, and pungent were the principal aroma descriptors in alembic spirits. Moreover, significantly higher ethanol yields and ethanol strengths were obtained with the packed-column distillation system.

Table 10.4 Content of Main Compounds Present in Kiwi Distillates

	L1	Yeast	L2	Yeast
	Mean	SD	Mean	SD
Ethanol (% v/v)	41.5	0.9	42.3	1.6
Methanol	1236.6	76.4	1137.3	47.4
Ethyl acetate	46.4	8.9	33.0	0.3
Acetaldehyde	153.9	38.0	212.2	41.6
Σ total higher alcohols[a]	317.30	4.37	374.06	28.91
Ethyl lactate[a]	6.48	2.45	1.86	0.67
1-Hexanol[a]	6.27	0.55	4.13	0.22
Σ acetates of higher alcohols	0.05	0.01	0.04	0.02
Σ ethyl esters C6—C12	2.14	1.06	2.04	1.00
Σ ethyl esters C14—C18	0.44	0.42	0.44	0.75
Trans-3-hexen-1-ol	0.03	0.01	0.02	0.01
Cis-3-hexen-1-ol[a]	0.14	0.02	0.08	0.02
Trans-2-hexen-1-ol	0.03	0.01	0.01	0.01
2-Phenylethanol	0.76	0.11	1.09	0.43
Σ minor alcohols[a]	0.46	0.03	0.30	0.40
Σ monoterpenols[a]	1.61	0.20	1.07	0.22

SD, *standard deviation.*
[a]*indicates significant differences (P < .05) between yeast used.*
Adapted from López-Vázquez, C., Garcia-Llobodanin, L., Perez-Correa, J.R., Lopez, F., Blanco, P., Orriols, I., 2012. Aromatic characterization of pot distilled kiwi spirits. Journal of Agricultural and Food Chemistry 60, 2242–2247.

6.2 MELON BRANDY

Hernández-Gómez et al. (2003) used melon fruits (*Cucumis melo*) to develop a new melon spirit drink. They studied the influence of raw material to ferment (total mash, mash without skin and juice) and the distillation system used (column and copper pot). The melon wines were double distilled. The distillation methods resulted in differences, and the distillate obtained with the copper pot distillation was preferred. With regard to the type of substrate, the mash obtained with all the fruit gives a better yield in the process; it does not seem to be the ideal substrate due to the sluggish fermentation, high methanol content, and negative sensory characteristics; the one obtained from the melon juice was preferred. Nevertheless from an industrial standpoint, the distillate obtained from mash without skin substrate can be regarded as preferable, because it produces less waste with a lower environmental impact and it is not necessary to press the paste to obtain it (Hernández-Gómez et al., 2005a). Adjusting the pH brought about substantial decreases in the acetaldehyde and methanol content (see Fig. 10.11), a facet that will have to be taken into account in the case of methanol in view of the maximum limits set by the regulations (Hernández-Gómez et al., 2008). The maceration of the final double-distilled spirit did enhance the color and aroma attributes of the final product (Hernández-Gómez et al., 2005b; Briones et al., 2012).

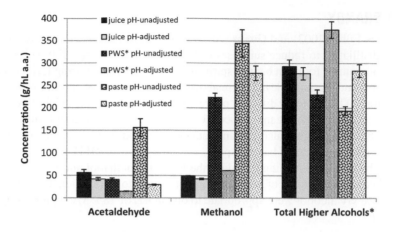

FIGURE 10.11

Effect of pH in acetaldehyde, methanol, and total higher alcohol content for different raw material fermented [juice, paste without skin (PWS), and paste].

Adapted from Hernández-Gómez, L.F., Úbeda, J., Briones, A., 2008. Characterisation of wines and distilled spirits from melon (Cucumis melo L.). International Journal of Food Science and Technology 43, 644–650.

6.3 ORANGE BRANDY

The high productivity of oranges, an appreciated fruit throughout the world, generates important postharvest losses. To reduce waste disposing costs and increase farmers' income, processed fruits can generate several industrial products such as jams, juices, wines, and spirits (Santos et al., 2013). Orange spirits have not been widely studied, except from some citrus-based drinks (Da Porto et al., 2002). The old study of Von Loesecke et al. (1936) described how to obtain orange and grapefruit spirits, as well as brandy, by distilling fermented orange and grapefruit juices. The brandies were subsequently aged in plain oak barrels. They also described how to obtain citrus liqueurs (citrus cordials) by adding citrus oils and sugar syrup to citrus spirits. Da Porto et al. (2002) characterized an orange brandy developed at industrial scale, using a distillation system formed by a boiler, a distillation column, and a rectification column. The system allowed working under reduced pressure, reaching a mean alcoholic strength of 50–60% v/v in the first distillation, and 75–80% v/v in the rectification column. Da Porto et al. (2006) studied the effect of the addition of different sweeteners in the aroma perception of orange spirits. They found that sucrose, glucose, and maple syrup, among the permitted sweeteners, and aspartame, among the unpermitted ones, increased the solubility of some important contributors to a desirable orange flavor. The increased solubility of such volatile compounds and their subsequent reduction in the headspace means that they will be less flavor-active in the aroma of orange spirits. Santos et al. (2013) found that a copper alembic distillate from orange wine had high concentrations of acetaldehyde, ethyl acetate, isoamyl alcohol, and 2-phenylethanol. Nevertheless, they showed that orange juice could be a good substrate for fermentation and distillation, and the sensory analysis performed revealed that the produced beverage had good acceptance by the tasters.

6.4 BANANA BRANDY

Banana (Guimarães-Filho, 2003) and Guava (Alves et al., 2008) have been used in Brazil to produce and characterize distillates, to reduce surpluses of these fruits. Guimarães-Filho (2003) analyzed the volatile organic compounds found in banana spirits that were generated during fermentation and distillation. High alcohols and methanol levels were above the maximum concentrations allowed by Brazilian legislation. Lara (2007) explored different pretreatments of banana pulp and fermentation conditions to decrease the viscosity of the must and reduce the formation of higher alcohols during the production of banana spirits; however, higher alcohols and total acidity continued to be outside the limits set by Brazilian legislation in the obtained banana spirits. Alvarenga et al. (2011) used different *S. cerevisiae* strains to ferment banana pulp, aiming to reduce the levels of methanol and higher alcohols. Teixeira et al. (2005) studied the conditions of banana pulp maceration to obtain liquors. Maceration times of two weeks with 95% v/v alcohol were appropriate to obtain 4 L of 18% v/v liquor, using 500 g of pulp and 350 g/L of sugar.

7. CONCLUSIONS

Alcoholic beverages derived from fruit present a wide research field, combining tradition with innovation, either using fruit not considered before or using traditional fruit in different ways. Better characterization and increased knowledge of so-called exotic fruit has already allowed the generation of aromatically appealing new products, especially liqueurs. In addition, through the incorporation of healthy compounds, the image of the alcoholic beverage products can be improved. Even though most exotic-fruit alcoholic beverages are consumed in the countries they are produced, globalization could boost their international trade, although it would be necessary to investigate how this trade would benefit developing countries.

REFERENCES

Alonso-Gonzalez, E., Torrado-Agrasar, A., Pastrana-Castro, L.M., Orriols-Fernandez, I., Pérez-Guerra, N., 2010. Production and characterization of distilled alcoholic beverages obtained by solid-state fermentation of black mulberry (*Morus nigra* L.) and black currant (*Ribes nigrum* L.). Journal of Agricultural and Food Chemistry 58, 2529–2535.

Alonso-Gonzalez, E., Torrado-Agrasar, A., Pastrana-Castro, L.M., Orriols-Fernandez, I., Pérez-Guerra, N., 2011. Solid-state fermentation of red raspberry (*Rubus ideaus* L.) and arbutus berry (*Arbutus unedo*, L.) and characterization of their distillates. Food Research International 44, 1419–1426.

Alvarenga, R.M., 2006. Efeito do tratamento enzimático da polpa na produção de aguardente de manga (Master thesis). Faculdade de Farmácia. Universidade Federal de Minas Gerais, Belo Horizonte, Brazil.

Alvarenga, R.M., Carrara, A.G., Silva, C.M., Oliveira, E.S., 2011. Potential application of *Saccharomyces cerevisiae* strains for the fermentation of banana pulp. African Journal of Biotechnology 10, 3608–3615.

Alves, J.G.L.F., Tavares, L.S., Andrade, C.J., Pereira, G.G., Duarte, F.C., Carneiro, J.D.S., Cardoso, M.G., December 2008. Desenvolvimento, avaliaçãoqualitativa, rendimento e custo de produção de aguardente de goiaba. Brazilian Journal of Food Technology VII BMCFB 2008, 64–68.

Arrieta-Garay, Y., Garcia-Llobodanin, L., Perez-Correa, J.R., López-Vázquez, C., Orriols, I., Lopez, F., 2013. Aromatically enhanced pear distillates from blanquilla and conference varieties using a packed column. Journal of Agricultural and Food Chemistry 61, 4936–4942.

Arrieta-Garay, Y., López-Vázquez, C., Blanco, P., Perez-Correa, J.R., Orriols, I., Lopez, F., 2014. Kiwi spirits with stronger floral and fruity characters were obtained with a packed column distillation system. Journal of the Institute of Brewing 120, 111–118.

Asquieri, E.R., Silva, A.G.M., Cândido, M.A., 2009. Aguardente de jabuticabaobtida da casca e borra afabricação de fermentado de jabuticaba. Ciência e Tecnologia de Alimentos 29, 896–904.

Balcerek, M., 2010. Carbonylcompounds in aroniaspirits. Polish Journal of Food and Nutrition Sciences 60, 243–249.

Balcerek, M., Pielech-Przybylska, K., Patelski, P., Sapinska, E., Ksiezopolska, M., 2013. The usefulness of intermediate products of plum processing for alcoholic fermentation and chemical composition of the obtained distillates. Journal of Food Science 78, S770–S776.

Balcerek, M., Szopa, J., 2012. Ethanol biosynthesis and hydrocyanic acid liberation during fruit mashes fermentation. Czech Journal of Food Sciences 30, 144–152.

Brandy, 2015. In: Encyclopedia Britannica Retrieved from: http://global.britannica.com/topic/brandy.

Briones, A., Ubeda-Iranzo, J., Hernández-Gómez, L., 2012. Spirits and liqueurs from melon fruits (*Cucumis melo* L.). In: Zereshki, S. (Ed.), Distillation – Advances from Modeling to Applications. InTech Books and Journal, Rijeka, Croatia, pp. 183–196.

Buglass, A.J., McKay, M., Lee, C.G., 2011a. Distilled spirits. In: Buglass, A.J. (Ed.), Handbook of Alcoholic Beverages: Technical, Analytical and Nutritional Aspects. John Wiley & Sons, Ltd., Chichester, UK, pp. 456–468.

Buglass, A.J., McKay, M., Lee, C.G., 2011b. Fruit spirits. In: Buglass, A.J. (Ed.), Handbook of Alcoholic Beverages: Technical, Analytical and Nutritional Aspects. John Wiley & Sons, Ltd., Chichester, UK, pp. 602–614.

Cigic, I.K., Zupancic-Kralj, L., 1999. Changes in odour of bartlett pear brandy influenced by sunlight irradiation. Chemosphere 38, 1299–1303.

Claus, M.J., Berglund, K.A., 2005. Fruit brandy production by batch column distillation with reflux. Journal of Food Process and Engineering 28, 53–67.

Coelho, M.I.S., Albuquerque, L.K.S., Mascarenhas, R.J., Coelho, M.C.S.C., Nunes, I.C., 2011. Elaboração de licores de umbucom diferentes álcoois. Revista Semiárido De Visu 1, 41–46.

Deák, E., Gyepes, A., ÉvaStefanovits-Bányai, E., Dernovics, M., 2010. Determination of ethyl carbamate in pálinka spirits by liquid chromatography–electrospray tandem mass spectrometry after derivatization. Food Research International 43, 2452–2455.

Duarte, W.F., Amorim, J.C., Lago, L.A., Dias, D.R., Schwan, R.F., 2011. Journal of Food Science 76, C782–C790.

Fundira, M., Blom, M., Pretorius, I.S., Van Rensburg, P., 2002. Comparison of commercial enzymes for the processing of marula pulp, wine, and spirits. Journal of Food Science 67, 2346–2351.

Galego, L.R., Da Silva, J.P., Almeida, V.R., Bronze, M.R., Boas, L.V., 2011. Preparation of novel distinct highly aromatic liquors using fruit distillates. International Journal of Food Science and Technology 46, 67–73.

García-Llobodanin, L., Achaerandio, I., Ferrando, M., Güell, C., López, F., 2007. Pear distillates from pear juice concentrate: effect of lees in the aromatic composition. Journal of Agricultural and Food Chemistry 55, 3462–3468.

García-Llobodanin, L., Senn, T., Ferrando, M., Güell, C., López, F., 2010. Influence of the fermentation pH on the final quality of blanquilla pear spirits. International Journal of Food Science and Technology 45, 839–848.

García-Llobodanin, L., Roca, J., López, J.R., Pérez-Correa, J.R., López, F., 2011. The lack of reproducibility of different distillation techniques and its impact on pear spirit composition. International Journal of Food Science and Technology 46, 1956–1963.

Genovese, A., Ugliano, M., Pessina, R., Gambuti, A., Piombino, P., Moio, L., 2004. Comparison of the aroma compounds in apricot (*Prunus armeniaca*, L. cv. Pellecchiella) and apple (*Malus pumila*, L. cv. Annurca) raw distillates. Italian Journal of Food Science 16, 185–196.

Gorjanović, S., Novaković, M.M., Vukosavljević, P.V., Pastor, F.T., Tešević, V.V., Sužnjević, D., 2010. Polarographic assay based on hydrogen peroxide scavenging in determination of antioxidant activity of strong alcohol beverages. Journal of Agricultural and Food Chemistry 58, 8400–8406.

Greger, V., Schieberle, P., 2007. Characterization of the key aroma compounds in apricots (*Prunus armeniaca*) by application of the molecular sensory science concept. Journal of Agricultural and Food Chemistry 55, 5221–5228.

Guimarães-Filho, O., 2003. Avaliação da produção artesanal da aguardente de banana utilizando Saccharomyces-cerevisiae CA-1174 (Doctoral thesis). de Doutorado. Universidade Federal de Lavras, Lavras, Brazil.

Hernández-Gómez, L.F., Úbeda, J., Briones, A., 2003. Melon fruit distillates: comparison of different distillation methods. Food Chemistry 82, 539–543.

Hernández-Gómez, L.F., Úbeda-Iranzo, J., García-Romero, E., Briones-Pérez, A., 2005a. Comparative production of different melon distillates: chemical and sensory analyses. Food Chemistry 90, 115–125.

Hernández-Gómez, L.F., Úbeda-Iranzo, J., Briones-Pérez, A., 2005b. Role of maceration in improving melon spirit. European Food Research and Technology 220, 55–62.

Hernández-Gómez, L.F., Úbeda, J., Briones, A., 2008. Characterisation of wines and distilled spirits from melon (*Cucumis melo* L.). International Journal of Food Science and Technology 43, 644–650.

Jernigan, D.H., 2009. The global alcohol industry: an overview. Addiction 104 (Suppl. 1), 6–12.

Joshi, V.K., Bhutani, V.P., Sharma, R.C., 1990. Effect of dilution and addition of nitrogen source on chemical, mineral and sensory qualities of wild apricot wine. American Journal of Enology and Viticulture 41, 229–231.

Joshi, V.K., Sadhu, D.K., 2000. Quality evaluation of naturally fermented alcoholic beverages, microbiological examinations of source of fermentation and ethanolic productivity of the isolates. Acta Alimentaria 29, 323–334.

Joshi, V.K., Sharma, S., 2004. Importance, nutritive value and medicinal contribution of wines. Beverage and Food World 31, 41–45.

Kister, H.Z., 1992. Distillation Design. McGraw-Hill, New York.

Lachenmeier, D.W., 2005. Rapid screening for ethyl carbamate in stone-fruit spirits using FTIR spectroscopy and chemometrics. Analytical and Bioanalytical Chemistry 382, 1407–1412.

Lara, C.A., 2007. Produção da aguardente de banana: emprego de enzimas pectinolíticas e efeito de fontes de nitrogênio e quantidade de inóculo na formação de álcoois superiores (Master thesis). Faculdade de Farmácia da Universidade Federal de Minas Gerais, Belo Horizonte, Brazil.

Ledauphin, J., Saint-Clair, J.F., Lablanquie, O., Guichard, H., Founier, N., Guichard, E., Barillier, D., 2004. Identification of trace volatile compounds in freshly distilled calvados and cognac using preparative separations coupled with gas chromatography–mass spectrometry. Journal of Agricultural and Food Chemistry 52, 5124–5134.

López-Vázquez, C., Garcia-Llobodanin, L., Perez-Correa, J.R., Lopez, F., Blanco, P., Orriols, I., 2012. Aromatic characterization of pot distilled kiwi spirits. Journal of Agricultural and Food Chemistry 60, 2242–2247.

Von Loesecke, H.W., Mottern, H.H., Pulley, G.N., 1936. Wines, brandies and cordials from citrus fruits. Industrial Engineering and Chemistry 28, 1224–1229.

Nikićević, N., 2005. Effects of some production factors on chemical composition and sensory qualities of Williams pear brandy. Journal of Agricultural Sciences 50, 193–206.

Nikićević, N., Veličković, M., Jadranin, M., Vučković, I., Novaković, M., Vujisić, L., Stanković, M., Urošević, I., Tešević, V., 2011. The effects of the cherry variety on the chemical and sensorial characteristics of cherry brandy. Journal of the Serbian Chemical Society 76, 1219–1228.

Da Porto, C., Bravin, M., Pizzale, L., Conte, L.S., Baseotto, S., 2002. Acquavited'arancia: sperimentazione di una tecnologia di produzione. Industrie delle Bevande 31, 1–4.

Da Porto, C., Cordaro, F., Marcassa, N., 2006. Effects of carbohydrate and noncarbohydrate sweeteners on the orange spirit volatile compounds. LWT – Food Science and Technology 39, 159–165.

"Paracelsus, Theophrastus Philippus Aureolus Bombastus von Hohenheim." Complete Dictionary of Scientific Biography, 2008. Retrieved from: http://www.encyclopedia.com/doc/1G2-2830903284.html.

Pischl, J., 2011. Distilling Fruit Brandy. Schiffer Pub., Cop, Atglen, PA.

Puskas, V., Miljic, U., Vasic, V., Jokic, A., Manovic, M., 2013. Influence of cold stabilisation and chill membrane filtration on volatile compounds of apricot brandy. Food and Bioproducts Processing 91, 348–351.

Reddy, L.V.A., Reddy, O.V.S., 2005. Production and characterization of wine from mango fruit (*Mangifera indica* L.). World Journal of Microbiology and Biotechnology 21, 1345–1350.

Regulation (EC) No 110/2008 of the European Parliament and of the Council of 15 January 2008 on the Definition, Description, Presentation, Labelling and the Protection of Geographical Indications of Spirit Drinks and Repealing Council Regulation (EEC) No 1576/89.

Rodríguez-Madrera, Picinelli-Lobo, A., Mangas-Alonso, J.J., 2010. Effect of cider maturation on the chemical and sensory characteristics of fresh cider spirits. Food Research International 43, 70–78.

Rodríguez-Madrera, R., Pando-Bedriñana, R., García-Hevia, A., Bueno-Arce, M., Suárez-Valles, B., 2013. Production of spirits from dry apple pomace and selected yeasts. Food and Bioproducts Processing 9, 623–631.

Santo, D.E., Galego, L., Gonçalves, T., Quintas, C., 2012. Yeast diversity in the Mediterranean strawberry tree (*Arbutus unedo* L.) fruits' fermentations. Food Research International 47, 45–50.

Santos, C.C.A.A., Duarte, W.F., Carreiro, S.C., Schwan, R.F., 2013. Inoculated fermentation of orange juice (*Citrus sinensis* L.) for production of a citric fruit spirit. Journal of the Institute of Brewing 119, 280–287.

Satora, P., Tuszynki, T., 2008. Chemical characteristics of Sliwowica Łacka and other plum brandies. Journal of the Science of Food and Agriculture 88, 167–174.

Satora, P., Tuszynski, T., 2010. Influence of indigenous yeasts on the fermentation and volatile profile of plum brandies. Food Microbiology 27, 418–424.

Schehl, B., Kachenmeier, D., Senn, T., Heinisch, J.J., 2005. Effect of the stone content on the quality of plum and cherry spirits produced from mash fermentations with commercial and laboratory yeast strains. Journal of Agricultural and Food Chemistry 53, 8230–8238.

Schehl, B., Senn, T., Lachenmeier, D.W., Rodicio, R., Heinisch, J.J., 2007. Contribution of the fermenting yeast strain to ethyl carbamate generation in stone fruit spirits. Applied Microbiology and Biotechnology 74, 843–850.

Sensidoni, A., Da Porto, C., Dalla Rosa, M., Testolin, R., 1997. Utilisation of reject kiwifruit fruit for alcoholic and non-alcoholic beverages. Acta Horticulturae 444, 663–670.

Small, R.W., Couturier, M., 2011. Beverage Basics: Understanding and Appreciating Wine, Beer, and Spirits. Wiley, Hoboken, NJ.

Soufleros, E.H., Pissa, I., Petridis, D., Lygerakis, M., Mermelas, K., Boukouvalas, G., Tsimitakis, E., 2001. Instrumental analysis of volatile and other compounds of Greek kiwi wine; sensory evaluation and optimisation of its composition. Food Chemistry 75, 487–500.

Spaho, N., Dürr, P., Grba, S., Velagić-Habul, E., Blesić, M., 2013. Effects of distillation cut on the distribution of higher alcohols and esters in brandy produced from three plum varieties. Journal of the Institute of Brewing 119, 48–56.

Tanner, H., Brunner, H.R., 1982. La distillation moderne des Fruits. Un guide pour les distillateurs. Heller, Schwä'bish Hall (RFA).

Teixeira, L.J.Q., Ramos, A.M., Chaves, J.B.P., Da Silva, P.H.A., Strongheta, P.C., 2005. Avaliaçao tecnológica da extraçao alcoólica no processamento de licor de banana. Boletim Centro de Pesquisa de Processamento de Alimentos 23, 329–346.

Tesevic, V., Nikicevic, N., Jovanovic, A., Djokovic, D., Vujisic, L., Vuckovic, I., Bonic, M., 2005. Volatile components of plum brandies. Food Technology and Biotechnology 43, 367–372.

Urosevic, I., Nikisevic, N., Stankovic, L., Andelkovic, B., Urosevic, T., Krstic, G., Tesevi, V., 2014. Influence of yeast and nutrients on the quality of apricot brandy. Journal of the Serbian Chemical Society 79, 1223–1234.

Versini, G., Franco, M.A., Moser, S., Barchetti, P., Manca, G., 2009. Characterisation of apple distillates from native varieties of Sardinia island and comparison with other Italian products. Food Chemistry 113, 1176–1183.

Versini, G., Franco, M.A., Moser, S., Manca, G., 2012. Characterisation of pear distillates from wild and cultivated varieties in Sardinia. International Journal of Food Science and Technology 47, 2519–2531.

Vidrih, R., Hribar, J., 1999. Synthesis of higher alcohols during cider processing. Food Chemistry 67, 287–294.

Vulić, T., Nikićević, N., Stanković, L., Veličković, M., Todosijević, M., Popović, B., Urošević, I., Stanković, M., Beraha, I., Tešević, V., 2012. Chemical and sensorial characteristics of fruit spirits produced from different black currant (*Ribes nigrum* L.) and red currant (*Ribes rubrum* L.) cultivars. Macedonian Journal of Chemistry and Chemical Engineering 31, 217–227.

Willner, B., Granvogl, M., Schieberle, P., 2013. Characterization of the key aroma compounds in bartlett pear brandies by means of the sensomics concept. Journal of Agricultural and Food Chemistry 61, 9583–9593.

Winchester Capital Research, 2012. Spirits Industry M&A Update: 2012. http://win-marine.com/uploads/WC%20 2012%20Spirits%20Industry%20Update.pdf.

Winchester Capital Research, 2015. The Spirits Industry 2015. http://winchestercapital.com/wp-content/uploads/ 2014/09/WC-Spirits-Industry-Report_2015-Final.pdf.

Zhang, H., Woodams, E.E., Hang, Y.D., 2011. Influence of pectinase treatment on fruit spirits from apple mash, juice and pomace. Process Biochemistry 46, 1909–1913.

WASTE FROM FRUIT WINE PRODUCTION

11

M.R. Kosseva
University of Nottingham Ningbo Campus, Ningbo, China

1. INTRODUCTION

The manufacture of the fruit wines can generate a vast amount of by-products, which comprise of various categories: waste from processing of fruit varieties, fruit- and wine-processing wastewaters, lees (fermentation sediment rich in yeast biomass), and solid waste like pomace, peels, seeds, and stems. Examples of such waste include apple, banana, berries, citrus, pineapple, and pear, among other residues remaining after industrial processing. Over 115 million tons of citrus fruit are produced annually, and about 30 million tons are processed industrially for juice production. After industrial processing, citrus peel waste accounts for almost 50% of the wet fruit mass. The annual world production of bananas, apples, and pears is approximately 107.1, 75.5, and 24.0 million tons, respectively, and 25–40% of this mass remains as waste after processing (Fig. 11.1) (FAOSTAT, 2013).

Fruit- and wine-processing residues are regarded as being food industry waste, so the definition of food waste is applied to these by-products.

1.1 DEFINING FOOD AND FRUIT WASTE

Kosseva (2013a) describes a number of different definitions of food waste with respect to the complexities of the food supply chains. In this chapter we use the waste definition proposed by the Waste and Resources Action Programme (WRAP) [food or drink products that are disposed of (includes all waste disposal and treatment methods) by manufacturers, packers/fillers, distributors, retailers, and consumers as a result of being damaged, reaching their end of life, being offcuts, or being deformed (outgraded) (WRAP, 2010)].

The World Resources Institute (WRI, 2015) refers to food waste as "food loss and waste or food, as well as associated inedible parts, removed from the food supply chain."

Because fruit waste is rich in sugars, organic acids, phenolic compounds, and other components, these forms of disposal may cause environmental pollution and loss of valuable components. Disposal of waste is also becoming gradually more expensive. For example, European Union (EU) landfill directives have caused landfill gate fees to increase in some cases because of land limitations and transport and labor costs (Lin et al., 2013). In America, the annual cost of apple pomace disposal alone reached US$10 million (Shalini and Gupta, 2010). Legislation has been used around the world to prevent, reduce, and manage waste (e.g., promoting recycling and energy recovery). Diverting waste from landfills is an important element in EU policy for improving the use of resources and reducing the environmental impacts of waste management, in particular, in pursuance of Directive 1999/31/EC on landfill

Science and Technology of Fruit Wine Production. http://dx.doi.org/10.1016/B978-0-12-800850-8.00011-9

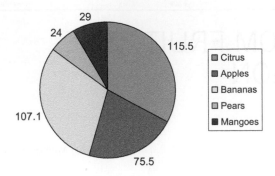

FIGURE 11.1

World annual of basic fruits production in millions of metric tons.

  FAOSTAT, 2013. Food and Agriculture Organization of the United Nations. Accessed December 2015 at: http://faostat.fao.org.

waste (hereinafter referred to as the Landfill Directive). According to this directive, member states must reduce the amount of biodegradable municipal waste going to landfill:

- to 35% of 1995 levels by 2016;
- to a 90% reduction in food waste going to landfills by 2020, in comparison with that produced in 2006 [based on Landfill Directive (65%), European Waste Framework Directive (15%), and future biowaste legislation following the EC communication on future steps in biowaste management in the EU (10%)].

The EU Landfill Directive is one of the stakeholder key drivers for waste minimization, management, and coproduct recovery in food processing (Waldron, 2007).

Increasing cost of waste and resultant legislation provide a strong incentive for its processing and appropriate re-use. Determining biowaste constituents is a first step to their valorization. Fruits are rich in antioxidants, pectin, fibers, carbohydrates, organic acids, mineral salts, food flavors, colorants, and so on. These compounds can be utilized as raw materials for secondary processes or as ingredients for new products. One example is production of ethanol from the fruit waste matrices, which are rich in reducing sugars. Another potential opportunity is to employ some of the soft fruit wastes as substrates for the fruit wine fermentation processes, converting food industry by-products to beneficial resources.

In general, fruit wine waste (FWW) is characterized by the heterogeneity of its sources, variations in its chemical composition, and microbial instability, which are not easy to control. Hazardous wastes may be occasionally generated by the application of pesticides, as well as contamination by toxic chemicals or pathogens. Fruit- and wine-processing wastewater is mainly characterized by a high degree of organic pollution, sometimes containing polyphenols and aromatics forming inhibitors, which constitute a serious environmental problem for soil, rivers, and groundwater. The great variety of components found in liquid and solid waste requires development of appropriate technologies to eliminate those that have harmful effects on the environment. In practical terms, it is essential to develop cost-effective usage for the final waste products.

Fruit- and wine-waste streams have a potential, which has been underestimated so far. Here we discuss the development of sustainable technologies from the organic waste steams, raising the potential to minimize waste quantities, and creating new opportunities for the market. We aim to characterize the wastes generated by the production of fruit wine, separating them into two main categories of liquid and solid wastes. The liquid stream consists mainly of the wastewater generated by the wine and distillery industries. It is characterized by high organic load, or high contents of chemical oxygen demand (COD) and biochemical oxygen demand (BOD). The difficulty in dealing with this fermentation wastewater is in the flows and loads of the waste (Kosseva, 2013c). It was calculated that a winery generates between 1.3 and 1.5 kg of residues per liter of grape wine made, 75% of which is winery wastewater (Lucas et al., 2010). A concise overview of treatment processes applied to winery wastewater is given with a focus on the usage of biological methods. The evaluation of organic solid waste generated from fruit wine production, the amount of unavoidable fruit waste quantities from food processing, and their characterization are considered. The principal methods for valorization of fruit wastes are emphasized with a focus on the elaboration of new products and materials as well as their applications.

There are various aspects that could influence the types of compounds present in FWW, the main being the type of wine (respectively, the fruit used), including the geographical location of the origin, the harvest period, and the characteristics of the products and their handling.

Finally, sustainability in the wine sector has been characterized focusing on the orchard and winery aspects. The life cycle assessment (LCA) method, based on the International Standards ISO 14040 series, is applied to verify the proposed production routes. LCA is also used to assess the environmental impact of fruit production as well as wine packaging.

2. UNAVOIDABLE SOLID FOOD AND FRUIT WASTE

"Unavoidable food waste" is classified as the waste arising from food and drink preparation that is not, and has not been, edible under normal circumstances (e.g., pineapple skin) (WRAP, 2010). Wastes from fruit processing, as a part of the entire food waste processing sector, such as pomace (skin, seed, and pulp), peels, seeds, and stems can be considered unavoidable waste. During processing, these plant parts are usually removed by peeling or are retained in the press residues (e.g., skins and seeds in fruit pomace). Depending on the raw material and the technologies applied, they emerge in large quantities and are often a considerable disposal problem for the food industry. For example, during cider and apple juice production, approximately 20% and 25%, respectively, of the raw material remains as pomace. Even higher proportions of by-products emerge from processing of some exotic fruits such as mangoes, for which the peels and seeds may amount to up to 60% of the total fruit weight.

Nonetheless, the by-products contain large amounts of secondary plant metabolites in concentrated form and represent promising sources of bioactive compounds, which can be used as components in functional food. Secondary plant metabolites such as polyphenols play an important role in the defense system of the plant, protecting it from biotic and abiotic stress. For instance, flavonoids act as UV-absorbing compounds and signal molecules. Phenolic compounds also show antimicrobial activity against plant pathogens. Because of their biological role in plants, secondary metabolites are located primarily in the outer layers of fruits and in the seeds (Schieber, 2009). Furthermore, a series of studies has proven that the extracts from tropical fruit waste components have higher polyphenol contents than the main soft tissue, thereby making them a potentially more prospective therapeutic source for further studies (Asyifah et al., 2014; Daud et al., 2010; Taing et al., 2012; Wilkinson et al., 2011). Among the bioactive phytochemical compounds discovered to be present as unavoidable waste components in the inedible part of the tropical fruits are quercetin, hydroxytyrosol, and oleuropein. The high amounts of phytochemicals present in these components are believed to be responsible for the antiobesity effect (Asyifah et al., 2014).

2.1 APPLE POMACE

The world production of apples is about 75.5 million tons per annum (FAOSTAT, 2013). About 20% of the apples are used to derive value-added products, of which 65% are processed into apple juice concentrate and the balance into other products that include packed natural ready-to-serve apple juice, apple cider, wine and vermouth, apple purees and jams, and dried apple products (Joshi et al., 1991; Joshi, 1997). Apple pomace is the main by-product of the apple cider- and juice-processing industries and accounts for about 25% of the original fruit mass at 85% (wb) moisture content (Sun et al., 2007). It demonstrates the highest values of BOD and COD and acidic pH 5.9 compared to other fruits like cherries and grapefruits: apple's BOD = 9.6 g O_2/L, COD = 18.7 g O_2/L, suspended solids (SS) = 0.45 g/L. The chemical composition of this pomace is illustrated in Table 11.1 (Vendruscolo et al., 2008). It is rich in pectin, crude fibers, proteins, vitamins, and minerals. Major components are simple sugars like glucose, fructose, and arabinose, whereas minor components are sucrose, galactose, and xylose. Therefore, several microorganisms can use apple residues as a substrate for growth (Kosseva, 2011).

Shalini and Gupta (2010) evaluated various applications of this residue, such as fuel production, food products, pectin extraction, cattle feed, biotransformation, source of fibers, and others. The study proposed manufacturing of a variety of products, which appears economically more feasible compared to the production of a single product, because of the large quantity of pomace (Mirabella et al., 2014). Various biotechnological applications of apple pomace, including relevant microorganisms and the fermentation processes applied, were also reviewed by Vendruscolo et al. (2008) and Kosseva (2013c).

Table 11.1 Chemical Composition of Apple Pomace

Composition	Albuquerque (2003)	Jin et al. (2002)	Joshi and Shandu (1996)	Villas-Boas and Esposito (2000)
Moisture (%)	79.2	5.8	3.97	80
Protein (%)	3.7	4.7	5.80	4.1
Lipids (%)	n.d.	4.2	3.90	n.d.
Fibers (%)	38.2	n.d.	14.70	40.3
Ash (%)	3.5	1.5	1.82	2.0
Carbohydrates (%)	59.8	83.8	48.00	n.d.
Reducing sugars (%)	10.8	n.d.	n.d.	15
Pectin (%)	7.7	n.d.	n.d.	5.5
pH	4.0	n.d.	4.20	n.d.
Titratable acidity (%)	0.13	n.d.	2.60	n.d.
Water activity	0.973	n.d.	n.d.	n.d.

Adapted from Vendruscolo, F., Albuquerque, P.M., Streit, F., Esposito, E., Ninow, J.L., 2008. Apple pomace: a versatile substrate for biotechnological applications. Critical Reviews in Biotechnology 28, 1–12.

2.2 MANGO PEELS

Mango fruit is cultivated in more than 100 countries of both tropical and subtropical regions. Mango production is highest in India, at 41% of the world's production, followed by China, Thailand, Mexico, Pakistan, Indonesia, the Philippines, Nigeria, and Brazil (Kim et al., 2009). The United Nations Food and Agriculture Organization (FAO) estimates that the mango harvest will be around 28.8 million tons in 2014, that is, 35% of the production of the world's tropical fruit. Sixty-nine percent of the total amount will be produced in Asia and the Pacific (India, China, Pakistan, Philippines, and Thailand), 14% in Latin America and the Caribbean (Brazil and Mexico), and 9% in Africa. Regarding mango production by developed countries (United States, Israel, and South Africa) it is estimated at 158,000 tons.

Depending on the cultivars and products made, the industrial by-products of mangoes, namely peels and seeds, represent 35–60% of the total weight of the fruit (Jahurul et al., 2015). The peel constitutes approximately 7–24% of the total weight of a mango fruit (Kim et al., 2009). Mango peels have become an attractive area for research because of their high content of valuable compounds, such as phytochemicals, polyphenols, carotenoids, enzymes, vitamin E, and vitamin C, which have predominant functional and antioxidant properties (Ajila et al., 2007). Moreover, Sogi et al. (2013) reported mango peels as a rich source of dietary fiber, cellulose, hemicellulose, lipids, proteins, enzymes, and pectin. Mango peels can be utilized for the production of valuable ingredients (i.e., dietary fiber and polyphenols) for various food applications, as has been reported by many researchers (Ajila et al., 2007, 2010; Aziz et al., 2012). Currently, mango peel flour is used as a functional ingredient in many food products, such as noodles, bread, sponge cakes, biscuits, and other bakery products (Aziz et al., 2012). Mango peels also contain fats (2.16–2.66%) (Ajila et al., 2007).

A number of studies on various tropical fruit by-products and their biological activities have highlighted their potential benefits. They can be used as supplements for the treatment of obesity, along with its related diseases such as diabetes (Asyifah et al., 2014). Taing et al. (2012) demonstrated experimentally the effect of mango peel extract on adipogenesis in 3T3-L1 cells. The authors also proved that the differences in phytochemical quantity and composition between species of the same fruit can result in different adipogenesis inhibition efficiencies.

An important finding concerning the storage of fruit wastes and variations in their bioactivity is correlated to the mango's heat tolerance. It fluctuates because of a number of factors including origin, species, fruit maturity, shape, size, and weight (Jacobi et al., 2001). The study of Kim et al. (2009) was focused on polyphenolic and antioxidant changes to mature, green mangoes due to different durations of hot water treatment (HWT) and their changes during short-term storage. Experimentally, fruit were immersed in 46.1°C water for from 70 to 110 min; half were evaluated within 2 h of treatment, and the remainder was evaluated after 4 days of storage at 25°C for changes in polyphenolics, antioxidant capacity, and fruit quality. Two major polyphenolics in mango, gallic acid and gallotannins, as well as total soluble phenolics, decreased as a result of prolonged HWT, whereas the antioxidant capacity remained unchanged in all heat-treated mangoes immediately after HWT.

2.2.1 Mango Seeds

Mango has a single large seed covered with a shell that has the kernel inside, which represents around 20% of the whole fruit (Solís-Fuentes and Durán-de-Bazúa, 2011). Mango seed kernel has several medicinal uses in various parts of the world (Raihana Noor, 2015). Fresh mango kernel is used in Fiji against dysentery and asthma, whereas the juice is applied as nasal drops to remove suffering from sinus problems. Dried seed powder is used to remove dandruff, and the kernel starch is eaten as a famine food in India. A hot water extract of the kernel is also recommended as an anthelmintic, aphrodisiac, and laxative tonic (Nithitanakool et al., 2013). Depending on the variety, mango seed kernel (on a dry weight basis) contains an average of 6.0% protein, 11% fat, 77% carbohydrate, 2.0% crude fiber, and 2.0% ash (Zein et al., 2005). Kittiphoom (2012) reported that mango seed kernel was high in potassium, magnesium, phosphorus, calcium, and sodium, so it could be further processed into functional food products. Being nutritious and nontoxic, it could directly substitute for any solid fat without adverse effects on the quality. The lipid composition of various mango kernel varieties has been a subject of many studies because of the potential application as a substitute for cocoa butter in the confectionery industry (Jahurul et al., 2014). The neutral lipids of different mango seed kernels vary from 95.2% to 96.2%, phospholipids from 2.7% to 3.3%, and glycolipids from 1.1% to 1.4% (Rashwan, 1990). Triglycerides constitute the major fraction of the neutral lipids and account for about 93.7–96.4%. The lipids of mango seed kernel consist of about 44–48% saturated fatty acids (majority stearic) and 52–56% unsaturated (majority oleic) (Raihana Noor, 2015). This review concluded that further research is needed to include not only the recovery of valuable compounds from mango waste, but also specific applications to ensure industrial exploitation and sustainability of the final product. Additionally, the sensorial and nutritional aspects of new food products containing mango seed fat from by-products have to be explored (Jahurul et al., 2015).

2.2.2 Mango Leaves

As a tropical fruit waste component, mango leaves have proven to be beneficial in alleviating diabetes and inhibiting triglyceride accumulation in 3T3-L1 cells. It is a prospective candidate for recycling and potential usage as therapeutics for related diseases, namely, diabetes and obesity. In a study conducted by Kumar et al. (2013) mangiferin, a bioactive phytochemical component, was extracted and isolated from stem bark of *Mangifera indica* and subjected to a glucose utilization test with 3T3-L1 cells. The results showed that treatment of the 3T3-L1 cells with mangiferin increased glucose utilization in a dose-dependent manner (Kumar et al., 2013).

2.3 CITRUS PEELS

The family of citrus fruits includes oranges, grapefruits, lemons, limes, mandarins, pomelos, and others, as the most plentiful fruits in the world. Although dry citrus peels are rich in pectin, cellulose, and hemicellulose, the disposal of the fresh peels is becoming a major problem for many manufacturers. Residues of citrus juice production contain several compounds, essentially, soluble sugars in water, fibers, organic acids, amino acids and proteins, minerals, oils and lipids, as well as flavonoids and vitamins (Mirabella et al., 2014). Because of their high water content (~80%), the residues are prone to microbial spoilage and need to be dried immediately after processing, which is an economically limiting factor. Various microbial transformations have been proposed for the use of this waste to produce valuable products, like biogas, ethanol, citric acid, chemicals, various enzymes (e.g., pectinase, phytase, xylanase), volatile flavoring compounds, fatty acids, and microbial biomass (Dhillon et al., 2010). In the food industry, two types of citrus by-products (lemon albedo and orange dietary fiber powder) at different concentrations were added to cooked and dry-cured sausages to increase their dietary fiber content (Fernández-López et al., 2004). Crizel de Moraes et al. (2013) characterized orange juice fiber by-products and proved the possibility of their application as a fat replacer in ice cream.

Citrus fruit waste can also serve as cattle feed, especially for ruminants. The significant by-products that can be used are fresh or dried pulp, citrus silage, citrus meal and fines, citrus molasses, citrus peel liquor, and citrus activated sludge. Bampidis and Robinson (2006) characterized the physical and nutrient composition, digestibility, fermentation, and effects on ruminants (weight and lactating production) of these feeds. The authors stated that these wastes could be effectively used as feedstuff in provisions that support growth and lactation in ruminants (Mirabella et al., 2014).

Citrus solid wastes are rich in fermentable soluble sugars such as glucose, fructose, and sucrose, along with structural cellulose and hemicellulose. The contents of fermentable sugars in the various citrus wastes range from 23.2% to 57.1%. Orange peel, mandarin peel, and grapefruit peel contain 53.2%, 57.1%, and 43.2%, respectively. Lemon peel and lime peel show moderate fermentable sugar levels of 31.2% and 23.2%, respectively (Choi et al., 2015). Therefore, they may be an excellent substrate for ethanol production, for example, from orange peel an ethanol yield coefficient of ~0.495 g/g and productivity of 4.85 g/L/h were achieved by Santi et al. (2014). From mandarin peel an ethanol concentration of 46.2 g/L and productivity of 3.85 g/L/h was obtained by Choi et al. (2013). Ethanol production in excess of 60 L per 1000 kg of lemon peel waste was gained both at the laboratory scale and in a 5-L bioreactor by Boluda-Aguilar and López-Gómez (2013).

Choi et al. (2015) developed a novel efficient technology for ethanol production: a D-limonene removal column (LRC) containing raw cotton and activated carbon, which acted as an adsorbent, successfully removed this inhibitor from the citrus waste. When the LRC was coupled with an immobilized cell reactor (ICR), yeast fermentation resulted in ethanol concentrations (14.4–29.5 g/L) and yields (90.2–93.1%) that were 12-fold greater than products from ICR fermentation alone. Traditionally, citrus peels are hydrolyzed by high-cost commercial enzymes, including pectinase, cellulase, and β-glucosidase, prior to the fermentation process taking place.

Another citrus waste, pomelo peel, is an attractive substrate for the production of polygalacturonase—a widely used pectinase in fruit juice/wine clarification and yield improvement. The pomelo peel is reported to contain 16.9% soluble sugars, 3.75% fibers (9.21% cellulose, 10.5% hemicelluloses, 0.84% lignin, and 42.5% pectin), 3.5% ash, 1.95% fat, and 6.5% protein (Cheong et al., 2011). Darah et al. (2013) proved the feasibility of producing polygalacturonase with an *Aspergillus niger* LFP-1 local strain (Universiti Sains Malaysia, Penang) from pomelo peel as a substrate in solid-state fermentation. The optimum conditions for production of polygalacturonase were found to be 5.0 g of substrate, a 107 spores/mL^3 solution, 30°C, and 5 days of cultivation under static conditions with addition of water at 1:1 (w/v) and 1.2% (w/w) ammonium nitrate.

As to the bioactive compounds contained in the same by-products, Ding et al. (2013) made an observation that the differences in phytochemical quantity and composition between species of the same fruit can produce different adipogenesis inhibition efficiencies in pomelo peels. Using an in vivo experimental model of obese C57BL/6 mice induced with a high-fat diet, they found that the peel crude extract was able to block body weight gain and lower the blood glucose level, serum total cholesterol, liver lipid levels, and serum insulin levels as well as improving glucose tolerance and insulin resistance.

2.4 BERRY PEELS, PULP, AND SEEDS

Global annual production of berries, including raspberry, blueberry, and cranberry, was about 1.5 million tons in 2013. Large amounts of solid wastes from berry processing usually originate from their pretreatment, washing and sorting, and they consist of damaged fruits, stems, and stalks. A major source of solid waste generation is the pressing process, in which peels, seeds, and pulps are separated from the fruit juice (Pap et al., 2004). Berries contain various bioactive components, such as phenolic phytochemicals (flavonoids, phenolic acids, polyphenols) and fibers (Häkkinen et al., 1999). Pap et al. (2004) used supercritical fluid extraction with natural CO_2 as an eco-friendly technology for the recovery of valuable compounds from berry waste.

Cranberry waste contains various phenolic compounds and anthocyanins. Woolford et al. (2013) provided methods of producing proanthocyanidin (PAC)-containing solutions, powders, and beverages from cranberry plant material. They proposed additional processing steps often needed to reduce the amount of sugars and/or acids present in extracts of cranberries, when it is desirable to obtain an extract or product that is relatively low in sugar and/or acids. PACs have antioxidant activity and interfere with adhesion of bacterial cells to epithelial cells. PACs are also thought to impart various health benefits. A number of processes are useful for isolating PACs from cranberries and other fruits. Additional methods for the selective capture and dry weight concentration of PACs can provide additional opportunities to prepare beverages and foods expected to have health benefits, as proposed in the aforementioned patent. The method includes providing a mixture of cranberry plant material (e.g., cranberry leaves and/or stems) and an aqueous medium, wherein the cranberry plant material is provided in the mixture in a range of about 1–12%, for example, 5–10% (on a w/w basis), and steeping the plant material in the aqueous medium with or without enzymes, or an organic solution (such as ethanol or propylene glycol solution) under conditions to extract the PACs. This creates a PAC-containing solution, and optionally the steeped cranberry plant material is separated from the PAC-containing solution. Combining the PAC-containing solution with one or more additional components, for example, water, a juice, a sweetener, a natural or artificial favor, or a tisane, can produce a healthy beverage.

Górecka et al. (2010) added raspberry pomace in a dried form as replacement for flour in cookies in the range from 25% to 50%. The use of raspberry pomace in cookies resulted in an increase in fiber contents, which did not affect the organoleptic properties of the product and was well accepted by consumers.

Bakowska-Barczak et al. (2009) evaluated the seeds from five black currant cultivars grown in western Canada for their oil content, fatty acid and triacylglycerol composition, and tocopherol and phytosterol profiles and contents. They also determined polyphenolic compounds and antioxidant activity in the seed extracts remaining after oil extraction. Canadian black currant seed oil was proven as a good source of essential fatty acids, tocopherols, and phytosterols.

The isolation and effective separation process of various chemicals from fruit waste components is essential for the effective development of this field. The presence of more advanced analytical devices such as prep–HPLC, supercritical fluid extraction technique, and liquid chromatography/mass spectrometry, for instance, will definitely facilitate this research. The subsequent steps would therefore require further elucidation of the chemical structures of the novel bioactive phytochemicals, whereas for the already identified bioactive chemicals, further investigation is required so that absolute confirmation of their respective bioactive capacity can be achieved (Asyifah et al., 2014).

2.5 COCONUT WASTE

2.5.1 Coconut Wine

Coconut wine is a popular drink in the Philippines, it is also known as tuba. In the northern part of the Philippines, coconut wine is called lambanog. It is made of pure nectar, milky white in color or almost colorless. Tuba is usually consumed fresh as it easily turns sour. If a more potent wine is desired, then often it can be distilled to bring out a stronger drink. In the Visayas Islands, particularly in Leyte, tuba is made of coconut sap mixed with barok (the bark of a red mangrove tree), which serves as a colorant and preservative that offsets fermentation. Coconut wine that is distilled for less than a year is called bahal. When it is distilled for 1 year or more, it is called bahalina, or mature drink. The longer it is aged, the finer and mellower the taste is (Jackson, 2010). Even so, coconut wine is yet to be recognized in other parts of the globe. The top world producer of coconuts is Indonesia at 31.2%, followed by the Philippines at 25.5% and India at 17%. The total world annual production of coconuts was about 62.14 million tons in 2012. Using this amount of coconuts, the waste generated annually could reach 12.2 million tons of coconut husk or coir (a coarse fiber removed from the fibrous outer shell–mesocarp part of the coconut fruit).

2.5.2 Coconut Coir

2.5.2.1 Case Study 1. Pretreatment and Enzymatic Hydrolysis of Coconut Coir

Coconut coir is composed of cellulose, lignin, pectin, and hemicellulose. The cellulose content in old coconuts is 41.7% (Fatmawati et al., 2013). The dried coconut coir was cut ($\pm 5 \times 5$ cm) and then milled using a disk mill at a speed of 5800 rpm. It was sieved to obtain the particle size of 70–100 mesh. Pretreatment and enzymatic hydrolysis are the key preliminary processes for producing biofuels from lignocellulosic wastes. So primarily, coconut coir was pretreated using various concentrations of NaOH (5–11%) solution in an autoclave at 121°C for 1 h to remove lignin and to increase enzyme accessibility to cellulose contents. After mapping of the pretreatment methods, among which microwave digestion reactor (MARS) and a sonication water bath with power inputs of 100 and 200 W were used. Alkaline–autoclave pretreatment was carried out at the lowest concentration of NaOH (5% wt) at 121°C for 15 min and a total holding time of 1 h. Ultrasound-assisted pretreatment in a water bath was chosen as an alternative to pretreatment in the autoclave. It was carried out in 5% wt NaOH at 50°C; the selected temperature also coincided with the temperature of the hydrolysis reaction. Two commercial enzymes, cellulase and β-galactosidase (Sigma–Aldrich), were used simultaneously as biocatalysts in Erlenmeyer flasks containing samples of coir and citrate buffer (100 mL) at pH 4.8 and temperature 50°C. Tetracycline antibiotic was added to the buffer to prevent bacterial contamination. Experiments were carried out in the rotating shaker at 50 rpm. The main products from hydrolysis were glucose and cellobiose. A spectrophotometric dinitrosalicylic method was applied to analyze the total concentration of reducing sugars (RSC) obtained after hydrolysis. Profiles of RSC versus time produced during the hydrolysis of samples containing various concentrations of coconut coir are shown in Fig. 11.2.

Maximum conversion of cellulose to sugar was achieved in the range of 40–60%, depending on the initial concentrations of the substrate used. With the same stock of coconut coir and an analogous procedure for alkaline pretreatment in an autoclave (11% wt NaOH), Fatmawati et al. (2013) achieved a maximum 7.57 g/L RSC after hydrolysis using identical enzymes. Our research group at the University of Nottingham Ningbo Campus found that autoclave pretreatment at 121°C can be successfully

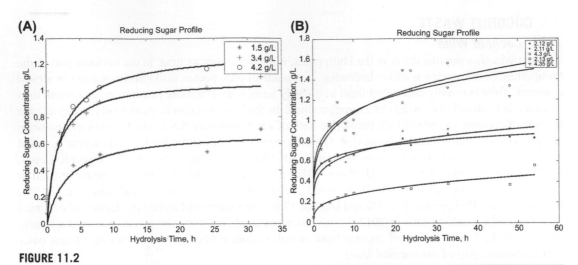

FIGURE 11.2

Kinetics of enzymatic hydrolysis. (A) Hydrolysis of coconut coir pretreated in an autoclave with NaOH. (B) Hydrolysis of coconut coir pretreated with NaOH via sonication at 100 W in a water bath.

substituted by treatment in ultrasonic bath at 50°C for the identical time of 1 h. One of the advantages of the ultrasound-assisted pretreatment is the lowest weight loss of the pretreated coir, which reached 48% w/w. When using an autoclave for pretreatment of biomass, the weight loss was much higher—about 66–67% of the initial dry weight of the samples. Fig. 11.3 illustrates raw coconut fibers as well as fibers after pretreatment in an autoclave (Fig. 11.3B) and fiber pretreated in an ultrasound water bath (Fig. 11.3C).

Vaithanomsat et al. (2011) investigated the possibility of ethanol production from young coconut husk—a lignocellulosic residue, containing 39.31% α-cellulose, 16.15% hemicellulose, 29.79% lignin, and 28.48% extractives. The pretreatment with NaOH solution resulted in higher α-cellulose (48.90%) and hemicellulose (22.04%), indicating the possibility of high ethanol conversion from coconut husk. The conversion of lignocellulosic biomass to ethanol usually employs three major steps: (1) pretreatment, to break down the lignin and open the cellulose structure; (2) hydrolysis, with a combination of enzymes to convert the cellulose to glucose; and (3) microbial fermentation of the glucose to ethanol. In the above study, the amounts of ethanol produced by the separate hydrolysis and fermentation (SHF) and simultaneous saccharification and fermentation (SSF) processes were compared. There were similar ethanol yields from SHF (21.21% based on pulp weight) and SSF (20.67% based on pulp weight). The ethanol yields were approximately 85% of the theoretical ethanol yield. However, optimization of the process at each stage should be further studied to develop the most suitable conversion technology (see Fig. 11.4).

Alternative attractive applications of dwarf–green coconut coir fibers were proposed by Esmeraldo et al. (2010), which include further reinforcement in composite materials as well as potential use as insulation for cables, electric motor wiring, small capacitors, and so on.

Two other vital applications were found for by-products of the coconut food industry. Naik et al. (2012) evaluated the by-products of the virgin coconut oil processing industry, such as coconut skim milk and insoluble protein, to obtain a value-added product, namely, coconut protein powder. This product had good emulsifying properties and had potential to find applications in emulsified foods as

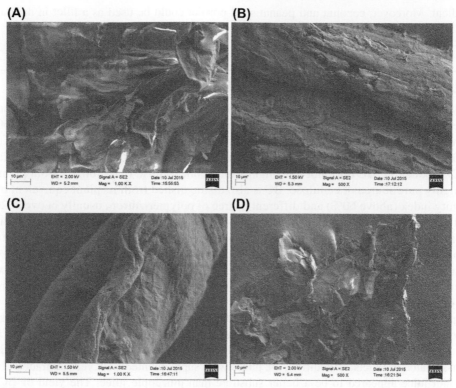

FIGURE 11.3

Scanning electron microscope micrographs. (A) Raw milled coconut coir (CC). (B) CC pretreated in an autoclave and hydrolyzed in a shaker. (C) CC pretreated in an ultrasound water bath (USB), hydrolyzed in a USB (at 100 W). (D) CC pretreated in a USB (200 W), hydrolyzed in a shaker.

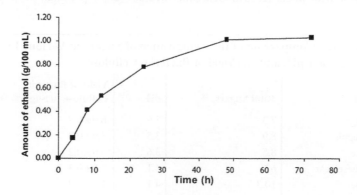

FIGURE 11.4

Ethanol produced by simultaneous saccharification and fermentation from young coconut husk pulp. The reaction contained 15 FPU Celluclast 1.5L and 15 IU Novozyme 188 per 1 g substrate and was carried out at 37°C.

Reproduced from Vaithanomsat, P., Apiwatanapiwat, W., Chumchuent, N., Kongtud, W., Sundhrarajun, S., 2011. The potential of coconut husk utilization for bioethanol production. Kasetsart Journal (Natural Science) 45, 159–164.

an ingredient. Moreover, coconut and peanut shell powder could be used as a filler in natural rubber (Sareena et al., 2012).

3. VALORIZATION OF FRUIT BY-PRODUCTS AND JUICES

A variety of fruit by-products, such as rotten fruits of plums, green grapes, pineapples, and apples (Jozala et al., 2015); orange peels (Kurosumi et al., 2009); coconut water (Kongruang, 2008); and juices from pineapple (Castro et al., 2011), orange, apple, Japanese pear, and grape (Kurosumi et al., 2009), were studied for the possibility of using them as carbon sources for bacterial cellulose (BC) production. BC does not have lignin and hemicellulose like biomass from plants, making it a highly pure source of cellulose. BC is also distinguished from its plant equivalent by a high crystallinity index (above 60%) and different degree of polymerization, usually between 2000 and 6000 (Mohite and Patil, 2014). It has been employed as a novel material in the food industry, as edible packing material, wound dressing materials, artificial skin, vascular grafts, scaffolds for tissue engineering, artificial blood vessels, medical pads, and dental implants (Shah et al., 2013). Industrial products such as sponges to collect leaking oil and materials for absorbing toxins and optoelectronics materials (liquid crystal displays) were also made because of its unique properties (Donini et al., 2010).

Kongruang (2008) carried out static batch fermentation of coconut and pineapple juices for BC production in 5-L fermenters at 30°C using three *Acetobacter* strains obtained from the Thailand Institute of Scientific and Technological Research, Bangkok. BCs produced from all strains were growth-associated products; in addition, the juices were supplemented with yeast extract and ethanol. The coconut juice appeared to be a better substrate than pineapple juice. The yield of BC on this supplemented medium was high, which made the production costs lower than expected.

Likewise, Kurosumi et al. (2009) employed various fruit juices, such as orange, pineapple, apple, Japanese pear, and grape, for BC production. Table 11.2 shows total sugars present in these juices, their pH, and the yield (in %) of the BC produced. The highest yield of BC was achieved from orange juice followed by Japanese pear juice. Hestrin–Schramm medium (2.0% peptone, 0.5% yeast extract, and

Table 11.2 Comparison of the Composition of Sugars in Various Fruit Juices, Their pH, and the Yield of Bacterial Cellulose

Fruit	Total Sugars, %	pH	Yield of Bacterial Cellulose to Sugars, %
Orange	7.3	3.9	6.9±0.2
Pineapple	8.9	3.5	3.9±0.3
Apple	8.5	3.6	3.9±0.2
Japanese pear	62.	4.1	4.8±0.3
Grape	10.3	4.1	1.4±0.2

Adapted from Kurosumi, A., Sasaki, C., Yamashita, Y., Nakamura, Y., 2009. Utilization of various fruit juices as carbon source for production of bacterial cellulose by Acetobacter xylinum *NBRC 13693. Carbohydrate Polymers 76, 333–335.*

0.12% citric acid) was added to the fruit juices, and the mixture was adjusted to pH 6 with disodium hydrogen phosphate buffer before the fermentation.

The authors proved that orange juice was a suitable medium for BC production. BC was also produced from the various components of oranges, such as peel and squeeze residue, and BC of 0.65 g (dry weight) was produced from 100 g of oranges, and the solid residue from the oranges was about 17.2 g (as shown in Fig. 11.5).

Jozala et al. (2015) evaluated BC production by *Gluconacetobacter xylinus* using rotten fruits composed of plums, green grapes, pineapples, and apples and milk whey as the culture medium without supplements. BC can be produced by employing a rotten fruit medium as a carbon source to achieve

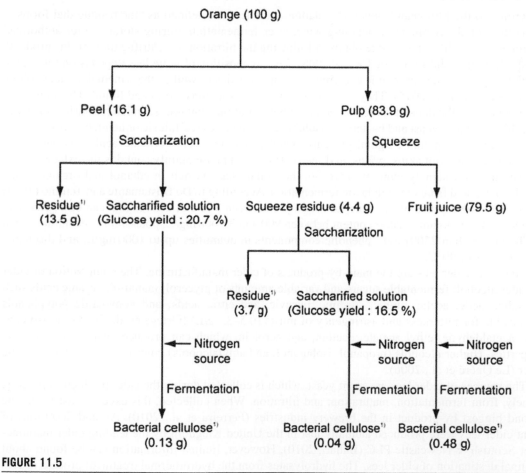

FIGURE 11.5

Bacterial cellulose production from various orange by-products. (1) Dry weight of residue and bacterial cellulose.

Reproduced from Kurosumi, A., Sasaki, C., Yamashita, Y., Nakamura, Y., 2009. Utilization of various fruit juices as carbon source for production of bacterial cellulose by Acetobacter xylinum NBRC 13693. Carbohydrate Polymers 76, 333–335.

high yields (the highest BC yield—60 mg/mL—was achieved with the rotten fruit culture). The ultimate time period to obtain the BC membranes appeared to be 96h. Biopolymer production can be optimized in relation to the cost-effectiveness of the medium in terms of cellulose yield and productivity (Panesar et al., 2012). Subsequently, the results provided a profitable alternative for generating valuable products from fruit waste.

To sum up, in this section examples are given to illustrate how to produce valuable materials with different possible applications by using wastes from the fruit and wine industries.

4. CIDER LEES

According to the European Council Regulation, wine lees are defined as "the residue that forms at the bottom of the recipients containing wine, after fermentation, during storage, after authorized treatments, as well as the residue obtained following the filtration or centrifugation of this product" (Pérez-Serradilla and Luque de Castro, 2008). The composition of wine lees depends on the region of their origin and the agronomic and climatic characteristics as well as the winemaking technology (Pérez-Bibbins et al., 2015). They consist mainly of two fractions: solid and liquid. The solid fraction contains all the deposits precipitated at the bottom of the containers, which primarily consist of microbial biomass (yeast and bacteria), insoluble carbohydrates (cellulosic or hemicellulosic materials), phenolic compounds, lignin, proteins, metals, inorganic salts, organic acid salts, and other materials such as fruit skins, grains, and seeds. The liquid phase mainly contains the exhausted fermentation broth, namely wine or cider. So, the liquid phase is rich in ethanol and organic acids, mainly lactic acid from the malolactic fermentation. According to De Bustamante and Temiño (1994) the main characteristics of the wine lees are a pH between 3 and 6, a COD that can be higher than 30,000 mg/L, an organic matter content between 900 and 35,000 mg/L, potassium concentration that can be higher than 2500 mg/L, phenolic components in quantities up to 1000 mg/L, and discharge temperatures of 90°C.

Likewise, cider lees are the main by-products of cider manufacturing. The composition of cider includes alcohol; fermentable sugars and variable amounts of glycerol; mannitol; organic acids such as malic, lactic, acetic, quinic, succinic, pyruvic, and citric acids; and nonvolatile polyphenols responsible for bitterness and astringency of musts (0.5–3.5 g/L) (Ortega et al., 2004). It also contains acetaldehyde, ethyl acetate, acetoin, and acrolein, which can produce changes to sensorial properties. Higher alcohols (propanol, isobutanol, and amylalcohols) contribute to the flavor of the cider (Le Quéré et al., 2006).

The sources of cider lees are spent yeast, which is collected during the cider production process, namely, from fermentation, maturation, and filtration. When collected, this excess yeast forms the second biggest by-product in the brewery industries (Ferreira et al., 2010). Around 8000 tons of spent cider lees were produced annually just in the United Kingdom by the leading cider manufacturer Scottish & Newcastle PLC (Bahari, 2010). However, limited information can be found about the final destination of cider lees. The hydrolysates from the hydrothermal treatment process of lees are a mixture of proteins and amino acids, which have been shown to be a useful source of nutrients for culture mediums as reported by Lamoolphak et al. (2007). Spent yeast (lees) from cider was used as animal feed previously, with a declining market recently, which resulted in additional costs for the drying and disposal of this waste.

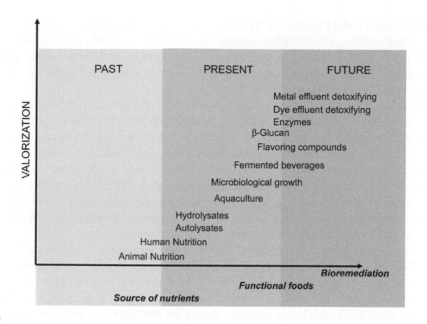

FIGURE 11.6

Schematic representation of the brewer's yeast biomass valorization at various times (Ferreira et al., 2010).

Yeast biomass is also known for its considerable concentrations of B-complex vitamins, nucleic acids, and the biologically active form of chromium known as glucose tolerance factor. The use of yeast biomass, especially in its dried form, has long been known in the food and flavor industries. For this purpose, yeast usually undergoes an autolysis reaction, which is carried out under controlled heating to kill the yeast without deactivating its enzymes. The main purpose of autolysis is to release the intracellular content of the yeast to the medium using the self-destructing mechanism of the cells and also to convert the complex structures of proteins to simpler amino acids and nucleotides, or other useful monomers. Furthermore, studies have reported that spent yeast biomass can be employed as a bioadsorbent to remove chemicals such as dyes (Yu et al., 2009), or metal ions (Cui et al., 2010). Fig. 11.6 shows how the valorization of yeast biomass has changed its role with time (Ferreira et al., 2010).

Research carried out at the University of Birmingham described the use of subcritical water-mediated hydrolysis for recovering intracellular yeast contents, phenolic compounds adsorbed on the cell walls, and yeast cell polymers (Bahari, 2010).

4.1 THE CHARACTERISTICS OF CRUDE CIDER LEES

Bahari (2010) studied spent cider lees comprising yeast cells (*Saccharomyces bayanus*), apple juice, apple residues, plant debris, organic acids (lactic acid), and additives that were introduced during the fermentation step. But the main constituent of the slurry was the yeast cells, which were made of the cell wall polysaccharide β-glucan and mannoproteins. The total dry weight of the slurry was calculated to be 11.5% (w/v). The pH of the samples was 3.6 ± 0.4. The acidic medium was a result of acidifying agents present upstream of the fermentation (mainly lactic acid). The protein concentration (~0.14 mg/mL or 1 μg/mg of dry yeast) was significantly lower than those observed in other studies [20–40 mg/mL (Illanes and Gorgollón, 1986) or 95 μg/mg of dry yeast (Lamoolphak et al., 2007)]. The initial concentration of protein in the washed samples was approximately 11 μg/mL. Total phenolic content of the crude sample was 680–730 mg/mL. Total organic content of the crude lees sample was about 50 mg/mL.

The results achieved by Bahari (2010) showed that subcritical water can completely hydrolyze yeast cells after 30 min at 200°C. Thus, β-glucan and mannoprotein in the cell wall were converted to glucose and mannose (monosaccharides), respectively, and their concentration was measured in both liquid and solid phases at different times and temperatures. Because of milder conditions needed for glucan conversion to sugars, the yeast biomass can be used for the production of intermediate bio-based chemicals using subcritical water. Subjecting the reaction medium to higher temperatures can lead the reaction to final derivatives such as organic acids. Comparing the kinetics of glucan conversion in subcritical water with cellulose conversion showed faster conversion in a less energy-intensive reaction and the feasibility of the production of bio-based chemicals via a biorefinery approach.

The type of phenolic compounds present in the lees depends on the origin of the fruit as well as the climate. Phenolic compounds contribute to color, flavor, astringency, bitterness, enzymatic or nonenzymatic browning, haze formation, and aging behavior (Pérez-Bibbins et al., 2015). Apple phenolics were identified in the original substrate and in the extracts, suggesting that they were carried over by the yeast cells after being adsorbed during the fermentation. Subcritical water was able to release these phenolics from yeast cells in levels comparable to organic solvent extraction. At the same time newly formed phenolic compounds were identified, which may have been formed or released from more condensed polymeric phenolics. The release of phenolic compounds resulted in a significant change in the antioxidant activity of the extracts at 225°C, which showed their potential as a source of highly active natural phenolics. Phenolic compounds, especially the identified individual phenolics, showed their highest concentration at 100°C, whereas the total phenolic content increased with temperature.

Comparing the yields of different compounds at different temperatures, it may be possible to design a sequential hydrolysis process in which different products are produced and separated from the reaction medium using selective separation techniques, and the remaining reactants can proceed to the next steps for other derivatives. Fabrication of different products can provide higher revenues for a single process and make the industrial applications of biomass conversion more appealing (Bahari, 2010).

4.2 VALORIZATION OF WINE LEES

A novel biorefinery concept based on grape wine lees valorization was developed at the Agricultural University of Athens (Fig. 11.7). The process started with the centrifugation or filtration of wine lees, to separate the liquid stream used for ethanol production via distillation. Ethanol can be used either to produce potable alcohol or as a platform chemical for the sustainable chemical industry. Alternatively, ethanol was also utilized as a carbon source for microbial fermentation aiming at polyhydroxybutyrate (PHB) production by the bacterial strain *Cupriavidus necator* NCIMB 12080 (Senior et al., 1986). The remaining liquid after ethanol extraction can be used in subsequent hydrolysis stages to increase the presence of nutrients.

After the distillation step, yeast cells were lysed and converted into a nutrient-rich supplement similar to yeast extract. Preliminary experiments with *C. necator* DSM 7237 and crude glycerol (e.g., from

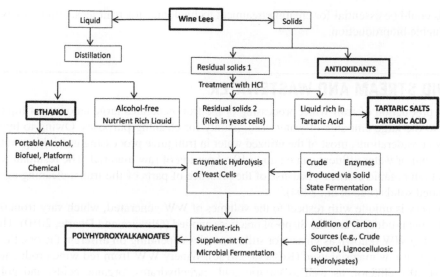

FIGURE 11.7

Valorization of wine lees.

Reproduced from Kachrimanidou, V., Kopsahelis, N., Webb, C., Koutinas, A.A., 2014. Bioenergy technology and food industry waste valorization for integrated production of polyhydroxyalkanoates. In: Gupta, V.K., et al. (Eds.), Bioenergy Research: Advances and Applications. Elsevier, Amsterdam, pp. 419–433.

biodiesel industries) as carbon source proved that PHB production was feasible using wine lees hydrolysates. However, supplementation with a low quantity of minerals was necessary, showing that this nutrient supplement was deficient in some minerals. Similar to the valorization concept for wine lees proposed by Kachrimanidou et al. (2014), the lees left after fruit winemaking can be treated to produce biofuels, antioxidants, and biopolymers. In turn, the supplements required should be adjusted based on the specific fruits used for the wine production and the composition of the fruit wine obtained. However, this research field is still in its infancy and the cost-competitiveness of the approach requires further elaboration.

Although various applications have been evaluated, including the recovery of ethanol by distillation, polyphenolic compounds, or salts, as fertilizers, as raw material for L-lactic acid production, or even for the production of biogas, waste lees appear to be an undervalued by-product up to now. The study of Pérez-Bibbins et al. (2015) aimed to evaluate lees as nutritional supplements for industrial fermentation processes considering that the residual content in lees of carbohydrates, nitrogen compounds, or essential vitamins makes this residue a promising supplement similar to yeast extract or corn steep liquor for formulating culture media. According to Pérez-Bibbins et al. (2015), lees are attractive as potential economic nutrient sources for subsequent larger-scale bioproduction of natural food additives such as xylitol, lactic acid, and citric acid. The preliminary calculated economic efficiency parameters identified lees as a lower cost and more effective nutrient source in comparison to corn steep liquor. These promising results need further confirmation. Additionally, the reasons for the poorer production observed in some cases with lees have to be clarified, especially when they are applied in high concentrations or as solid fractions. Expansion and combination of the studies on metals and phenolic compounds could be essential for creating treatment technologies for fruit lees, aiming to extend their use for valuable bioproduction.

5. LIQUID STREAM AND WASTEWATER

During the production of fruit wine, processing of the fruit generates large amounts of liquid effluent waste due to the large amounts of water used, mainly for cleaning purposes. Owing to hygienic and food safety considerations, most of the utilized water in fruit juice processing is drinking water quality and the amount of water effluent can be up to $10\,m^3$ per ton of raw material. The resultant wastewater (WW) has high organic contents, because of the presence of parts of the fruits, cleaning agents, salts, and suspended solids (Pap et al., 2004).

Each winery is unique with respect to the volumes of WW generated, which vary from 0.5 to 14 L per liter of wine produced, and the disposal practices applied (Oliveira and Duarte, 2010). The composition of winery WW (known as vinasse or stillage) varies depending on different factors, the main one being the type of raw material used (Kosseva, 2013c). Winery WW from red wine production usually contains 60–80% ethanol, as well as tartaric acid, carbohydrates, organic acids, and polyphenols, whereas vinasse from tropical fruit wine production also has sulfur compounds and a high sugar content (Montalvo et al., 2008). A comparison between parameters characterizing red wine WW and guava wine WW is shown in Table 11.3.

5.1 CHARACTERIZATION OF WINERY LIQUID EFFLUENTS

As illustrated in Tables 11.3–11.5, the aforementioned liquid effluents are acidic and have high contents of organic matter. The WWs from fruit-processing industries are slightly acidic in nature, which is due to the biological breakdown of fruits mainly under anaerobic conditions. Low pH can cause corrosion of the plant machinery and materials, can hamper the biological oxidation treatment, and can also cause poor settling of the primary sludge. Banana, apple, apricot, peach, and pear juice processing are the

Table 11.3 Comparison Between Parameters Characterizing Wastewater From Red Wine and Guava Wine (Montalvo et al., 2008)

Parameter	Red Wine Winery Wastewater	Tropical Fruit Wine Winery Wastewater
Soluble chemical oxygen demand (mg/L)	36,100 ± 1200	33,300 ± 1890
Total nitrogen (mg/L)	450 ± 25	515 ± 31
Total phosphorus (mg/L)	250 ± 11	287 ± 14
Sulfides (mg/L)	148 ± 1°	184 ± 12
Volatile fatty acids (g acetic acid/L)	7800 ± 189	6900 ± 160
Total polyphenols (mg/L as pyrogallol)	433 ± 45	233 ± 26
pH	4.1 ± 0.2	4.2 ± 0.3

examples exhibiting the highest pollution load from the fruit industry (Joshi, 2000). The characteristics of these fruit processing WWs are shown in Table 11.5.

The fruit pressing stage generates the highest amount of polluted WW characterized by greater BOD and SS concentrations. Segregation of different WW streams and assessment of their individual waste characteristics are the key factors for planning an efficiently designed WW from the resources recovery view point. Table 11.4 presents WW quantities and pollution loads generated in terms of pH, BOD, and settleable solids for various WW streams generated from different segments of a fruit juice processing plant. Data in the table indicate that the bottle cleaning procedure generates a high hydraulic load and a lower pollution load than the other streams. This stream could be treated with primary treatment and joined with the final outlet stream resulting after treatment for dilution, or it could be directly used for on-land application of irrigation.

Both dilute and concentrated vinasse can be used as organic fertilizer. In this case the agricultural soils are considered a land treatment system but will also benefit from nutrients present in the vinasse. However, possible adverse environmental impacts such as the enrichment of salt in the soil and nitrate leaching need to be considered. Sustainable management of vinasse spreading on agricultural fields therefore requires a precise understanding of C and N mineralization kinetics. The WWs, including

Table 11.4 Pollution Loads at Various Stages in Fruit Juice Factories (Joshi, 2000; Barnes et al., 1984)

Process Stage	pH	BOD (mg/L)	Settleable Solids (mg/L)	Wastewater (m³/m³)
Pressing	5.8–6.0	2850–2870	26.4–26.6	0.82–1.42
Container cleaning	7.0–9.3	730–810	4.8–36.0	0.015–0.019
Bottle cleaning	8.4–9.4	52–290	–	0.23–1.82
Filtration	5.9–6.9	–	12.0–20.4	0.005–0.013
Refining solids	–	67–500	–	0.06

BOD, *biochemical oxygen demand.*
Adapted from Joshi, C., 2000. Food processing waste treatment technology. In: Verma, L.R., Joshi, V.K. (Eds.), Postharvest Technology of Fruits and Vegetables, Vol. 1. Indus Publishing Co, New Delhi, p. 440.; Barnes, D., Forester, C.F., Hrudly, S.E., 1984. Survey in Industrial Waste Treatment. Food and Allied Industries, vol. 1. Pitman Pub, Co, London.

Table 11.5 Characteristics of Fruit Processing Wastewater (Joshi, 2000)

Fruit Processed	Wastewater (L per Ton of Raw Product)	BOD (kg per Ton of Raw Product)	TSS (kg per Ton Product)
Apples	10	9.0	2.2
Apricots	23	20	4.9
Peaches	13	17.5	4.3
Pears	15	25	–
Plums	10	5	1

BOD, *biochemical oxygen demand;* TSS, *total suspended solids.*

vinasses, from alcohol distilleries have different compositions and mineralization pathways, which could be helpful to understand their subsequent behaviors in soils (Parnaudeau et al., 2008).

In general, winery effluents are biodegradable and the ratio BOD_5/COD is higher is during the vintage period, because of the presence of molecules such as sugars and ethanol (Ganesh et al., 2010). Part of the organic load contained in the winery effluents is bio-recalcitrant and potentially toxic to various microorganisms and plant species. The COD concentration of grape winery effluents ranges from 320 to 49,105 mg/L (mean value: 11,886 mg/L), whereas the BOD_5 ranges from 203 to 22,418 mg/L (mean value: 6570 mg/L) (Ioannou et al., 2015).

5.2 METHODS OF WASTEWATER TREATMENT

Resource conservation is the most preferable approach in the food waste management hierarchy (Kosseva, 2013b), or "prevention is better than cure." The development of WW treatment strategies is usually guided by this preventive principle. Because the amount and quality of the effluent also greatly influence the economic feasibility of a company, efforts should be made to minimize the use of water and therefore to (World Bank, 1996):

- use dry methods such as vibration or air jets to clean raw fruit,
- separate and recirculate process WW,
- minimize the use of water for cleaning purposes,
- remove solid wastes without the use of water, and
- use countercurrent systems where washing is necessary (Pap et al., 2004).

The important factors that influence the selection of a specific WW treatment process are the nature of the product manufactured, the capacity of the manufacturing plant, the quantity of the WW generated, the availability and cost of land for the waste treatment plant, climatic factors (e.g., temperature) at the site of construction, regulatory requirements of pollution control authorities, and the existing status of water quality at the final disposal point at a natural stream or river. For example, a waste stabilization pond, oxidation ditches, and aerated lagoons are suitable where land is cheap and climate is moderately hot. Activated sludge and its modified system are energy intensive with a high cost. A fluidized-bed anaerobic system is suitable where land is scarce and costly. At times, the selection of treatment process is guided by the extent of treatment under environmental protection regulations. Secondary sludge may be dried on conventional sludge drying beds and used for composting or landfilled. Winery WW after treatment in a WW treatment plant is required to be disposed of in the environment as per the regulatory framework of the pollution control acts of the specific country (Joshi, 2000).

In Europe, legislative policies about winery WW treatment and disposal have not been published so far. The European Council Directive (91/271/EEC) regarding urban WW treatment sets maximum limits of 125 mg/L COD, 25 mg/L BOD_5, and 35 mg/L total suspended solids (TSS) for treated urban effluents that can be used for irrigation and can be discharged into water dams and other water bodies.

In contrast, California and Australia have established relevant guidelines regarding the management and discharge of winery WW (Ioannou et al., 2015). The California Regional Water Quality Control Board published in 2002 the "General waste discharge requirements for discharges of winery waste to land" (Order No. R1-2002-0012) (CRWQCB, 2002). The BOD and TSS of winery effluents, for example, should not be higher than 80 mg/L when the treated effluents are discharged to the land using spray irrigation, and not higher than 160 and 80 mg/L, correspondingly, when drip irrigation is applied. In Australia, the Agriculture and Resource Management Council of Australia and New Zealand and the Australian and New Zealand Environment and Conservation Council established in 1998 the "Effluent management guidelines for Australian wineries and distilleries" (ANZECC, 1998). The Australian limits for winery effluents for discharge into aquatic bodies are 15 mg/L BOD_5 and 50 mg/L TSS, and the limits for irrigation are 1500 kg BOD_5/ha/month and 704–2112 mg/L TDS (Kumar et al., 2009). Obviously, there are no COD limit values specified in both the Californian and the Australian guidelines.

Although there is an extremely limited number of publications on fruit wine WW, the treatment of grape winery WW has been broadly reviewed by Ioannou et al. (2015). The authors presented the state of the art of the processes currently applied and/or tested for the treatment of winery WW based on both bench- and pilot/industrial-scale processes. They compared numerous treatment technologies, which are currently available, for example, physical–chemical and biological processes, membrane filtration and separation processes, advanced oxidation processes, and combined biological with advanced oxidation processes. The physical–chemical processes (i.e., coagulation/flocculation and electrocoagulation) have been found to be effective for the pretreatment of winery WW, and more specifically, for lowering the TSS and the turbidity, as well as a part of the organic content, to levels that can facilitate further treatment by other biological, membrane filtration and separation, or advanced oxidation processes.

5.2.1 Biological Treatment Processes and Bioreactors

Biological WW treatment processes have been used for more than a century, starting with anaerobic treatment of the city WW in Exeter in the United Kingdom in the 1890s. The biological methods are considered eco-friendly and cost-effective, in most cases. Because the major part of winery WW is biodegradable, its biological treatment started with the conventional activated sludge process more than 20 years ago, achieving removal of COD up to 98%, reaching 85% for $P\text{-}PO_4$ elimination and 50% for BOD_5. Various types of bioreactors have been applied to both aerobic and anaerobic processes, including jet-loop reactor, sequencing batch reactor (SBR), fixed-bed biofilm reactor, air microbubble bioreactor, aerated submerged biofilter, rotating biological contactor, biological sand filter (BSF), membrane bioreactor (MBR), anaerobic digester, upflow anaerobic sludge blanket (UASB), anaerobic fluidized-bed reactor, upflow anaerobic filter, and anaerobic moving-bed biofilm reactor (Ioannou et al., 2015).

5.2.1.1 Aerobic Microbial Treatment Technologies

According to Fumi et al. (1995), the conventional activated sludge (CAS) treatment plant can withstand large variations in hydraulic and pollution loads. Other important factors that favor an economical operation of the plant, according to this study, were the small amount of sludge produced (0.065 kg TSS/kg COD); the absence of use of pH correctors, settling agents, and nutrients; the low level of labor required to manage the process; and the vertical design, which reduces the footprint area occupied. In the study of Petruccioli et al. (2000), the efficiency of the activated sludge process in removing the organic load from winery effluents was evaluated through the operation of (1) an air bubble column bioreactor (ABB), (2) a fluidized-bed bioreactor, and (3) a packed-bed bioreactor. According to the results, the highest efficiency was obtained by ABB (92.2% COD removal; influent value = 800–11,000 mg/L).

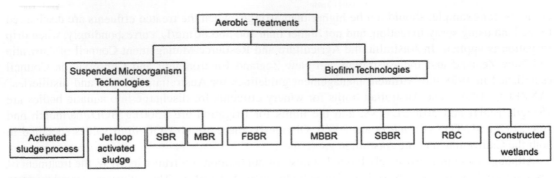

FIGURE 11.8

Aerobic technologies for treatment of wastewater. *FBBR*, fixed-bed biofilm reactor; *MBBR*, moving-bed biofilm reactor; *MBR*, membrane bioreactor; *RBC*, rotating biological contactor; *SBR*, sequencing batch reactor; *SBBR*, sequencing batch biofilm reactor.

Adapted from Lofrano, G., Meric, S., 2014. A comprehensive approach to winery wastewater treatment: a review of the state-of the-art. Desalination and Water Treatment, 1–18.

The performance (expressed as % COD removal) of the aforementioned aerobic biological processes for the treatment of winery WW from either real or simulated effluents is illustrated in Fig. 11.8. As shown, the most efficient aerobic biological process was found to be the MBR (or biocatalytic membrane reactor) (Bolzonella et al., 2010; Artiga et al., 2005, 2007; Valderrama et al., 2012), yielding the higher COD removal of 97% and almost complete removal of TSS (99%), thus proving very good commercial prospects.

MBRs have already been used in WW treatment for a number of years. The first full-scale commercial aerobic MBR process appeared in North America in the late 1970s, and then in Japan in the early 1980s. There are over 500 commercial MBRs in operation worldwide, and there are more under construction (Stephenson et al., 2000). However, over 98% of these systems couple the membrane separation process to an aerobic rather than to an anaerobic bioreactor.

The higher efficiency of the MBR can be explained by a more effective way to supply dissolved gases to microorganisms through membranes. The membrane's lumen can be pressurized with a gas, which diffuses through the membrane wall to a biofilm attached at the membrane's outer surface. When used to deliver air or oxygen, the process is often called a membrane aerated bioreactor (MABR) (Briddle et al., 1999; Casey et al., 2008; Stricker et al., 2011) or membrane biofilm reactor (MBfR) (Rittmann, 2006; Lofrano and Meric, 2014). Membranes are central to both the function and the cost of an MBR. A balance must be achieved between gas transfer properties, mechanical strength, chemical resistance, surface area, and cost (Semmens, 2005). MBRs may be based on either flat sheet or hollow-fiber membranes, although most studies have utilized hollow fibers because of their versatility, ability to provide high specific surface area, and superior biomass retention. One of the original goals of the MBR was to provide bubbleless aeration of WW for high-rate treatment or high-strength COD removal (Pankhania et al., 1999). MBRs have the potential to deliver oxygen at transfer efficiencies approaching 100% in the case of a dead-end operation (as shown in Case Study 2).

The membrane and module designs in the MBfRs have been considered in detail by Martin and Nerenberg (2012). For the membrane design the wall thickness, diameter, and length; material properties and gas transfer; and membranes and biofilm attachment are taken into account. A desirable module design maximizes mass transfer without requiring large amounts of mixing energy. The key considerations for MBR module design are dead-end or flow-through mode, orientation of membranes relative to the flow, membrane packing, and reactor configurations. The orientation of the membrane relative to flow affects the formation and thickness of the liquid diffusion layer (LDL). The LDL slows gas loss from the biofilm to the bulk liquid, which is especially beneficial when there are patchy biofilms or expensive gases being used. For the MBfR, the LDL resides on the outer edge of the biofilm, introducing resistance to gas transfer from the biofilm to the bulk. The LDL, located at the outer edge of the biofilm, helps to resist the loss of gas to the bulk liquid, thus contributing to higher gas utilization efficiencies. It is desirable to minimize the LDL, which decreases the flux of the target contaminant into the biofilm. Cross-flow, that is, flow perpendicular or transverse to the membrane axis, can deliver substrates to the biofilm advectively, minimizing the LDL (Motlagh et al., 2008; Wei et al., 2012).

Numerous MBR configurations have been evaluated, many of which mimic membrane filtration modules or previous bioreactor designs. For the parallel-flow, shell-and-tube reactor, membranes are potted at one end, either free or fixed at the distal end. They are fitted within an outer sleeve through which water flows parallel to the membranes (Martin and Nerenberg, 2012). Other reactors are continuously stirred tank reactors (CSTR) retrofitted with membranes. This type of bioreactor is suitable for control of pH through automatic dosing of acid/base or the addition/depletion of carbon dioxide. Control of pH is important in treatment processes, such as nitrification or denitrification, in which alkalinity is consumed or produced. Stricker et al. (2011) considered in-line pH control to be uneconomical and, instead, relied on the air sparging used for biomass control to accomplish acidification through the stripping of carbon dioxide. MBRs can be arranged in a variety of process configurations.

Compared to conventional suspended-growth systems (e.g., CAS, SBR), MBRs have the following advantages: (1) shorter hydraulic retention time (HRT), (2) less sludge production, (3) more stable operation, (4) reduced incidence of process upsets, and (5) simultaneous nitrification–denitrification. However, some of the disadvantages of the MBRs include (1) high capital costs for the membrane modules, (2) limited data on membrane life and thus a potentially high recurring cost of periodic membrane replacement, (3) higher energy costs due to membrane scouring compared to conventional suspended-growth processes, (4) potential membrane fouling that affects the ability to treat design flows, and (5) waste sludge from the membrane process that may be more difficult to dewater. The operational costs, based on chemical and energy consumption only, estimated by Valderrama et al. (2012) showed that for treatment of winery WW, CAS and MBR plants have very comparable operational costs equal to €0.38 and €0.40/m^3, respectively.

On the other hand, CASs and BSFs, when operating at optimum conditions, can also be considered two of the most effective aerobic biological processes, removing high rates of organic load from real winery effluents (up to 98%) (Ioannou et al., 2015).

5.2.1.1.1 Case Study 2. Application, Design, Long-Term Operation, and Performance of a Pilot-Plant MABR During Treatment of High-Strength Wastewater From Cider Production

Aiming to treat apple cider WW, Brindle et al. (1999) constructed a pilot-plant tubular reactor, which was held vertically, containing 4080 porous hollow fibers (Model MHF 200TL, Mitsubishi Rayon Co., Ltd., Tokyo). The Mitsubishi composite membrane was made of microporous polyethylene and dense polyurethane. Each fiber had an active length of 0.84 m, an outside diameter of 280 μm, with a 0.04- to 0.1-μm pore size range. The hollow fibers were grouped in a bundle with the oxygen supplied via a manifold at the base of the bundle. The fiber ends were individually sealed farthest from the oxygen source (dead-end configuration), so that each fiber could move independently, decreasing the LDL and controlling the biofilm accumulation. This configuration eliminated the option of purging the fibers and the motion could cause membrane breakage, especially if precipitation or mineral salts formed on the membrane surface (Lee and Rittmann, 2002). The specific membrane surface area was $447\,m^2/m^3$ with a void space of 97%. The combined volume of the reactor, recirculation line, and overflow vessel was 7.30 L with a reactor length, internal diameter, and working volume of 1.46 m, 0.077 m, and 6.75 L, respectively.

5.2.1.1.2 Experimental Conditions The influent to the MABR was alcoholic cider production WW collected at regular intervals and stored in a 1.0-m^3 container (Brindle et al., 1999). A submerged pump was used to mix the container contents before it was pumped to a 50-L feed tank, to prevent solid settlement and anaerobic degradation. Amendment of the WW with nitrogen in the form of NH_4HCO_3 and phosphorus as KH_2PO_4 and K_2HPO_4 was performed to give a total BOD_5 to N to P ratio of 100:5:1. A pH of about 7.6 was maintained with a feedback pH controller and chemical dosing unit using $NaHCO_3$ (2.0 M) as buffer. Temperature in the MABR was held at $25.7 \pm 0.5°C$ via an immersed heater in the feed tank and an insulation jacket around the reactor module. Pure oxygen (99.9%) was supplied from the gas cylinder to the lumen side of the fibers via an oxygen mass-flow controller. A regular cleaning procedure was applied to control biofilm development in the MABR module (Pankhania et al., 1999).

Pulse input tracer studies were performed with and without a biofilm attached to the membrane surface using NaCl and LiCl, respectively. Standard stimulus-response techniques were used to obtain residence time distribution (RTD) curves. The dimensionless residence time (h), number of CSTRs in series (N), and dispersion number (D/uL, where D is the diffusion coefficient of tracer in m^2/s, u is the liquid-phase velocity in m/s, and L is the length in m) were determined from the RTDs according to Levenspiel (1999). The results are presented in Table 11.6.

Influent and effluent total and dissolved COD (TCOD and DCOD), BOD_5, and total and suspended solids were analyzed according to standard methods (APHA, 1992).

Table 11.6 Experimental Values of the Residence Time Distribution Curves in an MABR With and Without the Presence of Biofilm on the Membrane

Qn (m^3/h)	Membrane Status	h	N	D/uL
0.78	Clean	1.10	1.0	0.59
	Biofilm	1.07	1.3	0.29
0.36	Clean	1.03	1.2	0.49
	Biofilm	0.95	1.5	0.21
0.18	Clean	0.93	2.9	0.35
	Biofilm	0.97	2.5	0.22
0.00	Clean	0.91	8.3	0.007
	Biofilm	–	–	–

D/uL, dispersion number; h, dimensionless mean residence time; MABR, membrane aerated bioreactor; N, number of continuously stirred tank reactors in series; Qn, recirculation flow rate.
Adapted from Brindle, K., Stephenson, T., Semmens, M.J., 1999. Pilot-plant treatment of a high strength brewery wastewater using a membrane-aeration bioreactor. Water Environment Research 71 (6), 1197–1203.

Table 11.7 Variations in the Operating Conditions and Parameters in an MABR During the CM and PF Trials

Parameter	Completely Mixed (CM)	Plug Flow (PF)
Influent TCOD	1742–3150 (mg/L)	1600–2798 (mg/L)
Influent DCOD	1422–2806 (mg/L)	1114–2345 (mg/L)
Influent suspended solids COD	217–1013 (mg/L)	285–508 (mg/L)
Organic loading rate	4.4–30 (kg COD/m^3/day)	7.5–34.9 (kg COD/m^3/day)
Hydraulic retention time	1.4–10.7 (h)	1.8–3.8 (h)
Recirculation flow rate	0.78 (m^3/h)	0.18–0.78 (m^3/h)
Bulk liquid phase velocity	0.047 (m/s)	0.022 (m/s)

COD, *chemical oxygen demand;* DCOD, *dissolved COD;* MABR, *membrane aerated bioreactor;* TCOD, *total COD.*
Adopted from Brindle, K., Stephenson, T., Semmens, M.J., 1999. Pilot-plant treatment of a high strength brewery wastewater using a membrane-aeration bioreactor. Water Environment Research 71 (6), 1197–1203.

Two trials (completely mixed, CM, and plug flow, PF) were completed to investigate the effects of organic loading rate (OLR) and liquid flow characteristics on MABR performance. The membrane-aeration-bioreactor operating conditions and WW parameters are presented in Table 11.7.

Oxygen was supplied to the MABR in excess of that required for complete oxidation of COD in WW, so the dissolved oxygen (DO) concentration in the bulk liquid remained greater than 1 mg/L through both trials. But it varied during the cleaning process. CM conditions prevailed at circulation flow rates of 0.78, 0.36, and 0.18 m^3/h, according to both the number of CSTRs in series ($N > 4$) and the dispersion model ($D/uL > 0.2$). At zero circulation, the MABR was operated as a PF system ($N > 4$).

Biofilm characteristics are usually influenced by the bulk liquid velocity and substrate loading during the startup conditions. In the present case study, process stabilization did not occur until day 10 of the CM trial compared to day 3 of the PF trial. Startup at a low liquid velocity with a short period of acclimation enabled the MABR process to operate at a much lower HRT and higher OLR from the start of the PF trial.

5.2.1.1.3 *Organic Removal*
The pilot-plant MABR was operated at five different OLRs during 82 days of the CM trial. The presence of a clearly visible biofilm from day 9 led to a 78% ± 6% removal efficiency of TCOD from day 10 to 35 at an ORL of 4.4 kg TCOD/m^3/day. From day 35 to 82, the removal efficiency of suspended solids was 69 ± 13% regardless of changes in SS loading. Recirculation flow rate was further reduced from 0.78 to 0.36 m^3/h from day 36.

During the PF trial, TCOD removal efficiencies and organic removal rates during four flow conditions averaged 77% at 23 kg TCOD/m^3/day, 88% at 28 kg TCOD/m^3/day, 73% at 26 kg TCOD/m^3/day, and 81% at 287 kg TCOD/m^3/day, respectively. Recirculation flow rate was further reduced to 0.18 m^3/h to give a liquid velocity of 0.11 m/s from day 42 to 46. Thus, the MABR operated under PF conditions from day 47 to 52. An average DCOD removal efficiency of 94% was achieved during this period, which was higher than that observed during CM operation. The COD saturation constant for mixed heterotrophic populations ranged from 5 to 30 mg/L. Regular monitoring of the on-site activated sludge process treating the same WW confirmed that nearly 200 mg/L of DCOD was non-biodegradable. The DCOD concentration during CM MABR operation averaged ~320 mg/L and was always greater than 170 mg/L. PF DCOD concentrations ranged from 2080 mg/L at the influent end of the module to ~170 mg/L at the effluent end. The concentration of biodegradable substrate could be rate limiting during CM MABR operation and near the effluent end of the module during PF operation.

The authors found that PF reactors performed slightly better than CM conditions because of the higher concentrations in the tank. However, the effect of the LDL could counter the advantage of having high organic concentrations in the module (Pankhania et al., 1999). Staging reactors in series allowed for well-mixed conditions within each individual module, whereas as a whole, the series acted as a PF reactor. An extra stage provided relief to shock loads, though during average loadings, the last reactor could be starved, causing patchy biofilms and loss of gas to the bulk liquid (Stricker et al., 2011). Moreover, the MABR allowed for the fine-tuning of the gas supply in each module. For removal of competing contaminants or secondary contaminants, modules in series may be ideal (Martin and Nerenberg, 2012).

5.2.1.2 Anaerobic Microbial Treatment Technologies

Natural zeolite from Chilean and Cuban deposits were used independently as bacterial support in the lab-scale anaerobic fluidized-bed reactor-treated winery WW from red grape wine and guava wine production at mesophilic temperature (35°C) (Table 11.3). The COD removal for OLR was achieved at the range 80–86% of up to 20 g COD/L/day, after which inhibition of the specific methanogenic activity was observed. pH and volatile fatty acids (VFAs) remained at values that are adequate for anaerobic processes even at high OLR values (around 20 g COD/L/day). In the bioreactor operating with tropical fruit wine WW, COD removal of up to 86% could be achieved at an OLR of up to 22 g COD/L/day. Biomass concentration at the support increased when OLR increased and a linear relationship between time (days) and biomass concentration was demonstrated (Montalvo et al., 2008).

According to the results reported in the literature, an anaerobic sequencing batch reactor (ASBR) can achieve a significant COD removal greater than 98% (influent COD = 8600 mg/L), with HRT of 2.2 days, and a specific OLR of 0.96 g COD/g VSS/day (Ruiz et al., 2001). In addition, this study showed that (1) the acidification of the organic matter and the methanization of the VFAs follow zero-order reactions, whereas (2) the effect on the gas production rate resulted in two level periods separated by a sharp break when the acidification stage was finished and only the breaking down of the VFAs continued. The type of sludge (e.g., granular, sewage) seeded in a UASB plays an important role in reducing bioreactor startup time and in increasing its efficiency, as well (Keyser et al., 2003; Fig. 11.9).

Anaerobic treatment technologies for wine WW are shown in Fig. 11.9. Considering biological treatment processes, the main factors affecting their efficiency are BOD, pH, temperature, TSS loading, sludge quality, mixed liquor suspended solids, DO concentration, C/N ratio, HRT, sludge retention time, needs of nutrients, growth kinetics, and sensitivity to the environmental conditions. As illustrated in Fig. 11.10, concerning anaerobic treatment, the ASBR and upflow sludge blanket filtration reactors were found to be the two most efficient processes for the treatment of winery WW, with a COD reduction up to 98% (Ruiz et al., 2001; Molina et al., 2007), followed by the laboratory-scale anaerobic MBR (97% COD removal) (Basset et al., 2014). However, for the biological treatment methods the effluent residual COD ranges from 132 up to 42,900 mg/L (Fig. 11.10), depending on the initial COD value of the influent used in each study, leading to the decision that further treatment is required before its disposal into the environment (if the qualitative limits of Directive 91/271/EEC concerning urban WW are considered).

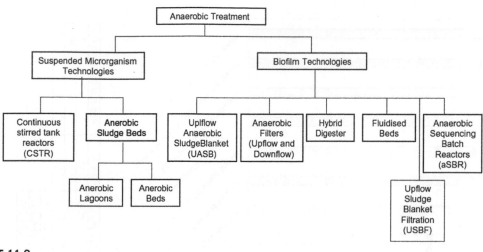

FIGURE 11.9

Anaerobic technologies for treatment of wastewater.

Adapted from Lofrano, G., Meric, S., 2014. A comprehensive approach to winery wastewater treatment: a review of the state-of-the-art. Desalination and Water Treatment, 1–18.

5.2.1.3 Combined Aerobic/Anaerobic Biological Systems

Anaerobic and aerobic biological treatments are often combined in brewery WW treatment. As shown in Fig. 11.11, there are essentially four types of integrated anaerobic–aerobic bioreactors (Chan et al., 2009). Some of the important advantages of combined aerobic/anaerobic treatment of brewery effluent over completely aerobic include a positive energy balance, reduced (bio)sludge production, and significantly low space requirements (Simate et al., 2011).

In recent years, attention has been focused on the use of membranes in conjunction with anaerobic reactors because of their inherent advantages over aerobic systems, such as low sludge production, net energy production, and a fully enclosed treatment tank (Hu and Stuckey, 2006). Hollow-fiber membranes (Mitsubishi Rayon) with an outside diameter of 540 μm and a wall thickness of 90 μm, and Kubota membranes (A4-size flat sheet membrane) were used in anaerobic MBRs. Both membranes were hydrophilic and were used to investigate the influence of membrane configuration on system performance over long-term operation. Each membrane module had 0.1 m^2 of total membrane surface area, with a pore size of 0.4 μm. Both Mitsubishi Rayon hollow-fiber and Kubota flat sheet membranes resulted in similar COD removal, but the transmembrane pressure across the hollow-fiber membranes at the sparging rate used, 5 L per minute, was higher than that of the flat sheets under similar conditions (Hu and Stuckey, 2006).

The combination of advanced oxidation processes (AOPs) and biological treatment (as pre- or posttreatment) can lead to a higher level of COD reduction than any single-stage treatment under the same operating conditions. Particularly, considering grape winery effluents, the use of AOPs as posttreatment of biological processes was found to be the most effective combination, because it permits almost complete purification, compared to AOPs used as pretreatment (Ioannou et al., 2015). AOPs include ozonation, Fenton oxidation, the application of H_2O_2 as an oxidant combined with light, the combination of

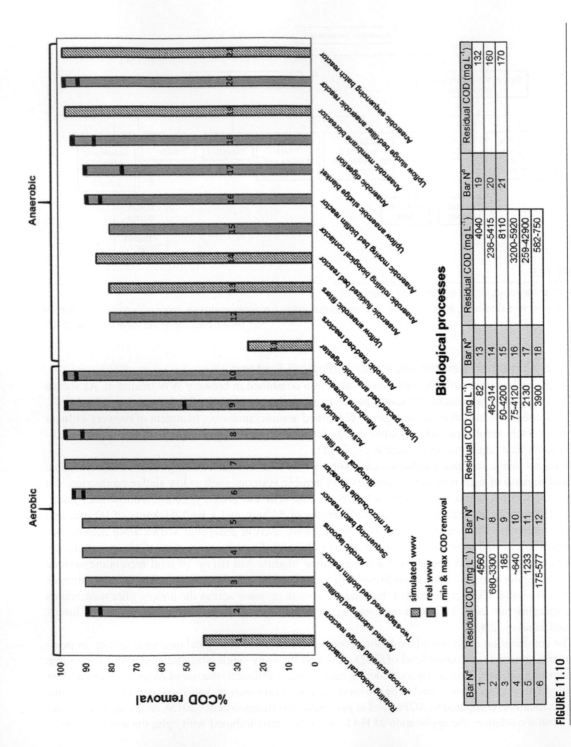

FIGURE 11.10

Organic load removal efficiency and effluent residual load of biological methods. *www*, wine wastewater.

Reproduced from Ioannou, L.A., Li Puma, G., Fatta-Kassinos, D., 2015. Treatment of winery wastewater by physicochemical, biological and advanced processes: a review. Journal of Hazardous Materials 286, 343–368.

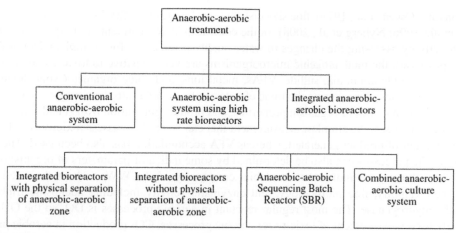

FIGURE 11.11

Types of combined anaerobic–aerobic systems for brewery wastewater treatment.

Adapted from Simate, G.S., Cluett, J., Iyuke, S.E., Musapatika, E.T., Ndlovu, S., Walubita, L.F., Alvarez, A.E., 2011. The treatment of brewery wastewater for reuse: state of the art. Desalination 273, 235–247; Driessen, W., Yspeert, P., Yspeert, P., Vereijken, T., May 24–26 2000. Compact Combined Anaerobic and Aerobic Process for the Treatment of Industrial Effluent, Environmental Forum, Columbia–Canada: Solutions to Environmental Problems in Latin America. Cartegena de Indias, Columbia.

UV-A radiation and O_3 in the presence of TiO_2 as a potential destructive technology, or wet air oxidation, catalytic wet air oxidation, wet peroxide oxidation, and catalytic wet peroxide oxidation. A new system for the treatment of winery WW (influent COD = 10,240 mg/L) was developed, which included electrochemical methods (electrooxidation, electrocoagulation using stainless steel, iron, and aluminum electrode sets) with simultaneous sonication and recirculation in a strong electromagnetic field (Orescanin et al., 2013).

The "best available technology not entailing excessive cost" (BATNEEC) has not been developed so far, because of the lack of information on specific cost analyses. Thus, accurate economic studies should be accomplished to determine whether the most efficient combined processes are cost-effective, environmentally sensitive, and technically reliable, allowing their full application at the industrial scale. In addition, the set of physicochemical and biological parameters used for the evaluation of the efficiency of the various processes, both individual or combined ones, should be expanded to include bioassays, for example, ecotoxicity and phytotoxicity tests. The development of alternative process combinations is essential, to increase the efficiency of removal of both the recalcitrant organic compounds and ecotoxicity, with a simultaneous reduction in the investment and operational costs (Ioannou et al., 2015).

6. ECOTOXICITY

Toxicity is the degree to which a toxic compound can harm an organism or a group of organisms. The level of damage can be measured directly by the deterioration of physiological functions of the organisms, such as oxygen uptake rate (OUR) (Lajoie et al., 2003; Andreottola et al., 2008), methane production rate (Owen et al., 1979), floc size distribution (Lajoie et al., 2002), and community changes (Madoni et al., 1996; Nyberg et al., 2008). In the context of WW treatment, toxicity can also be measured indirectly by assessing the changes in water quality parameters, for example, COD or BOD. In anaerobic processes, the methanogenic microorganisms are very sensitive to toxic compounds. Common toxicants include ammonia, sulfide, VFAs, metal ions, and some organics (Abouelenien et al., 2010). The approaches to measuring the toxic effects can be divided into three major categories: reactant consumption, indicator of organism properties, and product generation (Rozzi and Remigi, 2004). For anaerobic processes, the simplest measurement that can be made to assess toxicity is CH_4 production as the amount of product generated, whereas VFA accumulation has also been used. The activity or toxicity of the processes can also be determined by some indicator parameters of organisms natural to the system, that is, the indigenous microbial community of the WW, such as their cell density (growth), cell viability, floc morphology, and enzymatic activity including adenosine triphosphate (ATP) and dehydrogenase. The most regular reactant in aerobic processes is DO, so the OUR is the most widely used measurement of toxicity for aerobic processes. CO_2 production in aerobic processes has also been used for toxicity assessment. So, the respirometric methods, based on OUR and CO_2 production, are specific to aerobic processes. Changes in ammonium consumption, nitrite consumption or accumulation, and nitrate accumulation have all been used in toxicity assays for the nitrification process. Consequently, among the available toxicity assays for biological WWTPs are the OUR-based respiration inhibition test and the CH_4-based ATA assay, which are the most widely used in aerobic and anaerobic processes, respectively. Some of the biosensors developed since 2005 have also been shown to be effective for use in biological WWTPs, but most of them perform off-line and cannot be adapted to online monitoring tools, in order to provide an early warning to WWTP operators. For this reason, more research is needed to test the response of bioluminescence assays and whole-cell sensors based on specific organisms to various toxicants and to build a matrix of biosensors or a biosensor of multiple reporter organisms that can protect the integrity of WWTPs; also more advanced online data processing methods are needed to correctly interpret the results obtained from toxicity sensors in real WWTPs (Xiao et al., 2015).

7. SUSTAINABILITY IN THE WINEMAKING SECTOR

In the face of growing pressures upon the global population and the world's natural resources, many leaders in the beverage industry have been implementing a number of major sustainability initiatives. Sustainability is increasingly considered to be a key strategic priority, and many companies are setting sustainability targets hand in hand with the delivery of commercial objectives (e.g., improving energy efficiency and reducing carbon emissions). The initiatives includes efforts (1) to reduce carbon and water footprints, (2) to make innovations in the health and well-being sector, (3) to reduce and recycle packaging, and (4) to make improvements to the supply chain (Kosseva, 2013b). In line with the above initiatives in Europe, the United States, Australia, South Africa, and New Zealand, growing interest exists in the sustainability of wine production, which is focused primarily on environmental indicators, greenhouse gas (GHG) emissions, and LCA. Sustainability in the winemaking sector is based on two major objectives associated with: (1) the vineyard or orchards (with reference to fruit wines) and (2) the winery:

1. The first aspect includes soil management, water use, maintenance of biodiversity, and reduction of agrochemicals used in the vineyard or orchard (Chiusano et al., 2015).
2. The second includes WW treatment, GHG reduction, decreased or substitution of wine additives with more eco-friendly and natural substances (e.g., biopolymers), and packaging (e.g., lighter bottles, recycled paper) in the winery.

The importance of lowering the environmental impact during the production phase is also stated by the European Commission: "products of the future shall use less resources, have lower impacts and risks to the environment and prevent waste generation already at the conception stage" (European Commission, 2001).

7.1 LIFE CYCLE ASSESSMENT IN FRUIT PRODUCTION

Because LCA is a method of environmental analysis that assesses the environmental impact of products throughout their entire life from "the cradle to the grave," LCA approaches are now considered to form the basis for communicating the overall environmental performance of products (Ingwersen and Stevenson, 2012). The complexity of orchard systems affects the methods encompassed in applying LCA to agro-ecosystems (Notarnicola et al., 2012). As far as comparison between agricultural products is concerned, the standardization of methods used for characterization of sustainability is essential, or a reference framework for selection of the best settings for LCA applications in fruit production systems is required. There are already several frameworks for Environmental Product Declarations (EPDs) available and applicable to fruit production (Cerutti et al., 2014). One of the most widely used declaration systems to include the LCA of food products is the international EPD System, standardized as type III labeling (ISO 14025). It works with rules based on a hierarchical approach following the international standards ISO 9001, ISO 14001, ISO 14040, ISO 14044, and ISO 21930. As a consequence, the LCA approach is a mandatory procedure and reference is made to LCA-based information in the context of product category rules (PCRs), allowing transparency and comparability between EPDs. Another important international framework for EPDs is the product environmental footprint, which is an LCA-based method to calculate the environmental performance of a product. It was developed by the European Commission's Joint Research Centre and is based on existing methods that have been tested and used extensively (European Commission, 2013). In this framework product environmental footprint category rules (PEFCRs) are used; so far, the protocol has covered the testing phase and as of this writing there are no PEFCRs for fruit.

Thus far, various guidelines and approaches have been developed for harmonization of methods used to calculate the environmental impact of food production systems (Cerutti et al., 2014). A specific focus on the GHG emissions from the life cycle of goods and services, such as the Publicly Available Specification (PAS2050), was developed by the British Standards Institute and the Carbon Trust (BSI, 2011; Carbon Trust, 2007), the French Bilan Carbone (ADEME, 2010), and the GHG Protocol drawn up by the World Resources Institute and the World Business Council for Sustainable Development (WBCSD/WRI, 2009). Two specific ISO standards on product carbon footprints, ISO 14067 (ISO, 2013a), and water footprints, ISO 14046 (ISO, 2013b), have been developed. The main characteristic of most of these initiatives is the use of just one indicator (carbon or water). The first attempt to harmonize a method for the environmental assessment of food and drink products was the European Food Sustainable Consumption and Production Roundtable's Draft ENVIFOOD Protocol issued in 2013 (European Commission, 2012). As of this writing, it is going through a trial. The main objective is to provide guidance for assessments, which is instrumental for both the communication and the environmental improvement of business-to-business and business-to-consumer analyses (De Camillis et al., 2012). The ENVIFOOD Protocol includes a list of relevant impact categories for the environmental assessment of food and drink products and is expected to contain more detailed product footprint categories rules than the PCR in the EPD scheme. Fresh fruits in group 1 in the ENVIFOOD protocol are expected to be studied throughout the full life cycle, including the user phase, if relevant to the PCR. However, there is no specific guidance on accounting for the whole lifetime of orchard systems. Furthermore, according to the protocol, the impact categories relevant to agriculture and water consumption have to be reported separately. In response, Cerutti et al. (2014) developed a general orchard model, which included the whole lifetime of the system. This model will be helpful for initiatives to produce an international standardization of methods such as the ENVIFOOD protocol. The authors also presented practical recommendations on how to build the orchard system for LCA application avoiding over- or underestimations of the various orchard stages. As a result, a framework for selecting the best parameters for an LCA application in fruit production systems was proposed.

7.2 LIFE CYCLE ASSESSMENT OF WASTE MANAGEMENT IN CIDER PRODUCTION

Herrero et al. (2013) applied LCA with a cradle-to-grave perspective to a Spanish hypothetical factory producing sparkling cider. The full description of Case study 3 is provided in *Chapter 15: Life Cycle Assessment Focusing on Food Industry Wastes* of "Food Industry Waste: Assessment and Recuperation of Commodities" (Kosseva and Webb, 2013). The authors considered several subsystems, one of which was "waste management" (WM) involving the processes of composting, recycling, and landfill disposal. The environmental impact assessment results were strongly dependent on the treatment and valorization methods applied to the solid waste.

In conclusion, the management of solid wastes was a key factor in the environmental impacts resulting from cider production. Environmental improvement measures have to increase the amount of wastes that are valorized and recycled.

7.3 LIFE CYCLE ASSESSMENT OF WINE PACKAGING

The climate change impacts of packaging systems differ considerably from one another. Clearly (2013) used LCA input data from laboratory and field research, as well as questionnaires to the industry, to investigate the potential life cycle impacts of five types of 1-L wine packages. This study addressed: (1) lightweight single-use (LSU) glass bottles, (2) refillable glass (RFG) bottles, (3) polyethylene terephthalate (PET) containers, (4) conventional single-use (CSU) glass containers, and (5) aseptic cartons (ACs). The LCA results estimated using the ReCiPe impact assessment method addressed both the relative impact differences between the packaging systems and the cumulative impacts from the consumption of packages of multiple sizes and types within a municipality. Of the five types of packages evaluated, the RFG bottle and AC proved to have the lowest net end-point level environmental impacts, with impact reductions, relative to the CSU glass container, reaching as high as 87%. The PET containers were responsible for less than 45% of the CSU packaging emissions; additionally, the ACs and RFG containers were responsible for 19% and 14–17%, respectively. For all packages but RFG bottles, the production of the primary package generates the most GHG emissions over the life cycle. The proportional reduction in end-point level impacts from replacing CSU glass containers with LSU glass bottles predominantly reflects the mass reduction from the substitution. In comparison, the net impacts from CSU and LSU bottles were almost always more than double the magnitudes of those of the remaining container types. End-point level impacts for wine containers were almost identical for refillable bottles and ACs, with the exception of the damage to ecosystem quality impact, in which the RFG bottle had half the impact of the AC. ACs were responsible for greater ecosystem quality damage because of the land use required to supply the wood pulp for the paperboard input, an input not required for RFG bottles.

The substitution of lightweight and refillable packages for conventional glass containers has considerable potential to reduce the environmental impacts of wine packaging. The results of these LCA comparisons indicated that the net environmental burdens from each package life cycle broadly reflected the relative masses of the containers, with the exception of RFG bottles. Undertaking the packaging LCA at a municipal scale provides a more realistic appraisal of the potential environmental gains from package substitutions because, at this scale, one can address levels of product consumption, including variations in product types consumed (i.e., the sizes and types of packages). Together, the results from these LCAs can provide useful input for commercial procurement decisions, policy design, product design, and WM (Clearly, 2013).

8. CONCLUSIONS

1. Large amounts of unavoidable waste are generated by the fruit- and wine-processing industries. Depending on the raw materials and the technologies applied, solid wastes such as fruit skin, seed, pulp, peels, and stems often emerge as a considerable disposal problem for the beverage industry. FWWs can be characterized by the heterogeneity of their sources, variations in their chemical composition, and microbial instability. Rich in antioxidants, pectin, fibers, carbohydrates, organic acids, mineral salts, food flavors, and colorants, these compounds can be utilized as substrates for secondary processes or as ingredients for novel products. The most popular products derived from FWW are animal feed, biofuels, biopolymers, fertilizers, and ingredients for novel foodstuffs. An opportunity exists to employ some of the unavoidable soft fruit wastes as substrates for fruit wine fermentation processes, converting food industry by-products to beneficial resources.

2. Tropical fruits are especially favorable for wine production. Their unique flavor, attractive fragrance, and color make them highly desirable around the globe. Development of effective recycling techniques requires effective tapping of the maximum potential of the unavoidable tropical fruit wine waste components. For this purpose effective large-scale processing technologies have to be designed, including engineering of large-scale extraction techniques facilitating recovery of the desired materials, for example, bioactive compounds. Other relevant factors include finding the proper equipment and storage space for these phytochemical components, preventing their decay. Further study of chemical crystallization processes and their mathematical modeling will

also be required (Asyifah et al., 2014). Additional research requires the inclusion of specific applications to ensure industrial exploitation and sustainability of the final product, for example, functional ingredients with applications in food, nutraceutical, pharmaceutical, and cosmetic industries. Furthermore, the sensorial and nutritional aspects of the new food products derived from by-products have to be explored (Jahurul et al., 2015).

3. Yeast biomass, which is the main component of lees, has long been used in food and flavor industries in dried form. Although different applications include the recovery of ethanol, production of polyphenolic compounds, fertilizer, L-lactic acid production, and biogas, waste lees appear as an undervalued by-product so far. Subcritical water extraction was used to release the phenolics from yeast cells resulting from cider production at levels comparable to organic solvent extraction (Bahari, 2010). The release of phenolic compounds resulted in a significant change in the antioxidant activity of the extracts at high temperatures, showing their potentials as a source of highly active natural phenolics. Finally, the biorefinery approach for valorization of wine lees was proposed with production of polyhydroxyalkano-ates at the final stage. Synergy in the studies of metals and phenolic compounds could be essential for the design of lees treatment processes and the expansion of the derived bioproduct applications (Pérez-Bibbins et al., 2015).

4. A limited number of articles have been published about the treatment of WW from fruit wine production. Using the analogy with grape wine WW, one can conclude that application of combined methods like AOPs and biological treatment will lead to a higher level of COD reduction than any single-stage cleaning. The physical–chemical processes (coagulation or flocculation) have been found to be effective for the pretreatment of winery WW, particularly for lowering the TSS. The remaining organic contents can be further treated by other processes. The most efficient biological treatment technology includes the MBR, which has been already commercialized. Attention has focused on the use of membranes in conjunction with anaerobic systems. Integrated aerobic and anaerobic processing has been successfully used in brewery WW treatment. So far, the BATNEEC has not been developed because of the lack of information on detailed cost analyses. Accurate economic studies to determine whether the most efficient combined processes are cost-effective, environmentally sensitive, and technically reliable will allow their full application at the industrial scale (Ioannou et al., 2015).

5. Sustainability in the winemaking sector is based on the characterization of the environmental impact of orchard/vineyard and winery wastes. LCA was used to verify the proposed wine fabrication routes, weighing the environmental impact of fruit production as well as wine packaging. For the fabrication of cider, the emission of carbon dioxide produced during apple must fermentation and WW had a lower contribution to the total impact (under 5%). Thus, WM proved to be responsible for most of the total impact in the freshwater aquatic ecotoxicity category (85%), with management of solid waste as a key factor in the environmental impact resulting from cider-making (Herrero et al., 2013).

The scientific literature related to the application of LCA to orchard and fruit production is not particularly extensive, dated from 2005. Further research is expected to be conducted in particular among scientific communities in Asia, both as case studies and for the environmental evaluation of fruit commercialization, because of the high quantity of fruit produced and the new fruit wine belt formation in the Asian region.

REFERENCES

Abouelenien, F., Fujiwara, W., Namba, Y., Kosseva, M., Nishio, N., Nakashimada, Y., 2010. Improved methane fermentation of chicken manure via ammonia removal by biogas recycle. Bioresource Technology 101, 6368–6373.

ADEME, 2010. La méthode Bilan Carbone. Agence de L'Environnement et de la Maîtrise de l'Energie. Available from: www2.ademe.fr.

Ajila, C.M., Naidu, K.A., Bhat, S.G., Prasada Rao, U.J.S., 2007. Bioactive compounds and antioxidant potential of mango peel extract. Food Chemistry 105, 982–988.

Ajila, C.M., Rao, L.J., Rao, U.J.S.P., 2010. Characterization of bioactive compounds from raw and ripe *Mangifera indica* L. peel extracts. Food and Chemical Toxicology 48, 3406–3411.

Albuquerque, P.M., 2003. Estudo da producao de proteına microbiana a partir do bagaco de maca. UFSC, Florianopolis (Dissertation Master's degree in Food Engineering Foods).

Andreottola, G., Foladori, P., Ziglio, G., Cantaloni, C., Bruni, L., Cadonna, M., 2008. Methods for toxicity testing of xenobiotics in wastewater treatment plants and in receiving water bodies. In: Hlavinek, P., Bonacci, O., Marsalek, D.J., Mahrikova, I. (Eds.), Dangerous Pollutants in Urban Water Cycle. Springer, Netherlands, pp. 191–206.

APHA, 1992. Standard Methods for the Examination of Water and Wastewater. American Public Health Association.

Artiga, P., Ficara, E., Malpei, F., Garrido, J.M., Mendez, R., 2005. Treatment of two industrial wastewaters in a submerged membrane bioreactor. Desalination 179, 161–169.

Artiga, P., Carballa, M., Garrido, J.M., Mendez, R., 2007. Treatment of winery wastewaters in a membrane submerged bioreactor. Water Science & Technology 56, 63–69.

Asyifah, M.R., Lu, K., Ting, H.L., Zhang, D., 2014. Hidden potential of tropical fruit waste components as a useful source of remedy for obesity. Journal of Agricultural and Food Chemistry 62, 3505–3516.

Aziz, N.A.A., Wong, L.M., Bhat, R., Cheng, L.H., 2012. Evaluation of processed green and ripe mango peel and pulp flours (Mangifera indica var Chokanan) in term of chemical composition, antioxidant compounds and functional properties. Journal of the Science of Food and Agriculture 92, 557–563.

Effluent Management Guidelines for Australian Wineries and Distilleries, ANZECC (Agriculture and Resource Management Council of Australia and New Zealand), ARMCANZ (Australian and New Zealand Environment and Conservation Council), 1998. National Water Quality Management Strategy.

Bahari, A., 2010. Subcritical Water Mediated Hydrolysis of Cider Lees as a Route for Recovery of High Value Compounds (Ph.D. thesis). University of Birmingham, UK.

Bakowska-Barczak, A.M., Schieber, A., Kolodziejczyk, P., 2009. Characterization of Canadian black currant (*Ribes nigrum* L.) seed oils and residues. Journal of Agricultural and Food Chemistry 57, 11528–11536.

Bampidis, V.A., Robinson, P.H., 2006. Citrus by-products as ruminant feeds: a review. Animal Feed Science and Technology 128, 175–217.

Basset, N., López-Palau, S., Dosta, J., Mata-Álvarez, J., 2014. Comparison of aerobic granulation and anaerobic membrane bioreactor technologies for winery wastewater treatment. Water Science & Technology 69 (2), 320–327.

Barnes, D., Forester, C.F., Hrudly, S.E., 1984. Survey in Industrial Waste Treatment. Food and Allied Industries, vol. 1. Pitman Pub, Co, London.

Boluda-Aguilar, M., López-Gómez, A., 2013. Production of bioethanol by fermentation of lemon (*Citrus limon L.*) peel wastes pretreated with steam explosion. Industrial Crops and Products 41, 188–197.

Bolzonella, D., Fatone, F., Pavan, P., Cecchi, F., 2010. Application of a membrane bioreactor for winery wastewater treatment. Water Science & Technology 62, 2754–2759.

Brindle, K., Stephenson, T., Semmens, M.J., 1999. Pilot-plant treatment of a high strength brewery wastewater using a membrane-aeration bioreactor. Water Environment Research 71 (6), 1197–1203.

BSI, 2011. PAS 2050 Specification for the Assessment of the Life Cycle Greenhouse Gas Emissions of Goods and Services. British Standards, London, UK.

De Bustamante, I., Temiño, J., 1994. Vinasses purification model in carbonated materials by low-cost technologies: an example in the Llanura Manchega (Spain). Environmental Geology 24, 188–193.

Carbon Trust, 2007. Carbon Footprint Measurement Methodology. The Carbon Trust, London, UK.

California Regional Water Quality Control Board, 2002. ORDED No. R1-2002-0012, General Waste Discharge Requirements for Discharges of Winery Waste to Land.

Casey, E., Syron, E., Shanahan, J., Semmens, M., 2008. Comparative economic manalysis of full scale mabr configurations. In: IWA Membranes 2008. IWA, Amherst, MA.

Castro, C., Zuluaga, R., Putaux, J.L., Caro, G., Mondragon, I., Gañán, P., 2011. Structural characterization of bacterial cellulose produced by *Gluconacetobacter swingsii* sp. from Colombian agroindustrial wastes. Carbohydrate Polymers 84, 96–102.

Cerutti, A.K., Beccaro, G.L., Bruun, S., Bosco, S., Donno, D., Notarnicola, B., Bounous, G., 2014. LSA application in the fruit sector: state of the art and recommendations for environmental declarations of fruit products. Journal of Cleaner Production 73, 125–135.

Chan, Y.J., Chong, M.F., Law, C.L., Hassell, D.G., 2009. A review on anaerobic–aerobic treatment of industrial and municipal wastewater. Chemical Engineering Journal 155 (1–2), 1–18.

Cheong, M.W., Loke, X.Q., Liu, S.Q., Pramudya, K., Curran, P., Yu, B., 2011. Characterization of volatile compounds and aroma profiles of Malaysia pomelo [*Citrus grandis (L.) Osbeck*] blossom and peel. Journal of Essential Oil Research 23, 34–44.

Chiusano, L., Cerutti, A.K., Cravero, M.C., Bruun, S., Gerbi, V., 2015. An Industrial Ecology approach to solve wine surpluses problem: the case study of an Italian winery. Journal of Cleaner Production 91, 56–63.

Choi, I.S., Kim, J.-H., Wi, S.G., Kim, K.H., Bae, H.-J., 2013. Bioethanol production from mandarin (*Citrus unshiu*) peel waste using popping pretreatment. Applied Energy 102, 204–210.

Choi, I.S., Lee, Y.G., Khanal, S.K., Park, B.J., Bae, H.-J., 2015. A low-energy, cost-effective approach to fruit and citrus peel waste processing for bioethanol production. Applied Energy 140, 65–74.

Clearly, J., 2013. Life cycle assessments of wine and spirit packaging at the product and the municipal scale: a Toronto, Canada case study. Journal of Cleaner Production 44, 143–151.

Crizel de Moraes, T., Jablonski, A., Rios de Oliveira, A., Rech, R., Flôres Hickmann, S., 2013. Dietary fiber from orange byproducts as a potential fat replacer. LWT Food Science and Technology 53 (1), 9–14.

Cui, L., Wu, G., Jeong, T., 2010. Adsorption performance of nickel and cadmium ions onto brewer's yeast. The Canadian Journal of Chemical Engineering 88 (1), 109–115.

De Camillis, C., Bligny, J.C., Pennington, D., Pályi, B., 2012. Outcomes of the second workshop of the Food Sustainable Consumption and Production Round Table Working Group 1: deriving scientifically sound rules for a sector-specific environmental assessment methodology. The International Journal of Life Cycle Assessment 17, 511–515 Springer Berlin/Heidelberg.

Darah, I., Taufiq, M.M.J., Lim, S.H., 2013. Pomelo *Citrus grandis* (L.) Osbeck peel as an economical alternative substrate for fungal pectinase production. Food Science and Biotechnology 22 (6), 1683–1690.

Daud, N.H., et al., 2010. Mango extracts and the mango component mangiferin promote endothelial cell migration. Journal of Agricultural and Food Chemistry 58 (8), 5181–5186.

Dhillon, G.S., Brar, S.K., Verma, M., Tyagi, R.D., 2010. Recent advances in citric acid bio-production and recovery. Food and Bioprocess Technology 4, 505–529.

Ding, X., et al., 2013. Extracts of pomelo peels prevent high-fat dietinduced metabolic disorders in C57BL/6 mice through activating the PPARα and GLUT4 pathway. PLoS One 8 (10), 77915.

Donini, I.A.N., De Salvi, D.T.B., Fukumoto, F.K., Lustri, W.R., Barud, H.S., Marchetto, R., Messaddeq, Y., Ribeiro, S.J.L., 2010. Biosynthesis and recent advances in production of bacterial cellulose. Eclética Química 35, 165–178.

Driessen, W., Yspeert, P., Yspeert, P., Vereijken, T., May 24–26 2000. Compact Combined Anaerobic and Aerobic Process for the Treatment of Industrial Effluent, Environmental Forum, Columbia–Canada: Solutions to Environmental Problems in Latin America. Cartegena de Indias, Columbia.

Esmeraldo, M.A., Barreto, A.C.H., Freitas, J.E.B., Fechine, P.B.A., Sombra, A.S.B., Corradini, E., Mele, G., Maffezzoli, A., Mazzetto, S.E., 2010. Dwarf-green coconut fibers: a versatile natural renewable raw bioresource. Treatment, morphology, and physicochemical properties. BioResources 5 (4).

European Commission, 2001. Green Paper on Integrated Product Policy, 68 Final. COM. (Bruxelles, Belgium).

European Commission, November 2012. Environmental Assessment of Food and Drink Protocol. Draft Version 0.1 for Pilot Testing. European Food Sustainable Consumption and Production Roundtable.

European Commission, 2013. Product Environmental Footprint (PEF) Guide. European Commission. DG-JRC Ref. Ares (2012) 873782–17/07/2012.

European Council Directive (91/271/EEC). Concerning Urban Wastewater Treatment.

Fatmawati, A., Agustriyanto, R., Liasari, Y., 2013. Enzymatic hydrolysis of alkaline pretreated coconut coir. Bulletin of Chemical Reaction Engineering & Catalysis. http://bcrec.undip.ac.id.

Fernández-López, J., Fernández-Ginés, J.M., Aleson-Carbonell, L., Sendra, E., Sayas- Barberá, E., Pérez-Alvarez, J.A., 2004. Application of functional citrus by-products to meat products. Trends in Food Science & Technology 15, 176–185.

Ferreira, I., Pinho, O., Vieira, E., Tavarela, J.G., 2010. Brewer's *Saccharomyces* yeast biomass: characteristics and potential applications. Trends in Food Science & Technology 21 (2), 77–84.

FAOSTAT, 2013. Food and Agriculture Organization of the United Nations. Accessed December 2015 at: http://faostat.fao.org.

Fumi, M.D., Parodi, G., Parodi, E., Silva, A., Marchetti, R., 1995. Optimisation of long-term activated-sludge treatment of winery wastewater. Bioresource Technology 52 (1), 45–51.

Ganesh, R., Rajinikanth, R., Thanikal, J.V., Ramanujam, R.A., Torrijos, M., 2010. Anaerobic treatment of winery wastewater in fixed bed reactors. Bioprocess and Biosystems Engineering 33 (5), 619–628.

Górecka, D., Pacholek, B., Dziedzic, K., 2010. Raspberry pomace as a potential fiber source for cookies enrichment. Acta Scientiarum Polonorum, Technologia Alimentaria 9 (4), 451–462.

Häkkinen, S., Heinonen, M., Kärenlampi, S., Mykkänen, H., Ruuskanen, J., Törrönen, R., 1999. Screening of selected flavonoids and phenolic acids in 19 berries. Food Research International 32 (5), 345–353.

Herrero, M., Laca, A., Diaz, M., 2013. Life cycle assessment focusing on food industry wastes. In: Kosseva, M., Webb, C. (Eds.), Food Industry Wastes: Assessment and Recuperation of Commodities. Academic Press, Elsevier, San Diego, USA, pp. 268–280.

Hu, A.Y., Stuckey, D.C., 2006. Treatment of dilute wastewaters using a novel submerged anaerobic membrane bioreactor. Journal of Environmental Engineering 132 (2), 190–198.

Illanes, A., Gorgollón, Y., 1986. Kinetics of extraction of invertase from autolysed bakers' yeast cells. Enzyme and Microbial Technology 8 (2), 81–84.

Ingwersen, W., Stevenson, M.J., 2012. Can we compare the environmental performance of this product to that one? an update on the development of product category rules and future challenges toward alignment. Journal of Cleaner Production 24, 102–108.

Ioannou, L.A., Li Puma, G., Fatta-Kassinos, D., 2015. Treatment of winery wastewater by physicochemical, biological and advanced processes: a review. Journal of Hazardous Materials 286, 343–368.

ISO, 2013a. ISO/TS 14067 Greenhouse Gases Carbon Footprint of Products Requirements and Guidelines for Quantification and Communication.

ISO, 2013b. ISO/DIS 14046 Environmental Management Water Footprint Principles, Requirements and Guidelines.

Jackson, C., 2010. How to Make Coconut Wine or the Process of Making Tuba Wine. Food and Drink: Wine Spirits. http://EzineArticles.com/expert/C._Jackson/808631.

Jacobi, K.K., Macrae, E.A., Hetherington, E.H., 2001. Postharvest heat disinfestation treatments of mango fruit. Scientia Horticulturae 89, 171–193.

Jahurul, M.H., Zaidul, I.S.M., Norulaini, N.A.N., Sahena, F., Abedin, M.Z., Ghafoor, K., Mohd Omar, A.K., 2014. Characterization of crystallization and melting profiles of blends of mango seed fat and palm oil mid-fraction as cocoa butter replacers using differential scanning calorimetry and pulse nuclear magnetic resonance. Food Research International 55, 103–109.

Jahurul, M.H.A., Zaidul, I.S.M., Ghafoor, K., Al-Juhaimi, F.Y., Kar-Lin, N., Norulaini, N.A.N., Sahena, F., Mohd Omar, A.K., 2015. Mango (*Mangifera indica L.*) by-products and their valuable components: a review. Food Chemistry 183, 173–180.

Jin, H., Kim, H.S., Kim, S.K., Shin, M.K., Kim, J.H., Lee, J.W., 2002. Production of heteropolysaccharide-7 by *Beijerinckia indica* from agroindustrial byproducts. Enzyme and Microbial Technology 30, 822–827.

Joshi, V.K., Sandhu, D.K., Attri, B.L., Walia, R.K., 1991. Cider preparation from apple juice concentrate and its consumer acceptability. Indian Journal of Horticulture 48, 321–327.

Joshi, V.K., 1997. Fruit wines. In: Directorate of Extension Education, second ed. Dr. YS Parmar University of Horticulture and Forestry, Nauni, Solan (HP), p. 255.

Joshi, C., 2000. Food processing waste treatment technology. In: Verma, L.R., Joshi, V.K. (Eds.), Postharvest Technology of Fruits and Vegetables, Vol. 1. Indus Publishing Co, New Delhi, p. 440.

Joshi, V.K., Sandhu, D.K., 1996. Preparation and evaluation of an animal feed byproduct produced by solid state fermentation of apple pomace. Bioresource Technology 56, 251–255.

Jozala, A.F., Pértile, R.A.N., dos Santos, C.A., de Carvalho Santos-Ebinuma, V., Seckler, M.M., Gama, F.M., Pessoa Jr., A., 2015. Bacterial cellulose production by *Gluconacetobacter xylinus* by employing alternative culture media. Applied Microbiology and Biotechnology 99, 1181–1190.

Kachrimanidou, V., Kopsahelis, N., Webb, C., Koutinas, A.A., 2014. Bioenergy technology and food industry waste valorization for integrated production of polyhydroxyalkanoates. In: Gupta, V.K., et al. (Ed.), Bioenergy Research: Advances and Applications. Elsevier, Amsterdam, pp. 419–433.

Keyser, M., Witthuhn, R., Ronquest, L., Britz, T., 2003. Treatment of winery effluent with upflow anaerobic sludge blanket (UASB)-granular sludges enriched with *Enterobacter sakazakii*. Biotechnology Letters 25 (22), 1893–1898.

Kim, Y., Lounds-Singleton, A.J., Talcott, S.T., 2009. Antioxidant phytochemical and quality changes associated with hot water immersion treatment of mangoes (*Mangiferaindica L.*). Food Chemistry 115, 989–993.

Kittiphoom, S., 2012. Utilization of mango seed. International Food Research Journal 19, 1325–1335.

Kongruang, S., 2008. Bacterial cellulose production by *Acetobacter xylinum* strains from agricultural waste products. Applied Biochemistry and Biotechnology 148 (1–3), 245–256.

Kosseva, M.R., 2011. Wastes from agriculture, forestry and food processing; management and processing of food wastes. In: Moo-Young, M. (Ed.), Comprehensive Biotechnology, vol. 6. second ed. Elsevier, pp. 557–593.

Kosseva, M., 2013a. Recent EU legislation on management of wastes in the food industry. In: Kosseva, M., Webb, C. (Eds.), Food Industry Wastes: Assessment and Recuperation of Commodities. Academic Press, Elsevier, San Diego, USA, pp. 3–16.

Kosseva, M., 2013b. Development of green production strategies. In: Kosseva, M., Webb, C. (Eds.), Food Industry Wastes: Assessment and Recuperation of Commodities. Academic Press, Elsevier, San Diego, USA, pp. 17–36.

Kosseva, M., 2013c. Sources, characterization, and composition of food industry wastes. In: Kosseva, M., Webb, C. (Eds.), Food Industry Wastes: Assessment and Recuperation of Commodities. Academic Press, Elsevier, San Diego, USA, pp. 37–60.

Kumar, A., Arienzo, M., Quayle, W., Christen, E., Grocke, S., Fattore, A., Doan, H., Gonzago, D., Zandonna, R., Bartrop, K., 2009. Developing a Systematic Approach to Winery Wastewater Management, Report CSL05/02 (In: Final report to Grape and Wine Research and Development Corporation, CSIRO Land and Water Science Report Adelaide).

Kumar, B.D., et al., 2013. Effect of mangiferin and mahanimbine on glucose utilization in 3T3-L1 cells. Pharmacognosy Magazine 9 (33), 72–75.

Kurosumi, A., Sasaki, C., Yamashita, Y., Nakamura, Y., 2009. Utilization of various fruit juices as carbon source for production of bacterial cellulose by *Acetobacter xylinum* NBRC 13693. Carbohydrate Polymers 76, 333–335.

Lajoie, C.A., Lin, S.-C., Nguyen, H., Kelly, C.J., 2002. A toxicity testing protocol using a bioluminescent reporter bacterium from activated sludge. Journal of Microbiological Methods 50, 273–282.

Lajoie, C.A., Lin, S.C., Kelly, C.J., 2003. Comparison of bacterial bioluminescence with activated sludge oxygen uptake rates during zinc toxic shock loads in a wastewater treatment system. Journal of Environmental Engineering 129, 879–883.

Lamoolphak, W., Goto, M., Sasaki, M., 2007. Hydrothermal decomposition of yeast cells for production of proteins and amino acids. Journal of Hazardous Materials B137, 1643–1648.

Lee, K.C., Rittmann, B.E., 2002. Applying a novel autohydrogenotrophic hollow-fiber membrane biofilm reactor for denitrification of drinking water. Water Research 36 (8), 2040–2052.

Levenspiel, O., 1999. Chemical Reaction Engineering, third ed. J.Wiley & Sons, New York.

Lin, C.S.K., Pfaltzgraff, L.A., Herrero-Davila, L., Mubofu, E.B., Abderrahim, S., Clark, J.H., et al., 2013. Food waste as a valuable resource for the production of chemicals, materials and fuels. Current situation and global perspective. Energy & Environmental Science 6, 426–464.

Lofrano, G., Meric, S., 2014. A comprehensive approach to winery wastewater treatment: a review of the state-of the-art. Desalination and Water Treatment 1–18.

Lucas, M.S., Peres, J.A., Li Puma, G., 2010. Treatment of winery wastewater by ozone-based advanced oxidation processes (O_3, O_3/UV and O_3/UV/H_2O_2) in a pilot-scale bubble column reactor and process economics. Separation and Purification Technology 72 (3), 235–241.

Madoni, P., Davoli, D., Gorbi, G., Vescovi, L., 1996. Toxic effect of heavy metals on the activated sludge protozoan community. Water Research 30, 135–141.

Martin, K.J., Nerenberg, R., 2012. The membrane biofilm reactor (MBfR) for water and wastewater treatment: principles, applications, and recent developments. Bioresource Technology 122, 83–94.

Mirabella, N., Castellani, V., Sala, S., 2014. Current options for the valorization of food manufacturing waste: a review. Journal of Cleaner Production 65, 28–41.

Mohite, B.V., Patil, S.V., 2014. A novel biomaterial: bacterial cellulose and its new era applications. Biotechnology and Applied Biochemistry 61 (2), 101–110.

Molina, F., Ruiz-Filippi, G., Garcia, C., Roca, E., Lema, J., 2007. Winery effluent treatment at an anaerobic hybrid USBF pilot plant under normal and abnormal operation. Water Science & Technology 56, 25–31.

Montalvo, S., Guerrero, L., Borja, R., Cortes, I., Sanchez, E., Colmenarejo, M.F., 2008. Treatment of wastewater from red and tropical fruit wine production by zeolite anaerobic fluidized bed reactor. Journal of Environmental Science and Health, Part B 43, 437–442.

Motlagh, A.R.A., LaPara, T.M., Semmens, M.J., 2008. Ammonium removal in advective-flow membrane-aerated biofilm reactors (AF-MABRs). Journal of Membrane Science 319 (1–2), 76–81.

Naik, A., Raghavendra, S.N., Raghavarao, K.S.M.S., 2012. Production of coconut protein powder from coconut wet processing waste and its characterization. Applied Biochemistry and Biotechnology 167, 1290–1302.

Nithitanakool, S., Pithayanukul, P., Bourgeois, S., Fessi, H., Bavovada, R., 2013. The development, physicochemical characterization, and in vitro drug release studies of pectinate gel beads containing Thai mango seed kernel extract. Molecules 18, 6504–6520.

Notarnicola, B., Tassielli, G., Renzulli, P.A., 2012. Modeling the agri-food industry with life cycle assessment. In: Curran, M. (Ed.), Life Cycle Assessment Handbook: A Guide for Environmentally Sustainable Products. Wiley, Scrivener Publishing, Salem, MA.

Nyberg, L., Turco, R.F., Nies, L., 2008. Assessing the impact of nanomaterials on anaerobic microbial communities. Environmental Science & Technology 42, 1938–1943.

Oliveira, M., Duarte, E., 2010. Guidelines for the management of winery wastewaters. In: Treatment and Use of Non-conventional Organic Residues in Agriculture, RAMIRAN International Conference Lisboa, Portugal, 12–15 September, 2010.

Orescanin, V., Kollar, R., Nad, K., Mikelic, I.L., Gustek, S.F., 2013. Treatment of winery wastewater by electrochemical methods and advanced oxidation processes. Journal of Environmental Science and Health, Part A 48 (12), 1543–1547.

Ortega, R.M., López-Sobaler, A.M., Requejo, A.M., Andrés, P., 2004. La Composición de los Alimentos. In: Herramienta básica para la valoración nutricional, first ed. Tabla de composición de alimentos.

Owen, W.F., Stuckev, D.C., Healv, J.B., Young, L.Y., McCarty, P.L., 1979. Bioassay for monitoring biochemical methane potential and anaerobic toxicity. Water Research 13, 485–492.

Panesar, P.S., Chavan, Y., Chopra, H.K., Kennedy, J.F., 2012. Production of microbial cellulose: response surface methodology approach. Carbohydrate Polymers 87, 930–934.

Pankhania, M., Brindle, K., Stephenson, T., 1999. Membrane aeration bioreactors for wastewater treatment: completely mixed and plug-flow operation. Chemical Engineering Journal 73 (2), 131–136.

Pap, N., Pongrácz, E., Myllykoski, L., Keiski, R., 2004. Waste minimization and utilization in the food industry: processing of arctic berries, and extraction of valuable compounds from juice- processing by-products. In: Pongrácz, E. (Ed.), Proceedings of the Waste Minimization and Resources Use Optimization Conference. June 10th 2004. University of Oulu, Finland. Oulu University Press, Oulu, pp. 159–168.

Parnaudeau, V., Condom, N., Oliver, R., Cazevieille, P., Recous, S., 2008. Vinasse organic matter quality and mineralization potential, as influenced by raw material, fermentation and concentration processes. Bioresource Technology 99, 1553–1562.

Pérez-Bibbins, B., Torrado-Agrasar, A., Salgado, J.M., Pinheiro de Souza Oliveira, R., Domínguez, J.M., 2015. Potential of lees from wine, beer and cider manufacturing as a source of economic nutrients: an overview. Waste Management 40, 72–81.

Pérez-Serradilla, J.A., Luque de Castro, M.D., 2008. Role of lees in wine production: a review. Food Chemistry 111, 447–456.

Petruccioli, M., Duarte, J., Federici, F., 2000. High-rate aerobic treatment of winery wastewater using bioreactors with free and immobilized activated sludge. Journal of Bioscience and Bioengineering 90 (4), 381–386.

Le Quéré, J.M., Husson, F., Renard, C.M.G.C., Primault, J., 2006. French cider characterization by sensory, technological and chemical evaluations. LWT Food Science and Technology 39 (9), 1033–1044.

Raihana Noor, A.R., Marikkar, J.M.N., Amin, I., Shuhaimi, M., 2015. A review on food values of selected tropical fruits' seeds. International Journal of Food Properties 18 (11), 2380–2392.

Rashwan, M.R.A., 1990. Fatty acids composition, neutral lipids, and phospholipids fractionation in the kernel lipids of the mango varieties. Journal of Agricultural Science 21, 105–117.

Rittmann, B.E., 2006. The membrane biofilm reactors: the natural partnership of membranes and biofilm. Water Science & Technology 53 (3), 219–225.

Rozzi, A., Remigi, E., 2004. Methods of assessing microbial activity and inhibition under anaerobic conditions: a literature review. Reviews in Environmental Science and Bio/Technology 93, 93–115.

Ruiz, C., Torrijos, M., Sousbie, P., Martinez, J.L., Moletta, R., Delgenes, J., van Lier, J., Lubberding, H., 2001. Treatment of winery wastewater by an anaerobic sequencing batch reactor. In: Anaerobic Digestion: Concepts, Limits and Perspectives. 9th World Congress on Anaerobic Digestion, Antwerp, Belgium, pp. 219–224.

Santi, G., Crognale, S., D'Annibale, A., Petruccioli, M., Ruzzi, M., Valentini, R., Moresi, M., 2014. Orange peel pretreatment in a novel lab-scale direct steam-injection apparatus for ethanol production. Biomass Bioenergy 61, 146–156.

Sareena, C., Ramesan, M.T., Purushothaman, E., 2012. Utilization of peanut shell powder as a novel filler in natural rubber. Journal of Applied Polymer Science 125, 2322–2334.

Schieber, A., 2009. Nutraceuticals from By-Products of Plant Food Processing. Canadian Chemical News.

Semmens, M.J., 2005. Membrane Technology: Pilot Studies of Membrane-Aerated Bioreactors. Water Environment Research Foundation and the International Water Association.

Senior, P.J., Collins, S.H., Richardson, K.R., 1986. Copolymers of Poly(β-hydroxybutyric Acid) and Poly(β-hydroxyvaleric Acid) Are Produced by Culturing Alcohol-utlising Strains of Alcaligenes eutrophus on a Carbon Source Including Primary Alcohols Having an Odd Number of Carbon Atoms Such as Propan-1-ol. European Patent Office. Publication Number 0204442 A2.

Shah, N., Ul-Islam, M., Khattak, W.A., Park, J.K., 2013. Overview of bacterial cellulose composites: a multipurpose advanced material. Carbohydrate Polymers 98, 1585–1598.

Shalini, R., Gupta, D., 2010. Utilization of pomace from apple processing industries: a review. Journal of Food Science and Technology 47 (4), 365–371.

Simate, G.S., Cluett, J., Iyuke, S.E., Musapatika, E.T., Ndlovu, S., Walubita, L.F., Alvarez, A.E., 2011. The treatment of brewery wastewater for reuse: state of the art. Desalination 273, 235–247.

Sogi, D.S., Siddiq, M., Greiby, I., Dolan, K.D., 2013. Total phenolics, antioxidant activity, and functional properties of 'Tommy Atkins' mango peel and kernel as affected by drying methods. Food Chemistry 141, 2649–2655.

Solís-Fuentes, J.A., Durán-de-Bazúa, M.C., 2011. Mango (Mangifera indica L.) seed and its fats. In: Preedy, V., Watson, R.R., Patel, V.B. (Eds.), Nuts and Seeds in Health and Disease Prevention. Academic Press, San Diego, pp. 741–748.

Stephenson, T., Judd, S., Jefferson, B., Brindle, K., 2000. Membrane Bioreactors for Wastewater Treatment. IWA, London.

Stricker, A.-E., Lossing, H., Gibson, J.H., Hong, Y., Urbanic, J.C., 2011. Pilot scale testing of a new configuration of the membrane aerated biofilm reactor (MABR) to treat high-strength industrial sewage. Water Environment Research 83 (1), 3–14.

Sun, J., Hu, X., Zhao, G., Wu, J., Wang, Z., Chen, F., Liao, X., 2007. Characteristics of thin layer infrared drying of apple pomace with and without hot air pre-drying. Food Science and Technology International 13 (2), 91–97.

Taing, M.W., et al., 2012. Mango fruit peel and flesh extracts affect adipogenesis in 3T3-L1 cells. Food & Function 3 (8), 828–836.

Vaithanomsat, P., Apiwatanapiwat, W., Chumchuent, N., Kongtud, W., Sundhrarajun, S., 2011. The potential of coconut husk utilization for bioethanol production. Kasetsart Journal (Natural Science) 45, 159–164.

Valderrama, C., Ribera, G., Bahı, N., Rovira, M., Gimenez, T., Nomen, R., Lluch, S., Yuste, M., Martinez-Llado, X., 2012. Winery wastewater treatment for water reuse purpose: conventional activated sludge versus membrane bioreactor. Desalination 306, 1–7.

Vendruscolo, F., Albuquerque, P.M., Streit, F., Esposito, E., Ninow, J.L., 2008. Apple pomace: a versatile substrate for biotechnological applications. Critical Reviews in Biotechnology 28, 1–12.

Villas-Boas, S.G., Esposito, E., 2000. Bioconversao do bagaco de maca: enriquecimento nutricional utilizando fungos para produção de um alimento alternativo de alto valor agregado. Biotecnologia Ciência & Desenvolvimento 14, 38–42.

Waldron, K.W., 2007. Handbook of waste management and co-product recovery in food processing. In: Woodhead Publishing Series in Food ScienceTechnology and Nutrition No. 141, vol. 1.

WBCSD/WRI, 2009. The Greenhouse Gas Protocol. A Corporate Accounting and Reporting Standard. World Resources Institute-World Business Council for Sustainable Development, Washington, DC, USA.

Wei, X., Li, B., Zhao, S., Qiang, C., Zhang, H., Wang, S., 2012. COD and nitrogen removal in facilitated transfer membrane-aerated biofilm reactor (FT-MABR). Journal of Membrane Science 389, 257–264.

Wilkinson, A.S., et al., 2011. Bioactivity of mango flesh and peel extracts on peroxisome proliferator-activated receptor γ[PPARγ] activation and MCF-7 cell proliferation: fraction and fruit variability. Journal of Food Science 76 (1), H11–H18.

Woolford, et al., 2013. Patent US 2013/0090378 A1. Process for Extracting Compound from Cranberry Leaves and Related Products.

World Bank, 1996. Pollution Prevention and Abatement: Fruit and Vegetable Processing. Draft Technical Background Document. Environment Department, Washington, D.C.

WRAP, 2010. A Review of Waste Arisings in the Supply of Food and Drink to Households in the UK. (Banbury, UK).

WRI, March 20, 2015. FLW Protocol Accounting and Reporting Standard (FLW Standard) Draft. Retrieved from: http://www.wri.org/our-work/project/food-loss-waste-protocol/publications (23.10.15).

Xiao, Y., De Araujo, C., Sze, C.C., Stuckey, D., 2015. Toxicity measurement in biological wastewater treatment processes: a review. Journal of Hazardous Materials 286, 15–29.

Yu, J-xia, et al., 2009. Poly(Amic acid)-modified biomass of baker's yeast for enhancement adsorption of methylene blue and basic magenta. Applied Biochemistry and Biotechnology 160 (5), 1394–1406.

Zein, R.E., El-Bagoury, A.A., Kassab, H.E., 2005. Chemical and nutritional studies on mango seed kernel. Journal of Agricultural Science 30, 3285–3299.

BIOREFINERY CONCEPT APPLIED TO FRUIT WINE WASTES

12

M.-P. Zacharof

Swansea University Medical School, Swansea, United Kingdom

1. INTRODUCTION

The consumption of fruit wines is an emerging trend across the world. Although apple and pear cider, are well established alcoholic beverage alternatives to grape wine, other fruit wines are also becoming popular around the globe. Stone fruits such as plums, cherries, apricots; tropical fruits like mango and coconuts; and citrus fruits such as oranges, lemons, limes, or grapefruits are being used to produce wines of varying alcoholic content between 2% and 24% v/v.

The market size of fruit wines is rather difficult to estimate precisely mostly because these products are not classified as a distinct group. They are considered as sweet wines, namely dessert wines. It could be roughly estimated that their market size is about 2% of the total wine market which is estimated at about 2.8 billion cases of wines per year (Christ, 2014).

In Northern Europe, apple and pear cider are very popular, but they are considered of lesser quality when compared with grape wines. From a socioeconomic perspective, they are usually cheaper in price, of lower alcoholic content, and are consumed in a similar manner as beer, as a daily, casual alcoholic drink. On the other hand, fruit wines such as cherry wine are more sophisticated in taste, of varying, often higher, alcoholic content, and seen mostly as a luxury.

Certainly the production of fruit wines does generate a considerable amount of waste. There are not many reports regarding the size of production, which is often limited to small and medium enterprises rather than mass production. It is not straightforward to estimate the amount of waste generation regarding each fruit used for wine production. For example, for apple juice and cider generation, apple pomace, a mixture of peel, pulp, and seeds that represents about 25–35% of the processed apple mass, is reaching 1300 million kg (Carson et al., 1994) with disposal costs at about $10 million per year (Worall and Yang, 1992).

Such a large market demands a continuous production of the merchandise to be maintained. Consequently, intensive cultivation of land, harvesting of the goods, and fully organized manufacturing is implemented. Fruit winemaking is a timed multistage process producing a significant amount of predominantly organic waste. The winemaking industry, including fruit wines and grape-based wine alike, has been portrayed positively due to the socioeconomic and cultural benefits attributed. Regardless of the amounts of waste generated, the great use of water resources, and the large land usage, the industry has not been viewed negatively by the general public, encouraging its development and, therefore, the generation of higher amounts of waste. Much of the waste produced has been recycled as animal feed or fertilizer while still a large proportion is being landfilled.

Science and Technology of Fruit Wine Production. http://dx.doi.org/10.1016/B978-0-12-800850-8.00012-0

Waste can be seen as a virtually inexhaustible resource, being utilized in industrial markets to generate combined heat and power, fertilizers, chemicals, feeds, and food in the developed world (Appels et al., 2008; Frenkel et al., 2013). Within the next decade, driven largely by legislative, environmental, economic, and social drivers, these markets will be further developing. They will also be shifting into recovering chemicals and generating energy from waste to reduce the carbon footprint of their production. This will limit their manufacture by utilizing natural resources, achieving environmental sustainability and constituting waste safe for environmental discharge in the form of particle, nutrient free, and sterile effluents. Therefore, the utilization of waste as a valuable commodity and platform chemicals resource is an important step to the development and deployment of alternative sources for energy production (Hatti-Kaul et al., 2007; Zacharof et al., 2015, 2016).

Conventional treatment of waste is costly, demanding significant amounts of effort, resources, and energy for waste to be safely discharged into the environment (Østergaard, 2012). Currently living in a knowledge-driven economy, with growing awareness over environmental protection due to climate change and natural resources exhaustion; the need to recycle, reuse, and recover energy and valuable chemicals from waste and wastewater becomes apparent (The Biocomposites Centre Report, 2011).

Therefore, the overall aim of this chapter is to explore schemes that could be applied at an industrial scale to valorize fruit wine waste, introducing the concept of biorefinery, ergo the use of fruit wine waste as source of platform chemicals, fuel, heat, and energy.

1.1 GENERATION OF ENERGY AND PRODUCTS FROM ALTERNATIVE SOURCES: THE BIOREFINERY CONCEPT

Using agricultural goods for the production of other commodities is a well established concept. The use of plant based biomass as a raw material for the production of numerous products using complex physicochemical processing methods, a concept similar to a petroleum refinery, is a rather new idea, first initiated at the end of 20th century (Bustamante et al., 2005, 2008; Wadhwa et al., 2013).

This approach, though successful to an extent, has several disadvantages. Biomass is a rich source of lignin, carbohydrates, proteins, and fats, containing in smaller amounts vitamins, dyes, and flavors (de Villiers et al., 2012; Ioannou et al., 2014). It has to be intensively cultivated to produce considerable amounts of feedstock for the generation of fuels, chemicals, and power. This has led to land competition for crop development, potential shortage of feedstock to address needs in livestock feed, human food, and export market as well as possible water shortage.

Thus it has been a shift from a whole crop concept, where an entire crop of wheat, rye, corn, or triticale is used as feedstock, to a waste-based concept of mainly lignocellulosic feedstock, comprised of hard fibrous plant materials generated by municipal and industrial waste (Morley, 2008). This approach, although beneficial, has been difficult to apply due to the extensive demand in pretreatment (enzymatic hydrolysis or chemical digestion of cellulosic and hemicellulosic material).

On the other hand, several researchers (Tyagi and Lo, 2013; Zacharof and Lovitt, 2014a) have highlighted the importance of recycling all types of waste including domestic, municipal, industrial through bioconversion, i.e., applying a biorefinery (Fig. 12.1) concept but with waste as the main feedstock.

This approach has been supported by numerous governmental and nongovernmental bodies national and international but most importantly from the European Union (Legislative proposal

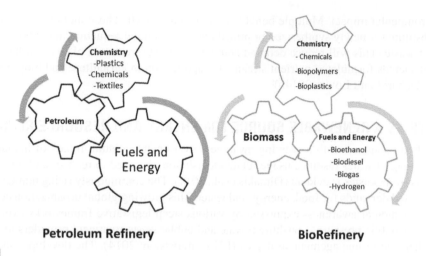

FIGURE 12.1

The petroleum refinery and the biorefinery concept.

02/07/2014 of EU Waste Framework Directive, 2008/98/EC). It has been suggested that the recycling and preparing for reuse of municipal waste is to be increased to 70% by 2030. In addition, landfilling will be phased out by 2025 for recyclables (i.e., plastics, paper, metals, glass, biowaste) corresponding to a maximum landfilling rate of 25%.

Among the various types of waste agricultural waste is a possible choice to be used as biorefinery feedstock. Agricultural waste contains various valuable chemicals, including carbohydrates, proteins, volatile fatty acids (VFA), nitrogen as ammonia, phosphate, and metals. Phosphate rock, is a nonrenewable natural resource, of critical importance because of its many applications including drinking water softening, feed and food additives, and fertilizers. Regardless of its production being carbon neutral, mining of phosphate is becoming more expensive and supply risks related to environmental and sociopolitical issues have risen. It has been reported that by 2035 the demand for phosphorus will outpace the supply. On the other hand, phosphorus removal from wastewater has to be improved as water discharge standards become more stringent, rising with the costs of wastewater treatment (Zacharof and Lovitt, 2013).

Other resources included in waste of agricultural sources such as ammonia, which has a current market value of $800/ton, has reached a global consumption of over 150 million tons. As well as being used heavily in fertilizers it is also an important component of various commercial and industrial products. It has a large production carbon footprint (common practice being 4 tons of CO_2 per ton of ammonia) as during its synthesis; methane is reformed to produce H_2 and CO_2. In addition, the disposal and return of ammonia to the atmosphere through nitrification and denitrification adds additional costs to wastewater treatment. Substantial value exists in the high contents of metal ions in agricultural waste and other types of industrial wastes.

Therefore, reclaiming these valuable chemicals into formulated feedstock suitable for biochemical conversion to industrially relevant products is an important step to take to improve sustainability and reduce environmental impact. Multiple benefits lie in this approach. These include recycled materials that will substitute for newly synthesized or mined materials and the reduction in the volume and concentration of waste. This will reduce demand and costs in waste treatment plants. Furthermore, valuable streams such as formulated nutrient streams for application in agriculture and bioprocessing will be created (Zacharof and Lovitt, 2014b).

1.2 WASTE AS A RENEWABLE SOURCE FOR ENERGY AND RESOURCE RECOVERY

In a low-carbon economy, with a growing awareness over environmental impact of human activities and strengthening of water resource usage related legislation, the need to recover and produce energy and chemicals from wastes is evident (Dimakis et al., 2011). The continuously rising human population results in elevated demands for food, energy, and water. This growing global urbanization coupled with elevated environmental awareness expressed by various steep legislative frameworks over waste disposal as well as public pressure is pushing private and public waste treatment providers to review and reengineer their waste management strategies (EU Commission, 2014). The development of a waste management methodology is of great interest to various groups such as contractors, engineering consultants, equipment providers, policy regulators (agencies, politicians, think tanks), and the general public, and it is to be decided upon the needs of the community in a microscale but also of the general good in a macroscale (Fig. 12.2; Tyagi and Lo, 2013).

Waste can be divided into numerous categories (Fig. 12.3) according to type, origin, or state of matter. Not all waste types are suitable to use as biorefinery feedstock, since several complications,

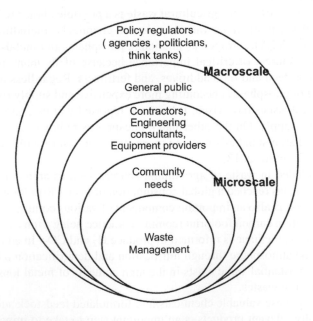

FIGURE 12.2

Decision-making process regarding waste management.

due to the complex physicochemical nature, might occur. Implications relevant to transportation or the need of extensive costly pretreatment might hinder the use of, for example, construction waste, where due to its heavily mixed nature and current ways of collection is unsuitable for such an approach (Wang et al., 2013).

Waste generated from the food, feed, and agricultural industries is possibly the best candidate for the biorefinery approach. Food production has become heavily industrialized and therefore regulated, generating tons of waste per annum (Fernando et al., 2006; Geoffrey et al., 2011).

The food industry is shifting toward the intensive production of ready to eat foods that are consumed in venues having fewer conventional methods of stabilizing food, therefore resulting in even larger amounts of waste (Jones et al., 2005). Apart from the directly occurring waste due to food processing (slaughterhouse, dairy, wheat and corn milling, sugar and starch processing, vegetative

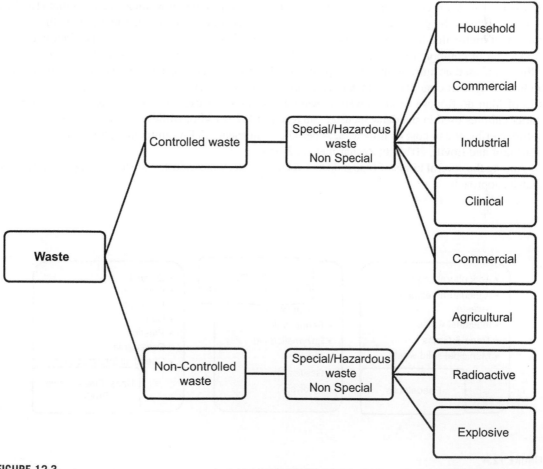

FIGURE 12.3

Waste categories and types (Tyagi and Lo, 2013).

processing, fish and poultry processing, alcoholic and nonalcoholic beverages, and soft drinks manufacturing and processing) the food industry is associated with agricultural waste (organic waste and agricultural residues) produced due to the intensive animal and crop farming to satisfy the food consumption demand, reaching 264,854 tons per annum (GOV.UK, 2011–2014) in United Kingdom alone. Agricultural waste is ranging third in terms of waste industry size, comparable only with municipal solid waste (Jefferson, 2008; Li and Yu, 2011), which imposes environmental constraints since conventional treatments, such as landfilling or landspreading, may cause eutrophication and land and water toxicity due to freely available nutrients and metals, spread in water and soil. There are also human health concerns due to land-related pathogenicity contained in the raw materials (Zacharof and Lovitt, 2014a,b).

Industrial wastewaters from food processing industries, wineries, breweries, and agricultural wastewater from animal confinements are suitable for biotechnological production of high-value and platform chemicals (Angenent et al., 2004; Lin and Tanaka, 2006); their effective formulation remains a necessity. The main goal of a biorefinery (Fig. 12.4) is to produce low-value high volume (LVHV) products to meet the global demand of energy, platform chemicals etc. at the same time with the production of high-value low volume products (HVLV) to enhance the profitability of the plant while the production of combined heat and power (CHP) can be used to reduce the cost of processing. These effluents, if used as nutrient media, are potentially highly profitable, especially when compared to the traditional synthetic media or that derived from food sources such as crops. For example, the cost per kilo of Man de Rogosa broth, a well-known nutrient medium used in research and development of starter cultures used in the dairy industry can reach $1311 per kilo, while a formulated waste deriving nutritive effluent can cost as little as $2.4 per kilo of nutrients (VFA, ammonia, phosphate) recovered (Zacharof and Lovitt, 2014a,b; Zacharof et al., 2015).

Among the several kinds of food industry–related waste, wine industry waste is of major interest for such an approach.

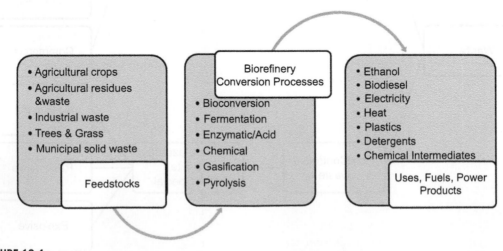

FIGURE 12.4

Use of waste streams within the biorefinery concept (Fernando et al., 2006).

1.3 BIOREFINERY FEEDSTOCK: THE FRUIT WINERY WASTE

1.3.1 The Fruit Winemaking Process

Fruit wines can be manufactured either of fruit juice concentrate or by the whole fruit, following a similar procedure of manufacturing as grape wines. Fruit wines have several types, being very versatile and suitable to accommodate the varying tastes of consumers. There are six popular types of fruit wines: the low alcohol content (2–7% v/v) known also as cider; the dry fruit wines of higher alcoholic content, usually 8.5–13.5% v/v alcohol; the sweet fruit wines usually manufactured of berry fruits of varying alcoholic content; the cryo-extracted fruit wines that are considered premium products; the fortified or "port-style" fruit wines that can have up to 24% v/v alcohol; and finally, the sparkling fruit wines of Charmat or Champenoise style, where carbon dioxide is injected to create the sparkling effect (Rivard, 2009).

Cider-style wines are commonly made of apple and pear fruits. Often apple is used as a base and then is blended with other fruits to create different flavors such as berry fruits, for example, strawberries or blackcurrant, or citrus fruits like lemon, lime, or orange. These wines are highly diluted to achieve a lower alcoholic content. They are produced in large volumes due to high popularity, especially in Western and Central Europe; they are also not highly taxed resulting into low prices and high consumption (Rivard, 2009).

Dry fruit wines have the highest appeal to the casual wine drinker, with a content of residual sugars (RS) lower than 30 g/L and an alcoholic content up to 13.5% v/v alcohol. Sweet fruit wines are produced mainly of berry fruits such as raspberries or currants, with an RS content higher that 40 g/L, while the total acidity is above 7–8 g/L (Rivard, 2009).

Cryo-extracted fruit wines, made of freezing juice of mostly tree fruits such as apples, cherries, apricots, and peaches, are high in RS (>140 g/L) while the totally acidity is about 9 g/L depending on the fruits.

Port-style fruit wines, usually made of raspberry, blackberry, or currants, have high alcoholic content up to 24%, with RS levels above 80 g/L. These wines can be easily transformed to sweet liquors by adding more alcohol. Charmat or Champenoise fruit wines are made by CO_2 injections and depending on the wine base can be made from dry to sweet (Rivard, 2009).

For the successful manufacturing of fruit wines it is essential to use fruits of high sugar content, low in acid, and ripped, with little bruises. In principle, grape and fruit winemaking is very similar except for certain variations that might be necessary depending on the type of the fruit.

In brief, winemaking is following a multiple step process: destemming, crushing, fermentation, pumping over, and pressing (Fig. 12.5).

The fruits are normally delivered to the distillery depending on the season that they most ripe, for example, apples are picked in autumn, while strawberries in midsummer. Destemming is the process of partial or total removal of stems from the fruit, is the next step in fruit wine formulation. Then the fruits can be crashed so pulp and juice are released. Crushing is done mechanically. The fruit comes through a pneumatic press and produces the must and the solid residues. Fermentation stage can be done either on the juice or on the solid parts depending on the type of wine desired and fruit used; the fermenting must is in contact with the seeds, skins, and sometimes even stems, while if juice is used there is not much involvement. During this process, conversion of fruit sugar into alcohol and carbon dioxide by yeasts takes place in a stainless steel, cement, or wooden fermentation tanks after pressing since the solids part should be in contact with the must to impart color, odor, and texture. During fermentation, continuous mixing is required since the fruit's solid parts have the tendency to surface. Continuous

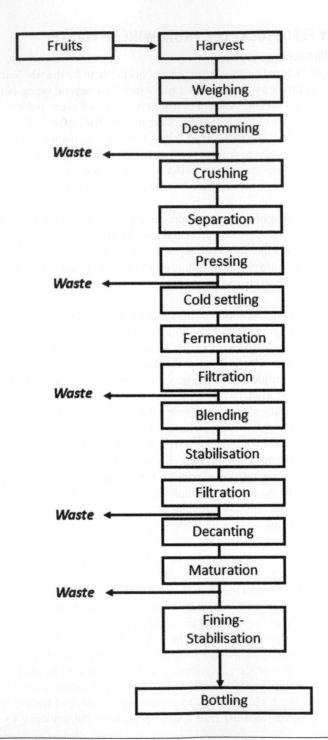

FIGURE 12.5

Waste generation during the winemaking process (Vlyssides et al., 2005).

mixing ensures the homogenous distribution of physicochemical conditions and yeasts. Instead of conventional fermentation, carbonic maceration is also an option where whole fruits are fermented in carbon dioxide prior to crushing.

After the fermentation, decanting takes place. During this process, the supernatant wine is separated from the produced wine lees and is fed by pumps to empty tanks that are filled 100% for further stabilization. The next stage is maturation, where decanted wine is kept in maximum capacity filled vessels. After maturation and stabilization, wine is filtered for quality improvement and then is decanted into empty tanks. After the desired timed period for settling, wine is bottled for transportation tanks and distributed to the contact points.

1.3.2 Generation of Fruit Winery Waste

Fruit winery waste can be divided into two main categories: solid and liquid waste generated during the collection of fruit, and liquid waste generated during the winemaking process (Fig. 12.5). Winery waste is varying in chemical composition and texture, depending on the fruits used. For example, apple processing waste is called apple pomace, which is a mixture of peels, seeds, and pieces of fruits. If dried, the apple pomace contains approximately 7.7% crude protein (CP) and 5% ether extract. On the other hand, banana processed solid waste is more complex, being composed of stalks, banana peels, leaves, and pseudo stems that contain 10–17% CP, 8% polyphenols, and a small percentage of tannins (leaves) and a further 3–5% CP (stems), while the waste is very rich in fiber. Ripe banana peels contain 8% CP, 6.2% EE, 13.8% soluble sugars, and 4.8% tannins and several trace elements such as iron, cooper, and zinc, and green peels have higher sugar content with up to 40% starch (Rivard, 2009; Wadhwa et al., 2013).

Winery waste is not only limited to solid waste generated at the first stages of fruit harvesting and initial stages of wine formulation. A fair amount of waste in the form of wastewater is generated during the later stages of processing including fermentation (vessels pre- and postwashing), storage and maturation (pre- and postwashing of storage tanks, pre- and postwashing of fermentation vessels, spillages), clarification (wastewater generated from filtration), decanting, and bottling (spillages and cleaning of vessels and bottles). Cleaning is not only done with water (cold or lukewarm) but also with solvents, detergents, and chemical agents such as sodium hydroxide. Each winery production step generates an amount of wastewater, into varying amounts as well as qualitative characteristics relevant to the process stage (Conradie et al., 2014).

The unregulated, unmonitored release of distillery wastewater to the soil and water streams can change their chemical and physical characteristics such as pH, conductivity, and color as well as having several other detrimental effects to the ecosystem. The high organic matter, indicated by biochemical oxygen demand (BOD), chemical oxygen demand (COD), and total organic carbon (TOC), results in a reduction of oxygen levels in the aquifer causing death of several aquatic organisms, generating odors due to the anaerobic decomposition (Conradie et al., 2014). High alkalinity or extreme acidity, indicated by the pH, affects the solubility of heavy metal content and is therefore constituting toxic water, which influences detrimentally both crops and aquatic organisms. The sodicity of soil, the high sodium content of soil, can cause disintegration of soil structure, resulting in surface crusting that causes low infiltration and hydraulic conductivity. On the other hand, high nutrient content such as nitrogen, potassium, and phosphorus leads to eutrophication and algal blooms while the drinking water, if containing nitrite and nitrate, can be highly toxic to humans. High ionic content or salinity, indicated by electrical conductivity (EC) and total dissolved solids (TDS) influences the palatability of water, its uptake from the crops as well as the well-being of aquatic organisms. High content in solids, indicated by total solids and total suspended solids (TS and TSS) can reduce light transmission, endangering the ecosystem's health and smothering its habitants (Conradie et al., 2014).

1.3.3 Applicability of Biorefinery Concept to Fruit Winery Waste

The biorefinery concept was introduced to tackle the global energy crisis and the climate change (global warming) attributed to the intensive industrialization across the globe. Energy production is among the most polluting processes, based majorly in nonrenewable sources such as coal, oil, and natural gas. The biorefinery concept was and still is majorly applied to cereals (crops such as wheat and corn) causing implications such as land competition, food shortages, and natural resources depletion such as water and soil nutrients.

The concept has been extended to the formulation of a bio-based economy that has been estimated to grow globally by 2020 to $250 billion ($77 billion at 2005, $125 billion at 2010) generating up to 380,000 jobs (120,000 at 2005, 190,000 at 2010). However, currently bio-based goods replace just 0.2% of petroleum-based goods, but alternatives exist for over 90% of them (Taylor, 2008; Perimenis et al., 2011). This prospect for scaling up has enlivened both supporters and critics of the technology (Petre, 2011; Richardson, 2012).

Economically, implementation of biorefineries on a large scale have not always been proven feasible due to the high cost of feedstock production and processing. Several attempts have been made to reduce the dependence on energy crops, involving the use of lignocellulosic (LCF) material; however, several complications regarding the cost of processing have arisen.

Waste can be seen as an inexhaustible resource due to its rich content in valuable nutrients, with agriculture waste (crops, plant and vegetation) related to food, feed, and beverage production becoming a strong nominee as biorefinery feedstock. Agricultural waste complex physicochemical nature might require pretreatment however in the case of winery waste due to its generation process this need is minimized.

Both the solid and the liquid winery waste can and has been used to an extent successfully as feedstock for the production of high-value chemicals (Fig. 12.6) either in a format of an LCF biorefinery (stalks, pomace, seeds), or as a conventional biorefinery where the effluent winery waste can be used as fermentation feedstock. In the case of fruit winery wastewater the high content is organic matter expressed by the COD (Fig. 12.7).

In a green concept biorefinery, fruit pomace, the solid residue that is left over after fruit processing for juice and wine, can be used as feedstock. Fermentation technology can be incorporated to produce a variety of high-value products, heat, and energy, while drying can help formulate pellets (Fig. 12.8).

In an LCF biorefinery the hard fibrous plant parts in the case of winery waste could be seeds and stalks, is fractionated, enzymatic, or chemical hydrolysis in three basic chemical parts, namely (1) hemicellulose, pentoses, 5-C polymers; (2) cellulose, hexoses, 6-C polymers; and (3) lignin, phenols (Fig. 12.9).

Further conversion of the produced fractions in useful chemicals such as ethanol, VFA, butanol, acetone, and others has been achieved in the small scale (Luguel, 2011; Sadhukhana et al., 2008). A biorefinery is a demanding capital investment and if based in one major conversion technology the cost of outputs for the consumer is increased. Therefore, several conversion technologies (thermochemical, biochemical, chemical, and biological) can be integrated (Figs. 12.8 and 12.9) so the biorefinery will be limited not only to the production of chemicals but also in the production of heat and electricity.

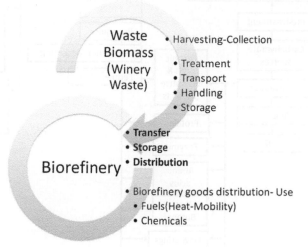

FIGURE 12.6

The biorefinery concept applied on the winery waste.

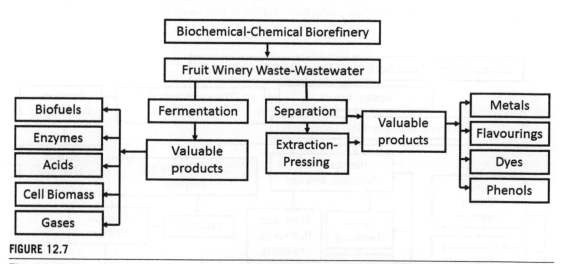

FIGURE 12.7

The chemical/biochemical biorefinery assortment applied to fruit wine waste (Fernando et al., 2006).

2. BIOTECHNOLOGICAL CONVERSION OF FRUIT WINE WASTE TO PLATFORM CHEMICALS AND ENERGY

2.1 CASE STUDIES

Limited studies have been conducted regarding the treatment of fruit wine waste. These attempts, mostly practiced in laboratory scale have had varying success rates while they have gone far beyond proving the concept, and most of them have shown highly promising results. In the following

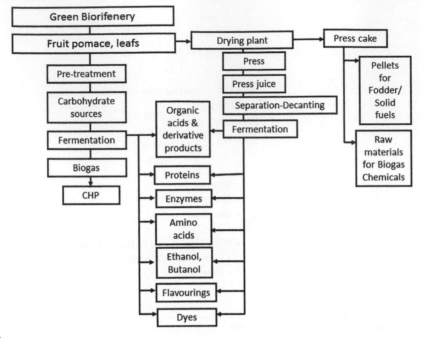

FIGURE 12.8

The green chemical/biochemical biorefinery assortment applied to wine waste (Fernando et al., 2006).

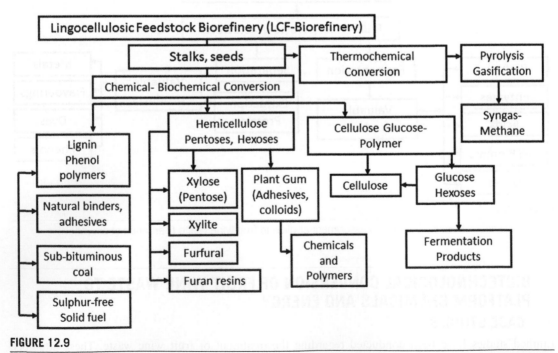

FIGURE 12.9

The lignocellulosic chemical/biochemical biorefinery assortment applied to fruit wine waste (Fernando et al., 2006).

paragraphs the case of apple pomace and mango waste will be discussed. Regardless of the great variety of fruits used in manufacturing of fruit wines, the apple pomace remains the most widely researched and documented.

2.2 THE USE OF APPLE POMACE AS FEEDSTOCK

Apple pomace is composed of peel, pulp, and seeds, and represents approximately 25% of the processed apple (Madrera et al., 2013). It has a high water content and is mainly composed of nonsoluble carbohydrates such as lignin, cellulose, and hemicellulose; monosaccharides such as glucose and fructose, and disaccharides like sucrose, along with minerals, proteins, and vitamins, constitute apple pomace as an ideal candidate for raw materials for bio-based value-added chemicals (Table 12.1). Its use includes the production of enzymes, organic acids, protein-enriched feeds, aroma compounds, natural antioxidants, and edible fibers (Vendruscolo et al., 2008).

Apple pomace has been used as a substrate of various microorganisms including bacteria, fungi, and yeast, in both solid and liquid fermentation arrangements (Joshi et al., 2011). For example, *Aspergillus niger* has been used to produce citric acid, pectolytic enzymes, polygalacturonase, and pectin methylesterase; *Trichoderma* spp. Have been used to produce phenolic compounds; *Candida utilis* has been used to produce single cell protein and feed enrichment (Vendruscolo et al., 2008).

Of the most important areas of apple pomace utilization is the production of enzymes, especially pectinases, for which it is an ideal substrate due to the high content of pectin. Enzymes such as hydrolytic depolymerases or polygalacturonases can be produced. These are extensively applied in applications in food and textile processing, wastewater treatments, and degumming of plant rough fibers (Favela-Torres et al., 2006).

Bioethanol is another important bio-based chemical that has been produced using apple pomace using mostly *Saccharomyces cerevisiae* with varying results ranging up to a 60% fermentation yield (Vendruscolo et al., 2008; Shalini and Gupta, 2010; Madrera et al., 2013). Citric acid production has also been tested on apple pomace using *A. niger* as well as fatty acids using *Thamnidium elegans* (Vendruscolo et al., 2008). Several studies have also been conducted for the production of biopolymers, pigments, and microbial biomass, especially, baker's yeast.

Table 12.1 Biological Conversion of Apple Pomace into High Value Products

Substrate	Process	Microorganism	Product	References
Apple pomace	Solid state fermentation (SSF)	*Aspergillus foetidus*	β-Glucosidase	Hang and Woodams (1994)
		Candida utilus	Lignocellulolytic enzymes	Villas-Boas et al. (2002)
	Submerged fermentation (SmF)		Single cell protein	Albuquerque (2003)
			Nutritional enrichment	Villas-Boas and Esposito (2000)
	Solid state fermentation (SSF)	*Aspergillus niger*	Pectolytic enzymes	Berovic and Ostroversnick (1997)
			Polygalacturonase	Hang and Woodams (1994)
			Citric acid	Shojaosadati and Babaeipour (2002)
			Pectin methylesterase	Joshi et al. (2006)
	Solid state fermentation (SSF)	*Saccharomyces cerevisiae*	Ethanol	Ngadi and Correa (1992)
			Animal feed	Joshi and Shandu (1996)

2.3 THE USE OF MANGO WASTE AS FEEDSTOCK

Mango (*Mangifera indica* L.) is of the most important, abundant, and popular fruits in Asia and Latin America, mainly Mexico. Currently, India holds the largest portion of production of mango with 54.2% of the world's total production (Santis Espinosa et al., 2014). Mango fruit has numerous varieties different to one another depending on the organoleptic properties such as flavor and aroma, shape, and agricultural properties (Santis Espinosa et al., 2014; Jaharul et al., 2015). It is perishable seasonal fruit that generates various products like juice, slices, jams, pickles, marmalades, and wine, among others (Varakumar et al., 2012).

Mango carbohydrate content favors the production of wine, generating waste—mainly peels. The main byproduct however, represents a major environmental constraint since its rapid decay becomes a source for insect multiplication and releases high amounts of organic matter. On the other hand, mango peel composition (cellulose, fibers, carbohydrates, phenolic compounds, and proteins) constitute an ideal candidate for biotechnological production of valuable chemicals. There are a limited number of studies, the most important are demonstrated here and in chapter 11 of this monograph (Varakumar et al., 2012; Santis Espinosa et al., 2014).

Dried mango peel is preferable due to its content of reduced sugars. Ethanol has been produced using the peel as substrate and *S. cerevisiae* as fermenting microorganism (Reddy et al., 2011) but with relatively low yields. Anaerobic digestion has also been proposed with various results (Devi and Nand, 1989; Madhukara et al., 1993); for example, $0.33\,m^3$ biogas/Kg total solids with 53% methane content at hydraulic retention time (HRT) of 15 days has been reported (Somayaji et al., 2001).

Mango peel has also been used for single cell protein production with *Pichia pinus* yeast (Rashad et al., 1990) with a maximum yield of 6.2 g/L. Lactic acid production has also been tested with *Lacobacillus casei* at a production of 63.33 g/L after pretreatment of the peel. Finally, pectinases using *Aspergillus foetidus* have been produced using solid state fermentation (Puligundla et al., 2014).

3. CONCLUSIONS

Fruit wine waste can be successfully used as feedstock in the biorefinery concept. The seasonal availability of the waste, however, does demand judicious handling and treatment to achieve economic feasibility and efficiency. Further research and practical experimentation is necessary since, in the case of wine distillery waste, limited studies have been conducted as well as life cycle analysis regarding full economic costing of the use of wine waste as a resource. The currently available results on the biotechnological applications of winery waste are a promising alternative to the current treatment techniques that are focusing on the waste remediation and treatment, rather than on resource recovery.

REFERENCES

Albuquerque, P.M., 2003. Estudo da producao de proteına microbiana a partir do bagaco de maca. Florianopolis: UFSC, 2003 Dissertation (Master's degree in Food Engineering). Departamento de Engenharia Quımica e Engenharia de Alimentos, Universidade Federal de Santa Catarina.

Appels, L., Baeyens, J., Degrève, J., Dewil, R., 2008. Principles and potential of the anaerobic digestion of waste-activated sludge. Progress in Energy and Combustion Science 34, 755–781.

Angenent, L.T., Karim, K., Muthanna, H.A.-D., Wrenn, B.A., Espinosa-Domiguez, R., 2004. Production of bio-energy and biochemicals from industrial and agricultural wastewater. Trends in Biotechnology 22, 477–485.

Berovic, M., Ostroversnik, H., 1997. Production of *Aspergillus niger* pectolytic enzymes by solid state bioprocessing of apple pomace. Journal of Biotechnology 53, 47–53.

Bustamante, M.A., Paredes, C., Moral, R., Moreno-Caselles, J., Perez-Espinosam, A., Perez-Murcia, M.D., 2005. Uses of winery and distillery effluents in agriculture: characterisation of nutrient and hazardous components. Water Science and Technology 51, 145–151.

Bustamante, M.A., Moral, R., Paredes, C., Perez-Espinosa, A., Moreno-Caselles, J., Perez-Murcia, M.D., 2008. Agrochemical characterisation of the solid by-products and residues from the winery and distillery industry. Waste Management 28, 372–380.

Carson, K.J., Collins, J.L., Penfield, M.P., 1994. Unrefined dried apple pomace as a potential food pomace as a potential food ingredient. Journal Food Science 59, 1213–1215.

Christ, K.L., 2014. Water management accounting and the wine supply chain: empirical evidence from Australia. The British Accounting Review 46, 379–396.

Commission, E.U., 2014. Directive of the European Parliament and of the Council Amending Directives 2008/98/EC on Waste, 94/62/EC on Packaging and Packaging Waste, 1999/31/EC on the Landfill of Waste, 2000/53/EC on End-of-Life Vehicles, 2006/66/EC on Batteries and Accumulators and Waste Batteries and Accumulators, and 2012/19/EU on Waste Electrical and Electronic Equipment.

Conradie, A., Sigge, G.O., Cloete, T.E., 2014. Influence of winemaking practices on the characteristics of winery wastewater and water usage of wineries. South African Society for Enology & Viticulture 35, 10–19.

de Villiers, A., Albertsa, P., Tredoux, A.G.J., Nieuwoudt, H.H., 2012. Analytical techniques for wine analysis: an African perspective; a review. Analytica Chimica Acta 730, 2–23.

Dimakis, A., et al., 2011. Methods and tools to evaluate the availability of renewable energy sources. Renewable Sustainability Energy Reviews 15, 1182–1200.

Devi, S.S., Nand, K., 1989. Microbial pretreatment of mango peel for biogas production. Journal Microbial Biotechnology 4, 110–115.

Favela-Torres, E., et al., 2006. Hydrolytic Depolymerising Pectinases. Food Technology and Biotechnology 44, 221–227.

Fernando, S., Adhikari, S., Chandrapal, C., Murali, N., 2006. Biorefineries: current status, challenges, and future direction. Energy & Fuels 20, 1727–1737.

Frenkel, V.S., Cummings, G., Maillacheruvu, K.Y., Tang, W.Z., 2013. Food-processing wastes. Water Environment Research 85, 1501–1514.

Geoffrey, S., Simate, G., Cluett, J., Lyuke, S.E., Musapatika, E.T., Ndlovu, S., Walubita, L.F., Alvarez, A.E., 2011. The treatment of brewery wastewater for reuse: state of the art. Desalination 273, 235–247.

GOV. AC.UK, 2011; 2012; 2013; 2014. The carbon plan: delivering our low carbon future presented to Parliament pursuant to Sections 12 and 14 of the Climate Change Act 2008 Amended 2nd December 2011 from the version laid before Parliament on 1st December 2011. In: Department of Energy & Climate Change (Ed.), HM Government, London- Waste Water Treatment in the United Kingdom – 2012 Implementation of the European Union Urban Waste Water Treatment Directive – 91/271/EEC, Department for Environment, Food and Rural Affairs, London.- Waste Management Plan for England. Department for Environment Food and Rural Affairs UK, London-2013., Review of Waste Policy and Legislation EU Waste Framework Directive 2008/98/EC, the Landfill Directive 1999//31/EC and the Packaging and Packaging Waste Directive 94/62/EC. Waste - Environment - European Commission. Brussels.

Hang, Y.D., Woodams, E.E., 1994. Apple pomace: a potential substrate for production of β-glucosidase by *Aspergillus foetidus*. LWT – Food Science and Technology 27, 587–589.

Hatti-Kaul, R., Tornvall, U., Gustafsson, L., Borjesson, P., 2007. Industrial biotechnology for the production of bio-based chemicals – a cradle-to-grave perspective. Trends in Biotechnology 25, 119–124.

Ioannou, L.A., Li Puma, G., Fatta-Kassinos, D., 2014. Treatment of winery wastewater by physicochemical, biological and advanced processes: a review. Journal of Hazardous Materials 286, 343–368.

Jahurul, M.H.A., Zaidul, I.S.M., Kashif, G., Al-Juhaimi, F.Y., et al., 2015. Mango (*Mangifera indica* L.) by-products and their valuable components: a review. Food Chemistry 183, 173–180.

Jefferson, J., 2008. Accelerating the transition to sustainable energy systems. Energy Policy 36, 4116–4125.

Jones, E., Salin, V., Williams, G.W., 2005. Nisin and the Market for Commercial Bacteriocins. Consumer and Product Research Report Texas Agribusiness Market Research Center (TAMRC), pp. 1–20.

Joshi, V.K., Parmar, M., Rana, N., 2011. Purification and characterization of pectinase produced from Apple pomace and evaluation of its efficacy in fruit juice extraction and clarification. Indian Journal of Natural Products and Resources 2, 189–197.

Joshi, V.K., Parmar, M., Rana, N.S., 2006. Pectin esterase production from apple pomace in solid-state and submerged fermentations. Food Technology and Biotechnology 44, 253–256.

Joshi, V.K., Sandhu, D.K., 1996. Preparation and evaluation of an animal feed byproduct produced by solid state fermentation of apple pomace. Bioresource Technology 56, 251–255.

Li, W.-W., Yu, H.-Q., 2011. From Wastewater to bioenergy and biochemicals via two-stage bioconversion processes: a future paradigm. Biotechnology Advances 29, 972–982.

Lin, Y., Tanaka, S., 2006. Ethanol fermentation from biomass resources: current state and prospects. Applied Microbiology & Biotechnology 69, 627–642.

Luguel, C., 2011. European Biorefinery Joint Strategic Research Roadmap for 2020 Strategic Targets for 2020 – Collaboration Initiative on Biorefineries, Europe.

Madhukara, K., Nand, K., Raju, N.R., Srilatha, H.R., 1993. Ensilage of mango peel for methane generation. Process Biochemistry 28, 119–123.

Madrera, R.R., Bedrinana, R.P., Hevia, A.G., Arce, M.B., Valles, B.S., 2013. Production of spirits from dry apple pomace and selected yeasts. Food and Bioproducts Processing 91, 623–631.

Morley, N., Bartlett, C., 2008. Mapping Waste in the Food Industry. Defra, Food and Drink Federation, UK.

Ngadi, M.O., Correia, L.R., 1992. Kinetics of solid state ethanol fermentation from apple pomace. Journal of Food Engineering 17, 97–116.

Østergaard, P.A., 2012. Comparing electricity, heat and biogas storages' impacts on renewable energy integration. Energy 37, 255–262.

Perimenis, A., Walimwipi, H., Zinoviev, S., Muller-Langer, F., Miertus, S., 2011. Development of a decision support tool for the assessment of biofuels. Energy Policy 39, 1782–1793.

Petre, M., 2011. Advances in Applied Biotechnology, first ed. InTech, Rijeka, Croatia.

Puligundla, P., Vijaya Sarathi, V., Obulam, R., Eun Oh, S., Mok, C., 2014. Biotechnological Potentialities and Valorization of Mango Peel Waste: A Review. Sains Malaysiana 1901–1906.

Rashad, M.M., Moharib, S.A., Jwanny, E.W., 1990. Yeast conversion of mango waste or methanol to single cell protein and other metabolites. Biological Waste 32, 277–284.

Reddy, L.V., Reddy, O.V.S., Wee, Y.J., 2011. Production of ethanol from mango (*Mangifera indica* L.) peel by *Saccharomyces cerevisiae* CFTRI101. African Journal of Biotechnology 10, 4183–4189.

Richardson, B., 2012. From a fossil-fuel to a biobased economy: the politics of industrial biotechnology. Environment and Planning C: Government and Policy 30, 282–296.

Rivard, D., 2009. The Ultimate Fruit Winemaker's Guide: The Complete Reference Manual for All Fruit Winemakers (Bacchus Enterprises Winemakers), second ed. CreateSpace Independent Publishing Platform.

Sadhukhana, J., Mustafa, M.A., Misailidis, N., Mateos-Salvadora, F., Dub, C., Campbell, G.M., 2008. Value analysis tool for feasibility studies of biorefineries integrated with value added production. Chemical Engineering Science 63, 503–519.

Santis Espinosa, L.F., Peréz-Sariñana, B.Y., Saldaña-Trinidad, S., 2014. Evaluation of agro-industrial wastes to produce bioethanol: case study - mango (*Mangifera indica* L.). Energy Procedia 57, 860–866.

Shalini, R., Gupta, D.K., 2010. Utilization of pomace from apple processing industries: a review. Journal Food Science Technology 47, 365–371.

Shojaosadati, S.A., Babaeipour, V., 2002. Citric acid production from apple pomace in multi-layer packed bed solid-state bioreactor. Process Biochemistry 37, 909–914.

Somayaji, D., Padshetty, N.S., Nand, K., 2001. Recycling of mango-peel waste for biogas production. Asian Journal of Microbiology Biotechnology & Environmental Sciences 3, 339–341.

Taylor, G., 2008. Biofuels and the biorefinery concept. Energy Policy 36, 4406–4409.

The Biocomposites Centre Report Centre, T.B., 2011. The Biorefining Opportunities in Wales: From Plants to Products, Bangor. .

Tyagi, V.-K., Lo, S.-L., 2013. Sludge : a waste or renewable source for energy and resources recovery? Renewable and Sustainable Energy Reviews 25, 708–728.

Varakumar, S., Naresh, K., Reddy, O.V.S., 2012. Preparation of mango (*Mangifera indica* L.) wine using a new yeast-mango-peel immobilised biocatalyst system. Czech Journal Food Science 30, 557–566.

Vendruscolo, F., Albuquerque, P.M., Streit, F., Esposito, E., Ninow, J.L., 2008. Apple pomace: a versatile substrate for biotechnological applications. Critical Reviews in Biotechnology 28, 1–12.

Villas-Boas, S.G., Esposito, E., 2000. Bioconversao do baga¸co dema¸ca: enriquecimento nutricional utilizando fungos para produ¸c¯ao de um alimento alternativo de alto valor agregado. Biotecnologia Ciencia e Desenvolvimento 14, 38–42.

Villas-Boas, S.G., Esposito, E., Mendon¸ca, M.M., 2002. Novel lignocellulolytic ability of *Candida utilis* during solid state cultivation on apple pomace. World Journal of Microbiology Biotechnology 18, 541–545.

Vlyssides, A.G., Barampouti, E.M., Mai, S., 2005. Wastewater characteristics from Greek wineries and distilleries. Water Science and Technology 51, 53–60.

Wadhwa, M., Bakshi, M.P.S., Makkar, H.P.S., 2013. Utilization of Fruit and Vegetable Wastes as Livestock Feed and as Substrates for Generation of Other Value-Added Products. Food and Agriculture Organization of the United Nations (FAO), Rome, pp. 1–56.

Wang, Z., Yu, H., Ma, J., Zheng, X., Wu, Z., 2013. Recent advances in membrane bio-technologies for sludge reduction and treatment. Biotechnology Advances 31, 1187–1199.

Worrall, J.J., Yang, C.S., 1992. Shiitake and oyster mushroom production on apple pomace and sawdust. Hortscience 27, 1113–1131.

Zacharof, M.-P., Lovitt, R.W., 2013. Complex effluent streams as a potential source of volatile fatty acids. Waste and Biomass Valorisation 4, 557–581.

Zacharof, M.-P., Lovitt, R.W., 2014a. The filtration characteristics of anaerobic digester effluents employing cross flow ceramic membrane microfiltration for nutrient recovery. Desalination 341, 27–37.

Zacharof, M.-P., Lovitt, R.W., 2014b. Recovery of volatile fatty acids (VFA) from complex waste effluents using membranes. Water Science and Technology 69, 495–503.

Zacharof, M.-P., Vouzelaud, C., Mandale, S.J., Robert, W., Lovitt, R.W., 2015. Valorization of spent anaerobic digester effluents through production of platform chemicals using *Clostridium butyricum*. Biomass and Bioenergy 81, 294–303.

Zacharof, M.-P., Mandale, S.J., Williams, P.M., Lovitt, R.W., 2016. Nanofiltration of treated digested agricultural wastewater for recovery of carboxylic acids. Journal of Cleaner Production 112, 4749–4761.

INNOVATIONS IN WINEMAKING

13

R.S. Jackson
Brock University, St. Catharines, ON, Canada

1. INTRODUCTION

Before the relevance of modern innovations can be fully appreciated, it is helpful to appreciate how wine's origin and early developments were dependent on ancient innovations. Grapes have been harvested since time immemorial. However, the nascence of wine appears to owe its origin partially to the development of pottery or some other sealable, impermeable containers. If grapes are crowded into a container of sufficient size, and left for even a short period, they begin to auto-ferment. This causes weakening of the skins, leading to rupture. Yeasts present on grape surfaces will begin to ferment the released juice and produce alcohol. Thus, without intention, wine begins its inception. When the savory nature of this salubrious accident was appreciated sufficiently to induce people to initiate the process intentionally is unknown. However, evidence that such a fermented beverage was being purposely produced goes back at least 7000 years (McGovern et al., 1996) in a jar found at Hijji Firuz Tepe, Turkey. This is a region conducive to wine's origin—the northern portions of the Fertile Crescent and adjacent southern Caucasus (modern-day northern Turkey, Iraq, and Iran). Here, the spread of agriculture as well as the native range of wild grapevines (*Vitis vinifera* f. *sylvestris*) and oak trees auspiciously overlapped. Evidence for wine production of similar antiquity has been unearthed in Georgia (Barnard et al., 2011). Evidence of an even older (9000 years ago) fermented beverage has been detected in China (McGovern et al., 2004). In this instance, it appears to have been produced from rice, honey, and fruit (hawthorn and/or grape).

To turn a seasonal, unstable, proto-wine, obtained by collecting fruit growing in the wild, into a beverage befitting the designation "wine" and available year-round, presupposed deliberate cultivation and a settled agricultural population. The vine only begins to yield fruit in reasonable quantity some years after planting—a situation ill-suited to a nomadic, hunter-gatherer culture. The initial proximity of oak trees may seem extraneous, until one appreciates that oak sap is likely the indigenous habitat of *Saccharomyces paradoxus*, the progenitor of *Saccharomyces cerevisiae*, the wine yeast (Phaff, 1986). It is required to metabolize all the fermentable sugars in grapes to generate a beverage with the potential for prolonged storage in a microbially stable condition.

Surprisingly, the typical epiphytic flora of grapes rarely includes *S. cerevisiae*, and when present, it is a minor component. The grape flora is composed primarily of members of *Klockera*, *Hansenula*, *Torulaspora*, *Candida*, *Pichia*, and *Cryptococcus*. When grapes break open, these indigenous yeasts would have begun to metabolize the sugars released by autofermented grapes. However, as they metabolize sugar, oxygen consumption would have demanded a switch to fermentative metabolism. Because ethanol is the principal organic by-product of yeast fermentative metabolism, to which they are sensitive, the growth of these epiphytic yeasts would have slowed and come to a halt. By comparison, *S. cerevisiae* is relatively alcohol insensitive. Thus, *S. cerevisiae* would have quickly come to dominate such spontaneous fermentations were it present. Without the presence of *S. cerevisiae*, such a proto-wine would rarely have reached alcohol values typical of most modern wines (11–13%). Without *S. cerevisiae*, such a beverage would have more likely resembled beer in its stability and alcohol content than wine as we know it. Thus, proximity to oak trees, or its bark, would have provided the opportunity for the transfer of *S. paradoxus* to grapes. *S. paradoxus* possesses many of the winemaking attributes of *S. cerevisiae*. The complete metabolism of grape fermentable sugars to alcohol possible with *S. cerevisiae* (or *S. paradoxus*) would have given the nascent wine some microbial stability. The natural acidity and phenolic content of grapes would have also favored spoilage resistance. That much of the acidity in grapes ripening in warm climates is tartaric acid, an acid not readily metabolized by bacteria or yeasts, is another inherent attribute of grapes favoring their transmutation into a relatively stable alcoholic Caucasus.

Presumably, similar events led to the discovery of cider, perry, and other fruit. However, because of their comparatively low levels of fermentable sugars (Joshi et al., 2004), the product would have developed comparatively low alcohol, unless an external source of sugars were added, such as honey. In contrast, wine grapes typically develop sugar contents above 20%. Only in cold regions, where the product might freeze, and the ice be removed, would fruit wines have achieved values > 10%. This situation would have remained largely unchanged until the precise role of simple (fermentable) sugars was discovered in the early 1800s, and abundant and inexpensive sources of sugar became readily available, to permit supplemented alcohol-producing ability.

As suggested earlier, grape cultivation likely began within the southern range of wild grapes in the Caucasus and Near East. Because of primitive transportation conditions, wine's availability outside this region would probably have been limited, even for the ruling classes of civilizations to the south (e.g., Sumeria). Subsequently, another alcoholic-containing beverage, made from grains (beer), came to be the staple tipple of the populace. Grains, being comparatively dry, were comparatively easy to store for subsequent use. However, grains are inherently devoid of fermentable sugars, most of their carbohydrates being stored as starch in the endosperm. Its hydrolysis is activated during a process termed malting, where grains are moistened and germination activates endogenous enzymes to break starches down into glucose. Subsequent mashing in water solubilizes the sugars and provides the liquid needed for yeast metabolism. However, even before the discovery of malting, honey could have supplied sufficient sugars to initiate a brief fermentation. Because of beer's poor preservative properties,[1] it would have been brewed locally and consumed shortly after production. In contrast, wine in sealed, impervious containers could be stored for several months. As such, it converted grapes, a readily perishable juicy fruit, into a beverage with potentially interesting physiological influences (loss of sobriety). Although grapes can be dried for long-term storage, there is no evidence that raisins were ever used for wine production.

Other than grapes, apples constitute the major fruit used in producing a wine-like beverage. Domesticated apples likely arose from *Malus sieversii* in Kazakhstan (Harris et al., 2002). From here, late Neolithic or early Bronze Age peoples carried seed east into central China and west into Europe. The origin of pears, the source of perry, is more complex, having independently been domesticated in China and the Caucasus. Other fruits have been fermented to produce wine-like beverages probably for centuries, but commercial production is more recent, and still comparatively limited in comparison to

---

[1]The addition of hops, as both a preservative and flavorant, began only millennia later in Central Europe.

wine, cider, or perry. Of these, possible kiwi-based wine is the most widely available, due to its extensive cultivation in many parts of the globe. Commercial production of any fruit-based wine almost automatically demands the occurrence of monoculture, to provide sufficient quantities of fruit to justify the purchase of the equipment required.

Once grape cultivation had begun, the opportunity was provided for noticing and selectively propagating vines that were more productive. The long canes of grape vines easily can trail down to the ground, where spontaneous rooting can occur. This would have facilitated the discovery and use of vegetative propagation. By selectively propagating the more fruitful vines, the practice would have resulted in roguing male (nonfruit-bearing) vines. This, in its turn, would have favored the selection of rare, functionally bisexual, self-fertile (more fruitful) plants. As male vines were eliminated, the remaining pure female vines would have become less fruitful, accentuating the comparative fruitfulness of self-fertile mutants. Subsequent selection of visibly distinctive variants would have begun the propagation of the variety of cultivars that now distinguishes modern viticulture.

Although the archeological record is basically silent on the origin of grape cultivars, genomic analysis has provided significant insights (see Jackson, 2014 for a review). It appears that viticulture and wine production spread from their origins in northern Turkey, Iraq, and Iran south into the Levant and Mesopotamia, regions distant from indigenous populations of wild vines, as well as to the north, into and beyond the Caucasus. Egyptian and Minoan use of wine suggests its quality was acceptable and adequate means for storage were available for at least a few years. However, written confirmation of wine quality is nonexistent until ancient Greek and Roman times. Some Roman wines are described with an enthusiasm resembling modern writings, with special vintages and provenances being highly prized. This situation was favored by technical advances in amphora production (formation of a glass-like inner lining; Koehler, 1986), and the use of a tight, gas- and water-impermeable closure (cork). As a consequence, fine wines had the potential of aging well for decades. In contrast, most wine had a short "shelf-life." To provide less-expensive amphorae with impermeability, they were coated with pitch (pine resin). This, and the occasional addition of pieces of pitch and various spices to the wine, obviated its developing a quality modern consumers would recognize or appreciate. In addition, the lowest wine grades were doctored with a host of ingredients, presumably to mask the wine's poor quality. Examples of additives recommended by Cato (234–149 BC) included seawater, boiled must, various herbs, honey, and vinegar. Although most unsavory by today's standards, the wine had the advantage of being potable, and when added to water made it relatively safe to drink. In ancient times, wine was typically diluted (~50:50). Drinking wine undiluted was generally frowned upon. The common habit of adding flavorants to wine, albeit helping to mask faults, was often used unscrupulously, even to the point of adding, often unknowingly, toxic substances. This practice was still surprisingly common in medieval times (Accum, 1820; Younger, 1966), for example, when lead salts were still being added to offset excessive acidity. More modern incidences (e.g., the addition of antifreeze and methanol) are just part of a, thankfully now rare but continuing, series of criminal wine doctorings (Hallgarten, 1987).

To most modern consumers, the addition of preservatives such as pitch would be anathema. However, there is still a market for such wines (e.g., Retsina). In addition, the practice of adding various agents, such as herbs and spices, to mask wine faults may be the origin of products such as vermouth. Alternatively, vermouth may have its origins in the use of wine as a solvent for medicinal herbs.

With the fall of the Roman Empire, wine production skills went into decline. Amphora production slowly fell into disuse, presumably because of reduced demand (decline in trade) and the technical skills required in their production. Wine increasingly came to be stored in oak cooperage, a vessel far less efficient in excluding oxygen and easily contaminated by spoilage microbes. Thus, most wine in medieval Europe rarely remained drinkable beyond the spring or early summer following production. Existent transport facilities would have further accentuated oxidation and spoilage. Consequently, few wines were exported, and what was transported was generally by boat, in small quantities, and to the few wealthy enough to afford its importation.

The use of oak cooperage during fermentation and storage also donated a distinctive flavor. Unlike that of pitch noted earlier, oak flavor has remained a desired attribute in many wines. This is principally with red wines. Their more intense flavors can more readily accommodate the presence of oaky attributes than more delicately flavored white wines. Equally, because most fruit wines have delicate flavors, they are seldom fermented or matured in oak. This may be beginning to change, notably with some ciders.

Following the collapse of the Roman Empire, knowledge of the benefits of wine's aging faded. Preconditions began to reappear near the end of the 1400s. At this time, the benefits of burning sulfur wicks in barrels, prior to adding wine, began to be recorded (Anonymous, 1986). This was some 350 years before the microbial nature of most wine spoilage was discovered. Nonetheless, the practice of barrel fumigation was slow to spread, possibly due to fears that it might poison the wine. Thus, the practice did not begin to be widely adopted until the mid-1600 in central Europe, and considerably later in more southern regions. Major improvements in glass making, during the mid-1600s, rapidly spread from England throughout Europe. Combined with the beginnings of industrialization, production costs fell and there was a slow shift in wine transport from barrel to glass. This change was itself dependent on the redeployment of cork as a closure. Thus, conditions were set for the rediscovery of the benefits of aging. The development of even stronger glass, combined with the use of cork, also permitted the evolution and spreading popularity of sparkling wines. The high internal pressures that developed during the gas entrapment during the second, in-bottle fermentation mandated strong bottles.

Although ciders produced in England and France are usually bubbly that may be so due to carbonation (a comparatively new innovation). In contrast, most premium sparkling ciders are made in a manner similar to that noted later for sparkling wine, and equally dependent on the availability of strong glass bottles.

Technological innovations, combined with improving economic conditions in northern and central Europe, also lead to the emergence of a sufficiently wealthy middle class to support increased wine importation. This, in its turn, provided an incentive for expanding and improving wine production. Regrettably, these conditions did not occur simultaneously in the warmer regions of southern Europe. Thus, their local cultivars and wines languished largely unknown, except locally—a situation that partially remains to this day. Only with developments in distillation did some southern wines achieve fame and a sizable export market. Adding distilled wine (brandy) provided them with sufficient stability to be transported to a welcoming public in the north. Distillation, although popular with medieval alchemists, was not applied to wine to any significant extent until the early 1600s. This innovation permitted the evolution of the major types of fortified wines, as well as brandy and other distilled beverages.

Changes in thinking, associated with the Renaissance, combining with economic improvements also fostered developments of science and related technological advances. Those in chemistry and microbiology were particularly applicable to improving grape and wine production—features that have subsequently been applied to fruit wines. Long-held traditional beliefs were either shown to be valid or disproven, and rough approximations replaced by objective measurements. For example, the precise relationship between sugar content and subsequent alcohol production and CO_2 release permitted the safer and more economical production of sparkling wine, avoiding the considerable losses previously associated with bottle explosion (primarily due to unpredictable excess carbon dioxide production). Also, knowledge of the role of microbes in fermentation and spoilage lead to marked improvements in wine quality. A better understanding of plant and yeast nutrition has also permitted improvements in fruit and wine quality. Current advancements continue at an ever increasing pace, resulting in universal enhancements in wine quality and increased control over their attributes. Independent developments, such as refrigeration, have significantly reduced one of the aspects that long delayed improvements in wine production in warm climates: the tendency for violent fermentation, overheating, and inactivation of the yeasts, resulting in fermentation terminating prematurely (becoming "stuck"). This left the wine sweet, low in alcohol content, and highly susceptible to spoilage.

The spread of these innovations has democratized wine production. Thus, quality wines can be produced almost worldwide. It is no longer the preserve of a few regions, blessed with conditions favoring the production and storage of wine, as well as proximity to markets willing and able to afford better wine. As markets have expanded globally, especially in recent years, new opportunities for producing good quality, mass-market wines have opened. In addition, the market for artisanal and fruit wine has expanded. This situation has encouraged experimentation, designed to enhance the product's distinctiveness and character. Examples are the transfer of techniques developed for ice wine production to producing ice cider, as well as the creation of a separate class of ciders: spiced apple wines. Equally, large-volume producers have the ability to produce wines tailored to specific consumer groups. There has also been a tendency for established premium wine producers to focus their production methodologies to match the preferences of influential critics. One clear indication of this trend is the increasing use of "flying" wine-makers–individuals with the skill and charisma to sell their expertise. Thus, all segments of wine production are being driven by innovations, and an increasing appreciation of the profits possible in satisfying the needs and desires of consumers.

Thus, in the past few decades, an industry, formerly hidebound by tradition, is now embracing innovation. Although certain aspects of individuality have been lost, the ideal of "fine wine for all on the supermarket shelf" is becoming increasing realized. As implied, the wine-maker is no longer just the equivalent of a midwife, present to avoid problems at the birth of a wine, but its designer, almost as potter molds clay.

Control begins, where it must, in the vineyard. The attributes of fruit, as it arrives at the cellar door, sets the outer parameters of the features ultimately possessed by the wine. Nonetheless, how the producer processes and ferments the fruit, and subsequently matures and ages the wine, largely defines the sensory properties it ultimately possesses. Examples of recent innovations include experimentation with specific strains of yeasts and lactic acid bacteria, their combination, or the use of local strains. Former regional and isolated procedures, such as *sur lies* maturation (Charpentier, 2010), *appassimento* (Paronetto and Dellaglio, 2011), *saignée* (Harbertson et al., 2009), and cold maceration (Allen, 2007) are also spreading across borders. Because scientific investigations on these procedures are now appearing in modern texts, scholarly, and trade journals, access is available to anyone desirous of learning their "secrets." Long gone are the days when critical knowledge was the prerogative of a few, or passed from father to son as proprietary information. Sensory studies have also allowed wine-makers to understand more the underpinnings of consumer preferences, both locally and internationally. Savvy marketers are also tapping into this database to direct their advertising to the best advantage.

Although the adoption of older techniques is being embraced as a means of adding distinctiveness, several newer techniques have not been as well received. Examples are cryoextraction, as the technical equivalent of natural freezing in producing ice wines; encapsulated yeasts in facilitation disgorging in sparkling wine production; spraying grapes with the spores of *Botrytis cinerea* and storage in climate-controlled warehouses for botrytized wine production; continuous versus batch fermentation; and genetically engineered yeasts. Ancient techniques are seen as natural and wholesome while modern equivalents are considered unnatural and contrived. The "natural product" image is often critical to the sales appeal of traditional agricultural products. However, with novel beverages, such as wines made from fruit such cagaita, jackfruit, and ginkgo, the benefits of modern procedures such yeast encapsulation (Oliveira et al., 2011; Singh and Sooch, 2009) are unlikely to deter sales. In addition, procedures such as thermovinification (Ribéreau-Gayon et al., 2006), originally designed to enhance anthocyanin extraction from grape cultivars low in pigmentation, has been applied in producing strawberry-based wines (Sharma et al., 2009).

Despite all this effort, it seems that most consumers are oblivious to most wine subtleties so diligently striven for by dedicated wine-makers. Most wine is too often consumed without thought, acting as little more than a liquid refreshment. Wine does act as a savory palate cleanser, but it has so much more to offer to those willing to take the time and effort to investigate. However, the wine-maker has little control over how their wine will be consumed. Innovation can improve the product but not the consumer. All the producer can do is keep tabs on changing preferences, and adjust accordingly, were possible and appropriate. Examples of recent shifts in consumer tastes have been an increase in the purchase of red wines, and during the summer months, rosé and fruit-flavored wines. The youngest segment of the consumer market is the most fickle in its preferences. The same may apply to fruit wines, which in most markets are niche products, often being sold directly to the consumer by the producer. Here, its image as an artisanal product, with distinct health benefits, is particularly value. As such, their sensory qualities are critical, as they may be consumed more by themselves than with a meal.

2. BASIC WINEMAKING

To further put modern innovations in perspective, it is essential to understand the basic steps in wine production. Technically, it begins when grapes, or their juice, reach the winery.

Upon arriving, extraneous material inadvertently harvested along with the fruit is removed. This is particularly necessary when grapes have been harvested mechanically. The grapes may also be inspected for immature, raisined, or diseased grapes, and these are removed. The grapes are then crushed, or in some cases, gently pressed whole, to release the juice.

If desired, a maceration phase (a period when the juice remains in contact with the pulp, seeds, and skins of the crushed fruit) is permitted before fermentation commences. Maceration may also continue throughout fermentation, notably in the production of red wines. Maceration facilitates the extraction of nutrients, flavorants, and other constituents located primarily in the grape skins and seeds (termed pomace). Crushed grapes and fermenting juice is termed must. Hydrolytic enzymes, released and/or activated during cell rupture, promote the liberation of constituents from the pomace. Prominent among these enzymes are pectinases. They assist in releasing cellular constituents by autolyzing cells not ruptured during crushing. Additional enzymes may synthesize or liberate flavorants, and degrade macromolecules, liberating nutrients supporting yeast and bacteria activity during fermentation.

Depending on the cultivar and desires of the wine-maker, different procedures are used in producing white, rosé, and red wines. These differences revolve primarily around the extraction of pigments and/or flavorants from the skins. With grapes, the most distinctive attributes (so essential to a wine's varietal aroma) are located in the skin, as well as most of the flavonoid phenolics. The situation is different with fruit wines, where distinctive varietal attributes have traditionally been less important in sales potential. Corresponding, there may be less concern about extracting constituents predominantly or exclusively localized in the skins. Nonetheless, the aroma profile of wines produced from different fruits can be sensorially marked (Fig. 13.1).

Another distinguishing characteristic separating most fruit from wine grapes is their significantly higher pectic content. Because of the tendency of pectins to cloud the finished product, pectins need to be removed, either by a process such as keeving (involving the action of naturally occurring pectinases in cider production) or by the addition of commercial pectin lyases and endopolygalaxturonases.

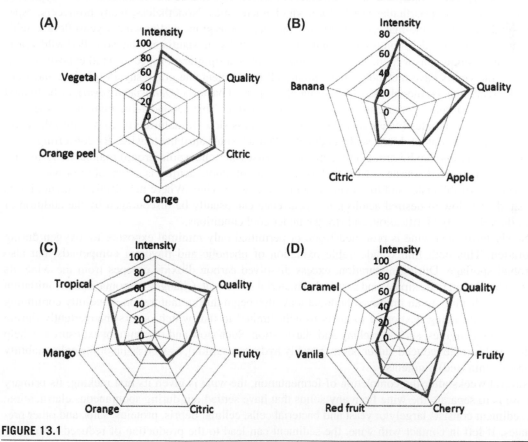

FIGURE 13.1

Aroma profile of several fruit wines: (A) orange, (B) banana, (C) mango, and (D) cherry.

From Coelho, E., Vilanova, M., Genisheva, Z., Oliveira, J.E., Teixeira, J.A., Domingues, L., 2015. Systematic approach for the development of fruit wines from industrially processed fruit concentrates, including optimization of fermentation parameters, chemical characterization and sensory evaluation. LWT – Food Science and Technology 62, 1043–1052; reproduced with permission.

Although demethylation releases more methanol than found in grape wines, methanol contents are typically below dangerous levels.

Typically fruit is pressed several times, with the various fractions either fermented separately, with possible subsequent blending, or combined before fermentation commences. In the case of cider, water may be added before subsequent pressing. Prefermentative juice clarification may be employed to reduce undesirable microbial contamination, but it also removes indigenous yeasts. Thus, cider juice is typically inoculated with cultures to initiate rapid fermentation. However, this reduction in microbial flora is desired in the production of French ciders. Here, slow fermentation is wanted, and further favored by having fermentation occur at a cool temperature. Sugar is typically added to the juice, depending on the desired alcohol content. Supplementation with ammonium nitrogen and vitamins may be needed to encourage complete fermentation.

Although the principal by-product of yeast fermentation is alcohol, yeast metabolism also supplies much of the typical bouquet and flavor attributes of wines. To enhance this feature, it is now typical to add an inoculum of a particular yeast of known characteristics. Nonetheless, many producers, especially small artisanal wine-makers, prefer to use the indigenous grape and/or winery yeast flora to initiate fermentation. This decision is based on the belief, partially substantiated, that so-called wild yeasts may donate a unique flavor to the beverage. Thus, it can be a significant aspect of market positioning.

When the yeasts have metabolized essentially all the fermentable grape sugars (glucose and fructose), the wine may be inoculated with a strain of lactic acid bacteria (*Oenococcus oeni*), or be treated in a manner fostering spontaneous malolactic fermentation. This second fermentation is especially desirable in cool climates (where malic acid metabolism is slow in maturing fruit) or fruit inherently high in malic acid content. Malic acid can donate a "hard" (sour) attribute to wine. By converting malic acid (possessing two acidic groups) to lactic acid (possessing one acidic group) malolactic fermentation reduces perceived sourness. Nonetheless, retention of ample acidity is essential to providing the wine with a "fresh" taste, and to prevent cloying in a sweet wine. Where malolactic fermentation is undesired (it is low in desired acidity), its occurrence can usually be discouraged by the addition of sulfur dioxide, early clarification, and storage under cool conditions.

Newly fermented wine is protected from, or permitted only minimal exposure to, oxygen during maturation. This both limits undesirable oxidation of phenolic and flavorant compounds, but also microbial spoilage. During maturation, excess dissolved carbon dioxide escapes from the wine, its yeasty odors begin to dissipate, and suspended material settles out. Changes in aroma, and the initiation of processes leading to an improved bouquet may also begin during maturation, hopefully continuing during in-bottle aging. Exposure to air is usually limited to that which occurs inadvertently during certain cellar activities, such as racking and clarification. Such slow and intermittent exposure can help oxidize undesirable reduced-sulfur odors (notably hydrogen sulfide), as well as favor the color stability of anthocyanin (red) pigments.

Several weeks after the completion of fermentation, the wine is given its first racking. Its primary purpose is to separate the wine from any solids that have settled out during spontaneous clarification. The sediment consists largely of yeast and bacterial cells, cellular debris, proteins, salts, and other precipitates. If left in contact with wine, the sediment can lead to the production of reduced-sulfur off-odors, as well as favor microbial spoilage (by supplying nutrients). Subsequent rackings separate the wine from additional precipitates encouraged by the use of clarifying agents.

Prior to bottling, the wine may be fined to remove dissolved proteins and other materials. Otherwise, these could generate cloudiness, especially if the wine is exposed to higher than normal temperatures. Fining may also be used to mollify the wine's taste by removing excess polyphenolics. Wines are also commonly chilled and filtered, to further enhance stability, notably due to salt crystallization. If separate wines are blended shortly before bottling, an extra period of maturation is usually required for their constituents to "harmonize" and reestablish equilibrium.

At bottling, white wines are generally given a small dose of sulfur dioxide to protect them against oxidation and microbial spoilage (about 0.8–1.5 mg/L free molecular SO_2). Sweet versions of wines are usually sterile-filtered as a further protection against microbial spoilage.

Before release, newly bottled wines are normally aged at the winery for several weeks, or occasionally several years. During the first few weeks, this allows acetaldehyde, potentially produced following bottling (as a consequence of oxygen uptake), to be assimilated (bind with other wine constituents, becoming nonvolatile). Thus, "bottle sickness," usually attributed to acetaldehyde, will have dissipated before the wine reaches the consumer.

3. INNOVATIONS IN THE VINEYARD/ORCHARD

Initially, advances leading to better fruit production were discovered empirically. This has changed radically in the past 150 years. Currently, an incredible range of scientific fields are being applied to improve fruit quality. Some of the more recent are noted later in this chapter.

That wine is "made in the vineyard" is an oft-quoted adage in marketing. Clearly, the fermented by-product of fruit can be no better than the potentials found at harvest. For optimal quality, all facets during the current and previous seasons must be ideal. As any aspect of wine production, the vineyard component is super complex and is far more prone to the fickle fluctuations of Mother Nature than its winery constituent. One of the central jobs of the grower is to diminish these vicissitudes, bringing to the cellar as much healthy, quality fruit as possible, respecting long-term vine/tree fruitfulness. It is a delicate tightrope that the grower must navigate each year.

In the incredibly diverse world of wine, there is one word that is everyone's lips: quality. Although it is the term to which everyone genuflects, there is no universally accepted definition, either for fruit or wine. Quality, like beauty, is in the eye of the beholder. Nonetheless, there is general agreement that when it comes to optimal fruit quality, it usually correlates with full maturity, typically declining thereafter. Nonetheless, this peak is climate and style dependent. For example, late harvesting is conducive to producing the best sweet wines, whereas moderate immaturity is preferable for producing most sparkling wines. This then begs the question of how one best defines and measures fruit maturity. Regrettably, this is no simple means by which to quantify maturity, depending as it does on the cultivar, provenance, and style.

Prior to the development of modern analytical methods, there were three principal indicators of fruit maturity: sweetness, sourness, and coloration. Assessment was based on experience. With developments in chemistry and physics, the assessment of fruit sugar, acid content, and coloration became comparatively easy and objective. Subsequent advances have confirmed the general correlation between these attributes and the development of desirable varietal flavors (Coelho et al., 2007). Regrettably, the correlation is not "perfect," and as suggested, can vary considerably with the cultivar, climatic conditions (including the soil), and wine style desired. Thus, much effort is still expended in developing even more precise physicochemical measures of grape quality, to gain better indicators of desirable ripeness. Once more is known about the precise origins of quality, similar techniques should be applicable to fruit wines.

Varietal flavor and pigmentation often do not develop until relatively late in ripening, reach a peak, and decline both quantitatively and qualitatively thereafter. The source of grape flavor is often exceptionally complex, both in its chemical diversity and presence in trace amounts. Flavorants often occur bound in nonvolatile complexes, being released only during fermentation, maturation, or subsequent aging. Although grape pigmentation is also chemically complex, more meaningful objective measurements are under development. This could further facilitate determining when best to harvest. An example is a handheld, fluorescence-based, optical sensor. It can measure fruit anthocyanin content under vineyard conditions (Ghozlen et al., 2010). In some cultivars, optimal maturity is signaled when anthocyanin content begins to decline. Regrettably, anthocyanin content does not necessarily correlate well with extraction (e.g., Rustioni et al., 2011), wine color, or its stability. Advances in near-infrared spectroscopy (NIRS) may also permit rapid (real-time), inexpensive, and accurate assessment of vineyard grape samples for sugar, acid, and color indicators (Cozzolino, 2015; Tuccio et al., 2011).

Spectroscopic measurements in the near-infrared (NIR) range (750–2500 nm) provide data on the relative proportions of —CH, —NH, and —OH bonds in opaque samples, thus, avoiding the time-consuming traditional sample adjustment. This feature depends on the unique vibrational properties of particular chemical bonds associated with specific light absorption. However, because of the large amount of data obtained, and various sources of error, multivariate statistical analysis is required for data interpretation, combined with frequent, on-site calibration with standard analytic techniques (Wilkes and Warner, 2014).

As noted, many aromatics accumulate in nonvolatile complexes. As glycosides, they accumulate more readily in cell vacuoles due to the greater water solubility of the complex. This has led to assessing glycosyl-glucose (G–G) content as a potential indicator of aromatic content (Williams and Francis, 1996). Regrettably, grape G–G content has often proven no more reliable as a predictor of wine flavor than grape anthocyanin content with wine coloration.

Subsequent research has also found several varietally important aromatics bound to glutathione. In addition, flavorants may accumulate as oxides (e.g., terpenes), exist weakly bound to other constituents, subsequently liberated by yeast action (e.g., linoleic and linolenic acids), or be released by acid hydrolysis during maturation and aging. To further understand these phenomena, most studies now focus on assessing the specific contents of individual compounds and their respective sensory significance.

Typically, the potential flavor impact of constituents is estimated by calculating its odor activity value (OAV). This is determined by dividing the concentration of the compound by its olfactory threshold. Although seemingly precise, aromatic compounds often interact synergistically with, or are masked by, other constituents (termed the matrix). Thus, a compound's threshold, and its OAV, is not fixed, and predicted olfactory significance may not be realized in practice. In addition, assessing the concentration of specific aromatics in juice or wine is far from easy, requiring extensive technical skill in the use of expensive equipment. Hopefully, technical advancement in vibrational spectroscopy (ultraviolet, visible, Fourier transforms infrared, or mid-infrared, near-infrared, and Raman spectroscopy) will bring aromatic assessment to within the reach of most wineries.

A subset of issues, relating to grape quality, is associated with variability within clusters, the vine (location within the canopy), and throughout the vineyard. Up to a point, limiting yield favors flavor development. However, as with other features, this property is not linear, and is cultivar and clone dependent (Fig. 13.2). Because limiting yield is best done as early, there is a need for simple predictors of grape yield. Hitherto, measurement of flower number and fruit set have been expensive and time-consuming. An innovative approach involves analyzing vineyard photographs with a cultivar-specific algorithm. It has the

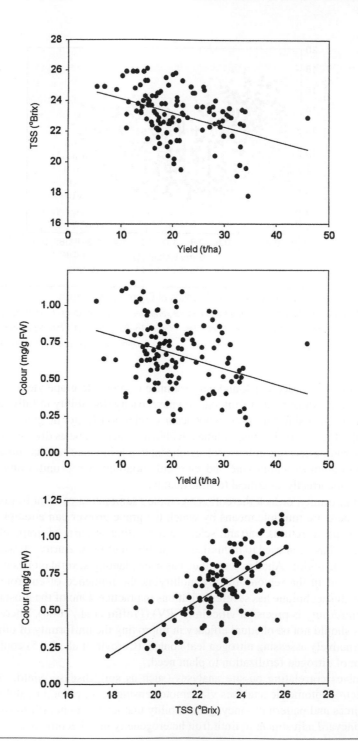

FIGURE 13.2

Variability between total soluble solids, yield, and color in "Shiraz" relative to yield and °Brix values. *FW*, fresh (berry) weight; *TSS*, total soluble solids, measured in Brix; *t/ha*, tons/hectare.

From Holzapfel, B.P., Rogiers, S.Y., Degaris, K.A., Small, G., 1999. Ripening grapes to specification: effect of yield on color development of Shiraz in the Riverina. Australian Grapegrower & Winemaker 428 (24), 26–28; reproduced by permission.

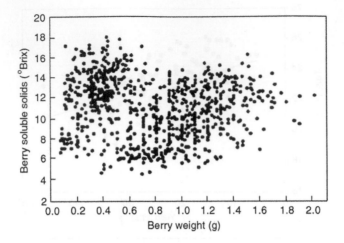

FIGURE 13.3

Variation in berry size in grape clusters and soluble solids of Chardonnay grapes.

From Trought, M., Tannock, S., 1996. Berry size and soluble solids variation. In: Henick-Kling, T., Wolf, T.E., Harkness, E.M. (Eds.), Proc. 4th Int. Symp. Cool Climate Vitic. Enol., Rochester, NY, July 16–20, 1996. NY State Agricultural Experimental Station, Geneva, New York, pp. V-70–73; reproduced by permission.

potential of providing the needed quick-and-easy estimate of probable grape yield (Diago et al., 2014). This may facilitate early adjustment of fruit load to vine capacity (the ability to fully ripen fruit). Regrettably, flower development and fertilization do not occur synchronously, nor do grapes in a cluster develop or mature uniformly (Fig. 13.3). The microclimate within a cluster can be as diverse as across the vine or throughout the vineyard. Thus, one of the most significant trends in viticulture has been the measurement of and practices to minimize within vine and vineyard variability. Crop uniformity is now becoming viewed, correctly or incorrectly, as critical to wine quality.

Because crop uniformity, or lack thereof, is a complex summation of what has occurred in current and past seasons, there are multiple means by which the grape grower can attempt to limit excessive variation. Where financial returns warrant, selective harvesting of parts of grape clusters, or distinct parcels of a vineyard, has been used to minimize variability in grape maturity. However, this is seldom an economically viable option. Another technique has been planting several clones of a cultivar, known to do particularly well in the region, thereby mollifying the influence of environmental variability. Although of value, this technique has severe limitations in practice. One of the most recent innovations to enhance fruit uniformity is precision viticulture (PV) (Proffitt et al., 2006). There is no reason why similar techniques should not be of equal efficacy in improving the uniformity of other fruit corps. For example, nondestructively assessing nitrogen leaf content (Cerovic et al., 2015) could permit the more precise adjustment of nitrogen fertilization to plant need.

PV often involves correlating on-site analyses (such as soil, disease, yield, composition), with broader scale spectrophotometric analyses via remote sensing (e.g., plane or satellite). The aim is to determine the sources and pattern of vineyard variability that lead to nonuniform ripening (Fig. 13.4). With such data, vineyard adjustment to limit fruit heterogeneity may become one of the principal tasks of grape growers.

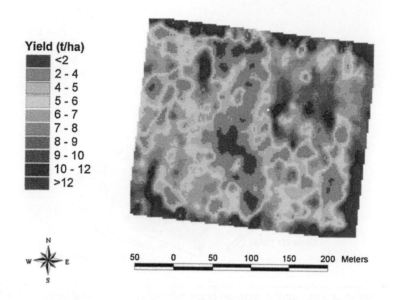

Yield (t/ha)
- <2
- 2 - 4
- 4 - 5
- 5 - 6
- 6 - 7
- 7 - 8
- 8 - 9
- 9 - 10
- 10 - 12
- >12

50   0   50   100   150   200   Meters

FIGURE 13.4

Variation in grape yield (1999), clearly showing the marked influence of microclimatic, terrestrial, and atmospheric conditions on vine yield.

From Bramley, R., Proffitt, T., 1999. Managing variability in viticultural production. Australian Grapegrower & Winemaker 427, 11–12, 15–16; reproduced by permission.

Studies to date indicate that considerable improvements in yield and maturity prediction are possible with data provided by PV (Proffitt and Malcolm, 2005). Developments in remote sensing are making the mapping of vineyard variability, down to individual vines, both feasible and increasingly affordable. Where considerable vineyard variability exists, and cannot be adequately addressed, it is important that maturity assessment be adjusted to reflect this reality. For example, if vines of high and low vigor were to represent 25% and 10% of the vineyard area, then approximately 25% of the fruit should be randomly selected from vigorous parcels, 10% from low vigor sites, and the remainder from intermediary sites. Adjusted sampling would more accurately represent the fruit quality of the crop as a whole. The same data could also be the basis for selectively timing harvest to reflect the local distribution of mature fruit.

Improvements in harvest timing could materialize one of the dreams of wine-makers: improved fruit quality (uniformity) at the winery door. Fruit of uniform maturity should yield wines with a more predictable character (Kontoudakis et al., 2011), allowing the wine-maker to adjust production procedures at an early date, to achieve producer and consumer preferences. Although the interrelationship between fruit chemical analysis and perceived wine quality is still illusive, progress is being made (Forde et al., 2011). However, caution is required. Wine quality is a multifactorial property (Fig. 13.5), not easily correlated with objective measurements (Bramley, 2005). In fact, variability in fruit maturity may not automatically be a negative attribute. It could accentuate flavor complexity, the *sine qua non* of quality. In addition, vineyard variability may partially compensate climatic vicissitudes—that is, not all vines being equally perturbed by climatic perturbations.

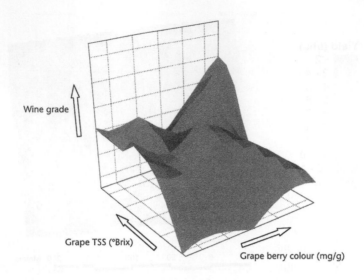

FIGURE 13.5

Schematic representation of a relationship between grape total soluble solids and berry color with the quality grading of the resultant wine.

From Gishen, M., Iland, P.G., Dambergs, R.G., Esler, M.B., Francis, I.L., Kambouris, A., Johnstone, R.S., Høj, P.B., 2002. Objective measures of grape and wine quality. In: Blair, R.J., Williams, P.J., Høj, P.B. (Eds.), 11th Aust. Wine Ind. Tech. Conf. October 7–11, 2001, Adelaide, South Australia. Winetitles, Adelaide, Australia, pp. 188–194; reproduced with permission.

Other improvements in fruit quality are likely to be achieved with remote sensing, such as by NIRS (de Bei et al., 2011) and chlorophyll fluorescence (Moya et al., 2004). These techniques can assess water-related photosynthetic efficiency, thereby giving the grower the potential to adjust irrigation to actual plant need, rather than by the current, indirect, and imprecise means (Escalona and Ribas-Carbó, 2010). Techniques such as regulated deficit irrigation and partial rootzone drying are of particular note in regard to irrigation efficiency. In addition, where irrigation is regularly required, it can be combined with simultaneous fertilization, a process termed fertigation. As water for agriculture becomes increasingly scarce and more expensive, such techniques may become of equal value with other vine and fruit-tree crops. Expertise in one agricultural endeavor often finds applicability in others.

Mechanical harvesting, initially employed to reduce production costs, has seen marked improvements in efficiency. This has been further improved by innovations in sorting. Special screens, or optical sorters, can automatically activate blowers over collector belts (Messenger, 2007), facilitating the removal of diseased fruit and extraneous material (leaves, twigs, and other plant and animal debris). Thus, essentially only healthy fruit passes on to the stemmer-crusher, masher, pitter, deseeder, or directly to the press. Further current developments in harvesting and sorting equipment are discussed in Christmann and Freund (2010) and Li et al. (2011).

Significant innovations in pest control have reduced the need for pesticide use, both synthetic and organic. This often involves the use of climate-based disease models. They direct pesticide application relative to need, avoiding older, wasteful, prophylactic application on a rigid schedule. Rotating selective and curative (systemic) pesticides also delays or avoids the development of pest resistance. Another class of compounds, such as chitosan and benzothiodiazole, do not affect pests directly, but activate inherent disease regulation in the host. A study by Vitalini et al. (2014) has shown that these agents (and possibly synthetic and organic pesticides) may have hitherto unknown effects on plant secondary metabolism and volatile production, influencing the sensory attributes of wines subsequently produced. In addition, the adoption of canopy adjustment techniques, producing more open canopies, has enhanced air flow around, and sun exposure to, fruit and the leaf canopy. These, combined with developments in nozzle design, have improved pesticide distribution, reducing application need and rates, minimizing run off, drift, and enhancing effectiveness.

The incorporation of genetic resistance into existing cultivars is possible, but often hindered or prevented by vocal minorities opposed to genetic engineering—the only means of achieving this goal in vegetatively reproduced crops. This is incomprehensible because genetic engineering is "environmentally friendly," drastically reducing the need for pesticide use. The potential of newer forms of genetic engineering (Dolgin, 2015), in avoiding the hypothetical fears of critics, should encourage the greater (rational) acceptance of genetically modified crops. That only multinational corporations are the principal producers of genetically modified organisms is no a priori reason for rejection.

The availability of the grape genome has increased the potential of using marker-assisted breeding in traditional techniques (Mackay et al., 2009). Regrettably, this applies only to the development of new cultivars. Traditional breeding techniques disrupt the genomic makeup of their parents, such that the new cultivar no longer possess the flavor characteristics on which varietal reputations in wine are based. Without the new variety possessing the name of any of its parents, the critical marketing value of consumer varietal recognition in wine sales would be lost.

Other innovations, useful in the biological control of arthropod pests, include a return to the use of indigenous ground covers. These provide potential sites for maintaining effective populations of natural pest-control agents. One of the more ingenious biological controls involves sheep grazing in weed control.

4. WINERY INNOVATIONS

In centuries past, crushing grapes under foot was standard practice. Although gentle, it was labor intensive, slow, and incomplete. Unbroken grapes in the ferment would undergo autofermentation (termed semicarbonic maceration). At a time when few wines aged well, it had the benefit of producing wines that were ready to drink shortly after production. Standard procedures often require several months to reach an equivalent state of drinkability.

The process of carbonic maceration is still used in the production of some wines, notably Beaujolais Nouveau. Traditionally this involves grape clusters piled in large but shallow, covered fermentors. Modern versions often use plastic covers and flush the grapes with carbon dioxide to encourage earlier autofermentation. Elsewhere, the carbonic maceration phase may occur in the containers in which the fruit is delivered to the winery (tightly wrapped in plastic), or occurs in specially designed chambers flushed with carbon dioxide (Càstino and Ubigli, 1984). After completion of the carbonic maceration phase, the grapes are pressed and fermented as per usual under the action of yeasts. The technique has been experimented with in producing a strawberry wine (Sharma et al., 2009).

With the development of mechanical crushers in the mid-1800s, the effectiveness of pressing increased. It also meant that the incidental occurrence of semicarbonic maceration became a thing of the past. Although crushers are not essential for soft fruit such as raspberries, the development of mechanical devices for mashing hard fruit greatly facilitated the production of wine from fruit such as apples or pears. Joint stemmer-crushers remove stems as well as other debris, minimizing the uptake of undesirable flavorants and contaminants.

Advancements in press design (beginning in the 1300s) have also increased the efficiency of juice extraction. The reduced pressure needed for juice extraction has also limited the extraction of undesirable oils and phenolics from the seeds and skins. For the production of white sparkling wines, the fruit is pressed whole (without prior crushing). This is especially important to limit the uptake of anthocyanins where red grapes are used.

Until comparatively recently, contact between the must and oxygen, during and immediately after crushing and pressing, was viewed as detrimental. It was considered to favor premature wine browning. Thus, the must was often "blanketed" with carbon dioxide or nitrogen. Nonetheless, the practice enhanced oxidative browning, by protecting readily oxidized phenolics from early oxidation, and their subsequent precipitation during fermentation and clarification. Thus, a return to older winemaking techniques has come into vogue. To enhance the removal of easily oxidized phenolics, some producers now purposely expose the juice or must to oxygen during maceration (a procedure termed hyperoxidation).

Another "innovation," being in reality a broader application of an old process, is cold presoaking (maceration) before the onset of fermentation. The procedure has been traditional in Burgundy, where crushing and fermentation in cold, unheated cellars delayed the beginning of fermentation. Incidentally, it facilitated the extraction, and possibly stability, of the limited pigmentation typical of Pinot noir grapes. It also appears to encourage the extraction of flavorants in several other grape varieties. In lieu of cold cellars, and with grapes harvested in warm climates, rapid cooling is frequently achieved by adding dry ice or liquid nitrogen. The technique is now being used experimentally with several of red cultivars, such as Syrah and Cabernet Sauvignon (Moss et al., 2013). The optimal parameters of the process are still being established for these and other cultivars. Its value, as an adjunct in fruit wine production, seems uninvestigated.

Another return to older procedures is spontaneous fermentation (or its modern equivalent: inoculation with a series of yeast strains and/or species). Initially, all wine fermentations were spontaneous, starting with the action of epiphytic yeasts or residuals on winery equipment. In most instances, fermentation subsequently came to be dominated by one or more strains of *S. cerevisiae*, typically derived from winery equipment. However, in the case with cider fermentations, other species appear to be selected, notably the related *Saccharomyces uvarum*, especially under cool fermentation temperatures.

The return to spontaneous fermentation is based on the view that indigenous yeasts donate that distinctive *je ne sais quoi* quality valuable in distinguishing local or small producers from the competition. For them, vintage, regional, and estate distinctiveness are often crucial to market recognition, and thereby, financial viability. At the moment, the effectiveness of these procedures is contentious, often depending on the region, the strains involved, and the season.

In the case of spontaneous French cider fermentations, there is an initial phase that permits the activity of oxidative or slowly fermentative yeasts, notably those belonging to *Metschnikovia*, *Hanseniaspora*, and *Candida*. During this phase little ethanol is produced, but numerous aromatic compounds are generated, providing the product with much of its traditional character. Subsequently, the alcohologenic phase commences with the action of *S. uvarum*, *Saccharomyces bayanus*, or both (Coton et al., 2016).

Where wine of low-alcohol content is desired, the use of non-*Saccharomyces* yeasts, susceptible to ethanol accumulation, may prove a desirable choice. Such yeasts may also possess additional advantages, such as *Williopsis saturnus* (formerly *Hansenula saturnus*; Li et al., 2012). Unlike *S. cerevisiae*, *W. saturnus* does not degrade the fruit terpenoids, a characteristic of ripe mangoes.

Alternatively, fermentation may be initiated with a mix of indigenous and/or commercial yeast strains and species. The result can be distinctly different from natural spontaneous fermentations, where the epiphytic yeast flora begins fermentation, only to be replaced and completed by *S. cerevisiae* (Heard and Fleet, 1985). In mixed, induced fermentation, all species are inoculated at the level (10^6–10^7 cells/mL). In this instance, all species may remain active to the end of fermentation (Chanprasartsuk et al., 2012).

Examples of the effects of mixed inoculation may include reduced alcohol production, and associated with enhanced glycerol accumulation (Giaramida et al., 2013). This reduces the burning aspect associated with high alcohol contents, while providing improved smoothness with the additional glycerol. Inoculation with *Metschnikowia pulcherrima*, followed by *S. cerevisiae,* can also reduce alcohol production, combined with other sensory benefits (Contreras et al., 2014). Joint inoculation has also been tried with fruit wines, apparently enhancing the desirable flavor profile of the wine (Sun et al., 2014). In addition, for specific wine styles, species closely related to *S. cerevisiae* may be preferable. For example, *S. bayanus* var. *bayanus* have advantages in the production of fine sherries and sparkling wines, whereas *S. bayanus* var. *uvarum* has benefits where cool fermentation is desired.

In contrast, inoculation with a single known strain is more predictable and can avoid several intractable fermentation problems. Correspondingly, attempts are underway to produce interspecific yeast strains with some of the sensory complexity claimed for spontaneous fermentations, but without the associated risks (see Bellon et al., 2011).

Inoculation involves the addition of a concentrated solution of active yeasts, and any nutrients considered necessary, to bring the population to about 10^6 to 10^7 cells/mL. This typically assures the rapid onset of fermentation and ethanol production. This, in turn, quickly limits the action of other yeast, as well as reducing the number of cell replications that occur during fermentation. Problems induced by killer-yeast factors, that only infect cells during multiplication, can usually be avoided by limiting the number of cell divisions with heavy yeast inoculation. The growing commercial availability of a diversity of yeast strains, with known attributes, has further encouraged the widespread adoption of induced fermentation. This practice has provided wine-makers with hitherto unknown control over flavor development. Breeding has also incorporated special attributes, such as limited H_2S production, resistance of killer factors, increased fruit ester production, and improved flocculation. Occasionally, strain improvement has involved overexpression of a gene, such as *STR3*. It increases the release of 3-mercaptohexan-1-ol from cysteinylated precursors in the grapes, improving the aroma of grapes containing this precursor (Holt et al., 2011). Another example of genetic modification has been the successful insertion of genes that can conduct malolactic fermentation in *S. cerevisiae*, avoiding the need to inoculate with *O. oeni* (Bony et al., 1997). Such strains are the mainstay of larger producers where it increases their ability to achieve brand consistency.

Until comparatively recently, malolactic fermentation by *O. oeni* tended to be both inconsistent and spontaneous. Successful development of a commercial dry inoculum has permitted both greater control on when malolactic fermentation occurs and its sensory consequences (besides deacidification). As with alcoholic (yeast) fermentation, spontaneous malolactic fermentation may one day be viewed as a means of encouraging regional or estate distinctiveness. Nonetheless, inconsistencies with spontaneous occurrence and sensory consequences favor the continued use of strains of known attributes. Inoculation with *O. oeni* also helps limit the action of undesirable lactic acid bacteria. Much of this benefit comes with the subsequent storage of the wine under conditions unfavorable to the activity of lactic acid bacteria.

The principal action of malolactic fermentation is deacidification—converting malic to lactic acid (see Reuss et al., 2010, regarding use in cider production). The conversion is desirable if the wine possesses high concentrations of malic acid (enhancing sourness), but is undesirable for those with insufficient acidity (lacking in taste vitality and reduced color and microbial stability). Nonetheless, much of the current discussion relating to malolactic fermentation focuses on flavor modification. With cider, it has been noticed to enhance its apple flavor, while reducing sourness, astringency, and yeastiness (Riekstina-Dolge et al., 2014). Marked improvement in the sensory qualities of mango wine were also found as a consequence of malolactic fermentation (Fig. 13.6). Although malolactic fermentation can

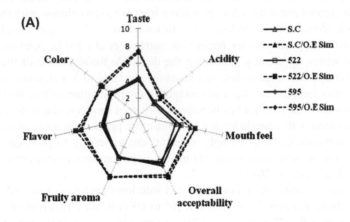

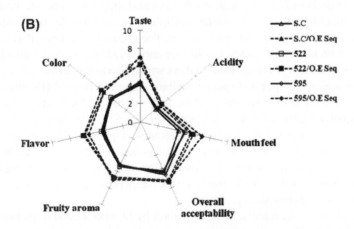

FIGURE 13.6

Influence of malolactic fermentation on the sensory profiles of mango wine fermented in simultaneous inoculations (A) and sequential inoculations (B). *522*, UCD 522—a University of California, Davis *S. cerevisiae* strain; *595*, UCD 595—a University of California, Davis *S. bayanus* strain; *S.C*, a strain of *Saccharomyces cerevisiae*; */O.E Seq*, inoculated sequentially with a *Oenococcus oeni* strain NCIM 2219 (originating from the National Collection for Industrial Microoganisms, Pune, India); */O.E Sim*, inoculated simultaneously with a *Oenococcus oeni* strain NCIM 2219.

From Varakumar, S., Naresh, K., Variar, P.S., Sharma, A., Reddy, O.V.S., 2013. Role of malolactic fermentation on the quality of mange (Mangifera indica L.) wine. Food Biotechnology 27, 119–136; reproduced with permission.

enhance flavor quality and complexity, it can also donate off-odors. Regrettably, the action of malolactic fermentation generates conditions that favor the action of spoilage lactic acid bacteria (*Pediococcus* and *Lactobacillus* spp.): increased pH.

The strain of *O. oeni*, and whether malolactic fermentation occurs during or after alcoholic fermentation, can profoundly influence flavor development. Similar findings have been found with cherry (Sun et al., 2013) and mango (Varakumar et al., 2013) wines. In addition, attributes such as apple and grapefruit-orange in Chardonnay wines, and strawberry-raspberry of Pinot noir wines, may be replaced by hazelnut, fresh bread and dried fruit aromas, and animal and vegetable notes, respectively (Sauvageot and Vivier, 1997). These influences are, in turn, affected by the wine's temperature, initial pH, and varietal origin. It is because of such variability that most wine-makers, if desiring malolactic fermentation, inoculate with commercial (freeze-dried) strains.

Alternatives to inoculation with *O. oeni* have been suggested, partially to avoid the occasional difficulty in completing malolactic fermentation. Systems based on encapsulation of the bacteria in a support system (Kosseva and Kennedy, 2004; Srivastava et al., 2013; Fig. 13.7), or enzyme immobilization have been proposed. Another route is inoculation with the yeast, *Schizosaccharomyces pombe* (Benito et al., 2013). It not only completely decarboxylates malic acid, reduces ethanol production, but also limits the accumulation of urea (and thereby ethyl carbamate generation). Despite these benefits, it has been rarely used due to its too frequent association with off-odor production. Immobilization (Silva et al., 2003), and delaying inoculation, until *S. cerevisiae* has been active for several days (Carre et al., 1983) appear to reduce the negative sensory effects of *Sz. pombe*.

As the abilities of analytic instruments have become more sophisticated, so have their potential to measure changes during fermentation. Their applicability under winery conditions has also improved as the robustness of the equipment has increased, and costs decreased. Although this availability has given the wine-maker unprecedented opportunities to follow the progress of fermentation, and institute quick corrective measures as needed, we are nowhere near "autopilot" winemaking. For that, constant and instantaneous physicochemical analysis would be required, as well as a better understanding of the intricacies of fermentation and the physicochemical parameters of wine quality.

Currently, censors located throughout the fermentor can record parameters such as sugar, nitrogen, and alcohol content. These data, supplied to computers, have facilitated the monitoring and regulation of fermentation in gargantuan cooperage. In addition, advances in spectroscopy and other analytic instruments are beginning to permit real-time analysis of multiple chemical parameters throughout fermentation (Cozzolino et al., 2011; Dubernet, 2010). Quantitative nuclear magnetic resonance (qNMR) is also being investigated as a potential procedure for profiling metabolite production during fermentation (López-Rituerto et al., 2009).

Until the development of modern refrigeration, temperature buildup during fermentation caused major problems in southern climes, and with large fermentors everywhere. Temperatures above 30°C progressively suppress and eventually terminate fermentation. With most larger fermentors now possessing some degree of temperature control, fermentation can be held to within a few degrees of any desired value.

For red wines, where a cap of seeds and skins accumulates on the surface of the must, various means are available to periodically, or continuously, transfer fermenting juice over and through the cap. This procedure, termed pumping over, facilitates approaching temperature equilibration throughout the tank (Schmid et al., 2009). Although temperature uniformity is now preferred, in the past, temperature

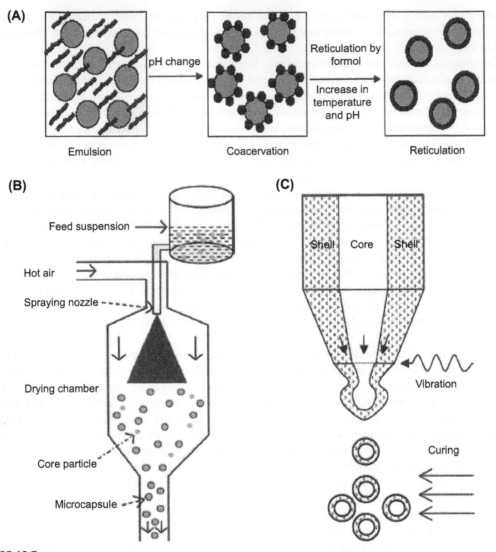

FIGURE 13.7

Microencapsulation techniques (A) coacervation; (B) spray drying; (C) extrusion.

From Srivastava, Y., Semwal, A.D., Sharma, G.K., 2013. Application of various chemical and mechanical microencapsulation techniques in food sector –a review. International Journal of Food and Fermentation Technology 3, 1–13; reproduced by permission.

differentials between the cap and fermenting juice could be up to 10°C (Guymon and Crowell, 1977). Whether this had some sensory benefit seems not to have been investigated. Temperature control has also provided one of the easier fermentation parameters to regulate, providing another aspect of control over the wine's flavor profile. This occurs not only by directly influencing yeast metabolism but also by affecting the species and strains of yeasts that conduct fermentation.

Fermentors have changed markedly, both in size, shape, and construction during the past half century. Stainless steel fermentors have become the industry standard, replacing large oak and redwood tanks and vats. Not only can stainless steel tanks be produced in almost any size or shape desired, they transfer heat rapidly (facilitating temperature control), are easy to clean and store empty, have the potential of being supplemental storage vessels, and are inert and impervious (protecting the wine's attributes). Nonetheless, oak fermentors have certain advantages for wine-makers desiring to produce particular wines. Recently, this has been almost exclusively associated with in-barrel fermentation. Innovations in the construction of large oak fermentors may encourage their redeployment. For example, modern oak tanks are designed for much easier emptying and cleaning between uses. They can be fitted with inserts that permit effective temperature control and facilitate pumping over. In addition, replacement of one of the staves with clear fiberglass permits direct visualization of fermentation from top to bottom.

Rotary fermentors are another comparatively recent innovation, designed to improve and automate color and flavor extraction. The horizontal position of the fermentor increases contact between the juice and the pomace. Rotating paddles, or jacket, help minimize the formation of a cap of seeds and skins, and stratification of temperature throughout the ferment. Alternatively, the fermentor may simply be gently rocked to reduce the extraction of solids or other constituents from the pomace.

Blending is most commonly thought of in terms of finished wines. Combining of the juice from several cultivars, prior to fermentation, is uncommon with grape-based wines, but seemingly common with fruit-based wines. Nonetheless, in the past, combining the juice or musts from different grape cultivars was common, if only due to the random interplanting of several cultivars throughout vineyards and their harvesting together. Although this practice is now rare, some producers are experimenting with intentionally blending musts from different cultivars before fermentation (termed cofermentation). There is evidence that the practice may enhance a wine's flavor (García-Carpintero et al., 2010), possibly beyond that of blending their respective finished wines. A more well-established practice is adding limited amounts of the juice from white grapes to the must of certain red cultivars. If the red cultivar is deficient in copigments (such as caftaric acid that promotes stable coloration), adding juice from a white cultivar amply supplied with copigments improves color stability (Lorenzo et al., 2005). Whether such a process would help color stability in some fruit wines is unknown. Another unique form of cofermentation is associated with producing some rosé wines (by drawing off a portion of the juice early during fermentation). The concentrated must that remains is added to must to enhance coloration. The technique, and its variations, is often referred to as *saignée*.

More commonly, blending occurs postfermentation. This may involve wines derived from separately fermenting different press fractions, from different fermentors (notably if fermentation occurs in-barrel), from fruit harvested at different times, or portions of a vineyard. This form of blending is typical where vineyard or orchard uniqueness (*terroir*) is a major element in marketing. In contrast, large wineries often blend wines produced from multiple sites, regions, or countries. To achieve the volumes and brand consistency required for world distribution, artful blending from different sources is necessary. Blending from different properties is also standard practice with particular wine styles, notably sparkling and fortified wines. For these wines, brand consistency is essential to their marketing and distribution strategy. In contrast to much of the mystique that surrounds the term *terroir*, and how it provides sensory uniqueness, research has shown that blending frequently diminishes flavor imbalances or deficiencies in the component wines, generating a more sensorially pleasing product (Singleton and Ough, 1962).

With the apparent shift in consumer demand to more flavorful wines, producers have often chosen to leave grapes on the vine longer than is traditional. This is based on the belief that somewhat overmature, partially dehydrated grapes supply enhanced flavor. This trend seems not to have infected the fruit wine industry. One of the regrettable consequences of this change has been harvesting grapes at elevated °Brix, and correspondingly the production of wines with atypically high alcohol contents. A similar result (regarding alcohol content) is a consequence of another, increasingly popular, supposedly, flavor enhancing/modifying technique, *appassimento*. In the process, fully mature clusters are laid on racks to partially dehydrate for several weeks under cool conditions. Amarones are the most well-known wines produced in this manner. Much of their distinctive flavor is (or was) derived from an unsuspected reactivation of dormant *Botrytis* infections in a portion of the grapes (Usseglio-Tomasset, 1986). Without the changes induced by this form of botrytization, *appassimento* seems to have little rationale other than marketing.

As noted, one of the more obvious consequences of both late harvesting and partial dehydration has been a pronounced increase in the alcohol content of table wines. Traditional values, previously within in the 11.5–12.5% range, are now often in the 13–14% range, and occasionally reach above 15%. Although accepted by the majority of consumers, possibly because they have little choice, most wine critics decry the change. Anecdotal reports, supported by some research, suggest that higher alcohol contents distort wine flavor-profiles, only occasionally enhancing fruit flavors (Heymann et al., 2013). Partial dealcoholization is a possible solution. However, older, heat-induced evaporative-dealcoholization procedures were usually aimed at producing fully dealcoholized wines. That the procedure generated a slightly baked or cooked odor, and a distinct change in flavor, seemed not to have been a significant commercial deterrent. Vacuum distillation avoids heat-induced flavor modification, but does involve flavor loss. Although flavorants lost can be trapped, and added back to the alcohol-reduced wine, it adds significantly to production costs. Alternative procedures include strip-column distillation, dialysis, pervaporation, spinning cone column, and carbon dioxide stripping (Wollan, 2010). All have, to varying degrees, the potential to reduce (modify) the wine's flavor.

A novel approach has been proposed by Kontoudakis et al. (2010). A portion of the crop is picked early, fermented, and treated with bentonite and charcoal to produce a flavorless, low-alcohol acidic wine. It is subsequently added to wine made from fully or overmature grapes, decreasing the blended wine's alcohol content. The resultant wine possesses a fuller, but still more traditional, flavor profile (but without an alcohol burn). The acidity supplied by the low-alcohol wine enhances color intensity and freshness. Nonetheless, returning to harvesting at more traditional °Brix values would avoid the need for dealcoholization.

Fruit wines are unlikely to experience problems with excessive alcohol content due to their inherent low contents of fermentable sugars. To reach traditional table wine alcohol contents sugar must be added. Thus, fruit wines are intrinsically endowed to fill a niche for low-alcohol wines.

Other techniques designed to improve flavor development promote the liberation of aromatics from nonvolatile complexes. Although the flavorant content increases during ripening, an increasing proportion of them becomes bound in nonvolatile complexes. This phenomenon affects many important varietal compounds, notably terpenes, norisoprenoids, and volatile phenols that bind with sugars as glycosides. These bonds tend to break spontaneously, but only slowly during wine aging. This is a desirable attribute in wines aged for many years, as it helps maintain the wine's flavor. However, this is of little value to wines consumed young, as is typical. To promote the earlier liberation of these flavorants, heating has been investigated to speed acid-induced hydrolysis. Regrettably, such heating also promotes the hydrolysis of desirable fruit esters, accentuates terpene oxidation, and enhances the production of methyl disulfide. To avoid these flavor distortions, enzymic hydrolysis is preferred (O'Kennedy and Canal-Llaubères, 2013a,b). Of these preparations, β-glycosidases have been the most studied. Because glycosides often involve sugars other than glucose, most commercial preparations possess some α-arabinosidase, α-rhanmosidase, β-xylanosidase, and β-apiosidase activities. Enzymatic preparations derived from filamentous fungi are preferable as they are relatively insensitive to the acidic conditions of wine. When used, these preparations are added at the end of fermentation; glucose (in juice/must) acts as an enzyme inhibitor. Enzyme inactivation can be quickly terminated, as and when desired, by adding bentonite. It absorbs and removes proteins. More efficient (but expensive) regulation can be achieved with enzymes immobilization on a column, through which the wine is slowly passed.

At least one study with β-glucosidase added during maceration was shown to significantly enhance the flavor of a mango wine (Fig. 13.8). This suggests that similar treatments may be equally beneficial with other fruit-based wines. Enzyme preparations may be also added to degrade problem-inducing polysaccharides, notably pectins. These can often be a particular problem with nonjuicy fruit such as kiwi (Vaidya et al., 2009).

Once fermentation has come to completion, and the wine separated from pomace with which it may have been fermented, the wine is matured for several months (to years) before bottling. One of the central purposes of maceration is to permit spontaneous clarification. Wine is initially cloudy after fermentation, possessing a variety of suspended solids. Given time, these usually precipitate on their own (such as yeasts and fruit cell particles), or after binding with suspended phenolics and/or proteins.

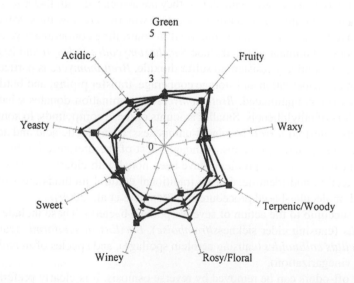

FIGURE 13.8

Aroma profile of mango wines: nonmacerated control (▲); macerated wine control (■); nonmacerated wine with enzyme treatment (◆); and macerated wine with enzyme treatment (x).

From Li, X., Lim, S.L., Yu, B., Curran, P., Liu, S-Q., 2013. Mango wine aroma enhancement by pulp contact and β-glucosidase. International Journal of Food Science & Technology 48, 2258–2266; reproduced with permission.

In addition, various potassium and calcium salts crystallize and precipitate. Although effective, these natural processes often occur more slowly than currently desired. Correspondingly, a wide range of techniques may be used to speed clarification.

Adding fining agents, either organic (e.g., tannin, protein, or their combination) or inorganic (e.g., bentonite), are traditional. Nonetheless, adjusting their addition is empirical and may accentuate cloudiness if overused. In addition, some fining agents can result in significant wine loss, notably bentonite. How significant the fining agents are in flavor loss is contentious. For a while, there was concern that gelatin derived from beef products, used as a fining agent, might contaminate the treated wine with prions of Bovine Spongiform Encephalophy. To counter this concern, plant-based proteins were studied as a substitute. Although effective, wheat-based protein fining agents may have their own problems, due to allergies associated with even trace amounts of residual glutens. Nonetheless, if clarification is to occur early, some form of fining or centrifugal-induced precipitation is required. Without these, producing commercially acceptable, stably clear wine would result is excessive, and thereby expensive, plugging of clarification filters. Cross-flow filters are a comparatively new alternative to the former use of throughput filters. The latter often involved the addition of filter aids, accentuating wine loss. Because wine in cross-flow filters passes horizontally through the filter tube, particulate matter, collected on the inside of the filter, is partially swept along with some of the wine. The majority of the wine passes through the filter wall.

When considered necessary, microbial contaminants and colloidal material can be removed by micro- or ultrafiltration. For sterile filtration, cutoff points of 0.45 μm have been traditional, but for greater protection, cutoff values of 0.3–0.2 μm are now recommended. Ultrafiltration techniques can be used to remove recalcitrant colloidal material, as well as pigments, tannins, and mannoproteins. However, ultrafiltration is usually employed only when clearly obligatory—they are associated with flavor loss.

To limit microbial contamination to a minimum, most wineries are now impressively hygienic. This is a significant change from the past, but is important in controlling contaminant yeasts, notably *Brettanomyces* spp. Other contaminant yeasts include *Saccharomycodes ludwigii* and *Zygosaccharomyces bailii*. They are all comparatively resistant to sulfur dioxide. *Brettanomyces* is particularly serious as it can easily contaminate fermentation and storage cooperage, transfer piping, and bottling lines, all sites difficult to sterilize once contaminated. *Brettanomyces* contamination donates what, to most people, are the repulsive odors of ethylphenols. Smaller amounts of these compounds, to some individuals, are considered as supply a desired *terroir* or winery uniqueness. Many lactic acid and acetic acid bacteria can also induce serious spoilage problems. Although most of the literature on these issues relates to wine, beer, and fruit juices, similar problems have been noted with ciders and presumably other fruit wines. Because detection and identification by traditional cultural methods are slow, most labs now employ more rapid, molecular-based procedures (Tessonnière et al., 2009).

Cider can also succumb to the action of several spoilage bacteria. These include various strains of *Zymomonas mobilis* (causing cider sickness/*framboise*), *Bacillus licheniformis* (causing ropy cider), strains of *Lactobacillus collinoides* (causing acrolein spoilage), and species of *Acetobacter* and *Gluconobacter* (causing vinegarization).

Although most off-odors can be removed by reverse osmosis, it is clearly preferable to avoid their occurrence. Thus, hygiene in all aspects of current wine production has become essential to minimize costly sterilization procedures and disruptive product returns.

Despite all the modern attention to hygiene, microbial spoilage remains a constant thread. This usually results from lapses in sterilization of winery equipment, but can also arise from contaminants arriving on grapes from the vineyard. Most grape epiphytic contaminants are unable to grow or metabolize in wine, especially in the absence of oxygen, but this does not apply to all. For this reason, at bottling, wines are often given a dose of sulfur dioxide. Not only is it antimicrobial, but it is also an efficient and useful antioxidant. Because some consumers are allergic to trace amounts of sulfites, there has been a concerted effort to find an effective alternative. To date, none has proven to possess sulfur dioxide's antimicrobial and antioxidant attributes, as well as its low human toxicity. Colloidal silver (Garde-Cerdán et al., 2014), high hydrostatic pressure (Buzrul, 2012), and ultraviolet exposure (Falguera et al., 2013) are possible substitutes in terms of microbial control, but their sensory consequences have yet to be adequately investigated. Sterile filtration and hot bottling are supplemental techniques used for wines, especially those susceptible to microbial spoilage, for example low-alcohol and sweet wines. Alternatives for the antioxidant attributes of sulfur dioxide may lie in the addition of a mixture of glutathione, caffeic acid, and gallic acid (Roussis et al., 2013), or chitosan (Chinnici et al., 2014).

Although oak cooperage was the traditional fermentation and storage container for centuries, stainless steel is now the industry standard. However, extended oak contact is likely to mask the delicate fragrance of most fruit wines and add undesired bitterness (Riekstina-Dolge et al., 2014), unless the barrels have housed wine several times previously (see Rous and Anderson, 1983). In addition, once barrels have become contaminated with spoilage microbes, their complete inactivation is difficult to impossible (Swaffield et al., 1997). These issues are avoided when maturation occurs in inert cooperage such as stainless steel.

Small amounts of oxygen are important in stabilizing the color of red wines. This typically occurs during their maturation in oak barrels, racking, or other procedures. However, whether contact with oak is beneficial for most fruit wines has as yet to be assessed adequately. Where limited oxygen exposure is desired, but without oak flavors, this can be achieved via the use of silicone-tube diffusers. The amount and rate of oxygen ingress can be regulated by adjusting the length, thickness, and gas pressure in the tubing. Other means include oxygen dispersed into wine via microporous diffusers. Amounts in the range of 5 mL O_2/liter/month (at 15–20°C) are roughly equivalent to barrel uptake, and lower than the maximal rate of oxygen consumption by red wine (Castellari et al., 2004). Treatment may last for as little as a few weeks to several months, depending on the cultivar, pH, and wine style desired (see Dykes and Kilmartin, 2005). An alternative technique involves the use of high-density polyethylene cooperage (del Alamo-Sanza et al., 2015). These are purportedly capable of being designed to provide a relatively precise oxygen diffusion rate. To have additional control over oxygen ingress, some cooperage can be fitted with fiber optics, connected to oxygen-measuring fluorescence spectroscopy.

In most instances, oxygen access is likely to be undesirable for fruit wines (as with most white wines)—oxygen increasing the likelihood of microbial spoilage and early browning. Only in fruit wines possessing anthocyanins, such as cherry, raspberry, strawberry, and pomegranate, might limited oxygen uptake during maturation be of value. However, this would be dependent on the copresence of appropriate flavonoid phenols to stabilize the fruit's anthocyanin pigments.

An older, but increasingly popular technique used in maturing white wines, involving a limited degree of oxidation, is *sur lies* maturation (Dubourdieu et al., 2000). Although used primarily to add flavors and extract mannoproteins from dead and dying yeast cells (lees), the procedure involves *battonage* (periodic stirring of the wine). During the process, small amounts of oxygen are absorbed, helping to limit the development of a highly reductive lees layer at the bottom of the cooperage. Thick lees layers are conducive to the generation of sulfur off-odors.

In some instances, oak flavor is desired, but at lower cost and without the maintenance or associated slight oxygen uptake that comes with barrel usage. One modern alternative involves the insertion of relatively thin oak slats (battens) in a stainless steel frame. The frame is submerged in the wine, usually maturing in stainless steel tanks. If the slats are presoaked in wine, air (and its oxygen content) in the slats is replaced by wine. A still simpler technique is the addition of presoaked oak chips. In this instance, care must be taken to ensure adequate exposure of the wine to the chips, while avoiding sufficient release of oak dust to clog drains, pumps, and filters during subsequent chip removal. This procedure has been investigated in cider production (Fan et al., 2006). The flavors donated by slats and chips can be adjusted, in a manner resembling that of oak barrels, by the choice of oak (American or European), degree of toasting, and shape (surface area to volume ratio). Where economy and foreshortening oak maturation is of particular importance, commercially available oak extracts or oak "flour" can be added. In the latter instance, addition is preferable during fermentation. This favors the deposition of the "flour" with the lees and subsequent removal during racking.

Recent innovations have not only touched aspects affecting wine fermentation and maturation, but also transport. The shift from using charcoal to coal at the beginning of the Industrial Revolution was critical to the production of strong glass. As already noted, strong glass permitted the insertion of tight-fitting cork closures. Bottle use also provided provenance and varietal identification (bottle shape, color, and labeling). Unlike barrels, bottled wine was easier to transport and, with cork closures, aged safely in an inert, essentially anoxygenic environment. In addition, the clarity of glass permitted visualization of certain faults. The addition of specific metal oxides during glass production generated of a range of distinctive colors. A less-expensive modern alternative, but surprisingly rarely used, is the application of a colored and/or textured coating. They provide improved abrasion and impact resistance, can protect against (absorb) ultraviolet radiation, and facilitate recycling (the coating melts off, leaving colorless glass for processing). Surprisingly, under certain circumstances, wine stored in clear or light-colored bottles can be less sensitive to browning than those in darker glass (Maury et al., 2010). This counter intuitive result may be due to the higher temperatures experienced by wines when exposed to sunshine in darker bottles. In addition, xanthylium-based pigments, potentially involved in browning, may be degraded more readily on the greater light exposure associated with clear or light-colored glass.

Although the advantages of glass apply as much today as in the past, some of its disadvantages—high production costs, breakability, weight, and disposal issues—have induced the introduction of innovative alternatives. One of the first and most successful to date has been the bag-in-box container. Because the bag collapses as the product is dispensed, the remaining wine remains under anaerobic conditions. This allows the consumer to periodically remove wine, without a marked loss in quality (aromatics do not escape into the headspace, as occurs when wine remains in a partially emptied bottle). The cost benefits and consumer convenience of bag-in-box containers are also credited with expanding wine consumption in areas without an established wine culture. Bag-in-box packaging is also ideally suited for restaurant use, especially in larger (10- to 20-L) sizes. The protective outer rigid housing also permits efficient stacking where storage space is limited. The major constraint to the more extensive use of the bag-in-box technology is wine shelf-life. Oxygen ingress around the tap often generates detectable deterioration within about 9 months, especially if the container is not kept under cool conditions. Thus, bag-in-box use is restricted to products not benefiting from aging, and with high turnover rates. For smaller volumes, 1 L cartons save both on production and shipping costs. As with bag-in-box packaging, there is abundant space for marketing information. Another alternative, based on soft-drink technology, is the aluminum can. None of these alternatives are elegant solutions, but for the sector for which they are aimed, refined presentation is not a significant issue.

One of the newest glass substitutes is polyethylene terephthalate (PET). It has many of the attributes of glass (moldability, transparency, impermeability), but is lighter, less breakable, easier to recycle, and less expensive to produce. With a gas barrier and/or incorporation of an oxygen scavenger, PET plastic bottles possess properties similar to glass in terms of wine stability and quality. Details on the oxygen permeability of PET and PET-flexible packaging films can be found in Lange and Wyser (2003). Another alternative is polylactic acid (PLA) bottles (Pati et al., 2010). Combined with an oxygen scavenger, it also appears acceptable as a glass substitute for wines with a short shelf-life. PLA has the additional advantage of being biodegradable.

Since the rediscovery of the advantages of cork as a bottle closure in the 1600s, cork has been used almost exclusively as the preferred bottle stopper, until recently. Wine possessing an off-odor, associated with cork closures, has been noted for more than a century. Nonetheless, cork-related problems took an apparent marked upturn some 30 to 40 years ago. Whether this was just coincidental, or due to changes in the production and processing of cork is uncertain, but probable. Most situations of "corked" wine are associated with above threshold amounts of 2,4,6-trichloroanisole (TCA). A similar odor is also generated by 2,4,6-tribromoanisole (TBA). Both compounds are the by-products of the fungal detoxification of industrial compounds—TCP and TBP, respectively. For the past half century they have been widely used as an insecticide or fire-retardant/wood preservative. Because the cork industry was slow to respond adequately to the growing number of complaints, many wineries began to question cork's continued use as a bottle closure. This spurred studies into cork substitutes, notably plastic. Corks made from polyethylene were the first to be used, but excessive gas permeability severely restricted wine shelf-life. Polyvinyl, or ethylene vinyl acetate, stoppers followed, both having enhanced gas impermeability. In the interim, innovations in cork production have largely removed or avoided odor faults due to contamination with TCA. Regardless, many wineries have shifted away from the use of any type of stopper and adopted the use of aluminum screw caps. The cap liner, based either on polyvinylidene chloride (Saran) or polytetrafluoroethylene (Teflon), can be manufactured to provide a specified degree of oxygen permeability, if desired (Karbowiak et al., 2010). In contrast, cork, being a natural product, can show considerable variability in oxygen permeability (Faria et al., 2011). Although this variability is likely the principal factor inducing early browning of white wines, some in the wine industry consider minimal and slow oxygen permeability essential to the optimal aging of red wines, and/or to limit the development of reduced-sulfur off-odors. Screw caps have also appealed to many consumers, due to the ease of bottle opening and reclosure. To counter issues concerning variability in the oxygen permeability of natural cork, the need for a corkscrew, and resealability, a bottle and closure producer have teamed up to produce a reinsertable T-cork. It is made of agglomerate cork with several curved grooves to match treads in the neck of a specially designed bottle (see www.helixconcept.com). Another novel closure system is based on a technique well-known in chemistry labs, as well as in closing wine decanters: the T-shaped ground-glass stopper. As with screw caps, the neck of the bottle must be especially molded to accept the stopper. It is only time before such technologies begin to influence the packaging of fruit wines, if they have not already.

Premium wines usually receive some aging in oak and in-bottle, before they are considered to have reached their sensory peak (plateau). Even standard wines require some maturation before they are bottled. Besides allowing excess dissolved carbon dioxide to escape, reduced-sulfur compounds to oxidize, and to permit clarification and fining, flavors mature and mellow, especially in red wines. Thus, with both standard and premium wines, stock is often tied up for months before being released. Correspondingly, there has long been interest in innovations that might hasten maturation and aging. Early investigations into accelerated aging (see Singleton, 1962) were not positive. Relatively short exposures to 45°C (Francis et al., 1993) have been reported to produce changes resembling those engendered by several years in-bottle. Alternating current (600 V/cm for 3 min) has also been reported to improve the balance and mouth-feel of Cabernet Sauvignon wines, reduce aldehyde and higher alcohol contents, and slightly increase ester content (Zeng et al., 2008). Other techniques investigated have included exposure to ultrasonic and gamma rays, nanogold photocatalysis, and high pressure (see Tao et al., 2014). Nonetheless, none of these treatments have achieved any industry acceptance to shortcut the development of the complexity ascribed to traditionally cellared wines.

5. SPARKLING WINES

As noted, the development of sparkling wine depended on innovations dating back to the late 1600s: improvements in glass manufacture and the reintroduction of cork stoppers. Without these discoveries, the safe accumulation of the standard 6 ATM of carbon dioxide would have been impossible. The incentive to produce effervescent wine appears to have arisen when barrels of still wine from Champagne began to referment in England, as temperatures warmed in the spring. The resultant frothy wine became highly popular with the, then, jet-set. Dom Perignon's contribution was not in this discovery, but instituting the blending of wines from different sites to both accentuate their individual desirable attributes and lessen their defects. He may also have been the first to use cork to close bottles of champagne. Later discoveries revealed the association between sugar content and carbon dioxide production. This permitted the addition of a precise amount of sugar to achieve the desired degree of effervescence. This avoided the hitherto frequent explosion of bottles in storage cellars—those having received too much sugar. Subsequently, a technique was developed to move yeast cells toward the neck of the bottle (for discharge) to achieve a crystal clear wine. At the same time, wine could be adjusted for sweetness, to fit the preferences of the time and particular clientele. Over the years, the degree of sweetening has diminished to the point that now most sparkling wines receive only about 1.5 g sucrose per 750 mL bottle. This is sufficient to avoid excessive sourness; sparkling wines are often higher in acidity than equivalent table wines. Modern innovations in sparkling wine production cover most aspects of production, from press to bottle closure.

Because red grapes are often used in producing white sparkling wines, the grapes need to be pressed whole (without prior stemming or crushing). This minimizes the extraction of pigments from the grapes, as well as minimizing the liberation of flavonoid phenolics. Both accentuate the tendency of the wine to gush upon opening. The grapes are also typically harvested prior to full maturity (and color development in red grapes). This decreases the likelihood of any *Botrytis*-infected grapes being harvested. *B. cinerea* typically does not parasitize unripe grapes, unless wet conditions occur for an extended period. In addition, incomplete ripeness provides the grapes with additional acidity, lower sugar content, and limited varietal flavors—all features typically desired in the base wines.

Early harvesting of apples is not required for producing sparkling cider as apples are less susceptible to spoilage as they mature than grapes, and any pigmentation extracted during mashing tends to precipitate during maturation. Nonetheless, there is concern about eliminating any damaged fruit, not only because of off-flavors that can be produced but also to avoid the presence of fungal toxins from secondary saprophytes, notably *Penicillium, Aspergillus*, and *Fusarium* spp.

Because the effervescence and foaming attributes of sparkling wine and ciders are central to their sensory appreciation, considerable effort has been directed into studying these phenomena (Liger-Belair et al., 2008; Pozo-Bayón et al., 2009). The release of mannoproteins during yeast autolysis is crucial to prolonged effervescence, whereas foam attributes, more important to cider quality, appear to be more associated with particular polysaccharides (Mangas et al., 1999), fatty acids (Cabrales et al., 2003), and polypeptides (Blanco-Gomis et al., 2007). The role of particular proteins is controversial. However, mannoprotein liberation is comparatively slow, tying up stock of prolonged periods. To accelerate this process, breeders have developed yeast strains that undergo more rapid autolysis (Lobo et al., 2005; Tabera et al., 2006).

Although bacterially induced malolactic fermentation usually occurs after alcoholic fermentation, and prior to the second, often in-bottle fermentation, experiments have been conducted with having joint alcohol and malolactic fermentations. Successful results involving inoculation with both *S. cerevisiae* and *Sz. pombe* (the latter being capable of malolactic fermentation) have been reported with plum wine (Bhardwaj and Joshi, 2009).

Riddling by hand, the movement of sedimented yeasts to the bottle neck traditionally took 3–8 weeks. When finding individuals willing to do this onerous task became increasingly difficult, engineers developed large, automated riddling machines. Amortized over several years, riddling machines are less expensive, economize on space usage, and minimize handling, as well as shortening the duration of riddling to about 7–10 days. However, further foreshortening the riddling time has proven counterproductive. It can increase the likelihood of a bentonite haze forming and can also interfere with mousse stability.

A minute amount of bentonite is usually added with the yeast inoculum and sugar before the onset of the second fermentation. It facilitates the formation of a tight yeast sediment, important in disgorging (the ejection of the yeast plug). Other innovations, designed to reduce the expense and complications of disgorging, have included incorporating the yeast in a stable gel matrix (immobilization; see Torresi et al., 2011), or in a membrane cartouche (Jallerat, 1990). Although succeeding in their primary intent, they appear not to generate the same sensory attributes as yeast cells free to disperse throughout the wine. Thus, neither of these techniques has achieved industry acceptance. The feasibility of adding magnetized nanoparticles to yeasts, permitting their rapid collection to the neck prior to disgorging, has been demonstrated (Berovic et al., 2014).

Bottles for the second, in-bottle fermentation used to be closed with cork, held in place with a metal clamp (agrafe). With the development of metal crown caps, their efficacy and lower price resulted in their replacing the cork/agrafe closure system. Some sparkling wines are now furnished with a crown cap for sale, but the association of crown caps with soft drinks does not provide the same "elegance" as popping the cork.

Manual disgorging is now largely reserved for demonstrations in champagne caves. It is unnecessarily expensive and dangerous, relative to the millions of bottle of sparkling wine produced annually. Disgorging is now automated, with the wine being first chilled to about 7°C, and the necks submerged in a −20°C ice bath. This quickly freezes the yeast sediment to the cap. This facilitates ejection of the yeast plug, minimizing wine loss due to gushing when the cap is removed. The wine volume is immediately adjusted to the appropriate value, associated with the addition of a dosage. The dosage is a mixture of aged wine, sulfur dioxide, and usually sugar. At this point, a specially designed cork is inserted and a retaining wire attached. The cork is typically composed of agglomerate cork, to which two disks (*rondelles*) of natural

FIGURE 13.9

The SPK Zork cork for sparkling wines that offers the option to seal the bottle effectively once it has been opened.

Photo courtesy of contacted Scholle Packaging.

cork are attached. The cork is positioned so that the rondelles will be in contact with the wine. This cork combination has replaced the former use of completely natural cork stoppers.

A new development in closure design employs a plastic resealable cap (Fig. 13.9). It fits most standard sparkling wine bottles and retains the outward appearance of a traditionally closed sparkling wine. The cap liner obviates the problem of gas permeability that previously limited plastic resealable corks to inexpensive petillant wines. The resealable cap also permits wine to be periodically poured over several days or weeks, without risking oxidation or significant carbon dioxide loss. Because carbon dioxide is heavier than air, escaping CO_2 expels any oxygen that enters the bottle during pouring. In addition, as sparkling wines contain a gaseous volume equivalent to six times the volume of the bottle, there is ample carbon dioxide to donate sufficient effervescence until the last glass is poured. Another alternative cork closure is a modified aluminum screw cap. It, however, requires the adoption of bottles specifically designed for their use.

Older innovations, designed to reduce production costs, include the transfer and Charmat (bulk) processes. The transfer process was developed to circumvent the expense of manual riddling and disgorging. It differed from the traditional method only by transferring wine from thousands of bottles, after the second fermentation, to a pressurized tank. From this tank the wine was filtered and transferred to bottle under isobaric pressure. However, the development of automated riddling machines eliminated this advantage. In contrast, the Charmat method still retains its appeal for sweet sparkling wines, designed to accentuate varietal character. It substitutes the usual second, in-bottle fermentation with a second fermentation in a sealed tank, capable of sustaining the high pressures that develop. Sweetness in the wine may be obtained by prematurely terminating fermentation. The system also works well with red sparkling wines, where the phenolic content tends to accentuate gushing and wine loss during disgorging. Riddling and disgorging to remove the yeasts are replaced by isobaric centrifugation or filtration. If an extended period of yeast contact period is desired for bouquet development and to enhance effervescence, the lees are intermittently stirred. Nonetheless, the latter can potentially be replaced by adding mannoproteins isolated from yeast autolyzate prior to bottling.

Carbonation, as with soft drinks, is also permitted in producing effervescent wines. In this instance, the saturation is more on the level of what is called petillance, that is about 5 g/L (~2 Atm). Sparkling ciders and perries, produced by the traditional (champagne) method, often possesses about 10 g/L (~3.5–4 Atm) CO_2.

6. FORTIFIED WINES

The production of fortified wines in southern Europe is no mere accident. As already noted, grapes cultivated in these regions tended to arrive at the winery warm to hot, low in acidity, and high in sugar content. Without the benefit of modern refrigeration, fermentation was often fast and furious, often terminating prematurely, leaving the wine sweet and tasting flat. If consumed shortly thereafter, this was not a problem. However, before the widespread adoption of sulfur dioxide, many southern European wines rapidly succumbed to the action of spoilage yeasts and bacteria. This tendency to spoil was accentuated if the wine were transported in-barrel, the only means at the time. The solution discovered, following the perfection of distillation, was the addition of brandy. Raising the alcohol level to above 18% effectively inhibited microbial spoilage, permitting the stable transport of wine long distances. Subsequent developments in different regions lead to the evolution of the three major modern styles of fortified wines: sherry, port, and madeira.

For sherries, the deployment of fractional (*solera*) blending was a major innovation. It involved periodic and sequentially blending of portions of wine, from one set of barrels (a *criadera*) to another, older *criadera*. The process averaged out vintage differences and gave rise to a uniform product—ideal for producing brand-named products.

Fino-style sherries develop in association with the growth of a surface yeast velum (*flor*). Fortification to 15–15.5% alcohol favors a change in yeast cell-wall chemistry, allowing the yeast to rise and float to the surface. Exposure of the *flor* to air switches yeast metabolism from fermentation to respiration, protecting the underlying wine from oxidation. In contrast, if the wine is fortified to about 18% alcohol, no *flor* develops, and the wine oxidizes. With fractional blending, this wine gives rise to *oloroso*-style sherries. It had the distinct advantage that, being oxidized, it could be stored and consumed over an extended period, without risking spoilage or deterioration. Another style, with its own set of sensory attributes, arose when young wine, developing as a *fino* sherry, was fortified to 18% alcohol. This terminated flor development, and subsequent maturation was similar to that of an *oloroso* sherry. Thus arose *amontillado* sherries.

Recent innovations in sherry production have primarily concentrated on increasing the precision with which base wine evolution can be directed, as desired by the winery (*bodega*) and consumer demand. Other innovations have been aimed at shortening and improving the conditions traditionally used in partially dehydrating grapes prior to fermentation. Additional research is aimed at reducing the skill and time involved in sherry maturation, and the associated labor and storage costs connected with holding vast volumes of wine for extended periods.

Ports also owe much of their stylistic evolution to the addition of fortifying wine spirits (unmatured brandy), but this time it is supplied partway through fermentation. For red ports, fortification occurs about halfway through fermentation, raising the final alcohol content to between 18% and 20%. As a consequence, fermentation stops, leaving about half the original grape sugars present. The resulting sweetness has a major ameliorating effect on the wine's potential bitterness and astringency. Depending on the quality of the wine, it may be matured for various periods in large-volume oak tanks, or in smaller volume barrels (pipes) prior to bottling. Wine, considered to be of well-above average quality, is stored in special oak cooperage (pipes) and labeled with the vintage year. In contrast, ruby and tawny ports (those aged in large-volume oak tanks) are not vintage designated. Blending of wines from different sites and vintages is employed to achieve brand consistency. The evolution of vintage-style ports particularly depended on the preexistence of glass bottles and cork stoppers. Such ports were one of the first modern wines for which the advantages of considerable in-bottle aging was observed. This innovative "discovery" appears to have precipitated a revolution in aging table wines. Heretofore, most table wines were best before the next vintage, older wines being sold at a discount.

Madeira is the third fortified wine style in worldwide distribution. It comes from the island of the same name, off the western coast of North Africa in the Atlantic Ocean. Its origin also involved a serendipitous discovery, based on an observation associated with the wine's export (in-barrel). Ships transporting the wine to North America often went via the West Indies. This exposed the wine to heat for a considerable period. It was observed that the attributes of the wine were significantly enhanced. This led to the innovation of heating the wine (*estufagem*) in Madeira prior to shipping. As in port production, fermentation is stopped prematurely by the addition of brandy. That different grape varieties had their fermentation terminated at different stages led to the development of a range of styles based on sweetness, frequently denoted by the name of the grape variety used. Other than sweetness, the principal attribute that distinguishes madeiras is a consequence of heating. Heating donates an oxidized, baked character that, along with its alcohol content, provides madeiras with possibly the longest aging potential of any wine. For most modern madeiras, heating involves slowly heating of the wine in stainless steel tanks up to 45–50°C, for a minimum of about three months, followed by gradual cooling. For superior wines, heating often occurs in oak *butts* in the top floors of warehouses. Storage in this location can last for upwards of eight years, and exposes the wine to alternating cycles of heat and cold.

Vermouths are a separate class of fortified wines distinguished by their being flavored with proprietary combinations of herbs and spices. Many fruit wine–based vermouths are produced and are the closest of any fortified fruit-based wine to an equivalent grape-based fortified wine (Panesar et al., 2011). Pommeau is another fruit-based fortified product, being a blending of a cider brandy (Calvados) and unfermented apple juice (about 1:2). The mixture, at about 16–18% alcohol, is aged for about 2.5 years in oak cooperage before bottling.

Although brandy is technically not wine, it is the distilled by-product of wine. As such, there are an incredible diversity of brandies, either based on wine (cognacs and armagnacs), cider (calvados), or almost any other fruit-based wine. As a distilled product, the technical aspects of its production are highly specialized. Grape and wine production techniques are clearly important to the quality of the brandy, but the desirable attributes of the base wines are slightly different from those for table wines. These differences relate specifically to the physicochemical nature of distillation and the products maturation. These are, however, beyond the scope of this chapter (see Bertrand, 2003; Jackson, 2014).

7. SENSORY EVALUATION

The scientific investigation of the sensory attributes of wine is comparatively recent, following in the footsteps of developments in the food industry. Although commonly called wine tasting, sensory evaluation involves much more than just tasting.

Most consumer tastings are designed to enhance the appreciation of wine and promote sales. Ideally this assesses the color, taste, mouth-feel, and fragrance of the wine. Such assessment has always been to some degree associated with attempts to verbalize the wine's fragrance, with stress on the aromas donated by the grape. Other attributes intriguing to the aficionado are aspects that hint to the wine's provenance, style, age, and more nebulous characteristics such as complexity, balance, development, duration, and quality. What is often poorly understood is how the tasting conditions can influence and distort these perceptions. One of the principal functions of a critical analysis is to, as far as possible, avoid the environment from biasing the pure sensory attributes of the wine. This also involves divorcing analysis from personal preferences.

One of the most important discoveries in sensory perception is that all chemical and visual senses are integrated in the cerebral orbitofrontal cortex (Rolls et al., 2010). Memory patterns develop around this sensory integration, possibly similar to how the brain constructs memories of faces and events. The orbitofrontal cortex also compares current sensations with past experience. Because vision often takes precedence over the chemical senses in humans, color has the potential to significantly influence a person's evaluation of wine, even when it conflicts with sensations derived from the tongue and nose (Österbauer et al., 2005). To avoid this phenomenon, wines ideally should be assessed in black glasses, or under conditions where the lighting falsifies the wine's color. With these tasting design changes, the influence of wine color can be eliminated.

Suggestion is another potent psychological factor that can influence perception, be it verbal or facial expression. To limit its effect, sensory researchers usually use rooms where each panel member is visually isolated and communication forbidden. In addition, the bottles from which the wines are dispensed are hidden. To ease data collection, sensory responses are often entered via a touchscreen on a computer monitor. This simplifies data collection and permits almost instantaneous analysis.

Depending on the purpose of the tasting, panel members may be specifically trained and assessed for the skills needed. Assessors are often used to describe or differentiate between wine based on selected sensory criteria, for which analytic instruments are currently unavailable. However, for some simpler tasks, such as fault detection, instruments termed electronic (e−) noses and tongues may be substituted, due to their superior consistency and immediate availability as needed. In contrast, several panel members are required to obtain similar, statistically valid data. Despite training and selection, individual tasters vary in their ability to both detect and identify aromatic and gustatory compounds (Tempère et al., 2011). Such idiosyncracies can also vary from day-to-day. Correspondingly, any one member is unlikely to provide consistently reliable data.

The presentation of sensory data in a clear and readily understandable manner has always been difficult. A numerical score, so popular at the retail level, has no value in wine assessment. It subjugates too many, potentially opposed, sensory responses into a single value. Considerably more informative, but still for amateurs, is a response chart (Fig. 13.10). It permits visual representation of descriptive responses on a graph that denotes changing perceptions and relative intensity throughout the tasting.

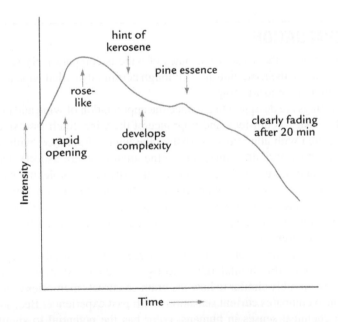

FIGURE 13.10

Graphic representation of the development of a wine's fragrance during a tasting. Specific observations can be applied directly to the point on the graph where the perception was detected.

From Jackson, R.S., 2009. Wine Tasting: A Professional Handbook, second ed. Academic Press, London, UK; reproduced with permission.

For scientific purposes, the development of spider charts (Fig. 13.8) was a significant advancement. It permits not only visual representation of the intensities of different sensory attributes but also their statistical significance. An alternative, but more complex representation, involves principal component analysis. It can graphically group sets of attributes that represent the main sources of variation among the products as well as panel members.

In the process of training, especially where a sensory profile or discrimination is desired, representative samples of each term employed are prepared and are made available during each assessment. The descriptors are usually chosen by the researchers, but may be developed in conjunction with the panel members during training. For discrimination purposes the least number of terms required is preferred. This simplifies the task for the assessor, minimizing sensory fatigue that could reduce discriminatory ability.

To further reduce sensory fatigue, the interval between sample presentation is extended as the detail required per sample is increased. Palate fatigue, associated with adaptation, has usually been thought to be reduced by providing water and/or cubes of bread between samples. While long-established, research has shown that they are relatively ineffective. In contrast, a solution of 1% pectin, or unsalted crackers, appear to be more effective palate cleansers (Ross et al., 2007).

One of the most important roles of tasting is in assessing the effect of different production techniques on the sensory attributes. Determining the statistical significance of detectable differences is important, but even more valuable is measuring the relative importance of the sensory attributes detected. In this regard, descriptive sensory analysis is used to rank individually the perceived intensity of aspects considered of importance. To obtain additional clarification, time intensity analysis can be used to express the temporal dynamics of how different sensory properties change during detection (Cliff and Heymann, 1993). However, focusing individually on single perceptions can potentially exaggerate their significance, through synergy with other attributes not raked (termed halo-dumping). To attempt to counteract this phenomenon, temporal dominance of sensations was developed (Pineau et al., 2009), a serious version of the amateur's version represented in Fig. 13.10. It provides information on the sequence of how a range of attributes express themselves, in relative dominance, over the course of a tasting. These and future developments in sensory techniques may provide researchers with better means of correlating wine chemistry with wine flavor, as well as wine-makers with useful information on how viticultural and enologic changes affect sensory attributes important to the consumer.

Although most sensory research uses trained panels, the data obtained does not necessarily, or even likely, validly represent consumer perception or preferences. Training and testing not only improves the ability of researchers to detect those people best able to perform the desired assessment(s) (Tempere et al., 2012), but can also modify the sensory skills of the tasters. Preference modification is especially prevalent as members develop extensive wine experience. Thus, to better understand purchaser preferences, selecting consumers likely to buy the product is essential. Tasting conducted by a random sample of consumers in a wine shop is not necessary likely to provide the data desired. To obtain more realistic information, the tasting should be done under situations approximating actual consumption, even if this can only be achieved by asking the consumers to mentally visualize a home setting. While it may provide information about consumer preferences, it may not equally provide realistic data on purchase intention—the most important data marketers want. Purchase is largely based on price, color, variety, and provenance, and for aficionados, the rarely valid correlation between reputation and quality. It is also important that any questions posed to consumers be carefully worded so as not to be "leading," unwittingly directing responses one way or the other. Although the sensory descriptions provided on back labels are intended to provide the consumer with useful sensory information, they are notoriously imprecise.

Another moot point is how researchers should assess consumer-derived data. Consumers often use terms inconsistently and in divergent ways. Thus, any simple summation and standard analysis of the data is futile. One potential statistical tool, specifically designed for this situation, is Procrustes analysis (Dijksterhuis, 1996). It mathematically "massages" the data, adjusting the measurements from all respondents to within a similar range before analysis. Although extensively used, whether its assumption— that all consumers perceive the same phenomena, only differing in the scale range used—appears highly dubious. Most data indicate that individuals perceive differently, not only quantitatively but also qualitatively. Thus, the data obtained by Procrustes analysis may provide data that is statistically significant, but irrelevant.

8. AUTHENTICITY

Part of a product's consumer appeal is its authenticity as a "natural" product. Attributes such as regional appellation, estate bottling, varietal origin, vintage date, and producer name are all expected to be guaranteed—the principal criteria by which most wine is chosen, besides price. However, how the wine will taste is only imperfectly known until the bottle is opened, hours to decades later. Thus, the consumer must put faith in the information presented on the label. Loss of such faith is almost lethal to any wine or region. Correspondingly, the wine industry is especially concerned in wanting the data noted on the label to be valid in all regards. These features are particularly pertinent when the product is exported thousands of miles and commands exorbitant prices. Thus, assuring the product's authenticity is critical to the exclusivity it provides the owner (as an original work of art). Although this is currently not as critical for most fruit wines, produced and sold locally and without appellation control laws regulating their production, the means of establishing provenance and adherence to production regulations, noted later, are the same as those needed to confirm the same attributes for fruit wines.

Despite public acceptance of what is written on the label, consumers are usually unaware of technical aspects behind what is noted. This is especially so for many European appellations. Regulations that affect features such as the cultivars used, geographic boundaries, regulations relative to grape cultivation and yield per hectare, and fermentation, clarification, and maturation conditions, often vary from region to region. Regulation of such features in New World countries is more liberal, with vineyard and winery innovation not only permitted but encouraged.

Consumers understandably consider only what is explicitly stated on the label. Nonetheless, they may misinterpret the degree of its precision. For example, that only one grape variety is mentioned on the label does not necessarily mean that it was the only cultivar used. Although this may be true, often small amounts of other cultivars may have been added. This depends on the laws in the producing region, which often vary from region to region. The primary intent of mentioning a varietal name on the label is to denote the source of the principal sensory attributes of the wine. Where other, unlisted cultivars are used, they are usually added to improve the flavor balance of the wine—the goal being to provide the consumer with the best possible sensory experience. This is especially so with the increasing popularity of blends, where the varietal origin may not even be mentioned. High quality, restrained price, and good marketing can be very rewarding for both consumer and producer.

Historically, verifying compliance with production regulations was difficult, with the detection of fraudulent activity often depending on anonymous tips or on-site inspection. Innovations in analytic chemistry have immensely expanded the ability of investigators to assess conformity with appellation regulations, as well as affirm varietal, geographic, and vintage designations. The arsenal of techniques now includes scintillation counters, mass spectrometry, nuclear magnetic resonance, and the multiple forms of gas and liquid chromatography. Because of the huge amount of data potentially derived, advanced statistical analysis such as principal component analysis may be required to isolate pertinent information. Furthermore, because of the incredible range of adulteration possible, it is almost necessary to suspect a priori any duplicity involved.

For provenance validation, variation in the distribution of hydrogen (H), carbon (C), and oxygen (O) isotopes are the primary data employed (Raco et al., 2015). However, other isotopes, ions, and mineral differences between one site and another may be useful in specific cases (Kelly et al., 2005). Because of differences in how cultivars concentrate these indicators, authentic samples of the wines and vintages must be available for comparison. Other aspects of modern authentication are discussed in Arvanitoyannis (2010).

Verification that no artificial flavors or illegal additives have been used can also be assessed. Most artificial flavors and additives do not possess the isotopic signature of grapes grown in the designated region. Isolation of suspect additions can be checked for their isotopic signature.

With innovation in modern surveillance techniques, it is becoming increasingly difficult for fraudulent wine to be produced without detection, to the point that it is almost more expensive to produce an undetectable fraudulent wine than an authentic one. This may not stop a fraudster, but should discourage legitimate producers from illegal tinkering with their wines. Detection and publication of the results usually will result in bankruptcy for the individual, and serious financial loss to the region's producers.

One of the newest techniques concerning provenance verification involves track-and-trace technologies (Kumar et al., 2009; Vierra, 2013). At its simplest, this involves an ID mark (code) that may be placed on the back label, under the closure, or embossed on the bottle—each code being unique to a particular producer. The mark can be read by anyone interested, using a scanner and the appropriate software. More complex verification marking permits the purchaser, or other interested parties, to know where the wine has been since shipment from the winery. Other versions, occasionally placed on the box in which the wine is transported, allow the wholesaler or retailer to determine the temperature range over which the wine has been subjected during shipment and storage. However, this only detects problems after the fact. Proposals to have detectors send warning signals, in time for shippers to take corrective measures, are under investigation (Lam et al., 2013).

9. FUTURE PROSPECTS

Gazing into the proverbial crystal ball is one of the best ways of visualizing what will not happen. Nonetheless, if current trends can be considered indicators of the future, then it is clear fruit-based wines will garner a larger portion of the wine market, especially in regions where grape culture is not propitious. The promotion of fruit wines as a safe and wholesome beverage is justified due to their nutraceutical content, notably phenolics (Dey et al., 2009; Rupasinghe and Clegg, 2007; Fig. 13.11). That fruit wines are often of lower alcohol content than most standard table wines might reduce the

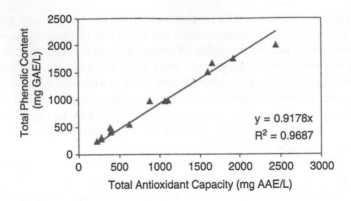

FIGURE 13.11

Relationship between total phenolic content and total antioxidant capacity in major fruit wines.

From Rupasinghe, H.P.V., Clegg, S., 2007. Total antioxidant capacity, total phenolic content, mineral elements, and histamine concentrations in wines of different fruit sources. Journal of Food Composition and Analysis 20, 133–137; reproduced with permission.

negative effects of excessive alcohol consumption (assuming the consumer is so inclined). In addition, low-alcohol content is no longer a limiting factor due to poor microbial stability. Sterile filtration, pasteurization, screw caps, newer antioxidants, and cool dark storage can extend stability for years. New packaging also facilitates efficient storage and use, avoiding the need to consume the product shortly after opening.

A distinct advantage of most fruit wines relates to limited production regulations, a feature that too often stifles innovation. For example, fruit wines can be commercially produced from concentrates (Coelho et al., 2015). This facilitates year-round production, without being limited by the seasonality of fresh fruit. It also frees the producer to experiment with fruit blends not permissible with grape-based wines, an example being wine made from grapes and janun (*Syzygium cuminii* L. Skeels; Chaudhary et al., 2014). In addition, chaptalization (increasing the sugar content of the must) is not simply restricted to sugar, as in the case of grape-based wines. For example, honey has been used as a sugar source in the production of apricot wine (Abrol and Joshi, 2011). This freedom to experiment permits easier and more rapid adjustment to perceived consumer demand. It also allows time to discover those conditions and cultivars most adapted for transformation into wine (Towantakavanit et al., 2011).

Conversion into wine also serves one of the original advantages of wine production: conversion of readily perishable fruit into a product with a prolonged shelf-life, and new and flavorful attributes. This transformation can also apply to parts of a fruit traditionally a waste product, such as the pulp from cocoa pods (*Theobroma cacao* L.; Dias et al., 2007) or cashew apple (*Anacardium occidentale* Linn.; Attri, 2009). The fruit used also does not have to possess the unblemished appearance increasingly demanded of fresh fruit in the modern marketplace.

Increased affluence may be expected to increase use by those unaccustomed to wine, especially when they come to appreciate its potential to enhance the pleasures of the table. Not only can wine add to the social enjoyment of eating with family and friends but also can ameliorate food flavors. What is surprising about this last aspect is that the improvement is largely due to reducing the more unpleasant aspects of the food and the wine. Thus, as in mathematics, two negatives generate a positive. I'm all for those types of positives!

REFERENCES

Abrol, G.S., Joshi, V.K., 2011. Effect of different initial TSS level on physico-chemical and sensory quality of wild apricot mead. International Journal of Food and Fermentation Technology 1, 221–229.

Accum, F., 1820. Adulteration of wine. In: A Treatise on Adulterations of Food and Culinary Poisons, second ed. Longman, Hurst, Rees, Orme, and Brown, London, UK, pp. 92–105.

Allen, D., 2007. Prefermentative cryomaceration. Australian & New Zealand Grapegrower & Winemaker 523, 59–64.

Anonymous, 1986. The history of wine: sulfurous acid – used in wineries for 500 years. German Wine Reviews 2, 16–18.

Arvanitoyannis, L.S., 2010. Wine authenticity, traceability and safety monitoring. In: Reynolds, A.G. (Ed.), Managing Wine Quality. Viticulture and Wine Quality, vol. 1. Woodhead Publishing Ltd, Cambridge, UK, pp. 218–272.

Attri, B.L., 2009. Effect of initial sugar concentration on the physico-chemical characteristics and sensory qualities of cashew apple wine. Natural Product Radiance 8, 374–379.

Barnard, H., Dooley, A.N., Areshian, G., Gasparyan, B., Faull, K.F., 2011. Chemical evidence for wine production around 4000 BCE in the Late Chalcolithic Near Eastern highlands. Journal of Archaeological Science 38, 977–984.

Bhardwaj, J.C., Joshi, V.K., 2009. Effect of cultivar, addition of yeast type, extract and form of yeast culture on foaming characteristics, secondary fermentation and quality of sparkling plum wine. Natural Product Radiance 8, 452–464.

Bellon, J.R., Eglinton, J.M., Siebert, T.E., Pollnitz, A.P., Rose, L., de Barros Lopes, M., Chambers, P.J., 2011. Newly generated interspecific wine yeast hybrids introduce flavour and aroma diversity to wines. Applied Microbiology and Biotechnology 91, 603–612.

Benito, S., Plaomero, F., Morata, A., Calderon, F., Palermo, D., Suarez-Lepe, J., 2013. Physiological features of *Schizosaccharomyces pombe* on interest in the making of white wines. European Food Research and Technology 236, 29–36.

Berovic, M., Berlot, M., Kralj, S., Makovec, D., 2014. A new method for the rapid separation of magnetized yeast in sparkling wine. Biochemical Engineering Journal 88, 77–84.

Bertrand, A., 2003. Brandy and cognac: armagnac, brandy and cognac and their manufacture. In: Caballero, B., Trugo, L.C., Finglas, P.M. (Eds.), Encyclopedia of Food Sciences and Nutrition, second ed. Academic Press, Oxford, UK, pp. 584–601.

Blanco-Gomis, D., Mangas-Alonso, J.J., Junco-Corujedo, S., Gutiérrez-Álvarez, M.D., 2007. Cider proteins and foam characteristics: a contribution to their characterization. Journal of Agricultural and Food Chemistry 55, 2526–2531.

Bony, M., Bidart, F., Camarasa, C., Ansanay, V., Dulau, L., Barre, P., Dequin, S., 1997. Metabolic analysis of *Saccharomyces cerevisiae* strains engineered for malolactic fermentation. FEBS Letters 410, 452–456.

Bramley, R.G.V., 2005. Understanding variability in winegrape production systems. 2. Within vineyard variation in quality over several vintages. Australian Journal of Grape and Wine Research 11, 33–42.

Bramley, R., Proffitt, T., 1999. Managing variability in viticultural production. Australian Grapegrower & Winemaker 427, 11–12 15–16.

Buzrul, S., 2012. High hydrostatic pressure treatment of beer and wine. A review. Innovative Food Science and Emerging Technologies 13, 1–12.

Cabrales, I.M., Abrodo, P.A., Blanco-Gomes, D., 2003. Influence of fatty acids on foaming properties of cider. Journal of Agricultural and Food Chemistry 51, 6314–6316.

Carre, E., Lafon-Lafourcade, S., Bertrand, A., 1983. Désacidification biologique des vins blancs secs par fermentation de l'acide malique par les levures. Connaissances de la Vigne et du Vin 17, 43–53.

Càstino, M., Ubigli, M., 1984. Prove di macerazione carbonica con uve Barbera. Vini Italia 26, 7–23.

Chanprasartsuk, O.-O., Pheanudomkitlert, K., Toonwai, D., 2012. Pineapple wine fermentation with yeasts isolated from fruit as single and mixed starter cultures. Asian Journal of Food and Agro-Industry 5, 104–111.

Chaudhary, C., Yadav, B.S., Grewal, R.B., 2014. Preparation of red wine by blending the grape (*Vitis vinifera* L.) and Janum (*Syzygium cuminii* L. Skeels) juices before fermentation. International Journal of Agriculture Food Science & Technology 5, 239–248.

Charpentier, C., 2010. Ageing on lees (*sur lies*) and the use of speciality inactive yeasts during wine fermentation. In: Reynolds, A.G. (Ed.), Managing Wine Quality. Oenology and Wine Quality, vol. 2. Woodhead Publishing Ltd, Cambridge, UK, pp. 164–187.

Castellari, M., Simonato, B., Tornielli, G.B., Spinelli, P., Ferrarini, R., 2004. Effects of different enological treatments on dissolved oxygen in wines. Italian Journal of Food Science 16, 387–397.

Cerovic, Z.G., Ghozlen, N.B., Milhade, C., Obert, M., Debuisson, S., Le Moigne, M., 2015. Nondestructive diagnostic test for nitrogen nutrition of grapevine (*Vitis vinifera* L.) based on Dualex leaf-clip measurements in the field. Journal of Agricultural and Food Chemistry 63, 3669–3680.

Chinnici, F., Natali, N., Riponi, C., 2014. Efficacy of chitosan in inhibiting the oxidation of (+)-catechin in white wine model solutions. Journal of Agricultural and Food Chemistry 62, 9868–9875.

Christmann, M., Freund, M., 2010. Advances in grape processing equipment. In: Reynolds, A.G. (Ed.), Managing Wine Quality. Viticulture and Wine Quality, vol. 1. Woodhead Publishing Ltd, Cambridge, UK, pp. 547–558.

Cliff, M.A., Heymann, H., 1993. Development and use of time-intensity methodology for sensory evaluation: a review. Food Research International 26, 375–385.

Coelho, E., Rocha, S.M., Barros, A.S., Delgadillo, I., Coimbra, M.A., 2007. Screening of variety- and pre-fermentation -related volatile compounds during ripening of white grapes to define their evolution profile. Analytica Chimica Acta 597, 257–264.

Coelho, E., Vilanova, M., Genisheva, Z., Oliveira, J.E., Teixeira, J.A., Domingues, L., 2015. Systematic approach for the development of fruit wines from industrially processed fruit concentrates, including optimization of fermentation parameters, chemical characterization and sensory evaluation. LWT – Food Science and Technology 62, 1043–1052.

Coton, E., Coton, M., Guichard, H., 2016. Cider: the product and its manufacture. In: Caballero, B., Finglas, P., Toldrá, F. (Eds.), Encyclopedia of Food and Health. Elsevier (in press).

Contreras, A., Hidalgo, C., Henschke, P.A., Chambers, P.J., Curtin, C., Varela, C., 2014. Evaluation of non-*Saccharomyces* yeasts for the reduction of alcohol content in wine. Applied and Environmental Microbiology 80, 1670–1678.

Cozzolino, D., 2015. Sample presentation, source of error and future perspectives on the application of vibrational spectroscopy in the wine industry. Journal of the Science of Food and Agriculture 95, 861–868.

Cozzolino, D., Cynkar, W., Shah, N., Smith, P., 2011. Technical solutions for analysis of grape juice, must, and wine: the role of infrared spectroscopy and chemometrics. Analytical and Bioanalytical Chemistry 40, 1475–1484.

de Bei, R., Cozzolino, D., Sullivan, W., Cynkar, W., Fuentes, S., Dambergs, R., Pech, J., Tyerman, S., 2011. Nondestructive measurement of grapevine water potential using near infrared spectroscopy. Australian Journal of Grape and Wine Research 17, 1078–1086.

del Alamo-Sanza, M., Laurie, V.F., Nevares, I., 2015. Wine evolution and spatial distribution of oxygen during storage in high-density polyethylene tanks. Journal of the Science of Food and Agriculture 95, 1313–1320.

Dey, G., Negi, B., Gandhi, A., 2009. Can fruit wines be considered as function food? – An overview. Natural Product Radiance 8, 314–322.

Diago, M.P., Sanz-Garcia, A., Millan, B., Blasco, J., Tardaguila, J., 2014. Assessment of flower number per inflorescence in grapevine by image analysis under field conditions. Journal of the Science of Food and Agriculture 94, 1981–1987.

Dias, D.R., Schwan, R.F., Freire, E.S., dos Santos Serôdio, R., 2007. Elaboration of a fruit wine from cocoa (*Theobroma cacao* L.) pulp. International Journal of Food Science & Technology 42, 319–329.

Dijksterhuis, G., 1996. Procrustes Analysis in sensory research. In: Naes, T., Risvik, E. (Eds.), Multivariate Analysis of Data in Sensory Science. Data Handling in Science and Technology, vol. 16. Elsevier Science Publ, Amsterdam, pp. 185–220.

Dolgin, E., 2015. Safety boost for GM organisms. Nature 517, 423.

Dubernet, M., 2010. Automatic analysers in oenology. In: Moreno-Arribas, M., Polo, C. (Eds.), Wine Chemistry and Biochemistry. Springer Verlag, New York/Heidelberg, pp. 649–676.

Dubourdieu, D., Moine-Ledoux, V., Lavigne-Cruège, V., Blanchard, L., Tominaga, T., 2000. Recent advances in white wine aging: the key role of lees. In: Proc. ASEV 50th Anniv. Ann. Meeting, Seattle, WA. June 19–23, 2000. American Society for Enology andViticulture, Davis, CA, pp. 345–352.

Dykes, S.I., Kilmartin, P.A., 2005. Effect of oxygenation dosage rate on the chemical and sensory properties of Cabernet Sauvignon. American Journal of Enology and Viticulture 56, 291A–292A.

Escalona, J.M., Ribas-Carbó, M., 2010. Methodologies for measurement of water flow in grapevines. In: Delrot, S., Medrano, H., Or, E., et al. (Eds.), Methodologies and Results in Grapevine Research. Springer, Dordrecht, The Netherlands, pp. 57–69.

Falguera, V., Forns, M., Ibarz, A., 2013. UV-vis irradiation: an alternative to reduce SO_2 in white wine. LWT – Food Science and Technology 51, 59–64.

Fan, W., Xu, Y., Yu, A., 2006. Influence of oak chips geographical origin, toast level, dosage and aging time on volatile compounds of apple cider. Journal of the Institute of Brewing 112, 255–263.

Faria, D.P., Fonseca, A.L., Pereira, H., Teodoro, O.M., 2011. Permeability of cork to gases. Journal of Agricultural and Food Chemistry 59, 3590–3597.

Forde, C.G., Cox, A., Williams, E.R., Boss, P.K., 2011. Associations between the sensory attributes and volatile composition of Cabernet Sauvignon wines and the volatile composition of the grapes used for their production. Journal of Agricultural and Food Chemistry 59, 2573–2583.

Francis, I.L., Leino, M., Sefton, M.A., Williams, P.J., 1993. Thermal processing of Chardonnay and Semillon juice and wine – sensory and chemical changes. In: Stockley, C.S., Johnstone, R.S., Leske, P.A., Lee, T.H. (Eds.), Proc. 8th Aust. Wine Ind. Tech. Conf. Oct. 25–29, 1992, Melbourne, Australia. Winetitles, Adelaide, Australia, pp. 158–160.

Garde-Cerdán, T., López, R., Garijo, P., González-Arenzana, L., Gutiérrez, A.R., López-Alfara, I., Santamaría, P., 2014. Application of colloidal silver versus sulfur diocide during vinification and storage of Tempranillo red wines. Australian Journal of Grape and Wine Research 20, 51–61.

García-Carpintero, E.G., Sánchez-Palomo, E., González Viñas, M.A., 2010. Influence of co-winemaking technique in sensory characteristics of new Spanish red wines. Food Quality and Preference 21, 705–710.

Ghozlen, B.N., Cerovic, Z.G., Germain, C., Toutain, S., Latouche, G., 2010. Non-destructive optical monitoring of grape maturation by proximal sensing. Sensors 10, 10040–10068.

Giaramida, P., Ponticello, G., Di Maio, S., Squadrito, M., Genna, G., Barone, E., Scacco, A., Corona, O., Amore, G., di Stefano, R., Oliva, D., 2013. *Candida zemplinina* for production of wines with less alcohol and more glycerol. South African Journal for Enology and Viticulture 34, 204–211.

Gishen, M., Iland, P.G., Dambergs, R.G., Esler, M.B., Francis, I.L., Kambouris, A., Johnstone, R.S., Høj, P.B., 2002. Objective measures of grape and wine quality. In: Blair, R.J., Williams, P.J., Høj, P.B. (Eds.), 11th Aust. Wine Ind. Tech. Conf. October 7–11, 2001, Adelaide, South Australia. Winetitles, Adelaide, Australia, pp. 188–194.

Guymon, J.F., Crowell, E.A., 1977. The nature and cause of cap-liquid temperature differences during wine fermentation. American Journal of Enology and Viticulture 28, 74–78.

Hallgarten, F.L., 1987. Wine Scandal. Weidenfeld and Nicolson, London, UK.

Harbertson, J.F., Mireles, M.S., Harwood, E.D., Weller, K.M., Ross, C.F., 2009. Chemical and sensory effects of saignée, water addition, and extended maceration on high Brix must. American Journal of Enology and Viticulture 60, 450–460.

Harris, S.A., Robinson, J.P., Juniper, B.E., 2002. Genetic clues to the origin of the apple. Trends in Genetics 18, 426–430.

Heard, S.J., Fleet, G.H., 1985. Growth of natural yeast flora during the fermentation of inoculated wines. Applied and Environmental Microbiology 50, 727–728.

Heymann, H., LiCalzi, M., Conversano, M.R., Bauer, A., Skogerson, K., Matthews, M., 2013. Effects of extended grape ripening with and without must and wine alcohol manipulations on Cabernet Sauvignon wine sensory characteristics. South African Journal for Enology and Viticulture 34, 86–99.

Holzapfel, B.P., Rogiers, S.Y., Degaris, K.A., Small, G., 1999. Ripening grapes to specification: effect of yield on colour development of Shiraz in the Riverina. Australian Grapegrower & Winemaker 428 (24), 26–28.

Holt, S., Cordente, A.G., Williams, S.J., Capone, D.L., Jitjaroen, W., Menz, I.R., Curtin, C., Anderson, P.A., 2011. Engineering *Saccharomyces cerevisiae* to release 3-mercaptohexan-1-ol during fermentation through overexpression of an *S. cerevisiae* gene, *STR3*, for improvement of wine aroma. Applied and Environmental Microbiology 77, 3626–3632.

Jackson, R.S., 2009. Wine Tasting: A Professional Handbook, second ed. Academic Press, London, UK.

Jackson, R.S., 2014. Wine Science: Principles and Application, fourth ed. Academic Press, San Diego, CA.

Jallerat, E., 1990. Les nouvelles techniques de tirage. III. Le minifermenteur. Un nouveau dispositif pour la prise de mousse en bouteilles. Vigneron de Champenois 10, 9–24.

Joshi, V.J., Sharma, S., Bhushan, S., Attri, D., 2004. Fruit-based alcoholic beverages. In: Pandey, A. (Ed.), Concise Encyclopedia of Bioresource Technology. Food Products Press, New York, pp. 335–345.

Karbowiak, T., Gougeon, R.D., Alinc, J.-B., Brachais, L., Debeaufort, F., Voilley, A., Chassagne, D., 2010. Wine oxidation and the role of cork. Critical Reviews in Food Science and Nutrition 50, 20–52.

Kelly, S., Heaton, K., Hoogewerff, J., 2005. Tracing the geographical origin of food: the application of multi-element and multi-isotope analysis. Trends in Food Science & Technology 16, 555–567.

Koehler, C.G., 1986. Handling of Greek transport amphoras. In: Empereur, J.-Y., Garlan, Y. (Eds.), Recherches sur les Amphores Greques. École française d'Anthènes, Paris, France, pp. 49–67. Bull. Correspondance Hellénique, supp. 13.

Kontoudakis, N., Esteruelas, M., Fort, F., Canals, J.M., and Zamora, F., 2010. Comparison of methods for estimating phenolic maturity in grapes: correlation between predicted and obtained parameters. Analytica Chimica Acta 660, 127–133.

Kontoudakis, N., Esteruelas, M., Fort, F., Canals, J.M., Zamora, F., 2011. Use of unripe grapes harvested during cluster thinning as a method for reducing alcohol content and pH of wine. Australian Journal of Grape and Wine Research 17, 230–238.

Kosseva, M.R., Kennedy, J.F., 2004. Encapsulated lactic acid bacteria for control of malolactic fermentation in wine. Artificial Cells, Blood Substitutes and Immobilization Biotechnology 32, 55–65.

Kumar, P., Reinitz, H.W., Simunovic, J., Sandeep, K.P., Franzon, P.D., 2009. Overview of RFID technology and its applications in the food industry. Journal of Food Science 74, R101–R106.

Lam, H.Y., Choy, K.L., Ho, G.T.S., Kwong, C.K., Lee, C.K.M., 2013. A real-time risk control and monitoring system for incident handling in wine storage. Expert Systems With Applications 40, 3665–3678.

Lange, J., Wyser, Y., 2003. Recent innovation in barrier technologies for plastic packaging – a review. Packaging Technology & Science 16, 149–158.

Li, P., Lee, S.-H., Hsu, H.-Y., 2011. Review on fruit harvesting method for potential use of automatic fruit harvesting systems. Proceed Engine 23, 351–366.

Li, X., Yu, B., Curran, P., Liu, S.-Q., 2012. Impact of two *Williopsis* yeast strains on the volatile composition of mango wine. International Journal of Food Science & Technology 47, 808–815.

Li, X., Lim, S.L., Yu, B., Curran, P., Liu, S.-Q., 2013. Mango wine aroma enhancement by pulp contact and β-glucosidase. International Journal of Food Science & Technology 48, 2258–2266.

Liger-Belair, G., Polidori, G., Jeandet, P., 2008. Recent advances in the science of champagne bubbles. Chemical Society Reviews 37, 2490–2511.

Lobo, A.P., Tascón, N.F., Madrera, R.R., Valles, B.S., 2005. Sensory and foaming properties of sparkling cider. Journal of Agricultural and Food Chemistry 53, 10051–10056.

López-Rituerto, E., Cabredo, S., López, M., Avenoza, A., Busto, J.H., Peregrina, J.M., 2009. A thorough study on the use of quantitative ^{1}H NMR in Rioja red wine fermentation processes. Journal of Agricultural and Food Chemistry 57, 2112–2118.

Lorenzo, C., Pardo, F., Zalacain, A., Alonzo, G.L., Salinas, M.R., 2005. Effect of red grapes co-winemaking in polyphenols and color of wines. Journal of Agricultural and Food Chemistry 53, 7609–7616.

Mackay, T.C., Stone, E.A., Ayroles, J.F., 2009. The genetics of quantitative traits: challenges and prospects. Nature Reviews Genetics 10, 565–577.

Mangas, J.J., Moreno, J., Rodríguez, R., Picinelli, A., Suárez, B., 1999. Analysis of polysaccharides in cider: their effect on sensory foaming properties. Journal of Agricultural and Food Chemistry 47, 152–156.

Maury, C., Clark, A.C., Scollary, G.R., 2010. Determination of the impact of bottle colour and phenolic concentration on pigment development in white wine stored under external conditions. Analytica Chimica Acta 660, 81–86.

McGovern, P.E., Glusker, D.L., Exner, L.J., Voigt, M.M., 1996. Neolithic resinated wine. Nature 381, 480–481.

McGovern, P.E., Zhang, J., Tang, J., Zhang, Z., Hall, G.,R., Moreau, R.A., Nuñez, A., Butrym, E.D., Richards, M.P., Wang, C.-S., Cheng, G., Zhao, Z., Wang, C., 2004. Fermented beverages of pre- and proto-historic China. Proceedings of the National Academy of the United States of America 101, 17593–17598.

Messenger, S., 2007. New MOG removal system makes the grade. Australian & New Zealand Grapegrower & Winemaker 516, 49–50.

Moss, R., Daniels, K., Shasky, J., 2013. Effect of cold soak on the phenolic extraction of Syrah. Australian & New Zealand Grapegrower & Winemaker 592, 59–62.

Moya, I., Camenen, L., Evain, S., Goulas, Y., Cerovic, Z.G., Latouche, G., et al., 2004. A new instrument for passive remote sensing. 1. Measurements of sunlight induced chlorophyll fluorescence. Remote Sensing of Environment 91, 186–197.

O'Kennedy, K., Canal-Llaubères, R.-M., 2013a. The A–Z of wime enzymes: Part 1. Australian & New Zealand Grapegrower & Winemaker 589, 57–61.

O'Kennedy, K., Canal-Llaubères, R.-M., 2013b. The A–Z of wine enzymes: Part 2. Australian & New Zealand Grapegrower & Winemaker 590, 42–46.

Oliveira, M.E.S., Pantoja, L., Cuarte, W.F., Collela, C.F., Valarelli, L.T., Schwan, R.F., Dias, D.R., 2011. Fruit wine produced from cagaita (*Eugenia dysenterica* DC) by both free and immobilized yeast cell fermentation. Food Research International 44, 2391–2400.

Österbauer, R.A., Matthews, P.M., Jenkinson, M., Beckmann, C.F., Hansen, P.C., Calvert, G.A., 2005. Color of scents: chromatic stimuli modulate odor responses in the human brain. Journal of Neurophysiology 93, 3434–3441.

Panesar, P.S., Joshi, V.K., Panesar, R., Abrol, G.S., 2011. Vermouth: technology of production and quality characteristics. Advances in Food and Nutrition Research 63, 251–284.

Paronetto, L., Dellaglio, F., 2011. Amarone: a modern wine coming from an ancient production technology. Advances in Food and Nutrition Research 63, 285–306.

Pati, S., Mentana, A., La Notte, E., Del Nobile, M.A., 2010. Biodegradable poly-lactic acid package for the storage of carbonic maceration wine. LWT – Food Science and Technology 43, 1573–1579.

Phaff, H.J., 1986. Ecology of yeasts with actual and potential value in biotechnology. Microbial Ecology 12, 31–42.

Pineau, N., Schlich, P., Cordelle, S., Mathonnière, C., Issanchou, S., Imbert, A., Rogeaux, M., Eteévant, P., Koster, E., 2009. Temporal dominance of sensations: construction of the TDS curves and comparison with time-intensity. Food Quality and Preference 20, 450–455.

Pozo-Bayón, M.Á., Martínez-Rodríguez, A., Pueyo, E., Moreno-Arribas, M.V., 2009. Chemical and biochemical features involved in sparkling wine production: from a tradition to an improved winemaking technology. Trends in Food Science & Technology 20, 289–299.

Proffitt, T., Malcolm, A., 2005. Zonal vineyard management through airborne remote sensing. Australian & New Zealand Grapegrower & Winemaker 502, 22–24 25–26.

Proffitt, T., Bramley, R., Lamb, D., Winter, E., 2006. Precision Viticulture. Winetitles Pty Ltd, Adelaide, Aust.

Raco, B., Dotsika, E., Poutoukis, D., Battaglini, R., Chantzi, P., 2015. O-H-C isotope ratio determination in wine in order to be used as a fingerprint of its regional origin. Food Chemistry 168, 588–594.

Reuss, R.M., Stratton, J.E., Smith, D.A., Read, P.E., Cuppett, S.L., Parkhurst, A.M., 2010. Malolactic fermentation as a technique for the deacidification of hard apple cider. Journal of Food Science and Technology 75, C74–C78.

Ribereau-Gayon, P., Doneche, B., Dubourdieu, D., Lonvaud, A., 2006. Handbook of Enology vol. I: The Microbiology of Wine and Vinifications, second ed. John Wiley and Sons, Chichester, UK.

Riekstina-Dolge, R., Kruma, Z., Cinknanis, I., Straumite, E., Sabovics, M., Tomsone, L., 2014. Influence of *Oenococcus oeni* and oak chips on the chemical composition and sensory properties of cider. In: Straumite, E. (Ed.), Conference Proceedings of the 9th Baltic Conference on Food Science and Technology "Food for Consumer Well-Being. FOODBALT 2014, Jelgava, Latvia pp. 178–183.

Rolls, E.T., Critchley, H.D., Verhagen, J.V., Kadohisa, M., 2010. The representation of information about taste and odor in the orbitofrontal cortex. Chemosensory Perception 3, 16–33.

Ross, C.F., Hinken, C., Weller, K., 2007. Efficacy of palate cleansers for reduction of astringency carryover during repeated ingestions of red wine. Journal of Sensory Studies 22, 293–312.

Rous, C., Alderson, B., 1983. Phenolic extraction curves for white wine aged in French and American oak barrels. American Journal of Enology and Viticulture 34, 211–215.

Roussis, I.G., Patrianakou, M., Drossiadis, A., 2013. Protection of aroma volatiles in a red wine with low sulphur dioxide by a mixture of glutathione, caffeic acid and gallic acid. South African Journal for Enology and Viticulture 34, 262–265.

Rupasinghe, H.P.V., Clegg, S., 2007. Total antioxidant capacity, total phenolic content, mineral elements, and histamine concentrations in wines of different fruit sources. Journal of Food Composition and Analysis 20, 133–137.

Rustioni, L., Rossoni, M., Calatrioni, M., Failla, O., 2011. Influence of bunch exposure on anthocyanins extractability from grapes skins (*Vitis vinifera* L.). Vitis 50, 137–143.

Sauvageot, F., Vivier, P., 1997. Effects of malolactic fermentation on sensory properties of four Burgundy wines. American Journal of Enology and Viticulture 48, 187–192.

Schmid, F., Schadt, J., Jiranek, V., Block, D.E., 2009. Formation of temperature gradients in large- and small-scale red wine fermentations during cap management. Australian Journal of Grape and Wine Research 15, 249–255.

Sharma, S., Joshi, V.K., Abrol, G., 2009. An overview on strawberry [*Fragaria* x *ananassa* (Weston) Duchesne ex Rozier] wine production technology, composition, maturation and quality evaluation. Natural Product Research 8, 356–365.

Silva, S., Ramón-Portugal, F., Andrade, P., Abreu, S., de Fatima Texeira, M., Strehaiano, P., 2003. Malic acid consumption by dry immobilized cells of *Schizosaccharomyces pombe*. American Journal of Enology and Viticulture 54, 50–55.

Singh, R.S., Sooch, B.S., 2009. High cell density reactors in production of fruit wine with special reference to cider – an overview. Natural Product Radiance 8, 323–333.

Singleton, V.L., 1962. Aging of wines and other spiritous products, acceleration by physical treatments. Hilgardia 32, 319–373.

Singleton, V.L., Ough, C.S., 1962. Complexity of flavor and blending of wines. Journal of Food Science 12, 189–196.

Srivastava, Y., Semwal, A.D., Sharma, G.K., 2013. Application of various chemical and mechanical microencapsulation techniques in food sector –a review. International Journal of Food and Fermentation Technology 3, 1–13.

Sun, S.Y., Gong, H.S., Zhao, K., Wang, X.L., Wang, W., Zhao, X.H., Yu, B., Wang, H.X., 2013. Co-inoculation of yeast and lactic acid bacteria to improve cherry wines sensory quality. International Journal of Food Science & Technology 48, 1783–1790.

Sun, S.Y., Gong, H.S., Jiang, X.M., Zhao, Y.P., 2014. Selected non-*Saccharomyces* wine yeasts in controlled multistarter fermentations with *Saccharomyces cerevisiae* on alcoholic fermentation behaviour and wine aroma of cherry wines. Food Microbiology 44, 15–23.

Swaffield, C.H., Scott, J.A., Jarvis, B., 1997. Observations on the microbial ecology of traditional alcoholic cider storage vats. Food Microbiology 14, 353–361.

Tabera, L., Munoz, R., Gonzalez, R., 2006. Deletion of *BCY1* from the *Saccharomyces cerevisiae* genome is semidominant and induces autolytic phenotypes suitable for improvement of sparkling wines. Applied and Environmental Microbiology 72, 2351–2358.

Tao, Y., García, J.F., Sun, D.-W., 2014. Advances in wine aging technologies for enhancing wine quality and accelerating wine aging process. Critical Reviews in Food Science and Nutrition 54, 817–835.

Tempère, S., Cuzange, E., Malik, J., Cougeant, J.C., de Revel, G., Sicard, G., 2011. The training level of experts influences their detection thresholds for key wine compounds. Chemosensory Perception 4, 99–115.

Tempère, S., Cuzange, E., Bougeant, J.C., de Revel, G., Sicard, G., 2012. Explicit sensory training improves the olfactory sensitivity of wine experts. Chemosensory Perception 5, 205–213.

Tessonnière, H., Vidal, S., Barnavon, L., Alexandre, H., Remize, F., 2009. Design and performance testing of a real-time PCR assay for sensitive and reliable direct quantification of *Brettanomyces* in wine. International Journal of Food Microbiology 129, 237–243.

Torresi, A., Frangipane, M.T., Anelli, G., 2011. Biotechnologies in sparkling wine production. Interesting approaches for quality improvement: a review. Food Chemistry 129, 1232–1241.

Towantakavanit, K., Park, Y.S., Gorinstein, S., 2011. Quality properties of wine from Korean kiwifruit new cultivars. Food Research International 44, 1364–1372.

Trought, M., Tannock, S., 1996. Berry size and soluble solids variation. In: Henick-Kling, T., Wolf, T.E., Harkness, E.M. (Eds.), Proc. 4th Int. Symp. Cool Climate Vitic. Enol., Rochester, NY, July 16–20, 1996. NY State Agricultural Experimental Station, Geneva, New York, pp. V-70–73.

Tuccio, L., Remorini, D., Pinelli, P., Fierini, E., Tonutti, P., Scalabrelli, G., et al., 2011. Rapid and non-destructive method to assess in the vineyard grape berry anthocyanins under different seasonal and water conditions. Australian Journal of Grape and Wine Research 17, 181–189.

Usseglio-Tomasset, L., 1986. Riattivazione della fermentazione e prevenzione degli arresti fermentativi mediante l'impiego di pareti cellulari di lievito. Enotecnico 1, 53–57.

Vaidya, D., Vaidya, M., Sharma, S., Ghanshayam, 2009. Enzymatic treatment of juice extraction and preparation and preliminary evaluation of Kiwifruit wine. Natural Product Radiance 8, 380–385.

Varakumar, S., Naresh, K., Variar, P.S., Sharma, A., Reddy, O.V.S., 2013. Role of malolactic fermentation on the quality of mange (*Mangifera indica* L.) wine. Food Biotechnology 27, 119–136.

Vierra, T., August 2013. The Proof Is in the Packaging. Wines Vines. p. 2 https://ipcybercrime.com/wp.../09/2013-08-IPC-Wines-and-Vines.pdf.

Vitalini, S., Ruggiero, A., Rapparini, F., Neri, L., Tonni, M., Iriti, M., 2014. The application of chitosan and benxothiadizole in vineyard (*Vitis vinifera* L. cv Groppello Gentile) changes the aromatic profile and sensory attributes of wine. Food Chemistry 162, 192–205.

Wilkes, E., Warner, L., 2014. Accurate mid-infrared analysis in wine production–fact or fable? Wine and Viticulture Journal 29 (3), 64–67.

Williams, P.J., Francis, I.L., 1996. Sensory analysis and quantitative determination of grape glycosides – the contribution of these data to winemaking and viticulture. In: Biotechnol. Improved Foods Flavors. Amer. Chemical Society, Washington, DC, pp. 124–133. ACS Symposium Series 637.

Wollan, D., 2010. Reducing wine alcohol: some myths busted. Australian & New Zealand Grapegrower & Winemaker 562, 54–59.

Younger, W., 1966. Gods, Men and Wine. George Rainbird, London, UK, pp. 258–312.

Zeng, X.A., Yu, S.J., Zhang, L., Chen, X.D., 2008. The effects of AC electric field on wine maturation. Innovative Food Science and Emerging Technologies 9, 463–468.

FURTHER READING

de Revel, G., Martin, N., Pripis-Nicolau, L., Lonvaud-Funel, A., Bertrand, A., 1999. Contribution to the knowledge of malolactic fermentation influence on wine aroma. Journal of Agricultural and Food Chemistry 47, 4003–4008.

Joshi, V.K., Sandhu, D.K., 2009. Flavour profiling of apple vermouth using descriptive analysis technique. Natural Product Research 8, 419–425.

Larpin, S., Sauvageot, N., Pichereau, V., Laplace, J.-M., Auffray, Y., 2002. Biosynthesis of exopolysaccharide by a *Bacillus licheniformis* strain isolated from ropy cider. International Journal of Food Microbiology 77, 1–9.

Valles, B.S., Bedriñana, R.P., Queipo, A.L., Alonso, J.J.M., 2008. Screening of cider yeasts for sparkling cider production (Champenoise method). Food Microbiology 25, 690–697.

TECHNICAL GUIDE FOR FRUIT WINE PRODUCTION

14

F. Matei

University of Agronomic Sciences and Veterinary Medicine of Bucharest, Bucharest, Romania

1. INTRODUCTION

Fruit wines are fermented alcoholic beverages made of fruits other than grapes; they may also have additional flavors taken from fruits, flowers, and herbs. In the United Kingdom, fruit wine is commonly called *country wine*; the term should not be confounded with the French term *vin de pays*, which is grape wine. The basic process is similar to grape juice production and fermentation. Grape juice is naturally suited for making wine and needs little adjustment prior to fermentation, while other fruits almost always require easy adjustments. The following aspects should be taken into account: the amount of fruit needed per gallon/liter of obtained wine; the amount of available sugars and the juice's acidity should be tested and adjusted (to regulate sugar content, the fruit mash is generally topped up with water prior to fermentation to reduce the acidity to pleasant levels); the nutritional aspects required by the yeast employed during the fermentation (winemakers can counter this with the addition of nitrogen, phosphorus, and potassium, available commercially as yeast nutrient).

When labeling, fruit wines are usually referred to by their main ingredient (e.g., plum wine or elderberry wine) because the usual definition of wine states that it is made from fermented grape juice.

Generally, the fruit wines should be maturated at least six months before opening the first bottle and to be consumed within three to four years. Some authors reported the possibility of aging in barrel or with wooden chips (Joshi et al., 2011a,b).

A high variety of fruits is employed for winemaking worldwide, depending on the region and its specific climate; most fruits and berries have the potential to produce wine. In the United States and Canada the examples of fruits include strawberries, plums, watermelons, peaches, blackberries, gooseberries, persimmons, citrus fruits, etc., while in Europe the predominant fruit wines are made of apples and pears (Hohenstein, 2005; Rivard, 2009; Peak, 2013).

In Asia the fruit palette is quite different and includes tropical and subtropical fruits such as pineapples, mango, lychees, cocoa, figs, or less-known fruits such as sapota (*Achras sapota*; Panda, 2014a,b) or jackfruit (*Artocarpus heterophyllus* L). Recently (Zhang, 2013) in China, fruits wines made of dogwood, known also as Japanese cornel (*Cornus officinalis*), have been proposed. Some other minor fruits have been reported to be used successfully for winemaking in the Philippines (Sanchez, 1979): bignay (*Antidesma bunius)*, calumpit (*Terminalia edulis*), mansanitas (*Zizyphus jujuba*), and passion fruit (*Passiflora foetida*). Meanwhile, in Brazil, underused fruits are used for winemaking, as cacao, cupuassu (*Theobroma grandiflorum*), gabiroba (*Campomanesia* sp.), jaboticaba (*Plinia cauliflora*), cagaita (*Eugenia dysenterica* DC), and umbu/Brazil plum (*Spondias tuberosa*) because the major

Science and Technology of Fruit Wine Production. http://dx.doi.org/10.1016/B978-0-12-800850-8.00014-4

components found in these fruit wines (alcohols, monterpenics compounds, and ethyl esters) contributed to the formation of aromas which could be characterized as fruity, green apple, banana, sweet, citrus, citronella, vanilla, roses, and honey (Duarte, 2010).

In Africa, traditional fermented foods include fermented beverages from indigenous fruits such as sand apple and hacha (*Parinari curatellifolia*), mazhanje (*Uapaca kirkiana*), and masau (*Ziziphus mauritiana*; Nyanga, 2007). Also, palm wine is an alcoholic beverage produced and consumed in very large quantities in West Africa, and it is known throughout the major parts of Africa under various names, such as "mimbo" in Cameroon, "nsafufuo" in Ghana, and "emu" in Nigeria. It can be produced starting from oil palm (*Elaeis guineensis*), coconut palm (*Phoenix dactylifera*), date palm, nipa palm, kithul palm, and raffia palm (*Raphia hookeri*; Amoa-Awua, 2007; Stringini, 2009).

2. FRUIT WINE TYPES AND STYLES

The definition of fruit wine varies, but generally fruit wine is made from the juice of ripe fruit, fruit juices, or concentrate, without containing any grape products. The alcohol content in wine can vary. According to certain regulations, fruit wine has to be greater than a minimum of 7.1% alcohol. If greater than 14.9% alcohol by volume it can be labeled as a "dessert" wine or "aperitif." Light wine is considered that of 9% alcohol by volume or less (Rivard, 2009).

Beside the classic variety of wine, there are three other categories of fruit wines that can be produced (Rivard, 2009):

- **Iced fruit wine** is made from juice obtained by cyroextraction followed by the fermentation.
- **Fortified fruit wine** is obtained by adding alcohol, derived from the alcoholic fermentation of a food source and distilled to not less than 80% alcohol by volume. This type of wine shall have an actual content no less than 14% alcohol by volume.
- **Sparkling fruit wine's** effervescence derives from primary or secondary alcoholic fermentation in a closed vessel, or carbon dioxide is added to the wine to a minimum of 200 kPa at 10°C and has an actual alcoholic strength of not less than 8.5% by volume.

Depending on the application and target, many types of fruit wines can be produced.

- Low alcohol "cider style" (2–7% alcohol) is very popular in Europe; usually, they are made with an apple base and blended with other fruit. They are characterized by a high dilution ratio.
- Dry or "off-dry" fruit wines are usually between 8.5 and 13.5% alcohol with less than 30 g/L residual sugar, being very similar to the grape wines. As with grape wines, the key is to have a good balance between the alcohol, sugar, acidity, and flavor concentration.
- Sweet fruit wines have residual sugar higher than 40 g/L and tannic acid 7–8 g/L. The key is the balance between sugar and flavor intensity. The best sweet fruit wines are made of higher acid fruit such as raspberry or currants.
- Cryoextracted fruit wines are made by freezing juice and freeze-fractioning it off. Their residual sugar goes higher than 140 g/L and the tannic acid is about 9 g/L, depending on the fruit.
- Fortified or "Port-style" fruit wines can be fortified up to 24% alcohol. The residual sugar levels are higher than 80 g/L. The best fortified wines are made of raspberry, blackberry, or currants, which are fruits of high acidity.
- Sparkling fruit wine can be made by carbon dioxide injection, by Charmat method, or traditional (Champenoise) style.

3. METHODS FOR FRUIT WINE PRODUCTION

Grape and fruit winemaking are similar except with some variations based on the fruit used. Important distinctions between juice and wine are described as followed by Bates et al. (2001):

- Fruits do not necessarily have to be peeled, cored, deseeded, pressed, etc., immediately after crushing. Light colored fruits are best pressed soon after crushing and treated with about 100 ppm sulfur dioxide (SO_2) to prevent browning; clarification is not required until after the fermentation.
- The fermentation can be initiated and allowed to proceed for some time prior to pressing.
- In the case of colored fruits where pigment extraction into the must is desired, the initial phase of fermentation serves to extract color and soften the crushed material.
- In contrast to grape must, many fruits lack the nutrients necessary to sustain yeast growth. Thus yeast nutrients such as yeast extract or diammonium phosphate may be needed up to 0.1%. Crushed fruit will have more yeast nutrients than pressed juice, so addition may be unnecessary in the latter.
- Elevated temperatures higher than 30°C can improve some dessert wines if fortified with additional alcohol.
- The most common quality defects of fruit wines are excessive sweetness and oxidized flavor and color. Practice and attention to detail can easily avoid both.

Depending on the initial sources ("raw material") the fruit wine **production types**, according to Rivard (2009), are the following:

- Grape base plus fruit extract, known as "Arbor Mist style";
- Grape base plus juice, known as ready to drink (RTD) "Wild Vine Style";
- Fruit juice or concentrate, known as "Standard" fruit wine;
- Whole fruit, known as "Premium" fruit wine.

Main equipment to be used during the fruit wine production (Fig. 14.1) consists of:

- *Fruit handling*: bins, crates, conveyor, pails, or paddles;
- *Fruit movement*: tractors, pallets, forklifts;
- *Fruit crushers*: rollers, pulverizers;

FIGURE 14.1

Some equipment used during the fruit wine production (from left to right: fruit crusher, fruit pressing, plastic and stainless fermentation tanks, and bottling equipment).

- *Fruit pressing*: press or screw press, basket, bladder, membrane;
- *Pumping equipment*: impellors, centrifugal, piston, lobe;
- *Fermentation tanks*: totes, plastic, stainless, cement;
- *Conditioning equipment*: mainly filters (plate, cartridge, cross-flow);
- *Bottling*: wash, fill, cork, shrink/spin, label, case;
- *Cleaning/sanitizing*: hose, hot water, steam generator, power washer;
- *Cooling facilities*: air conditioning, reefer, glycol chillers.

The main steps of the fruit wines technology are the following: fresh or frozen fruits reception and preliminary preparation; fruit must extraction and preparation by crushing, pressing, clarifying, and amending; fruit must fermentation with or without starter microorganisms; fruit wine conditioning and conservation; aging fruit wine (Fig. 14.2).

Raw material for wine fruits are available all year long, either fresh or frozen pulp, and sometimes concentrate. The moment of fruit harvest and the raw material availability lead to the possibility to make three production cycles early in the temperate climate (Fig. 14.3).

The fruit selection is a very important step in the fruit wine production. The fruit should have high sugar and low acidity, which should be adjusted when needed. It is better to employ slightly overripe fruit. If the fruit are picked too early the wines will lack in the fruit's particular character. For example, a pear wine will taste more like an apple wine if the pears are allowed to reach full maturity (Kraus, 1998). If molds or little bruises are present, SO_2 should be employed as little as possible in the initial step. The best results in balancing the fruit wines are the employment of base wines made of apples or rhubarb.

Many fruit wine producers provide wines made of fruit juices or extracts blended with grape juice bases. Some combination between grape varieties and fruit are known in the US market as black raspberry with Merlot, exotic fruits with White Zinfandel, green apple or raspberry with Riesling, or mango with Symphony grape variety (made of Muscat of Alexandria and Grenache gris). The fruit flavor typically becomes dominant in fruit and grape blends, which can complement the grape component and even sometimes change the wine in an expected way. For instance, Solomons Island Winery has revealed that when adding mango on the base of Symphony wine a coconut flavor appears (according to *WineMaker* magazine, June/July 2007). When naming a blend with the names of the ingredients (or varietals), the majority ingredient is always listed first, the next major ingredient listed second, etc. A mustang-huckleberry-elderberry blend means the mustang is dominant, the huckleberry second, and the elderberry is minor.

There are two main rules to respect when blending: to make sure that all the wines added to the blend are stable (they have fermented to dryness or fermentation was arrested several months earlier) and have aged a few months before blending (Keller, 2005).

3.1 CRUSHING THE FRUIT AND MUST PREPARATION

The fruit used in winemaking can be classified, depending on their hardness, in four groups: soft, medium, hard, and citrus, which are considered as special groups in this respect (Sanchez, 1979). Medium-hard fruit contain large amounts of juice, which can be easier extracted after a suitable maceration by pressing. For the hard fruit, generally, the boiling process should be applied, but only with fruit with low pectic content, otherwise problems may arise during clarification and in the final aroma (Sanchez, 1979).

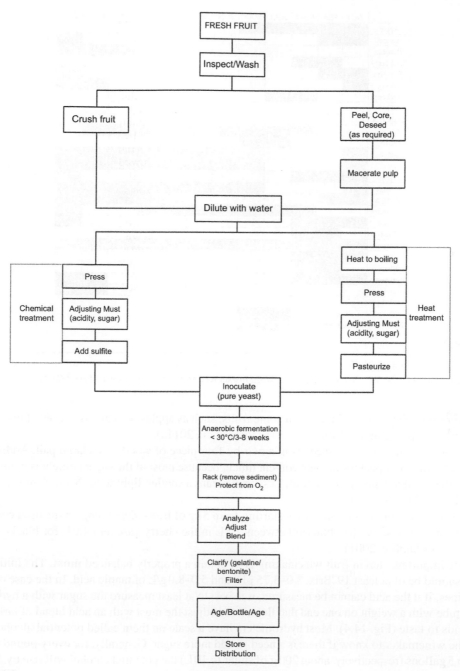

FIGURE 14.2

General flowsheet of fruit wine production.

Adapted after Sanchez, P., 1979. The prospects of fruit wines in Phillipines. Journal of Crop Science 4 (4), 183–190.

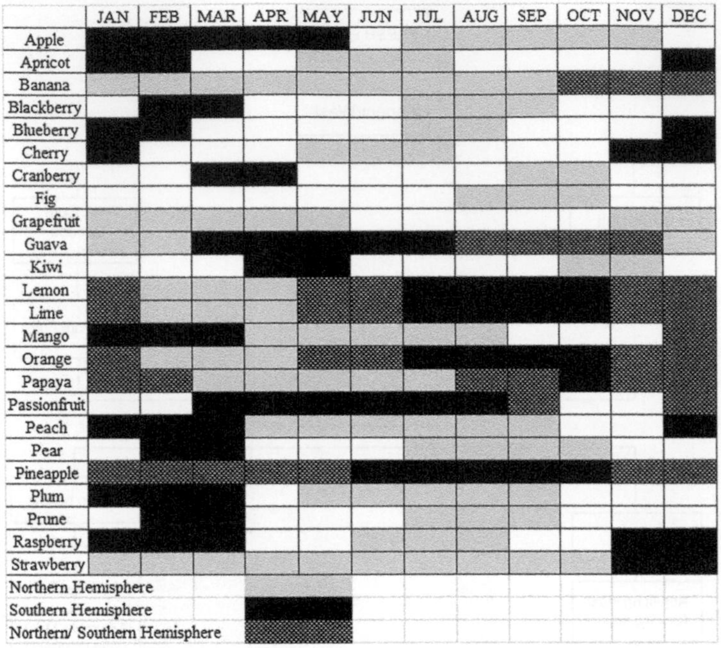

FIGURE 14.3

Harvest season of various fruits.

Adapted after Brandon, S.L., Ferreiro, J.D., 1998. World market for non-citrus juices. Journal of Food Quality 13 (6), 395–398.

On a commercial level, the juice from some fruits, such as apples or pears, is extracted first by grating followed by pressing in a hydraulic press (Joshi et al., 2011a).

At home the fruit can be crushed using a two-by-four piece of wood and a large pail. Adding more or less fruit will not impact the alcohol content much, because most of the sugar content is coming from the sugar that is added, not from the fruit. Less fruit means a subtler, light wine. More fruit means more intense flavor (Pomohaci et al., 2002).

Under homemaking conditions, when starting from 5 kg of fruits, depending on the used equipment and the fruit varieties, can be obtained between 2.5 L in the cherry case and 4.0 L for blackberries or cranberries (Delambre, 2001).

A very important step in fruit winemaking is to obtain a properly balanced **must**. The initial sugar content should be of at least 19°Brix, 3.0–3.75 pH, and 5.0–8.0 g/L of tannic acid. In the case of homemade wines, if it the acid cannot be measured, it is best to at least measure the sugar with a hydrometer (a glass tube with a weight on one end that floats) and adjust the must with an acid blend of tartaric and malic acids to taste (Fig. 14.4). Most hydrometers have a scale on them called potential alcohol which allows the winemaker to know if there is a need to add more sugar. Generally, for every pound of sugar added to 5 gallons (respectively, about 200 g of sugar to 10 L) the potential alcohol will rise by 1%. Different kinds of sugar may be added on homemade wines like cane sugar, corn sugar, beet sugar, brown sugar, rice sugar, fructose, and even malt sugar. Usually, for 4 kg (9 lbs) of added sugar, a few grams of citric acid should also be added. This helps the sugar go into the solution, as well as adding to the finish

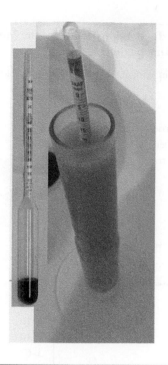

FIGURE 14.4

The use of a hydrometer for sugar content measurement.

of the wine. Adjusting the acid and the sugar will give more alcohol in the finished product, as well as added protection against premature aging.

Honey is an alternative. "Pyment" is a term used for fruit wines with honey added to them. Honey, in its simplest form, has the advantage of adding an "herbal" finish to a wine (Kraus, 1998).

3.1.1 Acidity Adjustment in Fruit Wines

Many kinds of fruit, like strawberries, cherries, pineapples, and raspberries, have a natural acid content which would be too high to produce a savory and pleasant fruit wine in undiluted form.

On homemade winemaking the taste will decide the "correct" acidity. Too low titratable acidity (TA) will taste flat and lifeless while too high will taste tart, sour, and/or bitter. When the wine tastes lively and crisp the acidity is just right. A TA of at least 0.55% contributes to the aseptic quality of the wine. As a rule, the pH should be measured. If it is above 4.55 the wine is susceptible to microorganisms. If it is below 2.5 the wine will be unpalatable. The perceived correctness for most palates is between 3.2 and 3.8 (Keller, 2014).

In Table 14.1 are presented the total acidity values (g/L) of the most used fruits in winemaking.

The dominant acid in apples is malic acid, but many other fruits are reported with citric acid content. In making acid adjustments to fruit wines, the winemaker has a choice of food-grade acids to use like pure tartaric, citric, and malic acids or blends of these may be purchased. Peak (2013) recommends a common recipe for an acid blend (by weight): 8 parts citric acid, 4 parts tartaric acid, and 1 part malic acid.

Table 14.1 Total Acidity Values of the Most Used Fruits in Winemaking

Fruits Used in Winemaking	Total Acidity (g/L)
Lemons	40–50
Cassis	25–35
Red gooseberries	21–30
White gooseberries	19–25
Blue gooseberries	15–20
Raspberries	12–21
Blackberries	12–20
North cherry	12–19
Orange	12–20
Plums	4–16
Blueberries	9–14
Elderberry	10–12
Quince	8–12
Peaches	7–10
Strawberries	6–10
Apples	6–10
Banana	3–6
Pears	3–5
Sweet cheery (bigarreaux)	3–5

One gram of malic acid contains acidity equal to 1.12 g of tartaric acid, and 1 g of citric acid represents the acidity of 1.17 g of tartaric. Although these acids are chemically different, as a practical matter, they may be fairly interchangeable. Peak (2013) provides some practical information for maximum precision. If the calculations call for adding 1 g of tartaric acid, but the malic acid will be used for flavor or fruit compatibility reasons, they should divide 1 by 1.12 and used just 0.9 g of malic in place of each gram of tartaric. Correspondingly, it should be 1/1.17 or 0.85 g of citric acid for each gram of tartaric. For the acid blend described earlier, this looks like $((8 \times 1.17) + (4 \times 1) + (1 \times 1.12))/13 = 1.11$ g effective tartaric acid addition for each gram of the blend (Peak, 2013).

The juice of highly acidic fruit such as some plum, apricot, or grape fruits can be corrected by dilution or by adding potassium or calcium carbonate. According to Joshi et al. (2011a) the carbonate should be added to the heated mixture 150–160°F to hasten the reaction to make calcium citrate less soluble. Ion exchange treatment can also be used for the same purpose (Joshi et al., 2011a).

In Table 14.2 there are presented the total acidity and alcohol desired for fruit wines on the European market (Dewez, 1998).

The acidity of the juice or must made directly from the fruit are at or above those usually desired in finished wine. Usually, the acidity level is decreased into an acceptable range by adding water, which also helps processing the pulp and the fermentation. For best results, an acid titration is necessary to determine the TA level of any juice or wine.

Table 14.2 Total Acidity and Alcohol Content Desired for Some Fruit Wines in Europe

Fruit Wine	Total Acidity (g/L)	Alcohol (% vol)
Cherry wine	6.5	12–13
Apple wine	5.5–6	12–13
Banana sweet wine	9.5	13–15
Strawberry wine	8	12–13
Plum wine	5.5–6	12–13
Redcurrant wine	8	12–15
Blackcurrant (cassis) wine	8	12–13

Dewez, B., 1998. Confrérie Temploutoise des fabricants de vins de fruits. Cours de Formation.

3.1.2 Adding Tannin or Phenolic Compounds in Fruits Wines

If the final wine is going to be darker in color, like blackberry wine, for example, it is recommended to add some additional tannin or phenolic compounds to the must before the fermentation. These will help the wine to have a pleasing mouthfeel and will also provide protection against excessive oxidation in the bottle. Tannins and phenolics can sometimes be purchased in home winemaking stores as commercial powders and potions. A most natural way to add them is to add grape juice or grape juice concentrate. When adding grape juices from the supermarket which are treated with high levels of potassium sorbate (as preservative) the yeast's growth is inhibited.

3.1.3 Adding Water

While grape wines consist of pure grape juice, fruit wines are usually diluted with water. The main reasons for adding water are the high acidity of some fruits, like gooseberries, or strong flavors, like elderberries, which can be astringent.

In many cases, the addition is needed because juice is difficult to extract directly from the fruit. This is generally the case with plum wine, for instance, where crushing and pressing is difficult and produces a low yield (Joshi et al., 2011a,b). In other cases, water may be added to reduce excessive acidity or tannins to palatable wine levels. Adding water means adding sugar as well to achieve a reasonable alcohol level.

Generally, the recipes to be followed advise a specific weight of whole fruit to be crushed, mashed, or cut and combined with water for a specified final volume. In nearly every such recipe, some acid will need to be added. Since there is no simple way to determine the TA of whole fruit, the winemaker can start with published values and make further adjustments later. Acid in grams per 100 mL (or in percent weight/volume) can be reasonably estimated as equal to grams per 100 g (or percent weight/weight) since water weighs 1 g per mL (Peak, 2013).

3.2 MUST FERMENTATION

Just as with grape wines, leaving the pulp with the juice for the first week or so for maceration and fermentation will also intensify the wine's body and character and deepen its color.

Before fermentation pectic enzyme may be added, which breaks down the pectin in the fruit. This helps the wine clear when it is done fermenting. Also, SO_2 may be added into the must in the concentration of 100–150 ppm as potassium metabisulfite (Josihi et al., 2011a). There is evidence that in the case of plum wine, the addition of sodium benzoate gave better quality wine than potassium metabisulfite. Home winemakers will find the pectin as powder or liquid and only a little of this enzyme is needed, following the instructions.

On an industrial level, the fruit wine fermentation can be made in different vessels of different capacities, made in stainless steel or plastic. The metal vessels are not recommended because they can interact with the fruit's acids. In homemade production, glass vessels of 5–10 L can be used, which have a fermentation plug attached (Fig. 14.5).

During the fermentation process, the container of fruit should be stored at room temperature, away from direct sunlight. Fermentation is better to be carried out at a temperature of 4–16°C (40–60°F). This is low for most kinds of fermentation but is beneficial for fruit wines, especially for cider and perries, as it leads to slower fermentation with less loss of delicate aromas. Temperatures higher than 26°C should be avoided because it causes loss of volatile components and alcohol (Joshi et al., 2011a,b). Depending on temperature, the fermentation can last from several days to a few weeks. The sugar content or Brix should be measured periodically to monitor the progress of fermentation. In the case of dried fruit wines, the fermentation is allowed to proceed until all of the sugar content is consumed completely (usually a reading of approximately 8°Brix).

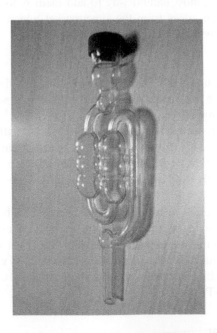

FIGURE 14.5

Fermentation plug (to be attached to a fruit wine fermentation vessel).

Some fruit wines, such as plum wine, can be prepared from plum fruit either with the fruit's skin or without skin, but the wines have considerable differences in physicochemical characteristics (Joshi et al., 2011a,b). As expected, after the fermentation with skin the color of the wines is more intense.

To avoid must spoilage with different microorganisms, potassium or sodium metabisulfite should be added before fermentation. Sanchez (1979) recommended 5 mL of 10% solution for one gallon of must. Dharmadhikari (2004) recommended the addition of SO_2 to must in the range 50–75 ppm. By sulfiting the fruit musts the glycerol formation will be improved and the vitamin C will be better preserved. Adding a high quantity of sulfite may lead to the must bleaching and the final product will become toxic (Sanchez, 1979).

3.3 YEAST INOCULATION

The fermentation process for elaboration of the beverage depends on the performance of yeast to convert the sugars into alcohol and esters. Different species of yeast that develop during fermentation determine the characteristics of the flavor and aroma of the final product. Due to the differences in fruit composition, yeasts strains used for fermentation have to adapt to different environments (e.g., sugar composition and concentrations, presence of organic acids, etc.). In addition, the applied yeast has to compete for sugar utilization with other microorganisms present in the mashes, e.g., other yeast species or bacteria, depending on the fruit of choice and varying climatic conditions. Natural grape and fruit fermentation involves a succession of yeasts, with *Saccharomyces cerevisiae* as the dominant species (Pretorius, 2000). The presence and permanence of different yeast species throughout fruit fermentation, and consequently, their influence on the final product, is determined by the fermentation conditions, such as inoculum of *S. cerevisiae* starter culture, the temperature of the fermentation, and the fruit juice composition.

Different non-*Saccharomyces* cultures have been isolated during the fruit fermentation (Pretorius, 2000; Chilaka et al., 2010; Matei et al., 2011). For example, from the indigenous fermentation gabiroba (*Campomanesia pubescens*) pulp have been isolated with species very similar to grape fermentation— respectively, *Candida*, *Issatchenkia*, and *Pichia* (Duarte, 2009).

A range of environmental factors influence the production of metabolites and survival of yeasts during industrial fermentations. The main factors are temperature, pH, sugar concentration, and acidity of fruit juice (substrate). In case of yeasts, temperature and tolerance of ethanol have an important influence on their performance.

It is very important to choose a yeast strain for a specific style of fruit wine. Fruit wines usually have a lower nitrogen content and the yeast cannot work properly; in this case, strains that work well in lower nutrient content and accentuate fresh aromatic qualities under less-than-ideal conditions should be employed (Parks, 2006).

Yeast starter culture can be purchased and used as slants or tablets, or in compressed form.

There are many commercial wine yeasts on the market, but for the fruit wines the list has been narrowed by the winemakers.

- *Lallemand 71B* may be used widely, for most off-dry fruit wines. It brings out the fresh fruitiness in most berry fruit wines.
- *Lallemand BA11* is excellent on tropical fruit wines. It helps develop the aroma and can increase mouthfeel.

- *Lallemand EC1118* is a highly used yeast, especially with wines with low pH or starting the fermentation at low temperatures. It is also recommended to ferment the fruit wines to a very dry level, or to make a sparkling wine.
- *Lallemand ICV-K1* is recommended for tree fruit wines such as apple or peach; it ferments well, no matter what the pH or temperature of the must is, and it emphasizes the freshness of the product.
- *Lallemand R2* is recommended for very sweet fruit wines, cryoextracted wines, and wines that need to ferment at low temperatures to retain the aromatic qualities.
- *Lallemand VIN13* is recommended for wines that need higher alcohol without fortification as it can ferment to almost 17% without any help. It gives good tropical notes and relatively clear flavors and can ferment under cooler temperatures.
- *Bio Springer CKS 102* is recommended for aromatic berry wines such as raspberry or delicate strawberry wines. Also works very well on tropical wines such as lychee, pineapple, and passion fruit.
- *Oenoferm Freddo* brings out the fresh fruit aroma.
- *Lesaffre UCD 595* is recommended by Joshi et al. (2011a,b) to obtain high-quality plum wines.

Yeast can be sprinkled directly on top of the must, hydrated separately, or added in a starter solution. The direct add into the must may be easiest, but it takes about two days to be certain the yeast was viable (Keller, 2010). Making a starter solution will help to know, within a reasonable time, if the yeast is viable before the must is ready for the yeast. The amount of active culture is added at a rate of 2–5% (v/v) to the must (Joshi et al., 2011a,b). However, the recommendations of the yeast producer should be followed.

Keller (2010) recommends some steps to follow to make a yeast starter: dried yeast should be added directly to 1 cup (240 mL) of 100°F (38°C) tap or spring water in a quart (liter) jar. After stirring gently, it should be covered and allowed to hydrate for at least 30 min, then should be checked to see if it is viable; it should be left for another 3–4 h for best results. During this time, allow the starter and must to adapt within 10°F (5°C) of one another, and then add to the starter 1/4 cup (60 mL) of strained must or white grape juice (not concentrate). The starter should be re-covered and set in a warm place for about 4 h left and added to another 1/4 cup (60 mL) of juice or strained must. Again should be covered and left for 4 h, then add it to the must or add another 1/2 cup (120 mL) of juice or strained must to really increase the yeast population.

Another good rehydration procedure was recommended by Monk (1986), which is given later in this chapter. Use water at 5 to 10 times the weight of the yeast. For example, for 500 g of dry yeast, use 3–5 L of water for rehydration. Rehydrate in warm water, 104° to 113°F (40–45°C). Slowly add yeast to water to obtain even hydration. Allow yeast to remain in warm water for 5–10 min before stirring (a longer duration will reduce yeast activity). The temperature difference between yeast starter and must should not be more than 18°F. To reduce the possibility of cold shock, gradually cool the starter, then add it to the must. The usual rate of inoculation is about 2 lbs./1000 gallons.

3.4 CLARIFYING/RACKING

Clarification of wines prior to bottling involves treatment with gelatin, albumin, isinglass, bentonite, potassium ferrocyanide, or salts. Alternative clarification procedures include chilling the wine prior to, or after, refining, and using microfiltration systems. A simple way to clarify wine is to add white gelatin (1 g per L of wine) to the fermented fruit solution, which is then allowed to stand in the refrigerator for 1 week, after which all of the suspended solids are precipitated and a clear transparent wine can be decanted from the top of the container. Following clarification, the wine will normally be flash pasteurized, hot-filled into bottles, or treated to give a residual SO_2 content of 100 ppm (Barbosa-Cánovas, 2003).

After fermentation, racking (decanting the top fluid portion from the residue, settled at the bottom) must be carried out at ambient temperature. The easiest way of racking is to siphon the clear wine through a plastic tube from the original container into a second clean container. Two or three rackings are usually done every 15–20 days; during the interracking period, no headspace is kept in the bottle or container—that is, the bottles are closed tightly to prevent the acetification of wine (Joshi et al., 2011a,b). Bentonite (0.04%) may be added before the final racking to remove the last remaining residues for clarification. After final racking, sodium metabisulfite (100 µg/mL as average) should be added as a preservative before bottling.

Malolactic fermentation in fruit wines is not desired since it can cause significant loss of acidity and may produce undesirable changes in flavor, resulting in high wine pH (Dharmadhikari, 2004). To discourage malolactic fermentation in fruit wines, the following provisions are suggested: maintain adequate free SO_2 levels (>0.8 molecular); store wines at cooler cellar temperatures; aim for lower pH (preferably <3.3) in the finished wine; use microorganism-proof bottling (0.45 µm filter); and stringent cleaning and sanitary measures during processing.

3.5 AGING FRUIT WINES

The fruit wines should be matured at least six months before opening the first bottle and be consumed within three or four years. Many fruit wines lack the natural acidity, alcohol content, tannin, and phenolic concentration of grape-based wines, all of which contribute appreciably to the longevity and "graceful" aging of the wine. The process of maturation is complex and the formation of esters takes place, thus improving the flavor of such beverages (Amerine et al., 1980).

As any wine, the fruit wines can be matured in an oak barrel or in other vessels by adding oak chips (Joshi et al., 2011a,b).

When aging the fruit wines in bottles, some conditions should be taken into account. The wine bottles should be stored on their side so the cork can remain moist. The wine will be kept away from sunlight as ultraviolet (UV) rays can penetrate the bottle and damage the contents. The wines should be stored in a place where the temperature is 55–75°F (13°C) and doesn't have a swing from high to low more than 10°F (~6°C) during the year.

3.6 BOTTLING FRUIT WINES

As with any product going to market, packaging and labeling are particularly important in the fruits winery industry. Generally, each country has strict regulations that apply to labeling, bottle size, and cork finish, including the packaging of waste directives, which will ensure that the environmental impact from cider and fruit wine producers is reduced even further. The US and European directives permit the active promotion of environmentally acceptable reusable packaging. In addition to protecting the product, packaging must not present a safety hazard or health risk to the consumer.

4. TRADITIONAL RECIPES OF FRUIT WINES

Fruit wines have been homemade for centuries, and it is recognized as a hobby for people having fruit trees or bushes in their yard. Worldwide, there are hundreds of local recipes for fruit wines, and the ingredients depend on the local fruits and the characteristics of the desired wine. People continue to follow their own traditions for preparing fruit wines, as well as accepting new fruits and recipes from all over the world.

For homemade wines, some guidelines have been given by Kraus (1998) on the amount of fruit to be used to produce a five-gallon batch: apricots 18 lbs.; blackberries 15 lbs.; blueberries 13 lbs.; currants 12 lbs.; elderberries 10 lbs.; gooseberries 11 lbs.; peaches 15 lbs.; pears 22 lbs.; persimmons 15 lbs.; pineapple 14 lbs.; plums 16 lbs.; raspberries 15 lbs.; strawberries 16 lbs.; watermelon 18 lbs. These recommendations can vary depending on the wine type to be obtained, sweet and high in alcohol like a dessert wine, or dry and crisp like a table wine (the difference is mainly the sugar concentration).

The homemade fruit wines require minimal equipment during the fermentation: a fermentation vessel linked by a plastic tube to another vessel with water, where the liberated CO_2 will be captured (Fig. 14.6).

The fruit wine production allows the producer to be very creative. After obtaining the wine there are many possibilities to blend it with other grape or fruit wines (strawberry or banana). Some local producers even add spices such as ginger or cinnamon, or oak chips for a barrel-aged effect and as a flavor and body enhancer. Moreover, the fruit wine may be fortified with brandy, which most commercial wineries use to fortify, or with vodka.

Depending on the types of fruit used in the homemade wines, the recipes and the process itself are slightly different. This is why in the following the traditional recipes have been presented by fruit categories, respectively, wines made from berry fruit (strawberry, blueberry, blackberry, etc.), stone fruit (apricot, peaches, plum, etc.), tropical and exotic fruit (banana, fig, etc.), citrus fruit (orange, lemon), and pome fruit (apple, pears).

FIGURE 14.6

Minimal equipment for homemade fruit wines.

4.1 BERRIES FRUITS WINES

4.1.1 Strawberry Wines

The relatively high sugar content of strawberries makes most strawberry varieties suitable for winemaking. For high-quality dessert wines, full-flavored, wild-type strawberries *Fragaria virginiana* and *Fragaria vesca* (known as Alpine strawberries) are preferred, but some other typical varieties are used, such as Mara de Bois (mainly in Europe), Purple Wonder, Albritton, Cardinal, Dunlap, Earliglow, Sparkle, Christine, Elegance, or Sweet Charlie. A mixture of strawberry varieties likely will produce better wine than a single variety (Kime, 1998). Joshi et al. (2005) have reported that the physicochemical characteristics of the cultivar Camarosa were rated superior to Chandler and Douglas cultivars. Wines from the Camarosa cultivar are found to have many desirable characteristics such as esters, optimum acidity, redder color units, alcohol, and total phenols, while Chandler cultivar had higher amount of ethyl alcohol, more phenols, and more anthocyanin than other cultivars (Sharma et al., 2009).

When preparing, strawberry wines may be used with ripe or overripe fruits, sometimes contaminated with molds. Overripe fruit with its higher anthocyanin and total phenolics gave wines better color than fully ripe fruit (Pilando et al., 1985). Because strawberry are acidic fruits, prior to the fermentation the acidity should be decreased to 0.8% by adding water and sugar; if the desired alcohol content is 10%, then more sugar should be added to the mix to bring the mixture to 20% Brix (Kime, 1998).

Jeong et al. (2006) found out that heating the must at 85°C during 10min improves the flavor compared with a 200ppm $K_2S_2O_5$ treatment. They propose to adjust the sugar to 16% of the soluble solids and to conduct the fermentation for 8days at 26°C.

Some other technology have been tested, such as thermovinification and carbonic maceration (Joshi et al., 2005; Sharma et al., 2009). Thermovinified wines had many desirable characteristics such as more total phenols, esters, and color with comparable amounts of higher alcohols, volatile acids, ethyl alcohol, sugars, and anthocyanin. The carbonic maceration resulted in wines with more alcohol, higher pH, lower acidity, lesser higher alcohol, and volatile acids than other wines.

In general, the maturation of strawberry wines improved the quality of wines and showed an effect on the contents, volatile acidity, higher alcohols, color, and esters (Joshi, 2006; Sharma et al., 2009). However, for some types the wines develop an orange color or an off-flavor (Kime, 1998).

There are many recipes worldwide for strawberry wines, depending on the kind of final product (sweet or semisweet) and on the desired quantity of the wine. Box 14.1–14.3 present some simple to more complex recipes.

A method for preparation of rose red strawberry wine has been described by Joshi et al. (2006); they propose the dilution of heated berries with 50% water at 60–65°C for 5–6min and raising the Total Soluble Solids (TSS) after crushing to 24°Brix produced wine with acceptable sensory quality.

For larger quantities Kraus (2000) has provided two homemade recipes.

4.1.2 Blueberry Wines

The wines made of blueberry and blackberry are characterized by high polyphenol content and antioxidant activities as compared to red grape wines (Ortiz et al., 2013). The blueberry wines have been recognized from the Middle Ages by the French monks as having a great health potential (*vin de myrtilles*).

BOX 14.1 RECIPE 1 FOR STRAWBERRY WINE (RECIPE FOR 1 GALLON/3.8 L, ACCORDING TO KELLER, 1998)

Ingredients:
   3 lbs. fresh strawberries
   2 lbs. granulated sugar
   2 tsp citric acid
   water to make 1 gallon
   wine yeast and nutrient
   Place all ingredients except yeast in crock; crush fruit with hands and cover with 5 pts. boiling water; stir with wooden paddle to dissolve sugar and simultaneously mash the strawberries. When cooled to 85°F, add yeast. Cover and stir daily. Strain on seventh day, transfer to secondary fermentation vessel, top up to 1 gallon, fit fermentation trap, and set aside. Rack after 30 days and again after additional 30 days. When become clear can be bottled. Allow to age 6 months to one year.

BOX 14.2 RECIPE 2 FOR STRAWBERRY WINE (RECIPE FOR 1 GALLON/3.8 L, ACCORDING TO KELLER, 1998)

Ingredients:
   3 1/2 lbs. fresh chopped strawberries
   1/4 lb. chopped golden raisins
   1/4 lb. chopped dates
   2 1/2 lbs. granulated sugar
   1 1/2 tsp. acid blend
   2 tsp. pectic enzyme
   1/4 tsp. grape tannin
   1 crushed Campden tablet
   yeast and yeast nutrient
   Place chopped fruit in nylon jelly bag, tied; place jelly bag and all other ingredients except Campden tablet, pectic enzyme, and yeast in crock and cover with 5 pts. boiling water; stir well to dissolve sugar and cover. After 2 h add crushed Campden tablet. After additional 10 h add pectic enzyme and 12 h later add yeast. Cover and stir daily. On seventh day remove jelly bag and hang over bowl to collect juice. Allow to drain thoroughly without squeezing. Pour all liquids into secondary fermentation vessel, top up to 1 gallon, fit fermentation trap, and set aside. Rack every 30 days. After third racking the wine will become clear and can be bottled. Allow to age at least 1 year.

Blueberry wines are made by fermenting fresh blueberry juice, fermenting heat-extracted blueberry juice, and carbonic maceration of crushed blueberries (Zee, 1973). Ripe blueberries can be crushed fresh or frozen, thawed, and crushed later. Fresh blueberries have a short shelf life, while the frozen fruit can be used at any time. Frozen fruit should thaw for 24 h before crushing. They aren't totally thawed when the crush starts. The Three Lakes Winery in Three Lakes, Wisconsin, recommends mixing the fruit with 70°F (21°C) water during the crush to help them run through the crusher (otherwise, the fruit gets thick and can gum up the machine). Then mix sugar with the remaining warm water and add this to the must. The goal is to get a must that is between 65° and 70°F (18–21°C) at the start of fermentation. Try to get 21°Brix in your prefermentation mix. The final wine will have an alcohol content of 11–12%.

Two recipes coming from the United States are presented in Box 14.4 and 14.5.

BOX 14.3 STRAWBERRY WINE RECIPES (RECIPE FOR 5 GALLONS/19 L, KRAUS, 2000)

Table strawberry wine (5 gallons)
12.5 lbs. strawberries
1/8 tsp. sodium bisulfite
pectic enzyme (as directed on package)
5 tsps. yeast nutrient
1 tsp. wine tannin
8 tsps. acid blend (0.60% tartaric)
8 lbs. sugar (1.078)
1 pkg. yeast

Dessert strawberry wine (5 gallons)
25 lbs. strawberries
¼ tsp. sodium bisulfite
pectic enzyme (as directed on package)
5 tsps. yeast nutrient
12 lbs. sugar (1.100)
1 pkg. yeast

If the strawberries are fresh, lightly rinse with water and allow to drain; then chop them up. If the strawberries have been frozen, thaw completely then mash them.

Take the chopped strawberries and put them into a primary vessel; add water to barely cover the strawberries. Add to this the sodium bisulfite, acid blend, and wine tannin; then add the pectic enzyme as directed on the package it came in. Let the mixture stand covered with a light towel for 24 h. It can be stirred from time to time. After waiting 24 h, it's time to dilute the winemaking liqueur with water to 5 gallons. Stir in the sugar until completely dissolved, then add the yeast nutrient and the wine yeast. Keep covered with a towel and allow to ferment. Around the seventh day of normal winemaking fermentation the activity will start to decrease. The specific gravity reading will usually be between 1.025 and 1.035 on a winemaking hydrometer. It is at this point when the wine is ready to be racked (siphoned) into a carboy leaving as much of the pulp and other sediment behind. At this point attach a wine airlock and allow the must to ferment until it has completely stopped, which will be about 4–6 weeks. At this point, rack one more time into a clean secondary container reattach the airlock to the carboy and allow to stand until the wine is completely clear. This will usually take an additional 1–2 weeks. For wines a little sweeter, now is the time to sweeten to taste. It can be sweetened with anything from table sugar to honey, but any time sugar is added to a finished wine either a stabilizer such as potassium sorbate must be added or it must be filtered with a pressurized-type filter system using "sterile" filter pads.

BOX 14.4 SUMMERTIME BLUEBERRY WINE (RECIPE FOR 5 GALLONS/19 L, FROM TAMUZZA VINEYARDS)

Ingredients:
15 lbs. (6.8 kg) blueberries
9 lbs. (4 kg) sugar
2–3 cups grape concentrate (optional; this will add to the wine's fruitiness)
2.5 tsp. acid blend
2.5 tsp. pectic enzyme
3 tsp. yeast nutrient
0.18 oz. (5 g) potassium metabisulfite (approximately 150 ppm SO_2)
2 tsp. potassium sorbate
1 tsp. tannin
yeast

Crush the blueberries. Add the water–sugar mixture and enough water to make 5 gallons (19 L). Add potassium metabisulfite. Cover and let sit for two days. Add sugar, if necessary, to reach specific gravity of 1.090. Add the tannin, acid blend, pectic enzyme, and yeast nutrient. Stir everything to blend. Maintain a constant fermentation temperature range between 70 and 75°F (21–24°C). Add yeast to the must. Stir the floating cap of fruit pulp into the fermenting must twice a day during fermentation. Fermentation will continue for approximately 14–21 days. Use a mesh bag to extract the juice from the blueberries in the must. Rack the remaining juice to a carboy, leaving the sediment (lees) behind. If possible, move the wine to a cooler place, like a basement, to clear. Rack the wine at least two more times before bottling it. Add another Campden tablet to the wine after each racking. The wine should age at least three months.

BOX 14.5 BLUEBERRY WINE RECIPE (RECIPE FOR 1 GALLON/3.8 L, ACCORDING TO KELLER, 2006)

Ingredients:

2 1/2 lb. (1.1 kg) blueberries

1 cup red grape concentrate

1 3/4 lb. (0.79 kg) granulated sugar

1/2 tsp. pectic enzyme

1 1/2 tsp. acid blend

1/2 tsp. yeast energizer

1 tsp. yeast nutrient

1/2 tsp. wine stabilizer

6 pts. (2.8 L) water

1 crushed Campden tablet

1 pkg. wine yeast

Wash and crush blueberries in a nylon straining bag and strain juice into the primary fermentation vessel. Tie the top of the nylon bag and place in primary fermentation vessel. Stir in all other ingredients except yeast, stabilizer, and red grape concentrate. Stir well to dissolve sugar, cover well, and set aside for 24 h. Add yeast, cover, and daily stir ingredients and press pulp in nylon bag to extract flavor. When specific gravity is 1.030 (about 5 days), strain juice from bag and siphon liquor off sediment into glass secondary fermentation vessel. Fit airlock. Rack in three weeks and again in two months. When wine is clear and stable, rack again and add stabilizer and red grape concentrate. Wait 3 weeks and bottle. Allow a year for aging.

4.1.3 Elderberry Wine

Elderberries make a rich, flavorful wine, but they have long been added to other fruit and berry wines, including grape, to add color, tannin, and complexity (Keller, 2009). Home winemakers may add elderberry in grape wine for a richer color and flavor (which is not allowed on industrial level). Dried elderberries do not leave the residue, described as "elderberry goo," that defies conventional clean-up methods.

Two species account for the majority of elderberry wine made in Europe and America. The European Elder (*Sambucus nigra*) grows all over in Europe, as well as in North Africa and in Asia. It has become naturalized in North America and the American Elder (*Sambucus canadensis*) grows throughout most of the United States and the eastern half of Canada.

According to Keller (2009) in the United States the most popular varieties, Adams No. 1 and Adams No. 2, are known as having good growth characteristics and large berries. They are followed by a newer variety, York, which is even more vigorous and yields bigger berries. Some other popular varieties have been reported, as Johns, Kent, Nova and Scotia. In Europe, the most cultivated varieties are Haschberg and Black Beauty.

Some caution has to be taken into account in the case of elderberry, because the stems, leaves, bark, roots, and all immature (green) berries are toxic.

Recipes of wild elderberry wines comes from the 16th century from the Scottish Highlands. Box 14.6 presents a recipe provided from the United States by Keller (2006).

4.1.4 Blackberry Wine

Blackberries are typically in season during late Summer to early Autumn months, and can be found in hedgerows all across the United States and Europe. Use fully ripe berries, fresh or frozen.

BOX 14.6 ELDERBERRY WINE (RECIPE FOR 1 GALLON/3.8 L, ACCORDING TO KELLER, 2006)

Ingredients:

- 4–5 1/4 oz. (0.11–0.15 g) dried elderberries
- 2 1/2 lbs. (1.1 kg) sugar
- 7 3/4 pts. (3.7 L) water
- 1 tsp. acid blend
- 1 tsp. yeast nutrient
- 1 crushed Campden tablet
- 1 pkg. of yeast

Boil the water with sugar; stir until the sugar is dissolved and the water is clear. Wash dried elderberries and put in nylon straining bag with several sanitized marbles for weight. Tie bag and put in primary. Pour boiling sugar water over elderberries and cover primary. When cool, stir in crushed Campden tablet, yeast nutrient, and acid blend until dissolved. Recover and set aside 12 h. Add activated yeast and ferment until specific gravity drops to 1.010, stirring and squeezing bag daily. Transfer liquid to secondary, fit airlock, and ferment to dryness. Rack every 30 days until wine clears.

BOX 14.7 RECIPE FOR BLACKBERRY WINE (RECIPE FOR 5 GALLONS, ADAPTED AFTER HERMITWOODS.COM)

Ingredients:

- 25 pounds of frozen blackberries
- 8.5–9.5 lb. cane or corn sugar
- 3 gallons of clean water
- 20 g of dry yeast
- 9.5 g of yeast nutrient added at yeast pitch and 3 days later

After mixing with sugar and water, allow the maceration for 4–6 days, then strain out berries just before bitter components appear; then ferment at 60°F; rack 2–3 times to clarify and, if possible, keep the wine under CO_2 at all times after fermentation is complete. Allow aging for 6 months and age to continue its improvement for 1–3 years depending on maceration time and/or use of oak/tannin additions.

Rommel et al. (1992) reported the most stable color from wine made of thawed fruit (Evergreen variety) by fermentation of pulp, depectinized juice, and high-temperature short-time (HTST)-treated.

Hermit Woods Vinery provides a recipe based on frozen blackberry (Box 14.7).

4.2 STONE FRUITS WINES

The category of stone fruits (drupes) includes apricots, peaches, nectarines, plums, and cherries. But some other fruits can be included in this group, like jujube, mango, sloe, or damson, cherry-plum, pluot, plumcot, and aprium. For winemaking fruits that are fully ripe and freshly picked are preferred because of their intense flavor (Keller, 2010).

When processing, it is important not to crack the stones or to remove them from the stone fruits. The reason is that most of the stones contain bitter components which will leech into the wine and some—including apricots, cherries, peaches, and plums—contain cyanogenic glycoside, which can transform into hydrogen cyanide (hydrogen cyanide is very toxic but present in extremely low concentrations in fruit stones).

The juices extracted from stone fruits lack the requisite sugar content or have poor fermentability (Joshi et al., 2011a,b). Most stone fruit wines lack body by themselves and require a second ingredient to increase the mouthfeel. Traditionally, golden or dark raisins were used for this purpose; an alternative is frozen grape concentrate. Bananas also contribute to final wine body and enhance the flavors, but can be added only on light colored fruit.

Keller (2010) provide an example of incorporating banana into a must by using one pound of ripe bananas per gallon (~120 g/L) of wine. The bananas should be peeled and sliced crosswise ¼ to ½ inch (0.64–1.3 cm) thick; the slices will be placed in a saucepan with 1 pt. of boiling water per pound of banana (~1.1 L/kg); should be simmered about 20 min and strained without squeezing; retain and cover liquid and set aside to cool; water should have reduced to about 12 oz. (360 mL).

4.2.1 Peach Wines

Compared to other stone fruits (plum or apricot), peaches have a lower acidity and it is much more pulpy, which requires them to be diluted with water (Joshi et al., 2011a,b).

The peach variety recommended to be used for wines are July Elberta, freestone Redhaven, Flavorcrest, Rich-a-haven, some old varieties such as Indian Blood and J.H. Hale or white-fleshed Georgia Belle (Keller, 2010; Joshi et al., 2011a,b; Davidovic, 2013). The peaches, no matter the variety, should be washed gently in a mild bleach solution (1 part bleach to 40 parts water) to remove dust, wild yeast, bacteria, and anything that may have been sprayed on the fruit. Rinse thoroughly twice to remove all traces of the bleach and pat dry. After bleaching the fruit will be halved and destoned and any brown spots will be cut out.

According to Joshi et al. (2011a,b) the method used for making peach wine consists of dilution of the pulp at a ratio of 1:1, increasing the initial TSS to 24°Brix, and adding pectinol and diammonium hydrogen phosphate (DAHP) at the rate of 0.5% and 0.1%, respectively; SO_2 will be added in 100 ppm.

Joshi and Shah (1998) have tested the effect of three different wood chips (*Quercus*, *Bombax*, and *Albezia)* treatment on the chemical composition and sensory qualities of peach wines. Sensory evaluation indicated that wines aged with wood chips were rated better than control and that wines treated with *Quercus* were the best.

A peach wine recipe mixed with grape concentrate or banana water provided by Keller (2010) is described in Box 14.8.

4.2.2 Nectarine Wines

A few nectarine varieties, both white-flesh or yellow-flesh, are proposed for winemaking, and they are Independence (bright red, freestone, firm yellow-flesh that is richly flavored, tangy and sweet), Merricrest (large, red-skinned, freestone, yellow-flesh and a rich, tangy flavor), Redgold (deep red, freestone, firm-golden flesh with a rich, satisfying flavor), White Freestone (red-blushed, white-flesh of excellent, sweet, juicy flavor and creamy texture), and one of best nectarine, the Snowqueen (freestone with a very large, light skin and a hint of russet blush, white-flesh, juicy, and very finely textured).

Usually the nectarine wines are made of mixture with grape concentrate or banana; famous recipes from France propose the mixture of rosé wines made of grapes with nectarine wines and strawberry wines. Box 14.9 presents a recipe proposed by Keller (2010) by adding grape concentrate or banana water.

BOX 14.8 PEACH WINE (RECIPE FOR 1 GALLON/3.8 L, ACCORDING TO KELLER, 2010)

Ingredients:

3 lbs. (1.4 kg) ripe peaches with skins (or 3.5 lbs./1.58 kg w/o skins)

12 oz. (0.34 kg) frozen white grape concentrate (or banana water)

1.75 lbs. (0.79 kg) granulated sugar

1.5 tsp. acid blend

0.5 tsp. pectic enzyme

0.25 tsp. tannin water to one gallon/3.8 L (about 3 qts./3 L)

1 crushed Campden tablet

1 tsp. yeast nutrient wine yeast

Boil the water, wash, rinse, and halve fruit, remove and discard stones, and tie fruit in nylon straining bag. Put bag in primary and mash and squeeze with hands until no solids remain except skins. When water boils, dissolve sugar in it. Pour over peaches. Add can of frozen white grape juice or banana water. When must cools, add acid blend, yeast nutrient, tannin, and finely crushed Campden tablet. Make up a starter solution for yeast. Cover primary and set aside 12 h. Stir in pectic enzyme and set aside another 12 h. Add yeast in starter solution and recover. Stir daily until vigorous fermentation begins to subside, then drip-drain pulp without squeezing. Siphon wine off sediments into secondary and fit airlock. Rack every 30 days until wine clears. Set aside two months and rack again into bottles; should be aged three months.

BOX 14.9 NECTARINE WINE (RECIPE FOR 1 GALLON/3.8 L, ACCORDING TO KELLER, 2010)

Ingredients:

4 lbs. (1.8 kg) nectarines

1.5 lbs. (0.68 kg) finely granulated sugar

12 oz. (0.34 kg) can of frozen white grape concentrate (or banana water)

1.5 tsp. acid blend

0.5 tsp. pectic enzyme

0.125 tsp. grape tannin

Water to 1 gallon/3.8 L (about 3 qts./3 L)

1 tsp. yeast nutrient

1 crushed Campden tablet

wine yeast

Boil the water, wash, destem, and destone the nectarines; cut the fruits into small pieces over a bowl, saving the juice. Pour into nylon straining bag and mash the fruit with hands, pour sugar and white grape concentrate or banana water over bag, and pour boiling water over all. Stir well with wooden spoon to dissolve sugar. When cool, add acid blend, tannin, yeast nutrient, and finely crushed Campden tablet. Cover and set aside 12 h. Add pectic enzyme, stir, recover, and set aside for an additional 12 h. Add yeast as starter solution, stir, and cover again. Gently squeeze bag twice daily to extract juice and stir. After five days of vigorous fermentation, drip-drain bag without squeezing, return drippings to primary, and discard pulp. Recover and let stand another week. Rack into secondary and fit airlock. After 30 days, rack again, top up, and refit airlock. Rack every 30 days until wine clears. Stabilize wine, add 1/4 cup simple syrup, wait 30 days, and rack into bottles. It is recommended to be aged 6–12 months.

4.2.3 Apricot Wines

The apricot is more fibrous and less juicy than most stone fruit, but that does not affect its flavor or sweetness. The juiciest varieties recommended to make fruit wine are the following: Harcot (sweet and richly flavored), Moongold (has large, plum-sized fruit that are very sweet and sprightly), Perfection (bright, yellow-orange skin and flesh and a delicious flavor), or Moorpark (rich flavor and aroma).

BOX 14.10 RECIPE FOR APRICOT WINE (RECIPE FOR 1 GALLON/3.8 L, ACCORDING TO KELLER, 2010)

Ingredients:

    4.5 lbs. (2.0 kg) chopped apricots

    12 oz. (0.34 kg) can of frozen white grape concentrate (or banana water)

    1.25 lbs. (0.57 kg) light brown sugar

    1.25 tsp. acid blend

    water to 1 gallon/3.8 L (about 3.25 qts./3 L)

    1 tsp. pectic enzyme

    0.25 tsp. grape tannin

    1 finely crushed Campden tablet

    1 tsp. yeast nutrient

    wine yeast

    Wash and chop fruit and tie in nylon straining bag; combine all ingredients in primary except pectic enzyme and yeast, stir to dissolve sugar, cover, and set in warm place for 12 h; add pectic enzyme, stir, cover, and set aside an additional 12 h. Add activated yeast in starter solution, cover, collapse bag, and stir daily until vigorous fermentation subsides. Drip-drain nylon bag into primary, pressing pulp lightly. Discard pulp, transfer wine to secondary and fit airlock. Rack, top up, and refit airlock after 30 days and again after another 60 days. When it becomes clear, it should be racked again and bottled. Should be aged one year.

Preparation of apricot wine from New Castle variety and wild apricot (*chulli*) has been reported in India by Joshi et al. (2011a,b).

A method for the preparation of wine from wild apricot has been developed by Joshi et al. (2011a,b) which consists of diluting the pulp at a ratio of 1:2, the addition of DAHP at 0.1% and 0.5% pectinol, TSS of 24°Brix. The wines prepared from 1:2 dilution were judged to be the best among those obtained from 1:1 diluted and undiluted pulps (Joshi et al., 1990). In another method, honey was used to ameliorate the wild apricot must instead of sugar. When adding wood chips, *Albezia* and *Quercus* produced wine of highest sensory qualities (Joshi et al., 2011a,b).

A homemade apricot recipe is presented in Box 14.10, adapted after Keller (2010).

4.2.4 Plum Wines

It is difficult to nominate any special plum varieties to be used for wine production because of the huge varieties existent worldwide; various cultivars of plums, which give the best wine, were evaluated by Joshi et al. (2011a,b). Generally, the variety to be employed should have sweet and juicy flesh and fine flavors.

Plum (as well as the apricot) are highly acidic and can reduce the fermentability of the juice, which can be corrected by dilution or by adding calculated amounts of potassium or calcium carbonate. In plum wine fermentation, the addition of sodium benzoate gave better quality wine than potassium metabisulfite; Joshi et al. (2011a,b) recommend 300 ppm; the addition of ammonium sulfate along with thiamine and biotin increased the rate of fermentation (Joshi et al., 2011a,b). According to Amerine et al. (1980) for every pound of plum, 1 L of water is added followed by the addition of starter culture. Because the fruits cannot be practically pressed before fermentation, the addition of pectolytic enzyme before fermentation facilitates processing by increasing the yield of juice and improving wine clarification.

> **BOX 14.11 PLUM WINE (RECIPE FOR 1 GALLON/3.8 L, ACCORDING TO KELLER, 2010)**
>
> Ingredients:
>
>   6 lbs. (2.7 kg) plums
>   1.25 lbs. (0.57 kg) fine granulated sugar
>   12 oz. (0.34 kg) can of frozen white or red grape concentrate
>   Water to 1 gallon/3.8 L (about 2.75–3 qts./2.5–3 L)
>   1.5 tsp. acid blend
>   1 tsp. pectic enzyme
>   1 crushed Campden tablet
>   0.75 tsp. yeast nutrient
>   0.25 tsp. yeast energizer
>   0.125 tsp. grape tannin
>   wine yeast
>
> Boil the water, wash the fruit, halve and remove stones, then chop fruit and tie in nylon straining bag in primary. Pour boiling water over fruit. Add sugar and stir well to dissolve. Cover and allow to cool to 70°F (21°C). Add acid blend, pectic enzyme, tannin, nutrient, and energizer, cover, and wait 12 h before adding yeast in a starter solution. Recover primary and allow to ferment 5–7 days, collapsing bag and stirring twice daily. Drip-drain bag without squeezing, transfer wine to secondary, and fit airlock. Rack after 30 days, adding finely crushed Campden tablet and topping up; refit airlock and repeat every 30 days (without adding Campden) until wine is brilliantly clear. Stabilize wine, sweeten if desired, wait additional 30 days, and rack into bottles. This wine can be sampled after only 6 months.

According to Vyas and Joshi (1982) a simple method to prepare plum wine consists in mixing 1:1 ratio fully ripened plums and water, adding 0.3% pectinol, 150 ppm SO_2, and add sugar to increase the TSS of the must to 24°Brix. After bottling and corking, they recommend pasteurization at 62°C for 20 min.

Homemade plum wine recipes recommend to add grape concentrate in the mixture. Box 14.11 is presents a recipe proposed by Keller (2010).

4.2.5 Cherry Wines

In the case of cherry fruits, the sour type of cultivars are more preferred (Joshi et al., 2011a,b). However, the cherry fruit has been found more suitable for preparation of dessert wine than a table wine. The alcohol content of such wines may range from 12% to 17%. A blend of current and table variety of cherries can also be used for winemaking.

There are a lot of cherry varieties and in winemaking can be used sweet cherries (known as *Prunus avium*) or tart cherry (know as *Prunus cerasus*). It is advisable to add in the mixtures 25–33% sweet cherries (Keller, 2010), but also sour juicy cherries may be added.

Fresh cherries can be dejuiced to make wine. In situations where fresh fruit is not available, cherry juice, syrup, or concentrate can also be used for winemaking. Using high-quality fresh fruit gives greater control of the quality of the wine and should be a preferred method. There are several ways in which cherries can be processed to obtain juice (Dharmadhikari, 2004).

Hot Pressing: Cherries should be washed, pitted, and heated in a stainless steel kettle to 140–150°F. The hot fruit should be pressed and the juice cooled to 50°F and stored for settling. Clear juice should then be siphoned off the sediment and filtered. The hot, pressed juice is darker and richer in color, but can have a canned cherry aroma instead of a fresh fruity aroma.

BOX 14.12 CHERRY WINE (RECIPE FOR 1 GALLON/3.8 L, ACCORDING TO KELLER, 2010)

Ingredients:

8 lbs. (3.6 kg) cherries

2.5 lbs. (1.1 kg) sugar

12 oz. (0.34 kg) can of frozen white or red grape concentrate (depends on juice color)

0.25 tsp. tannin

1 tsp. pectic enzyme

water to 1 gallon/3.8 L (about 3.25 qts./3 L)

1 crushed Campden tablet

1 tsp. yeast nutrient

0.25 tsp. yeast energizer

wine yeast

Boil the water, destem, wash, and crush the cherries in the primary without breaking any stones. Pour sugar over cherries. Pour the boiling water over the sugar and cherries and stir well to dissolve. Cover and set aside until cool. Add remaining ingredients and ferment 5 days. Strain juice into dark secondary, top up, attach an airlock, and discard pulp and stones. Rack every 30 days until wine become very clear, rack again, and add crushed Campden tablet. After aging two months it can be bottled. It is better to be stored in a dark place.

Cold Press: Cherries are washed, pitted, and pressed in a rack and cloth hydraulic press. The juice yield varies between 61% and 68%. The flavor is good, however, the color is not as dark as hot, pressed juice. To enhance the flavor, approximately 10% of the pits may be broken down while crushing the cherries (Joshi et al., 2011a,b).

Cold Pressing Thawed Fruit: Cherries are washed, pitted, frozen, and stored at 0°F or lower. When needed, the fruit is thawed. When the frozen fruit warms up to 45° to 50°F, it is pressed with a hydraulic press. The freezing and thawing action breaks the cells and helps in better extraction of color. The juice can be clarified by treating it with pectolytic enzyme and filtration. This process yields 60–75% juice of fresher flavor and richer color. This process yields better quality juice and, due to frozen storage, fruit can be processed at a later date at the winemaker's convenience.

The cherry wine does not require a long aging period (Joshi et al., 2011a,b).

Homemade recipes of cherry wine recommend the addition of grape concentrate (Box 14.12), as proposed by Keller (2010).

4.3 TROPICAL AND EXOTIC FRUIT WINES

Fruit wines are prepared from a variety of fruit sources whereas tropical fruit wines are mainly prepared from fruits including berries which are predominant in the tropical and subtropical regions (Chakraborty et al., 2014). The more common tropical fruits which can be used for winemaking are banana, carambola (star fruit), custard apple, cherimoya, guava, jujube, kiwi fruit, litchi, mango, melons, papaya, passion fruit, pineapple, tamarind, etc. The most popular on an indigenous level are the wines made of banana, mango, figs, and pomegranate. Their winemaking involves the same factors as making wine from fruit that is native to temperate regions. However, Keller (2014) has emphasized that the perception of people from temperate regions may not be as familiar with the inherent characteristics of the tropical fruit wines as they are with their regional favorites.

In the case of tropical fruits it is best to measure TA after the must has been constituted with crushed fruit for an hour or more—measuring after 10 h of pectic enzyme exposure is better. The principal acid in tropical fruit is citric acid, the rest is mostly malic and a very few succinic acid.

A few specific characteristics of tropical fruits should be taken into account when producing tropical fruit wines. Some tropical fruits contain enzymes that are uncommon to temperate zone fruit. For example, kiwi fruit are a commercial source for the proteolytic enzyme actinidin, useful as a meat tenderizer, but an allergen to a few (pineapple, mango, banana, and papaya also contain lesser amounts of actinidin). Figs contain ficin, an enzyme that aids digestion but also is mildly laxative. The peeling of the papaya is also a natural source of pectinase.

Generally, the tropical fruit wines do not develop a round body and require a body-building ingredient to increase the mouthfeel. Homemade recipes recommend the addition of frozen grape concentrate or banana water. Raisins may be added, but these can overwhelm the flavors of many tropical fruit. If using sultanas/raisins (golden or dark), 0.5 pound per gallon (0.23 kg per 3.8 L) of wine is usually sufficient and will also contribute 65% of their weight (5.2 ounces) as natural sugars to the must (Keller, 2014).

Usually, when incorporating banana into a must, it is recommended to use one pound of ripe bananas per gallon of wine (0.4 kg per 3.8 L). The bananas should be peeled off and sliced crosswise ¼- to ½-inch thick. The slices will be placed in 1 pt. of boiling water per pound of banana. Reduce the heat and simmer 20 min; strain without squeezing. Retain and cover liquid and set aside to cool. The water should have reduced to about 12 fluid ounces, but whatever the volume is this is the normal amount used in most tropical fruit wine recipes (Keller, 2014).

4.3.1 Banana Wine

Banana are consumed mainly as fresh as desert and its shelf life is very short, having a rapid rate of deterioration. This is why fermenting the banana juice may be an economical solution for overripe banana or surplus harvest.

Traditional recipes from Asia recommend as varieties Ae Ae (Hawaiian), Golden Aromatic (Chinese), Jamaican Red or Red Dwarf (Cuban), Mysore (Indian), Pysang Raja (Malaysian), or Raja Puri (Indian).

Akubor et al. (2003) recommend to adjust the sugar to 18°Brix, inoculate with 3% (V/V) yeast and hold at 30 ± 2°C for 14 days for a wine of about 5% (V/V) alcohol.

Polymeric carbohydrates like pectin and starch in banana cause the turbidity and viscosity of the wine and make the clarification process harder. Chersilip et al. (2008) has added to banana juice 0.05% (w/w) pectinase at 40°C for 2 h, followed by treating with 0.05% (w/w) of amylase at 50°C for 3 h and they have obtained a 2.7-fold increase in the amount of extracted juice and 39% more reducing sugar available for the yeast growth.

On an artisanal level, Keller (2006) proposes a recipe by adding also white grape concentrate (Box 14.13).

4.3.2 Mango Wines

Mango (*Mangifera indica* L.) is one of the most popular fruit choice in Asia. It contains a high concentration of sugar (16–18% w/v) and many acids with organoleptic properties, and also contains antioxidants like carotene. In India it has been developed as a research program to establish a coherent technology for mango wines as an alternative method for mango surplus (Reddy, 2005). Mangos

---

BOX 14.13 BANANA WINE (RECIPE FOR 1 GALLON/3.8 L, ACCORDING TO KELLER, 2006)

Ingredients:

4 ½ lbs. (2.0 kg) peeled bananas

½ lb. (0.23 kg) banana skins

1 ½ cups white grape concentrate

1 ¾ lbs. (0.79 kg) finely granulated sugar

2 tsps. citric acid

2 pinches grape tannin

6 1/2 pts. (3.1 L) water

1 tsp. yeast nutrient

1 pkg. wine yeast

Mash the bananas and finely chop the skins, placing both in primary fermenter. Boil the water and dissolve in it the sugar completely. Pour water over fruit and skins and cover primary fermenter. When cool, add all remaining ingredients except yeast and stir well to dissolve. Add activated yeast and recover primary. Ferment vigorously for two days and strain through muslin into secondary fermenter. Attach airlock and ferment to dryness. After 30 days, rack, top up, and refit airlock. Allow 90 days for wine to clear. If it does not clear on its own, add amylase according to its instructions. When clear, rack again, top up, and refit airlock. Age 2 months, stabilize, and refit airlock. After final 30 days, rack into bottles and allow 3 months for maturation.

---

needed for wine production should not be overripened, be free of mold and rot, and not badly damaged. Mango wine can be aged if any problems related to the "instability" of the wine are detected in time (Marshall, 2011).

Chakraborty et al. (2014) propose sugar raised to 20°Brix as a large-scale technology to add to the pulped fruits; usually 100 ppm SO_2 is used on the must, and pectinase enzyme (0.5%) is added to the pulp. The mango juice is fermented for 7–10 days at 22°C. After racking and filtration, the wine will be treated with bentonite and bottled with 100 ppm SO_2 as potassium metabisulfite. For making sweet wine, sugar should be added at the rate of 5 g/L.

According to Reddy (2005) approximately 500 mL of juice can be obtained from 1 kg of mangoes. To produce 1 L of wine, it requires about 1250 mL juice, as there would be fermentation and evaporation losses. The fruit should be washed, then peeled and pressed manually. In this recipe the final wine will reach 7–8.5% (w/v) alcohol. Reddy et al. (2009) recommend local Indian cultivars with the highest level in juice yield, like Raspur, Banganpalli or Alphonso, and Totapuri.

An artisanal mango wine recipe has been proposed by Garey (1996) as described in Box 14.14.

4.3.3 Fig Wine

In the case of figs, for winemaking only the sweetest, most tasty, almost overripe fruits should be used. Some traditional recipes recommend using the deep-frozen figs, which helps break down the tissue and get a better extraction of juices, flavors, and fermentable sugars. Keller (2006) has proposed an artisanal recipe starting from fresh figs (Box 14.15).

4.3.4 Kiwi Fruit Wine

The kiwi's flesh is firm until ripe and either bright green, yellowish-gold to brownish or even off-white. Flavor varies from sweet/tart to a sweet cross between banana and melon. Citric, galacturonic, lactic, and malic acids are the dominating organic acids in the kiwi fruit.

BOX 14.14 MANGO WINE (RECIPE FOR 1 GALLON/3.8 L, ADAPTED AFTER GAREY, 1996)

Ingredients:

3–4 lbs. fresh mango
1 lb. 13 oz. finely granulated sugar
7 1/4 pt. water
1 ½ tsp. acid blend
½ tsp. pectic enzyme
1 tsp. yeast nutrient
1/4 tsp. tannin
wine yeast

Boil the water, while peeling the mangos, cut the flesh away from the large seed, and slice and dice the flesh. Pour diced flesh in nylon straining bag, tie bag, and put in primary. Mash the flesh with the hands or a sterilized potato masher or piece of hardwood. Dissolve sugar in boiling water and pour over mashed fruit. Add acid blend, tannin, and yeast nutrient. Cover and allow to cool to room temperature. Add pectic enzyme, cover primary, and set aside for 12 h. Add yeast and recover the primary. Squeeze bag 2–3 times daily for 10 days. Drip-drain bag, squeeze gently to extract extra juice, and discard pulp (or use to make a "second wine"). Allow wine to settle overnight, then rack into secondary. Top up and fit airlock. Rack again after 30 days and again every two months for six months. Stabilize, sweeten to taste, wait 10 days, and rack into bottles. Should be aged one year.

BOX 14.15 FIG WINE (RECIPE FOR 1 GALLON/3.8 L, ACCORDING TO KELLER, 2006)

Ingredients:

4 1/2 lb. (2.0 kg) figs
6 1/2 pts. (3.1 L) water
1 3/4 lbs. (0.79 kg) granulated sugar.
3 tsp. acid blend
1 crushed Campden tablet
1 tsp. yeast nutrient
1 pkg. wine yeast

Cut off stems and chop figs. Place in large, fine mesh nylon straining bag, tie top, and put in primary fermentation vessel. Stir in all other ingredients except yeast. Check specific gravity (should be 1.085 to 1.095; if not, add up to 1/4 cup more sugar, stirring very well before rechecking gravity). Cover with sanitized cloth. Add activated yeast after 12 h and stir twice daily, pressing pulp lightly to aid extraction of juices. When specific gravity reaches 1.040 (3–5 days), hang bag over bowl to drain, lightly pressing to aid extraction. While pulp drains, siphon off liquid sediments into secondary. Add drained liquid and discard pulp. Fit airlock to secondary. Ferment to dryness (specific gravity 1.000 or lower—in about 3 weeks). Rack into clean secondary, top up to 1 gallon and reattach airlock. Rack again in 2 months. Rack again and bottle when clear. It will yield a dry wine. For a sweeter wine, it should be stabilized after last racking (but before bottling), then added 1/4 lb. dissolved sugar per gallon. Wait three weeks and bottle. Should be aged 3 months in bottle.

Kiwi fruits (*Actinidia chinensis*) are best for fresh consumption, but the surplus of nonattractive, small-sized kiwi fruit can be used for winemaking. It has been demonstrated that the yield of the juice is increased up to 75% in weight by using riper kiwi fruits and by processing them with pectolytic enzymes (Soufleros et al., 2001). A balanced kiwi wine will contain 10% vol. alcohol, more than 30 g/l sugars and 0.5 bar CO_2.

BOX 14.16 KIWI FRUIT WINE (RECIPE FOR 1 GALLON/3.8 L, ACCORDING TO KELLER, 2014)

Ingredients:

3–4 lbs. (1.3–1.8 kg) kiwi fruit

1.75 lbs. (0.80 kg) granulated sugar

12 fl. oz. (355 mL) banana water

7 pts. (3.3 L) water

0.75–1.0 tsp. acid blend

0.5 tsp. pectic enzyme

0.25 tsp. grape tannin (powder)

1 tsp. yeast nutrient

1 pkg. wine yeast

Add the granulated sugar to the water and bring the mixture to a boil, stirring occasionally until dissolved. Place the peeled, chopped fruit in nylon straining bag, tie closed, and put in the primary fermenter. Crush the bagged fruit with your hands. Add acid blend, tannin, and yeast nutrients to the primary and pour the sugar water over the fruit. Add the banana water, cover the primary, and set aside to cool. When it cools to 110°F (43°C) or below, add the pectic enzyme and stir. Cover and set aside 10–12 h. Add the yeast as a starter solution and recover the primary. Lift and dunk bag several times daily (do NOT squeeze) and stir. When vigorous (primary) fermentation slows, or on the seventh day, remove the bag and drip-drain without squeezing, returning the drippings to the primary. Allow the sediment to settle and rack into a secondary fermenter. Add 1/16 tsp. of potassium metabisulfite, stir, top up, and attach an airlock. Rack the wine again after three months, refit the airlock, and set it aside for two months or until the wine clears and the lees compact. The wine should be clear and completely dry. After 30 days, rack it into bottles and age for six months.

As a general operation on the commercial level, after washing and milling but before pressing 50 SO_2 mg/kg should be added and 100 mg kg pectic enzyme to help the pressing. After the pressing, it will be kept overnight at 45°C then filtered and the juice collected, which should be adjusted. Better fermentation should be conducted at 15°C for about 4 days (Joshi et al., 2011a).

An artisanal recipe has been proposed by Keller (2014) by adding banana water to the fresh kiwi fruit (Box 14.16).

4.3.5 Pomegranate Wines

Pomegranate fruit (*Punica granatum* L.) are usually used for fresh consumption as well as for the elaboration of juices, jams, and various other processed products. Nevertheless, large quantities of deteriorating secondary quality and overripe pomegranate fruits are often not desired by consumers and, hence, are wasted. Therefore, with the aim of minimizing production losses and generating more profits along with a sustainable use of wastes, new uses and methods for pomegranate processing have been developed. Two disadvantages should be taken into account: their relatively high price and the difficulty of peeling them and liberating the hundreds of seeds without damaging too many. The white pomegranate seeds that remain after the juice is squeezed are placed in a cold press to produce pomegranate oil, which is used as a food additive and helps reduce blood pressure in those suffering from heart disease or atherosclerosis. It also helps maintain balanced insulin levels in diabetes (Cohen, 2012).

To prepare a wine from pomegranate, the whole fruit should be pressed without crushing to avoid excessive astringency in the wine (Joshi et al., 2011a). The pomegranates only have a sugar concentration level of 12–16°Brix and for winemaking should increase the sugar to 22 and 26°Brix.

BOX 14.17 POMEGRANATE WINE (RECIPE FOR 1 GALLON/3.8 L, ACCORDING TO KELLER, 2002)

Ingredients:

    10–15 ripe pomegranates

    ½ lb. barley

    3 lb. granulated sugar

    1 lemon, juiced

    1 gallon water

    wine yeast and nutrient

    Peel the fruit and remove the seed-juice sacs from the bitter white membrane dividers. Ten fruit are sufficient if 5–6 inches in diameter; 15 are required for 3–4 inch diameters. Meanwhile, bring the water to boil with the barley in it. Simmer for about 5 min, then strain onto the pomegranate seeds, sugar, and lemon juice in the primary fermentation vessel. Stir well. When cool (70–75°F), add the activated yeast and nutrient. Cover and allow to ferment vigorously for five days, then strain into secondary fermentation jar and fit with fermentation trap. When wine clears, rack and bottle. May be aged between six months and one year.

A preservative potassium metabisulfite should be added in the must (2 lbs. per 1000 gallon). Usually the fermentation process takes longer and is done at a lower temperature than with grape juice (Cohen, 2012). To make sweet table wine, sugar is added to bring its TSS to 8–10°Brix after aging. The wine is flash pasteurized at 60°C bottled hot, and sealed (Joshi et al., 2011a).

Mena et al. (2012) have tested well-known varieties of pomegranate with good results, particularly Wonderful and Mollar de Elche for winemaking.

Keller (2002) has adapted a pomegranate wine recipe which is described in Box 14.17.

4.3.6 Cocoa Wines

Cocoa (*Theobroma cacao* L.) is known worldwide for its beans used in the manufacture of chocolate. The cocoa pulp is a substrate rich in nutrients, which can be used in industrial processes for byproduct manufacture. Reported fermentation trials on cocoa wines have been registered in Brazil (Dias, 2007).

The cocoa pulp has an average total sugar of 17%, but generally the cocoa must has a low pH. According to Dias (2007) before the beginning of the fermentative process, sucrose, pectinolytic enzymes, sulfite, and bentonite should be added to the cocoa pulp. To reach 22°Brix, 10 L of sucrose solution (350 g/L) had to be added to an equal volume of 10 L of cocoa pulp. The addition of the enzymatic complex Ultrazym AFP-L in the concentration of 0.7 ml/L favored the process of clarification of the wine of cocoa. To limit the bacterial growth, 100 mg of SO_2 should be added per liter of must (about 200 mg of $K_2S_2O_5$). To facilitate the nonfermentable solids to sediment, 1% of bentonite may be added to the must. The fermentation process takes around 50 days and after this period the beverage can be filtered and stored in glass bottles at 4°C.

4.4 CITRUS WINES

Botanists place citrus trees in the family *Rutaceae*. The fruits of citrus trees are actually a type of berry called a hesperidium. The citrus wines are mainly made from oranges, key limes, tangerines, and grapefruit. In their case only fresh juice will be employed. Another particularity of these wines is that they

---

BOX 14.18 ORANGE WINE (RECIPE FOR 1 GALLON/3.8 L)

Ingredients:
  1 pt. (~0.5 L) of freshly squeezed orange juice
  2 1/2 lbs. (1.1 kg) cane sugar
  6 1/2 pts. (3.1 L) water
  1/2 tsp. pectic enzyme
  1 tsp. yeast nutrient
  1 Campden tablet (potassium or sodium metabisulfite; crushed and dissolved)
  wine yeast

The juice will be obtained with a juice machine. The seeds or the peel are not allowed to get into the juice as this can make the wine bitter. Add water, nutrient, and Campden tablet to the juice. Stir in and completely dissolve the sugar. Cover the juice and let it set for a day. After about 18–24 h, dissolve the pectic enzyme per manufacturers' instructions and let it set for another 12 h. Add yeast and check the original gravity. Allow fermentation. It should be reached about 1.010 g/L residual sugar or below when primary fermentation is complete. After that may be added another Campden tablet. Rack the wine for secondary fermentation and allow it to ferment until dry. Wine should be racked once or twice during secondary fermentation. At this stage a little water may be added to bring the level back to a gallon, as well as a stabilizer and a clarifier. Allow it to age 4–6 months before bottling. If there is the desire to sweeten it to taste before bottling, should be added 1/2 tsp. of potassium sorbate to prevent fermentation from starting again. The wine should be aged in bottle for at least 6 months.

After Peragine, J., 2006. Citrus Wines. Wine Maker Magazine, p. 209.

---

can't be aged in oak and it is not allowed to induce malolactic fermentation. Some Florida recipes add chocolate or coffee in orange wine (Peragine, 2006).

In the case of citrus fruits the juice, squeezed together with peel, contain too much oil that consequently influences yeast fermentation. Some authors (Liou, 1986) have chosen to deacidify peeled tangerine orange juice with the addition of 10% raw peel juice to make tangerine orange wine. It was found that the quality of wine fermented with fruit juice without the peel was much better than that of wine fermented with the peel.

Presented in Box 14.18 and 14.19 and are two recipes for orange wine and grapefruit wine proposed by Peragine (2006) after visiting Florida Orange Groves Inc. and Winery, in Florida, United States. The ingredients allow for the production of 1 gallon (3.8 L) of wine.

4.5 POME FRUITS WINES

The most well-known and grown pome fruits are apples and pears, but this category also includes the quinces, and they all belong to the rose family (*Rosaceae*).

Modern archeology has proven that the Greeks and Romans knew about fermented drinks made from apples. When Romans invaded England around 55 BC, they found that cider was already being enjoyed by the locals there (Stewart, 2013). The technique of cider making has been perfected in Northern Spain, France, and Southern England. Following the invention of the screw press in the 13th century, larger scale production became possible and the growing of apple trees and making of cider spread rapidly from Western France.

According to European regulation, cider (made of apple) and perry (made of pears) are derived by the fermentation of the juices of apples or pears, respectively, without at any time adding distilled alcohol.

> **BOX 14.19 GRAPEFRUIT WINE (RECIPE FOR 1 GALLON/3.8 L)**
>
> Ingredients:
>
> 6 large grapefruit
>
> 3 lbs. (1.4 kg) cane sugar
>
> water (enough to make the final wine volume equal 1 gallon)
>
> 1/2 tsp. pectic enzyme
>
> 1/8 tsp. tannin
>
> 1 crushed Campden Tablet (potassium or sodium metabisulfite; crushed and dissolved)
>
> fining agent
>
> wine yeast (such as Sauternes)
>
> 1/2 tsp. potassium sorbate
>
> The juice will be obtained with a juice machine or by hand. The seeds or the peel are not allowed to get into the juice as this can make the wine bitter. Save a cleaned peel and juiced pulp from one of the grapefruit. Clean out the pith from the inside of the peel and add it to the juice along with half of the sugar, the Campden tablet, tannin, yeast nutrient, and water to make 1 gallon (3.8 L) of juice. Dissolve the sugar well in the mixture by stirring. Add the pulp making sure there is no pith or seeds on it. These can add bitterness and off flavors to the wine. Allow the juice to sit 12–18 h covered. Add the pectic enzyme per manufacturers' instructions. Allow the juice to sit for at least 12 more hours and then add the yeast. After 2 days of fermentation, add the remaining sugar and dissolve it well. Allow another four to 5 days of fermentation. After this, strain out the peel and pulp through a strainer and rack into secondary fermenter. Rack regularly and ferment for another 6 months. In this stage add water to bring it back to a gallon of wine. May be sweetened by adding a half cup of sugar to boiled water. Potassium sorbate must be added at this time to stabilize the wine and prevent refermentation. Bottle and allow to age at least 6 months.
>
> *After Peragine, J., 2006. Citrus Wines. Wine Maker Magazine, p. 209.*

4.5.1 Apple Wine (Cider)

Cider is a fermented alcoholic beverage made from the unfiltered juice of apples. Cider alcohol content varies, generally, between 3% and 8.5%, but some continental cider goes to 12% alcohol. In UK law, it must contain at least 35% apple juice (fresh or from concentrate). In the United States, there is a 50% minimum. In France, cider must be made solely from apples.

The flavor of cider varies, depending on the used apple variety. Ciders can be classified from dry to sweet. Their appearance ranges from cloudy with sediment to completely clear, and their color ranges from almost clear to amber to brown.

In all apple wine recipes the more acid and sour varieties are preferred. Sweeter eating varieties should be avoided. Winesap, McIntosh, Jonathans, and crab apples are best. In France the cider varieties are classified in five groups: sweet (Clos-Renaux), bittersweet (Bisquet), bitter (Cidor), acid (Judeline), and sour (Juliana) (IFPC, 2009). The apples' content in sugar allow to be obtained a maximum of 5–6% alcohol, and this is why, generally, sugar should be added.

On the commercial scale it is recommended, after washing and crushing the fruits, to add 50 ppm of sulfur dioxide and 10% water. Use of nitrogen source such as ammonium chloride or phosphate and pure yeast culture is required to produce a wine with 5.6–7.3% alcohol. The addition of DAHP improved the fermentability and made most of the physicochemical characteristics of apple wine desirable (Joshi et al., 2011a).

The cider can be still or sparkling. For the second one it is recommended not to add stabilizers (potassium sorbate and sodium metabisulfite) because it requires a secondary fermentation inside the bottles. The fermentation temperature is usually lower than in grape winemaking, respectively, between 7 and 14°C, just to keep the volatile substances of the final product. According to Mangas et al. (1994)

the main technological factors, aside from the fruit, of a good quality cider are the speed of pressing and the clarification system.

As a general remark, the cider can be aged from 6 months to 3 years; the aging depends mainly on the stabilization method (SO_2, refrigeration, or carbonation).

Homemade ciders are made in a simple way and Box 14.20 presents a French countryside recipe.

4.5.2 Pear Wine (Perry)

Perry (*poiré* in French) is an alcoholic beverage made from fermented pears, similar to the way cider is made from apples. It has been common for centuries in England; it is also made in parts of South Wales and France, especially Normandy and Anjou.

Traditional perry making is broadly similar to traditional cider making, in that the fruit is picked, crushed, and pressed to extract the juice, which is then fermented using the wild yeasts found on the fruit's skin. For higher alcohol content, sugar may be added.

In perry winemaking the pear variety is very important. In France, the most employed variety is Plant de Blanc.

On commercial scale, to prepare juice, the fruits are grated and pressed in rack and cloth press. The juice with original TSS of 14°Brix and 0.25% acidity is ameliorated to 21°Brix and 0.5% citric acid along with 100 ppm SO_2. The fermentation is carried out with a pure wine yeast culture. Adding pectic enzyme in the must will hasten the process of clarification (Joshi et al., 2011a).

Further is presented an easy homemade pear wine (Box 14.21) proposed by Combe (2013).

BOX 14.20 CIDER (FRENCH TRADITIONAL RECIPE FOR 1 L)

Ingredients:

    6 kg ripped apples

    0.2 L water

    200 g + 1.1 kg sugar

    1 kg yeast nutrient

    Chop the apples and boil them in water with 200 g sugar and the yeast nutrient added. After cooling, press, and in the obtained juice, add gradually (during 2–3 days) 1.1 kg of sugar in three doses and allow for fermentation. When starting to clarify, under sulfiting, rack the wine and bottle.

BOX 14.21 PERRY RECIPE (FOR 2 GALLON VESSEL)

Ingredients:

    3/4 gal. pears

    2 lbs. sugar

    1 1/4 gal. boiling water

    one packet or 1 tsp. of bread yeast

    First, the pears shout be cut up. Place the chopped fruit in the clean container, then put the sugar over the top. Pour the boiling water over the fruit and sugar and fill almost to the top, leaving a little room for foam during fermentation. Stir the mixture until the sugar is dissolved. Then fill a container about halfway or a little less with chopped fruit, add about a pound of sugar per gallon of container, then fill almost to the top with boiling water. After adding the water, cover the mixture and allow it to cool. Leave it overnight. When it is cool, add the yeast. After adding the yeast, gently stir or shake the must then put on the lid and/or airlock/balloon, which allow gas to escape but will keep organisms out. Put the wine in a place that is room temperature or slightly warmer and let it work for a few weeks. After 3–6 weeks it can be bottled and aged for one year.

After Combe, R., 2013. Easy Homemade Pear Wine Recipe. HubePags, Wine recipes. http://rickcombe.hubpages.com.

> **BOX 14.22 QUINCE WINE (RECIPE FOR 1 L)**
>
> Ingredients:
>   4 kg quinces
>   2 L of boiling water
>   100 g sugar
>   5 g citric acid
>   1 g nutritive salt
> The peeled and chopped quinces will be boiled in the water with the sugar, citric acid, and salt added. After cooling, yeast should be added; stir and keep 2 days. Then press, deburb, add a sugar solution (1.2 kg sugar/1 L water), and allow to ferment. After four to 6 weeks it can be clarified and bottled.
>
> *After Delambre, M., 2001. The Wine and the Cider. Ed. Alex 2000, Bucharest.*

4.5.3 Quince Wine

Quinces are hard fruits and should be very ripe before use in winemaking. Most of the varieties ripen in October (de Bourgeault, Champion, Géant de Vrania, de Leskowatz, Rea's Mammoth).

Box 14.22 presents a traditional homemade quince recipe.

5. THE FRUIT WINES IN THE MARKET

Around the world the making of "country" wine from fruit, flowers, and sometimes vegetables in the home is a popular hobby with a long tradition. It is generally accepted that fruit wines have been made since ~3500 BC, but according to the Belgian brotherhood—*Confrérie temploutoise des fabricants de vins de fruits*—the modern archeology has proven its existence before ~10,000 BC for the wines made of honey, wild apple, and pears or blackberries. Moran (2010) has emphasized that the part of the world that includes the Black and Caspian Seas along with the northern parts of what is today Iraq and Iran, is where scientists think fruit was first domesticated, and along with that, where the first wine was made.

The transition from homemade fruit wines to an industrial-level production has been slower than in the case of grape wine. Many people who have fruit trees or berry bushes in their back yard have turned those into wine instead of preparing jam or fruits pies. Larger quantities of fruit wines have been prepared mainly in farms having orchards and in the 18th century were reported in the United Kingdom as a custom to pay part of a farm laborer's wage in cider.

In terms of legislation, Europe was the first to adopt fiscal laws regarding the wines made of dried fruits, like the Belgian law from 1883. After World War I, the alcoholic beverages laws in Europe restricted the product to more than 15% alcohol, being subject to taxes; the exception was for the lower alcohol beverages like ciders and perries. During the 1970s and 1980s, professional and nonprofessional (amateur) fruit wine makers started to gain prominence.

After the financial crisis of 2007–2008, the fruit wine sector registered an interesting comeback. Wines made with fruit are gaining wider acceptance in the market (Rivard, 2009). Reports coming from 2010 place the fruit wine industry far away from the grape wine industry because fruit wines represent less than 4% of the overall production of world wine.

Canadian reports from 2008 have shown that there is a successful link between cottage wineries and tourism. Generally, a cottage winery producing fruit wine is linked to an already existing orchard that is in production. When selling fruit wines, there are two choices: onsite or by retailers. Onsite establishes a retail store at the orchard site; in this situation the location is key to attracting a customer base. Selling by retailers is when producers work with specific retailers to obtain shelf space or with chefs to promote the wines in restaurants. The major challenge in the last case is to convince a store merchant to accept the product on the shelf. Generally, if farm gate sales are successful, then they can move the product to retail outlets (AG Ventures, 2006).

The fruit wines can be sold in different ways, like in wineries, farmers' markets, local liquor stores or convenience stores, bars, festivals, and through the Internet and mail order services (Noller, 2009).

When selling fruit wine, one of the obstacles is the consumer perception that often equates fruit wines with "homemade" products, implying a lower quality product. This is why marketing initiatives, including an educational component as well as consistent standards of quality, will help change this perception. To help consumers appreciate and place value on fruit wines, it is important to conduct wine tastings, work with chefs, and attend special events (AG Ventures, 2006).

In the meantime, the increased consumer awareness of the health benefits of eating fruit could also be transferred to the fruit wine industry. Fruit contains many different dietary phytonutrients with strong antioxidant capacities such as flavonoids and phenolic acids, carotenoids, and vitamins. This antioxidant activity has been claimed to be beneficial in fighting heart disease, cancer, and the symptoms of aging. This is why several authors have reported their results on the antioxidant activity of several fruit wines (Table 14.3).

Generally, the wines prepared from berries such as elderberry, blueberry, raspberry, and cranberry had a greater total antioxidant capacity than wine prepared from pome fruits such as apple and pears. A recent example is presented by Panda (2014a,b) regarding the wine prepared from bael fruit pulp, which is a novel beverage rich in antioxidants with an alcoholic concentration of 7.87% (v/v); there is good prospects for commercialization of this wine as a medicinal wine. If such claims continue to gain mainstream acceptance they could, over time, lead to further increases in wine consumption by the general population.

In some European Union (EU) countries cider and fruit wines have enjoyed one of the fastest growth rates. Proving the growing interested in this industry, most of the European countries have

Table 14.3 Total Antioxidant Capacity and Total Phenolic Content of Different Fruit Wines

Fruit Wine	Total Antioxidant Capacity (mg AAE/L)	Total Phenolic Content (mg GAE/L)
Elderberry	1911	1753
Blueberry	1655	1676
Black currant	1595	1509
Cherry	1102	991
Raspberry	1067	977
Cranberry	875	971
Plum	618	555
Apple	404	451
Peach	395	418
Pear	271	310

Adapted after Vasantha Rupasinghe, H.P., Clegg, S., 2007. Total antioxidant capacity, total phenolic content, mineral elements, and histamine concentrations in wines of different fruit sources. Journal of Food Composition and Analysis 20, 133–137.

national stakeholders associated from 1968 in a professional organization, named European Cider and Fruit Wine Association known as AICV (in French *L'Association des Industries des Cidres et Vins de fruits de l'UE*). The association secretariat is set up in Brussels. The members of AICV commit themselves to produce cider and fruit wines on the basis of fermented fruit juice. This is manifested in the AICV Code of Practice. The affiliated members represent over 180 cider and fruit wine manufacturing companies in the EU. Most of them are relatively modest in size, although there are some large producers, mainly in England, France, Spain, Germany, Ireland, and Belgium. Over 4800 people are directly employed in the cider and fruit wine industries, but there are also many indirect jobs, mainly in the agricultural sector, through the production of apples and other fruits (http://www.aicv.org).

According to the regulations in Europe, fruit wines—"vins de fruits" in French—must be obtained by the fermentation of the juices of fruits other than grape. Fruit wines can be still or sparkling. Their alcoholic strength is permitted to be between 1.2% and 14% by volume. They can be fortified by adding distilled alcohol: in this case, the alcoholic strength may be as high as 22% by volume. The exceptions are cider (produced from apples and possibly a limited volume of pears) and perry (produced from pears and possibly a limited volume of apples) are derived by the fermentation of the juices of apples or pears without at any time adding distilled alcohol.

In France, Spain, and Belgium, cider is mainly consumed as a less alcoholic alternative to sparkling white wines. Furthermore, in France, cider is typically produced from bittersweet apples and remains firmly anchored in the country's culinary traditions where it also accompanies meals: the best chefs have no hesitation in using cider in their recipes. Sidra (name of the cider in Northern Spain) becomes more and more popular during springtime. Sweden started manufacturing cider only recently, but Finland has a long tradition of producing both cherry wines and ciders, which are more widely consumed than wines. In Germany, cider, or Apfelwein, is consumed; the product is dry, only slightly sparkling, and competes with local white wines. The Netherlands, Denmark, and Germany have a long tradition of fruit wines, mainly produced from local fruit (blackcurrants, redcurrants, strawberries, etc.). In the Eastern European countries, which have recently aligned legislation with EU laws, fruit wine is still local and made mainly from fruit berries (strawberries, blackcurrants, blackberries, raspberries) or pome fruits (apples, pears).

The total production of the members of the AICV in 2010 was over 14.5 million hectoliters, i.e., over 14 million hectoliters of cider and perry, and 0.5 million hectoliters of fruit wine. From this, 98% of total sales are made within the EU, while the rest goes outside the EU. Consumption in the EU is as follows: for cider and perry, 48% in the pubs, hotels, restaurants, and catering sector and 52% for home consumption (supermarkets, specialized shops, etc.), while for fruit wines there is home consumption only (http://www.aicv.org).

According to the *European Supermarket* trade magazine, in 2013 in the United Kingdom the cider market was growing constantly. This growth was driven by fruit, rising 35% to £245.7 million, resulting in an increase of 6% in its market share to 26%.

Fruits wine market trends are the following (Rivard, 2009): local first, 100-mile diet, all natural, low carbon footprint shoppers; hard ciders: huge growth, over 25% in the past years in the United States; fruit ciders: taking over the RTD market due to better quality perceptions; Ice fruit wines: on the Asian export markets; quality grape/fruit blends: increased consumption in the United Kingdom.

In the case of Canada, most of the fruit wineries are currently located in Ontario, British Columbia, and the Atlantic provinces and only recently in Alberta. There are a number of fruits native to Canada, which include cranberries, blueberries, strawberries, raspberries, blackberries, black raspberries, and saskatoons. The Canadian fruit industry has adapted to Canada's cool northern climate and short grow-ing season. The trend in Canada is to combine agriculture tourism components with wineries which may bring economic benefit (The Canadian Wine Industry, 2004). Imported wines are subject to import tariffs, excise tax, Goods and Services Taxes, levies, and provincial taxes. Promotion is mainly through producers' websites and fruit wine competitions.

According to ReportLinker.com, in China there were in 2013 about 100 fruit wine enterprises. Pro-duction locations spread all over China, but Henan, Shandong, Jiangsu, Anhui, and Guangdong prov-inces serve as the main areas. Apple wine processing regions are Shandong and Shaanxi provinces where the two large apple industrial bases are located. At the same time, with the growth of fruit pro-cessing technologies, some fruit wines have benefited from regional resources and are growing rapidly, such as Ningxia wolfberry wine and Zhejiang waxberry wine. From January to November in 2013, 61.2426 million kiloliters of beverage wines were produced in China, an increase of 4.29% year-on-year, of which fruit wines (the main kind is grape wine) were 1.0621 million kiloliters, only accounting for 1.73% of beverage wines. As for fruit wine production and consumption, the fruit wine industry shows good momentum in recent years along with the growth in living standards, national policy sup-port, and efforts from fruit wine producers. In 2013, China's fruit wine market scale reached CNY 112.5 billion; if the growth rate is about 13% in terms of retail market scale from 2014 to 2018, the retail market scale will exceed CNY 208.5 billion by 2018.

India has small, emerging domestic and export markets for fruit wines and large tropical fruit indus-tries. The industry produces fruit wines from tropical and temperate fruits, as well as grape wines and blends with fruit wine and juice. It should be taken into account that India has a legal drinking age of 25 and a Constitution that prohibits intoxicating drinks, so it does not have a tradition of alcoholic drinks.

A special case can be considered in Japan, having no long tradition in fruit wines production and consumption, but has an emerging fruit wine market which has stabilized (Noller, 2009). Key markets for fruit wine include young women and health conscious people. Domestic producers range from small wineries to large breweries diversifying into wine, using temperate and tropical fruits. Small quantities of tropical and temperate fruit wines are imported from a number of countries.

In Australia, fruit wine industry is emerging in major tropical fruit-growing regions of northern counties, using a wide range of tropical, exotic, and Australian native fruits. The industry has the poten-tial to emulate Australia's very large and successful grape wine industry in tropical regions. Reports (Noller, 2009) show that in Australia there are small, underdeveloped Australian markets of domestic and international tourists visiting tropical fruit wine–producing regions and people buying from local wineries; favorable demand trends and identifiable potential segments in the Australian mass market; potential segments in the Japanese market; and minor opportunities in a number of other countries.

5.1 NICHE MARKET SEGMENTS

Different studies (Noller, 2009; AICV) have highlighted that the main niche for fruit wines are young people, mostly women between 18 and 26 years of age, often with middle incomes, and often living outside large cities. The same studies noticed that the elderly are also a target because they are open to new experiences after tasting during their life a lot of grape wines.

Tourism remains one of the most important niches for the fruit wines market; the target are mainly tourists seeking organic and natural products. This sector perceives these wines as healthy and more "natural" than grape wines.

Organic food products offer new alternatives for fruit wines (Rivard, 2009) rising from the policies and public concern of sustainable agricultural development and its impact on human and animal health. The principle in organic production is the minimal inputs of chemicals; however, on an industrial scale, making fruit wines without sulfur dioxide is not a choice. For this reason there are clear regulations regarding the labeling, which should alert the consumer to the toxicity of this compound.

6. CONCLUSIONS

Fruit wines (*Vin de fruits* in French, *Vinos de frutas* in Spanish, or *Fruchtweinen* in German), known also as "country wines," have been homemade for centuries, being generally accepted that their history goes closer to 10,000 BC.

Fruit wines are fermented alcoholic beverages made of fruits other than grapes; they may also have additional flavors taken from fruits, flowers, and herbs.

Depending on the region and its specific climate are employed a high variety of fruits for winemaking and most fruits and berries have the potential to produce wine. In the United States and Canada the examples of fruits include berry and stone fruits (strawberries, plums, peaches, blackberries); in Europe, wines made of apples and pears are predominant; in Asia the wines are made tropical and subtropical fruits like banana, pomegranate, or kiwi. In Africa, traditional fermented foods include fermented beverages from indigenous fruits such as *P. curatellifolia* (sand apple, hacha), *U. kirkiana* (mazhanje), and *Z. mauritiana* (masau), as well as palm wine.

Fruit wines can be still or sparkling. Their alcoholic strength is permitted to be between 1.2% and 14% by volume. Different types of fruit wines are produced worldwide and include low alcohol "cider style," dry, or "off-dry" fruit wines (similar to grape wines), sweet fruit wine, cryoextracted fruit wines, fortified or "Port-style" fruit wines, and sparkling fruit wines.

Grape and fruit winemaking technology are similar except with some variations based on the fruit used. Grape juice is naturally suited for making wine and needs little adjustment prior to fermentation, while fruit other than grapes almost always requires adjustments.

The main steps of fruit wine technology are the following: fresh or frozen fruits reception and preliminary preparation; fruits musts extraction and preparation by crushing, pressing, clarifying, and amending; fruits musts fermentation with or without starter microorganisms; fruit wines conditioning and conservation; and aging fruit wines.

Fruit selection is a very important step in fruit wine production. The fruit should have high sugar and low acidity, which should be adjusted when needed. It is better to employ slightly overripe fruits. The amount of fruit needed per gallon/liter of obtained wine, the amount of available sugars, and the juice's acidity should be tested and adjusted.

Adding water in fruit wines is a must. The main reasons for adding water to cut down the high acidity of some fruits or to avoid a strong or astringent flavor. In many cases, the addition is needed because juice is difficult to extract directly from the fruit. Generally, the recipes to be followed advise a specific weight of whole fruit to be crushed, mashed, or cut and combined with water for a specified final volume. In nearly every such recipe, some acid will need to be added.

Before fermentation, pectic enzyme may be added, which breaks down the pectin in the fruit. This helps the wine clear when it is done fermenting. Fermentation is better to be carried out at a temperature of 4–16°C (40–60°F). This is low for most kinds of fermentation but is beneficial for fruit wines, as it leads to slower fermentation with less loss of delicate aromas. Temperatures higher than 26°C should be avoided because it causes loss of volatile components and alcohol. Depending on the temperature, the fermentation can last from several days to a few weeks. To avoid must spoilage with different microorganisms, potassium or sodium metabisulfite should be added before fermentation.

Yeast can be sprinkled directly on top of the must, hydrated separately, or added in a starter solution. In contrast to grape must, many fruits lack the nutrients necessary to sustain yeast growth. Thus, yeast nutrients such as yeast extract or diammonium phosphate may be needed up to 0.1%. Crushed fruit will have more yeast nutrients than pressed juice, so addition may be unnecessary.

Generally, the fruit wines should be matured at least six months before opening the first bottle and to be consumed within three or four years. The most common quality defects of fruit wines are excessive sweetness and oxidized flavor and color. When labeling, fruit wines are usually referred to by their main ingredient because the usual definition of wine states that it is made from fermented grape juice.

Making "country wines" is recognized as a hobby for people having fruit trees or bushes in their yard. On a homemade level, depending on the type of the fruits used, the recipes and the process itself are slightly different. The homemade fruit wines requires minimal equipment during the fermentation: a fermentation vessel linked by a plastic tube to another vessel with water where the liberated carbon dioxide will be captured.

Worldwide there are hundreds of local fruit wine recipes and the ingredients depend on the local fruits and the characteristics of the desired wine. People continue to follow their own traditions for preparing fruit wines, as well as accepting new fruits and recipes from all over the world.

The transition from homemade fruit wines to industrial-level production has been slower than in the case of grape wine. During the 1970s and 1980s, professional and nonprofessional (amateur) fruit wine makers started to gain in prominence. On the market, fruit wines have been produced on a large scale only in the past 40–50 years. After the financial crisis of 2007–2008, the fruit wine sector registered a clear comeback, and special efforts were made in this respect by marketing initiatives, including an educational component as well as developing consistent standards of quality for fruit wines.

More efforts should be made to help consumers find a place for fruit wines in their culture and to know how to appreciate these products. Meanwhile, consumers are already aware about the health benefits of eating fruits due to their antioxidant activity—a perception that could also be transferred to the fruit wine industry.

A niche market for fruit wines has been detected for young people, mostly female, and various tourism industries have identified the tourist seeking organic and natural products. Recent reports have shown that there is a successful link between cottage wineries and tourism.

REFERENCES

Akubor, P.I., Obio, S.O., Nwadomere, K.A., Obiomah, E., 2003. Production and quality evaluation of banana wine. Plant Foods for Human Nutrition 58 (3), 1–6.

Amerine, M.A., Berg, H.W., Kunkee, R.E., Qugh, C.S., Singleton, V.L., Webb, A.D., 1980. The Technology of Wine Making, fourth ed. AVI, Westport, CT.

Amoa-Awua, W.K., Sampson, E., Tano-Debrah, K., 2007. Growth of yeasts, lactic and acetic acid bacteria in palm wine during tapping and fermentation from felled oil palm (*Elaeis guineensis*) in Ghana. Journal of Applied Microbiology 102, 599–606.

AG Ventures, 2006. Commercial cottage wine industry: fruit wine. Agriculture Business Profiles Agdex 200/830–1.

An Introduction to the Fruit Wine Business in Alberta, 2008. Agri-Facts. AgDex 200/820–1.

Barbosa-Cánovas, G., Fernández-Molina, J., Alzamora, S., Tapia, S., López-Malo, A., Chanes, J.W., 2003. Handling and Preservation of Fruits and Vegetables by Combined Methods for Rural Areas. Technical Manual. FAO Agricultural Services Bulletin. 149.

Bates, R.P., Morris, J.R., Crandall, P.G., 2001. Principles and Practices of Small- and Medium-Scale Fruit Juice Processing. FAO Agricultural Services Bulletin. 146.

Brandon, S.L., Ferreiro, J.D., 1998. World market for non-citrus juices. Journal of Food Quality 13 (6), 395–398.

Chakraborty, K., Saha, J., Raychaudhuri, U., Runu Chakraborty, R., 2014. Tropical fruit wines: a mini review. Natural Products-An Indian Journal 10 (7), 219–228.

Cheirsilp, B., Umsakul, K., 2008. Processing of banana-based wine product using pectinase and α-amylase. Journal of Food Process Engineering 31 (1), 78–90.

Chilaka, C.A., Uchechukwu, N., Obidiegwu, J.E., Akpor, O.B., 2010. Evaluation of the efficiency of yeast isolates from palm wine in diverse fruit wine production. African Journal of Food Science 4 (12), 764–774.

Cohen, A., September 21, 2012. Israeli family takes pomegranate wine to heady heights. Haartez Journal.

Combe, R., 2013. Easy Homemade Pear Wine Recipe. HubePags. Wine recipes. http://rickcombe.hubpages.com.

Davidović, S.M., Veljović, M.S., Pantelić, M.M., Baošić, R.M., Natić, M.M., Dabić, D.Č., Pecić, S.P., Vukosavljević, P.V., 2013. Physicochemical, antioxidant and sensory properties of peach wine made from Redhaven cultivar. Journal of Agricultural and Food Chemistry 61 (6), 357–1363.

Delambre, M., 2001. The Wine and the Cider. Ed. Alex 2000, Bucharest.

Dewez, B., 1998. Confrérie Temploutoise des fabricants de vins de fruits. Cours de Formation.

Dias, D.R., Schwan, R.F., Freire, E.S., dos Santos Serodio, R., 2007. Elaboration of a fruit wine from cocoa (*Theobroma cacao* L.) pulp. International Journal of Food Science and Technology 42, 319–329.

Dharmadhikari, M., 2004. Wines from cherries and soft fruits. Vineyard & Vintage View, Mt. Grove, MO 19 (2), 9–15.

Duarte, W.F., Dias, D.R., de Melo Pereira, G.V., Gervásio, I.M., Schwan, R.F., 2009. Indigenous and inoculated yeast fermentation of gabiroba (*Campomanesia pubescens*) pulp for fruit wine production. Journal of Industrial Microbiology and Biotechnology 36, 557–569.

Duarte, W.F., Dias, D.R., Oliveira, J.M., Teixeira, J.A., de Alemeida e Silva, J.B., Schwan, R.F., 2010. Characterization of different fruit wines made from cacao, cupuassu, gabiroba, jaboticaba and umbu. LWT – Food Science and Technology 43, 1564–1572.

Garey, T., 1996. The Joy of Home Wine Making. Kindle Edition.

Hohenstein, J., 2005. Cottage Wine Industry, the World of Wine. Quality Production for a New Industry. http://hermitwoods.com/recipes/.

IFPC (Institut Francais de Production Cidricole), 2009. Pomme a Cidre. Les Variétés. Infos Techniques IFPC, France.

Jeong, E.-J., Kim, Y.-S., Jeong, Do-Y., Shin, D.-H., 2006. Yeast selection and comparison of sterilization method for making strawberry wine and changes of physicochemical characteristics during its fermentation. Korean Journal of Food Science and Technology 38 (5), 642–647.

Joshi, V.K., Attri, D., Singh, T.K., Abrol, G.S., 2011a. In: Handbook of Enology: Principles, Practices and Recent Innovations. Asiatech Publishers Inc.

Joshi, V.K., Rakesh, S., Ghanshyam, A., 2011b. Stone fruit: wine and brandy. In: Hue, et al. (Ed.), Handbook of Food and Beverage Fermentation Technology. CRC Press, Florida.

Joshi, V.K., Sharma, S., Kumar, K., 2006. Technology for production and evaluation of strawberry wine. Beverage and Food World 33 (1), 77–78.

Joshi, V.K., Sharma, S., Bhushan, S., 2005. Effect of method of preparation and cultivar on the quality of strawberry wine. Acta Alimentaria 34 (4), 339–353.

Joshi, V.K., Shah, P.K., 1998. Effect of wood treatment on chemical and sensory quality of peach wine during aging. Acta Alimentaria Budapest 27 (4), 307–318.

Joshi, V.K., Bhutani, V.P., Sharma, R.C., 1990. The effect of dilution and addition of nitrogen source on chemical, mineral, and sensory qualities of wild apricot wine. American Journal of Enology and Viticulture 41 (3), 229–231.

Kime, R., 1998. In: Pritts, M., Handley, D. (Eds.), Strawberry Wine Production, Strawberry Production Guide, NRAES-88 162.

Keller, J., 1998. http://winemaking.jackkeller.net/reques5.asp.

Keller, J., 2002. http://winemaking.jackkeller.net/pomegran.asp.

Keller, J., 2005. Grape/Non-grape Blending. Wine Maker Magazine, p. 326.

Keller, J., 2006. Grape/Can't-Miss Country Wine Recipes. Wine Maker Magazine, pp. 2–10.

Keller, J., 2009. Elderberry Wine. Wine Maker Magazine, p. 841.

Keller, J., 2010. Stone Fruit Wines. Wine Maker Magazine, p. 937.

Keller, J., 2014. A Taste of the Tropics. Wine Maker Magazine, p. 1422.

Kraus, E., 1998. There's More to Wine Than Just Grapes. Wine Maker Magazine, p. 688.

Kraus, E., 2000. Take a Sip of Strawberries. Wine Maker Magazine, p. 652.

Liou, G.D., 1986. Make an experiment of fermentation MikanBouya wine. Special Topics on Science & Technology of Alcoholic Beverages 8, 228–231.

Marshall, E., Mejia, D., 2011. Traditional Fermented Food and Beverages for Imporved Livelihoods. Diversification booklet number 21. Rural Infrastructure and Agro-Industries Division – FAO, Rome, Italy.

Mangas, J.J., Cabranes, C., Moreno, J., Gomis, D.B., 1994. Influence of cider-making technology on cider taste. LWT – Food Science and Technology 27 (6), 583–586.

Matei, F., Brinduse, E., Nicoale, G., Tudorache, A., Razvan, T., 2011. Yeast biodiversity evolution over decades in Dealu Mare-Valea Calugareasca vineyard. Romanian Biotechnological Letters 16 (No.1 S), 113–120.

Mena, P., Gironés-Vilaplana, A., Marti, N., García-Viguera, C., 2012. Pomegranate varietal wines: phytochemical composition and quality parameters. Food Chemistry 133 (1), 108–115.

Monk, P.R., 1986. Rehydration and propagation of active dry yeast. Australian Wine Industry Journal 3–5.

Moran, L., 2010. Fruit Wine Report in Vine and Wine Program of the AgInfo Network. http://www.aginfo.com/index.cfm/report/id/Vine-to-Wine-16558, June 07, 2010.

Noller, J., Wilson, B., 2009. Markets for Tropical Fruit Wine Products. RIRDC Publication No. 09/033.

Nyanga Loveness, K., Nout, M.J.R., Gadaga, T.H., Theelen, B., Boekhout, T., Zwietering, M.H., 2007. Yeasts and lactic acid bacteria microbiota from masau (*Ziziphus mauritiana*) fruits and their fermented fruit pulp in Zimbabwe. International Journal of Food Microbiology 120, 159–166.

Ortiz, J., Marín-Arroyo, M.-R., Noriega-Domínguez, M.-J., Navarro, M., Arozarena, I., 2013. Color, phenolics, and antioxidant activity of blackberry (*Rubus glaucus* Benth.), blueberry (*Vaccinium floribundum* Kunth.), and apple wines from Ecuador. Journal of Food Science 78 (7), C985–C993.

Pilando, L., Wrolstad, R.E., Heatherbell, D.A., 1985. Influence of fruit composition, maturity and mold contamination on the color and appearance of strawberry wine. Journal of Food Science 50 (4), 1121–1125.

Panda Sandeep, K., Sahu Umesh, C., Behera Sunil, K., Ray Ramesh, C., 2014a. Fermentation of sapota (*Achras sapota* Linn.) fruits to functional wine. Nutra Foods 13, 179–186.

Panda Sandeep, K., Sahu Umesh, C., Behera Sunil, K., Ray Ramesh, C., 2014b. Bio-processing of bael (*Aegle marmelos* L.) fruits into wine with antioxidants. Food Bioscience 5, 34–41.

Parks, B., 2006. Country Wine Yeast: Tips From the Pros. Wine Maker Magazine, p. 857.

Peragine, J., 2006. Citrus Wines. Wine Maker Magazine, p. 209.

Peak, B., 2013. Country Wine Acids: Techniques. Wine Maker Magazine, p. 1255.

Pomohaci, N., Cioltean, I., Vişan, L., Rădoi, F., 2002. Tuica (Romanian Plum Brandy) and Other Natural Spirits. Ed. Ceres, Bucureşti.

Pretorius, I.S., 2000. Tailoring wine yeast for the new millennium: novel approaches to the ancient art of winemaking. Yeast 16, 675–729.

Reddy, L.V.A., Reddy, O.V.S., 2009. Production, optimization and characterization of wine from Mango (*Mangifera indica* L.). Indian Journal of Natural Products and Resources 8 (4), 426–435.

Reddy, L.V.A., Reddy, O.V.S., 2005. Production and characterization of wine from mango fruit (*Mangifera indica* L.). World Journal of Microbiology & Biotechnology 21, 1345–1350.

Rivard, D., 2009. The Ultimate Fruit Winemarkers' Guide. Bacchus Enterprises Ltd.

Rommel, A., Wrolstad, R.E., Heatherbell, D.A., 1992. Blackberry juice and wine: processing and storage effects on anthocyanin composition, color and appearance. Journal of Food Science 57 (2), 385–391.

http://www.reportlinker.com/p02029306/Research-and-Development-Trend-of-China-Fruit-Wine-Market-2014-2018.html#utm_source=prnewswire&utm_medium=pr&utm_campaign=Wine.

Sanchez, P., 1979. The prospects of fruit wines in Phillipines. Journal of Crop Science 4 (4), 183–190.

Sharma, S., Joshi, V.K., Abrol, G., 2009. An overview on Strawberry [*Fragaria × ananassa* (Weston) Duchesne ex Rozier] wine production technology, composition, maturation and quality evaluation. Indian Journal of Natural Products and Resources 8 (4), 356–365.

Soufleros, E.H., Pissa, I., Petridis, D., Lygerakis, M., Mermelas, K., Boukouvalas, G., Tsimitakis, E., 2001. Instrumental analysis of volatile and other compounds of Greek kiwi wine; sensory evaluation and optimisation of its composition. Food Chemistry 75 (4), 487–500.

Stewart, A., 2013. The History of Cider Making. UTNE Redears-Arts.

Stringini, M., Comitini, F., Taccari, M., Ciani, 2009. Yeast diversity during tapping and fermentation of palm wine from Cameroon. Food Microbiology 26 (4), 415–420.

The Canadian Wine Industry, September 2004. Agriculture and Agri-Food Canada, p. 1035.

Vasantha Rupasinghe, H.P., Clegg, S., 2007. Total antioxidant capacity, total phenolic content, mineral elements, and histamine concentrations in wines of different fruit sources. Journal of Food Composition and Analysis 20, 133–137.

Vyas, K.K., Joshi, V.K., 1982. Plum wine making: standardization of a methodology. Indian Food Packer 36 (6), 80.

WineMaker magazine, June/July 2007. https://winemakermag.com/311-fruit-and-grape-blends-tips-from-the-pros.

Zee, J.A., Simard, R.E., 1973. Evolution of organic acids during the fermentation of lowbush blueberry in wine making. American Journal of Enology and Viticulture 24 (3), 86–90.

Zhang, Q.-A., Fan, X.-H., Zhao, Wu-Q., Wang, X.-Y., Liu, H.-Z., 2013. Evolution of some physicochemical properties in *Cornus officinalis* wine during fermentation and storage. European Food Research and Technology 237, 711–719.

Index

'*Note*: Page numbers followed by "f" indicate figures, "t" indicate tables and "b" indicate boxes.'

A

AAB. *See* Acetic acid bacteria (AAB)
ABB. *See* Air bubble column bioreactor (ABB)
Abiotic depletion (AD), 588
Absolute density, 236
Absorption rate, 198–199
Acetic acid, 92, 119–120
  fermentation, 132–133, 132f
Acetic acid bacteria (AAB), 92, 243
  fruit wine spoilage by, 142
Achras sapota Linn. *See* Sapota (*Achras sapota* Linn.)
Acid adjustment, 302
Acidification, 588
Acidity, 231–234
  adjustment in fruit wines, 669–670, 670t–671t
  total, 233, 233t
  volatile acidity, 233–234, 234f
ACs. *See* Aseptic cartons (ACs)
Actinidia chinensis. *See* Kiwi fruits (*Actinidia chinensis*)
AD. *See* Abiotic depletion (AD)
Adenosine triphosphate (ATP), 585–586
Adsorption method, 86–87
Advanced oxidation processes (AOPs), 583–585
Aeration, 275
Aerobic microbial treatment technologies, 577–582
  application, design, long-term operation, and performance of pilot-plant MABR, 579–580
  experimental conditions, 580–581
  organic removal, 581–582
  of wastewater, 578f
Aerobic/anaerobic biological systems, combined, 583–585
  types for brewery wastewater treatment, 585f
AF. *See* Alcoholic fermentation (AF)
AFL. *See* Aflatoxin (AFL)
Aflatoxin (AFL), 96
Aging
  of apple wine, 312
  of cider, 328–329
    biochemical changes during aging, 329
  vermouth, 494
Agricultural waste, 601
Agricultural wines, 463
  cocoa wine, 481–482
  mahua wines, 463–471
  mead wines, 471–474
  regulations for making, 482–483
  Rhododendron wine, 474–476

sweet potato wine, 477–478
tomato wine, 478–479
whey wines, 480–481
AICV, 696–698
Air bubble column bioreactor (ABB), 577
Alcoholic fermentation (AF), 6, 79, 106–107, 108f, 421
  yeast population development during, 93–94, 94f
  yeast strains used for fruit wine fermentation, 107t
Alcohols, 178–179, 196, 531
  beverages, 531
  removal from fruit wines, 268–269
  strength, 237
Aldehydes, 182–183
Alkalinity of ash, 234–235
Allicin, 403
Alpine strawberries, 677
Amelioration of cider juice, 324
American Elder (*Sambucus canadensis*), 680
Amines, 140–141, 140t
Amino acid
  composition of wine apples, 300
  utilization profile, 127–128
Ammonia, 601
Amphora production, 619–620
Anacardium occidentale. *See* Cashew tree (*Anacardium occidentale*)
Anaerobic microbial treatment technologies, 582
  of wastewater, 583f
Anaerobic sequencing batch reactor (ASBR), 582
Ananas comosus. *See* Pineapple (*Ananas comosus*)
Annurca apple (*Malus pumila* L.), 545
Anthocyanins, 189–191, 189t, 259, 418, 420f, 421–422
  anthocyanin-derived compounds, 189–191
  bioavailability, 198–200
    absorption rate and main forms detected, 198–199
    metabolic transformations and excretion, 199–200
  pigments, 389–390
Antiinflammatory effect, 210–211
Antimicrobials, 133
Antioxidant(s), 152–153, 195–196
  activity, 447
    of Mandarin wine, 434–435, 435f
  capacity of fruit wines, 151, 152t

Antioxidative effects, 206–212
  antiinflammatory effect, 210–211
  cumulative antioxidative effect, 211–212
  differences in antioxidative efficiency, 208–210
  direct antioxidative effect on targeted tissue, 210
  free radical depletion mechanisms, 207
  methods for estimation of antioxidative potential, 208
AOC. *See* Appelation Controle (AOC)
AOPs. *See* Advanced oxidation processes (AOPs)
Apigenin, 421
Apis mellifera (*A. mellifera*), 473
Apparent density, 236
appassimento technique, 638
Appelation Controle (AOC), 539–540
Apple, 13–15, 13f–15f, 296
  brandy, 539–541
    effect of year for apple native Sardinian varieties, 541f
    methanol content for different fermentation yeasts, 542f
    Sardinian varieties *vs.* Trentino varieties, 540f
  cultivar influence in methanol in cider, 331t
  eau de vie, 330
  juices
    ester concentrations, 334t
    fermentation process, 296
  pomace, 541, 560, 560t, 607
    use as feedstock, 611, 611t
Apple wine, 8, 15, 49, 296–312, 693–694, 694b
  aging/maturation, 312
  apple tea wine, 312–313
  bitterness, 300t
  choice of fruit, 296–297
  clarification, 312
  composition of fruit, 298–301, 298t
  fermentation
    behavior, 307f
    of must, 310–312
  inoculation with yeast culture, 309–310
  manufacturing, 297f
  with medicinal value, 313–315
  pectinesterase addition effect, 303f, 304t
  phenolic compounds, 300t
  physicochemical characteristics, 304f
    changes in, 308t
  processing apples for juice, 301–306
    adjusting sugar and acid, 302
    grinding, 301
    juice treatment, 301
    must clarification, 302
    nitrogen source, 302, 303f
    pressing, 301
    sulfur dioxide, 304
    temperature, 305
    yeast, 305–306

sensory qualities, 312
  SO_2 and temperature effect, 305t
Apricot brandy, 545–546
Apricot wine, 683–684, 684b. *See also* Cherry wines; Peach
    wine; Plum wine
  fruit and cultivar, 368
  of initial TSS level effect, 372t
  maturation, 368–370, 369t
  method, 368
  physicochemical composition, 370t
  wine preparation, 371f
Apricots (*Prunus armeniaca* L.), 15, 15f, 348, 545
Arbor Mist style, 665
Aroma compounds, 153–157, 154t, 156t, 534
Aroma-active compounds of orange wine, 417t
"Aromatized fortified wine", 487
Aronia berries, 548
ASBR. *See* Anaerobic sequencing batch
    reactor (ASBR)
Ascorbic acid, 384–386, 423
Aseptic cartons (ACs), 588–589
Asparagine, 300–301
Astringent compounds in apple fruit, 299–300
Atherosclerosis signaling pathways, influence on,
    205–206
ATP. *See* Adenosine triphosphate (ATP)
Australia, status of fruit wines, 57
Authenticity, winemaking, 652–653
Autolysis process, 511–512

B
Bacterial cellulose (BC), 568, 568t
  production, 569f
Bactericidal yeasts, 31–32
Bahal, 565
Bahalina, 565
Banana, 16, 16f
  brandy, 552
  wine, 53, 687, 688b
Barrels, 3–5
Bartlett, 315
BATNEEC. *See* Best available technology not entailing
    excessive cost (BATNEEC)
BC. *See* Bacterial cellulose (BC)
Benomyl fungicide, 363–365
Bentonite, 310–312, 675
Benzaldehyde, 534
Berries fruits wines. *See also* Citrus wines; Tropical fruit,
    wine from
  blackberry wine, 680–681
  blueberry wines, 677–678
  elderberry wine, 680
  strawberry wines, 677

Berry, 16–17, 16f, 382–383
  berry-based alcoholic beverages, 384
  blended passion fruit, 405–406
  brandies, 382
  chemical composition of berries, 548t
  fruit brandy, 547–549
  lychee wine, 403–404
  methods of preparation of table wine, 384–403
    blackberry jamun wine, 395–396
    fermented garlic beverage, 403
    persimmon wine, 401
    pumpkin wine, 392
    pumpkin-based herbal wine, 392–395
    red raspberry wine, 389–390
    red wine, 398–401
    sea buckthorn wine, 390–392
    strawberry wine, 384–387
  papaya wine, 404–405
  peels, pulp, and Seeds, 564
  production of berry wine
    composition and maturity of fruits, 384
    problems of, 383–384
    raw materials, 384
  types of berry fruits and developmental stages of
        strawberry, 383f
Berry wine, 177–178, 178t
  composition, 178–185
    alcohols, 178–179
    aldehydes, esters, and volatile constituents, 182–183
    carotenoids, 184
    dietary fiber, 185
    histamine presence, 185
    minerals, 184–185
    organic acids, 181–182, 181t
    sugar content, 180–181
    vitamins, 183–184
  enzymatic transformations, 196–197
  health benefit potential, 203–214
  nutritional facts, 195–196
  phenolic compounds, 186–195
Best available technology not entailing excessive cost
        (BATNEEC), 585
Bilberry anthocyanins, 199
Bio Springer CKS 102, 674
Bioavailability
  of anthocyanins, 198–200
  of flavonoids, 200–202
  of health benefit components of fruit wines, 197–202
Biochemical oxygen demand (BOD), 607–608
Bioconversion, 600
Bioethanol, 611
Biofilm characteristics, 581
Biological activity of nongrape fruit wines, 151–153

Biological sand filter (BSF), 577
Biological treatment
  of food waste, 609–611
  processes, 577–585
    aerobic microbial treatment technologies, 577–582, 578f
    anaerobic microbial treatment technologies, 582
    combined aerobic/anaerobic biological
        systems, 583–585
Biologically aged wines, 138–139
Biomass, 600
Biopolymer production, 569–570
Bioreactors, 577–585
  aerobic microbial treatment technologies, 577–582, 578f
  anaerobic microbial treatment technologies, 582
  combined aerobic/anaerobic biological systems, 583–585
Biorefinery, 600–602
  applicability to fruit winery waste, 608, 609f
  concept applicability to fruit winery waste, 608, 609f
  feedstock, 605–608
    fruit winemaking process, 605–607
    generation of fruit winery waste, 607–608
  use of waste streams within, 604f
Biotechnological conversion of FWW
  case studies, 609–611
  use of apple pomace as feedstock, 611, 611t
  use of mango waste as feedstock, 612
Biowaste constituents, 558
Bitterness of apple wine, 300t
Bittersweet, 300
Black raspberry (*Rubus occidentalis*), 382
Blackberry (*Rubus* sp.), 16–17, 382
Blackberry jamun wine, 395–396. *See also* Jamun wine
  dilution
    on physicochemical characteristics, 397t
    on sensory characteristics, 398t
  maturation on physicochemical characteristics, 398t
  preparation, 397f
  sensory characteristics, 396
Blackberry wine, 151, 203–204, 680–681
  recipe for, 681b
Blended passion fruit, 405–406
Blending in stone fruits, 352
Blueberry (*Vaccinium corymbosum*), 382
  wines, 677–678
    recipe, 680b
    summertime, 679b
BMR. *See* Membrane bioreactor (BMR)
BOD. *See* Biochemical oxygen demand (BOD)
Bottle fermentation, 509–511
  factors affecting, 510–511
    changes in titratable acidity and pH, 511f
    changes in TSS and ethanol concentration, 510f
  method, 502

Boukha, 10
Brandies, 382, 533
   distillation, 491
   factors affecting quality, 534
BSF. *See* Biological sand filter (BSF)
Budding, 94
Bulk method, 504

C

Caffeic acid, 418–421
Calvados, 539–540
Campomanesia pubescens. See Gabiroba (*Campomanesia pubescens*)
Canada, status of fruit wines, 56
Candida albicans (*C. albicans*), 214
Candida pulcherrima (*C. pulcherrima*), 356–357
CAR. *See* Carboxen (CAR)
Caramel, 491
Carbohydrates in apple fruit, 298
Carbon dioxide (CO_2), 487
Carbon metabolism, 109–125. *See also* Nitrogen metabolism
   citric acid cycle, 112–113
   decomposition of organic acids, 122–123
   end products of sugar metabolism, 116–117
   glycolysis, 109–111
   metabolic pathways of LAB, 124–125
   metabolism of pyruvate, 111, 112f
   regulation in yeast, 115–116
   types of fermentation, 113–115
   volatile and nonvolatile organic acids, 118–121
   yield of ethanol and by-products, 115
Carbonation, 647
Carbonic maceration process, 631
Carbonyl compounds in cider, 332
Carboxen (CAR), 241
Cardioprotective potential, 203–206
   enhancement in nitric oxide production, 206
   influence on atherosclerosis signaling pathways, 205–206
   inhibition of platelet aggregation, 204
   prevention of LDL oxidation, 204–205
   vasodilatory effect, 203–204
Carotenoid composition, 447
Carotenoids, 184
CAS. *See* Conventional activated sludge (CAS)
Cashew apple, 451–452, 452t, 453f
Cashew nut, 17, 17f
Cashew tree (*Anacardium occidentale*), 17
Casse, 337
Catechins, 201
CC. *See* Coconut coir (CC)
Cell immobilization technology, 449–450
Centrifugal pumps, 276

CFMF. *See* Cross-flow MF (CFMF)
Champagne, 487
Champagne cider, 318–319
Charmat process, 500–501, 504. *See also* Bulk method
   sparkling wine production by, 506f
Chemical engineering
   alternative methods for fruit juice extraction, 254–262
   membrane technologies applying to fruit winemaking, 263–275, 266f
   preservation processes applicable to wine production, 279–287
   racking process and transport of wine, 275–279
Chemical oxygen demand (COD), 607–608
Cherry, 17–18, 17f, 348
   brandy, 542–543
   Cherry Kijafa, 49
Cherry wines, 49, 685–686, 686b. *See also* Apricot wine; Peach wine; Plum wine
   fruit and cultivar, 370
   method, 370–376
   minerals in, 377t
China, status of fruit wines, 55
Chinese gooseberry. *See* Kiwifruit
Chinese plums. *See* Loquats
Chips, 3–5
Chlorophyll fluorescence, 630
Chokeberries, 548
Chromatographic analysis, 237–242. *See also* Microbiological analysis; Physicochemical analysis
   GC, 241–242
   liquid chromatography, 237–241
Chromatography, 237
Cider, 10, 15, 318–338
   cider-style wines, 605
   clarification, 328
   definition and characteristics, 318–319
   ester concentrations, 334t
   fermentation, 80
   flavor wheel, 335f
   flavor-affecting factors, 319, 319t
   lees, 570–574
      brewer's yeast biomass valorization, 571f
      characteristics of crude cider lees, 571–572
      valorization of wine lees, 572–574
   mean polyphenolic contents and antioxidant capacity, 336t
   production, 258f
      immobilized cells, 90
      LCA of waste management in, 588
      MW heating of fruit mash as pretreatment technique, 257–259
      regulating pressing conditions, 259–260
      treatment of high-strength wastewater from, 579–580

production technology, 322t
  aging/maturation and secondary fermentation, 328–329
  amelioration of juice, 324
  biochemical changes during aging, 329
  cider making methods, 320–321
  controlling microorganisms before fermentation, 323–324
  fermentation of must, 327–328
  final treatment and packaging, 329
  inoculation with yeast culture, 324–326
  milling and pressing, 321–323
  progress of lactic acid bacteria, 325t
  required raw materials, 321
  temperature, 326
  yeast, 326–327
quality
  chemical composition, 329–333
  sensory quality, 334–337
spoilage of cider, 337–338
Citric acid, 121, 121f, 134t, 232, 491
  cycle, 112–113
Citrus fruits, 33
  waste, 563
Citrus paradis. See Grapefruit (*Citrus paradis*)
Citrus peels, 562–563
Citrus sinensis. See Orange (*Citrus sinensis*)
Citrus solid wastes, 563
Citrus wines, 54, 410, 691–692. *See also* Berry; Tropical fruit, wine from
  comparison of flavor profiles of, 432f
  mandarin wine, 424–435
  orange wine, 410–424
Clementine Mandarin wine, 430
  volatile flavor of, 431t
Cocoa (*Theobroma cacao* L.), 481, 691
  composition of cocoa pulp, 482
  wine, 482
   alcoholic beverage production, 483f
Coconut coir (CC), 565–568
Coconut palm (*Cocos nucifera*), 454
Coconut waste. *See also* Fruit wine wastes (FWW)
  CC, 565–568
  coconut wine, 565
Coconut wine, 454, 565
Cocos nucifera. See Coconut palm (*Cocos nucifera*)
COD. *See* Chemical oxygen demand (COD)
Cognitive support, 213
Cold presoaking. *See* Maceration
Colored pigments, 477
Composition
  of apple fruit, 298–301, 298t
   astringent compounds, 299–300
   carbohydrates, 298
   nitrogenous compounds, 300–301

   organic acids, 299
   pectic substances, 299
   starch, 298–299
   water, 298
  berry wine, 178–185
   alcohols, 178–179
   aldehydes, esters, and volatile constituents, 182–183
   carotenoids, 184
   dietary fiber, 185
   histamine presence, 185
   minerals, 184–185
   organic acids, 181–182, 181t
   sugar content, 180–181
   vitamins, 183–184
  fruit wine, 178–185
   alcohols, 178–179
   aldehydes, esters, and volatile constituents, 182–183
   carotenoids, 184
   composition and nutritional significance, 143–157
   dietary fiber, 185
   histamine presence, 185
   minerals, 184–185
   organic acids, 181–182, 181t
   sugar content, 180–181
   vitamins, 183–184
  Mahua flower, 464–465, 465t
  of orange wine
   aroma-active compounds of, 417t
   composition of, 414–415, 415t
   flavor composition, 416
   organic acid composition of, 422t, 423–424
   phenolic composition of, 416–422
   sugar composition of, 422–423, 422t
  of pear fruit, 315–316
  sparkling plum wine, 507–509
  stone fruits wines, 349, 349t
  of volatiles in mango varieties, 448t
Continuous distillation systems, 536–537, 536f
Continuous-flow stirred-tank membrane reactor, 273–274
Continuously stirred tank reactors (CSTR), 579
Conventional activated sludge (CAS), 577
Conventional extraction methods, 255
Conventional single-use (CSU), 588–589
Cordials, 532–533
Corollas, 464–465
p-Coumaric acid, 8, 418–421
Counting cells technique, 93–94
Counting chamber, 93–94
Country wine, 663
CP. *See* Crude protein (CP)
Cranberry (*Vaccinium macrocarpon*), 382
  waste, 564

Crèmes, 533
Cross-flow MF (CFMF), 263
   wine components and sizes, 265t
Crude cider lees characteristics, 571–572
Crude enzymes, 274
Crude protein (CP), 607
Cryo-extracted fruit wines, 605
Cryoextraction, 621–622
CSTR. *See* Continuously stirred tank reactors (CSTR)
CSU. *See* Conventional single-use (CSU)
Cucumis melo. *See* Melon fruits (*Cucumis melo*)
Cucurbita moschata. *See* Pumpkin (*Cucurbita moschata*)
Cultivars, 315, 505
Cultural practices affecting wine quality, 13–25
   apple, 13–15, 13f–15f
   apricot, 15, 15f
   banana, 16, 16f
   berries, 16–17, 16f
   cashew nut, 17, 17f
   cherry, 17–18, 17f
   jamun, 18, 18f
   kiwifruit, 19–20, 19f
   loquat, 20–21, 20f–21f
   lychee, 20, 20f
   mandarin orange, 19, 19f
   mango, 21, 21f
   passion fruit, 21–22, 21f
   peach, 22, 22f
   pear, 22, 22f
   persimmon, 23, 23f
   plum, 23, 23f
   pomegranate, 24, 24f
   strawberry, 24–25, 24f
Cumulative antioxidative effect, 211–212
Cyanidin 3-glucoside, 389–390
Cyanogenic glycoside, 681

D

DAD. *See* Diode array detector (DAD)
DAHP. *See* Diammonium hydrogen phosphate (DAHP)
Darcy's Law, 263–264
DCOD. *See* Dissolved chemical oxygen demand (DCOD)
Deacidification, 634–635
Denaturing gradient gel electrophoresis (DGGE), 74
Densimetry, 236–237
   alcoholic strength, 237
   density and specific gravity, 236
Density in wines, 236
Dessert wines, 38
Destemming process, 605–607
DGGE. *See* Denaturing gradient gel electrophoresis (DGGE)
Diacetyl (2,3-butanedione), 155
Diafiltration mode, 269

Diammonium hydrogen phosphate (DAHP), 297f, 302, 306, 310–312, 350, 371f, 384, 682
Didymin, 421, 434
Dietary fiber, 185
Dihydroxyacetone phosphate, 117
Diode array detector (DAD), 240
Diospyros kaki. *See* Persimmon (*Diospyros kaki*)
2,2-Diphenylpicrylhydrazyl (DPPH), 394, 434, 478
   antioxidant activity, 402f
   assay, 335–337
Direct extraction method, 493
Discontinuous distillation systems, 534–536
Dissolved chemical oxygen demand (DCOD), 580
Dissolved oxygen (DO), 581
Distillation systems, 534–537
   continuous distillation systems, 536–537, 536f
   discontinuous distillation systems, 534–536
Distillation technique, 531
Distilled beverages, 531–532
Divinylbenzene (DVB), 241
DO. *See* Dissolved oxygen (DO)
DPPH. *See* 2,2-Diphenylpicrylhydrazyl (DPPH)
Drupes. *See* Stone fruits
Dry fruit wines, 605
Dry red table wine, 340–342
Dry wines, 10
DVB. *See* Divinylbenzene (DVB)
Dwarf–green coconut coir fibers, 566

E

EAE. *See* Ellagic acid equivalents (EAE)
EC. *See* Electrical conductivity (EC); Ethyl carbamate (EC)
EC regulations. *See* European Communities regulations (EC regulations)
Ecotoxicity, 585–586
Elderberry wine, 680, 681b
Electrical conductivity (EC), 607–608
Electron spin resonance spectroscopy (ESR), 208
Ellagic acid, 191
Ellagic acid equivalents (EAE), 187–188
EMR. *See* Enzyme membrane reactor (EMR)
Energy
   biotechnological conversion to, 609–612
   energy recovery, waste as renewable source for, 602–604
   production, 608
   and products generation from alternative sources, 600–602
Entrapment method, 86–87
ENVIFOOD Protocol, 587–588
Environmental Product Declarations (EPDs), 587
Enzyme membrane reactor (EMR), 271–272, 271f, 274
Enzyme(s), 111, 611
   bioreactor, 274
   enzymatic transformations of phenolic compounds, 196–197

treatment, 445
in winemaking, 133
EPDs. *See* Environmental Product Declarations (EPDs)
Epicatechin, 201
Eriobotrya japonica. See Loquat (*Eriobotrya japonica*)
ESR. *See* Electron spin resonance spectroscopy (ESR)
Esters, 182–183, 333
Ethanol, 413–415, 447, 494, 511, 617–618
 and by-products, yield of, 115
 changes in TSS and concentration, 510f
 in cider, 330
 concentrations, 85
 fermentation, 413f
Ethyl alcohol. *See* Ethanol
Ethyl carbamate (EC), 139–141, 141f, 242, 541–542
Ethyl esters, 543
European Communities regulations (EC regulations), 268
European Elder (*Sambucus nigra*), 680
European pears, 315
European Union (EU), 557–558, 696–697
 landfill directive, 558
Eutrophication (E), 588
Exotic fruit wines, 686–691. *See also* Berries fruits wines;
  Stone fruits wines
 banana wine, 687
 cocoa wines, 691
 fig wine, 688
 kiwi fruit wine, 688–690
 mango wines, 687–688
 pomegranate wines, 690–691

F

FAB. *See* Fermented alcoholic beverage (FAB)
FAN. *See* Free α-amino nitrogen (FAN)
FAO. *See* United Nations Food and Agriculture Organization (FAO)
Feedstock
 apple pomace use as, 611, 611t
 biorefinery, 605–608
 mango waste use as, 612
Feni beverage, 17, 54
Fenny, 17
Fermentation, 28–31, 35–36, 113–115, 194–195, 405, 620
 AF, 106–107, 108f
 of apple must, 310–312
  fermenter used for making wine, 311f
  wooden vat for, 310f
 bouquet, 135–137, 135f
 chemical changes
  biologically aged wines, 138–139
  sparkling wines, 137–138, 139t
 of cider must, 327–328

effects
 of cider in organic acids, 333t
 of cider in polyphenols, 333t
in-bottle, 645
initiation, 633
malolactic, 633, 634f
microbes in, 620–621
Neuberg's
 first form of fermentation, 113–114
 second form of fermentation, 114
 third form of fermentation, 114–115, 115f
of perry, 316–317
principal by-product of yeast, 624
spontaneous French cider, 632
in stone fruits, 351
temperature, 468
Fermented alcoholic beverage (FAB), 95
Fermented apple juice, 296
Fermented beverages, 664
Fermented garlic beverage, 403
Ferric reducing antioxidant power method (FRAP method), 208
Ferric-reducing ability of plasma (FRAP), 335–337, 423–424
Ferulic acid, 418–421, 433–434
FH. *See* Foam height (FH)
FID. *See* Flame ionization detection (FID)
Fig wine, 688, 689b
Film-forming yeasts, 337
Firethorn. *See* Pyracantha
Flame ionization detection (FID), 241
Flavanols, 194
Flavanones, 193, 420f, 421, 434
Flavones, 193
Flavonoids
 bioavailability, 200–202
  action in gastrointestinal tract, 202
  catechins, 201
  quercetin, 200
  resveratrol, 201–202
 compounds, 192–195
Flavonols, 259
Flavor composition
 of Mandarin wine, 430–431
 of orange wine, 416
Flavor compounds, 447
FLD. *See* Fluorimetric detection (FLD)
Flower wine, 463, 466
Fluorimetric detection (FLD), 96
Foam height (FH), 524
Foam stability height (FS), 524
Food
 industry, 603–604
 production, 603
 waste, 557–559

Fortified fruit wines, 487, 664
  vermouth, 487–500
    commercial production of, 500
    production from nongrape fruits, 494–500
    production technology, 488–494, 490f
    types, 488
Fortified wines, 10, 647–649
Fragaria × ananassa. See Strawberry (*Fragaria × ananassa*)
FRAP. *See* Ferric-reducing ability of plasma (FRAP)
FRAP method. *See* Ferric reducing antioxidant power method
    (FRAP method)
Free radical depletion mechanisms, 207
Free α-amino nitrogen (FAN), 131
Fresh apple juice, 301
Freshwater aquatic ecotoxicity (FWAE), 588
Friction head, 277
Fructose, 442
Fruit
  aggregate, and composite, 11f
  alcoholic beverages, 532–533
    legal restrictions of fruit spirit drinks types, 533t
  approximate composition, 14t
  classification, 12f
  composition and maturity, 34
  crushing, 666–671
    acidity adjustment in fruit wines, 669–670, 670t–671t
    adding tannin or phenolic compounds in fruits
      wines, 671
    adding water, 671
  cultivation practices and varieties, 10–25
    cultural practices affecting wine quality, 13–25
    nongrape fruits for winemaking, 11–13
  distillates, 533
  fruit-processing
    residues, 557
    wastewater, 558
  fruit-waste streams, 558–559
  LCA in fruit production, 587–588
  quality, 13
Fruit brandies
  alcoholic beverages, 531
  banana brandy, 552
  berry fruit brandy, 547–549
  distillation, 531, 534–537
  distilled beverages, 531–532
  factors affecting quality of brandies, 534
  fruit alcoholic beverages classification, 532–533
    legal restrictions of fruit spirit drinks types, 533t
  kiwi brandy, 549
  melon brandy, 550
  orange brandy, 551
  pome fruit brandy, 537–541
  stone fruit brandy, 541–547

Fruit juice
  alternative methods for extraction, 254–262
    case studies on cider production, 257–260
    microwave heating for improving extraction, 254–257
    PEF, 262
    US-assisted enzymatic extraction, 260–262
  clarification, 268
Fruit wine production, 557, 599, 605–607
  factors affecting yeast growth during fruit wine fermentation,
    84–85
  genetically modified microorganisms for, 96–98
    genetically engineered wine-yeast strains, 97–98
    legislation and consumer behavior against, 98
    methods for wine-yeast development, 97
  immobilized biocatalysts in winemaking, 86–90
  microbial biodiversity during fruit wine production,
    74–79
  microbiological analysis during fruit winemaking, 93–96
  microorganisms, 73–74
  MLF in fruit wines, 85–86
  nongrape wines, 73
  spoilage of fruit wines, 90–93
    fruit wine spoilage prevention, 93
    must spoilage, 91
    postfermentative spoilage, 91–92
    spoilage during MLF, 91
  waste generation, 606f
  yeast selection as starter cultures for, 79–83
Fruit wine wastes (FWW), 558. *See also* Coconut waste; Waste
    from fruit wine production
  applicability of biorefinery concept, 608, 609f
  biorefinery, 600–602, 601f
  biorefinery feedstock
    fruit winemaking process, 605–607
    generation, 607–608
  biotechnological conversion to platform chemicals and
    energy, 609–612
  production of fruit wines, 599
  treatment, 600
  valorization, 600
  virtually inexhaustible resource, 600
  waste as renewable source, 602–604
  waste categories and types, 603f
Fruit wine(s), 177, 178t, 227, 638
  alcohol removal from, 268–269
  analytical methods, 228t–229t
  bioavailability of health benefit components, 197–202
  chemical changes occurring fermentation, 137–139
  chemistry of wine spoilage, 142–143
  chemistry of winemaking, 107–133
  chromatographic analysis, 237–242
  classification of wine, 8
  composition, 178–185

alcohols, 178–179
aldehydes, esters, and volatile constituents, 182–183
carotenoids, 184
dietary fiber, 185
histamine presence, 185
minerals, 184–185
and nutritional significance of wine, 143–157
organic acids, 181–182, 181t
sugar content, 180–181
vitamins, 183–184
consumption, 599
enzymatic transformations, 196–197
enzymes in winemaking, 133
fermentation, 106–107
bouquet, 135–137, 135f
fermentation, 663
fruit cultivation practices and varieties, 10–25
health benefit potential, 203–214
HHP treatment, 279–281
on market, 695–699
fruit wine industry in Australia, 698
fruits wine market trends, 697
niche market segments, 698–699
obstacles in selling fruit wine, 696
total antioxidant capacity and total phenolic content, 696t
transition from homemade fruit wines to industrial-level
production, 695
methods for winemaking, 35–37
microbiological analysis, 242–244
mineral contents of, 7t
MLF, 133–135
nutritional facts, 195–196
phenolic compounds, 186–195
physicochemical analysis, 227–237, 229t–230t
production methods, 665–675
aging fruit wines, 675
bottling fruit wines, 675
clarifying/racking, 674–675
crushing fruit, 666–671
equipment, 665f
must fermentation, 671–673
must preparation, 666–671
yeast inoculation, 673–674
sensory analysis, 244–246
technology of production
aspects and problems in production of, 33
fruit composition and maturity, 34
microbiology of fermentation, 34–35
raw materials, 33–34
screening of varieties, 34
technology steps, 666
toxic metabolites of nitrogen metabolism, 139–141
traditional recipes of fruit wines, 676–695

types and styles, 664
types of, 9–10, 9f
use of antimicrobials, 133
yeast flavor compounds, 135–137
FS. *See* Foam stability height (FS)
Fuel alcohols, 116
FWAE. *See* Freshwater aquatic ecotoxicity (FWAE)
FWW. *See* Fruit wine wastes (FWW)

G
Gabiroba (*Campomanesia pubescens*), 74
GAE. *See* Gallic acid equivalents (GAE)
Gallic acid, 418
Gallic acid equivalents (GAE), 187
GAP. *See* General amino acid permease (GAP)
Garlic, 403
Gas chromatography (GC), 141, 309–310
analysis of volatile compounds, 241–242
ethyl carbamate, 242
methanol, 242
Gas chromatography flame ionization detection (GC—FID),
447
Gas chromatography mass spectrometry (GC—MS), 447
Gastrointestinal disorders prevention, 213
Gastrointestinal tract, action in, 202
GC. *See* Gas chromatography (GC)
GC/O. *See* GC/olfactometry (GC/O)
GC/olfactometry (GC/O), 416
GC—FID. *See* Gas chromatography flame ionization detection
(GC—FID)
GC—MS. *See* Gas chromatography mass spectrometry
(GC—MS)
GDHs. *See* Glutamate dehydrogenases (GDHs)
Gelatin, 35
General amino acid permease (GAP), 128–129
Genetic engineering in wine, 25–33
genetic modification of plants, 25–26
improvement of traits, 25
methods of genetic transformation, 25–26
genetically modified yeasts, 32–33
in plant modification, 27t
wine yeast, genetic engineering of, 26–32
Genetically modified microorganisms for fruit winemaking,
96–98
genetically engineered wine-yeast strains, 97–98
legislation and consumer behavior against, 98
methods for wine-yeast development, 97
Genetically modified organisms (GMOs), 28, 86
Genetically modified yeasts, 32–33
Genome shuffling, 97
G—G content. *See* Glycosyl-glucose content (G—G content)
GHG. *See* Greenhouse gas (GHG)
Global spirit market, 531–532, 532f

Global status, 46–57
country-wise status of fruit wines, 54–57
factors influencing, 46–48
global production of fruit wines, 49–54
red wines, 48t
types of wines, 47t
white wines, 48t
Global warming (GW), 588
Gluconacetobacter xylinus (*G. xylinus*), 569–570
Glucose, 442
Glutamate dehydrogenases (GDHs), 129–130
Glutathione (GSH), 502
Glyceraldehyde 3-phosphate, 109–111
Glycerol, 117, 118f, 179, 362, 447
Glycolysis, 109–111
Glycosyl-glucose content (G–G content), 626
GMOs. *See* Genetically modified organisms (GMOs)
Grape (*Vitis vinifera* L.), 398–399, 491
chemical analysis of, 399t
cultivation, 618
extraction of grape juice, 400f
pigmentation, 626
preparation of jamun–grape wine, 400f
production techniques, 649
red wine made by blending, 398–401
wines, 106, 146t, 147, 231
making, 665
minerals in, 377t
physicochemical characteristics, 365t, 368
technologies, 253
Grapefruit (*Citrus paradis*), 76
wine, 692, 693b
Green concept biorefinery, 608, 610f
Greenhouse gas (GHG), 586
Grinding, 301
GSH. *See* Glutathione (GSH)
GW. *See* Global warming (GW)

H

Hanseniaspora uvarum (*H. uvarum*), 356–357
HCN. *See* Hydrocyanic acid (HCN)
HDL. *See* High density lipoprotein (HDL)
Headspace (HS), 430
Health benefit potential
of different fruit and berry wines, 203–214
antioxidative effects, 206–212
cardioprotective potential, 203–206
cognitive support, 213
other, 214
prevention of gastrointestinal disorders, 213
types of cancers, 212
fruit wines, 106, 144, 151–152
phenolic compounds from fruit and berry wines with, 186–195

Hen egg white lysozyme (*HEL*1), 31
Herbs, 487–488, 492t–493t, 493, 498–499
in preparation of plum vermouth, 498t
Hesperidin, 421, 434
Heterofermentative pathways, 124–125
Heterogeneous bioreactor systems, 270–271
Hexanoic acid, 153–155
HHP treatment. *See* High hydrostatic pressure treatment (HHP treatment)
High density lipoprotein (HDL), 204
High hydrostatic pressure treatment (HHP treatment), 253
of fruit wine, 279–281
High-performance liquid chromatography (HPLC), 147–149, 150t, 237
analysis, 469
of control mahua wine, 469f
in lemon-treated mahua flower wine, 470f
High-pressure CO_2 sterilization, 281–283
mathematical models, 282t
High-pressure-treated foodstuffs, 279–280
High-temperature short-time (HTST), 681
High-value low volume (HVLV), 604
Higher alcohols, 116, 117f, 330, 447
"Hippocratic wine", 488
Hippophae rhamnoides L. *See* Sea buckthorn (*Hippophae rhamnoides* L.)
Histamine presence, 185
Hollow-fiber membranes, 583
Home-scale preparation, wine and ingredients for, 340t
Homofermentative pathways, 124–125
Homogeneous bioreactor systems, 270–271
Honey, 10, 473, 669
wine, 471
Hot water treatment (HWT), 561
HPLC. *See* High-performance liquid chromatography (HPLC)
HRT. *See* Hydraulic retention time (HRT)
HS. *See* Headspace (HS)
HT. *See* Human toxicity (HT)
HTST. *See* High-temperature short-time (HTST)
Human toxicity (HT), 588
HVLV. *See* High-value low volume (HVLV)
HWT. *See* Hot water treatment (HWT)
Hydraulic press, 301, 323f
Hydraulic retention time (HRT), 579, 612
Hydrocyanic acid (HCN), 544–545
Hydrodiffusion, 254
Hydrogen sulfide (H_2S), 130
Hydrometer for sugar content measurement, 669f
Hydrostatic pressure, 264
Hydroxybenzoic acids, 191, 418, 420f, 433
Hydroxycinnamic acids, 191, 418–421, 420f, 433–434

I

Iced fruit wine, 664
ICR. *See* Immobilized cell reactor (ICR)
ICV-GRE yeast, 305
Immersion (Im), 430
Immobilization studies, 449–450
Immobilized biocatalysts in winemaking, 86–90
    applications in enology, 88–90
        immobilized cells in cider production, 90
        immobilized yeast cells, 88–89
        MLF control by immobilized LAB, 89
        sparkling wines, 90
    methods of immobilization, 86–87, 88t
Immobilized cell reactor (ICR), 563
Immobilized cells
    in cider production, 90
    yeast cells, 88–89
India, status of fruit wines, 54–55
Industrial wastewaters, 604
Initial sugar concentration (ISC), 354
Inoculation, 413–414
    of apple wine with yeast culture, 309–310
    of cider with yeast culture, 324–326
    of wine yeast, 306
International Organization of Vine and Wine
        (OIV), 231
Ion-selective electrode, 235
Ipomoea batatas L. *See* Sweet potato (*Ipomoea batatas* L.)
ISC. *See* Initial sugar concentration (ISC)
Isoquercitrin, 193

J

Jabuticaba (*Myrciaria jabuticaba*), 549
Jamun (*Syzygium cumini* L.), 18, 18f, 395, 398–399
    physicochemical characteristics, 396t
Jamun wine, 53. *See also* Blackberry jamun wine
    chemical analysis of, 399t
    extraction of jamun juice, 399f
    preparation of jamun–grape wine, 400f
    red wine by blending of, 398–401
Japan, status of fruit wines, 56
Japanese plums. *See* Loquats

K

Kernel, 561–562
Killer yeast in wine fruits, 76–78
    killer test for yeast, 77f
    LAB, 78–79
Kiwi aroma, 549
Kiwi brandy, 549
Kiwi fruits (*Actinidia chinensis*), 19–20,
        19f, 689
    wine, 688–690, 690b
Kiwi wines, 54, 549

Kloeckera apiculata (*K. apiculata*), 356–357
KMS. *See* Potassium metabisulfite (KMS)

L

LAB. *See* Lactic acid bacteria (LAB)
Lactic acid, 120–121, 121f
    production, 612
Lactic acid bacteria (LAB), 78–79, 105, 108–109, 123f, 124,
        126f, 328–329, 456–457
    formation of ethanol, 124f
    metabolic pathways, 124–125
    MLF control, 89
    products in homolactic or heterolactic pathway, 125f
    progress, 325t
    wine spoilage by, 142–143
Lactobacillus fermentum (*L. fermentum*), 123
"*Lali Gurans*", 474
Lallemand 71*B*, 673
*Lallemand BA*11, 673
*Lallemand EC*1118, 674
*Lallemand ICV-K*1, 674
Lallemand R2, 674
*Lallemand VIN*13, 674
Lavoisier, Antoine, 6
LC. *See* Liquid chromatography (LC)
LCA. *See* Life cycle assessment (LCA)
*LCA*1. *See* *Leuconostoc carnosum* leucocin (*LCA*1)
LCF. *See* Lignocellulosic (LCF)
LCIA. *See* Life cycle impact assessment (LCIA)
LDL. *See* Liquid diffusion layer (LDL); Low density
        lipoprotein (LDL)
Lesaffre UCD 595, 674
Leuconostoc, 327–328
Leuconostoc carnosum leucocin (*LCA*1), 31
Life cycle assessment (LCA), 559
    in fruit production, 587–588
    of waste management in cider production, 588
    of wine packaging, 588–589
Life cycle impact assessment (LCIA), 588
Lightweight single-use (LSU), 588–589
Lignocellulosic (LCF), 608
    biorefinery, 608, 610f
D-Limonene removal column (LRC), 563
Lipids, 138
Liqueurs, 533
Liquid chromatography (LC), 96, 237–241
    analysis of organic acids, 239–241
    analysis of sugars, 237–239, 238t
Liquid diffusion layer (LDL), 578–579
Liquid stream, 574–585
    characterization of winery liquid effluents, 574–576
    methods of WW treatment, 576–585
        biological treatment processes, 577–585
        bioreactors, 577–585

Liquid vapor pressure, 278
Liquid waste, 607
Liquid–liquid extraction (LLE), 241
LLE. *See* Liquid–liquid extraction (LLE)
Loquat (*Eriobotrya japonica*), 20–21, 20f–21f, 338
    wine, 338
Low density lipoprotein (LDL), 203
    prevention of LDL oxidation, 204–205
Low value high volume (LVHV), 604
Low-acid cultivar, 321
LRC. *See* D-Limonene removal column (LRC)
LSU. *See* Lightweight single-use (LSU)
LVHV. *See* Low value high volume (LVHV)
Lychee, 20, 20f, 80
    wine, 403–404, 452–454

M

MABR. *See* Membrane aerated bioreactor (MABR)
Macerated fruit spirits, 533
Maceration, 493, 622, 632
Madeira-style wines, 38, 648–649
Madhuca indica (*M. indica*), 463
MAE. *See* Marine aquatic ecotoxicity (MAE)
MAF. *See* Maloalcoholic fermentation (MAF)
Mahua wines. *See also* Mead wines
    fermentation temperature, 468
        effect of, 468f
    flavor masking, 468–469
    mahua flower, 464
        collection and processing, 466
        composition, 464–465, 467f
        wine, 466
    mahua liquor, 465–466
    mahua tree, 463
    mahua vermouth, 470–471
    nutrients, addition of, 466–467
    phenolic profiling, 469
        HPLC analysis in lemon-treated mahua flower wine, 470f
        HPLC analysis of control mahua wine, 469f
    tannins, addition of, 468
Malic acid, 181, 232, 299, 302, 423
    reduction in must, 269–270
Maloalcoholic fermentation (MAF), 91
Malolactic acid bacteria, 33
Malolactic fermentation (MLF), 78, 85f, 133–135, 328
    in fruit wines, 85–86
        methods for cell immobilization, 87f
    MLF control by immobilized LAB, 89
    in sparkling wine production, 512–513
Malus pumila L. *See* Annurca apple (*Malus pumila* L.)
Mandarin orange, 19, 19f
Mandarin wine, 424–435. *See also* Orange wine
    antioxidant activity, 434–435, 435f
    blended kinnow–cane wine, 424

chemical composition of mandarin juices and, 425–428, 430t
flavor composition of, 430–431
phenolic composition, 431–434, 433t
physicochemical and sensory characteristics of kinnow wine, 429t
production, 425f
volatile flavor of Clementine, 431t
Mango (*Mangifera indica* L.), 21, 21f, 405, 495, 546, 612, 687–688
    leaves, 562
    peels, 561–562
        mango leaves, 562
        mango Seeds, 561–562
    Seeds, 561–562
    use of mango waste as feedstock, 612
Mango wine, 442–450, 687–688, 689b. *See also* Orange wine
    characterization, 446–447
    composition of volatiles in mango varieties, 448t
    cost economics, 450
    effect of storage, bottle color, and temperature, 447–449
    immobilization studies, 449–450, 450f
    optimization of production conditions, 445–446
        enzyme treatment, 445
        oxygen, 446
        pH, 446
        SO_2, 446
        temperature, 435
    physicochemical characteristics, 444t
        of mango pulp, 443t
    production, 443f–444f
    screening
        of mango varieties, 442
        of yeast strain, 445
    sensory evaluation, 449, 449t
Marine aquatic ecotoxicity (MAE), 588
Marula, 547
Mass per volume unit, 236
Mass spectrometry (MS), 241
Maturation, 37
    of apple wine, 312
    of apricot wine, 368–370, 369t
    of cider, 328–329
    of peach wine, 368
    of plum wine, 362
    in stone fruits, 351
Mature drink, 565
MBfR. *See* Membrane biofilm reactor (MBfR)
MBR. *See* Membrane reactor (MBR)
MC. *See* 3′-*O*-Methylcatechin (MC)
Mead wines, 471–474. *See also* Mahua wines
    method for mead production, 471–474
        physicochemical characteristics of mead, 473t
        types of mead preparation, 472t
Medicinal wines, 10

Medlar wine, 338–339
Medlars (*Mespilus germanica*), 338–339
Meijiu, 49
MeLo model system, 209
Melon brandy, 550
Melon fruits (*Cucumis melo*), 550
Membrane aerated bioreactor (MABR), 578
 pilot-plant MABR application, design, long-term operation,
  and performance, 579–580
 variations in the operating conditions and parameters in, 581t
Membrane biofilm reactor (MBfR), 578
Membrane bioreactor (BMR), 271–272, 577
 advantages, 579
 configurations, 579
 productivity, 274–275
Membrane reactor (MBR), 271–272
Membrane separation, 263, 271, 274, 287
Membrane technologies to fruit winemaking, 263–275, 266f
 alcohol removal from fruit wines, 268–269
 bioreactors for fruit wine processing, 270–274, 272f–273f
 clarification of fruit juice and wine, 268
 must correction, 266–268, 267f
 productivity of BMRs, 274–275
 reduction of malic acid in must, 269–270
Mesocarp, 295
Mespilus germanica. See Medlars (*Mespilus germanica*)
Metabolic transformations and excretion, 199–200
Methanol, 242
 apple cultivar influence, 331t
 in cider, 330
Methode ancestrale, 504
Méthode champenoise, 318–319, 502
Methyl alcohol (CH₃OH), 117, 118f
3'-*O*-Methylcatechin (MC), 201
MF. *See* Microfiltration (MF)
MFA. *See* Multiple factor analysis (MFA)
Microbial biodiversity during fruit wine production, 74–79
 killer yeast in wine fruits, 76–78
  killer test for yeast, 77f
  LAB, 78–79
  yeasts, 74–76
Microbial inactivation, 283–284
Microbial population level, 79
Microbiological analysis, 242–244. *See also* Chromatographic
  analysis; Physicochemical analysis
 classical techniques, 243–244
 during fruit winemaking, 93–96
  detecting spoiling microorganisms during, 94–95
  mycotoxins detection, 96
  yeast population development during alcoholic
   fermentation, 93–94, 94f
 molecular techniques, 244
Microbiology of fermentation, 34–35
Microbiota, 74, 79

Microencapsulation techniques, 636f
Microfiltration (MF), 253
Microorganism control before cider fermentation, 323–324
Microscopic techniques, 243–244
Microwaves (MWs), 253
 heating for improving extraction of fruit juice, 254–257
 heating of fruit mash as pretreatment technique, 257–259
 pasteurization of fruit juice, 284
Milling in cider, 321–323
Minerals, 143–144, 184–185
 present in select fruit wines, 145t
Mixed fruit wines, 340–342
 derivation of suppleness index, 341f
 wine and ingredients for home-scale preparation, 340t
MLF. *See* Malolactic fermentation (MLF)
Molecular techniques, 244
Monosaccharides, 611
Mousiness in cider, 337
MS. *See* Mass spectrometry (MS)
Multiple factor analysis (MFA), 335
Musanzeensis, 405
Must, 35–36, 302
 clarification, 302
 correction, 266–268, 267f
 fermentation, 671–673
  of apple, 310–312
  of cider, 327–328
  plug, 672f
  in stone fruits, 350–351
 preparation, 35, 666–671
  acidity adjustment in fruit wines, 669–670, 670t–671t
  adding tannin or phenolic compounds in fruits wines, 671
  adding water, 671
MWs. *See* Microwaves (MWs)
Mycotoxins detection, 96
Myrciaria jabuticaba. See Jabuticaba (*Myrciaria jabuticaba*)

N

Nanofiltration (NF), 263
Naringin, 421
Narirutin, 421, 434
National Association of Cider Makers, 315
Natural wines, 10
Natural zeolite, 582
Near-infrared (NIR), 626
Near-infrared spectroscopy (NIRS), 626
Nectarine wines, 682, 683b
Neohesperidin, 421
Net positive suction head (NPSH), 278
Neuberg's fermentation, 117
 first form of fermentation, 113–114
 second form of fermentation, 114
 third form of fermentation, 114–115, 115f
NF. *See* Nanofiltration (NF)

Niacin, 184
Niche market segments, 698–699
NIR. *See* Near-infrared (NIR)
NIRS. *See* Near-infrared spectroscopy (NIRS)
Nitric oxide production, 206
Nitrogen, 108, 607–608
   nitrogenous
      compounds in apple fruit, 300–301
      compounds in pear fruit, 316
      yeast food, 507
   source, 302, 303f
Nitrogen metabolism, 126–131. *See also* Carbon metabolism
   amino acid utilization profile, 127–128
   enological aspects, 130–131
   factors affecting nitrogen accumulation, 129–130
   products of nitrogenous compound degradation, 127f
   regulation of nitrogen transport, 129
   sources and supplements, 127
   toxic metabolites, 139–141
   uptake and transport of compounds, 128–129
Non-*Saccharomyces*
   species, 305
   yeast, 74, 75t, 242–243
Nongrape fruit wine industry, 46–57
   country-wise status of fruit wines, 54–57
      Australia, 57
      Canada, 56
      China, 55
      India, 54–55
      Japan, 56
      United Kingdom, 56
      United States, 56
   factors influencing, 46–48
   global production of fruit wines, 49–54
      apple wine/cider, 49
      banana wine, 53
      cherry wine, 49
      cider brands in countries, 53t
      citrus wine, 54
      fruits with growing countries, 50t–51t
      jamun wine, 53
      kiwi wine, 54
      pineapple wine, 49
      plum wine, 49
      strawberry wine, 53
      in various countries, 51t
   red wines, 48t
   types of wines, 47t
   white wines, 48t
Nongrape fruits
   apple vermouth, 494–495
   apricot vermouth, 499–500
   mango vermouth, 495
      physicochemical characteristics and sensory quality of, 497t
      spices for preparation of, 496t
   plum vermouth, 495–496
      physicochemical characteristics of, 498t
      spices and herbs in preparation of, 498t
   pomegranate vermouth, 496
   sand pear vermouth, 498–499
      physicochemical characteristics of, 499t
   tamarind vermouth, 499
   vermouth production from, 494–500
   for winemaking, 11–13
Nongrape wines, 39t–45t
Nonvolatile acids, 232–233
Nonvolatile organic acids, 118–121
NPSH. *See* Net positive suction head (NPSH)
Nutritional facts, 195–196
   alcohol, 196
   antioxidants, 195–196
   caloric value, 195

O

Oak chips, 340–342
OAV. *See* Odor activity value (OAV)
Oblačinska sour cherry, 372–374
Odor activity value (OAV), 155–157, 626
Off-flavors, 227–231
   chemistry of, 131–132
OIV. *See* International Organization of Vine and Wine (OIV)
OLD. *See* Ozone layer depletion (OLD)
OLR. *See* Organic loading rate (OLR)
Orange (*Citrus sinensis*), 76
   brandy, 551
Orange wine, 410–424, 692, 692b. *See also* Mandarin wine; Mango wine
   chemical composition of
      aroma-active compounds, 417t
      composition, 414–415, 415t
      flavor composition, 416
      organic acid composition, 422t, 423–424
      phenolic composition, 416–422
      sugar composition, 422–423, 422t
   orange winemaking, 412f
      alcoholic fermentation of the juice, 412–414
      ethyl alcohol fermentation, 413f
      extraction of juice, 411–412
      orange cultivars and properties, 411
   phenolic content of blonde and blood, 419t
   phenolic distribution of orange juices and wines, 418f
Orchard. *See* Vineyard
Organic acid(s), 181–182, 181t, 231–232, 232t
   analysis, 239–241
   in apple fruit, 299
   chromatogram, 239f
   cider fermentation effects, 333t

composition of orange wine, 422t, 423–424
decomposition, 122–123
fermentation, 92
in pear fruit, 316
Organic loading rate (OLR), 581
Organic removal, 581–582, 584f
OUR. *See* Oxygen uptake rate (OUR)
Oxidation, 259
Oxygen, 446
Oxygen uptake rate (OUR), 585–586
Ozone layer depletion (OLD), 588

P
PAC. *See* Proanthocyanidin (PAC)
Packaging of cider, 329
Palm wine, 455–457, 457f, 664
Papaya wine, 404–405
Partial dealcoholization, 638
PAS2050. *See* Publicly Available Specification2050 (PAS2050)
Passion fruit (*Passiflora edulis*), 21–22, 21f, 405
Pasteur effect, 116
Pasteurization, 37
   of fruit juice using MW, 284
   of perry, 318
   in stone fruits, 352
Patulin (PAT), 96
Paul's method, 235
PCA. *See* Principal component analysis (PCA)
PCR. *See* Polymerase chain reaction (PCR)
PCRs. *See* Product category rules (PCRs)
PDMS. *See* Polydimethylsiloxane (PDMS)
Peach (*Prunus persica*), 22, 22f, 348, 362–363
Peach wine, 362–368, 682, 683b. *See also* Apricot wine;
      Cherry wines; Plum wine
   comparison of sensory scores with grape wine, 367f
   fruit and cultivars, 363
   maturation, 368
   method of wine production, 363–367
   physicochemical characteristics, 365t, 368
   polyphenols, 367t
Pears (*Pyrus communis* L.), 22, 22f, 315
   brandy, 537–539
      effect of lees and copper, 538f
   wine, 315–318, 694, 694b
      composition of pear fruit, 315–316
      preparation, 317f
      process for making perry, 316–318
Pectate gel, 89
Pectic substances in apple fruit, 299
Pectin(s), 623–624, 687
   pectin-splitting enzyme, 384
Pectinesterase, 303f, 304t, 384
Pectinol, 35
Pediococcus acidilactici pediocin (*PED*1), 31

PEF. *See* Pulsed electric field (PEF)
PEFCRs. *See* Product environmental footprint category rules
      (PEFCRs)
Pellecchiella apricot (*Prunus armeniaca,* L), 545
Penicillium verrucosum (*P. verrucosum*), 96
Peroxidases (PODs), 196
Perry, 10, 315
   extraction of juice and preparation of must, 316
   fermentation, 316–317
   process for, 316–318
   quality, 318
   siphoning/racking/pasteurization, 318
Persimmon (*Diospyros kaki*), 23, 23f, 401
   additive concentration on chemical characteristics of, 403t
   flow sheet for production, 402f
   wine, 401
PET. *See* Polyethylene terephthalate (PET)
Petillance, 647
Petroleum refinery, 600, 601f
PFK. *See* Phosphofructokinase (PFK)
pH, 236, 446
PHB. *See* Polyhydroxybutyrate (PHB)
Phenolic characterization of fruit wines, 147–153,
      148t–149t
   antioxidant capacity of fruit wines, 151, 152t
   biological activity of nongrape fruit wines, 151–153
Phenolic composition
   Mandarin wine, 431–434, 433t
   of orange wine, 416–422
Phenolic compounds, 185, 268, 572
   concentration in cider juice, 332t
   enzymatic transformations, 196–197
   from fruit and berry wines, 186–195, 186t–187t
      acids, 191–192
      anthocyanins and anthocyanin-derived compounds,
         189–191, 189t
      flavonoid compounds, 192–195
   in fruits wines, 671
   in peach wine, 366
   in wine-apple juice, 300t
Phenolic profiling, 469
   HPLC analysis in lemon-treated mahua flower
      wine, 470f
   HPLC analysis of control mahua wine, 469f
Phenotypic approach, 244
Phosphofructokinase (PFK), 109–111
Phosphorus (P), 607–608
Photochemical oxidation (PO), 588
Physicochemical analysis, 227–237, 229t–230t. *See also*
      Chromatographic analysis; Microbiological analysis
   densimetry, 236–237
   potentiometry, 235–236
   regulations regarding fruit wine standards, 231t
   titrimetry, 231–235

Pilot-plant MABR, 581
    application, design, long-term operation, and performance,
        579–580
Pineapple (*Ananas comosus*), 76, 405, 451
    wine, 49, 451
PLA. *See* Polylactic acid (PLA)
Plants, genetic modification of, 25–26
    improvement of traits, 25
    methods of genetic transformation, 25–26
Platelet aggregation, inhibition of, 204
Platform chemicals, biotechnological conversion to, 609–612
Plating techniques, 243–244
Plum (*Prunus salicina* Linn), 23, 23f, 348, 495–496, 541–542
Plum brandy, 543–545
    chemical composition of analyzed plum spirit, 545t
    traditional plum spirit process, 544f
Plum jerkum, 23, 49
Plum juice process, 255
Plum wines, 11–13, 49, 673, 684–685, 685b. *See also* Apricot
        wine; Cherry wines; Peach wine
    chemical composition, 354t
    comparison of fermentation rates, 356f
    dilution effect of pulp/biological deacidification,
        357–359
    honey use on wine quality, 354
    initial sugar effects on wine quality, 354
    ISC effects on physicochemical characteristics, 355t
    low-alcoholic plum beverages, 362
    maturation, 362
    microbiology of fermentation, 356–357
    osmotic treatment effect of fruit on wine quality, 355
    physicochemical composition of fresh and fermented plum
        musts, 358t
    plum fruit cultivars, 352
    preparation methods, 352, 353f
    skin retention effect, 354
    use of preservatives, 355–356, 357f
    wine yeast strain effect, 357, 358t
    wood chips effects on physicochemical and sensory
        characteristics, 364t
PO. *See* Photochemical oxidation (PO)
PODs. *See* Peroxidases (PODs)
Polydimethylsiloxane (PDMS), 241
Polyethylene terephthalate (PET), 588–589, 643
Polyhydroxybutyrate (PHB), 572
Polylactic acid (PLA), 643
Polymerase chain reaction (PCR), 74, 95, 244
Polyphenol oxidase (PPO), 196, 254
Polyphenols, 105, 107, 133, 300, 447, 559–560
    in bittersweet English cider, 331f
    cider fermentation effects, 333t
    compounds, 213
    in peach wine, 367t
        comparison of sensory scores, 367f

Polypropylene (PP), 270
Polysaccharides, 268
Polytetrafluoroethylene, 643
Polyvinylidene chloride, 643
Pome, 295
Pome fruit brandy, 537–541. *See also* Stone fruit brandy
    apple brandy, 539–541
    pear brandy, 537–539
Pome fruit wines, 295, 692–695. *See also* Berries fruits wines;
        Stone fruit wines; Tropical fruit wines
    apple tea wine, 312–313
    apple wine, 296–315, 693–694
    cider, 318–338
    loquat wine, 338
    medlar wine, 338–339
    mixed fruit wines, 340–342
    name, botanical name, and family, 296t
    pear wine, 315–318, 694
    Pyracantha wine, 339
    quince wine, 339, 695, 695b
    toyon wine, 339
Pomegranate fruit (*Punica granatum* L.), 24, 24f, 496, 690
    wines, 690–691, 691b
Pomelo peel, 563
Pommeau, 649
Port-style fruit wines, 605
Positive-displacement pumps, 276
Potassium (K), 184, 607–608
Potassium metabisulfite (KMS), 350–351, 371f, 392, 507
Potential alcohol, 668–669
Potentiometry, 235–236
    pH, 236
PP. *See* Polypropylene (PP)
PPO. *See* Polyphenol oxidase (PPO)
Precision viticulture (PV), 628
"Premium" fruit wine, 665
Preservation processes to wine production, 279–287
    HHP treatment of fruit wine, 279–281
    high-pressure CO_2 sterilization, 281–283
    pasteurization of fruit juice using MW, 284
    PEF technology for wine preservation, 285–287, 285f
    ultrasound application in must treatment, 283–284
Pressing, 301
    in cider, 321–323
Principal component analysis (PCA), 306
Proanthocyanidin (PAC), 564
Procyanidin B1, 332
Procyanidin B2, 332
Product category rules (PCRs), 587
Product environmental footprint category rules
        (PEFCRs), 587
Protocatechuic acid, 433
Prunus armeniaca, L. *See* Pellecchiella apricot (*Prunus
        armeniaca,* L)

Prunus armeniaca L. *See* Wild apricot (*Prunus armeniaca* L)
Prunus armeniaca L. *See* Apricots (*Prunus armeniaca* L.)
Prunus avium. See Sweet cherries (*Prunus avium*)
Prunus cerasus L. *See* Tart cherry (*Prunus cerasus* L.)
Prunus ceresus. See Sour cherries (*Prunus ceresus*)
Prunus persica. See Peach (*Prunus persica*)
Prunus salicina Linn. *See* Plum (*Prunus salicina* Linn)
Publicly Available Specification2050 (PAS2050), 587–588
Pulsed electric field (PEF), 337–338
   effect on inactivation of PPO in apple cider, 338t
   methods, 286t
   technology, 253, 262, 285f
   for wine preservation, 285–287
Pumpkin (*Cucurbita moschata*), 392
   pumpkin-based herbal wine, 392–395
      physicochemical characteristics, 394t
      qualitative estimation, 395t
   wine, 392
      physicochemical characteristics, 394t
      preparation of, 393f
Pumps, 275–279, 278f
   centrifugal pumps, 276
   positive-displacement pumps, 276
   selection of, 277–279
Punica granatum L. *See* Pomegranate fruit (*Punica granatum* L.)
Punicine, 24
Purple sweet potato, 478
PV. *See* Precision viticulture (PV)
Pycnometry, 236
"Pyment", 669
Pyracantha, 339
   wine, 339
Pyrus nivalis. See Snow pears (*Pyrus nivalis*)
Pyrus pyrifolia. See Sand pear (*Pyrus pyrifolia*)
Pyruvate metabolism, 111, 112f

Q

QDA. *See* Quantitative descriptive analysis (QDA)
qNMR. *See* Quantitative nuclear magnetic resonance (qNMR)
QTL analysis. *See* Quantitative trait locus analysis (QTL analysis)
Quantitative descriptive analysis (QDA), 245, 306, 334
Quantitative nuclear magnetic resonance (qNMR), 635
Quantitative trait locus analysis (QTL analysis), 332
Quercetin, 200
   glycosides rutin, 193
Quince wine, 339, 695, 695b

R

Rack-and-frame hydraulic press, 257, 259f
Racking process(es), 253
   pumps, 275–279, 278f
   racking in stone fruits, 351

racking of perry, 318
   and transport of wine, 275–279
Rakia, 541–542
Rakija, 541–542
Raspberry (*R. idaeus*), 389
   wine, 189–190
Rate of deacidification (RDA), 513
Raw material, 666
RDA. *See* Rate of deacidification (RDA)
Reactive oxygen species (ROS), 206–207
Ready to drink (RTD), 665
Red raspberry (*Rubus idaeus*), 382
   wine, 389–390
Red wine, 204, 275, 398–401
   extraction
      of grape juice, 400f
      of jamun juice, 399f
      of juices, 401
      preparation of jamun–grape wine, 400f
Reducing sugars (RSC), 565
Refillable glass (RFG), 588–589
Refractive index (RI), 238
Remote sensing, 630
Renewable source for energy and resource recovery, 602–604
Residence time distribution (RTD), 580, 580t
Residual sugars (RS), 605
Resource conservation, 576
Resource recovery, waste as renewable source for, 602–604
Response surface methodology (RSM), 359, 361f
Restriction fragment length polymorphism (RFLP), 244
Resveratrol, 8, 201–202
Reverse osmosis (RO), 253, 263f, 269
RFG. *See* Refillable glass (RFG)
RFLP. *See* Restriction fragment length polymorphism (RFLP)
Rhododendron, 474
   flower, 475–476
      Rhododendron flower juice composition, 475t
   R. arboretum, 474–475
   tree, 474–475
   wine, 476
Rhone style, 340–342
RI. *See* Refractive index (RI)
Ripper's method, 235
RO. *See* Reverse osmosis (RO)
Ropiness, 92
ROS. *See* Reactive oxygen species (ROS)
Rotary fermentors, 637
RS. *See* Residual sugars (RS)
RSC. *See* Reducing sugars (RSC)
RSM. *See* Response surface methodology (RSM)
RTD. *See* Ready to drink (RTD); Residence time distribution (RTD)
Rubus idaeus. See Red raspberry (*Rubus idaeus*)

Rubus occidentalis. See Black raspberry (*Rubus occidentalis*)
Rubus sp. *See* Blackberry (*Rubus* sp.)

S

Saccharomyces
  multiplication and total soluble solids, 405f
  S. bayanus, 491
  S. cerevisiae, 1, 35, 74, 105, 114f, 122f, 179, 242–243, 281, 350, 541–542, 617, 673
  S. uvarum, 119, 305
  wine strains, 109
Saignée, 637
Sambucus canadensis. See American Elder (*Sambucus canadensis*)
Sambucus nigra. See European Elder (*Sambucus nigra*)
Sand pear (*Pyrus pyrifolia*), 498–499
Sapota (*Achras sapota* Linn.), 454–455
  physicochemical characteristics of sapota must and wine, 456t
  wine, 454–455, 455f
Saran. *See* Polyvinylidene chloride
SBR. *See* Sequencing batch reactor (SBR)
Schizosaccharomyces, 91
Schizosaccharomyces pombe (*S. pombe*), 357
  characteristics of plum musts fermented with, 359t–360t
Schnapps, 532–533
Screening
  mango varieties, 442
  varieties, 34
  yeast strain, 445
Sea buckthorn (*Hippophae rhamnoides* L.), 390–392
  wine, 390–392
    preparation of, 391f
Secondary fermentation methods, 509–512
  bottle fermentation, 509–511
  yeast autolysis, 511–512
Sensory
  analysis, 244–246, 245t, 651
  evaluation, 649–652
Separate hydrolysis and fermentation (SHF), 566
Sequencing batch reactor (SBR), 577
Sherry production, 648
SHF. *See* Separate hydrolysis and fermentation (SHF)
"Shotgun" approach, 28
Simultaneous saccharification and fermentation (SSF), 566, 567f
Sinapic acid, 418–421, 433–434
Siphoning, 36–37
  of perry, 318
  in stone fruits, 351
SLF. *See* Submerged liquid fermentation (SLF)
Slivovitz, 541–542
Snow pears (*Pyrus nivalis*), 22
SOD. *See* Superoxide dismutase (SOD)
Solid fruit waste, unavoidable, 559–568
Solid waste, 607
Solid-phase microextraction (SPME), 141, 241, 416

Solid-state fermentation (SSF), 547
Sonication treatment, 261
Sour cherries (*Prunus ceresus*), 542–543
Sparkling fruit wines, 487, 500–524, 664
  malolactic fermentation in sparkling wine production, 512–513
  nongrape wines, 502
  production of sparkling apple wine and cider, 520–524
  secondary fermentation methods, 509–512
    bottle fermentation, 509–511
    yeast autolysis, 511–512
  sparkling mead, 524
Sparkling plum wine
  base wine production, 505–507
    method, 505–507
  composition, 507–509
  production, 513–520
    effect of cultivar, 519t
    effect of treatment on fermentation behavior, 515f
    physicochemical characteristics of base wines, 516t
    sensory evaluation, 520
    spider web diagram, 518f
  sensory qualities, 509
Sparkling wine(s), 46, 90, 137–138, 139t, 500–502, 644–647
  SPK Zork cork for, 646f
  technology of production, 502–504
    bulk method, 504
    methode ancestrale, 504
    methode champenoise, 502
    steps in sparkling wine production, 503f
    transfer method, 504
Specific gravity, 236
Spices, 487, 493, 498–499
  for preparation of mango vermouth, 496t
  in preparation of plum vermouth, 498t
Spirits, 531, 534
SPME. *See* Solid-phase microextraction (SPME)
Spoilage of cider, 337–338
Spontaneous fermentation, 79
Spontaneous French cider fermentations, 632
SSF. *See* Simultaneous saccharification and fermentation (SSF); Solid-state fermentation (SSF)
"Standard" fruit wine, 665
Staphylococcus aureus (*S. aureus*), 214
Starch, 687
  in apple fruit, 298–299
Starter cultures, 79–83. *See also* Yeast(s)
Static head, 277
Sterols, 84
Stilbenoids, 194–195
Stone fruit brandy, 541–547. *See also* Pome fruit brandy
  apricot brandy, 545–546
  cherry brandy, 542–543
  other stone brandies, 546–547
  plum brandy, 543–545

Stone fruit wines, 348, 681–686. *See also* Berries fruits wines; Exotic fruit wines; Pome fruit wines
  apricot wines, 683–684
  cherry wines, 685–686
  composition of fruit, 349, 349t
  nectarine wines, 682
  peach wines, 682
  plum wines, 684–685
  production, 348
    blending, 352
    clarification, 351
    fermentation, 351
    maturation, 351
    pasteurization, 352
    preparation of must, 350–351
    problems, 350
    siphoning/racking, 351
    yeast starter culture preparation, 350
  table wine, 352–376
Stone fruits, 348, 541–542, 599
Storage sugar, 298–299
Strawberry (*Fragaria × ananassa*), 24–25, 24f, 382
Strawberry wine, 53, 241, 677. *See also* Mandarin wine; Mango wine; Orange wine
  comparative physicochemical and nutritional characteristics of berries, 385t
  cultivar on physicochemical characteristics of, 388t
  maturation of wine, 386–387
  method of wine preparation, 384–386
  phenolic compounds in, 388t
  preparation of, 387f
  recipes, 678b–679b
  total phenols, 388t
  treatment on flavor profiling, 389f
Submerged liquid fermentation (SLF), 547
Succinic acid, 119
Sucrose, 442
Sugar(s), 34, 384
  adjustment, 302
  analysis, 237–239, 238t
  chromatogram, 240f
  composition of orange wine, 422–423, 422t
  content, 180–181
  fermentation, 92
  metabolism, 116–117
  in pear fruit, 315–316
Sulfur, 84
  metabolism, 131–132
Sulfur dioxide (SO$_2$), 93, 133, 235, 279, 304, 412–413, 415, 446, 665
  effect in wine preparation, 305t
Sulfurous acid (H$_2$SO$_3$), 235
Sulfurous anhydride, 93
Superoxide dismutase (SOD), 196

Sweet cherries (*Prunus avium*), 18, 542–543, 685
Sweet cider, 318–319
  preparation, 320f
Sweet cultivar, 321
Sweet potato (*Ipomoea batatas* L.), 477
  nutritional value of, 477t
  wine, 477–478
    physicochemical properties of enzyme-treated, 479t
    preparation, 478f
Sweet wines, 10
Syzygium cumini L. *See* Jamun (*Syzygium cumini* L.)

T

TA. *See* Titratable acidity (TA)
TAA. *See* Total antioxidative activity (TAA)
Table wine(s)
  alcohol content, 638
  apricot wine, 368–370
  berry and other fruit wines
    blended passion fruit, 405–406
    lychee wine, 403–404
    methods of preparation, 384–403
    papaya wine, 404–405
    production of berry wine, 383–384
    types of berry fruits and developmental stages of strawberry, 383f
  cherry wines, 370–376
  citrus wines, 410
    mandarin wine, 424–435
    orange wine, 410–424
  peach wine, 362–368
  plum wine, 352–362
  pome fruit wines, 295
    apple tea wine, 312–313
    apple wine, 296–315
    cider, 318–338
    loquat wine, 338
    medlar wine, 338–339
    mixed fruit wines, 340–342
    name, botanical name, and family of pome fruits, 296tpear wine, 315–318
    Pyracantha wine, 339
    quince wine, 339
    toyon wine, 339
  stone fruit wines, 348
    composition of fruit, 349, 349t
    problems in wine production, 350
    production, 348, 350–352
  tropical fruit, wine from
    tropics, 441
    types of fruit wine, 442–450
TAC. *See* Total anthocyanin content (TAC)
Tamarind (*Tamarindus indica*), 499
Tamarindus indica. *See* Tamarind (*Tamarindus indica*)

Tangential flow filtration. *See* Cross-flow MF (CFMF)
Tank method. *See* Bulk method
Tannin(s), 24
  addition of, 468
  in cider, 330–332
  in fruits wines, 671
  in pear fruit, 316
Tart cherry (*Prunus cerasus* L.), 18, 685
TBA. *See* 2,4,6-Tribromoanisole (TBA)
TCA. *See* 2,4,6-Trichloroanisole (TCA)
TCA cycle. *See* Tricarboxylic acid cycle (TCA cycle)
TCOD. *See* Total chemical oxygen demand (TCOD)
TDS. *See* Total dissolved solids (TDS)
TE. *See* Terrestrial ecotoxicity (TE); Trolox equivalents (TE)
TEAC. *See* Trolox equivalent antioxidant capacity (TEAC)
Teflon. *See* Polytetrafluoroethylene
Temperature, 435
  in cider, 326
  in juice apple process, 305
  temperature-controlled fermenters, 374–375
Terrestrial ecotoxicity (TE), 588
"Terroir", 13, 637
Theobroma cacao L. *See* Cocoa (*Theobroma cacao* L.)
Thiamine, 184, 471
Tinjure Milke Jaljale (TMJ), 476
Tissue culture, 25–26
Titratable acidity (TA), 233, 669
Titrimetry, 231–235
  acidity, 231–234
  alkalinity of ash, 234–235
  sulfur dioxide, 235
TMJ. *See* Tinjure Milke Jaljale (TMJ)
TOC. *See* Total organic carbon (TOC)
Tomato, 478–479
  wine, 479
    preparation, 480f
Total acidity, 233, 233t
Total anthocyanin content (TAC), 190, 190t, 374–375
Total antioxidative activity (TAA), 187, 479
Total chemical oxygen demand (TCOD), 580
Total dissolved solids (TDS), 607–608
Total organic carbon (TOC), 607–608
Total phenolic content (TPC), 187, 188t, 366
Total solids (TS), 607–608
Total soluble solids (TSS), 297f, 306, 365, 371f, 389–390, 424, 442
  wood chip effect, 373t
Total soluble sugar (TSS), 470–471, 499
Total suspended solids (TSS), 576, 607–608
Toxic metabolites of nitrogen metabolism, 139–141
  amines, 140–141, 140t
  ethyl carbamate, 141
Toyon wine, 339

TPC. *See* Total phenolic content (TPC)
TPTZ. *See* 2,4,6-Tripyridyl-s-triazine (TPTZ)
Track-and-trace technologies, 653
Traditional recipes of fruit wines, 676–695
  berries fruits wines, 677–681
  citrus wines, 691–692
  exotic fruit wines, 686–691
  pome fruits wines, 692–695
  stone fruits wines, 681–686
  tropical fruit wines, 686–691
Transfer method, 504
2,4,6-Tribromoanisole (TBA), 643
Tricarboxylic acid cycle (TCA cycle), 112, 113f
2,4,6-Trichloroanisole (TCA), 643
Triose phosphate sugars, 109–111, 111f
2,4,6-Tripyridyl-s-triazine (TPTZ), 209
Trolox equivalent antioxidant capacity (TEAC), 208, 423–424
Trolox equivalents (TE), 209
Tropical fruit wines, 686–691. *See also* Berries fruits wines;
    Citrus wines; Pome fruits wines
  banana wine, 687
  cocoa wines, 691
  fig wine, 688
  kiwi fruit wine, 688–690
  mango wines, 687–688
  pomegranate wines, 690–691
  tropics, 441
  types
    cashew apple, 451–452, 452t
    coconut wine, 454
    lychee wine, 452–454
    mango wine, 442–450
    palm wine, 455–457
    pineapple wine, 451
    sapota wine, 454–455
TS. *See* Total solids (TS)
TSS. *See* Total soluble solids (TSS); Total soluble sugar (TSS);
    Total suspended solids (TSS)
Tuba, 565

U

UASB. *See* Upflow anaerobic sludge blanket (UASB)
UCD 505, 306, 357
UCD 522, 306
UCD 522, 357
UCD 595, 306, 357
UF. *See* Ultrafiltration (UF)
Ultrafiltration (UF), 253
Ultrasound (US), 253, 256f
  application in must treatment for microbial inactivation,
    283–284
  inactivation of microorganism via US and coventional
    heating, 283–284

Ultraviolet (UV), 239
  rays, 675
Umeshu, 49
Unavoidable food waste, 559
Unavoidable solid fruit waste, 559–568
  apple pomace, 560, 560t
  berry peels, pulp, and seeds, 564
  citrus peels, 562–563
  coconut waste, 565–568
  mango peels, 561–562
United Kingdom, status of fruit wines, 56
United Nations Food and Agriculture Organization (FAO), 561
United States, status of fruit wines, 56
Unsaturated fatty acids, 84
Upflow anaerobic sludge blanket (UASB), 577
Urea, 141
Urethane. *See* Ethyl carbamate
Urrac, 17
US. *See* Ultrasound (US)
US-assisted enzymatic extraction of fruit juice, 260–262
UV. *See* Ultraviolet (UV)

V

Vaccinium, 16
Vaccinium corymbosum. *See* Blueberry (*Vaccinium corymbosum*)
Vaccinium macrocarpon. *See* Cranberry (*Vaccinium macrocarpon*)
Valorization
  fruit by-products and juices, 568–570
    bacterial cellulose production, 569f
    brewer's yeast biomass valorization, 571f
  of wine lees, 572–574, 573f
Vanillic acid, 433
Vascular smooth muscle cells (VSMC), 203
Vasodilatory effect, 203–204
Vegetative bacteria, 281, 282f
Vermouth, 10, 38–46, 470–471, 487–500
  commercial production of, 500
    examples of, 501t
    nomenclature on color basis, 501t
  production from nongrape fruits, 494–500
  production technology, 488–494, 490f
    aging and finishing, 494
    base wine preparation, 491
    botanical extracts preparation, 491–493
    bottling, 494
    brandy distillation, 491
    fortification and blending, 494
    storage, 494
  types
    dry vermouth, 488, 489t
    sweet vermouth, 488, 489t
VFA. *See* Volatile fatty acid (VFA)
Vignoles wine, 305

Vineyard, 625–631
  availability of grape genome, 631
  biological control of arthropod pests, 631
  innovations in pest control, 630–631
  principal indicators of fruit maturity, 625
  relationship between grape total soluble solids and berry color, 630f
  spectroscopic measurements in NIR, 626
  variability between total soluble solids, yield, and color, 627f
  variation in berry size, 628f
  variation in grape yield, 629f
Vinification, 181, 194
  enzymatic transformations of phenolic compounds, 196–197
  factors, 107–108
Vintage cider. *See* White cider
Vitamins, 183–184
Viticulture, 619
Vitis vinifera (*V. vinifera*), 617
Vitis vinifera L. *See* Grape (*Vitis vinifera* L.)
Volatile acidity, 233–234, 234f
Volatile acids, 232–233
Volatile aroma compounds, 447
Volatile compounds, 153–157, 154t, 416, 430
  analysis, 241–242
Volatile constituents, 182–183
Volatile content of orange juice, 416
Volatile fatty acid (VFA), 582, 601
Volatile organic acids, 118–121
Volatile sulfur compounds (VSCs), 131
VSMC. *See* Vascular smooth muscle cells (VSMC)

W

Waste and Resources Action Programme (WRAP), 557
Waste(s)
  agricultural, 601
  biological treatment of food waste, 609–611
  citrus solid, 563
  decision-making process regarding waste management, 602f
  from fruit wine production
    cider lees, 570–574
    ecotoxicity, 585–586
    food waste, 557–559
    liquid stream and wastewater, 574–585
    sustainability in winemaking sector, 586–589
    unavoidable solid fruit waste, 559–568
    valorization of fruit by-products and juices, 568–570
    world annual fruit production, 558f
  LCA of waste management in cider production, 588
  liquid waste, 607
  mango waste use as feedstock, 612
  solid, 607
  unavoidable food, 559
  unavoidable solid fruit waste, 559–568

Wastewater (WW), 574
  characterization of winery liquid effluents, 574–576
  treatment methods, 576–585
    biological treatment processes, 577–585
    bioreactors, 577–585
Water in apple fruit, 298
WBCSD. *See* World Business Council for Sustainable
    Development (WBCSD)
Whey
  beer, 481
  composition, 480
  other fermented beverages, 480–481
  wine, 481
White cider, 318–319
Whole-fruit basis, 352
Wild apricot (*Prunus armeniaca* L), 499–500
Wild Vine Style, 665
Wild yeasts, 624
Wine, 1, 177, 410, 463, 487
  alcoholic fermentation, 6
  clarification of, 268
  composition and nutritional significance of
    aroma compounds, 153–157, 154t, 156t
    constituents in alcoholic beverages, 144t
    parameters, 143–147
    phenolic characterization of fruit wines, 147–153, 148t–149t
    volatile compounds, 153–157, 154t
  fermentation, 26
  as food and health benefits, 6–8
  genetic engineering in, 25–33
    genetic engineering of wine yeast, 26–32
    genetic modification of plants, 25–26
    genetically modified yeasts, 32–33
  history, 5–6
  LCA of packaging, 588–589
  lees, 570
    valorization, 572–574, 573f
  nongrape fruit wine industry, 46–57
  nongrape wines, 39t–45t
  origin of, 2–43
  preparation, 105–106
  preservation processes to production, 279–287
  production, 37–38, 649
  quality, 629
  salient features, 2t
  special wines, 38–46
    sparkling wine, 46
    vermouth, 38–46
  spoilage chemistry
    spoilage by acetic acid bacteria, 142
    spoilage by LAB, 142–143
  wine press, 4f
  wine-processing

residues, 557
  wastewater, 558
  wine-waste streams, 558–559
Wine yeast
  characteristics, 27t
  genetic engineering, 26–28
  methods and applications, 29f
  target genes for tailoring, 30f
  targets for, 28–32, 29f
    control of microbial spoilage, 31
    improvement in fermentation process, 28–31
    improvement in processing of wine, 31
    improvement in sensory qualities, 32
    improvement in wine wholesomeness, 31–32
  wine-yeast development, methods for, 97
Winemaking, 105, 242–243, 383–384
  aroma profile of fruit wines, 623f
  authenticity, 652–653
  bottling process, 625
  chemistry of, 107–133
    acetic acid fermentation, 132–133, 132f
    carbon metabolism, 109–125
    enzymes, 133
    metabolism of sulfur, 131–132
    nitrogen metabolism, 126–131
  cryoextraction, 621–622
  fermented wine protection, 624
  fortified wines, 647–649
  innovations in vineyard, 625–631
  maceration phase, 622
  methods
    clarification, 37
    fermentation, 35–36
    maturation, 37
    pasteurization, 37
    preparation of must, 35
    preparation of yeast starter culture, 35
    siphoning/racking, 36–37
  nongrape fruits for, 11–13
  pectins, 623–624
  percentage production of fruit crops, 12f
  process, 189–190, 195
  racking process, 624
  relationship between total phenolic content and total
      antioxidant capacity, 654f
  sensory evaluation, 649–652
  sparkling wines, 644–647
  sustainability in sector, 586–589
    LCA in fruit production, 587–588
    LCA of waste management in cider production, 588
    LCA of wine packaging, 588–589
  technological innovations, 620
  winery innovations, 631–644

Winery innovations, 631–644. *See also* Vineyard
   aroma profile of mango wines, 639f
   blending, 637
   cold presoaking, 632
   deacidification, 634–635
   fermentors, 637
   flavor development, 638–639
   influence of malolactic fermentation, 634f
   microencapsulation techniques, 636f
   oxygen, 641
   premium wines, 643–644
   process of carbonic maceration, 631
   spontaneous fermentation, 632
Winery liquid effluent characterization,
      574–576
   characteristics of fruit processing wastewater, 575t
   parameters characterizing wastewater from
         red wine, 574t
   pollution loads, 575t
World Business Council for Sustainable Development
      (WBCSD), 587–588
World Resources Institute (WRI), 557, 587–588
"Worm wood", 487
WRAP. *See* Waste and Resources Action Programme
      (WRAP)
WRI. *See* World Resources Institute (WRI)
WW. *See* Wastewater (WW)

Y

Yeast assimilable nitrogen (YAN), 471
Yeast(s), 3–5, 74–76, 94, 115, 118, 305–306, 617, 633

   autolysis, 511–512
      biochemical changes during, 512
   biomass, 571
   in cider, 326–327
   culture, 507
   factors affecting growth, 84–85
      acidity, 84
      aeration, 84
      ethanol concentrations, 85
      sugar concentrations, 84
      temperature, 84
   flavor compounds, 135–137
   growth, 116–117
   inoculation, 673–674
   non-*Saccharomyces* yeasts, 75t
   oxidative pathway, 119f
   regulation of carbon metabolism glycolysis, 115–116
   Saccharomyces yeasts, 76t
   selection as starter cultures for, 79–83
      inoculation with mixed yeast cultures, 80–81
      inoculation with pure yeast cultures, 81–82
      procedures for inoculums preparation for, 82–83
   starter culture
      for homemade fruit, 83b
      preparation in stone fruits, 350
      preparation of, 35
   strain, screening of, 445
   synthesis of malate, 120f

Z

Zygosaccharomyces bailii (*Z. bailii*), 122
Zymomonas anaerobia (*Z. anaerobia*), 337

Printed and bound by CPI Group (UK) Ltd, Croydon, CR0 4YY

03/10/2024

01040326-0002